Regulation of Plant Development

Fang Chang • Yingxiang Wang • Hong Ma
Editors

Regulation of Plant Development

Springer

Editors
Fang Chang
School of Life Sciences
Fudan University
Shanghai, Shanghai, China

Yingxiang Wang
College of Life Sciences
South China Agricultural University
Guangzhou, Guangdong, China

Hong Ma
Department of Biology, The Huck Institutes
of the Life Sciences
The Pennsylvania State University
University Park, PA, USA

ISBN 978-981-95-7032-4 ISBN 978-981-95-7033-1 (eBook)
https://doi.org/10.1007/978-981-95-7033-1

This Springer imprint is published by the registered company Springer Nature Singapore Pte Ltd.
The registered company address is: 152 Beach Road, #21-01/04 Gateway East, Singapore 189721, Singapore

Preface

Plant developmental biology is a multifaceted field that delves into the processes governing the growth, differentiation, and reproduction of plants. It seeks to understand how plants develop from the zygote into a complex multicellular organism, capable of responding and adapting to the environment. In the ever-evolving landscape of biological sciences, the understanding of plant development is not only crucial for unraveling the mysteries of the plant life cycle but also for addressing pressing global challenges, such as food security and ecological sustainability when facing climate changes. It is within this context that this comprehensive volume on the regulation of plant development emerges, offering a scholarly exploration of the fundamental molecular processes shaping the growth and development of plants.

The roots of plant developmental biology can be traced back to ancient times, when early naturalists and botanists observed and documented plant growth and development, including the effects of different environmental conditions. In particular, Darwin's experiments using dark-grown seedlings led to the subsequent discovery of auxin as a crucial plant hormone regulating plant development. Since the advent of modern experimental techniques, including the advances in microscopy, genetics, and molecular biology, plant developmental biology has flourished dramatically as a distinct scientific discipline. In the late nineteenth and early twentieth centuries, pioneering scientists such as Wilhelm Hofmeister, Julius von Sachs, and Edward Charles Jeffrey conducted studies on plant embryology, anatomy, and physiology, laid the groundwork for modern plant developmental biology. Their observations and experiments provided valuable insights into the processes of plant growth and development, setting the stage for more recent research in the field.

Some of the landmark discoveries in plant developmental biology were made since the mid-twentieth century with the identification of plant hormones, including auxins, gibberellins, cytokinins, ethylene, abscisic acid, brassinolide, and strigolactones, which play key roles in regulating diverse aspects of plant growth and development. This period also has seen the emergence of genetic studies in plants, facilitated by the development of model organisms such as *Arabidopsis thaliana*, rice, and others, which have collectively become a cornerstone of plant developmental research. In more recent decades from the late twentieth to century to the present day, advancements in molecular genetics and genomics have revolutionized the understanding of plant development. Key genes and signaling pathways have been discovered that control cell division, differentiation, patterning, cellular interactions, and other processes, uncovering the molecular mechanisms underlying plant development. Additionally, resources and techniques such as large-scale insertional mutant collections, RNA interference, and CRISPR-Cas9-based genome editing have enabled reverse genetic analyses, including precise manipulation of plant genomes, further unraveling the functions of genes and regulatory elements in plant development.

The significance of plant developmental biology extends far beyond the realm of basic science. Understanding how plants develop and respond to their environment has profound implications for agriculture, ecology, and environmental sustainability. Understanding the molecular mechanisms underlying plant growth and development can lead to the identification of novel strategies and targets for plant biotechnology and crop improvement, and development of resilient plant varieties capable of withstanding environmental stressors. The knowledge on plant development in combination with evolutionary developmental biology (evo-devo) can also facilitate the design of innovative strategies for ecosystem restoration and conservation.

The importance of plant developmental biology is amply demonstrated by numerous remarkable advancements that have reshaped the understanding of fundamental biological processes. Such progresses in gaining insights into the complex nature of plant development have greatly benefited from holistic approaches that integrate interdisciplinary insights and fosters collaboration across scientific disciplines. In this book, we have meticulously assembled a broad away of topics into 21 chapters, each dedicated to elucidating a specific aspect of plant development and its regulation. From the foundational principles of root and shoot development to the intricate processes of floral transition, sex determination, and fruit development, this book offers a comprehensive synthesis of current knowledge and emerging research trends in the field.

Chapters 1, 2, and 3 delve into the intricacies of root, shoot, and leaf development, respectively. Roots enable plants to anchor themselves in the soil and absorb water and nutrients essential for their survival. Root development is initiated in the root apical meristem and involves the formation of multiple tissues, including the epidermis where root hairs form via tip growth. Shoot development, on the other hand, governs the formation of stems, leaves, and flowers, which are essential for nutrient transport, photosynthesis, and reproduction, respectively. More specifically, leaf development starts with the initiation of the leaf primordium and leaf morphogenesis of leaf primordia, culminating in the formation of diverse leaf shapes and sizes adapted to various environmental conditions. Chapter 4 offers a distinct perspective on the process of organ regeneration, highlighting key regulatory processes, including hormonal actions and transcriptional control.

Chapters 5, 6, 7, and 8 explore the molecular mechanisms underlying the transition from vegetative to reproductive growth, the regulation of reproductive meristems, the specification of floral organs, and the evolutionary analysis of floral organ identity genes. The transition to flowering is a critical developmental switch in the life cycle of plants, marking the onset of reproductive development and the formation of flowers. Floral organ specification, meanwhile, involves the genetic and molecular processes that determine the identity and patterning of floral organs, including sepals, petals, stamens, and carpels. The evolutionary analysis of floral organ identity genes provides insights into the genetic basis of floral diversity across different plant species, shedding light on the mechanisms driving evolutionary change and adaptation in plants.

Chapters 9 and 10 focus on stamen and anther development and ovary and ovule development, respectively. Stamen and anther development, meanwhile, involves the differentiation and maturation of male reproductive structures responsible for producing and releasing pollen, the male gametophyte. Ovary and ovule development, on the other hand, governs the formation of female reproductive structures, ovaries and ovules, which house the female gametophyte and serve as the site of fertilization and subsequent seed development. Chapters 11, 12, and 13 explore meiosis, pollen development, and embryo sac development, respectively. Meiosis is a specialized cell division process that generates haploid gametes from diploid germ cells, promoting genetic diversity in part via recombination in sexually reproducing organisms. Pollen development, meanwhile, involves the formation and maturation of pollen grains, the male gametophytes responsible for delivering sperm cells to the female reproductive organs during pollination. Embryo sac development, on the other hand, governs the formation and maturation of the female gametophyte, which houses the egg cell and serves as the site of fertilization and seed development.

Chapters 14, 15, and 16 delve into pollination and fertilization, male fertility and hybrid sterility, and self-incompatibility, respectively. Pollination involves the transfer of pollen grains from the anther to the stigma at the top of the ovary and style, enabling fertilization and seed production. Fertilization, meanwhile, involves the fusion of a sperm and an egg cell to form a zygote, the first cell of the new plant embryo, and a second fusion of another sperm and the central cell for the formation of the endosperm. Male fertility and hybrid sterility, on the other hand, are genetic phenomena that influence the reproductive success and viability of hybrid offspring produced from crosses between different plant species or varieties. Self-incompatibility, meanwhile, is a genetic mechanism that prevents self-fertilization and promotes outcrossing in plants, thereby enhancing genetic diversity and reducing the risk of inbreeding depression.

Chapter 17 examines sex determination from a perspective of the sexual reproduction cycle from a comparison among plants, animals, and other eukaryotes and provides insights into the differences between sex determination of gametophytes and development of sexually different sporangia. Chapters 18 and 19 present development and regulations of the embryo and endosperm, respectively. Embryo developmental processes include early embryonic cell division, pattern formation of the embryo, and establishment of organ primordia. Chapter 18 also presents a discussion of the interplay between embryonic and post-embryonic developmental processes. Chapter 19 describes regulatory mechanisms of endosperm initiation and cellularization, specification of the specialized cell layer aleurone, accumulation of seed storage carbohydrates, proteins, and lipids, with a discussion of interactions with the embryo. Chapter 20 discusses gene functions for epigenetic regulation of maternal effects and imprinting during endosperm development, including key factors and relevant biological functions. Finally, Chap. 21 showcases fruit developmental regulation and the evolution of key fruit regulatory genes, focusing on comparative examination of *Arabidopsis*, rice, and tomato that supports the conservation of carpel identity genes. A role for developmental repatterning is also discussed for fruit diversification.

From the molecular intricacies of root and shoot development to the evolutionary analyses of floral organ identity genes and fruit developmental regulators, each chapter of this book represents a detailed examination of the multifaceted nature of plant development. More than a compilation of chapters, this volume is a testament to the collective endeavor of scholars and researchers dedicated to unraveling the mysteries of plant development and to the collaborative spirit of the scientific community with commitment to advancing knowledge for the betterment of society. As we embark on this journey of exploration and discovery, may this volume serve as a beacon of knowledge and inspiration for generations of scholars and researchers, propelling us toward a brighter and more sustainable future.

Shanghai, China — Fang Chang
Guangzhou, China — Yingxiang Wang
State College, PA, USA — Hong Ma
Beijing, China — Zhihong Xu

Contents

Chapter 1
Arabidopsis Root Development

Bingsheng Lv, Jiajia Liu, Huiyu Tian, Meng Li, Shuang Wu, Wei Xuan, Chuanyou Li, and Zhaojun Ding

Abstract The root is one of the most important organs for plants. It not only anchors plants to the soil, but also functions as the main organ for taking up water and nutrients from the soil environment. Our comprehension of root development initially stemmed from studies using classical genetic approaches and has been greatly enhanced by the advancements in modern molecular biology and genomics techniques. Unlike other plant organs, we now have a comprehensive map of individual cell types in roots, at different developmental stages, and in response to various stimuli. Consequently, root research has ascended to the forefront of plant development biology. This chapter delves into the exploration of root development using *Arabidopsis thaliana* as a model system. In *Arabidopsis*, the root system is mainly composed of the primary root, lateral roots, and root hairs. We review recent studies of the root apical meristem (RAM), discuss the lateral root development, and summarize the research progress about root epidermis—a pivotal interface facilitating root-soil interaction. Furthermore, we shed light on the pivotal role of plant hormones in modulating root development, responding to both internal developmental cues and external environmental signals.

B. Lv
College of Horticulture, Qingdao Agricultural University, Qingdao, Shandong, China

School of Life Sciences, Shandong University, Qingdao, Shandong, China

J. Liu · H. Tian · Z. Ding (✉)
School of Life Sciences, Shandong University, Qingdao, Shandong, China
e-mail: dingzhaojun@sdu.edu.cn

M. Li · S. Wu
Haixia Institute of Science and Technology, Horticultural Plant Biology and Metabolomics Center, Fujian Agriculture and Forestry University, Fuzhou, China

W. Xuan
State Key Laboratory of Crop Genetics & Germplasm Enhancement, Nanjing Agricultural University, Nanjing, China

C. Li
Institute of Genetics and Developmental Biology, Chinese Academy of Sciences, Beijing, China

F. Chang et al. (eds.), *Regulation of Plant Development*,
https://doi.org/10.1007/978-981-95-7033-1_1

Keywords Root development · Root apical meristem · Lateral root · Root hair · Hormonal regulation · Abiotic stress

1.1 Introduction

Developed in the early stage of the plant life cycle, roots evolved on land during the Devonian Period, which spanned from 416 million to 360 million years ago (Raven & Edwards, 2001). At the early stage of plant evolution, the primitive rooting structures appeared morphologically similar to the shoot system, suggesting that roots may have originated from an embryonic shoot dichotomy (Jiang & Feldman, 2005). The fully integrated vascular system of roots formed during evolution, enabled plants to acquire water and nutrients (Boyce, 2005). This evolutionary shift not only facilitated plant anchorage in soil but also established roots as the primary organ for water and nutrient absorption as plants adapted to terrestrial environments. The significance of the root system extends beyond mere anchorage and nutrient uptake; it also plays a fundamental role in interactions between plants and their environment. Therefore, ensuring appropriate root development, growth, and functionality holds immense importance for the survival of plants.

Angiosperm roots exhibit two distinct root systems, categorized by their developmental origin and branching pattern: the fibrous system and the taproot system. The fibrous root system is characteristic of monocot plants and serves as a morphological feature distinguishing them from dicot plants, which possess the taproot system. Both systems encompass the primary root, lateral roots (LRs), and adventitious roots. Roots that develop from the primary root are generally called lateral roots, while those that develop postembryonically from shoots, stems, or leaves are called adventitious roots. In the fibrous system, a dense network of slender roots—also known as crown roots—arises from the hypocotyl or stem. For instance, in maize, the fibrous root system encompasses brace or crown roots, seminal roots, and adventitious roots, besides the primary root and lateral roots. The taproot system is the system in which the primary root develops from a radicle and becomes dominant during the whole life cycle. The primary root holds pivotal importance throughout the entire lifespan of a plant within a taproot system. Generally, plants with a fibrous root system are shallow-rooted and are more susceptible to drought, but have the ability to respond quickly to fertilizer application. In contrast, plants with a taproot system are deep-rooted with better anchorage and allow the absorption of nutrients and water from deeper soil.

Our understanding of *Arabidopsis* root development was initially gained from classical genetic studies and has been further strengthened by analyses using molecular biology and genomics approaches. Numerous genes have been identified to play roles in a variety of regulatory processes of root growth and development. Unlike many other plant organs, we now possess an intricate map detailing the molecular landscape of individual cell types within roots across various developmental stages and in response to various cues. This exhaustive characterization has propelled root studies to the forefront of plant developmental biology. In this chapter, we summarize analyses of root development using *Arabidopsis*, whose root system includes the primary root, lateral roots, and root hairs. We delve into recent studies on root apical meristem (RAM), explore the intricacies of lateral root development. Furthermore, we offer a concise overview of the latest strides in understanding the root epidermis—a pivotal border tissue facilitating the crucial interaction between roots and soil. Additionally, we underscore the pivotal role of plant hormones in orchestrating root development, adeptly responding to both internal developmental signals and external environmental cues.

1.2 Root Apical Meristem

In dicots, the sustained growth of the primary root hinges on the Root Apical Meristem (RAM), a specialized structure situated at the root tip. This RAM is responsible for generating cells, known as stem cells, which significantly contribute to the development of nearly all root tissues postembryonically. Within the *Arabidopsis* RAM, a small cluster of infrequently dividing cells forms the quiescent center (QC), encircled by stem cells (Fig. 1.1). Stem cells positioned laterally and toward the shootward direction relative to the QC undergo divisions, yielding progeny that eventually differentiate into various root tissues, encompassing the epidermis, cortex, endodermis, lateral root cap, and vascular tissue. Stem cells located beneath the QC divide, producing cells earmarked for differentiation into the columella root cap (Dolan et al., 1993; Peret et al., 2009; Petricka et al., 2012). Recent advancements utilizing high-throughput single-cell RNA sequencing analyses have unveiled the remarkable heterogeneity of cells within the *Arabidopsis* root in terms of their transcriptomes. The spatial arrangement and temporal sequence of diverse

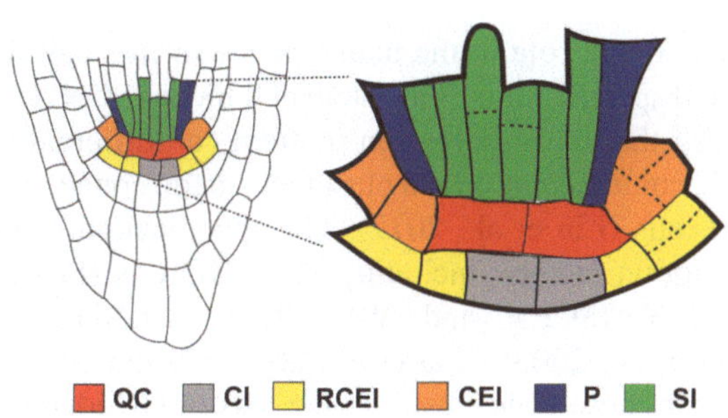

Fig. 1.1 The organization of the root apical meristem of *Arabidopsis thaliana* and the division planes of initial cells. QC quiescent center, CI columella cell initial, RCEI lateral root cap/epidermis cell initial, CEI cortex/endodermis cell initial, P pericycle, SI stele cell initial. Dotted lines represent the cell division direction

cell types across various developmental stages elucidate their hierarchical structures, facilitating the reconstruction of a continuous differentiation trajectory in root development (Zhang et al., 2019).

1.2.1 Stem Cell Niche

In the RAM, the centrally positioned QC is surrounded by four sets of stem cells—stele cell initial (SI), cortex/endodermis cell initial (CEI), lateral root cap/epidermis cell initial (RCEI), and columella cell initial (CI). Together, the QC and the surrounding stem cells compose the stem cell niche. These stem cells continually undergo asymmetric divisions, generating daughter cells that migrate away from the niche to commence their differentiation journey. Below the QC, the CIs typically divide only in a periclinal manner. Their offspring undergo rapid cell expansion, eventually differentiating into the central cells of the root cap. These specialized root cap central cells house amyloplasts (starch-filled plastids) and serve as gravity sensors. Conversely, the other three types of initials generally undergo both periclinal and anticlinal divisions. These divisions result in diverse cell lineages acquiring distinct identities (Dolan et al., 1993). The CEI initiates its division by an anticlinal split, resulting in a daughter cell in contact with the Quiescent Center (QC) retaining its stem-cell identity. Simultaneously, the other daughter cell undergoes a periclinal division, giving rise to two distinct cell layers—forming the endodermis and cortex within the ground tissue. The inner cell layer proceeds to specialize into the endodermis, while the outer layer matures into the cortex. The RCEI firstly undergoes an asymmetric periclinal division to give rise to one lateral root cap cell (the outmost/lower one) and the inner cell that retains the role of an initial, and that divides anticlinally to produce one initial cell (the lower one) and one epidermal cell (the upper one). Finally, several Sis undergo periclinal divisions, contributing cells to form the vascular tissues (xylem and phloem) along a singular central axis. The outermost SI cells divide transversely, contributing to the formation of the pericycle (Fig. 1.1).

The QC plays a crucial role in the maintenance of the undifferentiated state of stem cell initials and specification of the stem cell niche (SCN) (van den Berg et al., 1997). It's proposed that the suppression of stem cell differentiation is facilitated through symplastic communication via plasmodesmata, connecting the QC with the surrounding stem cells (Liu et al., 2017). The specification of the QC involves a number of transcription factors, including PLETHORAs (PLTs), SHORT ROOT (SHR)/SCARECROW (SCR), and WUSCHEL-RELATED HOMEOBOX 5 (WOX5) (Sabatini et al., 2003; Aida et al., 2004; Galinha et al., 2007). Loss-of-function mutations in these genes result in the loss of QC identity. For instance, in the *plt1/plt2/plt3* mutant, both the QC identity and the lateral root cap cell layer are lost, indicating a compromised formative cell division of RCEI stem cells (Galinha et al., 2007).

WOX5, a member of the WOX gene family, plays a role in maintaining root stem cells by regulating *CYCLING DOF FACTOR 4* (*CDF4*) and *CYCLIN D3;3* (*CYCD3;3*) (Sarkar et al., 2007; Forzani et al., 2014; Pi et al., 2015). In addition, the glutamine-rich SEUSS (SEU) protein facilitates the expression of *WOX5* via its physical interaction with both SCR and a methyltransferase SET DOMAIN GROUP 4 (SDG4), maintaining WOX5 activity (Zhai et al., 2020). It has also been shown that SCR and SHR control the formative cell divisions in CEI (Benfey et al., 1993; Scheres et al., 1995; Sabatini et al., 2003). Mutation of *SCR* or *SHR* results in a single ground tissue layer, while the overexpression of *SHR* triggers additional layer formation, indicating that *SHR* is both essential and sufficient for formative cell divisions (Helariutta et al., 2000). Interestingly, *SHR* is transcribed only in the stele, but the encoded protein moves to CEI and promotes the expression of *SCR*. SCR, in turn, limits radial SHR movement by binding and trapping SHR in the nucleus of endodermis (Cui et al., 2007). As a result, only CEI undergoes periclinal divisions to generate endodermis and cortex tissues (Helariutta et al., 2000; Gallagher et al., 2004; Heidstra et al., 2004).

Other transcription factors have also been reported to regulate the RAM and stem cell maintenance, such as FEZ and SOMBRERO (SMB). FEZ and SMB, both proteins belonging to the NAC transcription factor family, have been identified as direct regulators crucially involved in the division of columella stem cells (Willemsen et al., 2008). FEZ operates within stem cells, fostering cell division and stimulating the expression of *SMB* in subsequent daughter cells. SMB, in turn, downregulates *FEZ* expression, establishing a negative feedback loop crucial for tempering division and facilitating the differentiation of daughter cells (Willemsen et al., 2008). Meanwhile, the putative zinc finger transcription factor NO TRANSMITTING TRACT (NTT) has been demonstrated to orchestrate the initiation of the root meristem. Notably, this protein is not only essential but also adequate in conferring the identities of distal stem cells (stem cells below QC) in the root meristem (Crawford et al., 2015). Auxin can regulate root distal stem cell identity through dual regulation of the cell cycle factor DIMERIZATION PARTNER A (Dpa) at both transcriptional and posttranslational levels (Wang et al., 2023). A bHLH transcription factor named SPATULA (SPT) has been reported to regulate the size of RAM. The *spt* mutant has a larger RAM due to an increased size of QC and supernumerary divisions in initials (Makkena & Lamb, 2013). In an early study, the ETHYLENE RESPONSE FACTOR 115 (ERF115) was found to act as a rate-limiting factor in the division of QC (Heyman et al., 2013).The bHLH transcription factor UPBEAT1 (UPB1) is thought to keep the balance between cell proliferation and differentiation in the root meristem through regulating the reactive oxygen species (ROS) gradient between the RAM and the root differentiation zone above the RAM in *Arabidopsis* (Tsukagoshi et al., 2010). The maintenance of an optimal ROS seems also essential to control RAM identity. An increase in the rate of cell division in the QC was observed in the *app1* mutant, which has reduced ROS levels (Yu et al., 2016). On the contrary, the RAM defects in the *abo8–1* mutant were attributed to the increased ROS levels (Yang et al., 2014a).

1.2.2 Stele

The postembryonic vascular tissue in roots comprises distinct components: xylem, phloem, and procambium. Across all angiosperms, these tissues are encompassed by an outer layer of pericycle cells, collectively forming the stele (Bishopp et al., 2013). Beyond providing structural support, these tissues operate as a conductive network, facilitating the transport of water and nutrients. During primary root growth, xylem and phloem differentiate from the procambium. Once the primary vascular tissues are established, the thickening of stems and roots is orchestrated by cell division and differentiation from the cambium, derived from the procambium. In Arabidopsis, the development of secondary xylem progresses through two distinct phases. Initially, only vessel elements differentiate, while the remainder of the xylem cells retain their parenchyma state. Subsequently, in the later phase, both vessel elements and lignified fiber cells emerge from cambial cells, resulting in a stem anatomy akin to that of angiosperm tree wood (Chaffey et al., 2002). Within root structures, two types of xylem vessels, protoxylem and metaxylem, take shape. Protoxylem strands form on the periphery of the xylem axis, displaying lignin deposition in a spiral pattern and spiral cell wall thickenings. These cells differentiate early and tend to undergo cell death as the plant matures. In contrast, metaxylem strands develop centrally, showcasing lignin in a reticulate pattern and pitted cell wall thickenings. Differentiating later, these cells are larger in size and play a crucial role as the water-conducting vessels of the plant (Petricka et al., 2012).

Members of the class III homeo-domain leucine zipper (HD-ZIP III) transcription factor family exhibit expression within *Arabidopsis* vascular tissues. HD-ZIP IIIs play a crucial role in specifying the two xylem vessel types in a dosage-dependent manner: higher levels promote metaxylem formation, while lower levels favor protoxylem formation. Loss-of-function in all five HD-ZIP TFs in Arabidopsis root leads to a failure in xylem vessel development, underscoring the indispensability of HD-ZIP IIIs in defining xylem identity (Carlsbecker et al., 2010). The regulatory mechanism behind vascular cell division involves HD-ZIP III TFs positively regulating brassinosteroid biosynthesis-related genes, such as *CONSTITUTIVE PHOTOMORPHOGENIC DWARF* (*CPD*), in vascular cells. Elevated brassinosteroid levels suppress cytokinin response in vascular cells, thus restraining vascular cell division in the Root Apical Meristem (RAM) (Ohashi-Ito et al., 2023). NAC domain transcription factors VASCULAR-RELATED NAC-DOMAIN6 (VND6) and VND7 also play pivotal roles in xylem specification in Arabidopsis. VND6 promotes metaxylem differentiation, while VND7 induces protoxylem fate (Kubo et al., 2005). Further studies revealed that ectopic expression of VND6 leads to ectopic metaxylem formation, while ectopic expression of VND7 induces ectopic protoxylem (Ohashi-Ito & Fukuda, 2010; Yamaguchi et al., 2010). The interaction between auxin and cytokinin operates as a spatial mutually exclusive feedforward loop, crucial in delineating distinct hormonal response zones essential for vascular tissue formation (De Rybel et al., 2014). Modulators of cytokinin signaling, such as CLAVATA3/EMBRYO SURROUNDING REGION-related (CLE) and

FUMONISIN B1-RESISTANT 12 (FBR12), participate in protoxylem vessel formation. CLE10 inhibits protoxylem vessel formation by downregulating type-A ARRs, which negatively regulate cytokinin signal transduction, thereby activating the cytokinin signaling pathway (Kondo et al., 2011). Loss-of-function in FBR12, decreasing cytokinin sensitivity, results in the production of extra protoxylem vessels (Ren et al., 2013). It has been reported that perturbing phosphoinositide homeostasis differentially influences vascular tissue differentiation by modulating intracellular trafficking in Arabidopsis roots (Gujas et al., 2017).

Our comprehension of the molecular mechanisms governing phloem formation remains somewhat constrained. Phloem comprises sieve elements and companion cells, arising from the asymmetric divisions of phloem/procambial initial cells (Mahonen et al., 2000; Baum et al., 2002). Within this process, an MYB transcription factor known as ALTERED PHLOEM DEVELOPMENT (APL), which is expressed in developing phloem cells, not only fosters phloem development but also actively suppresses xylem identity. In *apl* mutants, the phloem identity becomes compromised, resulting in the absence of phloem sieve elements and companion cells. Intriguingly, these mutants exhibit the emergence of cells displaying xylem characteristics at the phloem poles instead of the typical phloem structure (Bonke et al., 2003).

1.2.3 Ground Tissue (Endodermis and Cortex)

In the *Arabidopsis* root, orchestrated cell division and specification within ground tissues play a pivotal role in establishing the radial pattern. During the early heart stage of embryos, the ground tissue founder cell undergoes asymmetric periclinal divisions, giving rise to the endodermis and cortex tissues (Scheres et al., 1994). Postembryogenesis, CEI cells embark on a series of asymmetric divisions, pivotal for root development. Anticlinal divisions ensure the self-renewal of CEI and the generation of CEI daughter cells (CEID). Subsequently, CEID experiences a periclinal division, orchestrating the formation of two distinct ground tissue layers: the endodermis and the cortex (Benfey et al., 1993; Dolan et al., 1993; Scheres et al., 1994, 1995; Helariutta et al., 2000). Around 1–2 weeks postseed germination, the endodermis undergoes another periclinal division, contributing to the creation of an additional cortex layer (Baum et al., 2002; Paquette & Benfey, 2005). This newly formed cortex layer is termed the middle cortex due to its spatial position. The emergence of the middle cortex layer signifies a more advanced phase in postembryonic root development (Baum et al., 2002; Paquette & Benfey, 2005).

Apart from their pivotal roles in regulating root stem cells, SHR and SCR play fundamental roles as determinants in ground tissue patterning during *Arabidopsis* root development (Benfey et al., 1993; Scheres et al., 1995). In *shr* and *scr* mutants, only a single ground tissue layer is observed between the epidermis and pericycle due to the failure of periclinal cell division in the ground tissue precursor. This underscores the indispensable role of SHR and SCR in orchestrating the periclinal asymmetric cell division, leading to the formation of two distinct cell files. In *scr* mutants, the single tissue layer displays a mixed identity, exhibiting characteristics of both cortex and endodermis, while in *shr* mutants, it predominantly exhibits cortex properties (Scheres et al., 1995; DiLaurenzio et al., 1996; Nakajima et al., 2001).

The overexpression of SHR not only triggers the formation of an additional endodermal layer but also signifies its role in promoting endodermis identity beyond asymmetric stem cell division (Helariutta et al., 2000). Intriguingly, while SHR RNA is absent in ground tissues and exclusively found in the stele (Helariutta et al., 2000; Nakajima et al., 2001), the SHR protein travels from the stele to the endodermis/cortex initials, illustrating its noncell-autonomous function (Nakajima et al., 2001). Upon entering endodermal cells, SCR sequesters SHR and restricts its location to the nucleus of endodermal cells, thus preventing its further movement (Cui et al., 2007). A related SCR homologue, SCARECROW-LIKE 23 (SCL23), interacts with SCR and SHR, inhibiting SHR's intercellular spread without nuclear retention (Long et al., 2015a). Nuclear-localized SHR plays a crucial role in the periclinal asymmetric divisions of CEID, and in patterning the ground tissue (Cui et al., 2007; Welch et al., 2007). In plants with reduced SCR expression (SCR RNAi), where residual SCR fails to capture all SHR proteins, additional endodermal layers emerge (Cui et al., 2007). Beyond SCR, the C2H2 transcription factor JACKDAW (JKD) and its close relatives participate in sequestering SHR within the nucleus (Long et al., 2015b). JKD, along with MAGPIE (MGP) and NUTCRACKER (NUC), collectively termed BIRDs, belong to the plant-specific INDETERMINATE DOMAIN C2H2 zinc finger protein family. They regulate CEID's asymmetric divisions and serve as determinants of CEI identity (Moreno-Risueno et al., 2015).

Auxin plays a pivotal role in embryo patterning, with the auxin effector ARF5 reported to exert both cell-autonomous and noncell-autonomous control over embryonic vascular tissue formation and root initiation (Moller et al., 2017). The HD-ZIP III transcription factor PHABULOSA (PHB) emerges as a key regulator governing middle cortex formation. It achieves this by activating the GA catabolic gene *GIBBERELLIN 2 OXIDASE 2* (*GA2ox2*) in the vascular tissue while controlling the accumulation of CYCD6;1 in the endodermis (Bertolotti et al., 2021). Moreover, a more recent finding highlights the role of NAC1, which genetically and physically interacts with SCR and SHR, in restricting excessive periclinal cell divisions within the endodermis. This fine-tuning is achieved by precisely regulating the spatiotemporal expression of *CYCD6;1*, ultimately contributing to the formation of middle cortex cells (Xie et al., 2023).

1.2.4 Exterior Surface of the Root

The lateral root cap and columella together form a protective layer that continually sheds as the root explores new areas. Along with the epidermis, these tissues compose the outermost surface of the root. The columella originates from the division of specific initials located beneath the QC, containing starch-filled plastids (amyloplasts) crucial for gravity sensing. Meanwhile, the epidermis and lateral root cap stem from a common set of initials. A specialized structure, the root hair, is produced by a subset of epidermal cells, vastly increasing the root's surface area for enhanced water and nutrient absorption. NAC domain transcription factors FEZ and SOMBRERO (SMB) play pivotal roles in the periclinal divisions of the columella and lateral root cap/epidermis initials (Willemsen et al., 2008). Mutations in *fez* result in reduced columella and lateral root cap layers, whereas *sombrero* mutants exhibit increased cell layers (Willemsen et al., 2008). *FEZ* expression in root cap stem cells autonomously drives periclinal cell division and activates *SMB* expression in the daughter cells. SMB, in turn, regulates FEZ activity negatively, thereby confining divisions to the initial cells (Willemsen et al., 2008). Research has uncovered that Class IIB NAC proteins control cell maturation in cells undergoing terminal differentiation, particularly those undergoing robust cell wall modifications (Bennett et al., 2010). Members of this family, including SOMBRERO (SMB), BEARSKIN1 (BRN1), and BEARSKIN2 (BRN2), regulate secondary cell wall synthesis in specialized cell types. Mutations in *smb-3* lead to continued division of lateral cap cells, hindering their detachment from the root. Additionally, in *brn1-1 brn2-1* double mutants, columella cells fail to detach. In triple mutants involving these three genes, cells across the cap fail to mature (Bennett et al., 2010).

1.3 Lateral Root Development

The root system's architecture is a fundamental aspect of root development and crucial for overall plant growth. A major determinant of root system architecture is the initiation of lateral roots or root branching. The adaptation of the plant's root architecture through the initiation and elongation of lateral roots is an important strategy used by plants to cope with the changing nutritional conditions by growing roots toward the source of nutrients (Malamy, 2005; Hodge et al., 2009). Lateral roots take shape within the differentiation zone, arising from mature pericycle cells adjacent to the xylem poles, known as the xylem pole pericycle. Among these cells, a specific subset—lateral root founder cells (LRFC)—undergoes a meticulously orchestrated series of divisions culminating in the formation of a lateral root primordium (LRP) (De Smet et al., 2006) (Fig. 1.2). The initial indication of a lateral root's emergence begins with the asymmetric and anticlinal divisions of these founder cells (Dubrovsky et al., 2001). A series of orchestrated periclinal and anticlinal cell divisions will eventually give rise to the organized LRP (Laskowski et al., 1995; Sussex et al., 1995; Malamy & Benfey, 1997). As the primordium emerges through the various cell layers of the parent root, this progression marks the lateral root emergence (Malamy & Benfey, 1997). To discern these stages, researchers have defined eight distinct phases based on specific anatomical characteristics and cell divisions during lateral root emergence (Malamy & Benfey, 1997). Once the lateral root has emerged, the RAM of the new lateral root is activated, and further growth occurs through divisions of initial cells in the new RAM.

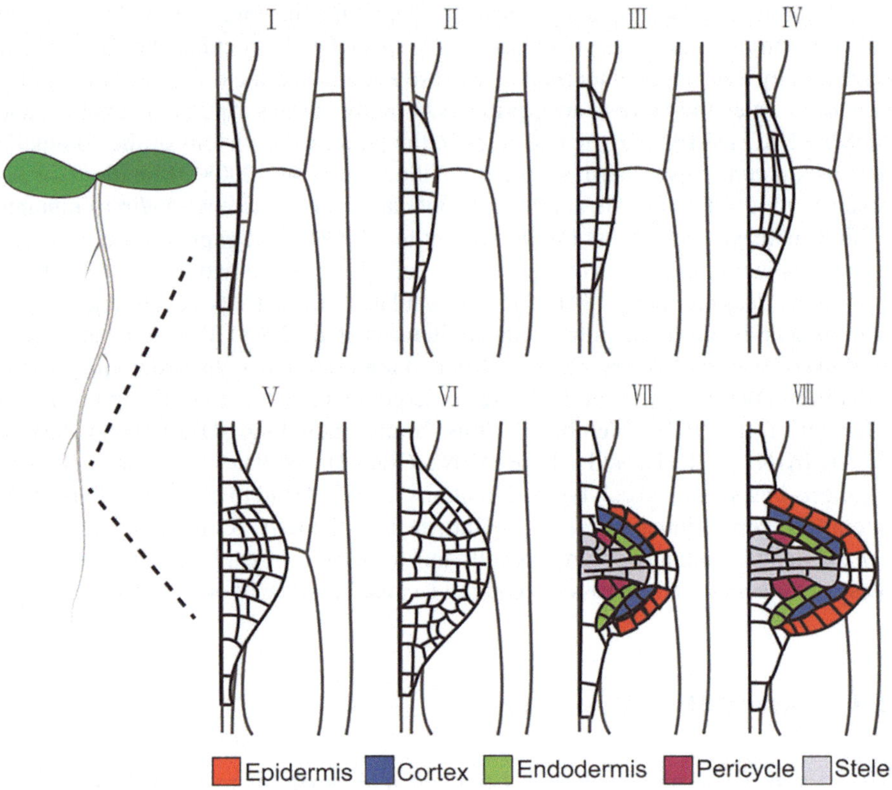

Fig. 1.2 Lateral root development

1.3.1 Regulation of Lateral Root Spacing

Although lateral roots are initiated from the pericycle, not all pericycle cells can be primed for lateral root initiation, suggesting that cells in the pericycle layer may belong to different subgroups. Exploring the positioning of lateral roots along the primary root has spurred the formulation of various hypotheses in recent times. An intriguing discovery involves the robust and periodic staining of the synthetic auxin-responsive reporter DR5:GUS within the protoxylem strands, specifically within an area termed the basal meristem—an intermediary region between the meristematic and elongation zones (Beemster et al., 2003)—which was linked to lateral root initiation (De Smet et al., 2007). This distinctive staining pattern has been linked to the initiation of lateral roots (De Smet et al., 2007). Researchers have postulated that this spatiotemporal distribution of auxin in the basal meristem might play a pivotal role in regulating lateral root production. It's proposed that auxin, within this region, acts as a primer, prepping specific subsets of pericycle cells to transform into founder cells, laying the groundwork for the emergence of lateral roots and establishing sites for potential lateral root branching (De Smet et al., 2007). Further experiments supported the idea that the auxin maxima in protoxylem cells result in the priming of neighboring pericycle cells that eventually initiate a lateral root in the more mature zone of the primary root (Overvoorde et al., 2010). However, an intriguing twist emerged from another study, challenging the notion that auxin alone determines prebranch sites. To unravel the amplified transcriptional reaction to auxin within the basal meristem, researchers fused the DR5 promoter with the Luciferase coding region (DR5:LUC), enabling real-time characterization of the reporter's behavior. Their observations unveiled a rhythmic pulsation of DR5:LUC expression recurring approximately every 6 hours within the basal meristem and the elongation zone, which was called the oscillation zone (OZ). This led to the hypothesis that the oscillatory pattern indicated by DR5:LUC might induce the formation of sites primed for lateral root emergence, evolving subsequently into fully developed lateral roots (Moreno-Risueno et al., 2010). The periodic establishment of these prebranch sites leads to a regularly spaced pattern of LRs along the primary root.

To determine whether prebranch site formation was related to changes in levels of auxin, Moreno-Risueno and colleagues engineered distinct auxin signaling reporters—merging the promoters of *IAA7*, *SLR/IAA14*, and *IAA19* with the luciferase reporter—to probe their responsiveness to externally applied auxin within the oscillation zone (OZ). Among these, the *pIAA19:LUC* reporter had a similar expression to exogenous auxin as DR5:LUC. Yet, unlike *DR5:LUC*, *pIAA19:LUC* didn't display oscillatory behavior. Intriguingly, attempts to induce prebranch sites using exogenous auxin, either alone or in conjunction with auxin transport inhibitors, failed to trigger their formation when not coinciding with a DR5 oscillation. This suggested that exogenous auxin was unlikely to be sufficient to generate prebranch sites (Moreno-Risueno et al., 2010). However, the knockout mutants of auxin receptor genes such as *TIR1* and *AFB2* exhibited weak DR5:LUC signal in the oscillation zone and few prebranch sites, suggesting that endogenous auxin signaling is required for the periodic prebranch site formation (Xuan et al., 2015).

In addition to the endogenous mechanism that controls the spacing of lateral roots under controlled growth conditions, lateral roots can also be induced by external factors such as gravistimulation (Richter et al., 2009). It has been proposed that lateral roots exhibit a preference for formation on the convex side of a bend, where the relocated PIN1 in a group of protoxylem cells drives auxin accumulation (Ditengou et al., 2008; Laskowski et al., 2008). Gravitropic stimuli have been observed to shorten the intervals between consecutive DR5:LUC oscillations, thus accelerating the production of prebranch sites (Moreno-Risueno et al., 2010). Adding complexity to this intricate process, a cyclic programmed cell death signal in the distal end of the lateral root cap (LRC) triggers the death of LRC cells. This event not only impacts the DR5 oscillation but also influences prebranch site formation by redistributing auxin from the dying cells back into the oscillation zone (Xuan et al., 2016). Moreover, carotenoid-derived molecules, like β-cyclocitral, have been identified as necessary for maintaining DR5 oscillation and facilitating prebranch site formation (Dickinson et al., 2021). Carotenoids, despite showing robust expression of the *PHYTOENE SYNTHASE* (*PSY*) gene in the root differentiation zone rather than in the oscillation zone, seem to act as mobile signals governing DR5 oscillation regulation. Meanwhile, genes involved in pectin distribution and esterification were also found to regulate prebranch site formation, indicating that the prebranch site formation also requires a balanced esterification level of pectin in the cell wall (Wachsman et al., 2020). Additionally, water availability is also known to regulate lateral root spacing. Loss of contact with water triggers the accumulation of endogenous abscisic acid in the roots, disrupting oscillations and inhibiting lateral root formation (Orman-Ligeza et al., 2018). Conversely, in the presence of water, ARF7 SUMOylation-dependent regulation of auxin response governs root branching patterns in response to water availability (Orosa-Puente et al., 2018). Beyond water, light signals have also been implicated in positively regulating prebranch site formation by activating the expression of light-responsive genes (e.g., Long hypocotyl1 (HY1) and HY5) and auxin signaling pathway genes (Duan et al., 2021).

1.3.2 Lateral Root Initiation

The LR initiation process begins before the division of the pericycle cells adjacent to the xylem poles (XPPs), which have to dedifferentiate or maintain the mitotic status once they begin to divide. XPP cells differ from other pericycle cells due to multiple distinguishing features, including size and various structural and ultrastructural characteristics (Beeckman et al., 2001; Parizot et al., 2008). Notably, research has revealed that XPP cells are tetraploid, a deviation from the diploid nature of all other pericycle cells (Beeckman et al., 2001). The mitotic cell cycle phase of XPP cells was investigated by the expression pattern of *CYCLIN A2;1* (*CYCA2;1*), whose transcriptional level rises during S phase and peaks at the end of G2 phase (Shaul et al., 1996). In the root, while the expression of CYCA2;1 diminishes beyond the root apical meristem, it reappears specifically in the XPP cells, indicating that the XPP cells are in G2 phase while other pericycle cells are in G1 phase (Beeckman et al., 2001). Notably, the expression of *CYCB1;1*, a marker linked to the G2 and M phases of the cell cycle, commonly serves as an indicator for the initiation of lateral roots (Himanen et al., 2002, 2004).

In most of the characterized plants, the initiation of lateral roots begins with the asymmetric division of founder cells (Casero et al., 1995; Malamy & Benfey, 1997). Auxin plays important roles in the regulation of cell division and initiation of lateral root (Himanen et al., 2002). External application of auxin prompts pericycle cells specifically positioned at the xylem poles—distinct from other pericycle cells—to participate in the formation of lateral root primordia (Himanen et al., 2004). The asymmetric division of founder cells is preceded by a significant event: the migration of nuclei from two adjacent founder cells toward their shared cell wall (De Smet et al., 2007). Auxin accumulation within these founder cells before the migration triggers the activation of ARF7 and ARF19, which appeared to be necessary for the coordinated nuclear migration (De Rybel et al., 2010). In response to auxin, the release of ARF7 and ARF19 from AUX/IAAs further leads to the activation of the members of the *LATERAL ORGAN BOUNDARIES DOMAIN/ASYMMETRIC LEAVES2-LIKE* (*LBD/ASL*) family, including *LBD16*, *18*, and *29* (Okushima et al., 2007). LBD18 interacts with LBD33 to activate the expression of E2Fa and regulate the cell cycle. This connection solidifies LBD18's role as a link between auxin signaling and the regulation of the cell cycle, leading to the initiation of lateral root formation (Berckmans et al., 2011). Besides auxin, MPK6 was reported to act downstream of the GOLVEN peptide signaling to restrict the number of initial asymmetric cell divisions for the initiation of lateral root (Fernandez et al., 2020).

1.3.3 Shaping of the New Primordium

Functional lateral root primordia (LRP) can only be formed when there is a combination of asymmetric cell division, cell specification, and correct hormonal balance. In *Arabidopsis*, even three founder cells can be sufficient for lateral root formation (Dubrovsky et al., 2001). The process of LRP formation and development was divided into eight stages according to a series of orchestrated periclinal and anticlinal cell divisions (Malamy & Benfey, 1997). Initially, the anticlinal divisions of the lateral root founder cells within the pericycle pave the way for the emergence of a stage I primordium (Fig. 1.2). These divisions occur rapidly, giving rise to approximately 8–10 short pericycle derivative cells in the longitudinal plane, which subsequently expand radially. As the process progresses to stage II, periclinal divisions lead to the formation of an outer layer (OL) and an inner layer (IL) within the primordium (Fig. 1.2). However, not all the founder cells participate in this division, and instead, most pericycle cells remain undivided. Consequently, the domed structure of the LRP begins to manifest as the OL and IL cells expand radially. Subsequent periclinal divisions in the OL cells result in the generation of a three-layered LRP by stage III (Fig. 1.2), maintaining the dome-like shape as some peripheral cells abstain from division. Advancing to stage IV, periclinal divisions occur in the IL cells, leading to the development of a four-layered LRP (Fig. 1.2). At this juncture, the LRP has penetrated the endodermal layer of the parental root. MPK14-mediated auxin signaling orchestrates the passage of the LRP through the endodermal layer by regulating ERF13-controlled biosynthesis of very long-chain fatty acids (VLCFA), crucial for lateral root emergence (VLCFA) (Lv et al., 2021). Another AP2/EREBP family transcription factor, PUCHI, directs the spatial expression of genes involved in VLCFA synthesis. This fine-tuned control contributes to regulating cell proliferation during the initial stages of LR development (Trinh et al., 2019). During the progression to stage V of LRP development (Fig. 1.2), a series of events unfold. Central cells within the OL undergo anticlinal division, resulting in the formation of four small cuboidal cells. Auxin and histone deacetylation affect *LATERAL ROOT PRIMORDIUM1* (*LRP1*) expression, acting downstream of LR, forming auxin response modules to negatively regulate LRP development at stage V by modulating auxin homeostasis (Singh et al., 2020). Concurrently, the cells adjacent to these cuboidal cells in the OL divide, increasing the OL cell count to 10–12 in the sectional plane through the central axis. Simultaneously, IL cells enlarge radially, contributing to the division and upward displacement of overlying layers. The advancing LRP, now pushing through the cortex of the primary root, marks this pivotal stage of its development. Advancing through subsequent stages, at stage VI, periclinal division within the OL gives rise to a new internal layer (Fig. 1.2). The LRP at this stage has penetrated the epidermis of the parental root. At stage VII, different cell types are being formed (Fig. 1.2). The LRP begins to resemble the mature root tip, containing three layers that could correspond to epidermis, cortex, and endodermis. By this time, the lateral root meristem has been established. As the LRP enlarges, it becomes more difficult to characterize the divisions. It appears that many of the cells of the LRP continue to undergo anticlinal divisions (Fig. 1.2). Upon emergence, the lateral root meristem is activated, signaling the commencement of further growth (Malamy & Benfey, 1997).

1.3.4 Emerging of Lateral Root

Lateral root initiation takes place in pericycle cells deep within the parental root, requiring new primordia to break through the overlying tissues before they emerge into the soil. In the case of *Arabidopsis*, this process involves breaking through three distinct tissues: the endodermis, cortex, and epidermis. Notably, the first barrier encountered by the developing lateral root is the endodermis, where a remarkable transformation takes place within the cells. The endodermal cells become progressively thinner, lose volume, up to the point where opposing plasma membranes fuse (Vermeer et al., 2014). Moreover, the Casparian strip, serving as a barrier between adjacent endodermal cells, locally degrades to facilitate a controlled disruption of the lignin network. This localized degradation allows for a confined opening within the lignin structure, creating a pathway through which the LRP can penetrate (Vilches-Barro & Maizel, 2015).

Once the LRP has crossed the endodermis, it must pass through the cortex and epidermis layers to emerge at the surface of the primary root. A crucial aspect of this process involves the separation of cells within the cortex and endodermis layers to create a pathway for the emerging LRP (Kumpf et al., 2013). This cellular separation hinges on the degradation of the pectin-rich middle lamella that typically binds adjacent cells together. Multiple enzymes participate in this cell wall remodeling. Notably, the pectin methylesterase PME1, alongside the pectin lyases PLA1 and PLA2, as well as the *PGAZAT/ADPG2*-encoded polygalacturonase, play pivotal roles in this intricate process (Laskowski et al., 2006; Gonzalez-Carranza et al., 2007a; Swarup et al., 2008). Upon demethylesterification by PME, the pectins within the middle lamellae are subject to degradation by pectin lyases. This degradation leads to the loosening of the cell wall between the LRP cells and the underlying tissues, facilitating the LRP's passage. Moreover, polygalacturonase also contributes to the breakdown of pectins. Activation of PGAZAT expression, crucial for this process, is instigated by auxin and the IDA-HAE signaling module, notably detected in cells of the cortex and epidermis surrounding the emerging LRP (Gonzalez-Carranza et al., 2007b; Kumpf et al., 2013). This orchestrated enzymatic activity and signaling pathway enable the necessary restructuring of the cell walls, allowing the lateral root primordium to successfully pass through these layers and emerge from the primary root's surface.

During the complex process of LR morphogenesis, auxin takes center stage by activating the MKK4/5–MPK3/6 cascade via transmembrane kinases 1/4 (TMK1/4), crucial for controlling cell division patterns (Huang et al., 2019). Furthermore, auxin's role extends to activating MPK3/6 in an IDA-HAE/HSL2-dependent manner, pivotal for regulating LR emergence by orchestrating pectin degradation and cell wall separation (Zhu et al., 2019). During the emergence of the LRP, the expression of LIKE AUX1 3 (LAX3) becomes noticeable within the outer layers of the primary root. LAX3, found in cortical and epidermal cells, encodes an auxin uptake transporter that facilitates the accumulation of auxin within these cell types (Swarup et al., 2008). This gene's activity in cortical and epidermal cells is further augmented by the buildup of auxin at the tip of the emerging LRP. Consequently, LAX3 plays a pivotal role in upregulating the expression of various cell wall remodeling enzymes essential for this process. Specifically, LAX3 triggers the increased expression of crucial cell wall-modifying enzymes such as AIR3 (a subtilisin-like protease), XTH23/XYLOGLUCAN ENDOTRANSGLYCOSYLASE6 (XTR6), and AtPLA2. These enzymes contribute to the restructuring of the cell wall, facilitating the successful emergence of the lateral root primordium (Swarup et al., 2008). The walls of the cortex and epidermis layers above the emerging LRP therefore start to soften, thereby creating an adequate room for LRP growth and emergence. The PR-1 homologue PRH1 was identified as a target of ARF7-mediated auxin signaling and participates in the regulation of LRP emergence through influencing the expression of *EXP*s that affect cell wall loosening (Zhang et al., 2020). Cell death in the layers overlying the LRP has also been shown recently to regulate LR emergence (Escamez et al., 2020).

1.4 Root Hair Development in *Arabidopsis*

Originating from a specialized group known as lateral root cap/epidermal stem cells, epidermal cells undergo a dynamic journey. Initially, these cells embark on rapid division within the meristem division zone. However, as they progress, they halt their division and swiftly elongate in alignment with root growth, ultimately reaching their final dimensions and forming what's known as the mature zone. It's within this mature zone that root hairs emerge, a characteristic feature of the epidermis (Dolan et al., 1993, 1994). Arabidopsis' root epidermis comprises two distinct cell types: root-hair cells and nonhair cells, each following a stereotypical pattern influenced by their specific positions. Root hairs play a multifaceted role, amplifying the root's surface area and contributing significantly to functions such as water uptake, nutrient absorption, anchorage, and fostering interactions with various microbes. This specialized structure is instrumental in aiding the plant's adaptation to its soil environment and optimizing its overall performance.

1.4.1 Root Hair Initiation and Tip Growth

Root hair initiation involves a fascinating process where a bulge forms at the RAM-proximal end of the cell, indicating the cell's polarization before hair outgrowth begins. Auxin likely plays a role in this intricate mechanism, considering its movement away from the meristem through epidermal cells. Mutants with impaired auxin signaling often display incorrect or absent hair development, underscoring its significance in this process (Masucci & Schiefelbein, 1994, 1996; Pitts et al., 1998). Numerous genes participate in the precise spatial control of hair outgrowth, evident from studies on mutants exhibiting hair emergence at wrong positions or multiple sites along epidermal cells. For instance, the *root hair defective 6* (*rhd6*) mutant showcases reduced hair production compared to the wild type, causing a basal shift in the site of hair emergence. This highlights RHD6's role in hair initiation and in specifying the emergence site along the apical-basal axis of the hair cell. Moreover, in the *rhd6* mutant, a higher frequency of epidermal cells produces multiple root hairs, suggesting RHD6's involvement in restricting the number of emerging sites as well as in bulge formation (Masucci & Schiefelbein, 1994). Jasmonate, another player in this process, modulates root hair development in a manner regulated by RHD6 and its interaction with RHD6 LIKE1 (RSL1). JASMONATE ZIM-DOMAIN (JAZ) proteins interfere with RHD6's transcriptional function and its interaction with RSL1, thus impacting Jasmonate-stimulated root hair development (Han et al., 2020). Additionally, the long noncoding RNA APOLO collaborates with the transcription factor WRKY42, triggering root hair cell expansion by activating *RHD6* transcriptional activity in response to cold (Moison et al., 2021). Meanwhile, the *tiny root hair 1* (*trh1*) mutant, which exhibits multiple hair initiation sites but fails to undergo tip growth, suggests that TRH1's wild-type activity is vital to maintain a single hair outgrowth site. TRH1, predicted to belong to the AtKT/AtKUP/HAK potassium transporter family, is also capable of catalyzing auxin efflux, supporting the notion that auxin is crucial in selecting the hair cell's outgrowth site (Rigas et al., 2001; Vicente-Agullo et al., 2004). Moreover, *Arabidopsis* mutant *supercentipede1* (*scn1*) develops ectopic sites of hair initiation in root-hair cells, implying that it restricts the number of sites of hair growth to a singularity (Carol et al., 2005). SCN1, a RhoGTPase GDP dissociation inhibitor (RhoGDI), spatially regulates hair growth sites. The presence of ROP2 at the root hair tip, depending on SCN1 function, further supports this model. Mutant versions of ROP2 lead to abnormal hair development, reinforcing the role of Rho GTPases in selecting the root hair initiation site (Carol et al., 2005; Jones et al., 2002).

1.4.2 Root Hairs and Nutrient Uptake

The scarcity of nutrients significantly hampers plant growth and global crop yields. In response to nutrient limitation, plants employ strategies such as augmenting their root surface area. The mere presence of root hairs, for instance, has been shown to amplify the total root surface area by two to fivefold (Gahoonia et al., 1997). Under nutrient stresses, there's an apparent increase in both the number and density of root hairs, suggesting that these structures play a pivotal role in enhancing the efficiency of nutrient ion uptake from the soil. Phosphorus, known for its extreme immobility in soil, often stands as a limiting factor for growth. The density of root hairs plays a direct role in determining the amount of phosphorus accessible to a plant for uptake and assimilation (Gahoonia & Nielsen, 2004). Consequently, the heightened density of root hairs serves as a critical adaptation for plants facing nutrient deficiencies, enabling them to access and utilize vital nutrients, particularly phosphorus, crucial for sustained growth and development.

In conditions where phosphorus is abundantly available, the shoot biomass of wild-type Arabidopsis, along with the *rhd6* mutant (characterized by reduced root hair number), and the *rhd2* mutant (developing very short hairs), showed no significant differences. This suggests that under plentiful phosphate conditions, the presence or absence of root hairs did not markedly impact shoot biomass. However, when subjected to low-phosphorus conditions, a distinct pattern emerged. Wild-type plants exhibited significantly greater shoot biomass compared to the *rhd6* mutant, which, in turn, displayed greater shoot biomass than the *rhd2* mutant (Bates & Lynch, 2000). These findings strongly imply that heightened root hair density and length contribute to an increased uptake of phosphate per unit root length when phosphorus availability is limited.

1.5 Hormonal Regulation of Root Growth

Plant hormones are crucial biotic factors regulating root growth (Fig. 1.3). Recent studies have unveiled a significant role of plant hormones in controlling cell division, cell differentiation, and cell growth in different root zones. However, various hormones govern organ growth by regulating specific growth processes such as cell proliferation, differentiation, or expansion in specific tissues. In this section, we will discuss the role of a few key plant hormones in root growth.

1.5.1 The Role of Auxin in Root Growth

The plant hormone auxin is known to be critical for root patterning, primary root growth, and root architecture (Tanaka et al., 2006; Vanneste & Friml, 2009). It plays a major role in regulating cell division and cell elongation to control root growth.

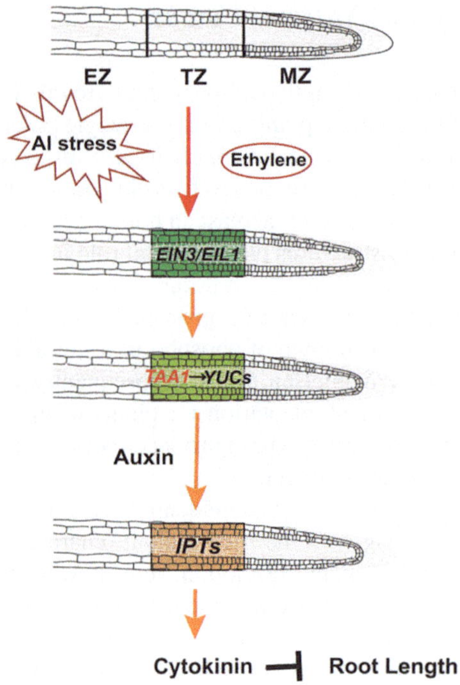

Fig. 1.3 Auxin maximum in root transition zone (TZ) mediates Aluminum stress to inhibit root growth in Arabidopsis. The root comprises three distinct zones: the proximal meristem zone (MZ), the transition zone (TZ), and the elongation/differentiation zone (EZ). Under aluminum stress, the ethylene signaling pathway, mediated by EIN3/EIL, triggers the local elevation of auxin levels in the root TZ through up-regulation of the expression of auxin biosynthetic genes such as *TAA1* and *YUC*s. This process leads to root-growth inhibition by subsequently enhancing the expression of cytokinin biosynthesis-related genes *IPT*s in the same region

Auxin is primarily synthesized in young leaves and cotyledons and is transported to the root sink through the phloem by the combined action of auxin influx carriers AUXIN RESISTANT 1/LIKE-AUXIN RESISTANT (AUX1/LAX) and auxin efflux carriers PIN-FORMED (PIN) (Swarup et al., 2001; Tanaka et al., 2006; Robert & Friml, 2009). An auxin maximum, which is generated by both polar auxin transport and local auxin biosynthesis, is essential for the maintenance of the QC identity (Grieneisen et al., 2007; Tian et al., 2014). Plants require an optimal auxin level in QC to maintain the QC identity, and either increasing or decreasing auxin levels both cause QC cell division and distal stem cell differentiation (Ding & Friml, 2010; Tian et al., 2014).

In the meristematic zone, auxin acts as a permissive signal for cell division and can regulate the cell cycle at both transcriptional and protein levels (Perrot-Rechenmann, 2010). Auxin has been shown to induce the expression of CYCLIN-DEPENDENT KINASE *CDKA;1*, which has a critical role in both G1 to S and G2 to M progressions in *Arabidopsis* seedlings (Hemerly et al., 1993; Ferreira et al.,

1994). Cyclin genes, such as *CYCB1* and *CYCA2*, are also potentially regulated by auxin (Roudier et al., 2003). Auxin positively regulates the transcription of DPa through ARF7 and ARF19, and enhances Dpa protein stability through MITOGEN-ACTIVATED PROTEIN KINASE 3 (MPK3)/MPK6-mediated phosphorylation to regulate distal stem cell maintenance (Wang et al., 2023). In addition to its effect on cell division, auxin is essential for root meristem maintenance by acting on the gradient of PLETHORA (PLTs) proteins, which belong to the AP2-domain transcription factor family and are essential for determining the root stem cell niche (Aida et al., 2004; Galinha et al., 2007).

Auxin tightly regulates lateral root formation. Gain-of-function mutations in genes such as *AXR5/IAA1, SHY2/IAA3, BDL/IAA12, SLR/IAA14, IAA18, IAA19*, and *IAA28* have been found to affect the capacity to form lateral roots (Tian & Reed, 1999; Rogg et al., 2001; Fukaki et al., 2002; Tatematsu et al., 2004; Yang et al., 2004; Uehara et al., 2008; Ploense et al., 2009). The *slr-1/iaa14* and *iaa28* gain-of-function mutant plants show the strongest reduction in the formation of lateral roots (Rogg et al., 2001; Fukaki et al., 2002). These results indicate that Aux/IAA proteins function as negative regulators of lateral root formation. The combination of forward and reverse genetic approaches indicates that ARF proteins also play important roles during lateral root formation (Tatematsu et al., 2004; Okushima et al., 2005). The *arf7* or *arf19* single mutant has a mild reduction in the number of formed lateral roots, whereas the *arf7/19* double mutant shows a dramatic delay in the formation of lateral roots (Okushima et al., 2005). ARF7 and ARF19 positively regulate many auxin-responsive genes such as *LATERAL ORGAN BOUNDARIES-DOMAIN (LBD)/ASYMMETRIC LEAVES2-like* (*ASL*) genes *LBD16/ASL18* and *LBD29/ASL16*, which are important for the initiation of lateral roots (Okushima et al., 2007). Overexpression of *LBD16* or *LBD29* partially rescued the lateral root phenotype of *arf7/19* mutant, indicating that LBD16 and LBD29 proteins regulate lateral root initiation downstream of ARF7/19 (Okushima et al., 2007). ARF5/MP was also shown to be involved in the formation of postembryonic lateral root (De Smet et al., 2010). Additionally, certain ARFs, such as ARF10 and ARF16, can function as negative regulators, as evidenced by the fact that the *arf10/16* double mutant exhibits an increased number of lateral roots compared to the wild type (Ulmasov et al., 1999; Wang et al., 2005).

Besides controlling cell division, exogenous application of auxin inhibits root growth by inhibiting cell elongation in all root cell layers. The inhibitory concentration of auxin varies according to the plant species and organs. Lower concentrations of auxin potentially promote the elongation of root cells, but its effect can be experimentally difficult to reveal. Notably, epidermal cells forming root hairs exhibit a distinct response to auxin compared to other root cells: while it inhibits the elongation of most root cells, it actually enhances the elongation of root hairs at equivalent auxin levels. Consequently, the majority of cellular reactions to auxin are presumably reliant on the dosage. Numerous early genes responsive to auxin have been initially identified in expanding tissues induced by auxin, highlighting their significant roles in this process (Perrot-Rechenmann, 2010).

Various gain-of-function mutants of the AUX/IAA repressor family (e.g., *axr5/iaa1*, *shy2/iaa3*, *axr2/iaa7*, or *axr3/iaa17*) show defects in cell elongation (Liscum & Reed, 2002; Mockaitis & Estelle, 2008), suggesting that the degradation of these repressors mediated by auxin is required for cell elongation. ABP1 has been established as a key player in auxin-mediated cell elongation. Deactivating ABP1 in roots triggers the elongation of basal meristematic cells and imparts resistance to IAA (Tromas et al., 2009), suggesting ABP1's involvement in modulating auxin sensitivity and positioning it as a negative regulator of root cell elongation. Beyond ABP1, various components of the TIR1/AFB signaling pathway, including TIR1 itself, have been identified as regulators of root growth. For instance, treatment with PEO-IAA, an auxin antagonist binding to TIR1 and obstructing its binding to Aux/IAA, has been observed to accelerate root growth (Fendrych et al., 2018). Furthermore, recent studies have shed light on AFB1's pivotal role in swiftly inhibiting root growth through membrane depolarization (Serre et al., 2021), underscoring its central significance in rapid auxin-mediated growth inhibition.

1.5.2 Ethylene Regulates Root Growth and Development

Ethylene, a gaseous plant hormone, profoundly influences various facets of plant growth and development, encompassing seed germination, hypocotyl and root elongation, floral transition, sex determination, fruit ripening, and responses to diverse biotic and abiotic stresses (Kendrick & Chang, 2008). Recently, there has been a growing focus on understanding the impact of ethylene on root growth and development. High levels of ethylene have been consistently linked to the inhibition of root elongation across different species (Le et al., 2001; Rahman et al., 2001; Swarup et al., 2002; Negi et al., 2010). Mutants displaying enhanced ethylene signaling or synthesis, such as *ctr1* or *eto1*, demonstrate a reduced rate of root elongation (Kieber et al., 1993). Conversely, mutants with diminished sensitivity to ethylene, like *etr1*, *ein2*, *ein3*, and *eil1*, exhibit elevated primary root growth rates (Barry & Giovannoni, 2006; Ruzicka et al., 2007; Stepanova et al., 2007; Swarup et al., 2007; Negi et al., 2010). These consistent findings strongly support the notion that ethylene acts as an inhibitor of root growth.

Ethylene also plays an important role in root gravitropism and root waving. In *Arabidopsis*, ethylene treatment impedes the gravitropic response while intensifying the frequency and magnitude of root waving, an action independent of gravity's influence (Buer et al., 2003, 2006; Lewis et al., 2011). In *ein2* and *etr1* mutants, the inhibition of gravitropism by ethylene is abolished (Buer et al., 2006), suggesting that the ethylene signaling pathway is necessary for the effect of ethylene on root gravitropism. Additionally, ethylene is involved in lateral root formation (Ivanchenko et al., 2008; Negi et al., 2008). Plants treated with ethylene or ethylene-sensitive mutants exhibit reduced lateral root formation, while ethylene-insensitive mutants show elevated numbers of lateral roots (Negi et al., 2008; Strader et al., 2010). In contrast to the negative effects of ethylene on root growth and lateral root formation, the development of root hair is induced by ethylene (Kieber et al., 1993; Pitts et al., 1998; Schiefelbein, 2000). Moreover, ethylene influences the positioning of root hair outgrowth, promoting basal positioning upon application, while ethylene-insensitive mutants show a shift toward apical root hairs (Fischer et al., 2007).

1.5.3 Abscisic Acid in Root Growth and Development

Abscisic acid (ABA), derived from carotenoids and initially isolated from cotton plants, is synthesized in both roots and leaves (Liu & Carnsdagger, 1961; Taylor et al., 2000). Known for its involvement in various stress responses—such as heat, cold, drought, and other environmental stresses (Melcher et al., 2010)—ABA has demonstrated a complex role in regulating root growth. Experimental evidence from exogenous ABA applications has indicated that ABA typically inhibits root growth. Initially, this inhibition was attributed to reduced cell elongation (Pilet, 1975). ABA has been shown to induce the expression of *PIN2* and *AUX1*, facilitating auxin redistribution to epidermal and cortical cells, thereby promoting cell elongation (Xu et al., 2013). Notably, during hydrotropism, the root elongation zone demonstrates a unique ABA-dependent sensing and response mechanism, distinct from other tropisms (Dietrich et al., 2017). However, ABA's impact on root growth is multifaceted. It not only inhibits cell elongation but also affects cell division. Studies have linked ABA to the induction of KRP1 expression, a CDK inhibitor, leading to cell division inhibition (Wang et al., 1998). Additionally, ABA treatment reduces CYCB1 expression while maintaining CDK expression, suggesting that ABA regulates cell division through cyclins and CDK inhibitors rather than directly affecting CDKs (Xu et al., 2010). Moreover, ABA's inhibition of root growth might involve modulation of ethylene synthesis or signaling, as blocking ethylene signaling partially restores root growth in aba2 mutants (Cheng et al., 2002).

Despite abundant evidence supporting ABA's inhibitory role in root growth, certain ABA-insensitive or ABA-deficient mutants exhibit slowed growth (Cheng et al., 2002; Lin et al., 2007; Fujii & Zhu, 2009). This discrepancy suggests that while exogenous ABA application inhibits growth, internally produced ABA might have different functions under normal conditions. The observations of retarded growth in ABA-insensitive or ABA-deficient mutants imply the necessity of basal ABA levels for maintaining typical root growth. This dual role of ABA in root growth—acting as an inhibitor at higher concentrations while potentially promoting growth at lower levels—illustrates its concentration-dependent regulation of root development.

ABA is closely linked to maintaining the quiescent state of Quiescent Center (QC) cells and inhibiting cell differentiation within the Stem Cell Niche (SCN) (Han et al., 2010; Zhang et al., 2010). In the root meristem, ABA's control over stem cell differentiation is mediated through the activity of WOX5. Overexpression of *WOX5* leads to the formation of additional layers of undifferentiated stem cells within the columella root cap (Sarkar et al., 2007). When ABA is externally applied, it further amplifies the number of stem cell layers induced by WOX5 (Han et al., 2010). Moreover, early investigations suggest a potential involvement of MONOPTEROS (MP) in regulating stem cell differentiation influenced by ABA within Arabidopsis root meristems (Zhang et al., 2010).

1.5.4 Brassinosteroid Action in Root Development

Brassinosteroids (BRs) are a class of plant hormones involved in various developmental processes, including root development (Vert et al., 2005; Ye et al., 2010; Gudesblat et al., 2012a, b; Wei & Li, 2016). BR, detected by the BRI1 receptor, activates the transcription factors BES1 and BZR1, which subsequently govern the transcription of a large number of genes (Yin et al., 2002; Yu et al., 2011). Mutants impaired in either BR synthesis or signaling cause a short-root phenotype, implying a critical role of BR in root growth and development (Li et al., 1996; Mussig et al., 2003; Lv et al., 2018). Besides promoting cell elongation, BR that regulates root growth is also mediated through the regulation of the cell cycle. The effects of BR on the cell cycle have been analyzed in BR mutants and BR-treated plants using cell cycle reporter lines. In *bri1* mutants, the immunofluorescence analysis of KNOLLE and misexpression of *CYCB1;1* and *ICK2/KRP2* revealed a reduced number of cell plates in the meristem of the root, pointing to an arrested or slower cell cycle. Defective root apical meristems in the *bri1* mutant were recovered by overexpression of the *CYCD3;1* gene (Gonzalez-Garcia et al., 2011). These results indicated that increased BR signaling can lead to an acceleration of the cell cycle during primary root growth. A recent study demonstrated that BR signaling affects neither cell volume nor cell proliferation capacity through combined digital 3D single-cell shape analysis and single-cell mRNA sequencing. They found that BR signaling is essential for the precise orientation of cell division planes and the extent and timing of anisotropic cell expansion (Graeff et al., 2021).

However, excessive application of bioactive BR affected normal development of plants (Zhu et al., 2013). Therefore, a finely tuned cellular regulation of BR level is important for plant development. It has been found that BR deficiency elicited the expression of BR biosynthesis genes, while an increase in endogenous BR levels leads to a feedback repression of BR metabolic genes to maintain the homeostasis of BR (Tanaka et al., 2005). Recent studies have unveiled the extensive interplay between BR and other plant hormones like abscisic acid, gibberellins, auxin, ethylene, and cytokinin, elucidating their collaborative role in governing plant growth and development (Hu & Yu, 2014; Tong et al., 2014; Chaiwanon & Wang, 2015; Lv et al., 2018). For instance, the functional analysis of BREVIS RADIX (BRX) has linked BR synthesis to auxin signaling during root growth (Mouchel et al., 2006). The coordinated actions of BRX, MONOPTEROS (MP), and SHORT HYPOCOTYL 2 (SHY2) establish a feedback loop that intricately regulates auxin transport mediated by PIN3, crucially influencing root meristem size (Scacchi et al., 2010). This network sheds light on the sophisticated mechanisms controlling root development at the molecular level (Scacchi et al., 2010).

1.5.5 Cytokinins Regulated Root Development

Cytokinins are N^6-prenylated adenine derivatives that are involved in a wide range of developmental and physiological processes, including root growth, vascular differentiation, lateral root initiation, and response to various biotic and abiotic stresses (Frugier et al., 2008; Argueso et al., 2010; Perilli et al., 2010). Numerous studies have demonstrated that applying external cytokinins inhibits root elongation, whereas a decrease in internal cytokinin levels correlates with enhanced primary root elongation (Beemster & Baskin, 1998; Werner et al., 2003). These findings strongly suggest that cytokinins function as suppressors of root development.

Exogenous application of cytokinins inhibits root growth by reducing the size of the root meristem without altering cell division and stem cell activity in the meristem zone (Dello Ioio et al., 2007). In mutants impaired in cytokinin biosynthesis and signaling, meristems exhibit enlargement, thereby enhancing root growth (Dello Ioio et al., 2007, 2008). Manipulating cytokinin levels specifically within the vascular tissue at the transition zone through the overexpression of CKX, cytokinin oxidase/dehydrogenase enzymes, is adequate to impact meristem size. However, reducing cytokinin levels in other regions of the root tissue does not yield a similar effect (Dello Ioio et al., 2007). These findings strongly suggest that the cytokinin signaling pathway operates distinctly within the vascular tissue at the transition zone, playing a pivotal role in regulating root meristem size and root growth.

It has been demonstrated that root meristem size is controlled by an increase in the rate of cell differentiation relative to cell division. The balance between cell differentiation and cell division is maintained by the expression of *SHY2*, activated by both ARR1 and ARR12 transcription factors (Moubayidin et al., 2010). SHY2 expression peaks around the 5-day mark postgermination, coinciding with the maximum impact of the SHY2 protein in diminishing and prematurely establishing root meristem size. This suggests that a maximum of SHY2 is important to determine the final size of the root meristem (Moubayidin et al., 2010).

Furthermore, cytokinins have been found to play a key role in specifying the root stem cell niche during the early stages of *Arabidopsis* embryo development (Muller & Sheen, 2008). It has been observed that reduced activity of both ARR7 and ARR15, which are repressors of cytokinin signaling, leads to ectopic phosphorelay output in the basal daughter cell, thereby preventing the establishment of a normal embryonic root stem cell niche (Muller & Sheen, 2008).

Cytokinins have emerged as inhibitors of lateral root formation. Introducing external cytokinins inhibits the initiation of lateral roots in *Arabidopsis* (Li et al., 2006; Kuderova et al., 2008). Notably, plants overexpressing *CKX* and certain type-B *ARR* and *AHK* mutants display heightened formation of lateral roots. Conversely, mutations in type-A ARR genes correlate with a reduction in the number of lateral roots (Werner et al., 2001, 2003; To et al., 2004; Mason et al., 2005; Riefler et al., 2006). The xylem pole pericycle is identified as the specific site where cytokinins exert their inhibitory effect on lateral root initiation. *IPT*s expressed in these cells have been shown to impede lateral root growth, while enhancing this process can be achieved through the expression of the *CKX1* gene in xylem pole pericycle cells (Laplaze et al., 2007).

Actually, each hormone is not involved in root growth regulation independently; there are very complex hormone cross talks in plants. For instance, the interaction between auxin and cytokinin has been observed to control root meristem size, with auxin and cytokinin exhibiting antagonistic behavior (Dello Ioio et al., 2007). Cytokinin-responsive transcription factors like ARR1 and ARR12 activate the *SHY2* gene in the root apex transition zone (TZ). SHY2, in turn, negatively regulates *PIN* genes responsible for auxin transport facilitation, thus establishing connections with auxin signaling to modulate root growth (Dello Ioio et al., 2008; Moubayidin et al., 2010). Additionally, ethylene influences root growth by regulating auxin biosynthesis or PIN-dependent polar auxin transport (Ruzicka et al., 2007; Swarup et al., 2007). The plant stress hormone jasmonate (JA) controls lateral root formation by modulating ASA1-dependent auxin biosynthesis and PIN-dependent auxin transport (Sun et al., 2009). Furthermore, JA modulates the identity of the root stem cell niche through MYC2-mediated repression of *PLT*s, which play a crucial role in the interaction between jasmonate and auxin for regulating root stem cell niche maintenance (Chen et al., 2011).

1.6 Root Response to Abiotic Stress

Abiotic stresses like salinity, drought, flooding, and cold pose significant challenges to plant growth and metabolism, profoundly impacting overall plant health. Among the various organ-level responses to abiotic stress in plants, root-specific processes hold particular importance. In order to mitigate and recover from the damaging effects of such adverse environmental conditions on root growth, plants have developed adaptive strategies at cellular and metabolic levels. Next, we will discuss the development of the root under different stress conditions.

1.6.1 Salinity Effects on Root Growth

Soil salinity stands as a prominent limitation to plant proliferation in natural environments. This issue is particularly prevalent in arid and semi-arid regions compared to more humid areas, becoming an escalating concern in these dry climates. The primary impact of salinity stress is the inhibition of root development, impacting not only the growth but also the morphology, anatomy, and physiology of roots. Moreover, it interferes with the root's ability to uptake water and crucial minerals.

For numerous plant species, salinity stress severely hampers root elongation, impeding overall root system development (Cramer et al., 1987; Neumann et al., 1994; West et al., 2004).

Under salinity stress, the inhibition of root elongation is linked to its impact on various cellular processes: cell division rate, duration of cell growth, and cell expansion rate. Studies have shown that salinity stress inhibits root cell proliferation, reducing the overall number of cells in roots (Kurth et al., 1986; Samarajeewa et al., 1999). This stress can also trigger deformation and degradation of the nucleus in meristematic root cells, leading to cell death (Katsuhara & Kawasaki, 1996). There's a suggested correlation between cell cycle regulation and tolerance to salt stress. Salinity can diminish the activity of cyclin-dependent kinase (CDK) and mitotic cyclin CYCB1;2, essential for cell cycle progression. Decreased transcription levels of genes like CDC2a and CycA2;1 in the root vascular cylinder under salinity stress correspond with the inhibition of root growth (Burssens et al., 2000).

Salt stress doesn't just influence cell division; it impacts cell expansion too. While mature root cell size diminishes, leading to shorter roots, the duration of root cell expansion increases under salt stress. This means that, despite the reduced growth zone, stressed cells elongate for longer periods than nonstressed cells. Moreover, salinity stress alters the growth orientation of root cells, resulting in thicker but shorter roots (Zhong & Lauchli, 1993; Huang & Redmann, 1995; Franco et al., 2011).

Salinity stress also alters root tissue patterning, stimulating early development of root endodermis and inducing exodermis formation (Reinhardt & Rost, 1995). Its primary impact is on dividing and expanding cells within the root. Understanding the spatial and temporal dynamics of root cell growth under salinity stress will provide insight into the complex effects of this stress on overall plant behavior.

1.6.2 Root Development in Drying Soil

The ongoing climate change is characterized by increased temperatures and altered precipitation patterns. In addition, the frequency, intensity, and duration of extreme climatic events, such as droughts, floods, and storms, have increased in recent decades. Drought can occur in all climatic regions, including the high and low rainfall areas. In general, drought stress is mainly related to a decline in rainfall over a period of time, high wind speeds, high temperature, and low relative humidity (Mishra & Singh, 2010).

Plants employ different strategies to cope with drought stress, including responses aimed at reducing water loss, sustaining water uptake at low soil moisture levels, and minimizing the negative impact of drought on growth and reproductive fitness (Verslues et al., 2006; Lawlor, 2013). Among these adaptive responses, the plasticity in root growth emerges as pivotal during drought stress, facilitating sustained water and nutrient absorption (Kano et al., 2011). In certain instances, mild drought stress has been noted to promote root growth (Kano et al., 2011). Studies reveal a robust positive correlation between root length and water extraction, particularly in deep, moist soil layers (Kamoshita et al., 2004). Varieties of plants exhibiting resistance to drought, with more compact yet extensive root systems, have demonstrated higher yields compared to standard varieties under drought conditions (Manschadi et al., 2006). Augmented root growth and increased rooting depth stand as pivotal traits in the pursuit of a drought-resistant plant ideotype (Zhang et al., 2009; Tardieu, 2012). However, it's essential to note that increased root growth or deeper root systems may offer limited benefits when deep soil layers hold little water. Under such circumstances, heightened root growth can rapidly deplete the remaining extractable water, eventually leading to a significant inhibition in root growth (Kamoshita et al., 2004; Tardieu, 2012). Research suggests that while mild drought can stimulate total root biomass and length, severe drought conditions inhibit their growth (Creelman et al., 1990). Therefore, the growth of the root may respond differently to different degrees of drought stress, but the mechanisms underlying this complex interaction are still not well understood.

1.6.3 Root Growth Under Aluminum Stress

Aluminum (Al) is one of the most abundant metals in the Earth's crust. Al toxicity is a significant issue that limits agricultural production in acid soils. Acidic soils are widespread, especially in the tropical and subtropical regions (Kochian et al., 2004). It has been reported that up to 50% of potentially cultivated crops worldwide are negatively affected by acid soils (Von Uexkull & Mutert, 1995). Acid soils inhibit crop production in many developing countries where the production of food is crucial. Inhibition of root elongation is the first visible phenotype of Al stress. The root apex (root cap, meristem, distal transition zone, and elongation zone) is the major target site of Al toxicity (Ryan & Kochian, 1993; Matsumoto, 2000). The distal part of the root-apex transition zone (TZ), located between the apical meristem and the basal elongation region, is the most Al-sensitive part of the root (Sivaguru & Horst, 1998; Illes et al., 2006; Rangel et al., 2007; Kong et al., 2018). Many studies have shown that the root epidermis may rupture or crack when exposed to Al stress (Yamamoto et al., 2001). An increase in the diameter and a decrease in the cell length in the outer layer of the cortex in the elongation zone is caused by Al stress, consistent with the shortening of the root elongation zone with the formation of cracks (Sasaki et al., 1996). Under Al stress, a notable inhibition in the production and release of border cells was observed. Roots subjected to Al stress exhibited a significantly lower count of released border cells compared to roots grown without such stress (Zhu et al., 2003).

Hormone signaling plays a crucial role in regulating root growth under Al stress. The inhibition of root growth caused by Al stress is clearly regulated by auxin (Kollmeier et al., 2000; Sun et al., 2010). It has been revealed that root growth is inhibited by Al stress through the expression of auxin response factors (ARFs) such as ARF10 and ARF16 (Wang et al., 2005; Liu et al., 2007). It was also found that the signaling of auxin between the root-apex transition zone and elongation zone determines the extent of root growth inhibition caused by Al stress. The exogenous supply of the auxin indole-3-acetic acid (IAA) to the elongation zone significantly reduces the extent of root elongation inhibition caused by Al stress, but the same treatment given to the meristematic zone has no such effect (Kollmeier et al., 2000). This indicated that the transition zone of the root apex is the perception site of Al stress.

Early research on *Arabidopsis* roots exposed to Al stress found that it induces a localized enhancement of auxin signaling in the root-apex transition zone, which is dependent on TAA1 and YUCs, enzymes involved in auxin biosynthesis (Yang et al., 2014b; Liu et al., 2016). In response to Al stress, ethylene signaling locally upregulates the expression of *TAA1* and *YUC* genes, thus increasing auxin responses in the transition zone and resulting in the inhibition of root growth regulated by auxin through several ARFs (Yang et al., 2014b). Local regulation of auxin transport in the root-apex transition zone also mediates Al-induced root growth inhibition (Li et al., 2021). In addition, cytokinin has also been shown to enhance root-growth inhibition under Al stress through mediating Al-induced auxin signaling (Yang et al., 2017). Al stress induces a local up-regulation of cytokinin response in the root-apex transition zone (TZ) that is dependent on adenosine phosphate isopentenyl transferases encoding genes *IPT*s, which are specifically upregulated in the root-apex TZ through ARF7-mediated auxin signaling and ethylene signaling, which acts upstream of auxin under Al stress. In summary, Al rewires the cross talk between auxin and cytokinin, and auxin synergistically acts with cytokinin to regulate Al-induced root growth inhibition in *Arabidopsis* (Fig. 1.3) (Yang et al., 2017).

1.6.4 The Role of Nutrient Availability in Root Growth

Nutrients in the soil play a critical role in plant growth and productivity. The availability of nutrients in the soil solution has a significant impact on root growth. Nutrients such as nitrogen (N), phosphorus (P), iron (Fe), and sulfur (S) have been reported to influence postembryonic root developmental processes. Notably, nutrient deficiencies often trigger specific alterations in root growth, such as the induction of root hairs. Recent research highlights that a lack of phosphorus not only impacts root hair elongation but also amplifies the count of root hairs (Ma et al., 2001). Similarly, low iron availability elicits morphological changes in root hairs akin to those seen in phosphorus-deficient conditions (Schmidt et al., 2000).

Furthermore, nutrient availability affects both root elongation and lateral root formation. Although nitrogen is one of the most abundant elements on earth, its availability in the soil is low, and it is easily lost by leaching or by bacterial denitrification. Therefore, nitrogen is the most critically limiting element for plant growth (Vance, 2001). In *Arabidopsis*, increasing nitrogen availability reduces the primary root elongation, whereas the lateral roots show two contrasting responses to the high nitrate. High concentration of nitrate (10 mM) reduces the elongation of lateral root throughout the root system, whereas in plants grown on a low concentration of nitrate (10 μM), localized application of a high nitrate treatment can induce lateral root elongation (Zhang & Forde, 1998). A NO_3^--inducible *Arabidopsis* gene *ARABIDOPSIS NITRATE-REGULATED1* (*ANR1*) that regulates root system architecture in response to nitrogen was previously identified. The *ANR1* gene encodes a NO_3^--inducible MADS-box transcription factor, and repression of ANR1 had an altered sensitivity to nitrate and a lack of response to NO_3^--rich zones by lateral root proliferation, indicating that ANR1 is an essential determinant of developmental plasticity in *Arabidopsis* roots in response to nitrate (Zhang & Forde, 1998).

Phosphorus is the second most critical factor in determining root growth. Under phosphorus-limiting conditions, the number of lateral roots was up to five times greater than that in plants supplied with optimal phosphorus (Lopez-Bucio et al., 2002). Despite advancements in understanding gene expression during phosphorus deficiency, the signal transduction pathway components initiating the phosphorus starvation response remain relatively obscure. A conserved MYB transcription factor, PHOSPHATE RESPONSE REGULATOR1 (PHR1), has been identified as a crucial participant in phosphate starvation signaling. Mutations in *phr1* alter the root-to-shoot growth ratio under phosphorus scarcity and disrupt the expression of genes regulated by low phosphorus levels. Through in vitro DNA-binding assays like gel-retardation, PHR1 has demonstrated its capacity to interact with DNA sequences from promoters regulated by low phosphorus levels. This supports the notion of PHR1's involvement in the phosphorus starvation signaling cascade (Rubio et al., 2001).

Under sulfate starvation conditions in *Arabidopsis* seedlings, lateral roots tend to form closer to the root tip and have increased density (Lopez-Bucio et al., 2003). This increased growth of the root system under limited sulfate conditions associates with the transcriptional activation of the *NITRILASE3* (*NIT3*) gene, a member of the nitrilase gene family that encodes enzymes catalyzing the conversion of indole-3-acetonitrile to indole-3-acetic acid. Concurrently, roots subjected to sulfate deficiency showcase increased levels of auxin, suggesting a direct involvement of NIT3 in auxin synthesis and consequent root branching (Kutz et al., 2002).

1.7 Concluding Remarks

Over the past decade, significant strides have been made in the research on plant root systems. However, the current understanding of molecular regulatory mechanisms in root systems predominantly stems from studies conducted on model plants like *Arabidopsis thaliana*. Despite substantial advancements in understanding root systems in crops like rice and maize, there remains a considerable gap to fulfill the needs for future crop breeding based on "ideal root architectures." Given the pivotal significance of integrating "ideal plant architecture" and "ideal root architecture" in future crop breeding strategies, there's a pressing need to intensify research efforts focused on crop root systems. Understanding the molecular regulatory mechanisms underlying different root architectures in crops and deciphering their commonalities and disparities is particularly important. For example, unraveling the regulatory mechanisms governing adventitious and lateral root development in cereal crops is essential. These roots serve as primary organs for water and nutrient uptake in the later growth stages of cereal crops, and a deeper understanding of their molecular regulation is crucial for guiding crop breeding based on "ideal root architectures." Furthermore, the terminal growth of the primary root during late growth stages in cereal crops is a characteristic feature, accompanied by the emergence of lateral roots and various adventitious roots. Therefore, elucidating the mechanisms behind the terminal growth of the primary root in cereal crops during the later growth stages is vital. Investigating the regulatory mechanisms behind this process forms a crucial foundation for understanding crop root development.

References

Aida, M., Beis, D., Heidstra, R., Willemsen, V., Blilou, I., Galinha, C., Nussaume, L., Noh, Y. S., Amasino, R., & Scheres, B. (2004). The PLETHORA genes mediate patterning of the Arabidopsis root stem cell niche. *Cell, 119*, 109–120.

Argueso, C. T., Raines, T., & Kieber, J. J. (2010). Cytokinin signaling and transcriptional networks. *Current Opinion in Plant Biology, 13*, 533–539.

Barry, C. S., & Giovannoni, J. J. (2006). Ripening in the tomato Green-ripe mutant is inhibited by ectopic expression of a protein that disrupts ethylene signaling. *Proceedings of the National Academy of Sciences of the United States of America, 103*, 7923–7928.

Bates, T. R., & Lynch, J. P. (2000). The efficiency of Arabidopsis thaliana (Brassicaceae) root hairs in phosphorus acquisition. *American Journal of Botany, 87*, 964–970.

Baum, S. F., Dubrovsky, J. G., & Rost, T. L. (2002). Apical organization and maturation of the cortex and vascular cylinder in Arabidopsis thaliana (Brassicaceae) roots. *American Journal of Botany, 89*, 908–920.

Beeckman, T., Burssens, S., & Inze, D. (2001). The peri-cell-cycle in Arabidopsis. *Journal of Experimental Botany, 52*, 403–411.

Beemster, G. T., & Baskin, T. I. (1998). Analysis of cell division and elongation underlying the developmental acceleration of root growth in Arabidopsis thaliana. *Plant Physiology, 116*, 1515–1526.

Beemster, G. T. S., Fiorani, F., & Inze, D. (2003). Cell cycle: The key to plant growth control? *Trends in Plant Science, 8*, 154–158.

Benfey, P. N., Linstead, P. J., Roberts, K., Schiefelbein, J. W., Hauser, M. T., & Aeschbacher, R. A. (1993). Root development in Arabidopsis: Four mutants with dramatically altered root morphogenesis. *Development, 119*, 57–70.

Bennett, T., van den Toorn, A., Sanchez-Perez, G. F., Campilho, A., Willemsen, V., Snel, B., & Scheres, B. (2010). SOMBRERO, BEARSKIN1, and BEARSKIN2 regulate root cap maturation in Arabidopsis. *Plant Cell, 22*, 640–654.

Berckmans, B., Lammens, T., Van Den Daele, H., Magyar, Z., Bogre, L., & De Veylder, L. (2011). Light-dependent regulation of DEL1 is determined by the antagonistic action of E2Fb and E2Fc. *Plant Physiology, 157*, 1440–1451.

Bertolotti, G., Unterholzner, S. J., Scintu, D., Salvi, E., Svolacchia, N., Di Mambro, R., Ruta, V., Scaglia, F. L., Vittorioso, P., Sabatini, S., Costantino, P., & Dello Ioio, R. (2021). A PHABULOSA-controlled genetic pathway regulates ground tissue patterning in the root. *Current Biology, 31*, 420–426.e6.

Bishopp, A., El-Showk, S., & Helariutta, Y. (2013). Vascular development in Arabidopsis roots. In *Plant roots: The hidden half* (4th ed.). CRC Press.

Bonke, M., Thitamadee, S., Mahonen, A. P., Hauser, M. T., & Helariutta, Y. (2003). APL regulates vascular tissue identity in Arabidopsis. *Nature, 426*, 181–186.

Boyce, K. (2005). The evolutionary history of roots and leaves. In N. M. Holbrook & M. A. Zwieniecki (Eds.), *Vascular transport in plants* (pp. 479–500). Elsevier.

Buer, C. S., Wasteneys, G. O., & Masle, J. (2003). Ethylene modulates root-wave responses in Arabidopsis. *Plant Physiology, 132*, 1085–1096.

Buer, C. S., Sukumar, P., & Muday, G. K. (2006). Ethylene modulates flavonoid accumulation and gravitropic responses in roots of Arabidopsis. *Plant Physiology, 140*, 1384–1396.

Burssens, S., Himanen, K., van de Cotte, B., Beeckman, T., Van Montagu, M., Inze, D., & Verbruggen, N. (2000). Expression of cell cycle regulatory genes and morphological alterations in response to salt stress in Arabidopsis thaliana. *Planta, 211*, 632–640.

Carlsbecker, A., Lee, J. Y., Roberts, C. J., Dettmer, J., Lehesranta, S., Zhou, J., Lindgren, O., Moreno-Risueno, M. A., Vaten, A., Thitamadee, S., Campilho, A., Sebastian, J., Bowman, J. L., Helariutta, Y., & Benfey, P. N. (2010). Cell signalling by microRNA165/6 directs gene dose-dependent root cell fate. *Nature, 465*, 316–321.

Carol, R. J., Takeda, S., Linstead, P., Durrant, M. C., Kakesova, H., Derbyshire, P., Drea, S., Zarsky, V., & Dolan, L. (2005). A RhoGDP dissociation inhibitor spatially regulates growth in root hair cells. *Nature, 438*, 1013–1016.

Casero, P. J., Casimiro, I., & Lloret, P. G. (1995). Lateral root initiation by asymmetrical transverse divisions of pericycle cells in 4 plant-species - Raphanus-sativus, Helianthus annuus, Zea-mays, and Daucus-carota. *Protoplasma, 188*, 49–58.

Chaffey, N., Cholewa, E., Regan, S., & Sundberg, B. (2002). Secondary xylem development in Arabidopsis: A model for wood formation. *Physiologia Plantarum, 114*, 594–600.

Chaiwanon, J., & Wang, Z. Y. (2015). Spatiotemporal brassinosteroid signaling and antagonism with auxin pattern stem cell dynamics in Arabidopsis roots. *Current Biology, 25*, 1031–1042.

Chen, Q., Sun, J. Q., Zhai, Q. Z., Zhou, W. K., Qi, L. L., Xu, L., Wang, B., Chen, R., Jiang, H. L., Qi, J., Li, X. G., Palme, K., & Li, C. Y. (2011). The basic helix-loop-helix transcription factor MYC2 directly represses PLETHORA expression during Jasmonate-mediated modulation of the root stem cell niche in Arabidopsis. *Plant Cell, 23*, 3335–3352.

Cheng, W. H., Endo, A., Zhou, L., Penney, J., Chen, H. C., Arroyo, A., Leon, P., Nambara, E., Asami, T., Seo, M., Koshiba, T., & Sheen, J. (2002). A unique short-chain dehydrogenase/reductase in Arabidopsis glucose signaling and abscisic acid biosynthesis and functions. *Plant Cell, 14*, 2723–2743.

Cramer, G. R., Lynch, J., Lauchli, A., & Epstein, E. (1987). Influx of Na+, K+, and Ca-2+ into roots of salt-stressed cotton seedlings - effects of supplemental Ca-2+. *Plant Physiology, 83*, 510–516.

Crawford, B. C. W., Sewell, J., Golembeski, G., Roshan, C., Long, J. A., & Yanofsky, M. F. (2015). Genetic control of distal stem cell fate within root and embryonic meristems. *Science, 347*, 655–659.

Creelman, R. A., Mason, H. S., Bensen, R. J., Boyer, J. S., & Mullet, J. E. (1990). Water deficit and abscisic-acid cause differential inhibition of shoot versus root-growth in soybean seedlings - analysis of growth, sugar accumulation, and gene-expression. *Plant Physiology, 92*, 205–214.

Cui, H. C., Levesque, M. P., Vernoux, T., Jung, J. W., Paquette, A. J., Gallagher, K. L., Wang, J. Y., Blilou, I., Scheres, B., & Benfey, P. N. (2007). An evolutionarily conserved mechanism delimiting SHR movement defines a single layer of endodermis in plants. *Science, 316*, 421–425.

De Rybel, B., Vassileva, V., Parizot, B., Demeulenaere, M., Grunewald, W., Audenaert, D., Van Campenhout, J., Overvoorde, P., Jansen, L., Vanneste, S., Moller, B., Wilson, M., Holman, T., Van Isterdael, G., Brunoud, G., Vuylsteke, M., Vernoux, T., De Veylder, L., Inze, D., Weijers, D., Bennett, M. J., & Beeckman, T. (2010). A novel aux/IAA28 signaling cascade activates GATA23-dependent specification of lateral root founder cell identity. *Current Biology, 20*, 1697–1706.

De Rybel, B., Adibi, M., Breda, A. S., Wendrich, J. R., Smit, M. E., Novak, O., Yamaguchi, N., Yoshida, S., Van Isterdael, G., Palovaara, J., Nijsse, B., Boekschoten, M. V., Hooiveld, G., Beeckman, T., Wagner, D., Ljung, K., Fleck, C., & Weijers, D. (2014). Plant development. Integration of growth and patterning during vascular tissue formation in Arabidopsis. *Science, 345*, 1255215.

De Smet, I., Vanneste, S., Inze, D., & Beeckman, T. (2006). Lateral root initiation or the birth of a new meristem. *Plant Molecular Biology, 60*, 871–887.

De Smet, I., Tetsumura, T., De Rybel, B., Frey, N. F. D., Laplaze, L., Casimiro, I., Swarup, R., Naudts, M., Vanneste, S., Audenaert, D., Inze, D., Bennett, M. J., & Beeckman, T. (2007). Auxin-dependent regulation of lateral root positioning in the basal meristem of Arabidopsis. *Development, 134*, 681–690.

De Smet, I., Lau, S., Voss, U., Vanneste, S., Benjamins, R., Rademacher, E. H., Schlereth, A., De Rybel, B., Vassileva, V., Grunewald, W., Naudts, M., Levesque, M. P., Ehrismann, J. S., Inze, D., Luschnig, C., Benfey, P. N., Weijers, D., Van Montagu, M. C. E., Bennett, M. J., Jurgens, G., & Beeckman, T. (2010). Bimodular auxin response controls organogenesis in Arabidopsis. *Proceedings of the National Academy of Sciences of the United States of America, 107*, 2705–2710.

Dello Ioio, R., Linhares, F. S., Scacchi, E., Casamitjana-Martinez, E., Heidstra, R., Costantino, P., & Sabatini, S. (2007). Cytokinins determine Arabidopsis root-meristem size by controlling cell differentiation. *Current Biology, 17*, 678–682.

Dello Ioio, R., Nakamura, K., Moubayidin, L., Perilli, S., Taniguchi, M., Morita, M. T., Aoyama, T., Costantino, P., & Sabatini, S. (2008). A genetic framework for the control of cell division and differentiation in the root meristem. *Science, 322*, 1380–1384.

Dickinson, A. J., Zhang, J. Y., Luciano, M., Wachsman, G., Sandoval, E., Schnermann, M., Dinneny, J. R., & Benfey, P. N. (2021). A plant lipocalin promotes retinal-mediated oscillatory lateral root initiation. *Science, 373*, 1532–1536.

Dietrich, D., Pang, L., Kobayashi, A., Fozard, J. A., Boudolf, V., Bhosale, R., Antoni, R., Nguyen, T., Hiratsuka, S., Fujii, N., Miyazawa, Y., Bae, T. W., Wells, D. M., Owen, M. R., Band, L. R., Dyson, R. J., Jensen, O. E., King, J. R., Tracy, S. R., Sturrock, C. J., Mooney, S. J., Roberts, J. A., Bhalerao, R. P., Dinneny, J. R., Rodriguez, P. L., Nagatani, A., Hosokawa, Y., Baskin, T. I., Pridmore, T. P., De Veylder, L., Takahashi, H., & Bennett, M. J. (2017). Root hydrotropism is controlled via a cortex-specific growth mechanism. *Nature Plants, 3*, 17057.

DiLaurenzio, L., WysockaDiller, J., Malamy, J. E., Pysh, L., Helariutta, Y., Freshour, G., Hahn, M. G., Feldmann, K. A., & Benfey, P. N. (1996). The SCARECROW gene regulates an asymmetric cell division that is essential for generating the radial organization of the Arabidopsis root. *Cell, 86*, 423–433.

Ding, Z., & Friml, J. (2010). Auxin regulates distal stem cell differentiation in Arabidopsis roots. *Proceedings of the National Academy of Sciences of the United States of America, 107*, 12046–12051.

Ditengou, F. A., Teale, W. D., Kochersperger, P., Flittner, K. A., Kneuper, I., van der Graaff, E., Nziengui, H., Pinosa, F., Li, X., Nitschke, R., Laux, T., & Palme, K. (2008). Mechanical induction of lateral root initiation in Arabidopsis thaliana. *Proceedings of the National Academy of Sciences of the United States of America, 105*, 18818–18823.

Dolan, L., Janmaat, K., Willemsen, V., Linstead, P., Poethig, S., Roberts, K., & Scheres, B. (1993). Cellular-organization of the Arabidopsis-Thaliana root. *Development, 119*, 71–84.

Dolan, L., Duckett, C. M., Grierson, C., Linstead, P., Schneider, K., Lawson, E., Dean, C., Poethig, S., & Roberts, K. (1994). Clonal relationships and cell patterning in the root epidermis of Arabidopsis. *Development, 120*, 2465–2474.

Duan, X. L., Xu, S., Xie, Y. M., Li, L., Qi, W. C., Parizot, B., Zhang, Y. H., Chen, T., Han, Y., Van Breusegem, F., Beeckman, T., Shen, W. B., & Xuan, W. (2021). Periodic root branching is influenced by light through an HY1-HY5-auxin pathway. *Current Biology, 31*, 3834–3847.e5.

Dubrovsky, J. G., Rost, T. L., Colon-Carmona, A., & Doerner, P. (2001). Early primordium morphogenesis during lateral root initiation in Arabidopsis thaliana. *Planta, 214*, 30–36.

Escamez, S., Andre, D., Sztojka, B., Bollhoner, B., Hall, H., Berthet, B., Voss, U., Lers, A., Maizel, A., Andersson, M., Bennett, M., & Tuominen, H. (2020). Cell death in cells overlying lateral root primordia facilitates organ growth in Arabidopsis. *Current Biology, 30*(455–464), e457.

Fendrych, M., Akhmanova, M., Merrin, J., Glanc, M., Hagihara, S., Takahashi, K., Uchida, N., Torii, K. U., & Friml, J. (2018). Rapid and reversible root growth inhibition by TIR1 auxin signalling. *Nature Plants, 4*, 453–459.

Fernandez, A. I., Vangheluwe, N., Xu, K., Jourquin, J., Claus, L. A. N., Morales-Herrera, S., Parizot, B., De Gernier, H., Yu, Q. Z., Drozdzecki, A., Maruta, T., Hoogewijs, K., Vannecke, W., Peterson, B., Opdenacker, D., Madder, A., Nimchuk, Z. L., Russinova, E., & Beeckman, T. (2020). GOLVEN peptide signalling through RGI receptors and MPK6 restricts asymmetric cell division during lateral root initiation. *Nature Plants, 6*, 533–543.

Ferreira, P. C., Hemerly, A. S., Engler, J. D., van Montagu, M., Engler, G., & Inze, D. (1994). Developmental expression of the arabidopsis cyclin gene cyc1At. *Plant Cell, 6*, 1763–1774.

Fischer, U., Ikeda, Y., & Grebe, M. (2007). Planar polarity of root hair positioning in Arabidopsis. *Biochemical Society Transactions, 35*, 149–151.

Forzani, C., Aichinger, E., Sornay, E., Willemsen, V., Laux, T., Dewitte, W., & Murray, J. A. (2014). WOX5 suppresses CYCLIN D activity to establish quiescence at the center of the root stem cell niche. *Current Biology, 24*, 1939–1944.

Franco, J. A., Cros, V., Vicente, M. J., & Martinez-Sanchez, J. J. (2011). Effects of salinity on the germination, growth, and nitrate contents of purslane (Portulaca oleracea L.) cultivated under different climatic conditions. *Journal of Horticultural Science and Biotechnology, 86*, 1–6.

Frugier, F., Kosuta, S., Murray, J. D., Crespi, M., & Szczyglowski, K. (2008). Cytokinin: Secret agent of symbiosis. *Trends in Plant Science, 13*, 115–120.

Fujii, H., & Zhu, J. K. (2009). Arabidopsis mutant deficient in 3 abscisic acid-activated protein kinases reveals critical roles in growth, reproduction, and stress. *Proceedings of the National Academy of Sciences of the United States of America, 106*, 8380–8385.

Fukaki, H., Tameda, S., Masuda, H., & Tasaka, M. (2002). Lateral root formation is blocked by a gain-of-function mutation in the SOLITARY-ROOT/IAA14 gene of Arabidopsis. *The Plant Journal, 29*, 153–168.

Gahoonia, T. S., & Nielsen, N. E. (2004). Barley genotypes with long root hairs sustain high grain yields in low-P field. *Plant and Soil, 262*, 55–62.

Gahoonia, T. S., Care, D., & Nielsen, N. E. (1997). Root hairs and phosphorus acquisition of wheat and barley cultivars. *Plant and Soil, 191*, 181–188.

Galinha, C., Hofhuis, H., Luijten, M., Willemsen, V., Blilou, I., Heidstra, R., & Scheres, B. (2007). PLETHORA proteins as dose-dependent master regulators of Arabidopsis root development. *Nature, 449*, 1053–1057.

Gallagher, K. L., Paquette, A. J., Nakajima, K., & Benfey, P. N. (2004). Mechanisms regulating SHORT-ROOT intercellular movement. *Current Biology, 14*, 1847–1851.

Gonzalez-Carranza, Z. H., Elliott, K. A., & Roberts, J. A. (2007a). Expression of polygalacturonases and evidence to support their role during cell separation processes in Arabidopsis thaliana. *Journal of Experimental Botany, 58*, 3719–3730.

Gonzalez-Carranza, Z. H., Rompa, U., Peters, J. L., Bhatt, A. M., Wagstaff, C., Stead, A. D., & Roberts, J. A. (2007b). HAWAIIAN SKIRT: An F-box gene that regulates organ fusion and growth in Arabidopsis. *Plant Physiology, 144*, 1370–1382.

Gonzalez-Garcia, M. P., Vilarrasa-Blasi, J., Zhiponova, M., Divol, F., Mora-Garcia, S., Russinova, E., & Cano-Delgado, A. I. (2011). Brassinosteroids control meristem size by promoting cell cycle progression in Arabidopsis roots. *Development, 138*, 849–859.

Graeff, M., Rana, S., Wendrich, J. R., Dorier, J., Eekhout, T., Fandino, A. C. A., Guex, N., Bassel, G. W., De Rybel, B., & Hardtke, C. S. (2021). A single-cell morpho-transcriptomic map of brassinosteroid action in the root. *Molecular Plant, 14*, 1985–1999.

Grieneisen, V. A., Xu, J., Maree, A. F. M., Hogeweg, P., & Scheres, B. (2007). Auxin transport is sufficient to generate a maximum and gradient guiding root growth. *Nature, 449*, 1008–1013.

Gudesblat, G. E., Betti, C., & Russinova, E. (2012a). Brassinosteroids tailor stomatal production to different environments. *Trends in Plant Science, 17*, 685–687.

Gudesblat, G. E., Schneider-Pizon, J., Betti, C., Mayerhofer, J., Vanhoutte, I., van Dongen, W., Boeren, S., Zhiponova, M., de Vries, S., Jonak, C., & Russinova, E. (2012b). SPEECHLESS integrates brassinosteroid and stomata signalling pathways. *Nature Cell Biology, 14*, 548–554.

Gujas, B., Cruz, T. M. D., Kastanaki, E., Vermeer, J. E. M., Munnik, T., & Rodriguez-Villalon, A. (2017). Perturbing phosphoinositide homeostasis oppositely affects vascular differentiation in Arabidopsis thaliana roots. *Development, 144*(19), 3578–3589.

Han, W., Zhang, H., & Wang, M. H. (2010). Fluridone affects quiescent centre division in the Arabidopsis thaliana root stem cell niche. *BMB Reports, 43*, 813–817.

Han, X., Zhang, M. H., Yang, M. L., & Hu, Y. R. (2020). Arabidopsis JAZ proteins interact with and suppress RHD6 transcription factor to regulate Jasmonate-stimulated root hair development. *Plant Cell, 32*, 1049–1062.

Heidstra, R., Welch, D., & Scheres, B. (2004). Mosaic analyses using marked activation and deletion clones dissect Arabidopsis SCARECROW action in asymmetric cell division. *Genes & Development, 18*, 1964–1969.

Helariutta, Y., Fukaki, H., Wysocka-Diller, J., Nakajima, K., Jung, J., Sena, G., Hauser, M. T., & Benfey, P. N. (2000). The SHORT-ROOT gene controls radial patterning of the Arabidopsis root through radial signaling. *Cell, 101*, 555–567.

Hemerly, A. S., Ferreira, P., de Almeida Engler, J., Van Montagu, M., Engler, G., & Inze, D. (1993). cdc2a expression in Arabidopsis is linked with competence for cell division. *Plant Cell, 5*, 1711–1723.

Heyman, J., Cools, T., Vandenbussche, F., Heyndrickx, K. S., Van Leene, J., Vercauteren, I., Vanderauwera, S., Vandepoele, K., De Jaeger, G., Van Der Straeten, D., & De Veylder, L. (2013). ERF115 controls root quiescent center cell division and stem cell replenishment. *Science, 342*, 860–863.

Himanen, K., Boucheron, E., Vanneste, S., Engler, J. D., Inze, D., & Beeckman, T. (2002). Auxin-mediated cell cycle activation during early lateral root initiation. *Plant Cell, 14*, 2339–2351.

Himanen, K., Vuylsteke, M., Vanneste, S., Vercruysse, S., Boucheron, E., Alard, P., Chriqui, D., Van Montagu, M., Inze, D., & Beeckman, T. (2004). Transcript profiling of early lateral root initiation. *Proceedings of the National Academy of Sciences of the United States of America, 101*, 5146–5151.

Hodge, A., Berta, G., Doussan, C., Merchan, F., & Crespi, M. (2009). Plant root growth, architecture and function. *Plant and Soil, 321*, 153–187.

Hu, Y., & Yu, D. (2014). BRASSINOSTEROID INSENSITIVE2 interacts with ABSCISIC ACID INSENSITIVE5 to mediate the antagonism of brassinosteroids to abscisic acid during seed germination in Arabidopsis. *Plant Cell, 26*, 4394–4408.

Huang, J., & Redmann, R. E. (1995). Responses of growth, morphology, and anatomy to salinity and calcium supply in cultivated and wild barley. *Canadian Journal of Botany-Revue Canadienne De Botanique, 73*, 1859–1866.

Huang, R., Zheng, R., He, J., Zhou, Z., Wang, J., Xiong, Y., & Xu, T. (2019). Noncanonical auxin signaling regulates cell division pattern during lateral root development. *Proceedings of the National Academy of Sciences of the United States of America, 116*, 21285–21290.

Illes, P., Schlicht, M., Pavlovkin, J., Lichtscheidl, I., Baluska, F., & Ovecka, M. (2006). Aluminium toxicity in plants: Internalization of aluminium into cells of the transition zone in Arabidopsis root apices related to changes in plasma membrane potential, endosomal behaviour, and nitric oxide production. *Journal of Experimental Botany, 57*, 4201–4213.

Ivanchenko, M. G., Muday, G. K., & Dubrovsky, J. G. (2008). Ethylene-auxin interactions regulate lateral root initiation and emergence in Arabidopsis thaliana. *The Plant Journal, 55*, 335–347.

Jiang, K., & Feldman, L. J. (2005). Regulation of root apical meristem development. *Annual Review of Cell and Developmental Biology, 21*, 485–509.

Jones, M. A., Shen, J. J., Fu, Y., Li, H., Yang, Z. B., & Grierson, C. S. (2002). The Arabidopsis Rop2 GTPase is a positive regulator of both root hair initiation and tip growth. *Plant Cell, 14*, 763–776.

Kamoshita, A., Rodriguez, R., Yamauchi, A., & Wade, L. J. (2004). Genotypic variation in response of rainfed lowland rice to prolonged drought and rewatering. *Plant Production Science, 7*, 406–420.

Kano, M., Inukai, Y., Kitano, H., & Yamauchi, A. (2011). Root plasticity as the key root trait for adaptation to various intensities of drought stress in rice. *Plant and Soil, 342*, 117–128.

Katsuhara, M., & Kawasaki, T. (1996). *Plant and Cell Physiology, 37*, 169–173.

Kendrick, M. D., & Chang, C. (2008). Ethylene signaling: New levels of complexity and regulation. *Current Opinion in Plant Biology, 11*, 479–485.

Kieber, J. J., Rothenberg, M., Roman, G., Feldmann, K. A., & Ecker, J. R. (1993). CTR1, a negative regulator of the ethylene response pathway in Arabidopsis, encodes a member of the raf family of protein kinases. *Cell, 72*, 427–441.

Kochian, L. V., Hoekenga, O. A., & Pineros, M. A. (2004). How do crop plants tolerate acid soils? - Mechanisms of aluminum tolerance and phosphorous efficiency. *Annual Review of Plant Biology, 55*, 459–493.

Kollmeier, M., Felle, H. H., & Horst, W. J. (2000). Genotypical differences in aluminum resistance of maize are expressed in the distal part of the transition zone. Is reduced basipetal auxin flow involved in inhibition of root elongation by aluminum? *Plant Physiology, 122*, 945–956.

Kondo, Y., Hirakawa, Y., Kieber, J. J., & Fukuda, H. (2011). CLE peptides can negatively regulate protoxylem vessel formation via cytokinin signaling. *Plant and Cell Physiology, 52*, 37–48.

Kong, X. P., Liu, G. C., Liu, J. J., & Ding, Z. J. (2018). The root transition zone: A hot spot for signal crosstalk. *Trends in Plant Science, 23*, 403–409.

Kubo, M., Udagawa, M., Nishikubo, N., Horiguchi, G., Yamaguchi, M., Ito, J., Mimura, T., Fukuda, H., & Demura, T. (2005). Transcription switches for protoxylem and metaxylem vessel formation. *Genes & Development, 19*, 1855–1860.

Kuderova, A., Urbankova, I., Valkova, M., Malbeck, J., Brzobohaty, B., Nemethova, D., & Hejatko, J. (2008). Effects of conditional IPT-dependent cytokinin overproduction on root architecture of Arabidopsis seedlings. *Plant and Cell Physiology, 49*, 570–582.

Kumpf, R. P., Shi, C. L., Larrieu, A., Sto, I. M., Butenko, M. A., Peret, B., Riiser, E. S., Bennett, M. J., & Aalen, R. B. (2013). Floral organ abscission peptide IDA and its HAE/HSL2 receptors control cell separation during lateral root emergence. *Proceedings of the National Academy of Sciences of the United States of America, 110*, 5235–5240.

Kurth, E., Cramer, G. R., Lauchli, A., & Epstein, E. (1986). Effects of NaCl and CaCl(2) on cell enlargement and cell production in cotton roots. *Plant Physiology, 82*, 1102–1106.

Kutz, A., Muller, A., Hennig, P., Kaiser, W. M., Piotrowski, M., & Weiler, E. W. (2002). A role for nitrilase 3 in the regulation of root morphology in sulphur-starving Arabidopsis thaliana. *The Plant Journal, 30*, 95–106.

Laplaze, L., Benkova, E., Casimiro, I., Maes, L., Vanneste, S., Swarup, R., Weijers, D., Calvo, V., Parizot, B., Herrera-Rodriguez, M. B., Offringa, R., Graham, N., Doumas, P., Friml, J., Bogusz, D., Beeckman, T., & Bennett, M. (2007). Cytokinins act directly on lateral root founder cells to inhibit root initiation. *Plant Cell, 19*, 3889–3900.

Laskowski, M. J., Williams, M. E., Nusbaum, H. C., & Sussex, I. M. (1995). Formation of lateral root-meristems is a 2-stage process. *Development, 121*, 3303–3310.

Laskowski, M., Biller, S., Stanley, K., Kajstura, T., & Prusty, R. (2006). Expression profiling of auxin-treated Arabidopsis roots: Toward a molecular analysis of lateral root emergence. *Plant and Cell Physiology, 47*, 788–792.

Laskowski, M., Grieneisen, V. A., Hofhuis, H., Hove, C. A., Hogeweg, P., Maree, A. F., & Scheres, B. (2008). Root system architecture from coupling cell shape to auxin transport. *PLoS Biology, 6*, e307.

Lawlor, D. W. (2013). Genetic engineering to improve plant performance under drought: Physiological evaluation of achievements, limitations, and possibilities. *Journal of Experimental Botany, 64*, 83–108.

Le, J., Vandenbussche, F., Van der Straeten, D., & Verbelen, J. P. (2001). In the early response of arabidopsis roots to ethylene, cell elongation is up- and down-regulated and uncoupled from differentiation. *Plant Physiology, 125*, 519–522.

Lewis, D. R., Ramirez, M. V., Miller, N. D., Vallabhaneni, P., Ray, W. K., Helm, R. F., Winkel, B. S., & Muday, G. K. (2011). Auxin and ethylene induce flavonol accumulation through distinct transcriptional networks. *Plant Physiology, 156*, 144–164.

Li, J., Nagpal, P., Vitart, V., McMorris, T. C., & Chory, J. (1996). A role for brassinosteroids in light-dependent development of Arabidopsis. *Science, 272*, 398–401.

Li, X., Mo, X., Shou, H., & Wu, P. (2006). Cytokinin-mediated cell cycling arrest of pericycle founder cells in lateral root initiation of Arabidopsis. *Plant and Cell Physiology, 47*, 1112–1123.

Li, C. L., Liu, G. C., Geng, X. Y., He, C. M., Quan, T. Y., Hayashi, K. I., De Smet, I., Robert, H. S., Ding, Z. J., & Yang, Z. B. (2021). Local regulation of auxin transport in root-apex transition zone mediates aluminium-induced Arabidopsis root-growth inhibition. *The Plant Journal, 108*, 55–66.

Lin, P. C., Hwang, S. G., Endo, A., Okamoto, M., Koshiba, T., & Cheng, W. H. (2007). Ectopic expression of ABSCISIC ACID 2/GLUCOSE INSENSITIVE 1 in Arabidopsis promotes seed dormancy and stress tolerance. *Plant Physiology, 143*, 745–758.

Liscum, E., & Reed, J. W. (2002). Genetics of Aux/IAA and ARF action in plant growth and development. *Plant Molecular Biology, 49*, 387–400.

Liu, W. C., & Carnsdagger, H. R. (1961). Isolation of Abscisin, an abscission accelerating substance. *Science, 134*, 384–385.

Liu, P. P., Montgomery, T. A., Fahlgren, N., Kasschau, K. D., Nonogaki, H., & Carrington, J. C. (2007). Repression of AUXIN RESPONSE FACTOR10 by microRNA160 is critical for seed germination and post-germination stages. *The Plant Journal, 52*, 133–146.

Liu, G., Gao, S., Tian, H., Wu, W., Robert, H. S., Ding, Z. (2016). Local transcriptional control of YUCCA regulates auxin promoted root-growth inhibition in response to aluminium stress in Arabidopsis. *PLoS Genetics, 12*, e1006360.

Liu, Y. T., Xu, M. Z., Liang, N. S., Zheng, Y. H., Yu, Q. Z., & Wu, S. (2017). Symplastic communication spatially directs local auxin biosynthesis to maintain root stem cell niche in Arabidopsis. *Proceedings of the National Academy of Sciences of the United States of America, 114*, 4005–4010.

Long, Y., Goedhart, J., Schneijderberg, M., Terpstra, I., Shimotohno, A., Bouchet, B. P., Akhmanova, A., Gadella, T. W., Jr., Heidstra, R., Scheres, B., & Blilou, I. (2015a). SCARECROW-LIKE23 and SCARECROW jointly specify endodermal cell fate but distinctly control SHORT-ROOT movement. *The Plant Journal, 84*, 773–784.

Long, Y. C., Smet, W., Cruz-Ramirez, A., Castelijns, B., de Jonge, W., Mahonen, A. P., Bouchet, B. P., Perez, G. S., Akhmanova, A., Scheres, B., & Blilou, I. (2015b). Arabidopsis BIRD zinc finger proteins jointly stabilize tissue boundaries by confining the cell fate regulator SHORT-ROOT and contributing to fate specification. *Plant Cell, 27*, 1185–1199.

Lopez-Bucio, J., Hernandez-Abreu, E., Sanchez-Calderon, L., Nieto-Jacobo, M. F., Simpson, J., & Herrera-Estrella, L. (2002). Phosphate availability alters architecture and causes changes in hormone sensitivity in the Arabidopsis root system. *Plant Physiology, 129*, 244–256.

Lopez-Bucio, J., Cruz-Ramirez, A., & Herrera-Estrella, L. (2003). The role of nutrient availability in regulating root architecture. *Current Opinion in Plant Biology, 6*, 280–287.

Lv, B. S., Tian, H. Y., Zhang, F., Liu, J. J., Lu, S. C., Bai, M. Y., Li, C. Y., & Ding, Z. J. (2018). Brassinosteroids regulate root growth by controlling reactive oxygen species homeostasis and dual effect on ethylene synthesis in Arabidopsis. *PLoS Genetics, 14*(1), e1007144.

Lv, B., Wei, K., Hu, K., Tian, T., Zhang, F., Yu, Z., Zhang, D., Su, Y., Sang, Y., Zhang, X., & Ding, Z. (2021). MPK14-mediated auxin signaling controls lateral root development via ERF13-regulated very-long-chain fatty acid biosynthesis. *Molecular Plant, 14*, 285–297.

Ma, Z., Bielenberg, D. G., Brown, K. M., & Lynch, J. P. (2001). Regulation of root hair density by phosphorus availability in Arabidopsis thaliana. *Plant, Cell and Environment, 24*, 459–467.

Mahonen, A. P., Bonke, M., Kauppinen, L., Riikonen, M., Benfey, P. N., & Helariutta, Y. (2000). A novel two-component hybrid molecule regulates vascular morphogenesis of the Arabidopsis root. *Genes & Development, 14*, 2938–2943.

Makkena, S., & Lamb, R. S. (2013). The bHLH transcription factor SPATULA regulates root growth by controlling the size of the root meristem. *BMC Plant Biology, 13*, 1.

Malamy, J. E. (2005). Intrinsic and environmental response pathways that regulate root system architecture. *Plant, Cell and Environment, 28*, 67–77.

Malamy, J. E., & Benfey, P. N. (1997). Organization and cell differentiation in lateral roots of Arabidopsis thaliana. *Development, 124*, 33–44.

Manschadi, A. M., Christopher, J., Devoil, P., & Hammer, G. L. (2006). The role of root architectural traits in adaptation of wheat to water-limited environments. *Functional Plant Biology, 33*, 823–837.

Mason, M. G., Mathews, D. E., Argyros, D. A., Maxwell, B. B., Kieber, J. J., Alonso, J. M., Ecker, J. R., & Schaller, G. E. (2005). Multiple type-B response regulators mediate cytokinin signal transduction in Arabidopsis. *Plant Cell, 17*, 3007–3018.

Masucci, J. D., & Schiefelbein, J. W. (1994). The Rhd6 mutation of Arabidopsis-Thaliana alters root-hair initiation through an auxin-associated and ethylene-associated process. *Plant Physiology, 106*, 1335–1346.

Masucci, J. D., & Schiefelbein, J. W. (1996). Hormones act downstream of TTG and GL2 to promote root hair outgrowth during epidermis development in the Arabidopsis root. *Plant Cell, 8*, 1505–1517.

Matsumoto, H. (2000). Cell biology of aluminum toxicity and tolerance in higher plants. *International Review of Cytology - a Survey of Cell Biology, 200*(200), 1–46.

Melcher, K., Zhou, X. E., & Xu, H. E. (2010). Thirsty plants and beyond: Structural mechanisms of abscisic acid perception and signaling. *Current Opinion in Structural Biology, 20*, 722–729.

Mishra, A. K., & Singh, V. P. (2010). A review of drought concepts. *Journal of Hydrology, 391*, 204–216.

Mockaitis, K., & Estelle, M. (2008). Auxin receptors and plant development: A new signaling paradigm. *Annual Review of Cell and Developmental Biology, 24*, 55–80.

Moison, M., Pacheco, J. M., Lucero, L., Fonouni-Farde, C., Rodríguez-Melo, J., Mansilla, N., Christ, A., Bazin, J., Benhamed, M., Ibañez, F., Crespi, M., Estevez, J. M., & Ariel, F. (2021). The lncRNA interacts with the transcription factor WRKY42 to trigger root hair cell expansion in response to cold. *Molecular Plant, 14*, 937–948.

Moller, B. K., ten Hove, C. A., Xiang, D. Q., Williams, N., Lopez, L. G., Yoshida, S., Smit, M., Datla, R., & Weijers, D. (2017). Auxin response cell-autonomously controls ground tissue initiation in the early Arabidopsis embryo. *Proceedings of the National Academy of Sciences of the United States of America, 114*, E2533–E2539.

Moreno-Risueno, M. A., Van Norman, J. M., Moreno, A., Zhang, J. Y., Ahnert, S. E., & Benfey, P. N. (2010). Oscillating gene expression determines competence for periodic Arabidopsis root branching. *Science, 329*, 1306–1311.

Moreno-Risueno, M. A., Sozzani, R., Yardimci, G. G., Petricka, J. J., Vernoux, T., Blilou, I., Alonso, J., Winter, C. M., Ohler, U., Scheres, B., & Benfey, P. N. (2015). Transcriptional control of tissue formation throughout root development. *Science, 350*, 426–430.

Moubayidin, L., Perilli, S., Dello Ioio, R., Di Mambro, R., Costantino, P., & Sabatini, S. (2010). The rate of cell differentiation controls the Arabidopsis root meristem growth phase. *Current Biology, 20*, 1138–1143.

Mouchel, C. F., Osmont, K. S., & Hardtke, C. S. (2006). BRX mediates feedback between brassinosteroid levels and auxin signalling in root growth. *Nature, 443*, 458–461.

Muller, B., & Sheen, J. (2008). Cytokinin and auxin interaction in root stem-cell specification during early embryogenesis. *Nature, 453*, 1094–U1097.

Mussig, C., Shin, G. H., & Altmann, T. (2003). Brassinosteroids promote root growth in Arabidopsis. *Plant Physiology, 133*, 1261–1271.

Nakajima, K., Sena, G., Nawy, T., & Benfey, P. N. (2001). Intercellular movement of the putative transcription factor SHR in root patterning. *Nature, 413*, 307–311.

Negi, S., Ivanchenko, M. G., & Muday, G. K. (2008). Ethylene regulates lateral root formation and auxin transport in Arabidopsis thaliana. *The Plant Journal, 55*, 175–187.

Negi, S., Sukumar, P., Liu, X., Cohen, J. D., & Muday, G. K. (2010). Genetic dissection of the role of ethylene in regulating auxin-dependent lateral and adventitious root formation in tomato. *The Plant Journal, 61*, 3–15.

Neumann, P. M., Azaizeh, H., & Leon, D. (1994). Hardening of root cell-walls - A growth-inhibitory response to salinity stress. *Plant, Cell and Environment, 17*, 303–309.

Ohashi-Ito, K., & Fukuda, H. (2010). Transcriptional regulation of vascular cell fates. *Current Opinion in Plant Biology, 13*, 670–676.

Ohashi-Ito, K., Iwamoto, K., Yamagami, A., Nakano, T., & Fukuda, H. (2023). HD-ZIP III-dependent local promotion of brassinosteroid synthesis suppresses vascular cell division in Arabidopsis root apical meristem. *Proceedings of the National Academy of Sciences of the United States of America, 120*, e2216632120.

Okushima, Y., Overvoorde, P. J., Arima, K., Alonso, J. M., Chan, A., Chang, C., Ecker, J. R., Hughes, B., Lui, A., Nguyen, D., Onodera, C., Quach, H., Smith, A., Yu, G. X., & Theologis, A. (2005). Functional genomic analysis of the AUXIN RESPONSE FACTOR gene family members in Arabidopsis thaliana: Unique and overlapping functions of ARF7 and ARF19. *Plant Cell, 17*, 444–463.

Okushima, Y., Fukaki, H., Onoda, M., Theologis, A., & Tasaka, M. (2007). ARF7 and ARF19 regulate lateral root formation via direct activation of LBD/ASL genes in Arabidopsis. *Plant Cell, 19*, 118–130.

Orman-Ligeza, B., Morris, E. C., Parizot, B., Lavigne, T., Babé, A., Ligeza, A., Klein, S., Sturrock, C., Xuan, W., Novák, O., Ljung, K., Fernandez, M. A., Rodriguez, P. L., Dodd, I. C., De Smet, I., Chaumont, F., Batoko, H., Périlleux, C., Lynch, J. P., Bennett, M. J., Beeckman, T., & Draye, X. (2018). The xerobranching response represses lateral root formation when roots are not in contact with water. *Current Biology, 28*, 3165–3173.e5.

Orosa-Puente, B., Leftley, N., von Wangenheim, D., Banda, J., Srivastava, A. K., Hill, K., Truskina, J., Bhosale, R., Morris, E., Srivastava, M., Kümpers, B., Goh, T., Fukaki, H., Vermeer, J. E. M., Vernoux, T., Dinneny, J. R., French, A. P., Bishopp, A., Sadanandom, A., & Bennett, M. J. (2018). Root branching toward water involves posttranslational modification of transcription factor ARF7. *Science, 362*, 1407–1410.

Overvoorde, P., Fukaki, H., & Beeckman, T. (2010). Auxin control of root development. *Cold Spring Harbor Perspectives in Biology, 2*, a001537.

Paquette, A. J., & Benfey, P. N. (2005). Maturation of the ground tissue of the root is regulated by gibberellin and SCARECROW and requires SHORT-ROOT. *Plant Physiology, 138*, 636–640.

Parizot, B., Laplaze, L., Ricaud, L., Boucheron-Dubuisson, E., Bayle, V., Bonke, M., De Smet, I., Poethig, S. R., Helariutta, Y., Haseloff, J., Chriqui, D., Beeckman, T., & Nussaume, L. (2008). Diarch symmetry of the vascular bundle in Arabidopsis root encompasses the pericycle and is reflected in distich lateral root initiation. *Plant Physiology, 146*, 140–148.

Peret, B., De Rybel, B., Casimiro, I., Benkova, E., Swarup, R., Laplaze, L., Beeckman, T., & Bennett, M. J. (2009). Arabidopsis lateral root development: An emerging story. *Trends in Plant Science, 14*, 399–408.

Perilli, S., Moubayidin, L., & Sabatini, S. (2010). The molecular basis of cytokinin function. *Current Opinion in Plant Biology, 13*, 21–26.

Perrot-Rechenmann, C. (2010). Cellular responses to auxin: Division versus expansion. *Cold Spring Harbor Perspectives in Biology, 2*, a001446.

Petricka, J. J., Winter, C. M., & Benfey, P. N. (2012). Control of Arabidopsis root development. *Annual Review of Plant Biology, 63*(63), 563–590.

Pi, L., Aichinger, E., van der Graaff, E., Llavata-Peris, C. I., Weijers, D., Hennig, L., Groot, E., & Laux, T. (2015). Organizer-derived WOX5 signal maintains root columella stem cells through chromatin-mediated repression of CDF4 expression. *Developmental Cell, 33*, 576–588.

Pilet, P. E. (1975). Abscisic acid as a root growth inhibitor: Physiological analyses. *Planta, 122*, 299–302.

Pitts, R. J., Cernac, A., & Estelle, M. (1998). Auxin and ethylene promote root hair elongation in Arabidopsis. *The Plant Journal, 16*, 553–560.

Ploense, S. E., Wu, M. F., Nagpal, P., & Reed, J. W. (2009). A gain-of-function mutation in IAA18 alters Arabidopsis embryonic apical patterning. *Development, 136*, 1509–1517.

Rahman, A., Amakawa, T., Goto, N., & Tsurumi, S. (2001). Auxin is a positive regulator for ethylene-mediated response in the growth of Arabidopsis roots. *Plant and Cell Physiology, 42*, 301–307.

Rangel, A. F., Rao, I. M., & Horst, W. J. (2007). Spatial aluminium sensitivity of root apices of two common bean (Phaseolus vulgaris L.) genotypes with contrasting aluminium resistance. *Journal of Experimental Botany, 58*, 3895–3904.

Raven, J. A., & Edwards, D. (2001). Roots: Evolutionary origins and biogeochemical significance. *Journal of Experimental Botany, 52*, 381–401.

Reinhardt, D. H., & Rost, T. L. (1995). Salinity accelerates endodermal development and induces an exodermis in cotton seedling roots. *Environmental and Experimental Botany, 35*, 563–574.

Ren, B., Chen, Q. G., Hong, S. L., Zhao, W. M., Feng, J., Feng, H. Z., & Zuo, J. R. (2013). The Arabidopsis eukaryotic translation initiation factor eIF5A-2 regulates root protoxylem development by modulating cytokinin signaling. *Plant Cell, 25*, 3841–3857.

Richter, G. L., Monshausen, G. B., Krol, A., & Gilroy, S. (2009). Mechanical stimuli modulate lateral root organogenesis. *Plant Physiology, 151*, 1855–1866.

Riefler, M., Novak, O., Strnad, M., & Schmulling, T. (2006). Arabidopsis cytokinin receptor mutants reveal functions in shoot growth, leaf senescence, seed size, germination, root development, and cytokinin metabolism. *Plant Cell, 18*, 40–54.

Rigas, S., Debrosses, G., Haralampidis, K., Vicente-Agullo, F., Feldmann, K. A., Grabov, A., Dolan, L., & Hatzopoulos, P. (2001). TRH1 encodes a potassium transporter required for tip growth in Arabidopsis root hairs. *Plant Cell, 13*, 139–151.

Robert, H. S., & Friml, J. (2009). Auxin and other signals on the move in plants. *Nature Chemical Biology, 5*, 325–332.

Rogg, L. E., Lasswell, J., & Bartel, B. (2001). A gain-of-function mutation in IAA28 suppresses lateral root development. *Plant Cell, 13*, 465–480.

Roudier, F., Fedorova, E., Lebris, M., Lecomte, P., Gyorgyey, J., Vaubert, D., Horvath, G., Abad, P., Kondorosi, A., & Kondorosi, E. (2003). The Medicago species A2-type cyclin is auxin regulated and involved in meristem formation but dispensable for endoreduplication-associated developmental programs. *Plant Physiology, 131*, 1091–1103.

Rubio, V., Linhares, F., Solano, R., Martin, A. C., Iglesias, J., Leyva, A., & Paz-Ares, J. (2001). A conserved MYB transcription factor involved in phosphate starvation signaling both in vascular plants and in unicellular algae. *Genes & Development, 15*, 2122–2133.

Ruzicka, K., Ljung, K., Vanneste, S., Podhorska, R., Beeckman, T., Friml, J., & Benkova, E. (2007). Ethylene regulates root growth through effects on auxin biosynthesis and transport-dependent auxin distribution. *Plant Cell, 19*, 2197–2212.

Ryan, P. R., & Kochian, L. V. (1993). Interaction between aluminum toxicity and calcium-uptake at the root apex in near-isogenic lines of wheat (Triticum-Aestivum L) differing in aluminum tolerance. *Plant Physiology, 102*, 975–982.

Sabatini, S., Heidstra, R., Wildwater, M., & Scheres, B. (2003). SCARECROW is involved in positioning the stem cell niche in the Arabidopsis root meristem. *Genes & Development, 17*, 354–358.

Samarajeewa, P. K., Barrero, R. A., Umeda-Hara, C., Kawai, M., & Uchimiya, H. (1999). Cortical cell death, cell proliferation, macromolecular movements and rTip1 expression pattern in roots of rice (Oryza sativa L,) under NaCl stress. *Planta, 207*, 354–361.

Sarkar, A. K., Luijten, M., Miyashima, S., Lenhard, M., Hashimoto, T., Nakajima, K., Scheres, B., Heidstra, R., & Laux, T. (2007). Conserved factors regulate signalling in Arabidopsis thaliana shoot and root stem cell organizers. *Nature, 446*, 811–814.

Sasaki, M., Yamamoto, Y., & Matsumoto, H. (1996). Lignin deposition induced by aluminum in wheat (Triticum aestivum) roots. *Physiologia Plantarum, 96*, 193–198.

Scacchi, E., Salinas, P., Gujas, B., Santuari, L., Krogan, N., Ragni, L., Berleth, T., & Hardtke, C. S. (2010). Spatio-temporal sequence of cross-regulatory events in root meristem growth. *Proceedings of the National Academy of Sciences of the United States of America, 107*, 22734–22739.

Scheres, B., Wolkenfelt, H., Willemsen, V., Terlouw, M., Lawson, E., Dean, C., & Weisbeek, P. (1994). Embryonic origin of the Arabidopsis primary root and root-meristem initials. *Development, 120*, 2475–2487.

Scheres, B., Dilaurenzio, L., Willemsen, V., Hauser, M. T., Janmaat, K., Weisbeek, P., & Benfey, P. N. (1995). Mutations affecting the radial organization of the Arabidopsis root display specific defects throughout the embryonic axis. *Development, 121*, 53–62.

Schiefelbein, J. W. (2000). Constructing a plant cell. The genetic control of root hair development. *Plant Physiology, 124*, 1525–1531.

Schmidt, W., Tittel, J., & Schikora, A. (2000). Role of hormones in the induction of iron deficiency responses in Arabidopsis roots. *Plant Physiology, 122*, 1109–1118.

Serre, N. B. C., Kralík, D., Yun, P., Slouka, Z., Shabala, S., & Fendrych, M. (2021). AFB1 controls rapid auxin signalling through membrane depolarization in root. *Nature Plants, 7*, 1229–1238.

Shaul, O., Mironov, V., Burssens, S., VanMontagu, M., & Inze, D. (1996). Two Arabidopsis cyclin promoters mediate distinctive transcriptional oscillation in synchronized tobacco BY-2 cells. *Proceedings of the National Academy of Sciences of the United States of America, 93*, 4868–4872.

Singh, S., Yadav, S., Singh, A., Mahima, M., Singh, A., Gautam, V., & Sarkar, A. K. (2020). Auxin signaling modulates LATERAL ROOT PRIMORDIUM1 (LRP1) expression during lateral root development in Arabidopsis. *The Plant Journal, 101*, 87–100.

Sivaguru, M., & Horst, W. J. (1998). The distal part of the transition zone is the most aluminum-sensitive apical root zone of maize. *Plant Physiology, 116*, 155–163.

Stepanova, A. N., Yun, J., Likhacheva, A. V., & Alonso, J. M. (2007). Multilevel interactions between ethylene and auxin in Arabidopsis roots. *Plant Cell, 19*, 2169–2185.

Strader, L. C., Chen, G. L., & Bartel, B. (2010). Ethylene directs auxin to control root cell expansion. *The Plant Journal, 64*, 874–884.

Sun, J. Q., Xu, Y. X., Ye, S. Q., Jiang, H. L., Chen, Q., Liu, F., Zhou, W. K., Chen, R., Li, X. G., Tietz, O., Wu, X. Y., Cohen, J. D., Palme, K., & Li, C. Y. (2009). Arabidopsis ASA1 is important for jasmonate-mediated regulation of auxin biosynthesis and transport during lateral root formation. *Plant Cell, 21*, 1495–1511.

Sun, P., Tian, Q. Y., Chen, J., & Zhang, W. H. (2010). Aluminium-induced inhibition of root elongation in Arabidopsis is mediated by ethylene and auxin. *Journal of Experimental Botany, 61*, 347–356.

Sussex, I. M., Godoy, J. A., Kerk, N. M., Laskowski, M. J., Nusbaum, H. C., Welsch, J. A., & Williams, M. E. (1995). Cellular and molecular events in a newly organizing lateral root-meristem. *Philosophical Transactions of the Royal Society of London. Series B, Biological Sciences, 350*, 39–43.

Swarup, R., Friml, J., Marchant, A., Ljung, K., Sandberg, G., Palme, K., & Bennett, M. (2001). Localization of the auxin permease AUX1 suggests two functionally distinct hormone transport pathways operate in the Arabidopsis root apex. *Genes & Development, 15*, 2648–2653.

Swarup, R., Parry, G., Graham, N., Allen, T., & Bennett, M. (2002). Auxin cross-talk: Integration of signalling pathways to control plant development. *Plant Molecular Biology, 49*, 411–426.

Swarup, R., Perry, P., Hagenbeek, D., Van Der Straeten, D., Beemster, G. T., Sandberg, G., Bhalerao, R., Ljung, K., & Bennett, M. J. (2007). Ethylene upregulates auxin biosynthesis in Arabidopsis seedlings to enhance inhibition of root cell elongation. *Plant Cell, 19*, 2186–2196.

Swarup, K., Benkova, E., Swarup, R., Casimiro, I., Peret, B., Yang, Y., Parry, G., Nielsen, E., De Smet, I., Vanneste, S., Levesque, M. P., Carrier, D., James, N., Calvo, V., Ljung, K., Kramer, E., Roberts, R., Graham, N., Marillonnet, S., Patel, K., Jones, J. D. G., Taylor, C. G., Schachtman, D. P., May, S., Sandberg, G., Benfey, P., Friml, J., Kerr, I., Beeckman, T., Laplaze, L., & Bennett, M. J. (2008). The auxin influx carrier LAX3 promotes lateral root emergence. *Nature Cell Biology, 10*, 946–954.

Tanaka, K., Asami, T., Yoshida, S., Nakamura, Y., Matsuo, T., & Okamoto, S. (2005). Brassinosteroid homeostasis in Arabidopsis is ensured by feedback expressions of multiple genes involved in its metabolism. *Plant Physiology, 138*, 1117–1125.

Tanaka, H., Dhonukshe, P., Brewer, P. B., & Friml, J. (2006). Spatiotemporal asymmetric auxin distribution: A means to coordinate plant development. *Cellular and Molecular Life Sciences, 63*, 2738–2754.

Tardieu, F. (2012). Any trait or trait-related allele can confer drought tolerance: Just design the right drought scenario. *Journal of Experimental Botany, 63*, 25–31.

Tatematsu, K., Kumagai, S., Muto, H., Sato, A., Watahiki, M. K., Harper, R. M., Liscum, E., & Yamamoto, K. T. (2004). MASSUGU2 encodes Aux/IAA19, an auxin-regulated protein that functions together with the transcriptional activator NPH4/ARF7 to regulate differential growth responses of hypocotyl and formation of lateral roots in Arabidopsis thaliana. *Plant Cell, 16*, 379–393.

Taylor, I. B., Burbidge, A., & Thompson, A. J. (2000). Control of abscisic acid synthesis. *Journal of Experimental Botany, 51*, 1563–1574.

Tian, Q., & Reed, J. W. (1999). Control of auxin-regulated root development by the Arabidopsis thaliana SHY2/IAA3 gene. *Development, 126*, 711–721.

Tian, H., Wabnik, K., Niu, T., Li, H., Yu, Q., Pollmann, S., Vanneste, S., Govaerts, W., Rolcik, J., Geisler, M., Friml, J., & Ding, Z. (2014). WOX5-IAA17 feedback circuit-mediated cellular auxin response is crucial for the patterning of root stem cell niches in Arabidopsis. *Molecular Plant, 7*, 277–289.

To, J. P., Haberer, G., Ferreira, F. J., Deruere, J., Mason, M. G., Schaller, G. E., Alonso, J. M., Ecker, J. R., & Kieber, J. J. (2004). Type-A Arabidopsis response regulators are partially redundant negative regulators of cytokinin signaling. *Plant Cell, 16*, 658–671.

Tong, H., Xiao, Y., Liu, D., Gao, S., Liu, L., Yin, Y., Jin, Y., Qian, Q., & Chu, C. (2014). Brassinosteroid regulates cell elongation by modulating gibberellin metabolism in rice. *Plant Cell, 26*, 4376–4393.

Trinh, D. C., Lavenus, J., Goh, T., Boutte, Y., Drogue, Q., Vaissayre, V., Tellier, F., Lucas, M., Voss, U., Gantet, P., Faure, J. D., Dussert, S., Fukaki, H., Bennet, M. J., Laplaze, L., & Guyomarc'h, S. (2019). PUCHI regulates very long chain fatty acid biosynthesis during lateral root and callus formation. *Proceedings of the National Academy of Sciences of the United States of America, 116*, 14325–14330.

Tromas, A., Braun, N., Muller, P., Khodus, T., Paponov, I. A., Palme, K., Ljung, K., Lee, J. Y., Benfey, P., Murray, J. A., Scheres, B., & Perrot-Rechenmann, C. (2009). The AUXIN-BINDING PROTEIN 1 is required for differential auxin responses mediating root growth. *PLoS One, 4*, e6648.

Tsukagoshi, H., Busch, W., & Benfey, P. N. (2010). Transcriptional regulation of ROS controls transition from proliferation to differentiation in the root. *Cell, 143*, 606–616.

Uehara, T., Okushima, Y., Mimura, T., Tasaka, M., & Fukaki, H. (2008). Domain II mutations in CRANE/IAA18 suppress lateral root formation and affect shoot development in Arabidopsis thaliana. *Plant and Cell Physiology, 49*, 1025–1038.

Ulmasov, T., Hagen, G., & Guilfoyle, T. J. (1999). Activation and repression of transcription by auxin-response factors. *Proceedings of the National Academy of Sciences of the United States of America, 96*, 5844–5849.

van den Berg, C., Willemsen, V., Hendriks, G., Weisbeek, P., & Scheres, B. (1997). Short-range control of cell differentiation in the Arabidopsis root meristem. *Nature, 390*, 287–289.

Vance, C. P. (2001). Symbiotic nitrogen fixation and phosphorus acquisition. Plant nutrition in a world of declining renewable resources. *Plant Physiology, 127*, 390–397.

Vanneste, S., & Friml, J. (2009). Auxin: A trigger for change in plant development. *Cell, 136*, 1005–1016.

Vermeer, J. E. M., von Wangenheim, D., Barberon, M., Lee, Y., Stelzer, E. H. K., Maizel, A., & Geldner, N. (2014). A spatial accommodation by neighboring cells is required for organ initiation in Arabidopsis. *Science, 343*, 178–183.

Verslues, P. E., Agarwal, M., Katiyar-Agarwal, S., Zhu, J., & Zhu, J. K. (2006). Methods and concepts in quantifying resistance to drought, salt and freezing, abiotic stresses that affect plant water status. (vol 45, pg 523, 2006). *The Plant Journal, 46*, 1092–1092.

Vert, G., Nemhauser, J. L., Geldner, N., Hong, F., & Chory, J. (2005). Molecular mechanisms of steroid hormone signaling in plants. *Annual Review of Cell and Developmental Biology, 21*, 177–201.

Vicente-Agullo, F., Rigas, S., Desbrosses, G., Dolan, L., Hatzopoulos, P., & Grabov, A. (2004). Potassium carrier TRH1 is required for auxin transport in Arabidopsis roots. *The Plant Journal, 40*, 523–535.

Vilches-Barro, A., & Maizel, A. (2015). Talking through walls: Mechanisms of lateral root emergence in Arabidopsis thaliana. *Current Opinion in Plant Biology, 23*, 31–38.

Von Uexkull, H. R., & Mutert, E. (1995). Global extent, development and economic impact of acid soils. In R. A. Date, N. J. Grundon, G. E. Raymet, & M. E. Probert (Eds.), *Plant-soil interactions at low pH: Principles and management* (pp. 5–19). Kluwer Academic Publishers.

Wachsman, G., Zhang, J. Y., Moreno-Risueno, M. A., Anderson, C. T., & Benfey, P. N. (2020). Cell wall remodeling and vesicle trafficking mediate the root clock in Arabidopsis. *Science, 370*, 819–823.

Wang, H., Qi, Q., Schorr, P., Cutler, A. J., Crosby, W. L., & Fowke, L. C. (1998). ICK1, a cyclin-dependent protein kinase inhibitor from Arabidopsis thaliana interacts with both Cdc2a and CycD3, and its expression is induced by abscisic acid. *The Plant Journal, 15*, 501–510.

Wang, J. W., Wang, L. J., Mao, Y. B., Cai, W. J., Xue, H. W., & Chen, X. Y. (2005). Control of root cap formation by MicroRNA-targeted auxin response factors in Arabidopsis. *Plant Cell, 17*, 2204–2216.

Wang, J. X., Li, X. X., Chen, X. L., Tang, W. X., Yu, Z. P., Xu, T. D., Tian, H. Y., & Ding, Z. J. (2023). Dual regulations of cell cycle regulator DPa by auxin in root distal stem cell maintenance. *Proceedings of the National Academy of Sciences of the United States of America, 120*, e2218503120.

Wei, Z., & Li, J. (2016). Brassinosteroids regulate root growth, development, and Symbiosis. *Molecular Plant, 9*, 86–100.

Welch, D., Hassan, H., Blilou, I., Immink, R., Heidstra, R., & Scheres, B. (2007). Arabidopsis JACKDAW and MAGPIE zinc finger proteins delimit asymmetric cell division and stabilize tissue boundaries by restricting SHORT-ROOT action. *Genes & Development, 21*, 2196–2204.

Werner, T., Motyka, V., Strnad, M., & Schmulling, T. (2001). Regulation of plant growth by cytokinin. *Proceedings of the National Academy of Sciences of the United States of America, 98*, 10487–10492.

Werner, T., Motyka, V., Laucou, V., Smets, R., Van Onckelen, H., & Schmulling, T. (2003). Cytokinin-deficient transgenic Arabidopsis plants show multiple developmental alterations indicating opposite functions of cytokinins in the regulation of shoot and root meristem activity. *Plant Cell, 15*, 2532–2550.

West, G., Inze, D., & Beemster, G. T. S. (2004). Cell cycle modulation in the response of the primary root of Arabidopsis to salt stress. *Plant Physiology, 135*, 1050–1058.

Willemsen, V., Bauch, M., Bennett, T., Campilho, A., Wolkenfelt, H., Xu, J., Haseloff, J., & Scheres, B. (2008). The NAC domain transcription factors FEZ and SOMBRERO control the orientation of cell division plane in Arabidopsis root stem cells. *Developmental Cell, 15*, 913–922.

Xie, C. T., Li, C. L., Wang, F. X., Zhang, F., Liu, J. J., Wang, J. X., Zhang, X. S., Kong, X. P., & Ding, Z. J. (2023). NAC1 regulates root ground tissue maturation by coordinating with the SCR/SHR-CYCD6;1 module in Arabidopsis. *Molecular Plant, 16*, 709–725.

Xu, J., Gao, G. L., Du, J. J., Guo, Y., & Yang, C. W. (2010). Cell cycle modulation in response of the primary root of Arabidopsis to aba. *Pakistan Journal of Botany, 42*, 2703–2710.

Xu, W. F., Jia, L. G., Shi, W. M., Liang, J. S., Zhou, F., Li, Q. F., & Zhang, J. H. (2013). Abscisic acid accumulation modulates auxin transport in the root tip to enhance proton secretion for maintaining root growth under moderate water stress. *The New Phytologist, 197*, 139–150.

Xuan, W., Audenaert, D., Parizot, B., Möller, B. K., Njo, M. F., De Rybel, B., De Rop, G., Van Isterdael, G., Mähönen, A. P., Vanneste, S., & Beeckman, T. (2015). Root cap-derived auxin pre-patterns the longitudinal axis of the root. *Current Biology, 25*, 1381–1388.

Xuan, W., Band, L. R., Kumpf, R. P., Van Damme, D., Parizot, B., De Rop, G., Opdenacker, D., Möller, B. K., Skorzinski, N., Njo, M. F., De Rybel, B., Audenaert, D., Nowack, M. K., Vanneste, S., & Beeckman, T. (2016). Cyclic programmed cell death stimulates hormone signaling and root development in Arabidopsis. *Science, 351*, 384–387.

Yamaguchi, M., Goue, N., Igarashi, H., Ohtani, M., Nakano, Y., Mortimer, J. C., Nishikubo, N., Kubo, M., Katayama, Y., Kakegawa, K., Dupree, P., & Demura, T. (2010). VASCULAR-RELATED NAC-DOMAIN6 and VASCULAR-RELATED NAC-DOMAIN7 effectively induce transdifferentiation into xylem vessel elements under control of an induction system. *Plant Physiology, 153*, 906–914.

Yamamoto, Y., Kobayashi, Y., & Matsumoto, H. (2001). Lipid peroxidation is an early symptom triggered by aluminum, but not the primary cause of elongation inhibition in pea roots. *Plant Physiology, 125*, 199–208.

Yang, X. Q., Lee, S., So, J. H., Dharmasiri, S., Dharmasiri, N., Ge, L., Jensen, C., Hangarter, R., Hobbie, L., & Estelle, M. (2004). The IAA1 protein is encoded by AXR5 and is a substrate of SCFTIR1. *The Plant Journal, 40*, 772–782.

Yang, L., Zhang, J., He, J., Qin, Y., Hua, D., Duan, Y., Chen, Z., & Gong, Z. (2014a). ABA-mediated ROS in mitochondria regulate root meristem activity by controlling PLETHORA expression in Arabidopsis. *PLoS Genetics, 10*, e1004791.

Yang, Z. B., Geng, X., He, C., Zhang, F., Wang, R., Horst, W. J., & Ding, Z. (2014b). TAA1-regulated local auxin biosynthesis in the root-apex transition zone mediates the aluminum-induced inhibition of root growth in Arabidopsis. *Plant Cell, 26*, 2889–2904.

Yang, Z. B., Liu, G. C., Liu, J. J., Zhang, B., Meng, W. J., Muller, B., Hayashi, K., Zhang, X. S., Zhao, Z., De Smet, I., & Ding, Z. J. (2017). Synergistic action of auxin and cytokinin mediates aluminum-induced root growth inhibition in Arabidopsis. *EMBO Reports, 18*, 1213–1230.

Ye, Q., Zhu, W., Li, L., Zhang, S., Yin, Y., Ma, H., & Wang, X. (2010). Brassinosteroids control male fertility by regulating the expression of key genes involved in Arabidopsis anther and pollen development. *Proceedings of the National Academy of Sciences of the United States of America, 107*, 6100–6105.

Yin, Y., Wang, Z. Y., Mora-Garcia, S., Li, J., Yoshida, S., Asami, T., & Chory, J. (2002). BES1 accumulates in the nucleus in response to brassinosteroids to regulate gene expression and promote stem elongation. *Cell, 109*, 181–191.

Yu, X., Li, L., Zola, J., Aluru, M., Ye, H., Foudree, A., Guo, H., Anderson, S., Aluru, S., Liu, P., Rodermel, S., & Yin, Y. (2011). A brassinosteroid transcriptional network revealed by genome-wide identification of BESI target genes in Arabidopsis thaliana. *The Plant Journal, 65*, 634–646.

Yu, Q., Tian, H., Yue, K., Liu, J., Zhang, B., Li, X., & Ding, Z. (2016). A P-loop NTPase regulates quiescent center cell division and distal stem cell identity through the regulation of ROS homeostasis in Arabidopsis root. *PLoS Genetics, 12*(9), e1006175.

Zhai, H. W., Zhang, X. Y., You, Y. R., Lin, L. H., Zhou, W. K., & Li, C. Y. (2020). SEUSS integrates transcriptional and epigenetic control of root stem cell organizer specification. *The EMBO Journal, 39*, e105047.

Zhang, H. M., & Forde, B. G. (1998). An Arabidopsis MADS box gene that controls nutrient-induced changes in root architecture. *Science, 279*, 407–409.

Zhang, H., Xue, Y. G., Wang, Z. Q., Yang, J., & Zhang, J. H. (2009). Morphological and physiological traits of roots and their relationships with shoot growth in "super" rice. *Field Crops Research, 113*, 31–40.

Zhang, H. M., Han, W., De Smet, I., Talboys, P., Loya, R., Hassan, A., Rong, H. L., Jurgens, G., Knox, J. P., & Wang, M. H. (2010). ABA promotes quiescence of the quiescent centre and suppresses stem cell differentiation in the Arabidopsis primary root meristem. *The Plant Journal, 64*, 764–774.

Zhang, T. Q., Xu, Z. G., Shang, G. D., & Wang, J. W. (2019). A single-cell RNA sequencing profiles the developmental landscape of root. *Molecular Plant, 12*, 648–660.

Zhang, F., Tao, W., Sun, R., Wang, J., Li, C., Kong, X., Tian, H., & Ding, Z. (2020). PRH1 mediates ARF7-LBD dependent auxin signaling to regulate lateral root development in Arabidopsis thaliana. *PLoS Genetics, 16*, e1008044.

Zhong, H. L., & Lauchli, A. (1993). Spatial and temporal aspects of growth in the primary root of cotton seedlings - effects of NaCl and $CaCl_2$. *Journal of Experimental Botany, 44*, 763–771.

Zhu, M. Y., Ahn, S., & Matsumoto, H. (2003). Inhibition of growth and development of root border cells in wheat by Al. *Physiologia Plantarum, 117*, 359–367.

Zhu, W., Wang, H., Fujioka, S., Zhou, T., Tian, H., Tian, W., & Wang, X. (2013). Homeostasis of brassinosteroids regulated by DRL1, a putative acyltransferase in Arabidopsis. *Molecular Plant, 6*, 546–558.

Zhu, Q., Shao, Y., Ge, S., Zhang, M., Zhang, T., Hu, X., Liu, Y., Walker, J., Zhang, S., & Xu, J. (2019). A MAPK cascade downstream of IDA-HAE/HSL2 ligand-receptor pair in lateral root emergence. *Nature Plants, 5*, 414–423.

Chapter 2
Shoot Development

Ying Hua Su, Li Ping Tang, and Xian Sheng Zhang

Abstract Plants can continuously and repetitively give rise to lateral organs through their meristems. The root apical meristem (RAM) generates the underground parts of plants, whereas the shoot apical meristem (SAM) produces the aboveground parts. The SAM originates from the embryo and gives rise to the aerial parts of plants. Within the SAM, a small number of stem cells confined to the upper region of the central zone (CZ) serve as the ultimate source of all shoot cells. Shoot stem cells have a strong ability to renew themselves and to generate daughter cells to produce new tissues. The plasticity of shoot stem cells allows plants to adapt their shape in response to developmental, physical, and environmental cues. The balance between the proliferation and differentiation of stem cells determines their populations in the SAM. The elaborate genetic network that regulates the initiation and maintenance of shoot stem cells has been a major area of interest in plant developmental research for decades.

Keywords Shoot · Shoot apical meristem · Embryonic shoot meristem · Organogenesis · Phyllotaxis · Axillary meristem

2.1 Introduction

In higher plants, post-embryonic development is characterized by the continuous generation of organs from the flanks of apical meristems. The shoot apical meristem (SAM) is initially formed during embryogenesis and continuously produces cells that will become incorporated into all the aerial organs and tissues of the plant, such as internodes, leaves, axillary shoot meristems, and flowers (Steeves & Sussex, 1989; Kerstetter & Hake, 1997). The aerial part of the plant can be divided into several iterative units; each unit comprises a leaf (or leaves) located at a node with a stem

Y. H. Su · L. P. Tang · X. S. Zhang (✉)
National Key Laboratory of Wheat Improvement, College of Life Sciences,
Shandong Agricultural University, Taian, Shandong, China
e-mail: zhangxs@sdau.edu.cn

F. Chang et al. (eds.), *Regulation of Plant Development*,
https://doi.org/10.1007/978-981-95-7033-1_2

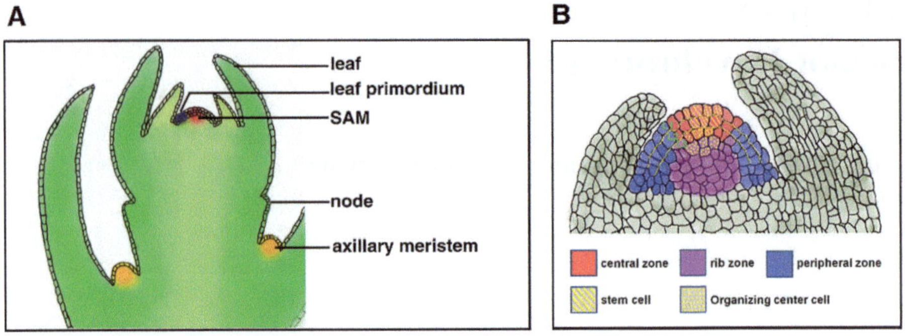

Fig. 2.1 Organization of *Arabidopsis* shoot apical meristem (SAM). (**a**) Schematic representation of shoot structure. The aerial part of the plant can be divided into several iterative units, each composed of a leaf (or leaves) located at a node with a stem segment, a subtending internode, and an axillary meristem in the leaf axil—the junction of leaf and stem. (**b**) Organization of shoot apical meristem: zones and layers. SAM can be divided into three distinct zones based on cytoplasmic densities and cell division rates: central zone, peripheral zone, and rib zone, which might represent a functional subdivision of the SAM. The central zone is thought to harbor stem cells. A small group of cells (organizing center) is positioned at the top of the rib zone, which acts as a niche to sustain the overlying stem cell reservoir. SAM is also organized into three distinct cell layers: L1 (outermost cell layer), followed by L2, L3, and pith, which connects SAM with the vasculature

segment, a subtending internode, and an axillary meristem (AM) in the leaf axil—the junction of the leaf and stem (Fig. 2.1a; Carles & Fletcher, 2003). In *Arabidopsis*, the number of rosette leaves depends on ambient environmental conditions that influence the time to flowering. After vegetative growth to form the stem and leaves, the SAM and AMs turn into inflorescence meristems that generate flowers.

The SAM harbors a collection of pluripotent stem cells (Aichinger et al., 2012). As the stem cells divide, some daughter cells are displaced toward the peripheral regions of the SAM to produce lateral organs such as leaves, while others are retained at the shoot apex to replenish the reservoir from which stem cells are maintained and used for further growth. Proper SAM function requires the maintenance of the delicate balance between the creation of new meristematic cells by stem cell division at its center and the continuous production of lateral organs by stem cell differentiation from its flanks (Lee & Clark, 2013). The SAM is located at the shoot apex, and leaves are produced from its peripheral regions during the vegetative developmental stage, whereas a floral meristem forms and gives rise to flowers from its own stem cell population during the reproductive developmental stage. To maintain the balance between indeterminate and determinate cells in the dynamic structure of the SAM, coordination among complicated signaling networks consisting of many genes, proteins, and plant hormones that govern the establishment and organization of the SAM is required. In this section, we discuss the molecular mechanisms of SAM development, mainly in *Arabidopsis*, focusing on five aspects: SAM organization; SAM formation during embryogenesis; maintenance of an active SAM during postembryonic growth; organogenesis from the SAM; and finally, the formation of AMs.

2.2 Organization of the Shoot Apical Meristem: Zones and Layers

Histological analyses have shown that the SAM can be divided into three distinct zones based on cytoplasmic densities and cell division rates: the central zone (CZ), the peripheral zone (PZ), and the rib zone (RZ). These zones may represent a functional subdivision of the SAM (Fig. 2.1b; Steeves & Sussex, 1989). The CZ at the summit of the SAM, where cells self-maintain and divide relatively infrequently, provides initials for the surrounding PZ. The CZ is thought to harbor stem cells, which maintain the CZ itself as well as replenish both the PZ and the RZ. A small group of cells called the organizing center (OC) is positioned at the top of the RZ, which acts as a niche that sustains the overlying stem cell reservoir (Carles & Fletcher, 2003). The CZ is surrounded by a more rapidly dividing cell population in the PZ at the meristem flanks. Lateral organs are initiated from cells recruited from the PZ (Steeves & Sussex, 1989), where cells divide rapidly and move downward into the RZ, which produces internal stem tissues. The relative stages of lateral primordium development are described in terms of plastochron (P), whereby the latest emerging primordium is termed P1, the next oldest primordium is termed P2, and so on. The region within the meristem from which the next lateral organ primordium will form is termed P0, and all these primordia are organized in a spiral pattern in the PZ (Doerner, 1999). Thus, the behavior of cells in the SAM is governed in a position-dependent manner.

From the outside to the inside, the SAM is organized into three distinct cell layers that contribute to different tissues of the shoot (Fig. 2.1b; Satina et al., 1940). The outermost cell layer is L1, followed by L2, L3, and then the pith, which connects the SAM with the vasculature. In *Arabidopsis*, cells in L1 give rise to the epidermis of shoots, leaves, and flowers; L2 cells form the ground tissues and germ cells, while derivatives of the L3 subepidermal layer form the vascular tissues of the stem and the most internal tissues of leaves and flowers. These three layers are morphologically defined, and cells rarely cross over among the different layers because they divide anticlinally. The cells within each layer are derived from stem cells in that layer (Stewart & Dermen, 1970). However, the development of each cell layer is flexible and can adapt to changes in cell proliferation that occur in other layers (Szymkowiak & Sussex, 1996).

Unlike animal cells, plant cells cannot move freely. They must receive positional information from their neighboring cells via intercellular communication. In the SAM, stem cells in the CZ prevent PZ cells from producing new stem cell reservoirs to generate new organs. Cells in the L1 layer inhibit the differentiation of L2 cells and regulate their division patterns (Reinhardt et al., 2003). Therefore, the dynamic properties of the SAM are maintained by the coordination of cellular behavior between different zones and cell layers (Fletcher, 2002).

2.3 Formation of the Shoot Apical Meristem During Embryogenesis

In *Arabidopsis*, the embryonic SAM forms gradually during embryogenesis. Beginning at the globular stage, the cells that form the embryonic SAM become confined to the most apical and central portion of the embryo (Fig. 2.2; Barton & Poethig, 1993). Additional rounds of asymmetric division of the apical cells produce the heart-stage embryo, wherein the cotyledon primordia develop from the apical, peripheral region of the embryo. Stem cells in the embryonic SAM are first detectable in the apical cells between the two cotyledon primordia (Fletcher et al., 1999). The embryonic SAM is eminent in the torpedo-stage embryo, with a characteristic three-layered SAM structure (Barton & Poethig, 1993). As the embryo matures, the SAM becomes morphologically delineated by further anticlinal divisions in the upper two layers of cells and mixed-plane divisions in the underlying cells of the embryo (Laux et al., 1996).

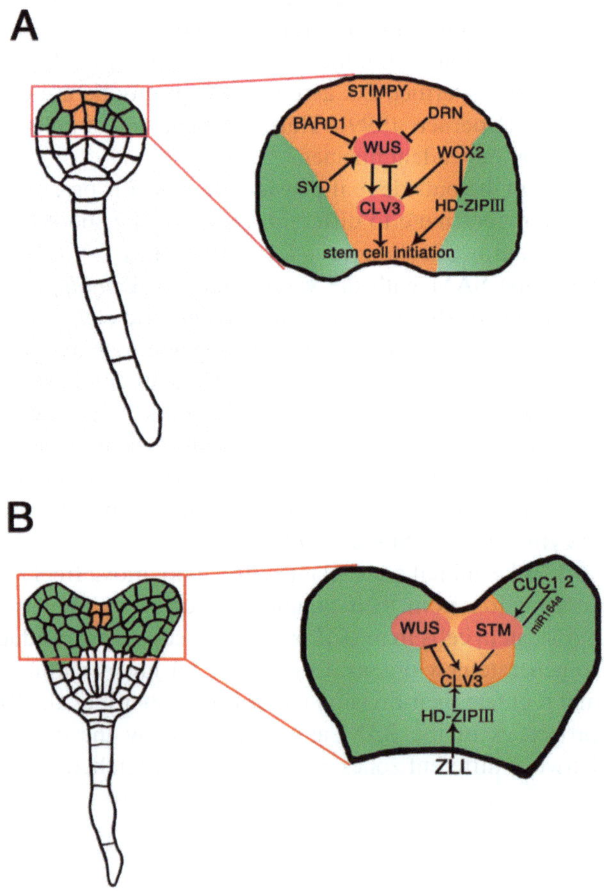

Fig. 2.2 Formation of shoot apical meristem during embryogenesis. Gene regulation in embryonic shoot apical meristem formation at globular embryo stage (**a**) and heart embryo stage (**b**). Different colors represent unique cell types. Green, presumptive cotyledon primordia; orange, embryonic shoot apical meristem

The SAM is gradually patterned during embryogenesis via the successive activation of meristem-restricted genes. During embryonic SAM initiation, the first molecular marker is *WUSCHEL* (*WUS*), which encodes a WOX family homeodomain transcription factor in *Arabidopsis*. Mutants lacking *WUS* function are unable to maintain a functional embryonic meristem, which leads to misspecification of stem cells and premature termination of the SAM (Laux et al., 1996). The expression of *WUS* begins in the four inner cells of the apical embryo region at the 16-cell stage of embryogenesis (Mayer et al., 1998). Its expression remains confined to a subset of cells in deeper OC regions during later stages of development. Therefore, *WUS* may act to specify the precursor cells of the OC during embryogenesis. The mechanisms underlying the initiation and early confinement of *WUS* expression in the embryonic SAM are still unclear.

The plant hormone cytokinin activates *WUS* expression during postembryonic SAM development (Gordon et al., 2009; Meng et al., 2017; Yang et al., 2021). Because WUS inhibits the expression of several negative regulators of cytokinin signaling, a positive feedback mechanism involving *WUS* and cytokinin signaling may operate to maintain SAM integrity (Leibfried et al., 2005). This crosstalk may already operate during embryogenesis. Cytokinin may contribute to the initial establishment of the OC by inducing *WUS* expression during early embryogenesis. In some mutants, the expression pattern of *WUS* is altered in embryos. For example, in the early globular embryo of mutants lacking *DORNRÖSCHEN* (*DRN*), *WUS* expression is shifted laterally into the apical regions (Chandler et al., 2011). *DRN* is expressed in the one-cell embryo of the embryo proper, suggesting that it is required to position *WUS* expression in the embryonic SAM. Two factors that interact directly with the *WUS* promoter are SPLAYED (SYD) and BRCA1-associated RING domain 1 (BARD1) (Kwon et al., 2005; Han et al., 2008). SYD, an SNF2 chromatin remodeling ATPase, binds to the promoter close to the transcriptional start site of *WUS*, and is required for its efficient transcription (Kwon et al., 2005). In *syd* mutants, the SAMs are reduced in size and *WUS* transcription levels are greatly reduced. The expression of *WUS* is also regulated by BARD1, which has previously been implicated in DNA repair (Han et al., 2008). Transcripts of *WUS* accumulate to much higher levels in *bard1* mutants than in wild type, and are located mainly in the outermost cell layers instead of OC cells. The overexpression of *BARD1* leads to reduced *WUS* transcript levels and the same phenotype as that of *wus* mutants. Therefore, BARD1 is required to repress *WUS* and confine its expression to the OC of the SAM. The BARD1 protein also binds to the *WUS* promoter and interacts with SYD (Han et al., 2008). Both *SYD* and *BARD1* are expressed during early embryogenesis, suggesting that both of these factors are involved in the *WUS* expression pattern and embryonic SAM establishment in early-stage embryos. The WOX family homeodomain transcription factor STIMPY/WOX9 is also involved in stem cell regulation. Mutations in *STIMPY/WOX9* cause premature termination of the SAM, the same phenotype as that of the *wus* mutant (Wu et al., 2005). *STIMPY* is required to maintain *WUS* expression in the SAM. Sucrose can rescue weak *stimpy* mutants, probably through its effect on cell division. Embryo development in the strong *stimpy* mutant is arrested at the early globular stage. It is likely that *STIMPY* plays essential roles in regulating early embryogenesis besides embryonic SAM formation (Wu et al., 2007).

A recent study showed that *WUS* is expressed during embryogenesis but is dispensable for stem cell initiation from the 16-cell stage to the globular stage (Zhang et al., 2017). Rather, the *WUS* paralog *WOX2*, the early embryo patterning factor, together with its redundant paralogs *WOX1*, *WOX3*, and *WOX5*, is required for the initiation of the stem cell program in this process. The *WOX2* and *WUS* expression domains overlap at the 16-cell stage (Mayer et al., 1998; Haecker et al., 2004). *WOX2* expression is then restricted to the apical domain of the embryo, which will give rise to the shoot, before its expression ceases. The *wox1,2,3,5* quadruple mutants are unable to form a shoot (Breuninger et al., 2008). *WOX2* regulates cell division patterns in the apical domain that give rise to the three-layered shoot meristem. After apical and basal embryo lineages are established, shoot apex fate is regulated by the expression of the *HOMEODOMAIN-LEUCINE ZIPPER CLASS III* (*HD-ZIP III*) transcription factor genes *PHABULOSA* (*PHB*), *PHAVOLUTA* (*PHV*), and *REVOLUTA* (*REV*) (Prigge et al., 2005; Byrne, 2006). Ectopic expression of *REV* in the root pole is sufficient to convert the root into a second shoot structure (Smith & Long, 2010). *WOX2* upregulates these *HD-ZIP III* transcription factors and initiates a regulatory cascade that establishes the stem cells in the embryonic SAM. During this process, the feedback signal *CLAVATA3* (*CLV3*) is activated; *CLV3* encodes a signaling peptide that restricts SAM activity (Fig. 2.2a).

The apical half of the globular embryo becomes partitioned into two cotyledon primordia and a central stripe that develops into the embryonic SAM. The interplay between *CUP-SHAPED COTYLEDON* (*CUC*) and *SHOOTMERISTEMLESS* (*STM*) is involved in specifying the central region as the embryonic SAM at the globular-embryo stage. Both *CUC* and *STM* are expressed during early embryogenesis and are required to establish and maintain the embryonic SAM. *STM* encodes a class I *KNOTTED-LIKE HOMEOBOX* (*KNOX1*) transcription factor. The *stm* mutant is unable to maintain a functional meristem during embryogenesis (Long et al., 1996; Long & Barton, 1998). *STM* is initially expressed in a few cells at the top of the globular embryo, implying that the presumptive SAM cells are specified at the embryo apex (Long & Barton, 1998). Then, *STM* is expressed throughout the vegetative and inflorescence SAMs, but not in the organ primordia. The expression pattern of *STM* in embryos and the phenotypes of *stm* mutants indicate that it functions in embryonic meristem formation. At the globular embryo stage, the closely related *CUC* genes *CUC1*, *CUC2*, and *CUC3* are first expressed in a narrow stripe at the apex of the embryo between the presumptive cotyledon primordia, such that the expression domains of *STM* and *CUCs* almost overlap at this stage of embryogenesis (Long et al., 1996; Aida et al., 1999; Takada et al., 2001). The single mutants of *cuc* display mildly defective phenotypes, whereas *cuc1 cuc2* double mutants exhibit fused cotyledons and lack a functional SAM (Aida et al., 1999; Takada et al., 2001; Vroemen et al., 2003). CUC1 and CUC2 are involved in inducing *STM* expression (Aida et al., 1999). *CUCs* function as positive regulators of the SAM, because overexpression of *CUC1* induces the formation of ectopic meristems (Takada et al., 2001). Recent studies have revealed that *STM* can, in turn, regulate

CUC expression in postembryonic tissue (Spinelli et al., 2011; Scofield et al., 2018). At the late stage of embryogenesis, *STM* is expressed throughout the SAM, but *CUC* expression is confined to the boundary region between the SAM and the two cotyledon primordia, a pattern complementary to that of *STM* expression (Aida et al., 1999; Takada et al., 2001; Vroemen et al., 2003). *STM* indirectly represses *CUC1* and *CUC2* expression by activating *MIR164a*, which targets *CUC* mRNA, suggesting the potential for negative feedback loops between *STM* and *CUC* (Fig. 2.2b; Spinelli et al., 2011). Therefore, precise spatiotemporal expression patterns and interactions of *STM*, *MIR164a*, and *CUC* during embryogenesis are required for embryonic SAM initiation.

By the early heart embryo stage, *CLV3* and *CLV1* mRNAs can be detected in the presumptive SAM between the developing cotyledons (Clark et al., 1996; Fletcher et al., 1999). *clv1* and *clv3* mutants display an increased expression domain of *WUS* from the heart stage onwards during embryogenesis. Therefore, CLV3 and CLV1 transmit a signaling cascade that restricts *WUS* expression to the OC during embryonic SAM initiation (Schoof et al., 2000). However, *CLV3* is not expressed in most mature *wus* embryos (Brand et al., 2002), indicating that *WUS* is required to maintain *CLV3* expression during embryogenesis. WUS specifies stem cell identity in a non-cell autonomous manner by promoting the expression of CLV3, which directly binds to the plasma membrane-localized receptor-like kinase, CLV1 (Fiers et al., 2005; Ito et al., 2006; Kondo et al., 2006; Ogawa et al., 2008; Rojo et al., 2002; Trotochaud et al., 1999). The onset of *CLV3* expression, the function of WUS, and the initiation of WUS/CLV3 feedback regulation are essential for maintaining the stem cell population in the embryonic SAM (Mayer et al., 1998; Schoof et al., 2000). Maintenance of *CLV1* expression requires STM function (Long & Barton, 1998), whereas maintenance of *CLV3* expression depends on both WUS and STM (Fig. 2.2b; Brand et al., 2002). In summary, patterning of the embryonic SAM is a gradual and dynamic process that is essential for the successful establishment of the OC and stem cells.

Mutation of *ZWILLE* (*ZLL*), which is a member of the ARGONAUTE (AGO) family, leads to an empty apex lacking a shoot meristem, as in *wus* seedlings, or a shoot apex with a differentiated leaf (Mcconnell & Barton, 1995; Moussian et al., 1998). Stem cell-specific expression of *CLV3* in *zll* embryos is initiated but not maintained, despite increased *WUS* transcription (Tucker et al., 2008). Furthermore, overexpression of *WUS*, which normally promotes stem cell identity, is ineffective in *zll* mutants. These findings indicate that *ZLL* is required to potentiate *WUS*-dependent signaling from the OC to maintain stem cell identity and *CLV3* expression. Transcripts of *ZLL* and ZLL protein are first detected throughout the proembryo between the 2-cell and 8-cell stages and subsequently become restricted to the vascular primordium, but are barely detectable in the shoot meristem until the embryo reaches maturity (Lynn et al., 1999; Moussian et al., 1998; Tucker et al., 2008). ZLL function in the vascular primordium but not in the stem cells is sufficient for stem cell maintenance (Tucker et al., 2008). Thus, ZLL acts to maintain the stem cells from the neighboring vascular primordium, indicating a non-cell-autonomous mechanism by which the vasculature sustains the overlying shoot meristem. In embryos of *zll* mutants, miR165/166 accumulate in the SAM and repress the expression of their targets; the *HD-ZIP III* genes *REV* and *PHB*, and Arabidopsis *HOMEOBOX GENE 8* (*AtHB15*) (Mallory et al., 2009). Overexpression of miRNA-resistant versions of *REV* or *PHB* and reduced miRNA165/166 levels can rescue shoot meristem development in *zll* mutants. ZLL appears to be less efficient than AGO1, the major AGO protein that degrades *HD-ZIPIII* mRNA. Therefore, accumulation of *HD-ZIPIII* mRNA in the shoot meristem region and maintenance of stem cells can be induced by ZLL sequestering miR165/166 away from AGO1 (Fig. 2.2b; Mallory et al., 2009). In the *zll* embryo, higher levels of miR165/166 can bind to AGO1, resulting in reduced *HD-ZIPIII* transcript levels and termination of the SAM and *CLV3* expression (Zhu et al., 2011; Boscá et al., 2011). Furthermore, auxin accumulation and auxin responses are enhanced in embryos of *zll* mutants, and transcription of *AUXIN RESPONSE FACTOR 2* (*ARF2*) is upregulated. Downregulation of *ARF2* can significantly rescue stem cell maintenance in *zll* mutants. Therefore, maintenance of stem cells in the embryonic SAM by ZLL involves the negative regulation of auxin signaling (Roodbarkelari et al., 2015).

2.4 Maintenance of the Shoot Apical Meristem During Postembryonic Development

2.4.1 Regulation of WUS*, the Master Regulator of the Shoot Apical Meristem*

The expression of *WUS* is maintained and confined to the OC cells (Laux et al., 1996; Mayer et al., 1998). This precise expression pattern is regulated by various transcription factors and chromatin remodeling factors, in addition to CLV3 peptide signaling (Kwon et al., 2005; Han et al., 2008; Yadav et al., 2011). *STIMPY* is a positive regulator of *WUS* expression not only during embryogenesis but also in postembryonic SAM maintenance (Wu et al., 2005). The transcriptional regulators OBERON1 (OBE1) and OBE2 have been implicated in maintaining stem cell activity in the SAM, and they function redundantly in the positive regulation of *WUS* and *CLV3* expression (Saiga et al., 2008). The double mutant *obe1 obe2*, but not the single mutants, displays an arrested SAM, like that of the *wus* mutant, with dramatically reduced expression of both *WUS* and *CLV3*. This implies that *OBE* factors are required to establish an appropriate SAM state via regulation of meristem maintenance genes (Saiga et al., 2008; Miwa et al., 2009). MERISTEM DEFECTIVE (MDF) has a putative role in transcriptional regulation or RNA processing. The *mdf* mutants show termination of the SAM and reduced *WUS* expression. Therefore, MDF may positively regulate *WUS* expression in the maintenance of the SAM (Casson et al., 2009). In contrast, overexpression of *MDF* causes ectopic meristem activity. APETALA 2 (AP2) plays an essential role in stem cell maintenance in the SAM, in addition to its roles as a floral homeotic protein in the ABC model of floral development (Wurschum et al., 2006). Mutations of *AP2* lead to premature termination of the SAM, like in *wus* mutants (Wurschum et al., 2006). AP2 regulates SAM maintenance through positively regulating WUS expression or negatively regulating CLV3 signaling. Another regulator in the WUS-mediated regulatory circuit is the basic helix-loop-helix transcription factor HECATE1 (HEC1), which activates stem cell proliferation in the SAM (Schuster et al., 2014). WUS directly represses the expression of *HEC1* in OC cells. Misexpression of *HEC1* in OC results in the loss of stem cell production. However, *HEC1* is expressed in the stem cells of the CZ, where it positively regulates cell proliferation by activating genes involved in the cell cycle and by suppressing *CLV3* expression.

Remodeling of chromatin organization is important in the activation and repression of *WUS* expression for SAM maintenance. *FASCIATED1* (*FAS1*) and (*FAS2*) encode subunits of CHROMATIN ASSEMBLY FACTOR-1 (CAF-1). The *fas1* and *fas2* mutants exhibit a defective SAM with expanded *WUS* expression (Kaya et al., 2001; Guyomarc'h et al., 2005). *BRUSHY1* (*BRU1*), which encodes a novel nuclear protein with two predicted protein–protein interaction domains, is associated with the regulation of *WUS* (Takeda et al., 2004). The *bru1* mutant phenotype is similar to that of *fas1* and *fas2*, suggesting that BRU1 and CAF-1 have similar roles in stabilizing chromatin structure and possibly in suppressing *WUS* expression (Takeda et al., 2004; Shen & Xu, 2009). The chromatin remodeling factor SYD specifically interacts with the *WUS* promoter to activate *WUS* expression for SAM maintenance (Guyomarc'h et al., 2005; Kwon et al., 2005). This was the first evidence of a chromatin remodeling factor directly regulating *WUS* expression. BARD1 has been implicated in DNA repair, but it also functions in SAM maintenance by confining *WUS* expression to the OC (Han et al., 2008). BARD1 directly interacts with the *WUS* promoter (Han et al., 2008; Shen & Xu, 2009) and also interacts with SYD to inhibit the chromatin remodeling process that is essential for *WUS* expression (Han et al., 2008). MGOUN1 (MGO1), which acts synergistically with chromatin remodeling factors, functions as a negative regulator of stem cell maintenance in the SAM. Genetic interaction studies have shown that the functions of WUS and MGO1 converge at a common downstream process in stem cell regulation, probably by controlling WUS-mediated gene repression via chromatin remodeling (Graf et al., 2010).

There are several factors that interact with WUS in the SAM, including transcriptional regulators in the HAIRY MERISTEM (HAM) family. These interactions are important for driving downstream transcriptional programs and in promoting shoot stem cell proliferation (Zhou et al., 2015). In the SAMs, WUS activates *CLV3* only in the absence of HAM, and in the AMs, an apical-basal gradient of HAM defines the patterning of the *CLV3* expression domain, suggesting that HAM TFs inhibit WUS from activating *CLV3* outside the SC domain (Zhou et al., 2018). WUS may also function as a repressor by recruiting TOPLESS (TPL) transcriptional corepressors to prevent the differentiation of stem cells (Kieffer et al., 2006). The carboxy terminus part of WUS interacts with the WD40 domain of the TPL, indicating that WUS may function in a repressor complex. A conservative strategy that WUS-mediated was revealed in plants to protect stem cells from virus invasion (Wu et al.,

2020). WUS can respond to cucumber mosaic virus (CMV) infection and repress virus accumulation in both the CZ and PZ of the meristem by repressing the expression of plant S-adenosyl-L-methionine–dependent methyltransferases, which are involved in ribosomal RNA processing and ribosome stability (Wu et al., 2020).

2.4.2 CLV3 Perception and Signaling Control the Number of Shoot Stem Cells

The first morphological meristem mutants identified in *Arabidopsis* were the *clv* mutants. A signal cascade essential for regulating the number of stem cells in the SAM has been described based on analyses of *clv* mutants. The *clv* mutants display an enlarged SAM and overproliferation of stem cells in the CZ, and they produce more organs than do wild-type plants (Leyser & Furner, 1992; Fletcher et al., 1999). The enlarged meristem of *clv3* mutants is caused by expanded *WUS* expression. The CLV3 peptide secreted from the CZ restricts *WUS* expression in cells of the OC (Fletcher et al., 1999; Rojo et al., 2002). It is perceived by many receptors, including plasma membrane-localized receptor-like kinases (RLKs), receptor-like proteins (RLPs), and receptor-like cytoplasmic kinases (RLCKs) (Brand et al., 2000; Fiers et al., 2005; Hobe et al., 2003; Müller et al., 2008; Schoof et al., 2000). Among these CLV3 receptors is a leucine-rich repeat (LRR) receptor kinase encoded by the RLK gene *CLV1*, which is expressed in the CZ of the SAM, including in OC cells and the L3 layer of stem cells (Fig. 2.1b; Leyser & Furner, 1992). The *clv1* mutant displays an enlarged SAM phenotype and was one of the first meristem mutants to be identified. CLV1 has three paralogs encoded by the *BARELY ANY MERISTEM1* (*BAM1*), *BAM2*, and *BAM3* genes, with their primary expression occurring on the periphery of the SAM (DeYoung et al., 2006; Nimchuk et al., 2015). As their name implies, the three *BAM* genes collaborate to facilitate the maintenance of stem cells. (DeYoung et al., 2006). Indeed, similar to CLV1, BAM1 and BAM2 can directly bind CLV3 peptides (Deyoung & Clark, 2008; Shinohara & Matsubayashi, 2015). There is also evidence indicating that CLV1 activity in the RZ suppresses the transcription of *BAM* genes, independently of its function in regulating WUS transcription, such that in *clv1* mutants, ectopic *BAM* expression within the inner regions of the SAM serves as a partial compensation for the lack of CLV1 (Nimchuk et al., 2015; Nimchuk, 2017). The RLP gene *CLV2* encodes an LRR-receptor-like transmembrane protein that lacks an intracellular kinase domain. Morphological analyses of *clv1* and *clv2* mutants have shown that the double mutant but not the single mutants mimic the nature and severity of the *clv3* phenotype, indicating at least partially redundant roles for the two receptors (Jeong et al., 1999; Kayes & Clark, 1998). The RLCK gene *CORYNE* (*CRN*) was identified as a co-receptor of *CLV2* (Müller et al., 2008). CLV2 interacts with CRN to form a CLV3-receptor complex for stem cell maintenance that is independent of CLV1. In contrast to CLV1, CLV2 and CRN are expressed not only in the CZ but also throughout the entire SAM;

therefore, they could function outside of central CLV3 perception (Müller et al., 2008). These three receptors constitute the two main signaling pathways of CLV3 peptide perception in the SAM: one signaling through CLV1 and the second independent signaling pathway through the CLV2–CRN heterodimer (Bleckmann et al., 2010; Guo et al., 2010; Müller et al., 2008). Another receptor kinase, RECEPTOR-LIKE PROTEIN KINASE 2 (RPK2), functions as a suppressor of the CLV3 peptide-mediated SAM-arrested phenotype (Kinoshita et al., 2010). Phenotypes of *rpk2* are additive to *clv1* and *clv2*, suggesting that RPK2 may function independently of the CLV1 and CLV2-CRN receptor pathways in the transmission of CLV3 signaling. A group of Leucine-Rich Repeat Receptor-Like Kinases (LRR-RLKs), known as CLAVATA3 INSENSITIVE RECEPTOR KINASES (CIKs), have essential roles in the regulation of stem cell homeostasis mediated by CLV3. The *cik1 2 3 4* quadruple mutant exhibits a significantly enlarged SAM, resembling *clv* mutants. Genetic analyses and biochemical assays demonstrated that CIKs function as co-receptors of CLV1, CLV2/CRN, and RPK2, facilitating CLV3 signaling through phosphorylation (Hu et al., 2018).

2.4.3 The WUS–CLV3 Pathway Is a Central Feedback Loop Essential for Shoot Meristem Maintenance

Many factors influence and control shoot meristem maintenance, as described above. The meristem homeostasis pathway explored in the greatest detail is the WUS–CLV3 pathway (Fig. 2.3a). This signaling pathway has evolved as the central regulatory pathway coordinating stem cell proliferation and differentiation. The coordination is achieved via an autoregulatory negative-feedback loop comprising the stem cell-promoting transcription factor WUS and the differentiation-promoting peptide CLV3 (Brand et al., 2000; Schoof et al., 2000). The essential nature of this pathway is exemplified by the phenotypes of mutants with disruptions in this pathway, ranging from shoots with no stem cells to shoots with only stem cells. This pathway appears to be central for signaling between the stem cells and the underlying OC.

WUS moves between cells through plasmodesmata to maintain the stem cell activity in the SAM (Yadav et al., 2011; Daum et al., 2014). *WUS* is expressed in OC cells just beneath the stem cell domain of the CZ, and moves to the CZ cells to promote the initiation and maintenance of the overlying stem cells by stimulating the activity of CLV3 therein. CLV3 is a small secreted peptide that serves as a stem cell marker (Mayer et al., 1998; Fletcher et al., 1999; Brand et al., 2000). To test the significance of plasmodesmata-mediated WUS movement, it was artificially restricted by fusing tandem repeats of green fluorescent protein (GFP) to increase its molecular weight, by fusing nuclear localization signals (NLS) to target it to the nucleus, or by artificially blocking the plasmodesmata pores (Yadav et al., 2011; Daum et al., 2014). In each case, restriction of WUS transport failed to completely

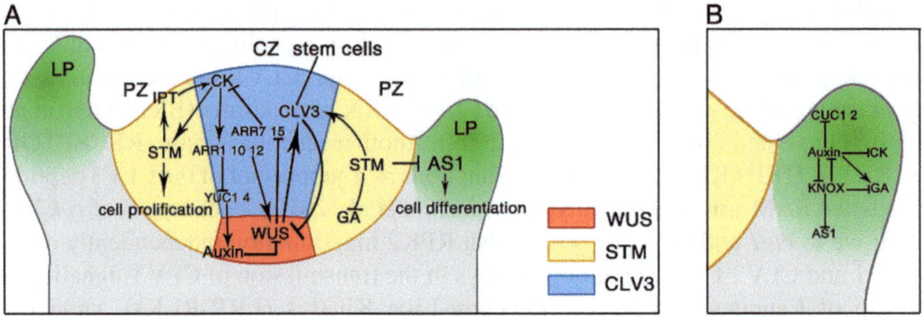

Fig. 2.3 Molecular mechanisms of regulation of shoot meristem maintenance and leaf primordia initiation. (**a**) Regulatory pathways controlling shoot meristem maintenance. The central regulatory pathway is an autoregulatory negative-feedback loop comprising stem cell-promoting transcription factor WUS and differentiation-promoting peptide CLV3. The parallel pathway to WUS–CLV3 pathway is principally mediated by the homeobox gene *STM*. Cytokinin positively regulates the maintenance of shoot stem cell activity. Expression of *STM* induces the synthesis of cytokinin, which positively regulates *STM* to maintain stem cell activity in the SAM. Feedback between WUS and the cytokinin signaling pathway produces accurate positioning of the stem cell niche. (**b**) In leaf primordia initiation, high auxin concentration activates gibberellin but represses cytokinin biosynthesis. Expression of *KNOX* and *CUC* is repressed in the incipient organ primordium by an auxin-dependent mechanism. KNOX proteins may inhibit auxin transport, indicating feedback regulation between auxin and KNOX. *AS1* is expressed in lateral organ primordia cells in a pattern complementary to that of *KNOX*

rescue *wus* mutant phenotypes, suggesting that WUS transport is crucial for maintenance of the SAM. WUS migrates from the OC into superficial cell layers of the CZ to activate *CLV3* transcription by binding to regulatory regions in the *CLV3* promoter (Yadav et al., 2011). Recently, it was shown that WUS binds to the same group of *cis*-elements in the *CLV3*-regulatory region to activate and repress *CLV3* expression in a concentration-dependent manner (Perales et al., 2016; Sloan et al., 2020). As a monomer at low levels, WUS activates the transcription of *CLV3*, but as a homodimer at higher levels, WUS may fail to engage the activation complex, thereby repressing *CLV3* transcription (Perales et al., 2016; Sloan et al., 2020; Wang & Jiao, 2023). In addition, CLV3 signaling attenuates the nuclear export of WUS protein from the nucleus into the cytosol, which limits its movement from the L3 cells into the outer cell layers of the SAM and also prevents its lateral dispersion (Plong et al., 2021).

The transcriptional regulation of *WUS* is limited by the *CLV3* signaling pathway. *clv3* mutants have opposite phenotypes to those of *wus* mutants, in that *clv3* mutants have enlarged meristems that specifically accumulate stem cells (Clark et al., 1993, 1995; Kayes & Clark, 1998). *WUS* expression expands into the apical layers of stem cells in *clv3* mutants so that *WUS* is expressed in both the apical and basal L3 daughter cells. The expanded OC represented by *WUS* expression leads to ectopic stem cell specification in the overlying cell layers, which explains the *clv3* phenotypes (Schoof et al., 2000; Brand et al., 2000). Overexpression of *CLV3* causes premature termination of the SAM (Mayer et al., 1998; Brand et al., 2000; Miwa et al., 2009). Thus, CLV3 acts as a signal to repress *WUS* expression. Accordingly, repression of *WUS* by CLV3 results in fewer stem cells being maintained, and, ultimately, in reduced *CLV3* production (Brand et al., 2000; Schoof et al., 2000). This feedback loop enables the stem cell compartment and the OC domain to maintain their size by adjusting relative to each other. In the model plant *Arabidopsis*, meristem growth in *wus* mutants comes to a halt during vegetative development, leading to the failure to produce flowers and fruits. Conversely, in *clv3* mutants, meristems undergo significant enlargement, resulting in the formation of an increased number of floral organs. However, the *clv3 wus* double mutants display no discernible differences from *wus* single mutants, indicating an epistatic relationship between WUS and CLV3 (Schoof et al., 2000). Research has further affirmed the conservation of this relationship through mutagenesis of tomato orthologs (Rodríguez-Leal et al., 2017, 2019; Wang et al., 2021). In tomatoes, there is a homologous gene to *CLV3* called *SlCLE9*, which can also interact with WUS. While null mutants of *Slcle9* are indistinguishable from wild-type plants, *Slclv3* is strongly enhanced by *Slcle9*, demonstrating a canonical unequal redundancy (Briggs et al., 2006) epistatic relationship between these paralogs (Kwon et al., 2022). It was shown that the regulation of *SlWUS* by SlCLE9 and SlCLV3 exhibits a dose-dependent effect. That is, once the expression of *SlCLE9* falls below or rises above a critical threshold, the regulatory effect on *SlWUS* by SlCLV3 becomes dominant. Within this range, SlCLE9 and SlCLV3 work together in a synergistic manner to inhibit the *SlWUS* gene (Aguirre et al., 2023).

2.4.4 The KNOX Pathway Maintains Shoot Meristem from Organogenesis

Genetic analyses have identified a pathway parallel to the WUS-CLV3 pathway that is essential for stem-cell initiation and maintenance in the SAM. The pathway is principally mediated by the homeobox gene *STM* (Fig. 2.3a; Long et al., 1996). *STM* is expressed in both the CZ and the PZ of the SAM but is repressed in organ primordia, indicating its role in maintaining cells in an undifferentiated state. Weak loss-of-function *STM* mutations (*stm* mutants) lack self-renewing stem cell populations in the SAM, which is prone to premature termination (Clark et al., 1996; Endrizzi et al., 1996). STM was reported to be a second candidate regulator of *CLV3* expression in the SAM, besides WUS (Brand et al., 2002). The expression level of *CLV3* is reduced in the SAM of embryos and seedlings of *stm* mutants. Although *STM* is unable to induce *CLV3* transcription and stem cell activity in the absence of *WUS* activity (Lenhard et al., 2002), its activity is necessary to maintain a high level of *CLV3* expression in the SAM (Brand et al., 2002). Co-expression of *WUS* and *STM* produces synergistic effects in conferring ectopic stem cell identity on the hypocotyl or cotyledons (Brand et al., 2002). STM interacts with WUS, enhancing WUS binding to the *CLV3* promoter. Additionally, STM has the capability to directly activate *CLV3* transcription, and its own expression is triggered by WUS. The interaction between WUS and STM induces a robust CLV3 signal, leading to the negative regulation of *WUS* transcription. This mechanism plays a crucial role in modulating the pool size of available stem cells (Su et al., 2020). In contrast, STM and WUS function in different pathways in shoot meristem regulation (Lenhard et al., 2002). WUS can activate stem cell specification in the absence of STM activity. Although CLV1::WUS in *stm* mutants can induce adventitious shoot meristem formation, the meristems grow much more slowly than the corresponding meristems in CLV1::WUS lines (Schoof et al., 2000; Lenhard et al., 2002). Conversely, STM suppresses stem cell differentiation and promotes cell division independently of WUS, by preventing the expression of the organ-formation factors *ASYMMETRIC LEAVES1* (*AS1*) and *AS2* in the center of the SAM (Fig. 2.3a; Byrne et al., 2002; Scofield and Murray, 2006). Whether and how the regulatory pathways are defined by the cooperation of WUS and STM in the activation of stem cell identity remains unclear.

The STM protein initiates self-activation by binding to the *STM* promoter. This process is facilitated by the BELL1-Like homeodomain family-binding proteins associated with STM, with a prominent role played by ARABIDOPSIS THALIANA HOMEOBOX GENE1 (ATH1). ATH1–STM heterodimers play a crucial role in sustaining STM expression, where ATH1 is responsible for DNA binding, and STM takes charge of transcriptional activation (Cao et al., 2020). Recent research has revealed that the STM protein has the capability to form nuclear condensates. Two classes of proteins, namely BELL and Mediator complex subunit MED8, have been identified interacting with the STM protein, collectively contributing to the formation of these condensates. The formation of STM protein condensates has been shown to enhance its own transcriptional regulatory activity, thereby contributing to the maintenance of SAM activities. Furthermore, it has been discovered that salt treatment enhances the formation of STM protein condensates. This augmentation results in increased meristem activity, thereby imparting plant resilience to salt stress (Cao et al., 2023).

Three other related KNOX proteins, KNAT1/BP (KNOTTED1-LIKE IN *ARABIDOPSIS* THALIANA/BREVIPEDICELLUS), KNAT2, and KNAT6, are expressed in the SAM of *Arabidopsis*. These KNAT proteins act redundantly with STM and restrict the expression of *AS1* and *AS2* to leaf primordia, thereby preventing stem cells from differentiating into leaves within the SAM (Fig. 2.3a; Byrne et al., 2002; Scofield and Murray, 2006). In turn, AS1 and AS2 repress the expression of *KNOXs* in developing organ primordia. This occurs as a result of their interaction with the histone chaperone HIRA to form a transcriptional repression complex (Phelps-Durr et al., 2005; Ha et al., 2010).

2.4.5 Cytokinin Is Involved in Shoot Meristem Maintenance

The central regions of the shoot meristem contain a population of undifferentiated stem cells that are the source of new organ primordia at the peripheral regions. The results of early studies confirmed that cytokinin promotes the proliferation of undifferentiated stem cells in the SAM. Individually overexpressing six different members of the *cytokinin oxidase/dehydrogenase* (*CKX*) gene family, which encode enzymes that irreversibly catalyze active cytokinin degradation, leads to a SAM with a reduced size (Werner et al., 2003). In contrast, the *Arabidopsis ckx3 ckx5* double mutants have larger inflorescence meristems due to increased numbers of stem cells and delayed cell differentiation (Bartrina et al., 2011). Mutation of all three genes encoding a subfamily of histidine kinases (CYTOKININ RESPONSE 1 (CRE1), *ARABIDOPSIS* HISTIDINE KINASE 2 (AHK2), and *ARABIDOPSIS* HISTIDINE KINASE 3 (AHK3) that function as cytokinin receptors results in a severe reduction in SAM size (Nishimura et al., 2004). Quadruple mutants of *Arabidopsis ISOPENTENYL TRANSFERASE1* (*IPT1*), *IPT3*, *IPT5*, and *IPT7*, which encode the enzyme catalyzing the first step of cytokinin biosynthesis, show significantly decreased cytokinin levels and reduced SAM size (Miyawaki et al., 2004, 2006). The fact that impaired cytokinin biosynthesis or signaling, as well as enhanced cytokinin degradation, confer a small-SAM phenotype indicates that cytokinin is a positive regulator of shoot stem cell activity. Recent studies have begun to uncover the mechanisms of cytokinin biosynthesis, distribution, degradation, and perception that are closely associated with SAM formation and maintenance.

The KNOX transcription factors, exemplified by KN1 (Vollbrecht et al., 1991) in maize and STM (Long et al., 1996) in *Arabidopsis*, are central players in the control of SAM activity. In *Arabidopsis*, *KNOXs* repress differentiation of stem cells and are required for the maintenance of stem cell activity in the SAM. Intriguingly, KNOX proteins activate cytokinin signaling by inducing cytokinin biosynthesis. Ectopic expression of *KNAT2*, *KN1*, and *STM* leads to increased *IPT7* expression and upregulated cytokinin responses (Yanai et al., 2005). In rice, transcription of *OsIPT2* and *OsIPT3* is induced by overexpression of the *KN1* homolog, *Oryza sativa* homeobox1 (*OSH1*) (Sentoku et al., 1999). Cytokinin promotes the expression of *STM* in *Arabidopsis*, setting up a positive feedback loop (Fig. 2.3a; Jasinski et al., 2005; Yanai et al., 2005). Overexpression of *IPT7* under control of the *STM* promoter in the SAM can partially rescue strong loss-of-function *stm* mutations. Isolation of a new cytokinin-deficient mutant, *lonely guy* (*log*) in rice, revealed a new aspect of strong links between cytokinin biosynthesis and shoot stem cell function. *LOG* encodes a phosphoribohydrolase that converts cytokinin from its inactive form into its active form (Kurakawa et al., 2007). *LOG* is specifically expressed at the tip of the first two or three layers of the SAM, where stem cells are located, suggesting that *LOG* may define the maintenance of stem cells. Rice *log* mutants have a perturbed SAM, markedly reduced meristem size, and reduced expression of *OSH1* (Matsuoka et al., 1993; Kurakawa et al., 2007). These results indicate that the expression of *KNOXs* in the SAM is sufficient to induce cytokinin synthesis, which in turn positively regulates *KNOXs* to maintain stem cell activity in the SAM.

A local increase in cytokinin induces *WUS* expression in OC cells (Gordon et al., 2009), indicating that cytokinin positively regulates *WUS* expression. Specific expression of *LOG* in CZ cells suggests that local cytokinin biosynthesis occurs in stem cells (Kurakawa et al., 2007; Perales & Reddy, 2012). This apically produced cytokinin in superficial layers of the CZ may diffuse into the OC to provide a positional cue for localized *WUS* expression. The cytokinin-degrading enzyme CKX3 is specifically expressed in OC cells of the *Arabidopsis* rib meristem (Bartrina et al., 2011). Furthermore, *ckx3 ckx5* double mutants exhibit higher levels of *WUS* expression in their larger SAM compared with wild type. In rice, a high level of *OsCKX2* expression reduces grain number, indicating a positive role of cytokinin in the formation of spikelet primordia in the meristem (Ashikari et al., 2005). Thus, the spatially controlled synthesis and degradation of cytokinin maintain *WUS* expression within the OC cells and stem cell activity in the SAM.

WUS is expressed in the OC cells of the internal layers and promotes stem cell identity in the overlying cells of the meristem. It also induces stem cells to express the peptide ligand CLV3, which feedback-inhibits *WUS* expression via its interaction with the receptor-like kinase CLV1 and maintains the stem cell niche in the SAM (Schoof et al., 2000). This WUS-CLV feedback loop interacts with the downstream response regulator of cytokinin signaling, which consists of both positive (B-type) and negative (A-type) Arabidopsis Response Regulator (ARR) transcription factors (Hwang et al., 2012). While B-type ARRs activate cytokinin signaling, A-type ARRs suppress it, which appears to limit the SAM size (Gordon et al., 2009). Constitutive expression of *OsARR6*, an A-type response regulator in rice, produces a smaller shoot meristem than that of wild-type plants (Hirose et al., 2008). WUS directly suppresses several negative A-type regulators (ARR5, ARR6, ARR7, and ARR15) of cytokinin signaling (Leibfried et al., 2005), thus enhancing downstream cytokinin responses. Thus, WUS might be necessary to provide restricted position information for cytokinin signaling in the SAM. WUS also reduces cytokinin biosynthesis via the downregulation of *LOG4* expression (Chickarmane et al., 2012). These results suggest that the positioning of *WUS* expression is crucial for the cell-type-specific control of cytokinin biosynthesis and signaling for SAM function. Overexpression of *ARR7* inhibits *WUS* expression, resulting in arrested SAM activity, whereas downregulation of *ARR7* and *ARR15* expression causes a decrease in *CLV3* levels, suggesting that they can maintain *CLV3* expression independently of *WUS* (Zhao et al., 2010). In addition, overexpression of *ARR7* and *ARR15* severely inhibits *WUS* expression and SAM regeneration from the root-induced callus, while the single mutants show accelerated SAM formation in vitro (Buechel et al., 2010). These results establish a model in which *WUS* expression increases in response to high cytokinin levels, thus increasing stem cell proliferation in both the vegetative and the reproductive SAMs (Bartrina et al., 2011). Recently, it has been shown that the type-B ARRs ARR1, ARR10, and ARR12, which are critical components of cytokinin signaling, directly activate *WUS* transcription by binding to its promoter. In parallel, these ARRs directly repress the expression of auxin biosynthetic genes *YUCCAs* (*YUCs*), which indirectly promote *WUS* induction (Fig. 2.3a; Meng et al., 2017; Wang et al., 2017; Xie et al., 2018). Therefore, the feedback between WUS and the cytokinin signaling pathway, together with WUS-induced *CLV3* expression, produces accurate positioning of the stem cell niche corresponding to the maximum cytokinin response (Chickarmane et al., 2012).

Besides WUS, another WUS-related homeodomain transcription factor, STIMPY, mediates cytokinin signaling and promotes cell proliferation within the *Arabidopsis* SAM (Wu et al., 2005; Skylar et al., 2010). *Stimpy* mutants have a small and differentiated SAM resembling that of the *wus* mutant, with a reduced number of stem cells (Wu et al., 2005). The triple mutants of cytokinin receptors (*ahk2 ahk3 cre1*) or B-type ARRs (*arr1 arr10 arr12*) show significant downregulation of *STIMPY* expression (Skylar et al., 2010), indicating that cytokinin induces *STIMPY* expression in the meristem. In the *stimpy* mutants, the expression of *ARR5* is reduced, suggesting that the cytokinin response is also suppressed by *STIMPY*. Therefore, cytokinin signaling is closely associated with the function of *STIMPY* in maintaining stem cell activity and preventing stem cell differentiation in the SAM. Research has expanded our understanding of cytokinin control in regulating growth within the SAM. Yang et al. discovered a long-sought molecular mechanism underlying the regulation of cell division by cytokinin in the SAM of *Arabidopsis* (Yang et al., 2021). They demonstrated that during the G2/M transition, elevated levels of cytokinin facilitate rapid nuclear trafficking of the Myb-domain protein 3R4 (MYB3R4), a transcriptional factor that activates mitosis gene expression. This mechanism enhances cell proliferation in the SAM (Yang et al., 2021).

2.4.6 Integration of Hormone Signaling in SAM Maintenance

The KNOX proteins are important for meristem development and have putative roles in simultaneously stimulating cytokinin accumulation and repressing gibberellin biosynthesis (Sakamoto et al., 2001; Hay et al., 2002; Chen et al., 2004). This antagonistic relationship between cytokinin and gibberellin is confirmed by the fact that mutations of *spindly* (*spy*) in *Arabidopsis* with constitutive gibberellin signaling enhance the phenotype of the weak *stm* allele, similar to the effect of reduced cytokinin perception on the weak *stm* phenotype (Jasinski et al., 2005). However, the combination of a *spy* mutation in a background overexpressing *CKX3* with reduced cytokinin levels results in strong defects in the SAM, compared with that of the *spy* mutant (Jasinski et al., 2005), indicating that increased cytokinin levels and reduced gibberellin signaling are critical for proper SAM function. Cytokinin also activates the expression of *GA2ox* to stimulate gibberellin deactivation at the base of the SAM. This avoids the inflow of active gibberellin from leaf primordia and maintains low gibberellin levels within the SAM (Jasinski et al., 2005; Wolters & Jürgens, 2009).

Auxin is biosynthesized by YUC enzymes in the central zone of the shoot meristem, and accumulates to an optimal level to stimulate downstream auxin-induced genes (Zhao et al., 2008). ARR7 and ARR15 are negative regulators of cytokinin signaling that have important functions in SAM maintenance (To et al., 2004). Although the expression of *ARR7* and *ARR15* is induced by cytokinin, a dramatic increase in their expression was observed in the SAM of *yuc* mutants, which lack essential local auxin biosynthesis (Zhao et al., 2010). *ARR7* and *ARR15* are ectopically expressed in the CZ and PZ of the SAM in mutants of MONOPTEROS (MP)/ARF5, a member of the auxin response factor (ARF) gene family. This result indicates that *ARR7* and *ARR15* are targets of *MP*-mediated auxin signaling in the regulation of shoot meristem function (Zhao et al., 2010). Because of their different effects on these type-A *ARRs*, auxin and cytokinin are able to act synergistically in the SAM. It seems that auxin and cytokinin signaling converge on *ARR7* and *ARR15* in the central regions of the meristem.

The determination of organ fate depends not only on the hormonal cues necessary to initiate the organogenesis program but also on the suppression of meristematic genes. This suppression enables the transformation of pluripotent meristematic cells from the PZ into organs (Shi & Vernoux, 2021). Auxin is also involved in this repression. MP/ARF5 indirectly inhibits *STM* through the action of its target *FILAMENTOUS FLOWER* (*FIL*). *STM* is also directly inhibited by two other members of the ARF transcription factor family, namely ARF3/ETTIN (ETT) and ARF4, which act as repressors. This inhibition is achieved through the recruitment of the histone deacetylase HDA19, which forms a corepressor complex with FIL, ETT, and ARF4 (Chung et al., 2019; Shi & Vernoux, 2021). Moreover, ETT/ARF3 is primarily induced at the periphery of the SAM and subsequently translocated into the OC cells. There, it directly suppresses WUS transcription and cytokinin activity, thereby restricting meristem size in a non-cell-autonomous manner (Zhang et al., 2022).

Ethylene signaling negatively regulates SAM function. The constitutive ethylene response caused by mutation of *CONSTITUTIVE TRIPLE RESPONSE1* (*CTR1*) results in a disrupted SAM structure with fewer cells in the meristem compared with wild type (Hamant et al., 2002). Ethylene signaling acts antagonistically with the activity of the *Arabidopsis KNOX* gene *KNAT2* in the SAM. The overexpression of *KNAT2* in transgenic tobacco plants induces a higher rate of shoot initiation in the leaves (Hamant et al., 2002). Exogenous treatment with the ethylene precursor severely inhibits the expression of *KNAT2* in the SAM and reduces the number of cells in the meristem, while the negative effect of ethylene precursor treatment is repressed by the induction of *KNAT2* (Hamant et al., 2002). Conversely, *KNAT2* expression is enhanced in the SAM of the ethylene-resistant *ethylene response1* (*etr1*) mutant, confirming the negative regulation of *KNAT2* by ethylene.

2.5 Organogenesis from the Shoot Apical Meristem

2.5.1 Leaf Primordia Initiation

Initiation of lateral organs from the PZ of the SAM requires a developmental switch from the pluripotent fate of meristem cells to the determinate fate of lateral organ cells. The pluripotent cell fate in the SAM is maintained by class I *KNOX* homeodomain transcription factors (Long et al., 1996). In *Arabidopsis*, the *KNOX* genes

STM, *KNAT1*, *KNAT2*, and *KNAT6* are expressed throughout the meristem but are down-regulated at the sites of leaf primordia (Belles-Boix et al., 2006; Lincoln et al., 1994; Long et al., 1996). Ectopic expression of *KNOX* genes in the leaf primordia marks the formation of the ectopic meristem (Chuck et al., 1996; Long et al., 1996). Thus, one of the earliest indications of proper organ differentiation is the repression of *KNOX* expression in the lateral organ primordia (Fig. 2.3b).

The MYB domain transcription factor *AS1* is expressed in lateral organ primordia cells in a pattern complementary to that of *KNOX* genes (Fig. 2.3b; Byrne et al., 2000). *STM* maintains the SAM integrity through repressing *AS1* (Byrne, 2005). In turn, *AS1* represses the expression of the *KNOX* genes *KNAT1*, *KNAT2*, and *KNAT6* in the developing leaf primordia (Ori et al., 2000; Byrne, 2005). The function of the plant-specific LATERAL ORGAN BOUNDARIES (LOB) domain protein AS2 in leaf primordia formation is similar to that of AS1 (Ikezaki et al., 2010). AS1 and AS2 physically interact to form a repressor complex that directly binds to specific promoter regions in *KNOX* genes (Guo et al., 2008). The AS1-AS2 repressor complex also interacts with the chromatin-remodeling protein HIRA and Polycomb-repressive complex2 to maintain a repressed chromatin state at the target loci in the *KNAT1* and *KNAT2* promoters during leaf initiation (Phelps-Durr et al., 2005; Guo et al., 2008; Lodha et al., 2013). Other chromatin-remodeling proteins PICKLE (PKL), SERRATE (SE), and RNA-DEPENDENT RNA POLYMERASE6 (RDR6) are also involved in the repression of *KNOX* genes in leaf primordia. Research has demonstrated that in rice, the polycomb repressive complex 2 (PRC2)-associated coiled-coil protein (PACP) has the capability to interact with both the core components of PRC2 and KNOX family proteins. This interaction plays a crucial role in maintaining the silencing of genes that promote cell differentiation in the shoot apex (Tan et al., 2022). Loss of function in PACP led to a reduction in genome-wide trimethylation of histone H3 at lysine 27 (H3K27me3) and diminished binding of the PRC2. Consequently, this resulted in differential gene expression patterns similar to those observed in plants where *PRC2* genes were knocked down. There is additional evidence suggesting that PACP is essential for the suppression of cell differentiation-promoting genes that are targeted by a rice KNOX1 protein in the SAM. PACP is involved in the recruitment or stabilization of PRC2 at genes targeted by KNOX/BELL transcription factors, ensuring the maintenance of H3K27me3 and gene repression in dividing cells of the shoot apex (Tan et al., 2022). These results suggest that epigenetic regulation to maintain the stable silencing of *KNOX* genes is required for differentiation of organ primordia (Li et al., 2005; Ori et al., 2000).

The AS1–AS2 repressor complex also binds to the JAGGED LATERAL ORGANS (JLO) protein to form a trimeric complex that is involved in regulating organ initiation (Rast & Simon, 2012). *JLO* is first expressed at leaf initiation sites and then resolves to the boundary of the leaf primordia during its outgrowth. *jlo* mutants display impaired leaf outgrowth, which is caused at least partly by the ectopic expression of *KNAT1* and *STM* at the base of leaf primordia (Rast & Simon, 2012). *JIO* regulates auxin distribution during organ initiation through activating PIN-mediated auxin efflux (Bureau et al., 2010; Rast & Simon, 2012; Žádníková & Simon, 2014).

BLADE-ON-PETIOLE 1 (*BOP1*) and *BOP2* are involved in cell differentiation during leaf primordia initiation (Ha et al., 2003, 2007). BOP1 and BOP2 likewise resolve to the boundary of the emerging leaf during its outgrowth (Ha et al., 2004; Hepworth et al., 2005; Norberg et al., 2005; Borghi et al., 2007). They negatively regulate the expression of *KNOX*, which confers meristem cell fate and positively regulates the *ASL*/*LBD* genes that promote lateral organ fate and polarity, such as *AS2* (Ha et al., 2003, 2007). BOP1 directly induces the expression of its target gene *AS2* as a transcriptional coactivator through binding to the *AS2* promoter (Jun et al., 2010). STM represses *BOP1* and *BOP2* expression to maintain meristem cell indeterminacy in embryonic SAM development. Conversely, BOP1 and BOP2 restrict *KNOX* expression to condition proper leaf differentiation (Jun et al., 2010).

2.5.2 Auxin Regulates Leaf Formation and Position-Phyllotaxis

Lateral organ initiation at the periphery of the SAM occurs after precise patterning known as phyllotaxis, which results from a self-organizing process of stem cell differentiation in the SAM (Adler et al., 1997). Auxin is a central regulator of SAM function, and is mainly associated with organ initiation and positioning at the SAM periphery. Auxin distribution is concentrated at the sites where organ primordia form in the SAM, as demonstrated by the auxin-inducible DR5 marker, indicating an activation of auxin signaling (Benková et al., 2003; Heisler et al., 2005; De Reuille et al., 2006; Smith et al., 2006). DR5 was only weakly detected in the CZ but showed rays of expression along the PZ, corresponding to incipient primordia (Ma et al., 2019; Galvan-Ampudia et al., 2020; Burian et al., 2022; Wang et al., 2022a, b). The inhibition of auxin signaling in the CZ is dependent on WUS, which broadly suppresses histone acetylation to hinder the expression of target loci, including *MP*/*ARF5* (Ma et al., 2019; Lopes et al., 2021). Through analyzing the distribution of signals of DII-VENUS, a recently developed perception sensor that is degraded in the presence of auxin, a low auxin concentration was detected in the region surrounding the organ primordium. Organ formation is inhibited in this auxin-depleted region (Vernoux et al., 2011; Ma et al., 2019; Galvan-Ampudia et al., 2020; Lopes et al., 2021). Therefore, polar auxin distribution in the SAM is necessary and sufficient to trigger organ initiation and provides precise instructions for organogenesis.

Polar auxin transport, which is mediated by carrier proteins, generates a controlled distribution of auxin at the SAM, resulting in higher auxin concentrations in specific cells that are required for organ cell differentiation. Several carrier proteins accounting for both cellular influx and efflux of auxin have been identified in *Arabidopsis*. AUXIN RESISTANT 1 (AUX1) is the only well-described auxin influx carrier (Bennett et al., 1996), whereas several members of the PIN-FORMED (PIN) family are known to mediate auxin efflux from cells (Leyser, 2005; Vieten et al., 2007; Friml, 2010). The efflux carrier PIN1 is localized in the L1 layer of the SAM, often on certain anticlinal sides of the cells (Gälweiler et al., 1998; De Reuille et al., 2006). This specific membrane localization of PIN1 provides directionality to auxin transport within the meristem L1 layer (Wisniewska et al., 2006). In the tomato vegetative shoot apex (Shi et al., 2017; Wang et al., 2022a, b) and *Arabidopsis* vegetative and inflorescence apices (Galvan-Ampudia et al., 2020; Burian et al., 2022), PIN1 polarity generally converges toward the CZ of the SAM from outside the SAM (Shi et al., 2017; Galvan-Ampudia et al., 2020; Burian et al., 2022; Wang et al., 2022a, b). Immunological analyses of PIN1 localization on the membranes have revealed that PIN1-mediated auxin transport is preferentially oriented towards the incipient organ primordia (Benková et al., 2003; Reinhardt et al., 2003). Within primordia, PIN1 is localized basally in the presumptive provascular cells, causing a downward flow of auxin from the SAM surface to deeper tissues (De Reuille et al., 2006). As a result, the organ primordia are transformed into auxin sinks, depleting auxin from surrounding cells where organ formation is inhibited. Inhibited auxin transport in the *pin1* mutant or by treatment with the transport inhibitor N-1-naphthylphthalamic acid (NPA) causes the disappearance of the phyllotactic pattern from the inflorescence meristem. This is due to the lack of the unequal distribution of auxin, as revealed by *DR5::GFP* signals in the peripheral regions (Reinhardt et al., 2000; Smith et al., 2006). However, leaf primordia also form in the vegetative SAM of *pin1* mutants, indicating that PIN1-independent mechanisms of organ formation also exist in the SAM. The PIN1 maxima are localized at the sites of organ primordia (Benková et al., 2003; Reinhardt et al., 2003; Galvan-Ampudia et al., 2020; Burian et al., 2022), implying that auxin positively regulates PIN1 expression in these peripheral regions of the SAM. Reversible phosphorylation of PIN proteins by the PINOID (PID) serine-threonine protein kinase plays a central role in the polar localization of PINs on the cell membranes (Friml et al., 2004; Michniewicz et al., 2007). Mutation of *PID* in *Arabidopsis* results in a *pin*-like inflorescence, due to an apical-to-basal shift in PIN1 polarity (Benjamins et al., 2001; Friml et al., 2004). Auxin polar transport is severely blocked in the *pin1 pid* double mutant, which fails to produce cotyledons and other lateral organs from the SAM (Furutani et al., 2004; Treml et al., 2005). The putative auxin influx transporter AUX1 is also mainly located in the L1 layer of the SAM (Reinhardt et al., 2003; De Reuille et al., 2006). In contrast to PIN1, it seems to be evenly distributed on cell membranes, suggesting that AUX1 helps to restrict auxin in the epidermis of the SAM (De Reuille et al., 2006). The AUX1 sequence is very similar to those of three putative auxin influx carriers in the *Arabidopsis* genome: LAX1, LAX2, and LAX3 (Like AUX1) (Parry et al., 2001; Parry et al., 2006; Swarup & Bennett, 2003). Mutations of *AUX1* or a single *LAX* do not cause an obvious phenotype in terms of SAM development. However, the *aux1 lax1 lax2 lax3* quadruple mutant shows irregular phyllotactic patterning in the SAM, with diffuse auxin distribution and disturbed PIN1 distribution in the SAM periphery (Bainbridge et al., 2008). These results indicate a role for auxin influx carriers in controlling polar auxin distribution and in interacting with PIN1. The *pin1 aux1* double mutant provides another clue about the role of active auxin influx in controlling the initiation of lateral organ primordia at the SAM

(Reinhardt et al., 2003). Localized auxin application to the *pin1 aux1* SAM does not induce single flowers on the apex, but rather fused-organ primordia that are wider than those of the single *pin1* mutant. In addition, treatment of the tomato SAM with auxin influx inhibitors leads to wider organ primordia than those of non-treated control plants (Stieger et al., 2002), suggesting that AUX1-mediated auxin influx is required for the restriction of organ boundaries (Reinhardt et al., 2003). These results suggest that AUX1 influx carriers may function to facilitate organ positioning, probably by stabilizing PIN1-mediated auxin maxima to form a regular phyllotactic pattern.

High levels of auxin in the primordia facilitate lateral organ initiation through downregulating the expression of *CUC* and *KNOX*, which are essential for meristem establishment and stem cell maintenance (Fig. 2.3b; Heisler et al., 2005; Hay et al., 2006). The expression regions of *STM* and *CUC2* are almost complementary to the regions of auxin maxima in the *Arabidopsis* inflorescence meristem (Heisler et al., 2005). Another *KNOX* gene, *KNAT1/BP*, is ectopically expressed in the leaf primordia of the *pin1* mutant, whereas its mutation moderately rescues the phenotypes of organ initiation in *pin1* or *pid* mutants (Hay et al., 2006). In the *pin1 pid* mutant, the expression of *STM* and *CUC1* is expanded into the incipient cotyledon primordia in the embryo or organ primordia in the SAM. Additional mutation of *STM* or *CUC1* in the *pin1 pid* mutant background can partially restore the formation of cotyledons and lateral organs in the SAM, suggesting that enhanced *STM* and *CUC1* expression in the incipient cotyledon primordia inhibits cotyledon formation in the *pin1 pid* mutant (Furutani et al., 2004). In maize, *KNOX* is expressed in the SAM, but is repressed in organ primordia in the peripheral regions (Scanlon, 2003). Inhibition of auxin transport by treatment with the auxin transport inhibitor NPA results in failure to downregulate maize *KNOX* expression at the SAM periphery, causing the arrest of organ initiation. Therefore, the expression of *KNOX* and *CUC* is repressed in the incipient organ primordium by an auxin-dependent mechanism. These results imply that auxin plays a role promoting stem cell differentiation during organogenesis and in repressing stem cell activity in the SAM periphery. Conversely, there is some evidence that KNOX inhibits auxin transport, indicating a feedback regulation between auxin polar transport and KNOX (Tsiantis et al., 1999; Treml et al., 2005).

The expression of *PIN1* is also controlled by the PLETHORA (PLT) family of transcription factors (Prasad et al., 2011). Mutation of *PLTs* reduces *PIN1* expression and alters auxin accumulation in the SAM periphery, resulting in abnormal phyllotaxy. Interestingly, *PLT*s have previously been shown to induce *PIN1* expression in root meristem maintenance (Blilou et al., 2005).

Although the role of polar auxin transport in organ initiation and phyllotactic patterning of the SAM has been extensively studied, there are probably other mechanisms that contribute to creating the polar auxin distribution in the SAM. Many studies have identified mutants in key auxin biosynthetic enzymes to highlight the importance of local auxin biosynthesis in the SAM. In *Arabidopsis*, auxin biosynthesis is regulated mainly through a two-step pathway involving the conversion of tryptophan to indole pyruvic acid by TRYPTOPHAN AMINOTRANSFERASE OF ARABIDOPSIS (TAA), and the subsequent conversion of indole pyruvic acid to

IAA by YUC flavin monooxygenase (Mashiguchi et al., 2011; Won et al., 2011). The importance of YUC-dependent local auxin biosynthesis in organ initiation has been revealed by analyses of multiple loss-of-function *yuc* mutants in *Arabidopsis* (Cheng et al., 2006, 2007). Although none of the single mutants show obvious phenotypes, several double and triple mutant combinations and the quadruple mutant of *yuc1 yuc2 yuc4 yuc6* display strong inflorescence meristem defects (Cheng et al., 2006). Furthermore, combining the *pin1* mutation with *yuc1 yuc4* mutations enhances the *pin1* phenotype, with no true leaves forming from the SAM. Combining the *aux1* mutation with *yuc1 yuc2 yuc4 yuc6* mutations produces a similar phenotype (Cheng et al., 2007), suggesting that the co-ordination of auxin transport and biosynthesis is important for regulating organ initiation (Weijers et al., 2005). Inhibition of both auxin biosynthesis and polar transport results in failed cell differentiation into leaves and flowers at the SAM periphery.

Exogenously applied auxin is not able to induce primordium initiation in the CZ (Reinhardt et al., 2003). In *pin1* mutants, cells at the SAM periphery can respond to exogenous auxin to initiate organs, whereas cells in the CZ cannot (Reinhardt et al., 2000). Although auxin also accumulates in the L1 layer of the meristem, organ primordia can only initiate at the sites of auxin accumulation in the PZ. Therefore, additional factors are required to control auxin perception in the PZ for organ initiation. One candidate is the *ARF* gene *MP/ARF5*, which is only expressed at the SAM periphery (Hardtke & Berleth, 1998). Mutation of *MP* induces a *pin1*-like phenotype and represses the competence of cells at the meristem periphery to respond to exogenous auxin for organ initiation (Hardtke & Berleth, 1998; Aida et al., 2002; Reinhardt et al., 2003). These results indicate that organ initiation at the SAM periphery at least partly depends on the spatial modulation of auxin signaling. In addition, the differential regulation of *CUC1* and *CUC2* by PIN1 and MP has been detected during early embryogenesis (Aida et al., 2002). *CUC1* expression is enhanced while *CUC2* expression is inhibited by mutations of both *PIN1* and *MP*. Therefore, *PIN1* and *MP* regulate SAM patterning partly through controlling the expression of *CUCs*, which encode regulators of cotyledon separation and SAM formation.

2.5.3 Establishment of Organ Boundaries Between Shoot Meristem and Leaves

As leaves initiate, morphological boundary regions form to delineate the leaves from the SAM. Many proteins are thought to be involved in organ boundary formation because their expression patterns are restricted to the boundary (Aida & Tasaka, 2006; Rast & Simon, 2008) and because disruptions of their expression lead to organ fusion, failure to initiate organs, and even termination of the SAM.

In addition to their function in establishing and maintaining the embryonic SAM, *CUCs* also function in regulating organ boundary specification (Fig. 2.4). The

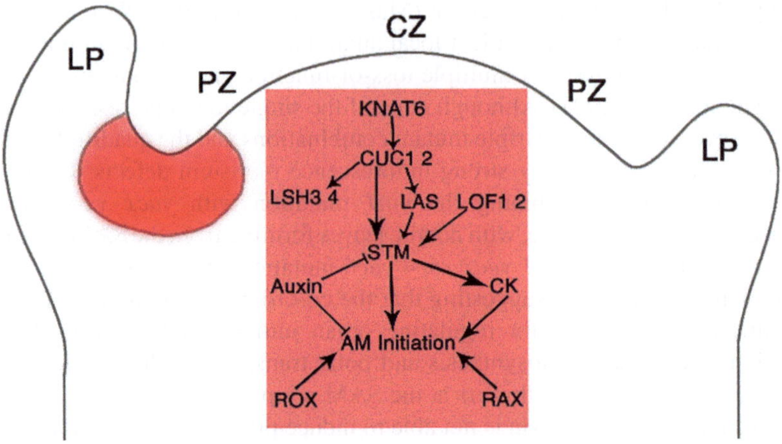

Fig. 2.4 Gene regulation in axillary meristem (AM) formation. *STM* is specifically expressed in leaf axils and is required for AM initiation. *CUC1* and *CUC2* are redundantly required for *STM* expression in AM formation. *KNAT6* is expressed at the boundary region between SAM and cotyledons, and this expression is required for the activities of *CUC1* and *CUC2*. CUC1 activates the expression of *LSH3* and *LSH4*. LOF1 and LOF2 function overlaps with CUCs in organ boundary specification by regulating *STM*. LAS acts downstream of CUC1 and CUC2, and functions in AM formation through maintaining meristematic competence of leaf axil cells via retaining *STM* expression. Low auxin level and high cytokinin level are required to avoid differentiation of these boundary cells and trigger AM initiation

expression of *CUCs* is restricted to the boundary regions via regulation by *STM* and auxin signaling (Aida et al., 2002; Spinelli et al., 2011; Scofield et al., 2018). This specific expression is required both to inhibit boundary cell proliferation for leaf separation from the SAM and to maintain the meristem. Another *KNOX* gene, *KNAT6*, is expressed at the boundary region between the SAM and the cotyledons, and its expression is required for the activities of CUC1 and CUC2 (Belles-Boix et al., 2006). CUC1 activates the expression of *LIGHT_DEPENDENT SHORT HYPOCOTYLS3* (*LSH3*) and *LSH4*, which are involved in suppressing differentiation in the boundary region (Takeda et al., 2011).

The functions of the MYB-domain transcription factors LATERAL ORGAN FUSION1 (LOF1) and LOF2 overlap with CUCs in organ boundary specification. They are expressed in the boundary region at the base of leaf primordia and contribute redundantly to the separation between leaves and the meristem (Lee et al., 2009). *LATERAL SUPPRESSOR* (*LAS*) encodes a GRAS family transcription factor and is expressed in the boundary region. LOF1 genetically interacts with CUC2 and CUC3, and is required for the transcription of *LOF2*, *CUC*, and *LAS* (Greb et al., 2003; Lee et al., 2009). LAS functions downstream of CUC1 and CUC2, and partially redundantly with CUC3 (Hibara et al., 2006). These genes act in a genetic regulatory hierarchy in the separation between leaves and the meristem (Fig. 2.4). Double mutants of *lof1stm-10* display more severe meristem defects than those of the *stm-10* single mutant (Lee et al., 2009), indicating that the boundary-expressed gene *LOF1* functions in meristem maintenance.

The boundary region between developing organs and the meristem is characterized by reduced growth to allow for the separation of distinct domains. The transcription factor *LOB*, which is a member of the LBD family in *Arabidopsis*, mediates reduced growth between developing organs and the meristem through negatively regulating brassinosteroid signaling (Singh & Savaldi-Goldstein, 2015). *lob* mutants show junctions between the main stem, axillary stem, and cauline leaves, and these boundary defects are mediated by brassinosteroid signaling. The gene encoding the brassinosteroid-inactivating enzyme BAS1 is a direct target of LOB (Bell et al., 2012). The expression of *LOB* is affected by brassinosteroid signaling (Arnaud & Laufs, 2013). The brassinosteroid signaling mutant *bzr1-d* shows organ separation defects similar to those of *lob* mutants and reduced expression of *CUCs* (Gendron et al., 2012). The expression of *LOB* is activated by *AS1*, *AS2*, *BOP1*, and *BOP2* in organ primordia (Byrne et al., 2002; Ha et al., 2007).

2.6 Axillary Meristem Formation

2.6.1 Boundary Zones Initiate Formation of Axillary Meristem

In addition to their function in separating lateral organs and the SAM, boundary zones can also initiate the formation of meristems. During embryogenesis, the first boundary zone is established between the two cotyledons, and subsequently initiates the formation of the embryonic SAM (Aida et al., 1999). During postembryonic shoot development, leaf primordia emerge from the flanks of the SAM, establishing new boundary zones with small, slowly dividing cells between the shoot apex and the developing organ (Barton, 2010). It is within these zones that the secondary meristems known as axillary meristems (AMs) initiate and outgrow to form buds with a number of nodes, which are responsible for producing all secondary axes, including inflorescence branches or flowers. The initiation of AMs is closely related to the formation of leaves and their separation from the SAM. The patterns of cell division and the absence of differentiation markers in the boundary between organs and the SAM indicate that the boundary at the leaf axil harbors pluripotent cells that retain competence for meristem formation (Rossmann et al., 2015). In *Arabidopsis*, AM initiation can first be detected in the axils of P16 leaf primordia, using the meristem marker *STM* (Greb et al., 2003). Recently, a live-cell imaging analysis showed that the initiation of AMs in the leaf axils requires the maintenance of the meristematic competence of leaf axil cells via the continuous expression of *STM*. Importantly, low levels of *STM* maintain a meristematic cell population while high levels of *STM* lead to AM initiation in the leaf axils (Shi et al., 2016). Hormone regulation and the gene network acting to maintain the pluripotent activity of the boundary cells in the leaf axils are essential for AM initiation (Fig. 2.4). The initiation of the AM is also strictly associated with leaf polarity, because AMs only initiate at the adaxial base of the leaf. Here, we focus on the gene regulatory network and hormone signals that control the developmental processes of AM initiation.

2.6.2 Regulation of Axillary Meristem Initiation

Key regulators that show boundary-specific expression patterns are involved in boundary formation. Their loss-of-function mutants display conspicuous organ fusions. Many of these genes affect the initiation of AMs. For example, *CUC1* and *CUC2* are required for the formation of the boundary region between cotyledons and the SAM (Aida et al., 1997, 1999). Besides their involvement in forming the boundary and the SAM, *CUCs* also play a very important role in AM initiation, as their mutations lead to defects in shoot branching in *Arabidopsis* and tomato (Hibara et al., 2006; Raman et al., 2008; Busch et al., 2011). The transcript levels of *CUC1* and *CUC2* are regulated by *miR164*, and are required for the formation of AMs at the leaf axil (Laufs et al., 2004; Raman et al., 2008). CUC2 and CUC3 directly activate the expression of *DA1*, which encodes a ubiquitin-dependent peptidase. Mutations in *UBIQUITIN-SPECIFIC PROTEASE15* (*UBP15*), which encodes a DA1 substrate, lead to repression of AM initiation (Li et al., 2020). Although CUC2 and CUC3 are required for AM initiation, excessively high expression of *CUC* genes can delay AM initiation, at least in cauline leaf axils (Nicolas et al., 2022). The NGATHA-LIKE transcription factors DEVELOPMENT-RELATED PcG TARGET IN THE APEX4 (DPA4) and SUPPRESSOR OF DA1-1 7 (SOD7) redundantly repress *CUC* expression in the leaf axil. *dpa4-2 sod7-2* double mutants display delayed AM initiation (Nicolas et al., 2022). Other genes involved in establishing the inter-cotyledon boundary and maintaining the SAM, such as *LOF1* and its paralog *LOF2*, contribute to modulating AM formation in *Arabidopsis* (Lee et al., 2009). Both *LOF1* and *LOF2* are specifically expressed in the adaxial leaf boundary, wherein the AMs initiate. The *lof1* single mutants and *lof1 lof2* double mutants both fail to form accessory shoot meristems and show reduced expression of the meristem determinant *STM* in the paraclade junctions. Analyses of the genetic interactions between *LOF1* and *CUCs* have revealed that *LOF1* also plays overlapping roles with *CUC2* and *CUC3* to control organ separation and AMs initiation (Lee et al., 2009). These data indicate that *LOF1* and *LOF2* function in both lateral organ separation and AM formation, at least partly through their interaction with *CUC2*, *CUC3*, and the *KNOX* gene *STM*. The expression of STM is sustained in a specific population of cells located in the central region of the leaf axils. Within the leaf axils, AMs originate specifically from cells expressing STM, with neighboring non-expressing cells not contributing to the initiation process (Wang et al., 2017; Xin et al., 2017; Gruel et al., 2018; Zhou et al., 2018). Another important regulator of AM formation in *Arabidopsis* is *LAS*, which is expressed in the boundary region (Greb et al., 2003). Mutations in *LAS* cause strong defects in AM formation, leading to the loss of side shoots during vegetative development. During new AM initiation with peak *STM* expression, *LAS* is co-expressed with *STM* in leaf axils. The expression of *STM* is lost in the leaf axil of *las* mutants, suggesting that *LAS* acts in AM formation through maintaining the meristematic competence of leaf axil cells via retaining *STM* expression (Greb et al., 2003; Schmitz & Theres, 2005). The expression of *LAS* in the *rev* mutants of *Arabidopsis* also revealed that *REV* expression in developing AMs is required for *LAS* activity during vegetative development (Greb et al., 2003). Genetic defects of *LAS* homologs alter plant architecture in monocots such as rice. *MONOCULM 1* (*MOC1*) in rice, which is an ortholog of *Arabidopsis LAS*, modulates AM formation in both vegetative and reproductive stages. Loss-of-function of *MOC1* leads to a lack of vegetative tillers, and reduced numbers of reproductive rachis branches and spikelets (Li et al., 2003).

The R2R3 MYB proteins REGULATOR OF AXILLARY MERISTEMS1 (RAX1), RAX2, and RAX3, which are specifically expressed in the boundary region between the SAM and leaf primordia, are partially redundant in the control of AM initiation during early vegetative development in *Arabidopsis* (Fig. 2.4; Keller et al., 2006; Müller et al., 2006). *RAX1* is expressed transiently in a small region at the center of the boundary between the SAM and developing leaf primordia. The initiation of AMs in the leaf axil is repressed in the *rax1 rax2 rax3* triple mutant. *EXCESSIVE BRANCHES1* (*EXB1*), encoding the transcription factor WRKY71, regulates the transcription of *RAXs* (Guo et al., 2015). The gain-of-function mutant *exb1-D* displays a bushy dwarf phenotype with massive AM initiation, suggesting that *EXB1* plays a positive role in AM initiation through regulating *RAX* expression.

In *Arabidopsis*, *REGULATOR OF AXILLARY MERISTEM FORMATION* (*ROX*) regulates AM formation during early vegetative shoot development (Yang et al., 2012). *ROX* is expressed in the adaxial boundary between the SAM and leaf primordia. Mutation of *ROX* leads to compromised AM and axillary bud formation during vegetative development, but no branching defects during reproductive development. *LAX PANICLE1* (*LAX1*) and *BARREN STALK1* (*BA1*) are key regulators of shoot and inflorescence branching in rice and maize, respectively (Komatsu et al., 2003; Gallavotti et al., 2004), and both genes are orthologs of *ROX* in *Arabidopsis*. In contrast to *rox*, *lax1* mutants in rice not only display defects in AM formation during vegetative development, but also show severely reduced formation of panicle branches during reproductive development (Oikawa & Kyozuka, 2009). The LAX1 protein moves from the boundary region towards the initiation sites of AMs to enhance cell proliferation. Similarly, mutation of maize *BA1* leads to compromised AM initiation during both vegetative and reproductive developmental stages (Ritter et al., 2002; Gallavotti et al., 2004).

2.6.3 Hormones Control Axillary Meristem Initiation

Besides numerous transcription factors and regulatory genes, hormones also modulate the specification of AMs. Similar to the SAM, formation of AMs at the boundary requires auxin and cytokinin (Fig. 2.4). Many studies in *Arabidopsis* have shown that a low auxin concentration is essential for the establishment and maintenance of meristematic competence in leaf axils, which is essential for AM initiation (Wang et al., 2014a, b). Treatment with chemical inhibitors of auxin polar transport or mutations in PIN1 or PID strongly inhibit AM initiation via the disruption of auxin minima in the leaf axil. Ectopic auxin synthesis in the leaf axil represses AM initiation, while downregulation of auxin signaling can at least partly rescue the AM initiation defects of the polar auxin transport mutant *pid* (Wang et al., 2014a). *STM*, which is a marker of meristematic cell competence in the leaf axil (Grbić & Bleecker, 2000; Long & Barton, 2000), is not expressed in *pid* mutants, indicating that a low-auxin environment is required to maintain *STM* expression (Wang et al., 2014a). Ectopic MP activity in the leaf axil is sufficient to inhibit STM maintenance and AM initiation (Guo et al., 2020). Furthermore, ectopic expression of the AUX/IAA protein BDL specifically in the boundary regions between cotyledons and the embryonic SAM to reduce auxin signaling in this area results in the formation of AMs in the axils of cotyledons, which never occurs in wild-type plants (Wang et al., 2014a). Therefore, a low level of auxin response in the boundary appears to be essential for AM formation. In contrast, a cytokinin pulse is required to precede AM formation in *Arabidopsis* leaf axils, following the auxin minimum (Wang et al., 2014b). *Arabidopsis* histidine kinase receptor mutants and B-type ARR transcription factor mutants show compromised AM initiation. The *Arabidopsis supershoot* mutants form more AMs in leaf axils because of increased cytokinin levels in the boundary regions, compared with wild type (Tantikanjana et al., 2001). In the *ipt3 ipt5 ipt7* triple mutant, AM initiation in the leaf axils is strongly repressed during early vegetative development, indicating that cytokinin biosynthesis is required for AM formation (Müller et al., 2015). Similarly, ectopic cytokinin biosynthesis in leaf axils can rescue AM initiation in the *Arabidopsis rax1* mutant, which cannot initiate AMs during early vegetative shoot development (Wang et al., 2014b). Because *STM* is specifically expressed in the leaf axils and is required for AM initiation, it may contribute to the cytokinin pulse based on its cytokinin-promoting function in SAM maintenance (Jasinski et al., 2005; Yanai et al., 2005). Therefore, both auxin and cytokinin strongly affect the AM initiation process. Before AMs initiate, boundary cells are kept in a meristematic state and maintain competence for later meristem formation. During this stage, low auxin levels and high cytokinin levels prevent the differentiation of these boundary cells and trigger AM initiation.

2.7 Concluding Remarks

The SAM of higher plants is a dynamic structure that initiates all of the aboveground organs and tissues, such as leaves, stems, and flowers. Within the SAM, unspecialized immature stem cells serve as the ultimate source of cells for future organ formation. Recent studies have identified a large and rapidly growing number of regulatory genes involved in the complex process of maintaining stem cell activity in the SAM. The functions of these genes in meristem maintenance have been identified by dissecting individual pathways. Such analyses have revealed many key processes, including transcriptional and epigenetic regulation and hormone signaling, which contribute to meristem maintenance and activity. More than a decade of research has demonstrated the existence of gene regulatory networks, in which each pathway is a part of a complex web of interacting effects. A significant challenge for future research is to determine how these different regulatory pathways are integrated to produce specific cellular outcomes, to mediate the complex growth patterns of SAMs, and to organize the initiation and maintenance of stem cell activity.

Molecular genetic studies have contributed much information about existing regulatory networks. More recently, live imaging technology has enabled high-resolution spatio-temporal analysis of gene expression patterns, which is useful to analyze the spatial parameters of hormonal, transcriptional, and epigenetic regulatory modules. Furthermore, computational methods for both image analysis and quantitative modeling will provide new insights into the complex and intricate models of stem-cell growth and organ morphogenesis in the SAM.

Acknowledgements This work was supported by the National Natural Science Foundation of China (32270378, 32070199) and the Natural Science Foundation of Shandong Province (ZR2022JQ12).

References

Adler, I., Barabe, D., & Jean, R. V. (1997). A history of the study of phyllotaxis. *Annals of Botany, 80*(3), 231–244.

Aguirre, L., Hendelman, A., Hutton, S. F., McCandlish, D. M., & Lippman, Z. B. (2023). Idiosyncratic and dose-dependent epistasis drives variation in tomato fruit size. *Science, 382*(6668), 315–320.

Aichinger, E., Kornet, N., Friedrich, T., & Laux, T. (2012). Plant stem cell niches. *Annual Review of Plant Biology, 63*(1), 615–636.

Aida, M., & Tasaka, M. (2006). Morphogenesis and patterning at the organ boundaries in the higher plant shoot apex. *Plant Molecular Biology, 60*(6), 915–928.

Aida, M., Ishida, T., Fukaki, H., Fujisawa, H., & Tasaka, M. (1997). Genes involved in organ separation in *Arabidopsis*: An analysis of the cup-shaped cotyledon mutant. *The Plant Cell, 9*(6), 841–857.

Aida, M., Ishida, T., & Tasaka, M. (1999). Shoot apical meristem and cotyledon formation during *Arabidopsis* embryogenesis: Interaction among the *CUP-SHAPED COTYLEDON* and *SHOOT MERISTEMLESS* genes. *Development, 126*(8), 1563–1570.

Aida, M., Vernoux, T., Furutani, M., Traas, J., & Tasaka, M. (2002). Roles of *PIN-FORMED1* and *MONOPTEROS* in pattern formation of the apical region of the *Arabidopsis* embryo. *Development, 129*(17), 3965–3974.

Arnaud, N., & Laufs, P. (2013). Plant development: Brassinosteroids go out of bounds. *Current Biology, 23*(4), 152–154.

Ashikari, M., Sakakibara, H., Lin, S., Yamamoto, T., & Matsuoka, M. (2005). Cytokinin oxidase regulates rice grain production. *Science, 309*(5735), 741–745.

Bainbridge, K., Guyomarc'h, S., Bayer, E., Swarup, R., & Kuhlemeier, C. (2008). Auxin influx carriers stabilize phyllotactic patterning. *Genes & Development, 22*(6), 810–823.

Barton, M. K. (2010). Twenty years on: The inner workings of the shoot apical meristem, a developmental dynamo. *Developmental Biology, 341*(1), 95–113.

Barton, M. K., & Poethig, R. S. (1993). Formation of the shoot apical meristem in *Arabidopsis thaliana*: An analysis of development in the wild type and in the shoot meristemless mutant. *Development, 119*(3), 823–831.

Bartrina, I., Otto, E., Strnad, M., Werner, T., & Schmülling, T. (2011). Cytokinin regulates the activity of reproductive meristems, flower organ size, ovule formation, and thus seed yield in *Arabidopsis thaliana*. *The Plant Cell, 23*(1), 69–80.

Bell, E. M., Lin, W. C., Husbands, A. Y., Yu, L., Jaganatha, V., Jablonska, B., Mangeon, A., Neff, M. M., Neff, T., & Springer, P. S. (2012). Arabidopsis lateral organ boundaries negatively regulates brassinosteroid accumulation to limit growth in organ boundaries. *Proceedings of the National Academy of Sciences, 109*(51), 21146–21151.

Belles-Boix, E., Hamant, O., Witiak, S. M., Morin, H., Traas, J., & Pautot, V. (2006). *KNAT6*: An Arabidopsis homeobox gene involved in meristem activity and organ separation. *The Plant Cell, 18*(8), 1900–1907.

Benjamins, R., Quint, A., Weijers, D., Hooykaas, P., & Offringa, R. (2001). The PINOID protein kinase regulates organ development in *Arabidopsis* by enhancing polar auxin transport. *Development, 128*(20), 4057–4067.

Benková, E., Michniewicz, M., Sauer, M., Teichmann, T., Seifertová, D., Jürgens, G., & Friml, J. (2003). Local, efflux-dependent auxin gradients as a common module for plant organ formation. *Cell, 115*(5), 591–602.

Bennett, M. J., Marchant, A., Green, H. G., May, S. T., & Feldmann, K. A. (1996). *Arabidopsis AUX1* gene: A permease-like regulator of root gravitropism. *Science, 273*(5277), 948–950.

Bleckmann, A., Weidtkamp-Peters, S., Seidel, C. A. M., & Simon, R. (2010). Stem cell signaling in *Arabidopsis* requires CRN to localize CLV2 to the plasma membrane. *Plant Physiology, 152*(1), 166–176.

Blilou, I., Xu, J., Wildwater, M., Willemsen, V., & Scheres, B. (2005). The PIN auxin efflux facilitator network controls growth and patterning in *Arabidopsis* roots. *Nature, 433*(7021), 39–44.

Borghi, L., Bureau, M., & Simon, R. (2007). *Arabidopsis JAGGED LATERAL ORGANS* is expressed in boundaries and coordinates KNOX and PIN activity. *The Plant Cell, 19*(6), 1795–1808.

Boscá, S., Knauer, S., & Laux, T. (2011). Embryonic development in *Arabidopsis thaliana*: From the zygote division to the shoot meristem. *Frontiers in Plant Science, 2*, 93.

Brand, U., Fletcher, J. C., Hobe, M., Meyerowitz, E. M., & Simon, R. (2000). Dependence of stem cell fate in *Arabidopsis* on a feedback loop regulated by CLV3 activity. *Science, 289*(5479), 617–619.

Brand, U., Grunewald, M., Hobe, M., & Simon, R. (2002). Regulation of *CLV3* expression by two homeobox genes in *Arabidopsis*. *Plant Physiology, 129*(2), 565–575.

Breuninger, H., Rikirsch, E., Hermann, M., Ueda, M., & Laux, T. (2008). Differential expression of *WOX* genes mediates apical-basal axis formation in the *Arabidopsis* embryo. *Developmental Cell, 14*(6), 867–876.

Briggs, G. C., Osmont, K. S., Shindo, C., Sibout, R., & Hardtke, C. S. (2006). *Trends in Plant Science, 11*(10), 492–498.

Buechel, S., Leibfried, A., To J P, Zhao, Z., Andersen, S. U., Kieber, J. J., & Lohmann, J. U. (2010). Role of A-type ARABIDOPSIS RESPONSE REGULATORS in meristem maintenance and regeneration. *European Journal of Cell Biology, 89*(2–3), 279–284.

Bureau, M., Rast, M. I., Illmer, J., & Simon, R. (2010). *JAGGED LATERAL ORGAN* (*JLO*) controls auxin dependent patterning during development of the *Arabidopsis* embryo and root. *Plant Molecular Biology, 74*(4–5), 479–491.

Burian, A., Paszkiewicz, G., Nguyen, K. T., Meda, S., Raczyńska-Szajgin, M., & Timmermans, M. C. P. (2022). Specification of leaf dorsiventrality via a prepatterned binary readout of a uniform auxin input. *Nature Plants, 8*(3), 269–280.

Busch, B. L., Schmitz, G., Rossmann, S., Piron, F., Ding, J., Bendahmane, A., & Theres, K. (2011). Shoot branching and leaf dissection in tomato are regulated by homologous gene modules. *The Plant Cell, 23*(10), 3595–3609.

Byrne, M. E. (2005). *Biochemical Society Transactions, 33*(6), 1499–1501.

Byrne, M. E. (2006). Shoot meristem function and leaf polarity: The role of class III HD–ZIP genes. *PLoS Genetics, 2*(6), e89.

Byrne, M. E., Barley, R., Curtis, M., Arroyo, J. M., & Martienssen, R. A. (2000). Asymmetric leaves1 mediates leaf patterning and stem cell function in *Arabidopsis*. *Nature, 408*(6815), 967–971.

Byrne, M. E., Simorowski, J., & Martienssen, R. A. (2002). ASYMMETRIC LEAVES1 reveals *knox* gene redundancy in *Arabidopsis*. *Development, 129*(8), 1957–1965.

Cao, X., Wang, J., Xiong, Y., Yang, H., Yang, M., Ye, P., Bencivenga, S., Sablowski, R., & Jiao, Y. (2020). A self-activation loop maintains meristematic cell fate for branching. *Current Biology, 30*(10), 1893–1904.e4.

Cao, X., Du, Q., Guo, Y., Wang, Y., & Jiao, Y. (2023). Condensation of STM is critical for shoot meristem maintenance and salt tolerance in *Arabidopsis*. *Molecular Plant, 16*(9), 1445–1459.

Carles, C. C., & Fletcher, J. C. (2003). Shoot apical meristem maintenance: The art of a dynamic balance. *Trends in Plant Science, 8*(8), 394–401.

Casson, S. A., Topping, J. F., & Lindsey, K. (2009). MERISTEM-DEFECTIVE, an RS domain protein, is required for the correct meristem patterning and function in *Arabidopsis*. *The Plant Journal, 57*(5), 857–869.

Chandler, J. W., Jacobs, B., Cole, M., Comelli, P., & Werr, W. (2011). *DORNRO¨SCHEN-LIKE* expression marks *Arabidopsis* floral organ founder cells and precedes auxin response maxima. *Plant Molecular Biology, 76*, 171–185.

Chen, M., Chory, J., & Fankhauser, C. (2004). Light signal transduction in higher plants. *Annual Review of Genetics, 38*, 87–117.

Cheng, Y., Dai, X., & Zhao, Y. (2006). Auxin biosynthesis by the YUCCA flavin monooxygenases controls the formation of floral organs and vascular tissues in *Arabidopsis*. *Genes & Development, 20*(13), 1790–1799.

Cheng, Y., Dai, X., & Zhao, Y. (2007). Auxin synthesized by the YUCCA flavin monooxygenases is essential for embryogenesis and leaf formation in *Arabidopsis*. *The Plant Cell, 19*(8), 2430–2439.

Chickarmane, V. S., Gordon, S. P., Tarr, P. T., Heisler, M. G., & Meyerowitz, E. M. (2012). Cytokinin signaling as a positional cue for patterning the apical–basal axis of the growing *Arabidopsis* shoot meristem. *Proceedings of the National Academy of Sciences, 109*(10), 4002–4007.

Chuck, G., Lincoln, C., & Hake, S. (1996). KNAT1 induces lobed leaves with ectopic meristems when overexpressed in *Arabidopsis*. *The Plant Cell, 8*(8), 1277–1289.

Chung, Y., Zhu, Y., Wu, M. F., Simonini, S., Kuhn, A., Armenta-Medina, A., Jin, R., Østergaard, L., Gillmor, C. S., & Wagner, D. (2019). Auxin Response Factors promote organogenesis by chromatin-mediated repression of the pluripotency gene *SHOOTMERISTEMLESS*. *Nature Communications, 10*(1), 886.

Clark, S. E., Running, M. P., & Meyerowitz, E. M. (1993). CLAVATA1, a regulator of meristem and flower development in *Arabidopsis*. *Development, 119*(2), 397–418.

Clark, S. E., Running, M. P., & Meyerowitz, E. M. (1995). CLAVATA3 is a specific regulator of shoot and floral meristem development affecting the same processes as CLAVATA1. *Development, 121*, 2057–2067.

Clark, S. E., Jacobsen, S. E., Levin, J. Z., & Meyerowitz, E. M. (1996). The CLAVATA and SHOOT MERISTEMLESS loci competitively regulate meristem activity in *Arabidopsis*. *Development, 122*(5), 1567–1575.

Daum, G., Medzihradszky, A., Suzaki, T., & Lohmann, J. U. (2014). A mechanistic framework for non-cell autonomous stem cell induction in *Arabidopsis*. *Proceedings of the National Academy of Sciences, 111*(40), 14619–14624.

De Reuille, P. B., Bohn-Courseau, I., Ljung, K., & Traas, J. (2006). Computer simulations reveal properties of the cell-cell signaling network at the shoot apex in *Arabidopsis*. *Proceedings of the National Academy of Sciences, 103*(5), 1627–1632.

Deyoung, B. J., & Clark, S. E. (2008). BAM receptors regulate stem cell specification and organ development through complex interactions with CLAVATA signaling. *Genetics, 180*(2), 895–904.

DeYoung, B. J., Bickle, K. L., Schrage, K. J., Muskett, P., Patel, K., & Clark, S. E. (2006). The CLAVATA1-related BAM1, BAM2 and BAM3 receptor kinase-like proteins are required for meristem function in *Arabidopsis*. *The Plant Journal, 45*(1), 1–16.

Doerner, P. (1999). Shoot meristems: Intercellular signals keep the balance. *Current Biology, 9*(10), 377–380.

Endrizzi, K., Moussian, B., HaeckerA, L. J. Z., & Laux, T. (1996). The *SHOOT MERISTEMLESS* gene is required for maintenance of undifferentiated cells in *Arabidopsis* shoot and floral meristems and acts at a different regulatory level than the meristem genes *WUSCHEL* and *ZWILLE*. *Plant Journal, 10*(6), 967–979.

Fiers, M., Golemiec, E., Xu, J., van der Geest, L., Heidstra, R., Stiekema, W., & Liu, C. M. (2005). The 14-amino acid CLV3, CLE19, and CLE40 peptides trigger consumption of the root meristem in *Arabidopsis* through a CLAVATA2-dependent pathway. *The Plant Cell, 17*(9), 2542–2553.

Fletcher, J. C. (2002). Shoot and floral meristem maintenance in *Arabidopsis*. *Annual Review of Plant Biology, 53*, 45–66.

Fletcher, J. C., Brand, U., Running, M. P., Simon, R., & Meyerowitz, E. M. (1999). Signaling of cell fate decisions by CLAVATA3 in *Arabidopsis* shoot meristems. *Science, 283*(5409), 1911–1914.

Friml, J. (2010). Subcellular trafficking of PIN auxin efflux carriers in auxin transport. *European Journal of Cell Biology, 89*(2–3), 231–235.

Friml, J., Yang, X., Michniewicz, M., Weijers, D., & Offringa, R. (2004). A PINOID-dependent binary switch in apical-basal PIN polar targeting directs auxin efflux. *Science, 306*(5697), 862–865.

Furutani, M., Vernoux, T., Traas, J., Kato, T., Tasaka, M., & Aida, M. (2004). *PIN-FORMED1* and *PINOID* regulate boundary formation and cotyledon development in *Arabidopsis* embryogenesis. *Development, 131*(20), 5021–5030.

Gallavotti, A., Zhao, Q., Kyozuka, J., Meeley, R. B., Ritter, M. K., Doebley, J. F., Pè, M. E., & Schmidt, R. J. (2004). The role of barren stalk1 in the architecture of maize. *Nature, 432*(7017), 630–635.

Galvan-Ampudia, C. S., Cerutti, G., Legrand, J., Brunoud, G., Martin-Arevalillo, R., Azais, R., Bayle, V., Moussu, S., Wenzl, C., Jaillais, Y., Lohmann, J. U., Godin, C., & Vernoux, T. (2020). Temporal integration of auxin information for the regulation of patterning. *eLife, 9*, e55832.

Gälweiler, L., Guan, C., Müller, A., Wisman, E., Mendgen, K., Yephremov, A., & Palme, K. (1998). Regulation of polar auxin transport by AtPIN1 in *Arabidopsis* vascular tissue. *Science, 282*(5397), 2226–2230.

Gendron, J. M., Liu, J. S., Fan, M., Bai, M. Y., Wenkel, S., & Wang, Z. Y. (2012). Brassinosteroids regulate organ boundary formation in the shoot apical meristem of *Arabidopsis*. *Proceedings of the National Academy of Sciences, 109*(51), 21152–21157.

Gordon, S. P., Chickarmane, V. S., Ohno, C., & Meyerowitz, E. M. (2009). Multiple feedback loops through cytokinin signaling control stem cell number within the *Arabidopsis* shoot meristem. *Proceedings of the National Academy of Sciences, 106*(38), 16529–16534.

Graf, P., Dolzblasz, A., Rschum, T. W., Lenhard, M., Pfreundt, U., & Laux, T. (2010). *MGOUN1* encodes an *Arabidopsis* type IB DNA topoisomerase required in stem cell regulation and to maintain developmentally regulated gene silencing. *The Plant Cell, 22*(3), 716–728.

Grbić, V., & Bleecker, A. B. (2000). Axillary meristem development in *Arabidopsis thaliana*. *The Plant Journal, 21*(2), 215–223.

Greb, T., Clarenz, O., Schafer, E., Muller, D., Herrero, R., Schmitz, G., & Theres, K. (2003). Molecular analysis of the *LATERAL SUPPRESSOR* gene in *Arabidopsis* reveals a conserved control mechanism for axillary meristem formation. *Genes & Development, 17*(9), 1175–1187.

Gruel, J., Deichmann, J., Landrein, B., Hitchcock, T., & Jönsson, H. (2018). The interaction of transcription factors controls the spatial layout of plant aerial stem cell niches. *npj Systems Biology and Applications, 4*, 36.

Guo, M., Thomas, J., Collins, G., & Timmermans, M. C. (2008). Direct repression of *KNOX* loci by the ASYMMETRIC LEAVES1 complex of *Arabidopsis*. *The Plant Cell, 20*(1), 48–58.

Guo, Y., Han, L., Hymes, M., Denver, R., & Clark, S. E. (2010). CLAVATA2 forms a distinct CLE-binding receptor complex regulating *Arabidopsis* stem cell specification. *The Plant Journal, 63*(6), 889–900.

Guo, D., Zhang, J., Wang, X., Han, X., Wei, B., & Qin, G. (2015). The WRKY transcription factor WRKY71/EXB1 controls shoot branching by transcriptionally regulating *RAX* genes in *Arabidopsis*. *The Plant Cell, 27*(11), 3112–3127.

Guo, Y., Tian, C., Jiao, Y., & Wang, Y. (2020). Multifaceted functions of auxin in vegetative axillary meristem initiation. *Journal of Genetics and Genomics, 47*(9), 591–594.

Guyomarc'h, S., Bertrand, C., Delarue, M., & Zhou, D. X. (2005). Regulation of meristem activity by chromatin remodelling. *Trends in Plant Science, 10*(7), 332–338.

Ha, C. M., Kim, G. T., Kim, B. C., Jun, J. H., & Nam, H. G. (2003). The *BLADE-ON-PETIOLE 1* gene controls leaf pattern formation through the modulation of meristematic activity in *Arabidopsis*. *Development, 130*(1), 161–172.

Ha, C. M., Jun, J. H., Nam, H. G., & Fletcher, J. C. (2004). *BLADE-ON-PETIOLE1* encodes a BTB/POZ domain protein required for leaf morphogenesis in *Arabidopsis thaliana*. *Plant and Cell Physiology, 45*(10), 1361–1370.

Ha, C. M., Jun, J. H., Nam, H. G., & Fletcher, J. C. (2007). *BLADE-ON-PETIOLE 1* and *2* control *Arabidopsis* lateral organ fate through regulation of LOB domain and adaxial-abaxial polarity genes. *The Plant Cell, 19*(6), 1809–1825.

Ha, C. M., Jun, J. H., & Fletcher, J. C. (2010). Shoot apical meristem form and function. *Current Topics in Developmental Biology, 91*, 103–140.

Haecker, A., Gross-Hardt, R., Geiges, B., Sarkar, A., Breuninger, H., Herrmann, M., & Laux, T. (2004). Expression dynamics of *WOX* genes mark cell fate decisions during early embryonic patterning in *Arabidopsis thaliana*. *Development, 131*(3), 657–668.

Hamant, O., Nogué, F., Belles-Boix, E., Jublot, D., Grandjean, O., Traas, J., & Pautot, V. (2002). The KNAT2 homeodomain protein interacts with ethylene and cytokinin signaling. *Plant Physiology, 130*(2), 657–665.

Han, I. S., Tseng, T. S., Eisinger, W., & Briggs, W. R. (2008). Phytochrome A regulates the intracellular distribution of phototropin 1-green fluorescent protein in *Arabidopsis thaliana*. *The Plant Cell, 20*(10), 2835–2847.

Hardtke, C. S., & Berleth, T. (1998). The *Arabidopsis* gene *MONOPTEROS* encodes a transcription factor mediating embryo axis formation and vascular development. *EMBO Journal, 17*(5), 1405–1411.

Hay, A., Kaur, H., Phillips, A., Hedden, P., Hake, S., & Tsiantis, M. (2002). The gibberellin pathway mediates KNOTTED1-type homeobox function in plants with different body plans. *Current Biology, 12*(18), 1557–1565.

Hay, A., Barkoulas, M., & Tsiantis, M. (2006). ASYMMETRIC LEAVES1 and auxin activities converge to repress *BREVIPEDICELLUS* expression and promote leaf development in *Arabidopsis*. *Development, 133*(20), 3955–3961.

Heisler, M. G., Ohno, C., Das, P., Sieber, P., Reddy, G. V., Long, J. A., & Meyerowitz, E. M. (2005). Patterns of auxin transport and gene expression during primordium development revealed by live imaging of the *Arabidopsis* inflorescence meristem. *Current Biology, 15*(21), 1899–1911.

Hepworth, S. R., Zhang, Y., McKim, S., Li, X., & Haughn, G. W. (2005). BLADE-ON-PETIOLE-dependent signaling controls leaf and floral patterning in *Arabidopsis*. *The Plant Cell, 17*(5), 1434–1448.

Hibara, K., Karim, M. R., Takada, S., Taoka, K., & Tasaka, M. (2006). *Arabidopsis* CUP-SHAPED COTYLEDON3 regulates postembryonic shoot meristem and organ boundary formation. *The Plant Cell, 18*(11), 2946–2957.

Hirose, N., Takei, K., Kuroha, T., Kamada-Nobusada, T., Hayashi, H., & Sakakibara, H. (2008). Regulation of cytokinin biosynthesis, compartmentalization and translocation. *Journal of Experimental Botany, 59*(1), 75–83.

Hobe, M., Müller, R., Grünewald, M., Brand, U., & Simon, R. (2003). Loss of CLE40, a protein functionally equivalent to the stem cell restricting signal CLV3, enhances root waving in *Arabidopsis*. *Development Genes and Evolution, 213*(8), 371–381.

Hu, C., Zhu, Y., Cui, Y., Cheng, K., Liang, W., Wei, Z., Zhu, M., Yin, H., Zeng, L., Xiao, Y., Lv, M., Yi, J., Hou, S., He, K., Li, J., & Gou, X. (2018). A group of receptor kinases are essential for CLAVATA signalling to maintain stem cell homeostasis. *Nature Plants, 4*(4), 205–211.

Hwang, I., Sheen, J., & Müller, B. (2012). Cytokinin signaling networks. *Annual Review of Plant Biology, 63*(1), 353–380.

Ikezaki, M., Kojima, M., Sakakibara, H., Kojima, S., Ueno, Y., Machida, C., & Machida, Y. (2010). Genetic networks regulated by *ASYMMETRIC LEAVES1* (*AS1*) and *AS2* in leaf development in *Arabidopsis thaliana*: *KNOX* genes control five morphological events. *The Plant Journal, 61*(1), 70–82.

Ito, Y., Nakanomyo, I., Motose, H., Iwamoto, K., Sawa, S., Dohmae, N., & Fukuda, H. (2006). Dodeca-CLE peptides as suppressors of plant stem cell differentiation. *Science, 313*(5788), 842–845.

Jasinski, S., Piazza, P., Craft, J., Hay, A., Woolley, L., & Tsiantis, M. (2005). *KNOX* action in *Arabidopsis* is mediated by coordinate regulation of cytokinin and gibberellin activities. *Current Biology, 15*(17), 1560–1565.

Jeong, S., Trotochaud, A. E., & Clark, S. E. (1999). The *Arabidopsis CLAVATA2* gene encodes a receptor-like protein required for the stability of the CLAVATA1 receptor-like kinase. *The Plant Cell, 11*(10), 1925–1934.

Jun, J. H., Ha, C. M., & Fletcher, J. C. (2010). BLADE-ON-PETIOLE1 co-ordinates organ determinacy and axial polarity in *Arabidopsis* by directly activating *ASYMMETRIC LEAVES2*. *The Plant Cell, 22*(1), 62–76.

Kaya, H., Shibahara, K. I., Taoka, K. I., Iwabuchi, M., Stillman, B., & Araki, T. (2001). *FASCIATA* genes for chromatin assembly factor-1 in *Arabidopsis* maintain the cellular organization of apical meristems. *Cell, 104*(1), 131–142.

Kayes, J. M., & Clark, S. E. (1998). CLAVATA2, a regulator of meristem and organ development in *Arabidopsis*. *Development, 125*(19), 3843–3851.

Keller, T., Abbott, J., Moritz, T., & Doerner, P. (2006). *Arabidopsis* REGULATOR OF AXILLARY MERISTEMS1 controls a leaf axil stem cell niche and modulates vegetative development. *The Plant Cell, 18*(3), 598–611.

Kerstetter, R. A., & Hake, S. (1997). Shoot meristem formation in vegetative development. *The Plant Cell, 9*(7), 1001–1010.

Kieffer, M., Stern, Y., Cook, H., Clerici, E., & Davies, B. (2006). Analysis of the transcription factor WUSCHEL and its functional homologue in Antirrhinum reveals a potential mechanism for their roles in meristem maintenance. *The Plant Cell, 18*(3), 560–573.

Kinoshita, A., Betsuyaku, S., Osakabe, Y., Mizuno, S., & Nagawa, S. (2010). RPK2 is an essential receptor-like kinase that transmits the CLV3 signal in *Arabidopsis*. *Development, 137*(22), 3911–3920.

Komatsu, K., Maekawa, M., Ujiie, S., Satake, Y., & Kyozuka, J. (2003). LAX and SPA: Major regulators of shoot branching in rice. *Proceedings of the National Academy of Sciences, 100*(20), 11765–11770.

Kondo, T., Sawa, S., Kinoshita, A., Mizuno, S., Kakimoto, T., Fukuda, H., & Sakagami, Y. (2006). A plant peptide encoded by CLV3 identified by in situ MALDI-TOF MS analysis. *Science, 313*(5788), 845–848.

Kurakawa, T., Ueda, N., Maekawa, M., Kobayashi, K., & Kyozuka, J. (2007). Direct control of shoot meristem activity by a cytokinin activating enzyme. *Nature, 445*(7128), 652–655.

Kwon, C. S., Chen, C., & Wagner, D. (2005). *WUSCHEL* is a primary target for transcriptional regulation by SPLAYED in dynamic control of stem cell fate in *Arabidopsis*. *Genes & Development, 19*(8), 992–1003.

Kwon, C.-T., Tang, L., Wang, X., Gentile, I., Hendelman, A., Robitaille, G., Van Eck, J., Xu, C., & Lippman, Z. B. (2022). Dynamic evolution of small signalling peptide compensation in plant stem cell control. *Nature Plants, 8*(4), 346–355.

Laufs, P., Peaucelle, A., Morin, H., & Traas, J. (2004). MicroRNA regulation of the *CUC* genes is required for boundary size control in *Arabidopsis* meristems. *Development, 131*(17), 4311–4322.

Laux, T., Mayer, K. F., Berger, J., & Jurgens, G. (1996). The *WUSCHEL* gene is required for shoot and floral meristem integrity in *Arabidopsis*. *Development, 122*, 87–96.

Lee, C., & Clark, S. E. (2013). Core pathways controlling shoot meristem maintenance. *Wiley Interdisciplinary Reviews. Developmental Biology, 2*(5), 671–684.

Lee, D. K., Geisler, M., & Springer, P. S. (2009). *LATERAL ORGAN FUSION1 and LATERAL ORGAN FUSION2* function in lateral organ separation and axillary meristem formation in *Arabidopsis*. *Development, 136*, 2423–2432.

Leibfried, A., To, J. P., Busch, W., Stehling, S., Kehle, A., Demar, M., Kieber, J. J., & Lohmann, J. U. (2005). *WUSCHEL* controls meristem function by direct regulation of cytokinin-inducible response regulators. *Nature, 438*(7071), 1172–1175.

Lenhard, M., Jurgens, G., & Laux, T. (2002). The *WUSCHEL* and *SHOOTMERISTEMLESS* genes fulfill complementary roles in *Arabidopsis* shoot meristem regulation. *Development, 129*, 3195–3206.

Leyser, O. (2005). Auxin distribution and plant pattern formation: How many angels can dance on the point of PIN? *Cell, 121*(6), 819–822.

Leyser, H. M. O., & Furner, I. J. (1992). Characterization of three shoot apical meristem mutants of *Arabidopsis thaliana*. *Development, 116*(2), 397–403.

Li, X., Qian, Q., Fu, Z., Wang, Y., Xiong, G., Zeng, D., Wang, X., Liu, X., Teng, S., Hiroshi, F., Yuan, M., Luo, D., Han, B., & Li, J. (2003). Control of tillering in rice. *Nature, 422*(6932), 618–621.

Li, H., Xu, L., Wang, H., Yuan, Z., Cao, X., Yang, Z., Zhang, D., Xu, Y., & Huang, H. (2005). The Putative RNA-dependent RNA polymerase RDR6 acts synergistically with ASYMMETRIC LEAVES1 and 2 to repress BREVIPEDICELLUS and MicroRNA165/166 in *Arabidopsis* leaf development. *The Plant Cell, 17*(8), 2157–2171.

Li, Y., Xia, T., Gao, F., & Li, Y. (2020). Control of plant branching by the CUC2/CUC3-DA1-UBP15 regulatory module. *The Plant Cell, 32*(6), 1919–1932.

Lincoln, C., Long, J., Yamaguchi, J., Serikawa, K., & Hake, S. (1994). Aknottedl-like homeobox gene in *Arabidopsis* is expressed in the vegetative meristem and dramatically alters L6af morphology when overexpressed in transgenic plants. *The Plant Cell, 6*(12), 1859–1876.

Lodha, M., Marco, C. F., & Timmermans, M. C. (2013). The ASYMMETRIC LEAVES complex maintains repression of *KNOX* homeobox genes via direct recruitment of Polycomb-repressive complex2. *Genes & Development, 27*(6), 596–601.

Long, J. A., & Barton, M. K. (1998). The development of apical embryonic pattern in *Arabidopsis*. *Development, 125*, 3027–3035.

Long, J., & Barton, M. K. (2000). Initiation of axillary and floral meristems in *Arabidopsis*. *Developmental Biology, 218*(2), 341–353.

Long, J., Moan, E., Medford, J., & Barton, M. (1996). A member of the KNOTTED class of homeodomain proteins encoded by the *STM* gene of *Arabidopsis*. *Nature, 379*(6560), 66–69.

Lopes, F. L., Galvan-Ampudia, C., & Landrein, B. (2021). WUSCHEL in the shoot apical meristem: Old player, new tricks. *Journal of Experimental Botany, 72*(5), 1527–1535.

Lynn, K., Fernandez, A., Aida, M., Sedbrook, J., Tasaka, M., Masson, P., & Barton, M. K. (1999). The *PINHEAD/ZWILLE* gene acts pleiotropically in *Arabidopsis* development and has overlapping functions with the *ARGONAUTE1* gene. *Development, 126*(3), 469–481.

Ma, Y., Miotk, A., Šutiković, Z., Ermakova, O., Wenzl, C., Medzihradszky, A., Gaillochet, C., Forner, J., Utan, G., Brackmann, K., Galván-Ampudia, C. S., Vernoux, T., Greb, T., & Lohmann, J. U. (2019). WUSCHEL acts as an auxin response rheostat to maintain apical stem cells in *Arabidopsis*. *Nature Communications, 10*(1), 5093.

Mallory, A. C., Hinze, A., Tucker, M. R., Bouché, N., Gasciolli, V., Elmayan, T., Lauressergues, D., Jauvion, V., Vaucheret, H., & Laux, T. (2009). Redundant and specific roles of the ARGONAUTE proteins AGO1 and ZLL in development and small RNA-directed gene silencing. *PLoS Genetics, 5*(9), e1000646.

Mashiguchi, K., Tanaka, K., Sakai, T., Sugawara, S., & Kawaide, H. (2011). The main auxin biosynthesis pathway in *Arabidopsis*. *Proceedings of the National Academy of Sciences, 108*(45), 18512–18517.

Matsuoka, M., Ichikawa, H., Saito, A., Tada, Y., Fujimura, T., & Kano-Murakami, Y. (1993). Expression of a rice homeobox gene causes altered morphology of transgenic plants. *The Plant Cell, 5*(9), 1039–1048.

Mayer, K. F., Schoof, H., Haecker, A., Lenhard, M., Jurgens, G., & Laux, T. (1998). Role of *WUSCHEL* in regulating stem cell fate in the *Arabidopsis* shoot meristem. *Cell, 95*(6), 805–815.

Mcconnell, J. R., & Barton, M. K. (1995). Effect of mutations in the pinhead gene of *Arabidopsis* on the formation of shoot apical meristems. *Developmental Genetics, 16*(4), 358–366.

Meng, W. J., Cheng, Z. J., Sang, Y. L., Zhang, M. M., Rong, X. F., Wang, Z. W., Tang, Y. Y., & Zhang, X. S. (2017). Type-B ARABIDOPSIS RESPONSE REGULATORs specify the shoot stem cell niche by dual regulation of *WUSCHEL*. *The Plant Cell, 29*(6), 1357–1372.

Michniewicz, M., Zago, M. K., Abas, L., Weijers, D., Schweighofer, A., Meskiene, I., Heisler, M. G., Ohno, C., Zhang, J., Huang, F., Schwab, R., Weigel, D., Meyerowitz, E. M., Luschnig, C., Offringa, R., & Friml, J. (2007). Antagonistic regulation of PIN phosphorylation by PP2A and PINOID directs auxin flux. *Cell, 130*(6), 1044–1056.

Miwa, H., Kinoshita, A., Fukuda, H., & Sawa, S. (2009). Plant meristems: CLAVATA3/ESR-related signaling in the shoot apical meristem and the root apical meristem. *Journal of Plant Research, 122*(1), 31–39.

Miyawaki, K., Matsumoto-Kitano, M., & Kakimoto, T. (2004). Expression of cytokinin biosynthetic isopentenyltransferase genes in *Arabidopsis*: Tissue specificity and regulation by auxin, cytokinin, and nitrate. *The Plant Journal, 37*(1), 128–138.

Miyawaki, K., Tarkowski, P., Matsumoto-Kitano, M., Kato, T., Sato, S., Tarkowska, D., Tabata, S., Sandberg, G., & Kakimoto, T. (2006). Roles of *Arabidopsis* ATP/ADP isopentenyltransferases and tRNA isopentenyltransferases in cytokinin biosynthesis. *Proceedings of the National Academy of Sciences, 103*(44), 16598–16603.

Moussian, B., Schoof, H., Haecker, A., Jürgens, G., & Laux, T. (1998). Role of the *ZWILLE* gene in the regulation of central shoot meristem cell fate during *Arabidopsis* embryogenesis. *EMBO Journal, 17*(6), 1799–1809.

Müller, D., Schmitz, G., & Theres, K. (2006). Blind homologous *R2R3 Myb*genes control the pattern of lateral meristem initiation in *Arabidopsis*. *The Plant Cell, 18*(3), 586–597.

Müller, R., Bleckmann, A., & Simon, R. (2008). The receptor kinase CORYNE of *Arabidopsis* transmits the stem cell–limiting signal CLAVATA3 independently of CLAVATA1. *The Plant Cell, 20*(4), 934–946.

Müller, D., Waldie, T., Miyawaki, K., To, J. P., Melnyk, C. W., Kieber, J. J., Kakimoto, T., & Leyser, O. (2015). Cytokinin is required for escape but not release from auxin mediated apical dominance. *The Plant Journal, 82*(5), 874–886.

Nicolas, A., Maugarny-Calès, A., Adroher, B., Chelysheva, L., Li, Y., Burguet, J., Bågman, A. M., Smit, M. E., Brady, S. M., Li, Y., & Laufs, P. (2022). *De novo* stem cell establishment in meristems requires repression of organ boundary cell fate. *The Plant Cell, 34*(12), 4738–4759.

Nimchuk, Z. L. (2017). CLAVATA1 controls distinct signaling outputs that buffer shoot stem cell proliferation through a two-step transcriptional compensation loop. *PLoS Genetics, 13*(3), e1006681.

Nimchuk, Z. L., Zhou, Y., Tarr, P. T., Peterson, B. A., & Meyerowitz, E. M. (2015). Plant stem cell maintenance by transcriptional cross-regulation of related receptor kinases. *Development, 142*(6), 1043–1049.

Nishimura, C., Ohashi, Y., Sato, S., Kato, T., Tabata, S., & Ueguchi, C. (2004). Histidine kinase homologs that act as cytokinin receptors possess overlapping functions in the regulation of shoot and root growth in *Arabidopsis*. *The Plant Cell, 16*(6), 1365–1377.

Norberg, M., Holmlund, M., & Nilsson, O. (2005). The *BLADE ON PETIOLE* genes act redundantly to control the growth and development of lateral organs. *Development, 132*(9), 2203–2213.

Ogawa, M., Shinohara, H., Sakagami, Y., & Matsubayashi, Y. (2008). *Arabidopsis* CLV3 peptide directly binds CLV1 ectodomain. *Science, 319*(5861), 294.

Oikawa, T., & Kyozuka, J. (2009). Two-step regulation of LAX PANICLE1 protein accumulation in axillary meristem formation in rice. *The Plant Cell, 21*(4), 1095–1108.

Ori, N., Eshed, Y., Chuck, G., Bowman, J. L., & Hake, S. (2000). Mechanisms that control *knox* gene expression in the *Arabidopsis* shoot. *Development, 127*(24), 5523–5532.

Parry, G., Marchant, A., May, S., Swarup, R., Swarup, K., James, N., Graham, N., Allen, T., Martucci, T., Yemm, A., Napier, R., Manning, K., King, G., & Bennett, M. (2001). Quick on the uptake: Characterization of a family of plant auxin influx carriers. *Journal of Plant Growth Regulation, 20*, 217–225.

Parry, G., Delbarre, A., Marchant, A., Swarup, R., Napier, R., Perrot-Rechenmann, C., & Bennett, M. J. (2006). Novel auxin transport inhibitors phenocopy the auxin influx carrier mutation aux1. *The Plant Journal, 25*(4), 399–406.

Perales, M., & Reddy, G. V. (2012). Stem cell maintenance in shoot apical meristems. *Current Opinion in Plant Biology, 15*(1), 10–16.

Perales, M., Rodriguez, K., Snipes, S., Yadav, R. K., Diaz-Mendoza, M., & Reddy, G. V. (2016). Threshold-dependent transcriptional discrimination underlies stem cell homeostasis. *Proceedings of the National Academy of Sciences, 113*(41), 6298–6306.

Phelps-Durr, T. L., Thomas, J., Vahab, P., & Timmermans, M. C. (2005). Maize rough sheath2 and its *Arabidopsis* orthologue ASYMMETRIC LEAVES1 interact with HIRA, a predicted histone chaperone, to maintain *knox* gene silencing and determinacy during organogenesis. *The Plant Cell, 17*(11), 2886–2898.

Plong, A., Rodriguez, K., Alber, M., Chen, W., & Reddy, G. V. (2021). CLAVATA3 mediated simultaneous control of transcriptional and post-translational processes provides robustness to the WUSCHEL gradient. *Nature Communications, 12*(1), 6361.

Prasad, K., Grigg, S. P., Barkoulas, M., Yadav, R. K., & Sanchez-Perez, G. F. (2011). *Arabidopsis* PLETHORA transcription factors control phyllotaxis. *Current Biology, 21*(13), 1123–1128.

Prigge, M. J., Otsuga, D., Alonso, J. M., & Clark, S. E. (2005). Class III homeodomain-leucine zipper gene family members have overlapping, antagonistic, and distinct roles in *Arabidopsis* development. *The Plant Cell, 17*(1), 61–76.

Raman, S., Greb, T., Peaucelle, A., Blein, T., Laufs, P., & Theres, K. (2008). Interplay of miR164, *CUP-SHAPED COTYLEDON* genes and *LATERAL SUPPRESSOR* controls axillary meristem formation in *Arabidopsis thaliana*. *The Plant Journal, 55*(1), 65–76.

Rast, M. I., & Simon, R. (2008). The meristem-to-organ boundary: More than an extremity of anything. *Current Opinion in Genetics & Development, 18*(4), 287–294.

Rast, M. I., & Simon, R. (2012). *Arabidopsis* JAGGED LATERAL ORGANS acts with ASYMMETRIC LEAVES2 to coordinate KNOX and PIN expression in shoot and root meristems. *The Plant Cell, 24*(7), 2917–2933.

Reinhardt, D., Mandel, T., & Kuhlemeier, C. (2000). Auxin regulates the initiation and radial position of plant lateral organs. *The Plant Cell, 12*(4), 507–518.

Reinhardt, D., Pesce, E. R., Stieger, P., Mandel, T., & Baltensperger, K. (2003). Regulation of phyllotaxis by polar auxin transport. *Nature, 426*(6964), 255–260.

Ritter, M. K., Padilla, C. M., & Schmidt, R. J. (2002). The maize mutant barren stalk1 is defective in axillary meristem development. *American Journal of Botany, 89*(2), 203–210.

Rodríguez-Leal, D., Lemmon, Z. H., Man, J., Bartlett, M. E., & Lippman, Z. B. (2017). Engineering quantitative trait variation for crop improvement by genome editing. *Cell, 171*(2), 470–480.e8.

Rodríguez-Leal, D., Xu, C., Kwon, C. T., Soyars, C., Demesa-Arevalo, E., Man, J., Liu, L., Lemmon, Z. H., Jones, D. S., Van Eck, J., Jackson, D. P., Bartlett, M. E., Nimchuk, Z. L., & Lippman, Z. B. (2019). Evolution of buffering in a genetic circuit controlling plant stem cell proliferation. *Nature Genetics, 51*(5), 786–792.

Rojo, E., Sharma, V. K., Kovaleva, V., Raikhel, N. V., & Fletcher, J. C. (2002). CLV3 is localized to the extracellular space, where it activates the *Arabidopsis* CLAVATA stem cell signaling pathway. *The Plant Cell, 14*(5), 969–977.

Roodbarkelari, F., Du, F., Truernit, E., & Laux, T. (2015). ZLL/AGO10 maintains shoot meristem stem cells during *Arabidopsis* embryogenesis by down-regulating ARF2-mediated auxin response. *BMC Biology, 13*(1), 1–12.

Rossmann, S., Kohlen, W., Hasson, A., & Theres, K. (2015). Lateral suppressor and Goblet act in hierarchical order to regulate ectopic meristem formation at the base of tomato leaflets. *The Plant Journal, 81*(6), 837–848.

Saiga, S., Furumizu, C., Yokoyama, R., Kurata, T., Sato, S., Kato, T., Tabata, S., Suzuki, M., & Komeda, Y. (2008). The *Arabidopsis OBERON1* and *OBERON2* genes encode plant homeodomain finger proteins and are required for apical meristem maintenance. *Development, 135*(10), 1751–1759.

Sakamoto, T., Kamiya, N., Ueguchi-Tanaka, M., Iwahori, S., & Matsuoka, M. (2001). KNOX homeodomain protein directly suppresses the expression of a gibberellin biosynthetic gene in the tobacco shoot apical meristem. *Genes & Development, 15*, 581–590.

Satina, S., Blakeslee, A. F., & Avery, A. G. (1940). Demonstration of the three germ layers in the shoot apex of Datura by means of induced polyploidy in periclinal chimeras. *American Journal of Botany, 27*, 895–905.

Scanlon, M. J. (2003). The polar auxin transport inhibitor N-1-naphthylphthalamic acid disrupts leaf initiation, KNOX protein regulation, and formation of leaf margins in maize. *Plant Physiology, 133*(2), 597–605.

Schmitz, G., & Theres, K. (2005). Shoot and inflorescence branching. *Current Opinion in Plant Biology, 8*(5), 506–511.

Schoof, H., Lenhard, M., Haecker, A., Mayer, K. F., Jürgens, G., & Laux, T. (2000). The stem cell population of *Arabidopsis* shoot meristems in maintained by a regulatory loop between the *CLAVATA* and *WUSCHEL* genes. *Cell, 100*(6), 635–644.

Schuster, C., Gaillochet, C., Medzihradszky, A., Busch, W., Daum, G., Krebs, M., Kehle, A., & Lohmann, J. U. (2014). A regulatory framework for shoot stem cell control integrating metabolic, transcriptional, and phytohormone signals. *Developmental Cell, 28*(4), 438–449.

Scofield, S., & Murrary, J. A. H. (2006). *KNOX* gene function in plant stem cell niches. *Plant Molecular Biology, 60*(6), 929–946.

Scofield, S., Murison, A., Jones, A., Fozard, J., Aida, M., Band, L. R., Bennett, M., & Murray, J. A. H. (2018). Coordination of meristem and boundary functions by transcription factors in the SHOOT MERISTEMLESS regulatory network. *Development, 145*(9), dev157081.

Sentoku, N., Sato, Y., Kurata, N., Ito, Y., Kitano, H., & Matsuoka, M. R. (1999). Expression of the rice *KN1-type homeobox* gene family during embryo, shoot, and flower development. *The Plant Cell, 11*, 1651–1663.

Shen, W. H., & Xu, L. (2009). Chromatin remodeling in stem cell maintenance in *Arabidopsis thaliana. Molecular Plant, 2*(4), 600–609.

Shi, B., & Vernoux, T. (2021). Hormonal control of cell identity and growth in the shoot apical meristem. *Current Opinion in Plant Biology, 65*, 102111.

Shi, B., Zhang, C., Tian, C., Wang, J., Wang, Q., Xu, T., Xu, Y., Ohno, C., & Sablowski, R. (2016). Two-step regulation of a meristematic cell population acting in shoot branching in *Arabidopsis. PLoS Genetics, 12*(7), e1006168.

Shi, J., Dong, J., Xue, J., Wang, H., Yang, Z., Jiao, Y., Xu, L., & Huang, H. (2017). Model for the role of auxin polar transport in patterning of the leaf adaxial-abaxial axis. *The Plant Journal, 92*(3), 469–480.

Shinohara, H., & Matsubayashi, Y. (2015). Reevaluation of the CLV3-receptor interaction in the shoot apical meristem: Dissection of the CLV3 signaling pathway from a direct ligand-binding point of view. *The Plant Journal, 82*(2), 328–336.
Singh, A. P., & Savaldi-Goldstein, S. (2015). Growth control: Brassinosteroid activity gets context. *Journal of Experimental Botany, 66*(4), 1123–1132.
Skylar, A., Hong, F. X., Chory, J., Weigel, D., & Wu, X. L. (2010). STIMPY mediates cytokinin signaling during shoot meristem establishment in *Arabidopsis* seedlings. *Development, 137*(4), 541–549.
Sloan, J., Hakenjos, J. P., Gebert, M., Ermakova, O., Gumiero, A., Stier, G., Wild, K., Sinning, I., & Lohmann, J. U. (2020). Structural basis for the complex DNA binding behavior of the plant stem cell regulator WUSCHEL. *Nature Communications, 11*(1), 2223.
Smith, Z. R., & Long, J. A. (2010). Control of *Arabidopsis* apical–basal embryo polarity by antagonistic transcription factors. *Nature, 464*(7287), 423–426.
Smith, R. S., Guyomarc'h, S., Mandel, T., Reinhardt, D., Kuhlemeier, C., & Prusinkiewicz, P. (2006). A plausible model of phyllotaxis. *Proceedings of the National Academy of Sciences, 103*(5), 1301–1306.
Spinelli, S. V., Martin, A. P., Viola, I. L., Gonzalez, D. H., & Palatnik, J. F. (2011). A mechanistic link between STM and CUC1 during *Arabidopsis*. *Development, 156*(4), 1894–1904.
Steeves, T., & Sussex, I. (1989). *Patterns in plant development*. Cambridge University Press.
Stewart, R. N., & Dermen, H. (1970). Determination of number and mitotic activity of shoot apical initial cells by analysis of mericlinal chimeras. *American Journal of Botany, 57*(7), 816–826.
Stieger, P. A., Reinhardt, D., & Kuhlemeier, C. (2002). The auxin influx carrier is essential for correct leaf positioning. *The Plant Journal, 32*(4), 509–517.
Su, Y. H., Zhou, C., Li, Y. J., Yu, Y., Tang, L. P., Zhang, W. J., Yao, W. J., Huang, R., Laux, T., & Zhang, X. S. (2020). Integration of pluripotency pathways regulates stem cell maintenance in the *Arabidopsis* shoot meristem. *Proceedings of the National Academy of Sciences of the United States of America, 117*(36), 22561–22571.
Swarup, R., & Bennett, M. (2003). Auxin transport: The fountain of life in plants? *Developmental Cell, 5*(6), 824–826.
Szymkowiak, E. J., & Sussex, I. M. (1996). What chimeras can tell us about plant development. *Annual Review of Plant Physiology, 47*, 351–376.
Takada, S., Hibara, K., Ishida, T., & Tasaka, M. (2001). The *CUP-SHAPED COTYLEDON1* gene of *Arabidopsis* regulates shoot apical meristem formation. *Development, 128*(7), 1127–1135.
Takeda, S., Tadele, Z., Hofmann, I., Probst, A. V., & Angelis, K. J. (2004). BRU1, a novel link between responses to DNA damage and epigenetic gene silencing in *Arabidopsis*. *Genes & Development, 18*(7), 782–793.
Takeda, S., Hanano, K., Kariya, A., Shimizu, S., Zhao, L., Matsui, M., Tasaka, M., & Aida, M. (2011). CUP-SHAPED COTYLEDON1 transcription factor activates the expression of *LSH4* and *LSH3*, two members of the *ALOG* gene family, in shoot organ boundary cells. *The Plant Journal, 66*(6), 1066–1077.
Tan, F. Q., Wang, W., Li, J., Lu, Y., Zhu, B., Hu, F., Li, Q., Zhao, Y., & Zhou, D. X. (2022). A coiled-coil protein associates Polycomb Repressive Complex 2 with KNOX/BELL transcription factors to maintain silencing of cell differentiation-promoting genes in the shoot apex. *The Plant Cell, 34*(8), 2969–2988.
Tantikanjana, T., Yong, J. W., Letham, D. S., Griffith, M., & Hussain, M. (2001). Control of axillary bud initiation and shoot architecture in *Arabidopsis* through the *SUPERSHOOT* gene. *Genes & Development, 15*(12), 1577–1588.
To, J. P., Haberer, G., Ferreira, F. J., Deruere, J., Mason, M. G., Schaller, G. E., Alonso, J. M., Ecker, J. R., & Kieber, J. J. (2004). Type-A *Arabidopsis* response regulators are partially redundant negative regulators of cytokinin signaling. *The Plant Cell, 16*(3), 658–671.
Treml, B. S., Winderl, S., Radykewicz, R., Herz, M., & Ruiz, R. A. (2005). The gene *ENHANCER OF PINOID* controls cotyledon development in the *Arabidopsis* embryo. *Development, 132*(18), 4063–4074.

Trotochaud, A. E., Hao, T., Wu, G., Yang, Z., & Clark, S. E. (1999). The CLAVATA1 receptor-like kinase requires CLAVATA3 for its assembly into a signaling complex that includes KAPP and a Rho-related protein. *The Plant Cell, 11*(3), 393–406.

Tsiantis, M., Brown, M. I., Skibinski, G., & Langdale, J. A. (1999). Disruption of auxin transport is associated with aberrant leaf development in maize. *Plant Physiology, 121*(4), 1163–1168.

Tucker, M. R., Hinze, A., Tucker, E. J., Takada, S., Jurgens, G., & Laux, T. (2008). Vascular signalling mediated by ZWILLE potentiates WUSCHEL function during shoot meristem stem cell development in the *Arabidopsis* embryo. *Development, 135*(17), 2839–2843.

Vernoux, T., Brunoud, G., Farcot, E., Morin, V., & Traas, J. (2011). The auxin signalling network translates dynamic input into robust patterning at the shoot apex. *Molecular Systems Biology, 7*(1), 508–508.

Vieten, A., Sauer, M., Brewer, P. B., & Friml, J. (2007). Molecular and cellular aspects of auxin-transport-mediated development. *Trends in Plant Science, 12*(4), 160–168.

Vollbrecht, E., Veit, B., Sinha, N., & Hake, S. (1991). The developmental gene *Knotted-1* is a member of a maize homeobox gene family. *Nature, 350*(6315), 241–243.

Vroemen, C. W., Mordhorst, A. P., Albrecht, C., Kwaaitaal, M. A., & de Vries, S. C. (2003). The *CUP-SHAPED COTYLEDON3* gene is required for boundary and shoot meristem formation in *Arabidopsis*. *The Plant Cell, 15*(7), 1563–1577.

Wang, Y., & Jiao, Y. (2023). Cell signaling in the shoot apical meristem. *Plant Physiology, 193*(1), 70–82.

Wang, Q., Kohlen, W., Rossmann, S., Vernoux, T., & Theres, K. (2014a). Auxin depletion from the leaf axil conditions competence for axillary meristem formation in *Arabidopsis* and tomato. *The Plant Cell, 26*(5), 2068–2079.

Wang, Y., Wang, J., Shi, B., Yu, T., Qi, J., Meyerowitz, E. M., & Jiao, Y. (2014b). The stem cell niche in leaf axils is established by auxin and cytokinin in *Arabidopsis*. *The Plant Cell, 26*(5), 2055–2067.

Wang, J., Tian, C., Zhang, C., Shi, B., Cao, X., Zhang, T. Q., Zhao, Z., Wang, J. W., & Jiao, Y. (2017). Cytokinin signaling activates *WUSCHEL* expression during axillary meristem initiation. *The Plant Cell, 29*(6), 1373–1387.

Wang, X., Aguirre, L., Rodríguez-Leal, D., Hendelman, A., Benoit, M., & Lippman, Z. B. (2021). Dissecting cis-regulatory control of quantitative trait variation in a plant stem cell circuit. *Nature Plants, 7*(4), 419–427.

Wang, Q., Marconi, M., Guan, C., Wabnik, K., & Jiao, Y. (2022a). Polar auxin transport modulates early leaf flattening. *Proceedings of the National Academy of Sciences of the United States of America, 119*(50), e2215569119.

Wang, W., Hu, C., Li, X., Zhu, Y., Tao, L., Cui, Y., Deng, D., Fan, X., Zhang, H., Li, J., Gou, X., & Yi, J. (2022b). Receptor-like cytoplasmic kinases PBL34/35/36 are required for CLE peptide-mediated signaling to maintain shoot apical meristem and root apical meristem homeostasis in *Arabidopsis*. *The Plant Cell, 34*(4), 1289–1307.

Weijers, D., Sauer, M., Meurette, O., Friml, J., Ljung, K., Sandberg, G., Hooykaas, P., & Offringa, R. (2005). Maintenance of embryonic auxin distribution for apical-basal patterning by PIN-FORMED-dependent auxin transport in *Arabidopsis*. *The Plant Cell, 17*(9), 2517–2526.

Werner, T., Motyka, V., Laucou, V., Smets, R., Van Onckelen, H., & Schmülling, T. (2003). Cytokinin deficient transgenic *Arabidopsis* plants show multiple developmental alterations indicating opposite functions of cytokinins in the regulation of shoot and root meristem activity. *The Plant Cell, 15*(11), 2532–2550.

Wisniewska, J., Xu, J., Seifertová, D., Brewer, P. B., & Friml, J. (2006). Polar PIN localization directs auxin flow in plants. *Science, 312*(5775), 883.

Wolters, H., & Jürgens, G. (2009). Survival of the flexible: Hormonal growth control and adaptation in plant development. *Nature Reviews Genetics, 10*(5), 305–317.

Won, C., Shen, X., Mashiguchi, K., Zheng, Z., & Zhao, Y. (2011). Conversion of tryptophan to indole-3-acetic acid by TRYPTOPHAN AMINOTRANSFERASES OF ARABIDOPSIS and YUCCAs in *Arabidopsis*. *Proceedings of the National Academy of Sciences, 108*(45), 18518–18523.

Wu, X., Dabi, T., & Weigel, D. (2005). Requirement of homeobox gene *STIMPY/WOX9* for *Arabidopsis* meristem growth and maintenance. *Current Biology, 15*(5), 436–440.

Wu, X. L., Chory, J., & Weigel, D. (2007). Combinations of WOX activities regulate tissue proliferation during *Arabidopsis* embryonic development. *Developmental Biology, 309*(2), 306–316.

Wu, H., Qu, X., Dong, Z., Luo, L., Shao, C., Forner, J., Lohmann, J. U., Su, M., Xu, M., Liu, X., Zhu, L., Zeng, J., Liu, S., Tian, Z., & Zhao, Z. (2020). WUSCHEL triggers innate antiviral immunity in plant stem cells. *Science, 370*(6513), 227–231.

Wurschum, T., Gross-Hardt, R., & Laux, T. (2006). *APETALA2* regulates the stem cell niche in the *Arabidopsis* shoot meristem. *The Plant Cell, 18*(2), 295–307.

Xie, M., Chen, H., Huang, L., O'Neil, R. C., Shokhirev, M. N., & Ecker, J. R. (2018). A B-ARR-mediated cytokinin transcriptional network directs hormone cross-regulation and shoot development. *Nature Communications, 9*(1), 1604.

Xin, W., Wang, Z., Liang, Y., Wang, Y., & Hu, Y. (2017). Dynamic expression reveals a two-step patterning of WUS and CLV3 during axillary shoot meristem formation in *Arabidopsis*. *Journal of Plant Physiology, 214*, 1–6.

Yadav, R. K., Perales, M., Gruel, J., Girke, T., Jönsson, H., & Reddy, G. V. (2011). WUSCHEL protein mediates stem cell homeostasis in the *Arabidopsis* shoot apex. *Genes & Development, 25*(19), 2025–2030.

Yanai, O., Shani, E., Dolezal, K., Tarkowski, P., Sablowski, R., Sandberg, G., Samach, A., & Ori, N. (2005). *Arabidopsis KNOXI* proteins activate cytokinin biosynthesis. *Current Biology, 15*(17), 1566–1571.

Yang, F., Wang, Q., Schmitz, G., Müller, D., & Theres, K. (2012). The bHLH protein ROX acts in concert with RAX1 and LAS to modulate axillary meristem formation in *Arabidopsis*. *The Plant Journal, 71*(1), 61–70.

Yang, W., Cortijo, S., Korsbo, N., Roszak, P., Schiessl, K., Gurzadyan, A., Wightman, R., Jönsson, H., & Meyerowitz, E. (2021). Molecular mechanism of cytokinin-activated cell division in *Arabidopsis*. *Science, 371*(6536), 1350–1355.

Žádníková, P., & Simon, R. (2014). How boundaries control plant development. *Current Opinion in Plant Biology, 17*, 116–125.

Zhang, Z., Tucker, E., Hermann, M., & Laux, T. (2017). A molecular framework for the embryonic initiation of shoot meristem stem cells. *Developmental Cell, 40*(3), 264–277.e4.

Zhang, K., Zhang, H., Pan, Y., Niu, Y., Guo, L., Ma, Y., Tian, S., Wei, J., Wang, C., Yang, X., Fu, Y., Qu, P., Liu, L., Zhang, Y., Sun, H., Bai, Z., Dong, J., Li, C., & Liu, X. (2022). Cell- and noncell-autonomous AUXIN RESPONSE FACTOR3 controls meristem proliferation and phyllotactic patterns. *Plant Physiology, 190*(4), 2335–2349.

Zhao, M., Morohashi, K., Hatlestad, G., Grotewold, E., & Lloyd, A. (2008). The TTG1-bHLH-MYB complex controls trichome cell fate and patterning through direct targeting of regulatory loci. *Development, 135*(11), 1991–1999.

Zhao, Z., Andersen, S. U., Ljung, K., Dolezal, K., & Lohmann, J. U. (2010). Hormonal control of the shoot stem-cell niche. *Nature, 465*(7301), 1089–1092.

Zhou, Y., Liu, X., Engstrom, E. M., Nimchuk, Z. L., & Pruneda-Paz, J. L. (2015). Control of plant stem cell function by conserved interacting transcriptional regulators. *Nature, 517*(7534), 377–380.

Zhou, Y., Yan, A., Han, H., Li, T., Geng, Y., & Liu, X. (2018). Meyerowitz EM. HAIRY MERISTEM with WUSCHEL confines *CLAVATA3* expression to the outer apical meristem layers. *Science, 361*(6401), 502–506.

Zhu, H., Hu, F., Wang, R., Zhou, X., Sze, S. H., Liou, L. W., Barefoot, A., Dickman, M., & Zhang, X. (2011). *Arabidopsis* argonaute10 specifically sequesters miR166/165 to regulate shoot apical meristem development. *Cell, 145*(2), 242–256.

Chapter 3
Leaf Development

Yuling Jiao and Ying Wang

Abstract For most plants, leaves function as solar panels, equipped with photosynthetic cells that transform carbon dioxide and water into carbohydrates and oxygen. To fulfill this function, the initiation and shape of leaves are precisely regulated in each species. In addition, many plant organs are derived from leaves. For example, all floral organs are specialized leaves, as first proposed by von Goethe in 1790. Thus, leaf development forms the developmental basis of many other plant organs. In this chapter, we will first review the basic structure of leaves. Next, we will review the molecular pathways underlying leaf development. We will discuss (1) how a leaf is specified in the shoot meristem, (2) how the leaf attains its flattened shape and polarity, and (3) how the leaf cell proliferation is activated and then terminated. In addition, we will review variations of leaf shape, in particular, compound leaf formation. Special emphasis is given to the recent findings of phytohormonal and mechanical regulation of leaf shape patterning.

Keywords Leaf · Morphogenesis · Patterning · Sepal · Organogenesis · Flatness

3.1 Introduction

A leaf is a typical organ of a vascular plant and is the primary lateral appendage of the stem (Fig. 3.1). A typical leaf is a thin, flattened organ with venation and is specialized for photosynthesis. Nevertheless, leaf development is highly flexible as

---

Y. Jiao
State Key Laboratory of Gene Function and Modulation Research, School of Life Sciences, Peking University, Beijing, China

Peking-Tsinghua Center for Life Sciences, Center for Quantitative Biology, Academy for Advanced Interdisciplinary Studies, Peking University, Beijing, China

Y. Wang (✉)
College of Life Sciences, Capital Normal University, Beijing, China
e-mail: ying.wang@cnu.edu.cn

F. Chang et al. (eds.), *Regulation of Plant Development*,
https://doi.org/10.1007/978-981-95-7033-1_3

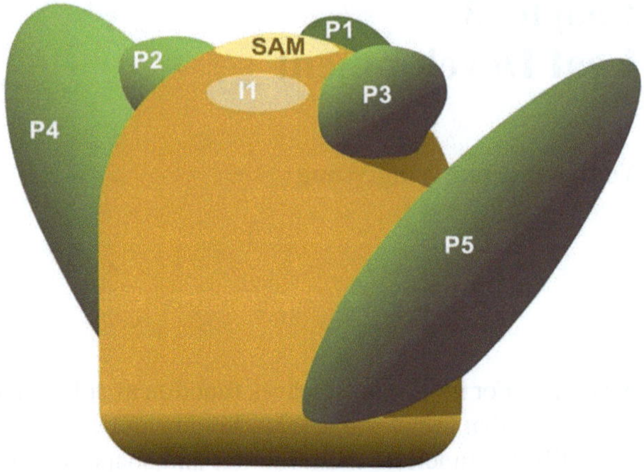

Fig. 3.1 Leaf formation at the shoot apex. The youngest leaf primordium, at the P1 stage, initiates as a radially symmetrical lateral appendage on the flanks of the SAM. During leaf development, a primordium elongates along the proximal-distal axis, forms the adaxial-abaxial axis, and flattens along the medio-lateral axis

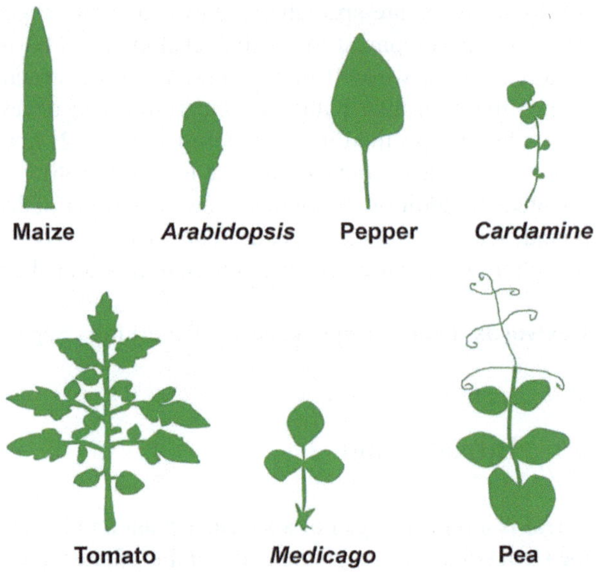

Fig. 3.2 Simple and compound leaves

reflected by the variable sizes and forms of leaves from different species and even from the same plant (Fig. 3.2).

Fossil evidence indicated that leaves originated from lateral branch structures. The development of determinate growth and a planar architecture transforms lateral branches into leaves (Boyce, 2010; Floyd & Bowman, 2010; Kaplan, 2001). Leaves evolved independently in different plant lineages (Tomescu, 2009), likely after the drop in carbon dioxide concentration in the atmosphere (Beerling et al., 2001). Ferns, gymnosperms, and angiosperms possess broad, flat leaves with intricate venation patterns, which are known as megaphylls. In contrast, clubmosses possess microphylls, which are smaller, simpler leaves featuring a single vein. These different leaf forms emerged independently in different lineages, although they share a similar planate shape (Tomescu, 2009). Leaves as the prototype can be further modified into diverse shapes to fulfill specialized functions. There are underground bulb scales, thick juicy leaves of succulent plants, and cataphyll leaves without major photosynthetic function. The floral organs, consisting of sepals, petals, stamens, and carpals, are all leaf derivatives and share a set of similar regulatory genes.

Leaves initiate at the flanks of the shoot apical meristem (SAM) (Fig. 3.1). Whereas an initial leaf primordium is a round-shaped bulge, it quickly acquires anisotropic development to form a flattened organ (Poethig & Sussex, 1985). There are three axes of symmetry, proximo-distal, adaxial-abaxial, and medio-lateral, established during leaf development. Leaves typically exhibit determinate growth, expanding and maturing into a particular form and then halting their development. In contrast, stems and roots display indeterminate growth, potentially continuing to elongate indefinitely providing they have sufficient resources. Leaf shape is highly diversified and can be roughly grouped into simple leaves and compound leaves (Fig. 3.2). These two groups actually represent extremes of a whole spectrum of leaf shapes.

Albeit highly diversified in shape, standard angiosperm leaves include, along the proximal-distal axis, stipules, a petiole, and a lamina. Stipules are small structures located on either side of the base of the petiole. Stipules can either be absent or noticeable, and they are only present in the mature leaves of a fraction of species. In the model plant *Arabidopsis*, stipules are formed early during leaf initiation but stop growing and eventually disappear when the leaf matures. Petioles connect the leaf to the plant mechanically and also enable transportation of water, salts, and sugars. However, petioles may also be absent so that the lamina directly connects to the stem. The lamina is typically flattened and is the major location for photosynthesis.

In a cross-section of a lamina, layers of cells are evident with parenchyma, i.e., ground tissues, termed mesophylls, located between the upper (adaxial) and lower (abaxial) epidermal layers. Mesophyll cells are usually responsible for photosynthesis. In many flowering plant species, there are two layers of mesophyll cells, the upper palisade layer and the lower spongy layer. The palisade mesophyll cells are vertically elongated, contain more chloroplasts, and are regularly arranged into one or more layers. All these characters make the palisade layer cells optimized for light harvesting. The spongy mesophyll cells are 'star-shaped', contain fewer chloroplasts, and are less tightly packed with large intercellular air spaces, which are connected to stomata in the epidermis. The spongy mesophyll cells are more suitable for gas exchange in respiration and photosynthesis. Because of the chlorophylls in chloroplasts in mesophyll cells, leaves are normally green with the upper side darker in color.

3.2 Leaf Initiation

Leaves originate from the periphery of the indeterminate shoot apical meristem (SAM). Initially established during embryogenesis, the SAM preserves a reservoir of undifferentiated cells at its center (Steeves & Sussex, 1989). Throughout post-embryonic development, the SAM orchestrates the production of stems, leaves, and floral organs in a specific sequence (Fig. 3.3).

In eudicots, the structure of the SAM is relatively consistent and can be delineated into an outer tunica layer and an inner corpus region. These regions exhibit distinct cellular behaviors: cells in the tunica predominantly divide perpendicular to the surface, or anticlinally, whereas those within the corpus divide without a preferential plane of cell division. This anticlinical division in the tunica results in a stratified structure where daughter cells remain in the same layer as their progenitors. The SAM can also be categorized into three functional domains: the central zone (CZ), the peripheral zone (PZ), and the rib meristem (RM) (Steeves & Sussex, 1989). The CZ, situated at the apex of the SAM, contains pluripotent stem cells that divide slowly to self-renew. Beneath the CZ lies the RM, which is composed of cells destined to form the internal stem tissues. In between the CZ and the RM is a small cluster of quiescent cells known as the organizing center (OC), which sustains the stem cell niche above. Some progeny from the CZ's stem cells are pushed into the surrounding PZ, where cells divide rapidly to initiate organ primordia. A distinct boundary separates the primordium from the rest of the SAM (Wang et al., 2016). During the vegetative phase of development, the SAM is responsible for the production of leaf primordia.

Leaf initiation is facilitated by the phytohormone auxin, as illustrated in Fig. 3.4. Experiments involving microapplications of auxin indicate that a localized auxin maximum is both necessary and sufficient to initiate primordium formation within the PZ of the SAM (Reinhardt et al., 2000). In normally growing plants, polar auxin

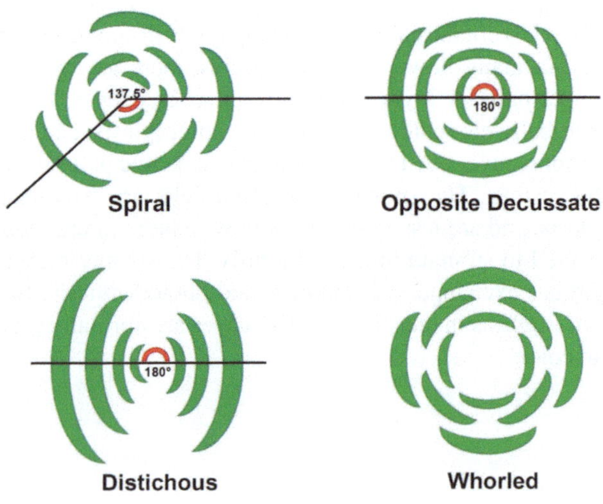

Fig. 3.3 Types of phyllotaxis. Spiral: one leaf per node with ~137.5° divergence angles. Opposite decussate: a pair of opposite leaves per node at 90° angles between pairs. Distichous: one leaf per node with 180° divergence angles. Whorled: three or more leaves per node

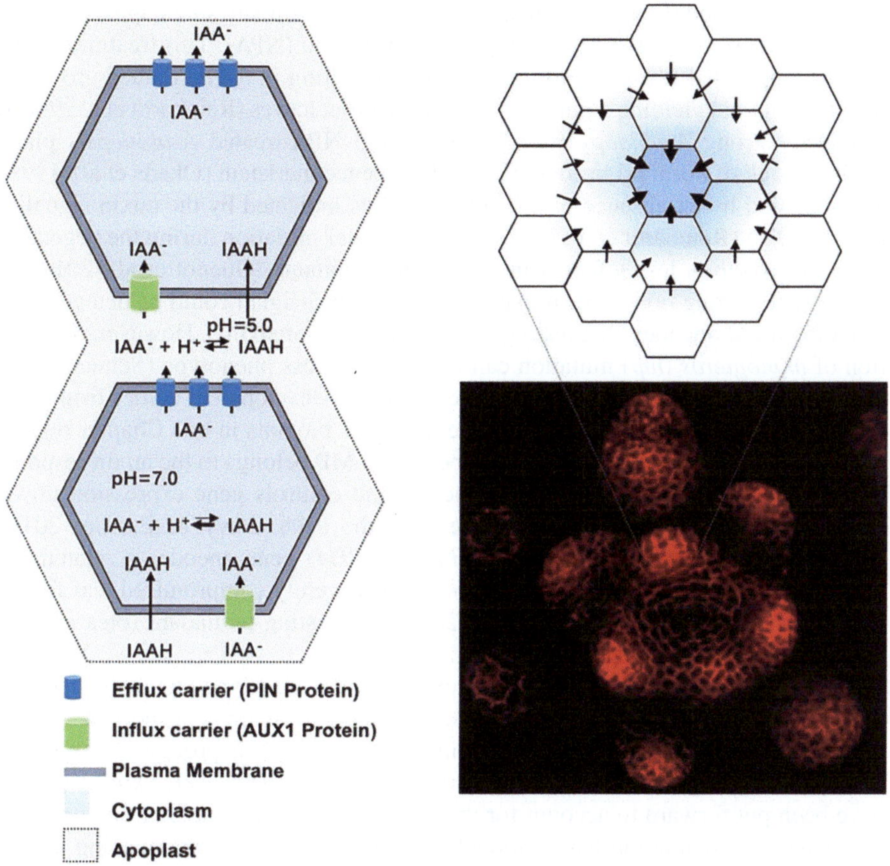

Fig. 3.4 Regulation of phyllotaxis by polar auxin transport. Top-down view of a SAM, showing the sites of developing primordia (P). PIN1 is localized toward the site of incipient primordia (P0), leading to high local concentrations of auxin. Schematic representation of polar auxin transport. According to the chemiosmotic model, the pH difference between the neutral cytoplasm and the acidic apoplast drives the polarized intercellular auxin movement. Protonated auxin (IAAH) forms in the acidic extracellular space apoplastic environment, and can passively diffuse through the lipidic plasma membrane or be transported by AUX1/LAX influx carrier into the cell. In the neutral cytosol, the deprotonated form (IAA^-) becomes dominant. IAA^- is less membrane permeable and requires efflux carriers, such as PINs. The subcellular PIN polarity determines the direction of intercellular auxin flow. In the SAM periphery, an auxin-PIN positive feedback loop generates auxin maxima, where leaf initiates

transport leads to the accumulation of auxin, creating local auxin maxima that designate the sites of incipient primordia. In *Arabidopsis*, the auxin efflux carrier PIN-FORMED1 (PIN1) is expressed in the epidermal layer and provascular cells of the SAM. The emergence of these local auxin maxima is a consequence of the coordinated and highly dynamic PIN1 polarity in each cell (Heisler et al., 2005; Reinhardt et al., 2003). The crucial roles of auxin and PIN1 in initiating leaf primordia have been demonstrated through the treatment of vegetative tomato shoot apices with the auxin transport inhibitor *N*-1-naphthylphthalamic acid (NPA). This treatment completely suppressed leaf primordium initiation, resulting in the continued growth of the meristem but yielding a pin-like structure without leaves (Reinhardt et al., 2000). Similarly, strong *Arabidopsis pin1* mutants and NPA-treated *Arabidopsis* plants exhibit a lack of floral primordia in the inflorescence meristem (Okada et al., 1991), accompanied by an absence of auxin maxima, as indicated by the auxin signaling reporter DR5 (Reinhardt et al., 2003). However, leaf initiation during the vegetative stage of growth is less affected in *Arabidopsis* mutants (Guenot et al., 2012). In *Arabidopsis* strong *pin1* mutant plants, a ring of DR5 signal could be detected surrounding the shoot apex, suggesting less focused auxin maxima. However, the addition of *monopteros* (*mp*) mutation can lead to a leafless phenotype (Schuetz et al., 2008). (Because much of our knowledge about leaf development comes from studies in the model plant *Arabidopsis*, gene names as mutants in this Chapter refer to those in *Arabidopsis* unless otherwise specified.) MP belongs to the auxin response factor (ARF) family of transcription factors and controls gene expression downstream of auxin signaling (Hardtke & Berleth, 1998; Lavy & Estelle, 2016). *ARF7/NONPHOTOTROPIC HYPOCOTYL4* (*NPH4*) gene encodes a related and redundant ARF. Double *mp nph4* mutants show severely compromised leaf formation (Guan et al., 2017; Krogan et al., 2016), suggesting redundant roles of these two ARFs in promoting leaf initiation.

It remains to be addressed how the auxin maxima are formed to enable leaf primordium initiation, although PIN1-mediated polar auxin transport is clearly involved. A comprehensive understanding of the molecular processes behind the self-organized regulation of PIN1 polarity is essential. However, various models have been put forward to account for the formation of auxin maxima. In particular, a mechanical regulation-based model assumes that cells experiencing higher mechanical strain have higher levels of membrane-localized PIN1 proteins (Nakayama et al., 2012). On the other hand, auxin levels positively correlate with cell expansion (Rayle & Cleland, 1992). A cell perceives its neighboring cell expansion to determine its own PIN1 protein membrane accumulation. This would cause the cell to export auxin (through PIN1) toward the expanding neighbor with higher auxin levels. As a result, the auxin flow will cause auxin accumulation against the concentration gradient and lead to a positive feedback to the cell expansion (Heisler et al., 2010) (Fig. 3.4).

Mechanical cues can also promote leaf primordium initiation. Ectopic manipulation of the cell wall of the SAM by wall-loosening agents, including expansin and pectin demethylesterifcase (PME), can trigger ectopic leaf and floral primordium initiation (Fleming et al., 1997; Peaucelle et al., 2008). It has been shown that pectin demethylesterification is associated with primordia formation and reduced wall elasticity (Peaucelle et al., 2011). Because auxin can promote pectin demethylesterification in the shoot apex (Braybrook & Peaucelle, 2013; Qi et al., 2017), auxin maximum may lead to cell wall loosening to promote leaf initiation. The primary cell wall can be considered as a hemicellulose and pectin matrix embedded with, and crosslinked by, rigid cellulose microfibrils. Auxin maximum also causes a local interference with the orientation of the rigid cellulose microfibrils, and thus anisotropy in wall stiffness, which can also promote primordium initiation (Sassi et al., 2014). On the other hand, cell wall loosening feeds back on polar auxin transport and further promotes auxin accumulation.

In addition to cell wall remodeling, an AP2 family transcription factor, LEAFLESS (LFS), is required for leaf initiation in tomato (Capua & Eshed, 2017). *LFS* expression is positively regulated by auxin, and *LFS* also promotes auxin maximum formation. Thus, *LFS* is part of a positive feedback loop that leads to auxin accumulation and auxin maximum formation. It remains to be explored how *LFS* promotes auxin accumulation and whether *LFS* affects wall remodeling. In *Arabidopsis*, there are two *LFS* homologs, named *DORNRÖSCHEN* (*DRN*) (also known as *ENHANCER OF SHOOT REGENERATION1*; *ESR1*) and *DRN-LIKE* (*DRNL*; also known as *ESR2*). Double *drn drnl* mutants lack cotyledons but can still form leaves (Chandler et al., 2007), suggesting the existence of additional regulators.

3.3 Phyllotaxis

The arrangement of leaf primordia at the shoot apex, termed phyllotaxis, reflects the site of leaf initiation and determines the pattern of leaves on a plant. The basic phyllotactic patterns include opposite, whorled, and alternate (spiral) (Fig. 3.3). Both *Arabidopsis* and tomato have spiral phyllotaxis. Phyllotaxis has attracted the attention of plant biologists, mathematicians, and artists. In 1868, Wilhelm Hofmeister noted that the initiation of leaves takes place at the region of the SAM that is farthest from the existing primordia. This observation led to the hypothesis that an existing leaf primordium generates an inhibition field (Snow & Snow, 1934). This theory is supported by microsurgical studies where separating an early leaf primordium from the SAM through a tangential cut alters the position of subsequent leaf primordia (Reinhardt et al., 2005; Snow & Snow, 1932).

Applications of the above-mentioned polar auxin transport-based positive feedback loop to a virtual SAM can recapitulate phyllotactic pattern formation. By hypothesizing that cells direct PIN1 toward the adjacent cells with a higher concentration of intracellular auxin (up-the-gradient) (Jönsson et al., 2006; Smith et al., 2006), computational models have successfully accounted for the self-organized patterns of phyllotaxis. Mechanics-based models provide additional molecular-level details to explain the up-the-gradient auxin transport, and can simulate phyllotaxis (Heisler et al., 2010).

The phyllotactic pattern is highly robust, with multiple feedback loops to ensure the precise site of leaf primordia initiation. Breaking these feedback loops may lead to divergence from, and/or higher dispersion of, the common distance and angles between two continuous leaf primordia. Among these regulatory components, MP is a vital component of the positive feedback loops. As a pivotal transcription factor that mediates the auxin-induced transcriptional response, *MP* undergoes auxin-dependent self-activation to control its own expression (Bhatia et al., 2016).

Additionally, auxin stimulates *PIN1* expression via MP (Krogan et al., 2016). Research has shown that localized MP activity directs PIN1 polarity in a non-cell autonomous manner, facilitating the formation of local auxin maxima and organogenesis. This non-cell autonomous auxin maxima formation may also hinge on flux-based PIN1 localization, where PIN1's presence on cell membranes correlates with the auxin flux rate across them (Mitchison, 1980; Stoma et al., 2008). Local auxin biosynthesis is also crucial for self-organized positive feedback loops. Several *AINTEGUMENTA-LIKE/PLETHORA* (*AIL/PLT*) genes are expressed in the shoot apex, encompassing the SAM and leaf primordia. *Arabidopsis ail5/plt5 ail6/plt3 ail7/ plt7* mutants have weakened phyllotaxis (Prasad et al., 2011), attributed to reduced auxin biosynthesis promoted by *AIL/PLT* (Pinon et al., 2013). Conversely, inhibitory domains of cytokinin signaling contribute to the robustness of phyllotactic patterning (Besnard et al., 2014). Auxin directly activates the expression of *ARABIDOPSIS HISTIDINE PHOSPHOTRANSFER PROTEIN 6* (*AHP6*), encoding a cytokinin signaling inhibitor. The intercellular movement of AHP6 protein further generates inhibitory fields, enhancing the robustness of phyllotaxis. Collectively, the interplay of positive and negative feedback loops contributes to the high robustness of the self-organized phyllotaxis.

3.4 Leaf Identity

After leaf initiation, a distinct set of genes defines the leaf fate and separates leaves from the SAM. The SAM function is promoted by KNOTTED-like TALE homeobox transcription factors, class 1 (KNOX1). KNOX1 is expressed throughout the SAM, and also in the leaf axils at the adaxial side, which faces the SAM (Hay & Tsiantis, 2010). In leaf primordia, *KNOX1* expression is inhibited, which requires a MYB family transcription factor called ARP proteins. Expression of the *ARP* genes, named after *ASYMMETRIC LEAVES1* (*AS1*) in *Arabidopsis*, *Rough sheath2* in maize, and *PHANTASTICA* in snapdragon, defines the leaf founder cell fate against the SAM cell fate. The expression of KNOX1 and ARP genes is antagonistic to each other. For example, expression of several *KNOX1* genes is maintained in leaf primordia of *Arabidopsis as1* mutants, leading to defective leaf patterning (Byrne et al., 2000; Ori et al., 2000). ARP proteins require partner proteins to inhibit *KNOX1* expression. In *Arabidopsis*, AS1 forms a repressive complex with AS2 (Xu et al., 2003), a LBD family transcription factor (Iwakawa et al., 2002; Xu et al., 2002), to inhibit *KNOX1* and other target gene expression (Guo et al., 2008; Lin et al., 2003). The AS1-AS2 repressive complex also interacts with epigenetic regulators, including the chromatin-remodeling factor HIRA, histone deacetylase HDA6, polycomb repressive complex 1 (PRC1), and/or PRC2 (Guo et al., 2008; Li et al., 2016; Lodha et al., 2013; Luo et al., 2012; Phelps-Durr et al., 2005), to form multimeric repressing complexes. In addition, AS2 interacts TCP family transcription factors to repress *KNOX1* gene expression (Li et al., 2012). All identified *ARP* mutants can still form leaves, suggesting their roles in leaf initiation are rather limited or redundant with additional factors.

Whereas the *KNOX1* genes are restricted to the SAM, the related class II *KNOX* (*KNOX2*) genes are specifically expressed in leaf primordia in *Arabidopsis* (Furumizu et al., 2015). The *KNOX2* gene functions are antagonistic to *KNOX1* genes and promote cell differentiation in leaves. Ectopic expression of *KNOX2* genes in the SAM, together with interacting partner BELL proteins, leads to the SAM termination. *KNOX2* genes are redundant in *Arabidopsis*, and loss-of-function phenotypes were only observed in high-order mutants. In addition to KNOX1 and KNOX2, a truncated version of *KNOX* genes, termed *KNOX-M*, exists in seed plants. These *KNOX* genes lack a homeodomain, which is responsible for protein-protein interaction (Magnani & Hake, 2008). In *Arabidopsis*, the KNOX-M gene *KNATM* is expressed in leaf primordia. *KNATM* overexpression experiments suggest its role in leaf proximal-distal patterning (Magnani & Hake, 2008). In tomato and *Medicago*, related *KNOX-M* genes regulate compound leaf patterning (Kimura et al., 2008; Peng et al., 2011). The roles of *KNOX2* and *KNOX-M* genes during the specification of leaf initiation remain to be completely understood.

Early leaf primordia also express additional genes, including several encoding transcription factors. *AINTEGUMENTA* (*ANT*), encoding an AP2 family transcription factor, is specifically expressed in leaf primordia but not the SAM (Scheres & Krizek, 2017). Several other *AIL/PLT* genes are also expressed in leaf primordia as well as the SAM. These *AIL/PLT* genes redundantly promote leaf cell division and differentiation (Scheres & Krizek, 2017). As mentioned above, these *AIL/PLT* genes also affect phyllotaxis. Whereas the SAM maintenance is compromised in *ant ail6/plt3 ail7/plt7* mutant plants, leaves still form (Mudunkothge & Krizek, 2012).

The YABBY (YAB) family of HMG family proteins is specifically expressed in leaf primordia (Goldshmidt et al., 2008; Siegfried et al., 1999). Whereas these *YAB* genes are expressed on the abaxial side (the side away from the SAM) of leaf primordia in *Arabidopsis* (Siegfried et al., 1999), it is expressed on the adaxial side (the side facing the SAM) in maize (Juarez et al., 2004) and has non-polar expression in rice (Yamaguchi et al., 2004). *YAB* genes are critical to specifying the leaf fate and to suppressing the SAM fate. In *yab* mutants, meristem marker gene expression is maintained or reactivated (Kumaran et al., 2002; Sarojam et al., 2010), leading to a hybrid leaf/meristem fate and partially indeterminate growth. In addition, the *YAB* genes also contribute to leaf polarity patterning and leaf blade expansion. At the molecular level, YAB regulates gene expression by forming protein complexes with LEUNIG proteins (Bonaccorso et al., 2012; Stahle et al., 2009).

3.5 Leaf Polarity—The Initial Pattern Setup

As previously stated, leaf tissues are arranged into a layered structure, and there is a clear distinction between the adaxial and abaxial sides (Fig. 3.5). This distinction is evident from the early stages of leaf initiation (Poethig & Sussex, 1985). Because of the very early morphological difference along the adaxial-abaxial axis, it has long been proposed that an incipient leaf primordium has been prepatterned into two

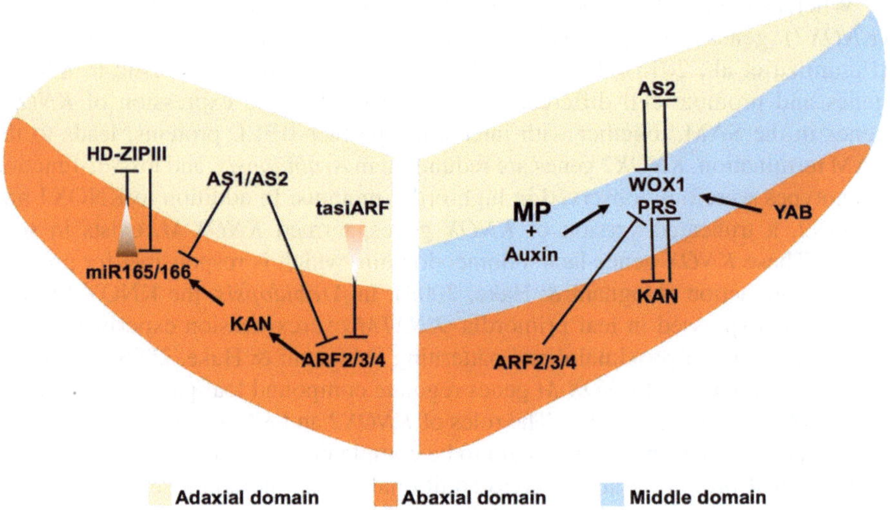

Fig. 3.5 Regulatory networks in the establishment of leaf polarity. Left: A variety of regulators play a role in establishing the adaxial-abaxial polarity of leaves; Right: The interplay between adaxial and abaxial factors aids in the determination and organization of the middle domain. AS: ASYMMETRIC LEAF; miR165/166: microRNA 165/166 targeting HD-ZIPIII; MP: MONOPTEROS; PRS: PRESSED FLOWER; tasiARFs: *trans*-acting short interfering RNA targeting ARF2/3/4; WOX1: WUSCHEL-RELATED HOMEOBOX1; YAB: YABBY; KAN: KANADI

domains (Hagemann & Gleissberg, 1996; Husbands et al., 2009). The adaxial domain may inherit positional information from the SAM. It was until recently that experimental evidence was provided to support this prepattern hypothesis. Time-lapse imaging revealed that genes involved in polarity determination display polarized expression patterns prior to the leaf being morphologically distinct from the surrounding PZ cells (Burian et al., 2022; Caggiano et al., 2017; Yu et al., 2017). A nascent leaf is prepatterned into an adaxial domain expressing the LBD gene *AS2* and HD-ZIPIII gene *REV*, and an abaxial domain expressing *KAN1*. The expression zones of *AS2* and *KAN1* are annular, encircling the SAM, indicating that *AS2* and *KAN1* expression does not forecast organogenesis but rather delineates developmental potential. Conversely, *REV* expression is heightened in an incipient primordium and also during the early stages of primordium formation. Therefore, *REV* expression is linked to both organogenesis and the determination of adaxial cell fate. The adaxial and abaxial domains are in a state of flux, indicating that these domains are not static during the early stages of leaf development (Yu et al., 2017). In fact, leaf cell gene expression still shifts until later stages (Tameshige et al., 2013).

The driving force for prepattern establishment remains elusive. A plausible candidate is the WUSCHEL protein (Yu et al., 2017). The exclusion of *AS2*, *KAN1*, and related *KAN* genes from the center of the SAM is critical. The OC-expressed *WUS* may establish the prepattern, as the WUS protein can diffuse to surrounding cells (Daum et al., 2014; Yadav et al., 2010). It has been shown that WUS can inhibit leaf gene expression, including *AS2* and *KAN1* (Yadav et al., 2013), making it an intriguing candidate for the initial setup of the prepattern.

Auxin maxima define new primordia (Reinhardt et al., 2000, 2003). The periodical auxin maxima, generated by the auxin efflux carrier PIN1, localize to the boundary between *REV* and *KAN1* expressing domains. The auxin maxima is not a spot and has its own shape: Auxin accumulates at the future leaf margins along the two lateral sides of a primordium (Shi et al., 2017; Wang et al., 2022). Auxin signaling activates the expression of *WUSCHEL-RELATED HOMEOBOX* (*WOX*) gene *PRS/WOX3* between the *REV* and *KAN1* domains at the P_1 stage (Caggiano et al., 2017; Yu et al., 2017), which designates the youngest leaf primordium. Later on, the homologous *WOX1* is expressed in the same middle domain (Caggiano et al., 2017; Yu et al., 2017). Both *PRS* and *WOX1* are directly activated by auxin through MP and related ARFs (Guan et al., 2017). PRS and WOX1 in turn promote auxin biosynthesis (Zhang et al., 2020), leading to further auxin accumulation in the leaf margins. The middle domain is critical for leaf flattening, which will be discussed below.

The division of the PZ into two distinct regions, which will eventually become the adaxial and abaxial domains, implies the existence of patterning signals emanating from the center of the SAM, specifically the CZ and the underlying OC. This notion has been proposed for quite some time, tracing back to the classic microsurgical experiments by Ian Sussex in the 1950s (Kuhlemeier & Timmermans, 2016; Reinhardt et al., 2005; Sussex, 1951). In potato, a microsurgical incision separating an emerging leaf primordium from the SAM leads to radially symmetric leaf formation (Sussex, 1951). The radially symmetric leaf, which is abaxialized and fails to develop a leaf blade, implies that the SAM also plays a role in promoting adaxial-abaxial patterning in leaves. This hypothetical meristem-derived signal is now called 'Sussex signal'. A more precise microsurgical incision using an infrared laser in tomato demonstrated that the Sussex signal is transmitted mainly through the epidermal (L1) layer (Reinhardt et al., 2005). On the other hand, tangential microsurgical incisions in the lateral regions of an I_1 or early P_1, separating it from older primordia but not the SAM, also lead to symmetric leaves in tomato and in *Epilobium hirsutum* (Shi et al., 2017; Snow & Snow, 1959). Thus, an adaxial-promoting signal may also come from older leaf primordia.

The Sussex signal can be largely attributed to polar auxin transport. As previously discussed, auxin moves from the adaxial side of an established leaf primordia back toward the SAM (Bayer et al., 2009; Heisler et al., 2005). Conversely, auxin accumulates at the margins of the leaf on both sides (Shi et al., 2017; Wang et al., 2022). This polar auxin transport results in a transient auxin minimum on the adaxial side (Qi et al., 2014) and auxin peaks along the leaf margins (Wang et al., 2022). Microsurgical incisions can induce ectopic auxin signaling maxima in the adaxial domain, but not along the leaf margins (Wang et al., 2022). Local inhibition of PIN1 activity is reminiscent of the microsurgical incision, suggesting that perturbing the flow of auxin leads to leaf abaxialization in tomato (Qi et al., 2014; Wang et al., 2022). Consistently, ectopic auxin application in tomato and ectopic auxin signaling on the adaxial side of *Arabidopsis* leaf primordia convert adaxial to abaxial cell fate.

Thus, auxin may constitute a polarizing signal, although it is not a simple adaxial-promoting signal from the SAM, that delivers polarity information. Together, the initial prepattern of KAN, AS2, and HD-ZIPIII expression is followed by auxin accumulation along the leaf margins to activate WOX gene expression in the middle domain. The interaction of the three domains ensures leaf blade formation.

3.6 Leaf Polarity—Pattern Maintenance

Although the early establishment of the adaxial-abaxial axis is not fully understood, over a decade of intensive molecular genetic research has uncovered a complex transcriptional regulatory network that includes genes promoting adaxial and abaxial identities, expressed either abaxially or adaxially (Barton, 2010; Husbands et al., 2009; Xu et al., 2007). These genes encode transcription factors and small RNAs, and their specific expression domains are crucial for asymmetric differentiation of adaxial and abaxial cells and the expansion of the leaf lamina. In the regulatory interplay, genes active in the abaxial domain inhibit those in the adaxial domain, and vice versa (Fig. 3.5).

The expression of the HD-ZIPIII transcription factors, including PHABULOSA, PHAVOLUTA, and REV, is restricted to the adaxial domain. These HD-ZIPIII genes are microRNA targets and are inhibited in the abaxial domain. Several members of the miR165/166 family are expressed in the epidermis of the abaxial side and diffuse throughout the entire abaxial domain (Chitwood et al., 2009; Tatematsu et al., 2015; Yao et al., 2009). MiR165/166 guides the cleavage of HD-ZIPIII mRNAs (Mallory et al., 2004). Dominant mutations in the HD-ZIPIII genes at the miR165/166 complementary site impair transcript cleavage and lead to ectopic HD-ZIPIII protein accumulation in the abaxial domain and adaxialized leaves.

In addition to miRNA, the transcription regulatory activity of HD-ZIPIII is also inhibited by a family of small leucine zipper-containing proteins, collectively termed LITTLE ZIPPER (ZPR) proteins (Kim et al., 2008; Wenkel et al., 2007). These *ZPR* genes are activated by HD-ZIPIII transcription factors (Wenkel et al., 2007), and the ZPR proteins engage with HD-ZIPIII proteins to form nonfunctional heterodimers through the ZIP motifs (Kim et al., 2008; Wenkel et al., 2007). Thus, the ZPR proteins establish a negative feedback mechanism to limit HD-ZIPIII activity within the adaxial domain, and also within the SAM.

The adaxial-expressed HD-ZIPIII transcription factors directly activate the expression of HD-ZIPII genes, which also encode transcription factors (Bou-Torrent et al., 2012; Turchi et al., 2013). These HD-ZIPII genes, including *HAT3*, *ATHB2*, and *ATHB4*, are similarly expressed in the adaxial domain, and their mutations lead to adaxial development defects. HD-ZIPII can form a regulatory protein complex with HD-ZIPIII transcription factors (Turchi et al., 2013). The HD-ZIPII/HD-ZIPIII protein complex directly represses the expression of several miR165/166 genes (Merelo et al., 2016). Thus, the HD-ZIPIII and miR165/166 form reciprocal inhibition of each other.

AS2 is another adaxial-expressed transcription factor. *AS2* expression is confined to the adaxial epidermis in the early stages of leaf development (Iwakawa et al., 2007). As mentioned above, the LOB domain transcription factor AS2 complexes with the MYB transcription factor AS1. At least in *Arabidopsis*, mutations in *as1* and *as2* lead to very similar phenotypes, although *AS1* is more broadly expressed than *AS2* (Byrne et al., 2000). As mutants of the *AS1* ortholog *PHAN* in *Antirrhinum* have clear leaf polarity defects (Waites & Hudson, 1995), the *Arabidopsis as1* and *as2* mutants lack obvious polarity phenotypes, such as radially symmetric or trumpet-like leaves. Instead, *as1* and *as2* mutants develop asymmetric leaves with reduced petioles.

A set of antagonistic transcription factors and small RNAs determines the abaxial identity. Prior to leaf initiation, KAN1 and potentially other KAN family transcription factors are already positioned within the abaxial domain, as evidenced by time-lapse live imaging (Burian et al., 2022; Caggiano et al., 2017; Yu et al., 2017). Loss-of-function mutations in both *KAN1* and *KAN2* result in adaxialization, whereas overexpression of *KAN1* or *KAN2* leads to abaxialization and termination of the SAM (Emery et al., 2003; Eshed et al., 2001; Kerstetter et al., 2001). Therefore, *KAN* genes promote the abaxial fate while inhibiting both the adaxial fate and the SAM fate. In *Arabidopsis*, three functionally redundant repressors, AUXIN RESPONSE FACTORs (ARFs), namely ETTIN (ETT, also known as ARF3), ARF4, and ARF2, are also expressed in the abaxial domain and enhance the abaxial identity (Guan et al., 2017; Pekker et al., 2005). Mutations in these *ARF* genes produce similar phenotypes to those observed in *kan* mutants. Indeed, ARF and KAN may form protein complexes (Kelley et al., 2012), as supported by genetic analysis. Although the expression levels of *ETT* and *ARF4* remain unchanged in *kan1 kan2* double mutants, *ETT* mutations have the ability to suppress the overexpression phenotype of *KAN1* in *Arabidopsis* (Pekker et al., 2005). Furthermore, *trans*-acting short interfering RNAs (*ta*-siRNAs), whose biogenesis is initiated by *miR390*, target *ETT*, *ARF4*, and *ARF2* (Allen et al., 2005; Fahlgren et al., 2006; Hunter et al., 2006; Marin et al., 2010). These *ta*-siRNAs are produced from the noncoding gene *TAS3*, which is cleaved by AGO7. Produced in the adaxial domain, these *ta*-siRNAs restrict their target ARFs (referred to collectively as tasiARFs) to the abaxial domain. The miR390-TAS3-ARF pathway is highly conserved across various species, including moss, grasses, and eudicots (Axtell & Bowman, 2008; Douglas et al., 2010; Nagasaki et al., 2007). Like miR165/166, the adaxially localized *ta*-siRNAs are mobile and precisely confine *tasiARF* expression in the abaxial domain (Skopelitis et al., 2017).

Further mutual inhibitions are present between regulators that govern the adaxial and abaxial domains. Specifically, the AS1-AS2 complex, which is positioned on the adaxial side, suppresses the epigenetic regulation, transcription, and post-transcriptional processing of *ARF3* and *ARF4* genes (Iwasaki et al., 2013). This complex also impedes the expression of *MiR166A* and *ETT* within the adaxial domain (Husbands et al., 2015). Furthermore, the AS1-AS2 complex interacts with the promoter regions of *TAS3A*, which encodes the precursor of *ta*-siARF, to enhance its transcription (Husbands et al., 2015). Conversely, in the abaxial domain,

KAN1 stimulates the expression of *MiR166A* and *MiR166F* (Merelo et al., 2013), creating a reinforcing feedback mechanism that consolidates the abaxial identity.

The transcription factor group known as YABBY (YAB) also plays a role in determining adaxial-abaxial polarity. In *Arabidopsis*, FILAMENTOUS FLOWER (FIL, also known as YAB1), YAB2, and YAB3 work in concert to enhance the abaxial identity (Sawa et al., 1999; Siegfried et al., 1999). These proteins are active on the abaxial side of lateral organs, and their simultaneous mutation leads to the development of adaxial characteristics (Sarojam et al., 2010; Stahle et al., 2009). YAB proteins dimerize with LEUNIG family transcriptional regulators to exert their function (Stahle et al., 2009). *YAB* expression stretches into the middle domain that lies between the adaxial and abaxial domains, a topic that will be delved into the subsequent section. Notably, YAB genes exhibit no polarization in rice (Yamaguchi et al., 2004), and are biased toward the adaxial domain in maize (Juarez et al., 2004), suggesting a shift in function over the course of evolution. It is even suggested that the primary role of YAB is to foster leaf development and repress the meristematic fate (Sarojam et al., 2010).

While most components of this ad-ab regulatory network are well-preserved among angiosperms, their individual impacts may differ. While mutations in the miR390-TAS3-ARF pathway result in minor leaf polarity defects in Arabidopsis, similar mutations in maize, rice, and tomato can result in pronounced polarity defects (Douglas et al., 2010; Nagasaki et al., 2007; Nogueira et al., 2007; Yifhar et al., 2012). The timing of *tasiARF* expression could account for these differences; *tasiARF* expression is abaxialized early in maize primordia but appears later in *Arabidopsis*. Likewise, mutants in the AS1-AS2 pathway show a spectrum of polarity phenotypes, as mentioned earlier.

Considering the mutual antagonistic interactions among adaxial, abaxial, and middle domain genes, an outstanding question pertains to the maintenance of these three domains within a leaf. Antagonistic interactions would commonly lead to unstable balance, and the diminishment of one domain. Computational simulation indicates that an upstream signal that activates both/all antagonistic domains could establish a robust balance (Guan et al., 2022). In leaf primordia, auxin acts as the upstream activator and functions through MP to activate not only *WOX1* and *PRS* in the middle domain (Guan et al., 2017), but also *REV* in the adaxial domain and *KAN1* in the abaxial domain (Guan et al., 2022). These interactions together stabilize the adaxial-abaxial axis, even after the leaf is pushed away from the SAM center.

3.7 Initial Leaf Shape Acquisition

Organ shape formation, a key aspect of morphogenesis, is not yet comprehended. Leaf and floral primordia originate from highly similar initial forms and even exhibit many comparable gene expression patterns. However, they eventually develop into distinct shapes. Leaf primordia develop into a flat, bilaterally symmetric structure, whereas floral primordia adopt a radial symmetry. Time-lapse imaging has revealed that in leaf primordia, the abaxial and the middle domains grow more rapidly than the adaxial domain (Peng et al., 2022). Computational models suggest that this differential growth rate is adequate to account for the transformation of the leaf primordium into a bilaterally symmetric shape. In contrast, floral primordia exhibit enhanced growth in the adaxial domain, which contributes to their axisymmetric form.

Transcription factors at the top of the regulatory hierarchy influence downstream targets to orchestrate local cell proliferation, expansion, and growth, resulting in three-dimensional shape alterations and the specification of functions. Compared to our knowledge of the interactions between adaxial and abaxial regulators, we have a limited understanding of their downstream effects. Nonetheless, recent genome-wide studies have begun to elucidate these mechanisms (Brandt et al., 2012; Huang et al., 2014; Merelo et al., 2013; Reinhart et al., 2013; Xie et al., 2015). Such genome-wide analyses imply that adaxial-promoting factors, such as HD-ZIPIII, and abaxial-promoting factors, like KAN, exert opposing effects on shared pathways.

In addition to other roles, the adaxial-abaxial polarity regulates the demethylesterification of pectins in the cell wall, which is associated with the mechanics of the cell wall (Qi et al., 2017). The abaxial domain contains methylesterified pectin and a reduced elastic modulus, while the remainder of the leaf primordium contains demethylesterified pectin and exhibits a higher elastic modulus. As the leaf grows, the pectin in the adaxial domain also undergoes demethylesterfication, resulting in a decrease in elastic modulus. The middle domain, however, retains methylesterfied pectin and maintains a low elastic modulus. Mathematical modeling suggests that mechanical heterogeneity alone can generate the asymmetry observed in leaf primordia, a prediction supported by experimental evidence. Floral primordia, on the other hand, display a uniform pattern of pectin demethylesterification, which aligns with floral growth dynamics (Peng et al., 2022). Thus, mechanical heterogeneity within tissues, regulated by polarity signals, may lead to the asymmetry in leaf shape.

3.8 Leaf Flattening

Leaf primordia sustain active cell proliferation in a highly coordinated spatial and temporal pattern. In contrast to the indeterminate growth of shoots, leaves show a determinate growth pattern, with cell proliferation being transiently maintained. Early research indicated that cell proliferation is concentrated in the leaf margin, and is referred to as the marginal blastozone or marginal leaf meristem (Avery, 1933; Hagemann & Gleissberg, 1996). However, subsequent findings suggest that the marginal region's contribution to the growth of the leaf blade is rather limited, and the expansion of the blade in angiosperms is primarily driven by active cell proliferation in a broader region known as the plate meristem, which spans the entire adaxial-abaxial junction (Donnelly et al., 1999; Schüepp, 1918; Waites &

Hudson, 1995). Molecular genetic studies have revealed a strong correlation between WOX genes and the activity of the blastozone.

In *Arabidopsis*, both *WOX1* and *PRESSED FLOWER* (*PRS*)/*WOX3* are expressed at the middle domain between the adaxial and abaxial domains, and function redundantly in promoting blade expansion. The double mutants of *wox1 prs* generate significantly narrower leaves than the wild type, suggesting that blade outgrowth along the medio-lateral axis is impaired (Nakata et al., 2012; Vandenbussche et al., 2009). Studies have identified the conserved roles of *WOX1* orthologs in controlling blade outgrowth across various eudicots, such as *LAM1* in *Nicotiana*, *MAEWEST* (*MAW*) in *Petunia*, *STENOFOLIA* (*STF*) in *Medicago*, and *LATHYROIDES* (*LATH*) in garden pea (Lin et al., 2013; McHale & Marcotrigiano, 1998; Tadege et al., 2011; Vandenbussche et al., 2009; Zhuang et al., 2012). Single mutants of each *WOX1* ortholog display a similar narrow-leaf phenotype to that of *wox1 prs* in *Arabidopsis*. Conversely, *PRS* orthologs regulate blade outgrowth in several monocots. For instance, the double mutants of maize *narrow sheath 1* and *2* (*ns1/2*) and rice *narrow leaf2* and *3* (*nal2/3*) show analogous leaf margin deficiencies (Ishiwata et al., 2013; Nardmann et al., 2004; Scanlon et al., 1996). *WOX1* and *PRS* are within the recently evolved *WUS* clade (van der Graaff et al., 2009). The slender-leaf phenotype of the *Nicotiana lam1* mutant can be restored by introducing the expression of *Arabidopsis WUS*-clade genes, including *WOX1* and *PRS*, in the middle domain, thereby proving that the function of *WUS*-clade genes is essential for the expansion of the leaf blade (Lin et al., 2013). However, the *WOX1* and *PRS* orthologs are expressed in different leaf domains in the monocots and eudicots mentioned earlier. *WOX1* orthologs are expressed throughout the entire middle domain, aligning with the hypothesized plate meristem region (Lin et al., 2013; Nakata et al., 2012; Tadege et al., 2011; Vandenbussche et al., 2009; Zhuang et al., 2012). Conversely, the expression of *PRS* orthologs is more confined to the marginal meristem (Ishiwata et al., 2013; Nakata et al., 2012; Nardmann et al., 2004). Consequently, the spatial variation in meristematic activities within leaves might result in distinct timings and positions for the lateral expansion of leaves in monocots and eudicots (Ishiwata et al., 2013).

Genetic and molecular data indicate that the adaxial-abaxial polarity is fundamental to the establishment of medio-lateral polarity and the growth of the lateral blade. The typical blade growth is anisotropic, with cell division predominantly occurring perpendicular to the medio-lateral axis, which results in a flattened leaf shape. In numerous mutants that exhibit adaxialized or abaxialized leaf phenotypes, the expansion of the leaf blade is markedly impaired. These leaves tend to display radial symmetry and isotropic growth across all directions, similar to the *Nicotiana lam1* mutant, where *WOX* function is inactive.

Based on the distinct expression patterns of *WOX1* and *PRS* in leaf tissues, a three-domain model has been put forward (Fig. 3.5), which includes a 'middle domain' positioned between the adaxial and abaxial domains (Nakata et al., 2012). Auxin plays a central role in the establishment of the middle domain by initially defining the adaxial and abaxial domains. MP, an ARF activator, is expressed in the adaxial domain (Guan et al., 2017), while the ARF repressors ARF2-4 are present in the abaxial domain (Pekker et al., 2005). Additionally, the abaxial domain is characterized by elevated auxin levels. These factors, along with the differential distribution of ARFs, result in an enhanced auxin response, particularly within the middle domain (Guan et al., 2017; Qi et al., 2014). Moreover, MP is a direct target of the adaxially expressed HD-ZIPIII proteins (Guan et al., 2022; Müller et al., 2016). Consequently, auxin serves as a positional signal for the differentiation of the middle domain sandwiched between the adaxial and abaxial domains.

Sophisticated regulatory mechanisms are in place to ensure the proper formation of the three domains within the leaf. The absence of *WOX1* and *PRS* expression leads to the ectopic expression of genes associated with adaxial and abaxial identities in the leaf margins, and a shift from the typical elongated marginal cell files to adaxialized cell types. This suggests that *WOX1* and *PRS* actively suppress adaxial and abaxial identities at the margins of the middle domain (McHale & Marcotrigiano, 1998; Nakata et al., 2012). *STF*, the *WOX1* ortholog in *Medicago*, directly inhibits the adaxial factor *AS2* in the leaf margin by recruiting TOPLESS family corepressors (Zhang et al., 2014). Conversely, *PRS* is suppressed by *AS2* within the adaxial domain (Alvarez et al., 2009). In the abaxial domain, the identity of the middle domain is suppressed by two sets of abaxial genes, *KAN1* and *2*, and repressors *ARF2*, *3*, and *4* (Guan et al., 2017; Nakata et al., 2012). *YAB* family members, which are expressed in both the middle and abaxial domains, can only activate *WOX1* expression in the middle domain, suggesting that *KAN* and *ARF* are epistatic to *YAB* in the regulation of *WOX* expression (Nakata et al., 2012).

WOX genes are known to stimulate cell proliferation. In mutants lacking *kan1 kan2* or *arf2 arf3 arf4*, *WOX1* and *PRS* are ectopically expressed in the abaxial domain of leaf primordia, resulting in the development of outgrowths on the abaxial leaf surface (Guan et al., 2017; Nakata et al., 2012). This alteration in morphology can be phenocopied by the overexpression of *WOX1* in the abaxial domain (Nakata et al., 2012). Indeed, *WOX1* and *PRS* enhance cell proliferation by activating the expression of the cytochrome P450 gene *CYP78A5/KLUH* in the leaf margins (Nakata et al., 2012). KLUH is responsible for extending cell proliferation. In *kluh* mutants, smaller organs are produced due to an early halt in cell proliferation, while the overexpression of *KLUH* leads to extended cell proliferation and the development of larger organs (Anastasiou et al., 2007). Furthermore, *WOX* genes may facilitate blade growth by modulating the balance of auxin and cytokinin in the leaves (Tadege et al., 2011). *WOX1* orthologs, *STF* in *Medicago* and *LAM1* in *Nicotiana*, have been shown to influence auxin levels, and the absence of blades in *lam1* leaves can be partially rescued by the combined application of auxin and cytokinin (Tadege et al., 2011). In *Arabidopsis*, it has been shown that WOX transcription factors activate the expression of *YUCCA1* (*YUC1*) that encode flavin monooxygenase-like enzymes (Zhang et al., 2020), *YUC* genes are essential for leaf margin development and blade expansion in *Arabidopsis* (Wang et al., 2011).

Leaf growth is highly directional, leading to the formation of thin, flat laminae. Computational simulation indicates that a bilateral symmetric organ has tensile stress pattern along the adaxial-abaxial axis (Zhao et al., 2020). Cortical microtubules align along the stress pattern and are indeed along the adaxial-abaxial axis.

Cortical microtubules further mediate cellulose deposition along the same orientation, leading to higher restriction of growth in this direction. Cell growth and division are both in the plain perpendicular to the adaxial-abaxial axis, leading to amplified flatness and blade formation. Because this directional growth strengthens the initial anisotropic stress pattern rather than reducing it, a positive feedback loop is established that continues blade expansion.

The microtubule-mediated mechanical feedback provides a parsimonious mechanism to explain leaf blade evolution (Jiao et al., 2021). Leaves evolved independently multiple times (Tomescu, 2009). Paleontological evidence suggests that the leaves of seed plants have evolved from ancestral shoot structures, and the existence of a leaf meristem that comes from a residual or modified apical or vascular meristem has long been debated (Foster, 1936; Hagemann & Gleissberg, 1996). In fact, different types of blades, such as unifacial leaves with blades along the adaxial-abaxial axis, rather than the medio-lateral axis, have been found (Yamaguchi et al., 2010). In the mechanical feedback theory, polar gene expression and/or phytohormone asymmetric accumulation generate bilateral flattened primordia (Peng et al., 2022), whose flattening is subsequently amplified by the microtubule-mediated mechanical feedback (Zhao et al., 2020). Once an initial axisymmetry breaking is established, the blade will form. In contrast to Zimmermann's telome theory, which is based on fossils and lacks an identified developmental mechanism (Beerling & Fleming, 2007), the mechanical feedback theory of leaf lamina formation is simple and supported by experimental evidence.

3.9 Transition from Morphogenesis to Differentiation

Following the initiation of the leaf blade, cellular proliferation and differentiation take place across the entire blade area, a process known as intercalary growth (Nakata et al., 2012). The transition from proliferation to differentiation is characterized by a basipetal gradient throughout leaf development in *Arabidopsis* and other species, where cell proliferation in the distal region of leaves halts initially, and the activity of cell proliferation gradually retreats toward the proximal region (Nath et al., 2003; Nelissen et al., 2012; Ori et al., 2007). Consistent with this pattern, leaf meristem activity, reflected in part by *PRS/WOX3* expression, is observed to progressively confine itself to the marginal region in immature leaves and subsequently to the proximal margin in more mature leaves (Alvarez et al., 2016). Nonetheless, the polarity of leaf growth can vary among species, with enhanced growth occurring either in the proximal or distal end, scattered without apparent polarity, or in a bidirectional manner (Fig. 3.6) (Das Gupta & Nath, 2015).

Two principal groups of miRNA/transcription factor modules exert a predominant influence on the maintenance of marginal meristems and the promotion of intercalary growth in a counteractive fashion. The class II TEOSINTE BRANCHED1/CYCLOIDEA/PROLIFERATING CELL FACTOR (TCP) transcription factor family suppresses the activity of the marginal meristem, thereby promoting the

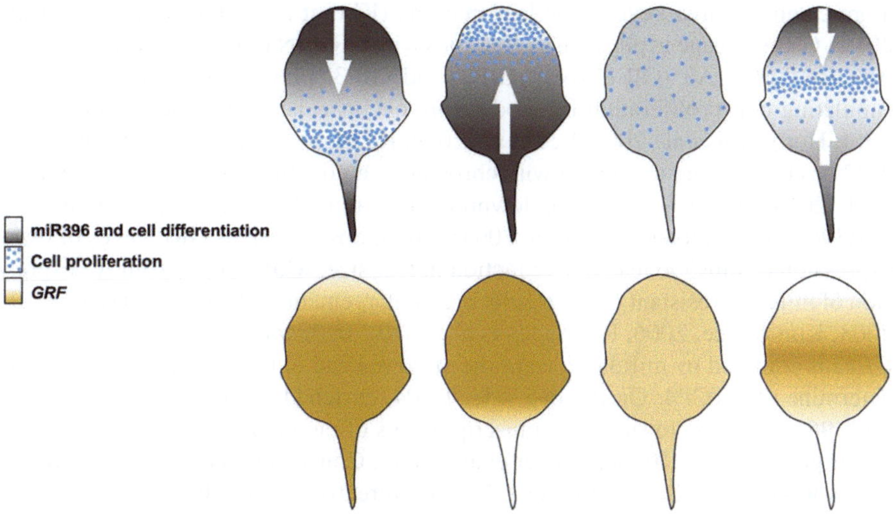

Fig. 3.6 Transition between proliferation and cell expansion. Expression gradient of GRF-targeting miR396, which inhibits cell proliferation and promotes cell differentiation, correlates with diverse leaf growth allometry, i.e., differential growth within a leaf. From left to right, shows positive allometry, negative allometry, isometry, and bidirectional allometry

transition from cellular proliferation to differentiation. When *TCPs* are inactivated in *Arabidopsis*, or their orthologs such as *CINCINNATA* (*CIN*) in *Antirrhinum* and *LANCEOLATE* (*LA*) in tomato are disabled, it leads to analogous excessive growth patterns in leaf margins, resulting in enlarged wrinkled leaves or leaflets with serrated margins (Efroni et al., 2008; Nath et al., 2003; Ori et al., 2007). The indeterminate cellular proliferation in the leaf margins of *tcp* mutants is probably due to the prolonged activity of *WOX1* and *PRS* genes, as well as the enduring mitotic activity triggered by the upregulation of cyclin genes like *CYCB* (Alvarez et al., 2016; Bresso et al., 2018). Meanwhile, the serrated marginal pattern is linked to the upregulation of several *CUC* genes, which are typically suppressed by *TCP*s in the leaves of wild-type plants (Koyama et al., 2017). The expression of class II *TCP*s is suppressed by miR319/*JAW*, which confines *TCP* expression to the more distal region of the leaves (Ori et al., 2007; Palatnik et al., 2003). Mutations that disrupt this regulation result in reduced leaf size with premature differentiation at the leaf margins and a loss of leaflet formation, likely due to the ectopic expression of *TCP*s in the proximal regions of the leaves (Efroni et al., 2008; Ori et al., 2007; Palatnik et al., 2003). Extended marginal growth and serrations are also noticeable when the function of four NGATHA (NGA) transcription factors is compromised, while diminished marginal growth is caused by the overexpression of *NGA3*, highlighting their overlapping roles in controlling the transition between leaf marginal expansion and differentiation (Alvarez et al., 2009; Trigueros et al., 2009).

In contrast, GROWTH REGULATING FACTORs (GRFs) prolong the shift from cellular proliferation to differentiation during intercalary growth, operating in a more extensive area beyond the leaf margins (Kim et al., 2003; Rodriguez et al., 2010). GRFs engage in physical interactions with GRF-INTERACTING FACTOR1/ANGUSTIFOLIA3 (GIF1/AN3), GIF2, and GIF3, playing overlapping roles in simulating cell proliferation by modulating cyclin gene levels (Debernardi et al., 2014; Horiguchi et al., 2005; Lee et al., 2009; Rodriguez et al., 2010). Additionally, AN3 has been found to interact with chromatin-remodeling complexes (Debernardi et al., 2014). Multiple *GRF*s are downregulated at the post-transcriptional level by miR396 (Jones-Rhoades & Bartel, 2004). Mutations in various *GRF* genes or overexpression of miR396 lead to a reduction in leaf size, while the constitutive expression of miR396-resistant *GRF*s results in an enlargement of the leaf area (Kim et al., 2003; Kim & Lee, 2006; Rodriguez et al., 2010). In *Arabidopsis*, the expression of *GRF*s is confined by miR396 toward the proximal end of the leaves, a pattern partly determined by TCP4. On the one side, *TCP4*, enriched distally, directly activates miR396 expression to suppress transcript levels of target *GRF*s in the distal region (Rodriguez et al., 2010; Schommer et al., 2014). Conversely, *TCP4* also suppresses the expression of *GIF1* and other *GRF*s not directly targeted by miR396 through unknown mechanisms (Rodriguez et al., 2010). Therefore, the miR319/*TCP* and miR396/*GRF* regulatory modules collaborate to regulate both marginal and overall leaf expansion by modulating cellular proliferation.

In addition to the control exerted by miRNA/transcript factor modules, hormonal regulation plays a role in intercalary growth. Gibberellin acids (GA) and brassinosteroids (BR) are two key hormones that stimulate leaf expansion by enhancing cell proliferation and expansion. Continuous activation of GA signaling or the overexpression of GA biosynthetic enzymes results in increased leaf size, while the inhibition of GA signaling leads to a decrease in leaf size (Achard et al., 2009; Huang et al., 1998). The beneficial effect of GA on cell proliferation may be mediated by the suppression of cell cycle inhibitors, such as KIP-RELATED PROTEIN 2 (KRP2) and SIAMESE (Achard et al., 2009). Elevated expression of BR receptor-encoding gene *BRI1* or the BR biosynthesis gene *DWARF4* also contributes to larger leaf size (Choe et al., 2001; Gonzalez et al., 2010; Zhiponova et al., 2013). In contrast, BR-deficient mutants exhibit smaller leaves and lower *CYCB* levels, indicating that BR influences the transition from mitosis (Zhiponova et al., 2013). Furthermore, BR signaling downregulates the expression of *PEAPOD1* (*PPD1*) and *PPD2*, which encode transcription factors that restrict the proliferation of meristemoid cells (Gonzalez et al., 2015). The influence of auxin on leaf growth is ambiguous with studies showing both positive and negative effects of auxin on cell proliferation and expansion in leaves (Saini et al., 2017; Wang & Guo, 2015). Given that auxin signaling is crucial for vascular development, which is responsible for nutrient transport to leaf cells, it might be challenging to distinguish the direct effects of auxin on cell proliferation and expansion from the indirect effects of metabolic changes.

3.10 Leaf Marginal Patterning—Simple and Compound Leaves

Despite differences in dimensions and shape, leaves are commonly categorized as either simple or compound, depending on the number of leaf blades. Simple leaves consist of a single unbroken blade with margins that may be smooth, serrated, or lobed. In contrast, compound leaves consist of multiple discrete blades known as leaflets, which are attached to a central axis named the rachis (Bar & Ori, 2015; Efroni et al., 2010; Runions et al., 2017). Genetic studies have revealed that the initiation of leaflet primordia in compound leaves resembles that of simple leaf primordia (Hasson et al., 2010).

The formation of leaflets represents a reiteration of the basic leaf formation program (Fig. 3.7). As mentioned above, continuous leaf development includes primordia specification, intercalary growth, and differentiation. The intercalary growth period, also termed primary morphogenesis, provides a transient stage that enables a partial reiteration of the leaf development program to form leaflets. Whereas this reiteration cycle is mostly absent in simple leaf species, it may occur one to multiple times in compound species, resulting in one to multiple orders of leaflets (Fig. 3.7). Unsurprisingly, many of the genes regulating leaf formation also play roles in leaflet formation.

By contrast, fronds of ferns, although they have leaf-like properties, are developmentally shoot-like (Sanders et al., 2011). Different from leaf primordia, fronds maintain apical initial cells that sequentially form pinnae ('leaflets') in a shoot-like manner, making frond development close to indeterminate. Thus, frond elaboration depends on indeterminate apical initials, whereas compound leaf elaboration depends on broadly but transiently existing undifferentiated leaf marginal cells during the intercalary growth phase (Fig. 3.7).

Extensive studies, mostly in tomato, *Cardamine hirsute*, *Medicago truncatula*, and pea, have identified many regulators of compound leaf development. Notably, genes enabling initiation and genes modulating the length of intercalary growth are commonly found regulating compound leaf morphology. We will elaborate on selected examples of both groups, and readers can refer to recent reviews for more comprehensive discussions (Bar & Ori, 2015; Blein et al., 2010; Koenig & Sinha, 2010).

The polar auxin transport-based positive feedback loop functions in leaf initiation at the SAM periphery, and also at the leaf lateral sides to promote leaflet initiation (Koenig et al., 2009). Genes involved in auxin transport and signaling have been found to regulate compound leaf development in several species (Ben-Gera et al., 2012, 2016; Bilsborough et al., 2011; Zhou et al., 2011). This auxin transport-based 'pattern generator' produces auxin maxima, which later become leaflets, and auxin minima, which become sinuses. *CUC* genes are also indispensable components of the auxin pattern generator. *CUC* genes are expressed in the sinus to suppress lamina growth (Berger et al., 2009; Blein et al., 2008; Busch et al., 2011). Periodical spatial patterning of *CUC* along the leaf margin is critical for the

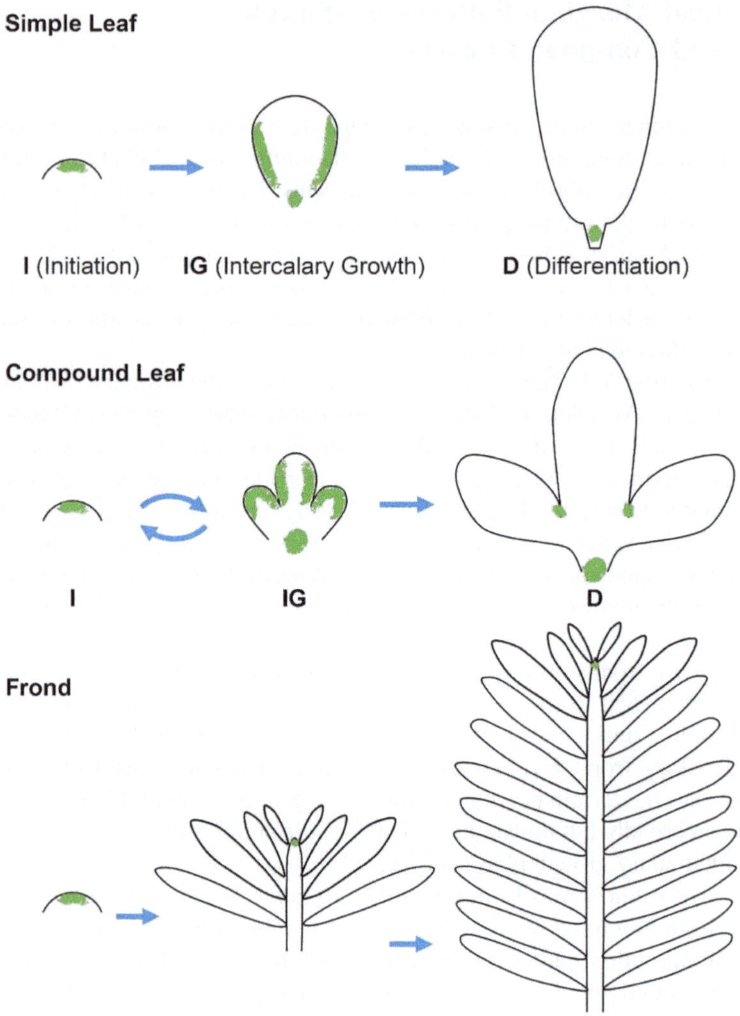

Fig. 3.7 Models showing simple leaf, compound leaf, and frond formation. During compound leaf development, leaflets initiate when the leaf blade is undergoing intercalary growth. Higher-order leaflets may continue to form on primary leaflets until leaf differentiation. Either a reduced intercalary phase or reduced activation of the blade/leaflet formation program may result in simple leaves. By contrast, fronds of ferns arise from a single apical initial, which is similar to a typical branch of seed plants. Meristematic cells are marked in green

specification of auxin maxima, which promote lamina growth and leaflet formation. In tomato, alterations in the expression patterns of the *CUC* gene *GOBLET*, whether diminished or enhanced, lead to leaflet fusion (Berger et al., 2009). The same suite of genes also functions in simple leaves to generate serrations, which can be considered as mini leaflets and lobes (Bilsborough et al., 2011).

In addition to leaflet initiation, modulation of the intercalary growth phase duration profoundly influences compound leaf patterning, with prolonged intercalary growth leading to higher orders of leaflets, and reduced intercalary growth results in reduced complexity. In several species, *KNOX1* genes promote leaf complexity (Chen et al., 1997; Hareven et al., 1996; Hay & Tsiantis, 2006; Kimura et al., 2008; Peng et al., 2011; Shani et al., 2009). Among others, *KNOX1* genes activate cytokinin signaling to promote the maintenance of prolonged intercalary growth (Shani et al., 2010). In peas and *Medicago*, orthologs of the *Arabidopsis* floral meristem identity gene *LEAFY* can also promote leaf intercalary growth (Hofer et al., 1997; Wang et al., 2008). Similarly, orthologs of the *Arabidopsis* floral patterning gene *APETALA1/FRUITFUL* in tomato promote intercalary growth and leaf complexity (Burko et al., 2013).

On the other hand, genes promoting differentiation terminate intercalary growth, and thus the leaflet patterning phase, to reduce leaf complexity. A notable example is the CIN-TCP family transcription factors, such as *LANCEOLATE* in tomato (Ori et al., 2007). Many of these CIN-TCP genes are miR319-targeting, and they promote gibberellin signaling to exert their function in differentiation promotion (Yanai et al., 2011).

Taken together, leaf development may utilize similar mechanisms in context-specific ways to obtain highly diversified leaf forms. Many of the parts, including transcription factors and hormones, are conserved in leaf development and leaflet development. On the other hand, some genes are re-purposed to alter leaf development, leading to leaf shape diversity. In addition to the above-mentioned examples, a homeodomain protein REDUCED COMPLEXITY (RCO) has evolved by duplication to enable leaflet formation in *Cardamine* but was lost in *Arabidopsis* with simple leaves (Vlad et al., 2014; Vuolo et al., 2016). Furthermore, diversification in RCO paralogs in the genus *Capsella* is temperature-responsive and may account for leaf shape plasticity under different temperatures (Sicard et al., 2014).

References

Achard, P., Gusti, A., Cheminant, S., Alioua, M., Dhondt, S., Coppens, F., Beemster, G. T., & Genschik, P. (2009). Gibberellin signaling controls cell proliferation rate in *Arabidopsis*. *Current Biology, 19*, 1188–1193.

Allen, E., Xie, Z., Gustafson, A. M., & Carrington, J. C. (2005). MicroRNA-directed phasing during trans-acting siRNA biogenesis in plants. *Cell, 121*, 207–221.

Alvarez, J. P., Goldshmidt, A., Efroni, I., Bowman, J. L., & Eshed, Y. (2009). The *NGATHA* distal organ development genes are essential for style specification in *Arabidopsis*. *Plant Cell, 21*, 1373–1393.

Alvarez, J. P., Furumizu, C., Efroni, I., Eshed, Y., & Bowman, J. L. (2016). Active suppression of a leaf meristem orchestrates determinate leaf growth. *eLife, 5*, e15023.

Anastasiou, E., Kenz, S., Gerstung, M., MacLean, D., Timmer, J., Fleck, C., & Lenhard, M. (2007). Control of plant organ size by KLUH/CYP78A5-dependent intercellular signaling. *Developmental Cell, 13*, 843–856.

Avery, G. S. (1933). Structure and development of the tobacco leaf. *American Journal of Botany, 20*, 565–592.

Axtell, M. J., & Bowman, J. L. (2008). Evolution of plant microRNAs and their targets. *Trends in Plant Science, 13*, 343–349.

Bar, M., & Ori, N. (2015). Compound leaf development in model plant species. *Current Opinion in Plant Biology, 23*, 61–69.

Barton, M. K. (2010). Twenty years on: The inner workings of the shoot apical meristem, a developmental dynamo. *Developmental Biology, 341*, 95–113.

Bayer, E. M., Smith, R. S., Mandel, T., Nakayama, N., Sauer, M., Prusinkiewicz, P., & Kuhlemeier, C. (2009). Integration of transport-based models for phyllotaxis and midvein formation. *Genes & Development, 23*, 373–384.

Beerling, D. J., & Fleming, A. J. (2007). Zimmermann's telome theory of megaphyll leaf evolution: A molecular and cellular critique. *Current Opinion in Plant Biology, 10*, 4–12.

Beerling, D. J., Osborne, C. P., & Chaloner, W. G. (2001). Evolution of leaf-form in land plants linked to atmospheric CO_2 decline in the Late Palaeozoic era. *Nature, 410*, 352–354.

Ben-Gera, H., Shwartz, I., Shao, M. R., Shani, E., Estelle, M., & Ori, N. (2012). *ENTIRE* and *GOBLET* promote leaflet development in tomato by modulating auxin response. *The Plant Journal, 70*, 903–915.

Ben-Gera, H., Dafna, A., Alvarez, J. P., Bar, M., Mauerer, M., & Ori, N. (2016). Auxin-mediated lamina growth in tomato leaves is restricted by two parallel mechanisms. *The Plant Journal, 86*, 443–457.

Berger, Y., Harpaz-Saad, S., Brand, A., Melnik, H., Sirding, N., Alvarez, J. P., Zinder, M., Samach, A., Eshed, Y., & Ori, N. (2009). The NAC-domain transcription factor *GOBLET* specifies leaflet boundaries in compound tomato leaves. *Development, 136*, 823–832.

Besnard, F., Refahi, Y., Morin, V., Marteaux, B., Brunoud, G., Chambrier, P., Rozier, F., Mirabet, V., Legrand, J., Laine, S., et al. (2014). Cytokinin signalling inhibitory fields provide robustness to phyllotaxis. *Nature, 505*, 417–421.

Bhatia, N., Bozorg, B., Larsson, A., Ohno, C., Jönsson, H., & Heisler, M. G. (2016). Auxin acts through MONOPTEROS to regulate plant cell polarity and pattern phyllotaxis. *Current Biology, 26*, 3202–3208.

Bilsborough, G. D., Runions, A., Barkoulas, M., Jenkins, H. W., Hasson, A., Galinha, C., Laufs, P., Hay, A., Prusinkiewicz, P., & Tsiantis, M. (2011). Model for the regulation of *Arabidopsis thaliana* leaf margin development. *Proceedings of the National Academy of Sciences of the United States of America, 108*, 3424–3429.

Blein, T., Pulido, A., Vialette-Guiraud, A., Nikovics, K., Morin, H., Hay, A., Johansen, I. E., Tsiantis, M., & Laufs, P. (2008). A conserved molecular framework for compound leaf development. *Science, 322*, 1835–1839.

Blein, T., Hasson, A., & Laufs, P. (2010). Leaf development: What it needs to be complex. *Current Opinion in Plant Biology, 13*, 75–82.

Bonaccorso, O., Lee, J. E., Puah, L., Scutt, C. P., & Golz, J. F. (2012). FILAMENTOUS FLOWER controls lateral organ development by acting as both an activator and a repressor. *BMC Plant Biology, 12*, 176.

Bou-Torrent, J., Salla-Martret, M., Brandt, R., Musielak, T., Palauqui, J. C., Martinez-Garcia, J. F., & Wenkel, S. (2012). ATHB4 and HAT3, two class II HD-ZIP transcription factors, control leaf development in *Arabidopsis*. *Plant Signaling & Behavior, 7*, 1382–1387.

Boyce, C. K. (2010). The evolution of plant development in a paleontological context. *Current Opinion in Plant Biology, 13*, 102–107.

Brandt, R., Salla-Martret, M., Bou-Torrent, J., Musielak, T., Stahl, M., Lanz, C., Ott, F., Schmid, M., Greb, T., Schwarz, M., et al. (2012). Genome-wide binding-site analysis of REVOLUTA reveals a link between leaf patterning and light-mediated growth responses. *The Plant Journal, 72*, 31–42.

Braybrook, S. A., & Peaucelle, A. (2013). Mechano-chemical aspects of organ formation in *Arabidopsis thaliana*: The relationship between auxin and pectin. *PLoS One, 8*, e57813.

Bresso, E. G., Chorostecki, U., Rodriguez, R. E., Palatnik, J. F., & Schommer, C. (2018). Spatial control of gene expression by miR319-regulated TCP transcription factors in leaf development. *Plant Physiology, 176*, 1694–1708.

Burian, A., Paszkiewicz, G., Nguyen, K. T., Meda, S., Raczynska-Szajgin, M., & Timmermans, M. C. P. (2022). Specification of leaf dorsiventrality via a prepatterned binary readout of a uniform auxin input. *Nature Plants, 8*, 269–280.

Burko, Y., Shleizer-Burko, S., Yanai, O., Shwartz, I., Zelnik, I. D., Jacob-Hirsch, J., Kela, I., Eshed-Williams, L., & Ori, N. (2013). A role for APETALA1/fruitfull transcription factors in tomato leaf development. *Plant Cell, 25*, 2070–2083.

Busch, B. L., Schmitz, G., Rossmann, S., Piron, F., Ding, J., Bendahmane, A., & Theres, K. (2011). Shoot branching and leaf dissection in tomato are regulated by homologous gene modules. *Plant Cell, 23*, 3595–3609.

Byrne, M. E., Barley, R., Curtis, M., Arroyo, J. M., Dunham, M., Hudson, A., & Martienssen, R. A. (2000). *Asymmetric leaves1* mediates leaf patterning and stem cell function in *Arabidopsis*. *Nature, 408*, 967–971.

Caggiano, M. P., Yu, X., Bhatia, N., Larsson, A., Ram, H., Ohno, C. K., Sappl, P., Meyerowitz, E. M., Jonsson, H., & Heisler, M. G. (2017). Cell type boundaries organize plant development. *eLife, 6*, e27421.

Capua, Y., & Eshed, Y. (2017). Coordination of auxin-triggered leaf initiation by tomato *LEAFLESS*. *Proceedings of the National Academy of Sciences of the United States of America, 114*, 3246–3251.

Chandler, J. W., Cole, M., Flier, A., Grewe, B., & Werr, W. (2007). The AP2 transcription factors DORNRÖSCHEN and DORNRÖSCHEN-LIKE redundantly control *Arabidopsis* embryo patterning via interaction with PHAVOLUTA. *Development, 134*, 1653–1662.

Chen, J. J., Janssen, B. J., Williams, A., & Sinha, N. (1997). A gene fusion at a homeobox locus: Alterations in leaf shape and implications for morphological evolution. *Plant Cell, 9*, 1289–1304.

Chitwood, D. H., Nogueira, F. T., Howell, M. D., Montgomery, T. A., Carrington, J. C., & Timmermans, M. C. (2009). Pattern formation via small RNA mobility. *Genes & Development, 23*, 549–554.

Choe, S., Fujioka, S., Noguchi, T., Takatsuto, S., Yoshida, S., & Feldmann, K. A. (2001). Overexpression of *DWARF4* in the brassinosteroid biosynthetic pathway results in increased vegetative growth and seed yield in *Arabidopsis*. *The Plant Journal, 26*, 573–582.

Das Gupta, M., & Nath, U. (2015). Divergence in patterns of leaf growth polarity is associated with the expression divergence of miR396. *Plant Cell, 27*, 2785–2799.

Daum, G., Medzihradszky, A., Suzaki, T., & Lohmann, J. U. (2014). A mechanistic framework for noncell autonomous stem cell induction in *Arabidopsis*. *Proceedings of the National Academy of Sciences of the United States of America, 111*, 14619–14624.

Debernardi, J. M., Mecchia, M. A., Vercruyssen, L., Smaczniak, C., Kaufmann, K., Inze, D., Rodriguez, R. E., & Palatnik, J. F. (2014). Post-transcriptional control of GRF transcription factors by microRNA miR396 and GIF co-activator affects leaf size and longevity. *The Plant Journal, 79*, 413–426.

Donnelly, P. M., Bonetta, D., Tsukaya, H., Dengler, R. E., & Dengler, N. G. (1999). Cell cycling and cell enlargement in developing leaves of *Arabidopsis*. *Developmental Biology, 215*, 407–419.

Douglas, R. N., Wiley, D., Sarkar, A., Springer, N., Timmermans, M. C., & Scanlon, M. J. (2010). ragged seedling2 Encodes an ARGONAUTE7-like protein required for mediolateral expansion, but not dorsiventrality, of maize leaves. *Plant Cell, 22*, 1441–1451.

Efroni, I., Blum, E., Goldshmidt, A., & Eshed, Y. (2008). A protracted and dynamic maturation schedule underlies *Arabidopsis* leaf development. *Plant Cell, 20*, 2293–2306.

Efroni, I., Eshed, Y., & Lifschitz, E. (2010). Morphogenesis of simple and compound leaves: A critical review. *Plant Cell, 22*, 1019–1032.

Emery, J. F., Floyd, S. K., Alvarez, J., Eshed, Y., Hawker, N. P., Izhaki, A., Baum, S. F., & Bowman, J. L. (2003). Radial patterning of *Arabidopsis* shoots by class III HD-ZIP and KANADI genes. *Current Biology, 13*, 1768–1774.

Eshed, Y., Baum, S. F., Perea, J. V., & Bowman, J. L. (2001). Establishment of polarity in lateral organs of plants. *Current Biology, 11*, 1251–1260.

Fahlgren, N., Montgomery, T. A., Howell, M. D., Allen, E., Dvorak, S. K., Alexander, A. L., & Carrington, J. C. (2006). Regulation of *AUXIN RESPONSE FACTOR3* by TAS3 ta-siRNA affects developmental timing and patterning in *Arabidopsis*. *Current Biology, 16*, 939–944.

Fleming, A. J., McQueenMason, S., Mandel, T., & Kuhlemeier, C. (1997). Induction of leaf primordia by the cell wall protein expansion. *Science, 276*, 1415–1418.

Floyd, S. K., & Bowman, J. L. (2010). Gene expression patterns in seed plant shoot meristems and leaves: Homoplasy or homology? *Journal of Plant Research, 123*, 43–55.

Foster, A. S. (1936). Leaf differentiation in angiosperms. *The Botanical Review, 2*, 349–372.

Furumizu, C., Alvarez, J. P., Sakakibara, K., & Bowman, J. L. (2015). Antagonistic roles for KNOX1 and KNOX2 genes in patterning the land plant body plan following an ancient gene duplication. *PLoS Genetics, 11*, e1004980.

Goldshmidt, A., Alvarez, J. P., Bowman, J. L., & Eshed, Y. (2008). Signals derived from *YABBY* gene activities in organ primordia regulate growth and partitioning of *Arabidopsis* shoot apical meristems. *Plant Cell, 20*, 1217–1230.

Gonzalez, N., De Bodt, S., Sulpice, R., Jikumaru, Y., Chae, E., Dhondt, S., Van Daele, T., De Milde, L., Weigel, D., Kamiya, Y., et al. (2010). Increased leaf size: Different means to an end. *Plant Physiology, 153*, 1261–1279.

Gonzalez, N., Pauwels, L., Baekelandt, A., De Milde, L., Van Leene, J., Besbrugge, N., Heyndrickx, K. S., Cuellar Perez, A., Durand, A. N., De Clercq, R., et al. (2015). A repressor protein complex regulates leaf growth in *Arabidopsis*. *Plant Cell, 27*, 2273–2287.

Guan, C., Wu, B., Yu, T., Wang, Q., Krogan, N. T., Liu, X., & Jiao, Y. (2017). Spatial auxin signaling controls leaf flattening in *Arabidopsis*. *Current Biology, 27*, 2940–2950.

Guan, C., Qiao, L., Xiong, Y., Zhang, L., & Jiao, Y. (2022). Coactivation of antagonistic genes stabilizes polarity patterning during shoot organogenesis. *Science Advances, 8*, eabn0368.

Guenot, B., Bayer, E., Kierzkowski, D., Smith, R. S., Mandel, T., Zadnikova, P., Benkova, E., & Kuhlemeier, C. (2012). PIN1-independent leaf initiation in *Arabidopsis*. *Plant Physiology, 159*, 1501–1510.

Guo, M., Thomas, J., Collins, G., & Timmermans, M. C. (2008). Direct repression of *KNOX* loci by the ASYMMETRIC LEAVES1 complex of *Arabidopsis*. *Plant Cell, 20*, 48–58.

Hagemann, W., & Gleissberg, S. (1996). Organogenetic capacity of leaves: The significance of marginal blastozones in angiosperms. *Plant Systematics and Evolution, 199*, 121–152.

Hardtke, C. S., & Berleth, T. (1998). The *Arabidopsis* gene *MONOPTEROS* encodes a transcription factor mediating embryo axis formation and vascular development. *The EMBO Journal, 17*, 1405–1411.

Hareven, D., Gutfinger, T., Parnis, A., Eshed, Y., & Lifschitz, E. (1996). The making of a compound leaf: Genetic manipulation of leaf architecture in tomato. *Cell, 84*, 735–744.

Hasson, A., Blein, T., & Laufs, P. (2010). Leaving the meristem behind: The genetic and molecular control of leaf patterning and morphogenesis. *Comptes Rendus Biologies, 333*, 350–360.

Hay, A., & Tsiantis, M. (2006). The genetic basis for differences in leaf form between *Arabidopsis thaliana* and its wild relative *Cardamine hirsuta*. *Nature Genetics, 38*, 942–947.

Hay, A., & Tsiantis, M. (2010). KNOX genes: Versatile regulators of plant development and diversity. *Development, 137*, 3153–3165.

Heisler, M. G., Ohno, C., Das, P., Sieber, P., Reddy, G. V., Long, J. A., & Meyerowitz, E. M. (2005). Patterns of auxin transport and gene expression during primordium development revealed by live imaging of the *Arabidopsis* inflorescence meristem. *Current Biology, 15*, 1899–1911.

Heisler, M. G., Hamant, O., Krupinski, P., Uyttewaal, M., Ohno, C., Jönsson, H., Traas, J., & Meyerowitz, E. M. (2010). Alignment between PIN1 polarity and microtubule orientation in the shoot apical meristem reveals a tight coupling between morphogenesis and auxin transport. *PLoS Biology, 8*, e1000516.

Hofer, J., Turner, L., Hellens, R., Ambrose, M., Matthews, P., Michael, A., & Ellis, N. (1997). *UNIFOLIATA* regulates leaf and flower morphogenesis in pea. *Current Biology, 7*, 581–587.

Horiguchi, G., Kim, G. T., & Tsukaya, H. (2005). The transcription factor AtGRF5 and the transcription coactivator AN3 regulate cell proliferation in leaf primordia of *Arabidopsis thaliana*. *The Plant Journal, 43*, 68–78.

Huang, S., Raman, A. S., Ream, J. E., Fujiwara, H., Cerny, R. E., & Brown, S. M. (1998). Overexpression of 20-oxidase confers a gibberellin-overproduction phenotype in *Arabidopsis*. *Plant Physiology, 118*, 773–781.

Huang, T., Harrar, Y., Lin, C., Reinhart, B., Newell, N. R., Talavera-Rauh, F., Hokin, S. A., Barton, M. K., & Kerstetter, R. A. (2014). *Arabidopsis* KANADI1 acts as a transcriptional repressor by interacting with a specific *cis*-element and regulates auxin biosynthesis, transport, and signaling in opposition to HD-ZIPIII factors. *Plant Cell, 26*, 246–262.

Hunter, C., Willmann, M. R., Wu, G., Yoshikawa, M., de la Luz Gutierrez-Nava, M., & Poethig, S. R. (2006). Trans-acting siRNA-mediated repression of *ETTIN* and *ARF4* regulates heteroblasty in *Arabidopsis*. *Development, 133*, 2973–2981.

Husbands, A. Y., Chitwood, D. H., Plavskin, Y., & Timmermans, M. C. (2009). Signals and prepatterns: New insights into organ polarity in plants. *Genes & Development, 23*, 1986–1997.

Husbands, A. Y., Benkovics, A. H., Nogueira, F. T., Lodha, M., & Timmermans, M. C. (2015). The ASYMMETRIC LEAVES complex employs multiple modes of regulation to affect adaxial-abaxial patterning and leaf complexity. *Plant Cell, 27*, 3321–3335.

Ishiwata, A., Ozawa, M., Nagasaki, H., Kato, M., Noda, Y., Yamaguchi, T., Nosaka, M., Shimizu-Sato, S., Nagasaki, A., Maekawa, M., et al. (2013). Two *WUSCHEL*-related homeobox genes, *narrow leaf2* and *narrow leaf3*, control leaf width in rice. *Plant & Cell Physiology, 54*, 779–792.

Iwakawa, H., Ueno, Y., Semiarti, E., Onouchi, H., Kojima, S., Tsukaya, H., Hasebe, M., Soma, T., Ikezaki, M., Machida, C., et al. (2002). The *ASYMMETRIC LEAVES2* gene of *Arabidopsis thaliana*, required for formation of a symmetric flat leaf lamina, encodes a member of a novel family of proteins characterized by cysteine repeats and a leucine zipper. *Plant & Cell Physiology, 43*, 467–478.

Iwakawa, H., Iwasaki, M., Kojima, S., Ueno, Y., Soma, T., Tanaka, H., Semiarti, E., Machida, Y., & Machida, C. (2007). Expression of the *ASYMMETRIC LEAVES2* gene in the adaxial domain of *Arabidopsis* leaves represses cell proliferation in this domain and is critical for the development of properly expanded leaves. *The Plant Journal, 51*, 173–184.

Iwasaki, M., Takahashi, H., Iwakawa, H., Nakagawa, A., Ishikawa, T., Tanaka, H., Matsumura, Y., Pekker, I., Eshed, Y., Vial-Pradel, S., et al. (2013). Dual regulation of *ETTIN* (*ARF3*) gene expression by AS1-AS2, which maintains the DNA methylation level, is involved in stabilization of leaf adaxial-abaxial partitioning in *Arabidopsis*. *Development, 140*, 1958–1969.

Jiao, Y., Du, F., & Traas, J. (2021). The mechanical feedback theory for leaf lamina formation. *Trends in Plant Science, 26*, 107–110.

Jones-Rhoades, M. W., & Bartel, D. P. (2004). Computational identification of plant microRNAs and their targets, including a stress-induced miRNA. *Molecular Cell, 14*, 787–799.

Jönsson, H., Heisler, M. G., Shapiro, B. E., Meyerowitz, E. M., & Mjolsness, E. (2006). An auxin-driven polarized transport model for phyllotaxis. *Proceedings of the National Academy of Sciences of the United States of America, 103*, 1633–1638.

Juarez, M. T., Twigg, R. W., & Timmermans, M. C. (2004). Specification of adaxial cell fate during maize leaf development. *Development, 131*, 4533–4544.

Kaplan, D. R. (2001). Fundamental concepts of leaf morphology and morphogenesis: A contribution to the interpretation of molecular genetic mutants. *International Journal of Plant Sciences, 162*, 465–474.

Kelley, D. R., Arreola, A., Gallagher, T. L., & Gasser, C. S. (2012). ETTIN (ARF3) physically interacts with KANADI proteins to form a functional complex essential for integument development and polarity determination in *Arabidopsis*. *Development, 139*, 1105–1109.

Kerstetter, R. A., Bollman, K., Taylor, R. A., Bomblies, K., & Poethig, R. S. (2001). *KANADI* regulates organ polarity in *Arabidopsis*. *Nature, 411*, 706–709.

Kim, J. H., & Lee, B. H. (2006). *GROWTH-REGULATING FACTOR4* of *Arabidopsis thaliana* is required for development of leaves, cotyledons, and shoot apical meristem. *Journal of Plant Biology, 49*, 463–468.

Kim, J. H., Choi, D., & Kende, H. (2003). The AtGRF family of putative transcription factors is involved in leaf and cotyledon growth in *Arabidopsis*. *The Plant Journal, 36*, 94–104.

Kim, Y. S., Kim, S. G., Lee, M., Lee, I., Park, H. Y., Seo, P. J., Jung, J. H., Kwon, E. J., Suh, S. W., Paek, K. H., et al. (2008). HD-ZIP III activity is modulated by competitive inhibitors via a feedback loop in *Arabidopsis* shoot apical meristem development. *Plant Cell, 20*, 920–933.

Kimura, S., Koenig, D., Kang, J., Yoong, F. Y., & Sinha, N. (2008). Natural variation in leaf morphology results from mutation of a novel KNOX gene. *Current Biology, 18*, 672–677.

Koenig, D., & Sinha, N. (2010). Evolution of leaf shape: A pattern emerges. *Current Topics in Developmental Biology, 91*, 169–183.

Koenig, D., Bayer, E., Kang, J., Kuhlemeier, C., & Sinha, N. (2009). Auxin patterns *Solanum lycopersicum* leaf morphogenesis. *Development, 136*, 2997–3006.

Koyama, T., Sato, F., & Ohme-Takagi, M. (2017). Roles of miR319 and TCP transcription factors in leaf development. *Plant Physiology, 175*, 874–885.

Krogan, N. T., Marcos, D., Weiner, A. I., & Berleth, T. (2016). The auxin response factor MONOPTEROS controls meristem function and organogenesis in both the shoot and root through the direct regulation of PIN genes. *The New Phytologist, 212*, 42–50.

Kuhlemeier, C., & Timmermans, M. C. P. (2016). The Sussex signal: Insights into leaf dorsiventrality. *Development, 143*, 3230–3237.

Kumaran, M. K., Bowman, J. L., & Sundaresan, V. (2002). *YABBY* polarity genes mediate the repression of *KNOX* homeobox genes in *Arabidopsis*. *Plant Cell, 14*, 2761–2770.

Lavy, M., & Estelle, M. (2016). Mechanisms of auxin signaling. *Development, 143*, 3226–3229.

Lee, B. H., Ko, J. H., Lee, S., Lee, Y., Pak, J. H., & Kim, J. H. (2009). The *Arabidopsis GRF-INTERACTING FACTOR* gene family performs an overlapping function in determining organ size as well as multiple developmental properties. *Plant Physiology, 151*, 655–668.

Li, Z., Li, B., Shen, W. H., Huang, H., & Dong, A. (2012). TCP transcription factors interact with AS2 in the repression of class-I *KNOX* genes in *Arabidopsis thaliana*. *The Plant Journal, 71*, 99–107.

Li, Z., Li, B., Liu, J., Guo, Z., Liu, Y., Li, Y., Shen, W. H., Huang, Y., Huang, H., Zhang, Y., et al. (2016). Transcription factors AS1 and AS2 interact with LHP1 to repress *KNOX* genes in *Arabidopsis*. *Journal of Integrative Plant Biology, 58*, 959–970.

Lin, W.-C., Shuai, B., & Springer, P. S. (2003). The *Arabidopsis* LATERAL ORGAN BOUNDARIES-domain gene *ASYMMETRIC LEAVES2* functions in the repression of *KNOX* gene expression and in adaxial-abaxial patterning. *Plant Cell, 15*, 2241–2252.

Lin, H., Niu, L., McHale, N. A., Ohme-Takagi, M., Mysore, K. S., & Tadege, M. (2013). Evolutionarily conserved repressive activity of WOX proteins mediates leaf blade outgrowth and floral organ development in plants. *Proceedings of the National Academy of Sciences of the United States of America, 110*, 366–371.

Lodha, M., Marco, C. F., & Timmermans, M. C. (2013). The ASYMMETRIC LEAVES complex maintains repression of *KNOX* homeobox genes via direct recruitment of Polycomb-repressive complex2. *Genes & Development, 27*, 596–601.

Luo, M., Yu, C. W., Chen, F. F., Zhao, L., Tian, G., Liu, X., Cui, Y., Yang, J. Y., & Wu, K. (2012). Histone deacetylase HDA6 is functionally associated with AS1 in repression of *KNOX* genes in *Arabidopsis*. *PLoS Genetics, 8*, e1003114.

Magnani, E., & Hake, S. (2008). *KNOX* lost the *OX*: The *Arabidopsis KNATM* gene defines a novel class of KNOX transcriptional regulators missing the homeodomain. *Plant Cell, 20*, 875–887.

Mallory, A. C., Reinhart, B. J., Jones-Rhoades, M. W., Tang, G., Zamore, P. D., Barton, M. K., & Bartel, D. P. (2004). MicroRNA control of *PHABULOSA* in leaf development: Importance of pairing to the microRNA 5' region. *The EMBO Journal, 23*, 3356–3364.

Marin, E., Jouannet, V., Herz, A., Lokerse, A. S., Weijers, D., Vaucheret, H., Nussaume, L., Crespi, M. D., & Maizel, A. (2010). miR390, Arabidopsis TAS3 tasiRNAs, and their AUXIN

RESPONSE FACTOR targets define an autoregulatory network quantitatively regulating lateral root growth. *Plant Cell, 22*, 1104–1117.
McHale, N. A., & Marcotrigiano, M. (1998). *LAM1* is required for dorsoventrality and lateral growth of the leaf blade in *Nicotiana*. *Development, 125*, 4235–4243.
Merelo, P., Xie, Y., Brand, L., Ott, F., Weigel, D., Bowman, J. L., Heisler, M. G., & Wenkel, S. (2013). Genome-wide identification of KANADI1 target genes. *PLoS One, 8*, e77341.
Merelo, P., Ram, H., Pia Caggiano, M., Ohno, C., Ott, F., Straub, D., Graeff, M., Cho, S. K., Yang, S. W., Wenkel, S., et al. (2016). Regulation of MIR165/166 by class II and class III homeodomain leucine zipper proteins establishes leaf polarity. *Proceedings of the National Academy of Sciences of the United States of America, 113*, 11973–11978.
Mitchison, G. J. (1980). A model for vein formation in higher plants. *Proceedings of the Royal Society B, 207*, 79–109.
Mudunkothge, J. S., & Krizek, B. A. (2012). Three *Arabidopsis AIL/PLT* genes act in combination to regulate shoot apical meristem function. *The Plant Journal, 71*, 108–121.
Müller, C. J., Valdés, A. E., Wang, G., Ramachandran, P., Beste, L., Uddenberg, D., & Carlsbecker, A. (2016). PHABULOSA mediates an auxin signaling loop to regulate vascular patterning in *Arabidopsis*. *Plant Physiology, 170*, 956–970.
Nagasaki, H., Itoh, J., Hayashi, K., Hibara, K., Satoh-Nagasawa, N., Nosaka, M., Mukouhata, M., Ashikari, M., Kitano, H., Matsuoka, M., et al. (2007). The small interfering RNA production pathway is required for shoot meristem initiation in rice. *Proceedings of the National Academy of Sciences of the United States of America, 104*, 14867–14871.
Nakata, M., Matsumoto, N., Tsugeki, R., Rikirsch, E., Laux, T., & Okada, K. (2012). Roles of the middle domain-specific *WUSCHEL-RELATED HOMEOBOX* genes in early development of leaves in *Arabidopsis*. *Plant Cell, 24*, 519–535.
Nakayama, N., Smith, R. S., Mandel, T., Robinson, S., Kimura, S., Boudaoud, A., & Kuhlemeier, C. (2012). Mechanical regulation of auxin-mediated growth. *Current Biology, 22*, 1468–1476.
Nardmann, J., Ji, J., Werr, W., & Scanlon, M. J. (2004). The maize duplicate genes *narrow sheath1* and *narrow sheath2* encode a conserved homeobox gene function in a lateral domain of shoot apical meristems. *Development, 131*, 2827–2839.
Nath, U., Crawford, B. C., Carpenter, R., & Coen, E. (2003). Genetic control of surface curvature. *Science, 299*, 1404–1407.
Nelissen, H., Rymen, B., Jikumaru, Y., Demuynck, K., Van Lijsebettens, M., Kamiya, Y., Inze, D., & Beemster, G. T. (2012). A local maximum in gibberellin levels regulates maize leaf growth by spatial control of cell division. *Current Biology, 22*, 1183–1187.
Nogueira, F. T., Madi, S., Chitwood, D. H., Juarez, M. T., & Timmermans, M. C. (2007). Two small regulatory RNAs establish opposing fates of a developmental axis. *Genes & Development, 21*, 750–755.
Okada, K., Ueda, J., Komaki, M. K., Bell, C. J., & Shimura, Y. (1991). Requirement of the auxin polar transport system in early stages of *Arabidopsis* floral bud formation. *Plant Cell, 3*, 677–684.
Ori, N., Eshed, Y., Chuck, G., Bowman, J. L., & Hake, S. (2000). Mechanisms that control knox gene expression in the *Arabidopsis* shoot. *Development, 127*, 5523–5532.
Ori, N., Cohen, A. R., Etzioni, A., Brand, A., Yanai, O., Shleizer, S., Menda, N., Amsellem, Z., Efroni, I., Pekker, I., et al. (2007). Regulation of *LANCEOLATE* by miR319 is required for compound-leaf development in tomato. *Nature Genetics, 39*, 787–791.
Palatnik, J. F., Allen, E., Wu, X., Schommer, C., Schwab, R., Carrington, J. C., & Weigel, D. (2003). Control of leaf morphogenesis by microRNAs. *Nature, 425*, 257–263.
Peaucelle, A., Louvet, R., Johansen, J. N., Höfte, H., Laufs, P., Pelloux, J., & Mouille, G. (2008). *Arabidopsis* phyllotaxis is controlled by the methyl-esterification status of cell-wall pectins. *Current Biology, 18*, 1943–1948.
Peaucelle, A., Braybrook, S. A., Le Guillou, L., Bron, E., Kuhlemeier, C., & Höfte, H. (2011). Pectin-induced changes in cell wall mechanics underlie organ initiation in *Arabidopsis*. *Current Biology, 21*, 1720–1726.

Pekker, I., Alvarez, J. P., & Eshed, Y. (2005). Auxin response factors mediate *Arabidopsis* organ asymmetry via modulation of KANADI activity. *Plant Cell, 17*, 2899–2910.

Peng, J., Yu, J., Wang, H., Guo, Y., Li, G., Bai, G., & Chen, R. (2011). Regulation of compound leaf development in *Medicago truncatula* by *fused compound leaf1*, a class M *KNOX* gene. *Plant Cell, 23*, 3929–3943.

Peng, Z., Alique, D., Xiong, Y., Hu, J., Cao, X., Lu, S., Long, M., Wang, Y., Wabnik, K., & Jiao, Y. (2022). Differential growth dynamics control aerial organ geometry. *Current Biology, 32*, 4854–4868.e4855.

Phelps-Durr, T. L., Thomas, J., Vahab, P., & Timmermans, M. C. (2005). Maize *rough sheath2* and its *Arabidopsis* orthologue *ASYMMETRIC LEAVES1* interact with HIRA, a predicted histone chaperone, to maintain *knox* gene silencing and determinacy during organogenesis. *Plant Cell, 17*, 2886–2898.

Pinon, V., Prasad, K., Grigg, S. P., Sanchez-Perez, G. F., & Scheres, B. (2013). Local auxin biosynthesis regulation by PLETHORA transcription factors controls phyllotaxis in *Arabidopsis*. *Proceedings of the National Academy of Sciences of the United States of America, 110*, 1107–1112.

Poethig, R. S., & Sussex, I. M. (1985). The developmental morphology and growth dynamics of the tobacco leaf. *Planta, 165*, 158–169.

Prasad, K., Grigg, S. P., Barkoulas, M., Yadav, R. K., Sanchez-Perez, G. F., Pinon, V., Blilou, I., Hofhuis, H., Dhonukshe, P., Galinha, C., et al. (2011). *Arabidopsis* PLETHORA transcription factors control phyllotaxis. *Current Biology, 21*, 1123–1128.

Qi, J., Wang, Y., Yu, T., Cunha, A., Wu, B., Vernoux, T., Meyerowitz, E., & Jiao, Y. (2014). Auxin depletion from leaf primordia contributes to organ patterning. *Proceedings of the National Academy of Sciences of the United States of America, 111*, 18769–18774.

Qi, J., Wu, B., Feng, S., Lü, S., Guan, C., Zhang, X., Qiu, D., Hu, Y., Zhou, Y., Li, C., et al. (2017). Mechanical regulation of organ asymmetry in leaves. *Nature Plants, 3*, 724–733.

Rayle, D. L., & Cleland, R. E. (1992). The Acid Growth Theory of auxin-induced cell elongation is alive and well. *Plant Physiology, 99*, 1271–1274.

Reinhardt, D., Mandel, T., & Kuhlemeier, C. (2000). Auxin regulates the initiation and radial position of plant lateral organs. *Plant Cell, 12*, 507–518.

Reinhardt, D., Pesce, E. R., Stieger, P., Mandel, T., Baltensperger, K., Bennett, M., Traas, J., Friml, J., & Kuhlemeier, C. (2003). Regulation of phyllotaxis by polar auxin transport. *Nature, 426*, 255–260.

Reinhardt, D., Frenz, M., Mandel, T., & Kuhlemeier, C. (2005). Microsurgical and laser ablation analysis of leaf positioning and dorsoventral patterning in tomato. *Development, 132*, 15–26.

Reinhart, B. J., Liu, T., Newell, N. R., Magnani, E., Huang, T., Kerstetter, R., Michaels, S., & Barton, M. K. (2013). Establishing a framework for the ad/abaxial regulatory network of *Arabidopsis*: Ascertaining targets of class III homeodomain leucine zipper and KANADI regulation. *Plant Cell, 25*, 3228–3249.

Rodriguez, R. E., Mecchia, M. A., Debernardi, J. M., Schommer, C., Weigel, D., & Palatnik, J. F. (2010). Control of cell proliferation in *Arabidopsis thaliana* by microRNA miR396. *Development, 137*, 103–112.

Runions, A., Tsiantis, M., & Prusinkiewicz, P. (2017). A common developmental program can produce diverse leaf shapes. *The New Phytologist, 216*, 401–418.

Saini, K., Markakis, M. N., Zdanio, M., Balcerowicz, D. M., Beeckman, T., De Veylder, L., Prinsen, E., Beemster, G. T. S., & Vissenberg, K. (2017). Alteration in auxin homeostasis and signaling by overexpression of PINOID kinase causes leaf growth defects in *Arabidopsis thaliana*. *Frontiers in Plant Science, 8*, 1009.

Sanders, H. L., Darrah, P. R., & Langdale, J. A. (2011). Sector analysis and predictive modelling reveal iterative shoot-like development in fern fronds. *Development, 138*, 2925–2934.

Sarojam, R., Sappl, P. G., Goldshmidt, A., Efroni, I., Floyd, S. K., Eshed, Y., & Bowman, J. L. (2010). Differentiating *Arabidopsis* shoots from leaves by combined YABBY activities. *Plant Cell, 22*, 2113–2130.

Sassi, M., Ali, O., Boudon, F., Cloarec, G., Abad, U., Cellier, C., Chen, X., Gilles, B., Milani, P., Friml, J., et al. (2014). An auxin-mediated shift toward growth isotropy promotes organ formation at the shoot meristem in *Arabidopsis*. *Current Biology, 24*, 2335–2342.
Sawa, S. S., Watanabe, K., Goto, K., Kanaya, E., Morita, E. H., & Okada, K. (1999). FILAMENTOUS FLOWER, a meristem and organ identity gene of Arabidopsis, encodes a protein with a zinc finger and HMG-related domain. *Genes & Development, 13*, 1079–1088.
Scanlon, M. J., Schneeberger, R. G., & Freeling, M. (1996). The maize mutant *narrow sheath* fails to establish leaf margin identity in a meristematic domain. *Development, 122*, 1683–1691.
Scheres, B., & Krizek, B. A. (2017). Coordination of growth in root and shoot apices by *AIL/PLT* transcription factors. *Current Opinion in Plant Biology, 41*, 95–101.
Schommer, C., Debernardi, J. M., Bresso, E. G., Rodriguez, R. E., & Palatnik, J. F. (2014). Repression of cell proliferation by miR319-regulated TCP4. *Molecular Plant, 7*, 1533–1544.
Schüepp, O. (1918). Zur Entwicklungsgeschichte des Blattes von *Acer pseudoplatanus* L. *Vierteljahressch Naturf Ges Zurich, 63*, 99–105.
Schuetz, M., Berleth, T., & Mattsson, J. (2008). Multiple MONOPTEROS-dependent pathways are involved in leaf initiation. *Plant Physiology, 148*, 870–880.
Shani, E., Burko, Y., Ben-Yaakov, L., Berger, Y., Amsellem, Z., Goldshmidt, A., Sharon, E., & Ori, N. (2009). Stage-specific regulation of *Solanum lycopersicum* leaf maturation by class 1 KNOTTED1-LIKE HOMEOBOX proteins. *Plant Cell, 21*, 3078–3092.
Shani, E., Ben-Gera, H., Shleizer-Burko, S., Burko, Y., Weiss, D., & Ori, N. (2010). Cytokinin regulates compound leaf development in tomato. *Plant Cell, 22*, 3206–3217.
Shi, J., Dong, J., Xue, J., Wang, H., Yang, Z., Jiao, Y., Xu, L., & Huang, H. (2017). Model for the role of auxin polar transport in patterning of the leaf adaxial-abaxial axis. *The Plant Journal, 92*, 469–480.
Sicard, A., Thamm, A., Marona, C., Lee, Y. W., Wahl, V., Stinchcombe, J. R., Wright, S. I., Kappel, C., & Lenhard, M. (2014). Repeated evolutionary changes of leaf morphology caused by mutations to a homeobox gene. *Current Biology, 24*, 1880–1886.
Siegfried, K. R., Eshed, Y., Baum, S. F., Otsuga, D., Drews, G. N., & Bowman, J. L. (1999). Members of the YABBY gene family specify abaxial cell fate in *Arabidopsis*. *Development, 126*, 4117–4128.
Skopelitis, D. S., Benkovics, A. H., Husbands, A. Y., & Timmermans, M. C. P. (2017). Boundary formation through a direct threshold-based readout of mobile small RNA gradients. *Developmental Cell, 43*, 265–273.
Smith, R. S., Guyomarc'h, S., Mandel, T., Reinhardt, D., Kuhlemeier, C., & Prusinkiewicz, P. (2006). A plausible model of phyllotaxis. *Proceedings of the National Academy of Sciences of the United States of America, 103*, 1301–1306.
Snow, M., & Snow, R. (1932). Experiments on phyllotaxis I - the effect of isolating a pininordium. *Philosophical Transactions of the Royal Society of London. Series B, Biological Sciences, 221*, 1–43.
Snow, M., & Snow, R. (1934). The interpretation of phyllotaxis. *Biological Reviews, 9*(1), 132–137.
Snow, M., & Snow, R. (1959). The dorsiventrality of leaf primordia. *The New Phytologist, 58*, 188–207.
Stahle, M. I., Kuehlich, J., Staron, L., von Arnim, A. G., & Golz, J. F. (2009). YABBYs and the transcriptional corepressors LEUNIG and LEUNIG_HOMOLOG maintain leaf polarity and meristem activity in *Arabidopsis*. *Plant Cell, 21*, 3105–3118.
Steeves, T. A., & Sussex, I. M. (1989). *Patterns in plant development* (2nd ed.). Cambridge University Press.
Stoma, S., Lucas, M., Chopard, J., Schaedel, M., Traas, J., & Godin, C. (2008). Flux-based transport enhancement as a plausible unifying mechanism for auxin transport in meristem development. *PLoS Computational Biology, 4*, e1000207.
Sussex, I. M. (1951). Experiments on the cause of dorsiventrality in leaves. *Nature, 167*, 651–652.

Tadege, M., Lin, H., Bedair, M., Berbel, A., Wen, J., Rojas, C. M., Niu, L., Tang, Y., Sumner, L., Ratet, P., et al. (2011). *STENOFOLIA* regulates blade outgrowth and leaf vascular patterning in *Medicago truncatula* and *Nicotiana sylvestris*. *Plant Cell, 23*, 2125–2142.
Tameshige, T., Fujita, H., Watanabe, K., Toyokura, K., Kondo, M., Tatematsu, K., Matsumoto, N., Tsugeki, R., Kawaguchi, M., Nishimura, M., et al. (2013). Pattern dynamics in adaxial-abaxial specific gene expression are modulated by a plastid retrograde signal during *Arabidopsis thaliana* leaf development. *PLoS Genetics, 9*, e1003655.
Tatematsu, K., Toyokura, K., Miyashima, S., Nakajima, K., & Okada, K. (2015). A molecular mechanism that confines the activity pattern of miR165 in *Arabidopsis* leaf primordia. *The Plant Journal, 82*, 596–608.
Tomescu, A. M. (2009). Megaphylls, microphylls and the evolution of leaf development. *Trends in Plant Science, 14*, 5–12.
Trigueros, M., Navarrete-Gomez, M., Sato, S., Christensen, S. K., Pelaz, S., Weigel, D., Yanofsky, M. F., & Ferrandiz, C. (2009). The *NGATHA* genes direct style development in the *Arabidopsis* gynoecium. *Plant Cell, 21*, 1394–1409.
Turchi, L., Carabelli, M., Ruzza, V., Possenti, M., Sassi, M., Penalosa, A., Sessa, G., Salvi, S., Forte, V., Morelli, G., et al. (2013). *Arabidopsis* HD-Zip II transcription factors control apical embryo development and meristem function. *Development, 140*, 2118–2129.
van der Graaff, E., Laux, T., & Rensing, S. A. (2009). The WUS homeobox-containing (WOX) protein family. *Genome Biology, 10*, 248.
Vandenbussche, M., Horstman, A., Zethof, J., Koes, R., Rijpkema, A. S., & Gerats, T. (2009). Differential recruitment of WOX transcription factors for lateral development and organ fusion in Petunia and *Arabidopsis*. *Plant Cell, 21*, 2269–2283.
Vlad, D., Kierzkowski, D., Rast, M. I., Vuolo, F., Dello Ioio, R., Galinha, C., Gan, X., Hajheidari, M., Hay, A., Smith, R. S., et al. (2014). Leaf shape evolution through duplication, regulatory diversification, and loss of a homeobox gene. *Science, 343*, 780–783.
Vuolo, F., Mentink, R. A., Hajheidari, M., Bailey, C. D., Filatov, D. A., & Tsiantis, M. (2016). Coupled enhancer and coding sequence evolution of a homeobox gene shaped leaf diversity. *Genes & Development, 30*, 2370–2375.
Waites, R., & Hudson, A. (1995). *phantastica*: A gene required for dorsoventrality of leaves in *Antirrhinum majus*. *Development, 121*, 2143–2154.
Wang, J.-J., & Guo, H.-S. (2015). Cleavage of *INDOLE-3-ACETIC ACID INDUCIBLE28* mRNA by microRNA847 upregulates auxin signaling to modulate cell proliferation and lateral organ growth in *Arabidopsis*. *Plant Cell, 27*, 574–590.
Wang, H., Chen, J., Wen, J., Tadege, M., Li, G., Liu, Y., Mysore, K. S., Ratet, P., & Chen, R. (2008). Control of compound leaf development by *FLORICAULA/LEAFY* ortholog *SINGLE LEAFLET1* in *Medicago truncatula*. *Plant Physiology, 146*, 1759–1772.
Wang, W., Xu, B., Wang, H., Li, J., Huang, H., & Xu, L. (2011). *YUCCA* genes are expressed in response to leaf adaxial-abaxial juxtaposition and are required for leaf margin development. *Plant Physiology, 157*, 1805–1819.
Wang, Q., Hasson, A., Rossmann, S., & Theres, K. (2016). *Divide et impera*: Boundaries shape the plant body and initiate new meristems. *The New Phytologist, 209*, 485–498.
Wang, Q., Marconi, M., Guan, C., Wabnik, K., & Jiao, Y. (2022). Polar auxin transport modulates early leaf flattening. *Proceedings of the National Academy of Sciences of the United States of America, 119*, e2215569119.
Wenkel, S., Emery, J., Hou, B. H., Evans, M. M., & Barton, M. K. (2007). A feedback regulatory module formed by LITTLE ZIPPER and HD-ZIPIII genes. *Plant Cell, 19*, 3379–3390.
Xie, Y., Straub, D., Eguen, T., Brandt, R., Stahl, M., Martinez-Garcia, J. F., & Wenkel, S. (2015). Meta-analysis of *Arabidopsis* KANADI1 direct target genes identifies a basic growth-promoting module acting upstream of hormonal signaling pathways. *Plant Physiology, 169*, 1240–1253.
Xu, Y., Sun, Y., Liang, W.-Q., & Huang, H. (2002). The *Arabidopsis AS2* gene encoding a predicted leucine-zipper protein is required for the leaf polarity formation. *Acta Botanica Sinica, 44*, 1194–1202.

Xu, L., Xu, Y., Dong, A., Sun, Y., Pi, L., Xu, Y., & Huang, H. (2003). Novel *as1* and *as2* defects in leaf adaxial-abaxial polarity reveal the requirement for *ASYMMETRIC LEAVES1* and *2* and *ERECTA* functions in specifying leaf adaxial identity. *Development, 130*, 4097–4107.

Xu, L., Yang, L., & Huang, H. (2007). Transcriptional, post-transcriptional and post-translational regulations of gene expression during leaf polarity formation. *Cell Research, 17*, 512–519.

Yadav, R. K., Tavakkoli, M., & Reddy, G. V. (2010). WUSCHEL mediates stem cell homeostasis by regulating stem cell number and patterns of cell division and differentiation of stem cell progenitors. *Development, 137*, 3581–3589.

Yadav, R. K., Perales, M., Gruel, J., Ohno, C., Heisler, M., Girke, T., Jönsson, H., & Reddy, G. V. (2013). Plant stem cell maintenance involves direct transcriptional repression of differentiation program. *Molecular Systems Biology, 9*, 654.

Yamaguchi, T., Nagasawa, N., Kawasaki, S., Matsuoka, M., Nagato, Y., & Hirano, H. Y. (2004). The YABBY gene *DROOPING LEAF* regulates carpel specification and midrib development in *Oryza sativa*. *Plant Cell, 16*, 500–509.

Yamaguchi, T., Yano, S., & Tsukaya, H. (2010). Genetic framework for flattened leaf blade formation in unifacial leaves of *Juncus prismatocarpus*. *Plant Cell, 22*, 2141–2155.

Yanai, O., Shani, E., Russ, D., & Ori, N. (2011). Gibberellin partly mediates *LANCEOLATE* activity in tomato. *The Plant Journal, 68*, 571–582.

Yao, X., Wang, H., Li, H., Yuan, Z., Li, F., Yang, L., & Huang, H. (2009). Two types of *cis*-acting elements control the abaxial epidermis-specific transcription of the *MIR165a* and *MIR166a* genes. *FEBS Letters, 583*, 3711–3717.

Yifhar, T., Pekker, I., Peled, D., Friedlander, G., Pistunov, A., Sabban, M., Wachsman, G., Alvarez, J. P., Amsellem, Z., & Eshed, Y. (2012). Failure of the tomato *trans*-acting short interfering RNA program to regulate *AUXIN RESPONSE FACTOR3* and *ARF4* underlies the wiry leaf syndrome. *Plant Cell, 24*, 3575–3589.

Yu, T., Guan, C., Wang, J., Sajjad, M., Ma, L., & Jiao, Y. (2017). Dynamic patterns of gene expression during leaf initiation. *Journal of Genetics and Genomics, 44*, 599–601.

Zhang, F., Wang, Y., Li, G., Tang, Y., Kramer, E. M., & Tadege, M. (2014). STENOFOLIA recruits TOPLESS to repress *ASYMMETRIC LEAVES2* at the leaf margin and promote leaf blade outgrowth in *Medicago truncatula*. *Plant Cell, 26*, 650–664.

Zhang, Z., Runions, A., Mentink, R. A., Kierzkowski, D., Karady, M., Hashemi, B., Huijser, P., Strauss, S., Gan, X., Ljung, K., et al. (2020). A WOX/auxin biosynthesis module controls growth to shape leaf form. *Current Biology, 30*, 4857–4868.

Zhao, F., Du, F., Oliveri, H., Zhou, L., Ali, O., Chen, W., Feng, S., Wang, Q., Lu, S., Long, M., et al. (2020). Microtubule-mediated wall anisotropy contributes to leaf blade flattening. *Current Biology, 30*, 3972–3985.e3976.

Zhiponova, M. K., Vanhoutte, I., Boudolf, V., Betti, C., Dhondt, S., Coppens, F., Mylle, E., Maes, S., Gonzalez-Garcia, M. P., Cano-Delgado, A. I., et al. (2013). Brassinosteroid production and signaling differentially control cell division and expansion in the leaf. *The New Phytologist, 197*, 490–502.

Zhou, C., Han, L., Hou, C., Metelli, A., Qi, L., Tadege, M., Mysore, K. S., & Wang, Z. Y. (2011). Developmental analysis of a *Medicago truncatula smooth leaf margin1* mutant reveals context-dependent effects on compound leaf development. *Plant Cell, 23*, 2106–2124.

Zhuang, L.-L., Ambrose, M., Rameau, C., Weng, L., Yang, J., Hu, X.-H., Luo, D., & Li, X. (2012). *LATHYROIDES*, encoding a WUSCHEL-related Homeobox1 transcription factor, controls organ lateral growth, and regulates tendril and dorsal petal identities in garden pea (*Pisum sativum* L.). *Molecular Plant, 5*, 1333–1345.

Chapter 4
De Novo Organ Regeneration in Vegetative Propagation

Ning Zhai and Lin Xu

Abstract Vegetative propagation is a plant reproductive process in which somatic cells can form a new plant without undergoing sexual reproduction. Vegetative propagation in seed plants can be achieved by either somatic embryogenesis or *de novo* organ regeneration. Here we review the basic concepts and recent advances in research on *de novo* organ regeneration, which refers to the organogenesis of adventitious roots and/or shoots after injury has occurred. *De novo* organ regeneration can be observed under natural conditions and is also exploited in many agricultural techniques such as cuttings and tissue culture. The *de novo* organ regeneration process can occur in either a direct or an indirect manner. Cuttings are an example of direct *de novo* organ regeneration, during which adventitious roots directly regenerate from the wound site after detachment of the stem or leaf explant. Indirect *de novo* organ regeneration occurs in tissue culture, where a high auxin concentration induces detached explants to form a group of pluripotent cells known as callus, from which adventitious roots or shoots can be induced by a low concentration of auxin or a high concentration of cytokinin, respectively. The *de novo* organ regeneration process involves wound signals, hormone actions, transcription factors, and fate transition of stem cells. Understanding the cellular and molecular framework of *de novo* organ regeneration will improve the efficiency and success of vegetative propagation.

Keywords Plant regeneration · Tissue culture · Callus · *De novo* organ regeneration · Plant stem cells

---

N. Zhai · L. Xu (✉)
National Key Laboratory of Plant Molecular Genetics, CAS Center for Excellence in Molecular Plant Sciences, Institute of Plant Physiology and Ecology, Chinese Academy of Sciences, Shanghai, China

Key Laboratory of Plant Carbon Capture, CAS, Shanghai, China
e-mail: xulin@cemps.ac.cn

F. Chang et al. (eds.), *Regulation of Plant Development*,
https://doi.org/10.1007/978-981-95-7033-1_4

4.1 Introduction to *De Novo* Organ Regeneration

Plants have powerful regenerative abilities that allow them to survive under severe natural conditions (Xu & Huang, 2014; Ikeuchi et al., 2016, 2019; Kareem et al., 2016a; Lup et al., 2016; Birnbaum & Sanchez Alvarado, 2008; Sugimoto et al., 2011). Regeneration is the process of recovery from injury. The cause of regeneration is damage to the organism, including wounding, severe stresses, or other disturbances.

The result of regeneration is recovery; for example, tissue repair to heal the wound and replace lost parts, and the formation of new organs to allow the plant to survive after damage.

In seed plants, the regeneration of adventitious roots and shoots is termed *de novo* organ regeneration (also known as *de novo* organogenesis), which includes *de novo* root regeneration (DNRR; also known as *de novo* root organogenesis) and *de novo* shoot regeneration (DNSR; also known as *de novo* shoot organogenesis) (Xu & Huang, 2014).

Under natural conditions, plants are frequently damaged during biotic and abiotic interactions. The regeneration of adventitious roots and shoots can promote plant growth to form or replenish adventitious branches in root and shoot systems. In such cases, adventitious roots and shoots can directly regenerate after tissue damage, via a process known as direct *de novo* organ regeneration that involves direct DNRR and direct DNSR. For example, cuttings of part of a shoot with a segment of stem and development leaves plus the shoot apical meristem (SAM) or lateral shoot meristems are commonly used for vegetative propagation of trees, and adventitious roots quickly regenerate from the wounding site of cuttings to ensure the survival of the wounded plants (direct DNRR, Fig. 4.1a). In many tree species, adventitious shoots regenerate after trimming the tree crown (direct DNSR, Fig. 4.1b).

Indirect *de novo* organ regeneration occurs in *in vitro* tissue culture. In tissue culture, adventitious roots and shoots do not directly regenerate from detached explants; instead, explants are first cultured on auxin-rich callus-inducing medium (CIM) to form a group of fast-dividing cells known as callus (Fig. 4.1c, d). The callus can then be cultured on root-inducing medium (RIM) with a low auxin level or no auxin to form adventitious roots (indirect DNRR, Fig. 4.1e). Alternatively, callus can be cultured on cytokinin-rich shoot-inducing medium (SIM) to induce the formation of adventitious shoots (indirect DNSR, Fig. 4.1f).

De novo organ regeneration in tissue culture has been studied for a long time (Sussex, 2008). Gottlieb Haberlandt was the first scientist to culture isolated somatic cells of higher plants *in vitro* (Haberlandt, 1902). Successes in the development of *in vitro* tissue culture were reported in 1939, and Philip White defined plant tissue culture as a system in which cells remained "undifferentiated" and showed "unlimited growth" (Gautheret, 1939; Nobécourt, 1939; White, 1939). In the 1950s, Folke Skoog and Carlos Miller found that different concentrations of auxin and cytokinin in the medium were the key to induce the formation of callus, roots, and shoots (Skoog & Miller, 1957). This discovery laid the foundation for the exploitation of *de novo* organ regeneration in tissue culture.

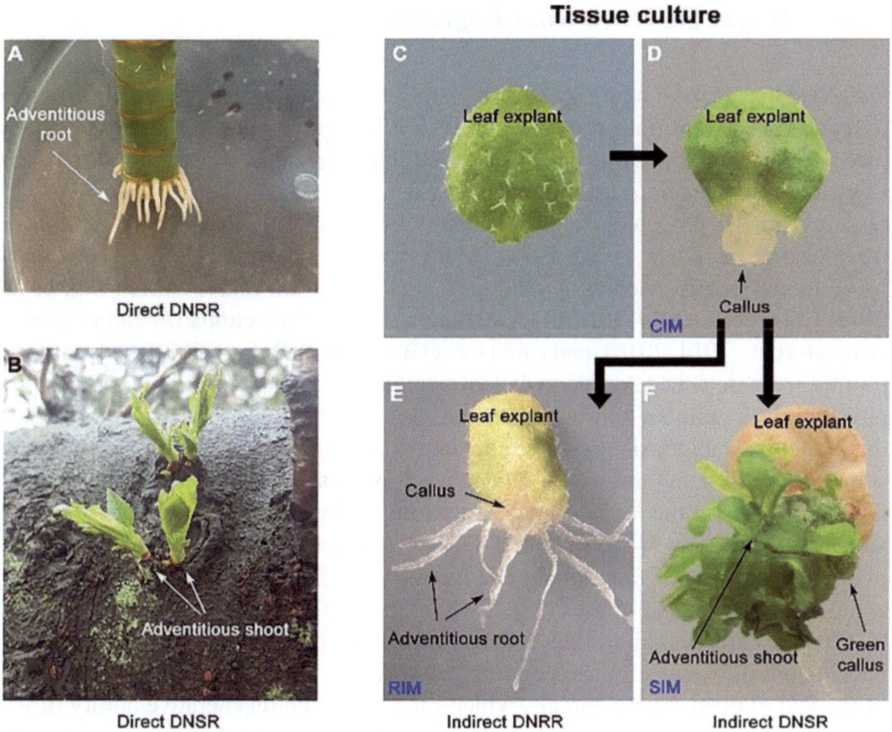

Fig. 4.1 *De novo* organ regeneration in seed plants. (**a**) Direct *de novo* root regeneration (DNRR) in a cutting (*Dracaena sanderiana*). (**b**) Direct *de novo* shoot regeneration (DNSR) on a tree (*Cerasus*). (**c–f**) Indirect DNRR and DNSR from Arabidopsis leaf explants in tissue culture

At the cellular level, *de novo* organ regeneration is based on the pluripotency of plant cells and reflects plant cell flexibility (Sugimoto et al., 2011; Xu & Huang, 2014). Here we summarize the current knowledge and recent progress in research on *de novo* organ regeneration, mainly focusing on Arabidopsis (*Arabidopsis thaliana*) and other model plants. This review updates our understanding of how plants regenerate new organs.

4.2 Direct *De Novo* Organ Regeneration

Direct DNRR and DNSR are important for plants to survive after tissue damage in natural conditions, and this regenerative ability has been exploited in many agricultural applications, such as cuttings (Morgan, 1900; Vöchting, 1885; Birnbaum & Sanchez Alvarado, 2008; Xu & Huang, 2014; Bellini et al., 2014; Druege et al., 2016; Ikeuchi et al., 2019).

4.2.1 Direct De Novo *Root Regeneration*

Detached leaves from many species (e.g., those in the Crassulaceae family) are able to regenerate adventitious roots for survival. When cuttings are taken from plants, a piece of the stem or a leaf is cut and inserted into wet soil or water to allow adventitious roots to regenerate from the wound site (Fig. 4.1a). There is a long history of research on the physiological mechanisms of the DNRR that occurs in cuttings (reviewed in de Klerk et al., 1999). Many rooting systems have been developed using model plants, such as Arabidopsis (Falasca et al., 2004; Correa Lda et al., 2012; Chen et al., 2014; Bustillo-Avendano et al., 2018), petunia (*Petunia hybrida*) (Druege et al., 2014, 2016), and some trees (Bannoud & Bellini, 2021). Adventitious rooting from detached Arabidopsis leaf explants is a simple system to analyze the physiological, cellular, and molecular mechanisms of direct DNRR (Fig. 4.2a). In this system, Arabidopsis leaves are detached and cultured on B5 medium without added hormones, and the whole regeneration process is dependent on the endogenous hormones in the leaf explants. Based on studies using this system, a cellular and molecular framework of direct DNRR has been proposed. This framework involves two types of functional cells (converter cells and regeneration-competent cells) and at least three phases (Xu, 2018) (Fig. 4.2b).

Phase I, early signaling: Upon detachment, leaf explants sense many early signals including wound signals, environmental signals, and the developmental status of the leaf explant. These signals, which may affect the regenerative ability of leaf explants, are sensed by converter cells, i.e., mesophyll cells, leaf margin cells, and some vascular cells. Wounding is the first event in regeneration. Wounding causes at least two consequences in the leaf explant. First, wounding prevents the exchange of many molecules (e.g., hormones, carbohydrates, and some metabolic products) between the explant and the mother plant. Second, wounding release many wound signals, which regulate regeneration in the explant. However, our knowledge of their signaling pathways is limited (Chen et al., 2016a). Some physical and chemical events occur upon wounding, such as changes in the plasma transmembrane potential and the reactive oxygen species (ROS) concentration. These changes are followed by the rapid accumulation of hormones such as jasmonates (JA) and ethylene. During adventitious rooting from petunia cuttings, JA was shown to be quickly accumulated after wounding and might promote the auxin peak at the wound site (Ahkami et al., 2009; Druege et al., 2016). Therefore, it is possible that JA may serve as one of the wound signals in cuttings. Recent studies in DNRR from Arabidopsis leaves showed that JA accumulation occurs after wounding within minutes, and the JA signaling pathway promotes auxin biosynthesis during DNRR (Zhang et al., 2019). Although many physical, chemical, and hormonal pathways, including JA and ethylene, have been reported to be involved in DNRR (Falasca et al., 2004; Lanteri et al., 2006; Lin et al., 2014; Bai et al., 2012; Liu et al., 2013, 2022; Druege et al., 2014, 2016; Veloccia et al., 2016; Ahkami et al., 2009; Lischweski et al., 2015; Fattorini et al., 2009; Gutierrez et al., 2012; da Costa et al., 2013; Zhang et al., 2019), how they serve as wound signals and how the wound

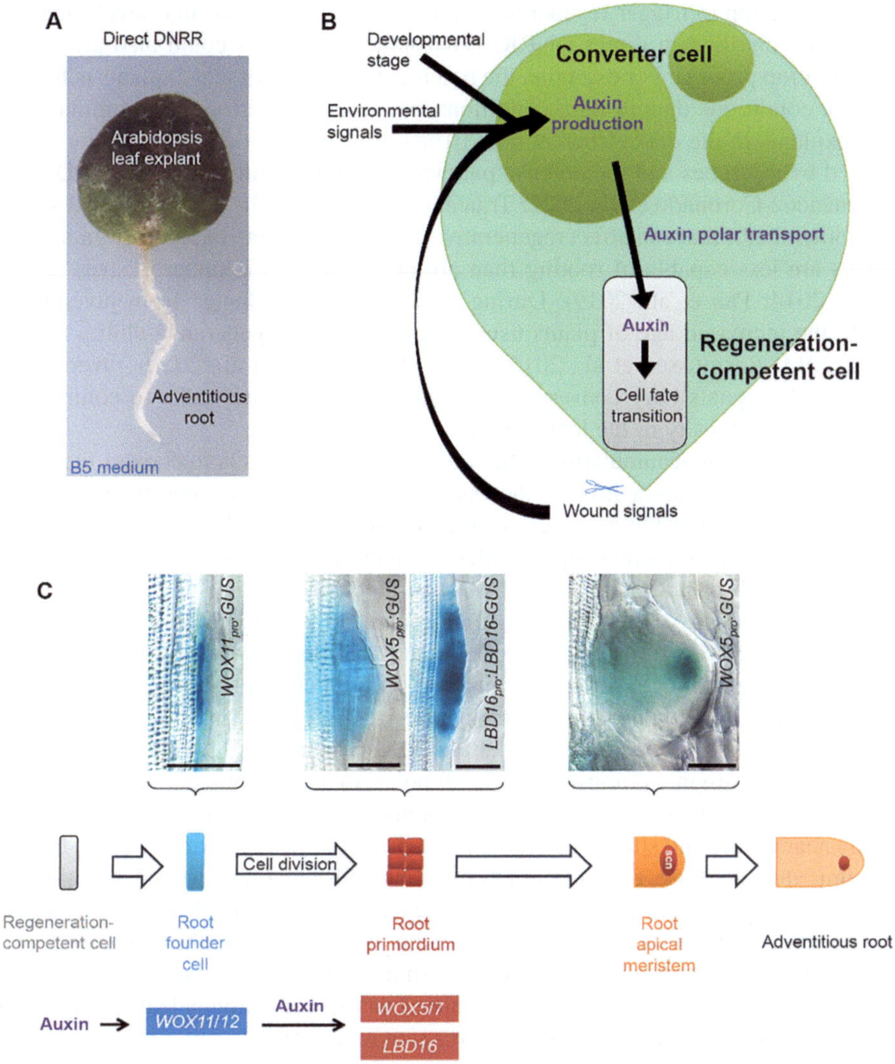

Fig. 4.2 Cellular and molecular framework of direct *de novo* root regeneration (DNRR) from Arabidopsis leaf explants. (**a**) Direct DNRR system from Arabidopsis leaf explant. (**b**) Schematic of direct DNRR process, involving converter cell and regeneration-competent cell. (**c**) Cell lineage and molecular markers for cell fate transition from the regeneration-competent cell to a root. GUS staining shows expression patterns of *WOX11* in the root founder cell, *WOX5* and *LBD16* in the root primordium, and *WOX5* in the stem cell niche (scn) of the root apical meristem. $WOX11_{pro}$:*GUS*, $WOX5_{pro}$:*GUS* and $LBD16_{pro}$:*LBD16-GUS* are in the wild-type Col-0 background. Scale bars, 50 μm in (**c**)

signals function to trigger regeneration at the cellular and molecular levels need to be further clarified in direct DNRR. Environmental signals could also affect the regeneration process. For example, the regenerative ability of leaf explants is higher in dark conditions than in light conditions, because auxin strongly accumulates in leaf explants in the dark (Chen et al., 2016b). Besides abiotic factors, biotic stress caused by microbes and the defense pathway could affect the efficiency of DNRR (Hernández-Coronado et al., 2022; Tran et al., 2023a, b). The developmental status is another factor that can affect regenerative ability. In general, explants from mature leaves are less capable of rooting than are explants from immature leaves (Chen et al., 2014; Pan et al., 2019). During vegetative phase change from juvenile to adult, the increased age of plants usually reduces their regenerative ability (Diaz-Sala, 2014; Rasmussen et al., 2015; Xu et al., 2016; Ye et al., 2020). Overall, all those early signals may converge as an input into converter cells to control the regenerative intensity of the leaf explant.

Phase II, auxin accumulation: The role of converter cells is to convert the input early signals into auxin flux as the output. Auxin is the key hormone in direct DNRR. Shortly after the discovery of the chemical nature of auxin (i.e., indole-3-acetic acid, IAA), scientists reported that the natural auxin and its synthetic forms, such as α-naphthaleneacetic acid (NAA) and indole-3-butyric acid (IBA), are able to promote rooting (Hitchcock & Zimmerman, 1936; Thimann & Went, 1934; Zimmerman & Wilcoxon, 1935). Therefore, auxin or its synthetic forms are usually included in the rooting powder to promote rooting in cuttings. During DNRR from Arabidopsis leaf explants, auxin is biosynthesized in converter cells guided by the early signals mentioned above (Liu et al., 2014; Chen et al., 2016b; Zhang et al., 2019). Then, auxin is transported in a polar manner from converter cells to regeneration-competent cells (i.e., procambium and some vascular parenchyma cells) near the wound site in the leaf explant (Liu et al., 2014). Auxin is then highly accumulated in those regeneration-competent cells near the wound site, where auxin cannot be further transported out of the leaf explant probably due to the physical barrier caused by the wounding. Blocking the auxin biosynthesis pathway and the auxin polar transport pathway can abolish the regeneration of adventitious roots from leaf explants, indicating that endogenous auxin production in leaf explants and transport to the would site are essential for root regeneration.

Phase III, cell fate transition: When auxin is transported into regeneration-competent cells, auxin triggers cell fate transition through its signaling pathway. The fate transition from a regeneration-competent cell to a mature root can be separated into four successive steps (Hu & Xu, 2016; Xu, 2018) (Fig. 4.2c). The first step of cell fate transition is priming of the regeneration-competent cell to become the root founder cell. At the molecular level, the auxin signaling pathway activates the expression of two partially redundant transcription factor genes; *WUSCHEL RELATED HOMEOBOX 11* and *12* (*WOX11/12*) (Liu et al., 2014) (Fig. 4.2c). This is a preparation step and no cell division occurs. The second step is initiation of the root primordium from the root founder cell by cell division. In this step, *WOX11/12*, together with the auxin signaling pathway, activate the expression of *WOX5/7*, *LATERAL ORGAN BOUNDARIES DOMAIN16* (*LBD16*), and *RGF1 INSENSITIVE*

1 and *2* (*RGF1/2*) (Liu et al., 2014; Hu & Xu, 2016; Zhang et al., 2023). Co-expression of *WOX5/7* and *LBD16* might serve as a molecular marker of root primordium cells (Fig. 4.2c). Auxin levels remain high in the root primordium cells. The third step is patterning of the root apical meristem (RAM) that arose from the root primordium by continuous cell division. In this step, *LBD16* expression decreases and *WOX5/7* expression is gradually restricted to the stem cell niche within the tip region of the RAM (Hu & Xu, 2016) (Fig. 4.2c). The root primordium cells continue to divide, and different functional regions gradually form. This could result from the increased cytokinin level, which restricts auxin distribution in the stem cell niche (Druege et al., 2016; de Klerk et al., 1999; Della Rovere et al., 2013, 2015). The fourth step is the emergence of the mature root tip with a mature RAM from the leaf explant (Chen et al., 2016c).

Overall, direct DNRR from Arabidopsis leaf explants is a process that requires the cooperation of converter cells and regeneration-competent cells. Converter cells sense multiple early signals and then produce an auxin flux towards the regeneration-competent cell for the fate transition to become a root.

The molecular pathway of DNRR from leaf explants is different from that in lateral root formation. When Arabidopsis is vertically grown on the medium, the lateral root formation does not require *WOX11* (Sheng et al., 2017). Thus, lateral roots are formed through the non-*WOX11*-mediated rooting pathway, in which an oscillatory root cap-derived auxin flux is transported into the xylem-pole pericycle cells and initiates the lateral root primordium via the auxin signaling pathway that involves IAA14 and AUXIN RESPONSE FACTOR7 and 19 (ARF7/19) (Peret et al., 2009; Lavenus et al., 2013; Moreno-Risueno et al., 2010; Van Norman et al., 2014; Xuan et al., 2015, 2016; Okushima et al., 2007; Lee et al., 2009; Goh et al., 2012; Fukaki et al., 2002). ARF7/19 directly activates the expression of *LBD16* and *RGI*s to initiate the lateral root primordium. In contrast to lateral rooting, the ARF6/8-WOX11 complex activates the expression of *LBD16* and *RGI*s to initiate the adventitious root primordium (Sheng et al., 2017; Zhang et al., 2023). Interestingly, when the primary root is wounded, it can initiate “adventitious lateral roots” via the *WOX11*-mediated rooting pathway, suggesting that the root system formation is highly plastic (Sheng et al., 2017).

4.2.2 *Direct* De Novo *Shoot Organogenesis*

Direct DNSR is commonly seen in tree stems upon wounding or stress (Fig. 4.1b). However, details of the cell lineages involved in direct DNSR and its underlying molecular mechanisms are largely unclear. Adventitious shoot formation from the stem is different from lateral shoot formation. First, lateral shoots usually initiate from the adaxial side of the leaf axillary, i.e., the junction between the leaf and the stem; while adventitious shoots usually form upon wounding or stress and are not restricted to a specific location. Second, lateral shoots bulge together with the epidermal cells, and thus, they do not burst out of the epidermis, while adventitious shoots seem to initiate from the vasculature and burst out of the epidermis when they emerge. A study on peach palm (*Bactris gasipaes* Kunth) shoot apex explants indicated that adventitious shoots are initiated from the pre-procambium (de Almeida et al., 2012). Third, the lateral shoot meristem is likely derived from a cell lineage originally from the peripheral zone of the SAM. Leaf axil cells express *SHOOT MERISTEMLESS* (*STM*) and maintain stem cell activity to initiate lateral shoots (known as the "detached meristem" model) (reviewed in Yang & Jiao, 2016; Wang & Jiao, 2018), while adventitious shoot formation from the stem seems to follow a *de novo* pattern.

A system in which shoots form directly from the Arabidopsis lateral root primordium has been established (Chatfield et al., 2013; Kareem et al., 2016b; Rosspopoff et al., 2017). Induction of shoots from the lateral root primordium requires supplementation with cytokinin, which activates the expression of *WUSCHEL* (*WUS*) to reprogram the lateral root primordium cells into shoot progenitor cells. Further analyses suggested that the lateral root primordium has higher competence for shoot formation at certain stages in response to cytokinin treatment. Interestingly, the newly formed shoot stem cells can revert to root stem cells under auxin treatment, suggesting that root–shoot reprogramming is relatively flexible and is controlled by cytokinin for shooting and by auxin for rooting (Rosspopoff et al., 2017).

Some seed plants could undergo DNSR in the absence of exogenous hormones. For example, in the leaf fragment of North American lake cress, *Rorippa aquatica*, both DNRR and DNSR occurred from the proximal side of the cut edge (Amano et al., 2020). In tomato, adventitious shoots could be regenerated from the wounded hypocotyls or stems (Plana et al., 2006; Larriba et al., 2022).

Further studies on direct DNSR based on an efficient system that mimics natural conditions without exogenous hormone treatment, e.g., adventitious shoot regeneration from tomato hypocotyls (Plana et al., 2006; Larriba et al., 2022), could improve our understanding of adventitious shooting and clarify the links between direct and indirect DNSR.

4.3 Indirect *De Novo* Organ Regeneration in Tissue Culture

In tissue culture, indirect *de novo* organ regeneration experiences at least two phases. In phase 1, i.e., pluripotency acquisition (also known as acquisition of competence for shooting or rooting), explants are first cultured on CIM to induce callus formation. In phase 2, i.e., shoot or root organogenesis, adventitious shoots or roots form from callus on SIM or RIM, respectively.

Callus contains pluripotent stem cells that are able to form either shoots or roots. The direction of the cell fate is determined by hormones: a high level of exogenous cytokinin leads to the shoot fate, while a decrease in exogenous auxin leads to the root fate.

4.3.1 Callus Formation

Many types of calli can form in tissue culture and also under natural conditions. The callus formed on CIM for *de novo* organ regeneration discussed here resembles the root primordium or the RAM. With the new understanding of the cellular features of callus, research on *de novo* organ regeneration has entered a new period.

The traditional view of callus formation was that differentiated somatic cells become callus via dedifferentiation stimulated by hormones (reviewed in de Klerk et al., 1997). In recent years, studies on callus at the cellular and molecular levels have shed light on the framework of callus initiation and revised the traditional view. Histological studies on Arabidopsis callus initiation indicated that callus initiates from xylem-pole pericycle cells in root explants (Che et al., 2007; Atta et al., 2009) and procambium and some vascular parenchyma cells (previously indicated as "pericycle-like cells") in leaf explants (Yu et al., 2010; Liu et al., 2014; Sugimoto et al., 2010; Hu et al., 2017). The xylem-pole pericycle cells are those that can initiate lateral roots. Procambium cells and some vascular parenchyma cells are adult stem cells in the vasculature that develop into veins or initiate adventitious roots in direct DNRR. Therefore, those adult stem cells (e.g., xylem-pole pericycle cells in roots and procambium and some vascular parenchyma cells in leaves) may serve as regeneration-competent cells, which are involved in lateral or adventitious rooting or vasculature development during normal plant development but are ready to initiate callus upon wounding and hormone treatment in tissue culture. Based on these findings, it is proposed that not all somatic cells but only regeneration-competent cells can initiate callus, and this fate transition is more like transdifferentiation (Sugimoto et al., 2011; Xu & Huang, 2014). Many differentiated somatic cells are not capable of forming callus, and this contradicts the traditional view that differentiated somatic cells can dedifferentiate to become competent for callus initiation.

An important study using genetic and transcriptome analyses demonstrated that callus formation follows the rooting pathway, regardless of the source of explant. This provided a novel understanding of the cellular and molecular features of callus (Sugimoto et al., 2010). At the chromatin level, the leaf-to-callus transition was shown to be a process of H3K27me3 (a repressive chromatin marker for gene silencing) removal from the genomic sites of root-related genes, thus allowing for the expression of root-related genes in callus cells. In addition, *LBD* genes, which are involved in root primordium initiation, were reported to be essential for callus formation. These findings confirmed that the rooting pathway is required for callus formation (He et al., 2012; Fan et al., 2012). Further studies using molecular markers suggested that the cell fate transition during callus initiation from leaf explants is similar to that in adventitious root regeneration from leaf explants (see direct DNRR above) (Liu et al., 2014). Single-cell RNA sequencing analysis revealed that there are three cell layers in callus and the middle layer with the quiescent center (QC)-like identity could be pluripotent stem cells for organ regeneration (Zhai & Xu, 2021; Zhai et al., 2023). This differs from the traditional view that callus is a group of dedifferentiated or undifferentiated cells (Sugimoto et al., 2011).

The cell fate transition during callus initiation is similar to the rooting process. In Arabidopsis roots or hypocotyls, callus could be initiated via either non-*WOX11*-mediated lateral rooting pathway or *WOX11*-mediated rooting pathway. In other explants of Arabidopsis, such as leaves, callus initiation usually follows the *WOX11*-mediated rooting pathway. Using callus initiation from hypocotyl explants as an example, there are several steps of cell fate transition (Fig. 4.3): The first step (the priming step) is from the regeneration-competent cell (i.e., pericycle, procambium, or some vascular parenchyma cells) to the callus founder cell. *WOX11* is activated during this step, and its expression marks the formation of the callus founder cell (Liu et al., 2014; Zhai et al., 2023). In the second step (the initiation step), the callus founder cell undergoes anticlinal cell division and fate transition to become the callus primordium. Since the callus primordium is a group of root primordium-like cells, coexpression of *LBD16* and *WOX5/7* can be considered as a marker of the callus primordium (Fan et al., 2012; Yu et al., 2017; Zhai & Xu, 2021; Zhai et al., 2023). In addition, *PLETHORA1/2* (*PLT1/2*), and *SCARECROW* (*SCR*) are expressed in the callus primordium. In the third step (the patterning step), the callus primordium undergoes both anticlinal and periclinal cell division and forms three cell layers in mature callus. The outer cell layer of callus resembles the epidermis and the lateral root cap of the RAM, the middle cell layer of callus has the QC-like feature, and the inner cell layer of callus resembles the vascular initial cells of the RAM (Zhai & Xu, 2021; Zhai et al., 2023). In this step, *LBD16* expression decreases, while *WOX5/7*, *PLT1/2*, and *SCR* are highly accumulated in the middle cell layer (Atta et al., 2009; Kim et al., 2018; Liu et al., 2018; Zhai & Xu, 2021; Zhai et al., 2023).

There are several molecular pathways involved in callus formation. During the transition from callus founder cells to callus primordium, *WOX11* activates the expression of *LBD16* and *WOX5/7* (Fan et al., 2012; Liu et al., 2014, 2018; Hu et al., 2017; Sheng et al., 2017). LBDs could form the protein complex with bZIP59 in callus initiation (Xu et al., 2018). *PLT3/5/7*, which are involved in both lateral rooting and adventitious rooting (Du & Scheres, 2017; Hofhuis et al., 2013; Krizek, 2009; Liu et al., 2022), are activated by the auxin signal pathway through WRKY23 and are expressed in all steps of callus formation (Zhai et al., 2023; Xu et al., 2023) (Fig. 4.3). *PLT3/5/7* promote the expression of *WOX5* and *PLT1/2* (Du & Scheres, 2017; Kareem et al., 2015; Zhai et al., 2023), while bHLH041 represses the expression of *WOX5* and *PLT1/2* (Xu et al., 2023). The pathway involving *SCR* and *SHOOT-ROOT* (*SHR*) are also critically required for the QC-like indentity sepcification in callus (Cui et al., 2007; Kim et al., 2018; Zhai et al., 2023). The very-long-chain fatty acids (VLCFA) biosynthesis pathway might also be involved in callus formation, because loss of VLCFA may lead to more callus formation on CIM (Shang et al., 2016). In addition, many chromatin factors and epigenetic pathways are also involved in callus formation, pluripotency acquisition, and organ regeneration in tissue culture (Li et al., 2011; Kim et al., 2018; Lee & Seo, 2018; Ishihara et al., 2019; Yan et al., 2020).

The cellular and molecular events leading to the acquisition of pluripotency in the callus include formation of the root primordium-like or RAM-like features, with

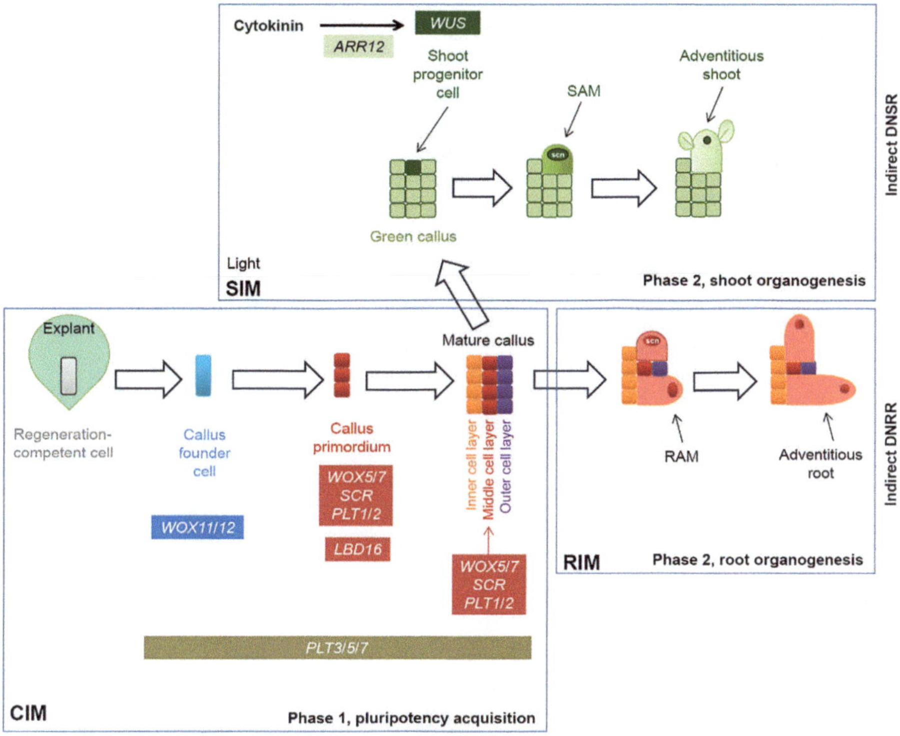

Fig. 4.3 Framework of indirect *de novo* organ regeneration in tissue culture. Cell lineage and molecular markers in indirect *de novo* organ regeneration in tissue culture, involving two phases. RP root primordium, RAM root apical meristem, SAM shoot apical meristem, scn stem cell niche

the pluripotent stem cells in the middle cell layer with QC-like identity. The root primordium, like callus, is able to directly form shoots upon cytokinin induction (Chatfield et al., 2013; Kareem et al., 2016b; Rosspopoff et al., 2017) (see direct DNSR above). Many rooting-related genes are essentially involved in the acquisition or maintenance of pluripotency. For examples, blocking of the *WOX11-LBD16* pathway, *PLT3/5/7* pathway, or the *SCR-SHR* pathway could result in the loss of pluripotency in the callus, therefore causing defects in shooting and/or rooting from callus (Liu et al., 2018; Kareem et al., 2015; Kim et al., 2018; Zhai & Xu, 2021; Zhai et al., 2023) (see the *lbd16* mutant as an example in Fig. 4.4).

The ability to form callus in tissue culture can differ among plant species. For example, the eudicot Arabidopsis is able to produce callus from the whole leaf at any developmental stage. However, mature leaves of monocot cereals (such as rice, maize, and wheat) are extremely insensitive to tissue culture conditions (Bhojwani et al., 1977; Cutler et al., 1991) (Fig. 4.5a, b). Only the very proximal region of cereal leaves is able to produce callus (Wernicke et al., 1981; Zamora & Scott, 1983; Wernicke & Milkovits, 1984; Cutler et al., 1991; Becher et al., 1992; Chen et al. 1995a, b; Ahmadabadi et al., 2007) (Fig. 4.5a). A recent study using rice and maize

Fig. 4.4 The *LBD16*-mediated rooting pathway is required for DNSR in tissue culture. (**a**, **b**) Shoot regeneration on green callus from hypocotyl explants of the wild-type Col-0 (**a**) and the *lbd16–2* mutant (**b**) cultured on CIM for 4 d and SIM for 14 d. Note that the *lbd16–2* green callus has defect in shoot organogenesis (Liu et al., 2018). Scale bars, 5 mm (**a**, **b**)

leaf and root explants showed that callus initiated from the bundle sheath and some immature vascular cells (the procambium) at the proximal region of leaves, and from phloem-pole pericycle cells in roots (Hu et al., 2017; Zeng et al., 2016). Therefore, the regeneration-competent cells could be partially different between eudicots and monocots. Interestingly, these diverse regeneration-competent cells use a common molecular mechanism, i.e., the *WOX11-WOX5* pathway, for callus initiation (Hu et al., 2017; Guo et al., 2018). *WOX11* is a marker of founder cells and *WOX5* is a marker of newly formed callus during Arabidopsis and rice callus initiation. In cereals, leaf maturation begins from the tip region and immature tissues remain only at the very base of the leaf. During leaf maturation, all the regeneration-competent cells differentiate into functional cells. For example, the bundle sheath in maize leaves differentiates into the Kranz anatomy for photosynthesis, and the outer bundle sheath in rice leaves differentiates into large parenchyma cells during leaf maturation (Zeng et al., 2016). Therefore, depletion of those cells during leaf

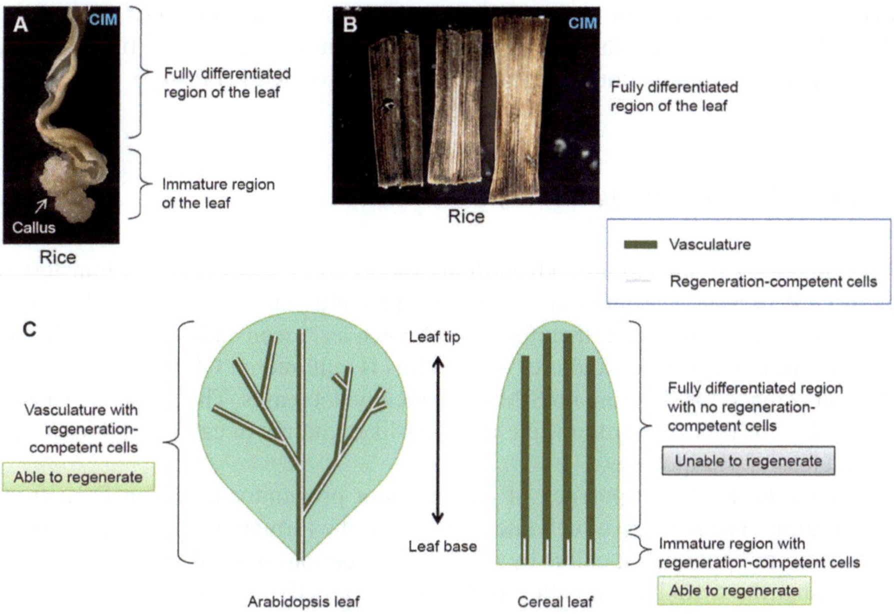

Fig. 4.5 Regenerative abilities in cereals and Arabidopsis. (**a, b**) Tissue culture of rice leaf on CIM for 2 weeks. Note that only the leaf base region (immature region) can produce callus (**a**), and the mature region is not able to regenerate (**a, b**). (**c**) Distribution of regeneration-competent cells (i.e., adult stem cells in vasculature) in the eudicot Arabidopsis and monocot cereals

maturation eliminates the regenerative ability of mature cereal leaves, and only the base region (the immature region) can produce callus (Hu et al., 2017) (Fig. 4.5c). Unlike cereals, the Arabidopsis leaf retains procambium cells during its whole life, and is therefore ready for callus induction at any stage (Hu et al., 2017) (Fig. 4.5c). Overall, the differentiation status and the distribution of regeneration-competent cells in explants are key factors in the application of tissue culture.

4.3.2 Root Organogenesis from Callus

Usually, when the callus continues to be cultured on CIM, partial differentiation occurs. Transcripts of many root meristem genes can be detected in diverse regions of the partially differentiated callus (Sugimoto et al., 2010; Zhai & Xu, 2021; Zhai et al., 2023). One hypothesis is that this represents a balance between two opposing forces: the endogenous developmental process that triggers differentiation directs the callus to become the root tip-like structure; and the effect of the high exogenous auxin level maintains the callus with the root primordium-like or RAM-like identity and prevents it from differentiation (Yu et al., 2017). When callus is transferred to RIM to decrease or eliminate auxin in the medium, there is little or no auxin to maintain root primordium-like or RAM-like cellular features of callus. Therefore, the callus mass could finish the differentiation to form the mature root tip (Yu et al., 2017), resulting in numerous roots form from callus on RIM (Fig. 4.3).

4.3.3 Shoot Organogenesis from Callus

In Arabidopsis tissue culture, adventitious shoots form when callus is transferred from CIM to SIM (Duclercq et al., 2011). The callus on SIM is cultured in light conditions and turns green; therefore, it is known as "green callus" (Figs. 4.1f and 4.3). In Arabidopsis tissue culture, when callus is cultured on CIM for a long time, the competence for shooting on SIM dramatically decreases. This might be caused by the loss of root primordium-like or RAM-like features in callus, leading to the partial loss of pluripotency.

On the basis of pluripotency of callus, shoot progenitor cells can form from green callus. The key step of cell fate transition is the formation of shoot progenitor cells from the pluripotent callus by cytokinin induction. Several studies have suggested that the cytokinin signaling pathway can directly activate *WUS* expression via B-type ARABIDOPSIS RESPONSE REGULATORs (ARRs), among which *ARR12* might be the major gene (Dai et al., 2017; Meng et al., 2017; Zhang et al., 2017). *WUS* is a key gene that marks and controls shoot progenitor cells (Fig. 4.3) (Zhang et al., 2017). Mutations in *WUS* result in a loss of shoot regeneration (Zhang et al., 2017), and ectopic expression of *WUS* is sufficient to induce shooting from the root tip (Gallois et al., 2004). In addition to *WUS* expression, *CUP SHAPED COTELYDON2* (*CUC2*) is also important for the establishment of shoot progenitor cells (Motte et al., 2011; Kareem et al., 2015; Gordon et al., 2007), probably by activating cell wall remodeling surrounding the shoot progenitor neighbor cells (Varapparambath et al., 2022).

Many genes are involved in the process of *WUS* activation by B-type ARRs. HD-ZIP III family transcription factors can form a protein complex with B-type ARRs and then participate in *WUS* activation (Zhang et al., 2017). WOUND INDUCED DEDIFFERENTIATION1 (WIND1) activates *ENHANCER OF SHOOT REGENERATION1* (*ESR1*) expression, and this pathway is involved in the formation of shoot progenitor cells (Iwase et al., 2011, 2017; Banno et al., 2001; Ikeda et al., 2006; Matsuo et al., 2011). In addition, the age of the explant may affect shoot regeneration via the microRNA156-targeted SQUAMOSA PROMOTER BINDING PROTEIN-LIKE (SPL) pathway, in which SPLs interact with B-type ARRs and regulate *WUS* activation (Zhang et al., 2015). Some epigenetic chromatin factors may also be involved in *WUS* activation (Li et al., 2011; Zhang et al., 2017; Ishihara et al., 2019; Wu et al., 2022). *WOX13* negatively regulates the specification of shoot progenitor cells through promoting non-meristematic cell fate (Ogura et al., 2023).

Shoot progenitor cells undergo cell division and enter the patterning step to form the SAM (Fig. 4.3a). In this step, *WUS* expression is gradually restricted to the organizing center (OC) of the shoot stem cell niche, *CLV3* expression is activated in the shoot stem cells above the OC (Gordon et al., 2007; Qiao et al., 2012; Cheng et al., 2013; Zhang et al., 2015). The patterning of the SAM from shoot progenitor cells might involve the spatial distribution pattern of auxin surrounding the *WUS*-expressing region (Qiao et al., 2012; Cheng et al., 2013). Shoot progenitor cells seem to contain the precursor of the shoot apical stem cell niche in the SAM, just like root primordium cells probably contain the precursor of the root apical stem cell niche in the RAM. Finally, the adventitious shoot with a mature SAM emerges from the green callus.

4.4 Conclusion and Perspectives

The underlying theme of *de novo* organ regeneration is cell fate transition guided by injury-induced endogenous hormones in the plant body or exogenous hormones in the medium. In plant development, auxin is essential for the maintenance of the QC in the root apical stem cell niche (Sabatini et al., 1999), while cytokinin is required to maintain or establish the OC in the shoot apical stem cell niche (Gordon et al., 2009; Leibfried et al., 2005; Sablowski, 2009; Zhao et al., 2010; Wang et al., 2017). Consistent with their roles in root and shoot development, auxin and cytokinin are the two core hormones directing *de novo* organ regeneration, controlling root/callus initiation and shoot initiation, respectively. At the cellular level, exogenous or endogenous auxin and cytokinin change the fate of stem cells. Regeneration-competent cells, root/callus founder cells, root/callus primordium cells, the middle cell layer of callus, the root apical stem cell niche, shoot progenitor cells, and the shoot apical stem cell niche could be regarded as different stem cells at different stages or with different functions controlled by auxin and cytokinin at different concentrations and ratios. At the molecular level, many key transcription factors are involved in controlling cell fate in response to hormones. For example, WOX family genes are involved in controlling different types of stem cells, e.g., *WOX11/12* in root/callus founder cells, *WOX5/7* in root/callus primordium cells, the middle cell layer of callus, and the root apical stem cell niche, and *WUS* in shoot progenitor cells and the shoot apical stem cell niche. The dynamic expression of these transcription factors is controlled by the level of auxin and/or cytokinin in the cell, e.g., *WOX11/12* and *WOX5/7* are controlled by a high auxin level and *WUS* is controlled by a high cytokinin level. Therefore, these transcription factors are a part of the machinery linking hormonal signaling to cell fate transition. Overall, the transition of cell fate is guided by hormonally controlled transcription factors to achieve root or shoot organogenesis during regeneration.

Based on current knowledge, many interesting questions remain to be answered. How do wound signals function together with other environmental and developmental signals? How do hormones, transcription factors, epigenetic chromatin factors, and other factors cooperate to regulate stem cell fate transition and stem cell division? Are there other pathways or different types of calli that can be induced to undergo organogenesis, e.g., *WIND1*-regulated wounding-induced callus (Iwase et al., 2011)? Understanding the physiological, cellular, and molecular mechanisms of *de novo* organ regeneration will advance new technologies in plant engineering.

4.5 Index

Callus A group of fast-dividing cells in tissue culture or on plants after injury. Different types of calli have different cellular features. In tissue culture of seed plants, root primordium- or RAM-like callus is usually required for *de novo* organ regeneration.

Callus-inducing medium (CIM) A type of medium in tissue culture usually with a high level of auxin and a low level of cytokinin to induce callus formation from explants.

Callus primordium During callus formation on CIM, callus founder cells undergo anticlinal cell division to form callus primordium, which has similar gene expression patterns compared with the root primordium. Callus primordium can further undergo patterning to form the mature callus with three cell layers.

Converter cells Cells that can convert early signals (wound signals, environmental signals, and developmental status) into auxin flux. During direct DNRR from Arabidopsis leaf explants, mesophyll cells, leaf margin cells, and some vascular cells may serve as converter cells.

***De novo* organ regeneration** Adventitious root and/or shoot organogenesis from damaged plant bodies or detached explants. This process can occur directly from plant bodies or indirectly from callus in tissue culture, and includes *de novo* root regeneration (DNRR) and *de novo* shoot regeneration (DNSR).

Green callus Callus cultured on SIM in light conditions. When root primordium- or RAM-like callus is transferred onto SIM and cultured in the light, it becomes green and can regenerate adventitious shoots.

Plant regeneration The process of recovery from injury in plants. Seed plants can usually undergo at least three types of regeneration: tissue regeneration (tissue repair); *de novo* organ regeneration (*de novo* organogenesis); and somatic embryogenesis.

Regeneration-competent cell Cell with potential competence to change its fate and undergo division for regeneration. Regeneration-competent cells for *de novo* organ regeneration might have adult stem cell features.

Root/callus founder cell Activated form of a regeneration-competent cell that is ready to undergo division to form the root or callus primordium. The root/callus founder cell is the source cell of the whole root/callus during *de novo* organ regeneration.

Root-inducing medium (RIM) A type of medium in tissue culture usually with a low level of auxin or without auxin to induce root organogenesis from callus.

Root primordium In *de novo* organ regeneration, the root primordium is a group of pluripotent cells that can both differentiate into a root or transdifferentiate into shoot progenitor cells for shoot formation. Co-expression of *WOX5/7* and *LBD16* is a molecular marker for the root primordium. The root primordium cell might be regarded as the rudimentary form or the precursor of the root tip stem cell niche.

Shoot-inducing medium (SIM) A type of medium in tissue culture usually with a high level of cytokinin and a low level of auxin to induce shoot organogenesis from callus.

Shoot progenitor cell Cell that is activated by cytokinin on SIM and is the source cell that differentiate and divide to become the SAM during DNSR. Shoot progenitor cell expresses *WUS* as a marker, and might be regarded as the rudimentary form or the precursor of the shoot apical stem cell niche.

Stem cell niche A group of highly organized cells that includes the organizer (QC in roots and OC in shoots) and its neighboring stem cells.

WOX family genes A group of transcription factor genes encoding homeodomains. These genes are usually expressed in different types of stem cells or stem cell niches.

Acknowledgments We thank Dr. Yuling Jiao for helpful discussion. We thank all members of the group for providing images and discussions. We apologize for references not cited due to space limitations. This work was supported by grants from the National Natural Science Foundation of China (32225007), the National Key R&D Program of China grant 2023YFE0101100, and the Strategic Priority Research Program of Chinese Academy of Sciences (Grant No. XDB0630000) to L.X., and the Young Scientists Fund of the National Natural Science Foundation of China (Grant No. 32300285) and China Postdoctoral Science Foundation (Grant No. 2023M733490) to N.Z.

References

Ahkami, A. H., Lischewski, S., Haensch, K. T., Porfirova, S., Hofmann, J., Rolletschek, H., Melzer, M., Franken, P., Hause, B., Druege, U., & Hajirezaei, M. R. (2009). Molecular physiology of adventitious root formation in *Petunia hybrida* cuttings: Involvement of wound response and primary metabolism. *The New Phytologist, 181*(3), 613–625.

Ahmadabadi, M., Ruf, S., & Bock, R. (2007). A leaf-based regeneration and transformation system for maize (*Zea mays* L.). *Transgenic Research, 16*(4), 437–448.

Amano, R., Nakayama, H., Momoi, R., Omata, E., Gunji, S., Takebayashi, Y., Kojima, M., Ikematsu, S., Ikeuchi, M., Iwase, A., Sakamoto, T., Kasahara, H., Sakakibara, H., Ferjani, A., & Kimura, S. (2020). Molecular basis for natural vegetative propagation via regeneration in north American Lake cress, *Rorippa aquatica* (Brassicaceae). *Plant & Cell Physiology, 61*(2), 353–369.

Atta, R., Laurens, L., Boucheron-Dubuisson, E., Guivarc'h, A., Carnero, E., Giraudat-Pautot, V., Rech, P., & Chriqui, D. (2009). Pluripotency of Arabidopsis xylem pericycle underlies shoot regeneration from root and hypocotyl explants grown *in vitro*. *The Plant Journal, 57*(4), 626–644.

Bai, X., Todd, C. D., Desikan, R., Yang, Y., & Hu, X. (2012). *N*-3-oxo-decanoyl-L-homoserine-lactone activates auxin-induced adventitious root formation via hydrogen peroxide- and nitric oxide-dependent cyclic GMP signaling in mung bean. *Plant Physiology, 158*(2), 725–736.

Banno, H., Ikeda, Y., Niu, Q. W., & Chua, N. H. (2001). Overexpression of Arabidopsis *ESR1* induces initiation of shoot regeneration. *Plant Cell, 13*(12), 2609–2618.

Bannoud, F., & Bellini, C. (2021). Adventitious rooting in *populus* species: Update and perspectives. *Frontiers in Plant Science, 12*, 668837.

Becher, T., Haberland, G., & Koop, H. U. (1992). Callus formation and plant regeneration in standard and microexplants from seedlings of barley (*Hordeum vulgare* L.). *Plant Cell Reports, 11*(1), 39–43.

Bellini, C., Pacurar, D. I., & Perrone, I. (2014). Adventitious roots and lateral roots: Similarities and differences. *Annual Review of Plant Biology, 65*, 639–666.

Bhojwani, S. S., Evans, P. K., & Cocking, E. C. (1977). Protoplast technology in relation to crop plants: Progress and problems. *Euphytica, 26*, 343–360.

Birnbaum, K. D., & Sanchez Alvarado, A. (2008). Slicing across kingdoms: Regeneration in plants and animals. *Cell, 132*(4), 697–710.

Bustillo-Avendano, E., Ibanez, S., Sanz, O., Sousa Barros, J. A., Gude, I., Perianez-Rodriguez, J., Micol, J. L., Del Pozo, J. C., Moreno-Risueno, M. A., & Perez-Perez, J. M. (2018). Regulation of hormonal control, cell reprogramming, and patterning during de novo root organogenesis. *Plant Physiology, 176*(2), 1709–1727.

Chatfield, S. P., Capron, R., Severino, A., Penttila, P. A., Alfred, S., Nahal, H., & Provart, N. J. (2013). Incipient stem cell niche conversion in tissue culture: Using a systems approach to probe early events in *WUSCHEL*-dependent conversion of lateral root primordia into shoot meristems. *The Plant Journal, 73*(5), 798–813.

Che, P., Lall, S., & Howell, S. H. (2007). Developmental steps in acquiring competence for shoot development in *Arabidopsis* tissue culture. *Planta, 226*(5), 1183–1194.

Chen, H., Xu, G., Loschke, D. C., Tomaska, L., & Rolfe, B. G. (1995a). Efficient callus formation and plant regeneration from leaves of oats (*Avena sativa* L.). *Plant Cell Reports, 14*(6), 393–397.

Chen, Z., Zhuge, Q., & Sundqvist, C. (1995b). Oat leaf base: Tissue with an efficient regeneration capacity. *Plant Cell Reports, 14*(6), 354–358.

Chen, X., Qu, Y., Sheng, L., Liu, J., Huang, H., & Xu, L. (2014). A simple method suitable to study *de novo* root organogenesis. *Frontiers in Plant Science, 5*, 208.

Chen, L., Sun, B., Xu, L., & Liu, W. (2016a). Wound signaling: The missing link in plant regeneration. *Plant Signaling & Behavior, 11*(10), e1238548.

Chen, L., Tong, J., Xiao, L., Ruan, Y., Liu, J., Zeng, M., Huang, H., Wang, J. W., & Xu, L. (2016b). *YUCCA*-mediated auxin biogenesis is required for cell fate transition occurring during *de novo* root organogenesis in arabidopsis. *Journal of Experimental Botany, 67*(14), 4273–4284.

Chen, X., Cheng, J., Chen, L., Zhang, G., Huang, H., Zhang, Y., & Xu, L. (2016c). Auxin-independent *NAC* pathway acts in response to explant-specific wounding and promotes root tip emergence during de novo root organogenesis in arabidopsis. *Plant Physiology, 170*(4), 2136–2145.

Cheng, Z. J., Wang, L., Sun, W., Zhang, Y., Zhou, C., Su, Y. H., Li, W., Sun, T. T., Zhao, X. Y., Li, X. G., Cheng, Y., Zhao, Y., Xie, Q., & Zhang, X. S. (2013). Pattern of auxin and cytokinin responses for shoot meristem induction results from the regulation of cytokinin biosynthesis by AUXIN RESPONSE FACTOR3. *Plant Physiology, 161*(1), 240–251.

Correa Lda, R., Troleis, J., Mastroberti, A. A., Mariath, J. E., & Fett-Neto, A. G. (2012). Distinct modes of adventitious rooting in *Arabidopsis thaliana*. *Plant Biology (Stuttgart, Germany), 14*(1), 100–109.

Cui, H., Levesque, M. P., Vernoux, T., Jung, J. W., Paquette, A. J., Gallagher, K. L., Wang, J. Y., Blilou, I., Scheres, B., & Benfey, P. N. (2007). An evolutionarily conserved mechanism delim-

iting SHR movement defines a single layer of endodermis in plants. *Science, 316*(5823), 421–425.
Cutler, A. J., Saleem, M., & Wang, H. (1991). Cereal protoplast recalcitrance. *In Vitro Cellular & Developmental Biology. Plant, 27*, 104–111.
da Costa, C. T., de Almeida, M. R., Ruedell, C. M., Schwambach, J., Maraschin, F. S., & Fett-Neto, A. G. (2013). When stress and development go hand in hand: Main hormonal controls of adventitious rooting in cuttings. *Frontiers in Plant Science, 4*, 133.
Dai, X., Liu, Z., Qiao, M., Li, J., Li, S., & Xiang, F. (2017). ARR12 promotes *de novo* shoot regeneration in *Arabidopsis thaliana* via activation of *WUSCHEL* expression. *Journal of Integrative Plant Biology, 59*(10), 747–758.
de Almeida, M., de Almeida, C. V., Mendes Graner, E., Ebling Brondani, G., & Fiori de Abreu-Tarazi, M. (2012). Pre-procambial cells are niches for pluripotent and totipotent stem-like cells for organogenesis and somatic embryogenesis in the peach palm: A histological study. *Plant Cell Reports, 31*(8), 1495–1515.
de Klerk, G.-J., Arnholdt-Schmitt, B., Lieberei, R., & Neumann, K.-H. (1997). Regeneration of roots, shoots, and embryos: Physiological, biochemical and molecular aspects. *Biologia Plantarum, 39*(1), 53–66.
de Klerk, G.-J., Van der Krieken, W., & De Jong, J. C. (1999). The formation of adventitious roots: New concepts, new possibilities. *In Vitro Cellular & Developmental Biology. Plant, 35*, 189–199.
Della Rovere, F., Fattorini, L., D'Angeli, S., Veloccia, A., Falasca, G., & Altamura, M. M. (2013). Auxin and cytokinin control formation of the quiescent Centre in the adventitious root apex of arabidopsis. *Annals of Botany, 112*(7), 1395–1407.
Della Rovere, F., Fattorini, L., D'Angeli, S., Veloccia, A., Del Duca, S., Cai, G., Falasca, G., & Altamura, M. M. (2015). Arabidopsis SHR and SCR transcription factors and AUX1 auxin influx carrier control the switch between adventitious rooting and xylogenesis *in planta* and in *in vitro* cultured thin cell layers. *Annals of Botany, 115*(4), 617–628.
Diaz-Sala, C. (2014). Direct reprogramming of adult somatic cells toward adventitious root formation in forest tree species: The effect of the juvenile-adult transition. *Frontiers in Plant Science, 5*, 310.
Druege, U., Franken, P., Lischewski, S., Ahkami, A. H., Zerche, S., Hause, B., & Hajirezaei, M. R. (2014). Transcriptomic analysis reveals ethylene as stimulator and auxin as regulator of adventitious root formation in petunia cuttings. *Frontiers in Plant Science, 5*, 494.
Druege, U., Franken, P., & Hajirezaei, M. R. (2016). Plant hormone homeostasis, signaling, and function during adventitious root formation in cuttings. *Frontiers in Plant Science, 7*, 381.
Du, Y., & Scheres, B. (2017). PLETHORA transcription factors orchestrate *de novo* organ patterning during *Arabidopsis* lateral root outgrowth. *Proceedings of the National Academy of Sciences of the United States of America, 114*(44), 11709–11714.
Duclercq, J., Sangwan-Norreel, B., Catterou, M., & Sangwan, R. S. (2011). *De novo* shoot organogenesis: From art to science. *Trends in Plant Science, 16*(11), 597–606.
Falasca, G., Zaghi, D., Possenti, M., & Altamura, M. M. (2004). Adventitious root formation in *Arabidopsis thaliana* thin cell layers. *Plant Cell Reports, 23*(1–2), 17–25.
Fan, M., Xu, C., Xu, K., & Hu, Y. (2012). LATERAL ORGAN BOUNDARIES DOMAIN transcription factors direct callus formation in *Arabidopsis* regeneration. *Cell Research, 22*(7), 1169–1180.
Fattorini, L., Falasca, G., Kevers, C., Rocca, L. M., Zadra, C., & Altamura, M. M. (2009). Adventitious rooting is enhanced by methyl jasmonate in tobacco thin cell layers. *Planta, 231*(1), 155–168.
Fukaki, H., Tameda, S., Masuda, H., & Tasaka, M. (2002). Lateral root formation is blocked by a gain-of-function mutation in the SOLITARY-ROOT/IAA14 gene of *Arabidopsis*. *The Plant Journal, 29*(2), 153–168.

Gallois, J. L., Nora, F. R., Mizukami, Y., & Sablowski, R. (2004). WUSCHEL induces shoot stem cell activity and developmental plasticity in the root meristem. *Genes & Development, 18*(4), 375–380.

Gautheret, R. J. (1939). Sur la possibilité de réaliser la culture indéfinie des tissus de tubercules de Carotte. *Comptes Rendus de l'Académie des Sciences, 208*, 118–130.

Goh, T., Joi, S., Mimura, T., & Fukaki, H. (2012). The establishment of asymmetry in *Arabidopsis* lateral root founder cells is regulated by LBD16/ASL18 and related LBD/ASL proteins. *Development, 139*(5), 883–893.

Gordon, S. P., Heisler, M. G., Reddy, G. V., Ohno, C., Das, P., & Meyerowitz, E. M. (2007). Pattern formation during *de novo* assembly of the *Arabidopsis* shoot meristem. *Development, 134*(19), 3539–3548.

Gordon, S. P., Chickarmane, V. S., Ohno, C., & Meyerowitz, E. M. (2009). Multiple feedback loops through cytokinin signaling control stem cell number within the *Arabidopsis* shoot meristem. *Proceedings of the National Academy of Sciences of the United States of America, 106*(38), 16529–16534.

Guo, F., Zhang, H. D., Liu, W., Hu, X. M., Han, N., Qian, Q., Xu, L., & Bian, H. W. (2018). Callus initiation from root explants employs different strategies in rice and Arabidopsis. *Plant & Cell Physiology, 59*(9), 1782–1789.

Gutierrez, L., Mongelard, G., Flokova, K., Pacurar, D. I., Novak, O., Staswick, P., Kowalczyk, M., Pacurar, M., Demailly, H., Geiss, G., & Bellini, C. (2012). Auxin controls *Arabidopsis* adventitious root initiation by regulating jasmonic acid homeostasis. *Plant Cell, 24*(6), 2515–2527.

Haberlandt, G. (1902). Kulturversuche mit isolierten Pflanzenzellen. *Sitz Akad Wiss Wien, 111*, 69–92.

He, C., Chen, X., Huang, H., & Xu, L. (2012). Reprogramming of H3K27me3 is critical for acquisition of pluripotency from cultured *Arabidopsis* tissues. *PLoS Genetics, 8*(8), e1002911.

Hernández-Coronado, M., Dias Araujo, P. C., Ip, P.-L., Nunes, C. O., Rahni, R., Wudick, M. M., Lizzio, M. A., Feijó, J. A., & Birnbaum, K. D. (2022). Plant glutamate receptors mediate a bet-hedging strategy between regeneration and defense. *Developmental Cell, 57*(4), 451–465.e456.

Hitchcock, A. E., & Zimmerman, P. W. (1936). Effect of the use of growth substances on the rooting response of cuttings. *Contributions from Boyce Thompson Institute, 8*, 63–79.

Hofhuis, H., Laskowski, M., Du, Y., Prasad, K., Grigg, S., Pinon, V., & Scheres, B. (2013). Phyllotaxis and rhizotaxis in *Arabidopsis* are modified by three PLETHORA transcription factors. *Current Biology, 23*(11), 956–962.

Hu, X., & Xu, L. (2016). Transcription factors WOX11/12 directly activate *WOX5/7* to promote root primordia initiation and organogenesis. *Plant Physiology, 172*(4), 2363–2373.

Hu, B., Zhang, G., Liu, W., Shi, J., Wang, H., Qi, M., Li, J., Qin, P., Ruan, Y., Huang, H., Zhang, Y., & Xu, L. (2017). Divergent regeneration-competent cells adopt a common mechanism for callus initiation in angiosperms. *Regeneration (Oxf), 4*(3), 132–139.

Ikeda, Y., Banno, H., Niu, Q. W., Howell, S. H., & Chua, N. H. (2006). The *ENHANCER OF SHOOT REGENERATION 2* gene in *Arabidopsis* regulates *CUP-SHAPED COTYLEDON 1* at the transcriptional level and controls cotyledon development. *Plant & Cell Physiology, 47*(11), 1443–1456.

Ikeuchi, M., Ogawa, Y., Iwase, A., & Sugimoto, K. (2016). Plant regeneration: Cellular origins and molecular mechanisms. *Development, 143*(9), 1442–1451.

Ikeuchi, M., Favero, D. S., Sakamoto, Y., Iwase, A., Coleman, D., Rymen, B., & Sugimoto, K. (2019). Molecular mechanisms of plant regeneration. *Annual Review of Plant Biology, 70*, 377–406.

Ishihara, H., Sugimoto, K., Tarr, P. T., Temman, H., Kadokura, S., Inui, Y., Sakamoto, T., Sasaki, T., Aida, M., Suzuki, T., Inagaki, S., Morohashi, K., Seki, M., Kakutani, T., Meyerowitz, E. M., & Matsunaga, S. (2019). Primed histone demethylation regulates shoot regenerative competency. *Nature Communications, 10*(1), 1786.

Iwase, A., Mitsuda, N., Koyama, T., Hiratsu, K., Kojima, M., Arai, T., Inoue, Y., Seki, M., Sakakibara, H., Sugimoto, K., & Ohme-Takagi, M. (2011). The AP2/ERF transcription factor WIND1 controls cell dedifferentiation in *Arabidopsis*. *Current Biology, 21*(6), 508–514.

Iwase, A., Harashima, H., Ikeuchi, M., Rymen, B., Ohnuma, M., Komaki, S., Morohashi, K., Kurata, T., Nakata, M., Ohme-Takagi, M., Grotewold, E., & Sugimoto, K. (2017). WIND1 promotes shoot regeneration through transcriptional activation of *ENHANCER OF SHOOT REGENERATION1* in Arabidopsis. *Plant Cell, 29*(1), 54–69.

Kareem, A., Durgaprasad, K., Sugimoto, K., Du, Y., Pulianmackal, A. J., Trivedi, Z. B., Abhayadev, P. V., Pinon, V., Meyerowitz, E. M., Scheres, B., & Prasad, K. (2015). *PLETHORA* genes control regeneration by a two-step mechanism. *Current Biology, 25*(8), 1017–1030.

Kareem, A., Radhakrishnan, D., Sondhi, Y., Aiyaz, M., Roy, M. V., Sugimoto, K., & Prasad, K. (2016a). *De novo* assembly of plant body plan: A step ahead of Deadpool. *Regeneration (Oxf), 3*(4), 182–197.

Kareem, A., Radhakrishnan, D., Wang, X., Bagavathiappan, S., Trivedi, Z. B., Sugimoto, K., Xu, J., Mähönen, A. P., & Prasad, K. (2016b). Protocol: A method to study the direct reprogramming of lateral root primordia to fertile shoots. *Plant Methods, 12*, 27.

Kim, J. Y., Yang, W., Forner, J., Lohmann, J. U., Noh, B., & Noh, Y. S. (2018). Epigenetic reprogramming by histone acetyltransferase HAG1/AtGCN5 is required for pluripotency acquisition in *Arabidopsis*. *The EMBO Journal, 37*(20), e98726.

Krizek, B. (2009). *AINTEGUMENTA* and *AINTEGUMENTA-LIKE6* act redundantly to regulate Arabidopsis floral growth and patterning. *Plant Physiology, 150*(4), 1916–1929.

Lanteri, M. L., Pagnussat, G. C., & Lamattina, L. (2006). Calcium and calcium-dependent protein kinases are involved in nitric oxide- and auxin-induced adventitious root formation in cucumber. *Journal of Experimental Botany, 57*(6), 1341–1351.

Larriba, E., Nicolás-Albujer, M., Sánchez-García, A. B., & Pérez-Pérez, J. M. (2022). Identification of transcriptional networks involved in de novo organ formation in tomato hypocotyl explants. *International Journal of Molecular Sciences, 23*(24), 16112.

Lavenus, J., Goh, T., Roberts, I., Guyomarc'h, S., Lucas, M., De Smet, I., Fukaki, H., Beeckman, T., Bennett, M., & Laplaze, L. (2013). Lateral root development in *Arabidopsis*: Fifty shades of auxin. *Trends in Plant Science, 18*(8), 450–458.

Lee, K., & Seo, P. J. (2018). Dynamic epigenetic changes during plant regeneration. *Trends in Plant Science, 23*(3), 235–247.

Lee, H. W., Kim, N. Y., Lee, D. J., & Kim, J. (2009). *LBD18/ASL20* regulates lateral root formation in combination with *LBD16/ASL18* downstream of *ARF7* and *ARF19* in Arabidopsis. *Plant Physiology, 151*(3), 1377–1389.

Leibfried, A., To JP, Busch, W., Stehling, S., Kehle, A., Demar, M., Kieber, J. J., & Lohmann, J. U. (2005). WUSCHEL controls meristem function by direct regulation of cytokinin-inducible response regulators. *Nature, 438*(7071), 1172–1175.

Li, W., Liu, H., Cheng, Z. J., Su, Y. H., Han, H. N., Zhang, Y., & Zhang, X. S. (2011). DNA methylation and histone modifications regulate *de novo* shoot regeneration in *Arabidopsis* by modulating *WUSCHEL* expression and auxin signaling. *PLoS Genetics, 7*(8), e1002243.

Lin, Y., Zhang, W., Qi, F., Cui, W., Xie, Y., & Shen, W. (2014). Hydrogen-rich water regulates cucumber adventitious root development in a heme oxygenase-1/carbon monoxide-dependent manner. *Journal of Plant Physiology, 171*(2), 1–8.

Lischweski, S., Muchow, A., Guthorl, D., & Hause, B. (2015). Jasmonates act positively in adventitious root formation in petunia cuttings. *BMC Plant Biology, 15*, 229.

Liu, R., Chen, S., Jiang, J., Zhu, L., Zheng, C., Han, S., Gu, J., Sun, J., Li, H., Wang, H., Song, A., & Chen, F. (2013). Proteomic changes in the base of chrysanthemum cuttings during adventitious root formation. *BMC Genomics, 14*, 919.

Liu, J., Sheng, L., Xu, Y., Li, J., Yang, Z., Huang, H., & Xu, L. (2014). *WOX11* and *12* are involved in the first-step cell fate transition during *de novo* root organogenesis in Arabidopsis. *Plant Cell, 26*(3), 1081–1093.

Liu, J., Hu, X., Qin, P., Prasad, K., Hu, Y., & Xu, L. (2018). The *WOX11-LBD16* pathway promotes pluripotency acquisition in callus cells during *de novo* shoot regeneration in tissue culture. *Plant & Cell Physiology, 59*(4), 734–743.

Liu, W., Zhang, Y., Fang, X., Tran, S., Zhai, N., Yang, Z., Guo, F., Chen, L., Yu, J., Ison, M. S., Zhang, T., Sun, L., Bian, H., Zhang, Y., Yang, L., & Xu, L. (2022). Transcriptional landscapes of *de novo* root regeneration from detached *Arabidopsis* leaves revealed by time-lapse and single-cell RNA sequencing analyses. *Plant Communications, 3*(4), 100306.

Lup, S. D., Tian, X., Xu, J., & Perez-Perez, J. M. (2016). Wound signaling of regenerative cell reprogramming. *Plant Science, 250*, 178–187.

Matsuo, N., Makino, M., & Banno, H. (2011). *Arabidopsis ENHANCER OF SHOOT REGENERATION* (*ESR*)*1* and *ESR2* regulate *in vitro* shoot regeneration and their expressions are differentially regulated. *Plant Science, 181*(1), 39–46.

Meng, W., Cheng, Z. J., Sang, Y. L., Zhang, M. M., Rong, X. F., Wang, Z. W., Tang, Y. Y., & Zhang, X. S. (2017). Type-B ARABIDOPSIS RESPONSE REGULATORs is critical to the specification of shoot stem cell niche by dual regulation of *WUSCHEL*. *Plant Cell, 29*(6), 1357–1372.

Moreno-Risueno, M. A., Van Norman, J. M., Moreno, A., Zhang, J., Ahnert, S. E., & Benfey, P. N. (2010). Oscillating gene expression determines competence for periodic *Arabidopsis* root branching. *Science, 329*(5997), 1306–1311.

Morgan, T. H. (1900). Regeneration in *Bipalium*. *Arch Entwick Mech, 9*, 563–586.

Motte, H., Verstraeten, I., Werbrouck, S., & Geelen, D. (2011). CUC2 as an early marker for regeneration competence in Arabidopsis root explants. *Journal of Plant Physiology, 168*(13), 1598–1601.

Nobécourt, P. (1939). Sur la pérennité et l'augmentation de volume des cultures de tissus végétaux. *Comptes Rendus Hebdomadaires de la Société de Biologie, 130*, 1270–1271.

Ogura, N., Sasagawa, Y., Ito, T., Tameshige, T., Kawai, S., Sano, M., Doll, Y., Iwase, A., Kawamura, A., Suzuki, T., Nikaido, I., Sugimoto, K., & Ikeuchi, M. (2023). WUSCHEL-RELATED HOMEOBOX 13 suppresses *de novo* shoot regeneration via cell fate control of pluripotent callus. *Science Advances, 9*(27), eadg6983.

Okushima, Y., Fukaki, H., Onoda, M., Theologis, A., & Tasaka, M. (2007). ARF7 and ARF19 regulate lateral root formation via direct activation of *LBD/ASL* genes in *Arabidopsis*. *Plant Cell, 19*(1), 118–130.

Pan, J., Zhao, F., Zhang, G., Pan, Y., Sun, L., Bao, N., Qin, P., Chen, L., Yu, J., Zhang, Y., & Xu, L. (2019). Control of *de novo* root regeneration efficiency by developmental status of *Arabidopsis* leaf explants. *Journal of Genetics and Genomics, 46*(3), 133–140.

Peret, B., De Rybel, B., Casimiro, I., Benkova, E., Swarup, R., Laplaze, L., Beeckman, T., & Bennett, M. J. (2009). *Arabidopsis* lateral root development: An emerging story. *Trends in Plant Science, 14*(7), 399–408.

Plana, D., Fuentes, A., Alvarez, M., Lara, R. M., Alvarez, F., & Pujol, M. (2006). A new approach for *in vitro* regeneration of tomato plants devoid of exogenous plant growth hormones. *Biotechnology Journal, 1*(10), 1153–1157.

Qiao, M., Zhao, Z., Song, Y., Liu, Z., Cao, L., Yu, Y., Li, S., & Xiang, F. (2012). Proper regeneration from *in vitro* cultured *Arabidopsis thaliana* requires the microRNA-directed action of an auxin response factor. *The Plant Journal, 71*(1), 14–22.

Rasmussen, A., Hosseini, S. A., Hajirezaei, M. R., Druege, U., & Geelen, D. (2015). Adventitious rooting declines with the vegetative to reproductive switch and involves a changed auxin homeostasis. *Journal of Experimental Botany, 66*(5), 1437–1452.

Rosspopoff, O., Chelysheva, L., Saffar, J., Lecorgne, L., Gey, D., Caillieux, E., Colot, V., Roudier, F., Hilson, P., Berthome, R., Da Costa, M., & Rech, P. (2017). Direct conversion of root primordium into shoot meristem relies on timing of stem cell niche development. *Development, 144*(7), 1187–1200.

Sabatini, S., Beis, D., Wolkenfelt, H., Murfett, J., Guilfoyle, T., Malamy, J., Benfey, P., Leyser, O., Bechtold, N., Weisbeek, P., & Scheres, B. (1999). An auxin-dependent distal organizer of pattern and polarity in the *Arabidopsis* root. *Cell, 99*(5), 463–472.

Sablowski, R. (2009). Cytokinin and *WUSCHEL* tie the knot around plant stem cells. *Proceedings of the National Academy of Sciences of the United States of America, 106*(38), 16016–16017.

Shang, B., Xu, C., Zhang, X., Cao, H., Xin, W., & Hu, Y. (2016). Very-long-chain fatty acids restrict regeneration capacity by confining pericycle competence for callus formation in *Arabidopsis*. *Proceedings of the National Academy of Sciences of the United States of America, 113*(18), 5101–5106.

Sheng, L., Hu, X., Du, Y., Zhang, G., Huang, H., Scheres, B., & Xu, L. (2017). Non-canonical WOX11-mediated root branching contributes to plasticity in *Arabidopsis* root system architecture. *Development, 144*(17), 3126–3133.

Skoog, F., & Miller, C. O. (1957). Chemical regulation of growth and organ formation in plant tissues cultured *in vitro*. *Symposia of the Society for Experimental Biology, 11*, 118–130.

Sugimoto, K., Jiao, Y., & Meyerowitz, E. M. (2010). *Arabidopsis* regeneration from multiple tissues occurs via a root development pathway. *Developmental Cell, 18*(3), 463–471.

Sugimoto, K., Gordon, S. P., & Meyerowitz, E. M. (2011). Regeneration in plants and animals: Dedifferentiation, transdifferentiation, or just differentiation? *Trends in Cell Biology, 21*(4), 212–218.

Sussex, I. M. (2008). The scientific roots of modern plant biotechnology. *Plant Cell, 20*(5), 1189–1198.

Thimann, K. V., & Went, E. W. (1934). On the chemical nature of the root forming hormone. *Proceedings of the Koninklijke Nederlandse Akademie van Wetenschappen. Series C, Biological and Medical Sciences, 37*, 456–459.

Tran, S., Ison, M., Ferreira Dias, N. C., Ortega, M. A., Chen, Y. S., Peper, A., Hu, L., Xu, D., Mozaffari, K., Severns, P. M., Yao, Y., Tsai, C. J., Teixeira, P., & Yang, L. (2023a). Endogenous salicylic acid suppresses *de novo* root regeneration from leaf explants. *PLoS Genetics, 19*(3), e1010636.

Tran, S., Stephanie Chen, Y. F., Xu, D., Ison, M., & Yang, L. (2023b). Microbial pattern recognition suppresses *de novo* organogenesis. *Development, 150*(9), dev201485.

Van Norman, J. M., Zhang, J., Cazzonelli, C. I., Pogson, B. J., Harrison, P. J., Bugg, T. D., Chan, K. X., Thompson, A. J., & Benfey, P. N. (2014). Periodic root branching in *Arabidopsis* requires synthesis of an uncharacterized carotenoid derivative. *Proceedings of the National Academy of Sciences of the United States of America, 111*(13), E1300–E1309.

Varapparambath, V., Mathew, M. M., Shanmukhan, A. P., Radhakrishnan, D., Kareem, A., Verma, S., Ramalho, J. J., Manoj, B., Vellandath, A. R., Aiyaz, M., Radha, R. K., Landge, A. N., Mähönen, A. P., Heisler, M. G., Weijers, D., & Prasad, K. (2022). Mechanical conflict caused by a cell-wall-loosening enzyme activates *de novo* shoot regeneration. *Developmental Cell, 57*(17), 2063–2080.e10.

Veloccia, A., Fattorini, L., Della Rovere, F., Sofo, A., D'Angeli, S., Betti, C., Falasca, G., & Altamura, M. M. (2016). Ethylene and auxin interaction in the control of adventitious rooting in *Arabidopsis thaliana*. *Journal of Experimental Botany, 67*(22), 6445–6458.

Vöchting, H. (1885). Über die regeneration der marchantieen. Jahrb f wiss Botanik XVI.

Wang, Y., & Jiao, Y. (2018). Axillary meristem initiation—A way to branch out. *Current Opinion in Plant Biology, 41*, 61–66.

Wang, J., Tian, C., Zhang, C., Shi, B., Cao, X., Zhang, T. Q., Zhao, Z., Wang, J. W., & Jiao, Y. (2017). Cytokinin signaling activates *WUSCHEL* expression during axillary meristem initiation. *Plant Cell, 29*(6), 1373–1387.

Wernicke, W., & Milkovits, L. (1984). Developmental gradients in wheat leaves – Response of leaf segments in different genotypes cultured *in vitro*. *Journal of Plant Physiology, 115*(1), 49–58.

Wernicke, W., Brettell, R., Wakizuka, T., & Potrykus, I. (1981). Adventitious embryoid and root formation from rice leaves. *Zeitschrift für Pflanzenphysiologie, 103*, 361–365.

White, P. R. (1939). Potentially unlimited growth of excised plant callus in an artificial nutrient. *American Journal of Botany, 26*, 59–64.

Wu, L. Y., Shang, G. D., Wang, F. X., Gao, J., Wan, M. C., Xu, Z. G., & Wang, J. W. (2022). Dynamic chromatin state profiling reveals regulatory roles of auxin and cytokinin in shoot regeneration. *Developmental Cell, 57*(4), 526–542.e7.

Xu, L. (2018). *De novo* root regeneration from leaf explants: Wounding, auxin, and cell fate transition. *Current Opinion in Plant Biology, 41*, 39–45.

Xu, L., & Huang, H. (2014). Genetic and epigenetic controls of plant regeneration. *Current Topics in Developmental Biology, 108*, 1–33.

Xu, M., Hu, T., Zhao, J., Park, M. Y., Earley, K. W., Wu, G., Yang, L., & Poethig, R. S. (2016). Developmental functions of miR156-regulated *SQUAMOSA PROMOTER BINDING PROTEIN-LIKE* (*SPL*) genes in *Arabidopsis thaliana*. *PLoS Genetics, 12*(8), e1006263.

Xu, C. Y., Cao, H. F., Zhang, Q. Q., Wang, H. Z., Xin, W., Xu, E. J., Zhang, S. Q., Yu, R. X., Yu, D. X., & Hu, Y. X. (2018). Control of auxin-induced callus formation by bZIP59-LBD complex in *Arabidopsis* regeneration. *Nature Plants, 4*(2), 108–115.

Xu, C. Y., Chang, P. J., Guo, S. Q., Yang, X. N., Liu, X. C., Sui, B. F., Yu, D. X., Xin, W., & Hu, Y. X. (2023). Transcriptional activation by WRKY23 and derepression by removal of bHLH041 coordinately establish callus pluripotency in *Arabidopsis* regeneration. *Plant Cell, 36*, 158. https://doi.org/10.1093/plcell/koad255

Xuan, W., Audenaert, D., Parizot, B., Moller, B. K., Njo, M. F., De Rybel, B., De Rop, G., Van Isterdael, G., Mahonen, A. P., Vanneste, S., & Beeckman, T. (2015). Root cap-derived auxin pre-patterns the longitudinal axis of the *Arabidopsis* root. *Current Biology, 25*(10), 1381–1388.

Xuan, W., Band, L. R., Kumpf, R. P., Van Damme, D., Parizot, B., De Rop, G., Opdenacker, D., Moller, B. K., Skorzinski, N., Njo, M. F., De Rybel, B., Audenaert, D., Nowack, M. K., Vanneste, S., & Beeckman, T. (2016). Cyclic programmed cell death stimulates hormone signaling and root development in *Arabidopsis*. *Science, 351*(6271), 384–387.

Yan, A., Borg, M., Berger, F., & Chen, Z. (2020). The atypical histone variant H3.15 promotes callus formation in *Arabidopsis thaliana*. *Development, 147*(11), dev184895.

Yang, M., & Jiao, Y. (2016). Regulation of axillary meristem initiation by transcription factors and plant hormones. *Frontiers in Plant Science, 7*, 183.

Ye, B. B., Shang, G. D., Pan, Y., Xu, Z. G., Zhou, C. M., Mao, Y. B., Bao, N., Sun, L., Xu, T., & Wang, J. W. (2020). AP2/ERF transcription factors integrate age and wound signals for root regeneration. *Plant Cell, 32*(1), 226–241.

Yu, Y., Feng, Z., Wang, G., Li, F., Du, X., & Zhu, J. (2010). Initiation of dedifferentiation and structural changes in *in vitro* cultured petiole of *Arabidopsis thaliana*. *Protoplasma, 241*(1–4), 75–81.

Yu, J., Liu, W., Liu, J., Qin, P., & Xu, L. (2017). Auxin control of root organogenesis from callus in tissue culture. *Frontiers in Plant Science, 8*, 1385.

Zamora, A. B., & Scott, K. J. (1983). Callus formation and plant regeneration from wheat leaves. *Plant Science Letters, 29*, 183–189.

Zeng, M., Hu, B., Li, J., Zhang, G., Ruan, Y., Huang, H., Wang, H., & Xu, L. (2016). Stem cell lineage in body layer specialization and vascular patterning of rice root and leaf. *Scientific Bulletin, 61*(11), 847–858.

Zhai, N., & Xu, L. (2021). Pluripotency acquisition in the middle cell layer of callus is required for organ regeneration. *Nature Plants, 7*(11), 1453–1460.

Zhai, N., Pan, X., Zeng, M., & Xu, L. (2023). Developmental trajectory of pluripotent stem cell establishment in *Arabidopsis* callus guided by a quiescent center-related gene network. *Development, 150*, dev200879.

Zhang, T. Q., Lian, H., Tang, H., Dolezal, K., Zhou, C. M., Yu, S., Chen, J. H., Chen, Q., Liu, H., Ljung, K., & Wang, J. W. (2015). An intrinsic microRNA timer regulates progressive decline in shoot regenerative capacity in plants. *Plant Cell, 27*(2), 349–360.

Zhang, T. Q., Lian, H., Zhou, C. M., Xu, L., Jiao, Y., & Wang, J. W. (2017). A two-step model for de novo activation of *WUSCHEL* during plant shoot regeneration. *Plant Cell, 29*(5), 1073–1087.

Zhang, G., Zhao, F., Chen, L., Pan, Y., Sun, L., Bao, N., Zhang, T., Cui, C., Qiu, Z., Zhang, Y., Yang, L., & Xu, L. (2019). Jasmonate-mediated wound signalling promotes plant regeneration. *Nature Plants, 5*(5), 491–497.

Zhang, T., Ge, Y., Cai, G., Pan, X., & Xu, L. (2023). WOX-ARF modules initiate different types of roots. *Cell Reports, 42*(8), 112966.

Zhao, Z., Andersen, S. U., Ljung, K., Dolezal, K., Miotk, A., Schultheiss, S. J., & Lohmann, J. U. (2010). Hormonal control of the shoot stem-cell niche. *Nature, 465*(7301), 1089–1092.

Zimmerman, W., & Wilcoxon, F. (1935). Several chemical growth substances which cause initiation of roots and other responses in plants. *Contributions from Boyce Thompson Institute, 7*, 209–217.

Chapter 5
Floral Transition

Lijing Xing, Yongyan Tang, Hanwen Deng, Peilei Cheng, Yunyuan Xu, and Kang Chong

Abstract Following seed germination, seedlings typically develop sufficient leaves to facilitate carbohydrate accumulation through photosynthesis, a stage known as vegetative growth. Throughout the vegetative growth phase, plants respond to external and internal signals that stimulate the initiation of floral primordia in the peripheral region of the shoot apical meristem, rather than leaf primordia. This intricate and orchestrated shift from vegetative to reproductive growth is referred to as floral transition. The upcoming chapter will primarily delve into the molecular mechanisms that govern the regulation of floral transition. First, the research results in *Arabidopsis* serve as a basis for briefly summarizing the six flowering pathways: age, autonomous, gibberellin(GA), sugar, photoperiod, and vernalization pathways. The following section will discuss the regulation of the photoperiod pathway in rice and soybean. The epigenetic regulation of *FLC* expression will be presented in the third section. In the last section, the principles and theories of vernalization will be elucidated using *Arabidopsis FLC* gene and wheat *VRN1* gene as case studies. The past research results of vernalization at the metabolic level will be reviewed briefly. The theory of vernalization perception will also be proposed. The role of protein phosphorylation and *O*-GlcNAc modification in wheat vernalization and the prolonged cold memory and resetting of vernalization requirements will also be summarized in this section.

Keywords Floral transition · Photoperiod · Vernalization · Epigenetic · Phosphorylation · *O*-GlcNAc

L. Xing · Y. Tang · H. Deng · P. Cheng · Y. Xu · K. Chong (✉)
The Key Laboratory of Plant Molecular Physiology, Institute of Botany, Chinese Academy of Sciences, Beijing, China
e-mail: chongk@ibcas.ac.cn

F. Chang et al. (eds.), *Regulation of Plant Development*,
https://doi.org/10.1007/978-981-95-7033-1_5

5.1 Flowering Regulation Pathways

In the past two decades, plenty of floral transition-associated mutants were acquired by large-scale artificial mutagenesis of the model plant *Arabidopsis*. Based on cloning and functional analyses of identified genes, scientists have clarified significant advances in flowering regulation mechanisms in plants. Flowering time pathways have been summarized to age, autonomous, gibberellin (GA), ambient temperature, sugar, photoperiod, and vernalization regulation. These pathways play roles not independently but coordinately regulate the floral meristem identity gene *LEAFY* (*LFY*) and MADS-box genes *APETALA1* (*AP1*) and *AGAMOUS LIKE 24* (*AGL24*) to promote flowering by acting on floral pathway integrators, *FLOWERING LOCUS T* (*FT*) and *SUPPRESSOR OF OVEREXPRESSION OF CONSTANS1*(*SOC1*). *FLOWERING LOCUS C* (*FLC*) encodes a MADS-box transcription factor and plays a negative role in flowering time regulation. *FLC* takes the central position in the flowering network and is always used as a marker to explain the flowering phenotype. Approximate 180 genes are implicated in flowering-time control in *Arabidopsis* (Fornara et al., 2010).

Development age, as an internal factor, is integrated with gibberellin, photoperiod, and ambient temperature through regulating *FT* transcription (Song et al., 2013). FT, a major component of florigen, is synthesized in the leaf vasculature and then moves to the shoot apical meristem by phloem translocation (Corbesier et al., 2007). FT interacts with bZIP protein FLOWERING LOCUS D (FD) to form a heterodimer to activate the MADS-box transcription factor SOC1. SOC1 interacts with AGL24 to directly promote the expression of *LFY*. In addition, another floral meristem identity gene *AP1* is a target of FT-FD complex (Wigge et al., 2005). Some transcriptional repressors, FLC, SHORT VEGETATIVE PHASE (SVP), TEMPRANILLO 1 (TEM1), and SCHLAFMU TZE (SMZ) repress *FT* expression by binding to *cis*-elements of *FT* during the juvenile development stage (Lee et al., 2007; Castillejo & Pelaz, 2008; Mathieu et al., 2009). The main pathways of flowering regulation were summarized in Fig. 5.1. It can be found that the floral meristem identity genes such as *LFY* and *AP1* are located at the downstream and are regulated by FLC, SOC1, and FT. The upstream regulators from six pathways directly or indirectly affect the expression of *FLC*, *SOC1*, *FT*, etc.

5.1.1 Aging Pathway

The microRNAs are one type of noncoding RNAs with a length of 21–24 nt and widespread occurrence in animal and plant cells (Bartel, 2009). More and more lines of evidence indicate that miRNAs play crucial roles in plant development and environmental adaptation. The microRNA156 (miR156) and its target genes *SQUAMOSA PROMOTER BINDING-LIKEs* (*SPLs*), encoding transcription factors, control flowering time under noninductive conditions, and this pathway was

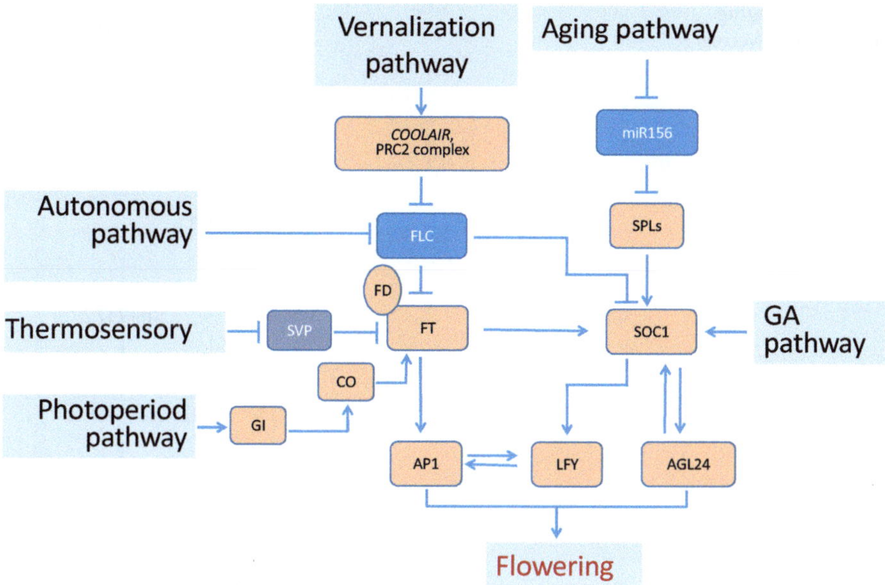

Fig. 5.1 Pathways of flowering regulation

defined as an aging pathway (Wang et al., 2009; Wang, 2014). In *Arabidopsis*, 11 members of the *SPL* family are targeted by miR156, which can be divided into two groups based on the size of their encoded products. One group, represented by *SPL3*, includes *SPL3*, *SPL4*, and *SPL5*, which encode smaller products. The other group, represented by *SPL9*, includes *SPL2*, *SPL6*, *SPL9*, *SPL10*, *SPL11*, *SPL13*, *SPL13-like*, and *SPL15* (Xing et al., 2010). The expression level of miR156 decreases With increasing plant age, the decrease in miR156 expression elevates *SPLs* transcription, thereby activating flower-promoting genes, including *AP1*, *SOC1*, *FRUITFULL* (*FUL*) and *LFY*, thereby inducing flowering in shoot apical meristem (SAM) (Wang, 2014). In addition to *Arabidopsis*, the expression pattern of miR156 is also observed in maize, rice, and other species (Chuck et al., 2007; Xie et al., 2012; Debernardi et al., 2022). The negative role of miR156 in regulating flowering time is conserved in plants (Xie et al., 2006; Zhou et al., 2013). miR159 has been reported to repress miR156 to regulate juvenile-to-adult transition in *Arabidopsis.* MYB33, a target of miR159, promotes *MIR156* transcription and mediate miR159-mediated repression of miR156 to prevent its hyperactivation, thus ensuring proper vegetative phase change (Guo et al., 2017) (Fig. 5.2).

miR172 is one of the earliest identified plant miRNAs. In dicotyledon and monocotyledon plants, the expression level of miR172 increases with plant development and accumulates in floral buds and leaves after flowering (Aukerman & Sakai, 2003). In leaves, SPLs directly activates miR172 (Wu et. al., 2009). Plants overexpressing miRNA172 exhibit an early flowering phenotype under both long day (LD) and short day (SD) conditions (Jung et al., 2007). Target genes of miR172 include

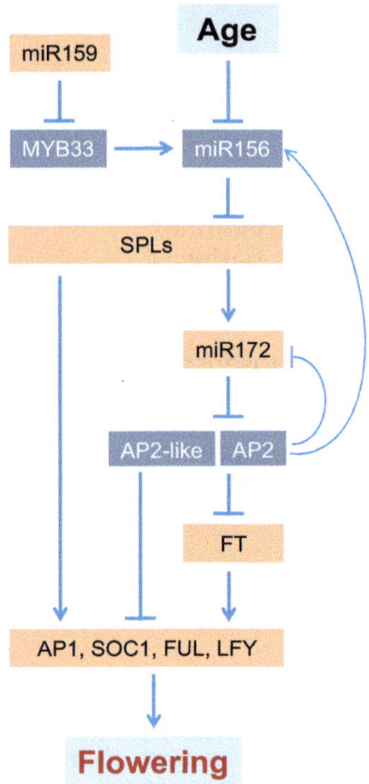

Fig. 5.2 The aging pathway of flowering regulation

APETALA2 (*AP2*) and *AP2*-like genes, such as *TARGET OF EAT1* (*TOE1*), *TOE2*, *TOE3*, *SCHNARCHZAPFEN* (*SNZ*), and *SMZ*, which encode plant-specific transcription factors regulated by development stages and repress floral transition (Jung et al., 2007; Mathieu et al., 2009; Zhu & Helliwell, 2011; Ó'Maoileidigh et al., 2021). AP2 directly represses the expression of the floral activator miR172 and promotes the expression of a floral repressor miRNA156 by binding to microRNA genes (Yant et al., 2010) (Fig. 5.2). Members of the *MIR172* family exhibit functional specificity and spatiotemporal expression differences in the regulation of flowering time. Some members promote *FT* expression in leaves under long-day conditions, while under short-day conditions, they facilitate flowering by activating the expression of *FUL* and *SOC1* in the shoot apical meristem (SAM) (Lian et al., 2021; Ó'Maoileidigh et al., 2021). The aging pathway integrates with gibberellin, vernalization anf photoperiod pathways to regulate flowering (Wang, 2014). DELLA proteins, key regulators of GA signaling, interact with SPLs to mediate the crosstalk between aging and GA pathways. The interaction between the DELLA protein RGA and SPL9 inhibits *FT* transcription by repressing miR172 in leaves, while in the shoot apex, it represses the transcription of *SOC1* and *FUL*, resulting in delayed flowering (Yu et al., 2012; Wang, 2014).

Elevated ambient temperatures typically induce the acceleration of flowering in most plant species (Jiang, 2023). Two ambient temperature sensors, FCA and SHORT VEGETATIVE PHASE (SVP), promote processing of *pri-MIR172* transcripts and represses miR172 transcription, respectively (Jung et al., 2012). Loss of function of SVP increases miR172a expression (Cho et al., 2012). Warm ambient temperatures promote flowering by relieving the transcriptional repression of *FT*, which is imposed by an inhibitory complex composed of the two crucial thermosensory factors SVP and FLM. In addition, higher temperatures activate *FT* expression through multiple pathways to accelerate flowering (Luo et al., 2025).

5.1.2 Autonomous Pathway

The pathway that promotes flowering in a manner independent of environment cues is called the autonomous pathway. The autonomous pathway genes include *FLOWERING CONTROL LOCUS A* (*FCA*), *FY*, *FLOWERING LOCUS PA* (*FPA*), *FLOWERING LOCUS K* (*FLK*), *FLOWERING LOCU D* (*FLD*), *LUMINIDEPENDENS* (*LD*), *FVE*, and *GLYCINE RICH PROTEIN 7* (*GRP7*) (Simpson & Dean, 2002). The mutants of these genes all exhibit elevated *FLC* expression and a delayed flowering phenotype under both LD and SD conditions. Vernalization can partially rescue the late flowering phenotype of mutants lacking these genes. Moreover, when crossed with the *flc* null mutant, the late flowering phenotype of these mutants can be converted, indicating that these genes accelerate flowering by repressing *FLC* transcription (Michwwaels & Amasino, 1999b, 2001). The repression of these genes on *FLC* expression is mediated by RNA-processing and histone modification.

Two genes encoding RNA-binding protein, *FCA* and *FPA*, are identified by screening flowering mutants (Macknight et al., 1997; Schomburg et al., 2001). They are involved in post-transcriptional processing by modulating poly(A) site selection of mRNA to regulate the protein level (Quesada et al., 2005; Hornyik et al., 2010a). FCA includes two RRM (RNA recognition motif) domains and a WW (because of two conserved tryptophan residues) domain mediating protein interaction. FY, an RNA 3′ processing factor, interacts with FCA to regulate *FCA* pre-mRNA through the selection of different poly(A) sites, and FCA/FY interaction is required for inhibiting *FLC* transcription (Simpson et al., 2003). *FPA* encodes an RNA-binding protein with three RRM domains, different from FCA. FPA does not need FY to assist in selecting RNA poly(A) sites. FCA and FPA are involved in activating the selection of proximal poly(A) of *FLC* antisense RNA, which results in the expression reduction of sense *FLC* RNA, and FLD also participates in this process (Hornyik et al., 2010a, b). The double mutants, *fca/fld* and *fpa/fld*, do not exhibit additive effects but show the flowering phenotype similar to the *fld* single mutant, indicating that FLD functions downstream of FCA and FPA (Baurle & Dean, 2008).

FLD encodes a lysine demethylase, mutation of *FLD* results in a severe late flowering phenotype, and *FLD* overexpression significantly accelerates floral transition in *Arabidopsis* (Sanda & Amasino, 1996). FLD is involved in *FLC* transcription regulation by regulating histone acetylation and methylation to alter chromatin status. In *fld*, histone H3 acetylation and H3K4me3 levels are elevated at the *FLC* locus, resulting in the activation of *FLC* transcription. The histone deacetylase HISTONE DEACETYLASE 6 (HDA6) interacts with FLD, mediating the crosstalk between histone deacetylation and demethylation. Loss function of *hda6* results in up-regulation of *FLC* and late-flowering phenotype (Yu et al., 2011). In addition, SIZ1-mediated sumoylation of FLD reduces acetylation of H4 in *FLC* chromatin to inhibit *FLC* transcription (Jin et al., 2008).

FVE encodes a homologue of mammalian retinoblastoma-associated protein (RbAp). As a component of the histone deacetylase complex (HDAC), FVE is involved in the deacetylation of *FLC* chromatin and accelerates floral transition (Jeon & Kim, 2011). MSI4/FVE assists the interaction between the CUL4-DDB1 E3 ligase and the CLF-PRC2 complexes to regulate H3K27me3 levels in *FLC* chromatin. MSI4/FVE mediates CUL4 binding to *FLC* chromatin and then influences flowering time (Pazhouhandeh et al., 2011).

FLK encodes an RNA-binding protein. The *flk* mutant shows a late flowering phenotype under LD and SD conditions, and vernalization treatment and exogenous spraying of GA3 can reverse the late flowering phenotype of the mutant. In *flk*, *FLC* transcription is increased, but *CONSTANS* (*CO*) and *SOC1* transcription are decreased (Lim et al., 2004). AtGRP7, a small glycine-rich RNA-binding protein, is also reported to be involved in flowering regulation in the autonomous pathway by suppressing *FLC* expression (Streitner et al., 2008). The jacalin-lectin encoding gene *AtJAC1* in *Arabidopsis* upregulates *FLC* expression and negatively regulates the floral transition via the autonomous pathway. AtJAC1 interacts with GRP7, affecting the binding and processing of the *FLC* antisense transcript (*COOLAIR*) by GRP7, thereby regulating *FLC* expression (Xiao et al., 2015).

GH1-HMGA gene family members were recently identified as autonomous pathway genes. GH1-HMGA family proteins alter *FLC* chromatin structure through binding to *FLC* flanking regions, inhibiting *FLC* gene loop formation to repress its transcription (Zhao et al., 2021a, b).

5.1.3 Gibberellin Pathway

Environmental cues and endogenous signaling are integrated to ensure flowering at the right time, and hormone action is very important in this process. A number of studies established the role of plant hormones in regulating flower induction in plants. It has been known that hormonal signaling plays roles in regulating the expression of *FLC*, *CO*, and *FT*. Almost all of the plant hormones including gibberellin (GA), brassinosteroids (BRs), auxin (indole-3-acetic acid, IAA), abscisic acid (ABA), ethylene (ET), jasmonate (JA), salicylic acid (SA), cytokinin (CKs), and nitric oxide (NO) are reported to be involved in flowering regulation. Among those hormones, the role of GA in flowering transition is the most studied (Conti, 2017).

Gibberellin promotes flowering by activating the expression of flowering-promoting factors such as *LFY*, *SOC1* and *FT* through GA biosynthetic, transport and signaling pathways (Bao et al., 2020). The mechanism of GA regulated floral transition has been well established (Fig. 5.3). The product encoded by *GA1* is necessary for the creation of ent-kaurene, which is the first intermediate in GA synthesis reaction. The severe GA-deficient mutant *ga1-3* is unable to flower unless treated with exogenous GA. Although another weak defective mutant of GA synthesis, *ga1-6*, can flower under SD conditions, but the flowering time of which is obviously delayed compared with the wild type. Similarly, other reported mutants with blocked GA signaling pathway show delayed flowering phenotype (Koornneef et al., 1991; Wilson et al., 1992). Studies on mutants with enhanced GA signaling also provide evidence that GA plays a positive role in regulating flowering time. For instance, *spindly* (*spy*) and *flowering promotive factor 1* (*fpf1*) mutants all flowered earlier than wild-type plants (Jacobsen & Olszewski, 1993; Kania et al., 1997).

Among the more than 100 gibberellins identified to date, only a few are biologically active (Hedden and Phillips 2000). Notably, GA4, one of the structurally similar molecules, shows the highest physiological activity and accumulates most abundantly in the shoot apex (Eriksson et al., 2006). Exogenous application of GA_4 can more effectively accelerate the flowering of GA-defective mutant *ga1*. At the time that GA_4 starts to rise in shoot apex, the transcription levels of feedback-regulated GA synthesis genes of *Gibberellin 20-oxidase* (*$GA20_{OX}$*) and *$GA3_{OX}$* are unaffected, but at this stage, the expression of *$GA2_{OX}$* is increased. Moreover, the application of GA_4 to a single leaf can induce flowering of plants grown under SD conditions, and labeled GA_4 can move from the leaf to the shoot apex. Together, these results indicate that GA_4 is a significant component of the floral signal (Eriksson et al., 2006).

GA signaling mediates flowering by regulating the degradation of DELLA proteins in plants. The late flowering phenotype of *ga1-3* can be recovered in the absence of *RGA*, *RGL1*, and *RGL2* (Tyler et al., 2004). Overexpression of any one of these regulators in *ga1* background can partly recover the late flowering phenotype (Moon et al., 2003). Exogenous GA3 treatment under short-day conditions can significantly promote SOC1 expression (Moon et al., 2003). In addition, exogenous application of GA promotes flowering by enhancing the promoter activity of *LFY* under SD (Blazquez & Weigel, 1999). A cis-element in the *LFY* promoter region,

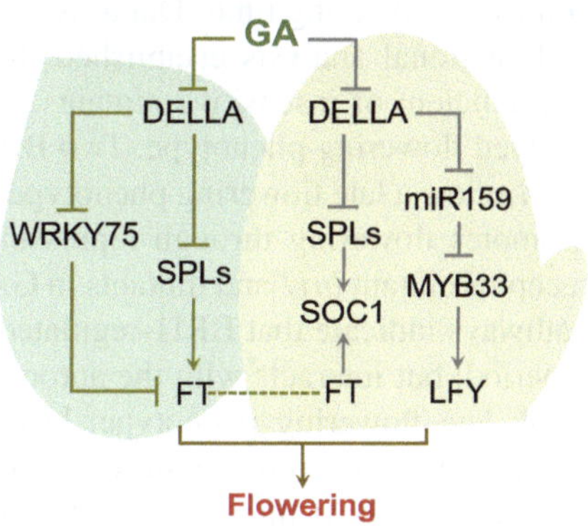

Fig. 5.3 GA pathway of flowering regulation in shoot apical meristems (SAMs) and leaves

distinct from the one responsive to photoperiod, is sufficient for GA-mediated upregulation of *LFY* expression (Blazquez & Weigel, 2000). In shoot apex, GA activates *SOC1* and *LFY* by repressing DELLA proteins (Mutasa-Gottgens & Hedden, 2009). SPLs meditae the repression of *SOC1* by DELLA, which integrates age amd GA pathways (Wang, 2014). The GA-dependent activation of *LFY* may be mediated through a GAMYB-like transcription factor, MYB33. It is shown that MYB33 binds to the GA response element of the *LFY* promoter region in vivo (Blazquez & Weigel, 2000; Gocal et al., 2001). GA positively regulates miR159 expression through repressing the degradation of DELLA proteins, and *MYB33* is the target of miR159. The *MYB33* transcription is decreased in *ga1*, which is consistent with the downregulation of *LFY* expression (Achard et al., 2004). In leaves, GA exerts regulatory function by enhancing the transcriptional activation of *FT* under long-day (LD) conditions, thereby facilitating flowering (Porri et al., 2012). The signal between DELLA and FT can be mediated by transcription factor WRKY75. Overexpression of WRKY75 obviously accelerates *Arabidopsis* flowering. WRKY75 directly binds to the promoter of *FT* and promotes its transcription. DELLA proteins RGL1 and GAI physically interact with WRKY75 and repress its transactivation activity. In the presence of GA, DELLA protein is degraded, and the repression of WRKY75 is released, which activates *FT* expression for promoting flowering (Zhang et al., 2018). The expression of *FT*, *SOC1*, and *LFY* mediated by DELLA in leaves and shoot apices is also regulated by multiple transcription factors, including SPLs (Bao et al., 2020).

Two GATA-like transcription factors, GNC (GATA, NITRATE-INDUCIBLE, CARBONMETABOLISM INVOLVED) and GNL (GNC-LIKE), are reported to be involved in GA-mediated flowering regulation. The mutants of *gnc*, *gnl*, and *gnc gnl* show an early flowering phenotype compared with wild-type plants. As target genes of PIF transcription factor, *GNC* and *GNL* are regulated by GA signaling in a DELLA-dependent pattern. *Gnc* and *gnl* can partially inhibit the defective flowering phenotype of *ga1*, indicating that *GNC*- and *GNL*-regulated floral transition is dependent on GA signaling (Richter et al., 2010). The rice early flowering mutant, *ef1*, has an enhanced GA response due to the loss of function of casein kinase I. EL1 phosphorylates DELLA protein Slender Rice 1 (SLR1) to increase its stability in vivo and in vitro, indicating that EL1 acts as a negative regulator in GA-mediated control of flowering time (Dai & Xue, 2010).

Functional analysis established the role of BR in regulating floral transition. BR-deficient or insensitive mutants all exhibit a prolonged vegetative stage and a delayed flowering phenotype. Two BR-deficient mutants, *de-etiolated-2* (*det2*) and *dwf4*, show a late flowering phenotype (Chory et al., 1991; Azpiroz et al., 1998). BR promotes flowering through repressing *FLC*. Analyses of double mutants of BR receptor mutant *bri1* and mutants in GA, photoperiod, vernalization, or autonomous pathways indicate that BRI1-regulated flowering does not depend on GA and photoperiod, but interacts with the autonomous pathway. The *bri1* single mutant has a weak late flowering phenotype; however, *bri1 ld*, *bri1 fca*, and *bri1 fri* double mutants show a severe late flowering phenotype, which can be alleviated by vernalization and RNAi interference of *FLC* expression, indicating that *BRI1* plays an assistant role in *FLC* repression of the autonomous pathway. In addition, BR is involved in regulating *FLC* chromatin status. The key transcription factor in the BR signaling pathway, bri1-ethyl methanesulfonate suppressor 1/brassinazole-resistant 1 (BES1/BZR1), interacts with early flowering 6 (ELF6) and RELATIVE OF EARLY FLOWERING 6 (REF6), which participate in chromatin remodeling and flowering time regulation (Noh et al., 2004; Domagalska et al., 2007).

IAA plays an important role in plant development processes. AUX/IAA acts as a repressor of auxin signaling to inhibit the function of auxin response factors (ARFs) in auxin signal transduction. Increased auxin results in AUX/IAA ubiquitination and degradation to release the repression of ARFs. Under SD conditions, the gain-of-function mutant *aux2-1* shows a late flowering phenotype, which can be recovered by exogenously applied GA. The expression levels of *GA20ox1* and *GA20ox2* in GA synthesis pathway are decreased in *aux2-1*. Overexpression of *LFY* and *SOC1* partly recovered the late flowering phenotype of *aux2-1*. Nevertheless, the mutation of members of the IAA receptor *TRANSPORT INHIBITOR RESISTANCE 1(TIR1)* family and *ARFs* does not result in a late flowering phenotype, demonstrating that IAA does not regulate flowering through the TIR1-ARF pathway. Some genes regulated by IAA participate in floral transition. AtFPF1 regulates flowering in a GA-dependent manner, and *AtFPF1* overexpression leads to early flowering (Kania et al., 1997). Transgenic plants overexpressing *OsRAA1*, the rice homolog of *AtFPF1*, also show an early flowering phenotype (Ge et al., 2004). In summary, IAA promotes floral transition, but the specific mechanism needs to be clarified.

Exogenous ABA inhibits flowering in *Arabidopsis*, and ABA synthesis-deficient mutant *aba1-1* and ABA signaling mutant *abi3-4* flowered earlier than the wild type (Kurup et al., 2000; Barrero et al., 2005; Razem et al., 2006). ABSCISIC ACID-INSENSITIVE 4 (ABI4) acts as a negative regulator of floral transition by promoting *FLC* expression (Shu et al., 2016).

Ethylene plays various roles in plant development and responses to both biotic and abiotic stress. Environmental stress stimulates ethylene production to activate ETR1-EIN3 signal pathway, resulting in decreased bioactive GA level, DELLA accumulation, and the inhibition of *LFY* and *SOC1* expression. Therefore, ethylene-regulated developmental stage transition from vegetative to reproductive growth is related to GA signaling (Achard et al., 2007). These results indicate that plant hormones play interconnected roles in regulating floral transition. Different pathways crosstalk with each other to form an integrated regulatory network and regulate floral transition in plants.

Epigenetic mechanisms also interact with plant hormone signals in floral transition. The activity of AUXIN RESPONSE FACTOR MONOPTEROS (MP/ARF5) is required for flower primordium initiation. Auxin treatment results in elevated H3K9ac in *LFY* chromatin, and high concentration of auxin induces MP to recruit SWI/SNF proteins to regulate the transcription of flowering initiation genes via chromatin remodeling (Wu et al., 2015). The loss of function of the chromatin remodeling enzyme POCKLE (PKL) results in delayed flowering time, a phenotype similar to that of mutants with defective GA biosynthesis and response. It has been proposed that DELLA proteins interact with PKL to recruit the PRC2 complex, which in turn regulates *FLC* transcription (Henderson et al., 2004; Park et al., 2017). The GA pathway interacts with other phytohormones to regulate flowering transition, in which DELLA proteins act as a key integration node of this hormonal crosstalk network and mediate the expression of *FT*, *LFY*, *SOC1* and *FLC* to modulate flowering (Bao et al., 2020).

5.1.4 Sugar Signaling of Floral Transition

Sugars, as primary carbon and energy sources, play crucial roles in the growth, development, and response to the environment in plants. Many studies established the role of sugar signaling in floral transition. Loss function of trehalose-6-phosphate synthase 1 (AtTPS1), a key enzyme in trehalose synthesis, results in delayed floral transition in *Arabidopsis*. The role of AtTPS1 in regulating flowering time does not depend on the vernalization and photoperiod pathways. The expression levels of *FLC*, *FT*, *CO*, *SOC1*, and *LFY* are not altered in *tps1*, and the functional mechanism of AtTPS1 in floral transition needs to be further studied (van Dijken et al., 2004). The induction of flowering is linked to the transport of leaf carbohydrates to the shoot apical meristem. Application of high concentrations of sucrose extends the vegetative phase and delays flowering, but exogenous supplementation of 1% sucrose can rescue the late flowering phenotype of plant mutants (Roldan et al., 1999). Another study indicated that the effect of sugars on floral transition is influenced by sugar concentration, the genetic background of the plant, and the timing of sugar application (Ohto et al., 2001).

Sugar functions in the juvenile-to-adult phase transition through repressing miR156. It is shown that the expression of miR156 responds to glucose, sucrose, and maltose treatment, and glucose plays a major role in this process (Yu et al., 2013). Sugar produced in leaves acts as a mobile signal to repress miR156 at transcription and post-transcription levels, which is conserved in plants. The transcription factor NF-YA8, predominantly expressed in juvenile leaves, has been identified as an activator of *MIR156* transcription. However, this activation is inhibited by sugar signals, indicating that sugars may regulate developmental phase transitions through modulation of miR156 levels (Chen et al., 2020).

Hexosamine biosynthetic pathway (HBP) is a glucose metabolic pathway that produces an amino sugar uridine diphosphate N-acetylglucosamine (UDP-GlcNAc), which is the donor of *O*-linked N-acetylglucosamine (*O*-GlcNAc) modification, and a very small proportion of glucose enters the HBP pathway in cells (Chatham et al., 2021). The posttranslational protein *O*-GlcNAc modification on serine and threonine residues is dynamically regulated by *O*-GlcNAc transferase (OGT) and *O*-GlcNAcase (OGA), which catalyze the addition and removal of *O*-GlcNAc, respectively (Love et al., 2003; Soesanto et al., 2008; Yang and Qian, 2017). A jacalin-like protein VER2 has been reported to play a role in the vernalization-induced floral transition and spikelet development, and exhibits a binding affinity for carbohydrates (Yong et al., 2003; Xing et al., 2009). The *Arabidopsis* homolog of OGT, *O*-GlcNAc transferase SECRET AGENT (SEC), modifies histone methyltransferase ARABIDOPSIS HOMOLOG OF TRITHORAX1 (ATX1) with *O*-GlcNAc. This modification leads to the activation of ATX1, enhancing the H3K4me3 levels at the *FLC* locus and ultimately repressing flowering (Xing et al., 2018). In winter wheat, the overexpression of leads to a decrease in glucose and sucrose concentrations in the leaves, while concurrently upregulating the expression of genes related to flowering, such as the vernalization genes *VRN1* and *VRN3* (Fan et al., 2021).

In summary, sugars play a crucial role in the regulation of flowering in plants by serving as energy sources, signaling molecules, and modulators of gene expression. The processes of sugar transport and metabolism are intricately connected to the timing of flowering, offering insights into how plants integrate environmental and internal cues to enhance reproductive success.

5.1.5 The Photoperiod and Vernalization Regulation Pathways

In addition to the endogenous factors mentioned above, two external environmental cues photoperiod and vernalization play critical roles in regulating floral transition in plants (Fig. 5.1). Photoperiod refers to the relative length of day and night, perceived by leaves to determine plant flowering time. In *Arabidopsis*, *CONSTANS* (*CO*) encodes a zinc finger-like transcription factor that plays a pivotal role in photoperiod-dependent flowering induction through its temporal and spatial expression patterns (Putterill et al., 1995). *CO* promotes flowering by activating the transcription of *FT* and *TSF*. The expression of *CO* is regulated by circadian rhythms and other regulators (Imaizumi et al., 2005; Sawa et al., 2007; Fornara et al., 2009). As a monocotyledon model plant, rice differs from *Arabidopsis* in photoperiod regulation. Rice is not strictly an SD plant, as it also flowers under LD conditions, although flowering is significantly delayed. Rice produces two types of florigens, Hd3a and RFT1, according to the day length. *Hd3a* RNAi transgenic plants exhibit a predominant delayed flowering phenotype under SD, but show similar flowering time to wild type plants under LD. LD induces the expression of *RFT1* in leaf phloem, and then the product is transported into shoot apex by long distance translocation system. Loss function of Hd3a and RFT1 blocks flowering in rice. The OsMADS50-Ehd1-RFT1 regulation pathway ensures the accomplishment of floral transition under noninducible LD conditions (Komiya et al., 2009).

Vernalization is a process by which prolonged cold exposure leads to competence to flower in winter-annual, biennial, and perennial plants (Woods et al., 2019). Most of the vernalization-requiring plants are LD plants, and LD and vernalization requirements permit plants to flower during spring but not prior to winter. The optimal temperatures for a vernalization response are typically at or above freezing. Studies on the model plant *Arabidopsis* clarify the molecular and epigenetic basis of vernalization, and *FRIGIDA* (*FRI*) and *FLOWERING LOCUS C* (*FLC*) are two key regulators (Clarke & Dean, 1994; Choi et al., 2011; Whittaker & Dean, 2017). *FRI* encodes a plant-specific protein which promotes *FLC* transcription (Lee & Amasino, 1995). *FLC* encodes a MADS-box transcription factor, which acts as a key negative regulator of flowering by repressing *FT* and *SOC1* expression (Michaels & Amasino, 1999a). Cereals exhibit different vernalization response mechanisms, indicating independent evolvement (Xu & Chong, 2018).

5.2 Photoperiod Pathway

The duration of day or night plays a crucial role in the growth and flowering of plants. The physiological responses of plants to relative lengths of light and dark periods are termed photoperiodism. Based on their specific day length needs for flowering, photoperiodic plants can be classified into three main categories: long-day plants, short-day plants, and day-neutral plants (Garner & Allard, 1920). Long-day plants, such as *Arabidopsis thaliana*, *Triticum aestivum* (wheat), *Hordeum vulgare* (barley), *Brassica napus* (rapeseed), and *Pisum sativum* (pea), usually flower during late spring or early summer, when the night length is shorter than their critical length, i.e., long day length promotes flowering of long-day plants. Short-day plants, such as *Oryza sativa* (rice), *Zea mays* (maize), *Glycine max* (soybean), *Nicotiana tabacum* (tobacco), and *Dendranthema morifolium* (chrysanthemum) generally flower during summer or fall, when the night lengths are longer than their critical length. Day-neutral plants can flower at any day-night period after reaching a certain developmental stage or age, or after some kinds of environmental stimuli, for example, low temperature (vernalization). This type of plant includes *Lycopersicon esculentum* (tomato), *Phaseolus vulgaris* (kidney bean), *Cucumis sativus* (cucumber), *Rosa chinensis* (Chinese rose), and *Solanum melongena* (eggplant).

In *Arabidopsis*, the photoperiod pathway includes two key components, CO and FLOWER LOCUS T (FT). CO activates *FT* expression to promote flowering under long-day condition. The expression of *CO* is tightly controlled by CYCLINGDOFFACTORs (CDFs), which bind to the *CO* promoter and inhibit its transcription. The protein abundance of CDFs is mainly controlled by FLAVIN-BINDING, KELCH REPEAT, F-BOX 1 (FKF1)-GIGANTEA (GI) complex under long-day condition. The detailed mechanisms can be found in Song's review (Song et al., 2015). In this section, we will focus on SD plants rice and soybean as examples to introduce photoperiodic flowering regulation.

5.2.1 Photoperiodic Flowering Regulation in Rice

5.2.1.1 Florigens in Rice

HEADING DATE 3a (Hd3a) and Rice FT1 (RFT1) are two rice homologs of *Arabidopsis* florigen FT (Tamaki et al., 2007; Komiya et al., 2009). They share 91% identity with each other and locate in the same region of chromosome 6. *Hd3a* and *RFT1* are mainly expressed in leaf vascular tissues and diurnally expressed with a peak before dawn under SD conditions. The expression of *Hd3a* and *RFT1* reaches a peak 30 days before flowering under SD conditions, but is barely detectable under LD conditions. Notably, the expression level of *Hd3a* is much higher than *RFT1* under SD conditions. Knockdown of *Hd3a* led to flowering 30 days later than wild-type under SD conditions, but flowers at the same time as wild-type under LD conditions. While in *RFT1* RNAi transgenic plants, the flowering time was only slightly delayed under SD conditions, but ~100 days later than wild type under LD conditions. However, overexpression of either *Hd3a* or *RFT1* caused early flowering under both SD and LD conditions. These results indicate that Hd3a and RFT1 have the same function in flowering regulation, but they preferentially work under different day length conditions. Hd3a and RFT1 are the major rice florigens under SD and LD, respectively (Kojima et al., 2002; Komiya et al., 2008; Matsubara et al., 2008).

Similar to FT in *Arabidopsis*, Hd3a and RFT1 are expressed in the leaf and then move into the shoot apical meristem (SAM) to induce rice flowering. Hd3a binds to the G-box factor 14-3-3c protein (GF14C) and transcription factor OsFD1 to form a florigen activation complex in the SAM (Tamaki et al., 2007; Komiya et al., 2009; Taoka et al., 2011). This complex activates the expression of *MADS14*, *MADS15*, *MADS18*, and *PAP2*, initiating floral meristem identity (Fig. 5.4). Knocking down *MADS14*, *MADS15*, *MADS18*, and *PAP2* individually did not affect flowering time. However, knocking down all three or four of these genes resulted in severe developmental defects in the inflorescence and a delay in flowering time (Kobayashi et al., 2012).

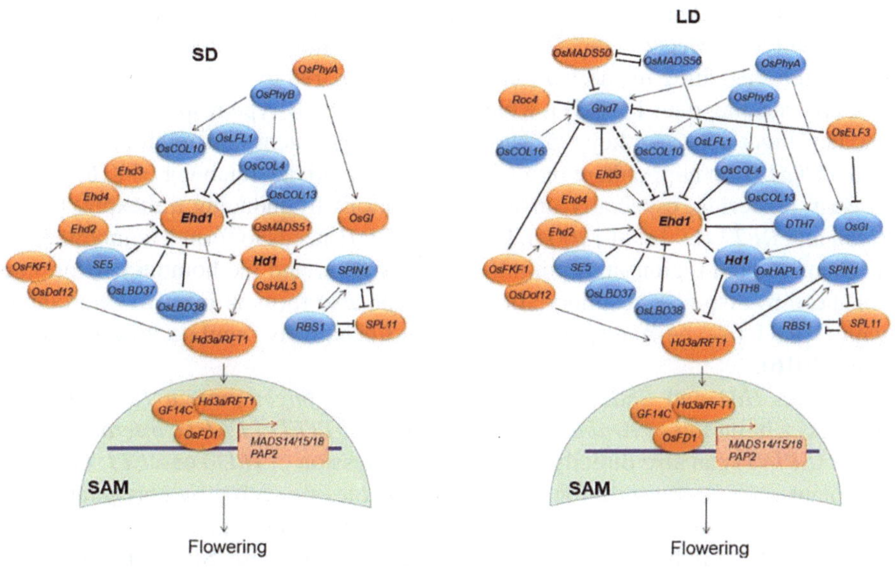

Fig. 5.4 Photoperiodic flowering regulation in rice. Photoperiodic flowering in rice is regulated differentially under SD and LD conditions. The florigens Hd3a and RFT1 play key roles in this process. *Ehd1,* the core regulator of *Hd3a* and *RFT1,* is strictly controlled by multiple upstream signals. Arrows represent activation, bars represent repression. Pink ovals signify positive regulators of flowering, light blue ovals indicate negative regulators of flowering

5.2.1.2 The Direct Upstream Regulators of Rice Florigen

HEADING DATA 1 (*Hd1*), the first identified rice flowering time QTL by the map-based cloning method, is one of the 16 *CO-Like* (*COL*) gene family members and can directly bind the promoters of *Hd3a* and *RFT1* to enhance their transcription (Kim et al., 2008b). The *Hd1* loss-of-function mutant flowers late under SD, but early under LD conditions, suggesting that *Hd1* is a flowering activator under SD but a flowering repressor under LD conditions. However, there is no significant difference in the transcription levels of *Hd1* between SD and LD conditions (Lin et al., 2000; Yano et al., 2000; Doi et al., 2004). Researchers found that the dual function of Hd1 is achieved by interacting with different proteins under different day-length conditions. Under SD, the positive regulation activity of Hd1 is dependent on putative blue-light sensor OsHAL3, which interacts with Hd1 and directly binds to the promoter of *Hd3a* to activate its expression (Su et al., 2016). Under LD, Hd1 forms a complex with a transcription factor, DAYS TO HEADING 8 (DTH8), to increase the H3K27 trimethylation level at *Hd3a* locus and thus repress *Hd3a* expression (Du et al., 2017). In addition, *Hd1* over-expressed plants (only over-expressed during the light period) delayed flowering under SD but not LD conditions by suppressing *Hd3a* expression in a light and PhyB-dependent manner (Ishikawa et al., 2011). It is likely that there are some unknown transcriptional suppressors interacting with Hd1 during the light period under SD conditions.

OsCO3 is also a COL family member. Just like *Hd1*, *OsCO3*-overexpressing results in delayed flowering under SD conditions by negatively regulating the expression of *Hd3a* and *RFT1* (Kim et al., 2008b; Ishikawa et al., 2011). However, the phenotypic effects of an *OsCO3* mutant have not been documented, leaving uncertainty regarding its in vivo functionality compared to *Hd1*.

Early heading date 1 (Ehd1), which belongs to the B-type response regulator, is another direct upstream key regulator of rice florigen genes. Introducing the functional *Ehd1* from Nipponbare into *Ehd1*-defected cultivar, Taichung 65 (T65), leads to early flowering, especially under SD conditions by inducing the expression of *Hd3a* and *RFT1* (Doi et al., 2004). Functional Ehd1 needs to form a homodimer via a 16-amino acid region in the middle of the protein, and the homodimerization is enhanced by Asp-63 phosphorylation. Ehd1 can also form an inactive heterodimer with OsRR1 (a type-A response regulator) to negatively regulate flowering (Cho et al., 2016).

Besides *Hd1* and *Ehd1*, *DNA binding with one finger protein 12* (*OsDof12*), which is a homolog of *Arabidopsis CDF*, and *SPOTTED LEAF11 interacting protein 1* (*SPIN1*) can also directly regulate the expression of *Hd3a* or *RFT1*. OsDof12 is a plant-specific transcription factor that promotes flowering by upregulating the expression of *Hd3a* and *OsMADS14* under LD conditions (Li et al., 2009). SPIN1, an RNA/DNA-binding protein, interacts with an E3 ligase SPOTTED LEAF11 (SPL11), which can monoubiquitinate SPIN1 and negatively regulate its activity. SPIN1 represses flowering by inhibiting the expression of *Hd1* and *Hd3a* under SD and LD conditions, respectively (Vega-Sanchez et al., 2008). SPIN1 can also interact with RNA-BINDING and SPIN1-interacting 1 (RBS1), which negatively regulates flowering under any conditions (Cai et al., 2014).

5.2.1.3 The Expression and Activity of *Hd1* and *Ehd1* Are Tightly Controlled

Hd1 and *Ehd1* are key regulators of rice flowering time. They integrate different types of environmental signals to regulate the expression of florigens and control rice flowering at the right time. Many upstream regulators of *Hd1* and *Ehd1* have been identified in recent decades. Some of them principally work under LD or SD conditions, while others function independent of day length.

Regulators principally work under LD conditions Rice is a short-day plant, and its flowering is suppressed under LD conditions. So there must be some suppressors that principally function under LD conditions to repress the expression of flowering activator *Ehd1* or induce the expression of flowering inhibitor *Hd1*. Many such suppressors have been identified in recent years, such as *Grain number, plant height, and heading date 7* (*Ghd7*), *DAYS TO HEADING 7* (*DTH7*), *DTH8*, *OsMADS50*, and *OsMADS56*.

The three traits, Grain number, plant height, and heading date (Ghd), are controlled by a single locus *Ghd7*, which encodes a nuclear-localized CCT domain containing protein. Enhanced expression of *Ghd7* causes late flowering and increased plant height and panicle size under LD conditions. Ghd7 can indirectly suppress *Ehd1* expression through *OsCOL10* and other unknown genes under LD conditions. OsCOL10, a COL family protein, is a negative regulator of flowering. Overexpression of *OsCOL10* results in delayed flowering under both SD and LD conditions (Xue et al., 2008; Tan et al., 2016). The upstream regulators of *Ghd7* are also reported, such as the positive regulator OsCOL16 and the negative regulator Rice outermost cell-specific gene 4 (Roc4) (Fig. 5.4). Overexpression of *OsCOL16* causes late flowering under both SD and LD conditions, while overexpression of *Roc4* results in early flowering only under LD conditions (Wei et al., 2016; Wu et al., 2017).
Days to heading 7 (*DTH7*), also known as *Ghd7.1*, *Hd2*, or *Pseudo-Response Regulator 37* (*OsPRR37*), is a major quantitative locus for photoperiod sensitivity and grain yield in rice. *DTH7* acts downstream of the photoreceptor gene *PhyB*, and suppresses flowering through repressing the expression of *Ehd1* under LD conditions (Yan et al., 2013; Gao et al., 2014). *DTH8*, encoding a CCAAT-box-binding transcription factor, is also a QTL locus that controls both flowering time and grain yield. Both rice varieties carrying nonfunctional DTH8 and the *dth8* mutant (*early flowering 8*, *ef8*) have weaker photoperiod sensitivity and early flowering phenotypes. Transferring functional DTH8 into the rice variety that carries nonfunctional DTH8 results in tall plant height, large panicles, and late flowering (Wei et al., 2010; Feng et al., 2014). DTH8 can form a complex with Hd1 and Heme Activator Protein-like 1 (OsHAPL1). The OsHAPL1-DTH8-Hd1 complex functions as a transcription

repressor and downregulates the expression of *Ehd1* and *Hd3a* to repress flowering under LD conditions (Du et al., 2017; Zhu et al., 2017).

OsMADS50 and OsMADS56 are two homologs of *Arabidopsis* Suppressor of Overexpression of Constans1 (SOC1). Knockout *OsMADS50* or overexpression of *OsMADS56* in rice causes late flowering under LD conditions (Lee et al., 2004; Ryu et al., 2009), while overexpression of *OsMADS50* results in extremely early flowering at the callus stage (Lee et al., 2004). OsMADS50 and OsMADS56 interact with each other and act antagonistically to regulate the expression of *O. sativa LEC2 and FUSCA3 Like 1* (*OsLFL1*), which is a repressor of *Ehd1* (Ryu et al., 2009).

Regulators principally work under SD conditions Rice flowers when the day length is shorter than its critical point. The relative duration of day and light is measured by the circadian clock. So the regulation of photoperiodic responses by circadian clocks is vital for plant flowering at the right time. *O. sativa* GIGANTEA (OsGI) is an orthologue of *Arabidopsis* GIGANTEA (GI), which is an important regulator of circadian rhythm and photoperiodic flowering (Izawa et al., 2011). The flowering time of the *osgi* mutant and *OsGI* RNAi transgenic plants is delayed under SD conditions, but there are no consistent changes among different reports under LD conditions (Lee & An, 2015). The expression levels of flowering activators, *Hd1*, *OsMADS51*, *Ehd1*, *Hd3a*, and *RFT1* are reduced in *osgi* mutant. Overexpression of *OsGI* in rice causes late flowering under both SD and LD conditions. Additionally, the expression of *Hd1* is increased while the expression of *Hd3a* is decreased in *OsGI* overexpression plants (Hayama et al., 2003; Lee & An, 2015). This flowering phenotype of *osgi* mutant and *OsGI*-overexpression plants under SD conditions is in accordance with *hd1* mutant and *Hd1*-overexpression plants, respectively. Thus, OsGI preferentially modulates *Hd3a* and *RFT1* expression through Hd1-mediated pathway under SD conditions.

OsMADS51, a type I MADS-box subfamily member, also preferentially promotes flowering under SD conditions (Kim et al., 2007). A mutant of *OsMADS51* flowers 2 weeks later than the wild type under SD conditions but has little change in flowering time under LD conditions. Overexpression of *OsMADS51* causes early flower under SD conditions by promoting the expression of *Ehd1* (Kim et al., 2007).

Regulators for general control of flowering time under all conditions Light is an important factor of photoperiod to decide the flowering time, and light signaling has been proven to be involved in flowering regulation. In plants, light is perceived by a series of photoreceptors, including phytochromes, cryptochromes, phototropins and so on. Mutation of the *PHOTOPERIOD SENSITIVITY5* (*SE5*), which is involved in the phytochrome chromophore biosynthesis pathway and encodes a heme oxygenase, results in early flowering under both SD and LD conditions by upregulating the expression of *Ehd1* (Izawa et al., 2000; Andres et al., 2009; Itoh et al., 2010). There are three phytochrome genes, *OsPhyA*, *OsPhyB*, and *OsPhyC*, in the rice genome. Knocking out either *PhyB* or *PhyC* causes early flowering under LD, while knocking out only *OsPhyA* has no effect on flowering time under LD conditions (Takano et al., 2005). However, knocking out *OsPhyA* and *OsPhyB* or

OsPhyA and *OsPhyC* together leads to double mutant flowering much earlier than the single mutants under LD conditions. This indicates that PhyA, PhyB, and PhyC inhibit flowering in a combinatorial manner. Under SD conditions, OsPhyA promotes flowering, while OsPhyB represses flowering (Takano et al., 2005). The mutant analyses suggested that PhyA can promote flowering through enhancing *OsGI* expression under SD conditions and delay flowering through enhancing *Ghd7* expression under LD conditions in the absence of OsPhyB (Lee et al., 2016). *OsPhyB* can inhibit flowering through activating the expression of COL family members, *OsCOL4*, *OsCOL10,* and *OsCOL13*, to suppress *Ehd1* transcription (Lee et al., 2010; Sheng et al., 2016; Tan et al., 2016). Cryptochrome OsCRY2 and the putative blue-light sensor OsHAL3 also participate in flowering time regulation. *OsCRY2* antisense transgenic plants flowered late under both SD and LD conditions (Singh et al., 2023). *OsHAL3* overexpression plants show an early flowering phenotype under SD conditions. OsHAL3 interacts with Hd1 and directly binds to the promoter of *Hd3a* to activate *Hd3a* expression under SD conditions (Hirose et al., 2006; Su et al., 2016). *O. sativa FLAVIN-BINDING, KELCH REPEAT, F-BOX 1* (*OsFKF1*), a homolog of *Arabidopsis FKF1,* is activated by blue light and promotes flowering by upregulating *Ehd2* and downregulating *Ehd7* to activate *Ehd1* expression under both SD and LD conditions. OsFKF1 can also directly activate the expression of *Ehd1* without moderating the expression of *Ehd2* and *Ghd7* under blue light (Han et al., 2015).

Numerous transcription factors have been identified to play a role in regulating rice flowering under both SD and LD conditions, such as *Ehd2*, *Ehd3*, *Ehd4*, *OsLFL1*, *O. sativa LATERAL ORGAN BOUNDARIES DOMAIN protein 37* (*OsLBD37*), and *OsLBD38*. Ehd2 and Ehd4 are zinc finger transcription factors that promote flowering by inducing the expression of *Ehd1*, and their mutants show late flowering under both SD and LD conditions (Matsubara et al., 2008; Gao et al., 2013). Ehd3 is a plant homeodomain (PHD) finger transcription factor, which can promote flowering through inhibiting the expression of *Ghd7* under LD conditions or directly activating the expression of *Ehd1* under SD conditions (Matsubara et al., 2011). OsLFL1 is a B3 DNA-binding domain-containing transcription factor that can repress the expression of *Ehd1* by directly binding to the RY *cis*-elements (CATGCATG) of *Ehd1* promoter. In the gain-of-function mutant W378, ectopic expression of *OsLFL1* results in a delay of flowering time by more than 1 month under both SD and LD conditions. By using the RNAi method to knock down the expression of *OsLFL1* in W378, the flowering time can be restored (Peng et al., 2007, 2008). Additionally, OsLBD37 and OsLBD38, plant-specific transcription factors, negatively regulate flowering by repressing the expression of *Hd3a* and *RFT1* through *Ehd1* under both SD and LD conditions. Overexpression of *OsLBD37* or *OsLBD38* in rice delays flowering time and increases yield (Li et al., 2017).

EARLY FLOWERING 3 (OsELF3), a homolog of *Arabidopsis* ELF3, is a positive regulator of *OsGI* under SD conditions and a negative regulator of *Ghd7* under LD conditions. The flowering time of *oself3* mutant and *OsELF3* RNAi transgenic plants is apparently delayed under both SD and LD conditions. Overexpression of

OsELF3 promotes flowering under LD conditions, but delays flowering under SD conditions (Saito et al., 2012; Yang et al., 2013).

The last decades saw great progress in the understanding of the genetic and molecular regulation of rice flowering due to genome sequencing and advanced gene mapping technology. Dozens of genes are identified to be involved in flowering regulation, exerting direct or indirect influence on the expression of *Ehd1* or *Hd1* to dictate the timing of rice flowering. However, the regulatory mechanisms of these genes and their response to photoperiod changes remain unclear. Thus, more flowering regulators need to be clarified for improving rice yield through controlling its flowering at the right time.

5.2.2 Photoperiodic Flowering Regulation in Soybean

Soybean is a significant leguminous crop that provides food and oil to humans. Soybean is a short-day plant and highly sensitive to photoperiod, meaning it flowers when the day length is shorter than its critical point. The flowering time of soybean has a significant impact on its growth habit, product yield, geographic evolutionary, and domestication. Over the past few decades, numerous genes involved in regulating soybean flowering have been identified and studied. In soybean, there are ten flowering time QTL loci known as E1-E9 and J. Among these loci, E1, E2, E3, E4, and E9 have been specifically identified through genetic mapping methods using experimental materials with natural variants (Cao et al., 2017). And the *E5* has been excluded from the flowering control QTL locus, as it may represent several identified QTL loci (Dissanayaka et al., 2016). *J* is the ortholog of Arabidopsis *ELF3* and plays a key role in the adaptation of plants in low-latitude regions. *J* promotes flowering by suppressing *E1*, a flowering repressor specific to legumes, which in turn releases the inhibition of *FT* expression by *E1* (Lu et al., 2017). *Tof11* and *Tof12* encode PRR (pseudo-response-regulator) proteins. Under LD conditions, *Tof11* and *Tof12* reduce the suppression of *E1* by inhibiting the circadian clock gene *LHY*, leading to the downregulation of *FT* expression, and ultimately delayed flowering. The gradual evolution of *Tof11* and *Tof12* promotes the adaptation of soybean to mid- and high-latitude regions (Lu et al., 2020). In addition, some orthologues of *Arabidopsis* flowering regulators have been identified from soybean by genome sequence comparison and transgenic analysis.

5.2.2.1 Florigens in Soybean

There are 24 phosphatidyl ethanolamine-binding protein (PEBP) family members in soybean, and 11 of them belong to the FT-like subfamily. These FT-like genes are further divided into two subgroups, the FTL subgroup and the TSF subgroup. *GmFT2a/GmFTL3*, *GmFT2b/GmFTL5*, *GmFT3a/GmFTL1*, *GmFT3b/GmFTL2*, *GmFT5a/GmFTL4*, *GmFT5b/GmFTL* and a truncated copy, *GmFTL7*, belong to the

FTL subgroup. *GmFT1a/GmTSF2*, *GmFT1b/GmTSF1*, *GmFT4/GmTSF4*, and *GmFT6/GmTSF3* belong to the TSF subgroup. Overexpression of FTL subgroup genes *GmFT2a*, *GmFT2b*, *GmFT3a*, *GmFT3b*, *GmFT5a*, and *GmFT5b* causes early flowering in *Arabidopsis* (Kong et al., 2010; Fan et al., 2014). In contrast, overexpression of *GmFT4*, one of the TSF subgroup genes, results in delayed flowering time under LD. The expression of flowering repressor *GmFT4* is upregulated under LD and downregulated under SD conditions. On the other hand, the expression of flowering activators, *GmFT2a* and *GmFT5a*, is preferably upregulated under SD conditions (Kong et al., 2010). Among the six putative soybean flowering activators, *GmFT2a* and *GmFT5a* have the highest expression levels in trifoliate leaf and are greatly induced under SD conditions; thus, GmFT2a and GmFT5a are the most likely florigens in soybean (Kong et al., 2010; Zhai et al., 2014; Cao et al., 2017). Both *GmFT2a* (*E9*) and *GmFT5a* are QTLs that exist in cultivated soybean varieties and control flowering time (Takeshima et al., 2016; Zhao et al., 2016). Overexpression of *GmFT2a* and *GmFT5a* in soybean cultivar Williams 82 promotes flowering under LD conditions (Nan et al., 2014). Unlike *Arabidopsis* and rice florigens encoding genes, which are only transcribed in leaf vascular bundles, the transcripts of *GmFT2a* are detected not only in vascular tissues but also in apical and lateral meristems of the soybean plant (Sun et al., 2011).

The FT/FD-AP1 pathway is also conserved in soybean. GmFT2a and GmFT5a interact with GmFDL19, a homolog of *Arabidopsis* FD, to activate the expression of floral identity genes such as *GmAP1*, *GmSOC1*, and *GmLFY*, promoting flowering in soybean in a similar manner to FT in *Arabidopsis* (Fig. 5.5). In addition to the physical interaction between GmFT2a and GmFDL19, GmFT2a can also activate the transcriptional of *GmFDL19*. GmFDL19 directly binds to the ACGT *cis*-elements of the *GmAP1* promoter to activate its expression. Overexpression of *GmFDL19* in soybean plants results in early flowering under LD conditions (Nan et al., 2014).

5.2.2.2 The Upstream Regulators of Soybean Florigen

Arabidopsis CO and rice *Hd1* are the direct and central upstream regulators of *FT* gene in the photoperiodic flowering regulation pathway. The CO-FT module also exists in soybean. There are 26 (13 pairs) CO homologs in soybean, and they are divided into three clades: 4 pairs in clade I, 3 pairs in clade II, and 6 pairs in clade III. GmCOL1a, GmCOL1b, GmCOL2a, and GmCOL2b, which exhibit the highest similarity with *Arabidopsis* CO and rice Hd1, belong to clade I. Ectopic expression of *GmCOL1a*, *GmCOL1b*, *GmCOL2a*, and *GmCOL2b* in the *Arabidopsis* mutant *co* can fully complement its late flowering phenotype. *GmCOL1a* and *GmCOL1b* have much higher expression than *GmCOL2a* and *GmCOL2b* under both SD and LD conditions. And *GmCOL1a/GmCOL1b* exhibit a strong diurnal rhythm and have a peak at dawn under SD conditions, which aligns with the expression pattern of GmFT5a. Thus, GmCOL1a and GmCOL1b might be the key flowering regulators in soybean (Wu et al., 2014). *gmcol1b* mutant shows an early flowering phenotype

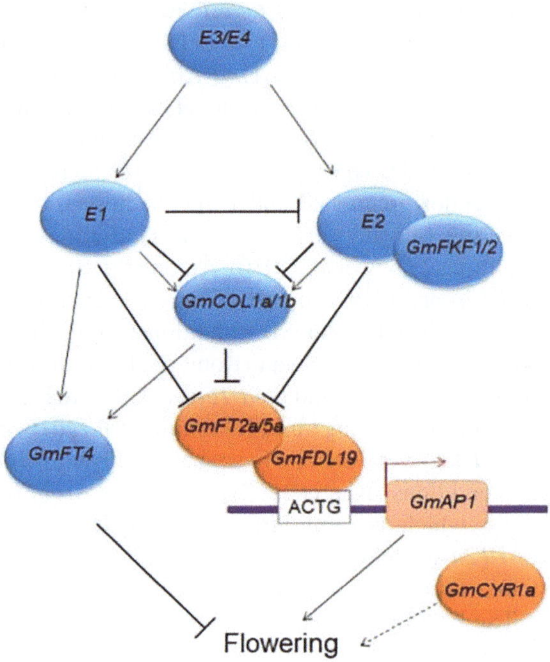

Fig. 5.5 Photoperiodic flowering regulation in soybean. *GmFT2a* and *GmFT5a* are the florigens in soybean. *E3/E4–E1/E2* module plays a central role in photoperiodic control of soybean flowering. Arrows indicate activation. Bars indicate repression. Pink ovals indicate positive regulators of flowering. Light blue ovals indicate negative regulators of flowering

under LD conditions, and *GmCOL1a* overexpression transgenic plants flowered later than the wild type under LD conditions. In *gmcol1b* mutant, the expressions of *GmFT2a* and *GmFT5a* are increased, while the expression of *GmFT4* is decreased under LD. Accordingly, in *GmCOL1a* overexpression plants, the expressions of *GmFT2a* and *GmFT5a* are decreased, while the expression of *GmFT4* is increased under LD conditions. Therefore, GmCOL1a and GmCOL1b function as repressors of *GmFT2a* and *GmFT5a*, and activators of *GmFT4*, ultimately inhibiting flower formation under LD conditions (Cao et al., 2015). Besides, the expression *of GmFT2a* and *GmFT5a* is upregulated in *GmCOL1a* overexpression plants under SD conditions, but no significant change in flowering time was observed (Ishikawa et al., 2011; Cao et al., 2015).

E1, encoding a nuclear-localized flowering repressor, was identified by the positional cloning method. Homologs of E1 are present in various legume species, but have not been discovered in *Arabidopsis* and rice. Plants with a dysfunctional allele of the *E1* locus and knock-out mutants all exhibit an early flowering phenotype (Xia et al., 2012). Two *E1* homologs, *E1La* and *E1Lb*, act as repressors of flowering in soybean (Xu et al., 2015). *E1*, *E1La*, and *E1Lb* are exclusively expressed during the light period, exhibiting two peaks under LD conditions. The induction of their expression after dawn requires a period of light before dusk on the previous day (Xu et al., 2015). *E1* and its homologs inhibit soybean flowering through repressing the expression of florigens *GmFT2a* and *GmFT5a* under LD conditions (Xu et al., 2015).

E2/GmGIa/GmGI3, a homolog of *Arabidopsis GI*, is a flowering QTL locus in soybean and was identified by a map-based cloning strategy (Watanabe et al., 2011). The *e2* mutant allele that harbors a premature stop codon of *E2* showed an early flowering phenotype (Watanabe et al., 2011). In *E2* loss-of-function genotype, the expression of *GmFT2a* is elevated; however, the expression of *GmFT5a* is not affected (Watanabe et al., 2011). Thus, it is possible that *E2* inhibits soybean flowering time through repressing *GmFT2a* expression under LD conditions (Watanabe et al., 2011). The expression of *GmCOL1a/GmCOL1b* is downregulated in *e1* and *e2* mutants, and upregulated in *E1* overexpression plants under LD conditions. The suppression of *GmFT2a* and *GmFT5a* by *E1* and *E2* is partially mediated by *GmCOL1a/GmCOL1b*. Interestingly, *GmCOL1a* can also repress the expression of *E1* and *E2* in a feedback manner. Additionally, *E1* also suppresses the expression of *E2* under LD conditions. Thus, *E1*, *E2*, and *GmCOL1a* form a negative feedback regulation loop (Cao et al., 2015).

GI interacts with FKF1 to control CDF1 stability for regulating *CO* transcription in *Arabidopsis* (Sawa et al., 2007). This pathway is also conserved in soybean. There are three homologs of *GI*, *GmGI1*, *GmGI2*, and *GmGI3/E2/GmGIa*, two homologs of *FKF1*, *GmFKF1* and *GmFKF2*, in soybean. GmGIs can also interact with GmFKFs and GmCDF1. Ectopic overexpression of *GmFKF1* and *GmFKF2* in *Arabidopsis* causes early flowering under SD (Li et al., 2013). Under LD conditions, E2-FKF1 complex interacts with J/ELF3, a key component of the evening complex EC, and promotes the degradation of J/ELF3. EC can bind to *E2* promoter to inhibit its transcription. These two processes jointly determine the flowering time of soybean by regulating *E1* expression. Under SD conditions, the function of E2 is suppressed, maintaining the inhibitory effect of EC on *E1* and resulting in early flowering of soybeans (Zhao et al., 2024).

E3 and *E4*, two loci responsible for controlling flowering, have been identified as phytochrome *GmPHYA3* and *GmPHYA2*, respectively (Liu et al., 2008; Watanabe et al., 2009). Both *E3* and *E4* were first identified by screening of photoperiod-insensitive plants using varying R:FR ratios of artificial light sources to extend the day length to 20 h. The *e3* recessive homozygote can initiate flowering under LD conditions with high R:FR ratios. The *e4* recessive homozygosity is necessary for the *e3* homozygote flowering under LD conditions with low R:FR ratios; however, *e4* alone cannot confer insensitivity to LD conditions with either high R:FR ratio or low R:FR ratio (Liu et al., 2008). In *e3, e4* single and double recessive mutants, the expression levels of *E1*, *E2*, *GmCOL1a*, *GmCOL1b, GmFT2a*, and *GmFT5a* are all downregulated (Kong et al., 2010; Xia et al., 2012; Cao et al., 2015). So *E3* and *E4* inhibit flowering under LD conditions through *E1*- and *E2*-mediated pathway.

The blue light receptor CYR2 regulates photoperiodic flowering in *Arabidopsis* and rice (Guo et al., 1998; Hirose et al., 2006). In soybean, there are six cryptochrome genes, *GmCYR1a-1d*, *GmCYR2a*, and *GmCYR2b*. Only *GmCYR1a* and *GmCYR2a* have been studied in detail. Both *GmCRY1a* and *GmCRY2a* participate in blue light inhibition of cell elongation, but only GmCRY2a is degraded under blue light through the 26S proteasome. However, GmCRY1a, but not GmCRY2a, can promote floral initiation in *Arabidopsis*. The diurnal expression pattern of

GmCRY1a correlates with photoperiodic flowering and latitudinal distribution of soybean cultivars (Zhang et al., 2008). So, *GmCRY1a* may be the blue light receptor that mediates photoperiodic flowering in soybean.

GmRAV is a homolog of the *Arabidopsis* AtRAV2, containing an AP2 domain and a B3 DNA-binding domain. *GmRAV* is rhythmically expressed and reaches a peak at 4 h after the beginning of a dark period under SD conditions, and the expression of *GmRAV* is higher under SD than LD before flowering. *GmRAV*-inhibition (*GmRAV-i*) plants exhibit an early flowering phenotype and have enhanced expression of *GmFT2a* and *GmFT5a*. In addition, *GmRAV-i* plants are less sensitive to sucrose in hypocotyl and root growth inhibition assays. Therefore, *GmRAV* is a negative flowering regulator under SD conditions, participating in both the photoperiodic flowering pathway and the sucrose-controlled flowering time pathway (Lu et al., 2014).

In the last decades, the flowering regulation network of soybean has been established mainly through traditional map-based cloning strategy and comparative studies between soybean and the model plants *Arabidopsis* and rice. The sequencing of soybean genome has greatly accelerated this process. Besides the conserved flowering regulation pathways, there are also some legume-unique pathways that do not exist in *Arabidopsis* and rice, such as the *E1*-dependent pathway. However, the specific regulation mechanisms remain unexplored. With the identification of more and more flowering regulators, the soybean-unique flowering pathway will be better understood.

5.3 Epigenetic Regulation of Floral Transition

Epigenetics refers to covalent modification of DNA, protein, or RNA, resulting in heritable changes of gene expression without alterations in the DNA sequence (Bollati & Baccarelli, 2010; Verma et al., 2023). Epigenetic regulation controls gene expression in different aspects: DNA methylation, RNA modifications, histone modifications, and nucleosome positioning. DNA methylation mainly occurs in the cytosine ring of CpG dinucleotides. Histones can be covalently modified by the addition of methyl, acetyl, phosphor or ubiquityl, or sumo groups. Besides, nucleosome positioning and microRNAs also contribute to the epigenetic regulation.

5.3.1 DNA Methylation

DNA methylation occurs in the cytosine ring of the CpG dinucleotide, playing a crucial role in gene transcription regulation and chromosomal integrity maintenance. It has been found that the demethylating agent 5-azacytidine can substitute for vernalization and accelerate flowering (Burn et al., 1993). The *decreased DNA methylation 1* (*ddm1*) mutant and *methyltransferase 1* (*MET1*) antisense *Arabidopsis* plants showed decreased DNA methylation and an early flowering phenotype (Finnegan et al., 1998). Tetrahydrofolate (THF), a group of essential B-complex vitamins, is a critical coenzyme in one-carbon transfer and crucial for DNA methylation. Wild-type *Arabidopsis* plants supplemented with exogenous THF or the high endogenous THF mutant, *dihydrofolate synthetase folypoly-Glu synthetase homolog b-3* (*atdfb-3*), exhibited a late flowering phenotype. Transcriptional array and qRT-PCR analysis indicated that the delayed flowering induced by THF may be attributed to the increased expression of the flowering repression gene *Flowering Wageningen* (*FWA*) (Wang et al., 2017).

The accumulation of THF reduces the DNA methylation level through MET1-mediated epigenetic modifications at CG and CHG sites. The HhaI-based cytosine methylation assay showed that the *FWA* promoter sequence in *atdfb-3* could not be detected, indicating that increased endogenous THF in *atdfb-3* abolished DNA methylation in the *FWA* promoter sequence. When the wild type was supplied with THF, the abundance of PCR product for *FWA* promoter was also lower than that from the wild type without THF (Wang et al., 2017). The DNA hypomethylation of *FWA* promoter sequence resulted in the ectopic *FWA* expression, and the accumulated FWA protein interacts with FT and inhibits its function to delay flowering transition (Ikeda et al., 2007). Methylcytosine-binding proteins VIM1, VIM2, and VIM3 collaboratively maintain the global CG methylation, and the late flowering phenotype of *vim1 vim2 vim3* triple mutant is also relevant to the DNA hypomethylation patterning of *FWA*. *FWA* is an important gene in epigenetic silencing to regulate the flowering time (Woo et al., 2008; Wang et al., 2017).

In rice, knockdown of S-adenosyl-l-methionine synthetase (*SAMS*) genes also causes alterations of DNA methylation, leading to late flowering. *SAMS* encoding S-adenosyl-l-methionine synthetases, which are methyl group donors involved in regulation of various cellular functions. RNA interference (RNAi) transgenic plants of *OsSAMS1*, *2*, and *3* showed late flowering phenotypes and a great reduction of the flowering key genes, including *Early heading date 1* (*Ehd1*), *Hd3a*, and *RFT1* (rice *FT*-like genes). Three types of cytosine DNA methyltransferases of METHYLTRANSFERASE 1 (MET1), CHROMOMETHYLASE 3 (CMT3), and DOMAINS REARRANGED METHYLTRANSFERASE 2 (DRM2) were greatly reduced in transgenic rice plants, so the SAM deficiency may inhibit DNA methyltransferases, which subsequently decreases DNA methylation at specific sites. The reduction of DNA methylation repressed the transcription of the *Ehd1*, *Hd3a*, and *RFT1* and led to a late-flowering phenotype in the RNAi transgenic rice plants (Li et al., 2011).

5.3.2 *Histone Modifications*

The nucleosome is the basic repeating structure unit of a chromosome, comprising an octamer core containing two sets of histones (H2A, H2B, H3, and H4) and 145–147 base pairs of DNA that coil around the octamer. Histone can be covalently

Table 5.1 The effects of histone modification on *FLC* chromatin

Methylation		
H3K4m	Activation	Liu et al. (2010)
H3K36me	Activation	Zhao et al. (2005)
H3K9me	Repression	Liu et al. (2010)
H3K27me	Repression	Liu et al. (2010)
H4R3sme2	Repression	Pei et al. (2007), Wang et al. (2007)
H4R3ame2	Repression	Niu et al. (2007)
Acetylation		
H3K9ac	Activation	Kim et al. (2013)
H3K27ac	Activation	Kim et al. (2013)
H3K9K14ac	Activation	Yu et al. (2011), Luo et al. (2015)
H3ac	Activation	Kang et al. (2015)
H4 ac	Activation	Bond et al. (2009)
H4K5ac	Activation	Xu et al. (2014a)

modified by a variety of groups, like methyl, acetyl, phosphor, or ubiquityl, which influence many cellular processes, including transcription, DNA repair, and replication. To date, a wide variety of modifications on histones have been identified, of which methylation and acetylation are two important forms (Huang et al., 2014; Lawrence et al., 2016). In *Arabidopsis*, histone modifications on *FLC* locus regulate its epigenetic activating and silencing (Table 5.1). In particular, the epigenetic regulation of *FLC* expression in response to vernalization is considered a model for understanding the epigenetic mechanisms of histone modification-regulated gene transcription (Xu & Chong, 2018; Luo & He, 2020).

5.3.2.1 Histone Methylation

Histone methylation with mono-, di-, or trimethyl can take place at multiple lysine or arginine sites (Fulton et al., 2018; Al Temimi et al., 2020). This modification specificity determines transcriptional activation or repression. The methylation on H3K4, H3K36, and H3K79 activates gene transcription, whereas H3K9 and H3K27 are associated with transcriptional repression (Sundar et al., 2013). The methylation is catalyzed by methyltransferase. Histone methyltransferases add methyl groups to specific histone Lys or Arg residues by using S-adenosylmethionine as a methyl donor. The methyltransferase can "write" the methyl marks on the histone. All of the H3K36-specific methyltransferases have the catalytic SET domain (SET domain: a conserved motif containing approximately 130 amino acids, which was originally identified in three *Drosophila* proteins: SU(var)3-9, Enhancer of zeste, and Trithorax) (Seni et al., 2023). In plants, the SET DOMAIN GROUP (SDG) proteins play important roles in multiple growth and development processes, including flowering, plant development, and stress response (Zhou et al., 2020; Zhang et al., 2021). JMJ14 is an H3K4 demethylase and functions as an eraser to prevent early flowering by repressing the expression of floral integrators *FT*, *AP*, *SOC1*, and *LFY* (Lu et al., 2010).

Polycomb repressive complex 2 (PRC2) is involved in maintaining gene repression by post-translationally modifying histones, specifically through trimethylation of H3K27 (Sharaf et al., 2022). The *Drosophila melanogaster* PRC2 consists of four subunits: Suz12, E(z), p55, and Esc (Bu et al., 2023). In *Arabidopsis*, Suz12 homologs include VRN2, EMF2, and FIS2; E(z) homologs are CURLYLEAF (CLF), MEDEA, and SWINGER; p55 and Esc homologs are MSI1 and FIE. Within PRC2, CLF functions as a writer, catalyzing the formation of H3K27me3 (Godwin & Farrona, 2022). The role of H3K27me3 in the vernalization response has been well established in *Arabidopsis*. Prior to cold exposure, the epigenome readers VALs recognize both the cold memory element (CME) and the histone mark H3K27me3 in *FLC* chromatin. Meanwhile, VALs also directly interact with LHP1 to recruit PRC2 for localizing PRC2 on the *FLC* chromatin. During vernalization, the plant homeodomain (PHD) protein VIN3 is induced and subsequently binds to PRC2 to form PHD (VIN3)-PRC2 complex. Elevated PRC2 activity results in H3K27me3 peak at the nucleation region. After vernalization, H3K27me3 spreads and covers the *FLC* locus to suppress its expression for flowering transition. At the embryo stage, transcription factor LEAFY COTYLEDON 1 (LEC1) promotes the enrichment of H3K36 methyltransferase EFS/SDG8 at the *FLC* to antagonize H3K27me3 and erases the vernalization memory. Here the LHP1 recognize H3K27me3 mark to play a reading function (Liu et al., 2010; Qüesta et al., 2016; Yuan et al., 2016; Tao et al., 2017; Yang et al., 2017).

Vernalization is not necessary for the flowering of rice, but its flowering time is still controlled by histone modifications. In rice, Hd1 represses the *Hd3a* expression and results in late flowering under LD (see Sect. 5.2). OsNF-YB11/DTH8, one of the NUCLEAR FACTOR Y transcription factors, directly interacts with Hd1 to form the DTH8-Hd1 complex. This complex is essential for Hd1 to suppress *Hd3a* expression. Under LD conditions, DTH8-Hd1 complex binds directly to the transcription start region of *Hd3a*, leading to the accumulation of repressive histone mark H3K27me3, which downregulates *Hd3a* expression and results in delayed flowering (Du et al., 2017). Rice *SDG701* encodes a H3K4-methyltransferase and promotes flowering under either LD or SD conditions. SDG701 specifically binds to chromatin regions of *Hd3a* and *RFT1* florigens to increase the deposition of H3K4me3, thereby enhancing the transcription of *Hd3a* and *RFT1* and promoting floral transition (Liu et al., 2017). The SDG708 protein is one of the H3K36-specific methyltransferases and promotes flowering by activating the expression of *EHD1*, *Hd3a*, and *RFT1* in rice. *SDG708* knockdown reduced the deposition of H3K36me1, H3K36me2, and H3K36me3 at *EHD1*, *Hd3a* and *RFT1* loci. Moreover, SDG708 is involved in diverse biological processes by catalyzing genome-wide modification levels of the mono-, di-, and trimethylation on H3K36 (Liu et al., 2016).

Another two rice H3K36 methyltransferases SDG724 and SDG725 promote flowering by affecting the levels of H3K36 methylation. SDG724 catalyzes H3K36me at *MADS50* and *RFT1* loci to promote the MADS50-Ehd1-RFT1 flowering pathway (Sun et al., 2012). SDG725 is involved in the promotion of BR-related gene expression to promote flowering (Sui et al., 2012). The histone marks must be recognized by special readers. MRG702 binds to H3K4me3 and H3K36me3, particularly in the nucleus, to promote flowering. *MRG702*-knockdown mutants displayed a late-flowering phenotype. Consistent with this phenotype, the transcript levels of flowering regulatory genes *Ehds*, *OsMADS50*, *OsMADS51*, *Hd3a*, and *RFT1* were downregulated in the mutants (Jin et al., 2015).

Apart from the modification on histone lysine, the histone arginine residue can also be modified with mono-methylation, asymmetric dimethylation, or symmetric dimethylation. The protein arginine methyltransferases are classified into two types, type I and type II. Type I enzymes catalyze the monomethylation and asymmetric dimethylation on Arg, and type II enzymes catalyze the monomethylation and symmetric dimethylation on Arg (Liu et al., 2010). The *Arabidopsis* histone arginine methyltransferase 5 (AtPRMT5, also named *Arabidopsis* Shk1-binding protein 1, SKB1) belongs to the type II class. AtPRMT5/SKB1 catalyzes histone H4R3 symmetric dimethylation (H4R3sme2) and represses *FLC* expression to promote the transition from vegetative growth to reproductive growth (Pei et al., 2007; Wang et al., 2007). Further analysis of the RNA-sequence revealed that deficiency in AtPRMT5 results in RNA splicing alterations, specifically affecting the splicing of pre-mRNA of *FLK*, a gene encoding a positive regulator of flowering by repressing *FLC*. In *atprmt5* mutant, there was an increase in unspliced *FLK* transcripts with intron1 retention, leading to a decrease in FLK protein levels. This ultimately resulted in the upregulation of FLC and delayed flowering (Deng et al., 2010). AtPRMT10 belongs to the type I group, which is responsible for H4R3 asymmetric dimethylation (H4R3ame2). The H4R3ame2 mark also represses *FLC* expression independent of AtPRMT5 (Niu et al., 2007). PRMT4 is another member of type I arginine methyltransferases. In *Arabidopsis*, AtPRMT4a and AtPRMT4b serve as major methyltransferases responsible for the formation of H3R17me2a. Although the *atprmt4a atprmt4b* double mutant displayed a late-flowering phenotype with elevated *FLC* level, the H3R17ame2 levels in *FLC* chromatin were not reduced, indicating that AtPRMT4a and AtPRMT4b do not regulate *FLC* transcription through the deposition of H3R17ame2 in *FLC* chromatin (Niu et al., 2008).

5.3.2.2 Histone Acetylation

Histone acetylation is also involved in epigenetic regulation of flowering, and this type of modification is regulated by histone acetyltransferases and histone deacetylases. *Arabidopsis* genome encodes 18 histone deacetylases (HDAs) (Hollender & Liu, 2008). In the *hda5-1* mutant, gene activation marks H3K9K14ac and H3K4me3 were upregulated in the promoter, the first exon, and the first intron regions of *FLC*, with similar results observed in *MAF1*. The upregulated H3K9K14ac and H3K4me3 lead to a late-flowering phenotype with increased transcription levels of flowering repressor genes *FLC* and *MAF1*. HDA5, HDA6, FVE, and FLD are present in the same complex to regulate flowering time (Luo et al., 2015). The deacetylases HDA5 and HDA6 and the demethylase FLD function as erasers of the activation marks of

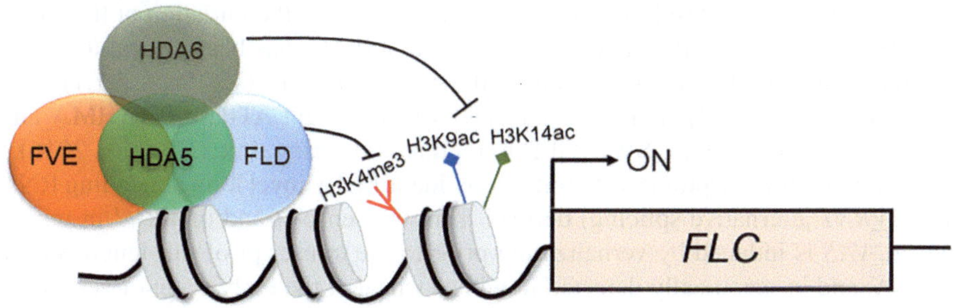

Fig. 5.6 The crosstalk between acetylation and methylation regulates flowering

H3K9ac, H3K14ac, and H3K4me3, respectively. The physical interaction between HDA6 and FLD mediates the crosstalk between histone acetylation and methylation (Fig. 5.6). For example, the modification levels of activation marks of H3K9K14ac and H3K4me3 were increased in *FLC*, *MAF4*, and *MAF5* loci and led to late flowering in *hda6* and *fld* mutants (Yu et al., 2011; Liu et al., 2014). In contrast, HDA5 binds to the *FLC* and *PAF1* loci to positively regulate flowering by repressing *FLC* expression (Luo et al., 2015).

The crosstalk between methylation and acetylation is also implied by H3K36me3 reader MRG1 and MRG2. MRGs bind to the H3K36me3 mark and recruit HAM1- and HAM2-containing complex to the target loci. HAMs are H4K5-specific histone acetyltransferases and can catalyze H4K5ac of *FLC* and *FT* chromatin to promote their expression (Xu et al., 2014b). HDA19 preferentially catalyzes deacetylation of H3K27, which is coordinated with CLF-PRC2-mediated H3K27me3 at *FLC* through the interaction between HDA9 and PRC2 partners, leading to *FLC* silencing in response to vernalization (Zeng et al., 2020).

Histone acetylation is influenced by special histone phosphorylation events. In *Arabidopsis*, the kinase MLK4 catalyzes the phosphorylation of H2A on serine 95 (H2AS95ph) and can be recruited to the *GI* promoter by transcription factor CCA1. CCA1 interacts with YAF9a, a co-subunit of SWR1 and NuA4 complex. This complex is responsible for histone H4 acetylation and H2A.Z deposition to regulate *GI* transcription. In the *mlk4* mutant, the levels of H2AS95ph, H4 acetylation, and H2A.Z accumulation are reduced, leading to repression of *GI* expression and a late flowering phenotype (Su et al., 2017).

5.3.3 *RNA Regulation*

5.3.3.1 Noncoding RNA

The noncoding RNA is also involved in epigenetic modification. The most common examples are long noncoding RNAs (lncRNAs) and microRNA (miRNAs). In *Arabidopsis*, vernalization induced the expression of lncRNAs COOLAIR,

COLDAIR, and COLDWRAP from *FLC*, which regulate the initiation of flowering by silencing *FLC* (see details in Sect. 5.4). The miR156 has been shown to be the master regulator of floral induction in the aging pathway (see Sect. 5.1.1). The expression of *MIR156A* is fine-tuned by DNA-dependent ATPase BRAHMA and histone methyltransferase SWINGER to regulate flowering time (Xu et al., 2016).

A splicing transcriptome screening assay identified a novel long noncoding RNA *VAS* (*VRN1* alternative splicing) derived from the sense strand of *VRN1* in winter wheat. *VAS* is induced by vernalization prior to the transcript of full-length *VRN1* mRNA, and is specifically detected in winter wheat at the early stage of vernalization. *VAS* associates with transcription factor TaRF2b to facilitate TaRF2b-TaRF2a complex binding to the Sp1 motif, breaking the gene loop between the promoter and the first intron to accelerate flowering through activating *VRN1* transcription (Xu et al., 2021).

5.3.3.2 RNA-Binding Proteins

RNA-binding proteins include glycine-rich RNA-binding protein (GRP) 1–7 and RZ-1A, RZ-1B, and RZ-1C in *Arabidopsis*. The RZ-1B and RZ-1C represent a unique group of GRPs and are required for normal plant development. RZ-1B and RZ-1C localize in nuclear speckles to modulate RNA splicing. *RZ-1C* overexpression plants displayed early flowering and reduced *FLC* mRNA level in *Arabidopsis*. The *FLC* transcript analysis indicated that RZ-1B and RZ-1C promote *FLC* pre-mRNA efficient splicing to repress *FLC* gene transcription (Wu et al., 2016).

In *Arabidopsis*, GRP7 is another RNA-binding protein that regulates flowering time in the autonomous pathway (Streitner et al., 2008). GRP7 directly binds to *FLC* antisense pre-mRNA at intron 6 to guide its process for producing two types of noncoding RNA transcripts, proximal and distal types of *COOLAIR*. The proximal *COOLLAIR* functions as a negative regulator for *FLC* mRNA maturation by histone methylation (see Sect. 5.4). In *grp7-1* mutant, the loss of function of GRP7 causes a decrease in proximal *COOLAIR* and an increase in *FLC* mRNA. *Arabidopsis* JACALIN-LECTIN LIKE 1 (AtJAC1) interacts with GRP7 in cytoplasm and inhibits the nuclear accumulation of GRP7 to weaken GRP7 function in nuclear. Therefore, both the *grp7-1* mutant and *AtJAC1*-overexpressing lines display late flowering phenotypes (Xiao et al., 2015).

5.3.3.3 mRNA Modification

RNA methylation mainly appears on cytosine and adenosine. The *N6*-methyladenosine (m6A) is the most common modification of RNA in eukaryotes, which is mediated by methyltransferases (Writer), and the methylation can be reversed by demethylases (Eraser). The m6A site is identified by m6A-binding proteins (Reader). FIP37 is a core component of *Arabidopsis* m6A methyltransferase complex and mediates the modification of shoot meristem genes to maintain meristem size (Shen et al., 2016). In *Arabidopsis*, ALKBH10B functions as an mRNA demethylase that is widely expressed in various organs. Overexpression of *ALKBH10B* decreases m6A levels, while deletion of *ALKBH10B* leads to an increase in total m6A levels in poly(A)$^+$-RNA. It is shown that the mRNA demethylation catalyzed by ALKBH10B stabilizes the mRNA of *FT*, *SPL3*, and *SPL9* to promote floral transition (Duan et al., 2017).

In summary, there are three levels of epigenetic control of gene expression: DNA modification, RNA modification and its splicing, and histone modification. These three systems collaborate in the regulation of gene expression, particularly in controlling *FLC*, the pivotal gene for flowering in *Arabidopsis*.

5.4 Vernalization Pathway

Flowering is precisely controlled by both genetic and external environmental factors. Vernalization is an important external environmental factor to induce flowering. Biennial plants (such as radish, cabbage, sugar beet, and hyacinth) and winter annual plants (such as winter wheat, and winter rye) flower in the spring or early summer after suffering a cold winter. The phenomenon in which prolonged exposure to low temperatures promotes flowering is known as vernalization. Based on the demand for vernalization, these plants can be divided into winter, semiwinter, and spring ecotypes.

5.4.1 Concept and Characteristics

The phenomenon of vernalization was first scientifically elucidated by the German botanist G. Gassner in 1918 through controlled temperature experiments conducted on cereal plants. In 1928, the Soviet agronomist Lysenko proposed the concept of “vernalization,” presenting it as a practical agricultural technique aimed at promoting earlier flowering and enhancing crop yields, particularly in winter cereals. Subsequently, scholars from various countries have conducted extensive research on vernalization in plants (Chouard, 1960). In winter wheat, vernalization is a necessary process and significantly contributes to the floral transition (Fig. 5.7).

There are four key characteristics of vernalization. First, vernalization is a slow process. The vernalization duration should be 0–3 °C for more than 35 days for winter wheat, 0–12 °C for 15–35 days for semiwinter wheat. Second, the effects of uncompleted vernalization can be erased by high temperature (~35 °C in winter wheat). Early studies have shown that the process of vernalization can be divided into three time phases: induction stage, acceleration stage, and stabilization stage (Yong et al., 2003). If the vernalization treatment only reached the level of the induction stage or/and the acceleration stage, winter wheat may not blossom normally, and the molecular memory of vernalization from these two stages can be

Fig. 5.7 The effect of vernalization duration on flowering time of winter wheat. Plants of the winter wheat cultivar Kenong 199 (KN199) were subjected to vernalization at 4 °C for 0, 7, 14, 21, 28, and 35 days, respectively. Following vernalization, the plants were transferred to natural environmental conditions and grown for 11 weeks

erased by devernalization (a short period of high temperature, for winter wheat, 35 °C for 5 days). However, if the cold treatment duration is sufficient and the stabilization stage is fully completed, the molecular memory of vernalization cannot be erased by devernalization until the next generation (Dennis & Peacock, 2009). Third, the perception site for vernalization signals is the shoot tip. According to grafting experiments, the classical florigen hypothesis posits that the shoot tip is responsible for perceiving vernalization signals (Curtis & Chang, 1930). However, due to the challenge of separating shoot apical meristems from young leaves in grafting experiments, later studies have proposed that young leaves also contribute to the perception of vernalization signals in plants (Crosthwaite & Jenkins et al., 1993; Yong et al., 2003). In winter wheat, in situ RNA hybridization analysis showed that the transcription of a vernalization response gene *VER2* is initially detected in immature leaves to accelerate flowering, indicating the potential role of young leaves in perceiving vernalization signal (Yong et al., 2003) (Fig. 5.8). Fourth, plants have evolved the ability to remember vernalization. Vernalized plants keep vernalization status until flowering in spring or early summer through rounds of mitotic division, which is cold vernalization "memory." The vernalization requirement is reset in the next generation, and the offspring still need vernalization for flowering. The "cold memory" and resetting of vernalization requirements involve epigenetic regulation.

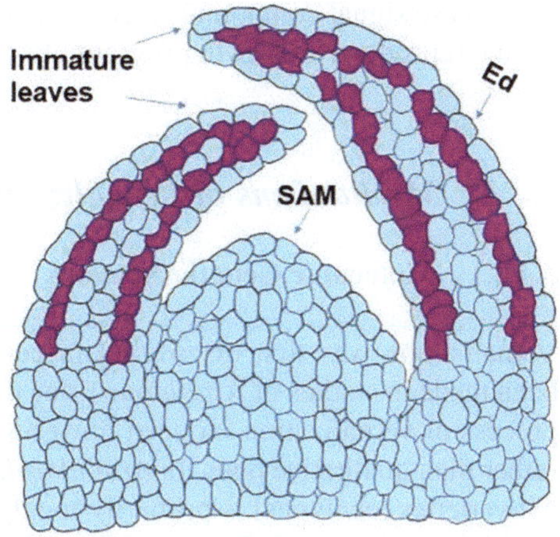

Fig. 5.8 A schematic representation illustrating the induction of *VER2* expression by vernalization in the immature leaves of winter wheat. Wheat plants were vernalized for 2 weeks after germination and then were collected for in situ hybridization analysis. Purple color indicates in situ signals

5.4.2 The Metabolic Dynamic During Vernalization

Vernalization is a complex physiological phenomenon. In the 1980s, it was found that vernalization treatment changed the activity of various enzymes in winter wheat. And then the respiratory metabolism and nucleic acid metabolism were changed obviously. According to metabolic changes, the five stages of winter wheat vernalization were put forward: oxidative phosphorylation stage, dehydrogenase enhanced stage, nucleic acid metabolism, protein metabolism, and induced stable stage. Based on the vernalized time, three phases can be divided: induction stage, acceleration stage, and stabilization stage (Yong et al., 2003).

In winter wheat, vernalization signals are activated in the induction phase. And if the vernalization process is interrupted at this stage, wheat plants cannot flower normally. In the acceleration phase, the effect of vernalization signals is accumulated (such as the translation of new proteins and the occurrence of protein modifications). If the vernalization process is stopped at this stage, the weaker winter wheat varieties could complete their life cycle. Moreover, the vernalization effects occurring before and during the acceleration stage can be erased by devernalization. If the duration of vernalization is sufficient to complete the stabilization stage, the vernalization status cannot be reversed by devernalization treatment (Li et al., 1987; Yong et al., 2003).

The initial step in the response to vernalization is the sensing of long-term cold exposure, which triggers a chain of biochemical reactions within cells. During the early stage of vernalization, environmental low temperature triggers responses to cold stress (Xiao et al., 2022). The subsequent accumulation of low temperature signals induces the synthesis of some compounds and/or the expression of vernalization genes, which cause accumulation of the vernalization signal molecules.

Once these signal compounds reach a specific threshold, shoot meristem cells initiate the reproductive growth program.

5.4.3 Mechanisms of Vernalization Response in Arabidopsis

5.4.3.1 Molecular Genetic Network

Vernalization was first discovered in cereals, but significant advancements have been made in studies of *Arabidopsis* due to the abundance of available mutants. The process of vernalization-induced flowering initiation involves quantitative changes rather than qualitative ones. In *Arabidopsis*, the vernalization requirement of different varieties (ecotypes) is primarily regulated by two genes, *FLC* and FRI (Shindo et al., 2005; Zhang & Jiménez-Gómez, 2020). There are dominant alleles of *FRI* and *FLC*, which establish overwintering growth habit in winter annuals of *Arabidopsis* (Amasino, 2010). *FLC* is the key negative regulator encoding a MADS-box transcription factor and inhibits flowering by directly suppressing the transcription of floral promoters of *FT* and *SOC1* in *Arabidopsis* (Michaels & Amasino, 1999a). There are five members of the *MADS BOX AFFECTING FLOWERING* (*MAF*) family homologous to *FLC* in *Arabidopsis* (Ratcliffe et al., 2003; Gu et al., 2013). *FLC* transcription is mainly regulated by *FRI*, *VRN1*, and *VRN2* (Xu & Chong, 2018). *FRI* encodes a plant-specific scaffold protein activating the transcription of *FLC* to establish vernalization requirement (Michaels et al., 2004). Null mutations of *FRI* and naturally occurring allelic variation in *FLC* establish rapid flowering summer-annual behavior (Michaels et al., 2003). *VRN1* and *VRN2* are not involved in the downregulation of *FLC* at the beginning stage of vernalization but are essential for maintaining the stable inhibition state of *FLC* after vernalization is accomplished (Levy et al., 2002). Different histone modification patterns mediated by protein complexes at the *FLC* locus maintain its epigenetic transcription activation and silencing states in response to vernalization (Xu & Chong, 2018; Luo & He, 2020).

Before vernalization, FRI associates with other proteins to form a transcription activator complex (FRIc) (Choi et al., 2011). FRIc together with chromatin modifiers, including COMPASS-like, HISTONE ACETYLTRANSFERASE OF THE MYST FAMILY1 (HAM1), H3K36 methyltransferase EARLY FLOWERING IN SHORT DAYS (EFS), H2A.Z-deposition complex SWR1c, H2B ubiquitination ligase UBC1, RNA Pol II-associated factor 1 complex (PAF1c) and the nuclear pre-mRNA processing factor CBC, form a FRI super complex (FRIsc) to establish active chromatin modifications of H3K36me3 and H3K4me3 (Fig. 5.9), and to form a gene loop to promote *FLC* transcription, fast elongation and mature *FLC* mRNA production (Hepworth & Dean, 2015; Li et al., 2016, 2018). The FRIsc-mediated active transcription state of *FLC* prevents floral transition and maintains vegetative growth before vernalization. Cold exposure induces the formation of FRI nuclear condensates, which is promoted by *COOLAIR*. FRI nuclear condensates inhibit

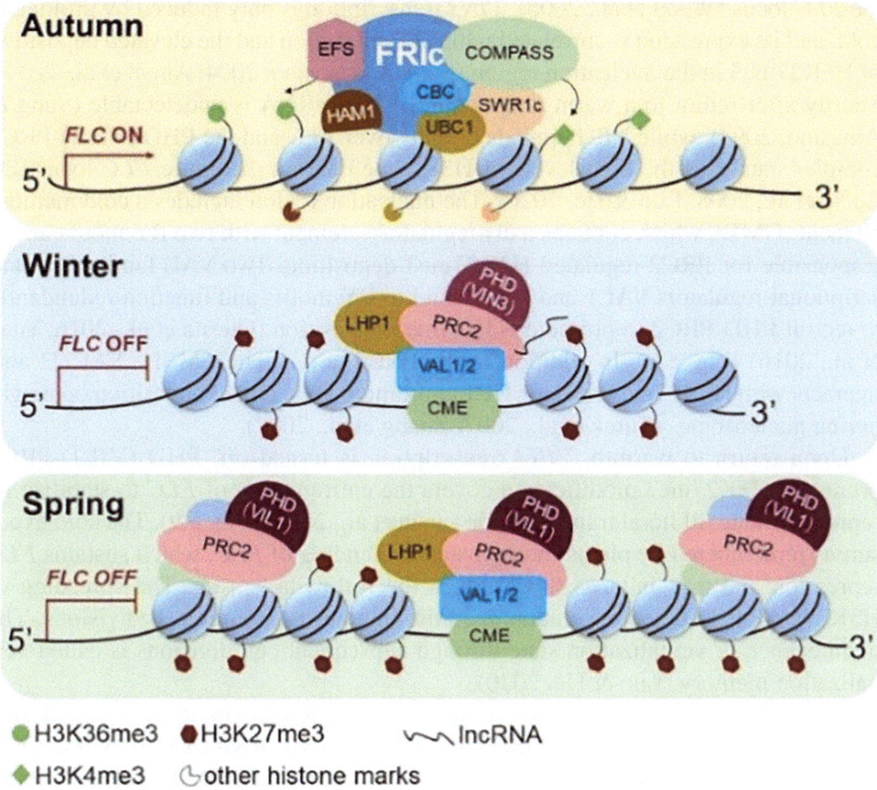

Fig. 5.9 The regulation mechanism of *FLC* expression during vernalization

FLC transcription by preventing FRI from associating with its promoter. Fluctuations in autumnal temperatures reversibly regulate the formation of FRI condensates, a mechanism that enables plants to adapt to varying thermal conditions in autumn (Zhu et al., 2021).

5.4.3.2 Cold Memory and Resetting of Vernalization Requirement

The Polycomb repressive complex 2 (PRC2) is evolutionarily conserved from animals to plants, and includes histone H3K27 methyltransferase and three structural subunits (Bieluszewski et al., 2021). During vernalization, prolonged cold exposure induces the repressive chromatin modification of H3K27me3 to be deposited on the nucleation region within *FLC* to turn off its transcription and establish the vernalization state (De Lucia et al., 2008; Coustham et al., 2012). A plant-specific protein with plant homeodomain (PHD), VERNALIZATION INSENSITIVE 3 (VIN3), together with another PHD protein VIL1(VIN3-LIKE 1)/VRN5, and other components cooperate with PRC2 to form PHD (VIN3)-PRC2 complex to trigger H3K27me3 at the *FLC* locus (Wood et al., 2006). *VIN3* transcription is only induced by prolonged cold, and its expression is correlated with *FLC* inhibition and the elevated deposition of H3K27me3 in the nucleation region (Sung & Amasino, 2004; Angel et al., 2011). Shortly after return to a warm temperature, *VIN3* mRNA is undetectable (Sung & Amasino, 2004), while VIL1 spreads broadly over *FLC* and the PHD (VIL1)-PRC2 complex catalyzes the high levels of H3K27me3 across the entire *FLC* locus (De Lucia et al., 2008; Luo & He, 2020). The nucleation region includes a cold memory element (CME), which contains a cis-regulatory element with two RY motifs and is responsible for PRC2-regulated H3K27me3 deposition. Two VAL family B3 transcriptional regulators VAL1 and VAL2 bind to RY motifs and function redundantly to recruit PHD-PRC2 to promote H3K27me3 deposition (Qüesta et al., 2016; Yuan et al., 2016) (Fig. 5.9). In addition to the PRC2 core subunit MSI1, VAL1/2 also interacts with H3K27me3 reader LHP1 to promote H3K27me3 deposition on neighboring nucleosomes (Turck et al., 2007; Zhang et al., 2007).

Upon return to warmth, *VIN3* transcription is turned off. PHD (VIL1)-PRC2 promoted H3K27me3 modification covers the entire region of *FLC* to stabilize its repressive state till floral transition (De Lucia et al., 2008) (Fig. 5.9). The cold exposure is remembered by plants through stable silencing of *FLC*, which sustains *FLC* repression and thus induces flowering in the following spring. The spreading of H3K27me3 at *FLC* locus is maintained throughout the mitotic cell divisions. The maintenance of vernalization state through subsequent cell divisions is called vernalization memory (Luo & He, 2020).

In *Arabidopsis*, three types of long noncoding RNAs (lncRNAs), COOLAIR (COLD INDUCED LONG ANTISENSE INTRAGENIC RNA), COLDAIR (COLD ASSISTED INTRONIC NONCODING RNA), and COLDWRAP (cold of winter-induced noncoding RNA from the promoter), are involved in vernalization-mediated *FLC* repression. At the beginning of vernalization (V0-V10), low temperature can induce COOLAIR peaks at 10 days of cold. COOLAIR is transcribed from the 3′ noncoding region of *FLC* and covers the whole coding region of *FLC* reversely. The COOLAIR is spliced into a short transcript by FCA containing RNA-binding protein complex. The short COOLAIR guides the H3K4 demethylase FLD to locate on *FLC* locus for decreasing the active mark level (Hepworth & Dean, 2015). COOLAIR-regulated repression of *FLC* is promoted by natural temperature fluctuation and is parallel with the role of PRC2 during vernalization (Zhu et al., 2021; Nielsen et al., 2024). COLDAIR is induced by cold exposure and transcribes from the first intron of the *FLC* in the sense direction, and its expression reaches the maximum level at 20 days of cold (V10-V20). COLDAIR interacts with and recruits the PRC2 complex to establish a stable repressed chromatin state at *FLC* (Heo & Sung, 2011) (Fig. 5.9). COLDWRAP is transcribed in a sense direction from the promoter region of *FLC*. Unlike the transient expression and association of COLDAIR with PRC2, COLDWRAP associates with PRC2 throughout cold exposure (Heo & Sung, 2011; Kim & Sung, 2017).

As mentioned above, it is the vernalization memory that gives plants the ability to flower in the spring of the coming year after a long winter cold treatment. How is the vernalization memory erased after flowering in the next generation? A seed-specific transcription factor LEAFY COTYLEDON1 (LEC1) functions in removing vernalization memory and reactivating the expression of *FLC* in the proembryo. In *Arabidopsis*, *LEC1* gene can be reactivated in early embryonic development and gathers H3K36 methyltransferase EFS and the chromatin-remodeling complex SWR1c to increase the levels of H3K4me3 and H3K36me3, which reactivate *FLC* expression. In the meantime, the repressive mark H3K27me3 is gradually eliminated by the EARLY FLOWERING 6 (ELF6) demethylase, thereby erasing the vernalization memory and maintaining the *FLC* chromatin in an active status in the period of embryonic development and postembryonic phases (Kim et al., 2009; He, 2012; Crevillen et al., 2014; Tao et al., 2017; Yang et al., 2017).

5.4.3.3 Long-Term Cold Sensing in *Arabidopsis*

Vernalization-induced VIN3 protein accumulation is slow and progressive, which is regarded as a sensing of long-term winter cold signaling to ensure adequate cold exposure duration, preventing flowering in late fall or winter. *VIN3* transcription activation is regulated by an NAC family transcription factor NTL8. In the gain-of-function mutant *nlt8-D*, overexpression of NTL8 constantly activates *VIN3* transcription and makes plants bypass vernalization requirements. Long-term cold does not alter NTL8 transcription, but slows down the growth rate of plants and results in decreased dilution rate of NTL8 protein concentrations, which can be regarded as information of long-term thermosensory to activate *VIN3* expression to silence *FLC* (Zhao et al., 2020).

5.4.4 *Regulation of Vernalization Response in Winter Wheat*

5.4.4.1 Genetic Network in Winter Wheat

In winter wheat, vernalization is regulated by *VRN1*, *VRN2*, and *VRN3*. Here, the wheat *VRNs* genes just have the same name as that of *Arabidopsis* but are first named in wheat and have completely different sequences and functions. *VRN1* encodes a MADS-box transcription factor, which is an ortholog of AP1. Long-term low temperature induces the expression of *VRN1* to accelerate flowering (Yan et al., 2003). *VRN2* encodes a negative flowering regulator with a conserved CCT domain to inhibit floral transition before vernalization in wheat (Yan et al., 2004), and *VRN3* encodes an ortholog of the *Arabidopsis* FT (Yan et al., 2006). Before vernalization, *VRN1* has a low transcription level because of the lack of cold exposure, while *VRN2* maintains active expression to suppress *VRN3* transcription in leaves. With the extension of winter low temperature, *VRN1* expression is gradually upregulated, while *VRN2* transcription is gradually downregulated. Meanwhile, *VRN2* is directly inhibited by *VRN1*, leading to the release of *VRN3* suppression (Chen & Dubcovsky, 2012; Oliver et al., 2013). VRN3, induced by VRN1 in leaves, translocates to the

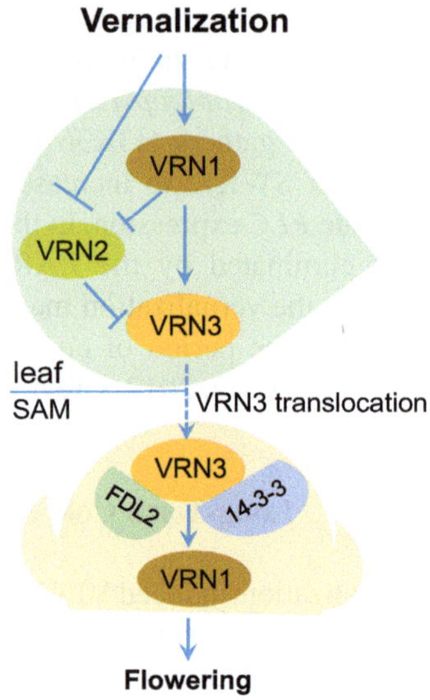

Fig. 5.10 The genetic relationship among major vernalization response genes in winter wheat

shoot apical meristem (SAM), where it forms a florigen activation complex (FAC) with FLOWERING LOCUS D-like 2 (FDL2) and 14-3-3C. This complex further induces *VRN1* expression to promote floral transition (Li & Dubcovsky, 2008; Distelfeld et al., 2009; Li et al., 2015) (Fig. 5.10). The early activation of *VRN1* can also be triggered by an SVP-like MADS transcription factor, which is rapidly upregulated and binds to *VRN1* promoter during short-term cold exposure (Kane et al., 2007; Xie et al., 2021).

Based on the requirements for duration of cold exposure, wheat cultivars can be divided into spring, facultative, and strong winter varieties. Differences in the vernalization requirements of wheat varieties are mainly determined by the natural variation of *VRN1*, *VRN2*, and *VRN3* (Yan et al., 2003; Xu & Chong, 2018; Luo & He, 2020). The main sources of differences in vernalization requirements are from the genetic variation at the *VRN1* locus (Fu et al., 2005; Kippes et al., 2015). Loss of function of *VRN2* or a mutation in the conserved domain of VRN2 results in the shift of winter growth habit to spring growth habit (Yan et al., 2004; Cockram et al., 2007). Allelic variation of *VRN3* results in significant differences in flowering time and alters the vernalization requirements of wheat and barley (Yan et al., 2006). In spring wheat, *VRN1* is continuously activated and thus maintains a high expression level due to mutations in the promoter and first intron in *VRN1*, which eliminates the vernalization requirements (Yan et al., 2003).

The ortholog of *FLC* in monocots, known as *ODDSOC2* (*OS2*), has been studied and characterized. *BdODDSOC2* negatively regulates flowering, and vernalization suppresses its transcription in *Brachypodium*. *TaODDSOC2* (*TaOS2*) shows similar expression patterns with *BdODDSOC2* during vernalization response, and the transcription inhibition of *TaOS2* by prolonged cold is independent on *VRN1* (Greenup et al., 2010; Sharma et al., 2017). In winter wheat, the miR172-*AP2L* (*TaTOE1*) module is involved in *VRN3* activation by vernalization in a manner not dependent on miR156. miR172 promotes flowering by inhibiting *AP2L*, which negatively regulates *VRN3*. Vernalization-induced *VRN1* upregulates miR172 and releases the suppression of *VRN3* in leaves (Debernardi et al., 2022).

5.4.4.2 *O*-GlcNAcylation and Phosphorylation in Vernalization Response

Research on the physiological effects of carbohydrates on the vernalization-induced flowering transition in cereals dates back to the 1940s. Early studies showed that sucrose and glucose promote vernalization, with their effects being most pronounced in the initial stages (Purvis, 1944).

Both *O*-GlcNAcylation and phosphorylation are reversible post-translational modifications. The cross-talk between *O*-GlcAcylation and phosphorylation participates in multiple processes, including protein interaction, localization and degradation, regulation of activity, and cell signaling transduction (Yang et al., 2017; Hart et al., 2011). *O*-GlcNAcylation and phosphorylation modifications of proteins can occur on the same or adjacent serine or threonine residues. These two modifications can compete with each other and form an either-or state, which is called "yin-yang" relationship (Slawson et al., 2006). If competitive cross-talk occurs on the same amino acid residue, this interaction is referred to as in situ cross-talk. Bioinformatics predictions for in situ cross-talk revealed three potential motifs associated with process and molecular functions. The motif Pxx[S] is significantly enriched in the endomembrane system organization and GTPase binding, Txxx[S] enriched in cytoskeleton organization, and [T]xxxxxxxxxP is related to the regulation of cell cycle (Yao et al., 2015).

VER2, a jacalin-like lectin encoding gene (Yong et al., 2003), is initially discovered through the mRNA differential display method, using subtractive cDNA cloning with unvernalized and vernalized wheat plants (Chong et al., 1994). VER2 exhibits high affinity for GlcNAc (N-acetylglucosamine) and galactose, and accumulates in the nucleus in response to vernalization. Prolonged cold exposure induces an overall increase in global *O*-GlcNAcylation levels in winter wheat (Xing et al., 2009). An *O*-GlcNAc transferase, TaOGT1, has been identified as a critical regulator of heading date in winter wheat. TaOGT1 activity was significantly positively correlated with *VRN1* and *VRN3* expression levels (Fan et al., 2021). During the vernalization process in winter wheat, proteins undergo dynamic *O*-GlcNAcylation and phosphorylation modifications, including GLYCINE-RICH RNA-BINDING PROTEIN2 (GRP2) (Xu et al., 2019). The *O*-GlcNAc modification of GRP2 is catalyzed by TaOGT1 (Fan et al., 2021). Mutation of *O*-GlcNAcylated sites on GRP2 alters its binding to a RIP-3 region within the first intron of *VRN1* (Xu et al., 2019). The RIP-3 region within *VRN-D4*, a duplicated copy of *VRN-A1* located on chromosome 5D, contains three close single-nucleotide polymorphisms (SNPs) in ancient wheat subspecies from South Asia. Variations in these SNPs influence the binding of GRP2 to RIP-3 (Kippes et al., 2015). These findings indicate a potential role of *O*-GlcNAc signaling in the regulation of flowering time mediated by vernalization.

Post-translational modification of key regulators plays a critical role in the transient cold signal transduction in plants (Ding et al., 2019). During vernalization in wheat, the phosphorylation modification levels of the key glucose metabolism pathway enzyme fructose-1,6-bisphosphate aldolase 10 (FBA10/HtL1) gradually decrease, while the *O*-GlcNAcylation modification levels, catalyzed by TaOGT1, gradually increase. This switch in modification patterns enhances the stability and enzymatic activity of FBA10/HtL1, thereby increasing the histone acetylation levels of H3K27ac and H3K14ac at the *VRN1* locus, activating its transcription and promoting flowering (Yang et al., 2025). The crosstalk between *O*-GlcNAcylation and phosphorylation, two key protein post-translational modifications, suggests their potential role in sensing and counting vernalization duration in winter wheat. In metazoans, the epigenetic "writers" and "erasers" of histone codes can interact with OGT and OGA, bridging the connection between *O*-GlcNAc signaling and epigenetic regulation of gene expression (Lewis & Hanover, 2014). Moreover, based on proteomic analysis to identify *O*-GlcNAc- and phosphorylation-modified proteins, histones were identified as being *O*-GlcNAc-modified in response to vernalization in winter wheat (Xu et al., 2019). Understanding the cross-talk between *O*-GlcNAc signaling and epigenetic regulation of vernalization response remains a key challenge for future research.

5.4.4.3 Memory of Winter Cold and Resetting of Vernalization Requirement

Before cold exposure, a repressive histone mark, H3K27me3, deposits on the promoter and the critical region within the first intron to maintain the silencing state of *VRN1* (Oliver et al., 2009). In wheat and barley, vernalization induces an increase in the activation marks of H3K4me3 and H3K36me3 and decrease in the repression mark H3K27me3 at *VRN1* locus, suggesting that *VRN1* is a target gene of epigenetic modification in response to vernalization (Oliver et al., 2009; Diallo et al., 2012). In cereals, the epigenetic modification-regulated flowering competence is also possibly obtained at *VRN3* and *ODDSOC2* loci during the vernalization process (Nelson et al., 2017).

Recently, the mechanisms of cold memory and vernalization requirement resetting have been reported in winter wheat. Prolonged cold exposure induces *VRN1* activation and *VRN2* repression, which are maintained during the later stages of growth and development to establish vernalization memory, ensuring timely spring flowering (Niu et al., 2024). The activation of *VRN1* is transmitted to early embryos and maintained by two active histone marks, H3K4me3 and H3K36me3. *VRN2* keeps silenced during seed development but is reactivated by light during seed germination. The chromatin-mediated stable transcription silencing of *VRN1* is maintained by polycomb repression-mediated H3K27me3 at the *VRN1* locus. TaRVR1, a homolog of BdRVR1 with highly conserved BAH domain, reads and maintains the H3K27me3 marks, mediating embryo resetting of *VRN1* (Niu et al., 2024).

5.5 Future Perspectives

Previous studies have comprehensively elucidated the molecular regulatory networks that govern floral transition in plants. Future research requires extending beyond the analysis of signaling pathways to actively harness the vast genetic diversity present in nature, particularly natural genetic variations. Identifying elite alleles for key agronomic traits, including reduced photoperiod sensitivity, optimized gibberellin response, and lowered vernalization requirements, is crucial for enabling the development of climate-resilient and adaptable crop varieties.

Flowering genes demonstrate pleiotropic effects, concurrently impacting plant height, yield, and stress resistance. Consequently, one important direction for future research will be to elucidate the mechanisms by which key natural variations regulate flowering time while simultaneously influencing other agronomic traits. By constructing polygenic interaction networks, researchers can assess the cumulative effects of integrating multiple advantageous alleles within the same genetic background, thus providing valuable insights for molecular design breeding.

References

Achard, P., Herr, A., Baulcombe, D. C., & Harberd, N. P. (2004). Modulation of floral development by a gibberellin-regulated microRNA. *Development, 131*, 3357–3365.

Achard, P., Baghour, M., Chapple, A., Hedden, P., Van Der Straeten, D., Genschik, P., Moritz, T., & Harberd, N. P. (2007). The plant stress hormone ethylene controls floral transition via DELLA-dependent regulation of floral meristem-identity genes. *Proceedings of the National Academy of Sciences of the United States of America, 104*, 6484–6489.

Al Temimi, A. H. K., Martin, M., Meng, Q., Lenstra, D. C., Qian, P., Guo, H., Weinhold, E., & Mecinović, J. (2020). Lysine ethylation by histone lysine methyltransferases. *Chembiochem, 21*, 392–400.

Amasino, R. (2010). Seasonal and developmental timing of flowering. *The Plant Journal, 61*, 1001–1013.

Andres, F., Galbraith, D. W., Talon, M., & Domingo, C. (2009). Analysis of *PHOTOPERIOD SENSITIVITY5* sheds light on the role of phytochromes in photoperiodic flowering in rice. *Plant Physiology, 151*, 681–690.

Angel, A., Song, J., Dean, C., & Howard, M. (2011). A Polycomb-based switch underlying quantitative epigenetic memory. *Nature, 476*, 105–108.

Aukerman, M. J., & Sakai, H. (2003). Regulation of flowering time and floral organ identity by a microRNA and its *APETALA2*-like target genes. *Plant Cell, 15*, 2730–2741.

Azpiroz, R., Wu, Y. W., LoCascio, J. C., & Feldmann, K. A. (1998). An *Arabidopsis* brassinosteroid-dependent mutant is blocked in cell elongation. *Plant Cell, 10*, 219–230.

Barrero, J. M., Piqueras, P., Gonzalez-Guzman, M., Serrano, R., Rodriguez, P. L., Ponce, M. R., & Micol, J. L. (2005). A mutational analysis of the *ABA1* gene of *Arabidopsis thaliana* highlights the involvement of ABA in vegetative development. *Journal of Experimental Botany, 56*, 2071–2083.

Bartel, D. P. (2009). MicroRNAs: Target recognition and regulatory functions. *Cell, 136*, 215–233.

Baurle, I., & Dean, C. (2008). Differential interactions of the autonomous pathway RRM proteins and chromatin regulators in the silencing of *Arabidopsis* targets. *PLoS One, 3*, e2733.

Bieluszewski, T., Xiao, J., Yang, Y., & Wagner, D. (2021). PRC2 activity, recruitment, and silencing: A comparative perspective. *Trends in Plant Science, 26*, 1186–1198.

Blazquez, M. A., & Weigel, D. (1999). Independent regulation of flowering by phytochrome B and gibberellins in *Arabidopsis*. *Plant Physiology, 120*, 1025–1032.

Blazquez, M. A., & Weigel, D. (2000). Integration of floral inductive signals in *Arabidopsis*. *Nature, 404*, 889–892.

Bollati, V., & Baccarelli, A. (2010). Environmental epigenetics. *Heredity (Edinb), 105*, 105–112.

Bond, D. M., Dennis, E. S., Pogson, B. J., & Finnegan, E. J. (2009). Histone acetylation, VERNALIZATION INSENSITIVE 3, FLOWERING LOCUS C, and the vernalization response. *Molecular Plant, 2*, 724–737.

Bu, S., Lau, S. S. Y., Yong, W. L., Zhang, H., Thiagarajan, S., Bashirullah, A., & Yu, F. (2023). Polycomb group genes are required for neuronal pruning in drosophila. *BMC Biology, 21*, 33.

Burn, J. E., Bagnall, D. J., Metzger, J. D., Dennis, E. S., & Peacock, W. J. (1993). DNA methylation, vernalization, and the initiation of flowering. *Proceedings of the National Academy of Sciences of the United States of America, 90*, 287.

Bao, S., Hua, C., Shen, L., & Yu, H. (2020). New insights into gibberellin signaling in regulating flowering in Arabidopsis. *Journal of Integrative Plant Biology, 62*(1), 118–131. https://doi.org/10.1111/jipb.v62.1, https://doi.org/10.1111/jipb.12892

Cai, Y., Vega-Sanchez, M. E., Park, C. H., Bellizzi, M., Guo, Z., & Wang, G.-L. (2014). RBS1, an RNA binding protein, interacts with SPIN1 and is involved in flowering time control in rice. *PLoS One, 9*, e87258.

Cao, D., Li, Y., Lu, S., Wang, J., Nan, H., Li, X., Shi, D., Fang, C., Zhai, H., Yuan, X., Anai, T., Xia, Z., Liu, B., & Kong, F. (2015). *GmCOL1a* and *GmCOL1b* function as flowering repressors in soybean under long-day conditions. *Plant & Cell Physiology, 56*, 2409–2422.

Cao, D., Takeshima, R., Zhao, C., Liu, B., Jun, A., & Kong, F. (2017). Molecular mechanisms of flowering under long days and stem growth habit in soybean. *Journal of Experimental Botany, 68*, 1873–1884.

Castillejo, C., & Pelaz, S. (2008). The balance between CONSTANS and TEMPRANILLO activities determines *FT* expression to trigger flowering. *Current Biology, 18*, 1338–1343.

Chatham, J. C., Zhang, J., & Wende, A. R. (2021). Role of *O*-linked N-acetylglucosamine protein modification in cellular (patho)physiology. *Physiological Reviews, 101*, 427–493.

Chen, A., & Dubcovsky, J. (2012). Wheat TILLING mutants show that the vernalization gene *VRN1* down-regulates the flowering repressor *VRN2* in leaves but is not essential for flowering. *PLoS Genetics, 8*, e1003134.

Chen, Q. S., Gan, S., & Li, G. (2020). *Arabidopsis* NUCLEAR FACTOR Y A8 inhibits the juvenile-to-adult transition by activating transcription of MIR156s. *Journal of Experimental Botany, 71*, 4890–4902.

Cheng, H., Qin, L., Lee, S., Fu, X., Richards, D. E., Cao, D., Luo, D., Harberd, N. P., & Peng, J. (2004). Gibberellin regulates *Arabidopsis* floral development via suppression of DELLA protein function. *Development, 131*, 1055–1064.

Cho, H. J., Kim, J. J., Lee, J. H., Kim, W., Jung, J. H., Park, C. M., & Ahn, J. H. (2012). SHORT VEGETATIVE PHASE (SVP) protein negatively regulates miR172 transcription via direct binding to the pri-miR172a promoter in *Arabidopsis*. *FEBS Letters, 586*, 2332–2337.

Cho, L.-H., Yoon, J., Pasriga, R., & An, G. (2016). Homodimerization of Ehd1 is required to induce flowering in rice. *Plant Physiology, 170*, 2159–2171.

Choi, K., Kim, J., Hwang, H. J., Kim, S., Park, C., Kim, S. Y., & Lee, I. (2011). The FRIGIDA complex activates transcription of *FLC*, a strong flowering repressor in *Arabidopsis*, by recruiting chromatin modification factors. *Plant Cell, 23*, 289–303.

Chong, K., Wang, L. P., Tan, K. H., Huang, H. L., & Liang, H. G. (1994). Molecular cloning and characterization of vernalization-related (ver) genes in winter wheat. *Physiologia Plantarum, 92*(3), 511–515. https://doi.org/10.1111/ppl.1994.92.issue-3, https://doi.org/10.1111/j.1399-3054.1994.tb08844.x

Chory, J., Nagpal, P., & Peto, C. A. (1991). Phenotypic and genetic analysis of *det2*, a new mutant that affects light-regulated seedling development in *Arabidopsis*. *Plant Cell, 3*, 445–459.

Chouard, P. (1960). Vernalization and its relations to dormancy. *Annual Review of Plant Physiology, 11*, 191–238.

Chuck, G., Cigan, A. M., Saeteurn, K., & Hake, S. (2007). The heterochronic maize mutant *Corngrass1* results from overexpression of a tandem microRNA. *Nature Genetics, 39*, 544–549.

Clarke, J. H., & Dean, C. (1994). Mapping *FRI*, a locus controlling flowering time and vernalization response in *Arabidopsis thaliana*. *Molecular & General Genetics, 242*, 81–89.

Cockram, J., Chiapparino, E., Taylor, S. A., Stamati, K., Donini, P., Laurie, D. A., & O'Sullivan, D. M. (2007). Haplotype analysis of vernalization loci in European barley germplasm reveals novel *VRN-H1* alleles and a predominant winter *VRN-H1/VRN-H2* multi-locus haplotype. *Theoretical and Applied Genetics, 115*, 993–1001.

Conti, L. (2017). Hormonal control of the floral transition: Can one catch them all? *Developmental Biology, 430*, 288–301.

Corbesier, L., Vincent, C., Jang, S., Fornara, F., Fan, Q., Searle, I., Giakountis, A., Farrona, S., Gissot, L., Turnbull, C., & Coupland, G. (2007). FT protein movement contributes to long-distance signaling in floral induction of *Arabidopsis*. *Science, 316*, 1030–1033.

Coustham, V., Li, P., Strange, A., Lister, C., Song, J., & Dean, C. (2012). Quantitative modulation of polycomb silencing underlies natural variation in vernalization. *Science, 337*, 584–587.

Crevillen, P., Yang, H., Cui, X., Greeff, C., Trick, M., Qiu, Q., Cao, X., & Dean, C. (2014). Epigenetic reprogramming that prevents transgenerational inheritance of the vernalized state. *Nature, 515*, 587–590.

Crosthwaite, S. K., & Jenkins, G. I. (1993). The role of leaves in the perception of vernalizing temperatures in sugar beet. *Journal of Experimental Botany, 44*(4), 801–806. https://doi.org/10.1093/jxb/44.4.801

Curtis, O. F., & Chang, H. T. (1930). The relative effectiveness of the temperature of the crown as contrasted with that of the rest of the plant upon the flowering of celery plants. *American Journal of Botany, 17*, 047–1048.

Dai, C., & Xue, H. W. (2010). Rice *early flowering1*, a CKI, phosphorylates DELLA protein SLR1 to negatively regulate gibberellin signalling. *The EMBO Journal, 29*, 1916–1927.

De Lucia, F., Crevillen, P., Jones, A. M., Greb, T., & Dean, C. (2008). A PHD-polycomb repressive complex 2 triggers the epigenetic silencing of *FLC* during vernalization. *Proceedings of the National Academy of Sciences of the United States of America, 105*, 16831–16836.

Debernardi, J. M., Woods, D. P., Li, K., Li, C., & Dubcovsky, J. (2022). *MiR172-APETALA2-like* genes integrate vernalization and plant age to control flowering time in wheat. *PLoS Genetics, 18*, e1010157.

Deng, X., Gu, L., Liu, C., Lu, T., Lu, F., Lu, Z., Cui, P., Pei, Y., Wang, B., Hu, S., & Cao, X. (2010). Arginine methylation mediated by the *Arabidopsis* homolog of PRMT5 is essential for proper pre-mRNA splicing. *Proceedings of the National Academy of Sciences of the United States of America, 107*, 19114–19119.

Dennis, E. S., & Peacock, W. J. (2009). Vernalization in cereals. *Journal of Biology, 8*, 57.

Diallo, A. O., Ali-Benali, M. A., Badawi, M., Houde, M., & Sarhan, F. (2012). Expression of vernalization responsive genes in wheat is associated with histone H3 trimethylation. *Molecular Genetics and Genomics, 287*, 575–590.

Ding, Y., Shi, Y., & Yang, S. (2019). Advances and challenges in uncovering cold tolerance regulatory mechanisms in plants. *The New Phytologist, 222*, 1690–1704.

Dissanayaka, A., Rodriguez, T. O., Di, S., Yan, F., Githiri, S. M., Rodas, F. R., Abe, J., & Takahashi, R. (2016). Quantitative trait locus mapping of soybean maturity gene *E5*. *Breeding Science, 66*, 407–415.

Distelfeld, A., Li, C., & Dubcovsky, J. (2009). Regulation of flowering in temperate cereals. *Current Opinion in Plant Biology, 12*, 178–184.

Doi, K., Izawa, T., Fuse, T., Yamanouchi, U., Kubo, T., Shimatani, Z., Yano, M., & Yoshimura, A. (2004). *Ehd1*, a B-type response regulator in rice, confers short-day promotion of flowering and controls *FT-like* gene expression independently of *Hd1l*. *Genes & Development, 18*, 926–936.

Domagalska, M. A., Schomburg, F. M., Amasino, R. M., Vierstra, R. D., Nagy, F., & Davis, S. J. (2007). Attenuation of brassinosteroid signaling enhances *FLC* expression and delays flowering. *Development, 134*, 2841–2850.

Du, A., Tian, W., Wei, M., Yan, W., He, H., Zhou, D., Huang, X., Li, S., & Ouyang, X. (2017). The DTH8-Hd1 module mediates day length-dependent regulation of rice flowering. *Molecular Plant, 10*, 948–961.

Duan, H. C., Wei, L. H., Zhang, C., Wang, Y., Chen, L., Lu, Z., Chen, P. R., He, C., & Jia, G. (2017). ALKBH10B is an RNA *N* (6)-methyladenosine demethylase affecting *Arabidopsis* floral transition. *Plant Cell, 29*, 2995–3011.

Eriksson, S., Bohlenius, H., Moritz, T., & Nilsson, O. (2006). GA4 is the active gibberellin in the regulation of *LEAFY* transcription and *Arabidopsis* floral initiation. *Plant Cell, 18*, 2172–2181.

Fan, C., Hu, R., Zhang, X., Wang, X., Zhang, W., Zhang, Q., Ma, J., & Fu, Y.-F. (2014). Conserved CO-FT regulons contribute to the photoperiod flowering control in soybean. *BMC Plant Biology, 14*, 9.

Fan, M., Miao, F., Jia, H., Li, G., Powers, C., Nagarajan, R., Alderman, P. D., Carver, B. F., Ma, Z., & Yan, L. (2021). *O*-linked N-acetylglucosamine transferase is involved in fine regulation of flowering time in winter wheat. *Nature Communications, 12*, 2303.

Feng, Z., Zhang, L., Yang, C., Wu, T., Lv, J., Chen, Y., Liu, X., Liu, S., Jiang, L., & Wan, J. (2014). *EF8* is involved in photoperiodic flowering pathway and chlorophyll biogenesis in rice. *Plant Cell Reports, 33*, 2003–2014.

Finnegan, E. J., Genger, R. K., Kovac, K., Peacock, W. J., & Dennis, E. S. (1998). DNA methylation and the promotion of flowering by vernalization. *Proceedings of the National Academy of Sciences of the United States of America, 95*, 5824–5829.

Fornara, F., Panigrahi, K. C., Gissot, L., Sauerbrunn, N., Ruhl, M., Jarillo, J. A., & Coupland, G. (2009). *Arabidopsis* DOF transcription factors act redundantly to reduce *CONSTANS* expression and are essential for a photoperiodic flowering response. *Developmental Cell, 17*, 75–86.

Fornara, F., de Montaigu, A., & Coupland, G. (2010). Snapshot: Control of flowering in *Arabidopsis*. *Cell, 141*, 550.

Fu, D., Szucs, P., Yan, L., Helguera, M., Skinner, J. S., von Zitzewitz, J., Hayes, P. M., & Dubcovsky, J. (2005). Large deletions within the first intron in *VRN-1* are associated with spring growth habit in barley and wheat. *Molecular Genetics and Genomics, 273*, 54–65.

Fulton, M. D., Brown, T., & Zheng, Y. G. (2018). Mechanisms and inhibitors of histone arginine methylation. *Chemical Record, 18*, 1792–1807.

Gao, H., Zheng, X.-M., Fei, G., Chen, J., Jin, M., Ren, Y., Wu, W., Zhou, K., Sheng, P., Zhou, F., Jiang, L., Wang, J., Zhang, X., Guo, X., Wang, J.-L., Cheng, Z., Wu, C., Wang, H., & Wan, J.-M. (2013). *Ehd4* encodes a novel and *Oryza*-genus-specific regulator of photoperiodic flowering in rice. *PLoS Genetics, 9*, e1003281.

Gao, H., Jin, M., Zheng, X. M., Chen, J., Yuan, D., Xin, Y., Wang, M., Huang, D., Zhang, Z., Zhou, K., Sheng, P., Ma, J., Ma, W., Deng, H., Jiang, L., Liu, S., Wang, H., Wu, C., Yuan, L., & Wan, J. (2014). *Days to heading 7*, a major quantitative locus determining photoperiod sensitivity and regional adaptation in rice. *Proceedings of the National Academy of Sciences of the United States of America, 111*, 16337–16342.

Garner, W. W., & Allard, H. A. (1920). Effect of the relative length of day and night and other factors of the environment on growth and reproduction in plants. *Journal of Agricultural Research, 18*, 0553–0606.

Ge, L., Chen, H., Jiang, J.-F., Zhao, Y., Xu, M. L., Xu, Y. Y., Tan, K. H., Xu, Z. H., & Chong, K. (2004). Overexpression of *OsRAA1* causes pleiotropic phenotypes in transgenic rice plants, including altered leaf, flower and root development and root response to gravity. *Plant Physiology, 135*, 1502–1513.

Gocal, G. F., Sheldon, C. C., Gubler, F., Moritz, T., Bagnall, D. J., MacMillan, C. P., Li, S. F., Parish, R. W., Dennis, E. S., Weigel, D., & King, R. W. (2001). *GAMYB*-like genes, flowering, and gibberellin signaling in *Arabidopsis*. *Plant Physiology, 127*, 1682–1693.

Godwin, J., & Farrona, S. (2022). The importance of networking: Plant polycomb repressive complex 2 and its interactors. *Epigenomes, 6*, 8.

Greenup, A. G., Sasani, S., Oliver, S. N., Talbot, M. J., Dennis, E. S., Hemming, M. N., & Trevaskis, B. (2010). ODDSOC2 is a MADS box floral repressor that is down-regulated by vernalization in temperate cereals. *Plant Physiology, 153*, 1062–1073.

Gu, X., Le, C., Wang, Y., Li, Z., Jiang, D., Wang, Y., & He, Y. (2013). *Arabidopsis* FLC clade members form flowering-repressor complexes coordinating responses to endogenous and environmental cues. *Nature Communications, 4*, 1947.

Guo, H., Yang, H., Mockler, T. C., & Lin, C. (1998). Regulation of flowering time by *Arabidopsis* photoreceptors. *Science, 279*, 1360–1363.

Guo, C., Xu, Y., Shi, M., Lai, Y., Wu, X., Wang, H., Zhu, Z., Poethig, R. S., & Wu, G. (2017). Repression of miR156 by miR159 regulates the timing of the juvenile-to-adult transition in *Arabidopsis*. *Plant Cell, 29*, 1293–1304.

Han, S.-H., Yoo, S.-C., Lee, B.-D., An, G., & Paek, N.-C. (2015). Rice FLAVIN-BINDING, KELCH REPEAT, F-BOX 1 (OsFKF1) promotes flowering independent of photoperiod. *Plant, Cell & Environment, 38*, 2527–2540.

Hart G. W., S. Chad, R. C. Genaro. And L. Olof. 2011. Cross talk between O-GlcNAcylation and phosphorylation: roles in signaling, transcription, and chronic disease. Annual Review of Biochemistry 80:825–858.

Hayama, R., Yokoi, S., Tamaki, S., Yano, M., & Shimamoto, K. (2003). Adaptation of photoperiodic control pathways produces short-day flowering in rice. *Nature, 422*, 719–722.

He, Y. H. (2012). Chromatin regulation of flowering. *Trends in Plant Science, 17*, 556–562.

Hedden, P., & Phillips, A. L. (2000). Gibberellin metabolism: New insights revealed by the genes. *Trends in Plant Science, 5*(12), 523–530. https://doi.org/10.1016/S1360-1385(00)01790-8

Henderson, J. T., Li, H. C., Rider, S. D., Mordhorst, A. P., Romero-Severson, J., Cheng, J. C., Robey, J., Sung, Z. R., de Vries, S. C., & Ogas, J. (2004). *PICKLE* acts throughout the plant to repress expression of embryonic traits and may play a role in gibberellin-dependent responses. *Plant Physiology, 134*, 995–1005.

Heo, J. B., & Sung, S. (2011). Vernalization-mediated epigenetic silencing by a long intronic noncoding RNA. *Science, 331*, 76–79.

Hepworth, J., & Dean, C. (2015). Flowering locus C's lessons: Conserved chromatin switches underpinning developmental timing and adaptation. *Plant Physiology, 168*, 1237–1245.

Hirose, F., Shinomura, T., Tanabata, T., Shimada, H., & Takano, M. (2006). Involvement of rice cryptochromes in de-etiolation responses and flowering. *Plant & Cell Physiology, 47*, 915–925.

Hollender, C., & Liu, Z. (2008). Histone deacetylase genes in *Arabidopsis* development. *Journal of Integrative Plant Biology, 50*, 875–885.

Hornyik, C., Duc, C., Rataj, K., Terzi, L. C., & Simpson, G. G. (2010a). Alternative polyadenylation of antisense RNAs and flowering time control. *Biochemical Society Transactions, 38*, 1077–1081.

Hornyik, C., Terzi, L. C., & Simpson, G. G. (2010b). The spen family protein FPA controls alternative cleavage and polyadenylation of RNA. *Developmental Cell, 18*, 203–213.

Huang, H., Sabari, B. R., Garcia, B. A., Allis, C. D., & Zhao, Y. (2014). Snapshot: Histone modifications. *Cell, 159*, 458–458.

Ikeda, Y., Kobayashi, Y., Yamaguchi, A., Abe, M., & Araki, T. (2007). Molecular basis of late-flowering phenotype caused by dominant epi-alleles of the *FWA* locus in *Arabidopsis*. *Plant & Cell Physiology, 48*, 205–220.

Imaizumi, T., Schultz, T. F., Harmon, F. G., Ho, L. A., & Kay, S. A. (2005). FKF1 F-box protein mediates cyclic degradation of a repressor of *CONSTANS* in *Arabidopsis*. *Science, 309*, 293–297.

Ishikawa, R., Aoki, M., Kurotani, K., Yokoi, S., Shinomura, T., Takano, M., & Shimamoto, K. (2011). Phytochrome B regulates *heading date 1* (*Hd1*)-mediated expression of rice florigen *Hd3a* and critical day length in rice. *Molecular Genetics and Genomics, 285*, 461–470.

Itoh, H., Nonoue, Y., Yano, M., & Izawa, T. (2010). A pair of floral regulators sets critical day length for *Hd3a* florigen expression in rice. *Nature Genetics, 42*, 635–638.

Izawa, T., Oikawa, T., Tokutomi, S., Okuno, K., & Shimamoto, K. (2000). Phytochromes confer the photoperiodic control of flowering in rice (a short-day plant). *The Plant Journal, 22*, 391–399.

Izawa, T., Mihara, M., Suzuki, Y., Gupta, M., Itoh, H., Nagano, A. J., Motoyama, R., Sawada, Y., Yano, M., Hirai, M. Y., Makino, A., & Nagamura, Y. (2011). Os-GIGANTEA confers robust diurnal rhythms on the global transcriptome of rice in the field. *Plant Cell, 23*, 1741–1755.

Jacobsen, S. E., & Olszewski, N. E. (1993). Mutations at the *SPINDLY* locus of *Arabidopsis* alter gibberellin signal transduction. *Plant Cell, 5*, 887–896.

Jeon, J., & Kim, J. (2011). FVE, an *Arabidopsis* homologue of the retinoblastoma-associated protein that regulates flowering time and cold response, binds to chromatin as a large multiprotein complex. *Molecules and Cells, 32*, 227–234.

Jiang, D. H. (2023). Complex regulation of flowering by high temperatures. *Plant, Cell & Environment, 46*, 1423–1426.

Jin, J. B., Jin, Y. H., Lee, J., Miura, K., Yoo, C. Y., Kim, W. Y., Van Oosten, M., Hyun, Y., Somers, D. E., Lee, I., Yun, D. J., Bressan, R. A., & Hasegawa, P. M. (2008). The SUMO E3 ligase, *AtSIZ1*, regulates flowering by controlling a salicylic acid-mediated floral promotion pathway and through affects on *FLC* chromatin structure. *The Plant Journal, 53*, 530–540.

Jin, J., Shi, J., Liu, B., Liu, Y., Huang, Y., Yu, Y., & Dong, A. (2015). MORF-RELATED GENE702, a reader protein of trimethylated histone H3 lysine 4 and histone H3 lysine 36, is involved in brassinosteroid-regulated growth and flowering time control in rice. *Plant Physiology, 168*, 1275–1285.

Jung, J. H., Seo, Y. H., Seo, P. J., Reyes, J. L., Yun, J., Chua, N. H., & Park, C. M. (2007). The *GIGANTEA*-regulated microRNA172 mediates photoperiodic flowering independent of *CONSTANS* in *Arabidopsis*. *Plant Cell, 19*, 2736–2748.

Jung, J. H., Seo, P. J., Ahn, J. H., & Park, C. M. (2012). *Arabidopsis* RNA-binding protein FCA regulates microRNA172 processing in thermosensory flowering. *The Journal of Biological Chemistry, 287*, 16007–16016.

Kane, N. A., Agharbaoui, Z., Diallo, A. O., Adam, H., Tominaga, Y., Ouellet, F., & Sarhan, F. (2007). TaVRT2 represses transcription of the wheat vernalization gene *TaVRN1*. *The Plant Journal, 51*, 670–680.

Kang, M. J., Jin, H. S., Noh, Y. S., & Noh, B. (2015). Repression of flowering under a noninductive photoperiod by the HDA9-AGL19-FT module in *Arabidopsis*. *The New Phytologist, 206*, 281–294.

Kania, T., Russenberger, D., Peng, S., Apel, K., & Melzer, S. (1997). *FPF1* promotes flowering in *Arabidopsis*. *Plant Cell, 9*, 1327–1338.

Kim, D. H., & Sung, S. (2017). Vernalization-triggered intragenic chromatin loop formation by long noncoding RNAs. *Developmental Cell, 40*, 302–312.e304.

Kim, S. L., Lee, S., Kim, H. J., Nam, H. G., & An, G. (2007). *OsMADS51* is a short-day flowering promoter that functions upstream of *Ehd1*, *OsMADS14*, and *Hd3a*. *Plant Physiology, 145*, 1484–1494.

Kim, S.-K., Yun, C.-H., Lee, J. H., Jang, Y. H., Park, H.-Y., & Kim, J.-K. (2008b). *OsCO3*, a *CONSTANS-LIKE* gene, controls flowering by negatively regulating the expression of *FT*-like genes under SD conditions in rice. *Planta, 228*, 355–365.

Kim, D.-H., Doyle, M. R., Sung, S., & Amasino, R. M. (2009). Vernalization: Winter and the timing of flowering in plants. *Annual Review of Cell and Developmental Biology, 25*, 277–299.

Kim, W., Latrasse, D., Servet, C., & Zhou, D. X. (2013). *Arabidopsis* histone deacetylase HDA9 regulates flowering time through repression of *AGL19*. *Biochemical and Biophysical Research Communications, 432*, 394–398.

Kippes, N., Debernardi, J. M., Vasquez-Gross, H. A., Akpinar, B. A., Budak, H., Kato, K., Chao, S., Akhunov, E., & Dubcovsky, J. (2015). Identification of the *VERNALIZATION 4* gene reveals the origin of spring growth habit in ancient wheats from South Asia. *Proceedings of the National Academy of Sciences of the United States of America, 112*, E5401–E5410.

Kobayashi, K., Yasuno, N., Sato, Y., Yoda, M., Yamazaki, R., Kimizu, M., Yoshida, H., Nagamura, Y., & Kyozuka, J. (2012). Inflorescence meristem identity in rice is specified by overlapping functions of three *AP1/FUL*-like MADS box genes and *PAP2*, a *SEPALLATA* MADS box gene. *Plant Cell, 24*, 1848–1859.

Kojima, S., Takahashi, Y., Kobayashi, Y., Monna, L., Sasaki, T., Araki, T., & Yano, M. (2002). *Hd3a*, a rice ortholog of the *Arabidopsis FT* gene, promotes transition to flowering downstream of *Hd1* under short-day conditions. *Plant & Cell Physiology, 43*, 1096–1105.

Komiya, R., Ikegami, A., Tamaki, S., Yokoi, S., & Shimamoto, K. (2008). *Hd3a* and *RFT1* are essential for flowering in rice. *Development, 135*, 767–774.

Komiya, R., Yokoi, S., & Shimamoto, K. (2009). A gene network for long-day flowering activates *RFT1* encoding a mobile flowering signal in rice. *Development, 136*, 3443–3450.

Kong, F., Liu, B., Xia, Z., Sato, S., Kim, B. M., Watanabe, S., Yamada, T., Tabata, S., Kanazawa, A., Harada, K., & Abe, J. (2010). Two coordinately regulated homologs of *FLOWERING LOCUS T* are involved in the control of photoperiodic flowering in soybean. *Plant Physiology, 154*, 1220–1231.

Koornneef, M., Hanhart, C. J., & van der Veen, J. H. (1991). A genetic and physiological analysis of late flowering mutants in *Arabidopsis thaliana*. *Molecular & General Genetics, 229*, 57–66.

Kurup, S., Jones, H. D., & Holdsworth, M. J. (2000). Interactions of the developmental regulator ABI3 with proteins identified from developing *Arabidopsis* seeds. *The Plant Journal, 21*, 143–155.

Lawrence, M., Daujat, S., & Schneider, R. (2016). Lateral thinking: How histone modifications regulate gene expression. *Trends in Genetics, 32*, 42–56.

Lee, I., & Amasino, R. M. (1995). Effect of vernalization, photoperiod, and light quality on the flowering phenotype of *Arabidopsis* plants containing the *FRIGIDA* gene. *Plant Physiology, 108*, 157–162.

Lee, Y.-S., & An, G. (2015). *OsGI* controls flowering time by modulating rhythmic flowering time regulators preferentially under short day in rice. *Journal Of Plant Biology, 58*, 137–145.

Lee, S. Y., Kim, J., Han, J. J., Han, M. J., & An, G. H. (2004). Functional analyses of the flowering time gene *OsMADS50*, the putative *SUPPRESSOR OF OVEREXPRESSION OFCO 1/AGAMOUS-LIKE 20* (*SOC1/AGL20*) ortholog in rice. *The Plant Journal, 38*, 754–764.

Lee, J. H., Yoo, S. J., Park, S. H., Hwang, I., Lee, J. S., & Ahn, J. H. (2007). Role of *SVP* in the control of flowering time by ambient temperature in *Arabidopsis*. *Genes & Development, 21*, 397–402.

Lee, Y. S., Jeong, D. H., Lee, D. Y., Yi, J., Ryu, C. H., Kim, S. L., Jeong, H. J., Choi, S. C., Jin, P., Yang, J., Cho, L. H., Choi, H., & An, G. (2010). *OsCOL4* is a constitutive flowering repressor upstream of *Ehd1* and downstream of *OsphyB*. *The Plant Journal, 63*, 18–30.

Lee, Y.-S., Yi, J., & An, G. (2016). *OsPhyA* modulates rice flowering time mainly through *OsGI* under short days and *Ghd7* under long days in the absence of phytochrome B. *Plant Molecular Biology, 91*, 413–427.

Levy, Y. Y., Mesnage, S., Mylne, J. S., Gendall, A. R., & Dean, C. (2002). Multiple roles of *Arabidopsis VRN1* in vernalization and flowering time control. *Science, 297*, 243–246.

Lewis, B. A., & Hanover, J. A. (2014). *O*-GlcNAc and the epigenetic regulation of gene expression. *The Journal of Biological Chemistry, 289*, 34440–34448.

Li, C., & Dubcovsky, J. (2008). Wheat FT protein regulates *VRN1* transcription through interactions with FDL2. *The Plant Journal, 55*, 543–554.

Li, X., Hao, N., & Tan, K. (1987). Effect of devernalization on soluble protein component in winter wheat seedling and subsequent plant developmental state. *Acta Botanica Sinica, 3*, 015.

Li, D. J., Yang, C. H., Li, X. B., Gan, Q., Zhao, X. F., & Zhu, L. H. (2009). Functional characterization of rice *OsDof12*. *Planta, 229*, 1159–1169.

Li, W. X., Han, Y. Y., Tao, F., & Chong, K. (2011). Knockdown of *SAMS* genes encoding S-adenosyl-L-methionine synthetases causes methylation alterations of DNAs and histones and leads to late flowering in rice. *Journal of Plant Physiology, 168*, 1837–1843.

Li, F., Zhang, X. M., Hu, R. B., Wu, F. Q., Ma, J. H., Meng, Y., & Fu, Y. F. (2013). Identification and molecular characterization of *FKF1* and *GI* homologous genes in soybean. *PLoS One, 8*, e79036.

Li, C., Lin, H., & Dubcovsky, J. (2015). Factorial combinations of protein interactions generate a multiplicity of florigen activation complexes in wheat and barley. *The Plant Journal, 84*, 70–82.

Li, Z., Jiang, D., Fu, X., Luo, X., Liu, R., & He, Y. (2016). Coupling of histone methylation and RNA processing by the nuclear mRNA cap-binding complex. *Nature Plants, 2*, 16015.

Li, C., Zhu, S., Zhang, H., Chen, L., Cai, M., Wang, J., Chai, J., Wu, F., Cheng, Z., Guo, X., Zhang, X., & Wan, J. (2017). OsLBD37 and OsLBD38, two class II type LBD proteins, are involved in the regulation of heading date by controlling the expression of *Ehd1* in rice. *Biochemical and Biophysical Research Communications, 486*, 720–725.

Li, Z., Jiang, D., & He, Y. (2018). FRIGIDA establishes a local chromosomal environment for *FLOWERING LOCUS C* mRNA production. *Nature Plants, 4*, 836–846.

Lian, H., Wang, L., Ma, N., Zhou, C. M., Han, L., Zhang, T. Q., & Wang, J. W. (2021). Redundant and specific roles of individual MIR172 genes in plant development. *PLoS Biology, 19*(2), e3001044. https://doi.org/10.1371/journal.pbio.3001044

Lim, M. H., Kim, J., Kim, Y. S., Chung, K. S., Seo, Y. H., Lee, I., Kim, J., Hong, C. B., Kim, H. J., & Park, C. M. (2004). A new *Arabidopsis* gene, *FLK*, encodes an RNA binding protein with K homology motifs and regulates flowering time via *FLOWERING LOCUS C*. *Plant Cell, 16*, 731–740.

Lin, H. X., Yamamoto, T., Sasaki, T., & Yano, M. (2000). Characterization and detection of epistatic interactions of 3 QTLs, *Hd1*, *Hd2*, and *Hd3*, controlling heading date in rice using nearly isogenic lines. *Theoretical and Applied Genetics, 101*, 1021–1028.

Liu, B., Kanazawa, A., Matsumura, H., Takahashi, R., Harada, K., & Abe, J. (2008). Genetic redundancy in soybean photoresponses associated with duplication of the phytochrome A gene. *Genetics, 180*, 995–1007.

Liu, C., Lu, F., Cui, X., & Cao, X. (2010). Histone methylation in higher plants. *Annual Review of Plant Biology, 61*, 395–420.

Liu, X., Yang, S., Zhao, M., Luo, M., Yu, C. W., Chen, C. Y., Tai, R., & Wu, K. (2014). Transcriptional repression by histone deacetylases in plants. *Molecular Plant, 7*, 764–772.

Liu, B., Wei, G., Shi, J., Jin, J., Shen, T., Ni, T., Shen, W. H., Yu, Y., & Dong, A. (2016). SET DOMAIN GROUP 708, a histone H3 lysine 36-specific methyltransferase, controls flowering time in rice (*Oryza sativa*). *The New Phytologist, 210*, 577–588.

Liu, K., Yu, Y., Dong, A., & Shen, W. H. (2017). *SET DOMAIN GROUP701* encodes a H3K4-methyltransferase and regulates multiple key processes of rice plant development. *The New Phytologist, 215*, 609–623.

Love, D. C., Kochan, J., Cathey, R. L., Shin, S. H., & Hanover, J. A. (2003). Mitochondrial and nucleocytoplasmic targeting of *O*-linked GlcNAc transferase. *Journal of Cell Science, 116*, 647–654.

Lu, F., Cui, X., Zhang, S., Liu, C., & Cao, X. (2010). JMJ14 is an H3K4 demethylase regulating flowering time in *Arabidopsis*. *Cell Research, 20*, 387–390.

Lu, Q., Zhao, L., Li, D., Hao, D., Zhan, Y., & Li, W. (2014). A *GMRAV* ortholog is involved in photoperiod and sucrose control of flowering time in soybean. *PLoS One, 9*, e89145.

Lu, S. J., Zhao, X. H., Hu, Y. L., Liu, S. L., Nan, H. Y., Li, X. M., Fang, C., Cao, D., Shi, X. Y., Kong, L. P., Su, T., Zhang, F. G., Li, S. C., Wang, Z., Yuan, X. H., Cober, E. R., Weller, J. L., Liu, B. H., Hou, X. L., Tian, Z. X., & Kong, F. J. (2017). Natural variation at the soybean J locus improves adaptation to the tropics and enhances yield. *Nature Genetics, 5*, 773–779.

Lu, S. J., Dong, L. D., Fang, C., Liu, S. L., Kong, L. P., Cheng, Q., Chen, L. Y., Su, T., Nan, H. Y., Zhang, D., Zhang, L., Wang, Z. J., Yang, Y. Q., Yu, D. Y., Liu, X. L., Yang, Q. Y., Lin, X. Y., Tang, Y., Zhao, X. H., Yang, X. Q., Tian, C. G., Xie, Q. G., Li, X., Yuan, X. H., Tian, Z. X., Liu,

B. H., Weller, J. L., & Kong, F. J. (2020). Stepwise selection on homeologous PRR genes controlling flowering and maturity during soybean domestication. *Nature Genetics, 4*, 428–436.

Luo, X., & He, Y. (2020). Experiencing winter for spring flowering: A molecular epigenetic perspective on vernalization. *Journal of Integrative Plant Biology, 62*, 104–117.

Luo, M., Tai, R., Yu, C. W., Yang, S., Chen, C. Y., Lin, W. D., Schmidt, W., & Wu, K. (2015). Regulation of flowering time by the histone deacetylase HDA5 in *Arabidopsis*. *The Plant Journal, 82*, 925–936.

Luo, X., Liu, X. L., Zhang, N., Song, C. Y., & He, Y. H. (2025). Molecular mechanisms of temperature-mediated flowering regulation: From Arabidopsis to short-day crops. *Plant, Cell & Environment, 48*, 7020–7037.

Macknight, R., Bancroft, I., Page, T., Lister, C., Schmidt, R., Love, K., Westphal, L., Murphy, G., Sherson, S., Cobbett, C., & Dean, C. (1997). *FCA*, a gene controlling flowering time in *Arabidopsis*, encodes a protein containing RNA-binding domains. *Cell, 89*, 737–745.

Mathieu, J., Yant, L. J., Murdter, F., Kuttner, F., & Schmid, M. (2009). Repression of flowering by the miR172 target SMZ. *PLoS Biology, 7*, e1000148.

Matsubara, K., Yamanouchi, U., Wang, Z.-X., Minobe, Y., Izawa, T., & Yano, M. (2008). *Ehd2*, a rice ortholog of the maize *INDETERMINATE1* gene, promotes flowering by up-regulating *Ehd1*. *Plant Physiology, 148*, 1425–1435.

Matsubara, K., Yamanouchi, U., Nonoue, Y., Sugimoto, K., Wang, Z.-X., Minobe, Y., & Yano, M. (2011). *Ehd3*, encoding a plant homeodomain finger-containing protein, is a critical promoter of rice flowering. *The Plant Journal, 66*, 603–612.

Michaels, S. D., & Amasino, R. M. (1999a). *FLOWERING LOCUS C* encodes a novel MADS domain protein that acts as a repressor of flowering. *Plant Cell, 11*, 949–956.

Michaels, S. D., & Amasino, R. M. (1999b). The gibberellic acid biosynthesis mutant *ga1-3* of *Arabidopsis thaliana* is responsive to vernalization. *Developmental Genetics, 25*, 194–198.

Michaels, S. D., & Amasino, R. M. (2001). Loss of *FLOWERING LOCUS C* activity eliminates the late-flowering phenotype of *FRIGIDA* and autonomous pathway mutations but not responsiveness to vernalization. *Plant Cell, 13*, 935–941.

Michaels, S. D., He, Y., Scortecci, K. C., & Amasino, R. M. (2003). Attenuation of FLOWERING LOCUS C activity as a mechanism for the evolution of summer-annual flowering behavior in *Arabidopsis*. *Proceedings of the National Academy of Sciences of the United States of America, 100*, 10102–10107.

Michaels, S. D., Bezerra, I. C., & Amasino, R. M. (2004). *FRIGIDA*-related genes are required for the winter-annual habit in *Arabidopsis*. *Proceedings of the National Academy of Sciences of the United States of America, 101*, 3281–3285.

Moon, J., Suh, S. S., Lee, H., Choi, K. R., Hong, C. B., Paek, N. C., Kim, S. G., & Lee, I. (2003). The *SOC1* MADS-box gene integrates vernalization and gibberellin signals for flowering in *Arabidopsis*. *The Plant Journal, 35*, 613–623.

Mutasa-Gottgens, E., & Hedden, P. (2009). Gibberellin as a factor in floral regulatory networks. *Journal of Experimental Botany, 60*, 1979–1989.

Nan, H., Cao, D., Zhang, D., Li, Y., Lu, S., Tang, L., Yuan, X., Liu, B., & Kong, F. (2014). GmFT2a and GmFT5a redundantly and differentially regulate flowering through interaction with and upregulation of the bZIP transcription factor GmFDL19 in soybean. *PLoS One, 9*, e97669.

Nelson, M. N., Książkiewicz, M., Rychel, S., Besharat, N., Taylor, C. M., Wyrwa, K., Jost, R., Erskine, W., Cowling, W. A., Berger, J. D., Batley, J., Weller, J. L., Naganowska, B., & Wolko, B. (2017). The loss of vernalization requirement in narrow-leafed lupin is associated with a deletion in the promoter and de-repressed expression of a *flowering locus T* (*FT*) homologue. *The New Phytologist, 213*, 220–232.

Nielsen, M., Menon, G., Zhao, Y., Mateo-Bonmati, E., Wolff, P., Zhou, S., Howard, M., & Dean, C. (2024). COOLAIR and PRC2 function in parallel to silence FLC during vernalization. *Proceedings of the National Academy of Sciences of the United States of America, 121*, e2311474121.

Niu, L., Lu, F., Pei, Y., Liu, C., & Cao, X. (2007). Regulation of flowering time by the protein arginine methyltransferase AtPRMT10. *EMBO Reports, 8*, 1190–1195.

Niu, L., Zhang, Y., Pei, Y., Liu, C., & Cao, X. (2008). Redundant requirement for a pair of PROTEIN ARGININE METHYLTRANSFERASE4 homologs for the proper regulation of *Arabidopsis* flowering time. *Plant Physiology, 148*, 490–503.

Niu, D., Gao, Z., Cui, B., Zhang, Y., & He, Y. (2024). A molecular mechanism for embryonic resetting of winter memory and restoration of winter annual growth habit in wheat. *Nature Plants, 1*, 37–52.

Noh, B., Lee, S. H., Kim, H. J., Yi, G., Shin, E. A., Lee, M., Jung, K. J., Doyle, M. R., Amasino, R. M., & Noh, Y. S. (2004). Divergent roles of a pair of homologous jumonji/zinc-finger-class transcription factor proteins in the regulation of *Arabidopsis* flowering time. *Plant Cell, 16*, 2601–2613.

Ohto, M., Onai, K., Furukawa, Y., Aoki, E., Araki, T., & Nakamura, K. (2001). Effects of sugar on vegetative development and floral transition in *Arabidopsis*. *Plant Physiology, 127*, 252–261.

Oliver, S. N., Finnegan, E. J., Dennis, E. S., Peacock, W. J., & Trevaskis, B. (2009). Vernalization-induced flowering in cereals is associated with changes in histone methylation at the *VERNALIZATION1* gene. *Proceedings of the National Academy of Sciences of the United States of America, 106*, 8386–8391.

Oliver, S. N., Deng, W., Casao, M. C., & Trevaskis, B. (2013). Low temperatures induce rapid changes in chromatin state and transcript levels of the cereal *VERNALIZATION1* gene. *Journal of Experimental Botany, 64*, 2413–2422.

Ó'Maoiléidigh, D. S., van Driel, A. D., Singh, A., Sang, Q., Le Bec, N., Vincent, C., ... & Coupland, G. (2021). Systematic analyses of the MIR172 family members of Arabidopsis define their distinct roles in regulation of APETALA2 during floral transition. *PLoS Biology, 19*(2), e3001043. https://doi.org/10.1371/journal.pbio.3001043

Park, J., Oh, D. H., Dassanayake, M., Nguyen, K. T., Ogas, J., Choi, G., & Sun, T. P. (2017). Gibberellin signaling requires chromatin remodeler PICKLE to promote vegetative growth and phase transitions. *Plant Physiology, 173*, 1463–1474.

Pazhouhandeh, M., Molinier, J., Berr, A., & Genschik, P. (2011). MSI4/FVE interacts with CUL4-DDB1 and a PRC2-like complex to control epigenetic regulation of flowering time in *Arabidopsis*. *Proceedings of the National Academy of Sciences of the United States of America, 108*, 3430–3435.

Pei, Y., Niu, L., Lu, F., Liu, C., Zhai, J., Kong, X., & Cao, X. (2007). Mutations in the type II protein arginine methyltransferase AtPRMT5 result in pleiotropic developmental defects in *Arabidopsis*. *Plant Physiology, 144*, 1913–1923.

Peng, L. T., Shi, Z. Y., Li, L., Shen, G. Z., & Zhang, J. L. (2007). Ectopic expression of *OsLFL1* in rice represses *Ehd1* by binding on its promoter. *Biochemical and Biophysical Research Communications, 360*, 251–256.

Peng, L. T., Shi, Z. Y., Li, L., Shen, G. Z., & Zhang, J. L. (2008). Overexpression of transcription factor *OsLFL1* delays flowering time in *Oryza sativa*. *Journal of Plant Physiology, 165*, 876–885.

Porri, A.,Torti, S., Romera-Branchat, M., Coupland, G. (2012). Spatially distinct regulatory roles for gibberellins in the promotion of flowering of Arabidopsis under long photoperiods. *Development, 139*, 2198–209.

Purvis, O. N. (1944). Studies in the vernalisation of cereals: VIII. The role of carbohydrate and nitrogen supply in the vernalisation of excised embryos of 'Petkus'. *Winter Rye, 8*, 285–314.

Putterill, J., Robson, F., Lee, K., Simon, R., & Coupland, G. (1995). The *CONSTANS* gene of *Arabidopsis* promotes flowering and encodes a protein showing similarities to zinc finger transcription factors. *Cell, 80*, 847–857.

Quesada, V., Dean, C., & Simpson, G. G. (2005). Regulated RNA processing in the control of *Arabidopsis* flowering. *The International Journal of Developmental Biology, 49*, 773–780.

Qüesta, J. I., Song, J., Geraldo, N., An, H., & Dean, C. (2016). *Arabidopsis* transcriptional repressor VAL1 triggers Polycomb silencing at *FLC* during vernalization. *Science, 353*, 485–488.

Ratcliffe, O. J., Kumimoto, R. W., Wong, B. J., & Riechmann, J. L. (2003). Analysis of the Arabidopsis MADS AFFECTING FLOWERING gene family: MAF2 prevents vernalization by short periods of cold. *Plant Cell, 15*, 1159–1169.

Razem, F. A., Baron, K., & Hill, R. D. (2006). Turning on gibberellin and abscisic acid signaling. *Current Opinion in Plant Biology, 9*, 454–459.

Richter, R., Behringer, C., Muller, I. K., & Schwechheimer, C. (2010). The GATA-type transcription factors GNC and GNL/CGA1 repress gibberellin signaling downstream from DELLA proteins and PHYTOCHROME-INTERACTING FACTORS. *Genes & Development, 24*, 2093–2104.

Roldan, M., Gomez-Mena, C., Ruiz-Garcia, L., Salinas, J., & Martinez-Zapater, J. M. (1999). Sucrose availability on the aerial part of the plant promotes morphogenesis and flowering of *Arabidopsis* in the dark. *The Plant Journal, 20*, 581–590.

Ryu, C.-H., Lee, S., Cho, L.-H., Kim, S. L., Lee, Y.-S., Choi, S. C., Jeong, H. J., Yi, J., Park, S. J., Han, C.-D., & An, G. (2009). *OsMADS50* and *OsMADS56* function antagonistically in regulating long day (LD)-dependent flowering in rice. *Plant, Cell & Environment, 32*, 1412–1427.

Saito, H., Ogiso-Tanaka, E., Okumoto, Y., Yoshitake, Y., Izumi, H., Yokoo, T., Matsubara, K., Hori, K., Yano, M., Inoue, H., & Tanisaka, T. (2012). *Ef7* encodes an ELF3-like protein and promotes rice flowering by negatively regulating the floral repressor gene *Ghd7* under both short- and long-day conditions. *Plant & Cell Physiology, 53*, 717–728.

Sanda, S. L., & Amasino, R. M. (1996). Ecotype-specific expression of a flowering mutant phenotype in *Arabidopsis thaliana. Plant Physiology, 111*, 641–644.

Sawa, M., Nusinow, D. A., Kay, S. A., & Imaizumi, T. (2007). FKF1 and GIGANTEA complex formation is required for day-length measurement in *Arabidopsis*. *Science, 318*, 261–265.

Schomburg, F. M., Patton, D. A., Meinke, D. W., & Amasino, R. M. (2001). *FPA*, a gene involved in floral induction in *Arabidopsis*, encodes a protein containing RNA-recognition motifs. *Plant Cell, 13*, 1427–1436.

Seni, S., Singh, R. K., & Prasad, M. (2023). Dynamics of epigenetic control in plants via SET domain containing proteins: Structural and functional insights. *Biochimica et Biophysica Acta, Gene Regulatory Mechanisms, 1866*, 194966.

Sharaf, A., Vijayanathan, M., Oborník, M., & Mozgová, I. (2022). Phylogenetic profiling resolves early emergence of PRC2 and illuminates its functional core. *Life Science Alliance, 5*, e202101271.

Sharma, N., Ruelens, P., D'Hauw, M., Maggen, T., Dochy, N., Torfs, S., Kaufmann, K., Rohde, A., & Geuten, K. (2017). A flowering locus C homolog is a vernalization-regulated repressor in *Brachypodium* and is cold regulated in wheat. *Plant Physiology, 173*, 1301–1315.

Shen, L. S., Liang, Z., Gu, X. F., Chen, Y., Teo, Z. W. N., Hou, X. L., Cai, W. M., Dedon, P. C., Liu, L., & Yu, H. (2016). *N*-6-methyladenosine RNA modification regulates shoot stem cell fate in *Arabidopsis*. *Developmental Cell, 38*, 186–200.

Sheng, P., Wu, F., Tan, J., Zhang, H., Ma, W., Chen, L., Wang, J., Wang, J., Zhu, S., Guo, X., Wang, J., Zhang, X., Cheng, Z., Bao, Y., Wu, C., Liu, X., & Wan, J. (2016). A *CONSTANS-like* transcriptional activator, *OsCOL13*, functions as a negative regulator of flowering downstream of *OsphyB* and upstream of *Ehd1* in rice. *Plant Molecular Biology, 92*, 209–222.

Shindo, C., Aranzana, M. J., Lister, C., Baxter, C., Nicholls, C., Nordborg, M., & Dean, C. (2005). Role of *FRIGIDA* and *FLOWERING LOCUS C* in determining variation in flowering time of *Arabidopsis*. *Plant Physiology, 138*, 1163–1173.

Shu, K., Chen, Q., Wu, Y., Liu, R., Zhang, H., Wang, S., Tang, S., Yang, W., & Xie, Q. (2016). ABSCISIC ACID-INSENSITIVE 4 negatively regulates flowering through directly promoting *Arabidopsis FLOWERING LOCUS C* transcription. *Journal of Experimental Botany, 67*, 195–205.

Simpson, G. G., & Dean, C. (2002). *Arabidopsis*, the rosetta stone of flowering time? *Science, 296*, 285–289.

Simpson, G. G., Dijkwel, P. P., Quesada, V., Henderson, I., & Dean, C. (2003). FY is an RNA 3' end-processing factor that interacts with FCA to control the *Arabidopsis* floral transition. *Cell, 113*, 777–787.

Singh, S., Vergish, S., Jain, N., Sharma, A. K., Khurana, P., & Khurana, J. P. (2023). OsCRY2 and OsFBO10 co-regulate photomorphogenesis and photoperiodic flowering in *indica* rice. *Plant Science, 330*, 111631.

Slawson, C., Housley, M. P., & Hart, G. W. (2006). O-GlcNAc cycling: How a single sugar post-translational modification is changing the way we think about signaling networks. *Journal of Cellular Biochemistry, 97*(1), 71–83. https://doi.org/10.1002/(ISSN)1097-4644, https://doi.org/10.1002/jcb.v97:1, https://doi.org/10.1002/jcb.20676

Soesanto, Y. A., Luo, B., Jones, D., Taylor, R., Gabrielsen, J. S., Parker, G., & McClain, D. A. (2008). Regulation of Akt signaling by *O*-GlcNAc in euglycemia. *American Journal of Physiology. Endocrinology and Metabolism, 295*, E974–E980.

Song, Y. H., Ito, S., & Imaizumi, T. (2013). Flowering time regulation: Photoperiod- and temperature-sensing in leaves. *Trends in Plant Science, 18*, 575–583.

Song, Y. H., Shim, J. S., Kinmonth-Schultz, H. A., & Imaizumi, T. (2015). Photoperiodic flowering: Time measurement mechanisms in leaves. *Annual Review of Plant Biology, 66*, 441–464.

Streitner, C., Danisman, S., Wehrle, F., Schoning, J. C., Alfano, J. R., & Staiger, D. (2008). The small glycine-rich RNA binding protein AtGRP7 promotes floral transition in *Arabidopsis thaliana. The Plant Journal, 56*, 239–250.

Su, L., Shan, J.-X., Gao, J.-P., & Lin, H.-X. (2016). OsHAL3, a blue light-responsive protein, interacts with the floral regulator Hd1 to activate flowering in rice. *Molecular Plant, 9*, 233–244.

Su, Y., Wang, S., Zhang, F., Zheng, H., Liu, Y., Huang, T., & Ding, Y. (2017). Phosphorylation of histone H2A at serine 95: A plant-specific mark involved in flowering time regulation and H2A.Z deposition. *Plant Cell, 29*, 2197–2213.

Sui, P., Jin, J., Ye, S., Mu, C., Gao, J., Feng, H., Shen, W. H., Yu, Y., & Dong, A. (2012). H3K36 methylation is critical for brassinosteroid-regulated plant growth and development in rice. *The Plant Journal, 70*, 340–347.

Sun, H., Jia, Z., Cao, D., Jiang, B., Wu, C., Hou, W., Liu, Y., Fei, Z., Zhao, D., & Han, T. (2011). *GmFT2a*, a soybean homolog of *FLOWERING LOCUS T*, is involved in flowering transition and maintenance. *PLoS One, 6*, e29238.

Sun, C., Fang, J., Zhao, T., Xu, B., Zhang, F., Liu, L., Tang, J., Zhang, G., Deng, X., & Chen, F. (2012). The histone methyltransferase SDG724 mediates H3K36me2/3 deposition at *MADS50* and *RFT1* and promotes flowering in rice. *Plant Cell, 24*, 3235–3247.

Sundar, I. K., Yao, H., & Rahman, I. (2013). Oxidative stress and chromatin remodeling in chronic obstructive pulmonary disease and smoking-related diseases. *Antioxidants & Redox Signaling, 18*, 1956–1971.

Sung, S., & Amasino, R. M. (2004). Vernalization in *Arabidopsis thaliana* is mediated by the PHD finger protein VIN3. *Nature, 427*, 159–164.

Takano, M., Inagaki, N., Xie, X. Z., Yuzurihara, N., Hihara, F., Ishizuka, T., Yano, M., Nishimura, M., Miyao, A., Hirochika, H., & Shinomura, T. (2005). Distinct and cooperative functions of phytochromes A, B, and C in the control of deetiolation and flowering in rice. *Plant Cell, 17*, 3311–3325.

Takeshima, R., Hayashi, T., Zhu, J., Zhao, C., Xu, M., Yamaguchi, N., Sayama, T., Ishimoto, M., Kong, L., Shi, X., Liu, B., Tian, Z., Yamada, T., Kong, F., & Abe, J. (2016). A soybean quantitative trait locus that promotes flowering under long days is identified as *FT5a*, a *FLOWERING LOCUS T* ortholog. *Journal of Experimental Botany, 67*, 5247–5258.

Tamaki, S., Matsuo, S., Wong, H. L., Yokoi, S., & Shimamoto, K. (2007). Hd3a protein is a mobile flowering signal in rice. *Science, 316*, 1033–1036.

Tan, J., Jin, M., Wang, J., Wu, F., Sheng, P., Cheng, Z., Wang, J., Zheng, X., Chen, L., Wang, M., Zhu, S., Guo, X., Zhang, X., Liu, X., Wang, C., Wang, H., Wu, C., & Wan, J. (2016). *OsCOL10*, a *CONSTANS-like* gene, functions as a flowering time repressor downstream of *Ghd7* in rice. *Plant & Cell Physiology, 57*, 798–812.

Tao, Z., Shen, L., Gu, X., Wang, Y., Yu, H., & He, Y. (2017). Embryonic epigenetic reprogramming by a pioneer transcription factor in plants. *Nature, 551*, 124–128.

Taoka, K., Ohki, I., Tsuji, H., Furuita, K., Hayashi, K., Yanase, T., Yamaguchi, M., Nakashima, C., Purwestri, Y. A., Tamaki, S., Ogaki, Y., Shimada, C., Nakagawa, A., Kojima, C., & Shimamoto, K. (2011). 14-3-3 proteins act as intracellular receptors for rice Hd3a florigen. *Nature, 476*, 332–335.

Turck, F., Roudier, F., Farrona, S., Martin-Magniette, M. L., Guillaume, E., Buisine, N., Gagnot, S., Martienssen, R. A., Coupland, G., & Colot, V. (2007). *Arabidopsis* TFL2/LHP1 specifically associates with genes marked by trimethylation of histone H3 lysine 27. *PLoS Genetics, 3*, e86.

Tyler, L.,Thomas, S. G., Hu, J. H.,Dill, A., Alonso, J. M., Ecker, J. R., & Sun, T. P. (2004). Della proteins and gibberellin-regulated seed germination and floral development in Arabidopsis. *Plant Physiology, 135*, 1008–1019.

van Dijken, A. J., Schluepmann, H., & Smeekens, S. C. (2004). *Arabidopsis* trehalose-6-phosphate synthase 1 is essential for normal vegetative growth and transition to flowering. *Plant Physiology, 135*, 969–977.

Vega-Sanchez, M. E., Zeng, L., Chen, S., Leung, H., & Wang, G.-L. (2008). SPIN1, a K homology domain protein negatively regulated and ubiquitinated by the E3 ubiquitin ligase SPL11, is involved in flowering time control in rice. *Plant Cell, 20*, 1456–1469.

Verma, M., McKay, J., & Verma, D. (2023). Role of epigenetics in innate lymphoid cells. *Epigenomics, 15*, 615–618.

Wang, J. W. (2014). Regulation of flowering time by the miR156-mediated age pathway. *Journal of Experimental Botany, 65*, 4723–4730.

Wang, X., Zhang, Y., Ma, Q., Zhang, Z., Xue, Y., Bao, S., & Chong, K. (2007). SKB1-mediated symmetric dimethylation of histone H4R3 controls flowering time in *Arabidopsis*. *The EMBO Journal, 26*, 1934–1941.

Wang, J. W., Czech, B., & Weigel, D. (2009). miR156-regulated SPL transcription factors define an endogenous flowering pathway in *Arabidopsis thaliana*. *Cell, 138*, 738–749.

Wang, L., Kong, D., Lv, Q., Niu, G. Q., Han, T., Zhao, X., Meng, S., Cheng, Q., Guo, S., & Du, J. (2017). Tetrahydrofolate modulates floral transition through epigenetic silencing. *Plant Physiology, 174*, 1274.

Watanabe, S., Hideshima, R., Xia, Z. J., Tsubokura, Y., Sato, S., Nakamoto, Y., Yamanaka, N., Takahashi, R., Ishimoto, M., Anai, T., Tabata, S., & Harada, K. (2009). Map-based cloning of the gene associated with the soybean maturity locus *E3*. *Genetics, 182*, 1251–1262.

Watanabe, S., Xia, Z., Hideshima, R., Tsubokura, Y., Sato, S., Yamanaka, N., Takahashi, R., Anai, T., Tabata, S., Kitamura, K., & Harada, K. (2011). A map-based cloning strategy employing a residual heterozygous line reveals that the *GIGANTEA* gene is involved in soybean maturity and flowering. *Genetics, 188*, 395–U260.

Wei, X., Xu, J., Guo, H., Jiang, L., Chen, S., Yu, C., Zhou, Z., Hu, P., Zhai, H., & Wan, J. (2010). *DTH8* suppresses flowering in rice, influencing plant height and yield potential simultaneously. *Plant Physiology, 153*, 1747–1758.

Wei, J., Choi, H., Jin, P., Wu, Y., Yoon, J., Lee, Y.-S., Quan, T., & An, G. (2016). *GL2*-type homeobox gene *Roc4* in rice promotes flowering time preferentially under long days by repressing *Ghd7*. *Plant Science, 252*, 133–143.

Whittaker, C., & Dean, C. (2017). The *FLC* locus: A platform for discoveries in epigenetics and adaptation. *Annual Review of Cell and Developmental Biology, 33*, 555–575.

Wigge, P. A., Kim, M. C., Jaeger, K. E., Busch, W., Schmid, M., Lohmann, J. U., & Weigel, D. (2005). Integration of spatial and temporal information during floral induction in *Arabidopsis*. *Science, 309*, 1056–1059.

Wilson, R. N., Heckman, J. W., & Somerville, C. R. (1992). Gibberellin is required for flowering in *Arabidopsis thaliana* under short days. *Plant Physiology, 100*, 403–408.

Woo, H. R., Dittmer, T. A., & Richards, E. J. (2008). Three SRA-domain methylcytosine-binding proteins cooperate to maintain global CpG methylation and epigenetic silencing in *Arabidopsis*. *PLoS Genetics, 4*, e1000156.

Wood, C. C., Robertson, M., Tanner, G., Peacock, W. J., Dennis, E. S., & Helliwell, C. A. (2006). The *Arabidopsis thaliana* vernalization response requires a polycomb-like protein complex that also includes VERNALIZATION INSENSITIVE 3. *Proceedings of the National Academy of Sciences of the United States of America, 103*, 14631–14636.

Woods, D., Dong, Y. X., Bouche, F., Bednarek, R., Rowe, M., Ream, T., & Amasino, R. (2019). A florigen paralog is required for short-day vernalization in a pooid grass. *eLife, 8*, e42153.

Wu, G., Park, M. Y., Conway, S. R., Wang, J. W., Weigel, D., & Poethig, R. S. (2009). The sequential action of miR156 and miR172 regulates developmental timing in Arabidopsis. *Cell, 138*(4), 750–759. https://doi.org/10.1016/j.cell.2009.06.031

Wu, F., Price, B. W., Haider, W., Seufferheld, G., Nelson, R., & Hanzawa, Y. (2014). Functional and evolutionary characterization of the *CONSTANS* gene family in short-day photoperiodic flowering in soybean. *PLoS One, 9*, e85754.

Wu, M. F., Yamaguchi, N., Xiao, J., Bargmann, B., Estelle, M., Sang, Y., & Wagner, D. (2015). Auxin-regulated chromatin switch directs acquisition of flower primordium founder fate. *eLlife, 4*, e09269.

Wu, Z., Zhu, D., Lin, X., Miao, J., Gu, L., Deng, X., Yang, Q., Sun, K., Zhu, D., & Cao, X. (2016). RNA binding proteins RZ-1B and RZ-1C play critical roles in regulating pre-mRNA splicing and gene expression during development in *Arabidopsis*. *Plant Cell, 28*, 55–73.

Wu, W., Zheng, X.-M., Chen, D., Zhang, Y., Ma, W., Zhang, H., Sun, L., Yang, Z., Zhao, C., Zhan, X., Shen, X., Yu, P., Fu, Y., Zhu, S., Cao, L., & Cheng, S. (2017). *OsCOL16*, encoding a CONSTANS-like protein, represses flowering by up regulating *Ghd7* expression in rice. *Plant Science, 260*, 60–69.

Xia, Z., Watanabe, S., Yamada, T., Tsubokura, Y., Nakashima, H., Zhai, H., Anai, T., Sato, S., Yamazaki, T., Lu, S., Wu, H., Tabata, S., & Harada, K. (2012). Positional cloning and characterization reveal the molecular basis for soybean maturity locus *E1* that regulates photoperiodic flowering. *Proceedings of the National Academy of Sciences of the United States of America, 109*, E2155–E2164.

Xiao, J., Li, C., Xu, S., Xing, L., Xu, Y., & Chong, K. (2015). JACALIN-LECTIN LIKE1 regulates the nuclear accumulation of GLYCINE-RICH RNA-BINDING PROTEIN7, influencing the RNA processing of *FLOWERING LOCUS C* antisense transcripts and flowering time in *Arabidopsis*. *Plant Physiology, 169*, 2102–2117.

Xiao, J., Liu, B., Yao, Y., Guo, Z., Jia, H., Kong, L., Zhang, A., Ma, W., Ni, Z., Xu, S., Lu, F., Jiao, Y., Yang, W., Lin, X., Sun, S., Lu, Z., Gao, L., Zhao, G., Cao, S., Chen, Q., Zhang, K., Wang, M., Wang, M., Hu, Z., Guo, W., Li, G., Ma, X., Li, J., Han, F., Fu, X., Ma, Z., Wang, D., Zhang, X., Ling, H. Q., Xia, G., Tong, Y., Liu, Z., He, Z., Jia, J., & Chong, K. (2022). Wheat genomic study for genetic improvement of traits in China. *Science China Life Sciences, 65*, 1718–1775.

Xie, K., Wu, C., & Xiong, L. (2006). Genomic organization, differential expression, and interaction of SQUAMOSA promoter-binding-like transcription factors and microRNA156 in rice. *Plant Physiology, 142*, 280–293.

Xie, K., Shen, J., Hou, X., Yao, J., Li, X., Xiao, J., & Xiong, L. (2012). Gradual increase of miR156 regulates temporal expression changes of numerous genes during leaf development in rice. *Plant Physiology, 158*, 1382–1394.

Xie, L., Zhang, Y., Wang, K., Luo, X., Xu, D., Tian, X., Li, L., Ye, X., Xia, X., Li, W., Yan, L., & Cao, S. (2021). *TaVrt2*, an *SVP-like* gene, cooperates with *TaVrn1* to regulate vernalization-induced flowering in wheat. *The New Phytologist, 231*, 834–848.

Xing, L., Li, J., Xu, Y., Xu, Z., & Chong, K. (2009). Phosphorylation modification of wheat lectin VER2 is associated with vernalization-induced *O*-GlcNAc signaling and intracellular motility. *PLoS One, 4*, e4854.

Xing, S., Salinas, M., Höhmann, S., Berndtgen, R., & Huijser, P. (2010). miR156-targeted and nontargeted SBP-box transcription factors act in concert to secure male fertility in Arabidopsis. *The Plant Cell, 22*(12), 3935–3950. https://doi.org/10.1105/tpc.110.079343

Xing, L., Liu, Y., Xu, S., Xiao, J., Wang, B., Deng, H., Lu, Z., Xu, Y., & Chong, K. (2018). Arabidopsis O-GlcNAc transferase SEC activates histone methyltransferase ATX1 to regulate flowering. *The EMBO Journal, 37*, e98115.

Xu, S., & Chong, K. (2018). Remembering winter through vernalisation. *Nature Plants, 4*, 997–1009.

Xu, M. Y., Zhang, L., Li, W. W., Hu, X. L., Wang, M. B., Fan, Y. L., Zhang, C. Y., & Wang, L. (2014a). Stress-induced early flowering is mediated by miR169 in *Arabidopsis thaliana. Journal of Experimental Botany, 65*, 89.

Xu, Y., Gan, E. S., Zhou, J., Wee, W. Y., Zhang, X., & Ito, T. (2014b). *Arabidopsis* MRG domain proteins bridge two histone modifications to elevate expression of flowering genes. *Nucleic Acids Research, 42*, 10960–10974.

Xu, M. L., Yamagishi, N., Zhao, C., Takeshima, R., Kasai, M., Watanabe, S., Kanazawa, A., Yoshikawa, N., Liu, B. H., Yamada, T., & Abe, J. (2015). The soybean-specific maturity gene *E1* family of floral repressors controls night-break responses through down-regulation of *FLOWERING LOCUS T* orthologs. *Plant Physiology, 168*, 1735–U1080.

Xu, Y., Guo, C., Zhou, B., Li, C., Wang, H., Zheng, B., Ding, H., Zhu, Z., Peragine, A., & Cui, Y. (2016). Regulation of vegetative phase change by SWI2/SNF2 chromatin remodeling ATPase BRAHMA. *Plant Physiology, 172*, 2416–2428.

Xu, S., Xiao, J., Yin, F., Guo, X., Xing, L., Xu, Y., & Chong, K. (2019). The protein modifications of *O*-GlcNAcylation and phosphorylation mediate vernalization response for flowering in winter wheat. *Plant Physiology, 180*, 1436–1449.

Xu, S., Dong, Q., Deng, M., Lin, D., Xiao, J., Cheng, P., Xing, L., Niu, Y., Gao, C., Zhang, W., Xu, Y., & Chong, K. (2021). The vernalization-induced long non-coding RNA *VAS* functions with the transcription factor TaRF2b to promote *TaVRN1* expression for flowering in hexaploid wheat. *Molecular Plant, 14*, 1525–1538.

Xue, W. Y., Xing, Y. Z., Weng, X. Y., Zhao, Y., Tang, W. J., Wang, L., Zhou, H. J., Yu, S. B., Xu, C. G., Li, X. H., & Zhang, Q. F. (2008). Natural variation in *Ghd7* is an important regulator of heading date and yield potential in rice. *Nature Genetics, 40*, 761–767.

Yan, L., Loukoianov, A., Tranquilli, G., Helguera, M., Fahima, T., & Dubcovsky, J. (2003). Positional cloning of the wheat vernalization gene *VRN1. Proceedings of the National Academy of Sciences of the United States of America, 100*, 6263–6268.

Yan, L., Loukoianov, A., Blechl, A., Tranquilli, G., Ramakrishna, W., SanMiguel, P., Bennetzen, J. L., Echenique, V., & Dubcovsky, J. (2004). The wheat *VRN2* gene is a flowering repressor down-regulated by vernalization. *Science, 303*, 1640–1644.

Yan, L., Fu, D., Li, C., Blechl, A., Tranquilli, G., Bonafede, M., Sanchez, A., Valarik, M., Yasuda, S., & Dubcovsky, J. (2006). The wheat and barley vernalization gene *VRN3* is an orthologue of *FT. Proceedings of the National Academy of Sciences of the United States of America, 103*, 19581–19586.

Yan, W. H., Liu, H. Y., Zhou, X. C., Li, Q. P., Zhang, J., Lu, L., Liu, T. M., Liu, H. J., Zhang, C. J., Zhang, Z. Y., Shen, G. J., Yao, W., Chen, H. X., Yu, S. B., Xie, W. B., & Xing, Y. Z. (2013). Natural variation in *Ghd7.1* plays an important role in grain yield and adaptation in rice. *Cell Research, 23*, 969–971.

Yang, X. Y., & Qian, K. (2017). Protein *O*-GlcNAcylation: Emerging mechanisms and functions. Nature reviews. *Molecular Cell Biology, 18*, 452–465.

Yang, Y., Peng, Q., Chen, G.-X., Li, X.-H., & Wu, C.-Y. (2013). *OsELF3* is involved in circadian clock regulation for promoting flowering under long-day conditions in rice. *Molecular Plant, 6*, 202–215.

Yang, H., Berry, S., Olsson, T. S. G., Hartley, M., Howard, M., & Dean, C. (2017). Distinct phases of Polycomb silencing to hold epigenetic memory of cold in *Arabidopsis. Science, 357*, 1142–1145.

Yang, P., Liu, Y., Dong, Q., Miao, Y., Zhang, J., Xu, S., ... & Chong, K. (2025). O-GlcNAc and phosphorylation modifications on HtL1/FBA10 regulate wheat vernalization for flowering. *Nature Communications*. https://doi.org/10.1038/s41467-025-67734-0

Yano, M., Katayose, Y., Ashikari, M., Yamanouchi, U., Monna, L., Fuse, T., Baba, T., Yamamoto, K., Umehara, Y., Nagamura, Y., & Sasaki, T. (2000). *Hd1*, a major photoperiod sensitivity quantitative trait locus in rice, is closely related to the arabidopsis flowering time gene *CONSTANS. Plant Cell, 12*, 2473–2483.

Yant, L., Mathieu, J., Dinh, T. T., Ott, F., Lanz, C., Wollmann, H., Chen, X., & Schmid, M. (2010). Orchestration of the floral transition and floral development in *Arabidopsis* by the bifunctional transcription factor APETALA2. *Plant Cell, 22*, 2156–2170.

Yao, H., Li, A., & Wang, M. (2015). Systematic analysis and prediction of in situ cross talk of O-GlcNAcylation and phosphorylation. *BioMed Research International, 2015*(100):1–12 279823. https://doi.org/10.1155/2015/279823

Yong, W., Xu, Y., Xu, W., Wang, X., Li, N., Wu, J., Liang, T., Chong, K., Xu, Z., & Tan, K. (2003). Vernalization-induced flowering in wheat is mediated by a lectin-like gene *VER2*. *Planta, 217*, 261–270.

Yu, C. W., Liu, X., Luo, M., Chen, C., Lin, X., Tian, G., Lu, Q., Cui, Y., & Wu, K. (2011). HISTONE DEACETYLASE6 interacts with FLOWERING LOCUS D and regulates flowering in *Arabidopsis*. *Plant Physiology, 156*, 173–184.

Yu, S., Galvao, V. C., Zhang, Y. C., Horrer, D., Zhang, T. Q., Hao, Y. H., Feng, Y. Q., Wang, S., Schmid, M., & Wang, J. W. (2012). Gibberellin regulates the *Arabidopsis* floral transition through miR156-targeted SQUAMOSA promoter binding-like transcription factors. *Plant Cell, 24*, 3320–3332.

Yu, S., Cao, L., Zhou, C. M., Zhang, T. Q., Lian, H., Sun, Y., Wu, J., Huang, J., Wang, G., & Wang, J. W. (2013). Sugar is an endogenous cue for juvenile-to-adult phase transition in plants. *eLife, 2*, e00269.

Yuan, W., Luo, X., Li, Z., Yang, W., Wang, Y., Liu, R., Du, J., & He, Y. (2016). A cis cold memory element and a trans epigenome reader mediate Polycomb silencing of *FLC* by vernalization in *Arabidopsis*. *Nature Genetics, 48*, 1527–1534.

Zeng, X. L., Gao, Z., Jiang, C., Yang, Y. P., Liu, R. Y., & He, Y. H. (2020). HISTONE DEACETYLASE 9 functions with polycomb silencing to repress *FLOWERING LOCUS C* expression. *Plant Physiology, 1*, 555–565.

Zhai, H., Lu, S., Liang, S., Wu, H., Zhang, X., Liu, B., Kong, F., Yuan, X., Li, J., & Xia, Z. (2014). *GmFT4*, a homolog of *FLOWERING LOCUS T*, is positively regulated by *E1* and functions as a flowering repressor in soybean. *PLoS One, 9*, e89030.

Zhang, L., & Jiménez-Gómez, J. M. (2020). Functional analysis of *FRIGIDA* using naturally occurring variation in *Arabidopsis thaliana*. *The Plant Journal, 103*, 154–165.

Zhang, X., Germann, S., Blus, B. J., Khorasanizadeh, S., Gaudin, V., & Jacobsen, S. E. (2007). The *Arabidopsis* LHP1 protein colocalizes with histone H3 Lys27 trimethylation. *Nature Structural & Molecular Biology, 14*, 869–871.

Zhang, Q., Li, H., Li, R., Hu, R., Fan, C., Chen, F., Wang, Z., Liu, X., Fu, Y., & Lin, C. (2008). Association of the circadian rhythmic expression of *GmCRY1a* with a latitudinal cline in photoperiodic flowering of soybean. *Proceedings of the National Academy of Sciences of the United States of America, 105*, 21028–21033.

Zhang, L., Chen, L., & Yu, D. (2018). Transcription factor WRKY75 interacts with DELLA proteins to affect flowering. *Plant Physiology, 176*, 790–803.

Zhang, S., Hao, H., Liu, X., Li, Y., Ma, X., Liu, W., Zheng, R., Liang, S., & Luan, W. (2021). SDG712, a putative H3K9-specific methyltransferase encoding gene, delays flowering through repressing the expression of florigen genes in rice. *Rice (N Y), 14*, 73.

Zhao, Z., Yu, Y., Meyer, D., Wu, C., & Shen, W. H. (2005). Prevention of early flowering by expression of *FLOWERING LOCUS C* requires methylation of histone H3 K36. *Nature Cell Biology, 7*, 1256–1260.

Zhao, C., Takeshima, R., Zhu, J., Xu, M., Sato, M., Watanabe, S., Kanazawa, A., Liu, B., Kong, F., Yamada, T., & Abe, J. (2016). A recessive allele for delayed flowering at the soybean maturity locus *E9* is a leaky allele of *FT2a*, a *FLOWERING LOCUS T* ortholog. *BMC Plant Biology, 16*, 20.

Zhao, Y., Antoniou-Kourounioti, R. L., Calder, G., Dean, C., & Howard, M. (2020). Temperature-dependent growth contributes to long-term cold sensing. *Nature, 583*, 825–829.

Zhao, B., Xi, Y., Kim, J., & Sung, S. (2021a). Chromatin architectural proteins regulate flowering time by precluding gene looping. *Science Advances, 7*, eabg3097.

Zhao, Y., Zhu, P., Hepworth, J., Bloomer, R., Antoniou-Kourounioti, R. L., Doughty, J., Heckmann, A., Xu, C., Yang, H., & Dean, C. (2021b). Natural temperature fluctuations promote COOLAIR regulation of *FLC*. *Genes & Development, 35*, 888–898.

Zhao, X. H., Li, H. Y., Wang, L. S., Wang, J. H., Huang, Z. R., Du, H. P., Li, Y. R., Yang, J. H., He, M. L., Cheng, Q., Lin, X. Y., Liu, B. H., & Kong, F. J. (2024). A critical suppression feedback loop determines soybean photoperiod sensitivity. *Developmental Cell, 59*, 1–14.

Zhou, C. M., Zhang, T. Q., Wang, X., Yu, S., Lian, H., Tang, H., Feng, Z. Y., Zozomova-Lihova, J., & Wang, J. W. (2013). Molecular basis of age-dependent vernalization in *Cardamine flexuosa*. *Science, 340*, 1097–1100.

Zhou, H., Liu, Y., Liang, Y., Zhou, D., Li, S., Lin, S., Dong, H., & Huang, L. (2020). The function of histone lysine methylation related SET domain group proteins in plants. *Protein Science, 29*, 1120–1137.

Zhu, Q. H., & Helliwell, C. A. (2011). Regulation of flowering time and floral patterning by miR172. *Journal of Experimental Botany, 62*, 487–495.

Zhu, S., Wang, J., Cai, M., Zhang, H., Wu, F., Xu, Y., Li, C., Cheng, Z., Zhang, X., Guo, X., Sheng, P., Wu, M., Wang, J., Lei, C., Wang, J., Zhao, Z., Wu, C., Wang, H., & Wan, J. (2017). The OsHAPL1-DTH8-Hd1 complex functions as the transcription regulator to repress heading date in rice. *Journal of Experimental Botany, 68*, 552–567.

Zhu, P., Lister, C., & Dean, C. (2021). Cold-induced Arabidopsis FRIGIDA nuclear condensates for FLC repression. *Nature, 599*, 657–661.

Chapter 6
Reproductive Meristems and Regulatory Genes

Dazhong Zhao, Weimin Ni, and Hong Ma

Abstract Reproductive meristems, emerging after the transition from vegetative to reproductive development (the flowering transition), generate floral organs and other meristems, thereby impacting the inflorescence architecture and numbers and arrangement of floral organs. In this chapter, we provide a comprehensive overview of the intricate process of the formation of reproductive meristems, including the inflorescence meristem (IM) and floral meristem (FM), and the function of FM in producing floral organ primordia. The IM is formed from the conversion of the shoot apical meristem (SAM), as a consequence of the flowering transition, and then generates multiple FMs, from which floral organ primordia are derived. We outline the regulation of IM formation by multiple pathways, such as floral repression, autonomous promotion, vernalization, photoperiod, and gibberellin signaling, as well as the *TFL1* gene. We then highlight the genes specifying FM identity and the genetic regulation of FM determinacy. Finally, we discuss the molecular mechanisms underlying the maintenance of reproductive meristem identities, the control of meristem growth and size, and the determination of floral organ numbers.

Keywords Inflorescence meristem · Floral meristem · Meristem identity genes · Meristem size

D. Zhao
Department of Biological Sciences, University of Wisconsin-Milwaukee, Milwaukee, WI, USA

W. Ni
Sutro Biopharma Inc., South San Francisco, CA, USA

H. Ma (✉)
Department of Biology, The Huck Institutes of the Life Sciences, 510D Mueller Laboratory, The Pennsylvania State University, University Park, PA, USA
e-mail: hxm16@psu.edu

F. Chang et al. (eds.), *Regulation of Plant Development*,
https://doi.org/10.1007/978-981-95-7033-1_6

6.1 Introduction

Reproductive development in flowering plants is a long and complex process characterized by the formation of reproductive meristems, including the inflorescence meristem (IM) and floral meristem (FM), which then generate floral organ primordia for the subsequent formation of floral organs (Zhao et al., 2001b). The transition from vegetative development to reproductive development, also called the flowering transition, leads to conversion from a vegetative shoot apical meristem (SAM) to an IM. The IM subsequently produces a number of FMs from which individual flowers are formed. The size, shape, and duration of activity of IM can affect the number and positions of FMs, which in turn determine the number and patterns (phyllotaxy) of flowers in an inflorescence (Kitagawa & Jackson, 2019). The IM can be indeterminate or determinate if, respectively, it produces an indefinite number of FMs or a limited and often specified number of FMs. During much of reproductive development, the indeterminate IMs remain active, while FMs generate floral organs, leading to the continuous production of FMs and flowers; in contrast, the determinate IMs have a limited duration of activity and are arrested after the formation of a certain number of flowers. Indeterminate IMs in many plants such as *Arabidopsis thaliana* and *Antirrhinum majus* (snapdragon) generate a raceme inflorescence with a spiral of FMs on their periphery in a temporal series of flowers from the bottom to the apex along the inflorescence stem. In some other species, like celery and onion, multiple FMs are produced at approximately the same time, forming umbel inflorescences with a cluster at the apex. In plants like maize and other grasses, the primary IM yields one or more additional levels of specialized reproductive/inflorescence meristems, the last of which are the spikelet meristems and initiate the FMs (Tanaka et al., 2013). In members of the sunflower family (Asteraceae), the IM is laterally expanded into a disc shape, with many FMs, resulting in the head (capitulum) type of inflorescence (Elomaa et al., 2018). In most Asteraceae members, the head inflorescence has peripheral ray flowers with zygomorphic corolla for attracting pollinators and multiple interior disk flowers for reproduction. In some plants with determinate IMs, such as tomato and tobacco, the last flower occupying the apex of the inflorescence is called the terminal flower (Pnueli et al., 2001). In some other plants, like tulip and magnolia, the floral transition leads to the conversion of SAM to FM for a single flower.

In the FM, cell division and differentiation result in the formation of floral organ primordia, which then develop into floral organs, whose types and numbers are characteristic of the species or groups of species. On the periphery of the flower are sterile perianth organs surrounding the interior reproductive organs. In eudicots, the largest monophyletic group of flowering plants, flowers generally have two distinct types of perianth organs arranged in concentric rings, called whorls; in these plants, the organs in the outermost whorl are sepals, with one or more whorls of petals interior to sepals (e.g., Bowman, 1997). In basal angiosperm species, the perianth organs are often numerous and arranged in a tight spiral (Soltis et al., 2007). In grass flowers, perianth organs have distinctive morphologies and unique names, such as lemma, palea, and lodicule (Kellogg, 2015). These outer grass floral organs share some features and regulatory genes as perianth organs in eudicots and other monocots, suggesting that palea and lodicule might be homologous, respectively, to outer and inner perianth organs of other monocots and eudicots (Schrager-Lavelle et al., 2017). The interior whorls of a flower are occupied by reproductive organs. The male and female reproductive organs are called the stamen and pistil, respectively. The stamen consists of a filament and an anther, which contains the male reproductive cells; the pistil can be divided into the stigma, style, and ovary, which houses the female reproductive cells. The reproductive cells undergo meiosis to produce microspores in the anther and megaspores in the ovary. The microspores and megaspores then develop into pollen grains and embryo sacs, which are male and female gametophytes, respectively. Most flowering plants produce perfect flowers that have both male and female organs within a single flower. Flowers that have only male or female organs are called imperfect flowers; monoecious species, such as maize, produce male and female flowers in the same plant, whereas dioecious species, such as asparagus, form male and female flowers on separate plants. The number, type, size, shape, and relative position of floral organs vary greatly among flowering plants, which are important characters for the classical systematics of flowering plants.

Since the mid-1980s, molecular genetic analyses and transcriptomics studies in *Arabidopsis* have generated valuable insights into molecular mechanisms underlying many aspects of reproductive development in this plant (Ma, 1994, 2005; Ge et al., 2010; Ma & Sundaresan, 2010). Additional research in *Antirrhinum* and various other plants found that many regulators critical for early reproductive development are conserved in these species (Schwarz-Sommer et al., 1990; Coen & Meyerowitz, 1991). At the same time, evidence suggests divergence in gene functions among more distant species (Soltis et al., 2007). Other comparative studies are beginning to allow glimpses into the conservation and divergence in other aspects of reproductive development (Soltis et al., 2007; Schrager-Lavelle et al., 2017; Wang et al., 2021; Zhang et al., 2024). In this chapter, we focus on molecular genetic analyses of reproductive meristems in *Arabidopsis*, with some brief relevant information from other species. Unless otherwise noted, the genes discussed here are from *Arabidopsis*.

6.2 Regulation of Inflorescence Meristem Formation

6.2.1 Regulation of Inflorescence Identity by Flowering Pathways

The transition from vegetative to reproductive development is a major switch in the life cycle of flowering plants (Simpson et al., 1999; Zhao et al. 2001b; see Chap. 5). During this transition, dramatic physiological and molecular changes caused by internal signals and environmental cues, such as temperature, light, and day length, occur in the SAM, resulting in the formation of IM that preludes the reproductive development of angiosperms. Genetic analyses in the past decades have identified many key genes that function in negative regulatory pathways to repress flowering and genes in an autonomous pathway to promote flowering (Araki, 2001; Mouradov et al., 2002). The repression pathways are an endogenous mechanism that allows flowering only when the plant has reached a particular age or size, whereas the autonomous promotion pathway positively regulates flowering as the plant develops (Quiroz et al., 2021). Furthermore, the photoperiod pathway controls flowering induction by day length, while the vernalization promotion pathway allows or accelerates flowering in response to an extended period of cold temperature (Kim et al., 2008; Osnato et al., 2021). Moreover, these pathways can interact with each other, mediating responses to environmental factors. The effects of environmental conditions vary with both the species and genotype of individuals. In many plants including *Arabidopsis*, when positive regulators of flowering overcome the effects of inhibitors of flowering, the IM identity is acquired; in other words, the plant is developmentally competent to flower (Chavez-Hernandez et al., 2022). To date, at least 30 genes that are involved in the flowering transition have been cloned and functionally characterized in *Arabidopsis* (Fig. 6.1; Table 6.1).

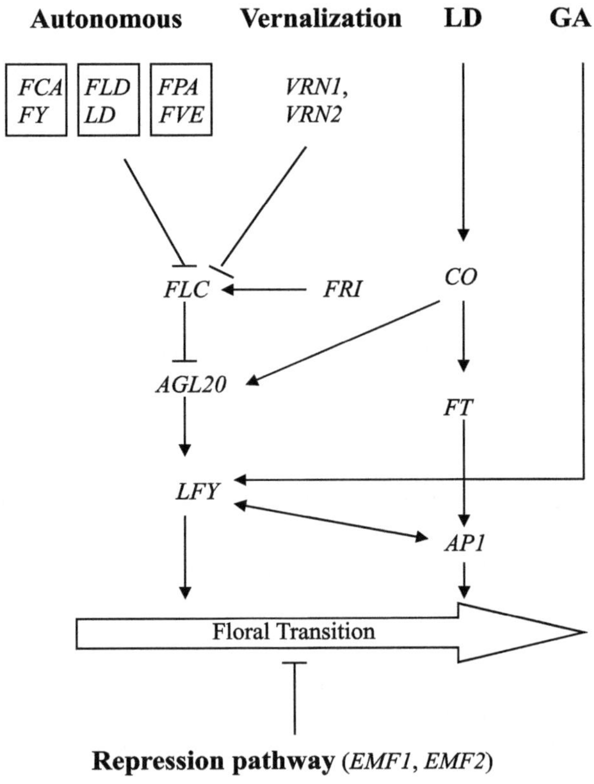

Fig. 6.1 The genetic network controlling the flowering transition. Four positive pathways are indicated in boldface at the top and one negative pathway at the bottom. Genes in each branch of the autonomous pathways are enclosed in a box. Arrows represent positive genetic interactions, whereas the lines ending with a short bar indicate negative interactions

Table 6.1 *Arabidopsis* genes affecting flowering time

Gene	Full name	Mutant phenotype[a]	Protein structure and function
AGL20/SOC1	*AGAMOUS-LIKE 20 SUPPRESSOR OF OVEREXPRESSION OF CO1*	Late flowering	MADS-domain protein
CCA1	*CIRCADIAN CLOCK ASSOCIATED1*	Defective circadian clock	MYB-domain protein
CO	*CONSTANS*	Late flowering under LD	Transcription factor, Zn finger
CRY1	*CRYPTOCHROMES 1*	Defective circadian clock	Blue light receptor
CRY2	*CRYPTOCHROMES 2*	Delayed flowering under LD	Blue light receptor
EBS	*EARLY BOLTING IN SHORT DAYS*	Early bolting in SD	
ELF1	*EARLY FLOWER1*	Early flowering	
ELF2	*EARLY FLOWER2*	Early flowering	
ELF3	*EARLY FLOWER3*	Early flowering	
EMF1	*EMBRYONIC FLOWER1*	Flowering upon germination	Novel protein
EMF2	*EMBRYONIC FLOWER2*	Flowering upon germination	Zinc-finger, putative Polycomb Group protein
FCA		Late flowering	RNA-binding protein
FKF1		Defective circadian clock and flowering time	LOV, F-box, kelch repeats
FLC	*FLOWERING LOCUS C*	Dominant allele causing late flowering	MADS-domain protein
FLD		Late flowering	
FPA		Late flowering	RNA-binding protein
FRI	*FRIGIDA*	Late flowering	
FT	*FLOWERING LUCUS T*	Late flowering	Homolog of TFL1
FVE		Late flowering	
FWA		Late flowering	Homeodomain protein
FY		Late flowering	
LD	*LUMINIDEPENDENS*	Late flowering	Transcription factor with a homeodomain
PHYA		Reduced flowering promotion	Red/far-red light receptor
PHYB		Early flowering	Red light receptor

(continued)

Table 6.1 (continued)

Gene	Full name	Mutant phenotype[a]	Protein structure and function
PHYD		Enhancing early flowering of phyB mutant	Red light receptor
PHYE		Early flowering in *phyA phyB* double mutant	Red light receptor
TFL1	*TERMINAL FLOWER1*	Early flowering	Phosphatidylethanolamine-binding protein
VRN1	*VERNALIZATION1*	Less sensitive to vernalization	
VRN2	*VERNALIZATION2*	Less sensitive to vernalization	Zinc-finger, putative Polycomb Group protein
ZTL	*ZEITLUPE*	Defective circadian clock and flowering time	LOV, F-box, kelch repeats

Abbreviations: *SD* short day, *LD* long day
[a]Unless otherwise indicated, the phenotypes are due to recessive mutations

Chapter 5 presents details of several molecular genetic pathways regulating the floral transition, including the autonomous, the photoperiod, the vernalization, the GA, the ambient temperature, and aging pathways. Thus, this chapter will focus on the floral repression pathway, with a summary of other pathways controlling the switch to reproduction development, followed by a discussion of the integration of pathways that regulate the inflorescence and floral meristem genes.

6.2.1.1 The Floral Repression Pathway

During vegetative development, it is beneficial for the plant to prevent premature flowering and allow sufficient vegetative development. In *Arabidopsis*, genes repressing flowering were uncovered by genetic analyses of early-flowering mutants (Fig. 6.1; Table 6.1). The *EMBRYONIC FLOWER* (*EMF1* and *EMF2*) genes have the most dramatic effects on repressing flowering (Aubert et al., 2001; Yoshida et al., 2001). Strong *emf* mutations, such as *emf1–1*, lead to flowering immediately following seed germination, bypassing the entire vegetative phase and formation of IM (Sung et al., 1992). Weak *emf* mutants precociously produce an inflorescence with a few floral buds and sessile leaves. *EMF1* encodes a plant-specific novel protein. In *Drosophila* and mammals, Polycomb group (PcG) proteins form complexes to regulate chromatin structure and stabilize gene expression patterns of homeotic genes that regulate development (Kassis & Brown, 2013). EMF1, HETEROCHROMATIN PROTEIN1 (LHP1), and a trimethyl histone H3 lysine-4 (H3K4) demethylase may constitute a PcG complex to regulate expression of

FLOWERING LOCUS T (*FT*) via remodeling the chromatin structure (Huang et al., 2019). The *EMF2* gene encodes a zinc-finger protein that shares sequence similarity to a member of the PcG protein family (Yoshida et al., 2001). Among PcG proteins, polycomb repressive complex 1 (PRC1) and 2 (PRC2) are critical for establishing a repressive chromatin state, thus inhibiting the expression of residing target genes. EMF2, together with two other PRC2 components, CURLY LEAF (CLF) and MULTICOPY SUPPRESSOR OF IRA1 (MSI1), form a PRC2 complex that interacts with RPA2A [(a subunit of the conserved heterotrimeric replication protein A (RPA)] to downregulate the expression of *FT*, thereby inhibiting flowering transition (Godwin & Farrona, 2022).

Another important gene that represses the flowering transition in *Arabidopsis* is *TERMINAL FLOWER1* (*TFL1*). The *TFL1* gene encodes a protein that is similar to animal phosphatidylethanolamine-binding proteins (PEBP) and is related to FT, the florigen (Ho & Weigel, 2014; Yang et al., 2023). Mutations in *TFL1* accelerate the transition from vegetative to reproductive development, resulting in the production of terminal flower(s) at the apex. However, after the floral transition, *TFL1* promotes the IM identity at the central portion of SAM by inhibiting the expression of FM identity genes, including *LEAFY* (*LFY*) and *APETALA1* (*AP1*) (Hanano & Goto, 2011; Goretti et al., 2020). TFL1 acts as a transcriptional coregulator to modulate the expression of thousands of genes, such as those important for flowering transition (Goretti et al., 2020; Cerise et al., 2023). Other genes involved in repressing early flowering include *EARLY BOLTING IN SHORT DAYS* (*EBS*) and *EARLY FLOWER1*, *2*, *3* (*ELF1*, *ELF2*, and *ELF3*) (Zagotta et al., 1992; Gómez-Mena et al., 2001; Hicks et al., 2001). In addition, *FLOWERING LOCUS C* (*FLC*) and *FRIGIDA* (*FRI*) were identified by genetic analysis of naturally occurring races, or ecotypes, of *Arabidopsis* (Clarke & Dean, 1994; Michaels & Amasino, 2001; Wang et al., 2006). *FLC* encodes a MADS-domain protein that represses flowering, and *FRI* enhances the expression of *FLC* (see Chap. 5). Although genetic studies suggest that many genes are involved in the repression or delay of flowering and promotion of vegetative development, understanding their relationships and molecular nature requires further study.

6.2.1.2 Key Genes in the Autonomous Promotion and Vernalization Pathways

In angiosperms, floral transition can occur after sufficient vegetative development regardless of environmental conditions due to the autonomous promotion pathway. In *Arabidopsis*, genes in the autonomous pathway were first identified by late-flowering mutations under both long- and short-day conditions, such as *FCA*, *FPA*, *FLD*, and *LD* (Simpson et al., 1999; Zhao et al. 2001b) (Fig. 6.1; Table 6.1; also see Chap. 5). Some double mutants exhibit more drastic flowering delays than the single mutants, suggesting that those genes act in different branches of this pathway, as also supported by early flowering caused by increased expression of *FCA* or *FPA*. The *LD* gene encodes a putative transcription factor with a homeodomain, and

LD interacts with FLD, an H3K4me demethylase that regulates the expression of *FLC* via removal of H3K4me2 in its promoter region.

Plants adapted temperate regions tend to flower in the spring following a long winter; such an extended cold period which can trigger or accelerate flowering via a process is called vernalization. Genetic and molecular analyses have identified key *Arabidopsis* genes for vernalization (Fig. 6.1; Table 6.1), including *VERNALIZATION1* (*VRN1*) and *VRN2* (Chandler et al., 1996), which are required for vernalization-mediated early flowering of *fca* mutants. *VRN1* encodes a DNA-binding protein affecting H3K9me3 of *FLC*, while *VRN2* encodes a PcG protein increasing H3K27me3 and H3K9me3 at *FLC* (Gendall et al., 2001; Levy et al., 2002). FLC is a MADS-box transcription factor and a repressor of flowering. Other proteins with roles in histone methylation also contribute to the regulation of *FLC* expression. The continued silencing of *FLC* after cold exposure is achieved through the maintenance of H3K27me3 and H3K9me3. The expression of *FLC* is reactivated during reproduction and embryogenesis due to the function of EARLY FLOWERING 6 (ELF6), which can demethylate H3K27me3 (Finnegan et al., 2020; Nakamura et al., 2017). Other factors for activating *FLC* expression include LEAFY COTYLEDON 1 (LEC1), LEC2, and FUSCA3 (FUS3) (Tao et al., 2017; Chen & Du, 2022).

In wheat (*Triticum aestivum*), the vernalization response is mainly controlled by three genes: *Ta-VRN1* (*VERNALIZATION 1*) encoding a MADS-box transcription factor similar to AP1, FRUITFULL (FUL), and CAULIFLOWER (CAL), *Ta-VRN2* encoding a zinc finger protein, and *Ta-VRN3* encoding FT1, a homolog of the *Arabidopsis FT* gene (Yan et al., 2003, 2004, 2006). Vernalization in winter induces the expression of *Ta-VRN1* but decreases the expression of *Ta-VRN2*. *Ta*-VRN1 promotes the expression of *Ta-VRN3* in leaves; the resulting *Ta*-VRN3 protein moves to the SAM where it enhances the expression of *Ta-VRN1*. Therefore, actions of the *Ta*-VRN1-*Ta*-VRN3 regulatory loop eventually cause flowering of wheat in spring. In addition, the noncoding RNA *TaVRN1 ALTERNATIVE SPLICING* (*VAS*), encoded by the *TaVRN1* sense strand, promotes the binding of the *Ta*RF2b-*Ta*RF2a transcription factor complex to the *TaVRN1* promoter, thereby activating its expression and inducing the flowering of winter wheat (Xu et al., 2021). Furthermore, vernalization induces the expression of another wheat gene, *VERNALIZATION-RELATED 2* (*VER2*), which promotes flowering and encodes a jacalin lectin protein binding to O-GlcNAc-modified proteins (Yong et al., 2003). VER2 can be phosphorylated, and the phosphorylated VER2 interacts with O-GlcNAc-modified GLYCINE-RICH RNA-BINDING PROTEIN 2 (*Ta*-GRP2); *Ta*-GRP2 is then relocated from the nucleus to the cytoplasm, which elevates the mRNA level of *Ta-VRN1*, thus triggering flowering (Xing et al., 2009; Xiao et al., 2014; Xu et al., 2019).

6.2.1.3 The Photoperiod Pathway and Regulation of Meristem Identity Genes

Photoperiod, an indicator of the season, is another important environmental factor that regulates flowering time. According to the way they respond to photoperiod, plants can be classified into three groups: long-day plants, short-day plants, and day-neutral plants. Obligate long-day or short-day plants require appropriate photoperiod conditions to flower and will remain vegetative until exposed to the necessary day length, while facultative plants will flower earlier under their favorable day-length conditions. For example, *Arabidopsis* is a facultative long-day plant; thus, it flowers earlier under long days than short days (Corbesier et al., 1996). Several *Arabidopsis* genes were identified as components of a long-day flowering promotion pathway by mutations that delay flowering under long days but have little or no effect under short days (Koornneef et al., 1991; Putterill et al., 1996; Fowler et al., 1999; Putterill, 2001; Mouradov et al., 2002) (Fig. 6.1; Table 6.1).

Photoperiod is perceived by leaves, and such perception requires photoreceptors, including phytochromes for red and far-red light and cryptochromes for sensing blue light (Mockler et al., 2003). In *Arabidopsis*, phytochromes are encoded by five genes: *PHYA* to *PHYE*. Genetic analyses indicate that *PHYA* and *PHYB* play major roles in the flowering transition, respectively, as positive and negative regulators. *PHYD* and *PHYE* may also act to repress flowering, as the *phyd* and *phye* mutations enhance the phenotype of the *phyB* mutant. Two *Arabidopsis* genes encode the blue light receptor cryptochromes (*CRY1* and *CRY2*) and affect the timing of flowering. Mutations in *CRY2* (also isolated as *fha*) cause delayed flowering under long-day conditions, whereas overexpression of *CRY2* results in accelerated flowering under long days but not short days, suggesting that *CRY2* is a key component of the long-day photoperiod pathway (Putterill, 2001; Guo et al., 1998). *CRY1* plays an essential role in the blue-light regulation of the circadian clock, which sets the timing of cellular activities in phase with the day-night cycle via a feedback mechanism (Yu et al., 2010). This feedback involves other clock genes, *ZEITLUPE* (*ZTL*) and *FLAVIN-BINDING KELCH REPEAT F-BOX 1* (*FKF1*), which encode proteins containing a blue-light responsive PAS/LOV domain, an F-box, and kelch repeats (Somers et al., 2000; Sawa et al., 2007). Another pair of genes, *LATE ELONGATED HYPOCOTYL* (*LHY*) and *CIRCADIAN CLOCK ASSOCIATED1* (*CCA1*), encode MYB-domain proteins that likely form a feedback loop to regulate gene expression (Lu et al., 2009).

The *CONSTANS* (*CO*) gene encodes a B-Box transcription factor and integrates external and internal cues to control flowering in the photoperiod pathway (Putterill et al., 1996; Yoo et al., 2005). *CO* expression is regulated by upstream components in the photoperiod pathway, including both photoreceptors and the circadian clock. Overexpression of *CO*, bypassing the regulation of these upstream factors, leads to early flowering under both short and long days. Specifically, blue light stabilizes photoreceptors FKF1 and ZTL, which promote the degradation of a repressor of *CO* expression, CYCLING DOF FACTOR (CDF), thereby resulting in increased levels of the *CO* mRNA. Furthermore, the CO protein is stabilized by FKF1 and FK

506-BINDING PROTEIN 12 (FKBP12), but the degradation of CO is facilitated by PHYB, HOS1, and ELF3. The *CO* homolog *HEADING DATE 1 (HD1)* in rice, a short-day plant, is also involved in the photoperiod response (Yano et al., 2000). CO activates the expression of *FT*, which is a major part of the previously proposed florigen (Wigge, 2011). FT is a small protein with ~175 amino acids and probably binds to phosphatidylethanolamine. The FT protein moves to the SAM, where it interacts with the basic leucine zipper domain transcription factor FD and 14-3-3 proteins. The FT-FD complex regulates the expression of numerous genes essential for IM and FM identity (see Sect. 6.3). In the SAM, TFL1 plays a critical role in maintaining the indeterminate IM by preventing the conversion of FM from IM. FD binds to G- and C-boxes in the second exon of the *LFY* gene. TFL1-FD inhibits the expression of *LFY* in the IM, thus preventing termination of IM, while in the FM, FT-FD activates the expression of *LFY*, which promotes FM identity.

6.2.1.4 Regulation of Flowering by the Phytohormone Gibberellin

Both internal signals and external cues that regulate flowering may affect the activity of plant hormones. The hormone gibberellin (GA) has been found to promote flowering in many plant species, with the exogenous application of GA being commonly effective in accelerating flowering time in long-day plants (Vince-Prue, 1985). In *Arabidopsis*, the application of GA accelerates the flowering of several late-flowering mutants, such as *fca*, *fd*, *co*, and *ft*, under long days, suggesting that GA functions via mechanisms separate from the autonomous and photoperiod pathways (Chandler & Dean, 1994). GA acts in part by transcriptional activation of the floral meristem identity gene *LFY* (see Sect. 6.3). Moreover, in the SAM, GA promotes the expression of *SUPPRESSOR OF OVEREXPRESSION OF CO1* (*SOC1*), which encodes a MADS-box transcription factor and positively regulates flowering (Hyun et al., 2016; Bao et al., 2020; Zhang et al., 2023). In *Arabidopsis*, five DELLA proteins are known to be important for mediating GA-dependent regulation of gene expression (Locascio et al., 2013; Yoshida et al., 2014). DELLAs interact with a zinc-finger transcription factor called GAI-ASSOCIATED FACTOR1 (GAF1); in maize, *INDETERMINATE1* (*ID1*) is a *GAF1* homolog and a positive regulator of flowering (Colasanti et al., 1998; Kozaki et al., 2004). In response to GAs, GAF1 promotes expression of *FT* and *SOC1* via directly downregulating expression of *ELF3*, *TEMPRANILLO1* (*TEM1*), *TEM2*, and *SHORT VEGETATIVE PHASE* (*SVP*) by binding to their promoters together with a DELLA protein.

6.2.1.5 Integration of Flowering Pathways

The existence of multiple pathways regulating the floral transition implies a need to integrate outputs from these pathways to control flowering time. Several genes downstream of two or more pathways may play vital roles in such convergence (Fig. 6.1). One of these genes is *FLC*, which acts downstream of both vernalization and autonomous promotion pathways (Michaels & Amasino, 2001; Searle et al., 2006). *FLC* encodes a MADS-domain repressor of the flowering transition, and its mRNA is downregulated by both autonomous and vernalization pathways (Michaels & Amasino, 2001). In *Brassica napus*, analyses of five *FLC*-related genes (*BnFLC1–5*) revealed that vernalization dramatically reduces the level of *BnFLC* mRNA and promotes flowering in the winter cultivar Colombus of *B. napus* (Tadege et al., 2001), similar to results from *Arabidopsis*.

FLC directly represses the expression of *AGMOUS-LIKE 20* (*AGL20*)/SOC1 via binding to CArG-box [CC(A/T)6GG] motifs in its promoter region (Searle et al., 2006). Thus, *SOC1* is regulated by both the autonomous and vernalization pathways via FLC. Furthermore, *SOC1* acts downstream of *CO* and the photoperiod pathway, as the *soc1* mutation suppresses the early-flowering phenotype of plants overexpressing *CO* (Samach et al., 2000). Hence, *SOC1* serves as an integrator of at least three flowering pathways to promote flowering, via activating floral meristem genes, such as *LFY* (Lee & Lee, 2010; Immink et al., 2012).

The photoperiod pathway and GA pathway also share redundant functions in flowering transition, as long-day conditions can bypass the requirement for GA. As mentioned above, GA promotes the expression of *SOC1*. Promoter fusion studies with the *LFY* gene show that *cis* elements responsible for the GA pathway are distinct from those responsive to the long-day pathway. Therefore, the transcriptional regulation of *LFY* is what integrates the GA and long-day pathways (Blazquez & Weigel, 2000). Genetic and molecular studies suggest that *FT* is another important integrator that functions downstream of floral repression, photoperiod, vernalization, and GA pathways in parallel with *SOC1* to promote the flowering transition (Wigge, 2011).

6.2.2 Genes Promoting Inflorescence Identity: TFL1 *and Its Homologs*

To continuously produce FMs over time, the IM needs to be maintained for a certain period of time. In *Arabidopsis*, *TFL1* plays a role in maintaining the indeterminate IM (Fig. 6.2) (Freytes et al., 2021; Wickland & Hanzawa, 2015; Shannon & Meeks-Wagner, 1991; Hanano & Goto, 2011; Yao et al., 2023). Mutations in the *TFL1* gene cause the production of a terminal flower at the apex of the primary inflorescence along with a few lateral flowers, indicating that the mutant IM produces a limited number of FMs, thus becoming determinate. Mutant plants with weak *tfl1* alleles form more lateral flowers before the inflorescence is terminated, suggesting that the effect of *TFL1* is quantitative. Normally, in *Arabidopsis*, branch inflorescences develop after the primary inflorescence; however, axillary meristems in the *tfl1–1* mutant produce a single flower instead of an entire branch inflorescence. Therefore, as the plant develops, the effect of *tfl1* mutations becomes greater, resulting in the conversion of the branch IM to one floral meristem. The *TFL1* gene is expressed in

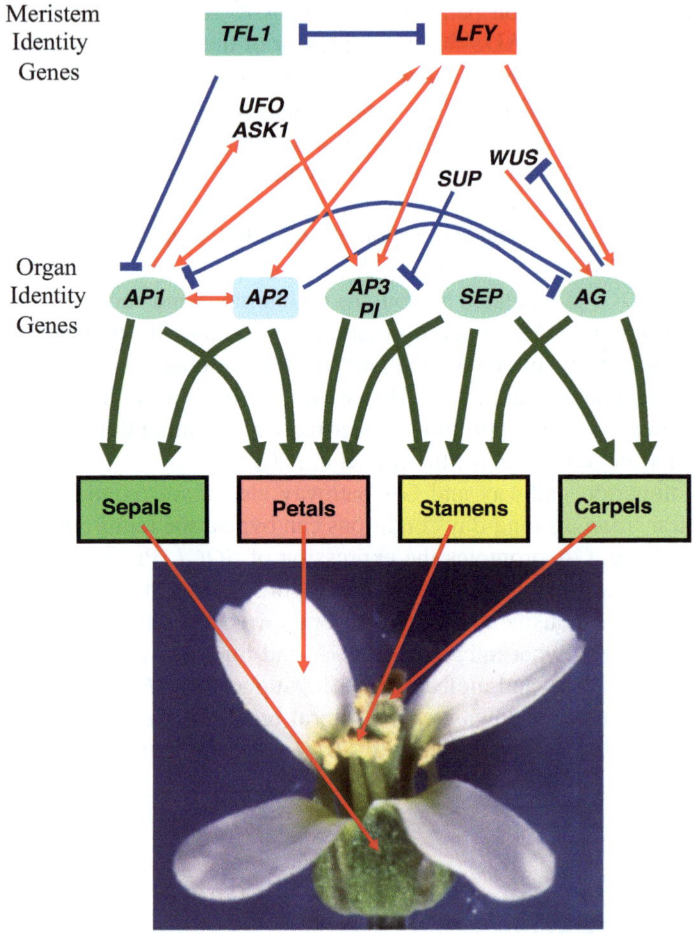

Fig. 6.2 Genes controlling floral meristem and organ identities. Genes controlling meristem identities are shown at the top. They interact with intermediate genes, which, together with the meristem identity genes, control the organ identity genes. The ultimate consequence of these gene functions is the specification of four organ identities, as shown in the photograph at the bottom

the central region of IM, which is consistent with its role in maintaining the IM identity. Because *TFL1* also represses flowering time, it can be considered a negative regulator of flower development both before and after the flowering transition.

In *Antirrhinum*, the *centroradialis* (*cen*) mutant has an inflorescence structure similar to that of *tfl1* mutants (Bradley et al., 1996, 1997). Therefore, the *TFL1* and *CEN* genes are hypothesized to have similar functions in maintaining IM. It is possible that in plants with determinate inflorescences, activities of genes like *TFL1* and *CEN* are absent or reduced. Also, plants that produce single flowers might have a stronger combination of positive regulators of flower development with the absence of TFL1/CEN-like activities.

Another gene homologous to *TFL1* and *CEN* is the tomato *SELF-PRUNING* (*SP*) gene that regulates shoot architecture (Pnueli et al., 1998). These regulatory genes form the *CETS* (*CEN*, *TFL1*, and *SP*) family, encoding putative membrane-associated proteins that are similar to animal phosphatidylethanolamine-binding proteins (see Sect. 6.2.1.1). CETS proteins potentially serve as adapter proteins that interact with a variety of proteins, thereby participating in signal transduction, transcriptional regulation, and other processes.

6.3 Genes Specifying Floral Meristem Identity

Molecular genetic studies have identified multiple key genes that control FM identity in *Arabidopsis* and *Antirrhinum* (Fig. 6.2, Table 6.2). In *Arabidopsis*, the *LFY* gene plays a central role in specifying the FM identity (Simon et al., 1996; Weigel et al., 1992; Blazquez et al., 1997; Busch et al., 1999; Zhao et al., 2001a; Yamaguchi, 2021; Huala & Sussex, 1992). Mutations in *LFY* result in the conversion of early flowers into shoot-like structures, while those flowers that arise later exhibit characteristics resembling those of inflorescences, such as partial indeterminacy and a spiral arrangement of organs. The *LFY* gene encodes a plant-specific helix-turn-helix transcription factor that is conserved in seed plants. The spatiotemporal expression patterns in IM and FM are critical for the functions of *LFY* in determining the floral fate, flower primordium founder fate, and floral organ fate. FD directly regulates the expression of *LFY* by binding to its promoter (Jaeger et al., 2013; Zhu et al., 2020; Abe et al., 2005). In the IM, when partnering with TFL1, FD inhibits the expression of *LFY*, thus preventing the termination of IM. Together with FT, FD activates the expression of *LFY* in the regions of IM where FMs will be formed before stage 1 of flower development (see below). AUXIN RESPONSE FACTOR5 (ARF5)/MONOPTEROS (MP) and SWI-SNF ATPase chromatin remodeling factors SPLAYED (SYD) and BRAHMA (BRM) also regulate *LFY* in floral primordia in later stages of flower development (Wagner & Meyerowitz, 2002; Yamaguchi et al., 2013; Wu et al., 2012).

The *Arabidopsis AP1* gene is another important gene promoting FM identity (Simon et al., 1996; Gustafson-Brown et al., 1994; Mandel et al., 1992; Ng & Yanofsky, 2001; Han et al., 2014). *ap1* mutants show a similar, but weak, phenotype to that of *lfy* mutants. The *ap1* mutant plants produce shoot-like structures instead of flowers in the early part of the inflorescence. Subsequently, the *ap1* flowers form axillary flowers, resulting in a short determinate reproductive shoot. The *AP1* gene is expressed in the entire FM at stages 1 and 2 of flower development. At stage 3, *AP1* expression is limited to the outer regions of FM corresponding to whorls 1 and 2, which is consistent with its role in controlling floral organ identity in the perianth. *AP1* encodes a MADS-domain protein that can bind to DNA sequences containing a CArG box [CC(A/T)6GG].

The function of *AP1* is partially redundant with two other genes, *CAL* and *FUL*, which are also expressed in the FM (Ferrandiz et al., 2000; Goslin et al., 2017; Ye

Table 6.2 *Arabidopsis* genes specifying floral meristem identity and organ identity

Gene	Full name	Mutant phenotype[a]	Protein structure and function
AG	*AGAMOUS*	Stamen and carpel identity defect; indeterminate (double) flower	MADS-domain protein, transcription factor
ANT	*AINTEGUMENTA*	Enhance petal defect of *ap2*–1 mutant	AP2 homolog, transcription factor
AP1	*APETALA1*	Inflorescence transformation defect, sepal and petal identity defect	MADS-domain protein, transcription factor
AP2	*APETALA2*	Sepal and petal identity defect	AP2-repeat, transcription factor
AP3	*APETALA3*	Petal and stamen identity defect	MADS-box, transcription factor
ASK1	*ARABIDOPSIS SKP1-LIKE1*	Abnormal petal	Skp1 homology, SCF ubiquitin ligase subunit
CAL	*CAULIFLOWER*	Enhancing *ap1* phenotypes	MADS-domain protein, transcription factor
CLF	*CURLY LEAF*	Floral organ defects	SET domain, Polycomb Group protein
CLV1	*CLAVATA1*	Extra-large SAM, indeterminate floral meristem, increased FON	Receptor-linked kinase
CLV2	*CLAVATA2*	Extra-large SAM, indeterminate floral meristem, increased FON	Receptor-like protein
CLV3	*CLAVATA3*	Extra-large SAM, indeterminate floral meristem, increased FON	Peptide ligand
FIL	*FILAMENTOUS FLOWER*	Enhance *ap1* floral meristem defect, filamentous structure instead of a flower	Transcription factor
FUL	*FRUITFUL*	Enhance *ap1 cal* meristem identity defect	MADS-domain protein, transcription factor
HEN1	*HUA ENHANCER1*	Enhance *hua1 hua2* stamen and carpel identity defects	Nuclear protein
HEN2	*HUA ENHANCER2*	Enhance *hua1 hua2* stamen and carpel identity defects	Similar to DExH-box RNA helicases
HUA1		Enhance *ag-4* phenotypes	CCCH zinc-finger, single-stranded RNA- and DNA-binding protein
HUA2		Enhance *ag-4* phenotypes	Novel transcription factor
LFY	*LEAFY*	Inflorescence transformation defect	Transcription factor

(continued)

Table 6.2 (continued)

Gene	Full name	Mutant phenotype[a]	Protein structure and function
LUG	*LEUNIG*	Abnormal sepal and petal	WD-40 repeat, transcriptional corepressor factor
PI	*PISTILLATA*	Petal and stamen identity defect	MADS-domain protein, transcription factor
SAP	*STERILE APETALA*	Sepal and petal identity defect	Novel transcription regulators
SEP1/ AGL2	*SEPALLATA1*	All floral organs are sepals in *sep1 sep2 sep3* triple mutant	MADS-domain protein, transcription factor
SEP2/ AGL4	*SEPALLATA2*	All floral organs are sepals in *sep1 sep2 sep3* triple mutant	MADS-domain protein, transcription factor
SEP3/ AGL9	*SEPALLATA3*	All floral organs are sepals in *sep1 sep2 sep3* triple mutant	MADS-domain protein, transcription factor
	SEUSS	Partial sepal and petal identity defect	Novel protein, two glutamine-rich domains, putative transcription coregulator
SHP1/ AGL1	*SHATTERPROOF1*	*shp1 shp2* eliminate remaining carpel identity in *ap2 ag* mutant	MADS-domain protein, transcription factor
SHP2/ AGL5	*SHATTERPROOF2*	*shp1 shp2* eliminate remaining carpel identity in *ap2 ag* mutant	MADS-domain protein, transcription factor
SUP	*SUPERMAN*	Extra whorls of stamens at the expense of the carpel	Zinc-finger protein
TFL1	*TERMINAL FLOWER1*	Terminal flower	Phosphatidylethanolamine-binding protein
UFO	*UNUSUAL FLORAL ORGAN*	Floral meristem defect, organ identity defect in the petal and stamens	F-box protein, SCF subunit
WUS	*WUSCHEL*	Ectopic expression cause ectopic formation of stamens and carpels	Homeodomain

Abbreviations: *FON* flower organ number, *SAM* shoot apical meristem
[a]Unless otherwise indicated, the phenotypes are due to recessive mutations

et al., 2016). Although a *cal* mutation alone does not cause any defect in FM identity, it greatly enhances the phenotype of the *ap1* mutant, resulting in an altered inflorescence meristem that produces a massive number of meristems, resembling a miniature cauliflower. Even though *ful* single mutants do not exhibit abnormalities in the FM either, the *ap1 cal ful* triple mutant displays a complete "nonflowering" phenotype. Both *CAL* and *FUL* genes encode MADS-domain proteins that are highly similar to AP1.

Furthermore, *lfy ap1* double mutants have much more severe phenotypes than either *lfy* or *ap1* single mutants, suggesting that *LFY* and *AP1* are also partially redundant (Grandi et al., 2012; Goslin et al., 2017). Several lines of evidence support the idea that LFY positively regulates the expression of *AP1* (Liljegren et al., 1999; Wagner et al., 1999; Mandel & Yanofsky, 1995; Jin et al., 2021). *AP1* expression is delayed and reduced in *lfy* mutants, while the constitutive expression of *LFY* results in the precocious expression of *AP1*. Nevertheless, the *AP1* expression is not abolished in *lfy* mutants, suggesting that other factors are also involved in activating *AP1*.

Conversely, AP1 has a positive effect on *LFY* expression. Moreover, *AP1*, *CAL*, and *FUL* together provide a much stronger upregulation of *LFY* expression than any one or two of these genes. The functional redundancy and the positive relationship between FM genes *LFY* and *AP1/CAL/FUL* are further supported by the fact that overexpression of either *LFY* or *AP1* is sufficient to specify FM identity. It is possible that these interactions form positive feedback loops that act together to ensure the sharp transition to reproductive development under favorable conditions.

Functions of *LFY* and *AP1* are opposite to those of *TFL1*, suggesting a genetic interaction between these genes (Goretti et al., 2020; Parcy et al., 2002; Liljegren et al., 1999). In the *tfl1* mutant, the region expressing both *LFY* and *AP1* expands to the shoot apex where *TFL1* is expressed, implying that *TFL1* represses the expression of both *LFY* and *AP1*. This negative interaction is mutual because overexpression of *LFY* or *AP1* represses the expression of *TFL1*. Therefore, the elegantly functional balance between *TFL1* and *LFY/AP1* not only maintains IM but also allows the formation of FMs.

It has been shown that LFY can access the closed chromatin, suggesting that it acts as a pioneer transcription factor (Yamaguchi, 2021; Jin et al., 2021; Lai et al., 2021). The N-terminal region of LFY contains a SAM (Sterile Alpha Motif) domain for oligomerization and a C-terminal DNA-binding domain (DBD). LFY regulates downstream gene expression by affecting chromatin structure via its ability to bind to the nucleosomal DNA at the promoter of *AP1* through its DBD. LFY displaces H1 linker histones and then recruits the SWI-SNF ATPase chromatin remodeling factors SYD and BRM, thereby opening the chromatin at regulatory regions of *AP1* (Bezhani et al., 2007; Jin et al., 2021; Pastore et al., 2011). The opened chromatin then allows binding of other transcription factors, such as FD, FT, LATE MERISTEM IDENTITY 2 (LMI2), and SQUAMOSA PROMOTER BINDING PROTEIN-LIKE (SPL), in the *AP1* regulatory regions; therefore, the expression of the FM identity gene *AP1* is activated by LFY.

In addition, LFY regulates the expression of *AP1* by a GA-dependent mechanism (Kinoshita et al., 2020; Fukazawa et al., 2021; Yamaguchi et al., 2014). In the FM, FT-FD-activated LFY induces the expression of *EUI-LIKE P450 A1* (*ELA1*) that encodes a GA catabolic enzyme. ELA1 and another GA catabolic enzyme, GIBBERELLIN 2-OXIDASE 4 (GA2ox4), decrease GA levels in the FM, resulting in the accumulation of GA-sensitive DELLA proteins. DELLA is then recruited by SPL transcription factors to the *AP1* promoter; the consequent activation of *AP1* expression then promotes the floral fate in the FM. As discussed above, LFY, FD,

FT, LMI2, and SPL directly activate the expression of *AP1*. On the other hand, AP1 upregulates the expression of *LFY* and directly represses the expression of *FD*, *SPL*, and *TFL1*. Moreover, AP1 upregulates *GIBBERELLIN 3-OXIDASE 1* (*GA3ox1*) and *GIBBERELLIN 2-OXIDASE 1* (*GA2ox1*), which encode GA anabolic and catabolic enzymes, respectively. These results suggest that GA homeostasis controlled by AP1 might maintain its own expression.

Several other genes also contribute to FM specification in *Arabidopsis*. Mutations in *APETALA2* (*AP2*) enhance both *ap1* and *lfy* phenotypes, suggesting a functional redundancy of *AP2* with *AP1* and *LFY* (Drews et al., 1991; Sang et al., 2022; Okamuro et al., 1997). *AP2* is expressed during vegetative development, and in FM, it encodes a putative transcription factor with two AP2 domains, which are present in many plant regulatory proteins (Wang et al., 2019). Mutations in either *UNUSUAL FLORAL ORGANS* (*UFO*) or *FILAMENTOUS FLOWER* (*FIL*) enhance the *ap1* mutant phenotype in FM specification (Wilkinson & Haughn, 1995; Rieu et al., 2023; Sawa et al., 1999; Ng & Yanofsky, 2001). In addition, both *ufo* and *fil* mutants produce unusual filamentous structures, which are possibly due to a failure of FM development. UFO interacts with LFY genetically and physically (Zhao et al., 2001a; Chae et al., 2008). *UFO* encodes an F-box protein, which might play a role in regulating protein degradation (see Sect. 6.6.2) (Ingram et al., 1995; Durfee et al., 2003), while the *FIL* gene is a member of the *YABBY* gene family, encoding a protein with a zinc finger and HMG-related domains (Boter et al., 2015).

Several genes in flowering pathways not only control IM identity but also play a role in FM identity, although the mechanisms of much of such regulation remain unclear. For example, CO has been demonstrated to regulate the expression of *LFY*, *AP1*, and *TFL1*, providing a link between the long-day pathway and FM identity, possibly via *SOC1* (Simon et al., 1996; Liu et al., 2007). In addition, as mentioned before, the GA pathway also regulates the expression of *LFY* and *AP1* (Kinoshita et al., 2020; Fukazawa et al., 2021; Yamaguchi et al., 2014). Furthermore, genetic studies suggest that two flowering time genes, *FT* and *FWA*, act in parallel with *LFY* to regulate the expression of *AP1* at the late stage (Ruiz-Garcia et al., 1997). This parallel gene regulation might be one of the reasons for the less severe flower defects observed in late-stage *lfy* mutants.

6.4 Genes Controlling Floral Meristem Determinacy

In many plant species, like *Arabidopsis*, FM is determinate, leading to the production of four whorls of floral organs. Molecular genetic studies show that the *AGAMOUS* (*AG*) gene plays a crucial role in controlling FM determinacy (Fig. 6.2, Table 6.2) (Yanofsky et al., 1990; Mizukami & Ma, 1997; Parcy et al., 2002). In *ag* mutants, flowers are indeterminate and produce indefinite numbers of sepals and petals, resulting in the formation of a "double flower." Furthermore, the ectopic expression of *AG* at the inflorescence apex causes the conversion of IM to FM, supporting its role in promoting meristem determinacy (Mizukami & Ma, 1997).

The mechanism of how *AG* regulates flower determinacy has been unraveled from analyses of *LFY*, *AG*, and *WUSCHEL* (*WUS*) genes (Lenhard et al., 2001; Liu et al., 2011; Sun et al., 2019; Guo et al., 2018; Lohmann et al., 2001; Roeder & Yanofsky, 2001). *WUS*, which encodes a homeobox transcription factor, is expressed in the center of both SAM and FM, and it is required to maintain stem cell populations in SAM and FM (see Sect. 6.6.1). The functional difference of *WUS* in SAM and FM is that it activates *AG* expression in conjunction with *LFY* at the center of FM at stage 3 of flower development but not in the SAM. AG in turn negatively regulates the expression of *WUS*. The feedback loop regulation between *AG* and *WUS* terminates the FM activity, causing the determinacy of FM.

Other genes that contribute to control of the FM determinacy include *CLAVATA1* (*CLV1*), *CLAVATA3* (*CLV3*), and *SUPERMAN* (*SUP*) genes (see Sect. 6.6.1) (Clark et al., 1997; Reddy & Meyerowitz, 2005; Schlegel et al., 2021; Fletcher, 2001; Brand et al., 2000; Jacobsen & Meyerowitz, 1997; Uemura et al., 2018). Mutations in these genes lead to an increased number of whorls of floral organs. In *clv1* and *clv3* mutant flowers, additional carpel-like organs are found in the interior of the fourth whorl carpels. CLV3 signaling affects the formation of floral primordia in an auxin-dependent manner (John et al., 2023). *SUP* encodes a C2H2-type zinc finger transcription factor. In *sup* mutants, FM is enlarged, and its termination is delayed (Jacobsen & Meyerowitz, 1997). Consequently, *sup* flowers produce several extra whorls of stamens in the center at the expense of carpels (Bowman et al., 1992).

The trithorax group protein (trxG) ULTRAPETALA1 (ULT1), the bZIP transcription factor PERIANTHIA (PAN), REBELOTE (RBL), and SQUINT (SQN) also activate the expression of *AG* (Prunet et al., 2008; Maier et al., 2009; Chuang & Meyerowitz, 2000). Early analyses of the spatial pattern of *AG* expression indicate that AP2 is required to limit *AG* expression to the inner two whorls, whereas AG is required for stamen and carpel identities and for floral meristem determinacy (Drews et al., 1991). Additionally, in the inner two whorls AP2 antagonizes AG by positively regulating expression of *WUS* and FM activity (Huang et al., 2017). Moreover, the tetramerization of AG and SEPALLATA3 (SEP3) is necessary for FM determinacy via regulating target genes *KNUCKLES* (*KNU*) and *CRABS CLAW* (*CRC*) (Honma & Goto, 2001; Lai et al., 2020; Yamaguchi et al., 2017).

Furthermore, auxin and cytokinin are involved in the control of FM determinacy. AG and AP2 upregulate the expression of *ETTIN* (*ETT*, also called *ARF3*) (Liu et al., 2014). ARF3 downregulates the expression of several genes related to cytokinin signaling and biosynthesis, such as *ARABIDOPSIS HISTIDINE KINASE 4* (*AHK4*) that encodes a cytokinin-binding receptor, *ISOPENTENYLTRANSFERASE* (*IPT*) family genes, and *LONELY GUY* (*LOG*) family genes that encode cytokinin biosynthesis enzymes (Zhang et al., 2018). Thus, AG and AP2 promote FM determinacy via increasing auxin activity but decreasing cytokinin activity. In addition, SUP, together with the PcG factors CURLY LEAF (CLF) and TFL2, represses the expression of the auxin biosynthesis genes *YUC1/4* to control FM determinacy (Xu et al., 2018). Furthermore, at stage 6 of flower development, *CRC*, which encodes a YABBY family transcription factor, controls FM determinacy possibly via modulating auxin homeostasis (Yamaguchi et al., 2017, 2018). *crc knu* double mutants have a strong FM indeterminacy defect. CRC directly upregulates the expression of *TORNADO2* (*TRN2*), which encodes a tetraspanin family gene involved in polar auxin transport. *YUC4* is also a target gene of CRC.

6.5 Maintenance of Reproductive Meristem Identities

In some plant species and genotypes, such as *Impatiens* and soybean, transferring them from inductive to noninductive environmental conditions can revert reproductive development to vegetative shoot programs (Tooke et al., 2005; Washburn & Thomas, 2000). Such reversions occur as either flower reversion or inflorescence reversion. Flower reversion happens when a flower that has already produced some floral organs subsequently produces leaves or a vegetative shoot. In inflorescence reversion, vegetative development is resumed after the inflorescence has produced floral meristems/flowers. In late-flowering soybean varieties, both flower reversion and inflorescence reversion occur when plants are transferred to long-day conditions after a 13-day short-day treatment. Interestingly, even after the inflorescence has terminated, long days can cause the abscission of almost all flowers and young fruits; the plants then resume vegetative growth like noninduced plants. Flowering reversion suggests that the flowering induction is reversible and that the maintenance of IM and FM identity needs proper environmental conditions.

In *Arabidopsis*, both *LFY* and *AG* are required for the maintenance of FM identity, as flowers of *lfy*/*LFY* heterozygous plants or *ag* mutants can be reverted to shoots under some specific conditions, such as short days or darkness for 24 h (Mizukami & Ma, 1997; Parcy et al., 2002; Okamuro et al., 1996; Ordidge et al., 2005; Tooke et al., 2005). The *lfy*/*LFY ag*/*ag* plants show a more severe defect in flower reversion than either *lfy*/*LFY* or *ag*/*ag* plants, suggesting that LFY contributes to the maintenance of FM identity separately from its regulation of the *AG* expression. In addition, the *TFL1* gene is ectopically expressed in flowers of transgenic plants with reduced *LFY* function, further suggesting that the function of *LFY* in hindering reversion from a FM to an IM is partially mediated by repressing the expression of *TFL1*.

Although *LFY* and its *Antirrhinum* homolog *FLORICAULA* (*FLO*) are important for promoting FM identity in *Arabidopsis* and *Antirrhinum*, respectively, such a function might not always be conserved for their homologs in other plant species (Coen et al., 1990). In *Impatiens*, the expression of the *LFY*/*FLO* homolog is not associated with the establishment or maintenance of FM (Pouteau et al., 1997). Similarly, the *Impatiens* homolog of *AP1* is not required for maintaining the floral state. Further loss-of-function analyses are needed to definitively rule out their role in controlling FM identity in *Impatiens*. In maize, flowering signals dependent on the *ID1* gene are required not only for floral induction but also for the maintenance of the reproductive state within the SAM (Colasanti et al., 1998; Kozaki et al., 2004).

6.6 Meristem Growth and Size and Floral Organ Number

6.6.1 Genetic Control of Meristem Growth and Maintenance

The SAM originates from the embryo, and its size is maintained in a precise manner throughout the life of a plant. In eudicots, a SAM typically consists of three layers of stem cells (L1–L3) at the apex of the meristem; stem cells (SC) in the center of the apex and the underlying organizing center (OC) constitute the central zone (CZ), which is flanked laterally by the peripheral zone (PZ) and by the rib zone (RZ) beneath (Fletcher, 2018; Holt et al., 2014). Cells in the PZ form FM in IM when floral transition completes. The size and integrity of the SAM are controlled by a balance between the proliferation and differentiation of a pool of stem cells at the center of the shoot apex. In *Arabidopsis*, *wus* mutants lack SAM and FM; *WUS* is expressed in cells of the OC at SAM; WUS proteins then move up to stem cells where WUS directly activates the expression of *CLV3* that encodes a small, secreted peptide (Mayer et al., 1998; Laux et al., 1996; Leibfried et al., 2005; Yadav et al., 2011). CLV3 elicits a putative receptor complex consisting of leucine-rich repeat receptor-like kinases CLV1 and BARELY ANY MERISTEMS (BAMs), CLV2 (a leucine-rich repeat protein), CORYNE (CRN, a novel membrane-associated receptor kinase), and CLAVATA3 INSENSITIVE RECEPTOR KINASE (CIKs), which in turn precisely controls the expression domain of *WUS* in the OC (Clark et al., 1997; Fletcher et al., 1999; Kondo et al., 2006; Rojo et al., 2002; Jeong et al., 1999; Muller et al., 2008; DeYoung et al., 2006; Hu et al., 2018; Brand et al., 2000; Schoof et al., 2000). WUS works with HAIRY MERISTEM (HAM) transcriptional regulators to control stem cell production and ensure that *CLV3* is exclusively expressed in the outermost apical layers of the SAM (Zhou et al., 2015b). Functional disruptions of genes in the CLV signaling pathway cause the enlarged SAM. Therefore, a negative feedback loop containing WUS and the CLV signaling pathway regulates SAM maintenance. Orthologous genes in the WUS-CLV pathway also exist in other plant species, including *Brassica*, tomato, rice, and maize (Kieffer et al., 2006; Fan et al., 2014; Chu et al., 2006; Suzaki et al., 2004, 2006; Bommert et al., 2005; Munos et al., 2011; Tanaka et al., 2015; Rodriguez-Leal et al., 2019). Another important gene that functions to maintain the SAM in *Arabidopsis* is the *SHOOT MERISTEMLESS* (*STM*) gene; *stm* mutants lack SAM or produce SAM with a reduced size, suggesting that *STM* prevents SAM cells from differentiation (Roth et al., 2018; Clark et al., 1996; Endrizzi et al., 1996; Balkunde et al., 2017). *STM* encodes a class I KNOX transcription factor, which not only directly upregulates the expression of *CLV3* but also facilitates WUS to bind to the *CLV3* promoter. WUS in turn activates the expression of *STM*.

It is believed that cytokinin promotes cell proliferation but represses cell differentiation in SAM, since overexpression of *ARABIDOPSIS RESPONSE REGULATOR 7* (*ARR7*) that encodes a cytokinin response antagonistic factor results in a strong *wus* phenotype, whereas increasing cytokinin levels leads to enlarged IM and FM (Leibfried et al., 2005). WUS directly downregulates the expression of *ARR7* and *ARR15* (Zhao et al., 2010). Overexpression of *STM* increased cytokinin levels, and cytokinin partially rescues the *stm* phenotype, suggesting that STM inhibits SAM cell differentiation via mediating cytokinin signaling (Scofield et al., 2013).

Additional factors independent of WUS are also involved in the control of FM determinacy. HD-zip III transcription factors, including PHABULOSA (PHB), PHAVOLUTA (PHV), and CORONA (CNA), inhibit stem cell activity in SAM (Turchi et al., 2013; Green et al., 2005; Prigge et al., 2005). miR165/166 negatively regulates the expression of *PHB*, *PHV*, and *CAN* at the posttranscriptional level (Zhou et al., 2015a; Williams et al., 2005). Overexpression of miR166 causes the enlargement of SAM due to the reduction in *PHB*, *PHV*, and *CAN* mRNA levels. ARGONAUTE 10 (AGO10)/ZWILLE (ZLL) prevents miR165/166 from being accumulated in SAM by competing with AGO1, which assists miR165/166 in binding to *PHB*, *PHV*, and *CAN* mRMAs for cleavage. REVOLUTA (REV), another HD-zip III transcription factor, promotes IM identity, possibly by antagonizing the function of PHB/PHV/CAN.

6.6.2 Control of Floral Organ Number

The number and type of floral organs vary greatly among flowering plants; however, little is known about the genetic basis for such differences. Flowers in *Arabidopsis* show fixed numbers and positions of organs: four sepals, four petals, five or six stamens, and two carpels in invariant positions. A number of genes have been found to affect the number of floral organs. The genes discussed above that regulate FM size also affect floral organ numbers in a relatively nonspecific fashion. Mutations in the *CLV1*, *CLV2*, *CLV3*, *WIGGUM* (*WIG*), and *ULTRAPETALA* (*ULT*) genes increase the number of floral organs or whorls, which is consistent with the enlargement of FM size caused by these mutations (Clark et al., 1997; Fletcher et al., 1999; Rojo et al., 2002; Jeong et al., 1999; Ziegelhoffer et al., 2000; Running et al., 1998; Fletcher, 2001; Carles et al., 2004). In contrast, *stm* and *wus* mutants, which have decreased FMs, produce flowers with a reduced number of floral organs in the inner whorls (Mayer et al., 1998; Laux et al., 1996; Roth et al., 2018; Clark et al., 1996; Endrizzi et al., 1996; Balkunde et al., 2017). These observations suggest that in a plant species, one factor controlling the floral organ number might be required for maintaining a fixed number of cells that make a floral organ primordium in the FM.

Other genes seem to regulate floral organ numbers without changing meristem size appreciably. The *PAN* gene controls the tetramerous floral pattern in *Arabidopsis*; *pan* mutants produce flowers with five sepals, five petals, and five stamens instead of the normally four sepals, four petals, and six stamens (Running & Meyerowitz, 1996). *PAN* encodes a TGA bZIP protein, suggesting that it is a transcription factor (Chuang et al., 1999). The PAN protein is detected throughout FM and all floral organ primordia; therefore, its function is not to define the boundary between floral organ primordia. It was proposed that *PAN* may regulate the number of organ primordia by interacting with another protein or that its activity may be regulated posttranslationally.

The floral organ identity genes *APETALA3* (*AP3*) and *PISTILLATA* (*PI*) also affect organ primordium initiation, since mutations in *AP3* and *PI* cause a slight reduction in floral organ number interior to the outer-whorl sepals (Krizek & Meyerowitz, 1996; Jack et al., 1992, 1994; Goto & Meyerowitz, 1994; Hu et al., 2023). The function of these genes in regulating floral organ number is further supported by the fact that they are expressed before initiation of the inner three whorls and that ectopic expression of *AP3* and *PI* results in extra whorls of stamens. As mentioned earlier, flowers of *sup* mutants produce extra whorls of stamens. *AP3* and *PI* expression was found to expand toward the center in *sup* mutant flowers, supporting the idea that *AP3* and *PI* affect cell proliferation and the initiation of floral organ primordia.

Mutations in the *UFO* gene also cause a reduction in or defects in floral organs in whorls two and three (Wilkinson & Haughn, 1995; Rieu et al., 2023). Similarly, a null mutation in the *ARABIDOPSIS SKP1-LIKE 1* (*ASK1*) gene also has a mild effect on reducing petal and stamen numbers (Zhao et al., 1999). Both SKP1 and F-box proteins are components of the SCF ubiquitin ligases that control the ubiquitination of specific proteins for their degradation (Deshaies, 1999; Jackson & Eldridge, 2002). In yeasts and humans, several SCFs have been shown to mediate the ubiquitination of the cell cycle and transcriptional regulators. Molecular genetic studies show that UFO interacts with ASK1, suggesting that UFO and ASK1 may regulate cell division in FM via ubiquitin-mediated degradation of target proteins (Ni et al., 2004; Zhao et al. 2001a). Furthermore, ASK1 and UFO also contribute to the control of floral organ number via the regulation of *AP3* and *PI* expression. Independent of ubiquitination, UFO may interact with LFY to form a transcriptional complex that regulates the expression of *AP3* and *PI* (Chae et al., 2008; Rieu et al., 2023; Dolde et al., 2023).

In addition, auxin signaling also appears to be important for the initiation and position of lateral organ primordia in the shoot. Auxin has long been thought to work as a morphogen, as it can maintain its gradient via polar transport (Yu et al., 2022). Some studies suggest that auxin might also control the initiation and position of floral organ primordia because the disruption of genes involved in auxin synthesis (e.g., *YUCCA*), polar transport (e.g., *PIN* and *PINOID*), or signaling perception (e.g., *AUX/IAA* and *ARF*) affects floral organ number (Cheng et al., 2006; Cance et al., 2022; Blilou et al., 2005; Christensen et al., 2000; Michniewicz et al., 2007; Huang et al., 2022). One line of evidence supporting this hypothesis is that the *ETT* gene is functionally redundant with the *PAN* gene (Sessions et al., 1997). *ett* mutants have an increased number of sepals and petals and a decreased number of stamens. *ETT* encodes ARF3, which regulates the expression of auxin-responsive genes upon auxin signaling. In petunia *floozy* (*fzy*) mutants, the formation of floral organ primordia in the outer three floral whorls is blocked at the early stage (Tobena-Santamaria et al., 2002). The *FZY* gene encodes a flavin mono-oxygenase-like protein that is a likely functional homolog of the *YUCCA* gene in *Arabidopsis* (Zhao et al., 2001c). In *Arabidopsis*, *yuc1 yuc2 yuc4 yuc6* quadruple mutants form IMs but not FMs, while flowers from *yuc1 yuc4* double mutants vary with reduced numbers of all four types of floral organs (Cheng et al., 2006). The fact that the *FZY* gene is expressed at the flanks of initiating petal and stamen primordia and that the ectopic *FZY* expression results in excessive auxin accumulation supports a role of *FZY* in floral organ formation via contributing to local auxin synthesis (Tobena-Santamaria et al., 2002). In *Arabidopsis petal loss* (*ptl*) mutant plants, petals are often absent; deformed sepals are sometimes fused (Griffith et al., 1999). *PTL* is expressed in sepal boundaries and encodes a plant-specific trihelix transcription activator. PTL may upregulate the expression of *UFO* to control petal development (Takeda et al., 2022). Mutations in the auxin influx carrier gene *AUX1* enhance the *ptl* phenotype. Expression of the auxin-inducible reporter *DR5rev:GFP-ER* is abnormal in petal-initiation zones in *ptl* flowers, while interaction between PTL and the glutaredoxin protein ROXY1 is essential for maintaining normal auxin activity in petal-initiation zones, suggesting that PTL controls petal initiation via possibly regulating auxin signaling (Lampugnani et al., 2012; Quon et al., 2017).

MicroRNA is also involved in controlling floral organ number. mir164 and its target genes *CUP SHAPED COTYLEDON1* (*CUC1*) and *CUC2* regulate floral organ number and floral whorl boundary (Laufs et al., 2004; Huang et al., 2012). The *early extra petals1* (*eep1*), a mutant of the *MIR164C* gene, produces extra petals, while the dominant *cuc1–1D cuc2–1D* double mutant flowers form more sepals, petals, and stamens than those of the wild type (Baker et al., 2005). Mutations in the *RABBIT EARS* (*RBE*) gene that encodes a C2H2 zinc finger transcriptional repressor cause the loss or malformation of petals (Li et al., 2016; Takeda et al., 2004). RBE directly represses the expression of *MIR164c* (Huang et al., 2012). Furthermore, the F-box gene *HAWAIIAN SKIRT* (*HWS*) controls floral organ number by affecting the expression of *MIR164*, *CUC1*, and *CUC2* in flowers (Gonzalez-Carranza et al., 2017). In *Arabidopsis*, a loss-of-function mutant of the *MICRORNA160a* gene, known as *floral organs in carpels* (*foc*), produces variable numbers of floral organs (Liu et al., 2010).

References

Abe, M., Kobayashi, Y., Yamamoto, S., Daimon, Y., Yamaguchi, A., Ikeda, Y., Ichinoki, H., Notaguchi, M., Goto, K., & Araki, T. (2005). FD, a bZIP protein mediating signals from the floral pathway integrator FT at the shoot apex. *Science, 309*, 1052–1056. https://doi.org/10.1126/science.1115983

Araki, T. (2001). Transition from vegetative to reproductive phase. *Current Opinion in Plant Biology, 4*, 63–68.

Aubert, D., Chen, L., Moon, Y. H., Martin, D., Castle, L. A., Yang, C. H., & Sung, Z. R. (2001). EMF1, a novel protein involved in the control of shoot architecture and flowering in Arabidopsis. *The Plant Cell, 13*, 1865–1875.

Baker, C. C., Sieber, P., Wellmer, F., & Meyerowitz, E. M. (2005). The early extra petals1 mutant uncovers a role for microRNA miR164c in regulating petal number in *Arabidopsis*. *Current Biology, 15*, 303–315. https://doi.org/10.1016/j.cub.2005.02.017

Balkunde, R., Kitagawa, M., Xu, X. M., Wang, J., & Jackson, D. (2017). SHOOT MERISTEMLESS trafficking controls axillary meristem formation, meristem size and organ boundaries in *Arabidopsis*. *The Plant Journal, 90*, 435–446. https://doi.org/10.1111/tpj.13504

Bao, S., Hua, C., & Shen, L. (2020). Yu H new insights into gibberellin signaling in regulating flowering in *Arabidopsis*. *Journal of Integrative Plant Biology, 62*, 118–131. https://doi.org/10.1111/jipb.12892

Bezhani, S., Winter, C., Hershman, S., Wagner, J. D., Kennedy, J. F., Kwon, C. S., Pfluger, J., Su, Y., & Wagner, D. (2007). Unique, shared, and redundant roles for the Arabidopsis SWI/SNF chromatin remodeling ATPases BRAHMA and SPLAYED. *The Plant Cell, 19*, 403–416. https://doi.org/10.1105/tpc.106.048272

Blazquez, M. A., & Weigel, D. (2000). Integration of floral inductive signals in *Arabidopsis*. *Nature, 404*, 889–892.

Blazquez, M. A., Soowal, L. N., Lee, I., & Weigel, D. (1997). *LEAFY* expression and flower initiation in *Arabidopsis*. *Development, 124*, 3835–3844.

Blilou, I., Xu, J., Wildwater, M., Willemsen, V., Paponov, I., Friml, J., Heidstra, R., Aida, M., Palme, K., & Scheres, B. (2005). The PIN auxin efflux facilitator network controls growth and patterning in *Arabidopsis* roots. *Nature, 433*, 39–44. https://doi.org/10.1038/nature03184

Bommert, P., Lunde, C., Nardmann, J., Vollbrecht, E., Running, M., Jackson, D., Hake, S., & Werr, W. (2005). Thick tassel dwarf1 encodes a putative maize ortholog of the *Arabidopsis* CLAVATA1 leucine-rich repeat receptor-like kinase. *Development, 132*, 1235–1245. https://doi.org/10.1242/dev.01671

Boter, M., Golz, J. F., Gimenez-Ibanez, S., Fernandez-Barbero, G., Franco-Zorrilla, J. M., & Solano, R. (2015). FILAMENTOUS FLOWER is a direct target of JAZ3 and modulates responses to jasmonate. *The Plant Cell, 27*, 3160–3174. https://doi.org/10.1105/tpc.15.00220

Bowman, J. L. (1997). Evolutionary conservation of angiosperm flower development at the molecular and genetic levels. *Journal of Biosciences, 22*, 515–527. https://www.ias.ac.in/article/fulltext/jbsc/022/04/0515-0527

Bowman, J. L., Sakai, H., Jack, T., Weigel, D., Mayer, U., & Meyerowitz, E. M. (1992). *SUPERMAN*, a regulator of floral homeotic genes in *Arabidopsis*. *Development, 114*, 599–615.

Bradley, D., Carpenter, R., Copsey, L., Vincent, C., Rothstein, S., & Coen, E. (1996). Control of inflorescence architecture in Antirrhinum. *Nature, 379*, 791–797.

Bradley, D., Ratcliffe, O., Vincent, C., Carpenter, R., & Coen, E. (1997). Inflorescence commitment and architecture in *Arabidopsis*. *Science, 275*, 80–83.

Brand, U., Fletcher, J. C., Hobe, M., Meyerowitz, E. M., & Simon, R. (2000). Dependence of stem cell fate in *Arabidopsis* on a feedback loop regulated by CLV3 activity. *Science, 289*, 617–619.

Busch, M. A., Bomblies, K., & Weigel, D. (1999). Activation of a floral homeotic gene in *Arabidopsis*. *Science, 285*, 585–587.

Cance, C., Martin-Arevalillo, R., Boubekeur, K., & Dumas, R. (2022). Auxin response factors are keys to the many auxin doors. *The New Phytologist, 235*, 402–419. https://doi.org/10.1111/nph.18159

Carles, C. C., Lertpiriyapong, K., Reville, K., & Fletcher, J. C. (2004). The ULTRAPETALA1 gene functions early in *Arabidopsis* development to restrict shoot apical meristem activity and acts through WUSCHEL to regulate floral meristem determinacy. *Genetics, 167*, 1893–1903. https://doi.org/10.1534/genetics.104.028787

Cerise, M., Falavigna, V. S., Rodrıguez-Maroto, G., Signol, A., Severing, E., Gao, H., van Driel, A., Vincent, C., Wilkens, S., Iacobini, F. R., Formosa-Jordan, P., Alice Pajoro, A., & Coupland, G. (2023). Two modes of gene regulation by *TFL1* mediate its dual function in flowering time and shoot determinacy of *Arabidopsis*. *Development, 150*, dev202089. https://doi.org/10.1242/dev.202089

Chae, E., Tan, Q. K., Hill, T. A., & Irish, V. F. (2008). An Arabidopsis F-box protein acts as a transcriptional co-factor to regulate floral development. *Development, 135*, 1235–1245.

Chandler, J., & Dean, C. (1994). Factors influencing the vernalization response and flowering time of late flowering mutants of *Arabidopsis thaliana* (L.) Heynh. *Journal of Experimental Botany, 45*, 1279–1288.

Chandler, J., Wilson, A., & Dean, C. (1996). *Arabidopsis* mutants showing an altered response to vernalization. *The Plant Journal, 10*, 637–644.

Chavez-Hernandez, E. C., Quiroz, S., Garcia-Ponce, B., & Alvarez-Buylla, E. R. (2022). The flowering transition pathways converge into a complex gene regulatory network that underlies the phase changes of the shoot apical meristem in *Arabidopsis thaliana. Frontiers in Plant Science, 13*, 852047. https://doi.org/10.3389/fpls.2022.852047

Chen, C., & Du, X. (2022). LEAFY COTYLEDONs: Connecting different stages of plant development. *Frontiers in Plant Science, 13*, 916831. https://doi.org/10.3389/fpls.2022.916831

Cheng, Y., Dai, X., & Zhao, Y. (2006). Auxin biosynthesis by the YUCCA flavin monooxygenases controls the formation of floral organs and vascular tissues in *Arabidopsis. Genes & Development, 20*, 1790–1799. https://doi.org/10.1101/gad.1415106

Christensen, S. K., Dagenais, N., Chory, J., & Weigel, D. (2000). Regulation of auxin response by the protein kinase PINOID. *Cell, 100*, 469–478. https://doi.org/10.1016/s0092-8674(00)80682-0

Chu, H., Qian, Q., Liang, W., Yin, C., Tan, H., Yao, X., Yuan, Z., Yang, J., Huang, H., Luo, D., Ma, H., & Zhang, D. (2006). The *floral organ Number4* gene encoding a putative ortholog of *Arabidopsis CLAVATA3* regulates apical meristem size in rice. *Plant Physiology, 142*, 1039–1052. https://doi.org/10.1104/pp.106.086736

Chuang, C. F., & Meyerowitz, E. M. (2000). Specific and heritable genetic interference by double-stranded RNA in *Arabidopsis thaliana. Proceedings of the National Academy of Sciences of the United States of America, 97*, 4985–4990.

Chuang, C. F., Running, M. P., Williams, R. W., & Meyerowitz, E. M. (1999). The *PERIANTHIA* gene encodes a bZIP protein involved in the determination of floral organ number in *Arabidopsis thaliana. Genes & Development, 13*, 334–344.

Clark, S. E., Jacobsen, S. E., Levin, J. Z., & Meyerowitz, E. M. (1996). The CLAVATA and SHOOT MERISTEMLESS loci competitively regulate meristem activity in *Arabidopsis. Development, 122*, 1567–1575.

Clark, S. E., Williams, R. W., & Meyerowitz, E. M. (1997). The *CLAVATA1* gene encodes a putative receptor kinase that controls shoot and floral meristem size in *Arabidopsis. Cell, 89*, 575–585.

Clarke, J. H., & Dean, C. (1994). Mapping of *FRI*, a locus controling flowering time and vernalization response in *Arabidopsis thaliana. Molecular & General Genetics, 242*, 81–89.

Coen, E. S., & Meyerowitz, E. M. (1991). The war of the whorls: Genetic interactions controlling flower development. *Nature, 353*, 31–37.

Coen, E. S., Romero, J. M., Doyle, S., Elliott, R., Murphy, G., & Carpenter, R. (1990). Floricaula: A homeotic gene required for flower development in *Antirrhinum majus. Cell, 63*, 1311–1322.

Colasanti, J., Yuan, Z., & Sundaresan, V. (1998). The indeterminate gene encodes a zinc finger protein and regulates a leaf-generated signal required for the transition to flowering in maize. *Cell, 93*, 593–603. https://doi.org/10.1016/s0092-8674(00)81188-5

Corbesier, L., Gadisseur, I., Silvestre, G., Jacqmard, A., & Bernier, G. (1996). Design in *Arabidopsis thaliana* of a synchronous system of floral induction by one long day. *The Plant Journal, 9*, 947–952.

Deshaies, R. J. (1999). SCF and Cullin/ring H2-based ubiquitin ligases. *Annual Review of Cell and Developmental Biology, 15*, 435–467. https://doi.org/10.1146/annurev.cellbio.15.1.435

DeYoung, B. J., Bickle, K. L., Schrage, K. J., Muskett, P., Patel, K., & Clark, S. E. (2006). The CLAVATA1-related BAM1, BAM2 and BAM3 receptor kinase-like proteins are required for meristem function in *Arabidopsis. The Plant Journal, 45*, 1–16. https://doi.org/10.1111/j.1365-313X.2005.02592.x

Dolde, U., Muzzopappa, F., Delesalle, C., Neveu, J., Erdel, F., & Vert, G. (2023). LEAFY homeostasis is regulated via ubiquitin-dependent degradation and sequestration in cytoplasmic condensates. *iScience, 26*, 106880.

Drews, G. N., Bowman, J. L., & Meyerowitz, E. M. (1991). Negative regulation of the *Arabidopsis* homeotic gene *AGAMOUS* by the *APETALA2* product. *Cell, 65*, 991–1002.

Durfee, T., Roe, J. L., Sessions, R. A., Inouye, C., Serikawa, K., Feldmann, K. A., Weigel, D., & Zambryski, P. C. (2003). The F-box-containing protein UFO and AGAMOUS participate in antagonistic pathways governing early petal development in *Arabidopsis*. *Proceedings of the National Academy of Sciences of the United States of America, 100*, 8571–8576.

Elomaa, P., Zhao, Y., & Zhang, T. (2018). Flower heads in Asteraceae—Recruitment of conserved developmental regulators to control the flower-like inflorescence architecture. *Horticulture Research, 5*, 36. https://www.nature.com/articles/s41438-018-0056-8

Endrizzi, K., Moussian, B., Haecker, A., Levin, J. Z., & Laux, T. (1996). The *SHOOT MERISTEMLESS* gene is required for maintenance of undifferentiated cells in *Arabidopsis* shoot and floral meristems and acts at a different regulatory level than the meristem genes *WUSCHEL* and *ZWILLE*. *The Plant Journal, 10*, 967–979.

Fan, C., Wu, Y., Yang, Q., Yang, Y., Meng, Q., Zhang, K., Li, J., Wang, J., & Zhou, Y. (2014). A novel single-nucleotide mutation in a *CLAVATA3* gene homolog controls a multilocular silique trait in *Brassica rapa* L. *Molecular Plant, 7*, 1788–1792. https://doi.org/10.1093/mp/ssu090

Ferrandiz, C., Gu, Q., Martienssen, R., & Yanofsky, M. F. (2000). Redundant regulation of meristem identity and plant architecture by *FRUITFULL, APETALA1* and *CAULIFLOWER*. *Development, 127*, 725–734.

Finnegan, E. J., Robertson, M., & Helliwell, C. A. (2020). Resetting *FLOWERING LOCUS C* expression after vernalizationis just activation in the early embryo by a different name. *Frontiers in Plant Science, 11*, 620155. https://doi.org/10.3389/fpls.2020.620155

Fletcher, J. C. (2001). The *ULTRAPETALA* gene controls shoot and floral meristem size in *Arabidopsis*. *Development, 128*, 1323–1333.

Fletcher, J. C. (2018). The CLV-WUS stem cell signaling pathway: A roadmap to crop yield optimization. *Plants (Basel), 7*, 87. https://doi.org/10.3390/plants7040087

Fletcher, J. C., Brand, U., Running, M. P., Simon, R., & Meyerowitz, E. M. (1999). Signaling of cell fate decisions by *CLAVATA3* in *Arabidopsis* shoot meristems. *Science, 283*, 1911–1914.

Fowler, S., Lee, K., Onouchi, H., Samach, A., Richardson, K., Morris, B., & Coupland, G. (1999). *GIGANTEA*: A circadian clock-controlled gene that regulates photoperiodic flowering in *Arabidopsis* and encodes a protein with several possible membrane-spanning domains. *The EMBO Journal, 18*, 4679–4688.

Freytes, S. N., Canelo, M., & Cerdan, P. D. (2021). Regulation of flowering time: When and where? *Current Opinion in Plant Biology, 63*, 102049. https://doi.org/10.1016/j.pbi.2021.102049

Fukazawa, J., Ohashi, Y., Takahashi, R., Nakai, K., & Takahashi, Y. (2021). DELLA degradation by gibberellin promotes flowering via GAF1-TPR-dependent repression of floral repressors in *Arabidopsis*. *The Plant Cell, 33*, 2258–2272. https://doi.org/10.1093/plcell/koab102

Ge, X., Chang, F., & Ma, H. (2010). Signaling and transcriptional control of reproductive development in *Arabidopsis*. *Current Biology, 20*, R988–R997.

Gendall, A. R., Levy, Y. Y., Wilson, A. & Dean, C. (2001). The VERNALIZATION 2 gene mediates the epigenetic regulation of vernalization in Arabidopsis. *Cell, 107*, 525–535. https://www.cell.com/cell/fulltext/S0092-8674(01)00573-6

Godwin, J., & Farrona, S. (2022). The importance of networking: Plant Polycomb repressive complex 2 and its interactors. *Epigenomes, 6*, 8. https://www.ncbi.nlm.nih.gov/pmc/articles/PMC8948837/

Gómez-Mena, C., Manuel Piñeiro, M., Franco-Zorrilla, J. M., Salinas, J., Coupland, G., & Martínez-Zapater, J. M. (2001). *Early bolting in short days*: An *Arabidopsis* mutation that causes early flowering and partially suppresses the floral phenotype of *leafy*. *The Plant Cell, 13*, 1011–1024.

Gonzalez-Carranza, Z. H., Zhang, X., Peters, J. L., Boltz, V., Szecsi, J., Bendahmane, M., & Roberts, J. A. (2017). *HAWAIIAN SKIRT* controls size and floral organ number by modulating *CUC1* and *CUC2* expression. *PLoS One, 12*, e0185106. https://doi.org/10.1371/journal.pone.0185106

Goretti, D., Silvestre, M., Collani, S., Langenecker, T., Mendez, C., Madueno, F., & Schmid, M. (2020). TERMINAL FLOWER1 functions as a mobile transcriptional cofactor in the shoot apical meristem. *Plant Physiology, 182*, 2081–2095. https://doi.org/10.1104/pp.19.00867

Goslin, K., Zheng, B., Serrano-Mislata, A., Rae, L., Ryan, P. T., Kwasniewska, K., Thomson, B., O'Maoileidigh, D. S., Madueno, F., Wellmer, F., & Graciet, E. (2017). Transcription factor interplay between LEAFY and APETALA1/CAULIFLOWER during floral initiation. *Plant Physiology, 174*, 1097–1109. https://doi.org/10.1104/pp.17.00098

Goto, K., & Meyerowitz, E. (1994). Function and regulation of the *Arabidopsis* floral homeotic gene *PISTILLATA*. *Genes & Development, 8*, 1548–1560.

Grandi, V., Gregis, V., & Kater, M. M. (2012). Uncovering genetic and molecular interactions among floral meristem identity genes in *Arabidopsis thaliana*. *The Plant Journal, 69*, 881–893. https://doi.org/10.1111/j.1365-313X.2011.04840.x

Green, K. A., Prigge, M. J., Katzman, R. B., & Clark, S. E. (2005). CORONA, a member of the class III homeodomain leucine zipper gene family in Arabidopsis, regulates stem cell specification and organogenesis. *The Plant Cell, 17*, 691–704. https://doi.org/10.1105/tpc.104.026179

Griffith, M. E., da Silva, C. A., & Smyth, D. R. (1999). *PETAL LOSS* gene regulates initiation and orientation of second whorl organs in the *Arabidopsis* flower. *Development, 126*, 5635–5644.

Guo, H., Yang, H., Mockler, T. C., & Lin, C. (1998). Regulation of flowering time by *Arabidopsis* photoreceptors. *Science, 279*, 1360–1363.

Guo, L., Cao, X., Liu, Y., Li, J., Li, Y., Li, D., Zhang, K., Gao, C., Dong, A., & Liu, X. (2018). A chromatin loop represses *WUSCHEL* expression in *Arabidopsis*. *The Plant Journal, 94*, 1083–1097. https://doi.org/10.1111/tpj.13921

Gustafson-Brown, C., Savidge, B., & Yanofsky, M. F. (1994). Regulation of the *Arabidopsis* floral homeotic gene *APETALA1*. *Cell, 76*, 131–143.

Han, Y., Zhang, C., Yang, H., & Jiao, Y. (2014). Cytokinin pathway mediates *APETALA1* function in the establishment of determinate floral meristems in *Arabidopsis*. *Proceedings of the National Academy of Sciences of the United States of America, 111*, 6840–6845. https://doi.org/10.1073/pnas.1318532111

Hanano, S., & Goto, K. (2011). *Arabidopsis TERMINAL FLOWER1* is involved in the regulation of flowering time and inflorescence development through transcriptional repression. *The Plant Cell, 23*, 3172–3184. https://doi.org/10.1105/tpc.111.088641

Hicks, K. A., Albertson, T. M., & Wagner, D. R. (2001). *EARLY FLOWERING3* encodes a novel protein that regulates circadian clock function and FLOWERING in *Arabidopsis*. *The Plant Cell, 13*, 1281–1292.

Ho, W. W., & Weigel, D. (2014). Structural features determining flower-promoting activity of Arabidopsis FLOWERING LOCUS T. *The Plant Cell, 26*, 552–564. https://doi.org/10.1105/tpc.113.115220

Holt, A. L., van Haperen, J. M., Groot, E. P., & Laux, T. (2014). Signaling in shoot and flower meristems of *Arabidopsis thaliana*. *Current Opinion in Plant Biology, 17*, 96–102. https://doi.org/10.1016/j.pbi.2013.11.011

Honma, T., & Goto, K. (2001). Complexes of MADS-box proteins are sufficient to convert leaves into floral organs. *Nature, 409*, 525–529.

Hu, C., Zhu, Y., Cui, Y., Cheng, K., Liang, W., Wei, Z., Zhu, M., Yin, H., Zeng, L., Xiao, Y., Lv, M., Yi, J., Hou, S., He, K., Li, J., & Gou, X. (2018). A group of receptor kinases are essential for CLAVATA signalling to maintain stem cell homeostasis. *Nature Plants, 4*, 205–211. https://doi.org/10.1038/s41477-018-0123-z

Hu, T., Li, X., Du, L., Manuela, D., & Xu, M. (2023). LEAFY and APETALA1 down-regulate ZINC FINGER PROTEIN 1 and 8 to release their repression on class B and C floral homeotic genes. *Proceedings of the National Academy of Sciences of the United States of America, 120*, e2221181120. https://doi.org/10.1073/pnas.2221181120

Huala, E., & Sussex, I. M. (1992). *LEAFY* interacts with floral homeotic genes to regulate Arabidopsis floral development. *The Plant Cell, 4*, 901–913.

Huang, T., Lopez-Giraldez, F., Townsend, J. P., & Irish, V. F. (2012). RBE controls microRNA164 expression to effect floral organogenesis. *Development, 139*, 2161–2169. https://doi.org/10.1242/dev.075069

Huang, Z., Shi, T., Zheng, B., Yumul, R. E., Liu, X., You, C., Gao, Z., Xiao, L., & Chen, X. (2017). APETALA2 antagonizes the transcriptional activity of AGAMOUS in regulating floral stem cells in *Arabidopsis thaliana*. *The New Phytologist, 215*, 1197–1209.

Huang, Y., Jiang, L., Liu, B. Y., Tan, C. F., Chen, D. H., Shen, W. H., & Ruan, Y. (2019). Evolution and conservation of polycomb repressive complex 1 core components and putative associated factors in the green lineage. *BMC Genomics, 20*, 533. https://doi.org/10.1186/s12864-019-5905-9

Huang, J., Zhao, L., Malik, S., Gentile, B. R., Xiong, V., Arazi, T., Owen, H. A., Friml, J., & Zhao, D. (2022). Specification of female germline by microRNA orchestrated auxin signaling in *Arabidopsis*. *Nature Communications, 13*, 6960. https://doi.org/10.1038/s41467-022-34723-6

Hyun, Y., Richter, R., Vincent, C., Martinez-Gallegos, R., Porri, A., & Coupland, G. (2016). Multi-layered regulation of SPL15 and cooperation with SOC1 integrate endogenous flowering pathways at the Arabidopsis shoot meristem. *Developmental Cell, 37*, 254–266.

Immink, R. G., Pose, D., Ferrario, S., Ott, F., Kaufmann, K., Valentim, F. L., de Folter, S., van der Wal, F., van Dijk, A. D., Schmid, M., & Angenent, G. C. (2012). Characterization of SOC1's central role in flowering by the identification of its upstream and downstream regulators. *Plant Physiology, 160*, 433–449.

Ingram, G. C., Goodrich, J., Wilkinson, M. D., Simon, R., Haughn, G. W., & Coen, E. S. (1995). Parallels between *UNUSUAL FLORAL ORGANS* and *FIMBRIATA*, genes controlling flower development in Arabidopsis and Antirrhinum. *The Plant Cell, 7*, 1501–1510.

Jack, T., Brockman, L. L., & Meyerowitz, E. M. (1992). The homeotic gene *APETALA3* of *Arabidopsis thaliana* encodes a MADS box and is expressed in petals and stamens. *Cell, 68*, 683–697.

Jack, T., Fox, G. L., & Meyerowitz, E. M. (1994). *Arabidopsis* homeotic gene *APETALA3* ectopic expression: Transcriptional and posttranscriptional regulation determine floral organ identity. *Cell, 76*, 703–716.

Jackson, P. K., & Eldridge, A. G. (2002). The SCF ubiquitin ligase: An extended look. *Molecular Cell, 9*, 923–925. https://doi.org/10.1016/s1097-2765(02)00538-5

Jacobsen, S. E., & Meyerowitz, E. M. (1997). Hypermethylated *SUPERMAN* epigenetic alleles in arabidopsis. *Science, 277*, 1100–1103. https://doi.org/10.1126/science.277.5329.1100

Jaeger, K. E., Pullen, N., Lamzin, S., Morris, R. J., & Wigge, P. A. (2013). Interlocking feedback loops govern the dynamic behavior of the floral transition in Arabidopsis. *The Plant Cell, 25*, 820–833. https://doi.org/10.1105/tpc.113.109355

Jeong, S., Trotochaud, A. E., & Clark, S. E. (1999). The Arabidopsis *CLAVATA2* gene encodes a receptor-like protein required for the stability of the CLAVATA1 receptor-like kinase. *The Plant Cell, 11*, 1925–1934.

Jin, R., Klasfeld, S., Zhu, Y., Fernandez Garcia, M., Xiao, J., Han, S. K., Konkol, A., & Wagner, D. (2021). LEAFY is a pioneer transcription factor and licenses cell reprogramming to floral fate. *Nature Communications, 12*, 626. https://doi.org/10.1038/s41467-020-20883-w

John, A., Smith, E. S., Jones, D. S., Soyars, C. L., & Nimchuk, Z. L. (2023). A network of CLAVATA receptors buffers auxin-dependent meristem maintenance. *Nature Plants, 9*, 1306–1317. https://doi.org/10.1038/s41477-023-01485-y

Kassis, J. A., & Brown, J. L. (2013). Polycomb group response elements in Drosophila and vertebrates. *Advances in Genetics, 81*, 83–118. https://doi.org/10.1016/B978-0-12-407677-8.00003-8

Kellogg, E. A. (2015). Flowering plants, monocots, poaceae. In K. Kubitski (Ed.), *The families and genera of vascular plants* (Vol. 13, pp. 1–416). Springer.

Kieffer, M., Stern, Y., Cook, H., Clerici, E., Maulbetsch, C., Laux, T., & Davies, B. (2006). Analysis of the transcription factor WUSCHEL and its functional homologue in Antirrhinum reveals a potential mechanism for their roles in meristem maintenance. *The Plant Cell, 18*, 560–573. https://doi.org/10.1105/tpc.105.039107

Kim, S. Y., Yu, X., & Michaels, S. D. (2008). Regulation of *CONSTANS* and *FLOWERING LOCUS T* expression in response to changing light quality. *Plant Physiology, 148*, 269–279. https://doi.org/10.1104/pp.108.122606

Kinoshita, A., Vayssieres, A., Richter, R., Sang, Q., Roggen, A., van Driel, A. D., Smith, R. S., & Coupland, G. (2020). Regulation of shoot meristem shape by photoperiodic signaling and phytohormones during floral induction of *Arabidopsis*. *eLife, 9*, e60661. https://doi.org/10.7554/eLife.60661

Kitagawa, M., & Jackson, D. (2019). Control of meristem size. *Annual Review of Plant Biology, 70*, 269–291. https://doi.org/10.1146/annurev-arplant-042817-040549

Kondo, T., Sawa, S., Kinoshita, A., Mizuno, S., Kakimoto, T., Fukuda, H., & Sakagami, Y. (2006). A plant peptide encoded by *CLV3* identified by in situ MALDI-TOF MS analysis. *Science, 313*, 845–848. https://doi.org/10.1126/science.1128439

Koornneef, M., Hanhart, C. J., & van der Veen, J. H. (1991). A genetic and physiological analysis of late flowering mutants in *Arabidopsis thaliana*. *Molecular & General Genetics, 229*, 57–66.

Kozaki, A., Hake, S., & Colasanti, J. (2004). The maize ID1 flowering time regulator is a zinc finger protein with novel DNA binding properties. *Nucleic Acids Research, 32*, 1710–1720. https://doi.org/10.1093/nar/gkh337

Krizek, B. A., & Meyerowitz, E. M. (1996). The *Arabidopsis* homeotic genes *APETALA3* and *PISTILLATA* are sufficient to provide the B class organ identity function. *Development, 122*, 11–22.

Lai, X., Stigliani, A., Lucas, J., Hugouvieux, V., Parcy, F., & Zubieta, C. (2020). Genome-wide binding of SEPALLATA3 and AGAMOUS complexes determined by sequential DNA-affinity purification sequencing. *Nucleic Acids Research, 48*, 9637–9648. https://doi.org/10.1093/nar/gkaa729

Lai, X., Blanc-Mathieu, R., GrandVuillemin, L., Huang, Y., Stigliani, A., Lucas, J., Thevenon, E., Loue-Manifel, J., Turchi, L., Daher, H., Brun-Hernandez, E., Vachon, G., Latrasse, D., Benhamed, M., Dumas, R., Zubieta, C., & Parcy, F. (2021). The LEAFY floral regulator displays pioneer transcription factor properties. *Molecular Plant, 14*, 829–837. https://doi.org/10.1016/j.molp.2021.03.004

Lampugnani, E. R., Kilinc, A., & Smyth, D. R. (2012). PETAL LOSS is a boundary gene that inhibits growth between developing sepals in *Arabidopsis thaliana*. *The Plant Journal, 71*, 724–735. https://doi.org/10.1111/j.1365-313X.2012.05023.x

Laufs, P., Peaucelle, A., Morin, H., & Traas, J. (2004). MicroRNA regulation of the CUC genes is required for boundary size control in *Arabidopsis* meristems. *Development, 131*, 4311–4322. https://doi.org/10.1242/dev.01320

Laux, T., Mayer, K. F., Berger, J., & Jurgens, G. (1996). The *WUSCHEL* gene is required for shoot and floral meristem integrity in *Arabidopsis*. *Development, 122*, 87–96.

Lee, J., & Lee, I. (2010). Regulation and function of SOC1, a flowering pathway integrator. *Journal of Experimental Botany, 61*, 2247–2254.

Leibfried, A., To JP, Busch, W., Stehling, S., Kehle, A., Demar, M., Kieber, J. J., & Lohmann, J. U. (2005). WUSCHEL controls meristem function by direct regulation of cytokinin-inducible response regulators. *Nature, 438*, 1172–1175. https://doi.org/10.1038/nature04270

Lenhard, M., Bohnert, A., Jurgens, G., & Laux, T. (2001). Termination of stem cell maintenance in *Arabidopsis* floral meristems by interactions between *WUSCHEL* and *AGAMOUS*. *Cell, 105*, 805–814.

Levy, Y. Y., Mesnage, S., Mylne, J. S., Gendall, A. R., & Dean, C. (2002). Multipe roles of *Arabidopsis VRN1* in vernalization and flowering time control. *Science, 297*, 243–246.

Li, J., Wang, Y., Zhang, Y., Wang, W., Irish, V. F., & Huang, T. (2016). RABBIT EARS regulates the transcription of TCP4 during petal development in *Arabidopsis*. *Journal of Experimental Botany, 67*, 6473–6480. https://doi.org/10.1093/jxb/erw419

Liljegren, S. J., Gustafson-Brown, C., Pinyopich, A., Ditta, G. S., & Yanofsky, M. F. (1999). Interactions among *APETALA1*, *LEAFY*, and *TERMINAL FLOWER1* specify meristem fate. *The Plant Cell, 11*, 1007–1018.

Liu, C., Zhou, J., Bracha-Drori, K., Yalovsky, S., Ito, T., & Yu, H. (2007). Specification of Arabidopsis floral meristem identity by repression of flowering time genes. *Development, 134*, 1901–1910. https://doi.org/10.1242/dev.003103

Liu, X., Huang, J., Wang, Y., Khanna, K., Xie, Z., Owen, H. A., & Zhao, D. (2010). The role of *floral organs in carpels*, an *Arabidopsis* loss-of-function mutation in *MicroRNA160a*, in organogenesis and the mechanism regulating its expression. *The Plant Journal, 62*, 416–428.

Liu, X., Kim, Y. J., Muller, R., Yumul, R. E., Liu, C., Pan, Y., Cao, X., Goodrich, J., & Chen, X. (2011). *AGAMOUS* terminates floral stem cell maintenance in Arabidopsis by directly repressing *WUSCHEL* through recruitment of Polycomb group proteins. *The Plant Cell, 23*, 3654–3670. https://doi.org/10.1105/tpc.111.091538

Liu, X., Dinh, T. T., Li, D., Shi, B., Li, Y., Cao, X., Guo, L., Pan, Y., Jiao, Y., & Chen, X. (2014). *AUXIN RESPONSE FACTOR 3* integrates the functions of *AGAMOUS* and *APETALA2* in floral meristem determinacy. *The Plant Journal, 80*, 629–641. https://doi.org/10.1111/tpj.12658

Locascio, A., Blazquez, M. A., & Alabadi, D. (2013). Genomic analysis of DELLA protein activity. *Plant & Cell Physiology, 54*, 1229–1237.

Lohmann, J. U., Hong, R. L., Hobe, M., Busch, M. A., Parcy, F., Simon, R., & Weigel, D. (2001). A molecular link between stem cell regulation and floral patterning in *Arabidopsis*. *Cell, 105*, 793–803.

Lu, S. X., Knowles, S. M., Andronis, C., Ong, M. S., & Tobin, E. M. (2009). CIRCADIAN CLOCK ASSOCIATED1 and LATE ELONGATED HYPOCOTYL function synergistically in the circadian clock of *Arabidopsis*. *Plant Physiology, 150*, 834–843. https://doi.org/10.1104/pp.108.133272

Ma, H. (1994). The unfolding drama of flower development: Recent results from genetic and molecular analyses. *Genes & Development, 8*, 745–756.

Ma, H. (2005). Molecular genetic analysis of microsporogenesis and microgametogenesis. *Annual Review of Plant Biology, 56*, 393–434.

Ma, H., & Sundaresan, V. (2010). Gametophyte development in flowering plant. In M. Timmermann (Ed.), *Current topics in developmental biology* (Vol. 91, pp. 379–412). Academic Press.

Maier, A. T., Stehling-Sun, S., Wollmann, H., Demar, M., Hong, R. L., Haubeiss, S., Weigel, D., & Lohmann, J. U. (2009). Dual roles of the bZIP transcription factor PERIANTHIA in the control of floral architecture and homeotic gene expression. *Development, 136*, 1613–1620. https://doi.org/10.1242/dev.033647

Mandel, M. A., & Yanofsky, M. F. (1995). A gene triggering flower formation in *Arabidopsis*. *Nature, 377*, 522–524.

Mandel, M. A., Gustafson-Brown, C., Savidge, B., & Yanofsky, M. F. (1992). Molecular characterization of the *Arabidopsis* floral homeotic gene *APETALA1*. *Nature, 360*, 273–277.

Mayer, K. F., Schoof, H., Haecker, A., Lenhard, M., Jurgens, G., & Laux, T. (1998). Role of *WUSCHEL* in regulating stem cell fate in the *Arabidopsis* shoot meristem. *Cell, 95*, 805–815.

Michaels, S. D., & Amasino, R. M. (2001). Loss of *FLOWERING LOCUS C* activity eliminates the late-flowering phenotype of *FRIGIDA* and autonomous pathway mutations but not responsiveness to vernalization. *The Plant Cell, 13*, 935–942.

Michniewicz, M., Zago, M. K., Abas, L., Weijers, D., Schweighofer, A., Meskiene, I., Heisler, M. G., Ohno, C., Zhang, J., Huang, F., Schwab, R., Weigel, D., Meyerowitz, E. M., Luschnig, C., Offringa, R., & Friml, J. (2007). Antagonistic regulation of PIN phosphorylation by PP2A and PINOID directs auxin flux. *Cell, 130*, 1044–1056. https://doi.org/10.1016/j.cell.2007.07.033

Mizukami, Y., & Ma, H. (1997). Determination of Arabidopsis floral meristem identity by *AGAMOUS*. *The Plant Cell, 9*, 393–408.

Mockler, T., Yang, H., Yu, X., Parikh, D., Cheng, Y. C., Dolan, S., & Lin, C. (2003). Regulation of photoperiodic flowering by *Arabidopsis* photoreceptors. *Proceedings of the National*

Academy of Sciences of the United States of America, 100, 2140–2145. https://doi.org/10.1073/pnas.0437826100

Mouradov, A., Cremer, F., & Coupland, G. (2002). Control of flowering time: Interacting pathways as a basis for diversity. *The Plant Cell, 14*(Suppl), S111–S130. https://doi.org/10.1105/tpc.001362

Muller, R., Bleckmann, A., & Simon, R. (2008). The receptor kinase CORYNE of Arabidopsis transmits the stem cell-limiting signal CLAVATA3 independently of CLAVATA1. *The Plant Cell, 20*, 934–946. https://doi.org/10.1105/tpc.107.057547

Munos, S., Ranc, N., Botton, E., Berard, A., Rolland, S., Duffe, P., Carretero, Y., Le Paslier, M. C., Delalande, C., Bouzayen, M., Brunel, D., & Causse, M. (2011). Increase in tomato locule number is controlled by two single-nucleotide polymorphisms located near WUSCHEL. *Plant Physiology, 156*, 2244–2254. https://doi.org/10.1104/pp.111.173997

Nakamura, M., Mozgova, I., & Hennig, L. (2017). Lost memories of winter: Breaking the *FLC* silence. *Molecular Plant, 10*, 1477–1479. https://doi.org/10.1016/j.molp.2017.11.009

Ng, M., & Yanofsky, M. F. (2001). Activation of the Arabidopsis B class homeotic genes by *APETALA1*. *The Plant Cell, 13*, 739–753.

Ni, W., Xie, D., Hobbie, L., Feng, B., Zhao, D., Akkara, J., & Ma, H. (2004). Regulation of flower development in Arabidopsis by SCF complexes. *Plant Physiology, 134*, 1574–1585.

Okamuro, J. K., den Boer, B. G., Lotys-Prass, C., Szeto, W., & Jofuku, K. D. (1996). Flowers into shoots: Photo and hormonal control of a meristem identity switch in *Arabidopsis*. *Proceedings of the National Academy of Sciences of the United States of America, 93*, 13831–13836.

Okamuro, J. K., Caster, B., Villarroel, R., Van Montagu, M., & Jofuku, K. D. (1997). The AP2 domain of APETALA2 defines a large new family of DNA binding proteins in *Arabidopsis*. *Proceedings of the National Academy of Sciences of the United States of America, 94*, 7076–7081.

Ordidge, M., Chiurugwi, T., Tooke, F., & Battey, N. H. (2005). *LEAFY, TERMINAL FLOWER1* and *AGAMOUS* are functionally conserved but do not regulate terminal flowering and floral determinacy in *Impatiens balsamina*. *The Plant Journal, 44*, 985–1000. https://doi.org/10.1111/j.1365-313X.2005.02607.x

Osnato, M., Cota, I., Nebhnani, P., Cereijo, U., & Pelaz, S. (2021). Photoperiod control of plant growth: Flowering time genes beyond flowering. *Frontiers in Plant Science, 12*, 805635. https://doi.org/10.3389/fpls.2021.805635

Parcy, F., Bomblies, K., & Weigel, D. (2002). Interaction of *LEAFY, AGAMOUS* and *TERMINAL FLOWER1* in maintaining floral meristem identity in *Arabidopsis*. *Development, 129*, 2519–2527.

Pastore, J. J., Limpuangthip, A., Yamaguchi, N., Wu, M. F., Sang, Y., Han, S. K., Malaspina, L., Chavdaroff, N., Yamaguchi, A., & Wagner, D. (2011). LATE MERISTEM IDENTITY2 acts together with LEAFY to activate APETALA1. *Development, 138*, 3189–3198. https://doi.org/10.1242/dev.063073

Pnueli, L., Carmel-Goren, L., Hareven, D., Gutfinger, T., Alvarez, J., Ganal, M., Zamir, D., & Lifschitz, E. (1998). The *SELF-PRUNING* gene of tomato regulates vegetative to reproductive switching of sympodial meristems and is the ortholog of *CEN* and *TFL1*. *Development, 125*, 1979–1989.

Pnueli, L., Gutfinger, T., Hareven, D., Ben-Naim, O., Ron, N., Adir, N., & Lifschitz, E. (2001). Tomato SP-interacting proteins define a conserved signaling system that regulates shoot architecture and flowering. *The Plant Cell, 13*, 2687–2702. https://www.ncbi.nlm.nih.gov/pmc/articles/PMC139482

Pouteau, S., Nicholls, D., Tooke, F., Coen, E., & Battey, N. (1997). The induction and maintenance of flowering in *Impatiens*. *Development, 124*, 3343–3351.

Prigge, M. J., Otsuga, D., Alonso, J. M., Ecker, J. R., Drews, G. N., & Clark, S. E. (2005). Class III homeodomain-leucine zipper gene family members have overlapping, antagonistic, and distinct roles in Arabidopsis development. *The Plant Cell, 17*, 61–76. https://doi.org/10.1105/tpc.104.026161

Prunet, N., Morel, P., Thierry, A. M., Eshed, Y., Bowman, J. L., Negrutiu, I., & Trehin, C. (2008). *REBELOTE, SQUINT*, and *ULTRAPETALA1* function redundantly in the temporal regulation of floral meristem termination in *Arabidopsis thaliana. The Plant Cell, 20*, 901–919. https://doi.org/10.1105/tpc.107.053306

Putterill, J. (2001). Flowering in time: Genes controlling photoperiodic flowering in *Arabidopsis. Philosophical Transactions of the Royal Society of London, Series B, 356*, 1761–1767.

Putterill, J., Robson, F., Lee, K., Simon, R., & Coupland, G. (1996). The *CONSTANS* gene of *Arabidopsis* promotes flowering and encodes a protein showing similarities to zinc finger transcription factors. *Cell, 80*, 847–857.

Quiroz, S., Yustis, J. C., Chavez-Hernandez, E. C., Martinez, T., Sanchez, M. P., Garay-Arroyo, A., Alvarez-Buylla, E. R., & Garcia-Ponce, B. (2021). Beyond the genetic pathways, flowering regulation complexity in *Arabidopsis thaliana. International Journal of Molecular Sciences, 22*, 5716. https://doi.org/10.3390/ijms22115716

Quon, T., Lampugnani, E. R., & Smyth, D. R. (2017). *PETAL LOSS* and *ROXY1* interact to limit growth within and between sepals but to promote petal initiation in *Arabidopsis thaliana. Frontiers in Plant Science, 8*, 152. https://doi.org/10.3389/fpls.2017.00152

Reddy, G. V., & Meyerowitz, E. M. (2005). Stem-cell homeostasis and growth dynamics can be uncoupled in the *Arabidopsis* shoot apex. *Science, 310*, 663–667. https://doi.org/10.1126/science.1116261

Rieu, P., Turchi, L., Thevenon, E., Zarkadas, E., Nanao, M., Chahtane, H., Tichtinsky, G., Lucas, J., Blanc-Mathieu, R., Zubieta, C., Schoehn, G., & Parcy, F. (2023). The F-box protein UFO controls flower development by redirecting the master transcription factor LEAFY to new cis-elements. *Nature Plants, 9*, 315–329. https://doi.org/10.1038/s41477-022-01336-2

Rodriguez-Leal, D., Xu, C., Kwon, C. T., Soyars, C., Demesa-Arevalo, E., Man, J., Liu, L., Lemmon, Z. H., Jones, D. S., Van Eck, J., Jackson, D. P., Bartlett, M. E., Nimchuk, Z. L., & Lippman, Z. B. (2019). Evolution of buffering in a genetic circuit controlling plant stem cell proliferation. *Nature Genetics, 51*, 786–792. https://doi.org/10.1038/s41588-019-0389-8

Roeder, A. H., & Yanofsky, M. F. (2001). Unraveling the mystery of double flowers. *Developmental Cell, 1*, 4–6.

Rojo, E., Sharma, V. K., Kovaleva, V., Raikhel, N. V., & Fletcher, J. C. (2002). CLV3 is localized to the extracellular space, where it activates the Arabidopsis CLAVATA stem cell signaling pathway. *The Plant Cell, 14*, 969–977. https://doi.org/10.1105/tpc.002196

Roth, O., Alvarez, J. P., Levy, M., Bowman, J. L., Ori, N., & Shani, E. (2018). The KNOXI transcription factor SHOOT MERISTEMLESS regulates floral fate in Arabidopsis. *The Plant Cell, 30*, 1309–1321. https://doi.org/10.1105/tpc.18.00222

Ruiz-Garcia, L., Madueno, F., Wilkinson, M., Haughn, G., Salinas, J., & Martinez-Zapater, J. M. (1997). Different roles of flowering-time genes in the activation of floral initiation genes in Arabidopsis. *The Plant Cell, 9*, 1921–1934.

Running, M. P., & Meyerowitz, E. M. (1996). Mutations in the *PERIANTHIA* gene of *Arabidopsis* specifically alter floral organ number and initiation pattern. *Development, 122*, 1261–1269.

Running, M. P., Fletcher, J. C., & Meyerowitz, E. M. (1998). The WIGGUM gene is required for proper regulation of floral meristem size in *Arabidopsis. Development, 125*, 2545–2553.

Samach, A., Onouchi, H., Gold, S. E., Ditta, G. S., Schwarz-Sommer, Z., Yanofsky, M. F., & Coupland, G. (2000). Distinct roles of CONSTANS target genes in reproductive development of *Arabidopsis. Science, 288*, 1613–1616.

Sang, Q., Vayssieres, A., O'Maoileidigh, D. S., Yang, X., Vincent, C., Bertran Garcia, de Olalla, E., Cerise, M., Franzen, R., & Coupland, G. (2022). MicroRNA172 controls inflorescence meristem size through regulation of *APETALA2* in *Arabidopsis. The New Phytologist, 235*, 356–371. https://doi.org/10.1111/nph.18111

Sawa, S., Ito, T., Shimura, Y., & Okada, K. (1999). *FILAMENTOUS FLOWER* controls the formation and development of Arabidopsis inflorescences and floral meristems. *The Plant Cell, 11*, 69–86.

Sawa, M., Nusinow, D. A., Kay, S. A., & Imaizumi, T. (2007). FKF1 and GIGANTEA complex formation is required for day-length measurement in *Arabidopsis*. *Science, 318*, 261–265.

Schlegel, J., Denay, G., Wink, R., Pinto, K. G., Stahl, Y., Schmid, J., Blumke, P., & Simon, R. G. (2021). Control of Arabidopsis shoot stem cell homeostasis by two antagonistic CLE peptide signalling pathways. *eLife, 10*, e70934. https://doi.org/10.7554/eLife.70934

Schoof, H., Lenhard, M., Haecker, A., Mayer, K. F., Jurgens, G., & Laux, T. (2000). The stem cell population of *Arabidopsis* shoot meristems in maintained by a regulatory loop between the *CLAVATA* and *WUSCHEL* genes. *Cell, 100*, 635–644.

Schrager-Lavelle, A., Klein, H., Fisher, A., & Bartlett, M. (2017). Grass flowers: An untapped resource for floral evo-devo. *Journal of Systematics and Evolution, 55*, 525–541. https://doi.org/10.1111/jse.12251. https://onlinelibrary.wiley.com/doi/full/10.1111/jse.12251.

Schwarz-Sommer, Z., Huijser, P., Nacken, W., Saedler, H., & Sommer, H. (1990). Genetic control of flower development by homeotic genes in *Antirrhinum majus*. *Science, 250*, 931–936.

Scofield, S., Dewitte, W., Nieuwland, J., & Murray, J. A. H. (2013). The *Arabidopsis* homeobox gene SHOOT MERISTEMLESS has cellular and meristem-organisational roles with differential requirements for cytokinin and CYCD3 activity. *The Plant Journal, 75*, 53–66. https://doi.org/10.1111/tpj.12198

Searle, I., He, Y., Turck, F., Vincent, C., Fornara, F., Krober, S., Amasino, R. A., & Coupland, G. (2006). The transcription factor FLC confers a flowering response to vernalization by repressing meristem competence and systemic signaling in *Arabidopsis*. *Genes & Development, 20*, 898–912.

Sessions, A., Nemhauser, J. L., McColl, A., Roe, J. L., Feldmann, K. A., & Zambryski, P. C. (1997). *ETTIN* patterns the *Arabidopsis* floral meristem and reproductive organs. *Development, 124*, 4481–4491.

Shannon, S., & Meeks-Wagner, D. R. (1991). A mutation in the Arabidopsis *TFL1* gene affects inflorescence meristem development. *The Plant Cell, 3*, 877–892.

Simon, R., Igeno, M. I., & Coupland, G. (1996). Activation of floral meristem identity genes in *Arabidopsis*. *Nature, 384*, 59–62.

Simpson, G. G., Gendall, A. R., & Dean, C. (1999). When to switch to flowering. *Annual Review of Cell and Developmental Biology, 99*, 519–550.

Soltis, D. E., Chanderbali, A. S., Kim, S., Buzgo, M., & Soltis, P. S. (2007). The ABC model and its applicability to basal angiosperms. *Annals of Botany, 100*, 155–163. https://www.ncbi.nlm.nih.gov/pmc/articles/PMC2735328/

Somers, D. E., Schultz, T. F., Milnamow, M., & Kay, S. (2000). *ZEITLUPE* encodes a novel clock-associated PAS protein from *Arabidopsis*. *Cell, 101*, 319–329.

Sun, B., Zhou, Y., Cai, J., Shang, E., Yamaguchi, N., Xiao, J., Looi, L. S., Wee, W. Y., Gao, X., Wagner, D., & Ito, T. (2019). Integration of transcriptional repression and Polycomb-mediated silencing of *WUSCHEL* in floral meristems. *The Plant Cell, 31*, 1488–1505. https://doi.org/10.1105/tpc.18.00450

Sung, Z. R., Belachew, L., Bai, S., & Bertrand-Garcia, R. (1992). *EMF*, an *Arabidopsis* gene required for vegetative shoot development. *Science, 258*, 1645–1647.

Suzaki, T., Sato, M., Ashikari, M., Miyoshi, M., Nagato, Y., & Hirano, H. Y. (2004). The gene *FLORAL ORGAN NUMBER1* regulates floral meristem size in rice and encodes a leucine-rich repeat receptor kinase orthologous to *Arabidopsis* CLAVATA1. *Development, 131*, 5649–5657. https://doi.org/10.1242/dev.01441

Suzaki, T., Toriba, T., Fujimoto, M., Tsutsumi, N., Kitano, H., & Hirano, H. Y. (2006). Conservation and diversification of meristem maintenance mechanism in *Oryza sativa*: Function of the *FLORAL ORGAN NUMBER2* gene. *Plant & Cell Physiology, 47*, 1591–1602. https://doi.org/10.1093/pcp/pcl025

Tadege, M., Sheldon, C. C., Helliwell, C. A., Stoutjesdijk, P., Dennis, E. S., & Peacock, W. J. (2001). Control of flowering time by *FLC* orthologues in *Brassica napus*. *The Plant Journal, 28*, 545–553.

Takeda, S., Matsumoto, N., & Okada, K. (2004). *RABBIT EARS*, encoding a SUPERMAN-like zinc finger protein, regulates petal development in *Arabidopsis thaliana*. *Development, 131*, 425–434. https://doi.org/10.1242/dev.00938

Takeda, S., Hamamura, Y., Sakamoto, T., Kimura, S., Aida, M., & Higashiyama, T. (2022). Non-cell-autonomous regulation of petal initiation in *Arabidopsis thaliana*. *Development, 149*, dev200684. https://doi.org/10.1242/dev.200684

Tanaka, W., Pautler, M., David Jackson, D., & Hirano, H. Y. (2013). Grass meristems II: Inflorescence architecture, flower development and meristem fate. *Plant & Cell Physiology, 54*, 313–324. https://doi.org/10.1093/pcp/pct016

Tanaka, W., Ohmori, Y., Ushijima, T., Matsusaka, H., Matsushita, T., Kumamaru, T., Kawano, S., & Hirano, H. Y. (2015). Axillary meristem formation in rice requires the *WUSCHEL* Ortholog *TILLERS ABSENT1*. *The Plant Cell, 27*, 1173–1184. https://doi.org/10.1105/tpc.15.00074

Tao, Z., Shen, L., Gu, X., Wang, Y., Yu, H., & He, Y. (2017). Embryonic epigenetic reprogramming by a pioneer transcription factor in plants. *Nature, 551*, 124–128.

Tobena-Santamaria, R., Bliek, M., Ljung, K., Sandberg, G., Mol, J. N., Souer, E., & Koes, R. (2002). FLOOZY of Petunia is a flavin mono-oxygenase-like protein required for the specification of leaf and flower architecture. *Genes & Development, 16*, 753–763.

Tooke, F., Ordidge, M., Chiurugwi, T., & Battey, N. (2005). Mechanisms and function of flower and inflorescence reversion. *Journal of Experimental Botany, 56*, 2587–2599. https://doi.org/10.1093/jxb/eri254

Turchi, L., Carabelli, M., Ruzza, V., Possenti, M., Sassi, M., Penalosa, A., Sessa, G., Salvi, S., Forte, V., Morelli, G., & Ruberti, I. (2013). *Arabidopsis* HD-Zip II transcription factors control apical embryo development and meristem function. *Development, 140*, 2118–2129. https://doi.org/10.1242/dev.092833

Uemura, A., Yamaguchi, N., Xu, Y., Wee, W., Ichihashi, Y., Suzuki, T., Shibata, A., Shirasu, K., & Ito, T. (2018). Regulation of floral meristem activity through the interaction of *AGAMOUS*, *SUPERMAN*, and *CLAVATA3* in *Arabidopsis*. *Plant Reproduction, 31*, 89–105. https://doi.org/10.1007/s00497-017-0315-0

Vince-Prue, D. (1985). Photoperiod and hormones. In R. P. Pharis & D. M. Reid (Eds.), *Hormonal regulation of development III: Role of environmental factors* (pp. 308–364). Springer-Verlag.

Wagner, D., & Meyerowitz, E. M. (2002). *SPLAYED*, a novel SWI/SNF ATPase homolog, controls reproductive development in *Arabidopsis*. *Current Biology, 12*, 85–94.

Wagner, D., Sablowski, R. W., & Meyerowitz, E. M. (1999). Transcriptional activation of *APETALA1* by *LEAFY*. *Science, 285*, 582–584.

Wang, J., Tian, L., Lee, H.-S., & Chen, Z. J. (2006). Nonadditive regulation of *FRI* and *FLC* loci mediates flowering-time variation in *Arabidopsis* allopolyploids. *Genetics, 173*, 965–974. https://www.ncbi.nlm.nih.gov/pmc/articles/PMC1526503/

Wang, L., Ma, H., & Lin, J. (2019). Angiosperm-wide and family level analyses of *AP2/ERF* genes reveal differential retention and sequence divergence after whole-genome duplication. *Frontiers in Plant Science, 10*, 196.

Wang, C., Yang, X., & Li, G. (2021). Molecular insights into inflorescence meristem specification for yield potential in cereal crops. *International Journal of Molecular Sciences, 22*, 3508. https://doi.org/10.3390/ijms22073508

Washburn, C. F., & Thomas, J. F. (2000). Reversion of flowering in Glycine max (Fabaceae). *American Journal of Botany, 87*, 1425–1438.

Weigel, D., Alvarez, J., Smyth, D. R., Yanofsky, M. F., & Meyerowitz, E. M. (1992). *LEAFY* controls floral meristem identity in *Arabidopsis*. *Cell, 69*, 843–859.

Wickland, D. P., & Hanzawa, Y. (2015). The *FLOWERING LOCUS T/TERMINAL FLOWER 1* gene family: Functional evolution and molecular mechanisms. *Molecular Plant, 8*, 983–997. https://doi.org/10.1016/j.molp.2015.01.007

Wigge, P. (2011). FT, a mobile developmental siganl in plants. *Current Biology, 21*, R374–R378.

Wilkinson, M. D., & Haughn, G. W. (1995). *UNUSUAL FLORAL ORGANS* controls meristem identity and floral organ primordia fate in Arabidopsis. *The Plant Cell, 7*, 1485–1499.

Williams, L., Grigg, S. P., Xie, M., Christensen, S., & Fletcher, J. C. (2005). Regulation of *Arabidopsis* shoot apical meristem and lateral organ formation by microRNA miR166g and its AtHD-ZIP target genes. *Development, 132*, 3657–3668. https://doi.org/10.1242/dev.01942

Wu, M. F., Sang, Y., Bezhani, S., Yamaguchi, N., Han, S. K., Li, Z., Su, Y., Slewinski, T. L., & Wagner, D. (2012). SWI2/SNF2 chromatin remodeling ATPases overcome polycomb repression and control floral organ identity with the LEAFY and SEPALLATA3 transcription factors. *Proceedings of the National Academy of Sciences of the United States of America, 109*, 3576–3581. https://doi.org/10.1073/pnas.1113409109

Xiao, J., Xu, S., Li, C., Xu, Y., Xing, L., Niu, Y., Huan, Q., Tang, Y., Zhao, C., Wagner, D., Gao, C., & Chong, K. (2014). *O*-GlcNAc-mediated interaction between VER2 and TaGRP2 elicits *TaVRN1* mRNA accumulation during vernalization in winter wheat. *Nature Communications, 5*, 4572.

Xing, L., Li, J., Xu, Y., Xu, Z., & Chong, K. (2009). Phosphorylation modification of wheat lectin VER2 is associated with vernalization-induced O-GlcNAc signaling and intracellular motility. *PLoS One, 4*, e4854.

Xu, Y., Prunet, N., Gan, E. S., Wang, Y., Stewart, D., Wellmer, F., Huang, J., Yamaguchi, N., Tatsumi, Y., Kojima, M., Kiba, T., Sakakibara, H., Jack, T. P., Meyerowitz, E. M., & Ito, T. (2018). *SUPERMAN* regulates floral whorl boundaries through control of auxin biosynthesis. *The EMBO Journal, 37*, EMBJ201797499. https://doi.org/10.15252/embj.201797499

Xu, S., Xiao, J., Yin, F., Guo, X., Xing, L., Xu, Y., & Chong, K. (2019). The protein modifications of O-GlcNAcylation and phosphorylation mediate vernalization response for flowering in winter wheat. *Plant Physiology, 180*, 1436–1449.

Xu, S., Dong, Q., Deng, M., Lin, D., Xiao, J., Cheng, P., Xing, L., Niu, Y., Gao, C., Zhang, W., Xu, Y., & Chong, K. (2021). The vernalization-induced long non-coding RNA VAS functions with the transcription factor TaRF2b to promote *TaVRN1* expression for flowering in hexaploid wheat. *Molecular Plant, 14*, 1525–1538.

Yadav, R. K., Perales, M., Gruel, J., Girke, T., Jonsson, H., & Reddy, G. V. (2011). WUSCHEL protein movement mediates stem cell homeostasis in the *Arabidopsis* shoot apex. *Genes & Development, 25*, 2025–2030. https://doi.org/10.1101/gad.17258511

Yamaguchi, N. (2021). LEAFY, a pioneer transcription factor in plants: A mini-review. *Frontiers in Plant Science, 12*, 701406. https://doi.org/10.3389/fpls.2021.701406

Yamaguchi, N., Wu, M. F., Winter, C. M., Berns, M. C., Nole-Wilson, S., Yamaguchi, A., Coupland, G., Krizek, B. A., & Wagner, D. (2013). A molecular framework for auxin-mediated initiation of flower primordia. *Developmental Cell, 24*, 271–282. https://doi.org/10.1016/j.devcel.2012.12.017

Yamaguchi, N., Winter, C. M., Wu, M. F., Kanno, Y., Yamaguchi, A., Seo, M., & Wagner, D. (2014). Gibberellin acts positively then negatively to control onset of flower formation in *Arabidopsis*. *Science, 344*, 638–641. https://doi.org/10.1126/science.1250498

Yamaguchi, N., Huang, J., Xu, Y., Tanoi, K., & Ito, T. (2017). Fine-tuning of auxin homeostasis governs the transition from floral stem cell maintenance to gynoecium formation. *Nature Communications, 8*, 1125. https://doi.org/10.1038/s41467-017-01252-6

Yamaguchi, N., Huang, J., Tatsumi, Y., Abe, M., Sugano, S. S., Kojima, M., Takebayashi, Y., Kiba, T., Yokoyama, R., Nishitani, K., Sakakibara, H., & Ito, T. (2018). Chromatin-mediated feed-forward auxin biosynthesis in floral meristem determinacy. *Nature Communications, 9*, 5290. https://doi.org/10.1038/s41467-018-07763-0

Yan, L., Loukoianov, A., Tranquilli, G., Helguera, M., Fahima, T., & Dubcovsky, J. (2003). Positional cloning of the wheat vernalization gene VRN1. *Proceedings of the National Academy of Sciences of the United States of America, 100*, 6263–6268. https://doi.org/10.1073/pnas.0937399100

Yan, L., Loukoianov, A., Blechl, A., Tranquilli, G., Ramakrishna, W., SanMiguel, P., Bennetzen, J. L., Echenique, V., & Dubcovsky, J. (2004). The wheat VRN2 gene is a flowering repressor down-regulated by vernalization. *Science, 303*, 1640–1644. https://doi.org/10.1126/science.1094305

Yan, L., Fu, D., Li, C., Blechl, A., Tranquilli, G., Bonafede, M., Sanchez, A., Valarik, M., Yasuda, S., & Dubcovsky, J. (2006). The wheat and barley vernalization gene *VRN3* is an orthologue of *FT*. *Proceedings of the National Academy of Sciences of the United States of America, 103*, 19581–19586.

Yang, J., Ning, C., Liu, Z., Zheng, C., Mao, Y., Wu, Q., Wang, D., Liu, M., Zhou, S., Yang, L., He, L., Liu, Y., He, C., Chen, J., & Liu, J. (2023). Genome-wide characterization of *PEBP* gene family and functional analysis of *TERMINAL FLOWER 1* homologs in *Macadamia integrifolia*. *Plants (Basel)., 12*, 2692. https://doi.org/10.3390/plants12142692

Yano, M., Katayose, Y., Ashikari, M., Yamanouchi, U., Monna, L., Fuse, T., Baba, T., Yamanoto, K., Umehara, Y., Nagamura, Y., & Sasaki, T. (2000). *Hd1*, a major photoperiod sensitivity quantitative trait locus in rice, is closely related to the Arabidopsis flowering time gene *CONSTANS*. *The Plant Cell, 12*, 2473–2484.

Yanofsky, M. F., Ma, H., Bowman, J. L., Drews, G. N., Feldmann, K. A., & Meyerowitz, E. M. (1990). The protein encoded by the *Arabidopsis* homeotic gene *agamous* resembles transcription factors. *Nature, 346*, 35–39.

Yao, W. J., Wang, Y. P., Peng, J., Yin, P. P., Gao, H., Xu, L., Laux, T., Zhang, X. S., & Su, Y. H. (2023). The RPT2a-MET1 axis regulates *TERMINAL FLOWER1* to control inflorescence meristem indeterminacy in Arabidopsis. *The Plant Cell, 36*, 1718–1735. https://doi.org/10.1093/plcell/koad249

Ye, L., Wang, B., Zhang, W., Shan, H., & Kong, H. (2016). Gains and losses of cis-regulatory elements led to divergence of the *Arabidopsis APETALA1* and *CAULIFLOWER* duplicate genes in the time, space, and level of expression and regulation of one paralog by the other. *Plant Physiology, 171*, 1055–1069. https://doi.org/10.1104/pp.16.00320

Yong, W., Xu, Y., Xu, W., Wang, X., Li, N., Wu, J., Liang, T., Chong, K., Xu, Z., Tan, K., & Zhu, Z. (2003). Vernalization-induced flowering in wheat is mediated by a lectin-like gene *VER2*. *Planta, 217*, 261–270.

Yoo, S. K., Chung, K. S., Kim, J., Lee, J. H., Hong, S. M., Yoo, S. J., Yoo, S. Y., Lee, J. S., & Ahn, J. H. (2005). CONSTANS activates *SUPPRESSOR OF OVEREXPRESSION OF CONSTANS 1* through FLOWERING LOCUS T to promote flowering in *Arabidopsis*. *Plant Physiology, 139*, 770–778. https://doi.org/10.1104/pp.105.066928

Yoshida, N., Yanai, Y., Chen, L., Kato, Y., Hiratsuka, J., Miwa, T., Sung, Z. R., & Takahashi, S. (2001). EMBRYONIC FLOWER2, a novel polycomb group protein homolog, mediates shoot development and flowering in Arabidopsis. *The Plant Cell, 13*, 2471–2481.

Yoshida, H., Hirano, K., Sato, T., Mitsuda, N., Nomoto, M., Maeo, K., Koketsu, E., Mitani, R., Kawamura, M., Ishiguro, S., Tada, Y., Ohme-Takagi, M., Matsuoka, M., & Ueguchi-Tanaka, M. (2014). DELLA protein functions as a transcriptional activator through the DNA binding of the indeterminate domain family proteins. *Proceedings of the National Academy of Sciences of the United States of America, 111*, 7861–7866. https://doi.org/10.1073/pnas.1321669111. Epub 2014 May 12.

Yu, X., Liu, H., Klejnot, J., & Lin, C. (2010). The cryptochrome blue light receptors. *Arabidopsis Book, 8*, e0135. https://doi.org/10.1199/tab.0135

Yu, Z., Zhang, F., Friml, J., & Ding, Z. (2022). Auxin signaling: Research advances over the past 30 years. *Journal of Integrative Plant Biology, 64*, 371–392. https://doi.org/10.1111/jipb.13225

Zagotta, M. T., Shannon, S., Jacobs, C., & Meeks-Wagner, D. R. (1992). Early-flowering mutants of *Arabidopsis thaliana*. *Australian Journal of Plant Physiology, 19*, 411–418.

Zhang, K., Wang, R., Zi, H., Li, Y., Cao, X., Li, D., Guo, L., Tong, J., Pan, Y., Jiao, Y., Liu, R., Xiao, L., & Liu, X. (2018). AUXIN RESPONSE FACTOR3 regulates floral meristem determinacy by repressing cytokinin biosynthesis and signaling. *The Plant Cell, 30*, 324–346. https://doi.org/10.1105/tpc.17.00705

Zhang, C., Jian, M., Li, W., Yao, X., Tan, C., Qian, Q., Hu, Y., Liu, X., & Hou, X. (2023). Gibberellin signaling modulates flowering via the DELLA-BRAHMA-NF-YC module in *Arabidopsis*. *The Plant Cell, 35*, 3470–3484. https://doi.org/10.1093/plcell/koad166

Zhang, G., Yang, J., Zhang, C., Jiao, B., Panero, J. L., Cai, J., Zhang, Z.-R., Gao, L.-M., Gao, T., & Ma, H. (2024). Nuclear phylogenomics of Asteraceae with increased sampling provides new insights into convergent morphological and molecular evolution. *Plant Communications, 5*, 100851. https://doi.org/10.1016/j.xplc.2024.100851

Zhao, D., Yang, M., Solava, J., & Ma, H. (1999). The *ASK1* gene regulates development and interacts with the *UFO* gene to control floral organ identity in *Arabidopsis*. *Developmental Genetics, 25*, 209–223.

Zhao, D., Yu, Q., Chen, M., & Ma, H. (2001a). The *ASK1* gene regulates B function gene expression in cooperation with *UFO* and *LEAFY* in *Arabidopsis*. *Development, 128*, 2735–2746.

Zhao, D. Z., Yu, Q. L., Chen, C. B., & Ma, H. (2001b). Genetic control of reproductive meristems. In M. T. McManus & B. E. Veit (Eds.), *Meristematic tissues in plant growth and development* (pp. 89–141). Shefield Academic Press.

Zhao, Y., Christensen, S. K., Fankhauser, C., Cashman, J. R., Cohen, J. D., Weigel, D., & Chory, J. (2001c). A role for flavin monooxygenase-like enzymes in auxin biosynthesis. *Science, 291*, 306–309.

Zhao, Z., Andersen, S. U., Ljung, K., Dolezal, K., Miotk, A., Schultheiss, S. J., & Lohmann, J. U. (2010). Hormonal control of the shoot stem-cell niche. *Nature, 465*, 1089–1092. https://doi.org/10.1038/nature09126

Zhou, Y., Honda, M., Zhu, H., Zhang, Z., Guo, X., Li, T., Li, Z., Peng, X., Nakajima, K., Duan, L., & Zhang, X. (2015a). Spatiotemporal sequestration of miR165/166 by *Arabidopsis* Argonaute10 promotes shoot apical meristem maintenance. *Cell Reports, 10*, 1819–1827. https://doi.org/10.1016/j.celrep.2015.02.047

Zhou, Y., Liu, X., Engstrom, E. M., Nimchuk, Z. L., Pruneda-Paz, J. L., Tarr, P. T., Yan, A., Kay, S. A., & Meyerowitz, E. M. (2015b). Control of plant stem cell function by conserved interacting transcriptional regulators. *Nature, 517*, 377–380. https://doi.org/10.1038/nature13853

Zhu, Y., Klasfeld, S., Jeong, C. W., Jin, R., Goto, K., Yamaguchi, N., & Wagner, D. (2020). TERMINAL FLOWER 1-FD complex target genes and competition with FLOWERING LOCUS T. *Nature Communications, 11*, 5118. https://doi.org/10.1038/s41467-020-18782-1

Ziegelhoffer, E. C., Medrano, L. J., & Meyerowitz, E. M. (2000). Cloning of the *Arabidopsis WIGGUM* gene identifies a role for farnesylation in meristem development. *Proceedings of the National Academy of Sciences of the United States of America, 97*, 7633–7638.

Chapter 7
Floral Organ Specification

Zheng Yuan and Dabing Zhang

Abstract In higher plants, flowers comprise distinct nonreproductive and reproductive organs, playing a pivotal role in both seed production and plant propagation. Although flower shapes exhibit significant variation across angiosperms, the fundamental genetic components governing floral organ development remain evolutionarily conserved. Over the past three decades, extensive research has focused on the genetic basis, cellular and molecular events regulating floral initiation, determination, patterning, organ boundary formation, cell differentiation, and the growth and development of floral organs in Arabidopsis (*Arabidopsis thaliana*) and other plant model systems, including snapdragon (*Antirrhinum majus*), petunia (*Petunia hybrida*), rice (*Oryza sativa*), and maize (*Zea mays*). In this chapter, our objective is to update the conserved and diverse components that control flower specification in higher plants. We specifically focus on rice in comparison with maize and Arabidopsis. Additionally, we aim to provide an overview of regulatory factors and signaling pathways involved in the specification of floral organ patterning.

Keywords Floral meristem · Spikelet · Identity · Specification · Determinacy · Phytohormone

Z. Yuan (✉)
Joint International Research Laboratory of Metabolic & Developmental Sciences, School of Life Sciences and Biotechnology, Shanghai Jiao Tong University, Shanghai, China

Yazhou Bay Institute of Deepsea Science and Technology, Shanghai Jiao Tong University, Sanya, China
e-mail: zyuan@sjtu.edu.cn

D. Zhang
Joint International Research Laboratory of Metabolic & Developmental Sciences, School of Life Sciences and Biotechnology, Shanghai Jiao Tong University, Shanghai, China
e-mail: zhangdb@sjtu.edu.cn

F. Chang et al. (eds.), *Regulation of Plant Development*,
https://doi.org/10.1007/978-981-95-7033-1_7

7.1 Introduction

Despite structural variations in angiosperm flowers, a typical flower is composed of four organ types arranged in whorls. Nonreproductive organs, such as sepals and petals in Arabidopsis and dicots and lemma, palea, and lodicule in rice, maize, and other grasses, are in outer whorls, while reproductive organs (stamens and carpels) are in inner whorls. Deciphering the regulatory "code" underlying floral organ specification has been a centuries-old challenge. Johann Wolfgang von Goethe proposed over 200 years ago that floral organs were modified leaves (van Goethe, 1790). This hypothesis was subsequently confirmed through extensive genetic and molecular studies in Arabidopsis and other species (Pelaz et al., 2001; Theissen & Saedler, 2001). The four fundamental floral organ identity genes, A, B, C, and E, encoding key transcription factors (TFs), have been identified in Arabidopsis and snapdragon. Their combinatorial activities are necessary and sufficient for the transformation of the leaf into the flower, summarized as the modified (A)B(C) model (Causier et al., 2010; Bowman et al., 2012; Theissen et al., 2016).

In addition to genes specifying floral organ identity, the formation of complexes involving various floral homeotic proteins and their downstream targets throughout consecutive developmental stages has been explored using advanced approaches in transcriptomics, proteomics, and computational methods (Kaufmann et al., 2009, 2010; Wuest et al., 2012; ÓMaoiléidigh et al., 2013; Pajoro et al., 2014; Wellmer et al., 2014a; O'Maoileidigh et al., 2015; Ruelens et al., 2017; Smaczniak et al., 2017; Thomson & Wellmer, 2019; Chahtane et al., 2023). Nevertheless, whether these regulators of flower development exhibit functional conservation in other species or if the proposed ABC model is applicable across diverse angiosperm species remains largely unknown (Soltis & Soltis, 2014). In this chapter, we will delineate recent advancements in floral organ specification, focusing on rice and comparing it with Arabidopsis, maize, and other species. Specifically, we will discuss recent progress in understanding signaling pathways and gene regulatory networks controlled by the flower meristem (FM) and floral organ identity genes during floral initiation and patterning.

7.2 Floral Organ Development: A Case Study from Rice and Maize

Grasses, constituting one of the largest angiosperm families, encompass numerous economically significant crop species, including rice, maize, sorghum (*Sorghum bicolor*), and others (Grass Phylogeny Working Group, 2001). Spikelets are the fundamental structural units of grass flowers. A typical spikelet comprises a pair of glumes and one to multiple florets (Whipple & Schmidt, 2006; Zhang & Yuan, 2014; Yuan et al., 2020). Rice, a crucial cereal grass crop in Asian countries, has spikelet meristems (SMs) derived from axillary meristems of inflorescence meristems (IMs) or their axillary meristems (AMs), including primary branch meristems (pBMs) and secondary branch meristems (sBMs) (Fig. 7.1a). Subsequently, SMs initiate floret meristems (FMs), leading to the differentiation of nonreproductive organs (one lemma and one palea in the outermost whorl and two lodicules close to the lemma base) and reproductive organs (6 stamens and 1 pistil) in the floral center (Ciaffi et al., 2011; Yoshida & Nagato, 2011; Tanaka et al., 2013; Hirano et al., 2014; Kyozuka et al., 2014; Zhang & Yuan, 2014). A rice floret is distinctively subtended by two pairs of glume-like appendages, known as sterile lemmas/empty glumes and rudimentary glumes, respectively (Fig. 7.1a, b). Despite prolonged debate over the identity of the glumes and nonreproductive organs, recent evidence from our group and others demonstrates that a genuine rice flower comprises paleas, lodicules, stamens, and pistils, as discussed below. Meanwhile, the sterile lemmas were analogous to lemma-like organs (floret bract-like organs), and the rudimentary glumes are equivalent to spikelet bract-like organs (see Sects. 7.4 and 7.5.2) (Yoshida et al., 2009; Yun et al., 2013; Zhang et al., 2017a; Zong et al., 2022).

Unlike the bisexual floral structure observed in rice, maize exhibits a monoecious nature, producing two distinct types of flowers. The female ear and the male tassel are situated on the upper and lower parts of the plant, originating from shoot apical meristems (SAMs) and AMs, respectively (Tanaka et al., 2013). An immature maize spikelet comprises two bisexual florets, the upper and lower florets, each consisting of a lemma, a palea, two lodicules, three stamens, and one pistil polymerized from three carpels. This structure is enclosed by a pair of glumes (Fig. 7.1b). The key distinction in the development of immature tassels and ears lies in the fact that tassel spikelets derive from IMs and BMs, while no BMs form during ear spikelet development (Li & Liu, 2017). In the mature tassel spikelets, both florets become unisexual without a pistil, composed of three stamens surrounded by a lemma, a palea, and two lodicules on the lemma side. Conversely, in the mature ear spikelets, the stamen primordium and the lower floret undergo arrest, and only the upper floret develops, housing one ovule enclosed by two lodicules, a palea, and a lemma (Fig. 7.1c).

In general, the characteristics of nonreproductive organs in rice and maize exhibit distinctions from those observed in eudicots and even other grass species (Fig. 7.1b–d). These disparities have led to considerable controversies regarding biological questions surrounding their morphogenesis (Rudall & Bateman, 2004; Whipple & Schmidt, 2006; Lombardo & Yoshida, 2015). For instance, does the ancestor of the *Oryzeae* spikelet have three florets? How did the derived grass floret evolve from a typical monocot flower? What defines the identity of lemma, palea, and lodicules? Is the lemma akin to a bract-like organ, or is the palea comparable to a prophyll-like organ? Do the lemma and palea serve as equivalents to modified sepals/tepals, or are lodicules equivalent to petals? Is the palea an organ fused from two sepals/tepals? Over the past 30 years, researchers have presented extensive molecular evidence to address these questions (Bommert et al., 2005a; Whipple & Schmidt, 2006; Ciaffi et al., 2011; Yoshida, 2012; Zhang et al., 2013; Hirano et al., 2014; Lombardo & Yoshida, 2015; Dreni & Zhang, 2016; Dreni et al., 2023), which will be discussed in this chapter.

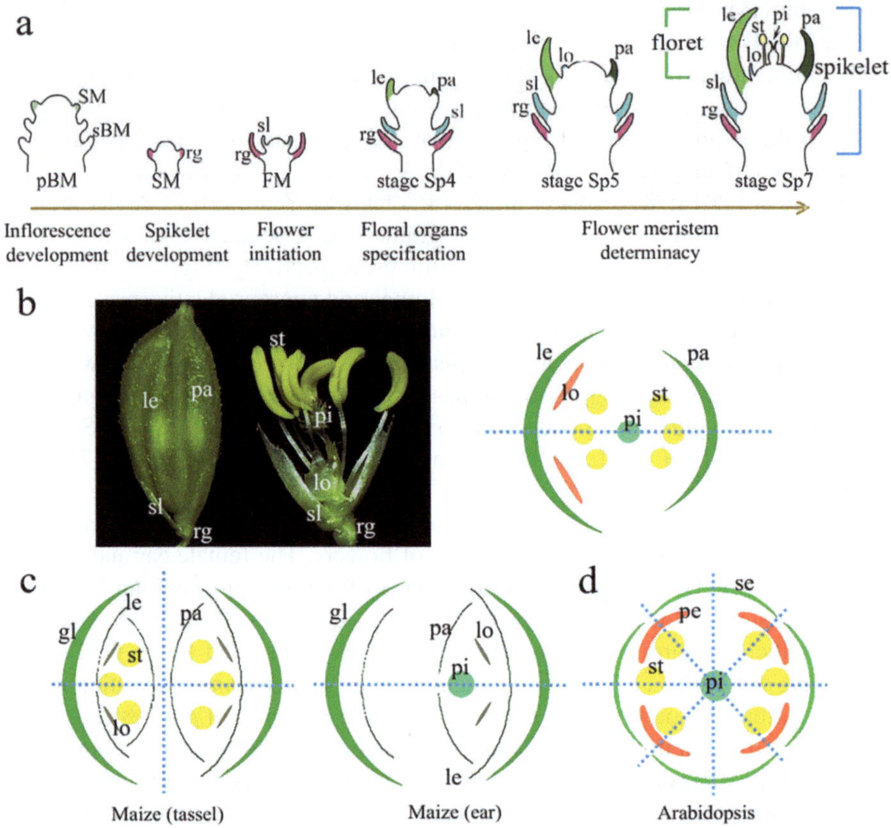

Fig. 7.1 Development and structure of the flowers in rice, maize, and Arabidopsis. (**a**) The diagram illustrates a series of meristem transition events in rice, encompassing the inflorescence meristem (IM), axillary meristem (AM), primary branch meristem (pBM), secondary branch meristem (sBM), spikelet meristem (SM), and floret meristem (FM). (**b–d**) The diagrams depict the structures of rice, maize spikelet and Arabidopsis flower. It is noteworthy that the rice spikelet comprises two whorls of glume-like organs and one floret, exhibiting bilaterally symmetric along the lemma-palea axis. The floret comprises one lemma, one palea, two lodicules, six stamens, and one pistil. The maize spikelet, also zygomorphic, produces two types of flowers, ear spikelet and tassel spikelet, in the upper and lower parts of the plant, respectively. Both the ear floret and tassel floret contain a lemma, a palea, and two lodicules. However, the tassel floret has three stamens without a pistil, while the ear floret has only one pistil. In contrast, the Arabidopsis flower is symmetrical and comprises four sepals, four petals, six stamens, and one pistil. Abbreviations: FM, flower meristem; gl, glume; le, lemma; lo, lodicule; pa, palea; pBM, primary branch meristem; pe, petal; pi, pistil; rg, rudimentary glume; sBM, secondary branch meristem; se, sepal; sl, sterile lemma; SM, spikelet meristem; st, stamen

7.3 Floral Meristem Fate Specification and Transition

Flower development involves a series of meristem fate transitions to generate lateral organs, including branches, bracts, and flowers (Chandler, 2011). Our understanding of the signaling and regulatory pathways governing FM fate specification and transition is primarily derived from research on eudicot model plants such as Arabidopsis, snapdragon, and petunia. In Arabidopsis, FM initiation occurs at the flank of the IM in an auxin-dependent manner (Cheng & Zhao, 2007; Chandler & Werr, 2014; Shi & Vernoux, 2022), and this developmental process is intricately regulated by a network that integrates environmental and hormonal signals, transcriptional regulators, and mechanical constraints (Chandler, 2012; Denay et al., 2017; Wagner, 2017; Wils & Kaufmann, 2017; Shi & Vernoux, 2022).

7.3.1 Auxin Activates Floral Meristem Initiation

During the initiation stage of flower development, genes associated with FM identity, such as *LEAFY* (*LFY*), *AINTEGUMENTA* (*ANT*), and *AINTEGUMENTA-LIKE 6* (*AIL6*), are directly activated by *AUXIN RESPONSE FACTOR* (*ARF*) TF, *MONOPTEROS*(*MP*)/*ARF5* (Yamaguchi et al., 2013). Furthermore, *LFY* triggers the expression of the MADS-box gene *APETALA1* (*AP1*), orchestrating feedback and feed-forward loops to control and sustain FM identity (Kaufmann et al., 2010; Pastore et al., 2011; Winter et al., 2011; Wils & Kaufmann, 2017). To fortify the commitment to FM identity, AP1 directly represses genes associated with flowering time and shoot identity in emerging floral primordia (Kaufmann et al., 2010). In developmental stages 1 and 2, the direct repression of *SEPALLATA 3* (*SEP3*) expression by flowering time genes, including *SHORT VEGETATIVE PHASE* (*SVP*), *SUPPRESSOR OF OVEREXPRESSION OF CONSTANS 1* (*SOC1*), and *AGAMOUS-LIKE 24* (*AGL24*), prevents the premature differentiation of floral meristems (Liu et al., 2009). Concurrently, the expression of meristem identity genes in the FM, such as *WUSCHEL* (*WUS*) and *CLAVATA* (*CLV*), establishes an organizing center and a stem cell niche (Denay et al., 2017). In early developmental stage 3, AP1 interacts with SEP3 as transcriptional activators to induce the expression of floral homeotic genes (Liu et al., 2009; Kaufmann et al., 2010). Analogous to other axillary meristems, the size of Arabidopsis FMs is determined by the WUS/CLV feedback loop (Pautler et al., 2013). Mutations in *CLV* genes lead to an increased number of flowers and floral organs (Clark et al., 1997; Fletcher et al., 1999; Jeong et al., 1999). The stem cell regulator *WUS* is essential for activating floral homeotic genes, in collaboration with *LFY*, ensuring the development of floral organs in the accurate number and pattern (Fig. 7.2a) (Lohmann et al., 2001). Consequently, the sustained activity of *LFY* serves diverse functions in repressing flowering and fostering the emergence and identity of FM (Denay et al., 2017).

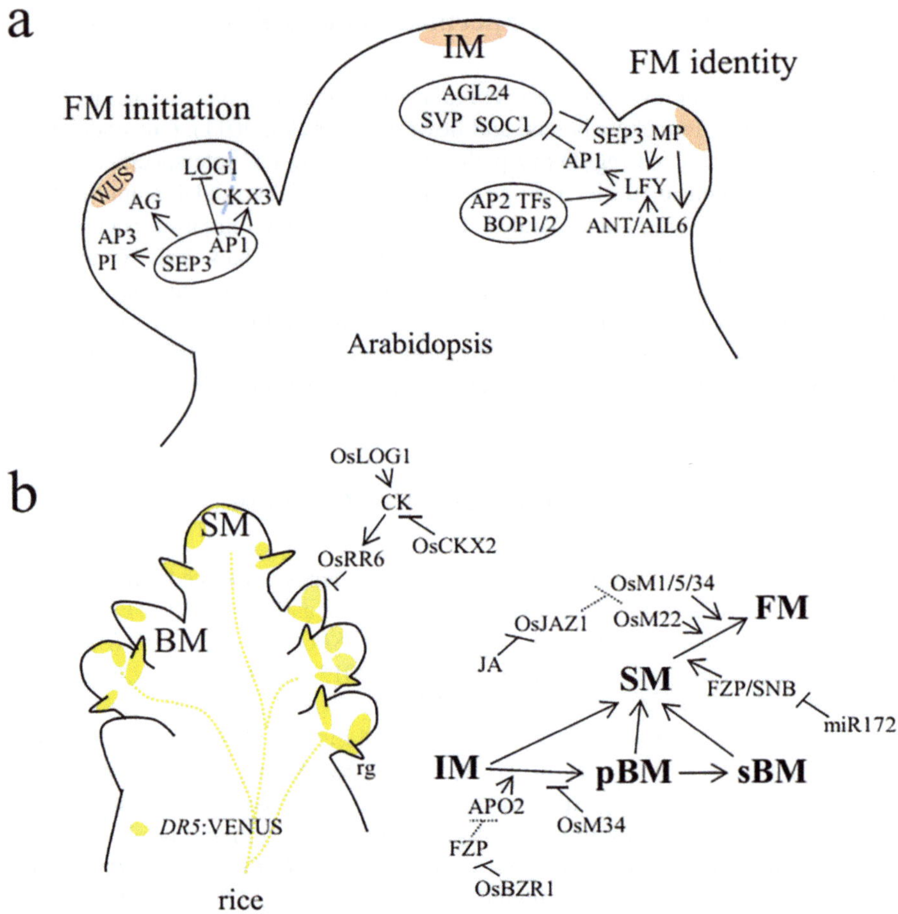

Fig. 7.2 Regulatory networks underlying FM identity and fate transition in Arabidopsis (**a**) and rice (**b**). In Arabidopsis, auxin maxima concurrently induce the expression of *MP* and *LFY* along with *AP2* TFs and respond to environmental signals to activate *AP1* expression. AP1, in turn, maintains FM identity by repressing the expression of genes related to flowering time and shoot identity. Furthermore, AP1, in association with SEP3, hinders the increase in CK in the sepal axil and promotes the expression of floral homeotic TFs to facilitate the initiation of floral organs. In rice and maize, the interplay between auxin and CK concentrations is pivotal for the initiation of BMs and SMs, while the transition from IMs to FMs involves a complex regulatory network encompassing *LFY*, *AP2*, MADS-box, and bHLH TFs. Jasmonates also contribute to the transition from SMs to FMs by activating the expression of MADS-box TFs such as *OsMADS1*. Abbreviations: BM, branch meristem; CK, cytokinin; FM, flower meristem; IM, inflorescence meristem; OsM, OsMADS; pBM, primary branch meristem; rg, rudimentary glume; sBM, secondary branch meristem; sl, sterile lemma; SM, spikelet meristem; TFs, transcription factors

In rice and maize, the transition from IMs to FMs is notably more intricate than that observed in Arabidopsis, involving a sequence of meristem identity transformations, including pBMs, sBMs, and SMs (Fig. 7.1a). Throughout this meristem transition, the fundamental roles of auxin in AM initiation for branch and spikelet formation have been elucidated (Fig. 7.2b) (McSteen, 2009, 2010; Zong et al., 2022). Employing auxin biosensor markers such as *DR5rev*-VENUS and *DII-VENUS* in rice and *DR5rev-mRFPer* and *pZmPIN1a-ZmPIN1a-YFP* in maize, researchers have demonstrated auxin maxima during spikelet initiation (Gallavotti et al., 2008a; Yang et al., 2017). Furthermore, mutations in genes associated with auxin biosynthesis and transport in maize, namely, *SPARSE INFLORESCENCE1* (*SPI1*, ortholog of *YUCCA*) and *BARREN INFLORESCENCE2* (*BIF2*, ortholog of *PINOID*), lead to barren inflorescence phenotypes, akin to pin-like inflorescence phenotypes observed in the auxin efflux carrier mutant of Arabidopsis, *pin1* (Gallavotti et al., 2008a, b). Conversely, mutants of the auxin influx transporter *ZmAux1* in maize and *SvAux1* in *Setaria viridis* exhibit disrupted AM initiation, decreased branching, and reduced spikelet numbers (Huang et al., 2017a). However, no conspicuous abnormal inflorescence development was reported in *Osaux1* and *Ataux1* mutants (Bainbridge et al., 2008; Yu et al., 2015; Zhao et al., 2015a), suggesting potential functional redundancy of AUX1 paralogs during inflorescence development in rice and Arabidopsis (McSteen, 2010). Indeed, abnormal IM phyllotaxy was solely observed in the combined mutation of *AtAUX1* and its paralogs *LAX1*/*LAX2*/*LAX3* (Bainbridge et al., 2008).

Nevertheless, the mutation of *FISH BONE*, the ortholog of TRYPTOPHAN AMINOTRANSFERASE OF ARABIDOPSIS, which encodes an enzyme involved in auxin biosynthesis, resulted in small inflorescences (Yoshikawa et al., 2014). Additionally, the RNAi-mediated expression reduction of the auxin efflux carrier-like gene *OsPIN5b* affects auxin homeostasis, transport, and distribution, leading to an increase in spikelet number and panicle length in rice (Lu et al., 2015a). Moreover, single-cell RNA sequencing (scRNA-seq) analysis in early rice inflorescence reveals the enrichment of auxin pathways in the transition from IMs to BMs. The new *Osaux1-1;4* allele exhibits fewer branches and spikelets (Zong et al., 2022), affirming the conserved role of auxin in rice inflorescence branching and flower initiation.

7.3.2 Cytokinins Promote Floral Meristem Outgrowth

The biosynthesis and signal transduction pathways of cytokinins actively contribute to the regulation of FM identity specification and outgrowth (Fig. 7.2a). In Arabidopsis, elevated concentrations of cytokinins activate the expression of the meristem marker gene *WUSCHEL* (*WUS*), facilitating AM initiation in shoot regeneration and leaf axils. The cytokinin-rich environment initially enhances *WUS* expression by removing repressive epigenetic marks. Subsequently, cytokinin-inducible B-type ARABIDOPSIS RESPONSE REGULATORs (ARRs) interact with HD ZIP III TFs, directly promoting *WUS* expression (Zhang et al., 2017b).

During FM initiation, the direct suppression of cytokinin accumulation, facilitated by AP1, acts by inhibiting the cytokinin biosynthetic gene *LONELY GUY1* (*LOG1*) and activating the cytokinin degradation gene *CYTOKININ OXIDASE/DEHYDROGENASE3* (*CKX3*) in the axils of sepals, thereby suppressing the formation of axillary secondary flowers (Han et al., 2014). This observation suggests potential evolutionary divergence in the regulatory pathways governing AM initiation during reproduction development. Currently, it remains unclear whether the miRNA165/166-HD-ZIP III TFs and the WUS pathway are independently recruited in the axils of sepals to suppress axillary secondary flower initiation. Recent research in rice, however, indicates that the miRNA165/166-HD-Zip III TFs do play a role in promoting axillary secondary flower formation by directly activating the expression of the FM identity gene *OSH1* (please also see Sect. 7.4.3) (Zhang et al., 2017a).

Nevertheless, the initiation of axillary branches and spikelets from inflorescence may necessitate a distinct regulatory pathway to modulate cytokinin activities. This includes the involvement of the cytokinin biosynthetic enzyme OsLOG1, the degradation enzyme OsCKX2, and genes in the cytokinin signaling pathway, such as A-type ARRs, *OsRR6* in rice, and *ABPH1* in maize (Zhang & Yuan, 2014). Therefore, although the cytokinin-ARR-HD-ZIP III regulatory pathway in the activation of meristem identity genes appears conserved across plant species, there are divergent pathways governing vegetative and reproductive development (Fig. 7.2). Expression profiling of 67 genes associated with cytokinin biosynthesis, degradation, and signaling suggests that subnetworks may play crucial roles in mediating dynamic cytokinin activity during rice panicle development (Yamburenko et al., 2017). Thus, the question of how plants employ distinct pathways to control cytokinin activity during AM and FM initiation remains open and merits further investigation.

7.3.3 Roles of LFY, AP2, *bHLH, and MADS-Box Transcription Factors in the Determinacy of Spikelet Meristems and the Initiation of Flower Meristems*

Unlike Arabidopsis, investigations into the roles of *LFY* orthologs in rice and maize have unveiled significant contributions of grass *LFY* in initiating lateral inflorescence primordia during early reproductive development (Wagner, 2017). In rice, RICE FLORICAULA LEAFY/ABERRANT PANICLE ORGANIZATION 2 (RFL/APO2) serves as the sole LFY ortholog, exhibiting exclusive expression in the apical meristem during the early stages of inflorescence development and subsequently in FMs and primordia of all floral organs at later stages (Kyozuka et al., 1998; Ikeda-Kawakatsu et al., 2012). The *apo2* mutant, characterized by a C-terminal-truncated *RFL*, displays reduced panicle size due to a defective precocious transition from IMs to SMs (Ikeda-Kawakatsu et al., 2012). A recent study further confirmed the precise spatial-temporal regulation of *RFL* expression in the far-upstream promoter region, highlighting the importance of potential cis elements in defining RFL's role in floral development (Lou et al., 2017).

Conversely, in maize, *Zea FLO/LFY* (*ZFL*) genes are initially expressed at the flank of IM, where spikelet pair meristems initiate. Subsequently, they exhibit spatial and temporal expression in SMs, the upper half of FMs, and all floral organ primordia. The tassel of the *zfl1 zfl2* double mutant also exhibits reduced branching (Bomblies et al., 2003). Despite *LFY* orthologs manifesting distinct functions in inflorescence development, a shared conserved role emerges during later flower development across plant species. Mutants such as *lfy*, *apo2*, *zfl1*, and *zfl2* exhibit similar defects in maintaining floral organ identity (Bomblies et al., 2003; Ikeda-Kawakatsu et al., 2012).

Using forward and reverse genetic analysis, a crucial AP2/ERF regulatory pathway has been identified for its pivotal role in facilitating the transition of FM fate in plants (Tanaka et al., 2013; Hirano et al., 2014). Investigations involving maize mutants, such as *ids1*, a *sister of indeterminate spikelet1* (*sid1*), *branched silkless* (*bd1*), *tasselseed4* (*ts4*), and *Tasselseed6* (*Ts6*), have indicated that *ids1/Ts6, sid1,* and *BD1* are affiliated with AP2/ERF TFs (Chuck et al., 1998, 2002, 2007, 2008). The expression of these genes is intricately regulated by *ts4*, a member of miR172 (Chuck et al., 2007). The dominant *Ts6* mutant, resembling *ts4*, exhibits a similar phenotype due to the repression release from miR172, leading to a widespread expression pattern of the IDS1 protein in SMs (Chuck et al., 2007). The characteristics of *ids1* spikelets, being indeterminate and producing additional florets (Chuck et al., 1998), and the production of repetitive glumes (bracts) in the *ids1 sid1* double mutant (Chuck et al., 2008), signify that *ids1* and *sid1* regulate SM determinacy and facilitate FM initiation. In rice, mutations of AP2/ERF TFs *SNB*, *OsIDS1*, *FRIZZY PANICLE* (*FZP*), and *MULTI-FLORET SPIKELET1* (*MFS1*) exhibit spikelet phenotypes similar to those of *ids1*. Additionally, the expression of *SNB* and *OsIDS1* is repressed by *OsmiR172d* (Zhu et al., 2009; Lee & An, 2012; Ren et al., 2013; Lee et al., 2014). In Arabidopsis, AP2/ERF TFs function redundantly during the transition from IMs to FMs. The study on *PUCHI*, an AP2/ERF ortholog, reveals its expression on the adaxial side at stage 1, interacting with BLADE-ON-PETIOLE1 (BOP1) and BOP2 in specifying FM identity by activating *LFY* and *AP1* (Hepworth et al., 2005; Karim et al., 2009; Xu et al., 2010). Flowers from the *bop1 bop2 puchi* triple mutants exhibit similar phenotypes, with partial FMs to IMs inversion, as seen in *lfy* mutants (Xu et al., 2010). Furthermore, when two other AP2/ERF orthologs, *DORNRÖSCHEN* (*DRN*) and *DORNRÖSCHEN-LIKE* (*DRNL*), are mutated and combined with *bop1 bop2 puchi* triple mutant, mutant FMs were completely converted into secondary inflorescence shoots (Chandler & Werr, 2017). Genetic analysis further demonstrates that these three closely related AP2/ERF TFs genetically interact with *BOP1/2* to regulate FM identity by activating *LFY* expression (Chandler & Werr, 2017). Recent work in rice also reveals that the AP2 homologous *FZP* represses *RFL/APO2* expression (Bai et al., 2016). A copy number variants-18bp (CNVs-18bp) in the promoter of *FZP* represses its expression levels, causing prolonged IM development and an increase in spikelet number. Expression levels of

FZP are tightly regulated by the transcription repressor OsBZR1 through the promoter sequence of CNVs-18bp (Bai et al., 2017). Similar results regarding the *AP2* TF *MORE SPIKELETS1* (*MOS1*) were obtained in Brachypodium (*Brachypodium distachyon*), showing that reduced expression of *MOS1* delayed the termination timing of IM development and produced more spikelets (Derbyshire & Byrne, 2013). Furthermore, studies in wheat and barley indicate that regulatory effectiveness between miR172 and AP2 TFs affects the size and density of spikes, heading time, and lodicule development, an important domestication site in wheat and barley breeding (Nair et al., 2010; Houston et al., 2013; Debernardi et al., 2017). Thus, plants might recruit diverse, balanced mechanisms independently governing AM and SM initiation, FM transition, and floral organ identity through controlling the activity of *AP2* genes. Genetic data in Arabidopsis demonstrate that the AP2 pathway acts upstream of *LFY* and auxin in FM initiation (Chandler & Werr, 2017), and the expression of *DRN, DRNL, and PUCHI* is also regulated by auxin at the transcriptional level (Hirota et al., 2007; Comelli et al., 2016). However, the molecular basis of *AP2* genes in promoting grass AM and SM initiation is still unclear (Fig. 7.2).

Of significance, a recently proposed model, known as the localized signaling center model, serves to elucidate the developmental mechanisms underlying inflorescence diversity. According to this model, inflorescence architecture genes exhibit differential expression in domains adjacent to AMs, exemplified by maize/rice *BA1*/*LAX1*, *BD1*/*FZP*, and *ids1*/*SNB*, offering positional information crucial for coordinating the determinacy of neighboring meristems (Whipple, 2017). Supporting this hypothesis, a functional analysis of maize *BIF1* and *BIF4* demonstrated that the auxin signaling pathway directly regulates the expression of *BARREN STALK1* (*BA1*), a bHLH TF, in the domain adjacent to AM (Galli et al., 2015). The disruption of AM outgrowth was similarly observed in *lax panicle 1* (*lax1*), a mutation affecting the *BA1* ortholog in rice (Oikawa & Kyozuka, 2009). These observations suggest that the involvement of bHLH TFs and the functional diversification of *LFY* in grass SM determinacy and AM initiation may be evolutionarily conserved and occur subsequent to the divergence of grass species. Therefore, the determinacy of SMs emerges as a critical factor in facilitating the transition from SMs to FMs in grasses.

The MADS-box TFs also play a role in SM determinacy and FM fate transition. In rice, the loss of function of *LEAFY HULL STERILE1* (*LHS1*)/*OsMADS1*, an E class gene, resulted in the production of extra lemma/palea organs or additional florets in spikelets, alongside abnormal lemma and palea identity (Prasad et al., 2001). Ectopic expression of *OsMADS1* led to a reduction in inflorescent height and the homeotic transformation of sterile lemmas into structures resembling lemma/palea organs (Prasad et al., 2001). Similarly, the loss of function of *OsMADS34*/*PANICLE PHYTOMER2* (*PAP2*), another E class gene, affected sterile lemma development and the formation of IMs (Gao et al., 2010; Kobayashi et al., 2010). Additionally, both *OsMADS1* and *OsMADS34* genetically interact with *AP1*/*FUL*-like and LOFSEP MADS box TFs to specify IM identity. This is evident from the severe perturbation in inflorescence development observed in *Osmads1*

Osmads15 and *Osmads5 Osmads34* double mutants and *Osmads34 Osmads14 Osmads15 Osmads18* quadruple knockdown mutants (Wang et al., 2010; Kobayashi et al., 2012; Zhu et al., 2022). Therefore, rice E class genes play diverse roles in regulating AM, SM, and FM activity throughout evolution. In tomato, *SEP4* family genes govern FM maturation and influence inflorescence branching in a dosage-dependent manner. This characteristic could be advantageous for tomato breeding endeavors aimed at enhancing fruit yield (Soyk et al., 2017). Investigating the specific functions of rice E genes during the reproductive stage could enhance the efficiency of high-yield breeding.

Genes belonging to the STMADS11-like clade have been identified as regulators in SM determinacy. In classic pod corn (*Zea mays* var. *tunicate*), a chromosomal rearrangement in the promoter of *ZMM19*, located at the *Tu1* locus, leads to the interruption of *ZMM19* and results in its ectopic expression in the inflorescence. This alteration causes distinct phenotypes, including elongated outer glumes covering the ear, lower florets in the ears, and more (Han et al., 2012; Wingen et al., 2012). The functional conservation of ZMM19 in SM determinacy has been confirmed in rice and barley. Overexpression of *OsMADS22* (the *ZMM19* ortholog in rice) resulted in spikelets with elongated glumes and extra florets (Sentoku et al., 2005), and the FM identity was repressed when the barley *ZMM19* ortholog *MADS1* (*BM1*) was ectopically expressed (Trevaskis et al., 2007). Thus, an intricate regulatory network involving AP2/ERF, bHLH, and MADS box TFs governs grass SM determinacy, necessitating further clarification of their relationships.

7.3.4 Jasmonates Facilitate the Developmental Transition from Spikelet Meristems to Floral Meristems

Jasmonates (JAs) are lipid-derived phytohormones playing diverse roles in stress response and development (Wasternack, 2007, 2014). While plants share conserved JA biosynthesis and signaling pathways, the roles of JA in plant reproductive development are diverse (Yuan & Zhang, 2015). In Arabidopsis, mutations in JA biosynthesis and signaling pathways affected flowering time, anthesis, and gynoecium abscission (Yuan & Zhang, 2015; Zhai et al., 2017). In maize, the JA concentration plays important roles in the sex determination of tassel florets (Acosta et al., 2009; Yan et al., 2012). In rice, the EXTRA GLUME 1 (EG1) belongs to class I of the phospholipase A1 family, homologous to Arabidopsis JA biosynthetic enzyme DEFECTIVE IN ANTHER DEHISCENCE1 (DAD1), involved in the release of a-linolenic acid (C18:3) from chloroplastic membrane lipids as metabolite precursors of JA biosynthesis (Ishiguro et al., 2001; Li et al., 2009). *eg1* mutants produced extra lemma and floret-like organs in spikelets (Li et al., 2009; Cai et al., 2014), indicating that JA is also required for rice SM determinacy. Indeed, the reduction in JA by the mutation of OsPEX5 and OsOPR7, two genes involved in JA biosynthesis, generated abnormal spikelets similar to *eg2* (You et al., 2019). The involvement of JA in FM fate transition has been further documented by the identification of the JA signaling pathway repressor, OsJAZ1/EG2, which interacted with the JA receptor OsCOI1b and responsive TF OsMYC2, a basic helix-loop-helix (bHLH) TF. Molecular experiments showed that at high JA content, OsJAZ1 was degraded through the 26S proteasome pathway, the repression of OsMYC2 was released, and the expression of E class genes such as *OsMADS1*, *OsMADS7,* and *OsMADS8* was activated to promote the transition from SMs to FMs (Cai et al., 2014). Dominant *eg2-1D* mutants with a single A to G substitution identified in the Jas domain of OsJAZ1 blocked the osjaz1-1D-OsCOI1b interaction, showing similar spikelet phenotypes to those in *eg1-1* mutants (Cai et al., 2014). Although OsMYC2 directly binds to the G-box motif (5′-CACGTG/CACATG-3′) in the promoter of *OsMADS1* (Cai et al., 2014; Uji et al., 2016), it is still unknown whether OsMYC2 or other JA responsive TFs are involved in the JA signaling pathway due to the absence of genetic mutants. In the *eg1-1 eg2*, *eg1 Osmads1-z,* and *Osmads1-z* mutants containing *OsJAZ1pro:Osjaz1-1D gDNA*, the genomic fragment amplified from *eg2-1D* mutants and spikelets grew reiterative glume-like organs, indicating that SM determinacy is compromised in these double mutants (Cai et al., 2014). Therefore, *OsMADS1* is a direct downstream gene of the JA signaling pathway in promoting the transition from SMs to FMs. However, the delayed transition in these double mutants was further exaggerated compared to *eg1-1*, *eg2-1D,* or *Osmads1-z* single mutants (Cai et al., 2014), suggesting that other unknown TFs might also be involved in the JA signaling pathway. On the other hand, over-expression of *mJAZ3*, *mJAZ6*, *mJAZ7,* and *mJAZ11*, carrying a mutated Jas domain, mimicked phenotypes from *eg2-1D* and generated similar phenotypes in the spikelets of *JAZ1pro:osjaz1-1D gDNA* transgenic lines, meaning members of the OsJAZ family other than OsJAZ1 are also involved in the JA signaling (Hori et al., 2014). A recent study indicated that OsJAZ6 exhibits increased sensitivity to JA signaling specifically during spikelet initiation (Cao et al., 2021). Therefore, a JA-MADS regulatory pathway could be employed in governing rice SM determinacy and FM fate transition (Fig. 7.2). The investigation of *NUMBER OF GRAINS 1* (*NOG1*), responsible for encoding enoyl-CoA hydratase/isomerase (ECH), a crucial enzyme in the fatty acid β-oxidation pathway, unveiled that a natural variation in the promoter of *NOG1* leads to a reduction in JA concentration and an increase in spikelet numbers (Huo et al., 2017). This implies the existence of an intricate regulatory network that likely impacts the spatial-temporal distribution of JA during inflorescence development, which could be further observed in vivo using the recently developed JA biosensor system, J6V-HM (Li et al., 2021b). Nevertheless, the questions persist regarding the universality of this pathway in other grass families.

7.4 Spikelet Development

A grass spikelet comprises one to 40 florets, depending on species (Bommert et al., 2005a). In rice spikelets, only one floret is enclosed by two whorls of glumes. One hypothesis posits that ancestral *Oryzeae* spikelet bore three florets, but only the terminal one persists, with the two lateral florets degenerating over evolution (Arber, 1934; Yoshida et al., 2009). Accordingly, the sterile lemmas are suggested to function as lemma-like organs analogous to floret bracts (please also see Sect. 7.5.2), and rudimentary glumes are morphologically categorized as spikelet bracts (Fig. 7.3).

7.4.1 Glume/Bract Suppression

In angiosperms, bracts are considered suppressed leaves that appear in the inflorescence or subtend flowers. While the reason for bract suppression or reduction remains unclear, one hypothetical explanation posits a selective advantage for pollination or growth competition leading to meristem size increase or proliferation after the floral transition (Whipple et al., 2010; Chandler, 2012). In Arabidopsis,

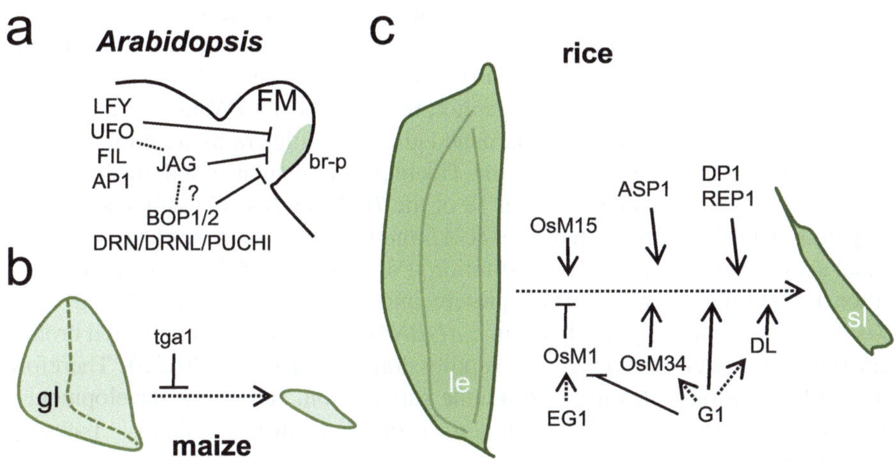

Fig. 7.3 Regulators in the suppression of Arabidopsis bract (**a**), maize glume (**b**), and rice sterile lemma (**c**). In Arabidopsis, the suppression of bract primordium in FM involves several TFs, including *LFY*, *UFO*, *FIL*, *AP1*, *JAG*, *BOP1/2*, and *DRN/DRNL/PUCHI*. However, the intricacies of the regulatory networks underlying this process remain a challenge to be addressed. In maize, TGA1 has been identified as a key regulator in the suppression of hardened glumes during the evolutionary transition from teosinte to maize. In rice, the development of sterile lemmas, considered a form of suppressed lemma, is positively regulated by *G1*, *OsMADS15*, *OsMADS34*, *DL*, *ASP1*, *DP1*, and *REP1*, while it is repressed by *OsMADS1* TFs. Abbreviations: br-p, bract primordium; gl, glum; FM, flower meristem; le, lemma; OsM, OsMADS; sl, sterile lemma

wild-type flowers lack visible bracts, and cryptic bracts are identified by a group of cells on the abaxial side of FMs (Sawa et al., 1999). Genes such as *LFY*, *UNUSUAL FLORAL ORGANS* (*UFO*), *FILAMENTOUS FLOWER* (*FIL*), *AP1*, *JAGGED* (*JAG*), *BOP1/2*, and *DRN/DRNL/PUCHI* have been found to regulate bract suppression, as evidenced by visible bracts and compromised flower development in corresponding mutants (Weigel et al., 1992; Sawa et al., 1999; Dinneny et al., 2004; Ohno et al., 2004; Hepworth et al., 2005; Norberg et al., 2005; Hepworth et al., 2006; Karim et al., 2009; Chandler & Werr, 2017). The question of why and how these genes suppress bract growth remains unaddressed. Genetic and molecular evidence indicates that *JAG* is necessary and sufficient for bract growth (Dinneny et al., 2004; Ohno et al., 2004), while *LFY* plays fundamental roles in suppressing bract growth and establishing FM identity (Chandler & Werr, 2017). Therefore, bract suppression is closely linked to FM identity in Arabidopsis (Chandler, 2012).

In the spikelet, florets are surrounded by two bract-like structures known as glumes. Unlike Arabidopsis, the gene homologs mentioned above do not appear to be involved in grass glume suppression, as normal glumes are consistently observed in corresponding mutants. Investigations into GATA zinc-finger proteins, TASSEL SHEATH1 (TSH1) in maize, NECK LEAF1 (NL1) in rice, THIRD OUTER GLUME (TRD) in barley, and HANABA TARANU (HAN) in Arabidopsis, have revealed the existence of a distinct Tsh1/NL1/TRD pathway for inflorescence bract suppression in the Poaceae (Whipple et al., 2010). However, glumes subtending spikelet pairs were still suppressed in *tsh1* mutants, indicating that an additional suppression pathway might participate in later grass flower development. This has been validated by detailed studies on the *teosinte glume architecture* (*tga1*) gene, which encodes an SBP TF and was originally identified as a domestication gene during maize evolution from its progenitor, teosinte species (Wang et al., 2005). In teosinte, the kernel is enclosed by hardened glumes, while grains are borne naked in the ear of maize. Research from the Doebley lab demonstrated that a single-nucleotide polymorphism at position 18 of the *TGA1* cDNA (G to C) among seven polymorphisms between the teosinte-*TGA1* and maize-*TGA1* genes leads to a functional alteration (Lys6-to-Asn6 substitution) (Wang et al., 2005). This single substitution transforms the maize-TGA protein into an active repressor (Wang et al., 2015b). Pleiotropic ear phenotypes in *tga1-RNAi* transgenic lines indicated broader functions of TGA1 beyond glume suppression (Wang et al., 2015b). Therefore, *TGA1* plays various roles in maize glume suppression, and glume development in rice and other grass species might evolve distinct regulatory pathways independently.

7.4.2 Sterile Lemma Development

Sterile lemmas grow between rudimentary glumes and the floret, a unique feature in the *Oryza* genus but not in maize and other grasses (Hirano et al., 2014). Several sterile lemma identity genes have been cloned to date, including *LONG STERILE*

LEMMA/ELONGATED EMPTY GLUME (*G1/ELE*) (Yoshida et al., 2009; Hong et al., 2010; Liu et al., 2016b), *OsMADS34/PAP2* (Gao et al., 2010; Kobayashi et al., 2010; Meng et al., 2017), *EG1* (Li et al., 2009), *ABERRANT SPIKELET AND PANICLE1* (*ASP1*) (Yoshida et al., 2012), *DEPRESSED PALEA1/PALEALESS1* (*DP1/PAL1*) (Jin et al., 2011), *NONSTOP GLUMES* (*NSG*) (Wang et al., 2013), and *FZP* (Ren et al., 2018). *G1* encodes a land plant-specific ALOG (Arabidopsis *LSH1* and *Oryza G1*) domain protein expressed in the inflorescence, sterile lemmas, and a portion of palea (Yoshida et al., 2009; Hong et al., 2010). Loss of function in *G1/ELE* results in spikelets with two larger/elongated sterile lemmas, resembling those of the wild rice species *Oryza. grandiglumis*, In *O. grandiglumis*, the *G1* locus exhibits amino acid substitution in the ALOG domain or nonsense mutations (Yoshida et al., 2009), suggesting a close association between the *G1* locus and sterile lemma evolution. Ectopic expression of the lemma identity gene *DROPPING LEAF* (*DL*) in *g1* sterile lemmas supports the notion that sterile lemmas are modified lemmas, with *G1* being a key regulator of sterile lemma development (Liu et al., 2016b). Down-regulation of another empty glume gene, *OsMADS34/PAP2,* in the *g1-6* allele indicates that *OsMADS34* acts downstream of *G1* to repress sterile lemma development (Liu et al., 2016b). However, sterile lemmas in *Osmads34-1 ele* double mutants developed even longer and wider than *Osmads34-1* and *ele* single mutants, suggesting that other genes might regulate sterile lemma development in parallel with the G1-OsMADS34 pathway. Consistently, the papery or leaf-like sterile lemmas of *Osmads34-1 Osmads15* double mutants appeared (Meng et al., 2017), indicating that *OsMADS15* is also a sterile lemma identity gene. In contrast to *OsMADS15* and *OsMADS34*, ectopic expression of the lemma and palea identity gene *OsMADS1* converts sterile lemmas into lemma-like structures (Wang et al., 2016), hinting that *OsMADS1* plays a pivotal role in establishing lemma identity in florets by repressing sterile lemma identity genes. Although *G1/ELE* was previously shown to regulate sterile lemma and lemma development independently of *OsMADS1* (Liu et al., 2016b), recent work demonstrates that G1 physically interacts with OsMADS1 and directly represses its expression in sterile lemma (Qin et al., 2025). These conflicting conclusions may reflect earlier technical limitations, underscoring the importance of mapping spatiotemporal expression patterns to define precise roles in floral organ specification. The roles of *EG1* and *ASP1* in sterile lemma development are still unclear, but they are presumed to control the transition of SMs to FMs (Yoshida et al., 2012; Cai et al., 2014). Therefore, abnormal sterile lemma development in *eg1* and *asp1* mutants probably resulted from compromised meristem transition. The palea identity genes, *DP1/PAL1* and *RETARDED PALEA1/SMALL DEGENERATIVE PALEA1* (*REP1/SDP1*), also synergistically regulate the asymmetrical development of sterile lemmas, as *dp1-2 sdp1* double mutants develop lemma-like sterile lemmas on the palea side (Jin et al., 2011). Moreover, the overexpression of *FZP* results in the transformation of sterile lemma into lemma (Ren et al., 2018). Therefore, the identity of sterile lemmas can be deduced as modified lemmas (Fig. 7.3), but the underlying regulatory network for these sterile lemma genes needs further elucidated.

7.4.3 *Hypothesis of Three-Florets-Spikelet Model in Grass Spikelet Formation*

Previous investigations into sterile lemma identity genes have led to the hypothesis that rice spikelets likely consist of three florets, with sterile lemmas being potential modified lemma structures (Yoshida et al., 2009). Direct support for this hypothesis comes from dominant *lf1* mutants, characterized by a single-nucleotide substitution (C to T) in the *miRNA165/166* binding site. These mutants exhibit a normal terminal floret along with rudimentary glumes and one or two lateral florets in the axils of sterile lemmas. The reciprocal expression patterns of *miRNA165/166* and *LF1*, encoding a class III homeodomain-leucine zipper (HD-ZIP III) protein, in the wild-type spikelet further underscore this hypothesis. Ectopic expression of the mutated *LF1* gene reproduces *lf1* phenotypes, confirming that misregulation of *LF1* in the axil of sterile lemmas triggers lateral floret development. Additionally, the meristem identity gene *OSH1* has been identified as a direct target of LF1. Notably, LF1 negatively regulates the MADS box TFs *OsMADS6* and *OsMADS15*, which play a pivotal role in the initiation of lateral FM in *lf1* mutants (Zhang et al., 2017a). This aligns with well-established studies on the *miRNA165/166*-HD-Zip III TF regulatory pathway in Arabidopsis and maize, focusing on leaf polarity and vascular development (please all see Sect. 7.3) (D'Ario et al., 2017). This finding further substantiates the proposal that sterile lemmas exhibit a closer structural resemblance to a leaf/bract configuration. Considering the relatively conserved regulatory network of HD-Zip III/cytokinin-WUS during shoot regeneration and AM initiation (Fig. 7.4) (Wang et al., 2017; Zhang et al., 2017b), it would be intriguing to explore potential interactions between the miRNA165/166-HD-Zip III pathway and the cytokinin pathway in repressing lateral FM initiation in Arabidopsis.

It is intriguing to observe that sterile lemmas in *lf1* mutants do not exhibit growth patterns similar to the lemma, suggesting an independence in sterile lemma development from subsequent FM initiation. A plausible explanation is that sterile lemma genes may be expressed normally in *lf1* mutants (Zhang et al., 2017a). Consequently, conducting genetic studies on *LF1* and sterile lemma genes or implementing the ectopic expression of *OsMADS1* in *lf1* mutants could contribute to a profound understanding of the sterile lemma identity. Moreover, insights from the miRNA165/166-LF1-OSH1 pathway suggest that lateral FMs may represent a de novo meristem rather than a detached one, with *OSH1* repression being dependent on the miRNA156/166-LF1 regulatory network. Comparable regulatory pathways are implicated in suppressing Arabidopsis AM initiation (Wang et al., 2017). The expression of AM identity gene *SHOOTMERISTEMLESS* (*STM*), an *OSH1* ortholog, is induced de novo during AM initiation (Wang et al., 2017).

In summary, the hypothesis proposing that ancient rice spikelet comprises three florets (Fig. 7.4), one apical floret, and two lateral florets enclosed by the rudimentary glumes (spikelet bracts) gains support from recent genetic and molecular data. The developmental suppression of two lateral florets by the miRNA165/166-HD-ZIP III pathway, a relatively conserved regulatory pathway governing leaf

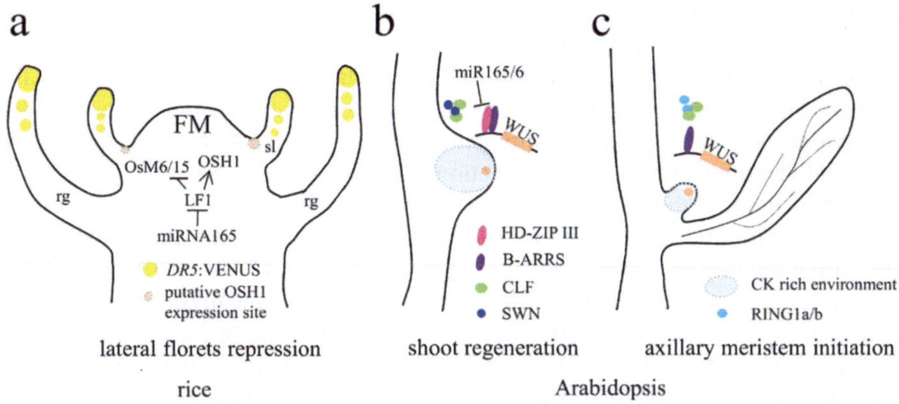

Fig. 7.4 A putative conserved miR165/6-HD ZIPIII regulatory network operates in rice to repress lateral floret development (**a**), in Arabidopsis during shoot regeneration (**b**), and in axillary meristem initiation (**c**). Within the sterile lemma axis, the miRNA165/166-HD ZIPIII (LF1) pathway represses the expression of the meristem identity gene, *OSH1*. This pathway is a conserved regulatory mechanism involved in the separation of leaf primordia from SAM, AM initiation, and shoot regeneration in Arabidopsis. Notably, recent discoveries indicate that a CK-rich environment de novo activates *WUS* expression, a meristem identity gene, through the HD-Zip III/cytokinin-WUS regulatory loop. Abbreviations: CK, cytokinin; FM, flower meristem; OsM, OsMADS; rg, rudimentary glume; sl, sterile lemma

primordia separation from SAM, leaf polarity establishment, and shoot regeneration (Wang et al., 2017; Zhang et al., 2017a, b), offers an explanation for the evolutionarily restrained development of these florets. Simultaneously, the G1-related pathway suppresses the development of two lateral lemmas (floret bracts with partial leaf identity) into sterile lemma structures (Yoshida et al., 2009). Therefore, the concept of lemma identity as a floret bract-like structure (refer to Sect. 7.5.2) necessitates further empirical examination. Additional open questions include: What distinguishes apical from lateral FMs? How is cytokinin buffered during inflorescence development? And how can cytokinin activity be precisely tuned to increase spikelet numbers in breeding? Although elevating floret number per spikelet has been proposed as a promising strategy for yield improvement (Ren et al., 2020; Yuan et al., 2020), the underlying regulatory mechanisms remain to be fully elucidated.

7.5 Specification and Development of Floral Organs

7.5.1 Flower Development in Arabidopsis

Following the establishment of FM identity, cells referred to as "founder" or "initial" cells undergo differentiation to form various floral organs, thereby defining floral organ identity (Chandler, 2011, 2012). In Arabidopsis, the classic ABC model emerged from genetic analysis of floral homeotic mutants, including *apetala1* (*ap1*), *apetala2* (*ap2*), *apetala3* (*ap3*), *pistillata* (*pi*), *agamous* (*ag*), and their combined double- and high-order mutants (Meyerowitz et al., 1989; Bowman et al., 1991b). This model posited that unique organ identity in each whorl is specified by combinations of floral organ identity genes: A class genes (*AP1* and *AP2*) specify the sepal identity in whorl 1, a combination of A class genes and B class genes (*AP3* and *PI*) determine the petal identity in whorl 2, B class and C class (*AG*) genes together control the stamen identity in whorl 3, and C class gene alone determines the carpel identity in whorl 4. Furthermore, A class and C class activities exhibit mutual repression; thus, C class activity persists throughout the flower in the absence of A activity, and vice versa (Bowman et al., 1991b; Coen & Meyerowitz, 1991; Rodríguez-Cazorla et al., 2020). The ABC model is expanded to the ABCD model, incorporating ovule identity genes from petunia and Arabidopsis (Angenent & Colombo, 1996). Further extension to the ABCDE model occurs with the identification of "class E genes" (*SEP* genes) in Arabidopsis, specifying that A+E class genes direct sepal development, A+B+E specify the petal identity, B+C+E direct stamen development, C+E determine carpel development, and C+D+E specify ovule development (Theissen, 2001).

All ABCDE genes, except the A function gene *AP2*, belong to the MADS-box TF family and contribute to various aspects of flower development, while AP2 is involved in sepal identity and FM specification (Wollmann et al., 2010; Prunet & Jack, 2014). The formation of multimeric complexes by DEFICIENS (DEF, AP3 ortholog), GLOBOSA (GLO, PI ortholog), and SQUAMOSA (SQUA, AP1 ortholog) in *A. majus* (Egea-Cortines et al., 1999), the identification of E (*SEP*) class genes (Pelaz et al., 2000), and experiments demonstrating that combined overexpression of B and E or B and A proteins can transform leaves into floral organs (Honma & Goto, 2001; Pelaz et al., 2001) collectively led to the proposal of the "floral quartet model" (FQM). This molecular model, the foundation of the ABCDE model, posits that MADS proteins assemble into larger complexes in a combinatorial fashion, rather than individual dimers, to control each type of floral organ identity (Theissen, 2001; Theissen & Saedler, 2001). Subsequent in vitro, in vivo, in planta, and in silico studies have provided substantial support for the crucial role of the ABCDE model or derived (A)B(C) model in regulating flower development across different species (Theissen & Melzer, 2007; Melzer & Theissen, 2009; Melzer et al., 2009; Smaczniak et al., 2012; Mendes et al., 2013; Jetha et al., 2014; Theissen et al., 2016). Comprehensive reviews have effectively summarized the historical evolution, modifications, and extensions of the original ABC model in several recent publications (Causier et al., 2010; Bowman et al., 2012; O'Maoileidigh et al., 2014; Prunet & Jack, 2014; Wellmer et al., 2014a, b; Sablowski, 2015; Stewart et al., 2016; Theissen et al., 2016; Yan et al., 2016; Thomson et al., 2017; Chongloi et al., 2019; Thomson & Wellmer, 2019; Chahtane et al., 2023; Dreni et al., 2023).

7.5.2 Flower Development in Rice and Maize

Extensive forward and reverse genetic analyses have provided compelling evidence supporting the applicability of the modified (A)B(C) model in rice and maize flower development. Additionally, specific TFs play crucial roles in ensuring distinct floral organ specification in monocots (Goto et al., 2001; Bommert et al., 2005a; Itoh et al., 2005; Whipple & Schmidt, 2006; Ciaffi et al., 2011; Yoshida & Nagato, 2011; Tanaka et al., 2013; Zhang et al., 2013; Hirano et al., 2014; Li & Liu, 2017). Notably, the elucidation of spikelet-specific organ development, focusing particularly on lemma and palea, has been primarily undertaken in rice (Lombardo & Yoshida, 2015).

7.5.2.1 Lemma and Palea Identity

As detailed in Sect. 7.4.2, current genetic and molecular investigations suggest a similarity between the rice lemma and sterile lemmas, portraying an identity akin to that of a floret bract-like organ. Consistently, mutations in genes associated with the transition from SMs to FMs frequently result in lemma structures exhibiting leafy characteristics (refer to Sect. 7.3). Conversely, mutations affecting genes related to FMs and floral organ identity manifest as phenotypes featuring defective palea but a normal lemma (refer to Sects. 7.5.2.2 and 7.5.2.3). In addition to the molecular evidence presented in Sect. 7.4.2, the specific expression of the *YABBY* family gene *DROOPING LEAF* (*DL*) in the midrib of leaves, at the base of the lemma and within carpel anlagen (Yamaguchi et al., 2004) provides further molecular insight, reinforcing the notion that lemma is likely a modified leaf/bract. This perspective aligns with our assertion that a complete rice flower comprises the palea, lodicule, stamen, and pistil. Our standpoint gains further validation through a recent analysis of rice flower developmental trajectories at the single-cell level (Zong et al., 2022).

In rice, *LHS1*/*OsMADS1* has undergone extensive study and has been classified as an SM and lemma/palea identity gene. *LHS1*/*OsMADS1* is expressed in the SMs, lemma, palea, and weakly in the pistil of rice flower (Prasad et al., 2001). In maize, *zmm8* and *zmm14*, homologs of *LHS1*/*OsMADS1*, are also expressed in SMs (Cacharron et al., 1999). Further expression analysis of *LHS1*/*OsMADS1* homologs in phylogenetically diverse grasses revealed its general role in specifying SMs and determining lemma and palea identity (Malcomber & Kellogg, 2004). Additionally, *LHS1*/*OsMADS1* participates in regulating floral organ development and floral determinacy (see Sect. 7.6). In lemma development, *LHS1*/*OsMADS1* promotes the differentiation of lemma-type cells, such as the silicified epidermal and sclerenchymal cells (Prasad et al., 2005; Wang et al., 2016), although the detailed mechanism remains unknown. Moreover, despite the similar cellular morphology of lemma and palea (Prasad et al., 2005; Yuan et al., 2009), ectopic expression of *LHS1*/*OsMADS1* is unable to convert the palea into a lemma-like structure (Prasad et al., 2001, 2005; Wang et al., 2016), suggesting the existence of specific regulators for palea morphogenesis.

The palea bears morphological resemblance to the lemma, yet distinctive features set it apart. Unlike the lemma, the palea exhibits two discernible cellular patterns and is smaller in size. Specifically, the palea possesses a marginal region of palea (MRP) with a smooth and unsilicified epidermis. Scanning electronic microscope (SEM) observations revealed that these cells are longer and larger than those in the body of the palea (BOP) and lack trichome-like bulges. Moreover, the palea is characterized by three vascular bundles, contrasting with five observed in the lemma (Prasad et al., 2005; Yuan et al., 2009). The reasons for these conspicuous differences remain unclear; however, several TFs have been identified with specific expression in the palea. *REP1/SDP1*, encoding a TCP family TF, exhibits special expression in the palea primordia, stamen, and vascular bundles in mature flowers (Yuan et al., 2009). Loss of *REP1* function results in the development of a partially lemma-like organ in the palea position, evident from the observation of five vascular bundles in the *rep1* mutant (Fig. 7.5a). In addition to its role in palea identity, *REP1* also regulates epidermal cell differentiation, as demonstrated by the silicified epidermis development on one side of MRP cells in transgenic lines with ectopic expression of *REP1/SDP1* (Yuan et al., 2009). The role of *REP1/SDP1* in palea identity and morphogenesis represents a conserved function, akin to the functions previously identified in *CYCLODEA* (*CYC*) and *DICHOTOMA* (*DICH*) in controlling floral asymmetry, as *rep1* flowers also exhibit partial loss of floral zygomorphy (see Sect. 7.9) (Yuan et al., 2009).

In addition, two MADS box TFs, *MOSAIC FLORAL ORGANS1* (*MFO1*)/*OsMADS6* and *CHIMERIC FLORAL ORGANS1/OsMADS32* (*CFO1/OsMADS32*), exhibit specific expression in the MRP. *OsMADS6*, belonging to *AGL6*-like MADS box genes, plays a role in positively regulating MRP development, as evidenced by the conversion of the palea identity into a lemma-like structure observed in *Osmads6* mutants (Fig. 7.5a) (Ohmori et al., 2009; Li et al., 2010; Li et al., 2011b). This suggests that *OsMADS6* is a key factor in promoting MRP development. Notably, this role may have evolved specifically after the divergence of *Streptochaeta* from the spikelet-bearing grasses, as indicated by the evolutionary expression pattern analysis of *AGL6*-like genes in grasses (Reinheimer & Kellogg, 2009). Similarly, *OsMADS32* also displays distinct expression in MRP, and *Osmads32* mutants exhibit a partial transformation of MRP into a lemma-like structure (Fig. 7.5a) (Sang et al., 2012; Wang et al., 2015c). A recent study demonstrates that mutual inhibition between the OsMADS6-OsMADS32 complex and REP1 regulates cell cycle progression and palea morphogenesis (Zhang et al., 2024). In the BOP, REP1 promotes the expression of S-CDK *OsCDKB1;1* while repressing *OsMADS6*, enabling M-CDK *OsCDKB2;1* expression. Elevated *OsCDKB1;1* and *OsCDKB2;1* levels facilitate normal cell proliferation and epidermal differentiation, forming a silicified enlarged BOP. In the MRP, OsMADS6 and OsMADS32 inhibit *REP1* and M-CDK *OsCDKB2;1* expression, reducing *OsCDKB1;1* and *OsCDKB2;1* levels to decrease cell division rate and promote epidermal cell elongation through endoreduplication. In the BOP of *rep1-1* mutants, *OsCDKB1;1* is downregulated while *OsMADS6* is upregulated, leading to *OsCDKB2;1* downregulation and subsequent epidermal cell endoreduplication, resembling MRP cell patterns. Conversely,

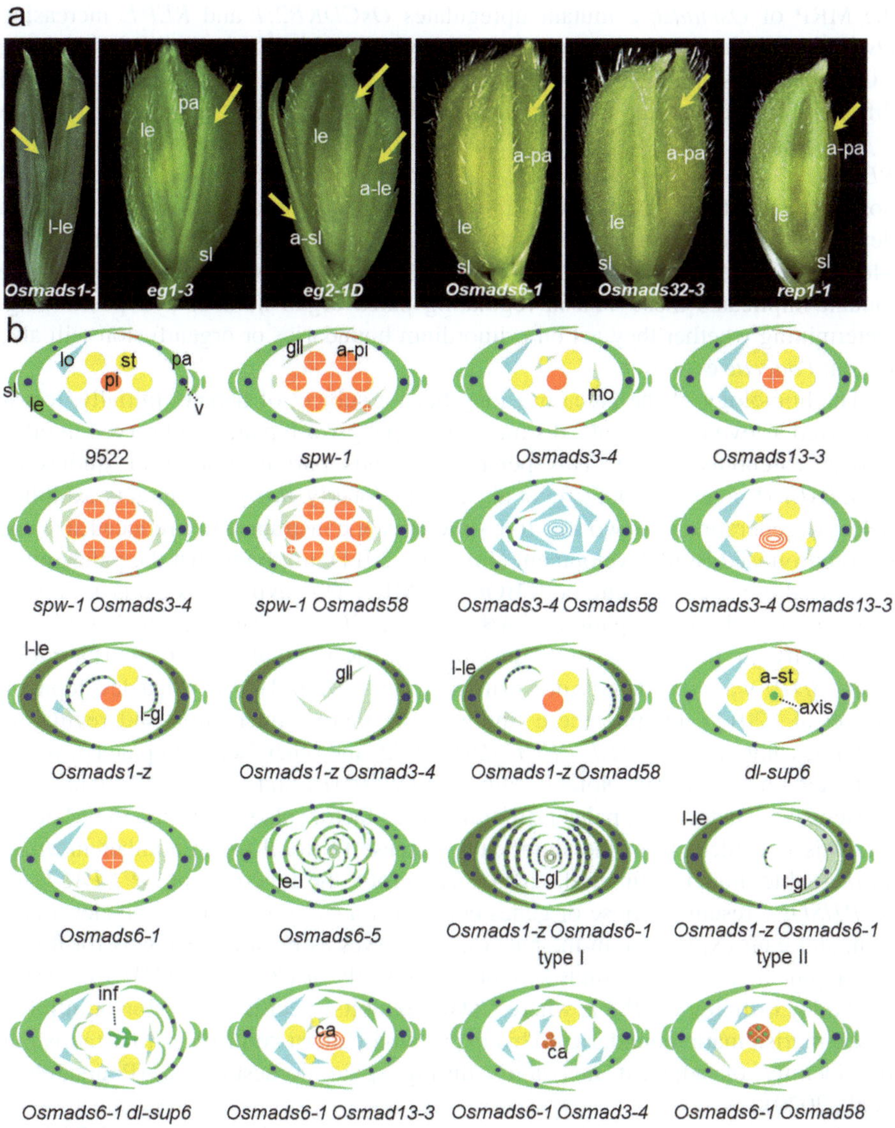

Fig. 7.5 Spikelet phenotypes observed in mutants *Osmads1-z*, *eg1-3*, *eg2-1D*, *Osmads6-1*, *Osmads32-3,* and *rep1-1* are depicted in panel (**a**), accompanied by schematic representations illustrating the spikelet structures in various mutants of floral homeotic TFs in panel (**b**). Note the conserved roles of B and C class genes in regulating reproductive development are evident, with both B and C function genes undergoing subfunctionalized during evolution. Additionally, distinct TFs such as *DL*, *OsMADS6*, *OsMADS32,* and *REP1* have been independently recruited to specify rice floral organ identity and morphogenesis, as well as FM activity and determinacy. Abbreviations: a-le, abnormal lemma like organ; a-pa, abnormal palea; a-pi, abnormal pistil; a-sl, abnormal steril lemma; a-st, abnormal stamen; gll, glume like organ; inf: inflorescence-like tissue; l-le, leafy like lemma; le, lemma; l-gl, leafy like glume; lle, leafy like lemma; lo, lodicule; mo, mosaic organ; pa, palea; pi, pistil; sl, sterile lemma; st, stamen; v, vascular

the MRP of *Osmads6-2* mutant upregulates *OsCDKB2;1* and *REP1*, increasing *OsCDKB1;1* expression to promote cell division and differentiation, resembling BOP cells. These findings reinforce that precise regulation of cell proliferation and differentiation is critical for organ morphology. In addition, genetic analyses in *rep1-1*, *Osmads6-2* (+/-) *rep1-1*, *Osmads6-2 rep1-1* combinatorial mutants and $REP1_{pro}$*::OsMADS6*$_{gDNA}$*/Osmads6-2* transgenic lines indicate that OsMADS6 acts dose dependently to trigger the mitosis-endoreduplication switch during palea development, underscoring the need to quantify spatiotemporal protein dynamics. Moreover, increased palea number in the *Osmads6-2 Osmads32-2 rep1-1* triple mutant implicates these TFs in regulating palea organ number and positioning. Determining whether they act on primordium boundaries or organ fusion will also inform *Poaceae* evolution.

The homology of the palea has long been a subject of debate. Initially, it was perceived as two fused tepals, distinct from tepal growth patterns observed in other grasses (Stebbins, 1956). This perspective finds partial support in studies on *DP1/PAL1* (Luo et al., 2005; Jin et al., 2011). Notably, *dp1/pal1* mutants exhibited paleas degenerating into two leaf-like structures or constrained to a reduced size due to compromised BOP development (Jin et al., 2011), implying that the palea might be a fused organ comprising both BOP and MRP. This hypothesis gains additional support from the investigation of *ABNORMAL HULL 2* (*AH2*), which functions specifically in promoting BOP development (Ren et al., 2019). Furthermore, *REP1/SDP1* was considered a downstream gene of *DP1/PAL1* in palea development (Jin et al., 2011). Despite these findings, the mutual interaction and regulatory dynamics among *DP1/PAL1*, *REP1/SDP1*, *AH2,* and other factors in palea development remain unexplored. Notably, although *DP1/PAL1* and *REP1/SDP1* expression is exclusively detected in palea primordia, these factors also contribute to FM activity. This is evident from the significant increase in floral organ number in *dp1-2 sdp1* double mutants (Jin et al., 2011). Therefore, the functions of *DP1/PAL1* and *REP1/SDP1* resemble those of genes essential for inflorescence or spikelet transition, and their expression in the palea likely serves as positional signals for flower development. This notion finds support in the study of *COMPOSITUM 1* (*COM1*), the barley ortholog of *REP1*, where *COM1* exhibits specific expression at inflorescence meristematic boundaries. This expression pattern contributes to establishing the identity of adjacent meristems during spike domestication (Poursarebani et al., 2020).

Alternatively, the palea might be analogous to a sepal-like organ, with *DP1/PAL1* and *REP1/SDP1* functioning akin to spikelet-specific (A) class genes, promoting flower development and specifying palea identity. Supporting the latter hypothesis, the downregulation of the *DEGENERATIVE PALEA/OsMADS15* (*DEP/OsMADS15*) gene, a homolog of the *APETALA1/FRUITFUL* (*AP1/FUL*)-like gene, in *dp1/pal1* mutants was observed (Jin et al., 2011). The defective palea in *dep/Osmads15* mutants lacking BOP structures mirrors the phenotypes in *dp1/pal1* spikelets (Wang et al., 2010). OsMADS15 interacts with its homolog OsMADS14 to specify palea identity, alongside their role as (A) function genes in spikelet development (Wu et al., 2017). Furthermore, the expression of *MFS1*, an AP2/ERF TF, in the palea and other floral organs, with a notable occurrence of degenerated BOP development in half of *mfs1* flowers (Ren et al., 2013). In contrast, mutation of *MFS2*, an MYB TF, leads to an increasing number of MRP-like structures (Li et al., 2020). These results suggest the involvement of a complex regulatory network in palea development, warranting further investigation.

7.5.2.2 Lodicule Development

The lodicule represents a unique floral organ to grasses, characterized by its transparent and scale-like morphology asymmetrically positioned near the lemma side (Fig. 7.1). During anthesis, lodicules undergo rapid swelling, facilitating the separation of the palea from the lemma. The biological function of lodicules is closely linked to their cellular constituents, comprising parenchyma cells and widely dispersed vasculature (Yadav et al., 2007). Considering the cellular composition, function, and molecular regulators involved in lodicule development, it is widely accepted that lodicules serve as analogs of petals (Yoshida, 2012). Genetic and molecular investigations have produced compelling evidence indicating that A+B+E genes play a pivotal role in specifying lodicule identity in both rice and maize, mirroring their counterparts' function in petal development in Arabidopsis (Yoshida, 2012).

It is noteworthy that the rice genome harbors two *PISTILLATA*-like B class genes (*OsMADS2* and *OsMADS4*), along with one *APETALA3*-like gene (*OsMADS16*/*SUPERWOMAN1*). The roles of these genes in lodicule and stamen development have undergone subfunctionalized over the course of evolution (Yao et al., 2008). *OsMADS2* is specifically dedicated to demarcating the boundary between the MRP and the lodicule, as well as regulating distal lodicule cells and vascular differentiation (Yadav et al., 2007). Additionally, *OsMADS2* plays a role in stamen dehiscence (Yao et al., 2008). The interaction between OsMADS2 and OsMADS4 is crucial for defining stamen and lodicule identity (Prasad & Vijayraghavan, 2003; Yao et al., 2008). However, single mutants of *Osmads16*/*spw1* exhibit flower phenotypes akin to those observed in *OsMADS2 OsMADS4* RNAi transgenic lines, with homeotic transformations resulting in lodicules resembling glume-like structures and stamens resembling carpel-like organs (Fig. 7.5b) (Nagasawa et al., 2003; Yao et al., 2008). This underscores the critical role of *OsMADS16*/*SPW1* in specifying lodicule and stamen development through formation of the B class protein complex. In alignment with its function in regulating lodicule development, weak allele mutations of *OsMADS16*/*SPW1*, known as *superwoman1-cleistogamy* (*spw1-cls*) and *spw1-cls2*, were identified through forward genetic screening for generating cleistogamous lines to facilitate gene containment (Yoshida et al., 2007; Lombardo et al., 2017). In *spw1-cls* mutants, the I45T mutation destabilized SPW1/OsMADS2 dimer formation (Yoshida et al., 2007), while the G27R mutation in *spw1-cls2* affected SPW1 binding activity (Lombardo et al., 2017). Although the abnormal lodicule development of *spw1-cls* was reversible under cold conditions, no such reversibility was observed in *spw1-cls2*

(Lombardo et al., 2017). Consequently, environmental temperature influences the stability of MADS-box protein complexes during flower development, yet the underlying molecular mechanism remains unclear.

Investigations in maize revealed that the PI-like genes *Zmm16* and *Zmm18/29* also undergo neo-functionalization or subfunctionalization. The individual RNAi transgenic lines of *Zmm18/29* exhibit no obvious floral phenotype, while *Zmm16* plays a role in lodicule identity and sex determination (Bartlett et al., 2015, 2016). The functional diversification of these maize PI-like genes may be associated with their distinct developmental expression pattern, or their alternative splicing pattern, influencing their protein-protein interactomes (Qi et al., 2021). In rice, transcriptional upregulation of *OsMADS16* in the *Osmads4* mutant could account for its normal lodicule development (Wang et al., 2024b), suggesting that OsMADS4 participates in distinct protein complexes relative to OsMADS2. These complexes may not only specify floral organs but also modulate target TF expression levels, ensuring stage specific quartet assembly at both transcriptional and protein levels. Supporting this model, *Osmads2-1 Osmads3-4* and *Osmads4-1 Osmads3-4* double mutants display distinct phenotypes: lodicules in *Osmads2-1 Osmads3-4* partially convert to lodicule-stamen chimeras, whereas *Osmads4-1 Osmads3-4* exhibits abnormal carpels in whorl 3 and lodicule-stigma or lodicule-stamen mosaics (Wang et al., 2024b). These findings underscore the need to map the spatiotemporal expression patterns of these proteins during floral organ specification.

Considering that lodicules in rice *bc* mutants (*spw1-1 Osmads3-4* and *spw1-1 Osmads58*) were transformed into glume-like organs with palea identity (Fig. 7.5b), a parallel phenotype observed in the maize *bc* mutant (*si1-R zag1-mum1*) (Ambrose et al., 2000; Yun et al., 2013), the authors posit that the palea is the putative counterpart of the sepal in Arabidopsis. This proposition aligns with findings from studies on palea identity genes (see Sect. 7.5.2.1). Ectopic *DL* expression in abnormal lodicules of *Osmads4-1 Osmads3-4* double mutants further suggests the rice lodicule is a bracteopetal (modified leaf) rather than an andropetal (stamen derived) organ (Wang et al., 2024b). Furthermore, CFO1/OsMADS32, encoding a MIKC^c-type MADS box protein absent in eudicots, participates in regulating lodicule development by interacting with OsMADS2 and OsMADS4 proteins via the K domain (Wang et al., 2015c). The loss of function of E class and AGL6 genes, such as *OsMADS1* and *OsMADS6*, results in abnormal lodicule development. However, considering that *OsMADS1* is not expressed in lodicules, the molecular mechanism governing the role of these regulators in lodicule identity remains unclear. One possibility is that NONSTOP GLUMES1 (NSG1), a JAG homolog, recruits the corepressor OsTPRs to silence *OsMADS1* in lodicules by reducing histone acetylation at its locus (Zhuang et al., 2020); however, the spatial control of NSG1 in the lodicule is still unknown.

7.5.2.3 Specification of Reproductive Organs

The field of plant reproductive development has generated extensive data, with recent years witnessing comprehensive reviews highlighting the pivotal roles played by genes within the *AGAMOUS* subfamily in the specification and development of reproductive organs (Dreni et al., 2013; Dreni & Kater, 2014). In angiosperms, the *AGAMOUS* subfamily consists of *AGAMOUS* and *AGL11* lineages. The *AGAMOUS* lineage is likely instrumental in specifying stamen and carpel identity, while the *AGL11* lineage is associated with ovule and seed development (Dreni et al., 2013). The involvement of the *AGAMOUS* subfamily in plant reproductive development is evolutionarily conserved. Several chapters in this book will delve into various aspects of reproductive organ development, including stamen and anther development (Ch10) and endosperm and embryo development (Ch15, Ch19, and Ch20). This section, however, provides a specific focus on the overview of reproductive organ specification in rice.

In rice, the *AGAMOUS* lineage encompasses the *OsMADS3* and *OsMADS58* genes. Similar to the floral phenotype observed in Arabidopsis *ag* mutants, the flowers of *Osmads3 Osmads58* double mutants exhibit a complete loss of reproductive organs (Fig. 7.5b) (Yanofsky et al., 1990; Dreni et al., 2011). However, these two genes have undergone subfunctionalized during evolution. Both *OsMADS3* and *OsMADS58* are expressed in stamen primordial, but only mutations in *OsMADS3* result in variable transformations of stamens to lodicules and male sterility (Yamaguchi et al., 2006; Dreni et al., 2011; Hu et al., 2011). The role of *OsMADS58* in carpel development is dependent on *OsMADS3*, *OsMADS6*, *OsMADS1*, or other unidentified floral meristem determinacy genes (Yamaguchi et al., 2006; Dreni et al., 2011; Li et al., 2011b; Hu et al., 2015). Conversely, the overexpression of *OsMADS3* converts lodicules into stamens without affecting lemma and palea development (Kyozuka & Shimamoto, 2002), while overexpression of *OsMADS58* induces carpelloid palea, abnormal lodicules, and stamens with carpel identity (Yan et al., 2015; Zheng et al., 2015). This affirms that the two C class orthologs in rice have undergone distinct specialization in the function of stamen and carpel specification during evolution. The subfunctionalization of OsMADS3 and OsMADS58 may stem from differences in their amino acid sequences or their slightly varied expression patterns (Dreni et al., 2013). Alternatively, the specific partners or regulators of these two genes might contribute to their function evolution.

There are two genes belonging to the *AGL11* lineage in the rice genome, namely, *OsMADS13* and *OsMADS21* (Dreni et al., 2013). However, only *OsMADS13* is involved in specifying ovule identity. The flowers of *Osmads13* mutants display a homeotic conversion of ovules to extra carpels (Fig. 7.5b) (Lopez-Dee et al., 1999; Dreni et al., 2007). Moreover, evidence suggests that rice ovules develop from the lateral region of the FM (also known as the ovule founder region, OFR) rather than from the carpel margin, as observed in Arabidopsis. *OsMADS13* is crucial for OFR differentiation but not for FM determinacy (Yamaki et al., 2011). The deformed FM inside the *Osmads13* carpels is proposed to originate from the OFR after the regular process of FM determination (Fig. 7.5b) (Yamaki et al., 2011). Therefore, the role of *OsMADS13* in ovule specification is highly conserved, especially compared to *STK/FBP11* clade genes, its orthologs in Arabidopsis and petunia (Angenent et al., 1995; Pinyopich et al., 2003). A comparative transcriptomic analysis between *Osmads1-z* and *Osmads13* mutants indicates that *OsMADS1* and *OsMADS13* may regulate shared developmental pathways in an antagonistic manner, with the latter acting as a repressor of carpel identity (Osnato et al., 2021).

Flowers of *Osmads3-3 Osmads58*, *Osmads3-4 Osmads58* double mutants (Fig. 7.5b) and *Osmads3-3 Osmads58 Osmads13* triple mutants exhibit a palea-like structure in the place of carpels (Dreni et al., 2011), not only indicating that the redundant function of *OsMADS3*, *OsMADS58*, and *OsMADS13* in carpel specification but also suggesting that FM determination occurs after carpel specification in rice, with the palea serving as sepal-like structures of these mutants. Despite the molecular differences observed in the palea-like structure of these mutants (Dreni et al., 2011), it is conceivable that additional regulators, in conjunction with these genes, are necessary for specifying palea identity and development in the place of carpels. Conversely, flowers of *Osmads13-3 Osmads3-4* double mutants exhibit indeterminate carpelloid structures in the flower center (Fig. 7.5b), indicating that *OsMADS13* also plays a role in floral meristem determinacy through synergetic interaction with *OsMADS3* (Li et al., 2011a) (see Sect. 7.6).

Moreover, the *LATERAL ORGAN BOUNDARIES DOMAIN/ASYMMETRIC LEAVES2-LIKE* (*LBD/ASL*) gene *OsIG1* plays distinct roles in inhibiting ovule primordium specification on the palea side (Zhang et al., 2015), suggesting that the rice flower development may involve a specialized regulatory network in FM determinacy and asymmetric morphogenesis (see Sect. 7.9). Therefore, there is an intriguing opportunity to further dissect the regulatory network of the *AGAMOUS* subfamily in rice. The results of such an investigation would contribute not only to understanding the evolutionary reasons for gene duplication and subfunctionalization but also to unraveling the molecular mechanism underlying the divergence of the *AGAMOUS* subfamily in reproductive organ specification and FM determinacy.

The rice *YABBY* gene family member *DL* is expressed in the carpel anlagen and primordia, and its involvement in carpel specification has been established (Yamaguchi et al., 2004). Null allele mutants of *DL* result in a complete homeotic transformation of carpels into stamens (Nagasawa et al., 2003; Yamaguchi et al., 2004). The Arabidopsis ortholog of *DL*, *CRABS CLAW* (*CRC*), is a direct target of AG (Gomez-Mena et al., 2005). CRC specifies carpel identity in an AG-independent pathway and regulates FM determinacy (Bowman & Smyth, 1999). Notably, mutations of *CRC* do not lead to the transformation of carpels into stamens (Alvarez & Smyth, 1999), suggesting that the *DL/CRC* subfamily plays a fundamental and conserved role in FM determinacy. However, *DL* acquires specific functions in carpel specification and leaf midrib development during grass evolution (Yamaguchi et al., 2004).

Genetic interactions between the *DL* and *AG* subfamilies in carpel specification and FM determinacy have been explored in rice. Rice *Osmads3-4 dl-sup6* double mutants exhibit supernumerary whorls of undifferentiated lodicule-like structures at the pistil position due to the ectopic expression of *SPW1/OsMADS16* in the inner whorls of flowers (Li et al., 2011a). This further confirms that *DL* represses *SPW1/OsMADS16* expression in the flower center, and *OsMADS3* and *DL* play redundant roles in rice FM determinacy. It has also been proven that *OsMADS13* and *DL* specify carpel/ovule and FM identity in the same genetic pathway (Li et al., 2011a). However, current expression pattern data do not reveal any direct regulatory relationship among the *DL*, *OsMADS3,* and *OsMADS13* genes (Li et al., 2011a). It appears that *DL* plays parallel roles in carpel specification with *OsMADS3* and in FM determinacy with *OsMADS3* and *OsMADS13*. The question of whether *DL* is directly regulated by *OsMADS3* or *OsMADS58* in carpel identity and FM determinacy remains unanswered.

7.6 Floral Meristem Maintenance and Determinacy

In contrast to the indeterminate growth of the shoot apical meristem (SAM), the Arabidopsis FM originates from the peripheral regions of the SAM and follows a determinate pattern. While the SAM continually generates lateral organs, including flowers, throughout the plant's life, the FM results in various floral organs in a stereotypical arrangement. The precise regulation of the position and number of each floral organ is a characteristic feature in different species (Sun & Ito, 2015; McSteen & Kellogg, 2022; Balanzà et al., 2023). However, as discussed below, despite extensive studies on FM determinacy, limited attention has been given to FM maintenance.

In Arabidopsis, the homeostasis between stem cell differentiation and self-maintenance is controlled by the CLV3-WUS negative feedback loop, a mechanism seemingly conserved in different stem cell niches and diverse plant species (Aichinger et al., 2012; Zhang & Yuan, 2014; Sun & Ito, 2015; Galli & Gallavotti, 2016; Somssich et al., 2016). Generally, the stem cell promoting TF *WUS* is expressed in the organizing center (OC) and migrates into the central zone (CZ) to positively regulate *CLV3* expression. CLV3, a stem cell proliferation gene, is expressed as a pre-pro peptide, which, in turn, is perceived as a signal by various receptor-like proteins, including CLV1. This perception restricts WUS activity in the underlying cells of the OC. Therefore, this feedback loop robustly maintains stem cell homeostasis and meristem size. Mutations in *CLV* genes lead to an increase in the number of floral organs (Clark et al., 1993, 1995, 1997), while mutations in *WUS* result in a decrease in floral organ numbers (Mayer et al., 1998).

In grasses, the CLV pathway appears relatively conserved, as evidenced by the analogous functions of *CLV* orthologs in maize and rice in regulating FM size. In maize, *THICK TASSEL DWARF1* (*TD1*) and *FASCIATED EAR2* (*FEA2*) encode orthologs of CLV1 and CLV2, respectively. Mutations in *td1* and *fea2* result in an increased floral organ number, along with inflorescence meristem enlargement/fasciation (Taguchi-Shiobara et al., 2001; Bommert et al., 2005b). Moreover, additive or synergistic phenotypes in *td1 fea2* double mutants, compared to the single mutants (Bommert et al., 2005b), suggest that *TD1* and *FEA2* may operate in distinct genetic pathways, unlike the shared pathway of *CLV1* and *CLV2* in Arabidopsis (Kayes & Clark, 1998).

In rice, mutations in *CLV1* and *CLV3* orthologs, *floral organ number 1* (*fon1*) and *fon2/4*, lead to an increased number of floral organs, especially stamens and pistils (Suzaki et al., 2004, 2006; Chu & Zhang, 2007). This trait, significant for agronomy, has found recent applications in hybrid rice breeding (Ye et al., 2017). Genetic studies reveal that *FON1* and *FON2/4* function in the same pathway, as the *fon1 fon2/4* double mutant mimics the phenotypes of each single mutant, and the overexpression of *FON2/4* is suppressed by the *fon1* mutant (Suzaki et al., 2006). Unlike the distinct cellular expression pattern of *CLV1*, *FON1* is uniformly expressed throughout the IM and FM, with no evident defects observed in the *fon1* inflorescence and vegetative development (Suzaki et al., 2004). This suggests the potential involvement of additional pathways in rice meristem cell development. Two additional signal peptides, FON2 SPARE1 (FOS1) and FON2-LIKE CLE PROTEIN1 (FCP1), have been identified, functioning parallel to FON2/4 in rice FM and SAM maintenance. Overexpression lines of *FCP1* and *FOS1* in a FON1-independent pathway show a reduced SAM size (Suzaki et al., 2008, 2009), indicating that rice stem cell maintenance involves a special regulatory pathway dependent on meristem type.

Similar results are observed in maize stem cell maintenance. The expression patterns of *TD1* and *FEA2* differ from those of *CLV1* and *CLV2* during reproductive development (Taguchi-Shiobara et al., 2001; Bommert et al., 2005b), suggesting the recruitment of diverse CLE signaling pathways during grass evolution for stem cell homeostasis. A new CLV-type receptor-like gene, *FEA3*, in maize further supports this notion. *FEA3* is expressed in and below the SAM CZ, the central region of ear SMs, and leaf primordia (Je et al., 2016). Unlike the *fea2* mutant, the *fea3* mutant is resistant to CLV3 and the ZmFCP1 peptide, suggesting that FEA3 functions as a receptor for ZmFCP1. Furthermore, Arabidopsis FEA3 RNAi plants, exhibiting a fascinated and enlarged SAM, also display resistance to CLE27 peptide treatment (Je et al., 2016). This implies the universality of organ-specific and redundant receptors in plant stem cell maintenance.

On the other hand, reports on genes with a WUS function in rice and maize are currently lacking. In rice, 13 *WUSCHEL-RELATED HOMEOBOX* (*WOX*) genes have been identified (Zhang et al., 2010). Despite the presence of the conserved WUS box domain in *OsWUS/MONOCULM 3* (*MOC3*)/*TILLERS ABSENT 1* (*TAB1*) and 13 WOX genes, which is implicated in regulating stem cell identity and size of Arabidopsis FM (Ikeda et al., 2009), knockout or knockdown of *OsWUS/MOC3/TAB1* and *OsWOX4* results in abnormal axillary meristems and SAM deficiency (Ohmori et al., 2013; Lu et al., 2015b; Tanaka et al., 2015; Tanaka & Hirano, 2020; Xia et al., 2020). The mutation of *FON2/4* rescues defective ovule development in *tab1* mutants (Tanaka et al., 2021), indicating the involvement of *OsWUS/MOC3/TAB1* in rice FM maintenance. Furthermore, mutations in *OsWOX9A/DWARF TILLER1* (*DWT1*) lead to tiller dwarfism (Wang et al., 2014; Fang et al., 2020) and defective floret development (Zong et al., 2022), suggesting subfunctionalization within the rice WOX family. Extensive investigations also reveal that certain *OsWOX* genes, like *OsWOX11*, are involved in the epigenetic regulation of cell proliferation in crown root meristems (Zhao et al., 2009; Zhao et al., 2015b; Chen et al., 2015;

Cheng et al., 2016, 2018; Jiang et al., 2017; Zhou et al., 2017; Zhang et al., 2018b; Geng et al., 2023; Xu et al., 2023). These studies imply that rice *WOX* homologous genes may evolve distinct functions in specific developmental stages.

In maize, there are two *WUS* homologs: *ZmWUS1* is expressed in the predicted OC of SAM and IM (Nardmann & Werr, 2006; Je et al., 2016), and *ZmWUS2* is expressed in a pattern similar to *TD1* in leaf primordia and the flank of SAM (Bommert et al., 2005b; Nardmann & Werr, 2006). This suggests that *TD1* and *ZmWUS2* might function in the same genetic pathway. Therefore, it appears that FM maintenance in grass likely evolves a distinct pathway independent of *WUS* genes. Further investigations to comprehend the functional redundancy among WOX genes and the feedback regulatory network of the CLV pathway underlying stem cell homeostasis in grass will be of considerable interest (McSteen & Kellogg, 2022).

Several additional pathways involved in meristem maintenance have been identified in both Arabidopsis and maize (Galli & Gallavotti, 2016). TFs, exemplified by the KNOTTED1-LIKE HOMEOBOX (KNOX) family genes, *SHOOT MERISTEMLESS* (*STM*) in Arabidopsis and *KNOTTED1* (*KN1*) in maize, intricately regulate meristem size in a parallel and complementary pattern with the CLV-WUS pathway (Lenhard et al., 2002; Bolduc et al., 2012). In maize, the *FEA4* gene, encoding a bZIP TF, interacts with ABERRENT PHYLLOTAXY2 (ABPH2) to control the rate of differentiation in PZ of IM. Genetic studies have revealed that *FEA4* negatively regulates *ABPH2* and operates distinctly WUS/CLV and STM/KN1 (Pautler et al., 2015; Yang et al., 2015), indicating the evolution of diverse regulatory pathways in buffering meristem activity.

Genetic studies investigating the combination of *clv* mutants and floral homeotic TF mutants have been conducted in Arabidopsis and rice. In Arabidopsis, the double mutants of *clv1* and *ap2*, *ap3*, *pi,* and *ag* exhibited additive phenotypes, and alterations in organ number were correlated with FM size (Clark et al., 1993). This implies that *CLV1* independently regulates FM development, irrespective of these floral homeotic TFs. In rice, combining *fon4* with mutants from C, D, and E class genes intensified abnormal phenotypes observed in single mutants, indicating that FON2/4 synergistically interacted with these TFs to regulate FM activity (Xu et al., 2017). Additional work on the isolation of *fon2*/*fon4* enhancers revealed that *FON2/4* synergistically interacted with *OsMADS3* to regulate pistil development and FM determinacy. *OsMADS3* might also regulate FM activity during early flower development (Yasui et al., 2017). Since no other mutants related to FM maintenance have been reported in rice, we hypothesize that rice likely evolved a specialized regulatory pathway to maintain stem cell activity. Mutations in key genes within this pathway might be either lethal or challenging to obtain through traditional genetic screening. Alternatively, single-gene mutations in this pathway might not manifest visible phenotypes due to gene redundancy. The use of inducible mutation or multiple mutation screening would be valuable for further elucidating the pathway involved in rice FM maintenance.

In Arabidopsis, the regulation of FM homeostasis by CLV-WUS is disrupted at floral stage 6 when carpel primordia arise, marking the onset of FM termination or floral determinacy (Sun et al., 2009; Gan et al., 2013; Sun & Ito, 2015). As detailed in Sect. 7.5.2.3, AG subfamily genes also crucially contribute to floral determinacy, ensuring proper reproductive development. The AG-WUS pathway precisely suppresses *WUS* expression to dictate the cell count for stamen and carpel development. Strong *ag* mutants, such as *ag-1*, exhibit a flower-in-flower phenotype, wherein a new flower bud replaces the carpel in the wt flower (Bowman et al., 1991a). Weaker alleles like *ag-4* display indeterminate sepal growth alongside defective floral organ identity (Sieburth et al., 1995). The *WUS*-dependent nature of indeterminate sepal growth in *ag* flowers is evident in the *ag-1 wus-1* double mutant, producing a precociously terminated flower with petals in the flower center (Laux et al., 1996). AG directly and indirectly represses *WUS* expression depending on developmental stage. At floral stage 3, when sepal primordia arise, AG recruits the Polycomb Group (PcG) complex to mildly repress *WUS* expression (Liu et al., 2011). Subsequently, AG actives the C2H2 zinc finger repressor protein KNUCKLES (KNU) to further repress *WUS* expression at stages 5–6, when petal and stamen primordia appear (Sun et al., 2009, 2014). At stage 6, KNU-mediated H3K27me3 deposition represses the auxin transporter *PIN1* and cytokinin biosynthesis gene *ISOPENTENYLTRANSFERASE7* (*IPT7*), modulating auxin and cytokinin activity to ensure timely floral meristem termination (Wang et al., 2024a). Moreover, KNU inhibits the expression of *CLV1* and *CLV3* through H3K27me3-mediated epigenetic mechanisms. It also interacts with WUS, disrupting the normal CLV-WUS feedback regulatory loop and controlling FM determinacy (Shang et al., 2021). Within the flower center, AP2 can activate *WUS* expression to antagonize AG activity (Fig. 7.6) (Huang et al., 2017b). Therefore, the temporal regulation of AG in directly and indirectly repressing *WUS* expression orchestrates the program of FM maintenance and termination. A KNU-independent regulatory pathway in floral determinacy has been identified. In this pathway, the AG target CRC controls auxin homeostasis, repressing *TORNADO2* (*TRN2*) and *YUC4* expression in the developing gynoecium (Fig. 7.6) (Walcher-Chevillet & Kramer, 2016; Yamaguchi et al., 2017; Yamaguchi et al., 2018). However, the specifics of how TRN2 regulates auxin homeostasis in a stage- and tissue-specific pattern remain unknown.

In rice, AG subfamily genes are also involved in FM termination and reproductive development, as evidenced by the abnormal development of flowers in mutants. Specifically, the *Osmads3-3 Osmads58*, *Osmads3-3 Osmads13* and *Osmads58 Osmads13* double mutants, *Osmads3-3 Osmads13/+ Osmads58/+* heterozygous, and *Osmads3-3 Osmads13 Osmads58* homozygous triple mutants exhibit indeterminate FM (Dreni et al., 2011). Notably, there is no report of *WUS*-like genes with comparable functions in rice and maize. The intriguing question of whether the AG-WUS pathway universally governs FM determinacy warrants further investigation.

B class genes also play important roles in AG-dependent FM maintenance and termination in Arabidopsis. The expression of *AG* driven by the AP3 promoter failed to rescue the indeterminate growth in *ag* mutants (Jack et al., 1997). Overexpression of *AP3* alone or in conjunction with *PI* led to a significant delay of FM termination, resulting in carpel-to-stamen transformation or the development of additional floral whorls, respectively (Jack et al., 1994; Krizek & Meyerowitz, 1996). The role of B

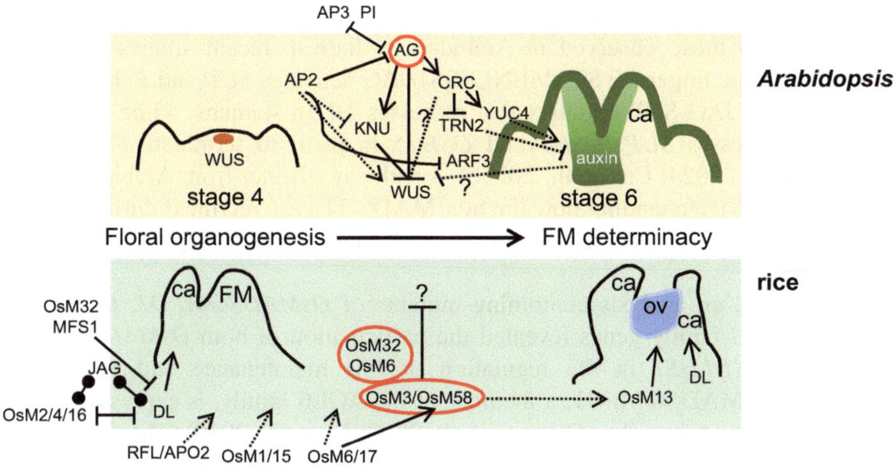

Fig. 7.6 Model for the role of C class genes in floral meristem determinacy in Arabidopsis and rice. C class genes play conserved functions in reproductive organ specification and FM determinacy. In Arabidopsis, AG activates KNU expression to repress WUS expression at floral stage 6, or it induces CRC expression to establish auxin maxima, promoting carpel development and terminal growth. However, AP2, AP3, and PI antagonize AG function to maintain FM activity. In rice, several TFs, such as *OsMADS32*, *MFS1*, *RFL/APO2*, *OsMADS1*, *OsMADS15*, *OsMADS6*, and *OsMADS17*, interacting with C class genes, *OsMADS3* and *OsMADS58*, contribute to FM determinacy through an unknown regulatory pathway. Abbreviations: ca, carpel; FM, floral meristem; OsM, OsMADS; ov, ovule

class proteins in FM maintenance is suggested to be related to the formation of stage-specific protein quartets during flower development (Prunet et al., 2009). Increased expression of AP3/PI proteins prompted a switch from AG/SEP-AG/SEP to AG/SEP-AP3/PI quartets, which was confirmed by in vitro EMSA experiments (Smaczniak et al., 2012). In addition, SUPERMAN (SUP), responsible for delineating the boundary between stamens and carpels in flowers, ensures the determinacy of FM by inhibiting the expression of B class genes (Prunet et al., 2017). Therefore, exploiting the specificities of different protein quartets aids in discerning the regulatory network in balancing FM maintenance and termination (Jetha et al., 2014; Puranik et al., 2014; Smaczniak et al., 2017). Notably, the structural change in the K domain of MADS proteins during evolution alters heterodimerization between MIKCC-type and MIKC*-type proteins, conferring distinct regulatory networks in development (Rümpler et al., 2023). In rice, the role of B class genes in FM termination appears conserved, as demonstrated by the presence of additional whorls of six glume-like structures in the position of stamens in *spw1-1 Osmads3-4* and *spw1-1 Osmads58* double mutants (Fig. 7.5b) (Yun et al., 2013). Furthermore, explorations into the rice gene SMALL REPRODUCTIVE ORGANS (SRO), an ortholog of SUP, disclose that the B class gene *OsMADS16/SPW1* is epistatic to *SRO*. Simultaneously, SRO collaboratively regulates reproductive organ specification and FM determinacy with C class genes OsMADS3 and OsMADS58 (Xu et al., 2022). Therefore, it appears that rice B and C class genes maintain FM size by suppressing FM activity, even though they may evolve distinct regulatory pathways separate from those observed in Arabidopsis. Indeed, recent studies show that rice C2H2- zinc finger TF STAMENLESS1/JAG activates *SUP* and *FON1* expression, forming JAG-SUP complex that represses *DL* in stamens, while a JAG-DL complex represses *SUP*, *OSH1* and *OSH15* in pistil to terminate FM activity (Zhuang et al., 2024; Cai et al., 2025b), a pathway distinct from Arabidopsis FM determinacy. Understanding how the non-MADS TFs are recruited during FM termination may uncover grass-specific mechanisms that regulate caryopsis development.

In contrast, an analysis combining mutants of *OsMADS6/MFO1*, *OsMADS1/LHS1,* and *AG* family genes revealed the participation of both *OsMADS6/MFO1* and *OsMADS1/LHS1* in the regulation of FM maintenance and termination (Fig. 7.6). OsMADS6/MFO1, a member of the AGL6 family, is expressed in FM and floral organ primordia (Ohmori et al., 2009; Li et al., 2010). The *mfo1-1* and *mfo1-2* weak mutants, as well as *Osmads6-1*, displayed abnormal carpel development and occasional formation of new immature spikelets (Ohmori et al., 2009; Li et al., 2010). The indeterminate FM phenotype was heightened in *Osmads6-1 dl-sup6* (an inflorescence-like structure in whorl 4), *Osmads6-1 osmad13-3* (higher-order carpel-like organs along the floral axis), *Osmads6-1 Osmads3-4* (elongated floral axis or an inflorescence-like structure in whorl 4), and *Osmads6-1 Osmads58* (two or three carpel/ovule-like structures) double mutants (Li et al., 2011b). However, the strong allele *Osmads6-5* exhibited indeterminate FM with reiterative lemma-like organs (Fig. 7.5b) (Duan et al., 2012), indicating that rice FM determinacy is regulated by distinct parallel pathways, including *AG*, *DL,* and *OsMADS6*-dependent pathways. Although *Osmads58* single mutants did not display obvious abnormal flower phenotypes, its expression was directly regulated by OsMADS6 (Li et al., 2011b). Therefore, OsMADS6 contributes to the FM context in both AG-independent and AG-dependent pathways. A recent study shows that OsMADS32 interacts synergistically with OsMADS6 and C class genes to regulate FM determinacy and the reversion from FM to IM in a dosage-dependent manner (Hu et al., 2021). Meanwhile, JAG represses *OsMADS6* in stamens but forms a complex with OsMADS6 and YABBY TFs in pistils to terminate FM activity (Cai et al., 2025a). The diverse indeterminate FM phenotypes described above also suggest the reversibility of transition signals among IM, SM, and FM, with FM identity intricately regulated in a developmental context-dependent manner.

Mutations in the *OsMADS1/LHS1* gene exhibit pleiotropic phenotypes, including abnormalities in floral organ identity and FM indeterminate growth (Jeon et al., 2000; Prasad et al., 2001, 2005; Hu et al., 2015). In contrast to the expression pattern of *OsMADS6/MFO1*, *OsMADS1/LHS1* is exclusively expressed in the FM during the early floral stage and is absent in sterile lemmas. Subsequently, it exhibits strong expression in the lemma and the palea and weak expression in the carpel (Prasad et al., 2001; Hu et al., 2015). Combined analysis of mutants from the AG subfamily genes with *Osmads1-z* reveals distinct phenotypes compared to those with *Osmads6* (Li et al., 2011b; Hu et al., 2015). In *Osmads1-z Osmads3-4* flowers, FM activity ceases earlier than in wt, resulting in extra glume-like structures at the flower center. Conversely, *Osmads1-z Osmads58* double mutants produce ectopic spikelet-like structures in the flower center, and *Osmads1-z Osmads13-3* double mutants display additive phenotypes compared to single mutants (Hu et al., 2015). These results not only provide genetic evidence of neofunctionalization occurring in rice C class genes but also suggest that *OsMADS1* and *OsMADS6* play different roles in regulating reproductive development and FM termination. Flowers of *Osmads6-1 Osmads1-z* and *mfo1-2 lhs1-2* with severely defective FM activity show extra spikelet(s) lacking inner floral organs or extra glume-like organs without inner floral organs (Ohmori et al., 2009; Li et al., 2010), suggesting a synergistic role of *OsMADS6* and *OsMADS1* in maintaining FM activity. RNAi transgenic lines of *OsMADS17*, another AGL6 gene, exhibit a normal flower phenotype, and transgenic *MFO1i+OM17i lhs1-2* triple mutant flowers mirror the phenotypes of *mfo1-2 lhs1-2* double mutants (Ohmori et al., 2009), suggesting that *OsMADS17* is likely one of the downstream genes of OsMADS1. This hypothesis is further confirmed by ChIP PCR analysis with OsMADS1 antibody, revealing enriched fragments in the promoter of *OsMADS17* (Hu et al., 2015). Furthermore, microarray analysis in *Osmads1-z* mutants demonstrates that *OsMADS1* activates the expression of floral homeotic TFs as a positive regulator, despite not being expressed in lodicules and stamens (Hu et al., 2015). Similar repression and derepression of *OsMADS6* expression in stamen and carpel primordia, respectively, are also observed (Ohmori et al., 2009), suggesting the existence of a negative feedback loop to repress *OsMADS1* and *OsMADS6* expression to maintain proper FM activity and to derepress their expression in carpels to cease FM activity. A study on *OsRAV11* revealed its role as the downstream of OsMADS1, contributing to the regulation of carpel development and FM determinacy (Osnato et al., 2020). However, the question of whether OsRAV11 functions synergistically with C class genes remains unanswered. Interestingly, the barley OsMADS1 ortholog, HvMADS1, modulates the cytokinin-degrading enzyme HvCKX3 in a temperature-dependent pattern, repressing the meristem cell cycle/division in barley inflorescence. Hence, *hvmads1* mutants grow branched spikes under higher temperatures (Li et al., 2021a). These results suggest that *OsMADS1* may regulate hormone homeostasis, and an unidentified TF-dependent regulatory network could potentially be involved in controlling rice FM maintenance and determinacy (Fig. 7.6).

7.7 Genetic and Molecular Models of Rice Floral Organ Patterning

As summarized in Sect. 7.5, since the classic genetic ABC model was proposed (Meyerowitz et al., 1989; Bowman et al., 1991b), extensive studies have been conducted to unravel the molecular basis of flower development in both eudicot and monocot model plants. The FQM and its modified versions have gained widespread usage in contemporary developmental and evolutionary biology, aiding in understanding the diversity of floral organs. In vitro, in vivo, bioinformatic, and protein crystal structure studies provide robust evidence that MADS box proteins form different dimers or tetramers (referred to as floral quartet-like protein complexes, FQCs) in a developmental context. These FQCs then selectively bind to the promoter of target genes, either activating or repressing their expression (Theissen et al., 2016; Käppel et al., 2023; Rümpler et al., 2023). In Arabidopsis, research endeavors have delved into uncovering the evolution of the FQCs, elucidating the specificity of different FQCs, and deciphering the molecular mechanisms through which these FQCs activate or repress their target genes (Muino et al., 2016; Ruelens et al., 2017; Smaczniak et al., 2017; Dreni et al., 2023; Goslin et al., 2023; Rümpler et al., 2023).

Both genetic and molecular evidence confirm the applicability of the modified genetic model (A)B(C) and the molecular model FQM to rice flowers, particularly in the specification of lodicule, stamen, carpel, and ovule development (refer to Sect. 7.5.2). The A function has been historically controversial due to the original definition limitations of A genes and varied opinions on outer sterile organs (Causier et al., 2010; Theissen et al., 2016). In rice, mutations of *AP1/FUL* orthologs, *Osmads14* and *Osmads15*, display pleiotropic flower defects, including smaller paleas and stamen-like papery lodicules. As the palea corresponds to a sepal homolog and lodicules are homologous to petals (see Sect. 7.5.2), single-mutant *Osmads14* or *Osmads15* phenotypes do not align with the A function of the classic ABC model, lacking observed ectopic carpel-like structures. However, *Osmads14 Osmads15*+ or *Osmads15 Osmads14/*+ heterozygous double mutants display the transformation of paleas to papery and carpel-like organs and homeotic transformation from lodicules to stamen-like organs. Furthermore, the knockdown of two other homologous genes, *OsMADS18* and *OsMAMS20,* in the *Osmads14 Osmads15* background results in the transformation of glumes and lemmas into leaf-like tissues, all paleas into carpel-like organs, and lodicules into papery or stamen-like structures (Wu et al., 2017), demonstrating the redundant role of rice *AP1/FUL*-like genes in specifying palea and lodicule identity, as well as FM activity. These observations suggest that rice A genes control both floral organ identity in the first two whorls and FM activity (Causier et al., 2010). Yeast two-hybrid (Y2H) assays indicate that OsMADS14 and OsMADS15 form homodimers with each other, not with OsMADS18. Protein interactions occur between OsMADS14 and OsMADS15, OsMADS14 and OsMADS18, and OsMADS15 and OsMADS18 (Wu et al., 2017), but their interactions with E class proteins, or palea-specific proteins, such as REP1/SDP1 or DP1/PAL1, have not been studied. The composition of the protein complex involving OsMADS20 remains unknown, and the possibility of OsMADS14/OsMADS15/OsMADS18 repressing the expression of *OsMADS58* in palea exists (Fig. 7.7).

In addition to the traditional ABCDE class proteins, other MADS proteins are implicated in controlling rice FM identity, often referred to as the "floral ground state." Notably, OsMADS6/MFO1 and OsMADS32/CFO1 play key roles in this process (Fig. 7.7) (Li et al., 2011b; Sang et al., 2012; Wang et al., 2015c; Dreni & Zhang, 2016; Hu et al., 2021). Based on the phylogenetic data and mutant analysis

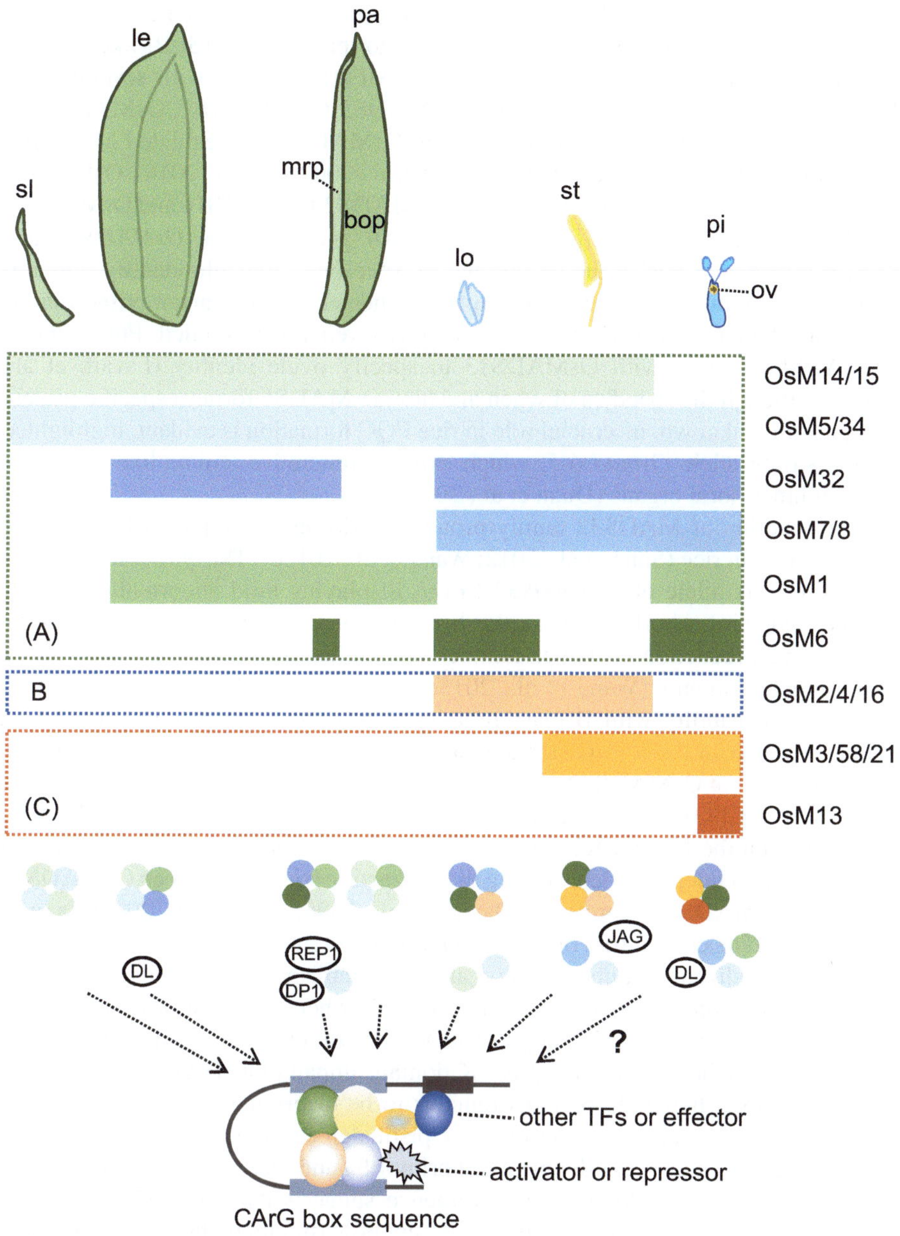

Fig. 7.7 The (A)B(C) model for floral organ specification in rice. The upper section of the figure represents floral organs in a rice floret. The middle portion outlines the expression pattern of OsMADS genes classified into (A), B, and (C) classes. The lower segment displays the fundamental specificity and mechanism of gene regulation by floral homeotic protein complexes. However, the specific tetramer complexes that form during rice flower development remain unknown. Abbreviations: bop, body of palea; gl, glume; le, lemma; lo, lodicule; mrp, marginal region of palea; OsM, OsMADS; ov, ovule; pa, palea; pi, pistil; sl, sterile lemma; st, stamen, TFs, transcription factors

in rice, maize, petunia, and Arabidopsis, AGL6 proteins have been classified as E function proteins, with an interactome similar to that of AGL2(SEP)-like proteins (Dreni & Zhang, 2016). In rice, OsMADS6/MFO1 interacts with several floral homeotic proteins, including OsMADS1, OsMADS5, OsMADS7/45, OsMADS8/24, OsMADS14, OsMADS15, OsMADS17, and OsMADS18, to regulate FM and floral organ identity (Moon et al., 1999; Ohmori et al., 2009; Li et al., 2010). OsMADS6 also forms ternary protein complexes with OsMADS2-OsMADS16 and OsMADS4-OsMADS16 heterodimers (Seok et al., 2010) and represses *OsMADS14* and *OsMADS15* expression in the second whorl, specifying lodicule identity (Ohmori et al., 2009). While *OsMADS6* transcripts are undetectable in stamen primordia, its physical interaction with C class proteins in rice remains unstudied. Furthermore, OsMADS6 interacts with OsMADS13 to specify ovule identity (Favaro et al., 2002). Although the specifics of when and how OsMADS6 forms a specific protein complex are unknown, its crucial role in rice FQC formation is evident, highlighted by the strong allele *Osmads6-5*, which exhibits reiterative lemma-like structures without inner floral organs (Duan et al., 2012).

The function of MADS32 family proteins in flower development has recently come to light in rice (Sang et al., 2012; Wang et al., 2015c). The *cfo1-3* mutant represents a weak allele of *OsMADS32/CFO1*, displaying mild abnormalities in the second and third whorls, such as the loss of lodicule identity, extra glume-like organs, and a reduction in stamen number, resembling the phenotypes observed in B class gene mutants (Wang et al., 2015c). Notably, CFO1-3 displayed reduced interaction capability with B class proteins, attributed to an extra 8 amino acid sequence between the K1 and K2 domains resulting from a C to A substitution in the third intron of OsMADS32. This substitution created a new 3′ splicing site in the third exon of CFO1-3 (Wang et al., 2015c). Through targeted deletion of the fragment between the K1 and K2 subdomains, this study confirms the significance of the K domain sequence in determining the interaction strength between OsMADS32 and B class MADS box proteins (Davies et al., 1996; Yang et al., 2003; Yang & Jack, 2004). The critical role of the K domain in protein interactions may be linked to nuclear localization, a phenomenon already established for Arabidopsis B class protein interactions (McGonigle et al., 1996). Further investigations will explore this functional significance through cellular localization and crystal protein structure analysis. Whether the disruptive K domain impairs OsMADS32 interactions with other floral homeotic proteins remains to be determined. Notably, while the *Osmads32* mutant in rice lacks obvious pistil defects, mutation of its sorghum ortholog results in adjacent double pistil primordia and increased grain number per panicle (Zhang et al., 2025). This divergence suggests that MADS32 may form distinct protein complexes or regulate different downstream pathways in FM determinacy across cereals—a possibility that warrants further study.

The interactions involving MADS box proteins and other TFs have not been thoroughly examined in rice. Despite *DL* often being utilized as a marker gene for carpel identity, its interactome remains unexplored. However, genetic analysis suggests that the repressing *DL* in palea, lodicule, and stamen primordia is essential for proper floral organ development, as evidenced by the observation of carpel-like structures in these positions (Yao et al., 2008; Sang et al., 2012; Wu et al., 2017). Conversely, the ectopic expression of *OsMADS58* can also induce the transformation of the palea into carpel-like structures (Yan et al., 2015; Zheng et al., 2015), suggesting the existence of a *DL*-independent pathway for carpel specification, and the relationship between DL and OsMADS58 requires further elucidation. In addition, the explored conserved role of *DRL*, a *DL* homolog, in leaf architecture in maize adds an intriguing dimension to our understanding (Strable et al., 2017). Mutants such as *drl1-R* and *drl1-Ds*, while maintaining carpel identity, exhibit abnormal phenotypes including reduced carpel wall growth, unfused carpel walls and silks, and protruding nucelli (Strable et al., 2017). Therefore, addressing questions about the potential subfunctionalization of maize YABBY TFs during genome duplication and whether rice *DL* evolved a specific function in specifying carpel identity adds an intriguing layer to the inquiry.

In Arabidopsis, in vitro experiments involving AP1, AP3, PI, AG, and SEP3 proteins have illustrated interactions with non-MADS proteins (Smaczniak et al., 2012). Flexible partners encompassed not only proteins known for their involvement in FM identity, such as BELLRINGER (BLR) and KNOTTED-LIKE 3 (KNAT3), and in early flower development, such as AUXIN RESPONSE FACTOR 2 (ARF2) and SQUAMOSA PROMOTER BINDING PROTEIN LIKE 8 (SPL8), but also contained transcriptional corepressors like SEUSS and LEUNIG-HOMOLOG (LUH) and chromatin remodeling and modifying factors such as the SWI/SNF ATPases BRAHMA (BRM) and SPLAYED (SYD), the CHD remodeler PICKLE (PKL), histone demethylase RELATIVE OF EARLY FLOWERING 6 (REF6), and ISWI-type remodelers CHROMATIN REMODELING 4 (CHR4), CHR11, and CHR17. Furthermore, the authors analyzed the DNA binding specificity of MADS-domain protein complexes with different compositions or concentrations, providing direct evidence that distinct protein complexes could coexist and compete for partner interaction and DNA-binding sites (Smaczniak et al., 2012). Combining information from DNA binding and gene expression data, the molecular mechanisms of these floral homeotic TFs are suggested to depend on domain variations (Lai et al., 2021) and their interaction partners (Heisler et al., 2022; Goslin et al., 2023; Rieu et al., 2023). One possible molecular module for MADS proteins has been proposed to act as "pioneer factors," modulating chromatin structure and accessibility in the promoters of target genes (Pajoro et al., 2014; Theissen et al., 2016). In rice, although there is evidence that rice A, B, C, D, E, and other MADS box proteins (OsMAD6 and OsMADS32) form homodimers or heterodimers in yeast, validation of whether and what kind of tetramer complexes form during rice flower development is needed through further in vitro and in vivo experiments (Fig. 7.7). Moreover, direct in vivo expression pattern analysis of these proteins is also essential for understanding their functions at the cellular level.

7.8 Transcriptional Regulatory Networks Controlling Rice Floral Organ Pattern

In Arabidopsis, a comprehensive transcriptional analysis of the floral homeotic TFs at the genomic level has been systematically conducted. This involved the use of microarrays, RNA sequencing, chromatin immunoprecipitation followed by deep sequencing (ChIP-seq), or hybridization to tiling array (ChIP-chip) assays. Additionally, bioinformatic analyses, coupled with inducible gene perturbation experiments, have contributed to a better understanding of gene regulatory networks governing flower development (Kaufmann et al., 2009, 2010; Yant et al., 2010; Winter et al., 2011; Wuest et al., 2012; ÓMaoiléidigh et al., 2013; O'Maoileidigh et al., 2015). These compiled datasets have accelerated insights into the regulatory networks controlling flower development (O'Maoileidigh et al., 2014; Stewart et al., 2016; Yan et al., 2016; Wils & Kaufmann, 2017; Thomson & Wellmer, 2019). Generally, these studies revealed that floral homeotic TFs target a broad spectrum of genes, including other TFs, as well as genes associated with cell growth, cell proliferation, and hormone signaling. The binding sites of SEP3, AP1, AP3/PI, and AG exhibit a high degree of correlation (Kaufmann et al., 2010; Wuest et al., 2012; Yan et al., 2016; Chen et al., 2018), confirming the formation of flexible protein complexes in vivo. These complexes regulate common targets, ensuring activity across diverse flower developmental processes, including floral organ initiation, cell specification, and organogenesis. In contrast, progress in deciphering regulatory networks of flower development in rice and other plants has been limited, with only a few target genes of rice floral homeotic TFs identified through a combination of microarray and ChIP-PCR analysis.

Genome-wide transcriptional analyses have been performed on various members of the rice MADS-box family (Arora et al., 2007), including E function genes *OsMADS1* (Khanday et al., 2013, 2016; Hu et al., 2015), *OsMADS34* (Meng et al., 2017) and *OsMADS6* (Li et al., 2011b), B class gene *OsMADS2* (Yadav et al., 2007), and C class gene *OsMADS3* (Hu et al., 2011). By using GeneChip@ Rice Genome Arrays, Arora et al. (2007) analyzed the expression profiling of 75 MADS-box genes across three vegetative and eleven reproductive developmental stages, along with three abiotic stress conditions (Arora et al., 2007), providing fundamental insights into expression patterns and potential biological functions. The rice genome encompasses five SEP MADS-box TFs, *OsMADS1/LHS1*, *OsMADS5*, *OsMADS7*, *OsMADS8,* and *OsMADS34/PAP2* (Malcomber & Kellogg, 2004; Zahn et al., 2005). The knockdown of *Osmads1/5/7/8* resulted in leaf-like floral organs, reminiscent of the role of Arabidopsis E genes in specifying floral organ identity and FM determinacy (Cui et al., 2010). In contrast to the highly redundant functions of Arabidopsis E class genes, single mutants, such as *Osmads1* or *Osmads34*, displayed distinct abnormalities in reproductive development. Current data suggest that *OsMADS34* acts earlier than other E class genes, influencing inflorescence structure and SM identity (Gao et al., 2010; Kobayashi et al., 2010). Transcriptomic profiling of <1 mm inflorescences identified potential target genes related to inflorescence development, including *RCN4*, a homolog of *TERMINAL FLOWER 1* (*TFL1*)-like group genes, and TFs regulating early spikelet development such as *OsMADS1*, *OsMADS5*, and *OsMADS6* (Meng et al., 2017). Subsequent investigations have elucidated that both OsMADS5 and OsMAD34 directly inhibit the expression of *RCN4*, thereby restricting inflorescence branching and facilitating the transition to SM identity (Zhu et al., 2022). However, the precise regulatory connections between *OsMADS34* and other genes remain unknown. Notably, a recent study revealed that *OsTAG10*, encoding a bZIP TF and regulating tapetum development, is a direct downstream target of *OsMADS8* (Zhang et al., 2018a), connecting floral organ specification to lateral stamen development. However, the precise role of *OsMADS8* in stamen development requires further clarification.

Several studies have been conducted to decipher the regulatory networks governed by the LOFSEP gene *OsMADS1*, employing microarray analysis on *Osmads1-z* inflorescences (Hu et al., 2015), panicles of RNAi plants, and ChIP-seq assay (Khanday et al., 2016). Khanday et al. (2013) unveiled OsMADS1's direct downstream targets, including auxin and cytokinin genes, and TFs such as *OsARF9/12/15/16/18/25*, *OsGH3*, *LONELY GUY* (*LOG*), *OsRR1/9*, *OsHB4*, *OsBLH1*, *OsKANADI2/4*, *OsETTIN2*, *OsMADS55*, and *OsMADS34*. The promoters of these genes harbored classic MADS protein binding sites and CArG elements (Khanday et al., 2013). In contrast to Khanday's report of approximately 8000 genes with expression changes in *OsMADS1* RNAi panicles (5–40 mm), Hu et al. observed 654 and 1659 genes with expression changes during the emergence of palea primordia (Sp4) and the appearance of carpel primordia (Sp7-Sp8), respectively (Hu et al., 2015). This discrepancy suggests that *OsMADS1* exerts distinct roles at different developmental stages. Subsequently, Vijayraghavan's group elucidated the DNA recognition properties of OsMADS1 through ChIP-seq assays, identifying a total of 3112 gene-associated binding sites (Khanday et al., 2016). Enrichment of motifs associated with other TFs, such as MYC, AP2/ERF, and bZIP, in OsMADS1 binding sites implies potential interactions forming protein complexes to regulate shared targets (Khanday et al., 2016). However, further efforts are required to explore their regulatory relationships between OsMADS1 and these TFs.

Using an Agilent 4 × 4 4K oligonucleotide DNA chip, Li et al. (2011b) conducted a microarray analysis on *Osmads6-1* mutant flowers at the stage of stamen primordia appearance (stage Sp6), revealing alterations in the expression of 59 genes, each exhibiting at least a twofold change (Li et al., 2011b). Among them, 26 genes displayed upregulated, while 33 exhibited downregulation. Notably, several floral homeotic genes experienced obvious downregulation in *Osmads6-1* flowers, underscoring the active role of *OsMADS6* in regulating rice flower development. Specifically, among these genes, *OsMADS58*, a C class gene, was identified as a direct target of OsMADS6. The transcriptional regulatory role of *OsMADS2* was also investigated using Agilent Technologies custom rice (22,000) microarray analysis in the 0.5–4 cm panicles of *dsRNAi-OsMADS2* lines, where lodicules underwent transformation into MRP-like structures (Yadav et al., 2007). Consistent with the abnormal phenotypes observed, the expression levels of genes related to cell proliferation, differentiation, and signaling factors were obviously altered in the *dsRNAi-Osmads2* lines.

7.9 Conclusion and Perspective

Over the last three decades, significant progress has been made in unraveling the genetic and molecular mechanisms that govern floral organ specification. However, a substantial portion of our understanding of flower development originates from the eudicot model plant Arabidopsis, whose flowers exhibit symmetry, regularity, and a relatively simple three-dimensional structure among angiosperms (Wellmer et al., 2014b). In contrast, grass flowers undergo complex and specialized developmental events (Soreng et al., 2017; Wang et al., 2022b). While the modified (A)B(C) model is partially applicable to elucidate grass floral organ specification (Hirano et al., 2014; Tanaka et al., 2014; Dreni et al., 2023), numerous unresolved questions persist.

Studies in Arabidopsis have demonstrated that the four whorls of floral organs originated from specific cell lineages within FM (Furner & Pumfrey, 1993; Bossinger & Smyth, 1996; Furner, 1996). Several groups of TFs have been identified as crucial in determining the position, boundaries, and patterning of floral organs (O'Maoileidigh et al., 2014; Wellmer et al., 2014a; Hepworth & Pautot, 2015; Yu & Huang, 2016). However, the mechanism by which floral organ founder cells attain their identity from undifferentiated FM remains elusive. Phytohormone distribution, especially the concentration maxima and polar transportation of auxin, plays a pivotal role in regulating the position, initiation, and arrangement of different floral organs in the confined space of the minute FM (Cheng & Zhao, 2007; Lampugnani et al., 2013). Nevertheless, marker genes or lines, except auxin, indicating their distribution during floral organ initiation, position, and patterning are currently lacking in grasses. Additionally, studies in Arabidopsis suggest that organ outgrowth represents a distinct developmental stage from the specification of organ identity (Wuest et al., 2012). Tissue-specific transcriptomics analysis in maize spikelet identified a boundary domain between upper and lower florets, resulting in reduced growth in the lower florets (Yang et al., 2022). Whether a similar mechanism governs the regulation of floral organ outgrowth and identity specification in grasses remains unknown. Furthermore, Arabidopsis floral homeotic MADS proteins exhibit active throughout flower morphogenesis, displaying dynamic changes in DNA occupancy at specific sites across different developmental stages. This suggests that they regulate target genes in a stage (and possibly tissue-)-specific manner (Pajoro et al., 2014). Therefore, the establishment of a marker line system, the dissection of the spatiotemporal resolution of the gene expression map, and the isolate of additional mutants are necessary to comprehensively understand the entire flower developmental process beyond floral organ identity specification.

Reverse genetic analysis has been employed to explore the conserved or diverse functions of TFs associated with innovations in floral patterns. As demonstrated in a recent study unveiling the potential functions of *YABBY* homologous genes in rice (Tanaka et al., 2012, 2017). In contrast to their Arabidopsis homologs, rice *YABBY* genes, *TONGARI-BOUSHI* (*TOB1*, *TOB2*, and *TOB3*), are expressed in bract and lateral organ primordia. Notably, *tob1* mutants exhibited flowers with pleiotropic phenotypes, including cone-shaped organs in place of the lemma or palea, depressed lemma and palea, and the development of two florets in a spikelet (Tanaka et al., 2012). Combining *tob1* with *tob2* and *tob3* resulted in the generation of pole-like branches devoid of spikelets, underscoring the role of rice TOB genes in regulating the maintenance and fate of all reproductive meristems (Tanaka et al., 2017). This suggests that genes expressed in the lateral organs may contribute to the developmental context for FM activity. However, intriguing questions persist regarding whether these TFs interact with FM regulators or floral homeotic TFs and how they initiate spikelet development.

In the past two decades, significant strides have been made in unraveling the molecular mechanisms underlying floral morphologies, particularly in *Antirrhinum* and Arabidopsis (Moyroud & Glover, 2017; Hong et al., 2018). Genetic and molecular evidence has established that two TCP family TFs, *CYC* and *DICH*, expressed specifically in the adaxial petal and stamen primordial, render flower zygomorphic development (Luo et al., 1996, 1999). Evolutionary data from various species have further demonstrated the independent recruitment of the CYC-dependent program to regulate bilateral flower symmetry in eudicots (Hileman, 2014a; Zhao et al., 2023). Unlike eudicot flowers, grass flowers lack distinct petal-like structures and exhibit a characteristic arrangement of floral organs in sets of three (Rudall & Bateman, 2004). While structural analysis suggests a pervasive transition from radial to bilateral flower symmetry among monocots (Rudall & Bateman, 2004; Hileman, 2014b), limited progress has been made in understanding these processes. Grass flowers, including rice, are categorized as zygomorphic along the lemma-palea axis, with the adaxial lodicule suppressed (Ambrose et al., 2000; Yuan et al., 2009). In rice, the *CYC* homolog, *REP1*, specially expressed in palea primordia, influences palea development, and overexpression of *REP1* results in an inner asymmetric growth pattern of the palea (Yuan et al., 2009), suggesting an independent role in specifying the palea pattern. The function of *CYC*-like genes in grass flower symmetry remains enigmatic. Additionally, studies in grass lodicule development have highlighted the regulatory role of B class genes, which are asymmetrically expressed in lodicules. In maize, the PI ortholog, *ZEA MAYS MADS16 (ZMM16)*/*STERILE TASSEL SILKY EAR1* (*STS1*), shows asymmetrical expression in the abaxial region of FMs, contributing to an unusual asymmetric program for flower patterning, distinct from the *CYC*/*DICH* expression pattern in snapdragon. Therefore, the floral organ identity gene, *STS*/*ZMM16*, links floral organ patterning and zygomorphy, although the detailed regulatory mechanism requires further exploration (Bartlett et al., 2015). Grass zygomorphy may also be influenced by the palea auxin maximum (palea inhibition zone), as supported by polar auxin distribution in the palea. However, the establishment of this polarity in maize flower zygomorphy remains unclear (Bartlett et al., 2015). In rice, polar auxin transport and high-level auxin maximum are observed in both lemma and palea primordia during early rice spikelet development (Yang et al., 2017). The correlation among space constraints, suppressed medial lodicules, and the palea inhibition zone warrants further investigation. The potential involvement of floral homeotic TFs in regulating rice flower patterning is also a captivating area for study. In *Osmads16*/*spw1* single mutants, extra palea/glume-like organs exhibit asymmetrical growth (Nagasawa et al., 2003), suggesting that the rice *AP3* class gene regulates flower patterning by restricting extra floral organs on the palea side. However, these extra organs might also result from abnormal FM identity. In *Osmads6-1 spw1-1* double mutants, more abnormal glume-like organs and inflorescence-like organs in the position of lodicules were observed (Li et al., 2011b), indicating that both *SPW* and *OsMADS6* jointly control FM and inner floral organ identity, besides their role in suppressing inflorescence reversion. The isolation of *Osmads2* and *Osmads4* mutants and their combination with different floral homeotic TF mutants will provide insights into the role of rice B class genes in flower zygomorphy. The availability of the clustered, regularly interspaced, short palindromic repeat (CRISPR/Cas9) system for targeted gene mutations facilitates such investigations (Ma et al., 2016). The discovery of a new symmetric plane in rice *two opposite lemma* (*tol*) spikelets, featuring a mirror image duplication of the half spikelet from the lemma side (Sugiyama et al., 2017), further underscores the existence of an unknown regulatory pathway controlling spikelet asymmetry.

Collected evidence suggests that plants have evolved diverse molecular mechanisms to regulate floral organ identity, reflecting the varied spatial restriction of genes involved in this process. In the classic ABC model for flower development, A class genes repress C functions in the outer whorls, specifying sepal and petal identity (Coen & Meyerowitz, 1991). However, increasing evidence in Arabidopsis indicates that the two A class genes, AP1 and AP2, play crucial roles in FM identity. Unlike *AP2*, *AP1* did not repress *AG* expression in the outer whorl, demonstrating a context-dependent genetic pathway (Prunet & Jack, 2014). Intriguingly, AP2 represses *AP3* and *SEP3* expression in whorl 1, and the reasons for this repression not occurring in whorl 2 (AP3 and SEP3) or whorl 3 (AG) remain unknown. The activity of *AP2* is dependent not only on the expression levels of *miR172* or *AG* but also on an unknown balancing mechanism (Prunet & Jack, 2014). Recent findings in petunia revealed that *AP2* TFs in petunia repressed B functions rather than C functions to regulate sepal, petal, and stamen identity (Morel et al., 2017). In rice and maize, *AP2* TFs regulate the transition from SMs to FMs and SM determinacy (Chuck et al., 1998, 2007, 2008; Lee et al., 2007; Lee & An, 2012). Although direct molecular data supporting the idea that rice A class genes may repress the expression of C class genes in outer whorls are lacking, abnormal transformation from the palea to carpel-like organs observed in specific transgenic plants suggests a role of rice *AP1/FUL* genes in repressing the expression of C class genes (Wu et al., 2017). Furthermore, studies in Arabidopsis have unveiled that certain MADS proteins, including SEP3 and AP1, harbor transcriptional activation domains. Other proteins, such as AG, can interact with transcriptional corepressors like SEUSS and LUH, or chromatin remodeling and modifying factors, to the expression of regulate target genes (Smaczniak et al., 2012). However, the intricate mechanisms governing the competition of different protein complexes competing to activate or repress target gene expression remain elusive. The roles of posttranslational modifications of floral homeotic TFs are also poorly understood. Therefore, our comprehension of the regulatory networks involving floral homeotic TFs is incomplete and demands further exploration. Recent advances in techniques equip the flower development community with robust tools to characterize spatial-temporal regulatory networks in flower development. These tools encompass genome-wide protein interactions (Struk et al., 2019; Yang et al., 2020), protein crystallography (Jumper et al., 2021; Varadi et al., 2021; Varadi et al., 2023), live imaging (Prunet, 2017; Colin et al., 2021; Silveira et al., 2022; Peng et al., 2023), computational modeling (Prusinkiewicz & Runions, 2012; Cieslak et al., 2022), machine learning (Kasianov et al., 2023), and innovative methods elucidating higher-order chromatin structure (e.g., Hi-C, ChIA-PET, HiChIP, and PLAC-seq) (Feng et al., 2014; Wang et al., 2015a, 2022a; Liu et al., 2016a; Zhu et al., 2023), exploring transcriptional regulation at the genomic scale (Zhao et al., 2019, 2020; Deng et al., 2023) and at single-cell resolution (Song et al., 2020; Marand et al., 2021; Xu et al., 2021; Nelms & Walbot, 2022; Neumann et al., 2022; Zong et al., 2022; Li et al., 2023).

Acknowledgments We acknowledge funding from the National Science Foundation (32470351, 32170322, 31470397 and 31671260), the National Key Research and Development Program of China (2022YFD1200102), Special Funds for Construction of Innovative Provinces in Hunan Province (2021NK1002), the Sino-Germany Mobility Program (M-0141) and the Innovative Research Team, Ministry of Education, 111 Project (B14016) to D. Z. and Z. Y.

References

Acosta, I. F., Laparra, H., Romero, S. P., Schmelz, E., Hamberg, M., Mottinger, J. P., Moreno, M. A., & Dellaporta, S. L. (2009). tasselseed1 is a lipoxygenase affecting jasmonic acid signaling in sex determination of maize. *Science, 323*, 262–265.

Aichinger, E., Kornet, N., Friedrich, T., & Laux, T. (2012). Plant stem cell niches. *Annual Review of Plant Biology, 63*, 615–636.

Alvarez, J., & Smyth, D. R. (1999). CRABS CLAW and SPATULA, two Arabidopsis genes that control carpel development in parallel with AGAMOUS. *Development, 126*, 2377–2386.

Ambrose, B. A., Lerner, D. R., Ciceri, P., Padilla, C. M., Yanofsky, M. F., & Schmidt, R. J. (2000). Molecular and genetic analyses of the *silky1* gene reveal conservation in floral organ specification between eudicots and monocots. *Molecular Cell, 5*, 569–579.

Angenent, G. C., & Colombo, L. (1996). Molecular control of ovule development. *Trends in Plant Science, 1*, 228–232.

Angenent, G. C., Franken, J., Busscher, M., van Dijken, A., van Went, J. L., Dons, H. J., & van Tunen, A. J. (1995). A novel class of MADS box genes is involved in ovule development in petunia. *Plant Cell, 7*, 1569–1582.

Arber, A. (1934). *The Gramineae: A study of cereal, bamboo, and grass*. Cambridge University Press.

Arora, R., Agarwal, P., Ray, S., Singh, A. K., Singh, V. P., Tyagi, A. K., & Kapoor, S. (2007). MADS-box gene family in rice: Genome-wide identification, organization and expression profiling during reproductive development and stress. *BMC Genomics, 8*, 242.

Bai, X., Huang, Y., Mao, D., Wen, M., Zhang, L., & Xing, Y. (2016). Regulatory role of FZP in the determination of panicle branching and spikelet formation in rice. *Scientific Reports, 6*, 19022.

Bai, X., Huang, Y., Hu, Y., Liu, H., Zhang, B., Smaczniak, C., Hu, G., Han, Z., & Xing, Y. (2017). Duplication of an upstream silencer of FZP increases grain yield in rice. *Nature Plants, 3*, 885–893.

Bainbridge, K., Guyomarc'h, S., Bayer, E., Swarup, R., Bennett, M., Mandel, T., & Kuhlemeier, C. (2008). Auxin influx carriers stabilize phyllotactic patterning. *Genes & Development, 22*, 810–823.

Balanzà, V., Merelo, P., & Ferrándiz, C. (2023). Flowering also has to end: Knowns and unknowns of reproductive arrest in monocarpic plants. *Journal of Experimental Botany, 74*, 3951–3960.

Bartlett, M. E., Williams, S. K., Taylor, Z., DeBlasio, S., Goldshmidt, A., Hall, D. H., Schmidt, R. J., Jackson, D. P., & Whipple, C. J. (2015). The maize PI/GLO ortholog Zmm16/sterile tassel silky ear1 interacts with the zygomorphy and sex determination pathways in flower development. *Plant Cell, 27*, 3081–3098.

Bartlett, M., Thompson, B., Brabazon, H., Del Gizzi, R., Zhang, T., & Whipple, C. (2016). Evolutionary dynamics of floral homeotic transcription factor protein-protein interactions. *Molecular Biology and Evolution, 33*, 1486–1501.

Bolduc, N., Yilmaz, A., Mejia-Guerra, M. K., Morohashi, K., O'Connor, D., Grotewold, E., & Hake, S. (2012). Unraveling the KNOTTED1 regulatory network in maize meristems. *Genes & Development, 26*, 1685–1690.

Bomblies, K., Wang, R. L., Ambrose, B. A., Schmidt, R. J., Meeley, R. B., & Doebley, J. (2003). Duplicate *FLORICAULA/LEAFY* homologs *zfl1* and *zfl2* control inflorescence architecture and flower patterning in maize. *Development, 130*, 2385–2395.

Bommert, P., Satoh-Nagasawa, N., Jackson, D., & Hirano, H. Y. (2005a). Genetics and evolution of inflorescence and flower development in grasses. *Plant & Cell Physiology, 46*, 69–78.

Bommert, P., Lunde, C., Nardmann, J., Vollbrecht, E., Running, M., Jackson, D., Hake, S., & Werr, W. (2005b). *thick tassel dwarf1* encodes a putative maize ortholog of the *Arabidopsis CLAVATA1* leucine-rich repeat receptor-like kinase. *Development, 132*, 1235–1245.

Bossinger, G., & Smyth, D. R. (1996). Initiation patterns of flower and floral organ development in Arabidopsis thaliana. *Development, 122*, 1093–1102.

Bowman, J. L., & Smyth, D. R. (1999). CRABS CLAW, a gene that regulates carpel and nectary development in Arabidopsis, encodes a novel protein with zinc finger and helix-loop-helix domains. *Development, 126*, 2387–2396.

Bowman, J. L., Drews, G. N., & Meyerowitz, E. M. (1991a). Expression of the Arabidopsis floral homeotic gene AGAMOUS is restricted to specific cell types late in flower development. *Plant Cell, 3*, 749–758.

Bowman, J. L., Smyth, D. R., & Meyerowitz, E. M. (1991b). Genetic interactions among floral homeotic genes of Arabidopsis. *Development, 112*, 1–20.

Bowman, J. L., Smyth, D. R., & Meyerowitz, E. M. (2012). The ABC model of flower development: Then and now. *Development, 139*, 4095–4098.

Cacharron, J., Saedler, H., & Theissen, G. (1999). Expression of MADS box genes ZMM8 and ZMM14 during inflorescence development of Zea mays discriminates between the upper and the lower floret of each spikelet. *Development Genes and Evolution, 209*, 411–420.

Cai, Q., Yuan, Z., Chen, M., Yin, C., Luo, Z., Zhao, X., Liang, W., Hu, J., & Zhang, D. (2014). Jasmonic acid regulates spikelet development in rice. *Nature Communications, 5*, 3476.

Cai, Z., Li, J., Su, Y., Zheng, L., Zhang, J., Zhu, M., Qiu, B., Kong, L., Ye, Y., Xue, Y., Wu, W., Duan, Y. (2025a) The MADS6 JAGGED and YABBY proteins synergistically determine floral organ development in rice. *Plant Physiology, 197*(3). https://doi.org/10.1093/plphys/kiaf076

Cai, Z., Su, Y., Kong, L., Ren, G., Zheng, L., Zhu, M., Qiu, B., Hong, Z., Ye, Y., Xue, Y., Wu, W., Duan, Y. (2025b) JAGGED and DROOPING LEAF synergistically establish sexual organs and terminate floral meristem in rice SUMMARY Significance Statement. *The Plant Journal, 123*(1). https://doi.org/10.1111/tpj.v123.1

Cao, L., Tian, J., Liu, Y., Chen, X., Li, S., Persson, S., Lu, D., Chen, M., Luo, Z., Zhang, D., & Yuan, Z. (2021). Ectopic expression of OsJAZ6, which interacts with OsJAZ1, alters JA signaling and spikelet development in rice. *The Plant Journal: For Cell and Molecular Biology, 108*, 1083–1096.

Causier, B., Schwarz-Sommer, Z., & Davies, B. (2010). Floral organ identity: 20 years of ABCs. *Seminars in Cell & Developmental Biology, 21*, 73–79.

Chahtane, H., Lai, X., Tichtinsky, G., Rieu, P., Arnoux-Courseaux, M., Cancé, C., Marondedze, C., & Parcy, F. (2023). Flower development in Arabidopsis. *Methods in Molecular Biology (Clifton, N.J.), 2686*, 3–38.

Chandler, J. W. (2011). Founder cell specification. *Trends in Plant Science, 16*, 607–613.

Chandler, J. W. (2012). Floral meristem initiation and emergence in plants. *Cellular and Molecular Life Sciences, 69*, 3807–3818.

Chandler, J. W., & Werr, W. (2014). Arabidopsis floral phytomer development: Auxin response relative to biphasic modes of organ initiation. *Journal of Experimental Botany, 65*, 3097–3110.

Chandler, J. W., & Werr, W. (2017). DORNROSCHEN, DORNROSCHEN-LIKE, and PUCHI redundantly control floral meristem identity and organ initiation in Arabidopsis. *Journal of Experimental Botany, 68*, 3457–3472.

Chen, G., Feng, H., Hu, Q., Qu, H., Chen, A., Yu, L., & Xu, G. (2015). Improving rice tolerance to potassium deficiency by enhancing OsHAK16p:WOX11-controlled root development. *Plant Biotechnology Journal, 13*, 833–848.

Chen, D., Yan, W., Fu, L. Y., & Kaufmann, K. (2018). Architecture of gene regulatory networks controlling flower development in Arabidopsis thaliana. *Nature Communications, 9*, 4534.

Cheng, Y. F., & Zhao, Y. D. (2007). A role for auxin in flower development. *Journal of Integrative Plant Biology, 49*, 99–104.

Cheng, S., Zhou, D. X., & Zhao, Y. (2016). WUSCHEL-related homeobox gene WOX11 increases rice drought resistance by controlling root hair formation and root system development. *Plant Signaling & Behavior, 11*, e1130198.

Cheng, S., Tan, F., Lu, Y., Liu, X., Li, T., Yuan, W., Zhao, Y., & Zhou, D. X. (2018). WOX11 recruits a histone H3K27me3 demethylase to promote gene expression during shoot development in rice. *Nucleic Acids Research, 46*, 2356–2369.

Chongloi, G. L., Prakash, S., & Vijayraghavan, U. (2019). Regulation of meristem maintenance and organ identity during rice reproductive development. *Journal of Experimental Botany, 70*, 1719–1736.

Chu, H., & Zhang, D. (2007). The shoot apical meristem size regulated by FON4 in rice. *Plant Signaling & Behavior, 2*, 115–116.

Chuck, G., Meeley, R. B., & Hake, S. (1998). The control of maize spikelet meristem fate by the *APETALA2*-like gene *indeterminate spikelet1*. *Genes & Development, 12*, 1145–1154.

Chuck, G., Muszynski, M., Kellogg, E., Hake, S., & Schmidt, R. J. (2002). The control of spikelet meristem identity by the *branched silkless1* gene in maize. *Science, 298*, 1238–1241.

Chuck, G., Meeley, R., Irish, E., Sakai, H., & Hake, S. (2007). The maize tasselseed4 microRNA controls sex determination and meristem cell fate by targeting Tasselseed6/indeterminate spikelet1. *Nature Genetics, 39*, 1517–1521.

Chuck, G., Meeley, R., & Hake, S. (2008). Floral meristem initiation and meristem cell fate are regulated by the maize *AP2* genes *ids1* and *sid1*. *Development, 135*, 3013–3019.

Ciaffi, M., Paolacci, A. R., Tanzarella, O. A., & Porceddu, E. (2011). Molecular aspects of flower development in grasses. *Sexual Plant Reproduction, 24*, 247–282.

Cieslak, M., Owens, A., & Prusinkiewicz, P. (2022). Computational models of auxin-driven patterning in shoots. *Cold Spring Harbor Perspectives in Biology, 14*, a040097.

Clark, S. E., Running, M. P., & Meyerowitz, E. M. (1993). *CLAVATA1*, a regulator of meristem and flower development in *Arabidopsis*. *Development, 119*, 397–418.

Clark, S., Running, M., & Meyerowitz, E. (1995). *CLAVATA3* is a specific regulator of shoot and floral meristem development affecting the same processes as *CLAVATA1*. *Development, 121*, 2057–2067.

Clark, S. E., Williams, R. W., & Meyerowitz, E. M. (1997). The *CLAVATA1* gene encodes a putative receptor kinase that controls shoot and floral meristem size in Arabidopsis. *Cell, 89*, 575.

Coen, E. S., & Meyerowitz, E. M. (1991). The war of the whorls: Genetic interactions controlling flower development. *Nature, 353*, 31–37.

Colin, L., Martin-Arevalillo, R., Bovio, S., Bauer, A., Vernoux, T., Caillaud, M.-C., Landrein, B., & Jaillais, Y. (2021). Imaging the living plant cell: From probes to quantification. *The Plant Cell, 34*, 247–272.

Comelli, P., Glowa, D., Chandler, J. W., & Werr, W. (2016). Founder-cell-specific transcription of the DORNROSCHEN-LIKE promoter and integration of the auxin response. *Journal of Experimental Botany, 67*, 143–155.

Cui, R., Han, J., Zhao, S., Su, K., Wu, F., Du, X., Xu, Q., Chong, K., Theissen, G., & Meng, Z. (2010). Functional conservation and diversification of class E floral homeotic genes in rice (Oryza sativa). *The Plant Journal: For Cell and Molecular Biology, 61*, 767–781.

D'Ario, M., Griffiths-Jones, S., & Kim, M. (2017). Small RNAs: Big impact on plant development. *Trends in Plant Science, 22*, 1056–1068.

Davies, B., Egea-Cortines, M., de Andrade Silva, E., Saedler, H., & Sommer, H. (1996). Multiple interactions amongst floral homeotic MADS box proteins. *The EMBO Journal, 15*, 4330–4343.

Debernardi, J. M., Lin, H., Chuck, G., Faris, J. D., & Dubcovsky, J. (2017). microRNA172 plays a crucial role in wheat spike morphogenesis and grain threshability. *Development, 144*, 1966–1975.

Denay, G., Chahtane, H., Tichtinsky, G., & Parcy, F. (2017). A flower is born: An update on Arabidopsis floral meristem formation. *Current Opinion in Plant Biology, 35*, 15–22.

Deng, L., Zhou, Q., Zhou, J., Zhang, Q., Jia, Z., Zhu, G., Cheng, S., Cheng, L., Yin, C., Yang, C., Shen, J., Nie, J., Zhu, J.-K., Li, G., & Zhao, L. (2023). 3D organization of regulatory elements for transcriptional regulation in Arabidopsis. *Genome Biology, 24*, 181.

Derbyshire, P., & Byrne, M. E. (2013). MORE SPIKELETS1 is required for spikelet fate in the inflorescence of Brachypodium. *Plant Physiology, 161*, 1291–1302.

Dinneny, J. R., Yadegari, R., Fischer, R. L., Yanofsky, M. F., & Weigel, D. (2004). The role of *JAGGED* in shaping lateral organs. *Development, 131*, 1101–1110.

Dreni, L., & Kater, M. M. (2014). MADS reloaded: Evolution of the AGAMOUS subfamily genes. *The New Phytologist, 201*, 717–732.

Dreni, L., & Zhang, D. (2016). Flower development: The evolutionary history and functions of the AGL6 subfamily MADS-box genes. *Journal of Experimental Botany, 67*, 1625–1638.

Dreni, L., Jacchia, S., Fornara, F., Fornari, M., Ouwerkerk, P. B., An, G., Colombo, L., & Kater, M. M. (2007). The D-lineage MADS-box gene *OsMADS13* controls ovule identity in rice. *The Plant Journal: For Cell and Molecular Biology, 52*, 690–699.

Dreni, L., Pilatone, A., Yun, D., Erreni, S., Pajoro, A., Caporali, E., Zhang, D., & Kater, M. M. (2011). Functional analysis of all AGAMOUS subfamily members in rice reveals their roles in reproductive organ identity determination and meristem determinacy. *Plant Cell, 23*, 2850–2863.

Dreni, L., Osnato, M., & Kater, M. M. (2013). The ins and outs of the rice AGAMOUS subfamily. *Molecular Plant, 6*, 650–664.

Dreni, L., Riechmann J. L., & Ferrándiz, C. (2023). The ABC of flower development in monocots: The model of rice spikelet. *Flower Development: Methods and Protocols, 2686*, 59–82. Springer US New York, NY.

Duan, Y., Xing, Z., Diao, Z., Xu, W., Li, S., Du, X., Wu, G., Wang, C., Lan, T., Meng, Z., Liu, H., Wang, F., Wu, W., & Xue, Y. (2012). Characterization of Osmads6-5, a null allele, reveals that OsMADS6 is a critical regulator for early flower development in rice (Oryza sativa L.). *Plant Molecular Biology, 80*, 429–442.

Egea-Cortines, M., Saedler, H., & Sommer, H. (1999). Ternary complex formation between the MADS-box proteins SQUAMOSA, DEFICIENS and GLOBOSA is involved in the control of floral architecture in Antirrhinum majus. *The EMBO Journal, 18*, 5370–5379.

Fang, F., Ye, S., Tang, J., Bennett, M. J., & Liang, W. (2020). DWT1/DWL2 act together with OsPIP5K1 to regulate plant uniform growth in rice. *The New Phytologist, 225*, 1234–1246.

Favaro, R., Immink, R. G., Ferioli, V., Bernasconi, B., Byzova, M., Angenent, G. C., Kater, M., & Colombo, L. (2002). Ovule-specific MADS-box proteins have conserved protein-protein interactions in monocot and dicot plants. *Molecular Genetics and Genomics: MGG, 268*, 152–159.

Feng, S., Cokus, S. J., Schubert, V., Zhai, J., Pellegrini, M., & Jacobsen, S. E. (2014). Genome-wide Hi-C analyses in wild-type and mutants reveal high-resolution chromatin interactions in Arabidopsis. *Molecular Cell, 55*, 694–707.

Fletcher, J. C., Brand, U., Running, M. P., Simon, R., & Meyerowitz, E. M. (1999). Signaling of cell fate decisions by *CLAVATA3* in *Arabidopsis* shoot meristems. *Science, 283*, 1911–1914.

Furner, I. J. (1996). Cell fate in the development of the Arabidopsis flower. *The Plant Journal: For Cell and Molecular Biology, 10*, 645–654.

Furner, I. J., & Pumfrey, J. E. (1993). Cell fate in the inflorescence meristem and floral buttress of Arabidopsis thaliana. *The Plant Journal, 4*, 917–931.

Gallavotti, A., Yang, Y., Schmidt, R. J., & Jackson, D. (2008a). The relationship between auxin transport and maize branching. *Plant Physiology, 147*, 1913–1923.

Gallavotti, A., Barazesh, S., Malcomber, S., Hall, D., Jackson, D., Schmidt, R. J., & McSteen, P. (2008b). *sparse inflorescence1* encodes a monocot-specific *YUCCA*-like gene required for vegetative and reproductive development in maize. *Proceedings of the National Academy of Sciences of the United States of America, 105*, 15196–15201.

Galli, M., & Gallavotti, A. (2016). Expanding the regulatory network for meristem size in plants. *Trends in Genetics: TIG, 32*, 372–383.

Galli, M., Liu, Q., Moss, B. L., Malcomber, S., Li, W., Gaines, C., Federici, S., Roshkovan, J., Meeley, R., Nemhauser, J. L., & Gallavotti, A. (2015). Auxin signaling modules regulate maize inflorescence architecture. *Proceedings of the National Academy of Sciences of the United States of America, 112*, 13372–13377.

Gan, E. S., Huang, J., & Ito, T. (2013). Functional roles of histone modification, chromatin remodeling and microRNAs in Arabidopsis flower development. *International Review of Cell and Molecular Biology, 305*, 115–161.

Gao, X., Liang, W., Yin, C., Ji, S., Wang, H., Su, X., Guo, C., Kong, H., Xue, H., & Zhang, D. (2010). The SEPALLATA-like gene OsMADS34 is required for rice inflorescence and spikelet development. *Plant Physiology, 153*, 728–740.

Geng, L., Li, Q., Jiao, L., Xiang, Y., Deng, Q., Zhou, D. X., & Zhao, Y. (2023). WOX11 and CRL1 act synergistically to promote crown root development by maintaining cytokinin homeostasis in rice. *The New Phytologist, 237*, 204–216.

Gomez-Mena, C., de Folter, S., Costa, M. M., Angenent, G. C., & Sablowski, R. (2005). Transcriptional program controlled by the floral homeotic gene AGAMOUS during early organogenesis. *Development, 132*, 429–438.

Goslin, K., Finocchio, A., & Wellmer, F. (2023). Floral homeotic factors: A question of specificity. *Plants (Basel), 12*, 1128.

Goto, K., Kyozuka, J., & Bowman, J. L. (2001). Turning floral organs into leaves, leaves into floral organs. *Current Opinion in Genetics & Development, 11*, 449–456.

Grass Phylogeny Working Group. (2001). Phylogeny and subfamilial classification of the grasses (Poaceae). *Annals of the Missouri Botanical Garden, 88*, 373–457.

Han, J. J., Jackson, D., & Martienssen, R. (2012). Pod corn is caused by rearrangement at the Tunicate1 locus. *Plant Cell, 24*, 2733–2744.

Han, Y., Zhang, C., Yang, H., & Jiao, Y. (2014). Cytokinin pathway mediates APETALA1 function in the establishment of determinate floral meristems in Arabidopsis. *Proceedings of the National Academy of Sciences of the United States of America, 111*, 6840–6845.

Heisler, M. G., Jönsson, H., Wenkel, S., & Kaufmann, K. (2022). Context-specific functions of transcription factors controlling plant development: From leaves to flowers. *Current Opinion in Plant Biology, 69*, 102262.

Hepworth, S. R., & Pautot, V. A. (2015). Beyond the divide: Boundaries for patterning and stem cell regulation in plants. *Frontiers in Plant Science, 6*, 1052.

Hepworth, S. R., Zhang, Y., McKim, S., Li, X., & Haughn, G. W. (2005). BLADE-ON-PETIOLE-dependent signaling controls leaf and floral patterning in Arabidopsis. *Plant Cell, 17*, 1434–1448.

Hepworth, S. R., Klenz, J. E., & Haughn, G. W. (2006). UFO in the Arabidopsis inflorescence apex is required for floral-meristem identity and bract suppression. *Planta, 223*, 769–778.

Hileman, L. C. (2014a). Bilateral flower symmetry—how, when and why? *Current Opinion in Plant Biology, 17*, 146–152.

Hileman, L. C. (2014b). Trends in flower symmetry evolution revealed through phylogenetic and developmental genetic advances. *Philosophical Transactions of the Royal Society B: Biological Sciences, 369*, 20130348.

Hirano, H. Y., Tanaka, W., & Toriba, T. (2014). Grass flower development. In J. L. Riechmann & F. Wellmer (Eds.), *Flower development: Methods and protocols* (pp. 57–84). Springer New York.

Hirota, A., Kato, T., Fukaki, H., Aida, M., & Tasaka, M. (2007). The auxin-regulated AP2/EREBP gene PUCHI is required for morphogenesis in the early lateral root primordium of Arabidopsis. *Plant Cell, 19*, 2156–2168.

Hong, L., Qian, Q., Zhu, K., Tang, D., Huang, Z., Gao, L., Li, M., Gu, M., & Cheng, Z. (2010). ELE restrains empty glumes from developing into lemmas. *Journal of Genetics and Genomics = Yi chuan xue bao, 37*, 101–115.

Hong, L., Dumond, M., Zhu, M., Tsugawa, S., Li, C. B., Boudaoud, A., Hamant, O., & Roeder, A. H. K. (2018). Heterogeneity and robustness in plant morphogenesis: From cells to organs. *Annual Review of Plant Biology, 69*, 469–495.

Honma, T., & Goto, K. (2001). Complexes of MADS-box proteins are sufficient to convert leaves into floral organs. *Nature, 409*, 525–529.

Hori, Y., Kurotani, K., Toda, Y., Hattori, T., & Takeda, S. (2014). Overexpression of the JAZ factors with mutated jas domains causes pleiotropic defects in rice spikelet development. *Plant Signaling & Behavior, 9*, e970414.

Houston, K., McKim, S. M., Comadran, J., Bonar, N., Druka, I., Uzrek, N., Cirillo, E., Guzy-Wrobelska, J., Collins, N. C., Halpin, C., Hansson, M., Dockter, C., Druka, A., & Waugh, R. (2013). Variation in the interaction between alleles of HvAPETALA2 and microRNA172 determines the density of grains on the barley inflorescence. *Proceedings of the National Academy of Sciences of the United States of America, 110*, 16675–16680.

Hu, L., Liang, W., Yin, C., Cui, X., Zong, J., Wang, X., Hu, J., & Zhang, D. (2011). Rice *MADS3* regulates ROS homeostasis during late anther development. *Plant Cell, 23*, 515–533.

Hu, Y., Liang, W., Yin, C., Yang, X., Ping, B., Li, A., Jia, R., Chen, M., Luo, Z., Cai, Q., Zhao, X., Zhang, D., & Yuan, Z. (2015). Interactions of OsMADS1 with floral homeotic genes in rice flower development. *Molecular Plant, 8*, 1366–1384.

Hu, Y., Wang, L., Jia, R., Liang, W., Zhang, X., Xu, J., Chen, X., Lu, D., Chen, M., Luo, Z., Xie, J., Cao, L., Xu, B., Yu, Y., Persson, S., Zhang, D., & Yuan, Z. (2021). Rice transcription factor MADS32 regulates floral patterning through interactions with multiple floral homeotic genes. *Journal of Experimental Botany, 72*, 2434–2449.

Huang, P., Jiang, H., Zhu, C., Barry, K., Jenkins, J., Sandor, L., Schmutz, J., Box, M. S., Kellogg, E. A., & Brutnell, T. P. (2017a). Sparse panicle1 is required for inflorescence development in Setaria viridis and maize. *Nature Plants, 3*, 17054.

Huang, Z., Shi, T., Zheng, B., Yumul, R. E., Liu, X., You, C., Gao, Z., Xiao, L., & Chen, X. (2017b). APETALA2 antagonizes the transcriptional activity of AGAMOUS in regulating floral stem cells in Arabidopsis thaliana. *The New Phytologist, 215*, 1197–1209.

Huo, X., Wu, S., Zhu, Z., Liu, F., Fu, Y., Cai, H., Sun, X., Gu, P., Xie, D., Tan, L., & Sun, C. (2017). NOG1 increases grain production in rice. *Nature Communications, 8*, 1497.

Ikeda, M., Mitsuda, N., & Ohme-Takagi, M. (2009). Arabidopsis WUSCHEL is a bifunctional transcription factor that acts as a repressor in stem cell regulation and as an activator in floral patterning. *Plant Cell, 21*, 3493–3505.

Ikeda-Kawakatsu, K., Maekawa, M., Izawa, T., Itoh, J., & Nagato, Y. (2012). ABERRANT PANICLE ORGANIZATION 2/RFL, the rice ortholog of Arabidopsis LEAFY, suppresses the transition from inflorescence meristem to floral meristem through interaction with APO1. *The Plant Journal: For Cell and Molecular Biology, 69*, 168–180.

Ishiguro, S., Kawai-Oda, A., Ueda, J., Nishida, I., & Okada, K. (2001). The DEFECTIVE IN ANTHER DEHISCIENCE gene encodes a novel phospholipase A1 catalyzing the initial step of jasmonic acid biosynthesis, which synchronizes pollen maturation, anther dehiscence, and flower opening in Arabidopsis. *Plant Cell, 13*, 2191–2209.

Itoh, J., Nonomura, K., Ikeda, K., Yamaki, S., Inukai, Y., Yamagishi, H., Kitano, H., & Nagato, Y. (2005). Rice plant development: From zygote to spikelet. *Plant & Cell Physiology, 46*, 23–47.

Jack, T., Fox, G. L., & Meyerowitz, E. M. (1994). Arabidopsis homeotic gene APETALA3 ectopic expression: Transcriptional and posttranscriptional regulation determine floral organ identity. *Cell, 76*, 703–716.

Jack, T., Sieburth, L., & Meyerowitz, E. (1997). Targeted misexpression of AGAMOUS in whorl 2 of Arabidopsis flowers. *The Plant Journal: For Cell and Molecular Biology, 11*, 825–839.

Je, B. I., Gruel, J., Lee, Y. K., Bommert, P., Arevalo, E. D., Eveland, A. L., Wu, Q., Goldshmidt, A., Meeley, R., Bartlett, M., Komatsu, M., Sakai, H., Jonsson, H., & Jackson, D. (2016). Signaling from maize organ primordia via FASCIATED EAR3 regulates stem cell proliferation and yield traits. *Nature Genetics, 48*, 785–791.

Jeon, J. S., Jang, S., Lee, S., Nam, J., Kim, C., Lee, S. H., Chung, Y. Y., Kim, S. R., Lee, Y. H., Cho, Y. G., & An, G. (2000). leafy hull sterile1 is a homeotic mutation in a rice MADS box gene affecting rice flower development. *Plant Cell, 12*, 871–884.

Jeong, S., Trotochaud, A. E., & Clark, S. E. (1999). The Arabidopsis *CLAVATA2* gene encodes a receptor-like protein required for the stability of the CLAVATA1 receptor-like kinase. *Plant Cell, 11*, 1925–1934.

Jetha, K., Theissen, G., & Melzer, R. (2014). Arabidopsis SEPALLATA proteins differ in cooperative DNA-binding during the formation of floral quartet-like complexes. *Nucleic Acids Research, 42*, 10927–10942.

Jiang, W., Zhou, S., Zhang, Q., Song, H., Zhou, D. X., & Zhao, Y. (2017). Transcriptional regulatory network of WOX11 is involved in the control of crown root development, cytokinin signals, and redox in rice. *Journal of Experimental Botany, 68*, 2787–2798.

Jin, Y., Luo, Q., Tong, H., Wang, A., Cheng, Z., Tang, J., Li, D., Zhao, X., Li, X., Wan, J., Jiao, Y., Chu, C., & Zhu, L. (2011). An AT-hook gene is required for palea formation and floral organ number control in rice. *Developmental Biology, 359*, 277–288.

Jumper, J., Evans, R., Pritzel, A., Green, T., Figurnov, M., Ronneberger, O., Tunyasuvunakool, K., Bates, R., Žídek, A., Potapenko, A., Bridgland, A., Meyer, C., Kohl, S. A. A., Ballard, A. J., Cowie, A., Romera-Paredes, B., Nikolov, S., Jain, R., Adler, J., Back, T., Petersen, S., Reiman, D., Clancy, E., Zielinski, M., Steinegger, M., Pacholska, M., Berghammer, T., Bodenstein, S., Silver, D., Vinyals, O., Senior, A. W., Kavukcuoglu, K., Kohli, P., & Hassabis, D. (2021). Highly accurate protein structure prediction with AlphaFold. *Nature, 596*, 583–589.

Käppel, S., Rümpler, F., & Theißen, G. (2023). Cracking the floral quartet code: How do multimers of MIKC(C)-type MADS-domain transcription factors recognize their target genes? *International Journal of Molecular Sciences, 24*, 8253.

Karim, M. R., Hirota, A., Kwiatkowska, D., Tasaka, M., & Aida, M. (2009). A role for Arabidopsis PUCHI in floral meristem identity and bract suppression. *Plant Cell, 21*, 1360–1372.

Kasianov, A. S., Klepikova, A. V., Mayorov, A. V., Buzanov, G. S., Logacheva, M. D., & Penin, A. A. (2023). Interspecific comparison of gene expression profiles using machine learning. *PLoS Computational Biology, 19*, e1010743.

Kaufmann, K., Muino, J. M., Jauregui, R., Airoldi, C. A., Smaczniak, C., Krajewski, P., & Angenent, G. C. (2009). Target genes of the MADS transcription factor SEPALLATA3: Integration of developmental and hormonal pathways in the Arabidopsis flower. *PLoS Biology, 7*, e1000090.

Kaufmann, K., Wellmer, F., Muino, J. M., Ferrier, T., Wuest, S. E., Kumar, V., Serrano-Mislata, A., Madueno, F., Krajewski, P., Meyerowitz, E. M., Angenent, G. C., & Riechmann, J. L. (2010). Orchestration of floral initiation by APETALA1. *Science, 328*, 85–89.

Kayes, J. M., & Clark, S. E. (1998). *CLAVATA2*, a regulator of meristem and organ development in Arabidopsis. *Development, 125*, 3843–3851.

Khanday, I., Yadav, S. R., & Vijayraghavan, U. (2013). Rice LHS1/OsMADS1 controls floret meristem specification by coordinated regulation of transcription factors and hormone signaling pathways. *Plant Physiology, 161*, 1970–1983.

Khanday, I., Das, S., Chongloi, G. L., Bansal, M., Grossniklaus, U., & Vijayraghavan, U. (2016). Genome-wide targets regulated by the OsMADS1 transcription factor reveals its DNA recognition properties. *Plant Physiology, 172*, 372–388.

Kobayashi, K., Maekawa, M., Miyao, A., Hirochika, H., & Kyozuka, J. (2010). PANICLE PHYTOMER2 (PAP2), encoding a SEPALLATA subfamily MADS-box protein, positively controls spikelet meristem identity in rice. *Plant & Cell Physiology, 51*, 47–57.

Kobayashi, K., Yasuno, N., Sato, Y., Yoda, M., Yamazaki, R., Kimizu, M., Yoshida, H., Nagamura, Y., & Kyozuka, J. (2012). Inflorescence meristem identity in rice is specified by overlapping functions of three AP1/FUL-like MADS box genes and *PAP2*, a SEPALLATA MADS box gene. *Plant Cell, 24*, 1848–1859.

Krizek, B. A., & Meyerowitz, E. M. (1996). The Arabidopsis homeotic genes APETALA3 and PISTILLATA are sufficient to provide the B class organ identity function. *Development, 122*, 11–22.

Kyozuka, J., & Shimamoto, K. (2002). Ectopic expression of *OsMADS3*, a rice ortholog of AGAMOUS, caused a homeotic transformation of lodicules to stamens in transgenic rice plants. *Plant & Cell Physiology, 43*, 130–135.

Kyozuka, J., Konishi, S., Nemoto, K., Izawa, T., & Shimamoto, K. (1998). Down-regulation of *RFL*, the *FLO/LFY* homolog of rice, accompanied with panicle branch initiation. *Proceedings of the National Academy of Sciences of the United States of America, 95*, 1979–1982.

Kyozuka, J., Tokunaga, H., & Yoshida, A. (2014). Control of grass inflorescence form by the fine-tuning of meristem phase change. *Current Opinion in Plant Biology, 17*, 110–115.

Lai, X., Vega-Léon, R., Hugouvieux, V., Blanc-Mathieu, R., van der Wal, F., Lucas, J., Silva, C. S., Jourdain, A., Muino, J. M., Nanao, M. H., Immink, R., Kaufmann, K., Parcy, F., Smaczniak, C., & Zubieta, C. (2021). The intervening domain is required for DNA-binding and functional identity of plant MADS transcription factors. *Nature Communications, 12*, 4760.

Lampugnani, E. R., Kilinc, A., & Smyth, D. R. (2013). Auxin controls petal initiation in Arabidopsis. *Development, 140*, 185–194.

Laux, T., Mayer, K. F., Berger, J., & Jurgens, G. (1996). The *WUSCHEL* gene is required for shoot and floral meristem integrity in Arabidopsis. *Development, 122*, 87–96.

Lee, D. Y., & An, G. (2012). Two AP2 family genes, *supernumerary bract* (*SNB*) and *Osindeterminate spikelet 1* (*OsIDS1*), synergistically control inflorescence architecture and floral meristem establishment in rice. *The Plant Journal: For Cell and Molecular Biology, 69*, 445–461.

Lee, D. Y., Lee, J., Moon, S., Park, S. Y., & An, G. (2007). The rice heterochronic gene SUPERNUMERARY BRACT regulates the transition from spikelet meristem to floral meristem. *The Plant Journal: For Cell and Molecular Biology, 49*, 64–78.

Lee, Y. S., Lee, D. Y., Cho, L. H., & An, G. (2014). Rice miR172 induces flowering by suppressing OsIDS1 and SNB, two AP2 genes that negatively regulate expression of Ehd1 and florigens. *Rice (New York, N.Y.), 7*, 31.

Lenhard, M., Jurgens, G., & Laux, T. (2002). The WUSCHEL and SHOOTMERISTEMLESS genes fulfil complementary roles in Arabidopsis shoot meristem regulation. *Development, 129*, 3195–3206.

Li, Q., & Liu, B. (2017). Genetic regulation of maize flower development and sex determination. *Planta, 245*, 1–14.

Li, H., Xue, D., Gao, Z., Yan, M., Xu, W., Xing, Z., Huang, D., Qian, Q., & Xue, Y. (2009). A putative lipase gene *EXTRA GLUME1* regulates both empty-glume fate and spikelet development in rice. *The Plant Journal: For Cell and Molecular Biology, 57*, 593–605.

Li, H., Liang, W., Jia, R., Yin, C., Zong, J., Kong, H., & Zhang, D. (2010). The AGL6-like gene *OsMADS6* regulates floral organ and meristem identities in rice. *Cell Research, 20*, 299–313.

Li, H., Liang, W., Yin, C., Zhu, L., & Zhang, D. (2011a). Genetic interaction of *OsMADS3, DROOPING LEAF*, and *OsMADS13* in specifying rice floral organ identities and meristem determinacy. *Plant Physiology, 156*, 263–274.

Li, H., Liang, W., Hu, Y., Zhu, L., Yin, C., Xu, J., Dreni, L., Kater, M. M., & Zhang, D. (2011b). Rice MADS6 interacts with the floral homeotic genes *SUPERWOMAN1*, *MADS3*, *MADS58*, *MADS13*, and *DROOPING LEAF* in specifying floral organ identities and meristem fate. *Plant Cell, 23*, 2536–2552.

Li, Y. F., Zeng, X. Q., Li, Y., Wang, L., Zhuang, H., Wang, Y., Tang, J., Wang, H. L., Xiong, M., Yang, F. Y., Yuan, X. Z., & He, G. H. (2020). MULTI-FLORET SPIKELET 2, a MYB transcription factor, determines spikelet meristem fate and floral organ identity in rice. *Plant Physiology, 184*, 988–1003.

Li, G., Kuijer, H. N. J., Yang, X., Liu, H., Shen, C., Shi, J., Betts, N., Tucker, M. R., Liang, W., Waugh, R., Burton, R. A., & Zhang, D. (2021a). MADS1 maintains barley spike morphology at high ambient temperatures. *Nature Plants, 7*, 1093–1107.

Li, S., Cao, L., Chen, X., Liu, Y., Persson, S., Hu, J., Chen, M., Chen, Z., Zhang, D., & Yuan, Z. (2021b). Synthetic biosensor for mapping dynamic responses and spatio-temporal distribution of jasmonate in rice. *Plant Biotechnology Journal, 19*, 2392–2394.

Li, C., Zhang, S., Yan, X., Cheng, P., & Yu, H. (2023). Single-nucleus sequencing deciphers developmental trajectories in rice pistils. *Developmental Cell, 58*, 694–708.e4.

Liu, C., Xi, W., Shen, L., Tan, C., & Yu, H. (2009). Regulation of floral patterning by flowering time genes. *Developmental Cell, 16*, 711–722.

Liu, X., Kim, Y. J., Muller, R., Yumul, R. E., Liu, C., Pan, Y., Cao, X., Goodrich, J., & Chen, X. (2011). AGAMOUS terminates floral stem cell maintenance in Arabidopsis by directly repressing WUSCHEL through recruitment of Polycomb Group proteins. *Plant Cell, 23*, 3654–3670.

Liu, C., Wang, C., Wang, G., Becker, C., Zaidem, M., & Weigel, D. (2016a). Genome-wide analysis of chromatin packing in Arabidopsis thaliana at single-gene resolution. *Genome Research, 26*, 1057–1068.

Liu, M., Li, H., Su, Y., Li, W., & Shi, C. (2016b). G1/ELE functions in the development of rice lemmas in addition to determining identities of empty glumes. *Frontiers in Plant Science, 7*, 1006.

Lohmann, J. U., Hong, R. L., Hobe, M., Busch, M. A., Parcy, F., Simon, R., & Weigel, D. (2001). A molecular link between stem cell regulation and floral patterning in Arabidopsis. *Cell, 105*, 793–803.

Lombardo, F., & Yoshida, H. (2015). Interpreting lemma and palea homologies: A point of view from rice floral mutants. *Frontiers in Plant Science, 6*, 61.

Lombardo, F., Kuroki, M., Yao, S. G., Shimizu, H., Ikegaya, T., Kimizu, M., Ohmori, S., Akiyama, T., Hayashi, T., Yamaguchi, T., Koike, S., Yatou, O., & Yoshida, H. (2017). The superwoman1-cleistogamy2 mutant is a novel resource for gene containment in rice. *Plant Biotechnology Journal, 15*, 97–106.

Lopez-Dee, Z. P., Wittich, P., Enrico Pe, M., Rigola, D., Del Buono, I., Gorla, M. S., Kater, M. M., & Colombo, L. (1999). OsMADS13, a novel rice MADS-box gene expressed during ovule development. *Developmental Genetics, 25*, 237–244.

Lou, S., Chen, S., Zhao, X., Chen, L., Zhang, J., Fu, H., Liu, Y. G., & Chen, Y. (2017). The far-upstream regulatory region of RFL is required for its precise spatial-temporal expression for floral development in rice. *Plant Molecular Biology, 93*, 185–195.

Lu, G., Coneva, V., Casaretto, J. A., Ying, S., Mahmood, K., Liu, F., Nambara, E., Bi, Y. M., & Rothstein, S. J. (2015a). OsPIN5b modulates rice (Oryza sativa) plant architecture and yield by changing auxin homeostasis, transport and distribution. *The Plant Journal: For Cell and Molecular Biology, 83*, 913–925.

Lu, Z., Shao, G., Xiong, J., Jiao, Y., Wang, J., Liu, G., Meng, X., Liang, Y., Xiong, G., Wang, Y., & Li, J. (2015b). MONOCULM 3, an ortholog of WUSCHEL in rice, is required for tiller bud formation. *Journal of Genetics and Genomics = Yi chuan xue bao, 42*, 71–78.

Luo, D., Carpenter, R., Vincent, C., Copsey, L., & Coen, E. (1996). Origin of floral asymmetry in Antirrhinum. *Nature, 383*, 794–799.

Luo, D., Carpenter, R., Copsey, L., Vincent, C., Clark, J., & Coen, E. (1999). Control of organ asymmetry in flowers of Antirrhinum. *Cell, 99*, 367–376.

Luo, Q., Zhou, K., Zhao, X., Zeng, Q., Xia, H., Zhai, W., Xu, J., Wu, X., Yang, H., & Zhu, L. (2005). Identification and fine mapping of a mutant gene for palealess spikelet in rice. *Planta, 221*, 222–230.

Ma, X., Zhu, Q., Chen, Y., & Liu, Y. G. (2016). CRISPR/Cas9 platforms for genome editing in plants: Developments and applications. *Molecular Plant, 9*, 961–974.

Malcomber, S. T., & Kellogg, E. A. (2004). Heterogeneous expression patterns and separate roles of the SEPALLATA gene LEAFY HULL STERILE1 in grasses. *Plant Cell, 16*, 1692–1706.

Marand, A. P., Chen, Z., Gallavotti, A., & Schmitz, R. J. (2021). A cis-regulatory atlas in maize at single-cell resolution. *Cell, 184*, 3041–3055.e21.

Mayer, K. F., Schoof, H., Haecker, A., Lenhard, M., Jurgens, G., & Laux, T. (1998). Role of WUSCHEL in regulating stem cell fate in the Arabidopsis shoot meristem. *Cell, 95*, 805–815.

McGonigle, B., Bouhidel, K., & Irish, V. F. (1996). Nuclear localization of the *Arabidopsis APETALA3* and *PISTILLATA* homeotic gene products depends on their simultaneous expression. *Genes & Development, 10*, 1812–1821.

McSteen, P. (2009). Hormonal regulation of branching in grasses. *Plant Physiology, 149*, 46–55.

McSteen, P. (2010). Auxin and monocot development. *Cold Spring Harbor Perspectives in Biology, 2*, a001479.

McSteen, P., & Kellogg, E. A. (2022). Molecular, cellular, and developmental foundations of grass diversity. *Science, 377*, 599–602.

Melzer, R., & Theissen, G. (2009). Reconstitution of 'floral quartets' in vitro involving class B and class E floral homeotic proteins. *Nucleic Acids Research, 37*, 2723–2736.

Melzer, R., Verelst, W., & Theissen, G. (2009). The class E floral homeotic protein SEPALLATA3 is sufficient to loop DNA in 'floral quartet'-like complexes in vitro. *Nucleic Acids Research, 37*, 144–157.

Mendes, M. A., Guerra, R. F., Berns, M. C., Manzo, C., Masiero, S., Finzi, L., Kater, M. M., & Colombo, L. (2013). MADS domain transcription factors mediate short-range DNA looping that is essential for target gene expression in Arabidopsis. *Plant Cell, 25*, 2560–2572.

Meng, Q., Li, X., Zhu, W., Yang, L., Liang, W., Dreni, L., & Zhang, D. (2017). Regulatory network and genetic interactions established by OsMADS34 in rice inflorescence and spikelet morphogenesis. *Journal of Integrative Plant Biology, 59*, 693–707.

Meyerowitz, E. M., Smyth, D. R., & Bowman, J. L. (1989). Abnormal flowers and pattern formation in floral. *Development, 106*, 209–217.

Moon, Y. H., Kang, H. G., Jung, J. Y., Jeon, J. S., Sung, S. K., & An, G. (1999). Determination of the motif responsible for interaction between the rice *APETALA1/AGAMOUS-LIKE9* family proteins using a yeast two-hybrid system. *Plant Physiology, 120*, 1193–1204.

Morel, P., Heijmans, K., Rozier, F., Zethof, J., Chamot, S., Bento, S. R., Vialette-Guiraud, A., Chambrier, P., Trehin, C., & Vandenbussche, M. (2017). Divergence of the Floral A-Function between an Asterid and a Rosid Species. *Plant Cell, 29*, 1605–1621.

Moyroud, E., & Glover, B. J. (2017). The evolution of diverse floral morphologies. *Current Biology: CB, 27*, R941–R951.

Muino, J. M., de Bruijn, S., Pajoro, A., Geuten, K., Vingron, M., Angenent, G. C., & Kaufmann, K. (2016). Evolution of DNA-binding sites of a floral master regulatory transcription factor. *Molecular Biology and Evolution, 33*, 185–200.

Nagasawa, N., Miyoshi, M., Sano, Y., Satoh, H., Hirano, H., Sakai, H., & Nagato, Y. (2003). *SUPERWOMAN1* and *DROOPING LEAF* genes control floral organ identity in rice. *Development, 130*, 705–718.

Nair, S. K., Wang, N., Turuspekov, Y., Pourkheirandish, M., Sinsuwongwat, S., Chen, G., Sameri, M., Tagiri, A., Honda, I., Watanabe, Y., Kanamori, H., Wicker, T., Stein, N., Nagamura, Y., Matsumoto, T., & Komatsuda, T. (2010). Cleistogamous flowering in barley arises from the suppression of microRNA-guided HvAP2 mRNA cleavage. *Proceedings of the National Academy of Sciences of the United States of America, 107*, 490–495.

Nardmann, J., & Werr, W. (2006). The shoot stem cell niche in angiosperms: Expression patterns of WUS orthologues in rice and maize imply major modifications in the course of mono- and dicot evolution. *Molecular Biology and Evolution, 23*, 2492–2504.

Nelms, B., & Walbot, V. (2022). Gametophyte genome activation occurs at pollen mitosis I in maize. *Science, 375*, 424–429.

Neumann, M., Xu, X., Smaczniak, C., Schumacher, J., Yan, W., Blüthgen, N., Greb, T., Jönsson, H., Traas, J., Kaufmann, K., & Muino, J. M. (2022). A 3D gene expression atlas of the floral meristem based on spatial reconstruction of single nucleus RNA sequencing data. *Nature Communications, 13*, 2838.

Norberg, M., Holmlund, M., & Nilsson, O. (2005). The BLADE ON PETIOLE genes act redundantly to control the growth and development of lateral organs. *Development, 132*, 2203–2213.

O'Maoileidigh, D. S., Graciet, E., & Wellmer, F. (2014). Gene networks controlling Arabidopsis thaliana flower development. *The New Phytologist, 201*, 16–30.

O'Maoileidigh, D. S., Thomson, B., Raganelli, A., Wuest, S. E., Ryan, P. T., Kwasniewska, K., Carles, C. C., Graciet, E., & Wellmer, F. (2015). Gene network analysis of Arabidopsis thaliana flower development through dynamic gene perturbations. *The Plant Journal: For Cell and Molecular Biology, 83*, 344–358.

Ohmori, S., Kimizu, M., Sugita, M., Miyao, A., Hirochika, H., Uchida, E., Nagato, Y., & Yoshida, H. (2009). *MOSAIC FLORAL ORGANS1*, an *AGL6*-like MADS box gene, regulates floral organ identity and meristem fate in rice. *Plant Cell, 21*, 3008–3025.

Ohmori, Y., Tanaka, W., Kojima, M., Sakakibara, H., & Hirano, H. Y. (2013). WUSCHEL-RELATED HOMEOBOX4 is involved in meristem maintenance and is negatively regulated by the CLE gene FCP1 in rice. *Plant Cell, 25*, 229–241.

Ohno, C. K., Reddy, G. V., Heisler, M. G., & Meyerowitz, E. M. (2004). The *Arabidopsis JAGGED* gene encodes a zinc finger protein that promotes leaf tissue development. *Development, 131*, 1111–1122.

Oikawa, T., & Kyozuka, J. (2009). Two-step regulation of LAX PANICLE1 protein accumulation in axillary meristem formation in rice. *Plant Cell, 21*, 1095–1108.

ÓMaoiléidigh, D. S., Wuest, S. E., Rae, L., Raganelli, A., Ryan, P. T., Kwasniewska, K., Das, P., Lohan, A. J., Loftus, B., Graciet, E., & Wellmer, F. (2013). Control of reproductive floral organ identity specification in Arabidopsis by the C function regulator AGAMOUS. *Plant Cell, 25*, 2482–2503.

Osnato, M., Matias-Hernandez, L., Aguilar-Jaramillo, A. E., Kater, M. M., & Pelaz, S. (2020). Genes of the RAV family control heading date and carpel development in rice. *Plant Physiology, 183*, 1663–1680.

Osnato, M., Lacchini, E., Pilatone, A., Dreni, L., Grioni, A., Chiara, M., Horner, D., Pelaz, S., & Kater, M. M. (2021). Transcriptome analysis reveals rice MADS13 as an important repressor of the carpel development pathway in ovules. *Journal of Experimental Botany, 72*, 398–414.

Pajoro, A., Madrigal, P., Muino, J. M., Matus, J. T., Jin, J., Mecchia, M. A., Debernardi, J. M., Palatnik, J. F., Balazadeh, S., Arif, M., O'Maoileidigh, D. S., Wellmer, F., Krajewski, P., Riechmann, J. L., Angenent, G. C., & Kaufmann, K. (2014). Dynamics of chromatin accessibility and gene regulation by MADS-domain transcription factors in flower development. *Genome Biology, 15*, R41.

Pastore, J. J., Limpuangthip, A., Yamaguchi, N., Wu, M. F., Sang, Y., Han, S. K., Malaspina, L., Chavdaroff, N., Yamaguchi, A., & Wagner, D. (2011). LATE MERISTEM IDENTITY2 acts together with LEAFY to activate APETALA1. *Development, 138*, 3189–3198.

Pautler, M., Tanaka, W., Hirano, H. Y., & Jackson, D. (2013). Grass meristems I: Shoot apical meristem maintenance, axillary meristem determinacy and the floral transition. *Plant & Cell Physiology, 54*, 302–312.

Pautler, M., Eveland, A. L., LaRue, T., Yang, F., Weeks, R., Lunde, C., Je, B. I., Meeley, R., Komatsu, M., Vollbrecht, E., Sakai, H., & Jackson, D. (2015). FASCIATED EAR4 encodes a bZIP transcription factor that regulates shoot meristem size in maize. *Plant Cell, 27*, 104–120.

Pelaz, S., Ditta, G. S., Baumann, E., Wisman, E., & Yanofsky, M. F. (2000). B and C floral organ identity functions require *SEPALLATA* MADS-box genes. *Nature, 405*, 200–203.

Pelaz, S., Tapia-Lopez, R., Alvarez-Buylla, E. R., & Yanofsky, M. F. (2001). Conversion of leaves into petals in Arabidopsis. *Current Biology: CB, 11*, 182–184.

Peng, Z., Jiao, Y., & Wang, Y. (2023). Live imaging of Arabidopsis shoot primordia via a confocal laser scanning microscope. *STAR Protocols, 4*, 102217.

Pinyopich, A., Ditta, G. S., Savidge, B., Liljegren, S. J., Baumann, E., Wisman, E., & Yanofsky, M. F. (2003). Assessing the redundancy of MADS-box genes during carpel and ovule development. *Nature, 424*, 85–88.

Poursarebani, N., Trautewig, C., Melzer, M., Nussbaumer, T., Lundqvist, U., Rutten, T., Schmutzer, T., Brandt, R., Himmelbach, A., Altschmied, L., Koppolu, R., Youssef, H. M., Sibout, R., Dalmais, M., Bendahmane, A., Stein, N., Xin, Z., & Schnurbusch, T. (2020). COMPOSITUM 1 contributes to the architectural simplification of barley inflorescence via meristem identity signals. *Nature Communications, 11*, 5138.

Prasad, K., & Vijayraghavan, U. (2003). Double-stranded RNA interference of a rice *PI/GLO* paralog, *OsMADS2*, uncovers its second-whorl-specific function in floral organ patterning. *Genetics, 165*, 2301–2305.

Prasad, K., Sriram, P., Kumar, C. S., Kushalappa, K., & Vijayraghavan, U. (2001). Ectopic expression of rice *OsMADS1* reveals a role in specifying the lemma and palea, grass floral organs analogous to sepals. *Development Genes and Evolution, 211*, 281–290.

Prasad, K., Parameswaran, S., & Vijayraghavan, U. (2005). *OsMADS1*, a rice MADS-box factor, controls differentiation of specific cell types in the lemma and palea and is an early-acting regulator of inner floral organs. *The Plant Journal: For Cell and Molecular Biology, 43*, 915–928.

Prunet, N. (2017). Live confocal imaging of developing Arabidopsis flowers. *Journal of Visualized Experiments*. https://doi.org/10.3791/55156

Prunet, N., & Jack, T. P. (2014). Flower development in Arabidopsis: There is more to it than learning your ABCs. In J. L. Riechmann & F. Wellmer (Eds.), *Flower development: Methods and protocols* (pp. 3–33). Springer New York.

Prunet, N., Morel, P., Negrutiu, I., & Trehin, C. (2009). Time to stop: Flower meristem termination. *Plant Physiology, 150*, 1764–1772.

Prunet, N., Yang, W., Das, P., Meyerowitz, E. M., & Jack, T. P. (2017). SUPERMAN prevents class B gene expression and promotes stem cell termination in the fourth whorl of Arabidopsis thaliana flowers. *Proceedings of the National Academy of Sciences of the United States of America, 114*, 7166–7171.

Prusinkiewicz, P., & Runions, A. (2012). Computational models of plant development and form. *The New Phytologist, 193*, 549–569.

Puranik, S., Acajjaoui, S., Conn, S., Costa, L., Conn, V., Vial, A., Marcellin, R., Melzer, R., Brown, E., Hart, D., Theissen, G., Silva, C. S., Parcy, F., Dumas, R., Nanao, M., & Zubieta, C. (2014). Structural basis for the oligomerization of the MADS domain transcription factor SEPALLATA3 in Arabidopsis. *Plant Cell, 26*, 3603–3615.

Qi, S., Chen, S., Wang, W., Li, W., Xu, X., Zhong, M., Lin, J., Yu, Y., Chen, Y., & Li, H. (2021). Alternative splicing and duplication of PI-like genes in maize. *Gene, 769*, 145064.

Qin, X., Sun, J., Li, R., Cen, W., Li, R., & Luo, J. (2025) G1-mediated OsMADS1 repression regulates sterile lemma development in rice. *Plant Cell Reports, 44*(10). https://doi.org/10.1007/s00299-025-03619-2

Reinheimer, R., & Kellogg, E. A. (2009). Evolution of AGL6-like MADS box genes in grasses (Poaceae): Ovule expression is ancient and palea expression is new. *Plant Cell, 21*, 2591–2605.

Ren, D., Li, Y., Zhao, F., Sang, X., Shi, J., Wang, N., Guo, S., Ling, Y., Zhang, C., Yang, Z., & He, G. (2013). MULTI-FLORET SPIKELET1, which encodes an AP2/ERF protein, determines spikelet meristem fate and sterile lemma identity in rice. *Plant Physiology, 162*, 872–884.

Ren, D., Hu, J., Xu, Q., Cui, Y., Zhang, Y., Zhou, T., Rao, Y., Xue, D., Zeng, D., Zhang, G., Gao, Z., Zhu, L., Shen, L., Chen, G., Guo, L., & Qian, Q. (2018). FZP determines grain size and sterile lemma fate in rice. *Journal of Experimental Botany, 69*, 4853–4866.

Ren, D., Cui, Y., Hu, H., Xu, Q., Rao, Y., Yu, X., Zhang, Y., Wang, Y., Peng, Y., Zeng, D., Hu, J., Zhang, G., Gao, Z., Zhu, L., Chen, G., Shen, L., Zhang, Q., Guo, L., & Qian, Q. (2019). AH2 encodes a MYB domain protein that determines hull fate and affects grain yield and quality in rice. *The Plant Journal: For Cell and Molecular Biology, 100*, 813–824.

Ren, D., Li, Y., He, G., & Qian, Q. (2020). Multifloret spikelet improves rice yield. *New Phytologist, 225*(6), 2301–2306. https://doi.org/10.1111/nph.v225.6; https://doi.org/10.1111/nph.16303

Rieu, P., Turchi, L., Thevenon, E., Zarkadas, E., Nanao, M., Chahtane, H., Tichtinsky, G., Lucas, J., Blanc-Mathieu, R., Zubieta, C., Schoehn, G., & Parcy, F. (2023). The F-box protein UFO controls flower development by redirecting the master transcription factor LEAFY to new cis-elements. *Nature Plants, 9*, 315–329.

Rodríguez-Cazorla, E., Ripoll, J. J., Ortuño-Miquel, S., Martínez-Laborda, A., & Vera, A. (2020). Dissection of the Arabidopsis HUA-PEP gene activity reveals that ovule fate specification requires restriction of the floral A-function. *The New Phytologist, 227*, 1222–1234.

Rudall, P. J., & Bateman, R. M. (2004). Evolution of zygomorphy in monocot flowers: Iterative patterns and developmental constraints. *New Phytologist, 162*, 25–44.

Ruelens, P., Zhang, Z., van Mourik, H., Maere, S., Kaufmann, K., & Geuten, K. (2017). The origin of floral organ identity quartets. *Plant Cell, 29*, 229–242.

Rümpler, F., Tessari, C., Gramzow, L., Gafert, C., Blohs, M., & Theißen, G. (2023). The origin of floral quartet formation-ancient exon duplications shaped the evolution of MIKC-type MADS-domain transcription factor interactions. *Molecular Biology and Evolution, 40*, msad088.

Sablowski, R. (2015). Control of patterning, growth, and differentiation by floral organ identity genes. *Journal of Experimental Botany, 66*, 1065–1073.

Sang, X., Li, Y., Luo, Z., Ren, D., Fang, L., Wang, N., Zhao, F., Ling, Y., Yang, Z., Liu, Y., & He, G. (2012). CHIMERIC FLORAL ORGANS1, encoding a monocot-specific MADS box protein, regulates floral organ identity in rice. *Plant Physiology, 160*, 788–807.

Sawa, S., Ito, T., Shimura, Y., & Okada, K. (1999). FILAMENTOUS FLOWER controls the formation and development of arabidopsis inflorescences and floral meristems. *Plant Cell, 11*, 69–86.

Sentoku, N., Kato, H., Kitano, H., & Imai, R. (2005). OsMADS22, an STMADS11-like MADS-box gene of rice, is expressed in non-vegetative tissues and its ectopic expression induces spikelet meristem indeterminacy. *Molecular Genetics and Genomics: MGG, 273*, 1–9.

Seok, H. Y., Park, H. Y., Park, J. I., Lee, Y. M., Lee, S. Y., An, G., & Moon, Y. H. (2010). Rice ternary MADS protein complexes containing class B MADS heterodimer. *Biochemical and Biophysical Research Communications, 401*, 598–604.

Shang, E., Wang, X., Li, T., Guo, F., Ito, T., & Sun, B. (2021). Robust control of floral meristem determinacy by position-specific multifunctions of KNUCKLES. *Proceedings of the National Academy of Sciences, 118*, e2102826118.

Shi, B., & Vernoux, T. (2022). Hormonal control of cell identity and growth in the shoot apical meristem. *Current Opinion in Plant Biology, 65*, 102111.

Sieburth, L. E., Running, M. P., & Meyerowitz, E. M. (1995). Genetic separation of third and fourth whorl functions of AGAMOUS. *Plant Cell, 7*, 1249–1258.

Silveira, S. R., Le Gloanec, C., Gómez-Felipe, A., Routier-Kierzkowska, A. L., & Kierzkowski, D. (2022). Live-imaging provides an atlas of cellular growth dynamics in the stamen. *Plant Physiology, 188*, 769–781.

Smaczniak, C., Immink, R. G., Muino, J. M., Blanvillain, R., Busscher, M., Busscher-Lange, J., Dinh, Q. D., Liu, S., Westphal, A. H., Boeren, S., Parcy, F., Xu, L., Carles, C. C., Angenent, G. C., & Kaufmann, K. (2012). Characterization of MADS-domain transcription factor complexes in Arabidopsis flower development. *Proceedings of the National Academy of Sciences of the United States of America, 109*, 1560–1565.

Smaczniak, C., Muino, J. M., Chen, D., Angenent, G. C., & Kaufmann, K. (2017). Differences in DNA binding specificity of floral homeotic protein complexes predict organ-specific target genes. *Plant Cell, 29*, 1822–1835.

Soltis, P. S., & Soltis, D. E. (2014). Flower diversity and angiosperm diversification. In J. L. Riechmann & F. Wellmer (Eds.), *Flower development: Methods and protocols* (pp. 85–102). Springer New York.

Somssich, M., Je, B. I., Simon, R., & Jackson, D. (2016). CLAVATA-WUSCHEL signaling in the shoot meristem. *Development, 143*, 3238–3248.

Song, Q., Ando, A., Jiang, N., Ikeda, Y., & Chen, Z. J. (2020). Single-cell RNA-seq analysis reveals ploidy-dependent and cell-specific transcriptome changes in Arabidopsis female gametophytes. *Genome Biology, 21*, 178.

Soreng, R. J., Peterson, P. M., Romaschenko, K., Davidse, G., Teisher, J. K., Clark, L. G., Barberá, P., Gillespie, L. J., & Zuloaga, F. O. (2017). A worldwide phylogenetic classification of the Poaceae (Gramineae) II: An update and a comparison of two 2015 classifications. *Journal of Systematics and Evolution, 55*, 259–290.

Soyk, S., Lemmon, Z. H., Oved, M., Fisher, J., Liberatore, K. L., Park, S. J., Goren, A., Jiang, K., Ramos, A., van der Knaap, E., Van Eck, J., Zamir, D., Eshed, Y., & Lippman, Z. B. (2017). Bypassing negative epistasis on yield in tomato imposed by a domestication gene. *Cell, 169*, 1142–1155.e12.

Stebbins, G. L. (1956). Cytogenetics and evolution of the grass family. *American Journal of Botany, 43*, 890–905.

Stewart, D., Graciet, E., & Wellmer, F. (2016). Molecular and regulatory mechanisms controlling floral organ development. *The FEBS Journal, 283*, 1823–1830.

Strable, J., Wallace, J. G., Unger-Wallace, E., Briggs, S., Bradbury, P. J., Buckler, E. S., & Vollbrecht, E. (2017). Maize YABBY genes drooping leaf1 and drooping leaf2 regulate plant architecture. *Plant Cell, 29*, 1622–1641.

Struk, S., Jacobs, A., Sánchez Martín-Fontecha, E., Gevaert, K., Cubas, P., & Goormachtig, S. (2019). Exploring the protein–protein interaction landscape in plants. *Plant, Cell & Environment, 42*, 387–409.

Sugiyama, S. H., Tanaka, W., & Hirano, H. Y. (2017). Polar patterning of the spikelet is disrupted in the two opposite lemma mutant in rice. *Genes & Genetic Systems, 91*, 193–200.

Sun, B., & Ito, T. (2015). Regulation of floral stem cell termination in Arabidopsis. *Frontiers in Plant Science, 6*, 17.

Sun, B., Xu, Y., Ng, K. H., & Ito, T. (2009). A timing mechanism for stem cell maintenance and differentiation in the Arabidopsis floral meristem. *Genes & Development, 23*, 1791–1804.

Sun, B., Looi, L. S., Guo, S., He, Z., Gan, E. S., Huang, J., Xu, Y., Wee, W. Y., & Ito, T. (2014). Timing mechanism dependent on cell division is invoked by Polycomb eviction in plant stem cells. *Science, 343*, 1248559.

Suzaki, T., Sato, M., Ashikari, M., Miyoshi, M., Nagato, Y., & Hirano, H. Y. (2004). The gene *FLORAL ORGAN NUMBER1* regulates floral meristem size in rice and encodes a leucine-rich repeat receptor kinase orthologous to *Arabidopsis* CLAVATA1. *Development, 131*, 5649–5657.

Suzaki, T., Toriba, T., Fujimoto, M., Tsutsumi, N., Kitano, H., & Hirano, H. Y. (2006). Conservation and diversification of meristem maintenance mechanism in *Oryza sativa*: Function of the *FLORAL ORGAN NUMBER2* gene. *Plant & Cell Physiology, 47*, 1591–1602.

Suzaki, T., Yoshida, A., & Hirano, H. Y. (2008). Functional diversification of CLAVATA3-related CLE proteins in meristem maintenance in rice. *Plant Cell, 20*, 2049–2058.

Suzaki, T., Ohneda, M., Toriba, T., Yoshida, A., & Hirano, H. Y. (2009). *FON2 SPARE1* redundantly regulates floral meristem maintenance with *FLORAL ORGAN NUMBER2* in rice. *PLoS Genetics, 5*, e1000693.

Taguchi-Shiobara, F., Yuan, Z., Hake, S., & Jackson, D. (2001). The *fasciated ear2* gene encodes a leucine-rich repeat receptor-like protein that regulates shoot meristem proliferation in maize. *Genes & Development, 15*, 2755–2766.

Tanaka, W., & Hirano, H. Y. (2020). Antagonistic action of TILLERS ABSENT1 and FLORAL ORGAN NUMBER2 regulates stem cell maintenance during axillary meristem development in rice. *The New Phytologist, 225*, 974–984.
Tanaka, W., Toriba, T., Ohmori, Y., Yoshida, A., Kawai, A., Mayama-Tsuchida, T., Ichikawa, H., Mitsuda, N., Ohme-Takagi, M., & Hirano, H. Y. (2012). The YABBY gene TONGARI-BOUSHI1 is involved in lateral organ development and maintenance of meristem organization in the rice spikelet. *Plant Cell, 24*, 80–95.
Tanaka, W., Pautler, M., Jackson, D., & Hirano, H. Y. (2013). Grass meristems II: Inflorescence architecture, flower development and meristem fate. *Plant & Cell Physiology, 54*, 313–324.
Tanaka, W., Toriba, T., & Hirano, H.-Y. (2014). Flower development in rice. In F. Fornara (Ed.), *The molecular genetics of floral transition and flower development* (pp. 221–262). Academic Press.
Tanaka, W., Ohmori, Y., Ushijima, T., Matsusaka, H., Matsushita, T., Kumamaru, T., Kawano, S., & Hirano, H. Y. (2015). Axillary meristem formation in rice requires the WUSCHEL Ortholog TILLERS ABSENT1. *Plant Cell, 27*, 1173–1184.
Tanaka, W., Toriba, T., & Hirano, H. Y. (2017). Three TOB1-related YABBY genes are required to maintain proper function of the spikelet and branch meristems in rice. *The New Phytologist, 215*, 825–839.
Tanaka, W., Ohmori, S., Kawakami, N., & Hirano, H. Y. (2021). Flower meristem maintenance by TILLERS ABSENT 1 is essential for ovule development in rice. *Development (Cambridge, England), 148*, dev199932.
Theissen, G. (2001). Development of floral organ identity: Stories from the MADS house. *Current Opinion in Plant Biology, 4*, 75–85.
Theissen, G., & Melzer, R. (2007). Molecular mechanisms underlying origin and diversification of the angiosperm flower. *Annals of Botany, 100*, 603–619.
Theissen, G., & Saedler, H. (2001). Plant biology. Floral quartets. *Nature, 409*, 469–471.
Theissen, G., Melzer, R., & Rumpler, F. (2016). MADS-domain transcription factors and the floral quartet model of flower development: Linking plant development and evolution. *Development, 143*, 3259–3271.
Thomson, B., & Wellmer, F. (2019). Molecular regulation of flower development. *Current Topics in Developmental Biology, 131*, 185–210.
Thomson, B., Zheng, B., & Wellmer, F. (2017). Floral organogenesis: When knowing your ABCs is not enough. *Plant Physiology, 173*, 56–64.
Trevaskis, B., Tadege, M., Hemming, M. N., Peacock, W. J., Dennis, E. S., & Sheldon, C. (2007). Short vegetative phase-like MADS-box genes inhibit floral meristem identity in barley. *Plant Physiology, 143*, 225–235.
Uji, Y., Taniguchi, S., Tamaoki, D., Shishido, H., Akimitsu, K., & Gomi, K. (2016). Overexpression of OsMYC2 results in the up-regulation of early JA-Rresponsive genes and bacterial blight resistance in rice. *Plant & Cell Physiology, 57*, 1814–1827.
van Goethe, J. W. (1790). *Versuch die Metamorphose der Pflanzen zu erklären*. Ettinger.
Varadi, M., Anyango, S., Deshpande, M., Nair, S., Natassia, C., Yordanova, G., Yuan, D., Stroe, O., Wood, G., Laydon, A., Žídek, A., Green, T., Tunyasuvunakool, K., Petersen, S., Jumper, J., Clancy, E., Green, R., Vora, A., Lutfi, M., Figurnov, M., Cowie, A., Hobbs, N., Kohli, P., Kleywegt, G., Birney, E., Hassabis, D., & Velankar, S. (2021). AlphaFold Protein Structure Database: Massively expanding the structural coverage of protein-sequence space with high-accuracy models. *Nucleic Acids Research, 50*, D439–D444.
Varadi, M., Bertoni, D., Magana, P., Paramval, U., Pidruchna, I., Radhakrishnan, M., Tsenkov, M., Nair, S., Mirdita, M., Yeo, J., Kovalevskiy, O., Tunyasuvunakool, K., Laydon, A., Žídek, A., Tomlinson, H., Hariharan, D., Abrahamson, J., Green, T., Jumper, J., Birney, E., Steinegger, M., Hassabis, D., & Velankar, S. (2023). AlphaFold protein structure database in 2024: Providing structure coverage for over 214 million protein sequences. *Nucleic Acids Research, 52*, D368.
Wagner, D. (2017). Key developmental transitions during flower morphogenesis and their regulation. *Current Opinion in Genetics & Development, 45*, 44–50.

Walcher-Chevillet, C. L., & Kramer, E. M. (2016). Breaking the mold: Understanding the evolution and development of lateral organs in diverse plant models. *Current Opinion in Genetics & Development, 39*, 79–84.

Wang, H., Nussbaum-Wagler, T., Li, B., Zhao, Q., Vigouroux, Y., Faller, M., Bomblies, K., Lukens, L., & Doebley, J. F. (2005). The origin of the naked grains of maize. *Nature, 436*, 714–719.

Wang, K., Tang, D., Hong, L., Xu, W., Huang, J., Li, M., Gu, M., Xue, Y., & Cheng, Z. (2010). *DEP* and *AFO* regulate reproductive habit in rice. *PLoS Genetics, 6*, e1000818.

Wang, N., Li, Y. F., Sang, X. C., Ling, Y. H., Zhao, F. M., Yang, Z. L., & He, G. H. (2013). nonstop glumes (nsg), a novel mutant affects spikelet development in rice. *Genes & Genomics, 35*, 149–157.

Wang, W., Li, G., Zhao, J., Chu, H., Lin, W., Zhang, D., Wang, Z., & Liang, W. (2014). Dwarf Tiller1, a Wuschel-related homeobox transcription factor, is required for tiller growth in rice. *PLoS Genetics, 10*, e1004154.

Wang, C., Liu, C., Roqueiro, D., Grimm, D., Schwab, R., Becker, C., Lanz, C., & Weigel, D. (2015a). Genome-wide analysis of local chromatin packing in Arabidopsis thaliana. *Genome Research, 25*, 246–256.

Wang, H., Studer, A. J., Zhao, Q., Meeley, R., & Doebley, J. F. (2015b). Evidence that the origin of naked kernels during maize domestication was caused by a single amino acid substitution in tga1. *Genetics, 200*, 965–974.

Wang, H., Zhang, L., Cai, Q., Hu, Y., Jin, Z., Zhao, X., Fan, W., Huang, Q., Luo, Z., Chen, M., Zhang, D., & Yuan, Z. (2015c). OsMADS32 interacts with PI-like proteins and regulates rice flower development. *Journal of Integrative Plant Biology, 57*, 504–513.

Wang, L., Zeng, X.-Q., Zhuang, H., Shen, Y.-L., Chen, H., Wang, Z.-W., Long, J.-C., Ling, Y.-H., He, G.-H., & Li, Y.-F. (2016). Ectopic expression of OsMADS1 caused dwarfism and spikelet alteration in rice. *Plant Growth Regulation, 81*, 433–442.

Wang, J., Tian, C., Zhang, C., Shi, B., Cao, X., Zhang, T. Q., Zhao, Z., Wang, J. W., & Jiao, Y. (2017). Cytokinin signaling activates WUSCHEL expression during axillary meristem initiation. *Plant Cell, 29*, 1373–1387.

Wang, G., Li, X., Shen, W., Li, M. W., Huang, M., Zhang, J., & Li, H. (2022a). The chromatin accessibility landscape of pistils and anthers in rice. *Plant Physiology, 190*, 2797–2811.

Wang, Y., Bi, X., & Zhong, J. (2022b). Revisiting the origin and identity specification of the spikelet: A structural innovation in grasses (Poaceae). *Plant Physiology, 190*, 60–71.

Wang, L., Li, Q. L., Hu, J. P., & Yuan, Z. (2024a) Neofunctionalization of B-class genes in regulating rice flower development. *Seed Biology, 3*(1), 0–0. https://doi.org/10.48130/seedbio-0024-0012

Wang, G., Wu, Z., & Sun, B. (2024b). KNUCKLES regulates floral meristem termination by controlling auxin distribution and cytokinin activity. *The Plant Cell 37*(1). https://doi.org/10.1093/plcell/koae312

Wasternack, C. (2007). Jasmonates: An update on biosynthesis, signal transduction and action in plant stress response, growth and development. *Annals of Botany, 100*, 681–697.

Wasternack, C. (2014). Action of jasmonates in plant stress responses and development—applied aspects. *Biotechnology Advances, 32*, 31–39.

Weigel, D., Alvarez, J., Smyth, D. R., Yanofsky, M. F., & Meyerowitz, E. M. (1992). LEAFY controls floral meristem identity in Arabidopsis. *Cell, 69*, 843–859.

Wellmer, F., Graciet, E., & Riechmann, J. L. (2014a). Specification of floral organs in Arabidopsis. *Journal of Experimental Botany, 65*, 1–9.

Wellmer, F., Bowman, J. L., Davies, B., Ferrandiz, C., Fletcher, J. C., Franks, R. G., Graciet, E., Gregis, V., Ito, T., Jack, T. P., Jiao, Y., Kater, M. M., Ma, H., Meyerowitz, E. M., Prunet, N., & Riechmann, J. L. (2014b). Flower development: open questions and future directions. In J. L. Riechmann & F. Wellmer (Eds.), *Flower development: Methods and protocols* (pp. 103–124). Springer New York.

Whipple, C. J. (2017). Grass inflorescence architecture and evolution: The origin of novel signaling centers. *The New Phytologist, 216*, 367–372.

Whipple, C. J., & Schmidt, R. J. (2006). Genetics of grass flower development. In J. H. L. P. S. S. D. E. Soltis & J. A. Callow (Eds.), *Developmental genetics of the flower* (pp. 385–424). Academic Press.

Whipple, C. J., Hall, D. H., DeBlasio, S., Taguchi-Shiobara, F., Schmidt, R. J., & Jackson, D. P. (2010). A conserved mechanism of bract suppression in the grass family. *Plant Cell, 22*, 565–578.

Wils, C. R., & Kaufmann, K. (2017). Gene-regulatory networks controlling inflorescence and flower development in Arabidopsis thaliana. *Biochimica et Biophysica Acta (BBA) – Gene Regulatory Mechanisms, 1860*, 95–105.

Wingen, L. U., Munster, T., Faigl, W., Deleu, W., Sommer, H., Saedler, H., & Theissen, G. (2012). Molecular genetic basis of pod corn (Tunicate maize). *Proceedings of the National Academy of Sciences of the United States of America, 109*, 7115–7120.

Winter, C. M., Austin, R. S., Blanvillain-Baufume, S., Reback, M. A., Monniaux, M., Wu, M. F., Sang, Y., Yamaguchi, A., Yamaguchi, N., Parker, J. E., Parcy, F., Jensen, S. T., Li, H., & Wagner, D. (2011). LEAFY target genes reveal floral regulatory logic, cis motifs, and a link to biotic stimulus response. *Developmental Cell, 20*, 430–443.

Wollmann, H., Mica, E., Todesco, M., Long, J. A., & Weigel, D. (2010). On reconciling the interactions between APETALA2, miR172 and AGAMOUS with the ABC model of flower development. *Development, 137*, 3633–3642.

Wu, F., Shi, X., Lin, X., Liu, Y., Chong, K., Theissen, G., & Meng, Z. (2017). The ABCs of flower development: Mutational analysis of AP1/FUL-like genes in rice provides evidence for a homeotic (A)-function in grasses. *The Plant Journal: For Cell and Molecular Biology, 89*, 310–324.

Wuest, S. E., O'Maoileidigh, D. S., Rae, L., Kwasniewska, K., Raganelli, A., Hanczaryk, K., Lohan, A. J., Loftus, B., Graciet, E., & Wellmer, F. (2012). Molecular basis for the specification of floral organs by APETALA3 and PISTILLATA. *Proceedings of the National Academy of Sciences of the United States of America, 109*, 13452–13457.

Xia, T., Chen, H., Dong, S., Ma, Z., Ren, H., Zhu, X., Fang, X., & Chen, F. (2020). OsWUS promotes tiller bud growth by establishing weak apical dominance in rice. *The Plant Journal: For Cell and Molecular Biology, 104*, 1635–1647.

Xu, M., Hu, T., McKim, S. M., Murmu, J., Haughn, G. W., & Hepworth, S. R. (2010). Arabidopsis BLADE-ON-PETIOLE1 and 2 promote floral meristem fate and determinacy in a previously undefined pathway targeting APETALA1 and AGAMOUS-LIKE24. *The Plant Journal: For Cell and Molecular Biology, 63*, 974–989.

Xu, W., Tao, J., Chen, M., Dreni, L., Luo, Z., Hu, Y., Liang, W., & Zhang, D. (2017). Interactions between FLORAL ORGAN NUMBER4 and floral homeotic genes in regulating rice flower development. *Journal of Experimental Botany, 68*, 483–498.

Xu, X., Crow, M., Rice, B. R., Li, F., Harris, B., Liu, L., Demesa-Arevalo, E., Lu, Z., Wang, L., Fox, N., Wang, X., Drenkow, J., Luo, A., Char, S. N., Yang, B., Sylvester, A. W., Gingeras, T. R., Schmitz, R. J., Ware, D., Lipka, A. E., Gillis, J., & Jackson, D. (2021). Single-cell RNA sequencing of developing maize ears facilitates functional analysis and trait candidate gene discovery. *Developmental Cell, 56*, 557–568.e6.

Xu, W., Zhu, W., Yang, L., Liang, W., Li, H., Yang, L., Chen, M., Luo, Z., Huang, G., Duan, L., Dreni, L., & Zhang, D. (2022). SMALL REPRODUCTIVE ORGANS, a SUPERMAN-like transcription factor, regulates stamen and pistil growth in rice. *The New Phytologist, 233*, 1701–1718.

Xu, Q., Wang, Y., Yue, Y., Chen, Z., Zhou, D. X., & Zhao, Y. (2023). Transcription factor WOX11 regulates protein translation via ribosome protein acetylation in rice roots. *Plant Physiology, 191*, 2224–2228.

Yadav, S. R., Prasad, K., & Vijayraghavan, U. (2007). Divergent regulatory *OsMADS2* functions control size, shape and differentiation of the highly derived rice floret second-whorl organ. *Genetics, 176*, 283–294.

Yamaguchi, T., Nagasawa, N., Kawasaki, S., Matsuoka, M., Nagato, Y., & Hirano, H. Y. (2004). The *YABBY* gene *DROOPING LEAF* regulates carpel specification and midrib development in *Oryza sativa. Plant Cell, 16*, 500–509.

Yamaguchi, T., Lee, D. Y., Miyao, A., Hirochika, H., An, G., & Hirano, H. Y. (2006). Functional diversification of the two C-class MADS box genes *OSMADS3* and *OSMADS58* in *Oryza sativa. Plant Cell, 18*, 15–28.

Yamaguchi, N., Wu, M. F., Winter, C. M., Berns, M. C., Nole-Wilson, S., Yamaguchi, A., Coupland, G., Krizek, B. A., & Wagner, D. (2013). A molecular framework for auxin-mediated initiation of flower primordia. *Developmental Cell, 24*, 271–282.

Yamaguchi, N., Huang, J., Xu, Y., Tanoi, K., & Ito, T. (2017). Fine-tuning of auxin homeostasis governs the transition from floral stem cell maintenance to gynoecium formation. *Nature Communications, 8*, 1125.

Yamaguchi, N., Huang, J., Tatsumi, Y., Abe, M., Sugano, S. S., Kojima, M., Takebayashi, Y., Kiba, T., Yokoyama, R., Nishitani, K., Sakakibara, H., & Ito, T. (2018). Chromatin-mediated feed-forward auxin biosynthesis in floral meristem determinacy. *Nature Communications, 9*, 5290.

Yamaki, S., Nagato, Y., Kurata, N., & Nonomura, K. (2011). Ovule is a lateral organ finally differentiated from the terminating floral meristem in rice. *Developmental Biology, 351*, 208–216.

Yamburenko, M. V., Kieber, J. J., & Schaller, G. E. (2017). Dynamic patterns of expression for genes regulating cytokinin metabolism and signaling during rice inflorescence development. *PLoS One, 12*, e0176060.

Yan, Y., Christensen, S., Isakeit, T., Engelberth, J., Meeley, R., Hayward, A., Emery, R. J., & Kolomiets, M. V. (2012). Disruption of *OPR7* and *OPR8* reveals the versatile functions of jasmonic acid in maize development and defense. *Plant Cell, 24*, 1420–1436.

Yan, D., Zhang, X., Zhang, L., Ye, S., Zeng, L., Liu, J., Li, Q., & He, Z. (2015). Curved chimeric palea 1 encoding an EMF1-like protein maintains epigenetic repression of OsMADS58 in rice palea development. *The Plant Journal: For Cell and Molecular Biology, 82*, 12–24.

Yan, W., Chen, D., & Kaufmann, K. (2016). Molecular mechanisms of floral organ specification by MADS domain proteins. *Current Opinion in Plant Biology, 29*, 154–162.

Yang, Y., & Jack, T. (2004). Defining subdomains of the K domain important for protein-protein interactions of plant MADS proteins. *Plant Molecular Biology, 55*, 45–59.

Yang, Y., Fanning, L., & Jack, T. (2003). The K domain mediates heterodimerization of the Arabidopsis floral organ identity proteins, APETALA3 and PISTILLATA. *The Plant Journal: For Cell and Molecular Biology, 33*, 47–59.

Yang, F., Bui, H. T., Pautler, M., Llaca, V., Johnston, R., Lee, B. H., Kolbe, A., Sakai, H., & Jackson, D. (2015). A maize glutaredoxin gene, abphyl2, regulates shoot meristem size and phyllotaxy. *Plant Cell, 27*, 121–131.

Yang, J., Yuan, Z., Meng, Q., Huang, G., Perin, C., Bureau, C., Meunier, A. C., Ingouff, M., Bennett, M. J., Liang, W., & Zhang, D. (2017). Dynamic regulation of auxin response during rice development revealed by newly established hormone biosensor markers. *Frontiers in Plant Science, 8*, 256.

Yang, X., Yang, S., Qi, H., Wang, T., Li, H., & Zhang, Z. (2020). PlaPPISite: A comprehensive resource for plant protein-protein interaction sites. *BMC Plant Biology, 20*, 61.

Yang, H., Nukunya, K., Ding, Q., & Thompson, B. E. (2022). Tissue-specific transcriptomics reveal functional differences in floral development. *Plant Physiology, 188*, 1158–1173.

Yanofsky, M. F., Ma, H., Bowman, J. L., Drews, G. N., Feldmann, K. A., & Meyerowitz, E. M. (1990). The protein encoded by the Arabidopsis homeotic gene agamous resembles transcription factors. *Nature, 346*, 35–39.

Yant, L., Mathieu, J., Dinh, T. T., Ott, F., Lanz, C., Wollmann, H., Chen, X., & Schmid, M. (2010). Orchestration of the floral transition and floral development in Arabidopsis by the bifunctional transcription factor APETALA2. *Plant Cell, 22*, 2156–2170.

Yao, S. G., Ohmori, S., Kimizu, M., & Yoshida, H. (2008). Unequal genetic redundancy of rice *PISTILLATA* orthologs, *OsMADS2* and *OsMADS4*, in lodicule and stamen development. *Plant & Cell Physiology, 49*, 853–857.

Yasui, Y., Tanaka, W., Sakamoto, T., Kurata, T., & Hirano, H. Y. (2017). Genetic enhancer analysis reveals that FLORAL ORGAN NUMBER2 and OsMADS3 co-operatively regulate maintenance and determinacy of the flower meristem in rice. *Plant & Cell Physiology, 58*, 893–903.

Ye, S., Yang, W., Zhai, R., Lu, Y., Wang, J., & Zhang, X. (2017). Mapping and application of the twin-grain1 gene in rice. *Planta, 245*, 707–716.

Yoshida, H. (2012). Is the lodicule a petal: Molecular evidence? *Plant Science: An International Journal of Experimental Plant Biology, 184*, 121–128.

Yoshida, H., & Nagato, Y. (2011). Flower development in rice. *Journal of Experimental Botany, 62*, 4719–4730.

Yoshida, H., Itoh, J., Ohmori, S., Miyoshi, K., Horigome, A., Uchida, E., Kimizu, M., Matsumura, Y., Kusaba, M., Satoh, H., & Nagato, Y. (2007). superwoman1-cleistogamy, a hopeful allele for gene containment in GM rice. *Plant Biotechnology Journal, 5*, 835–846.

Yoshida, A., Suzaki, T., Tanaka, W., & Hirano, H. Y. (2009). The homeotic gene *long sterile lemma* (*G1*) specifies sterile lemma identity in the rice spikelet. *Proceedings of the National Academy of Sciences of the United States of America, 106*, 20103–20108.

Yoshida, A., Ohmori, Y., Kitano, H., Taguchi-Shiobara, F., & Hirano, H. Y. (2012). Aberrant spikelet and panicle1, encoding a TOPLESS-related transcriptional co-repressor, is involved in the regulation of meristem fate in rice. *The Plant Journal: For Cell and Molecular Biology, 70*, 327–339.

Yoshikawa, T., Ito, M., Sumikura, T., Nakayama, A., Nishimura, T., Kitano, H., Yamaguchi, I., Koshiba, T., Hibara, K., Nagato, Y., & Itoh, J. (2014). The rice FISH BONE gene encodes a tryptophan aminotransferase, which affects pleiotropic auxin-related processes. *The Plant Journal: For Cell and Molecular Biology, 78*, 927–936.

You, X., Zhu, S., Zhang, W., Zhang, J., Wang, C., Jing, R., Chen, W., Wu, H., Cai, Y., Feng, Z., Hu, J., Yan, H., Kong, F., Zhang, H., Zheng, M., Ren, Y., Lin, Q., Cheng, Z., Zhang, X., Lei, C., Jiang, L., Wang, H., & Wan, J. (2019). OsPEX5 regulates rice spikelet development through modulating jasmonic acid biosynthesis. *The New Phytologist, 224*, 712–724.

Yu, H., & Huang, T. (2016). Molecular mechanisms of floral boundary formation in Arabidopsis. *International Journal of Molecular Sciences, 17*, 317.

Yu, C., Sun, C., Shen, C., Wang, S., Liu, F., Liu, Y., Chen, Y., Li, C., Qian, Q., Aryal, B., Geisler, M., Jiang de, A., & Qi, Y. (2015). The auxin transporter, OsAUX1, is involved in primary root and root hair elongation and in Cd stress responses in rice (Oryza sativa L.). *The Plant Journal: For Cell and Molecular Biology, 83*, 818–830.

Yuan, Z., & Zhang, D. (2015). Roles of jasmonate signalling in plant inflorescence and flower development. *Current Opinion in Plant Biology, 27*, 44–51.

Yuan, Z., Gao, S., Xue, D. W., Luo, D., Li, L. T., Ding, S. Y., Yao, X., Wilson, Z. A., Qian, Q., & Zhang, D. B. (2009). *RETARDED PALEA1* controls palea development and floral zygomorphy in rice. *Plant Physiology, 149*, 235–244.

Yuan, Z., Persson, S., & Zhang, D. (2020). Molecular and genetic pathways for optimizing spikelet development and grain yield. *aBIOTECH, 1*(4), 276–292. https://doi.org/10.1007/s42994-020-00026-x

Yun, D., Liang, W., Dreni, L., Yin, C., Zhou, Z., Kater, M. M., & Zhang, D. (2013). *OsMADS16* genetically interacts with *OsMADS3* and *OsMADS58* in specifying floral patterning in rice. *Molecular Plant, 6*, 743–756.

Zahn, L. M., Kong, H., Leebens-Mack, J. H., Kim, S., Soltis, P. S., Landherr, L. L., Soltis, D. E., Depamphilis, C. W., & Ma, H. (2005). The evolution of the SEPALLATA subfamily of MADS-box genes: A preangiosperm origin with multiple duplications throughout angiosperm history. *Genetics, 169*, 2209–2223.

Zhai, Q., Yan, C., Li, L., Xie, D., & Li, C. (2017). Jasmonates. In *Hormone metabolism and signaling in plants* (pp. 243–272). Academic Press.

Zhang, D., & Yuan, Z. (2014). Molecular control of grass inflorescence development. *Annual Review of Plant Biology, 65*, 553–578.

Zhang, X., Zong, J., Liu, J., Yin, J., & Zhang, D. (2010). Genome-wide analysis of WOX gene family in rice, sorghum, maize, Arabidopsis and poplar. *Journal of Integrative Plant Biology, 52*, 1016–1026.

Zhang, D., Yuan, Z., An, G., Dreni, L., Hu, J., & Kater, M. M. (2013). Panicle development. In Q. Zhang & R. A. Wing (Eds.), *Genetics and genomics of rice* (pp. 279–295). Springer.

Zhang, J., Tang, W., Huang, Y., Niu, X., Zhao, Y., Han, Y., & Liu, Y. (2015). Down-regulation of a LBD-like gene, OsIG1, leads to occurrence of unusual double ovules and developmental abnormalities of various floral organs and megagametophyte in rice. *Journal of Experimental Botany, 66*, 99–112.

Zhang, T., Li, Y., Ma, L., Sang, X., Ling, Y., Wang, Y., Yu, P., Zhuang, H., Huang, J., Wang, N., Zhao, F., Zhang, C., Yang, Z., Fang, L., & He, G. (2017a). LATERAL FLORET 1 induced the three-florets spikelet in rice. *Proceedings of the National Academy of Sciences of the United States of America, 114*, 9984–9989.

Zhang, T. Q., Lian, H., Zhou, C. M., Xu, L., Jiao, Y., & Wang, J. W. (2017b). A two-step model for de novo activation of WUSCHEL during plant shoot regeneration. *Plant Cell, 29*, 1073–1087.

Zhang, H., Xu, H., Feng, M., & Zhu, Y. (2018a). Suppression of OsMADS7 in rice endosperm stabilizes amylose content under high temperature stress. *Plant Biotechnology Journal, 16*, 18–26.

Zhang, T., Li, R., Xing, J., Yan, L., Wang, R., & Zhao, Y. (2018b). The YUCCA-auxin-WOX11 module controls crown root development in rice. *Frontiers in Plant Science, 9*, 523.

Zhang, X., Cai, Q., Li, L., Wang, L., Hu, Y., Chen, X., Zhang, D., Persson, S., Yuan, Z. (2024). OsMADS6-OsMADS32 and REP1 control palea cellular heterogeneity and morphogenesis in rice. *Developmental Cell, 59*(11), 1379–1395. https://doi.org/10.1016/j.devcel.2024.03.026

Zhang, D., Tang, S., Liu, F., Zhao, K., Li, C., Xia, R., Yu, F., Xie, Q., Xie, P. (2025). Natural variations in Multi-Grain Spikelet 1 enhance grain number in sorghum. *Journal of Integrative Plant Biology, 67*(6), 1441–1443. https://doi.org/10.1111/jipb.13871

Zhao, Y., Hu, Y., Dai, M., Huang, L., & Zhou, D. X. (2009). The WUSCHEL-related homeobox gene WOX11 is required to activate shoot-borne crown root development in rice. *Plant Cell, 21*, 736–748.

Zhao, H., Ma, T., Wang, X., Deng, Y., Ma, H., Zhang, R., & Zhao, J. (2015a). OsAUX1 controls lateral root initiation in rice (Oryza sativa L.). *Plant, Cell & Environment, 38*, 2208–2222.

Zhao, Y., Cheng, S., Song, Y., Huang, Y., Zhou, S., Liu, X., & Zhou, D. X. (2015b). The interaction between rice ERF3 and WOX11 promotes crown root development by regulating gene expression involved in cytokinin signaling. *Plant Cell, 27*, 2469–2483.

Zhao, L., Wang, S., Cao, Z., Ouyang, W., Zhang, Q., Xie, L., Zheng, R., Guo, M., Ma, M., Hu, Z., Sung, W. K., Zhang, Q., Li, G., & Li, X. (2019). Chromatin loops associated with active genes and heterochromatin shape rice genome architecture for transcriptional regulation. *Nature Communications, 10*, 3640.

Zhao, L., Xie, L., Zhang, Q., Ouyang, W., Deng, L., Guan, P., Ma, M., Li, Y., Zhang, Y., Xiao, Q., Zhang, J., Li, H., Wang, S., Man, J., Cao, Z., Zhang, Q., Zhang, Q., Li, G., & Li, X. (2020). Integrative analysis of reference epigenomes in 20 rice varieties. *Nature Communications, 11*, 2658.

Zhao, H., Liao, H., Li, S., Zhang, R., Dai, J., Ma, P., Wang, T., Wang, M., Yuan, Y., Fu, X., Cheng, J., Duan, X., Xie, Y., Zhang, P., Kong, H., & Shan, H. (2023). Delphinieae flowers originated from the rewiring of interactions between duplicated and diversified floral organ identity and symmetry genes. *Plant Cell, 35*, 994–1012.

Zheng, M., Wang, Y., Wang, Y., Wang, C., Ren, Y., Lv, J., Peng, C., Wu, T., Liu, K., Zhao, S., Liu, X., Guo, X., Jiang, L., Terzaghi, W., & Wan, J. (2015). DEFORMED FLORAL ORGAN1 (DFO1) regulates floral organ identity by epigenetically repressing the expression of OsMADS58 in rice (Oryza sativa). *The New Phytologist, 206*, 1476–1490.

Zhou, S., Jiang, W., Long, F., Cheng, S., Yang, W., Zhao, Y., & Zhou, D. X. (2017). Rice homeodomain protein WOX11 recruits a histone acetyltransferase complex to establish programs of cell proliferation of crown root meristem. *Plant Cell, 29*, 1088–1104.

Zhuang, H., Wang, H.L., Zhang, T., Zeng, X.Q., Chen, H., Wang, Z.W., Zhang, J., Zheng, H., Tang, J., Ling, Y.H., Yang, Z.L., He, G.H., Li, Y.F. (2020). NONSTOP GLUMES1 Encodes a C2H2 Zinc finger protein that regulates spikelet development in rice. *The Plant Cell, 32*(2), 392–413. https://doi.org/10.1105/tpc.19.00682

Zhuang, H., Li, Y.H., Zhao, X.Y., Zhi, J.Y., Chen, H., Lan, J.S., Luo, Z.J., Qu, Y.R., Tang, J., Peng, H.P., Li, T.Y. (2024). STAMENLESS1 activates SUPERWOMAN 1 and FLORAL ORGAN NUMBER 1 to control floral organ identities and meristem fate in rice SUMMARY. *The Plant Journal, 118*(3), 802–822. https://doi.org/10.1111/tpj.16637

Zhu, Q. H., Upadhyaya, N. M., Gubler, F., & Helliwell, C. A. (2009). Over-expression of miR172 causes loss of spikelet determinacy and floral organ abnormalities in rice (*Oryza sativa*). *BMC Plant Biology, 9*, 149.

Zhu, W., Yang, L., Wu, D., Meng, Q., Deng, X., Huang, G., Zhang, J., Chen, X., Ferrandiz, C., Liang, W., Dreni, L., & Zhang, D. (2022). Rice SEPALLATA genes OsMADS5 and OsMADS34 cooperate to limit inflorescence branching by repressing the TERMINAL FLOWER1-like gene RCN4. *The New Phytologist, 233*, 1682–1700.

Zhu, D., Wen, Y., Yao, W., Zheng, H., Zhou, S., Zhang, Q., Qu, L.-J., Chen, X., & Wu, Z. (2023). Distinct chromatin signatures in the Arabidopsis male gametophyte. *Nature Genetics, 55*, 706–720.

Zong, J., Wang, L., Zhu, L., Bian, L., Zhang, B., Chen, X., Huang, G., Zhang, X., Fan, J., Cao, L., Coupland, G., Liang, W., Zhang, D., & Yuan, Z. (2022). A rice single cell transcriptomic atlas defines the developmental trajectories of rice floret and inflorescence meristems. *The New Phytologist, 234*, 494–512.

Chapter 8
Floral Organ Identity Determination

Hongyan Shan and Hongzhi Kong

Abstract Flowers are unique to angiosperms. The origin and diversification of flowers have contributed to the great success of angiosperms in the terrestrial ecosystems. While a typical flower is composed of four types of organs: sepals, petals, stamens, and carpels, flowers in some angiosperms have evolved various other organ types, such as petaloid sepals, tepals, labella, lemmas, paleas, lodicules, and staminodes, which have greatly contributed to floral diversity. The elucidation of how different types of floral organs are specified is important for understanding the underlying mechanisms responsible for the origin and diversification of flowers. Since the proposal of the ABC model of flower development, the processes and mechanisms of floral organ identity determination have been investigated extensively in two pioneering model systems, *Arabidopsis thaliana* and *Antirrhinum majus*, as well as quite a few species from phylogenetically distant lineages. In the past 30 years, impressive progress has been made in identifying the patterns, mechanisms, and evolution of floral organ identity determination. In this chapter, we first briefly summarize the patterns of floral organ initiation and specification, and then provide a detailed overview of how floral organs of angiosperms and reproductive structures of gymnosperms are specified. Finally, we present current understanding of the evolutionary origins of the floral organ identity determination programs. We suggest that future research should focus on the regulatory evolution of floral MADS-box genes, functional and evolutionary analyses of non-MADS-box floral

H. Shan
State Key Laboratory of Plant Diversity and Specialty Crops, Chinese Academy of Sciences, Beijing, China

China National Botanical Garden, Beijing, China

H. Kong (✉)
State Key Laboratory of Plant Diversity and Specialty Crops, Chinese Academy of Sciences, Beijing, China

China National Botanical Garden, Beijing, China

University of Chinese Academy of Sciences, Beijing, China
e-mail: hzkong@ibcas.ac.cn

F. Chang et al. (eds.), *Regulation of Plant Development*,
https://doi.org/10.1007/978-981-95-7033-1_8

organ identity genes, and development of new model systems in basalmost angiosperms and gymnosperms.

Keywords Flower · Floral organ identity determination · Angiosperms · Gymnosperms · The origin of the flower

8.1 Introduction

Flowers are the most obvious distinguishing feature and one of the key innovations of angiosperms (i.e., flowering plants). Flowers of most angiosperms (particularly eudicots) have four types of organs—sepals, petals, stamens, and carpels—arranged from the outside to the center. In most cases, sepals are greenish and leaf-like, while petals are showy and colorful; together they are termed the perianth. However, floral organs are not limited only to these four basic types. In some angiosperms, flowers have evolved various other organ types, such as petaloid sepals in many eudicot species, undifferentiated perianth (i.e., tepals) in most basal angiosperms and nongrass monocots, lemmas, paleas, and lodicules in grasses (Poaceae), specialized labella in orchids, and staminodes in some lineages. The diversity of floral organ types has greatly contributed to the diversification of flowers. The elucidation of how different types of floral organs are specified is therefore key to understanding the origin and diversification of flowers.

In the past 30 years, the processes and mechanisms of floral organ identity specification have been investigated extensively in two pioneering model plants, *Arabidopsis thaliana* (Brassicaceae) and *Antirrhinum majus* (Plantaginaceae), as well as quite a few species from the Asteraceae, Solanaceae, Ranunculaceae, Papaveraceae, Poaceae, and Orchidaceae. The evolutionary patterns of floral organ identity genes have also been broadly explored in both angiosperms and gymnosperms. The major advances in these areas have not only provided a foundation for our understanding of the development and evolution of flowers, but also provided data essential for the growth of plant evolutionary developmental biology (evo-devo). In this chapter, we provide an overview of the patterns, mechanisms, and evolution of floral organ identity specification.

8.2 Patterns of Floral Organ Initiation and Specification

Soon after the formation of the floral meristem, floral organs are initiated in a centripetal sequence on the flank of the floral meristem. In this pattern, the outermost organ is initiated first and the innermost one is initiated last. However, flowers in some species deviate from a strict centripetal organ initiation, in which at least some organs are initiated in a centrifugal sequence, either within organ zones (e.g., *Dillenia alata* and *Couroupita guianensis*) or between organ zones (e.g., *Alisma plantago-aquatica* and *Lythrum salicaria*) (Endress, 2006; Rudall, 2010). Despite the frequent switches in initiation sequences of floral organs, the overall floral bauplan is conserved in flowering plants, i.e., nearly all angiosperm flowers have a perianth-stamen-carpel organization from the outermost to the innermost. Exceptions to this conserved floral architecture were only documented in a Mexican monocot species, *Lacandonia schismatica* (Triuridaceae) and a basalmost angiosperm species, *Trithuria submersa* (Hydatellaceae). Flowers in both species have central stamens surrounded by carpels and the stamens are formed before the carpels (Ambrose et al., 2006; Rudall et al., 2007).

Spatially, the initiation of floral organs follows either whorled or spiral patterns, which result in the formation of flowers with whorled or spiral phyllotaxy (Endress, 2006). In whorled flowers, organs present in a whorl are initiated concurrently, arranged equidistantly, and formed in a concentric circle, while organs between whorls appear with conspicuous time intervals (Endress, 2006). This kind of flower is frequently found in core eudicots, monocots, and some basal angiosperms. In spiral flowers, organs are initiated one by one at equal time intervals and regular divergence angles (137.5° or 99.5°), irrespective of organ types (Endress, 2006). Spiral flowers are mainly present in basal eudicots and basal angiosperms.

Floral organs can be initiated in various temporal and spatial patterns, but ontogenetic processes of individual organs are quite similar. Here we take the model species *A. thaliana* as an example. It has 16 floral organs, including four sepals, four petals, six stamens, and two carpels. By monitoring the expression of *DORNRÖSCHEN-LIKE* (*DRNL*), a founder cell identity gene, it has been clearly shown that each floral organ develops from a few undifferentiated founder cells of the floral meristem. These cells first proliferate to generate an anlage or preprimordium through cell division. Next the anlage further proliferates to produce an undifferentiated and developmentally recognizable primordium. Finally, the primordium differentiates into a mature floral organ. More importantly, the floral organ identity is specified during floral organogenesis, which is a key event that directs the developmental fate of a floral organ (Chandler, 2011; Chandler et al., 2011).

How floral organ identities are determined during floral organogenesis has been in debate for several decades. Before 1990s, two contrasting models were formulated to answer this question. The first one was the sequential model, which was proposed based on surgical experiments on floral buds. It states that sequential inductive signals are required to specify the formation of floral organs. For instance, signals from the outer whorl promote the initiation of the adjacent inner whorl (Cusick, 1956; Green, 1988; Hicks & Sussex, 1971; Hughen, 1980). In contrast, the spatial model predicts that positional signals laid down simultaneously in the meristem are required for the specification of floral organs (Holder, 1979). By specifically ablating petal and stamen primordia at an early developmental stage, Day et al. (1995) demonstrated that the sepal and carpel whorl can be formed normally without signals from adjacent whorls. This experiment provided direct evidence for the spatial model. The dispute between the two models remained unresolved until the proposal of the ABC model of flower development.

The ABC model was formulated based on extensive analyses of mutants affecting floral structures in *A. thaliana* and *A. majus* (Coen & Meyerowitz, 1991). It states that sepals are specified by the A function, petals by the combinational action of the A and B functions, stamens by the B and C functions, and carpels by the C function; the A and C functions are mutually repressive. In *A. thaliana*, the A function is conferred by *APETALA1* (*AP1*) and *APETALA2* (*AP2*), the B function by *APETALA3* (*AP3*) and *PISTILLATA* (*PI*), and the C function by *AGAMOUS* (*AG*), of which all but *AP2* are members of the MADS-box transcription factor family (Becker & Theißen, 2003). Later, with the characterization of *SEPALLATA* (*SEP*) genes affecting the identities of all floral organs, an E function was defined and the ABC model was extended to the ABCE model (Ditta et al., 2004; Pelaz et al., 2000; Theißen, 2001).

The ABCE model offered a new research paradigm for the study of the patterns of floral organ identity specification. According to this model, the identity determination of each floral organ type relies on genes expressed in the corresponding region. This fits well with the spatial model. However, based on the expression pattern of *FLORICAULA* (*FLO*), a floral meristem identity gene found in *A. majus*, Coen et al. (1990) proposed that floral organ identity genes are sequentially regulated by their upstream regulators. This fits well with the sequential model and is supported by later work in other species. For example, in *A. thaliana*, it has been shown that under the control of the FLO orthologous protein LEAFY (LFY) and other cofactors, *AP1* is expressed earliest in floral primordia at stage 1 of flower development, followed by *SEP3* at early stage 3, and B and C function genes at stage 3 (Liu & Mara, 2010). A study of a basal eudicot species, *Nigella damascena* (Ranunculaceae), showed that the MADS-box genes specifying floral organ identities constitute a relatively complex regulatory network, in which the changes in expression of early-acting genes can cause subsequent changes in late-acting genes (Wang et al., 2016). Therefore, in terms of the expression and regulation patterns of corresponding genes, the identities of floral organs are specified both temporally and spatially. In other words, both sequential and spatial models should be considered to explain the specification of floral organ identities.

However, the sequential regulation of the specification of floral organ identities is far from simple. Based on precisely timed activation and/or perturbation experiments of the *AP3*, *PI*, and *AG* genes in *A. thaliana*, it has been shown that stamen identity is specified relatively early at stages ~2–6, followed by carpel identity at stages 3–6, and petal identity later at stages ~8–10 (Ito et al., 2007; O'Maoileidigh et al., 2013; Wuest et al., 2012). Obviously, stamen and carpel identity specification is earlier than the formation of their recognizable primordia (stages 5 and 6, respectively), while petal identity specification is later than the formation of its recognizable primordium (stage 5) (Smyth et al., 1990). This indicates that the specification of floral organ identities is independent of the formation of floral primordia. Moreover, the timing of the specification of floral organ identities is not strictly synchronous with the initiation expression of floral ABC genes. In particular, there is a significant time gap between the B gene expression and the petal identity determination. However, whether this is caused by the delayed development of petals or by other causes is still unclear.

8.3 Mechanisms of Floral Organ Identity Determination

In this section, we provide detailed summaries of how the identities of different types of floral organs are specified. We begin with a discussion of the establishment of the floral context, and then focus on key genes responsible for the identities of four typical types of floral organs: sepals, petals, stamens, and carpels. We end this discussion by considering the molecular mechanisms responsible for the identities of special types of floral organs, including petaloid sepals, tepals, lemmas, paleas, lodicules, and staminodes. In addition, the evolutionary patterns of the responsible genes are also summarized in the wider context of angiosperm evolution. To make these complicated processes easier to understand, we present flowers and floral diagrams of eight representative species, and the corresponding models for their floral organ identity determination (Fig. 8.1).

8.3.1 Establishment of the Floral Context

While floral organ types vary among angiosperm species, a widely accepted notion is that each floral organ represents a modified leaf. The idea was initially put forward by Goethe in 1790, but was not confirmed until the discovery of the E function gene (Pelaz et al., 2001). Transgenic studies showed that vegetative leaves of *A. thaliana* were transformed into petaloid or staminoid organs when E function genes, together with B or B and C function genes, were simultaneously and ectopically expressed in the leaves (Goto et al., 2001; Honma & Goto, 2001; Pelaz et al., 2001). However, a similar conversion did not occur when only the B or C function genes were constitutively expressed (Causier et al., 2005; Honma & Goto, 2001; Krizek & Meyerowitz, 1996; Mizukami & Ma, 1992). This suggests that the E function genes are required for providing a floral context for the action of the ABC genes (Causier et al., 2010).

So far, E function genes have been identified in many angiosperm species. Four E function genes exist in the genome of *A. thaliana*: *SEP1*, *SEP2*, *SEP3*, and *SEP4* (Zahn et al., 2005). Except for *SEP3*, which is expressed in the inner three whorls, all other E function genes are expressed in all floral organs (Ditta et al., 2004; Huang et al., 1995; Ma et al., 1991; Mandel & Yanofsky, 1998). Inactivation of *SEP1–4* results in the conversion of all floral organs into leaf-like organs (Ditta et al., 2004). Similarly, multiple copies of *SEP*-like genes are present in other species. Defects affecting nearly all floral organs were observed in plants with mutant or silenced *SEP*-like genes in species such as *Petunia hybrida* and *Solanum lycopersicum*

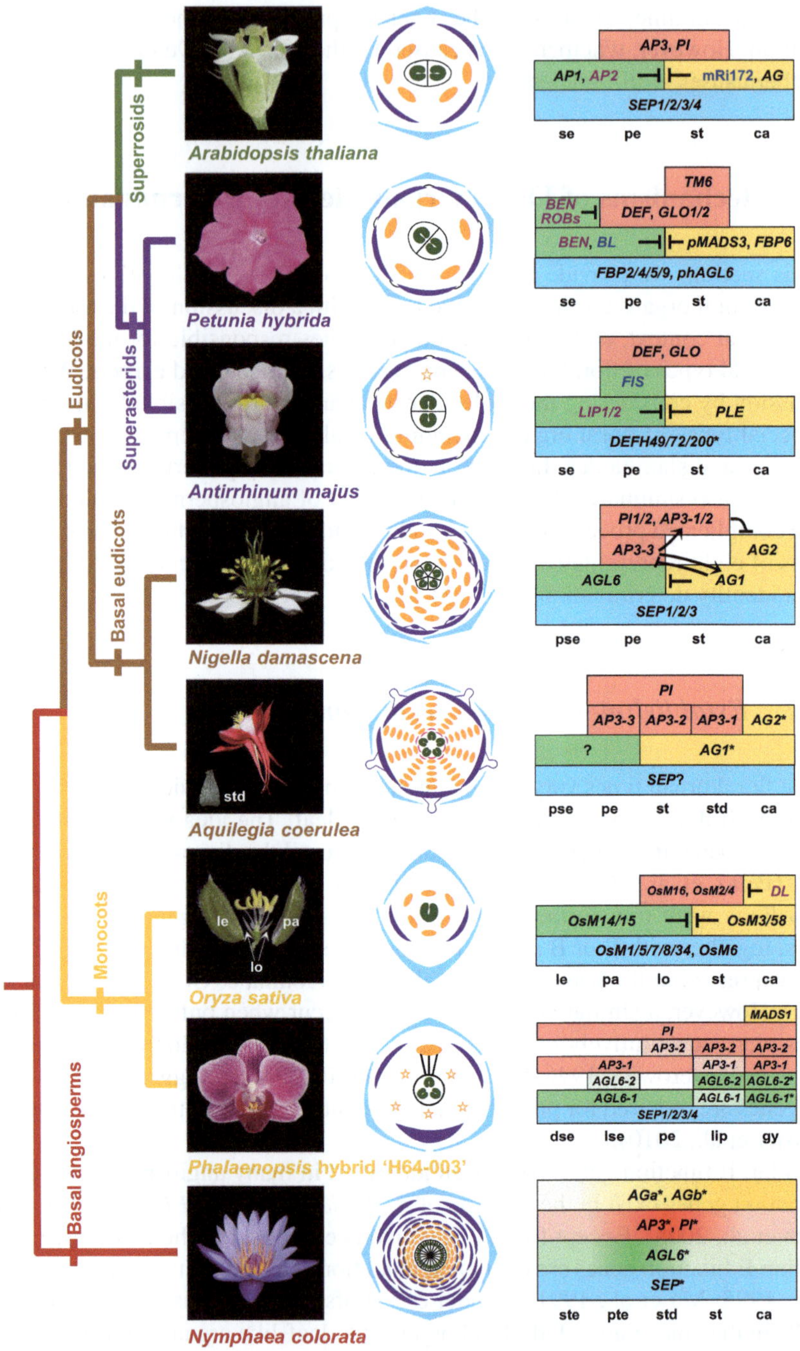

Fig. 8.1 Flowers, floral diagrams, and the corresponding models for floral organ identity determination in representative species of flowering plants. In the models, rectangles with different colors

(Solanaceae) (Ampomah-Dwamena et al., 2002; Morel et al., 2019; Pnueli et al., 1994; Vandenbussche et al., 2003), *Gerbera hybrida* (Asteraceae) (Zhang et al., 2017), and *Fragaria vesca* (Rosaceae) (Pi et al., 2021) in core eudicots, *N. damascena* and *Thalictrum thalictroides* (Ranunculaceae) in basal eudicots (Soza et al., 2016; Wang et al., 2016), as well as *Oryza sativa* (Poaceae) and *Phalaenopsis* and *Habenaria radiata* (Orchidaceae) in monocots (Cui et al., 2010; Pan et al., 2014; Mitoma & Kanno, 2018; Wu et al., 2018). These data indicate that *SEP*-like genes have a conserved function in the establishment of the floral context in flowering plants.

Phylogenetic analyses suggest that *SEP*-like genes are angiosperm specific and are closely related to *AP1* and *AGAMOUS-LIKE6* (*AGL6*) subfamily genes (Ruelens et al., 2013; Yu et al., 2016; Zahn et al., 2005; Zhao et al., 2017). Multiple independent gene duplication events that occurred on different evolutionary scales have been revealed in this subfamily. For instance, a gene duplication in the most recent common ancestor (MRCA) of all extant angiosperms gave rise to the *AGAMOUS-LIKE2/3/4* (*AGL2/3/4* or *SEP1/2/4*) and *AGAMOUS-LIKE9* (*AGL9* or *SEP3*) lineages. Moreover, gene duplications in basal core eudicots led to the generation of the *AGL2/4*, *AGL3*, and *FLORAL BINDING PROTEIN9* (*FBP9*) clades in the *AGL2/3/4* lineage, and gene duplications in the MRCA of all grasses produced three clades that include *Oryza sativa MADS1* (*OsMADS1*), *OsMADS5*, and *OsMADS34* in the *AGL2/3/4* lineage and two clades that include *OsMADS7* and *OsMADS8* in the *AGL9* lineage (Malcomber & Kellogg, 2005; Xu & Kong, 2007; Zahn et al., 2005). Extensive studies on sequence features, gene expression patterns, and functions have shown that despite functional redundancy, subfunctionalization and/or neofunctionalization have facilitated the persistence of multiple *SEP* paralogs in a genome (Malcomber & Kellogg, 2004, 2005; Soza et al., 2016; Wu et al., 2018; Zahn et al., 2005; Zhao et al., 2017).

Although *SEP*-like genes tend to diversify among paralogs, they still share common features. First, SEP-like proteins, which serve as "glues," can mediate the formation of different tetrameric complexes. The floral quartet model, which explains how the ABCE genes specify floral organ identities on the molecular level, predicts

Fig. 8.1 (continued) denote the functional domains of different floral organ identity genes, blue, E function; green, A function; vermillion, B function; and apricot, C function. The genes in black are floral MADS-box genes, those in purple are non-MADS-box transcription factors, and those in blue are microRNAs. Genes marked with stars indicate that only expression data are available. The question mark in the *Aquilegia coerulea* model indicates that no candidate genes have been characterized yet. Inferred positive and negative regulatory relationships between genes are indicated by arrows and bars, respectively. Abbreviations for floral organs in the models and color codes for floral organs in the floral diagrams: se, sepal (sky blue); pe, petal (purple); st, stamen (orange); ca, carpel (dark green); pse, petaloid sepal (sky blue); std, staminode (pink in *A. coerulea*, representing a novel organ type; and purle/orange in *Nymphaea colorata*, representing an intermediate structure between the petaloid tepal and the stamen); le, lemma (sky blue); pa, palea (sky blue); lo, lodicule (purple); dse, dorsal sepal (sky blue); lse, lateral sepal (sky blue); gy, gynostemium (a fused structure of a stamen and carpels in *Phalaenopsis* hybrid "H64-003"); ste, sepaloid tepal (sky blue); pte, petaloid tepal (purple)

that sepal, petal, stamen, and carpel identities in *A. thaliana* are specified by AP1-SEP-AP1-SEP, AP3-PI-AP1-SEP, AP3-PI-AG-SEP, and AG-SEP-AG-SEP combinations, respectively (Theißen, 2001). These interactions were later demonstrated by in vitro DNA-binding assays, yeast hybrid assays, and *in planta* detection (Immink et al., 2009; Melzer & Theißen, 2009; Rümpler et al., 2018; Smaczniak et al., 2012). Similar interaction patterns between SEP-like proteins and proteins encoded by other floral organ identity genes have been reported in a broad spectrum of angiosperm taxa, ranging from core eudicots, basal eudicots, and monocots to basalmost angiosperms (Li et al., 2015). This further confirms a conserved role for SEP-like proteins in mediating the formation of floral quartets in angiosperms.

Second, *SEP*-like genes of a species collectively show broad expression domains and high expression levels in all floral organs. This broad expression provides *SEP* genes with an opportunity to co-express and interact with their partners. In contrast, their relatively high expression levels are necessary for the normal formation of floral organs. For instance, plants constitutively expressing two *A. thaliana SEP* genes in a background overexpressing *AP1*, *AP3*, and *PI* result in a more complete transformation of rosette leaves into petals than plants containing only one *SEP* gene (Pelaz et al., 2001). Frequent observation of weak or no phenotypic changes in single mutants of *SEP*-like genes but severe phenotypic changes in multiple gene mutants also supports the idea that the relatively high level of expression of the *SEP* genes is essential for the specification of floral organ identities (Cui et al., 2010; Ditta et al., 2004; Soza et al., 2016; Vandenbussche et al., 2003; Wu et al., 2018; Zhang et al., 2017).

Third, *SEP*-like genes are also expressed in floral meristems of many angiosperm species. In *A. thaliana*, early expression of *SEP3* in floral meristems allows it to act as a cofactor of LFY and to directly activate B and C function genes; this is regarded as an important mechanism determining the appropriate timing of floral organ patterning (Castillejo et al., 2005; Kaufmann et al., 2009; Liu et al., 2009; Liu & Mara, 2010). Similarly, the *SEP1/2/4*-like genes in *O. sativa* (*OsMADS1/5/34*) are also expressed in floret meristems and positively regulate other floral organ identity genes, including B and C function genes. The regulatory link between them was determined based on the decrease in the expression of B and C genes in single, double, and triple mutants of *OsMADS1/5/34* (Wu et al., 2018). Whether B and C genes are regulated by *SEP*-like genes in other species is unknown. However, since *SEP*-like genes share conserved sequence features on the protein level as well as generally show a stable expression pattern in floral meristems, the regulatory link between *SEP*s and other floral organ identity genes may be evolutionarily conserved and may have been established early in the evolutionary history of angiosperms.

AGL6-like genes in several angiosperm species are thought to function redundantly with *SEP*-like genes in terms of their expression patterns, protein interaction behaviors, and functions. For example, *PhAGL6* in *P. hybrida* is broadly expressed in floral organs; its protein has similar interacting partners as the SEP-like protein FBP2. In sextuple mutants involving simultaneous mutations in *PhAGL6* and the E function genes (e.g., *FBP2*, *FBP4*, *FBP5*, and *FBP9*), all floral organs are green and densely covered by trichomes, exhibiting sepal/leaf-like characteristics (Morel et al., 2019). The phenotype is severer than did the *phagl6* mutant or *fbp* mutants (Morel et al., 2019; Rijpkema et al., 2009). Similarly, the *AGL6*-like gene in *O. sativa*, *OsMADS6* (also named *MOSAIC FLORAL ORGANS1*, *MFO1*), is expressed in floral meristems as well as all floral organs except the lemma and stamens. The protein encoded by *OsMADS6* can form complexes with AP3- and PI-like proteins (Seok et al., 2010). Mutations in *OsMADS6* result in the conversion of the palea and lodicules into lemma- and glume-like organs, respectively, defects in stamen and carpel development, and the partial loss of floral meristem determinacy (Li et al., 2010; Ohmori et al., 2009). Moreover, *OsMADS6* is required for the proper expression of B and C function genes (Li et al., 2011). Many of the roles of *PhAGL6* and *OsMADS6* resemble those of *SEP*-like genes in *P. hybrida* and *O. sativa*. However, to date only a few *AGL6*-like genes have been reported to function as E function genes. In most cases, the functions of *AGL6*-like genes are diverse, including the regulation of floral timing, ovule and seed development, male and female germline and gametophyte development, sepal and petal identity specification, and so on (Dreni & Zhang, 2016; Hsu et al., 2015; Wang et al., 2016; Zhao et al., 2023).

The significant similarities of *AGL6* and *SEP* subfamily genes obscure their evolutionary history and, in fact, different evolutionary scenarios have been proposed (Yu et al., 2016). A phylogenomic synteny study reported that they were generated by a whole genome duplication event (ζ) that occurred before the origin of seed plants. Subsequently, *AGL6*-like genes were retained in both angiosperms and gymnosperms, but the *SEP*-like genes were lost in gymnosperms and thus became angiosperm specific (Zhao et al., 2017). Based on the available functional data, it is clear that *SEP*- and *AGL6*-like genes experienced different evolutionary trajectories in angiosperms: *SEP*-like genes tend to have similar functions, while the *AGL6*-like genes have a wide range of functions (Dreni & Zhang, 2016). Thus, it is the *SEP*-like genes, rather than the *AGL6*-like genes, that function as the canonical E function to establish the floral context in angiosperms.

8.3.2 Identity Determination of Typical Floral Organs

8.3.2.1 Sepals

Sepals are the outermost part of flowers with dimorphic perianths, and enclose and protect inner floral organs. Sepals are generally greenish, leaf-like, and sessile and commonly have three vascular traces at a broad base (Endress, 2006; Ronse De Craene, 2008; Weberling, 1989). Leaf-like sepals are widely present in core eudicots and are also occasionally found in basal eudicots (e.g., Papaveraceae, Sabiaceae, and Nelumbonaceae), nongrass monocots (e.g., Zingiberales, Alismataceae, Bromeliaceae, Commelinales, and some Melanthiaceae), and basal angiosperms (e.g., Annonaceae, Aristolochiaceae, and Winteraceae), indicating their independent origins in flowering plants (Bartlett & Specht, 2010; Ronse De Craene & Brockington, 2013).

According to the ABCE model, sepal identity is specified by the A function (Coen & Meyerowitz, 1991). The model plant *A. thaliana* has two A function genes—*AP1* and *AP2*—that belong to unrelated gene families (Bowman et al., 1991, 1993; Bowman et al., 1989; Irish & Sussex, 1990; Jofuku et al., 1994; Kunst et al., 1989; Mandel et al., 1992). *AP1* is a member of the *AP1* MADS-box gene subfamily (Litt & Irish, 2003; Shan et al., 2007). It is strongly expressed in floral meristems and moderately expressed in the developing sepals and petals of stage 3–12 flowers (Ye et al., 2016). Loss of function of *AP1* causes the conversion of sepals into bracts, the absence of petals, and the concomitant formation of additional flowers in the axes of the bracts (Bowman et al., 1993; Irish & Sussex, 1990; Mandel et al., 1992). *AP2* is a member of the eu*AP2* lineage, which is a subgroup of the *AP2/ETHYLENE RESPONSE FACTOR* (*ERF*) transcription factor family (Kim et al., 2006). *AP2* mRNA is accumulated in all four whorls, but the activity is restricted to the outer two whorls due to the negative regulation of *AP2* by miR172 in stamens and carpels (Aukerman & Sakai, 2003; Chen, 2004; Jofuku et al., 1994). In the *AP2* loss-of-function mutant, sepals are transformed into carpeloid or leaf-like organs and petals are either missing or converted into stamen-like organs (Bowman et al., 1989, 1991; Jofuku et al., 1994; Komaki et al., 1988; Kunst et al., 1989). Despite the different phenotypic defects in *ap1* and *ap2* mutants, both genes contribute to sepal identity determination.

To date, multiple *AP1* homologs in different angiosperm species have been functionally characterized. Like *AP1* in *A. thaliana*, orthologs present in other core eudicots, such as *PROLIFERATING INFLORESCENCE MERISTEM* (*PIM*) in *Pisum sativum* (Fabaceae), *MtPIM* in *Medicago truncatula* (Fabaceae), *MACROCALYX* (*MC*) in *S. lycopersicum* (Solanaceae), and MADS-box gene 3 from *Physalis floridana* and *Physalis grisea* (Solanaceae) (*MPF3*), are also responsible for the specification of sepal identity. Flowers in plants with knockout or knockdown alleles of these genes show the transformation of sepals into leaf-like organs in addition to the defects in the transition of inflorescence meristems into floral meristems (Benlloch et al., 2006; He et al., 2023; Taylor et al., 2002; Vrebalov et al., 2002; Zhao et al., 2013). Interestingly, the *AP1*-like genes in basal eudicots, which are named *FRUITFULL* (*FUL*)-like genes, have been demonstrated by virus-induced gene silencing (VIGS) studies to have diverse functions. For instance, in two Papaveraceae species, *Papaver somniferum* and *Eschscholzia californica*, *FUL*-like genes specify sepal identity, and have roles in regulating axillary meristem growth, floral meristem identity specification, and fruit development (Pabón-Mora et al., 2012). However, *FUL*-like genes in *N. damascena* do not seem to be involved in flower development; they showed relatively low levels of expression in floral buds and no obvious phenotypic differences were found between VIGS-treated and wild type plants (Wang et al., 2016). The *FUL*-like gene in *Aquilegia coerulea* "Origami" (Ranunculaceae) mainly regulates leaf morphology rather than flower development (Pabón-Mora et al., 2013).

Phylogenetic analyses indicate that *AP1*-like genes are angiosperm specific (Ruelens et al., 2013; Yu et al., 2016; Zhao et al., 2017). Genes of this subfamily experienced two successive gene duplication events at the base of core eudicots, giving rise to three gene lineages: eu*AP1*, eu*FUL*, and *AGL79*. All core eudicot *AP1* genes mentioned above belong to the eu*AP1* lineage, which encode a diagnostic euAP1 motif at the C terminal region (Litt & Irish, 2003; Shan et al., 2007). In contrast, the genes present in the flowering plants outside core eudicots are referred to as *FUL*-like genes and include the abovementioned basal eudicot genes. Their protein products have a characteristic paleoAP1 motif at the C terminus (Litt & Irish, 2003; Shan et al., 2007). Based on the available functional and expression data for *AP1* and closely related subfamily genes, it has been hypothesized that the functions specifying both sepal and floral meristem identities are ancient and can be traced back to the MRCA of all *AP1*-, *SEP*-, and *AGL6*-like genes. During evolution, some *AP1*-like genes have independently lost the ability to specify sepal identity in some basal eudicot species (Pabón-Mora et al., 2013). However, since *SEP*-like genes are generally involved in the specification of floral meristem and all floral organ identities, and *AGL6*-like genes show diverse functions in different angiosperm species (as described in Sect. 8.3.1), an alternative scenario can be proposed: that the eu*AP1*-like genes gained the ability to specify sepal identity after gene duplication events at the base of core eudicots, while *FUL*-like genes in the Papaveraceae obtained this function independently. This later hypothesis is consistent with the independent origins of sepals across the flowering plants.

The situation is complicated for *AP2*-like genes that specify sepal identity. In *A. majus*, the recently duplicated *LIPLESS1* (*LIP1*) and *LIP2* are *A. thaliana AP2* orthologs (Keck et al., 2003). Mutation of either gene shows no obvious phenotypic changes, but the *lip1 lip2* double mutant has leaf-like organs in the first whorl, suggesting that these genes play a redundant role in sepal identity determination (Keck et al., 2003). In addition, four *AP2*-like genes—*BLIND ENHANCER* (*BEN*, previously named *AP2B*), *REPRESSOR OF B-FUNCTION1* (*ROB1*, previously named *AP2A*), *ROB2*, and *ROB3*—were found to be involved in sepal identity determination in *P. hybrida* (Maes et al., 2001; Morel et al., 2017). The *ben* single mutant has flowers with two petaloid sepals, while the *rob1 rob2 rob3* triple mutant produces flowers that have almost normal sepals. However, when both *BEN* and *ROB* genes are simultaneously mutated, flowers show a quasi-complete homeotic transformation of sepals into petals. Genetic and functional data not only suggest that *BEN* and *ROB* genes function redundantly in the restriction of the expression domain of B function genes, but also indicate that *BEN* plays the major role rather than the *ROB*s (Morel et al., 2017). Notably, in *P. grisea*, the *BEN* ortholog, *HUSKLESS* (*HU*), exhibits a similar function to *BEN*; inactivation of it results in the sepals with petaloid features (He et al., 2023). Interestingly, detailed phylogenetic analysis suggests that, while all abovementioned genes are eu*AP2* lineage members, the three *ROB*s are the counterparts of the *AP2* and *LIP* genes found in *A. thaliana* and *A. majus*, belonging to the *AP2/ TARGET OF EAT3* (*TOE3*) subclade, while *BEN* and *HU* belong to the *TOE1* subclade (Morel et al., 2017; Kerstens et al., 2020). These findings indicate that the eu*AP2*-like genes specify sepal identity in different manners in core eudicots.

Apart from the *AP1*- and *AP2*-like genes, a miR169 family gene in *P. hybrida* has also been reported to be involved in the specification of sepal identity (Morel et al., 2017). *BL* is a negative regulator of *NF-YA* genes that positively regulate the expression of C function genes. While the first whorl organs retain sepal identity in the *bl* single mutant, a complete conversion of sepals into carpeloid organs can be observed in flowers of the *bl ben* double mutant due to the expansion of C function genes. Obviously, this phenotype is more severe than those observed in the *bl* or *ben* single mutants, and is highly similar to that of the strongest *ap2* mutant found in *A. thaliana*. Therefore, *BL* is thought to act together with *BEN* to repress C function gene expression in the perianth of *P. hybrida*, fulfilling the A function (Morel et al., 2017).

Taken together, the available data have shown that the molecular mechanisms underlying sepal identity specification are not strictly conserved but are instead relatively flexible across the eudicots. The genetic programs have even diverged between Rosids and Asterids. More studies in other representative species are needed to further elucidate the function and evolution of genes regulating sepal identity, especially eu*AP2* and miR169.

8.3.2.2 Petals

Petals are showy and brightly colored floral organs positioned in the second whorl of flowers with a dimorphic perianth, which function to attract pollinators (Weberling, 1989). Traditional comparative morphological and anatomical studies suggest that petals have originated multiple times in different lineages, either from stamens (andropetals), or sepals/bracts (bractopetals) (Endress & Matthews, 2006; Ronse De Craene & Brockington, 2013). Andropetals usually have a narrow base with one vascular trace, contain specialized conical cells on the adaxial epidermis, and lack stomata. Moreover, their primordia show delayed development relative to stamen primordia (Kramer & Irish, 2000; Ronse De Craene & Brockington, 2013). Bracteopetals are similar to sepals or bracts in terms of their ontogeny and the number of vascular traces (Kramer & Irish, 2000; Ronse De Craene & Brockington, 2013). Petals are a characteristic feature of eudicots, although they are occasionally found in other clades. Petals show tremendous diversity in size, shape, structure, texture, and coloration (Endress & Matthews, 2006; Ronse De Craene & Brockington, 2013). However, extensive molecular genetic studies suggest that petals share a common identity specification program, suggestive of deep homology (Irish, 2009).

The B function of the ABCE model is of particular importance for petal identity specification. The genes responsible for the B function, such as *AP3* and *PI*, as well as *DEFICIENS* (*DEF*) and *GLOBOSA* (*GLO*), were first identified in *A. thaliana* and *A. majus*. *AP3* and *DEF* are orthologs that belong to the eu*AP3* lineage of the *AP3* subfamily (Kramer et al., 1998). The counterpart of *PI* is *GLO*, and both of them are members of the *PI* subfamily (Kramer et al., 1998). Despite the slight differences in expression between *AP3/DEF* and *PI/GLO*, these genes have overlapping expression domains in petals and stamens and maintain high levels of expression throughout flower development in both species. Mutations of these genes cause the homeotic transformation of petals into sepals and stamens into carpels (Bowman et al., 1989; Goto & Meyerowitz, 1994; Jack et al., 1992; Schwarz-Sommer et al., 1992; Sommer et al., 1990; Tröbner et al., 1992). Similarly, the replacement of petals by sepals was also found in loss-of-function mutants or knockdown plants of eu*AP3* and *PI* homologs in other core eudicot species, including representatives of the Solanaceae (i.e. *P. hybrida*, *Nicotiana benthamiana*, *S. lycopersicum*, and *Physalis*) and the Asteraceae (i.e., *G. hybrida*) (Broholm et al., 2010; de Martino et al., 2006; Guo et al., 2016; He et al., 2023; Liu et al., 2004; Vandenbussche et al., 2004; Zhang et al., 2014). These available data strongly indicate that eu*AP3*- and *PI*-like genes play a conserved role in specifying petal identity in core eudicots.

The petals of basal eudicots and basal angiosperms have been shown to have different evolutionary origins than those of core eudicots (Ronse De Craene & Brockington, 2013). However, the key genetic factors determining petal identity are also members of the *AP3* and *PI* subfamilies. In contrast to core eudicots, multiple independent gene duplication events have occurred in both *AP3*- and *PI*-like genes of basal eudicots (Kramer et al., 2003; Rasmussen et al., 2009). For instance, two successive duplications took place after the divergence of the Eupteleaceae from the other families of the Ranunculales, leading to the generation of three *AP3* lineages (i.e., *AP3-I*, *AP3-II*, and *AP3-III*). Many small-scale gene duplications have happened in different genera or species, resulting in an ongoing expansion of the *PI* lineage. Among the three *AP3* paralogs, only *AP3-III* genes show high and specific expression in petals of the examined species in the Ranunculaceae, Berberidaceae, and Lardizabalaceae (Duan et al., 2020; Hu et al., 2012; Sharma et al., 2011; Zhang et al., 2013). In contrast, the *PI*-like genes show relatively broader expression domains in floral organs (Kramer et al., 2003; Rasmussen et al., 2009). Mutation or downregulation of *AP3-III* and *PI*-like genes causes the conversion of petals into sepals in several emerging model species in the Ranunculaceae (*Aquilegia vulgaris*, *A. coerulea* "Origami," *N. damascena*, and *Delphinium ajacis*) and Papaveraceae (*P. somniferum* and *E. californica*) (Drea et al., 2007; Gonçalves et al., 2013; Lange et al., 2013; Sharma et al., 2011; Wang et al., 2016; Zhao et al., 2023). Moreover, the non-expression of *AP3-III* genes is closely associated with independent petal losses in the Ranunculaceae (Duan et al., 2020; Zhang et al., 2013). Similarly, *AP3*- and *PI*-like genes also show overlapping expression domains in the petals and stamens of basal angiosperms, including *Asimina longifolia* (Annonaceae), *Tradescantia reflexa* (Commelinaceae), *Saruma henryi* (Aristolochiaceae), and *H. radiata* (Orchidaceae) (Jaramillo & Kramer, 2004; Kim et al., 2005b, 2007; Ochiai et al., 2004). These lines of evidence not only strongly support the idea that that B class MADS-box genes are essential for petal identity determination in angiosperms but also suggest that the programs determining petal identity have evolved in parallel in different angiosperm lineages (Glover et al., 2015).

It should be pointed out that the A function also participates in the specification of petal identity. Mutations of the A function genes of *A. thaliana*, i.e., *AP1* and

AP2, result in the absence of petals or the transformation of petals into stamen-like organs (Bowman et al., 1989; Irish & Sussex, 1990; Kunst et al., 1989; Mandel et al., 1992). However, no such phenotypic change was observed in mutants of either *AP1* or *AP2* orthologs in other species (Litt, 2007; Morel et al., 2017). This led to a debate on whether *AP1* and *AP2* really were A function genes (Causier et al., 2010; Litt, 2007). By comparison and functional dissection of the regulatory differences of *AP1* and its closely related paralog, *CAULIFLOWER* (*CAL*), Ye et al. (2016) identified an autoregulation site—the CArG-box—in the promoter region of *AP1* but not in *CAL*. Furthermore, they revealed that this site confers the A function to *AP1* by maintaining its expression in developing sepals and petals. By further tracing the history of this site, they demonstrated that it was gained in the MRCA of *Arabidopsis* by modification of an existing sequence, suggesting that *AP1* has a recent origin as an A function gene. These findings, plus functional evidence that basal eudicot *AGL6*-like genes (*NdAGL6* of *N. damascena* and *DeajAGL6-1a/1b/2* of *D. ajacis*) are involved in sepal and petal identity determination, further support the notion that the A function genes have independent origins in eudicots (Wang et al., 2016; Zhao et al., 2023).

In addition to the classic A and B function genes, the abovementioned miR169 family genes are also involved in petal identity determination (Cartolano et al., 2007; Morel et al., 2017). In *P. hybrida* and *A. majus*, inactivation of the genes encoding miR169, i.e., *BL* and *FISTULATA* (*FIS*), results in the homeotic transformation of petals into stamens due to the outward expansion of the C function gene expression (Cartolano et al., 2007; Morel et al., 2017). However, the counterparts of *BL* and *FIS* in *A. thaliana* are not involved in petal identity determination, which suggests the functional divergence of miR169 genes between Rosid and Asterid species (Cartolano et al., 2007).

Despite the independent origins of petals in different angiosperm lineages, the available data suggest that the B function genes, including *AP3*- and *PI*-like genes, are necessary for petal identity determination, representing a deep homology in petal evolution. Other petal identity genes, such as *AP1*, *AP2*, *AGL6*, miR169, and (potentially) unknown regulators, may also be involved in a lineage- or species- specific manner.

8.3.2.3 Stamens

Stamens are the male reproductive organs of a flower and originated only once in angiosperms. A stamen is generally composed of an anther and a basal filament or stalk. An anther consists of two thecae and each theca consists of two pollen sacs (i.e., microsporangia), where pollen is produced and released. Filaments are slender in most cases. However, stamens of some basal angiosperms lack filaments, which have anthers being embedded in laminar structures. In addition, stamens of some flowering plants are morphologically complicated due to the fusion of organs. For example, fusion between stamens occurs in members of the Gesneriaceae, Scrophulariaceae, and Asterids; fusion of stamens with carpels is widely found in the Asclepiadoideae and Orchidaceae; and the adnation of filaments to the perianth occurs frequently in sympetal flowers in eudicots (D'Arcy & Keating, 1996).

Despite the diverse forms and configurations, stamen identity is defined by a common genetic program involving C and B function genes. The C function gene *AG* was first identified in *A. thaliana* (Yanofsky et al., 1990). *AG* mRNA starts to accumulate in floral primordia at stage 3—where stamens and carpels will later form—and is then concentrated in developing stamen and carpel primordia (Drews et al., 1991; Yanofsky et al., 1990). In the *ag* mutant, stamens and carpels are homeotically converted into petals and sepals, respectively. This is accompanied by a recurrence of these organs in the center of the flower, which suggests that *AG* not only plays a role in specifying stamen and carpel identities, but also controls floral determinacy (Bowman et al., 1989; Yanofsky et al., 1990). *AG* homologs have been functionally characterized in other phylogenetically distant species including *A. majus*, *P. hybrida*, *N. benthamiana*, *S. lycopersicum*, *M. truncatula*, and *G. hybrida* in core eudicots, *E. californica*, *P. somniferum*, *N. damascena*, and *T. thalictroides* in basal eudicots, and *O. sativa* and *Phalaenopsis equestris* in monocots. Loss of function or downregulation of *AG* genes in these species results in the loss of stamen and carpel identities, suggesting that their role in reproductive organ identity is functionally conserved (Causier et al., 2005; Davies et al., 1999; Dreni et al., 2011; Fourquin & Ferrandiz, 2012; Galimba et al., 2012; Hands et al., 2011; Heijmans et al., 2012; Hsieh et al., 2013; Pan et al., 2010; Serwatowska et al., 2014; Wang et al., 2016; Yellina et al., 2010; Yu et al., 1999).

Phylogenetic analyses indicate that the *AG* genes identified so far are members of the *AG* MADS-box gene subfamily (Kramer et al., 2004; Zahn et al., 2006). This subfamily experienced multiple independent duplications during angiosperm evolution. For example, a duplication event before the origin of core eudicots gave rise to the eu*AG* and *PLENA* (*PLE*) lineages; another duplication event—that possibly occurred at the base of the Ranunculales—generated the *AG1* and *AG2* clades; and the duplication events prior to the diversification of grasses produced the *OsMADS3*, *OsMADS58*, and *OsMADS13* lineages (Kramer et al., 2004; Zahn et al., 2006). Interestingly, by mapping gene expression data onto phylogenetic trees (Zahn et al., 2006) and referring to available functional data, it was revealed that the *AG*-like genes that specify stamen and carpel identities are not strictly orthologous among species. One classic example of this lack of orthology involves the genes *AG* and *PLE*, both of which are canonical C function genes required for the specification of stamen and carpel identities. However, they are actually not orthologous genes: *AG* is a member of the eu*AG* lineage, while *PLE* is a member of the *PLE* lineage (Causier et al., 2005). Similar random functional divergence has frequently occurred in this subfamily. Given that the protein-coding sequences of *AG*-like genes are highly conserved, it is possible that functional divergence between duplicate genes is mainly caused by the divergence of regulatory regions and thus expression patterns (Hong et al., 2003; Kramer et al., 2004; Zahn et al., 2006).

B function genes, including *AP3*- and *PI*-like genes, are also required for stamen identity determination. Loss of function or downregulation of either *AP3*- or *PI*-like genes in various species, from both eudicots and monocots, causes the homeotic transformation of stamens into carpeloid organs, as well as the conversion of petals or lodicules into sepals or paleas as described above (Ambrose et al., 2000; Broholm et al., 2010; de Martino et al., 2006; Drea et al., 2007; Guo et al., 2016; Kramer et al., 2007; Lange et al., 2013; Liu et al., 2004; Nagasawa et al., 2003; Otani et al., 2016; Rijpkema et al., 2006; Sharma & Kramer, 2013; Vandenbussche et al., 2004; Wang et al., 2016; Yao et al., 2008; Zhang et al., 2014; Zhao et al., 2023). Independent, large-scale gene duplications have also occurred in the *AP3* subfamily. These include the duplications that led to the eu*AP3* and *TOMATO MADS BOX GENE6* (*TM6*) lineages in core eudicots as well as those resulting in the *AP3-I*, *AP3-II*, and *AP3-III* clades in the Ranunculales. However, in contrast to the *AG* subfamily, the functional divergence pattern among paralogs of the *AP3* subfamily follows a general rule. In core eudicots, eu*AP3*-like genes tend to determine both stamen and petal identities, whereas paralogous *TM6*-like genes are required only for the specification of stamen identity (Broholm et al., 2010; Rijpkema et al., 2006; Vandenbussche et al., 2004). In basal eudicots, *AP3-III* genes specifically control the petal identity, while the other copies (i.e., *AP3-I* and/or *AP3-II*) mainly determine stamen identity instead (Drea et al., 2007; Sharma & Kramer, 2013; Wang et al., 2016).

It is notable that the miR172-eu*AP2* module also plays critical roles in determining stamen identity by defining the boundary between perianth and reproductive organs. As mentioned above, miR172 is a negative regulator of *AP2* and can prevent the accumulation of *AP2* mRNA and protein in the third and fourth whorls. In *A. thaliana* flowers expressing a miR172-resistant version of *AP2* or with reduced miR172 activity, the stamens are completely or partially converted into petals, due to strong *AP2* accumulation and decreased *AG* transcript levels in the third whorl (Wollmann et al., 2010). Similarly, in *Prunus persica* (Rosaceae), large fragment insertions or deletions disrupting a miR172 gene have been discovered to be responsible for the double-flower phenotype of some ornamentals, mainly represented by the replacement of stamens by petals (Cirilli et al., 2022). In addition, in several phylogenetically distant core eudicot taxa (e.g., *Dianthus*, *Petunia*, *Rosa*, and *Nicotiana*), it has been shown that independent mutations affecting the miR172 target site in the eu*AP2*-like *BEN* orthologs can also produce the double-flower trait (Gattolin et al., 2020). These findings strongly indicate that any mutations disrupting the miR172-mediated repression of eu*AP2* affect the identity of stamens and that this mechanism seems to be conserved at least in core eudicots.

8.3.2.4 Carpels

Carpels are the female reproductive organs of a flower and are the last organ to be formed by the floral meristem. A carpel is defined as a basic unit of the pistil, which is generally composed of ovary, style, and stigma. In basal angiosperms, either only one carpel is present in a flower, or the carpels are generally separate from each other (termed "apocarpy"). In contrast, in most angiosperms, such as eudicots and monocots, the carpels are congenitally united (termed "syncarpy"). As one of the key innovations in angiosperms, carpels confer many advantages on flowering plants, including enclosure and protection of early ovules and later seeds, improvement of the system for pollen capture and pollen tube guidance, and support for self- or inter-specific incompatibility (Endress, 2001b; Ferrándiz et al., 2010). As described above, carpel identity is specified by the *AG*-like genes in phylogenetically distant species.

In addition to *AG* subfamily genes, genes from other transcription factor families are also reported to be involved in the specification of carpel identity. *DROOPING LEAF* (*DL*) encodes a *CRABS CLAW* (*CRC*)-like gene of the *YABBY* gene family and was identified as a carpel identity gene in *O. sativa* (Yamaguchi et al., 2004). *DL* is expressed in the carpel anlagen and developing primordia. Mutation of *DL* shows a carpel-to-stamen transformation, indicating that *DL* is involved in the control of carpel identity by repressing the expression of the B function genes (Nagasawa et al., 2003; Yamaguchi et al., 2004). Identical expression patterns of *DL* orthologs have been discovered in other grass species, including *Zea mays*, *Triticum aestivum*, and *Sorghum bicolor*, suggestive of a potential functional conservation in grasses (Ishikawa et al., 2009). However, by looking at the expression patterns and functional data of *CRC*-like genes in the wider context of the angiosperm phylogeny, it becomes apparent that the carpel identity determination is a derived function of *CRC*-like genes in the Poaceae (Orashakova et al., 2009).

8.3.3 Identity Determination of Other Types of Floral Organs

8.3.3.1 Petaloid Sepals

In most cases, the outer sepals and inner petals are easily distinguished in a flower with a dimorphic perianth based on the features described above. However, the recognition of sepals can become difficult in cases where the sepals show petaloid rather than leaf-like characteristics. Petaloid sepals are frequently found in members of core eudicots (e.g., Saxifragaceae, Rhamnaceae, Balsaminaceae, Ericaceae, Plantaginaceae, and Caryophyllales) and basal eudicots (e.g., Ranunculaceae, Berberidaceae, Lardizabalaceae, and Menispermaceae) (Borchert et al., 2009; Geuten et al., 2006; Landis et al., 2012; Ronse De Craene & Brockington, 2013). The reconstruction of the ancestral character state suggests that petaloid sepals have evolved either independent of the petals or in response to a reduction or loss of petals (Ronse De Craene, 2008).

Given the general similarity of petals and petaloid sepals in appearance, it has been proposed that these two floral organ types share the same genetic program (i.e., have process homology). Moreover, the B function genes are the most potential regulators contributing to petaloidy (Albert et al., 1998; Kramer et al., 2003). However, gene expression analyses of several core eudicot species with petaloid sepals do not support this hypothesis. As shown in *Impatiens hawkeri* (Balsaminaceae), *Calluna vulgaris* (Ericaceae), *Sesuvium portulacastrum*

(Aizoaceae), and *Rhodochiton atrosanguineum* (Plantaginaceae), *AP3* and *PI* homologs are only weakly expressed, or not expressed at all, in petaloid sepals (Borchert et al., 2009; Brockington et al., 2012; Landis et al., 2012). Moreover, in contrast to the "true" petals, conical epidermal cells—a diagnostic trait—are absent in the petaloid sepals of these species (Borchert et al., 2009; Geuten et al., 2006; Landis et al., 2012).

The petaloid sepals of basal eudicot flowers are somehow different from those of core eudicots. They have conical epidermal cells on the adaxial surface, as shown in *Sinofranchetia chinensis* (Lardizabalaceae), as well as in *A. vulgaris*, *N. damascena*, and *D. ajacis* (Ranunculaceae) (Gonçalves et al., 2013; Hu et al., 2012; Kramer et al., 2007; Zhao et al., 2023). Moreover, among the B function genes, *PI*-like genes tend to be highly expressed in petaloid sepals of representative species in the Ranunculaceae, Berberidaceae, Lardizabalaceae, and Menispermaceae (Hu et al., 2012; Kramer et al., 2007; Shan et al., 2006; Sharma et al., 2011; Wang et al., 2016; Zhang et al., 2013; Zhao et al., 2023). However, unexpectedly, downregulation of *PI*-like genes of *A. vulgaris*, *N. damascena*, and *D. ajacis* result in no homeotic transformation of petaloid sepals (Kramer et al., 2007; Wang et al., 2016; Zhao et al., 2023). Again, these findings demonstrate that the B function genes are not essential for the specification of petaloid sepals in basal eudicots.

Notably, although the B function genes have no major role in the specification of petaloid sepals, a careful study revealed that the *PI*-like gene of *A. coerulea* "Origami," *AqPI*, is required for the production of anthocyanin in petaloid sepals. Knockdown of *AqPI* results in various degrees of color loss of petloid sepals (Sharma & Kramer, 2017). In addition, it has been reported that the *AGL6*-like genes found in *N. damascena* (*NdAGL6*) and *D. ajacis* (*DeajAGL-1a/1b/2*) are highly expressed in primordia of petaloid sepals and petals (Wang et al., 2016; Zhao et al., 2023). Downregulation of *NdAGL6* results in the conversion of these perianth organs into bract-like organs (Wang et al., 2016). Inspired by these studies, we propose that the independent recruitment of the pigmentation pathways and/or programs responsible for the conical cell differentiation (e.g., *MIXTA*) (Noda et al., 1994) into the sepal identity program may have been the key to the petaloidy of sepals.

8.3.3.2 Tepals

The term tepal is usually used for the simple, undifferentiated or only weakly differentiated perianth (Weberling, 1989). Typical tepals exist in flowers of nongrass monocots (e.g., Liliales and Asparagales) (Ronse De Craene & Brockington, 2013). Inspired by the ABC model, researchers have postulated that the undifferentiated petaloid tepals are generated by a spatial shift in B-class gene expression; this led to the formulation of the shifting boundary model (Bowman, 1997) and the sliding boundary model (Kramer et al., 2003). In the two decades since the articulation of these models, experimental data have suggested that the B-class genes are expressed in petaloid tepals of the examined species in the Liliales and Asparagales, which provides strong support for these models (Hsu et al., 2015; Kanno, 2016; Mondragon-Palomino & Theißen, 2011). Moreover, by investigating a tulip (*Tulipa gesneriana*) mutant—*viridiflora*, which has sepaloid rather than petaloid tepals—Hirai et al. (2010) found that the expression levels of two *AP3*-like genes are weaker than the expression levels found in wild type cultivars. A functional study in *Tricyrtis* sp. (Liliaceae) has also shown that repression of an *AP3*-like gene (*TrihDEFa*) expression results in the conversion of all petaloid tepals into greenish sepal-like organs (Otani et al., 2016). These data further support the shifting boundary or sliding boundary models.

It is also important to mention that one of the tepals in the inner perianth whorl of flowers of the Orchidaceae has evolved into novel forms, a structure known as the labellum ("lip"). The labellum acts as a main pollinator attractor and greatly contributes to Orchid family diversity. Several studies have shown that divergent expression patterns among the four *AP3*-like paralogs are closely associated with the differentiation of the labellum from the tepals (Chang et al., 2010; Mondragon-Palomino & Theißen, 2008). Comprehensive comparative expression and functional studies further demonstrate that labellum identity is determined by the L complex (i.e., OAP3-2/OAGL6-2/OPI), while the identity of other tepals is specified by the SP complex (i.e., OAP3-1/OAGL6-1 /OPI); this is the so-called P code model for perianth formation in orchids (Hsu et al., 2015). In addition, a later study refined the P code model and suggested that the lower part of the lateral sepal is regulated by a strong SP complex and a weak L' complex (OAP3-1/OAGL6-2/OPI); the petals are regulated by both SP and SP' (OAP3-2/ OAGL6-1/OPI) complexes; the labellum is primarily regulated by the L complex and, to a small degree, by the L'/SP/SP' complexes (Hsu et al., 2021).

In contrast to nongrass monocots, basal angiosperms generally have tepals that show a gradual morphological transition to adjacent floral parts. This pattern is seen in the Amborellaceae, as well as in some members of the Nymphaeaceae, Illiciaceae, Austrobaileyaceae, Schisandraceae, and Cabombaceae (Buzgo et al., 2004; Endress, 2001a; Kim et al., 2005b). Based on the floral developmental morphology of the basalmost angiosperm *Amborella trichopoda*, Buzgo et al. (2004) developed a "fading borders" model, which proposed that gradual morphological transitions between adjacent floral parts may have resulted from a gradient in the expression of floral organ identity genes, particularly B function genes. In support of this hypothesis, both *AP3*- and *PI*-like genes have been found to be expressed broadly in floral organs, including the tepals of many basal angiosperms (Kim et al., 2005a, b; Li et al., 2015). A comparative analysis of gene expression profiles was conducted using microarray data for the basal angiosperm species *Nuphar advena* (Nymphaeaceae) and *Persea americana* (Lauraceae). The results indicate that the transcriptional programs had broadly overlapping spatial domains across the morphologically intergrading organs in both species (Chanderbali et al., 2010). Similarly, the RNA-seq data of *Nymphaea colorata* also indicated that the ABCE homologs generally exhibit wide ranges of expression in floral organs (Zhang et al., 2020). These findings suggest that the "fading borders" transcriptional cascades may represent an ancestral state of floral organ identity determination and later be shaped into a strict pattern such as the ABCE model. Presumably, through canalization of gene expression patterns, angiosperm flowers have evolved from undifferentiated tepals into well-differentiated sepals and petals.

8.3.3.3 Lemmas, Paleas, and Lodicules

Flowers of Poaceae species possess a completely different perianth than other angiosperms. The perianth of Poaceae species is composed of three types of organs: a lemma, a palea, and two or three lodicules. The lemma is typically green, leaf- or bract-like. The palea is also bract-like, and is often thin and translucent. Lodicules are fleshy or scale-like and play a role in opening the flower by swelling to physically force apart the lemma and palea. Due to the fact that these organs have little morphological similarity to typical sepals and petals, whether these organs should be considered as parts of the perianth or as distinct, non-floral organs has been in debate (Kellogg, 2015). Accumulating genetic and functional studies of floral organ identity genes in the grass model species *O. sativa* and *Z. mays* not only demonstrated that these organs are indeed parts of the perianth, but also provided critical insights into how the identities of these organs are specified.

In *O. sativa*, the lemma and palea identities have been found to be specified by two *AP1* subfamily genes, *OsMADS14* and *OsMADS15*, which were generated by a gene duplication event before the divergence of the Poaceae (Shan et al., 2007). *OsMADS15* is mainly expressed in spikelet meristems, lemmas, and paleas, while *OsMADS14* is broadly expressed in vegetative tissues, spikelet meristems, and all floral organs (Preston & Kellogg, 2006, 2007; Wu et al., 2017). Lemma phenotypes are not affected in *osmads14* and *osmads15* single mutants or the *osmads14;osmads15/+* double mutants, but are transformed into leaf-like organs with stomata in the *osmads15;osmads14/+* double mutants. This suggests that *OsMADS14* and *OsMADS15* have redundant roles in specifying lemma identity, and that *OsMADS15* is a major contributor (Wu et al., 2017). In addition, the *OsMADS14* and *OsMADS15* single and double mutants produce carpels in the position of paleas, which indicates that *OsMADS14* and *OsMADS15* are also involved in the specification of palea identity by antagonizing the expression of C function genes (Wang et al., 2010; Wu et al., 2017).

Lodicule identity is determined by the combination of the activities of *AP3-*, *PI-*, and *AP1*-like genes. One *AP3*-like gene (*OsMADS16*, also named *SUPERWOMAN*, *SPW*) and two *PI*-like genes (*OsMADS2* and *OsMADS4*) exist in the genome of *O. sativa*. *OsMADS2* and *OsMADS4* were generated by a recent duplication event in the MRCA of all grasses (Xu & Kong, 2007). *OsMADS16* and *OsMADS2* are specifically expressed in lodicules and stamens, while *OsMADS4* has relatively broader expression domains in spikelet meristems and floral organs (Yadav et al., 2007). Yeast two-hybrid assays indicate that OsMADS16 is able to form heterodimers with both OsMADS2 and OsMADS4 (Yoshida et al., 2007). Loss of function of *OsMADS16* or concomitant suppression of *OsMADS2* and *OsMADS4* leads to the homeotic conversion of lodicules into palea-like organs (Nagasawa et al., 2003;

Yao et al., 2008). The *AP1*-like genes *OsMADS14* and *OsMADS15* specify lodicule identity by repressing the expression of C function genes, as revealed by their double mutant plants that produce flowers whose lodicules are replaced by stamens (Wu et al., 2017).

Despite the fact that detailed functional studies of floral organ identity genes are scarce in the Poaceae, available data suggest that *AP3* and *PI* homologs in *Z. mays* have similar roles in the specification of lodicule identity (Ambrose et al., 2000; Bartlett et al., 2015; Whipple et al., 2004). In other members of the Poaceae, excluding *Streptochaeta*, which has no lodicules, B-class genes also show a conserved expression pattern (Whipple et al., 2007). In addition, *OsMADS14* and *OsMADS15* orthologs have similar expression patterns in grass species (Preston & Kellogg, 2007; Wu et al., 2017). Based on these lines of evidence, it is highly likely that the programs determining the identities of perianth organs in grasses are conserved.

8.3.3.4 Staminodes

Staminodes are considered to be a novel organ type and are widely distributed taxonomically, having independently arisen in all major clades of flowering plants (Walker-Larsen & Harder, 2000). They are defined based on their resemblance to stamens or their positional association with stamens. A typical staminode looks like a stamen but is reduced to a greater or lesser extent and fails to produce viable male gametophytes. This type of staminode is often found in female flowers of both dioecious and monoecious species, and in many zygomorphic flowers such as those of the Fabales and Asterids. In some cases, staminodes are highly elaborate and specified, with flattened, petaloid, branched, or ciliate forms. Examples of such staminodes can be found in *Clematis* and *Aquilegia* (Ranunculaceae), *Alcea rosea* (Malvaceae), and *Parnassia palustris* (Celastraceae). This type of staminode generally has important ecological functions in pollinator attraction and/or the avoidance of self-pollination (Ronse De Craene & Smets, 2001; Walker-Larsen & Harder, 2000; Weberling, 1989).

So far, the molecular mechanism specifying staminode identity is largely unknown, and the only case study yet undertaken was on *Aquilegia*. The staminodes of this genus are membranous, antherless, and laterally expanded with ruffled margins. They are positioned between stamens and carpels. Functional studies of B-class MADS-box genes have indicated that one of the three *Aquilegia AP3* genes (*AP3-I*) as well as the *PI* gene are necessary for staminode identity specification. Silencing of these genes causes the transformation of the staminodes into carpeloid organs (Kramer et al., 2007; Sharma & Kramer, 2013). Based on these findings, Kramer and colleagues emphasized that the sub- and neo-functionalization of *AP3*-like genes has facilitated the evolution of the staminodes in the lineage leading to *Aquilegia*. However, staminodes have arisen independently throughout the angiosperms. Whether similar mechanisms are responsible for staminode identity determination in other clades awaits to be determined.

8.4 Reproductive Organ Identity Determination in Gymnosperms

Gymnosperms are the closest sister group to angiosperms. Living gymnosperms contain four different lineages: cycads, ginkgos, gnetophytes, and conifers (Wang & Ran, 2014). Their reproductive structures, called cones or strobili, are unisexual and are morphologically and structurally different from flowers. Generally, the male reproductive structure (i.e., pollen cone) is composed of microsporophylls, which harbor microsporangia producing pollens, and the female reproductive structure (i.e., seed cone) consists of megasporophylls with naked ovules (Simpson, 2010).

The formulation of the ABC and ABCE models has encouraged researchers to examine molecular mechanisms that may determine gymnosperm reproductive organ identities. So far, genes homologous to many floral MADS-box genes have been isolated in representative gymnosperm lineages, including *Picea*, *Pinus*, and *Cryptomeria* (Conifers) (Carlsbecker et al., 2013; Fukui et al., 2001; Mouradov et al., 1998, 1999; Rutledge et al., 1998; Sundström et al., 1999; Tandre et al., 1995), *Gnetum* and *Welwitschia* (gnetophytes) (Becker et al., 2000, 2003; Moyroud et al., 2017; Shindo et al., 1999; Winter et al., 1999), *Ginkgo biloba* (ginkgos) (Jager et al., 2003), and *Cycas edentata* (cycads) (Zhang et al., 2004). Extensive gene expression data have shown that homologs of B- and C-class genes have overlapping expression in male reproductive structures, while only homologs of C-class genes are expressed in female reproductive structures. Further, the introduction of gymnosperm B- and C-class genes into *A. thaliana* wild type or loss-of-function mutants result in phenotypic changes, which are partly similar to those caused by ectopic expression of endogenous genes (Rutledge et al., 1998; Sundström & Engstrom, 2002; Tandre et al., 1998; Winter et al., 2002; Zhang et al., 2004). In addition, a study on a *Picea abies* (var. *acrocona*) mutant showed that a homeotic transformation from vegetative shoot to female cone is closely related to the persistent activation of the C-class gene *DEFICIENS AGAMOUS-LIKE2* (*DAL2*) and a gymnosperm specific gene *DAL21* (Carlsbecker et al., 2013). By investigating protein–protein interactions among 12 highly expressed MADS-box genes in male and female cones of *Pinus tabuliformis*, Niu et al. (2022) proposed that the complex comprising B (DAL11 and DAL13), C (DAL2), and G (DAL14) class proteins, as well as DAL10 is responsible for the male identity, and the DAL2-DAL14-DAL10 complex for the female identity. These lines of evidence strongly indicate that the core components that specify reproductive organ identities are conserved between angiosperms and gymnosperms.

8.5 Evolutionary Origin of the Floral Organ Identity Determination Program

8.5.1 *Morphological Transitions During the Origin of the Flower*

The diversification of flowers has significantly contributed to the success of angiosperms in colonizing the terrestrial ecosystems. However, due to dramatic morphological differences and missing links between angiosperm flowers and gymnosperm reproductive organs, as well as the lack of informative fossil records, the origin of the flower remains enigmatic and has long been a focus of evolutionary botany. Over the past two centuries, based on different phylogenetic inferences of seed plant groups and evidence from paleontological, morphological, and anatomical studies, many theories have been proposed to explain the associated morphological changes before and after the origin of the flower. Representative well-known theories include the Euanthial theory (Arber & Parkin, 1907), the Telome theory (Wilson, 1942), and the Gonophyll theory (Melville, 1960). Although these theories were controversial in their predictions regarding the ancestor of angiosperms, the nature of the flower, and the origination process of the flower, they greatly stimulated the rise and rapid progress of molecular phylogenetics and evo-devo approaches for answering these questions. In recent years, the seed plant phylogeny has been further resolved and the molecular mechanisms underlying flower development and evolution have been explored in great detail. These advances have not only corrected the mistakes inherent in previous theories, but also brought new ideas into our understanding of the origin of the flower and floral organs.

Based on the phylogenetic framework of seed plants and available morphological and paleontological data, key changes during flower origin have been inferred. Now, it is generally believed that the flower did not originate abruptly but rather evolved gradually. The transition from unisexuality to bisexuality might have been the first step during the origin of the flower, followed by the compression of the reproductive axis, and the evolution of floral determinacy (Baum & Hileman, 2006; Specht & Bartlett, 2009; Theißen & Melzer, 2007). The floral organs also experienced multiple-step evolutionary modifications, which probably started long before angiosperms originated (Endress, 2001b). The temporal sequence of the evolution of floral organs was more or less the reverse of the sequence of floral development, which likely started with ovules, followed by stamens and carpels, and finally ended with undifferentiated tepals (Baum & Hileman, 2006; Endress, 2001b, 2006). The differentiation of the perianth into green sepals and showy petals possibly occurred before the divergence of core eudicots (Baum & Hileman, 2006; Ronse De Craene & Brockington, 2013).

By sampling 792 species representing 63 orders (98%) and 372 families (86%) of angiosperms, Sauquet et al. (2017) assembled the largest data set of floral traits yet seen and inferred the ancestral flower of extant angiosperms by using a model-based strategy. They proposed that the ancestral flower was most likely bisexual and contained three types of floral organs. They further predicted that it possessed more than ten undifferentiated tepals, more than ten stamens with introrse anthers, and five carpels, with tepals and stamens being arranged into whorls and carpels being arranged spirally (Sauquet et al., 2017). Despite some debates about floral phyllotaxy (Rümpler & Theißen, 2019), this finding not only addressed the uncertainty of previous studies in inferring the ancestral states of angiosperm flowers, but also provided an important base for further research on the mechanisms underpinning the origin of the flower.

8.5.2 *Molecular Changes During the Origin of the Flower*

8.5.2.1 Hypotheses Explaining the Origin of the Flower

One of the most important tasks in unravelling the mystery of the origin of the flower is to elucidate the molecular changes that occurred before and after the origin of the flower. In the past two decades, several famous hypotheses have been developed and provided detailed scenarios about how floral regulatory programs changed and how the first flower may have been formed. For instance, the mostly male theory (MMT) was the first molecular hypothesis regarding the origin of the flower. The MMT proposed that the first bisexual flower was derived from the male cones of the gymnosperm ancestor, which was caused by ectopic expression of ovule-promoting genes and the loss of the *LFY* paralog *NEEDLY* (*NLY*)—a female-specific gene; and thereby both stamens and carpels originated from microsporophylls (Albert et al., 2002; Frohlich & Parker, 2000). Two other hypotheses, the "out-of-male" and "out-of-female" (OOM/OOF) models suggested that homeotic transitions from microsporophylls to megasporophylls in male cones or vice versa in female cones caused by changes in B-class gene expression were responsible for the origin of the first bisexual flower. According to the OOM/OOF models, both microsporophylls and megasporophylls had the potential to evolve into stamens and/or carpels (Theißen et al., 2002). After the OOM model, Baum and Hileman (2006) further hypothesized that in the ancestor of angiosperms, the gradual increase in *LFY* expression along the reproductive axis resulted in differences in the expression levels of B- and C-class genes and thus promoted the conversion of the apical microsporophylls to megasporophylls in the male cones. They also postulated that the co-regulation of C-class genes and *WUSCHEL* (*WUS*) was a key determinant of the origin of undifferentiated tepals, while the cooption of *UNUSUAL FLORAL ORGAN* (*UFO*) for B-class gene activity coincided with the origin of a differentiated perianth (Baum & Hileman, 2006). Following Baum and Hileman's hypothesis, Theißen and Melzer (2007) added that the integration of *SEP*-like genes into the floral organ identity determination program led to the compression of the reproductive axis.

It should be noted that these models offer different explanations of how the first flower originated, but the hypothesized molecular events heavily relied on genetic and functional evidence from highly derived model plants, in particular *A. thaliana*. Whether these events occurred in the MRCA of extant angiosperms and contributed to the origin of the flower and floral organs need to be further evaluated. For instance, several studies have shown that the expression patterns of *LFY* and *NLY* in gymnosperms are not always male- or female-specific, which do not agree with the predictions of the MMT (Dornelas & Rodriguez, 2005; Moyroud et al., 2017; Vazquez-Lobo et al., 2007). In addition, the observation that no significant differences in gene components involved in the development of angiosperm flowers and gymnosperm male or female cones also does not favor the MMT (Tavares et al., 2010). However, by examining the regulatory evolution between LFY/NLY and B/C class genes, it has been suggested that LFY, rather than NLY, palys a major role in regulating B class genes in seed plants, which partially supports the MMT (Moyroud et al., 2017). More studies are required to refine the previous models or to propose new hypotheses.

8.5.2.2 Evolution of MADS-Box Genes and the Origin of the Flower

Despite the difficulties in exploring the molecular mechanisms responsible for the origin of the flower and testing these hypothetical mechanisms, valuable insights regarding the origin of the flower has been provided by research on the evolution of floral MADS-box genes. One of the most striking evolutionary changes in floral MADS-box genes is the increase in both gene type and copy number just before the origin of extant angiosperms (Fig. 8.2). Phylogenetic analyses indicate that all MADS-box genes involved in floral organ identity determination, including *AP1*-, *AP3*-, *PI*-, *AG*-, *SEP1/2/4*-, *SEP3*-, and *AGL6*-like gene, existed in the MRCA of extant angiosperms (Albert et al., 1998; Kramer et al., 1998, 2004; Ruelens et al., 2013; Shan et al., 2009; Zahn et al., 2005, 2006; Zhao et al., 2017). For simplicity, these genes are hereafter designated as A, B1, B2, C, E1, E2, and G, respectively. Interestingly, A, E1, E2, and G were derived from an ancestral gene that existed in the ancestor of seed plants. This ancestral gene experienced a tandem duplication event in the MRCA of seed plants and gave rise to the ancestral genes of E/G and A/S (*SUPPRESSOR OF OVEREXPRESSION OF CO1*, *SOC1*, also named *TM3*). The E/G–A/S cluster further duplicated through a whole genome duplication event (ζ), which generated two clusters. In the MRCA of seed plants, one cluster contained A and E and the other contained S and G (Fig. 8.2a). Following the whole genome duplication (ε) before the MRCA of extant angiosperms, E1, A, E2, and F (*FLOWERING LOCUS C*, *FLC*) were generated from one cluster and G and S from the other (Zhao et al., 2017). In addition, C and D (*AGL11*) may have been produced through the ε duplication before the origin of angiosperms, but the duplication generating B1 and B2 may be about 60 million years earlier (Ruelens et al., 2017) (Fig. 8.2b). In contrast, only three types of floral MADS-box genes (B', D', and G') were represented in the MRCA of gymnosperms, while the ancestral, tandemly arranged A–E genes were lost and the G' gene was duplicated (Shan et al., 2009; Zhao et al., 2017) (Fig. 8.2b).

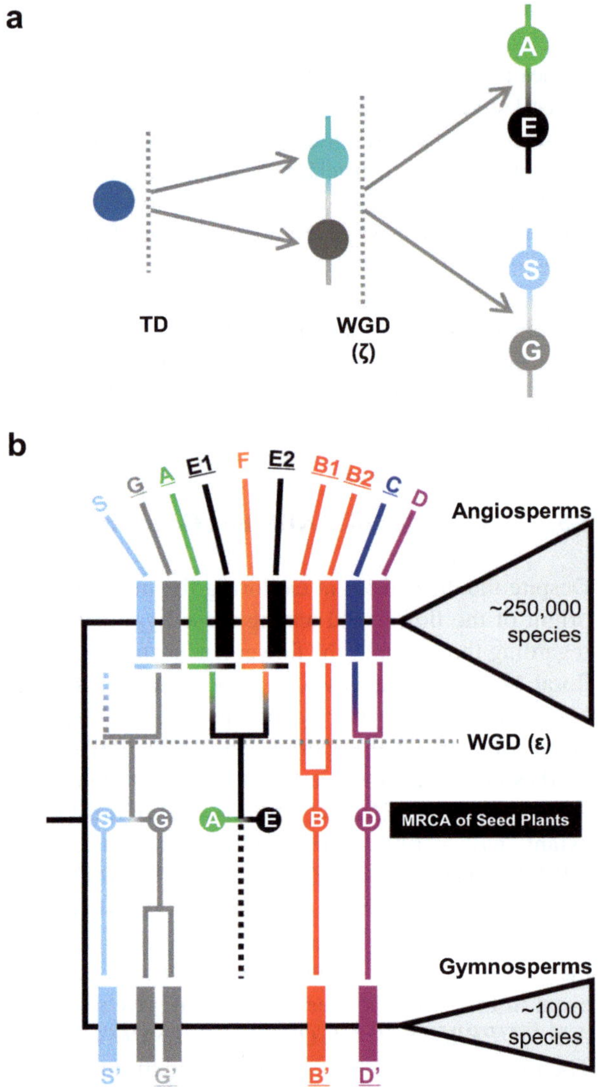

Fig. 8.2 Evolution of floral MADS-box genes. (**a**) Evolutionary origins of A, E, G, and S-class MADS-box genes. TD represents a tandem duplication event that occurred in the ancestor of seed plants. WGD (ζ) indicates the whole genome duplication that occurred in the most recent common ancestor (MRCA) of seed plants. (**b**) Evolutionary origins of floral organ identity MADS-box genes (underlined) in both angiosperms and gymnosperms. Colored bars represent the extant genes. Short lines underneath two bars stand for tandemly arranged genes. Dashed vertical lines denote gene losses. WGD (ε) indicates the whole genome duplication that occurred in the MRCA of angiosperms. In (**a**) and (**b**), the clustered two dots (e.g., A–E and S–G) indicate tandem duplicate genes

Like most duplicate genes, the angiosperm floral MADS-box genes tend to diverge in their coding and regulatory regions. For instance, changes in exon-intron structures have led to the formation of subfamily-specific motifs in different types of floral MADS-box genes. The ancestral B1 protein possessed both PI and paleoAP3 motifs, but the ancestral B2 protein had only a PI motif. The ancestral A, F, E1, and E2 proteins diverged more dramatically in their C-terminal regions relative to the closely related G protein (Kramer et al., 1998; Yu et al., 2016). Although the ancestral C and D genes showed less divergence in coding sequences, they were differentiated in expression patterns. The available data suggest that most C genes are expressed mainly in stamens and carpels and D genes in ovules, although rare exceptions do exist (Zahn et al., 2006).

Beyond these changes, it is notable that the interaction patterns of floral MADS-box proteins dramatically changed before and after the occurrence of extant angiosperms. By tracing the evolutionary histories of protein–protein interactions (PPIs) and investigating the interaction patterns of reconstructed ancestral proteins of floral MADS-box genes, it has been found that, in the MRCA of seed plants, the B′, D′, and G' proteins can form both homodimers and heterodimers (Li et al., 2015; Ruelens et al., 2017). Following duplications, the abilities of the proteins of the resultant genes to interact with each other, which were assumed to be retained, were actually lost in a non-random manner. In particular, almost all the PPIs that disappeared were related to B-class proteins; this may have facilitated the subsequent functional canalization of the *AP3* and *PI* genes and the evolution of the AP3-PI obligate heterodimer (Li et al., 2015; Ruelens et al., 2017). It has been proposed that the rapid canalization of the *AP3* and *PI* genes in both PPI and expression patterns reinforced their roles in specifying petal and stamen identities (Chanderbali et al., 2010; Li et al., 2015; Melzer et al., 2014).

Another important evolutionary event that occurred before the diversification of angiosperms was the integration of E-class proteins into the floral PPI network. Here they obtained the ability to mediate the formation of floral quaternary complexes (Li et al., 2015; Ruelens et al., 2017; Theißen & Melzer, 2007). The A and G-class proteins in the MRCA of extant angiosperms have been reported to have the same ability to interact with other proteins and to auto-activate as the ancestral E-class proteins (Li et al., 2015). However, when looking at the expression patterns of these genes across the seed plants, only the E-class genes have relatively broad expression domains in floral meristems and floral organs and maintain relatively stable expression levels throughout flower development (Becker et al., 2003; Carlsbecker et al., 2013; Li et al., 2015; Malcomber & Kellogg, 2005; Mouradov et al., 1999, 2017; Shan et al., 2007; Winter et al., 1999). Therefore, it is highly likely that the regulatory divergence between E-class genes and the other two classes of genes occurred at an early stage in their evolution, possibly prior to the origin of extant angiosperms. The integration of E genes into the floral PPI network not only facilitated the establishment of the floral context, but also promoted the evolution of bisexual flowers via the formation of different floral quaternary complexes along a reproductive axis (Li et al., 2015; Ruelens et al., 2017). Despite the fact that direct functional evidence *in planta* is still missing, these state-of-the-art studies have allowed us to move further toward understanding the molecular mechanisms responsible for the origin of the flower.

8.6 Outlook

After more than 30 years since the proposal of the classic ABC model, tremendous progress has been made in exploring the patterns, mechanisms, and evolution of floral organ identity determination. To date, it has become clear that: (i) floral organ identities are specified at different developmental stages, which are independent of the initiation processes of floral organ primordia; (ii) the floral MADS-box genes are central to the programs specifying floral organ identities; (iii) the establishment of the floral context mainly depends on E-class MADS-box genes; (iv) carpel and stamen identities are predominantly specified by the respective activities of C-class MADS-box genes and B- + C-class MADS-box genes, both of which are deeply conserved in angiosperms; (v) the identities of other types of floral organs are determined by group- or lineage-specific mechanisms; and (vi) abrupt increases in the number and in the type of floral MADS-box genes that occurred before the origin of extant angiosperms and the subsequent evolutionary changes in PPI and expression patterns have facilitated the origin of the perfect, bisexual flower.

Our current understanding of the floral organ identity determination remains incomplete. One of the major challenges remaining concerns how the floral MADS-box genes are precisely regulated. Data from loss-of-function mutants and computational modeling have shown that the appropriate expression of floral MADS-box genes is a prerequisite for conferring proper identities of floral organs at the right time and in the right place (Alvarez-Buylla et al., 2010). In *A. thaliana*, the initiation of expression of *AP1*, *AP3*, *AG*, and *SEP3* is mainly controlled by *LFY* and *UFO* or *WUS* (Lohmann et al., 2001; Rieu et al., 2023) and the maintenance of their subsequent expression mainly depends on autoregulation (Chen et al., 2018). From an evolutionary perspective, the regulatory link between *LFY* and B-class genes is evolutionarily conserved and may have existed in the MRCA of seed plants (Moyroud et al., 2017), while the autoregulation of *AP1* is a recently gained mechanism, which originated just before the divergence of the genus *Arabidopsis* (Ye et al., 2016). These findings clearly indicate that the links between floral MADS-box genes and their regulators in *A. thaliana* are composed of both evolutionarily conserved and newly recruited parts. It is now important to determine how floral MADS-box genes are regulated in other species as well as how the regulatory program originated and evolved. Such work may improve our understanding of the underlying mechanisms responsible for floral diversity of organ type and number as well as the genetic basis of the origin of the flower.

Other challenges concern non-MADS-box genes and less-studied basalmost angiosperms and gymnosperms. As we have reviewed in this chapter, non-MADS-box transcription factors and microRNAs are also required for the specification of floral organ identities in some species. However, it remains largely elusive whether these non-MADS-box genes and microRNAs play similar roles in other species or not. In addition, only expression models have been proposed to explain how floral organs in basalmost angiosperms and reproductive structures in gymnosperms are specified; this is primarily caused by the lack of suitable functional systems in these groups. In recent years, a promising technique for gene functional analysis (i.e., VIGS) has been successfully applied to several species of the Ranunculales and the functions of floral MADS-box genes have been studied in detail (Damerval & Becker, 2017). The novel data from these emerging model species have expanded our understanding of the conservation and divergence of the programs determining the identities of different types of floral organs (Shan et al., 2019). Given the important phylogenetic positions of basalmost angiosperms and gymnosperms, it is necessary to further develop functional techniques, especially VIGS or transgenic approaches, in these lineages. If these challenges are met, we may gain additional insight into patterns, mechanisms, and evolution of floral organ identity determination.

Acknowledgements We thank Boka Li for drawing floral diagrams, Xuehao Fu and Tianpeng Wang for editing the figures, and the Kong lab members for critical reading of the manuscript and for valuable discussion. We are also grateful to Zheng Meng, Xuelei Lin, Rui Zhang, Min Li, Xu Yao, Hong Liao, Ruidong Jia, and Lingling Ye for generously providing the beautiful flower photographs. This work was supported by National Natural Science Foundation of China (Grants nos. 32221001, 31970246, and 31930008).

References

Albert, V. A., Gustafsson, M. H. G., & Di Laurenzio, L. (1998). Ontogenetic systematics, molecular developmental genetics, and the angiosperm petal. In D. E. Soltis, P. S. Soltis, & J. J. Doyle (Eds.), *Molecular systematics of plants II: DNA sequencing* (pp. 349–374). Springer US. https://doi.org/10.1007/978-1-4615-5419-6_12

Albert, V. A., Oppenheimer, D. G., & Lindqvist, C. (2002). Pleiotropy, redundancy and the evolution of flowers. *Trends in Plant Science, 7*(7), 297–301. https://doi.org/10.1016/S1360-1385(02)02300-2

Alvarez-Buylla, E. R., Benítez, M., Corvera-Poiré, A., Chaos Cador, A., de Folter, S., Gamboa de Buen, A., Garay-Arroyo, A., García-Ponce, B., Jaimes-Miranda, F., Pérez-Ruiz, R. V., Piñeyro-Nelson, A., & Sánchez-Corrales, Y. E. (2010). Flower development. In *Arabidopsis Book* (Vol. 8, p. e0127). https://doi.org/10.1199/tab.0127

Ambrose, B. A., Lerner, D. R., Ciceri, P., Padilla, C. M., Yanofsky, M. F., & Schmidt, R. J. (2000). Molecular and genetic analyses of the *SILKY1* gene reveal conservation in floral organ specification between eudicots and monocots. *Molecular Cell, 5*(3), 569–579. https://doi.org/10.1016/S1097-2765(00)80450-5

Ambrose, B. A., Espinosa-Matías, S., Vázquez-Santana, S., Vergara-Silva, F., Martínez, E., Márquez-Guzmán, J., & Alvarez-Buylla, E. R. (2006). Comparative developmental series of the Mexican triurids support a euanthial interpretation for the unusual reproductive axes of *Lacandonia schismatica* (Triuridaceae). *American Journal of Botany, 93*(1), 15–35. https://doi.org/10.3732/ajb.93.1.15

Ampomah-Dwamena, C., Morris, B. A., Sutherland, P., Veit, B., & Yao, J. L. (2002). Down-regulation of *TM29*, a tomato *SEPALLATA* homolog, causes parthenocarpic fruit development and floral reversion. *Plant Physiology, 130*(2), 605–617. https://doi.org/10.1104/pp.005223

Arber, E. A., & Parkin, J. (1907). On the origin of angiosperms. *Linnean Society of London Zoology Transactions, 38*(263), 29–80. https://doi.org/10.1111/j.1095-8339.1907.tb01074.x

Aukerman, M. J., & Sakai, H. (2003). Regulation of flowering time and floral organ identity by a microRNA and its *APETALA2*-like target genes. *The Plant Cell, 15*(11), 2730–2741. https://doi.org/10.1105/tpc.016238

Bartlett, M. E., & Specht, C. D. (2010). Evidence for the involvement of *GLOBOSA*-like gene duplications and expression divergence in the evolution of floral morphology in the Zingiberales. *New Phytologist, 187*(2), 521–541. https://doi.org/10.1111/j.1469-8137.2010.03279.x

Bartlett, M. E., Williams, S. K., Taylor, Z., DeBlasio, S., Goldshmidt, A., Hall, D. H., Schmidt, R. J., Jackson, D. P., & Whipple, C. J. (2015). The maize *PI/GLO* ortholog *zmm16/sterile tassel silky ear1* interacts with the zygomorphy and sex determination pathways in flower development. *The Plant Cell, 27*(11), 3081–3098. https://doi.org/10.1105/tpc.15.00679

Baum, D. A., & Hileman, L. C. (2006). A developmental genetic model for the origin of the flower. In C. Ainsworth (Ed.), *Flowering and its manipulation, vol 20* (pp. 1–27). Blackwell publishing Ltd.

Becker, A., & Theißen, G. (2003). The major clades of MADS-box genes and their role in the development and evolution of flowering plants. *Molecular Phylogenetics and Evolution, 29*(3), 464–489. https://doi.org/10.1016/S1055-7903(03)00207-0

Becker, A., Winter, K. U., Meyer, B., Saedler, H., & Theißen, G. (2000). MADS-Box gene diversity in seed plants 300 million years ago. *Molecular Biology and Evolution, 17*(10), 1425–1434. https://doi.org/10.1093/oxfordjournals.molbev.a026243

Becker, A., Saedler, H., & Theißen, G. (2003). Distinct MADS-box gene expression patterns in the reproductive cones of the gymnosperm *Gnetum gnemon*. *Development Genes and Evolution, 213*(11), 567–572. https://doi.org/10.1007/s00427-003-0358-0

Benlloch, R., d'Erfurth, I., Ferrandiz, C., Cosson, V., Beltran, J. P., Canas, L. A., Kondorosi, A., Madueno, F., & Ratet, P. (2006). Isolation of *mtpim* proves *Tnt1* a useful reverse genetics tool in *Medicago truncatula* and uncovers new aspects of *AP1*-like functions in legumes. *Plant Physiology, 142*(3), 972–983. https://doi.org/10.1104/pp.106.083543

Borchert, T., Eckardt, K., Fuchs, J., Krueger, K., & Hohe, A. (2009). 'Who's who' in two different flower types of *Calluna vulgaris* (Ericaceae): Morphological and molecular analyses of flower organ identity. *BMC Plant Biology, 9*, 148. https://doi.org/10.1186/1471-2229-9-148

Bowman, J. L. (1997). Evolutionary conservation of angiosperm flower development at the molecular and genetic levels. *Journal of Biosciences, 22*(4), 515–527. https://doi.org/10.1007/Bf02703197

Bowman, J. L., Smyth, D. R., & Meyerowitz, E. M. (1989). Genes directing flower development in *Arabidopsis*. *The Plant Cell, 1*(1), 37–52. https://doi.org/10.1105/tpc.1.1.37

Bowman, J. L., Drews, G. N., & Meyerowitz, E. M. (1991). Expression of the *Arabidopsis* floral homeotic gene *AGAMOUS* is restricted to specific cell-types late in flower development. *The Plant Cell, 3*(8), 749–758. https://doi.org/10.1105/tpc.3.8.749

Bowman, J. L., Alvarez, J., Weigel, D., Meyerowitz, E. M., & Smyth, D. R. (1993). Control of flower development in *Arabidopsis thaliana* by *APETALA1* and interacting genes. *Development, 119*(3), 721–743. https://doi.org/10.1242/dev.119.3.721

Brockington, S. F., Rudall, P. J., Frohlich, M. W., Oppenheimer, D. G., Soltis, P. S., & Soltis, D. E. (2012). 'Living stones' reveal alternative petal identity programs within the core eudicots. *The Plant Journal, 69*(2), 193–203. https://doi.org/10.1111/j.1365-313X.2011.04797.x

Broholm, S. K., Pollanen, E., Ruokolainen, S., Tahtiharju, S., Kotilainen, M., Albert, V. A., Elomaa, P., & Teeri, T. H. (2010). Functional characterization of B class MADS-box transcription factors in *Gerbera hybrida*. *Journal of Experimental Botany, 61*(1), 75–85. https://doi.org/10.1093/jxb/erp279

Buzgo, M., Soltis, P. S., & Soltis, D. E. (2004). Floral developmental morphology of *Amborella trichopoda* (Amborellaceae). *International Journal of Plant Sciences, 165*(6), 925–947. https://doi.org/10.1086/424024

Carlsbecker, A., Sundström, J. F., Englund, M., Uddenberg, D., Izquierdo, L., Kvarnheden, A., Vergara-Silva, F., & Engstrom, P. (2013). Molecular control of normal and *acrocona* mutant

seed cone development in Norway spruce (*Picea abies*) and the evolution of conifer ovule-bearing organs. *New Phytologist, 200*(1), 261–275. https://doi.org/10.1111/nph.12360
Cartolano, M., Castillo, R., Efremova, N., Kuckenberg, M., Zethof, J., Gerats, T., Schwarz-Sommer, Z., & Vandenbussche, M. (2007). A conserved microRNA module exerts homeotic control over *Petunia hybrida* and *Antirrhinum majus* floral organ identity. *Nature Genetics, 39*(7), 901–905. https://doi.org/10.1038/ng2056
Castillejo, C., Romera-Branchat, M., & Pelaz, S. (2005). A new role of the *Arabidopsis SEPALLATA3* gene revealed by its constitutive expression. *The Plant Journal, 43*(4), 586–596. https://doi.org/10.1111/j.1365-313X.2005.02476.x
Causier, B., Castillo, R., Zhou, J., Ingram, R., Xue, Y., Schwarz-Sommer, Z., & Davies, B. (2005). Evolution in action: Following function in duplicated floral homeotic genes. *Current Biology, 15*(16), 1508–1512. https://doi.org/10.1016/j.cub.2005.07.063
Causier, B., Schwarz-Sommer, Z., & Davies, B. (2010). Floral organ identity: 20 years of ABCs. *Seminars in Cell & Developmental Biology, 21*(1), 73–79. https://doi.org/10.1016/j.semcdb.2009.10.005
Chanderbali, A. S., Yoo, M. J., Zahn, L. M., Brockington, S. F., Wall, P. K., Gitzendanner, M. A., Albert, V. A., Leebens-Mack, J., Altman, N. S., Ma, H., dePamphilis, C. W., Soltis, D. E., & Soltis, P. S. (2010). Conservation and canalization of gene expression during angiosperm diversification accompany the origin and evolution of the flower. *Proceedings of the National Academy of Sciences of the United States of America, 107*(52), 22570–22575. https://doi.org/10.1073/pnas.1013395108
Chandler, J. W. (2011). Founder cell specification. *Trends in Plant Science, 16*(11), 607–613. https://doi.org/10.1016/j.tplants.2011.08.005
Chandler, J. W., Jacobs, B., Cole, M., Comelli, P., & Werr, W. (2011). *DORNRÖSCHEN*-like expression marks *Arabidopsis* floral organ founder cells and precedes auxin response maxima. *Plant Molecular Biology, 76*(2), 171–185. https://doi.org/10.1007/s11103-011-9779-8
Chang, Y. Y., Kao, N. H., Li, J. Y., Hsu, W. H., Liang, Y. L., Wu, J. W., & Yang, C. H. (2010). Characterization of the possible roles for B class MADS box genes in regulation of perianth formation in orchid. *Plant Physiology, 152*(2), 837–853. https://doi.org/10.1104/pp.109.147116
Chen, X. M. (2004). A microRNA as a translational repressor of *APETALA2* in *Arabidopsis* flower development. *Science, 303*(5666), 2022–2025. https://doi.org/10.1126/science.1088060
Chen, D., Yan, W., Fu, L. Y., & Kaufmann, K. (2018). Architecture of gene regulatory networks controlling flower development in *Arabidopsis thaliana*. *Nature Communications, 9*, 4534. https://doi.org/10.1038/s41467-018-06772-3
Cirilli, M., Rossini, L., Chiozzotto, R., Baccichet, I., Florio, F. E., Mazzaglia, A., Turco, S., Bassi, D., & Gattolin, S. (2022). Less is more: Natural variation disrupting a miR172 gene at the *di* locus underlies the recessive double-flower trait in peach (*P. persica* L. Batsch). *BMC Plant Biology, 22*, 318. https://doi.org/10.1186/s12870-022-03691-w
Coen, E. S., & Meyerowitz, E. M. (1991). The war of the whorls: Genetic interactions controlling flower development. *Nature, 353*(6339), 31–37. https://doi.org/10.1038/353031a0
Coen, E. S., Romero, J. M., Doyle, S., Elliott, R., Murphy, G., & Carpenter, R. (1990). *FLORICAULA*: A homeotic gene required for flower development in *Antirrhinum majus*. *Cell, 63*(6), 1311–1322. https://doi.org/10.1016/0092-8674(90)90426-F
Cui, R., Han, J., Zhao, S., Su, K., Wu, F., Du, X., Xu, Q., Chong, K., Theißen, G., & Meng, Z. (2010). Functional conservation and diversification of class E floral homeotic genes in rice (*Oryza sativa*). *The Plant Journal, 61*(5), 767–781. https://doi.org/10.1111/j.1365-313X.2009.04101.x
Cusick, F. (1956). Studies of floral morphogenesis. I. Median bisections of flower primordia in *Primula bulleyana* Forrest. *Earth and Environmental Science Transactions of the Royal Society of Edinburgh, 63*(1), 153–166. https://doi.org/10.1017/S0080456800003057
Damerval, C., & Becker, A. (2017). Genetics of flower development in Ranunculales–a new, basal eudicot model order for studying flower evolution. *New Phytologist, 216*(2), 361–366. https://doi.org/10.1111/nph.14401

D'Arcy, W. G., & Keating, R. C. (1996). *The anther: Form, function and phylogeny*. Cambridge University Press.

Davies, B., Motte, P., Keck, E., Saedler, H., Sommer, H., & Schwarz-Sommer, Z. (1999). *PLENA* and *FARINELLI*: Redundancy and regulatory interactions between two *Antirrhinum* MADS-box factors controlling flower development. *The EMBO Journal, 18*(14), 4023–4034. https://doi.org/10.1093/emboj/18.14.4023

Day, C. D., Galgoci, B., & Irish, V. F. (1995). Genetic ablation of petal and stamen primordia to elucidate cell interactions during floral development. *Development, 121*(9), 2887–2895. https://doi.org/10.1242/dev.121.9.2887

de Martino, G., Pan, I., Emmanuel, E., Levy, A., & Irish, V. F. (2006). Functional analyses of two tomato *APETALA3* genes demonstrate diversification in their roles in regulating floral development. *The Plant Cell, 18*(8), 1833–1845. https://doi.org/10.1105/tpc.106.042978

Ditta, G., Pinyopich, A., Robles, P., Pelaz, S., & Yanofsky, M. F. (2004). The *SEP4* gene of *Arabidopsis thaliana* functions in floral organ and meristem identity. *Current Biology, 14*(21), 1935–1940. https://doi.org/10.1016/j.cub.2004.10.028

Dornelas, M. C., & Rodriguez, A. P. M. (2005). A *FLORICAULA/LEAFY* gene homolog is preferentially expressed in developing female cones of the tropical pine *Pinus caribaea* var. *caribaea*. *Genetics and Molecular Biology, 28*(2), 299–307. https://doi.org/10.1590/S1415-47572005000200021

Drea, S., Hileman, L. C., De Martino, G., & Irish, V. F. (2007). Functional analyses of genetic pathways controlling petal specification in poppy. *Development, 134*(23), 4157–4166. https://doi.org/10.1242/dev.013136

Dreni, L., & Zhang, D. (2016). Flower development: The evolutionary history and functions of the *AGL6* subfamily MADS-box genes. *Journal of Experimental Botany, 67*(6), 1625–1638. https://doi.org/10.1093/jxb/erw046

Dreni, L., Pilatone, A., Yun, D. P., Erreni, S., Pajoro, A., Caporali, E., Zhang, D. B., & Kater, M. M. (2011). Functional analysis of all *AGAMOUS* subfamily members in rice reveals their roles in reproductive organ identity determination and meristem determinacy. *The Plant Cell, 23*(8), 2850–2863. https://doi.org/10.1105/tpc.111.087007

Drews, G. N., Bowman, J. L., & Meyerowitz, E. M. (1991). Negative regulation of the *Arabidopsis* homeotic gene *AGAMOUS* by the APETALA2 product. *Cell, 65*(6), 991–1002. https://doi.org/10.1016/0092-8674(91)90551-9

Duan, X., Zhao, C., Jiang, Y., Zhang, R., Shan, H., & Kong, H. (2020). Parallel evolution of apetalous lineages within the buttercup family (Ranunculaceae): Outward expansion of *AGAMOUS1*, rather than disruption of *APETALA3-3*. *The Plant Journal, 104*(5), 1169–1181. https://doi.org/10.1111/tpj.14985

Endress, P. K. (2001a). The flowers in extant basal angiosperms and inferences on ancestral flowers. *International Journal of Plant Sciences, 162*(5), 1111–1140. https://doi.org/10.1086/321919

Endress, P. K. (2001b). Origins of flower morphology. *The Journal of Experimental Zoology, 291*(2), 105–115. https://doi.org/10.1002/jez.1063

Endress, P. K. (2006). Angiosperm floral evolution: Morphological developmental framework. In D. E. Soltis, J. H. LeebensMack, P. S. Soltis, & J. A. Callow (Eds.), *Advances in botanical research* (Vol. 44, pp. 1–61). Academic Press. https://doi.org/10.1016/S0065-2296(06)44001-5

Endress, P. K., & Matthews, M. L. (2006). Elaborate petals and staminodes in eudicots: Diversity, function, and evolution. *Organisms, Diversity and Evolution, 6*(4), 257–293. https://doi.org/10.1016/j.ode.2005.09.005

Ferrándiz, C., Fourquin, C., Prunet, N., Scutt, C. P., Sundberg, E., Trehin, C., & Vialette-Guiraud, A. C. M. (2010). Carpel Development. In J. Kader & M. Delseny (Eds.), *Advances in botanical research* (Vol. 55, pp. 1–73). Academic Press. https://doi.org/10.1016/B978-0-12-380868-4.00001-6

Fourquin, C., & Ferrandiz, C. (2012). Functional analyses of *AGAMOUS* family members in *Nicotiana benthamiana* clarify the evolution of early and late roles of C-function genes in eudicots. *The Plant Journal, 71*(6), 990–1001. https://doi.org/10.1111/j.1365-313X.2012.05046.x

Frohlich, M. W., & Parker, D. S. (2000). The mostly male theory of flower evolutionary origins: From genes to fossils. *Systematic Botany, 25*(2), 155–170. https://doi.org/10.2307/2666635

Fukui, M., Futamura, N., Mukai, Y., Wang, Y., Nagao, A., & Shinohara, K. (2001). Ancestral MADS box genes in Sugi, *Cryptomeria japonica* D. Don (Taxodiaceae), homologous to the B function genes in angiosperms. *Plant & Cell Physiology, 42*(6), 566–575. https://doi.org/10.1093/pcp/pce069

Galimba, K. D., Tolkin, T. R., Sullivan, A. M., Melzer, R., Theißen, G., & Di Stilio, V. S. (2012). Loss of deeply conserved C-class floral homeotic gene function and C- and E-class protein interaction in a double-flowered ranunculid mutant. *Proceedings of the National Academy of Sciences of the United States of America, 109*(34), E2267–E2275. https://doi.org/10.1073/pnas.1203686109

Gattolin, S., Cirilli, M., Chessa, S., Stella, A., Bassi, D., & Rossini, L. (2020). Mutations in orthologous *PETALOSA* TOE-type genes cause a dominant double-flower phenotype in phylogenetically distant eudicots. *Journal of Experimental Botany, 71*(9), 2585–2595. https://doi.org/10.1093/jxb/eraa032

Geuten, K., Becker, A., Kaufmann, K., Caris, P., Janssens, S., Viaene, T., Theißen, G., & Smets, E. (2006). Petaloidy and petal identity MADS-box genes in the balsaminoid genera *Impatiens* and *Marcgravia*. *The Plant Journal, 47*(4), 501–518. https://doi.org/10.1111/j.1365-313X.2006.02800.x

Glover, B. J., Airoldi, C. A., Brockington, S. F., Fernández-Mazuecos, M., Martínez-Pérez, C., Mellers, G., Moyroud, E., & Taylor, L. (2015). How have advances in comparative floral development influenced our understanding of floral evolution? *International Journal of Plant Sciences, 176*(4), 307–323. https://doi.org/10.1086/681562

Gonçalves, B., Nougue, O., Jabbour, F., Ridel, C., Morin, H., Laufs, P., Manicacci, D., & Damerval, C. (2013). An *APETALA3* homolog controls both petal identity and floral meristem patterning in *Nigella damascena* L. (Ranunculaceae). *The Plant Journal, 76*(2), 223–235. https://doi.org/10.1111/Tpj.12284

Goto, K., & Meyerowitz, E. M. (1994). Function and regulation of the *Arabidopsis* floral homeotic gene *PISTILLATA*. *Genes & Development, 8*(13), 1548–1560. https://doi.org/10.1101/gad.8.13.1548

Goto, K., Kyozuka, J., & Bowman, J. L. (2001). Turning floral organs into leaves, leaves into floral organs. *Current Opinion in Genetics & Development, 11*(4), 449–456. https://doi.org/10.1016/S0959-437x(00)00216-1

Green, P. B. (1988). A theory for inflorescence development and flower formation based on morphological and biophysical analysis in *Echeveria*. *Planta, 175*(2), 153–169. https://doi.org/10.1007/BF00392424

Guo, X. H., Hu, Z. L., Yin, W. C., Yu, X. H., Zhu, Z. G., Zhang, J. L., & Chen, G. P. (2016). The tomato floral homeotic protein *FBP1*-like gene, *SlGLO1*, plays key roles in petal and stamen development. *Scientific Reports, 6*, 20454. https://doi.org/10.1038/Srep20454

Hands, P., Vosnakis, N., Betts, D., Irish, V. F., & Drea, S. (2011). Alternate transcripts of a floral developmental regulator have both distinct and redundant functions in opium poppy. *Annals of Botany, 107*(9), 1557–1566. https://doi.org/10.1093/aob/mcr045

He, J., Alonge, M., Ramakrishnan, S., Benoit, M., Soyk, S., Reem, N. T., Hendelman, A., Van Eck, J., Schatz, M. C., & Lippman, Z. B. (2023). Establishing *Physalis* as a Solanaceae model system enables genetic reevaluation of the inflated calyx syndrome. *The Plant Cell, 35*(1), 351–368. https://doi.org/10.1093/plcell/koac305

Heijmans, K., Ament, K., Rijpkema, A. S., Zethof, J., Wolters-Arts, M., Gerats, T., & Vandenbussche, M. (2012). Redefining C and D in the petunia ABC. *The Plant Cell, 24*(6), 2305–2317. https://doi.org/10.1105/tpc.112.097030

Hicks, G., & Sussex, I. (1971). Organ regeneration in sterile culture after median bisection of the flower primordia of *Nicotiana tabacum*. *Botanical Gazette, 132*(4), 350–363. https://doi.org/10.1086/336602

Hirai, M., Ochiai, T., & Kanno, A. (2010). The expression of two *DEFICIENS*-like genes was reduced in the sepaloid tepals of viridiflora tulips. *Breeding Science, 60*(2), 110–120. https://doi.org/10.1270/jsbbs.60.110

Holder, N. (1979). Positional information and pattern formation in plant morphogenesis and a mechanism for the involvement of plant hormones. *Journal of Theoretical Biology, 77*(2), 195–212. https://doi.org/10.1016/0022-5193(79)90307-2

Hong, R. L., Hamaguchi, L., Busch, M. A., & Weigel, D. (2003). Regulatory elements of the floral homeotic gene *AGAMOUS* identified by phylogenetic footprinting and shadowing. *The Plant Cell, 15*(6), 1296–1309. https://doi.org/10.1105/tpc.009548

Honma, T., & Goto, K. (2001). Complexes of MADS-box proteins are sufficient to convert leaves into floral organs. *Nature, 409*(6819), 525–529. https://doi.org/10.1038/35054083

Hsieh, M. H., Pan, Z. J., Lai, P. H., Lu, H. C., Yeh, H. H., Hsu, C. C., Wu, W. L., Chung, M. C., Wang, S. S., Chen, W. H., & Chen, H. H. (2013). Virus-induced gene silencing unravels multiple transcription factors involved in floral growth and development in *Phalaenopsis* orchids. *Journal of Experimental Botany, 64*(12), 3869–3884. https://doi.org/10.1093/jxb/ert218

Hsu, H. F., Hsu, W. H., Lee, Y. I., Mao, W. T., Yang, J. Y., Li, J. Y., & Yang, C. H. (2015). Model for perianth formation in orchids. *Nature Plants, 1*(5), 15046. https://doi.org/10.1038/nplants.2015.46

Hsu, H. F., Chen, W. H., Shen, Y. H., Hsu, W. H., Mao, W. T., & Yang, C. H. (2021). Multifunctional evolution of B and *AGL6* MADS box genes in orchids. *Nature Communications, 12*, 902. https://doi.org/10.1038/s41467-021-21229-w

Hu, J., Zhang, J., Shan, H., & Chen, Z. (2012). Expression of floral MADS-box genes in *Sinofranchetia chinensis* (Lardizabalaceae): Implications for the nature of the nectar leaves. *Annals of Botany, 110*(1), 57–69. https://doi.org/10.1093/aob/mcs104

Huang, H., Tudor, M., Weiss, C. A., Hu, Y., & Ma, H. (1995). The *Arabidopsis* MADS-box gene *AGL3* is widely expressed and encodes a sequence-specific DNA-binding protein. *Plant Molecular Biology, 28*(3), 549–567. https://doi.org/10.1007/BF00020401

Hughen, A. M. (1980). The regulation of tobacco floral organ initiation. *Botanical Gazette, 141*(4), 389–395. https://doi.org/10.1086/337172

Immink, R. G., Tonaco, I. A., de Folter, S., Shchennikova, A., van Dijk, A. D., Busscher-Lange, J., Borst, J. W., & Angenent, G. C. (2009). *SEPALLATA3*: The 'glue' for MADS box transcription factor complex formation. *Genome Biology, 10*(2), R24. https://doi.org/10.1186/gb-2009-10-2-r24

Irish, V. F. (2009). Evolution of petal identity. *Journal of Experimental Botany, 60*(9), 2517–2527. https://doi.org/10.1093/jxb/erp159

Irish, V. F., & Sussex, I. M. (1990). Function of the *APETALA-1* gene during *Arabidopsis* floral development. *The Plant Cell, 2*(8), 741–753. https://doi.org/10.1105/tpc.2.8.741

Ishikawa, M., Ohmori, Y., Tanaka, W., Hirabayashi, C., Murai, K., Ogihara, Y., Yamaguchi, T., & Hirano, H. (2009). The spatial expression patterns of *DROOPING LEAF* orthologs suggest a conserved function in grasses. *Genes & Genetic Systems, 84*(2), 137–146. https://doi.org/10.1266/ggs.84.137

Ito, T., Ng, K. H., Lim, T. S., Yu, H., & Meyerowitz, E. M. (2007). The homeotic protein *AGAMOUS* controls late stamen development by regulating a jasmonate biosynthetic gene in *Arabidopsis*. *The Plant Cell, 19*(11), 3516–3529. https://doi.org/10.1105/tpc.107.055467

Jack, T., Brockman, L. L., & Meyerowitz, E. M. (1992). The homeotic gene *APETALA3* of *Arabidopsis thaliana* encodes a MADS-box and is expressed in petals and stamens. *Cell, 68*(4), 683–697. https://doi.org/10.1016/0092-8674(92)90144-2

Jager, M., Hassanin, A., Manuel, M., Le Guyader, H., & Deutsch, J. (2003). MADS-box genes in *Ginkgo biloba* and the evolution of the *AGAMOUS* family. *Molecular Biology and Evolution, 20*(5), 842–854. https://doi.org/10.1093/molbev/msg089

Jaramillo, M. A., & Kramer, E. M. (2004). *APETALA3* and *PISTILLATA* homologs exhibit novel expression patterns in the unique perianth of *Aristolochia* (Aristolochiaceae). *Evolution and Development, 6*(6), 449–458. https://doi.org/10.1111/j.1525-142X.2004.04053.x

Jofuku, K. D., den Boer, B. G. W., van Montagu, M., & Okamuro, J. K. (1994). Control of *Arabidopsis* flower and seed development by the homeotic gene *APETALA2*. *The Plant Cell, 6*(9), 1211–1225. https://doi.org/10.1105/tpc.6.9.1211

Kanno, A. (2016). Molecular mechanism regulating floral architecture in monocotyledonous ornamental plants. *Horticulture Journal, 85*(1), 8–22. https://doi.org/10.2503/hortj.MI-IR05

Kaufmann, K., Muino, J. M., Jauregui, R., Airoldi, C. A., Smaczniak, C., Krajewski, P., & Angenent, G. C. (2009). Target genes of the MADS transcription factor SEPALLATA3: Integration of developmental and hormonal pathways in the *Arabidopsis* flower. *PLoS Biology, 7*(4), e1000090. https://doi.org/10.1371/journal.pbio.1000090

Keck, E., McSteen, P., Carpenter, R., & Coen, E. (2003). Separation of genetic functions controlling organ identity in flowers. *The EMBO Journal, 22*(5), 1058–1066. https://doi.org/10.1093/emboj/cdg097

Kellogg, E. A. (2015). *Flowering plants. Monocots*. Springer US. https://doi.org/10.1007/978-3-319-15332-2

Kerstens, M. H., Schranz, M. E., & Bouwmeester, K. (2020). Phylogenomic analysis of the APETALA2 transcription factor subfamily across angiosperms reveals both deep conservation and lineage-specific patterns. *The Plant Journal, 103*(4), 1516–1524. https://doi.org/10.1111/tpj.14843

Kim, S., Koh, J., Ma, H., Hu, Y., Endress, P. K., Hauser, B. A., Buzgo, M., Soltis, P. S., & Soltis, D. E. (2005a). Sequence and expression studies of A-, B-, and E-class MADS-box homologues in *Eupomatia* (Eupomatiaceae): Support for the bracteate origin of the calyptra. *International Journal of Plant Sciences, 166*(2), 185–198. https://doi.org/10.1086/427479

Kim, S., Koh, J., Yoo, M. J., Kong, H., Hu, Y., Ma, H., Soltis, P. S., & Soltis, D. E. (2005b). Expression of floral MADS-box genes in basal angiosperms: Implications for the evolution of floral regulators. *The Plant Journal, 43*(5), 724–744. https://doi.org/10.1111/j.1365-313X.2005.02487.x

Kim, S., Soltis, P. S., Wall, K., & Soltis, D. E. (2006). Phylogeny and domain evolution in the *APETALA2*-like gene family. *Molecular Biology and Evolution, 23*(1), 107–120. https://doi.org/10.1093/molbev/msj014

Kim, S., Yun, P., Fukuda, T., Ochiai, T., Yokoyama, J., Kameya, T., & Kanno, A. (2007). Expression of a *DEFICIENS*-like gene correlates with the differentiation between sepal and petal in the orchid, *Habenaria radiata* (Orchidaceae). *Plant Science, 172*(2), 319–326. https://doi.org/10.1016/j.plantsci.2006.09.009

Komaki, M. K., Okada, K., Nishino, E., & Shimura, Y. (1988). Isolation and characterization of novel mutants of *Arabidopsis thaliana* defective in flower development. *Development, 104*(2), 195–203. https://doi.org/10.1242/dev.104.2.195

Kramer, E. M., & Irish, V. F. (2000). Evolution of the petal and stamen developmental programs: Evidence from comparative studies of the lower eudicots and basal angiosperms. *International Journal of Plant Sciences, 161*(S6), S29–S40. https://doi.org/10.1086/317576

Kramer, E. M., Dorit, R. L., & Irish, V. F. (1998). Molecular evolution of genes controlling petal and stamen development: Duplication and divergence within the *APETALA3* and *PISTILLATA* MADS-box gene lineages. *Genetics, 149*(2), 765–783. https://doi.org/10.1093/genetics/149.2.765

Kramer, E. M., Di Stilio, V. S., & Schluter, P. M. (2003). Complex patterns of gene duplication in the *APETALA3* and *PISTILLATA* lineages of the Ranunculaceae. *International Journal of Plant Sciences, 164*(1), 1–11. https://doi.org/10.1086/344694

Kramer, E. M., Jaramillo, M. A., & Di Stilio, V. S. (2004). Patterns of gene duplication and functional evolution during the diversification of the *AGAMOUS* subfamily of MADS box genes in angiosperms. *Genetics, 166*(2), 1011–1023. https://doi.org/10.1534/genetics.166.2.1011

Kramer, E. M., Holappa, L., Gould, B., Jaramillo, M. A., Setnikov, D., & Santiago, P. M. (2007). Elaboration of B gene function to include the identity of novel floral organs in the lower eudicot *Aquilegia*. *The Plant Cell, 19*(3), 750–766. https://doi.org/10.1105/tpc.107.050385

Krizek, B. A., & Meyerowitz, E. M. (1996). The *Arabidopsis* homeotic genes *APETALA3* and *PISTILLATA* are sufficient to provide the B class organ identity function. *Development, 122*(1), 11–22. https://doi.org/10.1242/dev.122.1.11

Kunst, L., Klenz, J. E., Martinezzapater, J., & Haughn, G. W. (1989). *AP2* gene determines the identity of perianth organs in flowers of *Arabidopsis thaliana*. *The Plant Cell, 1*(12), 1195–1208. https://doi.org/10.1105/tpc.1.12.1195

Landis, J. B., Barnett, L. L., & Hileman, L. C. (2012). Evolution of petaloid sepals independent of shifts in B-class MADS box gene expression. *Development Genes and Evolution, 222*(1), 19–28. https://doi.org/10.1007/s00427-011-0385-1

Lange, M., Orashakova, S., Lange, S., Melzer, R., Theißen, G., Smyth, D. R., & Becker, A. (2013). The *seirena* B class floral homeotic mutant of California Poppy (*Eschscholzia californica*) reveals a function of the enigmatic PI motif in the formation of specific multimeric MADS domain protein complexes. *The Plant Cell, 25*(2), 438–453. https://doi.org/10.1105/tpc.112.105809

Li, H., Liang, W., Jia, R., Yin, C., Zong, J., Kong, H., & Zhang, D. (2010). The *AGL6*-like gene *OsMADS6* regulates floral organ and meristem identities in rice. *Cell Research, 20*(3), 299–313. https://doi.org/10.1038/cr.2009.143

Li, H., Liang, W., Hu, Y., Zhu, L., Yin, C., Xu, J., Dreni, L., Kater, M. M., & Zhang, D. (2011). Rice *MADS6* interacts with the floral homeotic genes *SUPERWOMAN1*, *MADS3*, *MADS58*, *MADS13*, and *DROOPING LEAF* in specifying floral organ identities and meristem fate. *The Plant Cell, 23*(7), 2536–2552. https://doi.org/10.1105/tpc.111.087262

Li, L., Yu, X., Guo, C., Duan, X., Shan, H., Zhang, R., Xu, G., & Kong, H. (2015). Interactions among proteins of floral MADS-box genes in *Nuphar pumila* (Nymphaeaceae) and the most recent common ancestor of extant angiosperms help understand the underlying mechanisms of the origin of the flower. *Journal of Systematics and Evolution, 53*(4), 285–296. https://doi.org/10.1111/jse.12148

Litt, A. (2007). An evaluation of A-function: Evidence from the *APETALA1* and *APETALA2* gene lineages. *International Journal of Plant Sciences, 168*(1), 73–91. https://doi.org/10.1086/509662

Litt, A., & Irish, V. F. (2003). Duplication and diversification in the *APETALA1/FRUITFULL* floral homeotic gene lineage: Implications for the evolution of floral development. *Genetics, 165*(2), 821–833. https://doi.org/10.1093/genetics/165.2.821

Liu, Z., & Mara, C. (2010). Regulatory mechanisms for floral homeotic gene expression. *Seminars in Cell & Developmental Biology, 21*(1), 80–86. https://doi.org/10.1016/j.semcdb.2009.11.012

Liu, Y., Nakayama, N., Schiff, M., Litt, A., Irish, V. F., & Dinesh-Kumar, S. (2004). Virus induced gene silencing of a *DEFICIENS* ortholog in *Nicotiana benthamiana*. *Plant Molecular Biology, 54*(5), 701–711. https://doi.org/10.1023/B:PLAN.0000040899.53378.83

Liu, C., Xi, W., Shen, L., Tan, C., & Yu, H. (2009). Regulation of floral patterning by flowering time genes. *Developmental Cell, 16*(5), 711–722. https://doi.org/10.1016/j.devcel.2009.03.011

Lohmann, J. U., Hong, R. L., Hobe, M., Busch, M. A., Parcy, F., Simon, R., & Weigel, D. (2001). A molecular link between stem cell regulation and floral patterning in *Arabidopsis*. *Cell, 105*(6), 793–803. https://doi.org/10.1016/S0092-8674(01)00384-1

Ma, H., Yanofsky, M. F., & Meyerowitz, E. M. (1991). *AGL1-AGL6*, an *Arabidopsis* gene family with similarity to floral homeotic and transcription factor genes. *Genes & Development, 5*(3), 484–495. https://doi.org/10.1101/gad.5.3.484

Maes, T., Van de Steene, N., Zethof, J., Karimi, M., D'Hauw, M., Mares, G., Van Montagu, M., & Gerats, T. (2001). Petunia AP2-like genes and their role in flower and seed development. *The Plant Cell, 13*(2), 229–244. https://doi.org/10.1105/tpc.13.2.229

Malcomber, S. T., & Kellogg, E. A. (2004). Heterogeneous expression patterns and separate roles of the *SEPALLATA* gene *LEAFY HULL STERILE1* in grasses. *The Plant Cell, 16*(7), 1692–1706. https://doi.org/10.1105/tpc.021576

Malcomber, S. T., & Kellogg, E. A. (2005). *SEPALLATA* gene diversification: Brave new whorls. *Trends in Plant Science, 10*(9), 427–435. https://doi.org/10.1016/j.tplants.2005.07.008

Mandel, M. A., & Yanofsky, M. F. (1998). The *Arabidopsis AGL9* MADS box gene is expressed in young flower primordia. *Sexual Plant Reproduction, 11*(1), 22–28. https://doi.org/10.1007/s004970050116

Mandel, M. A., Gustafsonbrown, C., Savidge, B., & Yanofsky, M. F. (1992). Molecular characterization of the *Arabidopsis* floral homeotic gene *APETALA1*. *Nature, 360*(6401), 273–277. https://doi.org/10.1038/360273a0

Melville, R. (1960). A new theory of the angiosperm flower. *Nature, 188*(4744), 14–18. https://doi.org/10.1038/188014a0

Melzer, R., & Theißen, G. (2009). Reconstitution of 'floral quartets' in vitro involving class B and class E floral homeotic proteins. *Nucleic Acids Research, 37*(8), 2723–2736. https://doi.org/10.1093/nar/gkp129

Melzer, R., Harter, A., Rümpler, F., Kim, S., Soltis, P. S., Soltis, D. E., & Theißen, G. (2014). DEF- and GLO-like proteins may have lost most of their interaction partners during angiosperm evolution. *Annals of Botany, 114*(7), 1431–1443. https://doi.org/10.1093/aob/mcu094

Mitoma, M., & Kanno, A. (2018). The greenish flower phenotype of *Habenaria radiata* (Orchidaceae) is caused by a mutation in the *SEPALLATA*-like MADS-box gene *HrSEP-1*. *Frontiers in Plant Science, 9*, 831. https://doi.org/10.3389/fpls.2018.00831

Mizukami, Y., & Ma, H. (1992). Ectopic expression of the floral homeotic gene *AGAMOUS* in transgenic *Arabidopsis* plants alters floral organ identity. *Cell, 71*(1), 119–131. https://doi.org/10.1016/0092-8674(92)90271-D

Mondragon-Palomino, M., & Theißen, G. (2008). MADS about the evolution of orchid flowers. *Trends in Plant Science, 13*(2), 51–59. https://doi.org/10.1016/j.tplants.2007.11.007

Mondragon-Palomino, M., & Theißen, G. (2011). Conserved differential expression of paralogous *DEFICIENS*- and *GLOBOSA*-like MADS-box genes in the flowers of Orchidaceae: Refining the 'orchid code'. *The Plant Journal, 66*(6), 1008–1019. https://doi.org/10.1111/j.1365-313X.2011.04560.x

Morel, P., Heijmans, K., Rozier, F., Zethof, J., Chamot, S., Bento, S. R., Vialette-Guiraud, A., Chambrier, P., Trehin, C., & Vandenbussche, M. (2017). Divergence of the floral A-function between an Asterid and a Rosid species. *The Plant Cell, 29*(7), 1605–1621. https://doi.org/10.1105/tpc.17.00098

Morel, P., Chambrier, P., Boltz, V., Chamot, S., Rozier, F., Rodrigues Bento, S., Trehin, C., Monniaux, M., Zethof, J., & Vandenbussche, M. (2019). Divergent functional diversification patterns in the *SEP/AGL6/AP1* MADS-box transcription factor superclade. *The Plant Cell, 31*(12), 3033–3056. https://doi.org/10.1105/tpc.19.00162

Mouradov, A., Glassick, T. V., Hamdorf, B. A., Murphy, L. C., Marla, S. S., Yang, Y., & Teasdale, R. D. (1998). Family of MADS-Box genes expressed early in male and female reproductive structures of monterey pine. *Plant Physiology, 117*(1), 55–62. https://doi.org/10.1104/pp.117.1.55

Mouradov, A., Hamdorf, B., Teasdale, R. D., Kim, J. T., Winter, K. U., & Theißen, G. (1999). A *DEF/GLO*-like MADS-box gene from a gymnosperm: *Pinus radiata* contains an ortholog of angiosperm B class floral homeotic genes. *Developmental Genetics, 25*(3), 245–252. https://doi.org/10.1002/(SICI)1520-6408(1999)25:3<245::AID-DVG7>3.0.CO;2-N

Moyroud, E., Monniaux, M., Thévenon, E., Dumas, R., Scutt, C. P., Frohlich, M. W., & Parcy, F. (2017). A link between *LEAFY* and B-gene homologues in *Welwitschia mirabilis* sheds light on ancestral mechanisms prefiguring floral development. *New Phytologist, 216*(2), 469–481. https://doi.org/10.1111/nph.14483

Nagasawa, N., Miyoshi, M., Sano, Y., Satoh, H., Hirano, H., Sakai, H., & Nagato, Y. (2003). *SUPERWOMAN1* and *DROOPING LEAF* genes control floral organ identity in rice. *Development, 130*(4), 705–718. https://doi.org/10.1242/dev.00294

Niu, S., Li, J., Bo, W., Yang, W., Zuccolo, A., Giacomello, S., Chen, X., Han, F., Yang, J., Song, Y., Nie, Y., Zhou, B., Wang, P., Zuo, Q., Zhang, H., Ma, J., Wang, J., Wang, L., Zhu, Q., Zhao, H., Liu, Z., Zhang, X., Liu, T., Pei, S., Li, Z., Hu, Y., Yang, Y., Li, W., Zan, Y., Zhou, L., Lin, J., Yuan, T., Li, W., Li, Y., Wei, H., & Wu, H. X. (2022). The Chinese pine genome and methylome unveil key features of conifer evolution. *Cell, 185*(1), 204–217. https://doi.org/10.1016/j.cell.2021.12.006

Noda, K., Glover, B. J., Linstead, P., & Martin, C. (1994). Flower color intensity depends on specialized cell-shape controlled by a *MYB*-related transcription factor. *Nature, 369*(6482), 661–664. https://doi.org/10.1038/369661a0

Ochiai, T., Nakamura, T., Mashiko, Y., Fukuda, T., Yokoyama, J., Kanno, A., & Kameya, T. (2004). The differentiation of sepal and petal morphologies in Commelinaceae. *Gene, 343*(2), 253–262. https://doi.org/10.1016/j.gene.2004.08.032

Ohmori, S., Kimizu, M., Sugita, M., Miyao, A., Hirochika, H., Uchida, E., Nagato, Y., & Yoshida, H. (2009). *MOSAIC FLORAL ORGANS1*, an *AGL6*-like MADS box gene, regulates floral organ identity and meristem fate in rice. *The Plant Cell, 21*(10), 3008–3025. https://doi.org/10.1105/tpc.109.068742

O'Maoileidigh, D. S., Wuest, S. E., Rae, L., Raganelli, A., Ryan, P. T., Kwasniewska, K., Das, P., Lohan, A. J., Loftus, B., Graciet, E., & Wellmer, F. (2013). Control of reproductive floral organ identity specification in *Arabidopsis* by the C function regulator *AGAMOUS*. *The Plant Cell, 25*(7), 2482–2503. https://doi.org/10.1105/tpc.113.113209

Orashakova, S., Lange, M., Lange, S., Wege, S., & Becker, A. (2009). The *CRABS CLAW* ortholog from California poppy (*Eschscholzia californica*, Papaveraceae), *EcCRC*, is involved in floral meristem termination, gynoecium differentiation and ovule initiation. *The Plant Journal, 58*(4), 682–693. https://doi.org/10.1111/j.1365-313X.2009.03807.x

Otani, M., Sharifi, A., Kubota, S., Oizumi, K., Uetake, F., Hirai, M., Hoshino, Y., Kanno, A., & Nakano, M. (2016). Suppression of B function strongly supports the modified ABCE model in *Tricyrtis sp* (Liliaceae). *Scientific Reports, 6*, 24549. https://doi.org/10.1038/srep24549

Pabón-Mora, N., Ambrose, B. A., & Litt, A. (2012). Poppy *APETALA1/FRUITFULL* orthologs control flowering time, branching, perianth identity, and fruit development. *Plant Physiology, 158*(4), 1685–1704. https://doi.org/10.1104/pp.111.192104

Pabón-Mora, N., Sharma, B., Holappa, L. D., Kramer, E. M., & Litt, A. (2013). The *Aquilegia FRUITFULL*-like genes play key roles in leaf morphogenesis and inflorescence development. *The Plant Journal, 74*(2), 197–212. https://doi.org/10.1111/tpj.12113

Pan, I. L., McQuinn, R., Giovannoni, J. J., & Irish, V. F. (2010). Functional diversification of *AGAMOUS* lineage genes in regulating tomato flower and fruit development. *Journal of Experimental Botany, 61*(6), 1795–1806. https://doi.org/10.1093/jxb/erq046

Pan, Z. J., Chen, Y. Y., Du, J. S., Chen, Y. Y., Chung, M. C., Tsai, W. C., Wang, C. N., & Chen, H. H. (2014). Flower development of *Phalaenopsis* orchid involves functionally divergent *SEPALLATA*-like genes. *New Phytologist, 202*(3), 1024–1042. https://doi.org/10.1111/nph.12723

Pelaz, S., Ditta, G. S., Baumann, E., Wisman, E., & Yanofsky, M. F. (2000). B and C floral organ identity functions require *SEPALLATA* MADS-box genes. *Nature, 405*(6783), 200–203. https://doi.org/10.1038/35012103

Pelaz, S., Tapia-Lopez, R., Alvarez-Buylla, E. R., & Yanofsky, M. F. (2001). Conversion of leaves into petals in *Arabidopsis*. *Current Biology, 11*(3), 182–184. https://doi.org/10.1016/s0960-9822(01)00024-0

Pi, M., Hu, S., Cheng, L., Zhong, R., Cai, Z., Liu, Z., Yao, J. L., & Kang, C. (2021). The MADS-box gene *FveSEP3* plays essential roles in flower organogenesis and fruit development in woodland strawberry. *Horticulture Research, 8*, 247. https://doi.org/10.1038/s41438-021-00673-1

Pnueli, L., Hareven, D., Broday, L., Hurwitz, C., & Lifschitz, E. (1994). The *TM5* MADS box gene mediates organ differentiation in the three inner whorls of tomato flowers. *The Plant Cell, 6*(2), 175–186. https://doi.org/10.1105/tpc.6.2.175

Preston, J. C., & Kellogg, E. A. (2006). Reconstructing the evolutionary history of paralogous *APETALA1/FRUITFULL*-like genes in grasses (Poaceae). *Genetics, 174*(1), 421–437. https://doi.org/10.1534/genetics.106.057125

Preston, J. C., & Kellogg, E. A. (2007). Conservation and divergence of *APETALA1/FRUITFULL*-like gene function in grasses: Evidence from gene expression analyses. *The Plant Journal, 52*(1), 69–81. https://doi.org/10.1111/j.1365-313X.2007.03209.x

Rasmussen, D. A., Kramer, E. M., & Zimmer, E. A. (2009). One size fits all? Molecular evidence for a commonly inherited petal identity program in Ranunculales. *American Journal of Botany, 96*(1), 96–109. https://doi.org/10.3732/ajb.0800038

Rieu, P., Turchi, L., Thévenon, E., Zarkadas, E., Nanao, M., Chahtane, H., Tichtinsky, G., Lucas, J., Blanc-Mathieu, R., Zubieta, C., Schoehn, G., & Parcy, F. (2023). The F-box protein UFO controls flower development by redirecting the master transcription factor LEAFY to new *cis*-elements. *Nature Plants, 9*, 315–329. https://doi.org/10.1038/s41477-022-01336-2

Rijpkema, A. S., Royaert, S., Zethof, J., van der Weerden, G., Gerats, T., & Vandenbussche, M. (2006). Analysis of the *Petunia TM6* MADS box gene reveals functional divergence within the *DEF/AP3* lineage. *The Plant Cell, 18*(8), 1819–1832. https://doi.org/10.1105/tpc.106.042937

Rijpkema, A. S., Zethof, J., Gerats, T., & Vandenbussche, M. (2009). The *petunia AGL6* gene has a *SEPALLATA*-like function in floral patterning. *The Plant Journal, 60*(1), 1–9. https://doi.org/10.1111/j.1365-313X.2009.03917.x

Ronse De Craene, L. P. (2008). Homology and evolution of petals in the core eudicots. *Systematic Botany, 33*(2), 301–325. https://doi.org/10.1600/036364408784571680

Ronse De Craene, L. P., & Brockington, S. F. (2013). Origin and evolution of petals in angiosperms. *Plant Ecology and Evolution, 146*(1), 5–25. https://doi.org/10.5091/plecevo.2013.738

Ronse De Craene, L. P., & Smets, E. F. (2001). Staminodes: Their morphological and evolutionary significance. *The Botanical Review, 67*(3), 351–402. https://doi.org/10.1007/BF02858099

Rudall, P. J. (2010). All in a spin: Centrifugal organ formation and floral patterning. *Current Opinion in Plant Biology, 13*(1), 108–114. https://doi.org/10.1016/j.pbi.2009.09.019

Rudall, P. J., Sokoloff, D. D., Remizowa, M. V., Conran, J. G., Davis, J. I., Macfarlane, T. D., & Stevenson, D. W. (2007). Morphology of Hydatellaceae, an anomalous aquatic family recently recognized as an early-divergent angiosperm lineage. *American Journal of Botany, 94*(7), 1073–1092. https://doi.org/10.3732/ajb.94.7.1073

Ruelens, P., de Maagd, R. A., Proost, S., Theißen, G., Geuten, K., & Kaufmann, K. (2013). *FLOWERING LOCUS C* in monocots and the tandem origin of angiosperm-specific MADS-box genes. *Nature Communications, 4*, 2280. https://doi.org/10.1038/ncomms3280

Ruelens, P., Zhang, Z., van Mourik, H., Maere, S., Kaufmann, K., & Geuten, K. (2017). The origin of floral organ identity quartets. *The Plant Cell, 29*(2), 229–242. https://doi.org/10.1105/tpc.16.00366

Rümpler, F., & Theißen, G. (2019). Reconstructing the ancestral flower of extant angiosperms: The 'war of the whorls' is heating up. *Journal of Experimental Botany, 70*(10), 2615–2622. https://doi.org/10.1093/jxb/erz106

Rümpler, F., Theißen, G., & Melzer, R. (2018). A conserved leucine zipper-like motif accounts for strong tetramerization capabilities of SEPALLATA-like MADS-domain transcription factors. *Journal of Experimental Botany, 69*(8), 1943–1954. https://doi.org/10.1093/jxb/ery063

Rutledge, R., Regan, S., Nicolas, O., Fobert, P., Cote, C., Bosnich, W., Kauffeldt, C., Sunohara, G., Seguin, A., & Stewart, D. (1998). Characterization of an *AGAMOUS* homologue from the conifer black spruce (*Picea mariana*) that produces floral homeotic conversions when expressed in *Arabidopsis*. *The Plant Journal, 15*(5), 625–634. https://doi.org/10.1046/j.1365-313x.1998.00250.x

Sauquet, H., von Balthazar, M., Magallón, S., Doyle, J. A., Endress, P. K., Bailes, E. J., de Morais, E. B., Bull-Hereñu, K., Carrive, L., Chartier, M., Chomicki, G., Coiro, M., Cornette, R., Ottra, J. H. L., Epicoco, C., Foster, C. S. P., Jabbour, F., Haevermans, A., Haevermans, T. Little, S. A., Löfstrand, S., Luna, J. A., Massoni, J., Nadot, S., Pamperl, S., Prieu, C., Reyes, E., dos Santos, P., Schoonderwoerd, K. M., Sontag, S., Soulebeau, A., Staedler, Y., Tschan, G. F., Leung, A. W., & Schönenberger, J. (2017). The ancestral flower of angiosperms and its early diversification. *Nature Communications, 8*, 16047. https://doi.org/10.1038/ncomms16047

Schwarz-Sommer, Z., Hue, I., Huijser, P., Flor, P. J., Hansen, R., Tetens, F., Lonnig, W. E., Saedler, H., & Sommer, H. (1992). Characterization of the *Antirrhinum* floral homeotic MADS-box gene deficiens: Evidence for DNA binding and autoregulation of its persistent expression throughout flower development. *The EMBO Journal, 11*(1), 251–263. https://doi.org/10.1002/j.1460-2075.1992.tb05048.x

Seok, H. Y., Park, H. Y., Park, J. I., Lee, Y. M., Lee, S. Y., An, G., & Moon, Y. H. (2010). Rice ternary MADS protein complexes containing class B MADS heterodimer. *Biochemical and Biophysical Research Communications, 401*(4), 598–604. https://doi.org/10.1016/j.bbrc.2010.09.108

Serwatowska, J., Roque, E., Gomez-Mena, C., Constantin, G. D., Wen, J., Mysore, K. S., Lund, O. S., Johansen, E., Beltrán, J. P., & Canas, L. A. (2014). Two eu*AGAMOUS* genes control C-function in *Medicago truncatula*. *PLoS One, 9*(8), e103770. https://doi.org/10.1371/journal.pone.0103770

Shan, H., Su, K., Lu, W., Kong, H., Chen, Z., & Meng, Z. (2006). Conservation and divergence of candidate class B genes in *Akebia trifoliata* (Lardizabalaceae). *Development Genes and Evolution, 216*(12), 785–795. https://doi.org/10.1007/s00427-006-0107-2

Shan, H., Zhang, N., Liu, C., Xu, G., Zhang, J., Chen, Z., & Kong, H. (2007). Patterns of gene duplication and functional diversification during the evolution of the *AP1/SQUA* subfamily of plant MADS-box genes. *Molecular Phylogenetics and Evolution, 44*(1), 26–41. https://doi.org/10.1016/j.ympev.2007.02.016

Shan, H., Zahn, L., Guindon, S., Wall, P. K., Kong, H., Ma, H., DePamphilis, C. W., & Leebens-Mack, J. (2009). Evolution of plant MADS box transcription factors: Evidence for shifts in selection associated with early angiosperm diversification and concerted gene duplications. *Molecular Biology and Evolution, 26*(10), 2229–2244. https://doi.org/10.1093/molbev/msp129

Shan, H., Cheng, J., Zhang, R., Yao, X., & Kong, H. (2019). Developmental mechanisms involved in the diversification of flowers. *Nature Plants, 5*, 917–923. https://doi.org/10.1038/s41477-019-0498-5

Sharma, B., & Kramer, E. (2013). Sub- and neo-functionalization of *APETALA3* paralogs have contributed to the evolution of novel floral organ identity in *Aquilegia* (columbine, Ranunculaceae). *New Phytologist, 197*(3), 949–957. https://doi.org/10.1111/nph.12078

Sharma, B., & Kramer, E. M. (2017). *Aquilegia* B gene homologs promote petaloidy of the sepals and maintenance of the C domain boundary. *EvoDevo, 8*, 22. https://doi.org/10.1186/s13227-017-0085-7

Sharma, B., Guo, C. C., Kong, H. Z., & Kramer, E. M. (2011). Petal-specific subfunctionalization of an *APETALA3* paralog in the Ranunculales and its implications for petal evolution. *New Phytologist, 191*(3), 870–883. https://doi.org/10.1111/j.1469-8137.2011.03744.x

Shindo, S., Ito, M., Ueda, K., Kato, M., & Hasebe, M. (1999). Characterization of MADS genes in the gymnosperm *Gnetum parvifolium* and its implication on the evolution of reproductive organs in seed plants. *Evolution & Development, 1*(3), 180–190. https://doi.org/10.1046/j.1525-142x.1999.99024.x

Simpson, M. G. (2010). *Plant systematics*. Academic Press.

Smaczniak, C., Immink, R. G., Muino, J. M., Blanvillain, R., Busscher, M., Busscher-Lange, J., Dinh, Q. D., Liu, S., Westphal, A. H., Boeren, S., Parcy, F., Xu, L., Carles, C. C., Angenent, G. C., & Kaufmann, K. (2012). Characterization of MADS-domain transcription factor complexes in *Arabidopsis* flower development. *Proceedings of the National Academy of Sciences of the United States of America, 109*(5), 1560–1565. https://doi.org/10.1073/pnas.1112871109

Smyth, D. R., Bowman, J. L., & Meyerowitz, E. M. (1990). Early flower development in *Arabidopsis*. *The Plant Cell, 2*(8), 755–767. https://doi.org/10.1105/tpc.2.8.755

Sommer, H., Beltran, J.-P., Huijser, P., Pape, H., Lönnig, W., Saedler, H., & Schwarz-Sommer, Z. (1990). *DEFICIENS*, a homeotic gene involved in the control of flower morphogenesis in *Antirrhinum majus*: The protein shows homology to transcription factors. *The EMBO Journal, 9*(3), 605–613. https://doi.org/10.1002/j.1460-2075.1990.tb08152.x

Soza, V. L., Snelson, C. D., Hazelton, K. D. H., & Di Stilio, V. S. (2016). Partial redundancy and functional specialization of E-class *SEPALLATA* genes in an early-diverging eudicot. *Developmental Biology, 419*(1), 143–155. https://doi.org/10.1016/j.ydbio.2016.07.021

Specht, C. D., & Bartlett, M. E. (2009). Flower evolution: The origin and subsequent diversification of the angiosperm flower. *Annual Review of Ecology, Evolution, and Systematics, 40*, 217–243. https://doi.org/10.1146/annurev.ecolsys.110308.120203

Sundström, J., & Engstrom, P. (2002). Conifer reproductive development involves B-type MADS-box genes with distinct and different activities in male organ primordia. *The Plant Journal, 31*(2), 161–169. https://doi.org/10.1046/j.1365-313X.2002.01343.x

Sundström, J., Carlsbecker, A., Svensson, M. E., Svenson, M., Johanson, U., Theißen, G., & Engstrom, P. (1999). MADS-box genes active in developing pollen cones of Norway spruce (*Picea abies*) are homologous to the B-class floral homeotic genes in angiosperms. *Developmental Genetics, 25*(3), 253–266. https://doi.org/10.1002/(SICI)1520-6408(1999)25:3<253::AID-DVG8>3.0.CO;2-P

Tandre, K., Albert, V. A., Sundas, A., & Engstrom, P. (1995). Conifer homologs to genes that control floral development in angiosperms. *Plant Molecular Biology, 27*(1), 69–78. https://doi.org/10.1007/bf00019179

Tandre, K., Svenson, M., Svensson, M. E., & Engstrom, P. (1998). Conservation of gene structure and activity in the regulation of reproductive organ development of conifers and angiosperms. *The Plant Journal, 15*(5), 615–623. https://doi.org/10.1046/j.1365-313x.1998.00236.x

Tavares, R., Cagnon, M., Negrutiu, I., & Mouchiroud, D. (2010). Testing the recent theories for the origin of the hermaphrodite flower by comparison of the transcriptomes of gymnosperms and angiosperms. *BMC Evolutionary Biology, 10*, 240. https://doi.org/10.1186/1471-2148-10-240

Taylor, S. A., Hofer, J. M., Murfet, I. C., Sollinger, J. D., Singer, S. R., Knox, M. R., & Ellis, T. H. (2002). *PROLIFERATING INFLORESCENCE MERISTEM,* a MADS-box gene that regulates floral meristem identity in pea. *Plant Physiology, 129*(3), 1150–1159. https://doi.org/10.1104/pp.001677

Theißen, G. (2001). Development of floral organ identity: Stories from the MADS house. *Current Opinion in Plant Biology, 4*(1), 75–85. https://doi.org/10.1016/S1369-5266(00)00139-4

Theißen, G., & Melzer, R. (2007). Molecular mechanisms underlying origin and diversification of the angiosperm flower. *Annals of Botany, 100*(3), 603–619. https://doi.org/10.1093/aob/mcm143

Theißen, G., Becker, A., Winter, K., Munster, T., Kirchner, C., & Saedler, H. (2002). How the land plants learned their floral ABCs: The role of MADS-box genes in the evolutioanry origin of flowers. In Q. B. R. Cronk, R. M. Bateman, & J. Hawkins (Eds.), *Developmental genetics and plant evolution* (pp. 173–205). Taylor & Francis.

Tröbner, W., Ramirez, L., Motte, P., Hue, I., Huijser, P., Lönnig, W., Saedler, H., Sommer, H., & Schwarz-Sommer, Z. (1992). *GLOBOSA*: A homeotic gene which interacts with *DEFICIENS* in the control of *Antirrhinum* floral organogenesis. *The EMBO Journal, 11*(13), 4693–4704. https://doi.org/10.1002/j.1460-2075.1992.tb05574.x

Vandenbussche, M., Zethof, J., Souer, E., Koes, R., Tornielli, G. B., Pezzotti, M., Ferrario, S., Angenent, G. C., & Gerats, T. (2003). Toward the analysis of the petunia MADS box gene family by reverse and forward transposon insertion mutagenesis approaches: B, C, and D floral organ identity functions require *SEPALLATA*-like MADS box genes in *petunia*. *The Plant Cell, 15*(11), 2680–2693. https://doi.org/10.1105/tpc.017376

Vandenbussche, M., Zethof, J., Royaert, S., Weterings, K., & Gerats, T. (2004). The duplicated B-class heterodimer model: Whorl-specific effects and complex genetic interactions in *Petunia hybrida* flower development. *The Plant Cell, 16*(3), 741–754. https://doi.org/10.1105/tpc.019166

Vazquez-Lobo, A., Carlsbecker, A., Vergara-Silva, F., Alvarez-Buylla, E. R., Pinero, D., & Engstrom, P. (2007). Characterization of the expression patterns of *LEAFY/FLORICAULA* and *NEEDLY* orthologs in female and male cones of the conifer genera *Picea*, *Podocarpus*, and *Taxus*: Implications for current evo-devo hypotheses for gymnosperms. *Evolution & Development, 9*(5), 446–459. https://doi.org/10.1111/j.1525-142X.2007.00182.x

Vrebalov, J., Ruezinsky, D., Padmanabhan, V., White, R., Medrano, D., Drake, R., Schuch, W., & Giovannoni, J. (2002). A MADS-box gene necessary for fruit ripening at the tomato *Ripening inhibitor* (*Rin*) locus. *Science, 296*(5566), 343–346. https://doi.org/10.1126/science.1068181

Walker-Larsen, J., & Harder, L. D. (2000). The evolution of staminodes in angiosperms: Patterns of stamen reduction, loss, and functional re-invention. *American Journal of Botany, 87*(10), 1367–1384. https://doi.org/10.2307/2656866

Wang, X. Q., & Ran, J. H. (2014). Evolution and biogeography of gymnosperms. *Molecular Phylogenetics and Evolution, 75*, 24–40. https://doi.org/10.1016/j.ympev.2014.02.005

Wang, K., Tang, D., Hong, L., Xu, W., Huang, J., Li, M., Gu, M., Xue, Y., & Cheng, Z. (2010). *DEP* and *AFO* regulate reproductive habit in rice. *PLoS Genetics, 6*(1), e1000818. https://doi.org/10.1371/journal.pgen.1000818

Wang, P., Liao, H., Zhang, W., Yu, X., Zhang, R., Shan, H., Duan, X., Yao, X., & Kong, H. (2016). Flexibility in the structure of spiral flowers and its underlying mechanisms. *Nature Plants, 2*, 15188. https://doi.org/10.1038/nplants.2015.188

Weberling, F. (1989). *Morphology of flowers and inflorescences*. Cambridge Univercity Press.

Whipple, C. J., Ciceri, P., Padilla, C. M., Ambrose, B. A., Bandong, S. L., & Schmidt, R. J. (2004). Conservation of B-class floral homeotic gene function between maize and *Arabidopsis*. *Development, 131*(24), 6083–6091. https://doi.org/10.1242/dev.01523

Whipple, C. J., Zanis, M. J., Kellogg, E. A., & Schmidt, R. J. (2007). Conservation of B class gene expression in the second whorl of a basal grass and outgroups links the origin of lodicules and petals. *Proceedings of the National Academy of Sciences of the United States of America, 104*(3), 1081–1086. https://doi.org/10.1073/pnas.0606434104

Wilson, C. L. (1942). The telome theory and the origin of the stamen. *American Journal of Botany, 29*(9), 759–764. https://doi.org/10.2307/2437730

Winter, K. U., Becker, A., Munster, T., Kim, J. T., Saedler, H., & Theißen, G. (1999). MADS-box genes reveal that gnetophytes are more closely related to conifers than to flowering plants. *Proceedings of the National Academy of Sciences of the United States of America, 96*(13), 7342–7347. https://doi.org/10.1073/pnas.96.13.7342

Winter, K. U., Saedler, H., & Theißen, G. (2002). On the origin of class B floral homeotic genes: Functional substitution and dominant inhibition in *Arabidopsis* by expression of an orthologue from the gymnosperm *Gnetum*. *The Plant Journal, 31*(4), 457–475. https://doi.org/10.1046/j.1365-313X.2002.01375.x

Wollmann, H., Mica, E., Todesco, M., Long, J. A., & Weigel, D. (2010). On reconciling the interactions between *APETALA2*, miR172 and *AGAMOUS* with the ABC model of flower development. *Development, 137*(21), 3633–3642. https://doi.org/10.1242/dev.036673

Wu, F., Shi, X., Lin, X., Liu, Y., Chong, K., Theißen, G., & Meng, Z. (2017). The ABCs of flower development: Mutational analysis of *AP1/FUL*-like genes in rice provides evidence for a homeotic (A)-function in grasses. *The Plant Journal, 89*(2), 310–324. https://doi.org/10.1111/tpj.13386

Wu, D., Liang, W., Zhu, W., Chen, M., Ferrandiz, C., Burton, R. A., Dreni, L., & Zhang, D. (2018). Loss of *LOFSEP* transcription factor function converts spikelet to leaf-like structures in rice. *Plant Physiology, 176*(2), 1646–1664. https://doi.org/10.1104/pp.17.00704

Wuest, S. E., O'Maoileidigh, D. S., Rae, L., Kwasniewska, K., Raganelli, A., Hanczaryk, K., Lohan, A. J., Loftus, B., Graciet, E., & Wellmer, F. (2012). Molecular basis for the specification of floral organs by *APETALA3* and *PISTILLATA*. *Proceedings of the National Academy of Sciences of the United States of America, 109*(33), 13452–13457. https://doi.org/10.1073/pnas.1207075109

Xu, G., & Kong, H. (2007). Duplication and divergence of floral MADS-box genes in grasses: Evidence for the generation and modification of novel regulators. *Journal of Integrative Plant Biology, 49*(6), 927–939. https://doi.org/10.1111/j.1744-7909.2007.00502.x

Yadav, S. R., Prasad, K., & Vijayraghavan, U. (2007). Divergent regulatory *OsMADS2* functions control size, shape and differentiation of the highly derived rice floret second-whorl organ. *Genetics, 176*(1), 283–294. https://doi.org/10.1534/genetics.107.071746

Yamaguchi, T., Nagasawa, N., Kawasaki, S., Matsuoka, M., Nagato, Y., & Hirano, H. Y. (2004). The *YABBY* gene *DROOPING LEAF* regulates carpel specification and midrib development in *Oryza sativa*. *The Plant Cell, 16*(2), 500–509. https://doi.org/10.1105/tpc.018044

Yanofsky, M. F., Ma, H., Bowman, J. L., Drews, G. N., Feldmann, K. A., & Meyerowitz, E. M. (1990). The protein encoded by the *Arabidopsis h*omeotic gene *AGAMOUS* resembles transcription factors. *Nature, 346*(6279), 35–39. https://doi.org/10.1038/346035a0

Yao, S. G., Ohmori, S., Kimizu, M., & Yoshida, H. (2008). Unequal genetic redundancy of rice *PISTILLATA* orthologs, *OsMADS2* and *OsMADS4*, in lodicule and stamen development. *Plant & Cell Physiology, 49*(5), 853–857. https://doi.org/10.1093/pcp/pcn050

Ye, L. L., Wang, B., Zhang, W. G., Shan, H. Y., & Kong, H. Z. (2016). Gains and losses of *cis*-regulatory elements led to divergence of the *Arabidopsis APETALA1* and *CAULIFLOWER* duplicate genes in the time, space, and level of expression and regulation of one paralog by the other. *Plant Physiology, 171*(2), 1055–1069. https://doi.org/10.1104/pp.16.00320

Yellina, A. L., Orashakova, S., Lange, S., Erdmann, R., Leebens-Mack, J., & Becker, A. (2010). Floral homeotic C function genes repress specific B function genes in the carpel whorl of the basal eudicot California poppy (*Eschscholzia californica*). *EvoDevo, 1*, 13. https://doi.org/10.1186/2041-9139-1-13

Yoshida, H., Itoh, J. I., Ohmori, S., Miyoshi, K., Horigome, A., Uchida, E., Kimizu, M., Matsumura, Y., Kusaba, M., Satoh, H., & Nagato, Y. (2007). *Superwoman1-cleistogamy*, a hopeful allele for gene containment in GM rice. *Plant Biotechnology Journal, 5*(6), 835–846. https://doi.org/10.1111/j.1467-7652.2007.00291.x

Yu, D., Kotilainen, M., Pöllänen, E., Mehto, M., Elomaa, P., Helariutta, Y., Albert, V. A., & Teeri, T. H. (1999). Organ identity genes and modified patterns of flower development in *Gerbera hybrida* (Asteraceae). *The Plant Journal, 17*(1), 51–62. https://doi.org/10.1046/j.1365-313X.1999.00351.x

Yu, X. X., Duan, X. S., Zhang, R., Fu, X. H., Ye, L. L., Kong, H. Z., Xu, G. X., & Shan, H. Y. (2016). Prevalent exon-intron structural changes in the *APETALA1/FRUITFULL, SEPALLATA, AGAMOUS-LIKE6,* and *FLOWERING LOCUS C* MADS-Box gene subfamilies provide new insights into their evolution. *Frontiers in Plant Science, 7*, 598. https://doi.org/10.3389/fpls.2016.00598

Zahn, L. M., Kong, H., Leebens-Mack, J. H., Kim, S., Soltis, P. S., Landherr, L. L., Soltis, D. E., Depamphilis, C. W., & Ma, H. (2005). The evolution of the *SEPALLATA* subfamily of MADS-box genes: A preangiosperm origin with multiple duplications throughout angiosperm history. *Genetics, 169*(4), 2209–2223. https://doi.org/10.1534/genetics.104.037770

Zahn, L. M., Leebens-Mack, J. H., Arrington, J. M., Hu, Y., Landherr, L. L., dePamphilis, C. W., Becker, A., Theißen, G., & Ma, H. (2006). Conservation and divergence in the *AGAMOUS* subfamily of MADS-box genes: Evidence of independent sub- and neofunctionalization events. *Evolution & Development, 8*(1), 30–45. https://doi.org/10.1111/j.1525-142X.2006.05073.x

Zhang, P. Y., Tan, H. T. W., Pwee, K. H., & Kumar, P. P. (2004). Conservation of class C function of floral organ development during 300 million years of evolution from gymnosperms to angiosperms. *The Plant Journal, 37*(4), 566–577. https://doi.org/10.1046/j.1365-313X.2003.01983.x

Zhang, R., Guo, C., Zhang, W., Wang, P., Li, L., Duan, X., Du, Q., Zhao, L., Shan, H., Hodges, S. A., Kramer, E. M., Ren, Y., & Kong, H. (2013). Disruption of the petal identity gene *APETALA3-3* is highly correlated with loss of petals within the buttercup family (Ranunculaceae). *Proceedings of the National Academy of Sciences of the United States of America, 110*(13), 5074–5079. https://doi.org/10.1073/pnas.1219690110

Zhang, J. S., Li, Z., Zhao, J., Zhang, S., Quan, H., Zhao, M., & He, C. (2014). Deciphering the *Physalis floridana Double-Layered-Lantern1* mutant provides insights into functional divergence of the *GLOBOSA* duplicates within the Solanaceae. *Plant Physiology, 164*(2), 748–764. https://doi.org/10.1104/pp.113.233072

Zhang, T., Zhao, Y., Juntheikki, I., Mouhu, K., Broholm, S. K., Rijpkema, A. S., Kins, L., Lan, T., Albert, V. A., Teeri, T. H., & Elomaa, P. (2017). Dissecting functions of *SEPALLATA*-like MADS box genes in patterning of the pseudanthial inflorescence of *Gerbera hybrida*. *New Phytologist, 216*(3), 939–954. https://doi.org/10.1111/nph.14707

Zhang, L., Chen, F., Zhang, X., Li, Z., Zhao, Y., Lohaus, R., Chang, X., Dong, W., Ho, S. Y. W., Liu, X., Song, A., Chen, J., Guo, W., Wang, Z., Zhuang, Y., Wang, H., Chen, X., Hu, J., Liu, Y., Qin, Y., Wang, K., Dong, S., Liu, Y., Zhang, S., Yu, X., Wu, Q., Wang, L., Yan, X., Jiao, Y., Kong, H., Zhou, X., Yu, C., Chen, Y., Li, F., Wang, J., Chen, W., Chen, X., Jia, Q., Zhang, C., Jiang, Y., Zhang, W., Liu, G., Fu, J., Chen, F., Ma, H., Van de Peer, Y., & Tang, H. (2020). The water lily genome and the early evolution of flowering plants. *Nature, 577*(7788), 79–84. https://doi.org/10.1038/s41586-019-1852-5

Zhao, J., Tian, Y., Zhang, J. S., Zhao, M., Gong, P. C., Riss, S., Saedler, R., & He, C. Y. (2013). The euAP1 protein MPF3 represses *MPF2* to specify floral calyx identity and displays crucial roles in Chinese lantern development in *Physalis*. *The Plant Cell, 25*(6), 2002–2021. https://doi.org/10.1105/tpc.113.111757

Zhao, T., Holmer, R., de Bruijn, S., Angenent, G. C., van den Burg, H. A., & Schranz, M. E. (2017). Phylogenomic synteny network analysis of MADS-Box transcription factor genes reveals lineage-specific transpositions, ancient tandem duplications, and deep positional conservation. *The Plant Cell, 29*(6), 1278–1292. https://doi.org/10.1105/tpc.17.00312

Zhao, H., Liao, H., Li, S., Zhang, R., Dai, J., Ma, P., Wang, T., Wang, M., Yuan, Y., Fu, X., Cheng, J., Duan, X., Xie, Y., Zhang, P., Kong, H., & Shan, H. (2023). Delphinieae flowers originated from the rewiring of interactions between duplicated and diversified floral organ identity and symmetry genes. *The Plant Cell, 35*(3), 994–1012. https://doi.org/10.1093/plcell/koac368

Chapter 9
Stamen and Anther Development

Fang Chang and Hong Ma

Abstract Stamens are the male reproductive organs of angiosperms, and their proper development ensures male fertility and crop productivity. In addition, male sterile varieties are important for hybrid breeding and increased crop yield. Therefore, it is highly important for both basic plant development and crop breeding to understand the regulatory mechanisms of stamen development. Within the stamen, the anther contains reproductive cells and the surrounding somatic cells that support the formation and development of reproductive cells. The major developmental events and morphological changes of the anther are similar among different angiosperm species, including the establishment of anther polarity, anther morphogenesis to form the characteristic butterfly-like morphology of the cross section, meiosis to produce haploid microspores, pollen development, and release from the anther. There are great interests in investigating the molecular mechanisms underlying these processes using molecular genetics and cell biological approaches. In recent decades, breakthroughs have been made in understanding anther developmental processes due to the rapid advances in transcriptome sequencing, quantitative mass spectrometry, and microscopic technologies. This chapter presents a brief description of anther development and a detailed overview of advances in understanding the mechanisms underlying anther development in flowering plants, especially in model species such as *Arabidopsis*, rice, and maize. The emphasis here is on gene functions and molecular interactions, including feedback regulatory loops in both cell signaling and transcriptional control that are crucial for anther initiation, cell type specification, and tapetum development and function.

F. Chang
State Key Laboratory of Genetics and Development of Complex Phenotypes, Institute of Plant Biology, School of Life Sciences, Fudan University, Shanghai, China
e-mail: fangchang@fudan.edu.cn

H. Ma (✉)
Department of Biology, the Huck Institutes of the Life Sciences, The Pennsylvania State University, University Park, PA, USA
e-mail: hxm16@psu.edu

F. Chang et al. (eds.), *Regulation of Plant Development*,
https://doi.org/10.1007/978-981-95-7033-1_9

Keywords Stamen · Anther morphogenesis · Tapetum · Microspore · Cell-cell communications

9.1 Introduction

In flowering plants, the sexual organs, the stamens and the carpels, are composed primarily of somatic cells, and only a small population in each organ switches to male or female reproductive cells after organ identity is established (Bhatt et al., 2001). Stamens occupy positions in the third whorl (in some species further inner whorls) of the flower, interior to the perianth organs (sepals and petals), and surround the gynoecium; they are compound organs each consisting of an apical anther with four symmetrically organized lobes and a basal filament in many flowering plants including *Arabidopsis thaliana*, *Zea mays*, and rice (Goldberg et al., 1993; Sanders et al., 1999; Kelliher & Walbot, 2011; Zhang et al., 2011). The function of the anther is to house germ line cells and produce pollen grains, and the function of the filament is to support anther and pollen development by transporting signals and nutrients. The number of stamens varies among different angiosperms, ranging from a few to dozens. However, the general anther developmental stages and mature anther anatomy are similar in most angiosperms with multiple specialized cell types; the normal male fertility results from a series of developmental events including the anther initiation and morphogenesis, meiosis, postmeiotic gametophyte development, and anther dehiscence to release the pollen.

9.2 Anther Developmental Processes

Compared to the diversity of other floral organs, the organ shape and developmental steps of the anther are largely conserved among angiosperms (Goldberg et al., 1993; Sanders et al., 1999; Ma, 2005; Kelliher & Walbot, 2011; Zhang et al., 2011). The flower of the model plant *Arabidopsis* has six stamens, each with an anther and a filament. The anther contains four lobes, each with a group of reproductive pollen mother cells (PMCs; or meiotic cells), which are surrounded by four somatic cell layers: the tapetum, the middle layer, the endothecium, and the epidermis from the interior to the surface (Fig. 9.1). These somatic cell layers constitute the anther wall that provides developmental cues (Wilson & Zhang, 2009), structural and biochemical support for the meiocytes, and support microspore and pollen development by delivering nutrients and pollen coat materials. The filament has an epidermis and a vascular bundle, and the four-locular anther lobes are attached to the vascular bundle by connective tissues. The function of the filament is to connect the anther to the flower receptacle and supply signals, nutrients, and water to the growing anther.

Development of the *Arabidopsis* anther is divided into 14 stages according to distinctive tissue and cellular markers that could be examined in detail at the light

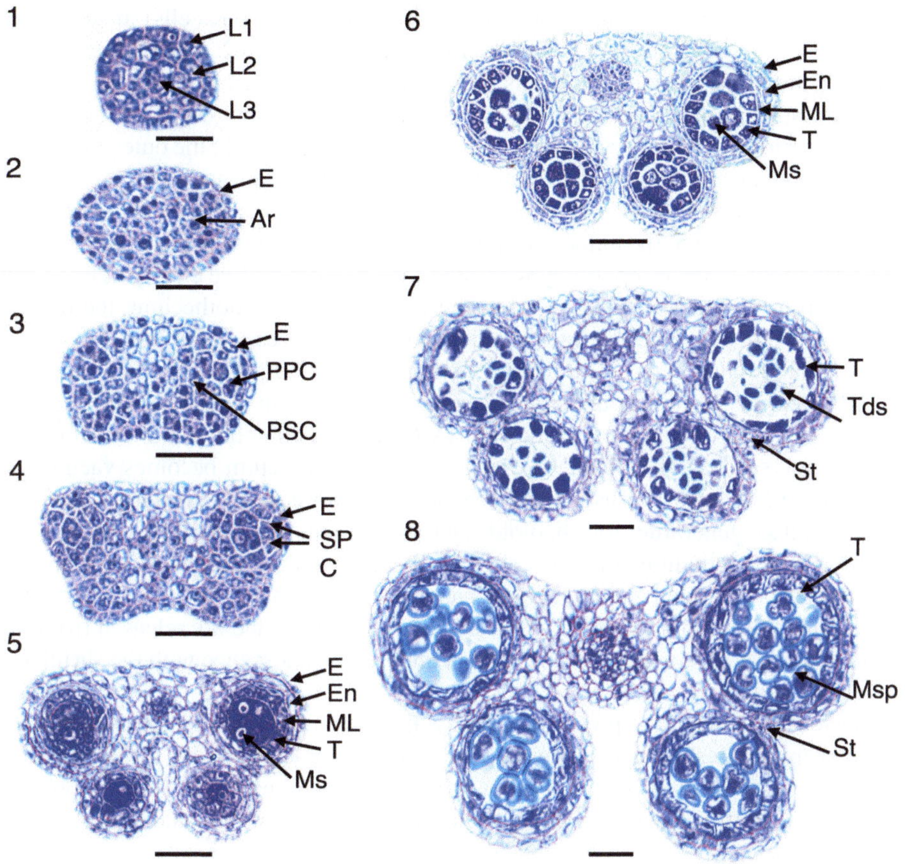

Fig. 9.1 Cell differentiation during anther development. Shown are the stages of wild-type anther development in *Arabidopsis thaliana*. The numbers indicate stages. Bar = 25 μm. L1, L2, L3, the anther cell layer 1, 2, 3, respectively; *Ar* archesporial cell, *E* epidermis, *En* endothecium, *PPC* primary parietal cell, *PSC* primary sporogenous cell, *SPC* secondary parietal cell, *ML* middle layer, *T* tapetum, *Ms.* microsporocytes, *Tds* tetrads, *St* stomium

microscope level. Stages 1–8 represent anther development phase one, during which anther tissues are formed as a result of cell division and differentiation (Fig. 9.1). Stages 9–14 represent anther developmental phase two, which includes the development of microspores inside the anther into pollen grains, with subsequent degeneration of anther tissues and the release of the pollen, along with dramatic elongation of the filament.

At stage 1, the round stamen primordium emerges with the L1, L2, and L3 cell layers derived from the floral meristem. During the subsequent stages 2–5, cells in the L1 layer undergo anticlinal divisions to form the anther epidermis. Cells in the L3 divide and differentiate to form largely the connective and vascular tissues. At the same time, cells of the L2 layer undergo a series of periclinal and anticlinal cell divisions and cell differentiation: firstly, along with the adaxial-abaxial anther polarity establishment, four clusters of archesporial cells (AR cells) arise in four "corners" of the L2 cell layer from periclinal division at stage 2; then periclinal division of the archesporial cells forms the primary parietal cell (PPC) layer and the primary sporogenous cell (PSC) layer, with further periclinal division of the PPC cells to form two secondary parietal cell (SPC) layers at stage 3; the outer SPC layer further divides and differentiates into the endothecium layer and the middle layer, while the inner SPC layer divides (anticlinally) and develops into the tapetum layer at stage 4; at stage 5, the anther morphogenesis is complete as marked by the characteristic four-lobed morphology with each lobe possessing five clearly defined cell layers from the surface to the interior: the epidermis, the endothecium, the middle layer, the tapetum, and the reproductive microsporocytes, which are also called microspore mother cells (MMCs) or pollen mother cells (PMCs) (Figs. 9.1 and 9.2).

At stage 6, the anther undergoes a general increase in size. The MMCs enter meiosis and thus are called meiotic cells (MCs), and their cell walls are surrounded by the polysaccharide callose. At the same time, the tapetum becomes vacuolated and the middle layer is crushed and degenerated. Subsequently, at stage 7, meiosis is completed, generating four haploid microspores that are transiently contained in a tetrad by the remaining meiotic cell wall. By stage 8, the callose wall surrounding the tetrads is degraded and microspores are released (Figs. 9.1 and 9.2). The microspores then develop into pollen grains during stages 9–12 and are released from the anther at stage 13. At stage 14, the stamen undergoes senescence with the shrinkage of anther cells.

Defects in premeiotic processes usually alter the cellular composition and function, therefore resulting in male sterility that is characterized by the absence of viable pollen. So far, dozens of genes with key roles in early anther development have been identified in model plants including *Arabidopsis*, rice, and maize.

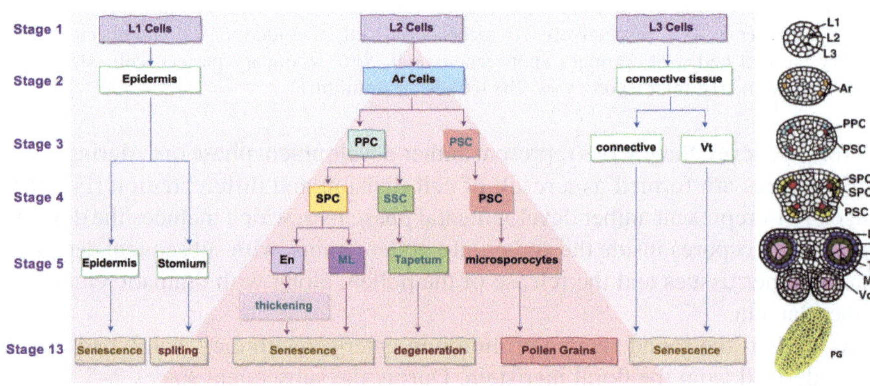

Fig. 9.2 Cell lineages and major events that occur during Arabidopsis anther development. Shown are the stages, timing events. *E* epidermis, *Ar* archesporial, *PPC* primary parietal cell, *PSC* primary sporogenous cell, *Vt* vascular tissue, *SPC* secondary parietal cell, *St* stomium, *En* endothecium, *ML* middle layer, *Ms*. microsporocytes, *PG* pollen grain

9.3 Regulation of the Early Anther Polarity

All lateral organs, including leaves and floral organs, develop from the flank of the meristem (Husbands et al., 2009). Furthermore, the establishment of adaxial-abaxial polarity is essential for the proper development of the lateral organs in plants; specifically for leaves, the establishment of the abaxial-adaxial polarity in the leaf primordia is the earliest and critical step for blade formation. Once the abaxial-adaxial polarity is established, this polarity is maintained and required for the specification of the developmental fate of leaf cells (Chitwood et al., 2009; Husbands et al., 2009; Toriba et al., 2011) (Fig. 9.3a). Floral organs are thought to be modified leaves, consistent with the sepal and petal being flat organs and morphologically similar to leaves and supported by leaf-like organs in mutant flowers lacking crucial floral organ identity genes (see Chap. 6 on control of floral organ identities). However, the stamen is morphologically different from the leaf. Firstly, the stamen has two parts, the proximal and radially shaped filament and the distal and bilaterally symmetric anther; secondly, the anther usually has four lobes positioned in two pairs (Fig. 9.3b). The two lobes of each pair are adjacent to each other and share a common dehiscence zone, the stomium, and each pair is a structure unit called a theca. The two thecae of the anther are both connected with the connective tissues (Fig. 9.3b).

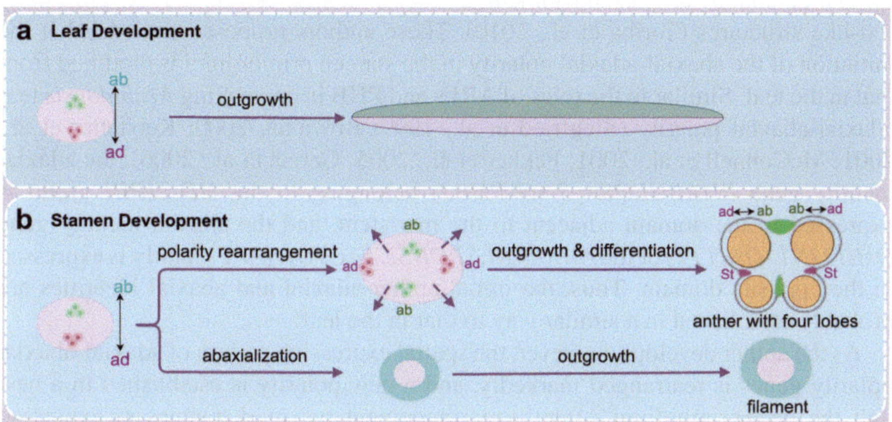

Fig. 9.3 A model showing the different developmental patterns of the leaf (a) and stamen (b, c). (**a**) In leaf development, lamina outgrowth occurs at the junction of the adaxial and abaxial domains. (**b**) A model of stamen patterning in Arabidopsis. In anther development (the top), the adaxial-abaxial polarity is rearranged, resulting in a different pattern; afterward, the boundary regions between the adaxial and abaxial domains grow outward and differentiate into anther lobes. In filament development (the bottom), the radially symmetrical structure is formed by abaxialization. The adaxial domains are shown in purple, and the abaxial domains are shown in green. Double-headed arrows present the axis of adaxial-abaxial polarity. Lilac arrows indicate the direction of outgrowth. *ab* abaxial side, *ad* adaxial side, *st* stomium. Figures are modified based on the references (Toriba et al., 2010; Walbot & Egger, 2016)

Therefore, the polarity establishment of stamen is more complicated than leaves, sepals, and petals (Fig. 9.3b).

Several *Arabidopsis* genes have been identified to play crucial roles in the establishment of leaf polarity, and their molecular functions have been characterized. For instance, the *HOMEODOMAIN-LEUCINE ZIPPERIII* (*HD-ZIPIII*) genes, such as *PHABULOSA* (*PHB*), are required for specification of adaxial cell fate, whereas *KANADI* (*KAN*) and *YABBY* genes are involved in specifying the abaxial cell fate (Siegfried et al., 1999; Bowman, 2000; Kerstetter et al., 2001; McConnell et al., 2001). In addition, small RNAs, such as miR166, act as the abaxial determinant by suppressing HD-ZIPIII function through post-transcriptional regulation (Emery et al., 2003; Kidner & Martienssen, 2004). *AUXIN RESPONSE FACTOR3* (*ARF3*)/*ETTIN*(*ETT*) and *ARF4* are required for the abaxial identity (Pekker et al., 2005; Garcia et al., 2006), and the TAS3 transacting small interfering small RNA2141 (ta-siRNA2141) and ta-siRNA2142 (also called tasiR-ARF) target ARF3/ETT and ARF4 and suppress the abaxial determination (Allen et al., 2005; Chitwood et al., 2007; Husbands et al., 2009; Voinnet, 2009).

It is generally accepted that the establishment of stamen polarity is regulated by an unusual mechanism different from that in the leaf: rearrangement of the adaxial-abaxial polarity in the anther and abaxialization in the filament (Fig. 9.3) (Toriba et al., 2010, 2011). The model of regulatory mechanism for the establishment of abaxial-adaxial polarity in stamen is mainly based on the study on the *rod-like lemma* (*rol*) mutant of rice, in which the stamen and lemma are defective in the abaxial-adaxial polarity establishment, and some stamens developed into pin-like or rod-like structures (Toriba et al., 2010). These authors proposed a model that the initiation of the abaxial-adaxial polarity in the stamen primordium is modified from that in the leaf. Similar to the roles of ARFs and PHB in controlling *Arabidopsis* leaf adaxial-abaxial polarity (Siegfried et al., 1999; Bowman, 2000; Kerstetter et al., 2001; McConnell et al., 2001; Pekker et al., 2005; Garcia et al., 2006), the adaxial identity gene *PHABULOSA 3* (*OsPHB3*) belonging to the *HD-ZIPIII* group is expressed in the domain adjacent to the meristem, and the abaxial identity gene *OsETTIN1* (*OsETT1*) of the *AUXIN RESPONSE FACTOR* (*ARF*) family is expressed in the opposite domain. Thus, the initial anther adaxial and abaxial identities are probably established in a similar way to that in the leaf.

As the anther develops, however, the spatial expression pattern of adaxial-abaxial polarity genes is rearranged markedly, and a new polarity is established in a new unit, the theca primordium. After the rearrangement, two marker genes are expressed in positions that crossed at right angles: *OsPHB3* is expressed in the two lateral regions of the anther primordium, whereas *OsETT1* is expressed in its original abaxial region and a region near the meristem where Os*PHB3* was expressed at the stamen primordium stage (Toriba et al., 2010, 2011). Thus, the new polarity of the anther is established and regulates further anther development. There are four boundary regions, each between two adjacent domains that express the adaxial or abaxial identity genes; these boundary regions undergo further outgrowth and subsequently differentiate into the four anther lobes. In comparison, the expression of *OsPHB3* is lost in the filament, suggesting the abaxialization for filament formation

(Toriba et al., 2010). Thus, the stamen has distinct polarity patterns in the anther and filament.

Prior work in *Arabidopsis* demonstrated that *ARF3/ETT* and *PBH* expression patterns are sufficiently similar to the patterns in rice (Sessions et al., 1997; Sessions, 1997; Dinneny et al., 2006). *ARF3* and *ARF4* are specifically expressed in the anther abaxial zone but not in the adaxial zone; on the other hand, *PHB* is expressed in the adaxial zone. In addition, *TAS3 ta-siRNAs* directly antagonize the *ARF* mRNAs and function to sharpen the boundaries between the abaxial and adaxial domains and establish the four-lobe architecture. These results support the idea that the regulatory mechanism underlying lobe development is conserved in the monocot rice and eudicot *Arabidopsis*.

9.4 Early Cell Division and Differentiation in Each Anther Lobe

As mentioned earlier, each anther lobe contains the reproductive cells surrounded by four somatic cell layers. The reproductive cells and the three inner somatic cell layers, tapetum, middle layer, and endothecium, are largely generated from AR cells in the L2 of the anther primordium (Sanders et al., 1999; Ma, 2005). Therefore, one of the most important events in early anther development is the formation of AR cells.

9.4.1 SPL *Regulates the Initiation of Cell Specification in Anther Lobes*

In *Arabidopsis*, the *SPOROCYTELESS/NOZZLE* (*SPL/NZZ*) (*SPL* for convenience) gene encoding a putative transcriptional regulator is essential for the early anther cell division and differentiation (Schiefthaler et al., 1999; Yang et al., 1999; Ito et al., 2004; Liu et al., 2009). *SPL* is a gene downstream of the MADS-box transcription factor AGAMOUS (AG), an essential component of C function required for stamen and carpel identity, and the loss of *SPL* function results in both male and female sterility. *SPL* is expressed in the L2 layer of the anther primordium and in PSCs as the cell layers are formed. In the *spl* mutant anther, neither the reproductive cells nor somatic cell layers develop normally, resulting in anthers without sporogenous cells and with only two undifferentiated somatic cell layers in each lobe (Yang et al., 1999; Liu et al., 2009). These suggest that *SPL* functions in both the specification of reproductive cell fate and differentiation of the inner somatic cell layers. Due to the lack of a molecular marker for the AR cell, it is uncertain whether the AR cells are formed in the *spl* mutant. Even so, it is likely that *SPL* promotes AR cell differentiation in the anther. In addition, studies in the ovule demonstrated that

SPL acts as a transcriptional suppressor of genes that are required for the ovule organ identity and thus behaves as an activator of the ovule cell specification (Wei et al., 2015). Interestingly, although *SPL* is essential for the development of both anthers and ovules, no *SPL* homologs have been detected in monocots, suggesting divergence of molecular programs between monocots and dicots for the regulation of early anther development. It is possible that the *SPL* function might not be part of the ancestral anther developmental pathway; alternatively, it might have been lost and replaced by other alternative functions and mechanisms.

9.4.2 *ROXY1/ROXY2 Are Required for Early Anther Cell Specification*

Glutaredoxins (GRXs) are small glutathione-dependent oxidoreductase that catalyzes the reduction of disulfide bonds of their substrate proteins and are involved in various cellular processes and responses to oxidative stress (Fernandes & Holmgren, 2004; Buchanan & Balmer, 2005). Plant GRXs include three types based on motif sequences, the CPYC-type, the CGFS type, and the CC-type (Lemaire, 2004; Ziemann et al., 2009). Evolutionary analyses suggested that the CC-type GRXs probably arose by diversification from the CPYC class at a time close to the colonization of land by plants (Ziemann et al., 2009). In a member of the early-divergent land plant lineage, mosses, *Physcomitrella patens*, only two CC-type GRXs are present. Then, this class dramatically expanded, especially in the angiosperms shortly after the divergence of monocots and eudicots; for instance, 21 CC-type GRXs are present in *Arabidopsis*, which likely have undergone functional specialization, including acquisition of new roles such as development and pathogenic response (Ziemann et al., 2009). Two of *Arabidopsis* CC-type GRX genes, *ROXY1* and *ROXY2,* are expressed at early anther stages in the tapetum and act redundantly to control early anther development (Xing & Zachgo, 2008; Wang et al., 2009). Loss of functions of both *ROXY1* and *ROXY2* result in defects in fertility, as the *roxy1 roxy2* double mutant flowers produced a reduced number of anthers that are sterile. At the anther stage 3, instead of the formation of the PPCs and PSCs in all four lobes as in the wild-type anthers, the *roxy1 roxy2* anthers lack PSCs on the adaxial side of the anther so that the differentiation of the PMCs and somatic cell layers failed to occur (Xing & Zachgo, 2008). Although differentiation of the PPCs and PSCs occurs in the abaxial lobes of the *roxy1 roxy2* anther and the PMCs and somatic cell layers are formed, these PMCs have abnormally low amounts of callose walls, adhere to each other, and subsequently degenerate, resulting in no pollen formation. In addition, the middle layer of the *roxy1 roxy2* anthers consists of two cell layers instead of one, and the tapetum grows hypertrophically and is eventually degraded (Xing & Zachgo, 2008; Xing et al., 2011). These results strongly suggest that *ROXY1* and *ROXY2* play an essential role in regulating early anther cell division, specification, and differentiation. Intriguingly, loss of function of a maize CC-type GRX gene, *MSCA1*, causes a similar phenotype of lacking sporogenous cells (Chaubal et al., 2003), suggesting that the role of ROXYs in early anther development might be conserved between monocots and eudicots.

The mechanistic basis of GRX activity has been intensively studied in *E. coli*, yeast, and mammals, and binding of the glutathione small tripeptide is required for their activity. The GRXs exert their activity through posttranslational reduction of disulfide bonds in their target proteins, thus altering the conformation and/or activity of their targets (Sun et al., 1998; Canales et al., 2002; Fernandes & Holmgren, 2004; Xing et al., 2006, 2011; Xing & Zachgo, 2008). The Arabidopsis ROXY1 and ROXY2 can interact with basic leucine-zipper (bZIP) transcription factors through a short pentapeptide located in the α5 helix, LVPLL (Li et al., 2011). This peptide resembles the α-helical L**LL (* any amino acid) motif known to mediate the interaction between transcriptional coactivators and ligand-bound nuclear receptors, as well as many other transcription factors in animals (Le Douarin et al., 1996; Heery et al., 1997; Voegel et al., 1998). ROXY1 interacts with PERIANTHIA (PAN) to regulate the correct petal numbers, and ROXY1 and ROXY2 both interact with TGA9 and TGA10 to regulate early anther development (Li et al., 2009; Murmu et al., 2010) (Fig. 9.4)

The similarity in the defective adaxial lobes of the *roxy1 roxy2* anthers to that of the *spl* anther suggests that the *ROXYs* and *SPL* genes are functionally related. This is supported by the observation that the anther of the *spl roxy1 roxy2* triple mutant resembles that of the *spl* single mutant; in addition, transcriptomic analysis showed that the *SPL* expression is reduced by approximately twofold in the *roxy1 roxy2* inflorescence (Xing & Zachgo, 2008). Xing and Zachgo considered the expression change of *SPL* in *roxy1 roxy2* is not large and proposed that the *ROXY1* and *ROXY2* probably act downstream of *SPL* (Xing & Zachgo, 2008). However, it is also possible that *ROXYs* act upstream of *SPL* through a positive feedback regulatory loop. As *SPL* homologs have not been identified in monocots such as rice and maize, whereas the function of ROXYs in regulating early anther development is likely more conserved, it will likely be informative to investigate whether these genes represented the ancestral function and how they function together or independently to control normal anther development. Further study on the SPL-regulated downstream transcriptional networks and ROXY1/2 target proteins and pathways will help to address these questions.

9.4.3 ER/ERL1/ERL2 and MPK3/6 Regulate Both Anther Lobe Formation and Anther Cell Differentiation

Normal anther development relies on the coordinated cell proliferation and differentiation; specifically, cell-to-cell communication mediated by leucine-rich repeat receptor-like protein kinases (LRR-RLKs) plays essential roles in developing anthers. Studies in *Arabidopsis* strongly suggested that the cytoplasmic MAP kinases (MPK) signaling pathways are also important for anther lobe formation and anther cell division and differentiation (Shpak et al., 2004; Hord et al., 2008).

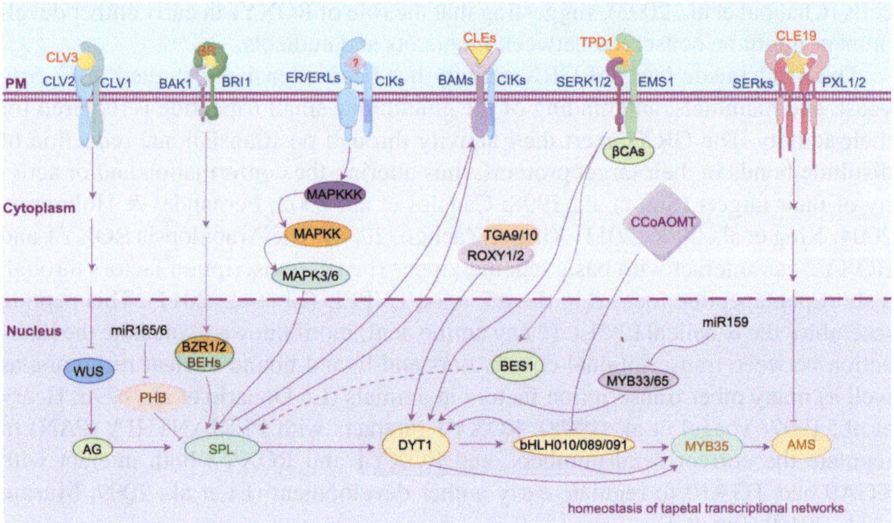

Fig. 9.4 Genetic interactions during anther development. Combined summary diagram showing the known genetic interactions between genes involved in stamen initiation and identity, anther morphogenesis, tapetum function, and pollen development. The genetic interactions are not necessarily direct. The CLVs-WUS-AG-SPL pathway promotes the initiation and identity of the anther, as well as the cell specification in anther lobes. The ER-MPK3/4 signaling cascade regulates both anther lobe formation and anther cell differentiation. The BAM1/2-SPL positive-negative feedback loops are required for the formation of PPC cells. The TPD1-EMS1-SERK1/2 ligand-receptor-co-receptor complex controls tapetum specification through BES1 and redundant transcription factors (TF). The DYT1-bHLH010/089/091-MYB35-AMS transcriptional pathway is crucial for tapetum function and pollen development. DYT1 and bHLH010/089/091 form Feed-forward and Positive Feedback Loops, and the CLE19-PXL1/2-SERKs ligand-receptor-co-receptor complex function as a "braking" system to suppress the expression of *AMS* and prevent harmful overexpression of the tapetum transcriptional network. Interactions between genes are shown by arrows for positive regulation, T-bars for negative regulation. Question marks indicate the unknown components in the pathways

The *Arabidopsis* ERECTA is an LRR-RLK that regulates organ shapes and inflorescence architecture. Landsberg *erecta*, one of the most widely used *Arabidopsis* ecotypes for molecular genetic studies, exhibits pleiotropic phenotypes, including compact inflorescence, blunt fruits, and short petioles, due to the *erecta* (*er*) mutation (Hwang et al., 1991; Anderson & Mulligan, 1992; Redel, 1992; Torii et al., 1996). In addition, ER has also been shown to function with two related LRR-RLKs, ER-Like 1 and 2 (ERL1, 2), to promote cell proliferation, control guard cell differentiation, direct ovule development, and regulate early anther development (Shpak et al., 2004, 2005; Hord et al., 2008). These three genes all possess 26 introns, and the ERL1 and ERL2 proteins share high overall sequence identity to ER (60% identity and 72% similarity), particularly in the LRR region and the kinase domain (Shpak et al., 2004). Phylogenetic analyses showed that ERL1 and ERL2 are the products of a recent duplication and are the closest paralogs of ER, and the three constitute a subfamily of LRR-RLKs (Shiu & Bleecker, 2001; Yin et al., 2002).

In floral organs, these three genes display overlapping expression patterns, with all three genes expressed in all flowering organs at early floral stages, but become differentially expressed in subsets of proliferating tissues: *ER* expressed largely in the mesocarp, *ERL1* predominantly in ovules, and *ERL2* in the style and ovules (Shpak et al., 2004). Although the *erl1* and *erl2* null single mutant alleles are distinguishable from wild-type plants, both *ERL1* and *ERL2* can rescue the *erecta* phenotype when expressed under the *ER* promoter and terminator, suggesting these receptor kinase proteins have similar functions in the regulation of organ size and flower development. More importantly, the *er-105 erl1-2 erl2-1* triple mutant showed dramatically defective anther development; however, the *er-105, erl1-2, erl2-1* single mutants, *er-105 erl1-2, er-105 erl2-1, erl1-2 erl2-1* double mutants, and *er-105 erl1-2 (+/−) erl2-1* mutant all produce anthers with a comparable size to wild type and with round and uniform pollen grains inside (Hord et al., 2008). These results further support the functional redundancy between these three genes in the regulation of anther development.

In the *er-105 erl1-2 erl2-1* triple mutant, many flowers produce no stamens, and other flowers produce fewer stamens with a range of anther defects (Hord et al., 2008). The most severe defect is filament-like stamens completely lacking an anther; some stamens have undifferentiated oval or kidney-shaped anthers. Only a few stamens produce viable pollen in one or two anther lobes at abaxial positions (Hord et al., 2008). These results indicated that the ER/ERL1/ERL2 RLKs function together to regulate both anther lobe formation and anther cell differentiation. These genes are probably required for the specification of the AR cells and/or promote the division of AR cells to form the multiple cell layers in each lobe.

In addition, the *mpk3 mpk6* double mutants defective in the *MPK3/6* MAP kinases genes show anther defects similar to those of the *er erl1 erl2* anthers, such as the formation of only two anther lobes sometimes, and abnormal cell patterning in the anther lobe (Shpak et al., 2004; Hord et al., 2008). A reasonable hypothesis is that the ER/ERL1/ERL2 RLKs may mediate signals necessary for anther lobe formation at the cell surface, and that the protein kinase cascade involving MPK3/MPK6 acts downstream of ER/ERL1/2. Both ER/ERLs and MPK3/6 are also key regulators of guard cell patterning and important for ovule development (Shpak et al., 2005; Pillitteri et al., 2007; Wang et al., 2008; Murphy et al., 2015); therefore, ER/ERL1/ERL2 RLKs and MPK3/6 might be parts of a regulatory pathway with important roles in several different developmental processes.

9.4.4 BAM1/BAM2 Are Required for the Formation of PPC Cells

The BARELY ANY MERISTEM1 (BAM1) and BAM2 are LRR-RLKs phylogenetically related to CLAVATA1 (CLV1), which is a key regulator of meristem size. BAM1/2 share high levels of amino acid sequence identity and form a four-gene monophyletic clade along with CLV1 and BAM3 (Shiu & Bleecker, 2001; DeYoung et al., 2006; Toriba et al., 2011). The *BAM1/2* genes are both expressed in the archesporial cells at anther stage 2 and in the PPCs and PSCs at anther stage 3, but only predominantly in the sporogenous cells at stage 4 and PMCs at stage 5. Their expression increased in the tapetum and decreased in the reproductive cells at stage 6 ~ 7, and no expression is detectable after stage 9 (DeYoung et al., 2006; Hord et al., 2006). The very similar expression patterns of *BAM1/2* suggest their potentially important but redundant role in anther development. Indeed, the *bam1/2* double mutant plants are male sterile due to defective early anther development. The wild-type anther forms the PPCs and PSCs in each lobe at stage 3, and sporogenous cells become visible at stage 4 at the center of each lobe while the PSCs divide to form the two SPC layers. However, in the *bam1/2* double mutant anthers, fewer and larger cells are produced in each lobe at stage 3, and the anther appears wider with no apparent cell layers at stage 4, compared to that of the wild type. At stage 5, the inner three somatic cell layers are replaced by cells resembling pollen mother cells (PMC) in the *bam1/2* double mutant anthers, although such PMC-like cells fail to complete meiosis (Hord et al., 2006). These together suggest that cell differentiation was affected in the *bam1/2* double mutant, probably since the earliest differentiation of the parietal cells.

As *BAM1/BAM2* are likely required in promoting the differentiation of parietal cells, and *SPL* was shown to be crucial for the initiation of anther cell differentiation, it is important to investigate the functional and genetic relationship between *BAM1/2* and *SPL*. Interestingly, compared with that in the wild-type anther, the expression of *SPL* expands to all subepidermal cells in the *bam1/2* double mutant anthers, indicating that the BAM1/2-dependent pathway normally limits the region of *SPL* expression. In turn, *BAM1/BAM2* expression is reduced in the *spl* mutant, suggesting SPL promotes *BAM1/2* expression (Hord et al., 2006). These observations together support a positive–negative feedback regulatory loop between BAM1/2 and SPL (Fig. 9.4). Through this regulatory loop, the balance between the reproductive cell and the somatic cell fates during anther development can be precisely controlled.

9.4.5 CIKs Functions Possibly as Co-receptors in Regulating Early Anther Development

Recent studies revealed that four CLAVATA3 INSENSITIVE RECEPTOR KINASES (CIK1/2/3/4) function with BAM1/2 and PRK2, possibly acting as co-receptors, to regulate early anther development in *Arabidopsis* (Cui et al., 2018). The *cik2/3*, *cik1/2/3*, and *cik1/2/3/4* mutant anthers showed increasingly severe anther defects similar to *bam1/2* mutants: 28% of the *cik2/3* double, half of the *cik1/2/3* triple, and most of the *cik1/2/3/4* quadruple anthers produce AR cells that fail to divide properly, resulting in defective cell fate determination and lack of proper PPCs. Therefore, the mutant anthers form excess microsporocytes and lack one to three somatic cell layers in each lobe. In addition, CIKs directly interact with and can be phosphorylated by BAM1/2. These results suggested that CIKs are redundantly required for cell fate determination in the early anther stage, possibly as co-receptors of BAM1/2. Moreover, only the *cik2/3* double mutant, but not the other double mutants of these *CIK* genes, showed defective anthers, and *cik1/2/3* showed much severe anther defects than *cik2/3/4*. These findings suggest that the *CIK1*, *CIK2*, *CIK3*, and *CIK4* genes may have partially distinctive functions in regulating anther development, possibly through interacting with both common and different RLKs to form multiple and diverse receptor-co-receptor complexes. Further studies are needed to reveal the redundant and distinctive functions of each CIK and uncover the diverse receptor-co-receptor complexes in the cell specification and differentiation aspects during early anther development.

9.4.6 Maize MAC1 Promotes the Differentiation of PPCs

MAIZE MULTIPLE ARCHESPORIAL CELLS 1 (MAC1) is a protein that is secreted by maize archesporial (AR) cells and stimulates the differentiation of PPCs into the endothecium cell layers and the secondary parietal cells (SPCs) (Wang et al., 2012). The fact that loss of MAC1 function causes excess AR cells in the maize anther suggests that MAC1 represses AR proliferation and/or inhibits the differentiation of PPCs. The observation that AR cells begin to proliferate earlier in the *mac1* mutant than in the wild type (Wang et al., 2012) also suggests that MAC1 acts on multiple cell types over an extended period during early lobe development. Other studies on MAC1 suggest that it also plays a role in the regulation of tapetum formation and differentiation (Sheridan et al., 1999; Al-Sady, 2008; Zhao et al., 2008), similar to the functions of MAC1 orthologs in both rice [*Oryza sativa TAPETUM DETERMINANT1-LIKE 1a* (*OsTDL1a*)] and *Arabidopsis* [TAPETUM DETERMINANT1 (TPD1)] (see next section for more detailed discussion).

9.5 The Development and Degeneration of Tapetum

As mentioned earlier, the tapetum is the innermost somatic anther cell layer that surrounds the microsporocytes or microspores directly. At the anther stage 5, tapetum cells are differentiated from the SPCs, which are in two layers with the inner layer forming the tapetum and the outer layer giving rise to the middle layer and the endothecium (Ma, 2005). Subsequently, the tapetal cells develop and undergo morphological changes, including characteristic changes of the organelles inside. The formation and differentiation of tapetum cells have been studied extensively in model plants like *Arabidopsis* and rice, as well as other species of the Brassicaceae family, including *Brassica* (Murgia et al., 1991; Owen & Makaroff, 1995; Piffanelli et al., 1997; Wang et al., 1997; Wu et al., 1997, 1999; Kim et al., 2001; Hsieh & Huang, 2005). Early after the formation of the tapetum cells, they possess abundant rough endoplasmic reticulum (RER) and secretory vesicles, apparently for active secretion of molecules into the locule for pollen development and pollen coat (exine) formation. At the mid-stage of anther development, the tapetum becomes a warehouse for temporary storage, as the tapetal cells are packed with two storage organelles, the elaioplasts and the tapetosomes. Elaioplasts are specialized plastids derived from protoplastids for temporary storage of steryl esters. In comparison, the tapetosomes are highly enriched with vesicles and oleosin-coated oil droplets, both of which are assembled in and then detached from the ER. During late anther development, the tapetal cells undergo programmed cell death (PCD) and discharge their steryl esters, oleosins, and alkanes onto the surface of pollen grains to form the pollen coat (Hsieh & Huang, 2005). Studies in recent decades have uncovered multiple genes and regulatory pathways essential for tapetum differentiation, development, or degeneration.

9.5.1 TPD1-EMS1/SERK1/2 Ligand-Receptor Complex Control Tapetum Specification

Similar to early anther development, cell-cell communication is also important for tapetum formation and differentiation, and a signaling pathway has been uncovered with an extracellular protein peptide signal and several LRR-RLKs serving as receptors and co-receptors, which mediate the communication between the tapetal precursor cells and the sporogenous cells (Canales et al., 2002; Zhao et al., 2002; Jia et al., 2008; Li et al., 2017). Members of this pathway include the secreted protein TAPETUM DETERMINANT1 (TPD1), the plasma membrane (PM)-distributed LRR-RLK EXCESS MICROSPOROCYTES1 (EMS1)/EXTRA SPOROGENOUS CELLS (EXS), and SOMATIC EMBRYOGENESIS RECEPTOR-LIKE KINASE 1 (SERK1) and SERK2 (Fig. 9.4). The *EMS1/EXS* gene is expressed in the tapetal precursors and developing tapetal cells, and mutants of this gene produce anthers lacking the tapetum while having extra PMCs (Canales et al., 2002; Zhao et al., 2002); for convenience, it is referred to as *EMS1* here. Nearly identical phenotypes of the absence of tapetal cells with extra PMCs have also been observed in the *tpd1* mutant and the *serk1/2* double mutant (Yang et al., 2003; Albrecht et al., 2005; Colcombet et al., 2005; Yang et al., 2005). Further cytological and biochemical studies on these genes and the corresponding proteins have revealed crucial support for their interactions: first, the highest *EMS1* expression is in the nascent tapetum, whereas that of TPD1 is in the PMCs; also TPD1 physically interacts with EMS1 and induces the phosphorylation of EMS1; in addition, EMS1 and SERK1/2 form a heteromeric receptor complex for the extracellular ligand TPD1; further, the transphosphorylation between SERK1 and EMS1 enhances EMS1 kinase activity (Albrecht et al., 2005; Li et al., 2017). These results together support the following model: the tapetum-promoting signal TPD1 is produced and secreted from the PMCs and perceived by EMS1 and the co-receptor SERK1/2, which are localized on the surface of tapetal precursors; the physical interaction of TPD1-EMS1/

SERK1/2 facilitates the communication between the anther somatic cell and PMCs by inducing the phosphorylation of EMS1 and subsequent transphosphorylation between EMS1 and SERK1/2. The signal transduction between the PMCs and the tapetal precursors promotes the tapetal fate and supports the completion of meiosis and subsequent development from microspores to pollen grains.

Besides the studies in *Arabidopsis*, TPD1 orthologs have been analyzed in rice (*OsTDL1a*) and maize (*MAC1*) (Sheridan et al., 1999; Zhao et al., 2008); in addition, the receptor of OsTDL1a in rice was identified as MSP1 (Zhao et al., 2008). Consistently, the *tdl1a, mac1*, and *msp1* mutants all showed generally similar phenotypes those of *ems1* and *tpd1* (Sheridan et al., 1999; Nonomura et al., 2003; Zhao et al., 2008; Yang et al., 2016), suggesting that the cell-cell communication mechanism for tapetum specification is conserved in flowering plants. However, there seems to be a difference between TPD1 in the eudicot model plant *Arabidopsis* and the homologs in the grasses rice and maize. TPD1 likely acts after the differentiation of SPCs when the cells of the endothecium and middle layers have already formed, and more specifically affects the tapetum. However, in rice and maize, these ligands are required at an earlier time and act over a longer period of time for the specification of the somatic cells (Wang et al., 2012), as mentioned in the above section. Although the receptor of MAC1 in maize has not been identified yet, it is possible that different RLKs might act at different stages to control the specification of SPCs and the tapetal cells.

The current knowledge on the downstream effectors of these ligand-receptor modules for tapetal cell specification is very limited, although *Arabidopsis* β-carbonic anhydrases (βCAs) have been reported as direct downstream targets of EMS1 (Huang et al., 2017). *βCAs* are expressed in the tapetum, and loss of *βCA* function causes tapetal cell differentiation defects, and overexpression of *βCAs* leads to anthers with extra tapetal cells. EMS1 can directly interact with and phosphorylate βCAs, and the phosphorylation modification of βCAs increases their activity to promote tapetal cell differentiation, possibly by regulating the pH of tapetal cell precursors.

9.5.2 Protein Kinase RPK2 Is Essential for the Formation of Tapetum and Middle Layer

Another RLK, RECEPTOR-LIKE PROTEIN KINASE2 (RPK2), is essential for the formation of both the tapetum and the middle layer (Mizuno et al., 2007). RNA in situ hybridization and RT-PCR analyses indicate that *RPK2* is highly expressed in the tapetum at anther stage 8 and 9, but not before stage 7 and after stage 10. The *rpk2* mutant plants are male sterile with multiple cellular defects, including hypertrophic tapetum, a lack of the middle layer, and collapse of anther lobes without dehiscence. In addition, the *rpk2* pollen mother cells could undergo meiosis and form tetrads, although their differentiation was subsequently defective. These results suggested that RPK2 positively regulates the differentiation of the inner

SPCs. The *bam1 bam2* mutant anthers were shown to lack the endothecium as well as the middle layer and tapetum (Hord et al., 2006), whereas the *rpk2* anthers showed failure to differentiate the middle layer (Mizuno et al., 2007) and the *tpd*, *ems1/exs,* and *serk1 serk2* mutant anthers lack the tapetum (Canales et al., 2002; Zhao et al., 2002; Yang et al., 2003; Albrecht et al., 2005; Colcombet et al., 2005). Therefore, it is likely that the RPK2- and EMS1/TPD1/SERK-mediated signaling processes function antagonistically to determine the fate of the inner parietal cell to differentiate into the middle layer or tapetum, respectively, downstream of BAM1/BAM2.

9.5.3 The DYT1-bHLH010/089/091-TDF1/MYB35-AMS *Genetic Pathway Controls Tapetum Development and Function*

Studies in the last two decades have uncovered a series of genes essential for tapetum development and function, especially those encoding transcription factors DYSFUNCTIONAL TAPTETUM1 (DYT1), bHLH010/089/091, MYB35/DEFECTIVE IN TAPETAL DEVELOPMENT AND FUNCTION1 (TDF1), ABORTED MICROSPORES (AMS), MYB103/MYB80/MS188, and MALE STERILITY1 (MS1); these form a regulatory pathway for tapetum development and functions that support the development of microspores (Fig. 9.4) (Wilson et al., 2001; Sorensen et al., 2003; Ma, 2005; Zhang et al., 2006; Wilson & Zhang, 2009; Ge et al., 2010; Xu et al., 2010, 2014; Chang et al., 2011; Zhu et al., 2011, 2015; Feng et al., 2012; Cui et al., 2016; Fu et al., 2020; Lai et al., 2022). Among these transcription factors, DYT1, bHLH010/089/091, MYB35/TDF1, and AMS are required for early tapetum development, and MYB103/MYB80/MS188 and MS1 are important for late tapetum function.

DYT1 is a basic helix-loop-helix (bHLH) transcription factor. It is regarded as a gatekeeper transcription factor in tapetal development in *Arabidopsis thaliana* (Zhang et al., 2006; Gu et al., 2014) and regulates over 1000 anther-expressed genes (Feng et al., 2012; Fu et al., 2020). The *dyt1* mutant exhibits abnormal anther morphology beginning at anther stage 4, with tapetal cells with excess and/or enlarged vacuoles. The meiocytes in the *dyt1* mutant can complete meiosis, but they lack the thick callose-rich cell wall and fail to complete meiotic cytokinesis (Zhang et al., 2006). Microarray- and RNA-seq-based transcriptomic analyses, respectively, revealed that the expression of 882 and 1124 genes exhibited over twofold changes in the *dyt1* anthers, including genes encoding transcription factors such as *MYB35*, *AMS*, and *MYB103/MYB80/MS188* (Feng et al., 2012; Fu et al., 2020), suggesting that the DYT1 acts as a "starter" of the tapetum transcriptional network. In addition, the rice homologs of *DYT1* [*UNDEVELOPED TAPETUM1* (*UDT1*)] and *AMS* [*TAPETUM DEGENERATION RETARDATION* (*TDR*)] have been identified and found to play similar roles, suggesting that the regulation of anther tapetum development and function is likely conserved in eudicots and monocots.

Using the in vitro SELEX (systematic evolution of ligands by exponential enrichment) approach, a consensus sequence TCACGTGA was found to bind to DYT1 strongly (Feng et al., 2012). However, only 12 genes differentially expressed in *dyt1* contain this consensus DYT1 binding site (Feng et al., 2012), suggesting these 12 genes are probably DYT1 direct targets and regulated by DYT1 homodimers. The fact that most of the DYT1 downstream genes lack the TCACGTGA DYT1 binding site suggests that some of these genes are likely regulated by protein complexes with DYT1 (such as heterodimers) and many others are not direct DYT1 targets. Further studies using a combination of ChIP-PCR, EMSA, and transcriptional activation assay revealed that DYT1-bHLH089 heterodimer directly recognizes and binds the CATGTG E-box and activates the expression of *MYB35*, *ATA20*, *EXL4*, *MEE48,* and 36 other genes with the CATGTG E-box in the promoter region (Cui et al., 2016; Fu et al., 2020).

bHLH010, bHLH089, and bHLH091 are highly similar in sequence and expression, and are redundantly required in the regulation of tapetum development and function (Zhu et al., 2015). The *bhlh010*, *bhlh089*, and *bhlh091* single mutants produce flowers with normal anther and pollen development; however, the corresponding double and triple mutants exhibit progressively more severe anther defects, including abnormal tapetum morphology, delayed callose degeneration, and aborted pollen development, indicating that they together are required for normal anther development (Zhu et al., 2015). Both the pollen exine and intine structures were severely defective in *bhlh010 bhlh089* mutant plants (Lai et al., 2022). Transcriptomic and metabolomic analyses revealed that bHLH010 and bHLH089 directly or/and indirectly regulate the expression of genes encoding metabolic enzymes and transporters related to pollen wall morphogenesis and the accumulation of metabolites involved in pollen wall morphogenesis, including polysaccharides, glyceryl derivatives, and flavonols. For instance, bHLH010 directly interacted with the two E-box sequences at the promoter of *CSLB03* and directly activated the expression of *CSLB03*, which is required for pollen development, and its single mutants show *bhlh010 bhlh089*-like pollen developmental defects (Lai et al., 2022).

The *Arabidopsis bHLH010*, *bHLH089*, and *bHLH091* genes are derived from the same ancestral gene, owing to two duplication events in the most recent common ancestor of Brassicaceae, with the first duplication generating *bHLH091* and the ancestor of *bHLH010* and *bHLH089*, and the second resulting in *bHLH010* and *bHLH089* (Zhu et al., 2015). These duplications are different from the independent gene duplication in Poaceae that generated the rice *ETERNAL TAPETUM1* (*EAT1*) and *TDR INTERACTING PROTEIN2* (*TIP2*) genes, which have distinct functions (Niu et al., 2013; Fu et al., 2014). Loss function mutant of either *EAT1* or *TIP2* is male sterile, indicating that they are each required for anther development. Mutation in *EAT1* causes defective programmed cell death of the tapetum, affecting a step downstream of *TDR* (the rice functional homolog of *AMS*) and directly regulating the expression of aspartic protease genes *OsAP25* and *OsAP37,* which can induce programmed cell death in both yeast and plants (Niu et al., 2013). In comparison, *TIP2* is required for early anther development by directly regulating the expression of *TDR* and *EAT1* (Fu et al., 2014). The functions of *EAT1* and *TIP2* are different, possibly because the duplication happened in the MRCA of grasses approximately

70 million years ago, allowing more time for *EAT1* and *TIP1* to diverge than the *bHLH010, bHLH089*, and *bHLH091* genes (Zhu et al., 2015).

The *MYB35* gene encodes an R2R3 MYB transcription factor. It is also named *DEFECTIVE IN TAPETAL DEVELOPMENT AND FUNCTION1 (TDF1)* because of its mutant exhibited irregular division and dysfunction of the tapetum (Zhu et al., 2008). *MYB35/TDF1* is highly expressed in the tapetum, meiocytes, and microspores during anther development, and acts downstream of DYT1 in the control of tapetum function and callose dissolution (Zhu et al., 2008; Feng et al., 2012). *MS1* (*MALE STERILITY1*) is expressed later than *MYB35/TDF1* and is dramatically reduced in the *myb35/tdf1* anther, whereas the expression of *MYB35/TDF1* is not changed in the *ms1* mutant. Therefore, *MYB35/TDF1* acts downstream of *DYT1* but upstream of *MS1* (Zhu et al., 2008, 2011; Feng et al., 2012). The binding of DYT1 to the E-box site in the *MYB35/TDF1* promoter is required, but not sufficient for the high-level *MYB35/TDF1* expression, which requires an interaction between DYT1 and other bHLH transcription factors, such as bHLH089 (Gu et al., 2014; Cui et al., 2016).

AMS is another member of the bHLH gene family and functions downstream of *DYT1* and *MYB35/TDF1*. AMS binds in vitro to a 6-bp consensus motif CANNTG, and directly regulates the expression of 23 genes implicated in callose dissolution, fatty acids elongation, formation of phenolic compounds, and lepidic transporter; moreover, the *ams* mutant exhibits premature degeneration of the tapetum, defective microspore release, lack of sporopollenin deposition, and a dramatic reduction of phenolic compounds and cutin monomers (Sorensen et al., 2003; Xu et al., 2010, 2014). Analysis of gene expression in the *dyt1*, *tdf1*, and *ams* mutants revealed that another member of the MYB family, named as MYB80 or MYB103 by different authors, is further downstream than *AMS* and is required for tapetum and pollen development (Phan et al., 2011).

The *MS1* gene was one of the earliest identified *Arabidopsis* genes important for tapetum development (Wilson et al., 2001; Ito & Shinozaki, 2002). *MS1* encodes a PHD-containing transcriptional regulator and is required for normal expression of tapetum genes essential for pollen wall formation, including *MYB99* (Ito et al., 2007). In the *ms1* mutant, the early stages of tapetum development and meiosis in PMCs occur normally, and the tapetum then becomes abnormally vacuolated, and the microspores and tapetum degenerated (Wilson et al., 2001; Ito & Shinozaki, 2002; Ariizumi et al., 2005). MS1 acts downstream of MYB103/MYB80, and is a direct regulator of *MYB99* (Ito et al., 2007), which promotes the expression of genes for the biosynthesis of pollen coat components. The rice homolog of *MS1* [*PERSISTENT TAPETAL CELL1* (*PTC1*)] has been shown to play similar roles.

9.6 The Regulation of Tapetum Homeostasis

The tapetum, positioned as the innermost layer of the four layers of anther wall cells and directly adjacent to the microspores and developing pollen, plays a vital role in providing essential precursors for pollen wall formation. Maintaining the homeostasis of the tapetal functions supporting pollen wall biogenesis is crucial for normal pollen fertility, yet the mechanisms governing this homeostasis were unclear previously. Recent investigations reveal that the tapetum "gatekeeper" transcription factor DYT1 engages in multiple feed-forward and feedback loops with downstream bHLH transcription factors to regulate this homeostasis.

9.6.1 The DYT1-bHLH010/089/091 Feed-Forward and Positive Feedback Loop

As a tapetum gatekeeper TF, *DYT1* is expressed at an early stage of anther development (late stage 4 to early stage 5; Zhang et al., 2006). The DYT1 homodimers show a limited nuclear distribution, with most of the proteins remaining in the cytoplasm (Cui et al., 2016). The basal level of DYT1 in the nucleus could activate low-level expression of *bHLH010/089/091*, whose protein products then physically interact with DYT1 through the C-terminal BIF (the bHLH protein interaction and function) domain and facilitate the nuclear localization of DYT1, so that the expression of DYT1-*bHLH010/089/091* downstream genes is increased.

In addition, different DYT1-bHLH heterodimers bind to the E-boxes of various downstream targets and activate the expression of specific downstream genes. These feed-forward loops ensure the temporal expression specificity of a large number of genes in the tapetum. Specifically, the DYT1-bHLH010/bHLH089/bHLH091 feed-forward regulatory loops control the complex and precise tapetal gene expression patterns in both *Arabidopsis* and rice (Niu et al., 2013; Zhu et al., 2015; Cui et al., 2016; Wang et al., 2017). First, DYT1 activates the expression of *bHLH010*, *bHLH089*, and *bHLH091*, whose protein products form heteropolymer with DYT1 to directly regulate the expression of downstream genes, including *MYB35/TDF1, ATA20, EXL4,* and *MEE48,* in a positive feed-forward loop (Cui et al., 2016; Fu et al., 2020). Furthermore, DYT1 and several other protein products of DYT1-regulated genes, including *MYB35/TDF1*, *AMS*, *MYB103*, and *MS1*, coregulate additional downstream genes; there are other possible feed-forward loops in anther development, although understanding of specific mechanisms for such regulatory loops will need further analyses (Fig. 9.5).

Moreover, feed-forward loops are also found among other tapetal transcription factors in *Arabidopsis* and rice. For instance, MYB35/TDF1 directly regulates the expression of *AMS* via an AACCT cis-element, and the *AMS* protein products interact with MYB35/TDF1 and, in a feed-forward loop, the AMS-MYB35/TDF1 complex activates *EXPB5* expression (Lou et al., 2018). Similarly, TIP2 directly regulates the expression of the *TDR* gene and interacts with the TDR protein in rice (Fu et al., 2014), suggesting a feed-forward loop with the TIP2 and TDR transcription factors. It is likely that the feed-forward loops among the transcription factors occur frequently in the tapetum and are important for the precise spatiotemporal expression of downstream anther genes to promote normal anther development under varying environmental conditions.

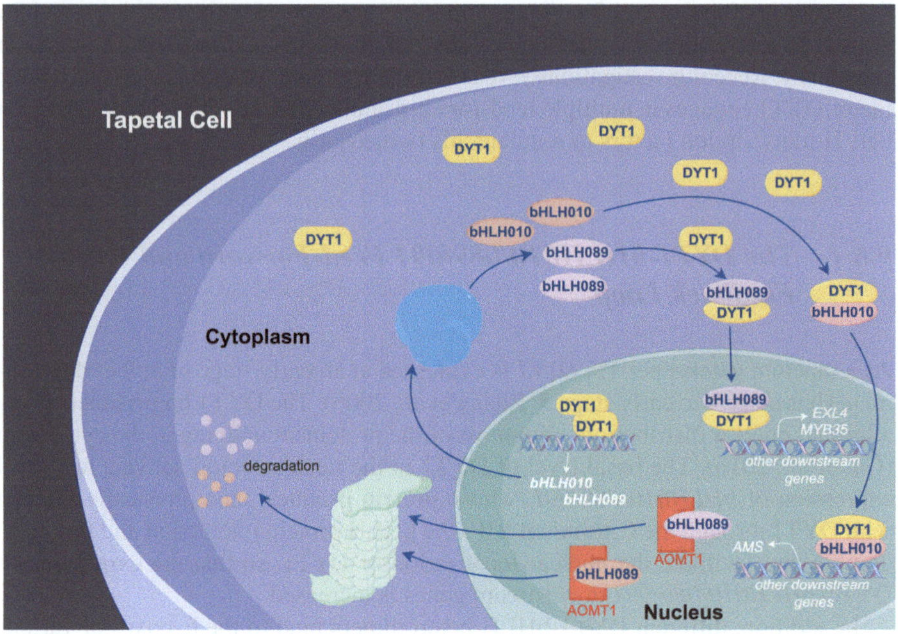

Fig. 9.5 A model showing regulatory loops in maintaining the homeostasis of DYT1-bHLH010/089/091 transcriptional network in anther tapetum. The tapetum gatekeeper transcription factor DYT1 exhibited a constrained nuclear localization, predominantly residing within the cytoplasm. Nuclear DYT1 at basal levels activates the modest expression of *bHLH010/089/091*. Subsequently, the protein products of *bHLH010/089/091* engage in physical interactions with DYT1, facilitating its nuclear translocation, thereby establishing a positive feedback regulatory loop. This loop ensures swift activation of an array of downstream genes. Within the nucleus, bHLH010/089/091 forms diverse DYT1-bHLH heteropolymers, which selectively bind to distinct E-box motifs of downstream targets and activate the expression of specific genes. These intricate feed-forward loops contribute to the temporal specificity of gene expression in the tapetum, orchestrated in a temporally dependent manner. Furthermore, CCoAOMT1 serves as a post-translational negative modulator of bHLH010 and bHLH089. Through direct interactions, CCoAOMT1 promotes the nuclear-cytoplasm translocation and subsequent degradation of these bHLH factors in the cytoplasm. Consequently, the homeostasis of tapetum function and pollen development is maintained

9.6.2 CCoAOMT1 Negatively Regulates bHLH010 and bHLH089

Given the presence of the “positive-feedback” and “feed-forward” regulatory loops promoting tapetum development and function, it is necessary to have a “braking” system to prevent harmful overexpression of the tapetum transcriptional network and maintain its function at a proper level. Caffeoyl coenzyme A O-methyltransferase 1 (CCoAOMT1) has been reported to function as a lignin synthetase in vascular bundles and is involved in various metabolite synthetic pathways (Fellenberg et al., 2012). In a recent investigation, CCoAOMT1 has been identified as a post-translational negative modulator of bHLH010 and bHLH089 (Lai et al., 2022, 2023). *CCoAOMT1* is highly expressed in reproductive tissues, specifically within the tapetal cells of anthers, and is widely distributed across the cell plasma membrane, endoplasmic reticulum (ER), and cell nucleus. Transgenic plants overexpressing *CCoAOMT1* exhibit phenotypic traits resembling those of *bhlh010 bhlh089* double mutant plants, including abbreviated siliques, reduced pollen grain numbers, predominantly shrunken and aborted, and an exceptionally low pollen germination rate. bHLH010/bHLH089 transcription factors exhibited distinctive nuclear localization, while CCoAOMT1 facilitates their nuclear export and subsequent degradation. The conserved methylated site K259 on bHLH010/089 is identified as a pivotal target of CCoAOMT1. Mutation at this site (bHLH089K259A) disrupts the negative regulatory influence of CCoAOMT1, resulting in substantial accumulation of the bHLH089K259A protein within the cell nucleus. Moreover, expression of mutated bHLH089 (bHLH089K259A) in *bhlh010 bhlh089* double mutant plants fails to effectively restore the wild-type phenotype, but instead induces phenotypic abnormalities resembling the overexpression of bHLH089. Significantly, the abnormal pollen development caused by *bHLH089* overexpression can be rescued through the concomitant overexpression of *CCoAOMT1*. These results collectively elucidate the intricate molecular mechanism by which CCoAOMT1 maintains homeostasis in pollen development by reducing the nuclear accumulation of bHLH transcription factors.

9.6.3 CLE Peptides Are Microspore-Derived "Brake" Signal to Restrict Tapetal Activity

Although previous studies have uncovered multiple molecular mechanisms by which tapetal cells maintain homeostasis in their transcriptional programs for pollen wall formation, largely through positive and negative feedback loops, a central question remained unresolved: how do developing pollen grains (microspores) communicate their developmental status back to the tapetum, and how is this information integrated into the tapetal transcriptional network?

This long-standing gap in understanding the bidirectional communication between reproductive cells and somatic tapetal cells was addressed in a breakthrough study in 2017 by a team at Fudan University. The researchers demonstrated that seven CLAVATA3/EMBRYO SURROUNDING REGION-related (CLE) peptides, produced by developing microspores, function redundantly as diffusible "brake" signals that prevent overactivation of the tapetal transcriptional machinery (Wang et al., 2017). Among these, CLE19 emerges as the predominant effector. In a follow-up study, the same group showed that CLE19 is perceived by a tapetum-localized receptor-co-receptor complex, composed of PXY-LIKE1/2 (PXL1/2) and SOMATIC EMBRYOGENESIS RECEPTOR KINASEs (SERKs). Signal perception triggers intracellular pathways that restrict the expression of key transcription factors, thereby fine-tuning tapetal gene expression in response to microspore-derived cues (Yu et al., 2023).

9.6.3.1 Overcoming Functional Redundancy: A Milestone Through Antagonistic Peptide Technology

The *CLE* gene family was named after the founding members *CLAVATA3 (CLV3)* and *EMBRYO SURROUNDING REGION (ESR)-RELATED*. This family comprises at least 32 members in *Arabidopsis thaliana* (Cock & McCormick, 2001; Oelkers et al., 2008; Fiers et al., 2007; Jun et al., 2010). CLE peptides are proteolytically processed from larger precursor proteins (Oelkers et al., 2008). The mature peptide hormones are usually 12 to 14 amino acids in length, and interact with membrane-localized leucine-rich repeat receptor-like kinases (LRR-RLKs) to form ligand-receptor modules that mediate intercellular communication and are key regulators of plant growth, development, and environmental responses (Yu et al., 2025).

Despite substantial progress in elucidating CLE peptide functions across various developmental contexts (Fiers et al., 2007; Fiume & Fletcher, 2012; Xu et al., 2015; Takahashi et al., 2018), their roles in male reproduction remain poorly understood. This is largely due to the extensive functional redundancy among CLE family members and the complexity of CLE-receptor interactions. A major conceptual advance came in 2013, when Song et al. demonstrated that a Gly-to-Tyr substitution at position six (G6T) in CLE peptides abolishes their ability to activate downstream signaling, and overexpression of such mutated peptides causes dominant-negative phenotypes (Song et al., 2013). This strategy provides a conceptual and experimental framework for using inhibitory CLE peptide variants (dominant-negative CLEs, DN-CLEs) to investigate CLE function while partially overcoming functional redundancy.

However, the reliability of DN-CLEs remained contentious until recent biochemical evidence validated their molecular mechanism. Yu et al. (2023) demonstrated that CLE19^{G6T} peptide retains the ability to bind its receptor PXL1 and promotes the formation of the receptor-co-receptor complex, while failing to activate receptor phosphorylation or initiate downstream signaling (Yu et al., 2023). These findings validate CLE19^{G6T} as an authentic dominant-negative variant and offer a clear mechanistic rationale for employing DN-CLE peptides in dissecting CLE signaling pathways.

9.6.3.2 Function of CLE19 Signaling in Tapetum-Microspore Communication

Recent studies revealed that seven CLE family genes, *CLE9, CLE16, CLE17, CLE19, CLE41, CLE42,* and *CLE45,* are expressed during anther development. Transgenic plants expressing dominant-negative constructs for these genes all exhibited abnormal anther morphology and pollen abortion, indicating that these CLE peptides collectively contribute to male fertility (Wang et al., 2017).

Among them, CLE19 plays a pivotal role as a negative regulator of tapetum function, particularly in sporopollenin biosynthesis and pollen exine formation. The wild-type pollen grains are covered by a well-organized network-like lacunae structure. In *cle19* mutants and *DN-CLE19* transgenic lines, pollen grains exhibited thicker exine walls with broader muri and smaller lacunae units, where most of the lacunae units were filled with excessive sporopollenin-like deposits. In contrast, *CLE19*-overexpression (*CLE19*-OX) lines produced pollen grains with reduced and disorganized exine networks, marked by a lack of the characteristic reticulate architecture formed by bacula and tectum (Wang et al., 2017). These observations highlight that both deficient and excessive CLE19 activity disrupt pollen exine formation, suggesting that the precise CLE19 dosage is essential for proper pollen wall patterning.

Further genetic and transcriptomic analyses revealed that *CLE19* acts as a negative regulator of the tapetal transcription factor *AMS* and approximately 280 of its downstream targets, including key regulators such as *MYB103/MYB80* and *MS1*. Loss of *CLE19* function (either in *cle19* mutants or *DN-CLE19* transgenic plants) led to hyperactivation of the AMS transcriptional program and excessive deposition of sporopollenin, resulting in abnormal pollen wall formation. Conversely, *CLE19* overexpression resulted in suppressed *AMS* expression and produced pollen grains with defective exine architecture (Wang et al., 2017). Notably, the *ams* (+/−) heterozygous genotype partially rescued the pollen exine defects induced by *DN-CLE19*, providing genetic evidence for CLE19-mediated modulation of *AMS* activity (Wang et al., 2017). Together, these findings support a model in which CLE19 peptide functions as a diffusible "brake" signal to fine-tune *AMS* expression within the tapetum to maintain pollen exine homeostasis and normal male fertility.

9.6.3.3 PXL1/2 and SERKs Function as Receptors and Co-receptors to Transduce CLE19 Signals into Tapetal Cells

Although CLE peptides are well-established mediators of intercellular communication in plants, the mechanism by which extracellular CLE signals are perceived and transduced into intracellular responses to regulate tapetal transcriptional networks remained unclear until the identification of the CLE19 receptor-co-receptor complex, comprising PXL1/2 and SERK1/2/3 (Yu et al., 2023).

During pollen development, CLE19 binds directly to the extracellular leucine-rich repeat (LRR) domain of PXL1, triggering PXL1 phosphorylation and promoting its interaction with SERK co-receptors. This ligand-receptor interaction activates a signaling cascade that transduces the CLE19 signals derived from developing microspores into intracellular responses within the tapetal cells. Loss of function T-DNA insertion mutants of PXL1 and PXL2, as well as transgenic plants expressing a dominant-negative PXL1 variant (lacking the intracellular kinase domain), exhibit defective pollen exine architecture and upregulation of exine-related transcription factors and biosynthetic genes—phenotypes that closely resemble those of *DN-CLE19* lines. Similarly, *serk1 serk2* double and *serk1 serk2 serk3* triple mutants display similar defects, underscoring the critical role of PXL1/PXL2–SERK1/2/3 as the functional receptor-co-receptor complex for CLE19 during pollen wall formation. Importantly, expression of the DN-PXL1 variant suppresses the pollen exine defects caused by CLE19 overexpression, confirming that PXL1 kinase activity is required for downstream signaling (Fig. 9.4). Collectively, these findings firmly support that PXL1 and SERKs act as the receptor and co-receptor complex responsible for CLE19 signal transduction in tapetal cells during pollen exine formation (Yu et al., 2023).

9.6.3.4 Evolutionary Relationships and Ligand Specificity of PXL Receptors

PXL1, PXL2, and PX Y belong to a subclass of leucine-rich repeat receptor-like kinases (LRR-RLKs). These three members were generated by two major duplication events: an ancient duplication before the divergence of monilophytes and seed plants generated the ancestral PXL and PXY clades, followed by a more recent duplication in early eudicots that gave rise to PXL1 and PXL2. These RLKs appear to function redundantly or synergistically in supporting normal pollen development (Yu et al., 2023).

Biochemical evidences show that both PXL1 and PXL2 directly bind CLE19, with dissociation constants (Kd) of approximately 346 nM and 714 nM, respectively. In contrast, PXY binds CLE19 only weakly, suggesting that PXL1 and PXL2 are the principal CLE19 receptors in anther tissue. Consistently, only PXL1 and PXL2, but not PXY, are capable of forming ligand-induced receptor-co-receptor complexes with SERK1/2/3 in response to CLE19, highlighting the functional divergence of PXY. One plausible explanation for this divergence is that PXY has evolved to recognize distinct CLE peptides that act redundantly with CLE19 in regulating pollen wall biogenesis.

Supporting this model, six additional CLE peptides, CLE9, CLE16, CLE17, CLE41, CLE42, and CLE45, are expressed in developing anthers and exhibit functional redundancy with CLE19 in controlling pollen development. Among these, CLE41 emerges as a strong candidate ligand for PXY, as it plays a dual role in anther development and vascular patterning. CLE41 binds to PXY with high affinity (with dissociation constants (Kd) of ~33 nM), whereas its binding to PXL1 and PXL2 is much weaker (Kd values of 2.1 μM and 9.9 μM, respectively) (Yu et al., 2023). This ligand-receptor specificity suggests that CLE41 preferentially activates PXY, whereas CLE19 predominantly signals through PXL1 and PXL2. These findings support the concept that CLE–RLK signaling modules are developmentally repurposed across distinct tissues, with context-dependent ligand-receptor interactions conferring distinct functional outcomes (Yu et al., 2023) (Fig. 9.6).

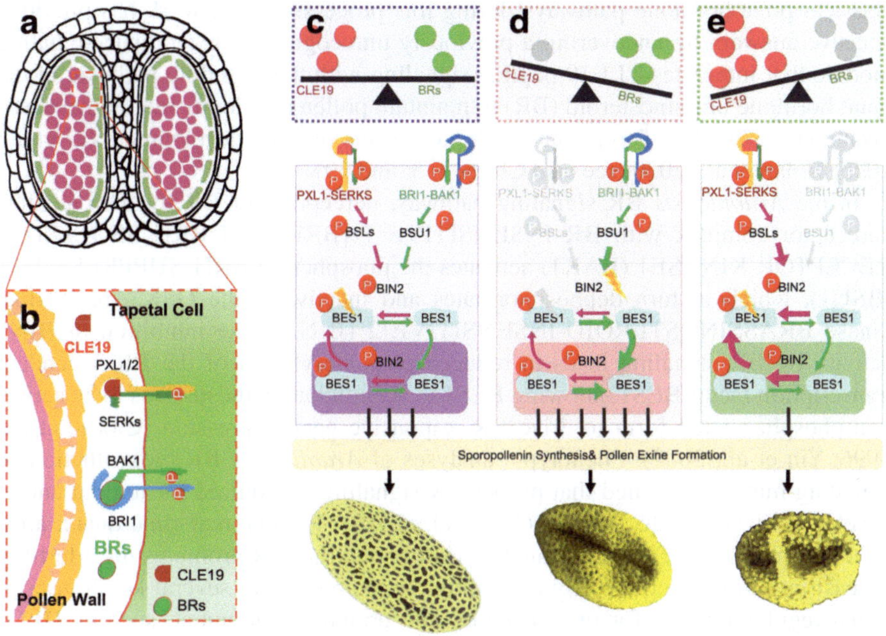

Fig. 9.6 A diagram illustrating the regulation of pollen exine developmental homeostasis by coordinated peptide and hormone signaling (Modified). Panel (**a**) shows a longitudinal-section of an *Arabidopsis* anther, highlighting pollen sacs containing developing pollen grains. Panel (**b**) depicts the interface between developing microspores and the surrounding tapetum, emphasizing the intercellular space where signaling molecules accumulate. CLE19 peptide secreted from microspores and BR signals originating from the tapetal layer are shown in this extracellular region. These signals are perceived by their respective receptor complexes at the tapetal plasma membrane, including the CLE19–PXL1–SERK and BR–BRI1–BAK1 modules. Panels (**c–e**) illustrate a conceptual model for maintenance of pollen exine developmental homeostasis. Panel (**c**) represents a balanced signaling state, in which CLE19 and BR signaling coordinately regulate tapetal transcriptional activity to support orderly pollen exine formation. CLE19 signaling constrains BR-induced transcriptional output through the PXL1–BSLs–BIN2–BES1 phosphorylation cascade. Panel (**d**) illustrates conditions of relatively low CLE19 signaling compared with BR activity, leading to enhanced BES1 nuclear accumulation and excessive or disorganized exine deposition. Panel (**e**) depicts conditions of greater CLE19 signaling relative to BR activity, in which increased BES1 phosphorylation and turnover impair tapetal transcriptional regulation and result in defective pollen exine formation. (Figures (**c–e**) are modified based on reference (Wang et al., 2025)

9.6.3.5 Antagonistic Signaling Between CLE19 and BR Maintains Pollen Developmental Homeostasis

A key unresolved question in CLE-mediated tapetum–microspore communication concerns how extracellular CLE19 peptide signals are transduced from the plasma membrane to the nucleus, and whether CLE peptide signaling intersects with classical phytohormone pathways during this process. Recent work has provided decisive answers by uncovering a previously unrecognized antagonistic signaling module that integrates CLE19 peptide signaling with the signaling pathway of the plant hormone brassinosteroid (BR) to maintain pollen developmental homeostasis (Wang et al., 2025). BR regulates multiple aspects of plant development (Li et al., 2025; Nolan et al., 2020; see also Chaps 3, 5, and 19).

In the *Arabidopsis* BR signaling pathway, perception of BR by the receptor-coreceptor complex with BR INSENSITIVE 1 (BRI1) and BRI1-SSOCIATED RECEPTOR KINASE1 (BAK1) activates the phosphatase BRI1 SUPPRESSOR 1 (BSU1), which in turn dephosphorylates and inactivates the GSK3/shaggy-like kinase BRASSINOSTEROID-INSENSITIVE 2 (BIN2). The inhibition of BIN2 activity by BR signaling leads to reduced phosphorylation of the BR-dependent transcription factor BES1, allowing BES1 to accumulate in the nucleus in its active, non-phosphorylated form to drive BR-responsive gene expression (Clouse et al., 1996; Yin et al., 2002). Phenotypic analyses of *Arabidopsis* BR biosynthetic and signaling mutants revealed that proper BR signaling is required for normal anther development and male fertility (Ye et al., 2010). Further real time-quantitative RT-PCR, microarray, and ChIP analyses supported that BR promotes male fertility at least in part through directly regulating the expression of several key transcriptional regulator genes for proper anther and pollen development, such as *DYT1*, *AMS*, *MYB103*, and *MS1* (Ye et al., 2010).

Importantly, CLE19 does not interfere with BR biosynthesis nor does it disrupt the formation or activity of the canonical BR ligand–receptor complex composed of BRI1 and BAK1. Instead, CLE19 antagonizes BR signaling downstream of membrane receptor activation, at the level of the shared intracellular BSLs–BIN2 signaling axis that controls the phosphorylation and activity of BES1 (Wang et al., 2025). CLE19 perception by its receptor PXL1 (Yu et al., 2023) actively recruits BSU1-LIKE 1/2/3 (BSL1/2/3) and BIN2. This recruitment enhances phosphorylation and activity of BIN2, as well as increased BES1 phosphorylation at specific serine residues (S219 and S223), promoting BES1 nuclear export and proteasomal degradation. Consequently, the transcriptional output of BES1 is attenuated, leading to reduced expression of core tapetal regulatory genes such as *MYB35/TDF1* and *AMS*, as well as their downstream gene networks (Wang et al., 2025). In this novel peptide–hormone antagonistic circuit, CLE19 suppresses BR signal output by modulating the phosphorylation status, subcellular localization, and stability of the key transcription factor BES1 through a shared intracellular BSL–BIN2 signaling module. Therefore, through this antagonistic yet coordinated regulation, CLE19 and BR act in opposition but converge on a common intracellular signaling axis to maintain BES1 nuclear abundance and activity within an optimal range. This finely tuned balance ensures stable transcriptional output in tapetal cells and safeguards pollen wall development and reproductive robustness.

9.7 Perspective

Molecular genetic studies in recent decades have identified numerous crucial regulatory molecules and signaling pathways that play important roles in the regulation of anther initiation, morphogenesis, and pollen development, providing a solid foundation in the understanding of these essential aspects of plant development and evolution and contributing to efforts to improve crops. However, many important questions still need to be addressed. For example: What are the earliest cues that trigger the spatiotemporal initiation of the anther primordia? How is the AR cell fate determined? What are the signals that mediate cell-cell communications governing tissue formation and specification during early anther stages? What are the receptors that perceive and mediate the CLE peptide signals in the tapetum? Further investigations using a combination of genetics, cell biology, biochemistry, and omics will likely lead to important discoveries. In addition, rapid developments of single-cell sequencing and advanced microscopy technologies to reveal cellular processes at a high resolution and in a dynamic manner will help to uncover cellular interactions from the cell surface to the nucleus in the developing anther.

Acknowledgments This work was supported by grants from the Ministry of Science and Technology, People's Republic of China (2023YFA0913500) and National Natural Science Foundation of China (32270347, 32470348). Figures 9.2, 9.3, 9.4, and 9.5 were drawn with Figdraw 2.0.

References

Albrecht, C., Russinova, E., Hecht, V., et al. (2005). The *Arabidopsis thaliana* SOMATIC EMBRYOGENESIS RECEPTOR-LIKE KINASES1 and 2 control male sporogenesis. *The Plant Cell, 17*, 3337–3349. https://doi.org/10.1105/tpc.105.036814

Allen, E., Xie, Z., Gustafson, A. M., & Carrington, J. C. (2005). microRNA-directed phasing during trans-acting siRNA biogenesis in plants. *Cell, 121*, 207–221. https://doi.org/10.1016/j.cell.2005.04.004

Al-Sady, B., Kikis, E. A., Monte, E., & Quail, P. H. (2008). Mechanistic duality of transcription factor function in phytochrome signaling. *Proceedings of the National Academy of Sciences of the United States of America, 105*(6), 2232–2237. https://doi.org/10.1073/pnas.0711675105. PMID: 18245378; PMCID: PMC2538903.

Anderson, M., & Mulligan, B. (1992). *Arabidopsis mutant collection methods in Arabidopsis research* (pp. 419–437). World Scientific. https://doi.org/10.1142/9789814439701_0018

Ariizumi, T., Hatakeyama, K., Hinata, K., et al. (2005). The HKM gene, which is identical to the MS1 gene of *Arabidopsis thaliana*, is essential for primexine formation and exine pattern formation. *Sexual Plant Reproduction, 18*, 1–7.

Bhatt, A. M., Canales, C., & Dickinson, H. G. (2001). Plant meiosis: The means to 1N. *Trends in Plant Science, 6*, 114–121. https://doi.org/10.1016/s1360-1385(00)01861-6

Bowman, J. L. (2000). The YABBY gene family and abaxial cell fate. *Current Opinion in Plant Biology, 3*, 17–22. https://doi.org/10.1016/s1369-5266(99)00035-7

Buchanan, B. B., & Balmer, Y. (2005). Redox regulation: A broadening horizon. *Annual Review of Plant Biology, 56*, 187–220. https://doi.org/10.1146/annurev.arplant.56.032604.144246

Canales, C., Bhatt, A. M., Scott, R., & Dickinson, H. (2002). EXS, a putative LRR receptor kinase, regulates male germline cell number and tapetal identity and promotes seed development in *Arabidopsis*. *Current Biology, 12*, 1718–1727. https://doi.org/10.1016/s0960-9822(02)01151-x

Chang, F., Wang, Y., Wang, S., & Ma, H. (2011). Molecular control of microsporogenesis in *Arabidopsis*. *Current Opinion in Plant Biology, 14*, 66–73. https://doi.org/10.1016/j.pbi.2010.11.001

Chaubal, R., Anderson, J. R., Trimnell, M. R., et al. (2003). The transformation of anthers in the msca1 mutant of maize. *Planta, 216*, 778–788. https://doi.org/10.1007/s00425-002-0929-8

Chitwood, D. H., Guo, M., Nogueira, F. T., & Timmermans, M. C. (2007). Establishing leaf polarity: The role of small RNAs and positional signals in the shoot apex. *Development, 134*, 813–823. https://doi.org/10.1242/dev.000497

Chitwood, D. H., Nogueira, F. T., Howell, M. D., et al. (2009). Pattern formation via small RNA mobility. *Genes & Development, 23*, 549–554. https://doi.org/10.1101/gad.1770009

Clouse, S. D., Langford, M., & McMorris, T. C. (1996). A brassinosteroid-insensitive mutant in Arabidopsis thaliana exhibits multiple defects in growth and development. *Plant Physiology, 111*(3), 671–678. https://doi.org/10.1104/pp.111.3.671

Cock, J. M., & McCormick, S. (2001). A large family of genes that share homology with CLAVATA3. *Plant Physiology, 126*, 939–942. https://doi.org/10.1104/pp.126.3.939

Colcombet, J., Boisson-Dernier, A., Ros-Palau, R., et al. (2005). Arabidopsis SOMATIC EMBRYOGENESIS RECEPTOR KINASES1 and 2 are essential for tapetum development and microspore maturation. *The Plant Cell, 17*, 3350–3361. https://doi.org/10.1105/tpc.105.036731

Cui, J., You, C., Zhu, E., et al. (2016). Feedback regulation of DYT1 by interactions with downstream bHLH factors promotes DYT1 nuclear localization and anther development. *The Plant Cell, 28*, 1078–1093. https://doi.org/10.1105/tpc.15.00986

Cui, Y., Hu, C., Zhu, Y., et al. (2018). CIK receptor kinases determine cell fate specification during early anther development in *Arabidopsis*. *The Plant Cell, 30*, 2383–2401. https://doi.org/10.1105/tpc.17.00586

DeYoung, B. J., Bickle, K. L., Schrage, K. J., et al. (2006). The CLAVATA1-related BAM1, BAM2 and BAM3 receptor kinase-like proteins are required for meristem function in *Arabidopsis*. *The Plant Journal, 45*, 1–16. https://doi.org/10.1111/j.1365-313X.2005.02592.x

Dinneny, J. R., Weigel, D., & Yanofsky, M. F. (2006). NUBBIN and JAGGED define stamen and carpel shape in *Arabidopsis*. *Development, 133*, 1645–1655. https://doi.org/10.1242/dev.02335

Emery, J. F., Floyd, S. K., Alvarez, J., et al. (2003). Radial patterning of *Arabidopsis* shoots by class III HD-ZIP and KANADI genes. *Current Biology, 13*, 1768–1774. https://doi.org/10.1016/j.cub.2003.09.035

Fellenberg, C., van Ohlen, M., Handrick, V., & Vogt, T. (2012). The role of CCoAOMT1 and COMT1 in *Arabidopsis* anthers. *Planta, 236*, 51–61. https://doi.org/10.1007/s00425-011-1586-6

Feng, B., Lu, D., Ma, X., et al. (2012). Regulation of the Arabidopsis anther transcriptome by DYT1 for pollen development. *The Plant Journal, 72*, 612–624. https://doi.org/10.1111/j.1365-313X.2012.05104.x

Fernandes, A. P., & Holmgren, A. (2004). Glutaredoxins: Glutathione-dependent redox enzymes with functions far beyond a simple thioredoxin backup system. *Antioxidants & Redox Signaling, 6*, 63–74. https://doi.org/10.1089/152308604771978354

Fiers, M., Ku, K. L., & Liu, C. M. (2007). CLE peptide ligands and their roles in establishing meristems. *Current Opinion in Plant Biology, 10*, 39–43. https://doi.org/10.1016/j.pbi.2006.11.003

Fiume, E., & Fletcher, J. C. (2012). Regulation of *Arabidopsis* embryo and endosperm development by the polypeptide signaling molecule CLE8. *The Plant Cell, 24*, 1000–1012. https://doi.org/10.1105/tpc.111.094839

Fu, Z., Yu, J., Cheng, X., et al. (2014). The rice basic helix-loop-helix transcription factor TDR INTERACTING PROTEIN2 is a central switch in early anther development. *The Plant Cell, 26*, 1512–1524. https://doi.org/10.1105/tpc.114.123745

Fu, Y., Li, M., Zhang, S., et al. (2020). Analyses of functional conservation and divergence reveal requirement of bHLH010/089/091 for pollen development at elevated temperature in *Arabidopsis*. *Journal of Genetics and Genomics, 47*, 477–492. https://doi.org/10.1016/j.jgg.2020.09.001

Garcia, D., Collier, S. A., Byrne, M. E., & Martienssen, R. A. (2006). Specification of leaf polarity in Arabidopsis via the trans-acting siRNA pathway. *Current Biology, 16*, 933–938. https://doi.org/10.1016/j.cub.2006.03.064

Ge, X., Chang, F., & Ma, H. (2010). Signaling and transcriptional control of reproductive development in Arabidopsis. *Current Biology, 20*, R988–R997. https://doi.org/10.1016/j.cub.2010.09.040

Goldberg, R. B., Beals, T. P., & Sanders, P. M. (1993). Anther development: Basic principles and practical applications. *The Plant Cell, 5*, 1217–1229.

Gu, J. N., Zhu, J., Yu, Y., et al. (2014). DYT1 directly regulates the expression of TDF1 for tapetum development and pollen wall formation in *Arabidopsis*. *The Plant Journal, 80*, 1005–1013. https://doi.org/10.1111/tpj.12694

Heery, D. M., Kalkhoven, E., Hoare, S., & Parker, M. G. (1997). A signature motif in transcriptional co-activators mediates binding to nuclear receptors. *Nature, 387*, 733–736. https://doi.org/10.1038/42750

Hord, C. L., Chen, C., Deyoung, B. J., et al. (2006). The BAM1/BAM2 receptor-like kinases are important regulators of *Arabidopsis* early anther development. *The Plant Cell, 18*, 1667–1680. https://doi.org/10.1105/tpc.105.036871

Hord, C. L., Sun, Y. J., Pillitteri, L. J., et al. (2008). Regulation of *Arabidopsis* early anther development by the mitogen-activated protein kinases, MPK3 and MPK6, and the ERECTA and related receptor-like kinases. *Molecular Plant, 1*, 645–658. https://doi.org/10.1093/mp/ssn029

Hsieh, K., & Huang, A. H. (2005). Lipid-rich tapetosomes in Brassica tapetum are composed of oleosin-coated oil droplets and vesicles, both assembled in and then detached from the endoplasmic reticulum. *The Plant Journal, 43*, 889–899. https://doi.org/10.1111/j.1365-313X.2005.02502.x

Huang, J., Li, Z., Biener, G., et al. (2017). Carbonic anhydrases function in anther cell differentiation downstream of the receptor-like kinase EMS1. *The Plant Cell, 29*, 1335–1356. https://doi.org/10.1105/tpc.16.00484

Husbands, A. Y., Chitwood, D. H., Plavskin, Y., & Timmermans, M. C. (2009). Signals and prepatterns: New insights into organ polarity in plants. *Genes & Development, 23*, 1986–1997. https://doi.org/10.1101/gad.1819909

Hwang, I., Kohchi, T., Hauge, B. M., et al. (1991). Identification and map position of YAC clones comprising one-third of the *Arabidopsis* genome. *The Plant Journal, 1*, 367–374. https://doi.org/10.1046/j.1365-313x.1991.t01-5-00999.x

Ito, T., & Shinozaki, K. (2002). The MALE STERILITY1 gene of *Arabidopsis*, encoding a nuclear protein with a PHD-finger motif, is expressed in tapetal cells and is required for pollen maturation. *Plant & Cell Physiology, 43*, 1285–1292. https://doi.org/10.1093/pcp/pcf154

Ito, T., Wellmer, F., Yu, H., et al. (2004). The homeotic protein AGAMOUS controls microsporogenesis by regulation of SPOROCYTELESS. *Nature, 430*, 356–360. https://doi.org/10.1038/nature02733

Ito, T., Nagata, N., Yoshiba, Y., et al. (2007). *Arabidopsis* MALE STERILITY1 encodes a PHD-type transcription factor and regulates pollen and tapetum development. *Plant Cell, 19*, 3549–3562. https://doi.org/10.1105/tpc.107.054536

Jia, G., Liu, X., Owen, H. A., & Zhao, D. (2008). Signaling of cell fate determination by the TPD1 small protein and EMS1 receptor kinase. *Proceedings of the National Academy of Sciences of the United States of America, 105*, 2220–2225. https://doi.org/10.1073/pnas.0708795105

Jun, J., Fiume, E., Roeder, A. H., et al. (2010). Comprehensive analysis of CLE polypeptide signaling gene expression and overexpression activity in *Arabidopsis*. *Plant Physiology, 154*, 1721–1736. https://doi.org/10.1104/pp.110.163683

Kelliher, T., & Walbot, V. (2011). Emergence and patterning of the five cell types of the Zea mays anther locule. *Developmental Biology, 350*, 32–49. https://doi.org/10.1016/j.ydbio.2010.11.005

Kerstetter, R. A., Bollman, K., Taylor, R. A., et al. (2001). KANADI regulates organ polarity in *Arabidopsis*. *Nature, 411*, 706–709. https://doi.org/10.1038/35079629

Kidner, C. A., & Martienssen, R. A. (2004). Spatially restricted microRNA directs leaf polarity through ARGONAUTE1. *Nature, 428*, 81–84. https://doi.org/10.1038/nature02366

Kim, H. U., Wu, S. S., Ratnayake, C., & Huang, A. H. (2001). *Brassica rapa* has three genes that encode proteins associated with different neutral lipids in plastids of specific tissues. *Plant Physiology, 126*, 330–341. https://doi.org/10.1104/pp.126.1.330

Lai, Z., Wang, J., Peng, S. Q., & Chang, F. (2022). bHLH010/089 transcription factors control pollen wall development via specific transcriptional and metabolic networks in *Arabidopsis thaliana*. *International Journal of Molecular Sciences, 23*(19), 11683. https://doi.org/10.3390/ijms231911683

Lai, Z., Wang, J., Fu, Y., et al. (2023). Revealing the role of CCoAOMT1: Fine-tuning bHLH transcription factors for optimal anther development. *Science China. Life Sciences, 67*(3), 565–578. https://doi.org/10.1007/s11427-023-2461-0

Le Douarin, B., Nielsen, A. L., Garnier, J. M., et al. (1996). A possible involvement of TIF1 alpha and TIF1 beta in the epigenetic control of transcription by nuclear receptors. *The EMBO Journal, 15*, 6701–6715.

Lemaire, S. D. (2004). The glutaredoxin family in oxygenic photosynthetic organisms. *Photosynthesis Research, 79*(3), 305–318. https://doi.org/10.1023/B:PRES.0000017174.60951.74. PMID: 16328797.

Li, S., Lauri, A., Ziemann, M., et al. (2009). Nuclear activity of ROXY1, a glutaredoxin interacting with TGA factors, is required for petal development in *Arabidopsis thaliana*. *The Plant Cell, 21*, 429–441. https://doi.org/10.1105/tpc.108.064477

Li, S., Gutsche, N., & Zachgo, S. (2011). The ROXY1 C-terminal L**LL motif is essential for the interaction with TGA transcription factors. *Plant Physiology, 157*, 2056–2068. https://doi.org/10.1104/pp.111.185199

Li, Z., Wang, Y., Huang, J., et al. (2017). Two SERK receptor-like kinases interact with EMS1 to control anther cell fate determination. *Plant Physiology, 173*, 326–337. https://doi.org/10.1104/pp.16.01219

Liu, X., Huang, J., Parameswaran, S., et al. (2009). The SPOROCYTELESS/NOZZLE gene is involved in controlling stamen identity in *Arabidopsis*. *Plant Physiology, 151*, 1401–1411. https://doi.org/10.1104/pp.109.145896

Li, X., Li, J., Zabed, H. M., Li, J., Xiong, M., Shi, H., & Li, J. (2025). Manipulating brassinosteroid signaling pathway to genetically improve horticultural plants. *aBIOTECH, 6*(2), 328–345. https://doi.org/10.1007/s42994-025-00201-y

Lou, Y., Zhou, H. S., Han, Y., et al. (2018). Positive regulation of AMS by TDF1 and the formation of a TDF1-AMS complex are required for anther development in *Arabidopsis thaliana*. *The New Phytologist, 217*, 378–391. https://doi.org/10.1111/nph.14790

Ma, H. (2005). Molecular genetic analyses of microsporogenesis and microgametogenesis in flowering plants. *Annual Review of Plant Biology, 56*, 393–434. https://doi.org/10.1146/annurev.arplant.55.031903.141717

McConnell, J. R., Emery, J., Eshed, Y., et al. (2001). Role of PHABULOSA and PHAVOLUTA in determining radial patterning in shoots. *Nature, 411*, 709–713. https://doi.org/10.1038/35079635

Mizuno, S., Osakabe, Y., Maruyama, K., et al. (2007). Receptor-like protein kinase 2 (RPK 2) is a novel factor controlling anther development in *Arabidopsis thaliana*. *The Plant Journal, 50*, 751–766. https://doi.org/10.1111/j.1365-313X.2007.03083.x

Murgia, M., Charzynska, M., Rougier, M., & Cresti, M. (1991). Secretory tapetum of *Brassica oleracea L.*: Polarity and ultrastructural features. *Sexual Plant Reproduction, 4*, 28–35.

Murmu, J., Bush, M. J., DeLong, C., et al. (2010). Arabidopsis basic leucine-zipper transcription factors TGA9 and TGA10 interact with floral glutaredoxins ROXY1 and ROXY2 and are redundantly required for anther development. *Plant Physiology, 154*, 1492–1504. https://doi.org/10.1104/pp.110.159111

Murphy, K. M., Egger, R. L., & Walbot, V. (2015). Chloroplasts in anther endothecium of *Zea mays* (*Poaceae*). *American Journal of Botany, 102*, 1931–1937. https://doi.org/10.3732/ajb.1500384

Niu, N., Liang, W., Yang, X., et al. (2013). EAT1 promotes tapetal cell death by regulating aspartic proteases during male reproductive development in rice. *Nature Communications, 4*, 1445. https://doi.org/10.1038/ncomms2396

Nolan, T. M., Vukasinovic, N., Liu, D., Russinova, E., & Yin, Y. (2020). Brassinosteroids: Multidimensional regulators of plant growth, development, and stress response. *Plant Cell, 32*(2), 295–318. https://doi.org/10.1105/tpc.19.00335

Nonomura, K., Miyoshi, K., Eiguchi, M., et al. (2003). The MSP1 gene is necessary to restrict the number of cells entering into male and female sporogenesis and to initiate anther wall formation in rice. *The Plant Cell, 15*, 1728–1739. https://doi.org/10.1105/tpc.012401

Oelkers, K., Goffard, N., Weiller, G. F., et al. (2008). Bioinformatic analysis of the CLE signaling peptide family. *BMC Plant Biology, 8*, 1. https://doi.org/10.1186/1471-2229-8-1

Owen, H. A., & Makaroff, C. A. (1995). Ultrastructure of microsporogenesis and microgametogenesis in *Arabidopsis thaliana* (L.) *Heynh*. Ecotype *Wassilewskija* (*Brassicaceae*). *Protoplasma, 185*, 7–21.

Pekker, I., Alvarez, J. P., & Eshed, Y. (2005). Auxin response factors mediate *Arabidopsis* organ asymmetry via modulation of KANADI activity. *The Plant Cell, 17*, 2899–2910. https://doi.org/10.1105/tpc.105.034876

Phan, H. A., Iacuone, S., Li, S. F., & Parish, R. W. (2011). The MYB80 transcription factor is required for pollen development and the regulation of tapetal programmed cell death in *Arabidopsis thaliana*. *The Plant Cell, 23*, 2209–2224. https://doi.org/10.1105/tpc.110.082651

Piffanelli, P., Ross, J. H., & Murphy, D. J. (1997). Intra- and extracellular lipid composition and associated gene expression patterns during pollen development in *Brassica napus*. *The Plant Journal, 11*, 549–562. https://doi.org/10.1046/j.1365-313x.1997.11030549.x

Pillitteri, L. J., Bemis, S. M., Shpak, E. D., & Torii, K. U. (2007). Haploinsufficiency after successive loss of signaling reveals a role for ERECTA-family genes in *Arabidopsis* ovule development. *Development, 134*, 3099–3109. https://doi.org/10.1242/dev.004788

Redel, J. P. (1992). *A note on Columbia wild type and Landsberg erecta*. World Scientific.

Sanders, P. M., Bui, A. Q., Weterings, K., et al. (1999). Anther developmental defects in *Arabidopsis thaliana* male-sterile mutants. *Sexual Plant Reproduction, 11*, 297–322. https://doi.org/10.1007/s004970050158

Schiefthaler, U., Balasubramanian, S., Sieber, P., et al. (1999). Molecular analysis of NOZZLE, a gene involved in pattern formation and early sporogenesis during sex organ development in *Arabidopsis thaliana*. *Proceedings of the National Academy of Sciences of the United States of America, 96*, 11664–11669. https://doi.org/10.1073/pnas.96.20.11664

Sessions, R. (1997). Arabidopsis (Brassicaceae) flower development and gynoecium patterning in wild type and Ettin mutants. *American Journal of Botany, 84*, 1179.

Sessions, A., Nemhauser, J. L., McColl, A., et al. (1997). ETTIN patterns the Arabidopsis floral meristem and reproductive organs. *Development, 124*, 4481–4491. https://doi.org/10.1242/dev.124.22.4481

Sheridan, W. F., Golubeva, E. A., Abrhamova, L. I., & Golubovskaya, I. N. (1999). The *mac1* mutation alters the developmental fate of the hypodermal cells and their cellular progeny in the maize anther. *Genetics, 153*, 933–941. https://doi.org/10.1093/genetics/153.2.933

Shiu, S. H., & Bleecker, A. B. (2001). Receptor-like kinases from *Arabidopsis* form a monophyletic gene family related to animal receptor kinases. *Proceedings of the National Academy of Sciences of the United States of America, 98*, 10763–10768.

Shpak, E. D., Berthiaume, C. T., Hill, E. J., & Torii, K. U. (2004). Synergistic interaction of three ERECTA-family receptor-like kinases controls *Arabidopsis* organ growth and flower development by promoting cell proliferation. *Development, 131*, 1491–1501. https://doi.org/10.1242/dev.01028

Shpak, E. D., McAbee, J. M., Pillitteri, L. J., & Torii, K. U. (2005). Stomatal patterning and differentiation by synergistic interactions of receptor kinases. *Science, 309*, 290–293. https://doi.org/10.1126/science.1109710

Siegfried, K. R., Eshed, Y., Baum, S. F., et al. (1999). Members of the YABBY gene family specify abaxial cell fate in *Arabidopsis. Development, 126*, 4117–4128. https://doi.org/10.1242/dev.126.18.4117

Song, X. F., Guo, P., Ren, S. C., et al. (2013). Antagonistic peptide technology for functional dissection of CLV3/ESR genes in *Arabidopsis. Plant Physiology, 161*, 1076–1085. https://doi.org/10.1104/pp.112.211029

Sorensen, A. M., Krober, S., Unte, U. S., et al. (2003). The *Arabidopsis* ABORTED MICROSPORES (AMS) gene encodes a MYC class transcription factor. *The Plant Journal, 33*, 413–423. https://doi.org/10.1046/j.1365-313x.2003.01644.x

Sun, C., Berardi, M. J., & Bushweller, J. H. (1998). The NMR solution structure of human glutaredoxin in the fully reduced form. *Journal of Molecular Biology, 280*, 687–701. https://doi.org/10.1006/jmbi.1998.1913

Takahashi, F., Suzuki, T., Osakabe, Y., et al. (2018). A small peptide modulates stomatal control via abscisic acid in long-distance signalling. *Nature, 556*, 235–238. https://doi.org/10.1038/s41586-018-0009-2

Toriba, T., Suzaki, T., Yamaguchi, T., et al. (2010). Distinct regulation of adaxial-abaxial polarity in anther patterning in rice. *The Plant Cell, 22*, 1452–1462. https://doi.org/10.1105/tpc.110.075291

Toriba, T., Ohmori, Y., & Hirano, H. Y. (2011). Common and distinct mechanisms underlying the establishment of adaxial and abaxial polarity in stamen and leaf development. *Plant Signaling & Behavior, 6*, 430–433. https://doi.org/10.4161/psb.6.3.14494

Torii, K. U., Mitsukawa, N., Oosumi, T., et al. (1996). The *Arabidopsis* ERECTA gene encodes a putative receptor protein kinase with extracellular leucine-rich repeats. *The Plant Cell, 8*, 735–746. https://doi.org/10.1105/tpc.8.4.735

Voegel, J. J., Heine, M. J., Tini, M., et al. (1998). The coactivator TIF2 contains three nuclear receptor-binding motifs and mediates transactivation through CBP binding-dependent and -independent pathways. *The EMBO Journal, 17*, 507–519. https://doi.org/10.1093/emboj/17.2.507

Voinnet, O. (2009). Origin, biogenesis, and activity of plant microRNAs. *Cell, 136*, 669–687. https://doi.org/10.1016/j.cell.2009.01.046

Walbot, V., & Egger, R. L. (2016). Pre-meiotic anther development: Cell fate specification and differentiation. *Annual Review of Plant Biology, 67*, 365–395. https://doi.org/10.1146/annurev-arplant-043015-111804

Wang, T. W., Balsamo, R. A., Ratnayake, C., et al. (1997). Identification, subcellular localization, and developmental studies of oleosins in the anther of *Brassica napus. The Plant Journal, 11*, 475–487. https://doi.org/10.1046/j.1365-313x.1997.11030475.x

Wang, H., Liu, Y., Bruffett, K., et al. (2008). Haplo-insufficiency of MPK3 in MPK6 mutant background uncovers a novel function of these two MAPKs in *Arabidopsis* ovule development. *The Plant Cell, 20*, 602–613. https://doi.org/10.1105/tpc.108.058032

Wang, Z., Xing, S., Birkenbihl, R. P., & Zachgo, S. (2009). Conserved functions of *Arabidopsis* and rice CC-type glutaredoxins in flower development and pathogen response. *Molecular Plant, 2*, 323–335. https://doi.org/10.1093/mp/ssn078

Wang, C. J., Nan, G. L., Kelliher, T., et al. (2012). Maize multiple archesporial cells 1 (mac1), an ortholog of rice TDL1A, modulates cell proliferation and identity in early anther development. *Development, 139*, 2594–2603. https://doi.org/10.1242/dev.077891

Wang, S., Lu, J., Song, X. F., et al. (2017). Cytological and transcriptomic analyses reveal important roles of CLE19 in pollen exine formation. *Plant Physiology, 175*, 1186–1202. https://doi.org/10.1104/pp.17.00439

Wang, S., Zhang, S., Yu, Y., Wang, J., Wang, J., Li, M., Lu, J., Ye, J., Li, H., Liu, Y., Zhao, Y., Song, W., Dong, J., Li, J., Liu, C., Ma, H., & Chang, F. (2025). CLE19 suppresses brassinosteroid signaling output via the BSL-BIN2 module to maintain BES1 activity and pollen exine patterning in Arabidopsis. *Journal of Integrative Plant Biology, 67*(12), 3216–3230. https://doi.org/10.1111/jipb.70024

Wei, B., Zhang, J., Pang, C., et al. (2015). The molecular mechanism of sporocyteless/nozzle in controlling *Arabidopsis* ovule development. *Cell Research, 25*, 121–134. https://doi.org/10.1038/cr.2014.145

Wilson, Z. A., & Zhang, D. B. (2009). From *Arabidopsis* to rice: Pathways in pollen development. *Journal of Experimental Botany, 60*, 1479–1492. https://doi.org/10.1093/jxb/erp095

Wilson, Z. A., Morroll, S. M., Dawson, J., et al. (2001). The *Arabidopsis* MALE STERILITY1 (MS1) gene is a transcriptional regulator of male gametogenesis, with homology to the PHD-finger family of transcription factors. *The Plant Journal, 28*, 27–39.

Wu, S. S., Platt, K. A., Ratnayake, C., et al. (1997). Isolation and characterization of neutral-lipid-containing organelles and globuli-filled plastids from *Brassica napus* tapetum. *Proceedings of the National Academy of Sciences of the United States of America, 94*, 12711–12716. https://doi.org/10.1073/pnas.94.23.12711

Wu, S. S., Moreau, R. A., Whitaker, B. D., & Huang, A. H. (1999). Steryl esters in the elaioplasts of the tapetum in developing Brassica anthers and their recovery on the pollen surface. *Lipids, 34*, 517–523. https://doi.org/10.1007/s11745-999-0393-5

Xing, S., & Zachgo, S. (2008). ROXY1 and ROXY2, two Arabidopsis glutaredoxin genes, are required for anther development. *The Plant Journal, 53*, 790–801. https://doi.org/10.1111/j.1365-313X.2007.03375.x

Xing, S., Lauri, A., & Zachgo, S. (2006). Redox regulation and flower development: A novel function for glutaredoxins. *Plant Biology (Stuttgart, Germany), 8*, 547–555. https://doi.org/10.1055/s-2006-924278

Xing, S., Salinas, M., & Huijser, P. (2011). New players unveiled in early anther development. *Plant Signaling & Behavior, 6*, 934–938. https://doi.org/10.4161/psb.6.7.15668

Xu, J., Yang, C., Yuan, Z., et al. (2010). The ABORTED MICROSPORES regulatory network is required for postmeiotic male reproductive development in Arabidopsis thaliana. *The Plant Cell, 22*, 91–107. https://doi.org/10.1105/tpc.109.071803

Xu, J., Ding, Z., Vizcay-Barrena, G., et al. (2014). ABORTED MICROSPORES acts as a master regulator of pollen wall formation in *Arabidopsis*. *The Plant Cell, 26*, 1544–1556. https://doi.org/10.1105/tpc.114.122986

Xu, T. T., Ren, S. C., Song, X. F., & Liu, C. M. (2015). CLE19 expressed in the embryo regulates both cotyledon establishment and endosperm development in *Arabidopsis*. *Journal of Experimental Botany, 66*, 5217–5227. https://doi.org/10.1093/jxb/erv293

Yang, W. C., Ye, D., Xu, J., & Sundaresan, V. (1999). The SPOROCYTELESS gene of Arabidopsis is required for initiation of sporogenesis and encodes a novel nuclear protein. *Genes and Development, 13*, 2108–2117.

Yang, S. L., Xie, L. F., Mao, H. Z., et al. (2003). Tapetum determinant1 is required for cell specialization in the *Arabidopsis* anther. *The Plant Cell, 15*, 2792–2804. https://doi.org/10.1105/tpc.016618

Yang, S. L., Jiang, L., Puah, C. S., et al. (2005). Overexpression of TAPETUM DETERMINANT1 alters the cell fates in the *Arabidopsis* carpel and tapetum via genetic interaction with excess microsporocytes1/extra sporogenous cells. *Plant Physiology, 139*, 186–191. https://doi.org/10.1104/pp.105.063529

Yang, L., Qian, X., Chen, M., et al. (2016). Regulatory role of OsTDL1A-MSP1 signaling in specifying anther cell identity in rice. *Plant Physiology, 171*(3), 2085–2100.

Ye, Q., Zhu, W., Li, L., Zhang, S., Yin, Y., Ma, H., & Wang, X. (2010). Brassinosteroids control male fertility by regulating the expression of key genes involved in Arabidopsis anther and pollen development. *Proceedings of the National Academy of Sciences of the United States of America, 107*(13), 6100–6105. https://doi.org/10.1073/pnas.0912333107

Yin, Y., Wu, D., & Chory, J. (2002). Plant receptor kinases: Systemin receptor identified. *Proceedings of the National Academy of Sciences of the United States of America, 99*, 9090–9092. https://doi.org/10.1073/pnas.152330799

Yin, Y., Wang, Z. Y., Mora-Garcia, S., Li, J., Yoshida, S., Asami, T., & Chory, J. (2002). BES1 accumulates in the nucleus in response to brassinosteroids to regulate gene expression and promote stem elongation. *Cell, 109*(2), 181–191. https://doi.org/10.1016/s0092-8674(02)00721-3

Yu, Y., Song, W., Zhai, N., Zhang, S., Wang, J., Wang, S., Liu, W., Huang, C. H., Ma, H., Chai, J., & Chang, F. (2023). PXL1 and SERKs act as receptor-coreceptor complexes for the CLE19 peptide to regulate pollen development. *Nature Communications, 14*(1), 3307. https://doi.org/10.1038/s41467-023-39074-4

Yu, Y., Zhang, S., & Chang, F. (2025). CLE peptides: Key regulators of plant development and stress adaption. *Seed Biology, 4*, e009. https://doi.org/10.48130/seedbio-0025-0009

Zhang, W., Sun, Y., Timofejeva, L., et al. (2006). Regulation of *Arabidopsis* tapetum development and function by DYSFUNCTIONAL TAPETUM1 (DYT1) encoding a putative bHLH transcription factor. *Development, 133*, 3085–3095. https://doi.org/10.1242/dev.02463

Zhang, D., Luo, X., & Zhu, L. (2011). Cytological analysis and genetic control of rice anther development. *Journal of Genetics and Genomics, 38*, 379–390. https://doi.org/10.1016/j.jgg.2011.08.001

Zhao, D. Z., Wang, G., Speal, B., & Ma, H. (2002). The EXCESS MICROSPOROCYTES1 gene encodes a putative leucine-rich repeat receptor protein kinase that controls somatic and reproductive cell fates in the *Arabidopsis* anther. *Genes and Development, 16*, 2021–2031.

Zhao, X., Justina, D. P., Rowena, O., et al. (2008). OsTDL1A binds to the LRR domain of rice receptor kinase MSP1, and is required to limit sporocyte numbers. *The Plant Journal, 54*, 375–387.

Zhu, J., Chen, H., Li, H., et al. (2008). Defective in Tapetal development and function 1 is essential for anther development and tapetal function for microspore maturation in *Arabidopsis*. *The Plant Journal, 55*, 266–277. https://doi.org/10.1111/j.1365-313X.2008.03500.x

Zhu, J., Lou, Y., Xu, X., & Yang, Z. N. (2011). A genetic pathway for tapetum development and function in *Arabidopsis*. *Journal of Integrative Plant Biology, 53*, 892–900. https://doi.org/10.1111/j.1744-7909.2011.01078.x

Zhu, E., You, C., Wang, S., et al. (2015). The DYT1-interacting proteins bHLH010, bHLH089 and bHLH091 are redundantly required for *Arabidopsis* anther development and transcriptome. *The Plant Journal, 83*, 976–990. https://doi.org/10.1111/tpj.12942

Ziemann, M., Bhave, M., & Zachgo, S. (2009). Origin and diversification of land plant CC-type glutaredoxins. *Genome Biology and Evolution, 1*, 265–277. https://doi.org/10.1093/gbe/evp025

Chapter 10
Ovary and Ovule Development

Heming Zhao and Yuan Qin

Abstract The gynoecium is the general term for the carpel of angiosperms, located at the center of most flowers, and consists of stigma, style, and ovary. It plays the important function of mediating pollination as well as protecting ovules. After a double fertilization event, the gynoecium gives rise to fruits and seeds with great importance for providing a key nutrient source for humans. It is necessary to deeply understand the molecular mechanisms that control the gynoecium development in angiosperm species. The gynoecium development in plant is highly complex and orderly processes and is controlled by robust regulatory networks, which are being elucidated with extensive studies. In the past decades, many genes important for gynoecium development has been identified functionally, such as the genes encoding key transcription factors. More recently, it is revealed that plant hormones particularly auxin, cytokinin, and brassinolide serve important roles in shaping the gynoecium through communication among them and interactions with transcription factors.

Ovule is initiated from the placenta, a structure developed from the lateral margins of the carpels, and it is composed of the funiculus, the nucellus, the integument, and the female gametophyte. Since the ovules develop into the seeds after fertilization, it is crucial to understand the molecular mechanisms that control ovule development as they ultimately determine the final yield in crop plants. Ovule development originates from a protrusion on the edges of the septum of the carpel. A single hypodermal cell at the tip of the ovule primordium is specified into an archesporial cell. The archespore directly differentiates into megaspore mother cell (MMC) that undergoes further meiosis to produce four haploid megaspores. In most flowering

H. Zhao
Center for Crop Biotechnology, College of Agriculture, Anhui Science and Technology University, Chuzhou, China

Y. Qin (✉)
Fujian Provincial Key Laboratory of Haixia Applied Plant Systems Biology, Key Lab of Genetics, Breeding and Multiple Utilization of Crops, Ministry of Education, State Key Laboratory of Ecological Pest Control for Fujian and Taiwan Crops, Fujian Agriculture and Forestry University, Fuzhou, China
e-mail: yuanqin@fafu.edu.cn

F. Chang et al. (eds.), *Regulation of Plant Development*,
https://doi.org/10.1007/978-981-95-7033-1_10

plants, three megaspores near micropylar undergo programmed cell death, and only the chalazal-most megaspore continues to develop into the functional megaspore, which undergoes three rounds of nuclear division to form a coenocytic, eight nucleated embryo sacs. Subsequently, cellularization, nuclear migration, and polar nuclear fusion take place to yield ultimately a seven-celled embryo sac composed of one egg cell, two synergids, one central cell, and three antipodals. At early stages of the ovule development, cell-cell communication between the somatic and reproductive cells involving complex epigenetic and signaling networks controls the initiation, differentiation, and the number of megasporocyte in ovule primordium. It has been shown that an asymmetric auxin gradient plays a key role in embryo sac polarity, gametophytic cell specification, and female gametophyte patterning. Moreover, evidences from many studies have demonstrated that RNA processing and some key components of small RNA pathways are involved to regulate female gametic cell fate and ovule development.

Keywords Angiosperm · Ovary · Ovule · Female gametogenesis · Germline cell fate

10.1 Introduction

Most flowering plants produce a pistil or a gynoecium at the center of their flowers, which is the female reproductive structure composed of stigma, style, and ovary. After a double fertilization event, the pistil gives rise to fruits and seeds with great importance for the plant by protecting the growth of a new generation, and provides a key nutrient source for humans. In addition, the gynoecium is a highly complex plant organ and a model system for studying plant morphogenesis and development. In this section, we describe the gynoecium structure and its developmental process. The focus then lies on genes and signaling networks that control carpel marginal meristem development and gynoecium morphogenesis. The role of hormones, particularly auxin and cytokinin, during gynoecium development is further discussed and highlighted.

Ovule, a structure initiated from the lateral margins of the carpels, consists of four parts, the funiculus, the nucellus, the integument, and the female gametophyte. The haploid female gametophyte is embedded in the sporophytic nucellus and integuments, which form an opening structure called micropyle where pollen tube enters. Female gametophyte produces the female gametes, representing the site of fertilization, and is an attractive system for studies of cell fate specification, cell-cell communication, and programmed cell death. Proper ovule and female gametophyte development are critical for reproductive success. Here we introduce the ovule development processes, which involves both sporophytic and gametophytic tissue growth, discuss the molecular control of the female gametophyte development and highlight the role of cell-cell communication that takes place throughout ovule development.

10.2 Gynoecium Structure

The gynoecium in flowering plant is the most complex organ with different tissues and cell types, and is the female reproductive structure made up of one, two, or more fused or unfused carpels (Sundberg & Ferrándiz, 2009). The carpels generally harbor and protect one or several ovules. The mature gynoecium plays a critical role during double-fertilization in plants, including inducing pollen germination, guiding the pollen tubes toward the ovules through the transmitting tract, and facilitating the fusion between the egg and the sperm and the embryo development. In angiosperms, along its apical-basal axis, the mature gynoecium consists of four distinct pattern elements: the stigmatic tissue located apically, which is required for the reception and germination of pollen grains; followed by the style, which contains the apical portion of the transmitting tract; ovary that is the majority of the apical-basal extent of the gynoecium, containing the ovules internally; and the gynophore, the basal most symmetrical internode structure that connect the gynoecium to the base of the flower (Fig. 10.1a).

The ovary of the gynoecium in *Arabidopsis* is composed of two congenitally fused carpels that arise from the floral meristem, but the style and the stigmatic papillae are postgenitally fused structures. In the transversal plane, the medial-lateral, adaxial-abaxial, and radial patterns are developed and distinct tissues differentiate sequentially in ovary. A mature ovary shows three well distinct territories along the medial-lateral axis: the valves, corresponding to the two carpel walls that enclose the majority of the ovary chamber, the valve margins, and the replum located at the medial plane of the flower.

Both the medial and lateral domains of the ovary contain morphologically and/or functionally distinct structures and give rise to distinct cell types from their adaxial and abaxial portions (Fig. 10.1b, c). At the adaxial side of the hollow ovary, a meristematic medial ridge is formed along each medial domain, which is also called the carpel margin meristem (CMM). As development continues, the CMM can further differentiate into the carpel marginal tissues: the ovules, septum, transmitting tract, placenta, style, and stigma. The adaxial section of the medial domain is formed into the ovules, the transmitting tract, and the septum that divides the ovary into two locules. The abaxial section of the medial domain is formed into the replum, and the lateral domains into the valves (Fig. 10.1b, c).

At stage 7 of the wild type gynoecium development, the gynoecial primordium is an open-ended cylinder with distinct lateral and medial planes. Between stages 7 and 9, the style cells and the stigmatic papillae first differentiate along the medial plane, and then differentiate along the lateral plane. By stage 9, the stigma, style, and the valves of the ovary become distinct from each other. *SPT* and *CRC* genes are key for the differentiation of style cells in the medial plane, mutations in any of them leads to loss of postgenital fusion of the septum (Alvarez & Smyth, 1999).

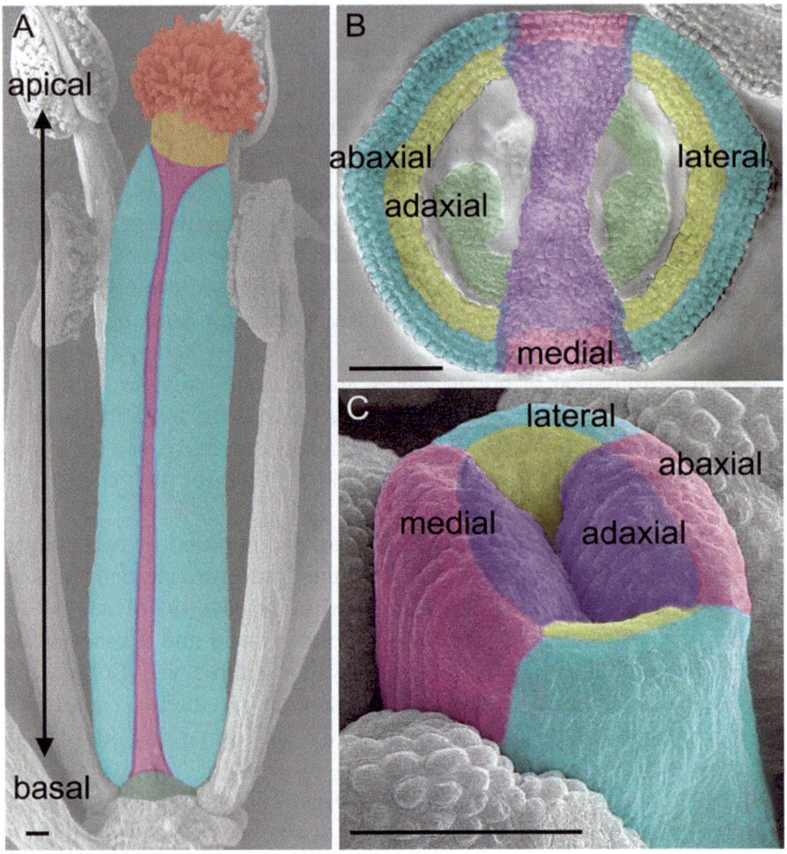

Fig. 10.1 Gynoecium structure in *Arabidopsis* (Larsson et al., 2013). (**a**) Mature gynoecium at floral stage 12; arrow indicates apical-basal axis. (**b**) Transverse section of a floral stage-11 gynoecium, indicating medial-lateral and abaxial-adaxial domains. (**c**) Young gynoecium at floral stage 7, indicating medial-lateral and abaxial-adaxial domains. (**a**) Red, stigma; mustard, style; blue and pink, ovary; green, gynophore. (**b**, **c**) Purple, medial adaxial domain; pink, medial abaxial domain; yellow, lateral adaxial domain; blue, lateral abaxial domain. (**b**) Green, ovules; purple, septum and transmitting tract. Bars, 50 μm. (Reprinted with permission from Oxford University Press)

10.3 Gynoecial Development

The development of flowers and gynoecial in *Arabidopsis* begins with the production of floral meristems and arises from the inflorescence meristem (Balanzá et al., 2006; Bowman & Smyth, 1999; Sundberg & Ferrándiz, 2009; Østergaard, 2009). After the formation of sepals, petals, and stamen primordia, the gynoecial primordia originate from the center of the flower from the remaining dome of floral meristem cells, which is at stage 6 of floral development (Smyth et al., 1990). Meanwhile, the basic patterns of medial-lateral, and inner (adaxial)-outer (abaxial) domains were established, leading to growth and differentiation of distinct tissues in the different regions. As is well-known, the gynoecium of *Arabidopsis* consists of two carpel organs that arise congenitally fused at the margins in this ring-like structure (Smyth et al., 1990). Two carpel walls form lateral domains and become valves. Internally, the margins of the fused carpels form a medial ridge, also known as the carpel margin meristem (CMM), which has meristematic activity for producing the marginal tissues.

The ring of gynoecium continues to develop into a tubular structure with an elliptical cross-section. In the early stages of gynoecial development, unique gene expression patterns can be used to distinguish the medial and lateral domains of the ovary, as well as the abaxial and adaxial positional identities (Fig. 10.1c). The cell proliferates in the medial domain and the two medial ridges of tissue begin to develop adaxially within the domain. The cells within the medial ridges retain a certain degree of meristematic potential, followed by the development of the ovules and the gynoecial septum (with associated transmitting tract) (Fig. 10.1b), and in addition likely conduce to the formation of styles and stigma tissues. As stigma cells and style start to develop at the top of the cylinder, the apical-basal patterning becomes apparent, and the cylinder closes at the top at floral stage 10. The cells in the abaxial portions of the medial domain produce the abaxial replum, while the cells in the lateral domains produce the valves. In the later stage of gynoecial development, the cells of the valve margin develop into the valve dehiscence zone required for seed propagation (Roeder & Yanofsky, 2006). At floral stage 12 (Fig. 10.1a), the gynoecia are at fully mature stage of development and ready to receive pollen.

10.4 Genes Involved in Carpel Marginal Meristem Development

The gynoecium is a highly complex reproductive organ, and its development requires the establishment of the three fundamental patterns: distinct apical-basal, medial-lateral and abaxial-adaxial pattern. Therefore, the development of gynoecium is a complex process that relies on the coordinated activity of a large number of genes, most of which have been identified in genetic and molecular studies of the model species *Arabidopsis thaliana* over the past few years (Balanzá et al., 2006; Østergaard, 2009; Roeder & Yanofsky, 2006).

CRABS CLAW (*CRC*) gene is early reported to play an important role for carpel morphogenesis (Alvarez & Smyth, 1999). The gynoecium in *crc* mutant is wider and shorter, its apex cannot fuse, and the number of style tissues decreases. *CRC* encodes a putative transcription factor belonging to the *YABBY* family and is essential for the development of both nectary and carpel (Alvarez & Smyth, 1999; Bowman & Smyth, 1999). Moreover, *CRC* also governs the abaxial-adaxial polarity establishment and cell fate determination in carpels (Eshed et al., 1999). *SPATULA* (*SPT*) encodes a putative *bHLH* family transcription factor, which is known to regulate the development of carpel marginal tissue in Arabidopsis. The gynoecium in *spt* mutant exhibits a slightly split apical region without the formation of transmitting tract, and the growth of the CMM and of the CMM-derived marginal tissues is also reduced. The double mutant *crc spt* shows a dramatically enhanced phenotype of almost completely unfused carpels and a dramatical reduction in marginal tissue formation (Alvarez & Smyth, 1999).

The MADS-box gene *AGAMOUS* (*AG*) is the key gene for the very early function in the whole gynoecium formation. The expression of *AG* could be detected from floral stage 3, and mutations in *AG* lead to a complete loss of gynoecium development. In the *ag* mutant, due to a loss of carpel identity and floral meristem determinacy, a new indeterminate flower is emerged to replace the gynoecium (Bowman et al., 1989). The *ag crc* double mutant is early reported to show an ectopic carpel development in the gynoecium organs (Alvarez & Smyth, 1999; Bowman & Smyth, 1999). Moreover, *AG* is involved to repress the transcription of the homeodomain transcription factor *WUSCHEL* (*WUS*), either directly or indirectly via *KNUCKLES* (*KNU*) that encodes a C2H2 zinc-finger transcription factor. Recently, it has been reported that *KNU* and *CRC* act together to repress the expression of *WUS*. The *knu crc* double mutant displays a dramatically enhanced phenotype of additional carpel whorls (Yamaguchi et al., 2017). These observations suggest that the adequate floral meristem termination is required for proper carpel marginal tissues development in gynoecia.

The other two redundant MADS-box genes, *SHATTERPROOF1* (*SHP1*) and *SHP2*, closely related to *AG*, are mainly involved in specifying valve margin identity (Liljegren et al., 2000). *SHP1* and *SHP2* redundantly control the differentiation of dehiscence zone, and the mature fruits in the *shp1 shp2* double mutant do not open (Liljegren et al., 2000). These genes are at the top of a series of transcription factors, including *INDEHISCENT* (*IND*) and *ALCATRAZ* (*ALC*) (Liljegren et al., 2004; Rajani & Sundaresan, 2001), which specify the identity of valve margin. *IND* takes responsibility for establishing a local minimum of auxin by coordinating auxin efflux from the separated layer cells, which is necessary for specifying the identity of these cells (Sorefan et al., 2009).

The *HECATE* genes (*HEC1*, *HEC2*, and *HEC3*) encode three putative transcription factors belonging to the bHLH family, it has been shown that they redundantly regulate the carpel development in *Arabidopsis*. The *hec1hec2hec3* triple mutant has a reduced stigma, a split style, and altered transmitting tract development, resulting in decreased seed set. Interestingly, the yeast two-hybrid experiments confirm that the three HEC proteins can interact with SPT to form dimers. The gyneocium in *hec1hec2hec3spt* quadruple mutant has a dramatically enhanced apical carpel fusion defect, no stigma, and reduced marginal tissue development (Schuster et al., 2015).

AINTEGUMENTA (*ANT*), a putative transcriptional regulator belonging to the *AP2* family, is required for the control of cell proliferation and play a pivotal role in carpel development, which is expressed specifically in the chalaza and in floral organ primordial (Elliott et al., 1996). The carpel development in *ant* single mutant shows only slight defects, and the development of gynoecium generally have fewer minor defects in stigmatic papillae and apical fusion (Elliott et al., 1996; Liu et al., 2000). However, the *crcant* double mutant has a completely split carpel, suggesting that the two genes redundantly control the carpel marginal tissues development in *Arabidopsis*. Moreover, *ANT*, *SEUSS* (*SEU*), *LEUNIG* (*LUG*), and *FILAMENTOUS FLOWER* (*FIL*) also exhibit partial functional redundancy in the marginal tissues differentiation (Azhakanandam et al., 2008; Liu et al., 2000; Nole-Wilson & Krizek, 2006; Franks et al., 2002; Sridhar et al., 2004). It has been suggested that these proteins can form a multimeric complex involved in carpel medial domain development (Azhakanandam et al., 2008; Nole-Wilson & Krizek, 2006; Nole-Wilson et al., 2010). Furthermore, the *crc ant shp1 shp2* quadruple mutant flowers have a decreased number of organs and the carpels lack almost completely the marginal tissues, suggesting that *CRC*, *ANT*, *SHP1*, and *SHP2* redundantly function in specifying carpel marginal tissues.

Among the transcriptional regulators directing carpel patterning, it has been shown that the *RAV* subfamily of the B3 transcription factor superfamily, the four *NGATHA* (*NGA*) genes, redundantly specify style development in a dosage-dependent manner. The *nga* quadruple mutants completely lack the development of style and stigma, forming a clearly undisturbed ovaries, and the apical region in gynoecium present an open structure with several protrusions of valve-like tissue (Alvarez et al., 2009; Trigueros et al., 2009). The other small family of unrelated factors is the SHI/STY family, which belongs to RING finger-like zinc finger motif transcription factors, and the mutations in the *SHI/STY* genes produce similar phenotypes with the *nga* quadruple mutants (Kuusk et al., 2006). *NGA* and *SHI/STY* genes also exhibit similar expression patterns and all interact with auxin, which include the apical gynoecium domain from stage 6 to stage 11–12 (Alvarez et al., 2009; Kuusk et al., 2002; Trigueros et al., 2009). The *nga3sty1* double mutant shows strong synergistic phenotypes similar to those observed in the *nga* quadruple mutant. Taken together, these data suggest that the *NGA* and *STY* factors might act cooperatively to promote carpel marginal tissues specification and formation in the apical gynoecium domain.

10.5 Auxin Is Required for Carpel Initiation and Gynoecium Development

It is well known that the plant hormone auxin controls many aspects of plant growth and development. More and more experimental evidences confirm that auxin plays an important role in gynoecium development, and the appropriate patterning relies on the establishment and maintenance of asymmetric auxin distribution, with the maximum of auxin at the apex (Balanzá et al., 2006; Larsson et al., 2014). It is widely reported that the mutations in the genes involved in biosynthesis, metabolism, polar transport, signaling, and response components of auxin pathway can lead to alterations in carpel development and gynoecium fundamental patterning, including a set of defects in carpel marginal tissues development. Mutations in the auxin biosynthesis *YUCCA* genes give rise to the defects in flower and gynoecium development. The ovules outside of the gynoecium and stigma-like tissues are observed in the *yuc1yuc4* double mutant, as well as in the *yuc1yuc2yuc4* and *yuc1yuc4yuc6* triple mutants (Cheng et al., 2006). Mutations in the auxin efflux transporter *PIN-FORMED* (*PIN*) genes give rise to abnormal flower and carpel development. The flower in *pin1* mutant displays an abnormal structure at the top of inflorescence axes, and the *pin3pin7* double mutant develop abnormal flower with missing carpels and stamens (Okada et al., 1991; Benková et al., 2003). *ETTIN* (*ETT*) encodes the auxin response factor 3 (ARF3), mutations in *ETT* lead to abnormal development in gynoecium, such as shortening of ovary region, increase in stylar and stigmatic regions, elongation of gynophore, and the abaxial (outward) proliferation of transmitting tract tissue (Cheng et al., 2013; Sessions & Zambryski, 1995). The *ETT* transcript were detected throughout the entire first stage floral meristems and subsequently in carpel primordia, suggesting that *ETT* participates in the regional identity and differentiation of the apical and basal boundaries in the gynoecium primordium (Sessions, 1997).

To a better understanding of auxin dynamics during early gynoecium development, expression patterns of markers and genes related to the auxin pathway were systematically observed. The combined use of synthetic auxin-responsive promoter fused with reporter genes facilitates the visualization of auxin biosynthesis, flow, and signaling sites, such as DR5::GUS/GFP constructs and highly active synthetic auxin response elements. It has been known that DR5rev:GFP strongly express at two lateral foci in the apical tips of the lateral domains at stage 7 in wild-type gynoecium (Fig. 10.2a, e). Starting at stage 8, two weaker foci of expression were detected at the apical tips of the medial domain (Fig. 10.2b, f). At early stage 9, the four DR5rev:GFP response foci begin to expand until DR5rev:GFP signal was seen as a ring encircling the apex of the gynoecium by stage 10 (Fig. 10.2c, d, g) (Larsson et al., 2013; Schuster et al., 2015; Girin et al., 2011), suggesting that the ring-like structure of auxin response in the apical region of the gynoecium may be necessary for the fusion of the two carpel apexes and the formation of style and stigma. The gynoecia in *hec1,2,3* triple mutant fail to establish this radial ring-shape DR5 expression pattern and retain the disturbed pattern with four prominent and isolated auxin signaling foci (Fig. 10.2j) (Schuster et al., 2015).

According to the localization analysis of the auxin efflux carriers PINs in the young developing gynoecium, it has been suggested that PIN1 is mainly necessary for polar auxin transport of the lateral region from the base to the gynoecium apex via the adaxial and abaxial epidermises, resulting in the two initial DR5 lateral foci (Larsson et al., 2014). This indicate that PIN1 mediated auxin flow is necessary for the differentiation and the outgrowth of the lateral domains. Moreover, it has also been proposed that the proper auxin drainage in the lateral domains play an important role in the correct apical-basal patterning of the gynoecium (Larsson et al., 2014). Interestingly, both PIN1 and PIN3 localization in the gynoecial apex and the medial region appear to be apolar, which can explain the formation of the ring-shape DR5 signal at the apex. The formation of ring-like auxin structure at the gynoecial

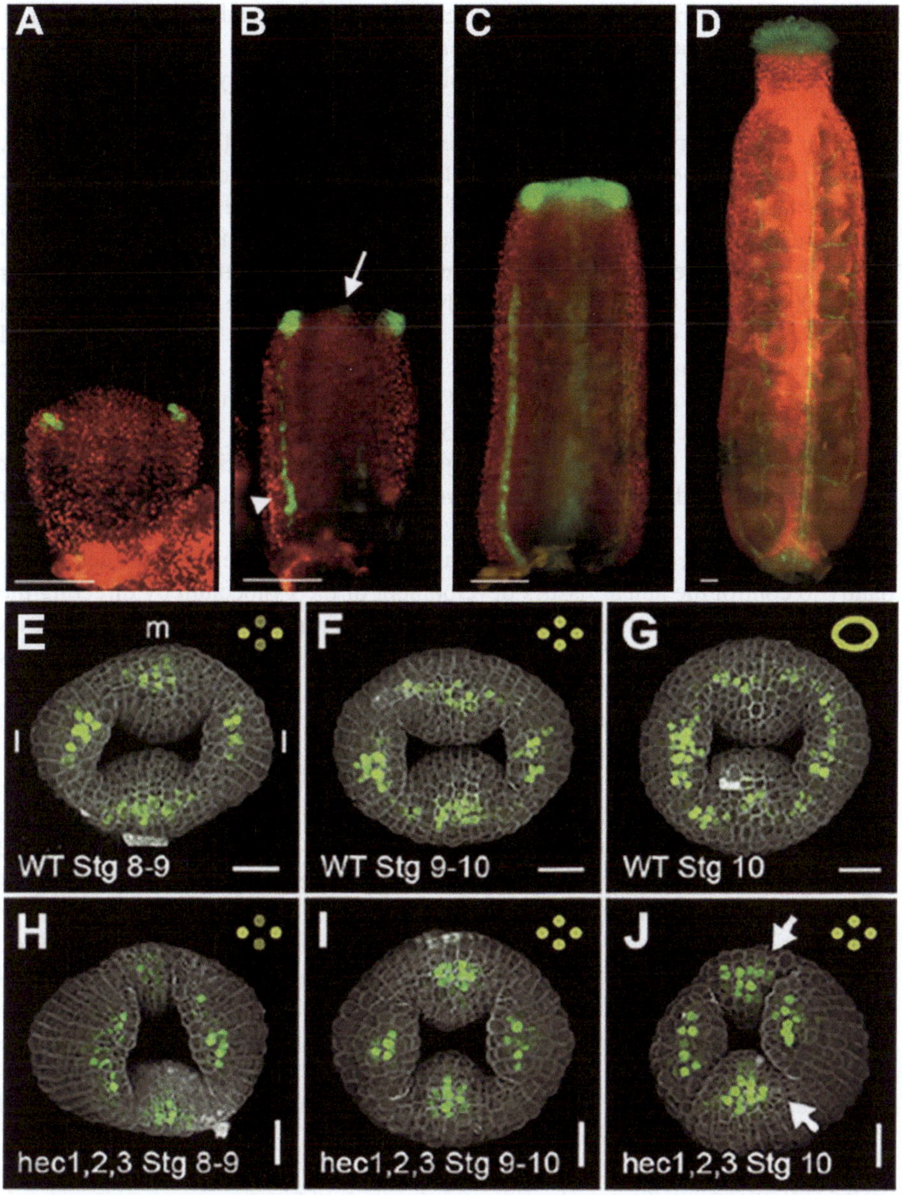

Fig. 10.2 Auxin location during gynoecium development (Larsson et al., 2013; Schuster et al., 2015). (**a**) *DR5rev* is expressed apically in two lateral foci in a stage-7 gynoecium. (**b**) Arrow indicates weak expression in apical medial tip, arrowhead indicate expression in pro-vasculature of a stage-8 gynoecium. (**c**) *DR5rev* expression encircles the whole apical part of a stage-9 gynoecium. (**d**) *DR5rev* is expressed in vasculature and in stigmatic papillae of mature stage-12 gynoecia. (**e–j**) pDR5:3xYFP-NLS expression in the developing gynoecium of wild type (**e–g**) and *hec1,2,3* mutants (**h–j**) at stage 8–9 (**e, h**), stage 9–10 (**f, i**), and stage 10 (**g, j**). In wild type, four regions of local auxin response are connected to a ring-like structure at the onset of stage 10 (**e–g**, insets), but *hec1,2,3* mutants fail to establish this radial symmetry (**h–j**). Arrows in **j** indicate retarded growth in the medial region. (Reprinted with permission from Oxford University Press and Company of Biologists)

apex might be essential for style and stigma development and the apical fusion of the two carpels. It has been proved that *SPT* can interact with *IND*, which can promote the apolar PIN localization at the apical region by directly repressing *PINOID* (*PID*) expression. The developing gynoecia in *spt* and *spt ind* double mutant fail to establish the two medial foci and subsequent ring-shaped DR5 signal element, resulting in the gynoecia with a split-style phenotype (Girin et al., 2011; Marsch-Martínez & de Folter, 2016). The apical fusion defect in the *sty1-1sty2-1* double mutant results from the reduced expression of auxin biosynthesis gene *YUC4* and can be restored by exogenous application of auxin (Eklund et al., 2010; Sohlberg et al., 2006; Ståldal et al., 2008). Similarly, the *hec1,2,3* triple mutant also shows a reduced expression of *YUC4*, *PIN1*, and *PIN3*, and *HEC1* is able to directly bind and activate *PIN1* and *PIN3* in the gynoecium (Schuster et al., 2015). Moreover, the *HEC* genes genetically interact with *SPT*, and the gynoecium in *hec1hec2hec3spt* quadruple mutant cannot form the DR5 signal ring-shape structure and shows enhanced unfusion in apical region, suggesting that HECs-SPT is important for auxin homeostasis during gynoecium development (Gremski et al., 2007; Schuster et al., 2015).

The NGA transcription factors activity have also been reported to act redundantly to control the apical tissues development of the gynoecium by directing the distribution of auxin at the distal end through repressing *PID* expression, as described for *IND* and *SPT*, and upregulating the expression of the biosynthesis genes *YUC2*, *YUC4*, *YUC8*, *TRYPTOPHAN AMINOTRANSFERASE OF ARABIDOPSIS1* (*TAA1*), and *AMIDASE1* (*AMI1*) (Martínez-Fernández et al., 2014; Steffen et al., 2007; Trigueros et al., 2009).

The SQUAMOSA PROMOTER BINDING PROTEIN-LIKE (SPL) transcription factors family has also been found that redundantly function in gynoecium development by activating auxin biosynthesis gene *YUC4* and reducing *HEC2* expression (Marsch-Martínez & de Folter, 2016; Xing et al., 2013). The *spl8* mutant shows abnormal style development, and the combined mutants with the *crc*, *spt*, or *ett* present an enhanced apical fusion defect in gynoecia. According to reports, *CRC* can fine-tune auxin homeostasis in the medial region of the developing gynoecium to produce an appropriate auxin maxima by directly inhibiting *TORNADO2* (*TRN2*). *KNU* and *CRC*, the two direct *AG* target genes, cooperatively repress *WUS* expression, and *WUS* expression is distinctly upregulated in floral meristem at developmental stage 6 in *kun crc* double mutant, which give rise to subsequent floral meristems activity and the indeterminate phenotypes in gynoecia (Yamaguchi et al., 2017). These data suggested that the termination of floral meristems mediated by *KNU* and *CRC* through auxin homeostasis is required for proper gynoecium development (Fig. 10.3).

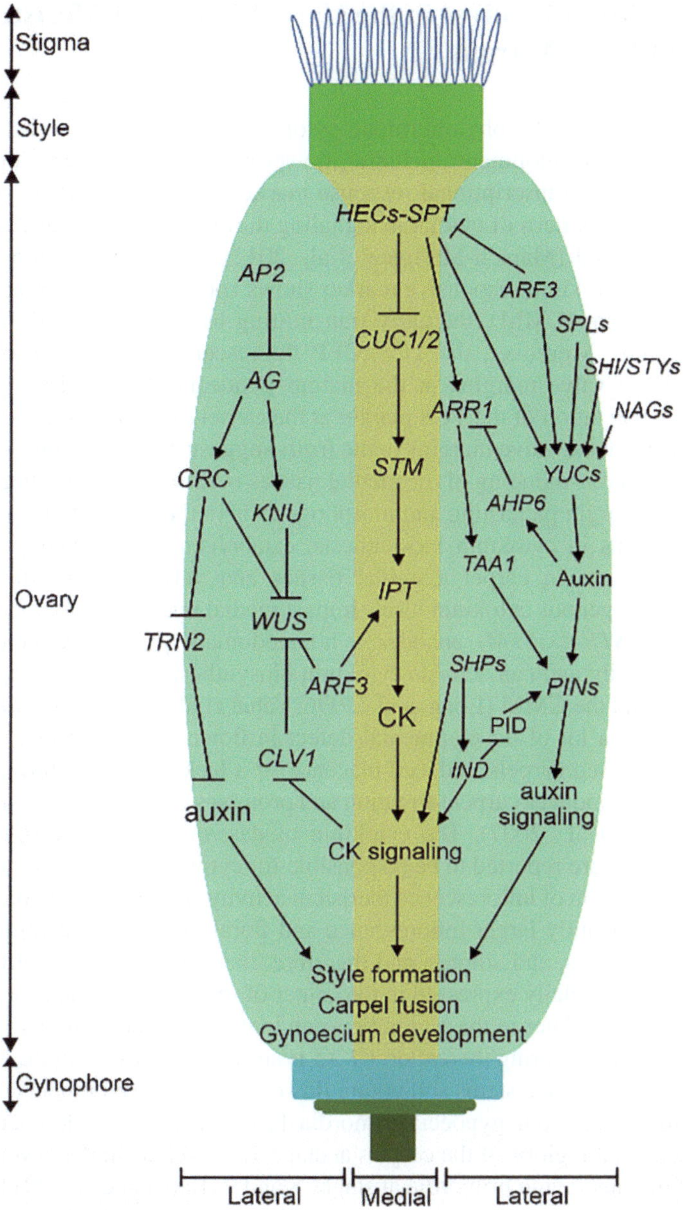

Fig. 10.3 Network model in gynoecium development integrating auxin and cytokinin pathways. This regulatory network integrates the results that auxin biosynthesis, transport, signaling, and cytokinin response are involved in style formation, carpel fusion, and gynoecium development. SPT, an important player of gynoecium development, enables cytokinin signaling in the medial domain of the young gynoecium by activating the transcription of type-B *ARR* genes, which proteins become active upon phosphorylation because of a phosphorelay cascade initiated when cytokinin is present, and then together activate auxin biosynthesis (*TAA1*) and transport important (*PIN*) for growth. It is likely that SPT also affects other components of the cytokinin signaling pathway. Black arrows indicate a positive regulation and a T-bar indicates a repression function, a broken black arrow indicates possible positive regulation; CK, cytokinin

10.6 Cytokinin Involves in the Carpel Marginal Meristem and Valve Margin

It is well known that the phytohormone cytokinin also has many crucial roles in plant growth and development, including flower and gynoecium development. By using the synthetic transcriptional response marker line TCS::GFP, the spatiotemporal localization pattern of cytokinin signaling during pistil and fruit development has been visualized (Marsch-Martínez et al., 2012). In young developing gynoecium, the *TCS* cytokinin responsive marker shows specific expression at the carpel marginal meristem (CMM), the early transmitting tract, and later in the external layer of replum. Moreover, the TCS::GFP fluorescence signal is also precisely localized at the valve margins in the mature gynoecia (Marsch-Martínez et al., 2012). The localization of the TCS marker at the carpel marginal tissues of the early gynoecium and at the valve margins in the fruits suggests that cytokinin is important for the proliferation-inducing of the medial tissues in the early developing gynoecia and the valve margin patterning and morphogenesis in the developing fruits.

The mutations in cytokinin biosynthesis, catabolism, and signaling genes can cause clear defects in carpel marginal tissues and valve margins development. Changes in endogenous cytokinin levels impact medial region development in fruits. *SHOOT MERISTEM* (*STM*) encodes a homeodomain protein belonging to the KNOTTED family and can activate cytokinin biosynthesis in the shoot apical meristem (SAM) and the CMM (Long et al., 1996; Yanai et al., 2005). The inducible *stm* mutants present a lot of developmental defect in flower and gynoecia, such as no gynoecium, unfused carpels, reduced placenta, or a lack of septum formation, suggesting that *STM* induces carpel formation and promotes the marginal tissue development (Scofield et al., 2007). The cytokinin oxidase/dehydrogenase (CKX) genes *CKX3* and *CKX5* are reported to be responsible for cytokinin catabolism and participate in the regulation of inflorescence meristem activity. The *ckx3 ckx5* double mutant generated significantly larger inflorescence and floral meristems, resulting in more ovules, more flowers, and siliques and therefore, increased seed set (Bartrina et al., 2011). *CKX3* is specially expressed in the center of the floral meristems in stage 2 to stage 5 flowers, and later in the region between long stamen primordia and gynoecia primordia and disappearing thereafter. *CKX5* is strongly expressed in the basal meristem of the inflorescence stems and young flowers, and as development continues, it is specifically expressed in gynoecia primordia during different developmental stages, later in the medial regions of the carpels at stage 7, as well as in the developing placentas and initiated ovule primordia at stages 8 and 9 (Bartrina et al., 2011).

It has been reported that the *TEOSINTE BRANCHED1/CYCLOIDEA/PROLIFERATING CELL FACTOR* (*TCP*)*14* and *TCP15* transcription factors, which mediate cytokinin responses, can be involved in gynoecium development. The mutations in the two genes lead to the excessive proliferation along the boundaries of the valves and the replum, resulting in the development of a protrusion topped with stigmatic papillae, which is similar to the phenotypes observed after cytokinin treatment (Kieffer et al., 2011; Steiner et al., 2012). Further study suggested that *TCP15* and related proteins regulate the normal development of the different tissues in the gynoecium through participating in interaction between cytokinin and auxin (Lucero et al., 2015; Uberti-Manassero et al., 2012).

The *CUP-SHAPED COTYLEDON* (*CUC*) genes *CUC1* and *CUC2*, belonging to the NAC transcription factor family, are expressed in the CMM and can promote the formation of the CMM and apical organ fusion in the gyneocium by activating the KNOX transcription factor *STM*. The gynoecia in the *cuc1 cuc2* double mutant display severe defects in carpel marginal tissues development and both septum growth and fusion. Furthermore, the transgenic plants carrying miR164-resistant forms of *CUC1* and *CUC2* show upregulated *STM* expression, which probably cause increased cytokinin levels, and give rise to the expansion of carpel margin structures, including a wider septum and repla. All these observations thus suggest that these two regulatory proteins play critical roles in gynoecium development through the control of meristematic activity, especially in the early medial tissue development (Kamiuchi et al., 2014; Nahar et al., 2012).

It was previously shown that the expression levels of *CUC1* and *CUC2* are ectopically elevated in *spt* mutant, the *cuc1* and *cuc2* mutations can rescue the split carpel phenotype of *spt*, suggesting that *SPT* controls carpel fusion in the apical part of the gynoecium by negatively regulating *CUC* genes expression (Nahar et al., 2012). Recently, there have been reports that *SPT* can emit cytokinin signaling in the medial domain. It is worth noting that the signal transduction of cytokinins is restricted to the medial domain by the cytokinin response repressor *ARABIDOPSIS HISTIDINE PHOSPHOTRANSFERASE 6* (*AHP6*) and/or ARR16 (a type-A ARR). In addition, SPT and cytokinin may jointly activate the expression of the auxin biosynthetic gene *TRYPTOPHAN AMINOTRANSFERASE OF ARABIDOPSIS 1* (*TAA1*) and the auxin efflux transporter *PIN3* (Reyes-Olalde et al., 2017).

The bHLH transcription factors *HEC1*, *HEC2*, and *HEC3* are proposed to participate in transmitting tract formation during carpel marginal tissue development by affecting not only auxin response but also cytokinin response. The cytokinin-treated *hec* and *spt* mutants show the extensive tissue proliferation and apical fusion defect during gynoecium development, and HEC1 can interact with SPT, suggesting that the HEC1-SPT module control marginal tissue development and carpel fusion by negatively modulating cytokinin signaling and restricting sensitivity to cytokinin during gynoecium development. Taken together, these data propose that *HEC1* and *SPT* function as local regulator of auxin and cytokinin responses to modulate gynoecium development (Schuster et al., 2015) (Fig. 10.3).

10.7 Ovule Development

Ovule is a female reproductive organ located within the carpel of a flower. In *Arabidopsis* and most flowering plants, ovule consists of four parts, namely, the embryo sac, the integument, the nucellus, and the funiculus that connect the ovule to placenta (Fig. 10.4a, b). The embryo sac is also commonly called the female

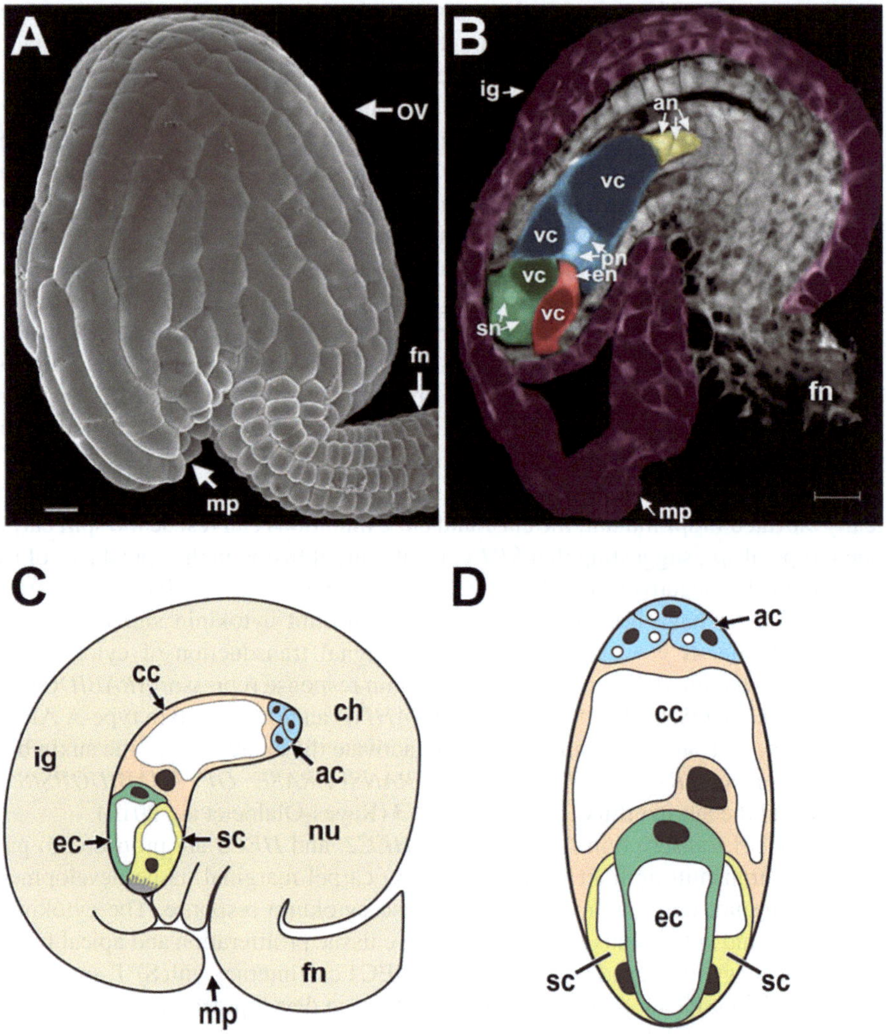

Fig. 10.4 The ovule and female gametophyte in Arabidopsis (Drews & Koltunow, 2011; Shi & Yang, 2011). (**a**) An electron scanning image of ovule. (**b**) A confocal optical image showing a typical ovule structure with a seven-celled and eight-nucleate embryo sac. (**c**) Cartoon of mature ovule. (**d**) Cartoon of mature female gametophyte. The mature female gametophyte in Arabidopsis is approximately 105 μm in length and approximately 25 μm in width. Abbreviations: ov, ovule; fn, funiculus; mp, micropyle; an, antipodal nucleus; cn: central nucleus; en, egg nucleus; pn, polar nucleus; sn, synergid nucleus; vc, vacuole; ig, integument; ac, antipodal cells; cc, central cell; ch, chalazal region of the ovule; ec, egg cell; sc, synergid cell. (Reprinted with permission from Elsevier)

gametophyte or megagametophyte, and is embedded in the integument and nucellus, which degenerates ultimately in seeds. The female gametophyte, a multicellular haploid structure, generally is composed of one egg cell, two synergid cells, one central cell, and three antipodal cells (Fig. 10.4c, d). After double fertilization, the egg cell and the central cell develop into an embryo and endosperm, respectively. The formation of female gametophyte is necessary for seed development during sexual reproduction in angiosperms. In the past 30 years, the development of female gametophyte in plants has become an excellent system to undertake fundamental questions in biology, such as cell specification, cell-cell interaction, and the developmental role of basic cellular machinery. Significant progress has been made in determining the genetic components that control gametogenesis.

Female gametophyte development includes many different patterns in angiosperms. Over 70% of flowering plants exhibit the polygonum-type pattern, including Arabidopsis, rice, maize, wheat, tomato, soybean, and Brassica, and so on. With exception of the specific developmental pattern, female gametophyte development includes two main phases: megasporogenesis and megagametogenesis (Fig. 10.5). Megasporogenesis in Arabidopsis includes three major events: megaspore mother cell formation, meiosis, and functional megaspore selection (Fig. 10.5a). Megagametogenesis comprises three identifiable events: three rounds of mitosis without cytokinesis, followed by nuclei cellularization and then cell specification and differentiation (Fig. 10.5c, d).

Ovule development in Arabidopsis begins with a finger-like projections on the edges of the placental tissue of the ovary (Fig. 10.5a). The protrusion (primordium) elongates and forms a finger-like structure (nucellus). Then, a sub-epidermal cell at the tip of the nucellus differentiates directly and forms an archesporial cell, which enters meiotic development to differentiate into the megaspore mother cell (MMC) (Fig. 10.5b). Compared with the somatic cells, the megaspore mother cells are larger in size, denser in cytoplasm, and larger in nucleus. The megasporocyte then undergoes meiosis to produce four one-nucleate, haploid megaspores. In most flowering plants, such as Arabidopsis and rice, micropylar megaspores undergo programmed cell death, and the single megaspore at chalazal end becomes functional and then undergo continually three rounds of mitosis without cytokinesis to form a huge eight-nucleate coenocyte with four nuclei at each pole (Fig. 10.5b, c). After the first mitosis, a central vacuole is formed between the two daughter nuclei that are separated to the poles and then enlarges in size. The cellularization process of the eight-nucleate coenocyte begin with the formation of phragmoplasts and cell plates between sister and non-sister nuclei during the third mitosis, and then the female gametophyte cells are quickly completely surrounded by cell walls. During the process of cellularization, two polar nuclei, one from each pole, migrate toward the center of the developing embryo sac and finally fuse to form a diploid central nucleus (Fig. 10.5d). The mature female gametophyte consists of seven cells with four cell types: three antipodal cells at the chalazal end, a diploid central cell, two synergids, and an egg cell at the micropylar end. The two nuclei from the egg and central cells lie very close to each other, which is important for double fertilization

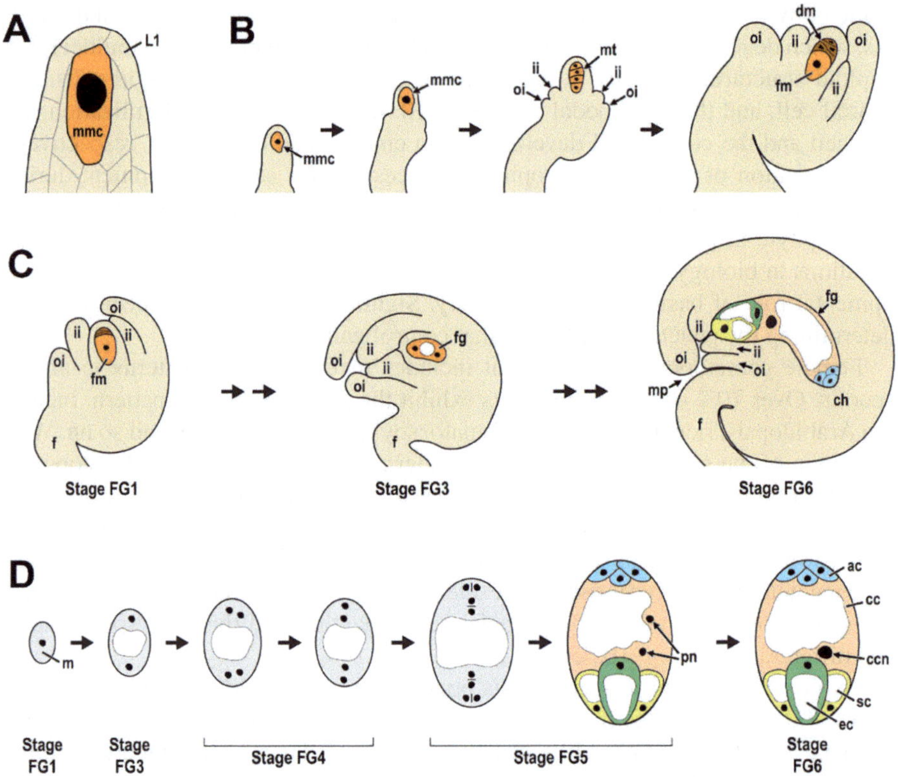

Fig. 10.5 Ovule development: Megasporogenesis and megagametogenesis in Arabidopsis (Drews & Koltunow, 2011). (**a**) Apical region of a finger-like ovule primordium. The megaspore mother cell differentiates from a sub-epidermal cell at the distal end of the ovule primordium. L1 is the outer layer of cells. (**b**) Steps of megasporogenesis. Ovule primordia arise as finger-like projections from the placenta. The megaspore mother cell undergoes meiosis and forms four megaspores. Three of the megaspores undergo cell death. The chalazal-most megaspore survives, becomes the functional megaspore, and undergoes megagametogenesis. (**c**) Steps of megagametogenesis emphasizing development within the ovule. (**d**) Stages of megagametogenesis (Christensen et al., 1998). The megaspore contains a single nucleus (stage FG1). This nucleus undergoes two rounds of mitosis, producing a four-nucleate coenocyte, with two nuclei at each pole separated by a large central vacuole (stage FG4). During a third mitosis, phragmoplasts and cell plates form between sister and non-sister nuclei and the nuclei become completely surrounded by cell walls (Stage FG5). During cellularization, the polar nuclei migrate toward the center of the female gametophyte and fuse before fertilization. These events produce a seven-celled structure consisting of three antipodal cells, one central cell, two synergid cells, and one egg cell. If the female gametophyte is not fertilized, the antipodal cells eventually degenerate (Stage FG7, not shown). Abbreviations: dm, degenerating megaspores; fm, functional megaspore; ii, inner integument; oi, outer integument; L1, L1 epidermal layer of the ovule primordium; mmc, megaspore mother cell; mt, meiotic tetrad; ac, antipodal cells; cc, central cell; ccn; central cell nucleus; ch, chalazal region of the ovule; ec, egg cell; f, funiculus; fg, female gametophyte; fm, functional megaspore; m, megaspore; mp, micropyle; oi, outer integument; pn, polar nuclei; sc, synergid cells. (Reprinted with permission from Elsevier)

because these two nuclei are fused with the two sperm nuclei, respectively (Fig. 10.5d).

Antipodal cells degenerate shortly before fertilization in *Arabidopsis* or undergo further mitosis as seen in maize and rice. The central cell, the egg, and two synergid cells are defined as a female germ unit-a functional unit that is able to attract a pollen tube, interact with the tube to trigger sperm release, and complete double fertilization.

At the same time, epidermal cells at the proximal third of the nucellus divide parallel to the long axis, forming two primordia, which become the inner and outer integuments, respectively (Fig. 10.5b). These surround the functional megaspore, which becomes the embryo sac, and form a narrow opening at the micropyle where the pollen tube enters after pollination.

As mentioned above, ovule development involves processes of sporophyte and gametophyte, and is an excellent system for studying the basic developmental mechanisms that govern germline formation, cell growth and division, and gametic cell fate specification.

10.8 The Transition of Somatic to Germline Cell Fate

In flowering plants, archesporial cells are the initial cells of the germline, which develop from the sub-epidermal L2 cell layer within the early ovule primordium. In most ovule primordia, just a single hypodermal L2 cell at the tip of the nucellus differentiates into the archesporial cell. However, about 5–6% of ovules in Arabidopsis produce additional enlarged cells (Armenta-Medina et al., 2011; Olmedo-Monfil et al., 2010). Extensive researches in Arabidopsis, rice, and maize have identified many genes that govern the development of megaspore mother cell, including the rice *MEIOSIS ARRESTED AT LEPTOTENE* (*MEL1*) gene (Nonomura et al., 2007), the maize *ARGONAUTE104* (*AGO104*) gene (Singh et al., 2011), *WUSCHEL* (*WUS*), *WINDHOSE1* (*WH1*), *WH2*, and *TORNADO2* (*TRN2*) genes (Lieber et al., 2011) (Fig. 10.6). Moreover, *MEIOSIS ARRESTED AT LEPTOTENE1* (*MEL1*), a germline-specific member of *ARGONAUTE* (*AGO*) genes family, is highly expressed in archesporial cells and sporogenous cells, and disappears when sporogenous cells undergo meiosis in anthers and ovules. The mutations in *MEL1* lead to ovule development arrested at different stages from pre-meiosis to tetrad. The germline cells in *mel1* exhibited uncondensed chromosomes and aberrant chromatin modification, indicating that *MEL1* acts on chromatin structures, similar to those of the PIWI subfamily of AGOs in *Drosophila*. This indicates that the *MEL1*, involving in small RNA-mediated gene silencing pathway, plays an important role in plant female germline development, most likely by controlling cell division of pre-meiotic germline cells, modification of meiotic chromosomes, and progression of meiosis, but not the initiation, establishment, and early mitotic division of germline cells (Nonomura et al., 2007).

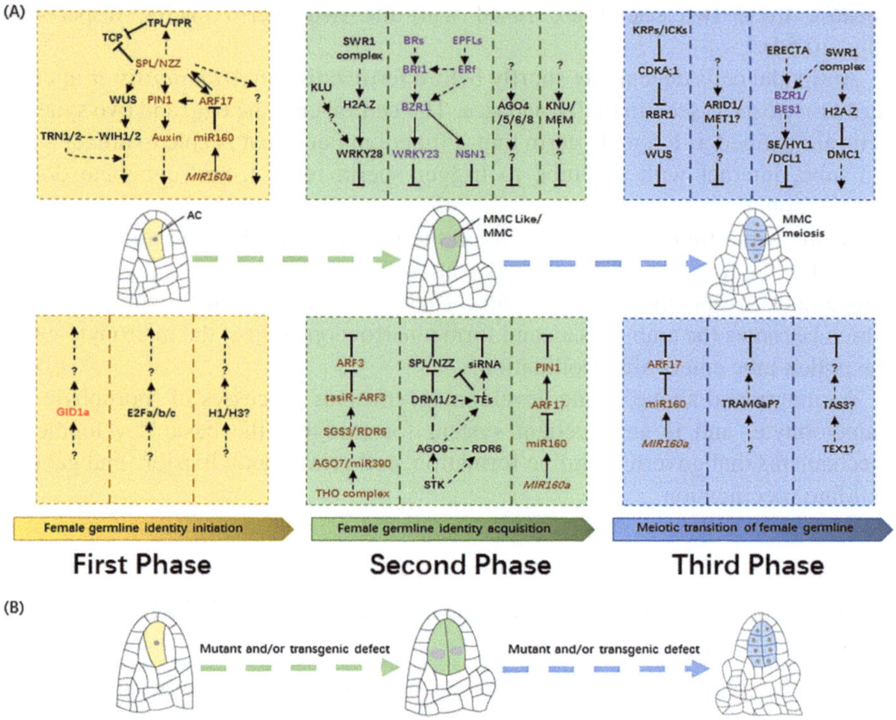

Fig. 10.6 Network model for megasporocyte specification. (**a**) The main gene regulatory network (GRN) restricting MMC formation to a single cell, such as SPL-WUS-WIH1/2 pathway. BRs/ BRI1-EPFLs/ERf-BZR1 pathway, STK-AGO9/RDR6-DRM1/2-SPL/NZZ pathway, and KRPs/ ICKs-CKA;1-RBR1-WUS pathway. Question marks indicate unknown factors/mechanisms that remain to be identified. The genes marked in red are involved in the GA signaling pathway, while those marked in brown are involved in the auxin signaling pathway, those in purple are involved in the BR signaling pathway and those in blue are involved in the cytokinin signaling pathway. (**b**) Mutants and/or transgenic defects corresponding to the three phases of MMC development in Arabidopsis

The archesporial cell morphologically differentiates from its surrounding somatic cells by its larger size and pronounced nucleus. The archesporial cell could differentiate into a megasporocyte (also termed megaspore mother cell, MMC) to initiate the female germline lineage. The *SPOROCYTELESS* gene (*SPL/NOZZLE*) has been early demonstrated to promote germline cell fate in *Arabidopsis*. *spl* mutants can form archesporial cells in both anther and ovule primordia, but the archesporial cells fail to develop further, which results in a complete lack of spore mother cell in male and female organs (Schiefthaler et al., 1999; Yang et al., 1999b). This suggests that *SPL* is necessary for promoting the formation of megaspore mother cell. *AGAMOUS* (*AG*), a floral homeotic regulator, has been shown to bind to the CArG-box in 3′ region of the *SPL* gene and activate it (Ito et al., 2004). Although it has been demonstrated that *SPL* is required to control male archesporial cell specification in *Arabidopsis*, whether *SPL* is also required for female archesporial cell specification remains unclear. *SPL/NZZ* encodes a nuclear protein that shares some similarities with MADS-box transcription factors and is expressed in a range of ovule tissues including the megaspore mother cell (Yang et al., 1999b). Until recently, it has been discovered that SPL recruit TOPLESS/TOPLESS-RELATED (TPL/TPR) co-repressors using its EAR motif at the C-terminal end, and using its N-terminal part to bind and inhibit the CINCINNATA (CIN)-like TEOSINTE BRANCHED1/CYCLOIDEA/PCF (TCP) transcription factors during ovule development, thereby acting as an adaptor-like transcriptional repressor (Wei et al., 2015). It was also proposed that *SPL/NZZ* may be involved in auxin homeostasis by repressing *YUCCA* genes in lateral organ development (Li et al., 2008). According to this hypothesis, both auxin and *Pin formed 1* (*PIN1*), which control polar auxin transport, have been observed in the unicellular layer of the nucellar epidermis, adjacent to the region where female gametophyte specification occurs (Benková et al., 2003; Pagnussat et al., 2009). A similar microenvironment related to auxin during megasporogenesis have also been found in two early-divergent angiosperms (*Annona cherimola* and *Persea americana*) and the evolutionarily derived *Arabidopsis thaliana* (Lora et al., 2017). It would be interesting to investigate how auxin is involved in germline formation. Furthermore, these findings indicate that the position of the single germline precursor is strictly restrained. In the early stage of ovule development, establishing an environment in the epidermal cell layer promotes the formation of female germline, which is crucial for the acquisition of female reproductive fate (Cai et al., 2022a, b).

Plant development is influenced by various signals, and phytohormones have received appreciable attention due to their involvement in various aspects of plant development, including those that are crucial for female germline specification (Cai et al., 2023a, b). Similar to auxins, a recent report has shown that there is a BR gradient in the ovule (Cai et al., 2022a). The analysis of both transcriptional and translational expression pattern of the ten known BR biosynthetic genes have shown that the BR biosynthetic mechanism is expressed exclusively in the sporophytic tissues of ovules, not in MMC (Cai et al., 2022a). Further study has found that BRs signaling affects the acquisition of female germline identity and meiotic transition of MMC through the *BRASSINOZOLE RESISTANT 1* (*BZR1*) family and their targeted genes in Arabidopsis (Cai et al., 2022a). Moreover, EPFL-ERf ligand-receptor pairs function upstream of the BZR1 family and synergistically regulate female germline specification by directly inducing the expression of a nucleolar GTP-binding protein, NUCLEOSTEMIN-LIKE 1 (NSN1) (Cai et al., 2023a).

In most flowering plant, only one megasporocyte is formed in one ovule. In the past decade, several studies have identified a group of genes that control the number of megasporocytes through genetic approaches in *Arabidopsis*, rice, and maize. Arabidopsis AGO9, which binds 24-nt small RNAs derived from transposable elements (TEs) and RNA-DEPENDENT RNA POLYMERASE 6 (RDR6) encoding an RNA-dependent RNA polymerase involved in the biogenesis of siRNAs restrict megasporocyte formation to a single cell in the ovule (Olmedo-Monfil et al., 2010). Loss-of-function mutations in *AGO9* or *RDR6* lead to supernumerary MMCs in ovule primordia (Olmedo-Monfil et al., 2010). The activities of these two genes are also involved in the epigenetic silencing that control the specification of megasporocyte which variated among different ecotypes in *Arabidopsis* (Rodríguez-Leal et al., 2015).

MULTIPLE ARCHESPORIAL CELLS 1 (*MAC1*) is reported to play important roles in the transition of the hypodermal cells from the somatic to germline cell fate in maize, and mutant *mac1* shows more than one archesporial cell and an excessive number of embryo sacs in ovules (Sheridan et al., 1996, 1999). Similarly, *MULTIPLE SPOROCYTE 1* (*MSP1*) in rice governs early sporogenic development and the *msp1* mutants likely lead to an excessive number of archesporial cells (Nonomura et al., 2003). *MSP1* gene is expressed in nucellar cells that surround germline cells but not in the developing germline cells, which indicates that *MSP1* plays key roles in determining archesporial cell fate by limiting the surrounding cells into germline cells (Nonomura et al., 2003). Similar to *MSP1*, the expression of *TDL1A* is detected exclusively in nucellar cells but not in germline cells. The knockdown expression of *OsTDL1A* by RNAi phenocopies the *msp1* phenotype in the archesporial cells fate (Zhao et al., 2008). *OsTDL1A* and *MSP1* are both expressed in young ovules, in all cells surrounding the archesporial cells. Rice *MSP1* encodes a leucine-rich repeat containing receptor-like kinase and *OsTDL1A* encodes a putative ligand for *MSP1*. OsTDL1A is shown to bind the extracellular domain of OsMSP1 by yeast two-hybrid and BiFC assays (Zhao et al., 2008). These data suggest that OsTDL1A and MSP1 form a ligand-receptor signaling system to control signaling between the archesporial cell and its neighboring cells and to specify germline cell fate and number during early ovule development, indicating that a lateral inhibition mechanism may act in controlling the number of germline cells.

Intercellular signaling among somatic cells to restrict MMC specification, without involving the cell that eventually becomes the megasporocyte, is beginning to be understood. It was recently showed that the epidermal layer (L1)-expressed TEX1 protein plays an important role in this process by promoting the biogenesis of TAS3-derived trans-acting small interfering RNAs (ta-siRNAs), which repress the expression of ARF3 (Su et al., 2017). Specifically, the expansion of *ARF3* expression into lateral epidermal cells from the medio domain of ovule primordia in a TAS3 ta-siRNA-insensitive mutant led to the formation of supernumerary MMCs (Su et al., 2017). In Arabidopsis, microRNA160 (miR160) targeted gene *ARF17* (*AUXIN RESPONSE FACTOR17*) genetically interacts with the *SPL/NZZ* function in promoting MMC specification. Moreover, miR160 and ARF17 define the expression domain of PIN1, which contributes to establishment of the local auxin maximum at the ovule apex (Huang et al., 2022). These results suggest that the small-RNA pathway involved intercellular signaling among somatic cells in the ovule primordia plays an important role in restricting the germline cell fate to a single cell. Recent study further demonstrated a cytochrome P450 gene *KLU* mediated mechanism in somatic cells surrounding the germline precursor to prevent ectopic germ-line formation. The inner integument expressed *KLU* non-cell-autonomously suppresses germline cell fate in somatic cells surrounding the megasporoctye through chromatin remodeling complex SWR1-mediated activation of transcription factor *WRKY28* expression (Zhao et al., 2018). *WRKY28* is exclusively detected in hypodermal somatic cells surrounding the germline cell and is a necessary component in repressing the somatic-to-germline cell fate transition (Zhao et al., 2018). This study shows how somatic cells cooperatively use chromatin remodeling to restrict germline cell specification to a single cell in the ovule primordia, in line with the finding that during megasporogenesis the somatic-to-germline cell fate transition is accompanied by large-scale chromatin reprogramming in megasporocyte (Baroux & Autran, 2015; She et al., 2013), as in microsporocyte (She & Baroux, 2015) and in primordial germ cells of animals (Hackett et al., 2012). With the except of the epigenetic factors, other components such as ER receptor-like kinase family also play an important role in limiting the fate of germline cell to the single MMC (Hou et al., 2021). Two key epigenetic factors, SWR1 and SDG2, cooperate with ER receptor kinase signaling pathway to govern female germline development in *Arabidopsis* (Huang et al., 2023).

In addition to the mechanism in which germline cell fate is repressed in somatic cells surrounding megasporocyte, a novel mechanism involved in cell cycle regulation is recently revealed to ensure a single megasporocyte formation per ovule by preventing over-proliferation of an already specified megasporocyte (Zhao et al., 2017). This study demonstrated that the number of megasporocyte and subsequent embryo sac in ovule is controlled by a group of redundantly acting cyclin-dependent kinase (CDK) inhibitors of the *KIP-RELATEDPROTEIN* (*KRP/ICK*) class (Zhao et al., 2017). *KRPs/ICKs* function by restricting CDKA;1-dependent inactivation of the Arabidopsis Retinoblastoma homolog RBR1, which directly represses the expression of the stem cell factor *WUSCHEL* (*WUS*) and regulates megasporocyte formation (Zhao et al., 2017).

10.9 The Regulation of MMC Meiosis and the Formation of the Functional Megaspore

The developed female megasporocyte produces four haploid megaspores through meiosis. Since meiosis is a fundamental process, the molecular mechanisms underlying this process should be similar in microsporogenesis and megasporogenesis. Previous studies have shown that many mutations in *Arabidopsis* that disrupt male meiosis also affect female meiosis (Bai et al., 1999; Couteau et al., 1999; Sebastian et al., 2009; Stacey et al., 2006). Therefore in plants, the understanding of meiosis is gender-biased, as most researches that have identified the expression and subcellular localization of meiotic proteins were performed on male meiocytes due to their large amount and easy accessibility compared with female meiocytes (Ronceret & Pawlowski, 2010).

However, genetic analysis suggests that several processes such as the rate of recombination are differentially regulated between female and male meiosis. For example, *MMD1* (Yang et al., 2003) and *ASK1* (Yang et al., 1999a) genes and different alleles of the meiotic genes *SWITCH1/DYAD* (Mercier et al., 2001) or

ACTIN-RELATED PROTEIN6 (Qin et al., 2014; Rosa et al., 2013) can differentially impact male and female meiosis. In addition, four new Arabidopsis *fm* mutants have been reported to affect specifically female meiosis (Zhao et al., 2014). These data imply that specific regulating mechanisms exist in male and female meiosis. The significant differences in the distribution and crossover rate along chromosomes during male and female meiosis further support this viewpoint (Giraut et al., 2011). A whole-mount immunolocalization method for analyzing the subcellular localization of proteins during female meiosis in Arabidopsis has been reported, providing a necessary technique for studying plant meiosis and allowing for the analysis of plant reproductive alternatives to sexual reproduction (Escobar-Guzmán et al., 2015).

In most flowering plants, of the four megaspores produced by female meiosis, only one close to the chalazal end participates in the formation of the final female gametophyte, while the other three undergo degeneration. The selection of functional megaspore is position dependent. Complex mechanisms must be evolved to define such precisely developmental control. It has been proposed that functional megaspores are selected by positive survival or protective signals secreted from nucellus cells at the chalazal end (Bell, 1996). Cytokinin is one of the sporophytic signals that was shown to play an important role in the specification of functional megaspore in Arabidopsis (Cheng et al., 2013). A sporophytic integument cells expressed protein, arabinogalactan protein 18 (AGP18) located in the abaxial region of the ovule, was also showed to be important for the selection and survival of megaspores in Arabidopsis (Demesa-Arévalo & Vielle-Calzada, 2013). Overexpression of *AGP18* in the ovule results in the abnormal maintenance of surviving megaspores that acquire a functional megaspore identity, suggesting that AGP18 positively promotes the selection of viable megaspores (Demesa-Arévalo & Vielle-Calzada, 2013). Although there are several pieces of evidence indicating that the communication between the sporophyte and the gametophyte prevails during megasporogenesis (Bencivenga et al., 2011; Olmedo-Monfil et al., 2010; Tucker et al., 2012), the position signals that determine the selection of functional megaspore are yet to be determined. The correlation between callose deposition and functional megaspore selection in several species including Arabidopsis and maize suggested that a physical isolation barrier is required for the selection of functional meiotic products (Russell, 1979; Webb & Gunning, 1990).

The polarity exists within the megasporocyte, for instance the plasmodesmata preferentially located at the MMC chalazal pole and dynamic distribution of the microtubule cytoskeleton has been observed in many species. In *Arabidopsis switch1 (swi1)/dyad* ovules, the polarity of megasporocytes was disrupted (Motamayor et al., 2000; Siddiqi et al., 2000). *SWI1* encodes a protein required for chromatid cohesion establishment and chromosome structure during meiosis (Agashe et al., 2002; Mercier et al., 2001). The *dyad* megasporocyte undergoes defective meiosis and produces two unreduced megaspores, of which the chalazal end but not the micropylar end diploid megaspore expresses functional megaspore specific markers (Ravi et al., 2008). This indicates that only the chalazal end megaspore is selected as functional megaspore and a position-dependent mechanism on work to decide the cell fate at this stage of ovule development. Similarly, a maize *ameiotic* (*am1*) mutant allele was also identified, in which megasporocyte underwent a mitotic division instead of meiosis, as in *swi1/dyad* (Pawlowski et al., 2009). The maize *AM1* encodes a plant-specific chromatin binding protein and shares 30% identity with SWI1, suggesting that SWI and AM1 are necessary for the switch between meiotic and mitotic division cycles. Other cell cycle regulators are also recently shown to involve in the selection of surviving megaspores. Mutation in all seven *ICK/KRP* genes, the plant CDK inhibitors, caused the number and position of surviving megaspores from one MMC were variable, leading to more than one embryo sac was formed in the septuple mutant ovule (Cao et al., 2018). Moreover, ICK4 protein was detected in the degenerative megaspores but absent in the functional megaspore. These data indicate that the ICKs/KRPs cell cycle regulators function redundantly in governing the selection of functional megaspore and the degeneration of non-functional megaspores, in addition to its role in defining one megasporocyte formation per ovule as supernumerary MMCs was formed in ick/krp mutants by over-proliferation of an already specified megasporocyte (Cao et al., 2018; Zhao et al., 2017). The functional dissection of SWI1, AM1, and ICKs/KRPs will reveal how the transition from meiosis to mitosis is regulated.

10.10 Control of the Mitotic Cycle in Female Gametogenesis

The functional megaspore rapidly expands, occupying the space left by the degenerating megaspores, and undergoes through two rounds of karyokinesis to form a four-nucleate coenocyte with two nuclei at each pole. During a third mitosis, cytokinesis takes place to form an eight-nucleate syncytial embryo sac. Then, the eight female gametophyte cells quickly become surrounded by cell walls. During cellularization, the polar nuclei migrate from each pole toward the center of the developing female gametophyte and subsequently fuse. These events result in the formation of a seven-celled female gametophyte, composed of one egg cell, two synergid cells, one central cell, and three antipodal cells (Figs. 10.5 and 10.6). In the past two decades, genetic studies have identified a lot of genes that regulate different stages of megagametogenesis. Comparative gene expression profiling also uncovered many genes that are involved in the female gametophyte development (Lê et al., 2005; Steffen et al., 2007; Weterings & Russell, 2004; Yang et al., 2006; Yu et al., 2005).

Little is known, however, about the molecular control of the initiation of female gametogenesis. The *Arabidopsis* AGL23, a type I MADS-box transcription factor expressed in the functional megaspore and developing female gametophyte, plays a very important role in megagametogenesis. Mutations in *AGL23* block the development of female gametophyte at the one-nucleate stage, indicating the role of AGL23 in early female gametogenesis (Colombo et al., 2008). The cell surface proteoglycan AGP18 was also reported to be essential for initiation of female gametogenesis in *Arabidopsis* (Coimbra et al., 2007). *AGP18* is expressed in the functional megaspore, but weakly detected in somatic nucellar cells. The functional megaspores in AGP18 RNAi-induced posttrancriptional silencing ovules fails to mitotically divide and initiate female gametogenesis. These data show that *AGP18* plays a key role during interactions between gametophytic and sporophytic cells in the young ovules and that these interactions are required for the initiation of the functional megaspore mitosis in Arabidopsis (Acosta-García & Vielle-Calzada, 2004; Coimbra et al., 2007). Mutations in the sporophytic genes *INNER NO OUTER* (*INO*) and *WUS* transcription factors resulting in compromised early stages of gametophyte development also suggested the importance of inter-regional signaling (Elliott et al., 1996; Lieber et al., 2011). Moreover, the small RNA-dependent pathway mediated by AGO5 independent of AGO9 acting in somatic ovule tissues has also been shown to be necessary for promoting the initiation of megagametogenesis in functional megaspore (Tucker et al., 2012).

It has been known that the anaphase-promoting complex or cyclosome (APC/C) plays a crucial role in regulating important transitions during mitotic progression. APC2, a subunit of APC/C, genetically interact with two other APC/C subunits APC11 and APC8/CDC23 to control female gametophytic cell cycle in *Arabidopsis*. Null mutation of APC2 causes megagametogenesis arrest (Capron et al., 2003). *Arabidopsis NOMEGA* encodes APC6/CDC16 subunit of APC/C, and *nomega* mutant embryo sacs fail to degrade Cyclin B, an important APC/C substrate, and lead to the arrest of female gametogenesis at the two-nucleate stage (Kwee & Sundaresan, 2003). These suggest that the APC/C ubiquitin-mediated proteolysis pathway is important for female gametophytic cell cycle control. The duplicated genes *RPT5a* and *RPT5b* encode *regulatory particle triple A ATPases* (*RPT*) of the 26S proteasome and are partially redundant in female gametophytes. The *rpt5a-4 rpt5b-1* double mutant exhibits a defect of the first or second mitosis of the megagametophyte that arrest embryo sac development at the one- or two-nucleate stage (Gallois et al., 2009). It has been reported that RHF1a and RHF2a, two RING-finger E3 ligases, play an important role in *Arabidopsis* gametogenesis. RHF1a can directly interact with a cyclin-dependent kinase inhibitor ICK4/KRP6 for proteasome to mediate degradation. The double mutation of them results in the interphase arrest of the mitotic cell cycle in the embryo sac development (Liu et al., 2008). These publications indicate that the ubiquitin/26S proteasome system participated in regulating cell cycle progression during female gametophyte development in plant.

The *PROLIFERA* (*PRL*) gene in *Arabidopsis* encodes DNA replication licensing factor subunit MCM7, a homologue of the yeast and mammalian MCM proteins, and functions in populations of dividing cells (Springer et al., 2000). Mutation in PRL results in failure of triggering mitosis in female germline cells and arrested megagametogenesis at the one-nucleate stage. It was shown that *SWA1* in *Arabidopsis* encodes a nucleolar WD40-containing protein in interphase cells, which is involved in the processing of pre-18S rRNA. The transcripts of *SWA1* are detected in dividing cells such as functional megaspores and the female gametophytic cells. *swa1* mutation brings about embryo sacs to arrest at two-, four-, or eight-nucleate stages during megagametogenesis. This suggests that SWA1 could mediate rRNA biogenesis to regulate the progression of the mitotic division cycles during megagametogenesis in plants (Shi et al., 2005).

The Arabidopsis retinoblastoma related (RBR) protein is known to act as a negative regulator of cell proliferation by inhibiting E2F transcription factors to control G1/S transition (Ebel et al., 2004). Overproliferation of megasporocytes and microsporocytes was observed in a temperature-sensitive *rbr1* mutant (Zhao et al., 2017). During megagametogenesis, the female gametophyte in *rbr1* mutant is also unable to prevent mitosis and undergoes excessive nuclear proliferation resulting in supernumerary nuclei with partially enclosed cell-wall-like structures at the micropylar end of the embryo sacs (Ebel et al., 2004). This suggests that cellularization and karyokinesis during megagametogenesis can be uncoupled, and these two events are regulated independently. The *rbr* mutant ovules fail to form gametophytic cells as indicated by no expression of female gametophyte marker genes, suggesting that cell cycle control is connected to cellular differentiation during megagametogenesis.

AtKin-1, a Kinesin-1 member microtubule-dependent motor protein playing a role in long-distance transport of organelles and vesicles in animal and fungi, contains a putative E2Fa binding sequence and also participated in the regulation of nuclear division cycles during megagametogenesis in *Arabidopsis* (Wang et al., 2014). In maize, *indeterminate gametophyte 1* (*IG1*) encodes a LATERAL ORGAN BOUNDARIES domain protein with high similarity to ASYMMETRIC LEAVES2 of *Arabidopsis* and limits the proliferative phase of female gametophyte development. Female gametophytes in *ig1* mutant exhibit a prolonged phase of free nuclear divisions with different cellular patterns, leading to a variety of abnormal embryo sacs, including two or more egg cells, three or more synergids, and two or more central cells with extra polar nuclei (Evans, 2007; Guo et al., 2004). While, its rice orthologue, OsIG1, is responsible for the initiation of ovule number control in addition to restricting the proliferative phase of female gametophyte development. OsIG1-RNAi ovules show double ovules with an extra egg cell in embryo sac (Zhang et al., 2015).

These studies collectively demonstrate that timely mitotic progression of the division cycle controlled by the conserved cell cycle machinery is crucial for megagametogenesis. While other regulatory mechanisms are also proved to play a role in gametophytic cell development in plants. The *CYTOKININ INDEPENDENT1* (*CKI1*) encodes a plant histidine kinase and is necessary for megagametophyte development. Knockout of *CKI1* also gives rise to the early abortion and degeneration of female gametophytic cells (Pischke et al., 2002). CKI1 promotes Arabidopsis female gametophyte development through activating the cytokinin signaling pathway via interacting with the Arabidopsis histidine phosphotransfer proteins AHP2, AHP3, and AHP5 (Deng et al., 2010; Liu et al., 2017). Chromatin regulation is also involved in the control of nuclear proliferation during megagametogenesis. It was reported that *Arabidopsis thaliana* chromatin-remodeling protein 11 (CHR11), a member of the ATP-dependent SWI2/SNF2 family, plays important roles for haploid nuclear proliferation during megagametogenesis. *CHR11*-RNAi lines exhibit defective female gametophytes arrested at the onset of megagametogenesis, which results in uncellularized embryo sac arrested at the one-, two-, four-, or eightnucleate stages (Huanca-Mamani et al., 2005). *HAM1* and *HAM2* genes, encoding histone acetyltransferase within the MYST subfamily, function redundantly in female gametophyte development. The *ham1ham2* double mutant arrest megagametogenesis at the one-nucleate stage (Latrasse et al., 2008), suggesting that hypoacetylation of histones in the functional megaspore is important for initiation of nuclear proliferation during megagametogenesis. Furthermore, a histone deacetylase AtHDA7 was demonstrated to be involved in nuclear proliferation during megagametogenesis. Silencing of *AtHDA7* blocks the entering of the third round of mitosis of female gametophyte and *Athda7* mutant ovules display degeneration of nuclei at the stage of four-nucleate female gametophyte (Cigliano et al., 2013), indicating the importance of histone deacetylation in promoting nuclear proliferation during megagametogenesis.

10.11 Cellularization and Polarity of the Female Gametophyte

During the process of megagametogenesis, cellularization of the syncytial embryo sac occurs after the two rounds of mitosis without cytokinesis, and upon the third round of mitosis, nuclear division is accompanied by the formation of phragmoplast and cell plate. During female gametophyte cellularization, mitosis ceases and cell growth attenuates, accompanied by gametophytic cell fate specification and cell differentiation. This process involves the precise regulation in multiple aspects, including cell wall formation, cell-cycle regulation, and cellular differentiation. Several genes implicating this process have been identified. The *GEMINI POLLEN 2* (*GEM2*) is demonstrated to play an important role in coordinating karyokinesis and cytokinesis during male and female gametogenesis. The predominant female defect in *gem2* is the lack of cellularization in the embryo sac during female gametogenesis (Park et al., 2004). Aberrant division and incomplete cytokinesis were also displayed in *gem2* pollen during male gametogenesis. Although cellularization is defective in most developing *gem2* female gametophyte, some partially cellularized *gem2* female gametophyte can further develop to form an embryo sac with a cell possessing characteristics of an egg cell (Park et al., 2004). This suggests that cellularization may reinforce cell fate specification that could have been established in the coenocytic embryo sac. Two closely related MYB transcription factors, *MYB64* and *MYB119*, are expressed in the nucleus of macrogametophyte during cellularization and cell differentiation. Most *myb64 myb119* double mutant gametophytes cannot undergo cellularization and cease nuclear division, resulting in the enlargement of coenocytic gametophytes with supernumerary nuclei (Rabiger & Drews, 2013). Female gametophyte marker genes were not expressed properly in *myb64 myb119* mutant gametophytes, indicating that cellular differentiation is also affected in *myb64 myb119* gametophytes. These data together suggest that cellularization and gamete specification are close-connected processes during megagametogenesis.

Cellularization of macrogametophyte is also controlled by genes involved in positioning phragmoplast and cell plate formation. Mutation in *GEM1/MOR1*, a microtubule-associated protein gene, and *TIO*, a phragmoplast associated protein kinase gene, results in defects in cell plate formation in somatic and reproductive cells (Oh et al., 2005; Twell et al., 2002). *gem1/mor1* and *tio* mutant ovules display uncellularized female gametophyte. In addition, loss of function of γ-tubulin encoding *TUBG1* and *2* genes, also cause uncellularized female gametophyte with aberrant positioning and number of nuclei (Pastuglia et al., 2006). A mitogen-activated protein kinase (MAPK) pathway is also involved in regulating cellularization of macrogametophyte. AtNACK1 and AtNACK2 are two kinesin-like proteins that directly activate the MAPKK kinase NPK1. *Atnack1 Atnack2* double mutant ovules exhibit mispositioning of nuclei and abnormal cell plate formation during megagametogenesis (Tanaka et al., 2004). Moreover, AtNACK2, AtMPK4, and AtMKK6 play a role in regulating meiotic cytokinesis during male gametogenesis (Zeng et al., 2011). The kinesin-MAPKKK pathway is also involved in the lateral expansion of phragmoplast and cell plate during sporophytic cytokinesis, indicating that the cellularization of the embryo sac is controlled by a conserved mechanism that regulates positioning phragmoplast and cell plate expansion in plants.

The macrogametophyte is a polarized structure, with the egg cell and synergid cells located at the micropylar end, and antipodal cells located at the chalazal end. The polarity of the female gametophyte is established during megasporogenesis. The polar distribution of cytoplasm and organelle along the chalazal-micropylar axis is examined in megaspore mother cell (Bajon et al., 1999). During function megaspore selection, only the chalazal end megaspore survives, whereas the other three degenerate (Fig. 10.5b). At four nucleate stage, the nuclei at the chalazal end are located one above the other with respect to the micropylar-chalazal axis, whereas the nuclei at the micropylar end are positioned side by side. During gametophytic cell differentiation, the micropylar end cells develop into egg apparatus, whereas the chalazal end of cell differentiates into antipodal cells (Fig. 10.5c). The gametophytic cells in the female gametophyte are also polarized at the sub-cellular level. The nuclei of the synergid and central cells are located at the micropylar end of the cell, whereas the nucleus of the egg cell is located at the chalazal end of the cell (Fig. 10.5d). The opposite polarity of the egg cell and central cell brings the two nuclei that will be fused with sperm nuclei respectively close to each other. The polarity of gametophytic cells may be important for double fertilization, as the sperm nuclei are released in synergids. The polarity of the female gametophyte is established through coordinated control of nuclear division and positioning, cellularization, and vacuole expansion. The migration and precise localization of the nuclei during megasporogenesis indicate that the cell fates of female gametophyte are programmed prior to cellularization.

Key factors related to polarity of the female gametophyte and gametophytic cell specification have been identified, and current data indicate that the proper positioning of nuclei is a prerequisite for gametophytic cell fate specification. The mispositioning of nuclei often causes altered cell identity. In the *eostre* mutant, nuclear migration abnormalities result in two egg cells and one synergid cell formation. The *eostre* phenotype is caused by ectopic expression of the homeodomain protein BEL1-like homeodomain 1 (BLH1) in the embryo sac, which interfere its interaction with the KNOX protein KNAT3 (Pagnussat et al., 2007). In the maize *indeterminate gametophyte 1* (*ig1*) mutant, additional free nuclear division occur, which is associated with extra egg cells and extra central cells whose fate also likely depend on their relative position in the female gametophyte (Evans, 2007). IG1 is a LATERAL ORGAN BOUNDARIES (LOB) domain protein and expressed in the functional megaspore and the subsequent egg cell and antipodal cells, but not in the synergid cells (Gallois et al., 2009). The asymmetrical expression of *IG1* in the embryo sac suggested the existence of embryo sac polarity.

An asymmetric auxin gradient in the syncytial female gametophyte was shown to play an essential role in gametic cell fate determination (Pagnussat et al., 2009). An auxin maximum is formed at the micropylar pole of the developing syncytial female gametophyte, as shown in the synthetic *DR5::GFP* or *DR5::GUS* reporter sensitive to auxin. Downregulating *AUXIN RESPONSE FACTOR* (*ARF*) genes expression within the female gametophyte simulates reduced concentration of auxin, which results in aberrant gametophytic cell identities at the micropylar end, showing synergid identity replaced with egg cell identity. The auxin biosynthesis YUCCA genes are preferentially expressed on the micropylar region of the nucellus and female gametophyte during early embryo sac development. Overexpression of the auxin biosynthetic *YUCCA1* gene in female gametophyte compromises gametophytic cell specification. These data suggest that auxin acts as a positional signal to direct the fate of the gametophytic cell along the micropyle-chalazal axis. Consistently, the auxin efflux facilitators PIN proteins are not expressed in the female gametophyte and PIN1 was expressed in the nucellus during the earliest stages of female gametophyte development (Pagnussat et al., 2009). This indicates that the auxin gradient established by a correlation between position-specific auxin biosynthesis and diffusion plays an important role in cell differentiation of female gametophyte.

A lateral inhibition mechanism is the second strategy that determines cell fate within female gametophyte, according to which female gametes inhibit gamete identity in adjacent accessory cells (Fig. 10.7). Three mutants, *lachesis* (*lis*), *clotho* (*clo*), and *atropos* (*ato*) that displayed altered expression of an egg cell marker, were identified (Gross-Hardt et al., 2007; Moll et al., 2008). In these mutant embryo sacs, the expression of the egg cell-specific marker is expanded to the synergids and central cell, while the antipodal cells express a central cell marker, indicating that accessory cells are recruited as gametic cells. The corresponding genes encode proteins that are components of the spliceosome, which may be involved in processing specific genes within the gametic cells to produce a signal molecule that inhibit gamete differentiation in the accessory cells (Gross-Hardt et al., 2007, Moll et al., 2008). This lateral inhibition mechanism involves in communication between the egg cell and the central cell, as well as cellular signaling between the gametic cell and accessory cells. The developmental plasticity of the gametophytic cells could serve as a backup mechanism to enhance the robustness of female gametophyte (Gross-Hardt et al., 2007; Lituiev & Grossniklaus, 2014; Tekleyohans et al., 2017).

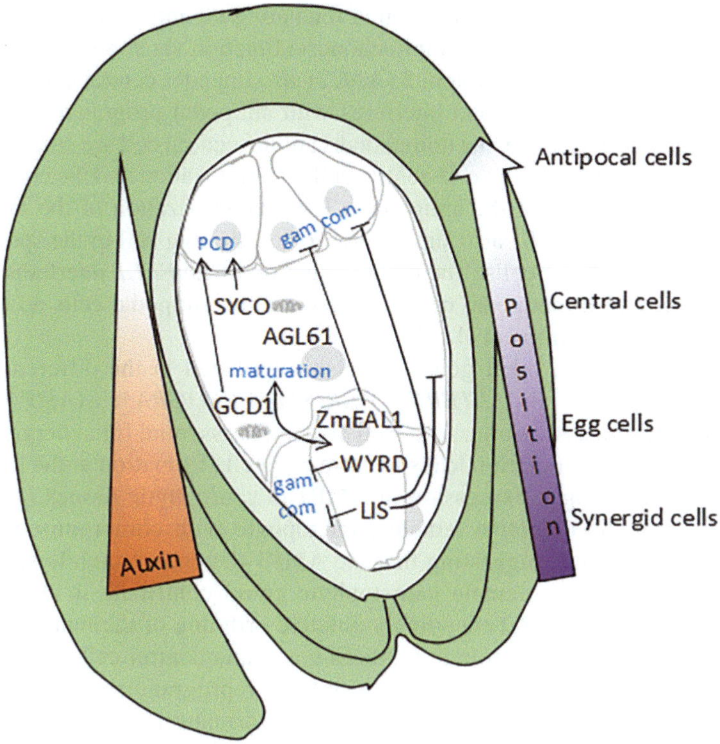

Fig. 10.7 Cell-cell communication involved in cell differentiation in female gametophyte. Auxin mediated positional signal is responsible for the proper differentiation of the cell within the female gametophyte. ZmEAL1, WYRD, and LIS involved lateral inhibition mechanism, in which female gametes repress gamete identity in adjacent accessory cells, is important for the cell fate determination within female gametophyte. GCD1-, AGL61-, and SYCO-mediated central cell communication with the egg cell, synergid cells, and antipodals also plays a role in maintenance of cell identity, gamete maturation, and antipodals degeneration. gam com.: gametic competence; PCD, programmed cell death

The characterization of other mutants further demonstrates the role of gametic cells in the specification of other cells in the female gametophyte. The *agamous-like61/diana* mutant ovules exhibit early degeneration of the central cell. After degeneration, the expression of synergid and egg cell markers spreads throughout the chalazal pole, associated with an expanded expression of antipodal markers to the micropylar pole (Bemer et al., 2008). This demonstrates an important major role for the central cell in determining cellular identity of adjacent cells. In addition, *fiona/syco-1* mutant exhibits affected integrity of the mitochondria of the central cell and defective fusion of polar nuclei. *SYCO* encoding a cysteinyl-tRNA synthetase is specifically expressed in the central cell, but mutation in *SYCO* increases the life span of the antipodal cells (Kägi et al., 2010), as observed in *lis* and *clotho* mutants (Gross-Hardt et al., 2007, Moll et al., 2008). GAMETE CELL DEFECTIVE 1 (GCD1), enriched in female gametophyte, plays an important role in central cell and egg cell maturation via its function in regulating the integrity of mitochondria. Targeted disruption of central cell mitochondria function via site-specific mutagenesis of the ATP/ADP carrier protein 2 (AAC2) also impedes central cell maturation, blocks egg cell maturation, and interferes with antipodal programmed cell death (Wu et al., 2012). Therefore, the mitochondria of the central cell are essential for the fusion of the polar nuclei, egg cell-central cell communication and the control of the programmed cell death of the antipodals. The characterization of the maize egg-cell-specific gene *ZmEAL1* also support the role of the egg cell in the specification of female gametophytic cells. The downregulation of *ZmEAL1* interferes with the expression of *IG1* in antipodal cells and induces the antipodal cells adopting the central cell identity (Krohn et al., 2012).

In addition, several other genes are also implicated in the differentiation of female gametophyte cells. *ALTERED MERISTEM PROGRAM 1* (*AMP1*), encoding an endoplasmic reticulum associated protein, is essential for synergid cell fate specification. *AMP1* mutation leads to extra egg cells formation at the expense of synergids. Interestingly, expression of AMP1 in sporophytic tissues is sufficient to complement the defective female gametophyte with compromised synergid identity in the mutant, suggesting that the AMP1-dependent mobile signals provided from outside the female gametophyte could contribute to its patterning (Kong et al., 2015). *WYRD* encoding a putative ortholog of the inner centromere protein (INCENP) plays a role in regulating egg and central cell fate and differentiation. Arabidopsis *wyrd* mutant exhibits a reprogramming of cell fate in female gametophyte with additional egg cells formation at the expense of the accessory synergids (Kirioukhova et al., 2011). Similarly, *rkd3* and *rkd4* mutants also display abnormal cell identity showing two egg cells and one synergid within one female gametophyte (Tedeschi et al., 2017). The five *RWP-RK domain-containing* (*RKD*) transcription factor genes encoded in Arabidopsis genome are expressed during different stages of female gametophyte development. Overexpression of *RKD1* and *RKD2* induces the formation of egg cell-like structure within the female gametophyte as well as in the sporophytic tissues (Koszegi et al., 2011; Tedeschi et al., 2017). Interestingly, the only *RKD* gene in liverwort was also demonstrated to function in germ cell differentiation and gametophyte–sporophyte transition (Koi et al., 2016; Rövekamp et al., 2016), indicating the conserved role of RKDs in the control of cell differentiation. *VERDANDI*, a putative plant-specific B3 superfamily transcription factor gene, acts downstream of the three MADS box genes SEEDSTICK (STK), SHATTERPROOF1 (SHP1), and SHP2 that redundantly regulate ovule development. Loss of *VERDANDI* can lead to compromised accessory cell (synergid and antipodal) specification, as well as defects in antipodal programmed cell death. An inversion of the expression of antipodal cell marker in the synergids and the expression of synergid marker in the antipodals was observed in *verdandi* mutant ovules (Matias-Hernandez et al., 2010).

10.12 Conclusions

Molecular mechanisms that control the gynoecium development and ovule development have been extensively studied in the past three decades. Significant progresses have been made in understanding the development of carpel marginal meristem, the definition of different carpel domains, patterning of gynoecium, megasporocyte specification, and intercellular interactions in female gametogenesis. The gynoecium is a highly complex female reproductive structure specifically used for fertilization and seed production. The ovule, located inside the carpel of a gynoecium, provides us an exciting system to investigate fundamental questions in biology as the developmental events require coordination development between the sporophyte and the female gametophyte. But accessibility of this reproductive tissue still remains a challenge. Emerging new techniques including live cell imaging, single-cell sequencing, and targeted gene editing will revolutionize the way through which we can reveal the genetic complexity and molecular mechanisms that govern ovary and ovule development at the cellular and molecular level. In the near future, significant progress is expected to be made in understanding this agronomically important miniature reproductive structure.

Acknowledgments This work was supported by the National Natural Science Foundation of China (NSFC; U1605212, 31761130074, 31970333, and 31571264), a Guangxi Distinguished Experts Fellowship, a Newton Advanced Fellowship (NA160391), and the Fujian Natural Science Foundation (2019J01425), the Natural Science Foundation of Anhui Province (2308085MC70), the key project of the Education Department of Anhui Province (2023AH051850), the Excellent Scientific Research and Innovation team of the Education Department of Anhui Province (2022AH010087), the Scientific and Technological Innovation Team of Anhui University of Science and Technology (2023KJCXTD001), and the Talent Introduction Project of Anhui Science and Technology University (NXYJ202104).

References

Acosta-García, G., & Vielle-Calzada, J. P. (2004). A classical arabinogalactan protein is essential for the initiation of female gametogenesis in Arabidopsis. *Plant Cell, 16*, 2614–2628.

Agashe, B., Prasad, C. K., & Siddiqi, I. (2002). Identification and analysis of DYAD: A gene required for meiotic chromosome organisation and female meiotic progression in Arabidopsis. *Development, 129*, 3935–3943.

Alvarez, J., & Smyth, D. R. (1999). CRABS CLAW and SPATULA, two Arabidopsis genes that control carpel development in parallel with AGAMOUS. *Development, 126*, 2377–2386.

Alvarez, J. P., Goldshmidt, A., Efroni, I., Bowman, J. L., & Eshed, Y. (2009). The NGATHA distal organ development genes are essential for style specification in Arabidopsis. *Plant Cell, 21*, 1373–1393.

Armenta-Medina, A., Demesa-Arévalo, E., & Vielle-Calzada, J. P. (2011). Epigenetic control of cell specification during female gametogenesis. *Sexual Plant Reproduction, 24*, 137–147.

Azhakanandam, S., Nole-Wilson, S., Bao, F., & Franks, R. G. (2008). SEUSS and AINTEGUMENTA mediate patterning and ovule initiation during gynoecium medial domain development. *Plant Physiology, 146*, 1165–1181.

Bai, X., Peirson, B. N., Dong, F., Xue, C., & Makaroff, C. A. (1999). Isolation and characterization of SYN1, a RAD21-like gene essential for meiosis in Arabidopsis. *Plant Cell, 11*, 417–430.

Bajon, C., Horlow, C., Motamayor, J. C., Sauvanet, A., & Robert, D. (1999). Megasporogenesis in Arabidopsis thaliana L.: An ultrastructural study. *Sexual Plant Reproduction, 12*, 99–109.

Balanzá, V., Navarrete, M., Trigueros, M., & Ferrándiz, C. (2006). Patterning the female side of Arabidopsis: The importance of hormones. *Journal of Experimental Botany, 57*, 3457–3469.

Baroux, C., & Autran, D. (2015). Chromatin dynamics during cellular differentiation in the female reproductive lineage of flowering plants. *The Plant Journal, 83*, 160–176.

Bartrina, I., Otto, E., Strnad, M., Werner, T., & Schmülling, T. (2011). Cytokinin regulates the activity of reproductive meristems, flower organ size, ovule formation, and thus seed yield in Arabidopsis thaliana. *Plant Cell, 23*, 69–80.

Bell, P. R. F. (1996). Megaspore abortion: A consequence of selective apoptosis? *International Journal of Plant Sciences, 157*, 1–7.

Bemer, M., Wolters-Arts, M., Grossniklaus, U., & Angenent, G. C. (2008). The MADS domain protein DIANA acts together with AGAMOUS-LIKE80 to specify the central cell in Arabidopsis ovules. *Plant Cell, 20*, 2088–2101.

Bencivenga, S., Colombo, L., & Masiero, S. (2011). Cross talk between the sporophyte and the megagametophyte during ovule development. *Sexual Plant Reproduction, 24*, 113–121.

Benková, E., Michniewicz, M., Sauer, M., et al. (2003). Local, efflux-dependent auxin gradients as a common module for plant organ formation. *Cell, 115*, 591–602.

Bowman, J. L., & Smyth, D. R. (1999). CRABS CLAW, a gene that regulates carpel and nectary development in Arabidopsis, encodes a novel protein with zinc finger and helix-loop-helix domains. *Development, 126*, 2387–2396.

Bowman, J. L., Smyth, D. R., & Meyerowitz, E. M. (1989). Genes directing flower development in Arabidopsis. *Plant Cell, 1*, 37–52.

Cai, H., Huang, Y., Liu, L., et al. (2023a). Signaling by the EPFL-ERECTA family coordinates female germline specification through the BZR1 family in Arabidopsis. *Plant Cell, 35*, 1455–1473.

Cai, H., Liu, L., Huang, Y., et al. (2022a). Brassinosteroid signaling regulates female germline specification in Arabidopsis. *Current Biology, 32*, 1102–14.e5.

Cai, H., Liu, L., Ma, S., Aslam, M., & Qin, Y. (2023b). Insights into the role of phytohormones in plant female germline cell specification. *Current Opinion in Plant Biology, 75*, 102439.

Cai, H., Ma, S., Su, H., Liu, K., Aslam, M., & Qin, Y. (2022b). Positional signals establishment in the regulation of female germline specification. *Seed Biology, 1*, 1–7.

Cao, L., Wang, S., Venglat, P., et al. (2018). Arabidopsis ICK/KRP cyclin-dependent kinase inhibitors function to ensure the formation of one megaspore mother cell and one functional megaspore per ovule. *PLoS Genetics, 14*, e1007230.

Capron, A., Serralbo, O., Fülöp, K., et al. (2003). The Arabidopsis anaphase-promoting complex or cyclosome: Molecular and genetic characterization of the APC2 subunit. *Plant Cell, 15*, 2370–2382.

Cheng, Y., Dai, X., & Zhao, Y. (2006). Auxin biosynthesis by the YUCCA flavin monooxygenases controls the formation of floral organs and vascular tissues in Arabidopsis. *Genes & Development, 20*, 1790–1799.

Cheng, Z. J., Wang, L., Sun, W., et al. (2013). Pattern of auxin and cytokinin responses for shoot meristem induction results from the regulation of cytokinin biosynthesis by AUXIN RESPONSE FACTOR3. *Plant Physiology, 161*, 240–251.

Christensen, C. A., Subramanian, S., & Drews, G. N. (1998). Identification of gametophytic mutations affecting female gametophyte development in Arabidopsis. *Developmental Biology, 202*, 136–151.

Cigliano, R. A., Cremona, G., Paparo, R., et al. (2013). Histone deacetylase AtHDA7 is required for female gametophyte and embryo development in Arabidopsis. *Plant Physiology, 163*, 431–440.

Coimbra, S., Almeida, J., Junqueira, V., Costa, M. L., & Pereira, L. G. (2007). Arabinogalactan proteins as molecular markers in Arabidopsis thaliana sexual reproduction. *Journal of Experimental Botany, 58*, 4027–4035.

Colombo, M., Masiero, S., Vanzulli, S., Lardelli, P., Kater, M. M., & Colombo, L. (2008). AGL23, a type I MADS-box gene that controls female gametophyte and embryo development in Arabidopsis. *The Plant Journal, 54*, 1037–1048.

Couteau, F., Belzile, F., Horlow, C., Grandjean, O., Vezon, D., & Doutriaux, M. P. (1999). Random chromosome segregation without meiotic arrest in both male and female meiocytes of a dmc1 mutant of Arabidopsis. *Plant Cell, 11*, 1623–1634.

Demesa-Arévalo, E., & Vielle-Calzada, J. P. (2013). The classical arabinogalactan protein AGP18 mediates megaspore selection in Arabidopsis. *Plant Cell, 25*, 1274–1287.

Deng, Y., Dong, H., Mu, J., et al. (2010). Arabidopsis histidine kinase CKI1 acts upstream of histidine phosphotransfer proteins to regulate female gametophyte development and vegetative growth. *Plant Cell, 22*, 1232–1248.

Drews, G. N., & Koltunow, A. M. (2011). The female gametophyte. *The Arabidopsis Book, 9*, e0155.

Ebel, C., Mariconti, L., & Gruissem, W. (2004). Plant retinoblastoma homologues control nuclear proliferation in the female gametophyte. *Nature, 429*, 776–780.

Eklund, D. M., Ståldal, V., Valsecchi, I., et al. (2010). The Arabidopsis thaliana STYLISH1 protein acts as a transcriptional activator regulating auxin biosynthesis. *Plant Cell, 22*, 349–363.

Elliott, R. C., Betzner, A. S., Huttner, E., et al. (1996). AINTEGUMENTA, an APETALA2-like gene of Arabidopsis with pleiotropic roles in ovule development and floral organ growth. *Plant Cell, 8*, 155–168.

Escobar-Guzmán, R., Rodríguez-Leal, D., Vielle-Calzada, J. P., & Ronceret, A. (2015). Whole-mount immunolocalization to study female meiosis in Arabidopsis. *Nature Protocols, 10*, 1535–1542.

Eshed, Y., Baum, S. F., & Bowman, J. L. (1999). Distinct mechanisms promote polarity establishment in carpels of Arabidopsis. *Cell, 99*, 199–209.

Evans, M. M. (2007). The indeterminate gametophyte1 gene of maize encodes a LOB domain protein required for embryo Sac and leaf development. *Plant Cell, 19*, 46–62.

Franks, R. G., Wang, C., Levin, J. Z., & Liu, Z. (2002). SEUSS, a member of a novel family of plant regulatory proteins, represses floral homeotic gene expression with LEUNIG. *Development, 129*, 253–263.

Gallois, J. L., Guyon-Debast, A., Lécureuil, A., et al. (2009). The Arabidopsis proteasome RPT5 subunits are essential for gametophyte development and show accession-dependent redundancy. *Plant Cell, 21*, 442–459.

Giraut, L., Falque, M., Drouaud, J., Pereira, L., Martin, O. C., & Mézard, C. (2011). Genome-wide crossover distribution in Arabidopsis thaliana meiosis reveals sex-specific patterns along chromosomes. *PLoS Genetics, 7*, e1002354.

Girin, T., Paicu, T., Stephenson, P., et al. (2011). INDEHISCENT and SPATULA interact to specify carpel and valve margin tissue and thus promote seed dispersal in Arabidopsis. *Plant Cell, 23*, 3641–3653.

Gremski, K., Ditta, G., & Yanofsky, M. F. (2007). The HECATE genes regulate female reproductive tract development in Arabidopsis thaliana. *Development, 134*, 3593–3601.

Gross-Hardt, R., Kägi, C., Baumann, N., et al. (2007). LACHESIS restricts gametic cell fate in the female gametophyte of Arabidopsis. *PLoS Biology, 5*, e47.

Guo, F., Huang, B. Q., Han, Y., & Zee, S. Y. (2004). Fertilization in maize indeterminate gametophyte1 mutant. *Protoplasma, 223*, 111–120.

Hackett, J. A., Zylicz, J. J., & Surani, M. A. (2012). Parallel mechanisms of epigenetic reprogramming in the germline. *Trends in Genetics, 28*, 164–174.

Hou, Z., Liu, Y., Zhang, M., et al. (2021). High-throughput single-cell transcriptomics reveals the female germline differentiation trajectory in Arabidopsis thaliana. *Communications Biology, 4*, 1149.

Huanca-Mamani, W., Garcia-Aguilar, M., León-Martínez, G., Grossniklaus, U., & Vielle-Calzada, J. P. (2005). CHR11, a chromatin-remodeling factor essential for nuclear proliferation during female gametogenesis in Arabidopsis thaliana. *Proceedings of the National Academy of Sciences of the United States of America, 102*, 17231–17236.

Huang, J., Zhao, L., Malik, S., et al. (2022). Specification of female germline by microRNA orchestrated auxin signaling in Arabidopsis. *Nature Communications, 13*, 6960.

Huang, Y., Liu, L., Chai, M., et al. (2023). Epigenetic regulation of female germline development through ERECTA signaling pathway. *The New Phytologist, 240*, 1015–1033.

Ito, T., Wellmer, F., Yu, H., et al. (2004). The homeotic protein AGAMOUS controls microsporogenesis by regulation of SPOROCYTELESS. *Nature, 430*, 356–360.

Kägi, C., Baumann, N., Nielsen, N., Stierhof, Y. D., & Gross-Hardt, R. (2010). The gametic central cell of Arabidopsis determines the lifespan of adjacent accessory cells. *Proceedings of the National Academy of Sciences of the United States of America, 107*, 22350–22355.

Kamiuchi, Y., Yamamoto, K., Furutani, M., Tasaka, M., & Aida, M. (2014). The CUC1 and CUC2 genes promote carpel margin meristem formation during Arabidopsis gynoecium development. *Frontiers in Plant Science, 5*, 165.

Kieffer, M., Master, V., Waites, R., & Davies, B. (2011). TCP14 and TCP15 affect internode length and leaf shape in Arabidopsis. *The Plant Journal, 68*, 147–158.

Kirioukhova, O., Johnston, A. J., Kleen, D., et al. (2011). Female gametophytic cell specification and seed development require the function of the putative Arabidopsis INCENP ortholog WYRD. *Development, 138*, 3409–3420.

Koi, S., Hisanaga, T., Sato, K., et al. (2016). An evolutionarily conserved plant RKD factor controls germ cell differentiation. *Current Biology, 26*, 1775–1781.

Kong, J., Lau, S., & Jürgens, G. (2015). Twin plants from supernumerary egg cells in Arabidopsis. *Current Biology, 25*, 225–230.

Koszegi, D., Johnston, A. J., Rutten, T., et al. (2011). Members of the RKD transcription factor family induce an egg cell-like gene expression program. *The Plant Journal, 67*, 280–291.

Krohn, N. G., Lausser, A., Juranić, M., & Dresselhaus, T. (2012). Egg cell signaling by the secreted peptide ZmEAL1 controls antipodal cell fate. *Developmental Cell, 23*, 219–225.

Kuusk, S., Sohlberg, J. J., Long, J. A., Fridborg, I., & Sundberg, E. (2002). STY1 and STY2 promote the formation of apical tissues during Arabidopsis gynoecium development. *Development, 129*, 4707–4717.

Kuusk, S., Sohlberg, J. J., Magnus Eklund, D., & Sundberg, E. (2006). Functionally redundant SHI family genes regulate Arabidopsis gynoecium development in a dose-dependent manner. *The Plant Journal, 47*, 99–111.

Kwee, H. S., & Sundaresan, V. (2003). The NOMEGA gene required for female gametophyte development encodes the putative APC6/CDC16 component of the Anaphase Promoting Complex in Arabidopsis. *The Plant Journal, 36*, 853–866.

Larsson, E., Franks, R. G., & Sundberg, E. (2013). Auxin and the Arabidopsis thaliana gynoecium. *Journal of Experimental Botany, 64*, 2619–2627.

Larsson, E., Roberts, C. J., Claes, A. R., Franks, R. G., & Sundberg, E. (2014). Polar auxin transport is essential for medial versus lateral tissue specification and vascular-mediated valve outgrowth in Arabidopsis gynoecia. *Plant Physiology, 166*, 1998–2012.

Latrasse, D., Benhamed, M., Henry, Y., et al. (2008). The MYST histone acetyltransferases are essential for gametophyte development in Arabidopsis. *BMC Plant Biology, 8*, 121.

Lê, Q., Gutièrrez-Marcos, J. F., Costa, L. M., et al. (2005). Construction and screening of subtracted cDNA libraries from limited populations of plant cells: A comparative analysis of gene expression between maize egg cells and central cells. *The Plant Journal, 44*, 167–178.

Li, L. C., Qin, G. J., Tsuge, T., et al. (2008). SPOROCYTELESS modulates YUCCA expression to regulate the development of lateral organs in Arabidopsis. *The New Phytologist, 179*, 751–764.

Lieber, D., Lora, J., Schrempp, S., Lenhard, M., & Laux, T. (2011). Arabidopsis WIH1 and WIH2 genes act in the transition from somatic to reproductive cell fate. *Current Biology, 21*, 1009–1017.

Liljegren, S. J., Ditta, G. S., Eshed, Y., Savidge, B., Bowman, J. L., & Yanofsky, M. F. (2000). SHATTERPROOF MADS-box genes control seed dispersal in Arabidopsis. *Nature, 404*, 766–770.

Liljegren, S. J., Roeder, A. H., Kempin, S. A., et al. (2004). Control of fruit patterning in Arabidopsis by INDEHISCENT. *Cell, 116*, 843–853.

Lituiev, D. S., & Grossniklaus, U. (2014). Patterning of the angiosperm female gametophyte through the prism of theoretical paradigms. *Biochemical Society Transactions, 42*, 332–339.

Liu, J., Zhang, Y., Qin, G., et al. (2008). Targeted degradation of the cyclin-dependent kinase inhibitor ICK4/KRP6 by RING-type E3 ligases is essential for mitotic cell cycle progression during Arabidopsis gametogenesis. *Plant Cell, 20*, 1538–1554.

Liu, Z., Franks, R. G., & Klink, V. P. (2000). Regulation of gynoecium marginal tissue formation by LEUNIG and AINTEGUMENTA. *Plant Cell, 12*, 1879–1892.

Liu, Z., Yuan, L., Song, X., Yu, X., & Sundaresan, V. (2017). AHP2, AHP3, and AHP5 act downstream of CKI1 in Arabidopsis female gametophyte development. *Journal of Experimental Botany, 68*, 3365–3373.

Long, J. A., Moan, E. I., Medford, J. I., & Barton, M. K. (1996). A member of the KNOTTED class of homeodomain proteins encoded by the STM gene of Arabidopsis. *Nature, 379*, 66–69.

Lora, J., Herrero, M., Tucker, M. R., & Hormaza, J. I. (2017). The transition from somatic to germline identity shows conserved and specialized features during angiosperm evolution. *The New Phytologist, 216*, 495–509.

Lucero, L. E., Uberti-Manassero, N. G., Arce, A. L., Colombatti, F., Alemano, S. G., & Gonzalez, D. H. (2015). TCP15 modulates cytokinin and auxin responses during gynoecium development in Arabidopsis. *The Plant Journal, 84*, 267–282.

Marsch-Martínez, N., & De Folter, S. (2016). Hormonal control of the development of the gynoecium. *Current Opinion in Plant Biology, 29*, 104–114.

Marsch-Martínez, N., Ramos-Cruz, D., Irepan Reyes-Olalde, J., Lozano-Sotomayor, P., Zúñiga-Mayo, V. M., & De Folter, S. (2012). The role of cytokinin during Arabidopsis gynoecia and fruit morphogenesis and patterning. *The Plant Journal, 72*, 222–234.

Martínez-Fernández, I., Sanchís, S., Marini, N., et al. (2014). The effect of NGATHA altered activity on auxin signaling pathways within the Arabidopsis gynoecium. *Frontiers in Plant Science, 5*, 210.

Matias-Hernandez, L., Battaglia, R., Galbiati, F., et al. (2010). VERDANDI is a direct target of the MADS domain ovule identity complex and affects embryo sac differentiation in Arabidopsis. *Plant Cell, 22*, 1702–1715.

Mercier, R., Vezon, D., Bullier, E., et al. (2001). SWITCH1 (SWI1): A novel protein required for the establishment of sister chromatid cohesion and for bivalent formation at meiosis. *Genes & Development, 15*, 1859–1871.

Moll, C., Von Lyncker, L., Zimmermann, S., et al. (2008). CLO/GFA1 and ATO are novel regulators of gametic cell fate in plants. *The Plant Journal, 56*, 913–921.

Motamayor, J. C., Vezon, D., Bajon, C., et al. (2000). Switch (swi1), an Arabidopsis thaliana mutant affected in the female meiotic switch. *Sexual Plant Reproduction, 12*, 209–218.

Nahar, M. A., Ishida, T., Smyth, D. R., Tasaka, M., & Aida, M. (2012). Interactions of CUP-SHAPED COTYLEDON and SPATULA genes control carpel margin development in Arabidopsis thaliana. *Plant & Cell Physiology, 53*, 1134–1143.

Nole-Wilson, S., & Krizek, B. A. (2006). AINTEGUMENTA contributes to organ polarity and regulates growth of lateral organs in combination with YABBY genes. *Plant Physiology, 141*, 977–987.

Nole-Wilson, S., Rueschhoff, E. E., Bhatti, H., & Franks, R. G. (2010). Synergistic disruptions in seuss cyp85A2 double mutants reveal a role for brassinolide synthesis during gynoecium and ovule development. *BMC Plant Biology, 10*, 198.

Nonomura, K., Miyoshi, K., Eiguchi, M., et al. (2003). The MSP1 gene is necessary to restrict the number of cells entering into male and female sporogenesis and to initiate anther wall formation in rice. *Plant Cell, 15*, 1728–1739.

Nonomura, K., Morohoshi, A., Nakano, M., et al. (2007). A germ cell specific gene of the ARGONAUTE family is essential for the progression of premeiotic mitosis and meiosis during sporogenesis in rice. *Plant Cell, 19*, 2583–2594.

Oh, S. A., Johnson, A., Smertenko, A., et al. (2005). A divergent cellular role for the FUSED kinase family in the plant-specific cytokinetic phragmoplast. *Current Biology, 15*, 2107–2111.

Okada, K., Ueda, J., Komaki, M. K., Bell, C. J., & Shimura, Y. (1991). Requirement of the auxin polar transport system in early stages of Arabidopsis floral bud formation. *Plant Cell, 3*, 677–684.

Olmedo-Monfil, V., Durán-Figueroa, N., Arteaga-Vázquez, M., et al. (2010). Control of female gamete formation by a small RNA pathway in Arabidopsis. *Nature, 464*, 628–632.

Østergaard, L. (2009). Don't "leaf" now. The making of a fruit. *Current Opinion in Plant Biology, 12*, 36–41.

Pagnussat, G. C., Alandete-Saez, M., Bowman, J. L., & Sundaresan, V. (2009). Auxin-dependent patterning and gamete specification in the Arabidopsis female gametophyte. *Science, 324*, 1684–1689.

Pagnussat, G. C., Yu, H. J., & Sundaresan, V. (2007). Cell-fate switch of synergid to egg cell in Arabidopsis eostre mutant embryo sacs arises from misexpression of the BEL1-like homeodomain gene BLH1. *Plant Cell, 19*, 3578–3592.

Park, S. K., Rahman, D., Oh, S. A., & Twell, D. (2004). Gemini pollen 2, a male and female gametophytic cytokinesis defective mutation. *Sexual Plant Reproduction, 17*, 63–70.

Pastuglia, M., Azimzadeh, J., Goussot, M., et al. (2006). Gamma-tubulin is essential for microtubule organization and development in Arabidopsis. *Plant Cell, 18*, 1412–1425.

Pawlowski, W. P., Wang, C. J., Golubovskaya, I. N., et al. (2009). Maize AMEIOTIC1 is essential for multiple early meiotic processes and likely required for the initiation of meiosis. *Proceedings of the National Academy of Sciences of the United States of America, 106*, 3603–3608.

Pischke, M. S., Jones, L. G., Otsuga, D., Fernandez, D. E., Drews, G. N., & Sussman, M. R. (2002). An Arabidopsis histidine kinase is essential for megagametogenesis. *Proceedings of the National Academy of Sciences of the United States of America, 99*, 15800–15805.

Qin, Y., Zhao, L., Skaggs, M. I., et al. (2014). ACTIN-RELATED PROTEIN6 regulates female meiosis by modulating meiotic gene expression in Arabidopsis. *Plant Cell, 26*, 1612–1628.

Rabiger, D. S., & Drews, G. N. (2013). MYB64 and MYB119 are required for cellularization and differentiation during female gametogenesis in Arabidopsis thaliana. *PLoS Genetics, 9*, e1003783.

Rajani, S., & Sundaresan, V. (2001). The Arabidopsis myc/bHLH gene ALCATRAZ enables cell separation in fruit dehiscence. *Current Biology, 11*, 1914–1922.

Ravi, M., Marimuthu, M. P., & Siddiqi, I. (2008). Gamete formation without meiosis in Arabidopsis. *Nature, 451*, 1121–1124.

Reyes-Olalde, J. I., Zúñiga-Mayo, V. M., Serwatowska, J., et al. (2017). The bHLH transcription factor SPATULA enables cytokinin signaling, and both activate auxin biosynthesis and transport genes at the medial domain of the gynoecium. *PLoS Genetics, 13*, e1006726.

Rodríguez-Leal, D., León-Martínez, G., Abad-Vivero, U., & Vielle-Calzada, J. P. (2015). Natural variation in epigenetic pathways affects the specification of female gamete precursors in Arabidopsis. *Plant Cell, 27*, 1034–1045.

Roeder, A. H., & Yanofsky, M. F. (2006). Fruit development in Arabidopsis. *The Arabidopsis Book, 4*, e0075.

Ronceret, A., & Pawlowski, W. P. (2010). Chromosome dynamics in meiotic prophase I in plants. *Cytogenetic and Genome Research, 129*, 173–183.

Rosa, M., Von Harder, M., Cigliano, R. A., Schlögelhofer, P., & Mittelsten Scheid, O. (2013). The Arabidopsis SWR1 chromatin-remodeling complex is important for DNA repair, somatic recombination, and meiosis. *Plant Cell, 25*, 1990–2001.

Rövekamp, M., Bowman, J. L., & Grossniklaus, U. (2016). Marchantia MpRKD regulates the gametophyte-sporophyte transition by keeping egg cells quiescent in the absence of fertilization. *Current Biology, 26*, 1782–1789.

Russell, S. D. (1979). Fine structure of megagametophyte development in Zea mays. *Botany, 57*, 1093–1110.

Schiefthaler, U., Balasubramanian, S., Sieber, P., Chevalier, D., Wisman, E., & Schneitz, K. (1999). Molecular analysis of NOZZLE, a gene involved in pattern formation and early sporogenesis during sex organ development in Arabidopsis thaliana. *Proceedings of the National Academy of Sciences of the United States of America, 96*, 11664–11669.

Schuster, C., Gaillochet, C., & Lohmann, J. U. (2015). Arabidopsis HECATE genes function in phytohormone control during gynoecium development. *Development, 142*, 3343–3350.

Scofield, S., Dewitte, W., & Murray, J. A. (2007). The KNOX gene SHOOT MERISTEMLESS is required for the development of reproductive meristematic tissues in Arabidopsis. *The Plant Journal, 50*, 767–781.

Sebastian, J., Ravi, M., Andreuzza, S., Panoli, A. P., Marimuthu, M. P., & Siddiqi, I. (2009). The plant adherin AtSCC2 is required for embryogenesis and sister-chromatid cohesion during meiosis in Arabidopsis. *The Plant Journal, 59*, 1–13.

Sessions, R. (1997). Arabidopsis (Brassicaceae) flower development and gynoecium patterning in wild type and Ettin mutants. *American Journal of Botany, 84*, 1179.

Sessions, R. A., & Zambryski, P. C. (1995). Arabidopsis gynoecium structure in the wild and in ettin mutants. *Development, 121*, 1519–1532.

She, W., & Baroux, C. (2015). Chromatin dynamics in pollen mother cells underpin a common scenario at the somatic-to-reproductive fate transition of both the male and female lineages in Arabidopsis. *Frontiers in Plant Science, 6*, 294.

She, W., Grimanelli, D., Rutowicz, K., et al. (2013). Chromatin reprogramming during the somatic-to-reproductive cell fate transition in plants. *Development, 140*, 4008–4019.

Sheridan, W. F., Avalkina, N. A., Shamrov, I., Batygina, T. B., & Golubovskaya, I. N. (1996). The mac1 gene: Controlling the commitment to the meiotic pathway in maize. *Genetics, 142*, 1009–1020.

Sheridan, W. F., Golubeva, E. A., Abrhamova, L. I., & Golubovskaya, I. N. (1999). The mac1 mutation alters the developmental fate of the hypodermal cells and their cellular progeny in the maize anther. *Genetics, 153*, 933–941.

Shi, D. Q., Liu, J., Xiang, Y. H., Ye, D., Sundaresan, V., & Yang, W. C. (2005). SLOW WALKER1, essential for gametogenesis in Arabidopsis, encodes a WD40 protein involved in 18S ribosomal RNA biogenesis. *Plant Cell, 17*, 2340–2354.

Shi, D. Q., & Yang, W. C. (2011). Ovule development in Arabidopsis: Progress and challenge. *Current Opinion in Plant Biology, 14*, 74–80.

Siddiqi, I., Ganesh, G., Grossniklaus, U., & Subbiah, V. (2000). The dyad gene is required for progression through female meiosis in Arabidopsis. *Development, 127*, 197–207.

Singh, M., Goel, S., Meeley, R. B., et al. (2011). Production of viable gametes without meiosis in maize deficient for an ARGONAUTE protein. *Plant Cell, 23*, 443–458.

Smyth, D. R., Bowman, J. L., & Meyerowitz, E. M. (1990). Early flower development in Arabidopsis. *Plant Cell, 2*, 755–767.

Sohlberg, J. J., Myrenås, M., Kuusk, S., et al. (2006). STY1 regulates auxin homeostasis and affects apical-basal patterning of the Arabidopsis gynoecium. *The Plant Journal, 47*, 112–123.

Sorefan, K., Girin, T., Liljegren, S. J., et al. (2009). A regulated auxin minimum is required for seed dispersal in Arabidopsis. *Nature, 459*, 583–586.

Springer, P. S., Holding, D. R., Groover, A., Yordan, C., & Martienssen, R. A. (2000). The essential Mcm7 protein PROLIFERA is localized to the nucleus of dividing cells during the G(1) phase and is required maternally for early Arabidopsis development. *Development, 127*, 1815–1822.

Sridhar, V. V., Surendrarao, A., Gonzalez, D., Conlan, R. S., & Liu, Z. (2004). Transcriptional repression of target genes by LEUNIG and SEUSS, two interacting regulatory proteins for Arabidopsis flower development. *Proceedings of the National Academy of Sciences of the United States of America, 101*, 11494–11499.

Stacey, N. J., Kuromori, T., Azumi, Y., et al. (2006). Arabidopsis SPO11-2 functions with SPO11-1 in meiotic recombination. *The Plant Journal, 48*, 206–216.

Ståldal, V., Sohlberg, J. J., Eklund, D. M., Ljung, K., & Sundberg, E. (2008). Auxin can act independently of CRC, LUG, SEU, SPT and STY1 in style development but not apical-basal patterning of the Arabidopsis gynoecium. *The New Phytologist, 180*, 798–808.

Steffen, J. G., Kang, I. H., Macfarlane, J., & Drews, G. N. (2007). Identification of genes expressed in the Arabidopsis female gametophyte. *The Plant Journal, 51*, 281–292.

Steiner, E., Efroni, I., Gopalraj, M., et al. (2012). The Arabidopsis O-linked N-acetylglucosamine transferase SPINDLY interacts with class I TCPs to facilitate cytokinin responses in leaves and flowers. *Plant Cell, 24*, 96–108.

Su, Z., Zhao, L., Zhao, Y., et al. (2017). The THO complex non-cell-autonomously represses female germline specification through the TAS3-ARF3 module. *Current Biology, 27*, 1597–609.e2.

Sundberg, E., & Ferrándiz, C. (2009). Gynoecium patterning in Arabidopsis: A basic plan behind a complex structure. *Annual Plant Reviews, 38*, 35–69.

Tanaka, H., Ishikawa, M., Kitamura, S., et al. (2004). The AtNACK1/HINKEL and STUD/TETRASPORE/AtNACK2 genes, which encode functionally redundant kinesins, are essential for cytokinesis in Arabidopsis. *Genes to Cells, 9*, 1199–1211.

Tedeschi, F., Rizzo, P., Rutten, T., Altschmied, L., & Bäumlein, H. (2017). RWP-RK domain-containing transcription factors control cell differentiation during female gametophyte development in Arabidopsis. *The New Phytologist, 213*, 1909–1924.

Tekleyohans, D. G., Nakel, T., & Groß-Hardt, R. (2017). Patterning the female gametophyte of flowering plants. *Plant Physiology, 173*, 122–129.

Trigueros, M., Navarrete-Gómez, M., Sato, S., et al. (2009). The NGATHA genes direct style development in the Arabidopsis gynoecium. *Plant Cell, 21*, 1394–1409.

Tucker, M. R., Okada, T., Hu, Y., Scholefield, A., Taylor, J. M., & Koltunow, A. M. (2012). Somatic small RNA pathways promote the mitotic events of megagametogenesis during female reproductive development in Arabidopsis. *Development, 139*, 1399–1404.

Twell, D., Park, S. K., Hawkins, T. J., et al. (2002). MOR1/GEM1 has an essential role in the plant-specific cytokinetic phragmoplast. *Nature Cell Biology, 4*, 711–714.

Uberti-Manassero, N. G., Lucero, L. E., Viola, I. L., Vegetti, A. C., & Gonzalez, D. H. (2012). The class I protein AtTCP15 modulates plant development through a pathway that overlaps with the one affected by CIN-like TCP proteins. *Journal of Experimental Botany, 63*, 809–823.

Wang, H., Liu, R., Wang, J., Wang, P., Shen, Y., & Liu, G. (2014). The Arabidopsis kinesin gene AtKin-1 plays a role in the nuclear division process during megagametogenesis. *Plant Cell Reports, 33*, 819–828.

Webb, M. C., & Gunning, B. E. S. (1990). Embryo sac development in Arabidopsis thaliana. *Sexual Plant Reproduction, 3*, 244–256.

Wei, B., Zhang, J., Pang, C., et al. (2015). The molecular mechanism of sporocyteless/nozzle in controlling Arabidopsis ovule development. *Cell Research, 25*, 121–134.

Weterings, K., & Russell, S. D. (2004). Experimental analysis of the fertilization process. *Plant Cell, 16*(Suppl), S107–S118.

Wu, J. J., Peng, X. B., Li, W. W., He, R., Xin, H. P., & Sun, M. X. (2012). Mitochondrial GCD1 dysfunction reveals reciprocal cell-to-cell signaling during the maturation of Arabidopsis female gametes. *Developmental Cell, 23*, 1043–1058.

Xing, S., Salinas, M., Garcia-Molina, A., Höhmann, S., Berndtgen, R., & Huijser, P. (2013). SPL8 and miR156-targeted SPL genes redundantly regulate Arabidopsis gynoecium differential patterning. *The Plant Journal, 75*, 566–577.

Yamaguchi, N., Huang, J., Xu, Y., Tanoi, K., & Ito, T. (2017). Fine-tuning of auxin homeostasis governs the transition from floral stem cell maintenance to gynoecium formation. *Nature Communications, 8*, 1125.

Yanai, O., Shani, E., Dolezal, K., et al. (2005). Arabidopsis KNOXI proteins activate cytokinin biosynthesis. *Current Biology, 15*, 1566–1571.

Yang, H., Kaur, N., Kiriakopolos, S., & Mccormick, S. (2006). EST generation and analyses towards identifying female gametophyte-specific genes in Zea mays L. *Planta, 224*, 1004–1014.

Yang, M., Hu, Y., Lodhi, M., Mccombie, W. R., & Ma, H. (1999a). The Arabidopsis SKP1-LIKE1 gene is essential for male meiosis and may control homologue separation. *Proceedings of the National Academy of Sciences of the United States of America, 96*, 11416–11421.
Yang, W. C., Ye, D., Xu, J., & Sundaresan, V. (1999b). The SPOROCYTELESS gene of Arabidopsis is required for initiation of sporogenesis and encodes a novel nuclear protein. *Genes & Development, 13*, 2108–2117.
Yang, X., Makaroff, C. A., & Ma, H. (2003). The Arabidopsis MALE MEIOCYTE DEATH1 gene encodes a PHD-finger protein that is required for male meiosis. *Plant Cell, 15*, 1281–1295.
Yu, H. J., Hogan, P., & Sundaresan, V. (2005). Analysis of the female gametophyte transcriptome of Arabidopsis by comparative expression profiling. *Plant Physiology, 139*, 1853–1869.
Zeng, Q., Chen, J. G., & Ellis, B. E. (2011). AtMPK4 is required for male-specific meiotic cytokinesis in Arabidopsis. *The Plant Journal, 67*, 895–906.
Zhang, J., Tang, W., Huang, Y., et al. (2015). Down-regulation of a LBD-like gene, OsIG1, leads to occurrence of unusual double ovules and developmental abnormalities of various floral organs and megagametophyte in rice. *Journal of Experimental Botany, 66*, 99–112.
Zhao, L., Cai, H., Su, Z., et al. (2018). KLU suppresses megasporocyte cell fate through SWR1-mediated activation of WRKY28 expression in Arabidopsis. *Proceedings of the National Academy of Sciences of the United States of America, 115*, E526–E535.
Zhao, L., He, J., Cai, H., et al. (2014). Comparative expression profiling reveals gene functions in female meiosis and gametophyte development in Arabidopsis. *The Plant Journal, 80*, 615–628.
Zhao, X., Bramsiepe, J., Van Durme, M., et al. (2017). RETINOBLASTOMA RELATED1 mediates germline entry in Arabidopsis. *Science, 356*, eaaf6532.
Zhao, X., De Palma, J., Oane, R., et al. (2008). OsTDL1A binds to the LRR domain of rice receptor kinase MSP1, and is required to limit sporocyte numbers. *The Plant Journal, 54*, 375–387.

Chapter 11
Regulatory Mechanism of Plant Meiosis

Jing Xu, Lian Zhou, Zhukuan Cheng, Hong Ma, and Yingxiang Wang

Abstract Meiosis is essential for sexual reproduction in eukaryotes and involves a specialized cell division that serves to halve the diploid chromosome number to the haploid level, ultimately producing sperm or egg cells (also called male or female gametes). In comparison to mitosis, which undertakes nuclear division once following a round of DNA replication, meiosis undergoes two rounds of nuclear division (meiosis I and II) following a single DNA replication. Meiosis I is unique and involves interactions between homologous chromosomes (homologues), including pairing, synapsis, and recombination, followed by their segregation. Meiosis II is similar to mitosis and results in the segregation of sister chromatids. Over the last few decades, molecular genetic studies have provided substantial understanding of the molecular mechanisms involved in the regulation of the meiotic process, using the model plant *Arabidopsis thaliana,* as well as the monocots rice, maize, and barley. This chapter will describe the fundamental processes of meiotic progression, with an emphasis on recent advances regarding the critical gene functions that mediate and regulate these processes.

Jing Xu and Lian Zhou contributed equally with all other contributors.

J. Xu · Y. Wang (✉)
Guangdong Basic Research Center of Excellence for Precise Breeding of Future Crops, Guangdong Laboratory for Lingnan Modern Agriculture, Guangdong Provincial Key Laboratory of Protein Function and Regulation in Agricultural Organisms, College of Life Sciences, South China Agricultural University, Guangzhou, China
e-mail: yx_wang@fudan.edu.cn

L. Zhou
Rice Research Institute, Guangdong Academy of Agricultural Sciences, Guangzhou, China

Z. Cheng
Jiangsu Key Laboratory of Crop Genomics and Molecular Breeding/Key Laboratory of Plant Functional Genomics of the Ministry of Education, Jiangsu Co-Innovation Center for Modern Production Technology of Grain Crops, Yangzhou University, Yangzhou, China

H. Ma
Department of Biology, The Huck Institutes of the Life Sciences, The Pennsylvania State University, University Park, PA, USA

F. Chang et al. (eds.), *Regulation of Plant Development*,
https://doi.org/10.1007/978-981-95-7033-1_11

Keywords Plants · Sexual reproduction · Meiosis · Homologous chromosomes · Molecular mechanism

11.1 Introduction

Angiosperms are the most common type of plants, making up over 90% of the extant plant species present on Earth. The life cycle of angiosperms is complex, beginning with seed germination and concluding with seed maturation (Fig. 11.1). An annual angiosperm plant undergoes two developmental phases during its life cycle, a vegetative phase and a reproductive phase. The vegetative phase progresses from seed germination through vegetative growth to adulthood, generating full size of vegetative organs, including roots and shoots, with stems and leaves. A combination of internal and external signals triggers a transition from the vegetative to reproductive phase and the conversion of vegetative shoot meristem to an inflorescence meristem. The subsequent formation of floral meristems then allows the initiation of floral organ primordia for further development of floral organs. Following the formation of the reproductive floral organs, stamens and carpels, there are three important stages in reproductive development: sporogenesis, gametogenesis, and embryogenesis (Fig. 11.1), with meiosis the most important and specific event occurring during late sporogenesis. Except for meiosis, the other processes in the life cycle are all completed by way of mitosis and cell differentiation.

Here we focus on molecular genetic studies of meiosis in *Arabidopsis thaliana,* with additional information from rice and, sometimes, other plants as noted. The structure of the flower has been described in detail in Chap. 8. Meiosis occurs in the male reproductive organs, the stamens (refer to Chap. 8 for details), and the female reproductive organs, the ovules (refer to Chap. 4 for details). Because the male reproductive tissue is more easily obtained compared with the female reproductive tissue, we mainly describe the progress in the understanding of the male meiosis. During the development of the Arabidopsis anther involving 15 stages based on various degrees of differentiation, meiosis takes place from stage 6 to stage 7 in the pollen mother cells, marking the transition from the diploid generation to the haploid generation (Sanders et al., 1999). At the completion of male meiosis in the diploid pollen mother cell, the four haploid microspores are enclosed in a tetrad. Subsequently, with the help of enzymes from the tapetum, the tetrad wall is degraded, thereby releasing the microspores (Parish & Li, 2010). Microspores (n) divide mitotically to form mature pollen grains. During this process, they will form bicellular pollen grains, each composed of two cells known as the generative cell and vegetative cell. The generative cell then divided into two sperm cells (Borg et al., 2009). Mature pollen grains generally have two sperm cells within the vegetative cell surrounded by a multiple-layered pollen wall, the outer exine and the inner intine. At this point, two pollen sacs of the mature anther fuse into a single pollen sac, which opens to release the pollen grains, thus preparing for subsequent pollination and fertilization. Within the ovary and in each ovule, a megaspore mother cell

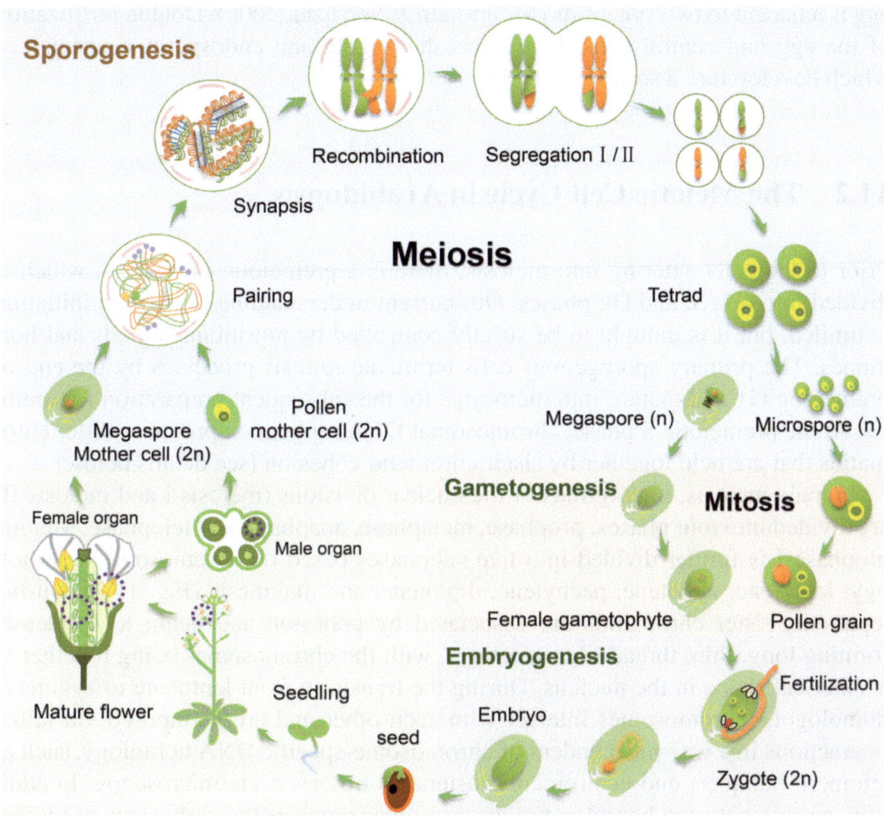

Fig. 11.1 The life cycle of Arabidopsis. The life cycle of Arabidopsis can be generally divided into two periods, the vegetative growth period and the reproductive growth period. The vegetative growth period begins at seed germination and proceeds through seedling growth with the development of full sized vegetative organs, taking about 1 month under standard conditions. The process from differentiation of reproductive organs to seed maturation is called the reproductive growth period, which contains three important stages: meiosis, gametogenesis, and embryogenesis. Meiosis is a specific cell division adopted by megaspore mother cells and pollen mother cells to give rise to megaspores and microspores. The diploid cells proceed through one round of DNA replication and two successive cell divisions, resulting in the production of four haploid cells. The haploid megaspores and microspores mature into female and male gametes (egg cell, pollen) via several rounds of independent mitosis. The female and male gametes combined to form a zygote via fertilization, which then developed into an embryo and mature seeds via cell division and differentiation

(2n) undergoes meiosis to produce four megaspores (n); however, only one megaspore (n) survives and goes through three rounds of mitosis to form an 8-nuclei (7-cell) embryo sac, which is also called the female gametophyte. Both poles of the embryo sac have four cells; one of each of these sets of four cells moves toward each other and then fuses to form the central cell nucleus. Of the other six cells, three cells at the one pole are antipodal cells, while at the other pole of the embryo sac, an egg is adjacent to two synergids (Mohrmann & Verrijzer, 2005). Double fertilization of the egg and central cells then forms the zygote and endosperm, respectively, which develop into a seed (Fig. 11.1).

11.2 The Meiotic Cell Cycle in Arabidopsis

Prior to the cells entering into meiosis, there is a premeiotic interphase, which is divided into G1, S, and G2 phases. Our current understanding of meiotic initiation is limited, but it is thought to be strictly controlled by nutritional signals and hormones. The primary sporogenous cells terminate mitosis processes by the end of premeiotic G1 and mature into meiocytes for the subsequent preparation for meiosis. At the premeiotic S phase, chromosomal DNA replication produces sister chromatids that are held together by sister chromatid cohesion (see details below).

In male meiosis, both rounds of the nuclear divisions (meiosis I and meiosis II) are divided into four phases: prophase, metaphase, anaphase, and telophase. Meiotic prophase I is further divided into five subphases based on chromosome morphology: leptotene, zygotene, pachytene, diplotene, and diakinesis (Fig. 11.2). During leptotene, sister chromatids are associated by cohesion and begin to condense, forming long, thin, thread-like structures, with the chromosomes being together in a spherical shape in the nucleus. During the transition from leptotene to zygotene, homologous chromosomes interact with each other and are juxtaposed via initial interactions in a way independent of chromosome-specific DNA homology, such as telomere bouquets and centromere clusters, as observed via microscopy. In addition, pairing between homologous chromosomes involves the early steps of recombination of homologous DNA sequences (see Sect. 11.4). At late zygotene, a stable complex proteinaceous structure consisting of lateral elements and central elements called the synaptonemal complex (SC) begins to form between paired chromosomes, which are normally homologous but sometimes are non-homologous chromosome pairs in yeast and maize mutants (Golubovskaya et al., 2011; Zickler et al., 1999). The SCs are initially formed regionally and gradually spread throughout the entire chromosomes. The fully synapsed homologues are seen as a thick thread-like structure, and this subphase is called pachytene and persists for a relatively long period. The next stage is diplotene, during which the disassembly of SC results in the separation of homologues except at the chiasma regions, which are formed due to the occurrence of homologous recombination while the sister chromatids are still associated. Subsequently, at diakinesis, tightly associated homologue pairs (bivalents) are highly compacted and appear as dots distributed randomly in the nucleus.

During metaphase I, each kinetochore of a bivalent is attached to spindle fibers from one of two opposite poles, while the spindle pulls the bivalents and aligns them at the equatorial plate. When all of the kinetochores are correctly attached to the spindle fibers, anaphase I begins. Homologues are then separated to the opposite poles. Telophase I is the stage at which separated chromosomes cluster together at the poles and begin to decondense. At prophase II, chromosomes condense again to

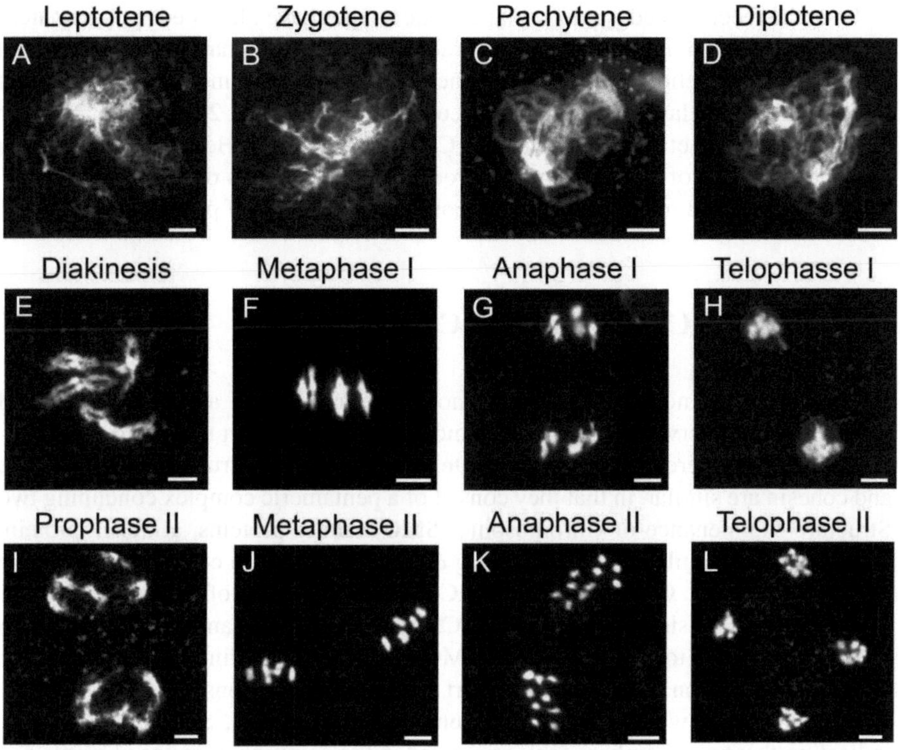

Fig. 11.2 The normal meiotic process in Arabidopsis. Images of DAPI-stained chromosome spreads are shown. (**a**) Leptotene: chromatin begins to condense, forming long, thin, thread-like structures similar to netting in the nucleus. (**b**) Zygotene: homologous chromosomes begin to search for each other and pair up through DNA homologue-dependent or independent approaches. The synaptonemal complex (SC) then begins to form to bring the homologues close to each other at some sites, and synapsis progresses along the chromosome. (**c**) Pachytene: fully synapsed chromosomes appeared as thick thread-like structures. (**d**) Diplotene: the SC is disassembled, and homologues are partially separated, except at the chiasma region. (**e**) Diakinesis: highly compacted bivalents look like thick dots distributed randomly in the nucleus. (**f**) Metaphase I: bivalents are pulled by spindles ejected from both poles to align along the equatorial plane. (**g**) Anaphase I: homologous chromosomes separate from each other. (**h**) Telophase I: the two sets of chromosomes reach to the poles of the cell. (**i**) Prophase II: the two sets of chromosomes begin to condense again. (**j**) Metaphase II: the two sets of chromosomes are pulled by spindles to line up along the new equatorial plates. (**k**) Anaphase II: sister chromatids separate and move toward opposite poles. (**l**) Telophase II: four daughter nuclei with a haploid set of five chromosomes are formed

form two clusters of five chromosomes each, with a band of organelles forming in between the clusters. At metaphase II, each of the two chromosome groups homologues' lines up in the middle of one of the meiosis II spindles, which are perpendicular to each other. The subsequent processes resemble that of mitosis, in which sister chromatids separate from each other and move to the opposite poles of the spindles, thus resulting in the formation of four haploid nuclei. Meiotic cytokinesis then yields four cells, the microspores of the tetrad, as the products of meiosis.

Over the recent decades, molecular genetic studies have identified approximately 100 genes involved in this process. Several excellent reviews have also summarized the molecular genetic studies of components and regulatory mechanisms involved in plant meiosis (Hamant et al., 2006; Luo et al., 2014; Ma, 2005, 2006; Mercier et al., 2015; Wang et al., 2014; Wang & Copenhaver, 2018). Here we mainly focus on the description of the fundamental processes and concepts of meiosis and highlight important and relatively recent discoveries in the field of plant meiosis.

11.3 Meiotic Condensins and Cohesins

Both mitosis and meiosis require chromosome condensation and sister chromatid cohesion, which serve to associate chromosomes and compact them into short condensed entities, thereby ensuring accurate segregation. The structures of condensin and cohesin are similar, in that they consist of a pentameric complex containing two Structure Maintenance of Chromosome (SMC) family proteins, a kleisin protein, and two HEAT-family proteins (Haering & Gruber, 2016). In eukaryotes, there are six SMC members, with SMC2 and SMC4 forming the core of condensins, SMC1 and SMC3 as cohesin subunits, and SMC5/6 assembled into another complex with a structure similar to the prokaryotic SMC complexes (Haering & Gruber, 2016). Molecular evolutionary analyses support orthologous relationships between plant and animal genes within each SMC member group (SMC1, SMC2, and others), indicating these six members each originated in the common ancestor of plants and animals (Surcel et al., 2008). In addition, in the SMC molecular phylogenetic tree, one subunit of both condensins (SMC4) and cohesins (SMC1) are sister clades, as are the other subunit of condensins (SMC2) and cohesins (SMC3), whereas both subunits of the SMC5-SMC6 complex are sisters (Surcel et al., 2008).

11.3.1 Condensins

In multicellular eukaryotes, there are two forms of the condensin complexes, condensin I and condensin II, both containing SMC2 and SMC4 as the backbone, but the β-kleisin and HEAT subunits in condensin I are CAP-H, CAP-D2, and CAP-G, with CAP-H2, CAP-D3, and CAP-G2 present in condensin II. In Arabidopsis, there are two highly similar homologues of the *SMC2* genes, *AtCAP-E1* and *AtCAP-E2* (Chromosome Associated Protein subunit E) (Siddiqui et al., 2003). The *AtCAP-E1* transcript comprises up to 85% of the total *SMC2* transcript, and mutation of each individual gene does not lead to any developmental defects during vegetative or reproductive stages. However, the double mutants are embryonically lethal, supporting the functional redundancy between these two genes (Siddiqui et al., 2003). The Arabidopsis *SMC4* homologue is called *AtCAP-C*, and the loss-of-function mutation causes embryonic lethality, suggesting its essential role in mitosis. Nevertheless, the *AtSMC4/Atsmc4* heterozygous plants showed a failure in homologue association and formation of univalents (unpaired chromosomes). Moreover, a reduction of *AtSMC4* expression during meiosis affected chromosome condensation and crossover frequency (Smith et al., 2014). Consistent with these observations, expression reduction of the other condensin subunits, such as *AtCAP-D3* and *AtCAP-D2*, also results in meiotic defects in chromosome condensation (Smith et al., 2014).

The compaction of chromosomes by the condensin complex is highly conserved across eukaryotic organisms, but the molecular mechanism underlying the regulation of the condensin subunits remains elusive. In particular, chromosome condensation in meiosis I can be divided into two phases: an early partial condensation that results in the formation of leptotene through pachytene chromosomes, and a later phase from late pachytene to metaphase I that generates the highly compacted short chromosomes. The early partially condensed chromosomes might allow both synapsis and recombination, which involve close interactions of DNA molecules of homologous chromosomes. However, little was known about the regulation of meiotic condensation until recently. In Arabidopsis, *AtDUET/MMD1* is a meiosis-specific gene that encodes a PHD domain (plant homeodomain) protein (Reddy et al., 2003; Wang et al., 2016c; Yang et al., 2003b). The *mmd1* mutant exhibits multiple meiotic defects, including reduced chromosome condensation from prophase I through metaphase I (Wang et al., 2016c). Expression of a truncated MMD1 without the PHD domain does not rescue the meiotic chromosome condensation defect in the *mmd1* mutant, suggesting that the MMD1 PHD domain is required for its function in meiotic chromosome condensation. In vitro peptide assays demonstrated that the MMD1 PHD can bind to modified H3K4, and transcriptome analysis found that hundreds of genes are affected in the *mmd1* mutant compared with wild type, including a set of condensin genes such as *CAP-D3*. Consistently, the *cap-d3* mutants also exhibit chromosome condensation defects similar to *mmd1*, revealing the novel finding that MMD1-dependent epigenetic regulation is required for sufficient condensin gene expression that promotes meiotic chromosome condensation (Wang et al., 2016c). Further studies showed that MMD1 binds to the histone demethylase JMJ16, which is expressed in both somatic cells and male meiocytes (Wang et al., 2020b). In male meiocytes the binding of MMD1 to JMJ16 alters the substrate selectivity of JMJ16 on histidine residues of the histone H3 tail and promotes the *CAP-D3* and *CAP-H* expression in an H3K9me3-dependent manner (Wang et al., 2020b). In addition, it has been shown that AtSMC4 acts as a corepressor to silence transposons in the heterochromatin region and to suppress hundreds of expressed genes on the euchromatic chromosome arms (Wang et al., 2017b), providing additional evidence that the condensin complex is involved in epigenetic regulatory mechanisms.

11.3.2 Cohesins

The cohesin subunits include the core components SMC1 and SMC3 and the α-kleisin and HEAT members SCC1/Rad21, SCC3, and PDS5 (Fig. 11.3). In Arabidopsis, there are four copies of α-kleisins, termed AtSYN1–4. The homozygous *Atsmc1*, *Atsmc3*, *Atsyn3*, and *Atscc3* null mutants are lethal (Schubert et al.,

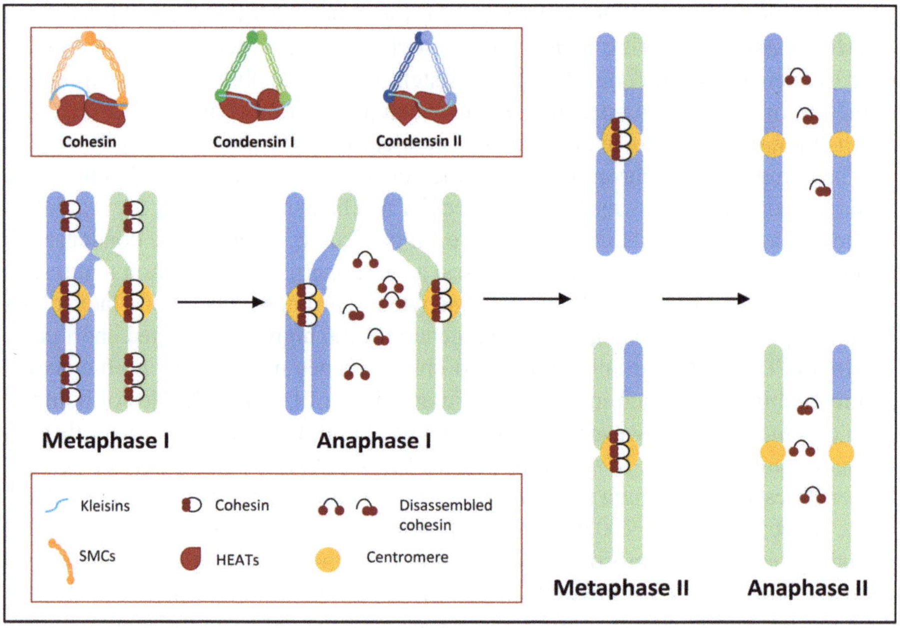

Fig. 11.3 The stepwise removal of cohesin during meiosis. Both cohesin and condensin must undergo timely removal to ensure the proper segregation of chromosomes. They both share a similar structure consisting of different SMC proteins, kleisins, and HEAT-family proteins. In meiosis, cohesin is loaded to hold sister chromatids during the meiotic S phase or earlier. At anaphase I, cohesion along the chromosome arms is removed by cleavage of the kleisin proteins to facilitate homologous chromosome segregation. The cohesion at the centromere is protected and not removed until anaphase II, thereby ensuring the correct segregation of sister chromatids

2009), indicating an essential role in mitotic cell division for these proteins. Unexpectedly, the transgenic plants with reduced expression of *AtSYN3* using an RNAi did not display a defect in sister chromatid cohesion, but rather the *AtSYN3* RNAi transgene was found to impact meiotic chromosome condensation, synapsis, and expression of a large number of genes (Yuan et al., 2012). These results suggest that SYN3 might have a function beyond its role in cohesion or that the residual SYN3 level, or another *SYN* gene, in the transgenic plants is sufficient for cohesion during meiosis. The Arabidopsis *SYN1*, rice *REC8*, and maize *AFD1* are the homologues of mammalian *REC8* and function in meiosis-specific cohesion (Chelysheva et al., 2005; Golubovskaya et al., 2006; Shao et al., 2011). The SYN1 protein shows a linear distribution on meiotic chromosomes from leptotene to pachytene (Cai et al., 2003). The *syn1* mutants in the examined plant species show normal vegetative development, but they are sterile, consistent with the role of this protein as a meiosis-specific cohesin. In Arabidopsis, the *Atsyn1* mutant exhibits premature segregation of sister chromatids and defects in chromosome axis formation in a way independent of chromosome pairing and recombination initiated by the DNA endonuclease AtSPO11 (Chelysheva et al., 2005). The *syn1/rec8* mutants in Arabidopsis and rice also show chromosome fragmentation and loss of monopolar orientation of the kinetochores during meiosis I. In addition, the SYN1/REC8 protein is also required for the assembly of SC in Arabidopsis and rice (Chelysheva et al., 2005; Shao et al., 2011).

In yeast, it was shown that cohesin is recruited to chromatin by the Scc2/Scc4 complex (Choudhary & Kupiec, 2023). In Arabidopsis, the homologous proteins of SCC2 and SCC4 interact with each other, consistent with results from analyses in other organisms indicating that these proteins form a complex (Ciosk et al., 2000). Studies in the budding yeast, *Xenopus laevis*, human, maize, and Arabidopsis further support the idea that interaction of SCC4 with the kinetochore component Ctf19 or the MCM2–7 complex promotes the recruitment of SCC2/SCC4 to the kinetochore, thus facilitating the assembly of cohesin proteins (Fernius & Marston, 2009; Fernius et al., 2013; Wang et al., 2020a; Zheng et al., 2018). A weak allele of *Atscc2* shows defects in meiotic-specific cohesin loading, homologue pairing, and synapsis. Further analysis demonstrates that the PHD motif in SCC2 could be functional in meiosis in land plants (Wang et al., 2020a). Due to the essential role of cohesion during cell division, the *scc2* and *scc4* null alleles are lethal, but the reduced expression of *AtSCC2* causes premature segregation of sister chromatids (Minina et al., 2017), suggesting a role in sister chromatid cohesion. In addition to their primary role in the establishment of cohesion, *AtSCC2* and *AtSCC4* were found to have a role in controlling cell fate during embryogenesis (Minina et al., 2017; Sebastian et al., 2009). Besides SCC2 and SCC4, the Arabidopsis SWITCH1 (SWI1) protein is also required for the establishment of sister chromatid cohesion (Mercier et al., 2001; Mercler et al., 2003). Sequence alignments show that SWI1 is likely a novel protein due to its low similarity to known proteins (Mercier et al., 2001). The *swi1* mutant meiocytes have 20 chromatids instead of 5 bivalents at metaphase I, indicating loss of cohesion at both the centromere and chromosome arms. In addition, SWI1 appears on chromosomes at an early stage before entering meiosis, consistent with a role in the establishment of sister chromatid cohesion. Further studies found that SWI1 is required for the formation of axial elements and initiation of meiotic recombination (Mercler et al., 2003). This idea is supported by the loss of signal for the recombinase RAD51 in *swi1* mutants and the ability of SWI1 to suppress the appearance of chromosome fragments in recombination-defective mutants (Mercier et al., 2001; Mercler et al., 2003). Additionally, rice OsAM1, a SWI1 homologue, has been identified and characterized (Che et al., 2011). Similar to AtSWI1, OsAM1 is localized on chromosomes at an early stage of meiosis. Unlike the meiotic defects in *Atswi1*, *Osam1* meiocytes are arrested at leptotene, suggesting a role in the leptotene-zygotene transition. Early meiotic events such as telomere bouquets and homologue pairing are also affected in the *Osam1* mutant, but the initiation of meiotic recombination seems to be unaffected in *Osam1,* as supported by the normal distribution of the recombination mark γH2AX. Moreover, the meiosis-specific cohesin REC8 could still be loaded onto meiotic chromosomes in *Osam1* (Che et al., 2011), suggesting a divergence of SWI1/AM1 beyond cohesion establishment in meiosis between dicots and monocots. Besides, the meiotic prophase I is regulated by both the ubiquitin-proteasome and the small ubiquitin-like modifier (SUMO) pathways for protein degradation (Rao et al., 2017). A recent study in budding yeast showed that the E3 SCF (SKP–Cullin–F-box) complex, Ufd4, and the cohesin component Pds5 regulate the length of meiotic chromosome axes (Yang et al., 2022). Cohesin-Pds5 recruits the proteasome onto chromosome axes to mediate axis-related protein ubiquitination and shortens meiotic chromosome axes; this mechanism might be conserved in multicellular eukaryotes (Yang et al., 2022).

11.4 Homologous Chromosome Pairing

Homologue pairing is a unique feature of meiosis (Zickler & Kleckner, 2015). Fundamentally, pairing requires the homologues to overcome spatial separation and to associate in close proximity. Currently, the mechanisms involved in controlling this process are largely unclear. There are two common viewpoints that support two potential mechanisms involved in this process. One is the DNA homology-independent mechanism, such as the formation of telomere bouquets and centromere clusters, and the other is a DNA homology-dependent mechanism, which mainly relies on the formation of double-strand breaks (DSB) and subsequent events during the early portion of meiotic recombination (Wang et al., 2014).

11.4.1 DNA Homology-Independent Pairing

During the early phase of meiotic prophase I or prior to the formation of DSBs, two kinds of dynamic chromosome features are recognized, namely the telomere bouquet and centromere clusters, which are thought to play a role in promoting homologue pairing. Telomere bouquet refers to the gathering of the ends of chromosomes at the nuclear envelope as observed in yeast, mammals, and plants and is thought to facilitate homologue pairing by bringing the homolog ends together (Scherthan, 2001). Differing from rice, maize, and rye, Arabidopsis does not have a typical telomere bouquet, but telomeres form a cluster in the nucleolus during premeiotic interphase (Armstrong et al., 2001). By leptotene, the paired telomeres are no longer associated with the nucleolus and become widely dispersed. At this point, the telomeres are loosely distributed in one hemisphere of the nucleus, while homologous chromosome synapsis is initiated during zygotene.

Several proteins are reported to be required for the attachment of the telomere to the nuclear envelope, including the SUN domain proteins. It has been shown that the N-terminus of SUN domain proteins interacts with telomere-binding proteins, while its C-terminus is located in the lumen between two layers of the nuclear envelope in yeast, mammals, and *C. elegans* (Ding et al., 2007; Penkner et al., 2007; Starr & Fridolfsson, 2010). Arabidopsis has two SUN domain proteins, AtSUN1 and AtSUN2, which are functionally redundant (Varas et al., 2015). The double mutants show multiple meiotic defects, such as delayed meiotic progression and failure of synapsis and recombination (Varas et al., 2015). How the SUN proteins are regulated during bouquet formation is unclear, although analyses in rice and maize support roles of possible protein ubiquitination or other forms of protein modification in bouquet formation (Jing et al., 2022; Orr et al., 2021; Zhang et al., 2017). A meiotic functional F-box protein, OsMOF, was first identified in rice and plays a role in the formation of the telomere bouquet and subsequent synapsis and recombination (Orr et al., 2021). Another F-box protein identified from rice, OsZYGO1, is also involved in bouquet formation (Zhang et al., 2017), and its close homolog in maize, the F-box protein Abnormal Chromosome Organization in Zygotene1 (ACOZ1), is essential for telomere bouquet formation (Jing et al., 2022). F-box proteins might be components of SCF E3 ubiquitin ligases, but whether these F-box proteins are involved in protein ubiquitination is not known. Nevertheless, in the *zygo1* mutants, a rice SUN homolog, OsSAD1, was found to have an abnormal distribution around the nuclear envelope, suggesting that OsZYGO1 mediates telomere bouquet formation, likely by affecting SUN protein localization (Zhang et al., 2017).

For centromere clusters, centric heterochromatin is important for mediating homologue pairing in Drosophila oocytes, thus ensuring the subsequent correct segregation of homologues, which lack recombinational crossovers (Dernburg et al., 1996). In Arabidopsis, the centromere regions do not pair at meiotic interphase or leptotene but associate pairwise from zygotene to pachytene due to homologue synapsis (Armstrong et al., 2001). In maize, centromere pairing occurs prior to the formation of the telomere bouquet and depends on conventional centromere activity in chromosome pairing and segregation, as observed of active centromeres and reactivated ones from previously inactive centromeres (Zhang et al., 2013). Furthermore, centromere coupling promotes the installation of the SC (Zhang et al., 2013). However, our current knowledge regarding DSB-independent pairing, such as the telomere bouquet and centromere clusters, remains largely vague, although these processes are thought to promote homology-dependent pairing. One possibility is that presynapsis colocalization of telomeres or centromeres increases the proximity of a chromosome with a DSB-strand and its homologue, thereby facilitating homologue pairing (Zickler & Kleckner, 2015).

11.4.2 DNA Homology-Dependent Pairing

DNA homology-dependent pairing during meiosis depends on the formation of programmed DSBs during the leptotene stage. DSBs and subsequent DNA single-end invasion into the homologous duplex and formation of a D-loop (see below in Sect. 11.6 on recombination for more information) result in close association of the two individual interacting DNA segments, thereby mediating the spatial co-alignment of homologous chromosomes. In particular, the DSB ends facilitate the search for and pairing of homologues. In species that form the telomere bouquet, this

DSB-mediated juxtaposition requires the association of telomeres with the nuclear periphery and local chromosome axis ensembles (Zickler & Kleckner, 2015). In both maize and Arabidopsis, the PHS1 protein is required for homologue pairing. The maize and Arabidopsis *pshl* mutants exhibit association between non-homologous chromosomes during meiosis (Khoo et al., 2012; Ronceret et al., 2009). Images obtained using 3-D microscopy coupled with FISH techniques imply that PHS1 may affect pairing of the gene-rich fraction of the genome rather than repetitive genome regions. The PHS1 protein is localized in the cytoplasm during leptotene and zygotene and controls nuclear import of RAD50 during meiosis. This provides a possible explanation for the defective homologue pairing and DSB repair in *phs1* mutants (Khoo et al., 2012; Ronceret et al., 2009). The maize F-box protein ZmACOZ1 also promotes homologous chromosome pairing (Guan et al., 2022). SPO11 (see details below) is a highly conserved DNA endonuclease required for meiotic DSB formation. Mutation of SPO11–1 in both rice and Arabidopsis influences homologue pairing, supporting the idea that DSB-dependent homologue identification is important for homologue pairing (Stacey et al., 2006; Yu et al., 2010). The mechanisms for the involvement of centromeres during homology-dependent homologue pairing are largely unknown. It is plausible that some DNA-RNA, RNA-RNA, or protein-DNA interactions are involved (Zhang et al., 2014a), but future studies are needed to test these ideas in plants.

11.5 Synapsis

Synapsis is one of the critical events and serves to maintain the pairing of homologues to each other during meiotic prophase I and requires the SC, a meiosis-specific protein structure (Fig. 11.4). The SC connects homologous chromosomes entirely during meiotic prophase I, and maintenance of this close attachment between homologues at early meiosis is essential for recombination and segregation events. The structure of the SC is evolutionarily conserved among eukaryotic organisms and appears similar to a sandwich, with two layers of lateral elements bridged closely by the transverse filaments and central elements in the middle (Page & Hawley, 2004). SC assembly begins in late leptotene and concludes at pachytene. The SC then disassembles when homologous recombination is completed at late pachytene. At early leptotene, the lateral elements are present as discontinuous axial elements (AEs) along each chromosome. These AEs rearrange along the two aligned homologues with evenly formed large chromatin loops. The AEs associate closely with the cohesin subunits REC8, SMC1, SMC3, and SCC3 (Page & Hawley, 2004). When the AE is formed continuously along the whole length of chromosome, it is called the lateral element. At this time, the homologues have been pulled adjacent to one another with a distance of less than 100 nm. Then, the transverse elements, compiled as antiparallel dimers, bridge the two lateral elements together, and the assembly of the SC is completed with the recruitment of other proteins to the central element (Page & Hawley, 2004).

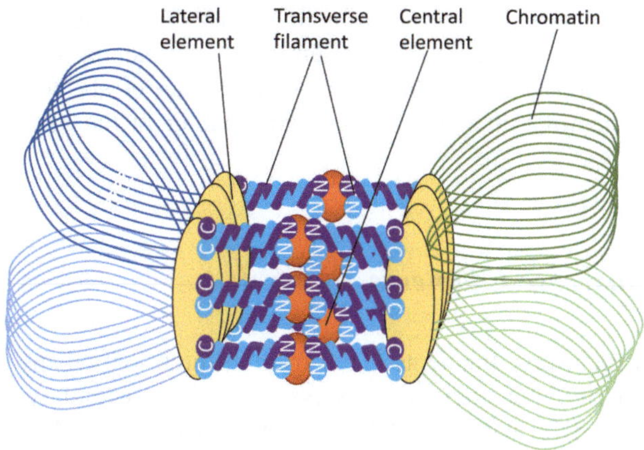

Fig. 11.4 Graphical structure of the synaptonemal complex (SC). The SC consists of lateral elements, transverse filaments, and central elements. The lateral elements are also known as axial elements and are arranged along the coiled chromosome. The transverse filaments (TF) are assembled as antiparallel dimers, combined with the central elements, to bridge the two sets of lateral elements along the homologous chromosomes

11.5.1 Lateral Elements

Several AE proteins have been identified in multiple organisms, including plants. Arabidopsis contains two known AE component proteins, ASY1 and ASY3, which are homologues of the rice proteins HOMOLOGOUS PAIRING ABERRATION IN RICE MEIOSIS 2 (PAIR2) and HOMOLOGOUS PAIRING ABERRATION IN RICE MEIOSIS 3 (PAIR3) (Caryl et al., 2000; Ferdous et al., 2012; Nonomura et al., 2004; Wang et al., 2011a; Yuan et al., 2009). The Arabidopsis ASY1 and rice PAIR2 are both homologues of the yeast HOP1 protein, the first identified lateral structural element of the SC (Caryl et al., 2000; Nonomura et al., 2004). Mutation of either *ASY1* or *PAIR2* results in the failure of synapsis, and the corresponding mutants present with all univalents instead of bivalents at diakinesis. The protein localization of OsPAIR2 is associated with axial elements from leptotene to zygotene (Nonomura et al., 2004). Afterward it is removed from AEs on the arm regions during the completion of synapsis, but it is maintained at centromeres until diakinesis. Furthermore, observation of normally assembled AEs in the *pair2* mutant supports the fact that OsPAIR2 is not required for AE formation (Nonomura et al., 2006), and the proper localization of OsPAIR2 depends on OsPAIR3, a homologue of AtASY3 in Arabidopsis (Ferdous et al., 2012; Yuan et al., 2009). Similarly, AtASY3 is also required for the proper localization of AtASY1 (Ferdous et al., 2012). Although OsPAIR3 and AtASY3 do not appear to be structural axial elements, mutation of both genes leads to the failure of assembly of lateral elements and synapsis. In addition, OsREC8, the Arabidopsis SYN1 homologue, is a meiosis-specific cohesin protein and is essential for the localization of OsPAIR3 (Wang et al., 2011a), suggesting that some cohesin proteins could also be axis-associated proteins, at least in part, thus participating in AE installation. SCF component SKP1 localizes to the lateral elements and is essential for maintaining synapsis and subsequent recombination in mouse male meiocytes (Guan et al., 2020). It would be meaningful to identify plant E3 ligases participating in the synapsis process in future studies.

11.5.2 Transverse Filaments

The identified transverse filaments (TFs) from different organisms show very low sequence similarity, but they do possess the same secondary structure. The Arabidopsis AtZYP1a and AtZYP1b, the barley homologue ZYP1, and the rice homologue ZEP1 are reported to be TFs (Higgins et al., 2005; Wang et al., 2010). Unlike rice ZEP, which occurs in a single copy, Arabidopsis possesses two copies (Higgins et al., 2005). These proteins have redundant functions in the SC. Mutation of these genes in both Arabidopsis and rice results in the failure of synapsis, supporting a basic role for these proteins in the SC. However, these proteins have a divergent role in crossover (CO) formation. In Arabidopsis, the *Atzyp1a Atzyp1b* mutants show a significant reduction of recombination frequency due to the formation of univalents, as opposed to the formation of bivalents that occurs in wild type (Higgins et al., 2005). A similar phenotype was observed in the barley homologue (Barakate et al., 2014). In contrast, in the absence of *OsZEP1* in rice, the number of crossovers is significantly increased (Wang et al., 2010). Immunostaining data has shown that OsZEP1 impacts the loading of OsPAIR2 and OsMER3 on the chromosome, but it possesses a negative effect in regard to the regulation of ZIP1 foci (Wang et al., 2015a). As such, it is still unclear if the role of ZEP1 in CO formation is similar between Arabidopsis and barley but in contrast to its role in rice. In transgenic plants with knockdown of meiocyte-specific DNA polymerase *POLD1,* little ZYP1 signal was shown in Arabidopsis male meiocytes, indicating *POLD1* is essential for synapsis (Wang et al., 2019). Also, a Drosophila SCF component F-box protein was found to regulate transverse protein C(3)G assembly and synapsis complex maintenance during female meiosis in a ubiquitin-dependent manner (Barbosa et al., 2021).

11.5.3 Central Elements

In addition to the TF elements, several other proteins are found in the SC central region. Central elements generally vary in number, localization, and function among species and include SYP-2, SYP-3, and SYP-4 in *C. elegans* (Schild-Prufert et al., 2011), as well as SCEP1 and SCEP2 in Arabidopsis (Vrielynck et al., 2023), and OsCRC1 and OsP31^{comet} in rice (Miao et al., 2013). Homodimerized SYP-1 proteins line up in a parallel fashion in the middle of the SC, similar to other transverse filaments, while SYP-2, SYP-3, and SYP-4 intermix with the SC via interactions among different protein regions (Schild-Prufert et al., 2011). Arabidopsis *SCEP1* and *SCEP2* encode two coiled-coil proteins and form a complex that colocalizes with ZYP1 in the central region of the SC. Mutation of either *SCEP1* or *SCEP2* causes a failure in TF formation and synapsis (Vrielynck et al., 2023). The rice *OsCRC1* gene encodes a conserved AAA-ATPase and is the homologue of *Pch2/thyroid receptor interacting protein 13* (*TRIP13*), whose mutation causes a pachytene checkpoint in budding yeast (Miao et al., 2013; San-Segundo & Roeder, 1999). OsP31^{comet} is a mitotic arrest-deficient 2 (Mad2)-interacting protein (Ji et al., 2016). Both CRC1 and P31^{comet} are located in the central regions of the SC and are co-associated with the ZEP1 protein. This is further supported by the direct interaction observed between CRC1 and the N-termi of ZEP1 in yeast two-hybrid assays. Interestingly, no interaction has been detected between P31^{comet} and ZEP1 (Ji et al., 2016), indicating a divergent function in regard to the SC between these two proteins. E3 ubiquitin ligase DESYNAPSIS1 (DSNP1) in rice has been identified to act in the synapsis process. The immunofluorescence analyses in the *dsnp1* mutant showed that the disassembly of the transverse filament protein ZEP1 was delayed, so it is possible that DSNP1 regulates SC assembly by mediating ZEP1 proper degradation during rice meiosis (Ren et al., 2021). Moreover, the normal recruitment of OsPAIR2 onto the chromosome requires OsCRC1 but not OsP31^{comet} (Miao et al., 2013), further supporting their different functions during SC installation. Together, these results provide evidence that CRC1 likely serves as a bridge to connect P31^{comet} with ZEP1 during SC installation, at least in rice.

11.6 Meiotic Recombination

In addition to pairing and synapsis, exchanging genetic information between homologues occurs during prophase I and is defined as homologous recombination. Crossovers, the products of recombination, in combination with sister chromatid cohesion, maintain homologue association (bivalent) after the disassembly of the SC. In plants like Arabidopsis and others, meiotic recombination initiates from the programmed formation of DSBs, which likely take place from leptotene to early zygotene as indicated by immunofluorescence assay using antibodies against the γ-H2AX (phosphorylated histone H2AX) and meiosis-specific recombinase DMC1 (Bishop et al., 1992; Lang et al., 2012; Turinetto & Giachino, 2015). For a relatively small subset of DSBs, following their formation, the subsequent procession of the DNA ends, facilitation of single-end invasion with the formation of the D-loop, and DNA synthesis result in the formation of a double Holliday junction (dHJ). Resolution of dHJs yields crossover (CO) with exchange of flanking sequences or noncrossover (NCO) events without such an exchange. Alternatively, a majority of DSBs are repaired as NCOs by way of the synthesis-dependent strand-annealing (SDSA) pathway (Fig. 11.5). Both COs and NCOs can be associated with gene conversion, the copying of sequence from one homologue to the other by DNA synthesis, if the two recombining homologous chromosomes exhibit sequence polymorphisms near the site of the DSB.

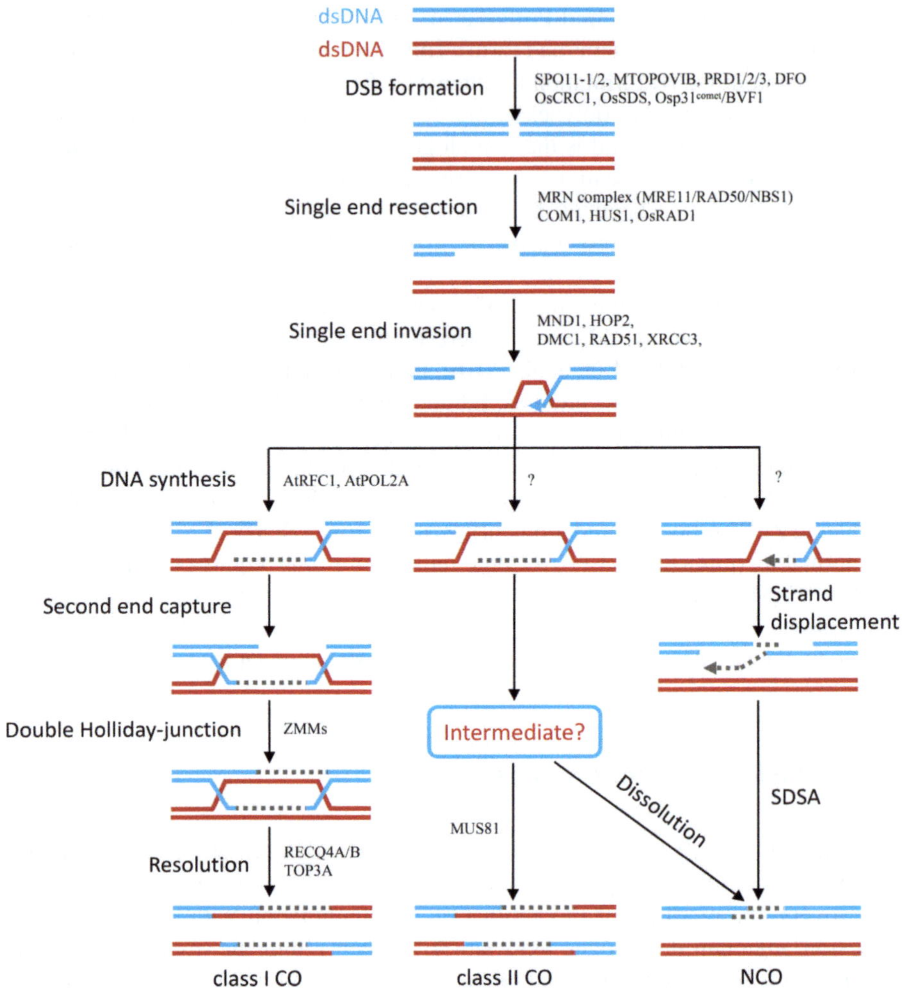

Fig. 11.5 A model for meiotic recombination. The blue lines and red lines independently represent the two strands of two non-sister chromatids, which belong to a pair of homologous chromosomes. Meiotic recombination begins with the formation of DSBs, which is facilitated by topoisomerases. The broken strand ends are then resected by an exonuclease to yield 3′ single-strand tails, which invade the non-sister chromatid to form a D-loop. Then, DNA repair can be processed via one of three pathways. The most typical repair pathway is the class I CO pathway, involving ZMM proteins and the formation of a double Holliday junction. The class II CO pathway is mediated by MUS81, but what type of intermediate is formed during this process is unclear. The majority of DSBs are repaired by synthesis-dependent strand annealing (SDSA) without yielding any crossover events

11.6.1 DSB Formation

The cleavage of DNA to produce DSBs is accomplished by a conserved topoisomerase-like protein, SPO11, similar to the A subunit of archaeal topoisomerase VI (TopVIA), which contains two subunits and usually functions as a heterotetramer with two each of A and B subunits (Keeney, 2008; Keeney et al., 1997). In Arabidopsis, there are three *SPO11* homologues (*SPO11–1*, *–2*, and *–3*), and *meiotic topoisomerase VIB–like* (*AtMTOPVIB*) is a single copy (Stacey et al., 2006). Among the three *SPO11* homologues, *AtSPO11–1* and *AtSPO11–2* play non-redundant roles in meiotic DSB formation; mutation of either *AtSPO11–1* or *AtSPO11–2* causes near absence of DSBs and great reduction or lack of meiotic homologous chromosome pairing, synapsis, and recombination. These meiotic defects result in generally 10 univalents and abnormal segregation of homologues, thus leading to pollen lethality and sterility. On the other hand, *AtSPO11–3* is involved in nuclear replication as a subunit of the AtTOPVI complex in somatic cells. Protein-protein interaction assays demonstrate that AtSPO11–1 and AtSPO11–2 interact with each other and also interact with AtMTOPVIB. Additional evidence has shown that formation of a heterodimer between AtSPO11–1 and AtSPO11–2 requires the direct interaction between AtSPO11–1, AtSPO11–2, and AtMTOPVIB (Stacey et al., 2006; Vrielynck et al., 2016). Mutation of *AtMTOPVIB* causes a similar defect in meiotic DSB formation to that of either *spo11–1* or *spo11–2*, suggesting that SPO11-MTOPVIB might form a heterotetramer during meiotic DSB formation (Vrielynck et al., 2016). In rice, there are five SPO11 homologues, though only OsSPO11–1 and OsSPO11–4 are reported to be involved in rice meiosis. Reduced expression of *OsSPO11–1* results in defects in homologous synapsis and recombination, thus leading to the formation of univalents (An et al., 2011; Yu et al., 2010). Rice also possesses a single copy of *OsMTOPVIB*, which is required for meiotic DSB formation (Fu et al., 2016; Tang et al., 2017; Xue et al., 2016). The observation that the transducer domain of OsMTOPVIB interacts with both OsSPO11–1 and OsSPO11–2 provides additional evidence that the SPO11–MTOPVIB complex is highly conserved and has an essential role in meiotic DSB formation. Furthermore, *OsMTOPVIB* is essential for converting an abnormal multipolar spindle into a bipolar spindle during rice metaphase I (Xue et al., 2019). A genome-wide analysis of small RNAs in Arabidopsis wild type and *Atspo11–1* meiocytes shows that *AtSPO11*-independent small RNAs are positively correlated with preferentially expressed meiotic genes, and *AtSPO11*-dependent ones exhibit enrichment of sequence motifs (CTT repeat and A-rich) that are associated with meiotic recombination hotspots (Huang et al., 2019).

In addition to the SPO11–MTOPVIB complex, there are several auxiliary proteins involved in DSB formation in plants, including Putative Recombination initiation Defect 1 (AtPRD1/ZmPRD1), AtPRD2, AtPRD3/OsPAIR1, DSB forming (AtDFO), SOLO DANCERS (OsSDS), and OsP31^{comet}/OsBVF1. AtPRD1 and AtPRD2 are the likely homologues of MEI1 and MEI4 in mice, respectively, and AtPRD3/OsPAIR1 and AtDFO are likely plant-specific proteins (Zhang et al.,

2012). It was reported that AtPRD1 probably functions as a bridge protein to link different heterodimers or form heterotetramers during meiotic DSB formation due to its direct interaction with multiple factors, including AtPRD3, AtDFO1, and the three SPO11 proteins (Tang et al., 2017). The ZmPRD1 interacts with ZmMTOPVIB, ZmPRD2, and ZmPRD3 to regulate meiotic DSB formation, homologous pairing, and recombination (Wang et al., 2022b), while unlike PRD1 in rice, the maize PRD1 is not required for meiotic DSB formation, suggesting a divergent role of PRD1 in rice meiosis (Shi et al., 2021). Another study in rice found that OsP31^{comet}/OsBVF1, similar to the MAD2-binding protein, is likely an SC component, but it also participates in DSB formation (Zhou et al., 2017). A similar case was also observed for the central element OsCRC1 (see above Sect. 11.5.3). Together, these data suggest that some SC-associated proteins are also important for DSB formation occurring during meiosis.

11.6.2 DSB End Processing

After the formation of DSBs, SPO11 bound to the 5′ ends needs to be removed by the MRX/N (MRE11-RAD50-XRS2/NBS1) complex and COM1/SAE2. The 5′ ends are further resected by EXO1 to yield a 3′ single-strand DNA overhang. In Arabidopsis, AtMRE11 and AtRAD50 are essential for mitotic and meiotic DSB repair, while MRE11 partially compensates for ATM deficiency during meiosis (Bleuyard et al., 2004; Samanic et al., 2013). The single mutant *Atnbs1* does not exhibit any defects in either mitotic or meiotic DNA repair, but the *Atnbs1–1 Atatm* double mutant demonstrates severe meiotic defects, revealing that AtNBS1 is also required for meiotic recombination (Waterworth et al., 2007). The Arabidopsis ATM and ATR proteins are important protein kinases that can transduce DNA-damage signaling to recruit DSB repair proteins for meiotic DSB repair (Amiard et al., 2010; Waterworth et al., 2007). Additionally, the Arabidopsis homologue of COM1 is also required for meiotic recombination. In the *Atcom1* mutant, overaccumulation of AtSPO11–1 and failure of AtRAD51 loading are observed, suggesting that AtCOM1 functions upstream of RAD51 (Uanschou et al., 2007). In rice, OsMRE11 and OsCOM1 are found to have similar functions in meiotic recombination. Moreover, the loading of OsMRE11 depends on the normal installation of SC AEs, while the loading of OsCOM1 depends on OsMRE11, but the underlying mechanisms involved in these processes are unclear (Ji et al., 2012, 2013). The COM1 homologue ZmCOM1 in maize is highly expressed in meiocytes and is required for the formation of telomere bouquet, homologous chromosome pairing, and synapsis, suggesting a conserved role of COM1 in homologous recombination in plants (Wang et al., 2018).

The single-strand DNA overhang is protected due to the binding of the replication protein A (RPA), which has a conserved function in binding to single-strand DNA (ssDNA) (Aklilu et al., 2014; Chang et al., 2009; Eschbach & Kobbe, 2014). In yeast and most mammals, RPA is a complex composed of three subunits: RPA1,

RPA2, and RPA3. Arabidopsis has five *RPA1* genes (*RPA1-a*, *b*, *c*, *d*, and *e*) and two of each *RPA2* and *RPA3* (Eschbach & Kobbe, 2014). *AtRPA1a* and *AtRPA1c* have roles in promoting homologous recombination, but *AtRPA1b*, *AtRPA1d*, and *AtRPA1e* are only involved in mitotic DSB repair (Aklilu et al., 2014; Chang et al., 2009). The rice genome contains three *RPA1* genes (*RPA1a*, *RPA1b*, and *RPA1c*), three *RPA2* genes (*RPA2a*, *RPA2b*, and *RPA2c*), and one *RPA3*, of which *OsRPA1a*, *OsRPA2c*, and *OsRPA1c* participate in meiotic recombination (Chang et al., 2009; Li et al., 2013), suggesting a conserved function for *RPAs* during meiotic recombination. However, whether *RPAs* in meiosis provide protection of ssDNA following the formation of DSBs still requires direct biochemical assessment.

11.6.3 Single-End Invasion

Following single-end procession, the replacement of RPAs on the 3′ overhanging ssDNA by the RecA recombinases, Disrupted Meiotic cDNA1 (DMC1) and RADiation sensitive51 (RAD51), facilitates invasion of the 3′ ssDNA end into the double helix between homologues to form a D-loop (Masson & West, 2001). In yeast, both DMC1- and RAD51-mediated strand invasion require the MND1-HOP2 complex, which is a V-shaped structure that possesses binding ability for both single- and double-stranded DNAs, thereby stabilizing the interaction between the single-strand end and the homologous double-stranded molecule. Arabidopsis also encodes a corresponding homologue, and results show that a high concentration of the MND1-HOP2 complex promotes the AtDMC1-mediated inter-homologue interaction (Uanschou et al., 2013).

In vertebrate animals and plants, the *RecA/RAD51* gene family is conserved and consists of seven members, including *DMC1*, *RAD51*, *RAD51B*, *RAD51C*, *RAD51D*, *XRCC2*, and *XRCC3* (Lin et al., 2006). In Arabidopsis, AtDMC1 is specifically required for meiotic recombination with inter-homologue bias, while AtRAD51 plays a supporting role in DMC1-dependent HR and also mediates HR between sister chromatids (Pradillo et al., 2012). In the *Atdmc1* mutant, all DSBs are thought to be repaired using sister chromatids as templates in a RAD51-dependent manner, while the *Atrad51* mutant shows severe chromosome fragmentation. Additional evidence supporting the divergent function of RAD51 and DMC1 comes from analysis following disruption of the RAD51 recombinase activity, but not its DNA binding or protein interacting properties (Da Ines et al., 2013). The results showed that the catalytically inactive RAD51 could rescue the *rad51* mutant phenotype in a DMC1-dependent manner, as the catalytically inactive RAD51 could not rescue the chromosome fragmentation phenotype of the *dmc1 rad51* double mutant. In vivo high-resolution visualization of DMC1 and RAD51 on meiotic ssDNA shows that DMC1 binds near the DSB sites to promote strand exchange between homologous chromosomes, while RAD51 binds farther away from the DSB sites (Hughes et al., 2019). Besides, there are more DNA-bound DMC1 foci than RAD51 foci on average (Hinch et al., 2020), and DMC1 foci are found further away from the synaptonemal complex axes relative to RAD51 foci (Bishop et al., 2020), indicating that distinct spatial localization on ssDNA of DMC1 and RAD51 implies their divergent function in meiotic recombination. Together, these results suggest that DMC1 is critical for meiosis, but that RAD51 serves an essential supporting role. Under certain conditions, DMC1 can replace RAD51 to conduct its recombinase activity. In contrast, DMC1 could inactivate RAD51 activity in mediating strand exchange through an unknown mechanism during Arabidopsis meiotic recombination (Henderson et al., 2022). In addition, other AtRAD51 paralogues were implicated in meiotic recombination; specifically, AtRAD51C can interact with both AtRAD51 and AtXRCC3, and the AtRAD51C-AtXRCC3 complex can facilitate the localization of AtRAD51 onto chromosomes (Su et al., 2017). These observations indicate that the role of AtRAD51C and AtXRCC3 in meiosis is dependent on AtRAD51; this idea is also consistent with the indistinguishable meiotic defects among the mutants of the three genes. The homologue of AtRAD51C in maize ZmRAD51C also functions in central element ZYP1 assembly, DSB repair, and homologous recombination (Jing et al., 2019). Another yeast factor, Rdh54/Tid1, is a regulator of RAD51-RAD54 and inhibits single-end invasion (Shah et al., 2020). In Arabidopsis, the homologue of RAD54 is indispensable for RAD51-mediated meiotic recombination when DMC1 is unavailable (Grelon et al., 2021).

In rice, there are two copies of *DMC1* homologues, *OsDMC1A* and *OsDMC1B*, which play redundant roles in meiosis. Every single mutant does not show any defect in fertility or meiosis (Wang et al., 2016b), but the *Osdmc1a Osdmc1b* mutant is completely sterile and shows defects in synapsis and recombination during meiosis. Immunolocalization assays show that normal OsRAD51C foci are formed in the *Osdmc1a Osdmc1b* mutant, but OsDMC1 is not loaded in the *Osrad51c* mutant, further suggesting that OsRAD51 functions upstream of OsDMC1. However, the *Osrad51c Osdmc1a Osdmc1b* triple mutant shows severe chromosome entanglement, greater than that of the *Osrad51c* single mutant, consistent with the idea in Arabidopsis that the function of RAD51 in meiosis is likely via the impact of RAD51 localization on chromosomes. Unexpectedly, unlike DMC1 in Arabidopsis and other organisms, the rice DMC1 is not required for early homologue pairing, suggesting a divergent role for DMC1 among these species. A recent study in Arabidopsis showed that RAD51 competitively interacts with the SMC5/6 complex to relieve the inhibitory effect of SMC5/6 on DMC1 localization on meiotic chromosomes, thereby promoting DMC1-mediated recombination (Chen et al., 2021).

After DMC1 completes its function in meiotic recombination, it is important to eliminate its function, possibly by removing the DMC1 protein; however, the factors and mechanism for this negative regulation were previously unknown. Very recently, Xu et al. reported the discovery of an E3 ubiquitin ligase $SCF^{RMF1/2}$ that is required for normal meiosis and demonstrated that $SCF^{RMF1/2}$ ubiquitinates AtDMC1 for degradation in 26S proteasome dependent manner (Xu et al., 2023).

11.6.4 Meiotic Recombination-Associated DNA Synthesis

Meiosis requires two steps of DNA synthesis. Premeiotic DNA replication makes double chromosomes called sister chromatids, and repair of meiotic DSBs requires recombination-associated DNA synthesis. Premeiotic DNA replication is thought to share the same DNA replication machinery as mitotic DNA replication, while the DSB-dependent DNA synthesis machinery employed during meiotic recombination remains largely unknown. According to the classical DSB repair model (Fig. 11.5), DNA leading strand synthesis is facilitated by single-end invasion and second-end capture for dHJ formation. Due to a lack of molecular genetic evidence, the role of DNA synthesis during meiotic recombination was previously not well understood. Recently, two review papers have summarized the current understanding of meiotic recombination-associated DNA synthesis and presented a new model (Fig. 11.5) (Huang et al., 2016; Wang & Copenhaver, 2018), which includes both DNA leading and lagging strand synthesis. The reason for including lagging strand synthesis in the meiotic recombination model is that genes encoding DNA polymerase α, δ, and replication factor C (RFC), which are required for DNA lagging strand replication, play an essential role in meiotic recombination. The corresponding mutants show SPO11–1-dependent chromosome fragmentation and reduced interference-sensitive meiotic crossover frequency (White et al., 2012). This finding argues that meiotic DSB repair requires DNA lagging strand synthesis. To test the role of leading strand synthesis during meiosis, a mutation of the *AtPOL2A* (the largest subunit of DNA polymerase Pol ε) exonuclease domain was assessed. This mutation causes the alternation of CO distribution but does not alter the number of COs (Huang et al., 2015). Meiocyte-specific knockdown of DNA polymerase δ *POL D1* in Arabidopsis causes reduction of type I CO formation, suggesting that POL2A coordinates with POLD1 in DNA synthesis for meiotic recombination (Wang et al., 2019). A similar mutation of the yeast homologue *Pol ε* has been shown to decrease enzymatic activity for leading strand elongation (Huang et al., 2015). Given the highly conserved nature of this domain in all organisms, the observed CO alternation in the Arabidopsis *pol2a* mutant could be due to reduced DNA elongation efficiency. Together, building on these results, an updated meiotic recombination model includes both DNA leading and lagging strand synthesis and also presents that DNA synthesis is important to differentiate meiotic recombination pathways. A recent study reported that AtPOL2A binds and recruits SU(VAR)3–9 homologue SUVH2 to the meiotic DSB hotspots; as SUVH2 is implicated in gene silencing, this finding suggests that AtPOL2A is likely involved in regulating meiotic DSB-associated gene silencing (Wang et al., 2022a).

11.6.5 Double Holliday Junction Resolution

In eukaryotes, resolution of dHJs is mediated by structure-specific endonucleases, including MMS and UV-sensitive protein 81 (MUS81), noncatalytic subunit essential meiotic endonuclease 1 (EME1), the resolvase synthetic lethal 1/4 (SLX1/4), and Holliday junction resolvase YEN1 (Gen endonuclease homolog1) (Schwartz & Heyer, 2011). In Arabidopsis, mutation of *AtMUS81* causes no apparent defects in plant fertility or meiosis but leads to the reduction of interference-insensitive COs (Berchowitz et al., 2007). Arabidopsis has two *EME1* homologues, AtEME1A and AtEME1B. Each of these can associate with AtMUS81 to form a functional endonuclease. Consistently, in vitro assays have demonstrated that both the AtMUS81-AtEME1A and AtMUS81-AtEME1B complexes are able to cleave 3′-flap structures, nicked HJs, and intact HJs, indicating that both the MUS-EME1 endonuclease complexes are required for resolution of HJs (Geuting et al., 2009; Wang et al., 2017a). The Yen1 (GEN1) resolvase resembles the *E. coli* RuvC, belonging to the large and well-characterized Rad2/XPG family of nucleases (Ip et al., 2008). The Arabidopsis GEN1 and SEND1 are the homologues of the rice GEN-L and SEND1 (Bauknecht & Kobbe, 2014) and possess the 5′-flap activities, but they are active on a broad range of substrates in contrast to EcRuvC, HsGEN1, ScYen1, and CeGEN1 (Bailly et al., 2010). Mutation of *AtGEN1* and *AtSEND1*, individually or together, shows no apparent effect on development, fertility, or sensitivity to DNA damage, but simultaneous loss of *AtMUS81* and *AtSEND1* causes severe developmental defects. In comparison to AtGEN1, AtSEND1 seems to have a minor role in the resolution of dHJs, but it acts as an essential backup to AtMUS81 for the resolution of toxic replication structures (Olivier et al., 2016). Similarly, loss of the rice *GEN1* leads to a reduction in chiasma frequency and persistent DSBs, while depletion of the rice *SEND1* has no effects on plant development (Wang et al., 2017a).

11.6.6 Three Types of Meiotic Recombination Pathways

In most organisms, meiotic crossovers occur via two genetic pathways, the interference-sensitive pathway (type I CO) and the interference-insensitive pathway (type II CO) (Paques & Haber, 1999). In Arabidopsis, formation of type I CO is dependent on the ZMM group of proteins (MutS homologue 4 (AtMSH4), AtMSH5, Meiotic recombination 3 (AtMER3), *Homo sapiens* Enhancer of Invasion 10 (AtHEI10), Zipper 4 (AtZIP4), AtSHOC1, PARTING DANCRES (AtPTD) (Wijeratne et al., 2006), MutL-homologue 1 (AtMLH1), and AtMLH3), in which AtMSH4 and AtMSH5 are two meiosis-specific members of the MutS mismatch repair family (Higgins et al., 2004, 2008b; Lin et al., 2007). These proteins interact with each other to form a heterocomplex, and the localization of AtMSH5 on chromosomes is dependent on AtMSH4 (Higgins et al., 2008b). Evidence was found in yeast suggesting that Msh4 is regulated by phosphorylation and proteasomal proteolysis (He et al., 2020). AtMLH1 and AtMLH3 are two MutL mismatch repair proteins and form a heterocomplex to impact meiotic recombination (Dion et al., 2007; Jackson et al., 2006). AtSHOC1, AtHEI10, and AtZIP4 are conserved homologues corresponding to yeast Zip2, Zip3, and Zip4 (Chelysheva et al., 2007, 2012; Macaisne et al., 2008). AtPTD exhibits sequence similarity to both XPF and ERCC1 proteins, and the *Atptd* mutant is defective in meiotic recombination, especially type I CO formation (Wijeratne et al., 2006). AtPTD-AtSHOC1 forms an XPF-ERCC1-like heterodimer that functions in type I CO formation (Macaisne et al., 2011). Each corresponding homologue is also found in rice, including OsMSH4, OsMSH5, OsMER3, OsZIP4, and OsHEI10 (Shen et al., 2012; Wang et al., 2016a; Zhang et al., 2014b), suggesting that the ZMM proteins are also evolutionarily and functionally conserved among plants. In rice, OsMSH4 functions upstream of the ZMM pathway and interacts with OsMSH5 to promote the recruitment of other ZMM proteins (Zhang et al., 2014b).

Differing from the type I CO pathway, which accounts for 80–90% of COs, type II COs only account for 10–20% of COs and are MUS81-dependent (Berchowitz et al., 2007; Wang et al., 2012a). Because the *msh4 mus81* double mutant still presents with 5–10% residual COs, this implies the existence of a potential alternative pathway for generating COs (Higgins et al., 2008a). Arabidopsis *FANCD2* encodes a homologue of the *Fanconi anemia D2* gene, whose defect has been reported to participate in Fanconi anemia (FA), a human autosomal recessive disorder; the *Atfancd2* mutant showed a reduction in CO numbers, which are MSH4- and MUS81-independent (Kurzbauer et al., 2018). Although the *msh4 mus81 fancd2* triple mutant exhibits a significant reduction in COs compared to that of the *msh4 mus81* double mutant, residual COs are still estimated to be around 5–10% (Kurzbauer et al., 2018). Future studies are still needed, however, to elucidate the molecular mechanism of *FANCD2* in meiotic recombination pathways.

11.6.7 Regulatory Mechanisms of Meiotic Recombination

The frequency and distribution of COs on chromosomes are uneven and are coordinately regulated. On chromosomes, the regions with the highest CO frequency are called hot spots, while the regions with the lowest CO frequency are called cold spots. At the genome-wide level, three regulatory mechanisms influence CO formation: crossover homeostasis, crossover interference, and crossover assurance (Chakraborty et al., 2017). Regarding individual chromosomes, the telomere, centromere, and heterochromatin regions with their highly repetitive sequences are special chromosome structures that have a role in suppressing CO formation. In contrast, chromosomal regions called recombination hot spots exhibit more frequent DSB formation and CO formation. Given that meiotic recombination takes place at the direct DNA level, including formation of DSBs and subsequent processes, the local chromatin structure also has a critical influence on meiotic recombination. As such, DNA modifications like DNA methylation, histone modification, and chromatin remodeling are thought to be involved in regulating CO formation.

11.6.7.1 Crossover Homeostasis, Assurance, and Interference

Results from several organisms have shown that the CO numbers per cell are stable even when the total number of DSBs within a cell changes, a phenomenon referred to as crossover homeostasis (Phadnis et al., 2011). Compared to the total number of DSBs, the number of COs is low. In Arabidopsis, there are 150–250 DSBs per meiocyte, but only 9–11 COs are produced (Ma, 2006). In yeast, CO homeostasis is mainly regulated by partner choice. The repair of hot-spot DSBs tends to use the inter-sister chromatids, while the repair of hotspot-poor DSBs likely uses the DMC1-dependent inter-homologue pathway (Hyppa & Smith, 2010). Although the phenomenon of CO homeostasis exists across species, the regulatory mechanisms involved in this process are likely different. For example, in maize, using the RAD51 foci as a DSB indicator, research has shown that the number of COs varies in relation to the number of DSBs among different inbred lines, suggesting that increasing DSBs number may be a potential way to increase recombination frequency, at least in maize inbred lines (Sidhu et al., 2015).

Crossover assurance refers to the fact that at least one CO event occurs between each pair of homologues, also known as obligate COs, which assures proper bivalent maintenance and alignment at metaphase I and homologue segregation during anaphase I. The obligate CO frequency varies greatly in different organisms and is lower in organisms that have strong genetic interference and higher in organisms that lack interference (Chakraborty et al., 2017). In yeast, there is a hypothesis that CO homeostasis, assurance, and interference are manifestations of a single underlying process (Wang et al., 2015b). Surprisingly, mutation of *FANCM* in Arabidopsis and rapeseeds causes the occurrence of extra COs that are MUS81-dependent and interference insensitive, implying the possibility of a relationship between assurance and interference (Lorenz & Whitby, 2012), as discussed below.

Crossover interference refers to the distribution of COs that are nonrandom across chromosomes, and the formation of an obligate CO inhibits other COs formed in an adjacent chromosomal interval (Bishop & Zickler, 2004). Interference may contribute to the dispersal of COs evenly along and between chromosomes. Generally, interference has a negative correlation with distance, but this is measured as the length along the chromosome axis rather than base pairs (Zickler & Kleckner, 2016). Crossover assurance is defined by the distribution of type I CO, while the distribution of type II CO among chromosomes is thought to be random at a genome scale. A recent study in Arabidopsis showed that FANCD2 (a promoter of type II CO), FANCM (an inhibitor of type II CO), FIGL1, as well as RMI1 are required for the proper distribution of type I CO among chromosomes, thereby maintaining CO assurance (Li et al., 2021). So far, our understanding of the underlying mechanisms in controlling crossover interference in plants is still limited.

11.6.7.2 Epigenetic Influence

Studies in yeast and animals have demonstrated that COs preferentially occur at hot spots associated with histone H3K4me3 modification (Borde et al., 2009; Brick et al., 2012; Kniewel & Keeney, 2009; Smagulova et al., 2011). In mouse, PRDM9 encodes an H3K4me3 methyltransferase and is a determinant of CO hot spots (Baudat et al., 2010; Parvanov et al., 2010). Similarly, the yeast Set1/COMPASS complex is also required to mediate H3K4 methylation and plays a role in the determination of DSB levels (Borde et al., 2009; Choi & Henderson, 2015). In Arabidopsis, meiotic recombination has been shown to be positively correlated with H3K4me3 and the histone variant H2A.Z (Choi et al., 2013; Drouaud et al., 2013; Liu et al., 2009; Shilo et al., 2015; Wijnker et al., 2013). As described above, Arabidopsis MMD1/DUET is a PHD domain protein, and this domain can bind to H3K4me2/3 (Andreuzza et al., 2015; Wang et al., 2016c; Yang et al., 2003a). The *mmd1* mutant also exhibits a reduction in CO frequency, suggesting a link between histone modification and COs. In addition, the histone acetyltransferase MCC1 has been reported to regulate CO distribution instead of total CO frequency (Perrella et al., 2010). Also, the Jmj-domain protein INCREASE IN BONSAI METHYLATION 1 (IBM1) is crucial for synapsis and recombination during male meiosis via regulation of the expression of meiosis-related genes; as the Jmj-domain can remove methyl groups from histone tails, it is likely IBM1 functions by affecting histone modification and, indirectly, DNA methylation (He et al., 2022). Moreover, in rice, Meiosis Arrested at Leptotene 1 (MEL1) is an AGO family member that can bind to 21-nt phasiRNAs and functions to reprogram histone H3 modifications, including H3K9me2, H3K9ac, and H3S10p, and loss of *MEL1* causes defects in meiotic recombination and synapsis (Komiya et al., 2014; Liu & Nonomura, 2016; Nonomura et al., 2007).

In addition to histone modification, DNA methylation is another important epigenetic marker that also plays an important role in shaping meiotic crossover patterns. Previous studies in Arabidopsis found that DNA methylation plays important roles in suppressing CO formation at centromeric and pericentromeric regions (Mirouze et al., 2012). In Arabidopsis, DNA methylation mainly appears in three DNA sequence contexts: CG, CHH, and CHG. RNA-directed DNA methylation (known as RdDM) is able to locally silence Arabidopsis crossover hot spots on euchromatin via targeted CG, CHG, and CHH DNA methylation (Yelina et al., 2015). AtMET1 is a DNA methyltransferase mainly conferring CG methylation, and mutation of *AtMET1* does not change the total number of COs, but it does alter the distribution of COs along the chromosome with a reduction of COs in the pericentromere region and increases in the centromere and chromosome arms (Mirouze et al., 2012). Meanwhile, DDM1 is an SWI2/SNF2-like chromatin-remodeling factor that regulates both CG and non-CG methylation. Mutation of *DDM1* increases COs in euchromatic regions but not heterochromatic regions (Melamed-Bessudo & Levy, 2012). In comparison to *met1*, CO frequency at pericentromeric regions is not altered in *ddm1* (Yelina et al., 2012). CG methylation is also maintained by VIM/ORTH, which acts at the DNA methylation-histone interface to maintain centromeric heterochromatin (Colome-Tatche et al., 2012; Woo et al., 2007). In Arabidopsis, CMT3 and DRM2 cytosine methyltransferases redundantly maintain non-CG methylation and mediate all CHH methylation in the genome (Cao et al., 2003; Stroud et al., 2014). In addition, there is evidence that associates CHG methylation with H3K9me2 (Du et al., 2012; West et al., 2014). A study has provided evidence that H3K9me2 and non-CG DNA methylation differentially mediate the formation of meiotic DSBs and COs in centromeric and pericentromeric heterochromatin, as supported by the mutation of *CMT3* required for the establishment of CHG DNA methylation, which results in the decondensed heterochromatin and progressively greater increases in CEN3 CO frequencies (Underwood et al., 2018). Taken together, these results support the idea that DNA methylation plays an important role in regulating pericentromeric COs.

11.7 Chromosome Segregation

Unlike mitosis, which only involves sister chromatid segregation during anaphase, meiosis involves two rounds of chromosome segregation. Homologous chromosomes are segregated at meiotic anaphase I, while sister chromatids are separated at anaphase II, similar to mitosis (Fig. 11.3). Homologous chromosome interactions during prophase I, including pairing, synapsis, and recombination, ensure the subsequent proper segregation of homologues at anaphase I. Furthermore, the formation of a mono-oriented kinetochore on the chromosome and its association with microtubules from one pole of the spindle is also important for pulling the bivalents from both poles and properly aligning them on the metaphase plate. Afterwards, the stepwise removal of sister chromatid cohesion along the chromosomal arms and centromeres, respectively, results in the separation of homologues and sister chromatids at anaphase I and II (Ma, 2006). Therefore, formation of mono-oriented sister chromatids of the same chromosome, stepwise removal of cohesion, and assembly of the meiotic spindle are of great importance for ensuring proper chromosome segregation.

11.7.1 Sister Chromatid Mono-Orientation

Homologue segregation during meiosis I requires two unique and conserved events: sister kinetochore mono-orientation and centromeric cohesin protection (detailed in the following section). The molecular mechanism of monopolar attachment has been well established in yeast and is mediated by a dedicated four-protein monopolin complex (i.e., Mam1, Csm1, Lrs4, and Hrr25/casein kinase 1) and the meiosis-specific protein Spo13 (Corbett & Harrison, 2012). Monopolin is thought to bind to the sister kinetochores and direct sister chromatid co-orientation in a cohesion-dependent manner (Corbett et al., 2010). Crystal structure analyses of the monopolin complex found that Csm1 and Lrs4 form a distinctive V-shaped clamp-like structure that serves to directly crosslink sister kinetochores; mutations of the monopolin complex subunits cause the mitosis-like attachment between spindles and kinetochores during meiosis I to form bi-oriented sister kinetochores from opposite spindle poles (Corbett et al., 2010). In plants, the mechanism of monopolar attachment is poorly understood, as no monopolin homologue has been identified. In addition to monopolin, cohesion is also important for the establishment of sister chromatid mono-orientation in several organisms, including plants, and the meiosis-specific cohesin SYN1/REC8 is also required for the establishment of a monopolar attachment (Chelysheva et al., 2005). The *syn1/rec8* mutant displays a mitosis-like bi-orientation of sister chromatids at metaphase I. In addition, the Arabidopsis cohesin subunit SCC3 has been demonstrated to have a role in monopolar attachment of sister kinetochores (Chelysheva et al., 2005).

The Aurora B kinase is mainly required for correcting improper chromosome attachment and also plays a role in ensuring sister chromatid mono-orientation during meiosis I in yeast, mouse, human as well as Arabidopsis (Meyer et al., 2013; Nikalayevich et al., 2022; Sakuno et al., 2011; Shrestha et al., 2017). The inactivation of the fission yeast *Aurora B* homologue *Ark1* causes a defect in sister chromatid mono-orientation in meiosis I (Sakuno et al., 2011). Also, Aurora B homologue Ipl1 and Mps1 kinase act sequentially to control meiotic kinetochore mono-orientation in budding yeast (Meyer et al., 2013). Arabidopsis has three *Aurora* homologues (*Aurora 1, 2, 3*), and every single mutant shows normal vegetative and reproductive development, whereas the double mutants affecting two of the three homologues are lethal (Demidov et al., 2005, 2014; Petrovska et al., 2012). Meiosis-specific reduction of *Aurora 1* expression causes defects in bivalent alignment and partial loss of bi-orientation of homologous chromosomes at metaphase I (Niu et al., 2015), suggesting a potential role of this protein in sister chromatid mono-orientation, but the underlying mechanism is unclear. In addition, the Arabidopsis Aurora 3 kinase was found to function like mammalian Aurora B in regulating meiotic chromosome segregation by controlling kinetochore function (Komaki et al., 2020).

Previous results showed that the spindle assembly checkpoint (SAC) functions in monitoring the interaction between kinetochores and spindle microtubules in mitosis (Musacchio & Salmon, 2007). The SAC targets cell division cycle 20 (CDC20), an activator of the anaphase-promoting complex/cyclosome (APC/C) E3 ubiquitin ligase, to negatively regulate the ability of CDC20, thus inhibiting APC/C-mediated ubiquitination of cyclin B and securin, which is important for mediating cohesion degradation; therefore, the correct attachment of the kinetochore-spindle ensures the metaphase I-anaphase I transition (Musacchio & Salmon, 2007). Arabidopsis has five copies of CDC20 (CDC20.1–20.5) (Niu et al., 2015). Among them, only AtCDC20.1 is involved in chromosome segregation during meiosis. The *Atcdc20.1* mutant exhibits a reduction in male fertility, asynchronous alignment of bivalents, and aberrant morphology of the spindle at metaphase I, thus leading to uneven segregation of homologues and sister chromatids at anaphase I and II, respectively (Niu et al., 2015). Estimation of meiotic progression suggests that the metaphase

I-anaphase I transition is significantly delayed. Furthermore, the distribution of histone H3 phosphorylation mediated by the Aurora kinase is altered in *Atcdc20.1*. Consistently, meiosis-specific reduction of *AtAurora1* expression causes similar meiotic chromosome segregation defects to those of *Atcdc20.1*. The Arabidopsis E3 ligase APC/C component AtAPC8 is also indispensable for kinetochore orientation in meiosis, probably through cooperating with AtCDC20.1 to regulate the Aurora kinase activity, which ensures the SAC-dependent meiotic chromosome segregation and normal progression of meiosis (Xu et al., 2019). Observation of three *CDC20* genes in rice with essential roles in faithful meiotic chromosome alignment and segregation (Lin et al., 2022), indicating their meiotic function is conserved in plants.

11.7.2 Cohesion Removal

Sister chromatid cohesion mediated by the cohesin complex is essential for the proper segregation of chromosomes during mitosis and meiosis. Unlike mitosis, which only involves sister chromatid segregation, meiosis involves two events of chromosome segregation, with separation of homologues at anaphase I and sister chromatids at anaphase II. These processes require the removal of cohesion in a stepwise manner (Marston, 2015). In higher eukaryotes, removal of arm cohesin during mitotic prophase requires Polo-like kinase, cyclin-dependent kinase, and WAPL (Shintomi & Hirano, 2009), while protection of centromeric cohesin involves the Shugoshin (Sgo1)-protein phosphatase 2A (PP2A) complex and WAPL inhibitor Switch1 (SWI1) (Liu et al., 2013; Yang et al., 2019). In meiosis, studies in flies have demonstrated that WAPL regulates cohesion removal by mediating the opening of the cohesin ring (Gandhi et al., 2006). In Arabidopsis, two *WAPL* copies are reported to have overlapping functions in cohesin removal (De et al., 2014). Mutation of either *AtWAPL1* or *AtWAPL2* does not cause obvious defects in vegetative growth or development, but a significant reduction in male and female fertility is apparent. The *Atwapl1 Atwapl2* mutant exhibits chromosome bridges, uneven chromosome segregation, and a delay in the decrease in the signal of meiosis-specific cohesin SYN1 (De et al., 2014), indicating a defect in cohesin removal. These observations also suggest that the mechanism involved in the removal of cohesin between mitosis and meiosis might be divergent. This idea is further supported by the inactivation of the two *WAPL* genes, which can suppress the embryo lethal phenotype caused by the complete loss of Establishment of cohesion 1/Chromosome transmission fidelity 7 (Eco1/Ctf7) acetyltransferase CTF7 (Bolanos-Villegas et al., 2013). Moreover, the *wapl1 wapl2 ctf7* triple homozygous mutant plants are fertile and exhibit a recovery from the defect in the removal of SYN1 in meiocytes. A previous study in Arabidopsis showed that phosphorylation of SWI by CDKs leads the phosphorylated SWI to be degraded by the 26S proteasome, thereby releasing its repression of WAPL to promote the meiotic prophase removal of cohesin (Yang et al., 2019).

Homologues of the Shugoshin (SGO) protein have been identified in several plant species, including Arabidopsis, rice, and maize (Hamant et al., 2005; Wang et al., 2011b; Zamariola et al., 2013). SGO proteins contain two conserved domains: an N-terminal coiled-coil domain, which is important for protein-protein interaction and homodimerization, and a C-terminal basic region required for localization of the protein to chromosomes. In the examined plant species mentioned above, the SGO proteins are shown to localize at centromeric regions and play an important role in the protection of centromeric cohesion; the *sgo* mutants exhibit premature separation of centromeric regions and, subsequently, improper chromosome segregation (Wang et al., 2011b; Zamariola et al., 2013). However, the functions of the SGO proteins are divergent among species. In maize, the centromeric localization of maize SGO1 is dependent on the meiosis-specific cohesin REC8, but the rice SGO1 demonstrates a role in the assembly and maintenance of SCs during early prophase I (Hamant et al., 2005; Wang et al., 2011b). Although the mechanisms for the regulation of SGO1 during meiosis are largely unclear, the rice gene *OsBRK1* has been implicated. *OsBRK1* encodes a Bub1-related kinase and is required for the centromeric recruitment of SGO1; the *brk1* mutant shows defects in spindle shape and improper homologue segregation, but the underlying role of BRK1 in mediating SGO1 localization still requires further study (Wang et al., 2012b).

PATRONUS 1 (*PANS1*) is also required for protecting centromeric cohesin at interkinesis and early meiosis II (Zamariola et al., 2014). Loss of *PANS1* function does not affect meiosis I but results in the formation of 10 single chromatids at each of two metaphase II plates. Additionally, no REC8/SYN1 signal is detected at metaphase II in chromatids in the *pans1* mutant. Protein interaction assays indicate that PANS1 interacts with components of the APC/C that regulate the metaphase to anaphase transition via ubiquitination of specific targets (Cromer et al., 2013). Recently, it was found that PANS1 and PANS2 function as plant securins and need to be degraded in a timely fashion to activate the protease separase for accurate segregation of meiotic homologous chromosomes (Cromer et al., 2013).

Besides, kinetochore proteins have been well studied in meiotic chromosome segregation. A study using mouse oocytes found that meiosis-specific kinetochore protein Meikin is cleaved by separase and APC/C-dependent ubiquitination and degradation to ensure faithful meiotic chromosome alignment and segregation (Maier et al., 2021). In the fission yeast, Aurora B kinase phosphorylates the condensin subunit Cnd2 and regulates telomere and chromosome arm segregation during meiosis I in a Rec8-independent manner (Berthezene et al., 2020), while Aurora B in mammalian oocytes phosphorylates Rec8 for cleavage and subsequent homologous chromosome segregation (Nikalayevich et al., 2022).

Arabidopsis ASK1 is a Skp1-like1 protein that is part of the SCF ubiquitin ligase, which targets a variety of proteins for degradation (Wang et al., 2004). The *Atask1* mutant is defective in chromosome segregation at anaphase I and exhibits an abnormal AtSYN1 distribution (Wang et al., 2004; Yang et al., 1999, 2006; Zhao et al., 2006), suggesting that ASK1 might mediate cohesion distribution via protein degradation, but the underlying mechanism of this process is unclear.

11.7.3 Assembly of Meiotic Spindle

In eukaryotes, chromosome alignment and segregation during cell division require the normal assembly of the microtubule-based spindle. During meiotic prophase I, microtubules begin to circle the nuclear envelope. Subsequently, the nuclear envelope breaks down, and the spindle structure is organized by the minus ends of the microtubules, while the plus ends of the microtubules attach to kinetochores, facilitating the pulling of the chromosomes by the spindle to the metaphase plate (McMichael & Bednarek, 2013). Different spindle-assembly-related structures, known as microtubule organizing centers (MTOCs), are found in different organisms, including the spindle pole body in budding yeast and centrosomes in animals (Bennabi et al., 2016). Although plant cells lack a centrosome, they possess plant-specific microtubule-based structures known as the phragmoplast and pro-prophase band (PPB), which are important for the formation of the cell wall between daughter nuclei and the position of the spindle following mitosis and meiosis (Liu & Lee, 2022).

Microtubules can bind to a class of motor proteins called kinesins, which are a large superfamily of proteins (kinesin-1 to kinesin-14 types) and possess a highly conserved ATPase/motor domain that can bind to and release microtubules to generate force and to move along microtubles. The kinesin motors are known to be involved in both mitosis and meiosis. The motor activity of most kinesins is plus-end directed, but the kinesin-14 subfamily is unique, as its motor domain moves in a minus-end direction (Gicking et al., 2018). Unlike animals, which have few kinesin-14 members (e.g., four in humans), land plants have multiple kinesin-14s, suggesting divergent functions in plant species. The Arabidopsis genome encodes 21 *kinesin-14* genes, and kinesin-14A has four close homologues: KIN14a (ATK1), KIN14b (ATK5), KIN14c (ATK2), and KIN14d (ATK3) (Chen et al., 2002; Quan et al., 2008; Richardson et al., 2006). ATK1 was the first identified plant kinesin-14 with a role in meiosis and has been implicated in spindle assembly and/or maintenance (Chen et al., 2002). The *atk1* mutant is male sterile and exhibits abnormal chromosome segregation and broad, unfocused, multi-axial spindles (Chen et al., 2002), indicating a role in spindle assembly. Similarly, maize has two Kinesin-14A paralogues, including *Divergent spindle-1 (Dv1)* (Higgins et al., 2016). The *dv1* mutant shows a similar phenotype to *atk1*, including reduced fertility, improper chromosome segregation, and broad-ended spindle bipolarity defects. The other maize Kinesin-14A paralogue is called *KINDR*, which is crucial for meiotic drive in maize (Dawe et al., 2018). The *kindr* mutant is defective in the activity of neocentromeres, which are ectopic centromeres that originate occasionally from noncentromeric regions of chromosomes (Dawe et al., 2018). Further results have demonstrated that KINDR is a minus-end-directed motor that interacts specifically with neocentromeres containing the 180 bp knob repeats. Additional evidence demonstrates that a mutant lacking neocentromere activity is a result of increased DNA methylation across the entire gene cluster (Dawe et al., 2018), but the exact mechanism underlying this phenotype is unclear. In contrast, the closest related paralogue of *ATK1*, *ATK5*, is not essential for either mitosis or meiosis, suggesting some form of functional redundancy. Indeed, the *atk1 atk5* double mutants are embryonic lethal. Nevertheless, the double heterozygote mutants show a reduction in fertility and meiotic defects, such as abnormal chromosome and spindle morphology, similar to the *atk1* single mutant (Quan et al., 2008), providing evidence supporting the redundant function in meiotic spindle assembly between these two proteins.

Another study showed that the BTB-domain protein MAB1 in maize may function as a substrate-specific adaptor for an E3 ligase that also contains Cullin3A to regulate meiotic spindle assembly and faithful chromosome segregation (Juranic et al., 2012). Moreover, PRD1 in rice also participates in meiotic bipolar spindle assembly (Shi et al., 2021).

11.8 Meiotic Cell-Cycle Control

The unique process of meiosis involves two successive nuclear divisions and chromosome segregation following a single round of DNA replication. As such, it is critical to accurately control the termination of meiosis at meiosis II instead of meiosis I, leading to a single copy of the haploid genome per nucleus. These processes rely on a number of cell-cycle regulators, including cyclin-dependent kinases and the anaphase-promoting complex/cyclosome (APC/C) (Wijnker & Schnittger, 2013).

In Arabidopsis, AtCDKA;1 is a homologue of the major cyclin-dependent kinase Cdc2+/Cdc28 of yeast (Dissmeyer et al., 2007), and phosphorylation of Thr-161 at the conserved T-loop is essential for its function. The *Atcdk1;1* mutant is completely sterile and shows premature cytokinesis after the first meiotic division (Dissmeyer et al., 2007), suggesting a role in meiotic cytokinesis. AtCDKA;1 also takes part in synapsis and bivalency formation. AtCDKA;1 colocalizes with ASY1 on chromosome axis and phosphorylates ASY1 to counteract the disassembly activity of the ATPase for meiotic chromosome axis assembly (Yang et al., 2020). AtTAM/CYCA1;2 is an A-type cyclin and forms a complex with AtCDKA;1 and is required for the transition from meiosis I to meiosis II; the *Attam* mutants fail to enter meiosis II and subsequently produce diploid spores and functional diploid gametes. Double mutants with *Atspo11–1* and *Atrec8* produce diploid gametes, likely via a mitotic-like division (d'Erfurth et al., 2010).

Arabidopsis OSD1/GIG1, an APC/C inhibitor, is a likely homologue of vertebrate Erp1/Emi2 and Mes1 of fission yeast (Cromer et al., 2012). Protein-protein interaction assays have shown that AtOSD1 interacts with a range of APC/C activators, including AtCDC20.1, AtCDC20.5, and the three CDH1 homologues. Mutation of *AtOSD1* leads to a premature exit from meiosis after meiotic I. AtOSD1 is phosphorylated by the AtTAM-AtCDKA;1 complex; meanwhile, the AtTAM-AtCDKA;1 complex can inhibit AtTDM1/AtMS5 activity at meiosis I to prevent premature meiotic exit (Cifuentes et al., 2016). Mutation of *AtTDM1/AtMS5* leads to an aberrant third division after meiosis II (Glover et al., 1998), suggesting that AtTDM1 functions as a safeguard to ensure meiotic termination at the end of meiosis

II. AtSMG7 is a protein involved in nonsense-mediated RNA decay (NMD) and has a conserved function. Mutation of *Atsmg7* causes an unusual cell-cycle arrest in meiotic anaphase II (Riehs et al., 2008). AtSMG7 may act downstream of APC/C^{Cdc20} to regulate the transcription of other meiosis-related genes, and genetic analysis shows that the epistatic relationship among these proteins (up to down) is AtOSD1, AtSMG7, AtTDM, and AtTAM (Riehs et al., 2008).

Arabidopsis thaliana Parallel Spindle 1 (*AtPS1*) encodes a protein with an N-terminal FHA domain and a C-terminal PINc domain; mutation of *AtPS1* leads to abnormal spindle orientation at meiosis II, producing diploid gametes and triploid plants (d'Erfurth et al., 2008). The identification of *AtPS1* may provide the basis of new strategies in plant breeding programs and progress in evolutionary studies. Similar to *Atps1*, *Atjason* produces more than 25% 2n pollen grains and triploid offspring (De Storme & Geelen, 2011). Cytological analysis indicates that the formation of parallel-arranged and fused spindles may be due to the 2n spores. Further analysis shows that the *AtPS1* transcript is reduced in *Atjason* mutants, indicating that AtJASON positively regulates the expression of *AtPS1*, affecting its participation in the proper organization of the spindle at meiosis II (De Storme & Geelen, 2011).

11.9 Perspective

Over the last several decades, molecular genetic studies have identified almost 100 meiotic genes in plants, mainly from Arabidopsis, rice, and maize. Such progress has greatly contributed to our understanding of the meiotic process, one of the most important events during sexual reproduction. With the discovery of more genes from different species, emerging evidence has shown both functional conservation and divergence of the paralogues across species. Some of the future research directions in the study of these processes are highlighted below.

The commonality and divergence of regulatory mechanisms during meiosis across organisms are not well characterized, although there are some specific examples of differences. In animals, PRDM9 possesses a long zinc finger domain and a SET domain that is proposed to trimethylate H3K4 and H3K36 (Powers et al., 2016; Smagulova et al., 2011). PRDM9 functions to recruit the topoisomerase complex to mediate meiotic DSB formation and/or CO hotspots. However, plants do not have a PRDM9-like homologue; thus, how the sites of DSBs are specified in plants is unclear. Similarly, the PHD domain protein MMD1 is plant specific; it plays a role in meiotic chromosome condensation via its binding to modified H3K4 and serves to enhance condensin gene expression (Wang et al., 2016c). Given that chromosome condensation is a widespread process, determining how the corresponding regulatory mechanisms have diverged between plants and animals is of great interest. In addition, several known meiotic proteins have divergent functions, even among different plant species. For example, the central element ZYP1 promotes recombination in Arabidopsis and barley, but its homologue ZEP1 limits recombination in rice. The recombination initiation factor PRD1 participates in meiotic bipolar spindle assembly in rice and Arabidopsis, but its homologue does not seem to have a similar function in maize (Wang et al., 2022b). Additionally, Arabidopsis SDS does not participate in DSB formation (De Muyt et al., 2009), but its rice homologue is required for DSB formation (Wu et al., 2015).

New techniques are also constantly being developed, including several powerful technologies established in recent years. First, observation of dynamic chromosome interactions in live cells has become possible via live cell imaging and super-resolution structured illumination microscopy (SIM); this approach has been successfully applied to observe chromosome pairing and spindle movements in maize and Arabidopsis (Sheehan & Pawlowski, 2009; Wang et al., 2022a). Second, single-cell sequencing has also been used to study meiotic recombination in a single tetrad, and single-cell RNA sequencing has been used to examine the transcriptome profiles during premeiotic and meiotic cellular processes (Li et al., 2015; Nelms & Walbot, 2019). This technology combined with ATAT-seq is used to unveil the distribution of the genomic cis-regulatory elements in maize at single-cell resolution (Marand et al., 2021), providing a means for further analyses of regulatory mechanisms of meiotic gene expression. Third, great advances in Cryo-EM technologies have made the analyses of the three-dimensional structure of a protein relatively easy. This will greatly help determining the precise mechanisms for protein interaction with other proteins or with nucleic acids at an atomic resolution, as was demonstrated with RAD51 (Xu et al., 2017).

Meiotic recombination is a fundamental process important for crop breeding. However, how to apply a meiotic gene identified from analyses in a model plant toward benefiting the growth of crops is a significant challenge. To this end, natural variation selection has identified HEI10, which functions to modify meiotic CO frequency in Arabidopsis, and MS5, which mediates early meiotic progression for hybrid vigor in *Brassica napus* (Serra et al., 2018; Ziolkowski et al., 2017). Moreover, identification of several anti-CO factors, such as TOP3α, RECQ4a/b, and FIGL1, has provided potential candidates for manipulating crop breeding by either genetic modification or natural hybridization (Girard et al., 2015; Seguela-Arnaud et al., 2015; Zeng et al., 2016). Future analyses might uncover other meiotic genes with potential impact in crop breeding.

References

Aklilu, B. B., Soderquist, R. S., & Culligan, K. M. (2014). Genetic analysis of the Replication Protein A large subunit family in Arabidopsis reveals unique and overlapping roles in DNA repair, meiosis and DNA replication. *Nucleic Acids Research, 42*, 3104–3118.

Amiard, S., Charbonnel, C., Allain, E., Depeiges, A., White, C. I., & Gallego, M. E. (2010). Distinct roles of the ATR kinase and the Mre11-Rad50-Nbs1 complex in the maintenance of chromosomal stability in Arabidopsis. *The Plant Cell, 22*, 3020–3033.

An, X. J., Deng, Z. Y., & Wang, T. (2011). OsSpo11-4, a rice homologue of the archaeal TopVIA protein, mediates double-strand DNA cleavage and interacts with OsTopVIB. *PLoS One, 6*, e20327.

Andreuzza, S., Nishal, B., Singh, A., & Siddiqi, I. (2015). The chromatin protein DUET/MMD1 controls expression of the meiotic gene *TDM1* during male meiosis in Arabidopsis. *PLoS Genetics, 11*, e1005396.

Armstrong, S. J., Franklin, F. C., & Jones, G. H. (2001). Nucleolus-associated telomere clustering and pairing precede meiotic chromosome synapsis in *Arabidopsis thaliana*. *Journal of Cell Science, 114*, 4207–4217.

Bailly, A. P., Freeman, A., Hall, J., Declais, A. C., Alpi, A., Lilley, D. M., Ahmed, S., & Gartner, A. (2010). The *Caenorhabditis elegans* homolog of Gen1/Yen1 resolvases links DNA damage signaling to DNA double-strand break repair. *PLoS Genetics, 6*, e1001025.

Barakate, A., Higgins, J. D., Vivera, S., Stephens, J., Perry, R. M., Ramsay, L., Colas, I., Oakey, H., Waugh, R., Franklin, F. C., Armstrong, S. J., & Halpin, C. (2014). The synaptonemal complex protein ZYP1 is required for imposition of meiotic crossovers in barley. *The Plant Cell, 26*, 729–740.

Barbosa, P., Zhaunova, L., Debilio, S., Steccanella, V., Kelly, V., Ly, T., & Ohkura, H. (2021). SCF-Fbxo42 promotes synaptonemal complex assembly by downregulating PP2A-B56. *The Journal of Cell Biology, 220*, e202009167.

Baudat, F., Buard, J., Grey, C., Fledel-Alon, A., Ober, C., Przeworski, M., Coop, G., & de Massy, B. (2010). PRDM9 is a major determinant of meiotic recombination hotspots in humans and mice. *Science, 327*, 836–840.

Bauknecht, M., & Kobbe, D. (2014). AtGEN1 and AtSEND1, two paralogs in Arabidopsis, possess Holliday junction resolvase activity. *Plant Physiology, 166*, 202–216.

Bennabi, I., Terret, M. E., & Verlhac, M. H. (2016). Meiotic spindle assembly and chromosome segregation in oocytes. *The Journal of Cell Biology, 215*, 611–619.

Berchowitz, L. E., Francis, K. E., Bey, A. L., & Copenhaver, G. P. (2007). The role of *AtMUS81* in interference-insensitive crossovers in *A. thaliana*. *PLoS Genetics, 3*, e132.

Berthezene, J., Reyes, C., Li, T., Coulon, S., Bernard, P., Gachet, Y., & Tournier, S. (2020). Aurora B and condensin are dispensable for chromosome arm and telomere separation during meiosis II. *Molecular Biology of the Cell, 31*, 889–905.

Bishop, D. K., & Zickler, D. (2004). Early decision; meiotic crossover interference prior to stable strand exchange and synapsis. *Cell, 117*, 9–15.

Bishop, D. K., Park, D., Xu, L., & Kleckner, N. (1992). *DMC1*: A meiosis-specific yeast homolog of *E. Coli recA* required for recombination, synaptonemal complex formation, and cell cycle progression. *Cell, 69*, 439–456.

Bishop, D. K., Slotman, J. A., Paul, M. W., Carofiglio, F., de Gruiter, H. M., Vergroesen, T., Koornneef, L., van Cappellen, W. A., Houtsmuller, A. B., & Baarends, W. M. (2020). Super-resolution imaging of RAD51 and DMC1 in DNA repair foci reveals dynamic distribution patterns in meiotic prophase. *PLoS Genetics, 16*, e1008595.

Bleuyard, J. Y., Gallego, M. E., & White, C. I. (2004). Meiotic defects in the Arabidopsis *rad50* mutant point to conservation of the MRX complex function in early stages of meiotic recombination. *Chromosoma, 113*, 197–203.

Bolanos-Villegas, P., Yang, X., Wang, H. J., Juan, C. T., Chuang, M. H., Makaroff, C. A., & Jauh, G. Y. (2013). Arabidopsis CHROMOSOME TRANSMISSION FIDELITY 7 (AtCTF7/ECO1) is required for DNA repair, mitosis and meiosis. *The Plant Journal, 75*, 927–940.

Borde, V., Robine, N., Lin, W., Bonfils, S., Geli, V., & Nicolas, A. (2009). Histone H3 lysine 4 trimethylation marks meiotic recombination initiation sites. *The EMBO Journal, 28*, 99–111.

Borg, M., Brownfield, L., & Twell, D. (2009). Male gametophyte development: A molecular perspective. *Journal of Experimental Botany, 60*, 1465–1478.

Brick, K., Smagulova, F., Khil, P., Camerini-Otero, R. D., & Petukhova, G. V. (2012). Genetic recombination is directed away from functional genomic elements in mice. *Nature, 485*, 642–645.

Cai, X., Dong, F., Edelmann, R. E., & Makaroff, C. A. (2003). The *Arabidopsis* SYN1 cohesin protein is required for sister chromatid arm cohesion and homologous chromosome pairing. *Journal of Cell Science, 116*, 2999–3007.

Cao, X., Aufsatz, W., Zilberman, D., Mette, M. F., Huang, M. S., Matzke, M., & Jacobsen, S. E. (2003). Role of the DRM and CMT3 methyltransferases in RNA-directed DNA methylation. *Current Biology, 13*, 2212–2217.

Caryl, A. P., Armstrong, S. J., Jones, G. H., & Franklin, F. C. H. (2000). A homologue of the yeast *HOP1* gene is inactivated in the Arabidopsis meiotic mutant *asy1*. *Chromosoma, 109*, 62–71.

Chakraborty, P., Pankajam, A. V., Lin, G., Dutta, A., Krishnaprasad, G. N., Tekkedil, M. M., Shinohara, A., Steinmetz, L. M., & Nishant, K. T. (2017). Modulating crossover frequency and interference for obligate crossovers in *Saccharomyces cerevisiae* meiosis. *G3: Genes, Genomes, Genetics, 7*, 1511–1524.

Chang, Y., Gong, L., Yuan, W., Li, X., Chen, G., Li, X., Zhang, Q., & Wu, C. (2009). Replication protein A (RPA1a) is required for meiotic and somatic DNA repair but is dispensable for DNA replication and homologous recombination in rice. *Plant Physiology, 151*, 2162–2173.

Che, L., Tang, D., Wang, K., Wang, M., Zhu, K., Yu, H., Gu, M., & Cheng, Z. (2011). OsAM1 is required for leptotene-zygotene transition in rice. *Cell Research, 21*, 654–665.

Chelysheva, L., Diallo, S., Vezon, D., Gendrot, G., Vrielynck, N., Belcram, K., Rocques, N., Marquez-Lema, A., Bhatt, A. M., Horlow, C., Mercier, R., Mezard, C., & Grelon, M. (2005). AtREC8 and AtSCC3 are essential to the monopolar orientation of the kinetochores during meiosis. *Journal of Cell Science, 118*, 4621–4632.

Chelysheva, L., Gendrot, G., Vezon, D., Doutriaux, M. P., Mercier, R., & Grelon, M. (2007). Zip4/Spo22 is required for class I CO formation but not for synapsis completion in *Arabidopsis thaliana*. *PLoS Genetics, 3*, e83.

Chelysheva, L., Vezon, D., Chambon, A., Gendrot, G., Pereira, L., Lemhemdi, A., Vrielynck, N., Le Guin, S., Novatchkova, M., & Grelon, M. (2012). The Arabidopsis HEI10 is a new ZMM protein related to Zip3. *PLoS Genetics, 8*, e1002799.

Chen, C., Marcus, A., Li, W., Hu, Y., Calzada, J. P., Grossniklaus, U., Cyr, R. J., & Ma, H. (2002). The Arabidopsis *ATK1* gene is required for spindle morphogenesis in male meiosis. *Development, 129*, 2401–2409.

Chen, H., He, C., Wang, C., Wang, X., Ruan, F., Yan, J., Yin, P., Wang, Y., & Yan, S. (2021). RAD51 supports DMC1 by inhibiting the SMC5/6 complex during meiosis. *The Plant Cell, 33*, 2869–2882.

Choi, K., & Henderson, I. R. (2015). Meiotic recombination hotspots–A comparative view. *The Plant Journal, 83*, 52–61.

Choi, K. H., Zhao, X. H., Kelly, K. A., Venn, O., Higgins, J. D., Yelina, N. E., Hardcastle, T. J., Ziolkowski, P. A., Copenhaver, G. P., Franklin, F. C. H., McVean, G., & Henderson, I. R. (2013). Arabidopsis meiotic crossover hot spots overlap with H2A.Z nucleosomes at gene promoters. *Nature Genetics, 45*, 1327–1336.

Choudhary, K., & Kupiec, M. (2023). The cohesin complex of yeasts: Sister chromatid cohesion and beyond. *FEMS Microbiology Reviews, 47*, fuac045.

Cifuentes, M., Jolivet, S., Cromer, L., Harashima, H., Bulankova, P., Renne, C., Crismani, W., Nomura, Y., Nakagami, H., Sugimoto, K., Schnittger, A., Riha, K., & Mercier, R. (2016). TDM1 regulation determines the number of meiotic divisions. *PLoS Genetics, 12*, e1005856.

Ciosk, R., Shirayama, M., Shevchenko, A., Tanaka, T., Toth, A., Shevchenko, A., & Nasmyth, K. (2000). Cohesin's binding to chromosomes depends on a separate complex consisting of Scc2 and Scc4 proteins. *Molecular Cell, 5*, 243–254.

Colome-Tatche, M., Cortijo, S., Wardenaar, R., Morgado, L., Lahouze, B., Sarazin, A., Etcheverry, M., Martin, A., Feng, S., Duvernois-Berthet, E., Labadie, K., Wincker, P., Jacobsen, S. E., Jansen, R. C., Colot, V., & Johannes, F. (2012). Features of the Arabidopsis recombination landscape resulting from the combined loss of sequence variation and DNA methylation. *Proceedings of the National Academy of Sciences of the United States of America, 109*, 16240–16245.

Corbett, K. D., & Harrison, S. C. (2012). Molecular architecture of the yeast monopolin complex. *Cell Reports, 1*, 583–589.

Corbett, K. D., Yip, C. K., Ee, L. S., Walz, T., Amon, A., & Harrison, S. C. (2010). The monopolin complex crosslinks kinetochore components to regulate chromosome-microtubule attachments. *Cell, 142*, 556–567.

Cromer, L., Heyman, J., Touati, S., Harashima, H., Araou, E., Girard, C., Horlow, C., Wassmann, K., Schnittger, A., De Veylder, L., & Mercier, R. (2012). OSD1 promotes meiotic progression via APC/C inhibition and forms a regulatory network with TDM and CYCA1;2/TAM. *PLoS Genetics, 8*, e1002865.

Cromer, L., Jolivet, S., Horlow, C., Chelysheva, L., Heyman, J., De Jaeger, G., Koncz, C., De Veylder, L., & Mercier, R. (2013). Centromeric cohesion is protected twice at meiosis, by SHUGOSHINs at anaphase I and by PATRONUS at interkinesis. *Current Biology, 23*, 2090–2099.

Da Ines, O., Degroote, F., Goubely, C., Amiard, S., Gallego, M. E., & White, C. I. (2013). Meiotic recombination in Arabidopsis is catalysed by DMC1, with RAD51 playing a supporting role. *PLoS Genetics, 9*, e1003787.

Dawe, R. K., Lowry, E. G., Gent, J. I., Stitzer, M. C., Swentowsky, K. W., Higgins, D. M., Ross-Ibarra, J., Wallace, J. G., Kanizay, L. B., Alabady, M., Qiu, W., Tseng, K. F., Wang, N., Gao, Z., Birchler, J. A., Harkess, A. E., Hodges, A. L., & Hiatt, E. N. (2018). A kinesin-14 motor activates neocentromeres to promote meiotic drive in maize. *Cell, 173*, 839–850. e18.

De Muyt, A., Pereira, L., Vezon, D., Chelysheva, L., Gendrot, G., Chambon, A., Laine-Choinard, S., Pelletier, G., Mercier, R., Nogue, F., & Grelon, M. (2009). A high throughput genetic screen identifies new early meiotic recombination functions in *Arabidopsis thaliana*. *PLoS Genetics, 5*, e1000654.

De Storme, N., & Geelen, D. (2011). The Arabidopsis mutant *jason* produces unreduced first division restitution male gametes through a parallel/fused spindle mechanism in meiosis II. *Plant Physiology, 155*, 1403–1415.

De, K., Sterle, L., Krueger, L., Yang, X., & Makaroff, C. A. (2014). *Arabidopsis thaliana* WAPL is essential for the prophase removal of cohesin during meiosis. *PLoS Genetics, 10*, e1004497.

Demidov, D., Van Damme, D., Geelen, D., Blattner, F. R., & Houben, A. (2005). Identification and dynamics of two classes of aurora-like kinases in Arabidopsis and other plants. *The Plant Cell, 17*, 836–848.

Demidov, D., Lermontova, I., Weiss, O., Fuchs, J., Rutten, T., Kumke, K., Sharbel, T. F., Van Damme, D., De Storme, N., Geelen, D., & Houben, A. (2014). Altered expression of Aurora kinases in Arabidopsis results in aneu- and polyploidization. *The Plant Journal, 80*, 449–461.

d'Erfurth, I., Jolivet, S., Froger, N., Catrice, O., Novatchkova, M., Simon, M., Jenczewski, E., & Mercier, R. (2008). Mutations in *AtPS1* (*Arabidopsis thaliana parallel spindle 1*) lead to the production of diploid pollen grains. *PLoS Genetics, 4*, e1000274.

d'Erfurth, I., Cromer, L., Jolivet, S., Girard, C., Horlow, C., Sun, Y., To, J. P., Berchowitz, L. E., Copenhaver, G. P., & Mercier, R. (2010). The cyclin-A CYCA1;2/TAM is required for the meiosis I to meiosis II transition and cooperates with OSD1 for the prophase to first meiotic division transition. *PLoS Genetics, 6*, e1000989.

Dernburg, A. F., Sedat, J. W., & Hawley, R. S. (1996). Direct evidence of a role for heterochromatin in meiotic chromosome segregation. *Cell, 86*, 135–146.

Ding, X., Xu, R., Yu, J., Xu, T., Zhuang, Y., & Han, M. (2007). SUN1 is required for telomere attachment to nuclear envelope and gametogenesis in mice. *Developmental Cell, 12*, 863–872.

Dion, E., Li, L. L., Jean, M., & Beizile, F. (2007). An Arabidopsis *MLH1* mutant exhibits reproductive defects and reveals a dual role for this gene in mitotic recombination. *The Plant Journal, 51*, 431–440.

Dissmeyer, N., Nowack, M. K., Pusch, S., Stals, H., Inze, D., Grini, P. E., & Schnittger, A. (2007). T-loop phosphorylation of Arabidopsis CDKA;1 is required for its function and can be partially substituted by an aspartate residue. *The Plant Cell, 19*, 972–985.

Drouaud, J., Khademian, H., Giraut, L., Zanni, V., Bellalou, S., Henderson, I. R., Falque, M., & Mezard, C. (2013). Contrasted patterns of crossover and non-crossover at *Arabidopsis thaliana* meiotic recombination hotspots. *PLoS Genetics, 9*, e1003922.

Du, J., Zhong, X., Bernatavichute, Y. V., Stroud, H., Feng, S., Caro, E., Vashisht, A. A., Terragni, J., Chin, H. G., Tu, A., Hetzel, J., Wohlschlegel, J. A., Pradhan, S., Patel, D. J., & Jacobsen, S. E. (2012). Dual binding of chromomethylase domains to H3K9me2-containing nucleosomes directs DNA methylation in plants. *Cell, 151*, 167–180.

Eschbach, V., & Kobbe, D. (2014). Different replication protein A complexes of *Arabidopsis thaliana* have different DNA-binding properties as a function of heterotrimer composition. *Plant & Cell Physiology, 55*, 1460–1472.

Ferdous, M., Higgins, J. D., Osman, K., Lambing, C., Roitinger, E., Mechtler, K., Armstrong, S. J., Perry, R., Pradillo, M., Cunado, N., & Franklin, F. C. (2012). Inter-homolog crossing-over and synapsis in Arabidopsis meiosis are dependent on the chromosome axis protein AtASY3. *PLoS Genetics, 8*, e1002507.

Fernius, J., & Marston, A. L. (2009). Establishment of cohesion at the pericentromere by the Ctf19 kinetochore subcomplex and the replication fork-associated factor, Csm3. *PLoS Genetics, 5*, e1000629.

Fernius, J., Nerusheva, O. O., Galander, S., Alves Fde, L., Rappsilber, J., & Marston, A. L. (2013). Cohesin-dependent association of scc2/4 with the centromere initiates pericentromeric cohesion establishment. *Current Biology, 23*, 599–606.

Fu, M., Wang, C., Xue, F. Y., Higgins, J., Chen, M. J., Zhang, D. B., & Liang, W. Q. (2016). The DNA topoisomerase VI-B subunit OsMTOPVIB is essential for meiotic recombination initiation in rice. *Molecular Plant, 9*, 1539–1541.

Gandhi, R., Gillespie, P. J., & Hirano, T. (2006). Human Wapl is a cohesin-binding protein that promotes sister-chromatid resolution in mitotic prophase. *Current Biology, 16*, 2406–2417.

Geuting, V., Kobbe, D., Hartung, F., Durr, J., Focke, M., & Puchta, H. (2009). Two distinct MUS81-EME1 complexes from Arabidopsis process Holliday junctions. *Plant Physiology, 150*, 1062–1071.

Gicking, A. M., Swentowsky, K. W., Dawe, R. K., & Qiu, W. (2018). Functional diversification of the kinesin-14 family in land plants. *FEBS Letters, 592*, 1918–1928.

Girard, C., Chelysheva, L., Choinard, S., Froger, N., Macaisne, N., Lemhemdi, A., Mazel, J., Crismani, W., & Mercier, R. (2015). AAA-ATPase FIDGETIN-LIKE 1 and helicase FANCM antagonize meiotic crossovers by distinct mechanisms. *PLoS Genetics, 11*, e1005369.

Glover, J., Grelon, M., Craig, S., Chaudhury, A., & Dennis, E. (1998). Cloning and characterization of *MS5* from Arabidopsis: A gene critical in male meiosis. *The Plant Journal, 15*, 345–356.

Golubovskaya, I. N., Hamant, O., Timofejeva, L., Wang, C.-J. R., Braun, D., Meeley, R., & Cande, W. Z. (2006). Alleles of *afd1* dissect REC8 functions during meiotic prophase I. *Journal of Cell Science, 119*, 3306–3315.

Golubovskaya, I. N., Wang, C. J. R., Timofejeva, L., & Cande, W. Z. (2011). Maize meiotic mutants with improper or non-homologous synapsis due to problems in pairing or synaptonemal complex formation. *Journal of Experimental Botany, 62*, 1533–1544.

Grelon, M., Hernandez Sanchez-Rebato, M., Bouatta, A. M., Gallego, M. E., White, C. I., & Da Ines, O. (2021). RAD54 is essential for RAD51-mediated repair of meiotic DSB in Arabidopsis. *PLoS Genetics, 17*, e1008919.

Guan, Y., Leu, N. A., Ma, J., Chmatal, L., Ruthel, G., Bloom, J. C., Lampson, M. A., Schimenti, J. C., Luo, M., & Wang, P. J. (2020). SKP1 drives the prophase I to metaphase I transition during male meiosis. *Science Advances, 6*, eaaz2129.

Guan, Y., Lin, H., Leu, N. A., Ruthel, G., Fuchs, S. Y., Busino, L., Luo, M., & Wang, P. J. (2022). SCF ubiquitin E3 ligase regulates DNA double-strand breaks in early meiotic recombination. *Nucleic Acids Research, 50*, 5129–5144.

Haering, C. H., & Gruber, S. (2016). Snapshot: SMC protein complexes part I. *Cell, 164*, 326. e1.

Hamant, O., Golubovskaya, I., Meeley, R., Fiume, E., Timofejeva, L., Schleiffer, A., Nasmyth, K., & Cande, W. Z. (2005). A REC8-dependent plant Shugoshin is required for maintenance of centromeric cohesion during meiosis and has no mitotic functions. *Current Biology, 15*, 948–954.

Hamant, O., Ma, H., & Cande, W. Z. (2006). Genetics of meiotic prophase I in plants. *Annual Review of Plant Biology, 57*, 267–302.

He, W., Rao, H., Tang, S., Bhagwat, N., Kulkarni, D. S., Ma, Y., Chang, M. A. W., Hall, C., Bragg, J. W., Manasca, H. S., Baker, C., Verhees, G. F., Ranjha, L., Chen, X., Hollingsworth, N. M., Cejka, P., & Hunter, N. (2020). Regulated proteolysis of mutSγ controls meiotic crossing over. *Molecular Cell, 78*, 168–183. e5.

He, C., Chen, Z., Zhao, Y., Yu, Y., Wang, H., Wang, C., Copenhaver, G. P., Qi, J., & Wang, Y. (2022). Histone demethylase IBM1-mediated meiocyte gene expression ensures meiotic chromosome synapsis and recombination. *PLoS Genetics, 18*, e1010041.

Henderson, I. R., Da Ines, O., Bazile, J., Gallego, M. E., & White, C. I. (2022). DMC1 attenuates RAD51-mediated recombination in Arabidopsis. *PLoS Genetics, 18*, e1010322.

Higgins, J. D., Armstrong, S. J., Franklin, F. C., & Jones, G. H. (2004). The Arabidopsis MutS homolog AtMSH4 functions at an early step in recombination: Evidence for two classes of recombination in Arabidopsis. *Genes & Development, 18*, 2557–2570.

Higgins, J. D., Sanchez-Moran, E., Armstrong, S. J., Jones, G. H., & Franklin, F. C. H. (2005). The Arabidopsis synaptonemal complex protein ZYP1 is required for chromosome synapsis and normal fidelity of crossing over. *Genes & Development, 19*, 2488–2500.

Higgins, J. D., Buckling, E. F., Franklin, F. C., & Jones, G. H. (2008a). Expression and functional analysis of AtMUS81 in Arabidopsis meiosis reveals a role in the second pathway of crossing-over. *The Plant Journal, 54*, 152–162.

Higgins, J. D., Vignard, J., Mercier, R., Pugh, A. G., Franklin, F. C., & Jones, G. H. (2008b). AtMSH5 partners AtMSH4 in the class I meiotic crossover pathway in *Arabidopsis thaliana*, but is not required for synapsis. *The Plant Journal, 55*, 28–39.

Higgins, D. M., Nannas, N. J., & Dawe, R. K. (2016). The maize *divergent spindle-1* (*dv1*) gene encodes a kinesin-14A motor protein required for meiotic spindle pole organization. *Frontiers in Plant Science, 7*, 1277.

Hinch, A. G., Becker, P. W., Li, T., Moralli, D., Zhang, G., Bycroft, C., Green, C., Keeney, S., Shi, Q., Davies, B., & Donnelly, P. (2020). The configuration of RPA, RAD51, and DMC1 binding in meiosis reveals the nature of critical recombination intermediates. *Molecular Cell, 79*, 689–701. e10.

Huang, J., Cheng, Z., Wang, C., Hong, Y., Su, H., Wang, J., Copenhaver, G. P., Ma, H., & Wang, Y. (2015). Formation of interference-sensitive meiotic cross-overs requires sufficient DNA leading-strand elongation. *Proceedings of the National Academy of Sciences of the United States of America, 112*, 12534–12539.

Huang, J. Y., Copenhaver, G. P., Ma, H., & Wang, Y. X. (2016). New insights into the role of DNA synthesis in meiotic recombination. *Scientific Bulletin, 61*, 1260–1269.

Huang, J., Wang, C., Wang, H., Lu, P., Zheng, B., Ma, H., Copenhaver, G. P., & Wang, Y. (2019). Meiocyte-specific and AtSPO11-1-dependent small RNAs and their association with meiotic gene expression and recombination. *The Plant Cell, 31*, 444–464.

Hughes, S. E., Hemenway, E., Guo, F., Yi, K., Yu, Z., & Hawley, R. S. (2019). The E3 ubiquitin ligase Sina regulates the assembly and disassembly of the synaptonemal complex in Drosophila females. *PLoS Genetics, 15*, e1008161.

Hyppa, R. W., & Smith, G. R. (2010). Crossover invariance determined by partner choice for meiotic DNA break repair. *Cell, 142*, 243–255.

Ip, S. C., Rass, U., Blanco, M. G., Flynn, H. R., Skehel, J. M., & West, S. C. (2008). Identification of Holliday junction resolvases from humans and yeast. *Nature, 456*, 357–361.

Jackson, N., Sanchez-Moran, E., Buckling, E., Armstrong, S. J., Jones, G. H., & Franklin, F. C. (2006). Reduced meiotic crossovers and delayed prophase I progression in AtMLH3-deficient Arabidopsis. *The EMBO Journal, 25*, 1315–1323.

Ji, J., Tang, D., Wang, K., Wang, M., Che, L., Li, M., & Cheng, Z. (2012). The role of OsCOM1 in homologous chromosome synapsis and recombination in rice meiosis. *The Plant Journal, 72*, 18–30.

Ji, J., Tang, D., Wang, M., Li, Y., Zhang, L., Wang, K., Li, M., & Cheng, Z. (2013). MRE11 is required for homologous synapsis and DSB processing in rice meiosis. *Chromosoma, 122*, 363–376.

Ji, J., Tang, D., Shen, Y., Xue, Z., Wang, H., Shi, W., Zhang, C., Du, G., Li, Y., & Cheng, Z. (2016). P31^{comet}, a member of the synaptonemal complex, participates in meiotic DSB formation in rice. *Proceedings of the National Academy of Sciences of the United States of America, 113*, 10577–10582.

Jing, J., Zhang, T., Wang, Y., Cui, Z., & He, Y. (2019). ZmRAD51C is essential for double-strand break repair and homologous recombination in maize meiosis. *International Journal of Molecular Sciences, 20*, 5513.

Jing, J., Wu, N., Xu, W., Wang, Y., Pawlowski, W. P., & He, Y. (2022). An F-box protein ACOZ1 functions in crossover formation by ensuring proper chromosome compaction during maize meiosis. *The New Phytologist, 235*, 157–172.

Juranic, M., Srilunchang, K. O., Krohn, N. G., Leljak-Levanic, D., Sprunck, S., & Dresselhaus, T. (2012). Germline-specific MATH-BTB substrate adaptor MAB1 regulates spindle length and nuclei identity in maize. *The Plant Cell, 24*, 4974–4991.

Keeney, S. (2008). Spo11 and the formation of DNA double-strand breaks in meiosis. *Genome Dynamics and Stability, 2*, 81–123.

Keeney, S., Giroux, C. N., & Kleckner, N. (1997). Meiosis-specific DNA double-strand breaks are catalyzed by Spo11, a member of a widely conserved protein family. *Cell, 88*, 375–384.

Khoo, K. H., Able, A. J., & Able, J. A. (2012). Poor homologous synapsis 1 interacts with chromatin but does not colocalise with ASYnapsis 1 during early meiosis in bread wheat. *International Journal of Plant Genomics, 2012*, 514398.

Kniewel, R., & Keeney, S. (2009). Histone methylation sets the stage for meiotic DNA breaks. *The EMBO Journal, 28*, 81–83.

Komaki, S., Takeuchi, H., Hamamura, Y., Heese, M., Hashimoto, T., & Schnittger, A. (2020). Functional analysis of the plant chromosomal passenger complex. *Plant Physiology, 183*, 1586–1599.

Komiya, R., Ohyanagi, H., Niihama, M., Watanabe, T., Nakano, M., Kurata, N., & Nonomura, K. (2014). Rice germline-specific Argonaute MEL1 protein binds to phasiRNAs generated from more than 700 lincRNAs. *The Plant Journal, 78*, 385–397.

Kurzbauer, M. T., Pradillo, M., Kerzendorfer, C., Sims, J., Ladurner, R., Oliver, C., Janisiw, M. P., Mosiolek, M., Schweizer, D., Copenhaver, G. P., & Schlogelhofer, P. (2018). *Arabidopsis thaliana* FANCD2 promotes meiotic crossover formation. *The Plant Cell, 30*, 415–428.

Lang, J., Smetana, O., Sanchez-Calderon, L., Lincker, F., Genestier, J., Schmit, A. C., Houlne, G., & Chaboute, M. E. (2012). Plant γH2AX foci are required for proper DNA DSB repair responses and colocalize with E2F factors. *The New Phytologist, 194*, 353–363.

Li, X., Chang, Y., Xin, X., Zhu, C., Li, X., Higgins, J. D., & Wu, C. (2013). Replication protein A2c coupled with replication protein A1c regulates crossover formation during meiosis in rice. *The Plant Cell, 25*, 3885–3899.

Li, X., Li, L., & Yan, J. (2015). Dissecting meiotic recombination based on tetrad analysis by single-microspore sequencing in maize. *Nature Communications, 6*, 6648.

Li, X., Zhang, J., Huang, J., Xu, J., Chen, Z., Copenhaver, G. P., & Wang, Y. (2021). Regulation of interference-sensitive crossover distribution ensures crossover assurance in Arabidopsis. *Proceedings of the National Academy of Sciences of the United States of America, 118*, e2107543118.

Lin, Z., Kong, H., Nei, M., & Ma, H. (2006). Origins and evolution of the *recA/RAD51* gene family: Evidence for ancient gene duplication and endosymbiotic gene transfer. *Proceedings of the National Academy of Sciences of the United States of America, 103*, 10328–10333.

Lin, Z., Nei, M., & Ma, H. (2007). The origins and early evolution of DNA mismatch repair genes-multiple horizontal gene transfers and co-evolution. *Nucleic Acids Research, 35*, 7591–7603.

Lin, Y. N., Jiang, C. K., Cheng, Z. K., Wang, D. H., Shen, L. P., Xu, C., Xu, Z. H., & Bai, S. N. (2022). Rice cell division cycle 20s are required for faithful chromosome segregation and cytokinesis during meiosis. *Plant Physiology, 188*, 1111–1128.

Liu, B., & Lee, Y. J. (2022). Spindle assembly and mitosis in plants. *Annual Review of Plant Biology, 73*, 227–254.

Liu, H., & Nonomura, K. I. (2016). A wide reprogramming of histone H3 modifications during male meiosis I in rice is dependent on the Argonaute protein MEL1. *Journal of Cell Science, 129*, 3553–3561.

Liu, S., Yeh, C. T., Ji, T., Ying, K., Wu, H., Tang, H. M., Fu, Y., Nettleton, D., & Schnable, P. S. (2009). Mu transposon insertion sites and meiotic recombination events co-localize with epigenetic marks for open chromatin across the maize genome. *PLoS Genetics, 5*, e1000733.

Liu, H., Jia, L., & Yu, H. (2013). Phospho-H2A and cohesin specify distinct tension-regulated Sgo1 pools at kinetochores and inner centromeres. *Current Biology, 23*, 1927–1933.

Lorenz, A., & Whitby, M. C. (2012). How not to get cross(ed): A novel role for FANCM orthologs in meiotic recombination. *Cell Cycle, 11*, 3347–3348.

Luo, Q., Li, Y., Shen, Y., & Cheng, Z. (2014). Ten years of gene discovery for meiotic event control in rice. *Journal of Genetics and Genomics, 41*, 125–137.

Ma, H. (2005). Molecular genetic analyses of microsporogenesis and microgametogenesis in flowering plants. *Annual Review of Plant Biology, 56*, 393–434.

Ma, H. (2006). A molecular portrait of Arabidopsis meiosis. *The Arabidopsis Book, 4*, e0095.

Macaisne, N., Novatchkova, M., Peirera, L., Vezon, D., Jolivet, S., Froger, N., Chelysheva, L., Grelon, M., & Mercier, R. (2008). SHOC1, an XPF endonuclease-related protein, is essential for the formation of class I meiotic crossovers. *Current Biology, 18*, 1432–1437.

Macaisne, N., Vignard, J., & Mercier, R. (2011). SHOC1 and PTD form an XPF-ERCC1-like complex that is required for formation of class I crossovers. *Journal of Cell Science, 124*, 2687–2691.

Maier, N. K., Ma, J., Lampson, M. A., & Cheeseman, I. M. (2021). Separase cleaves the kinetochore protein Meikin at the meiosis I/II transition. *Molecular Cell, 56*, 2192–2206. e8.

Marand, A. P., Chen, Z., Gallavotti, A., & Schmitz, R. J. (2021). A cis-regulatory atlas in maize at single-cell resolution. *Cell, 184*, 3041–3055. e21.

Marston, A. L. (2015). Shugoshins: Tension-sensitive pericentromeric adaptors safeguarding chromosome segregation. *Molecular and Cellular Biology, 35*, 634–648.

Masson, J. Y., & West, S. C. (2001). The Rad51 and Dmc1 recombinases: A non-identical twin relationship. *Trends in Biochemical Sciences, 26*, 131–136.

McMichael, C. M., & Bednarek, S. Y. (2013). Cytoskeletal and membrane dynamics during higher plant cytokinesis. *The New Phytologist, 197*, 1039–1057.

Melamed-Bessudo, C., & Levy, A. A. (2012). Deficiency in DNA methylation increases meiotic crossover rates in euchromatic but not in heterochromatic regions in Arabidopsis. *Proceedings of the National Academy of Sciences of the United States of America, 109*, E981–E988.

Mercier, R., Vezon, D., Bullier, E., Motamayor, J. C., Sellier, A., Lefevre, F., Pelletier, G., & Horlow, C. (2001). SWITCH1 (SWI1): A novel protein required for the establishment of sister chromatid cohesion and for bivalent formation at meiosis. *Genes & Development, 15*, 1859–1871.

Mercier, R., Mezard, C., Jenczewski, E., Macaisne, N., & Grelon, M. (2015). The molecular biology of meiosis in plants. *Annual Review of Plant Biology, 66*, 297–327.

Mercler, R., Armstrong, S. J., Horlow, C., Jackson, N. P., Makaroff, C. A., Vezon, D., Pelletier, G., Jones, G. H., & Franklin, F. C. H. (2003). The meiotic protein SWI1 is required for axial element formation and recombination initiation in *Arabidopsis*. *Development, 130*, 3309–3318.

Meyer, R. E., Kim, S., Obeso, D., Straight, P. D., Winey, M., & Dawson, D. S. (2013). Mps1 and Ipl1/Aurora B act sequentially to correctly orient chromosomes on the meiotic spindle of budding yeast. *Science, 339*, 1071–1074.

Miao, C., Tang, D., Zhang, H., Wang, M., Li, Y., Tang, S., Yu, H., Gu, M., & Cheng, Z. (2013). Central region component1, a novel synaptonemal complex component, is essential for meiotic recombination initiation in rice. *The Plant Cell, 25*, 2998–3009.

Minina, E. A., Reza, S. H., Gutierrez-Beltran, E., Elander, P. H., Bozhkov, P. V., & Moschou, P. N. (2017). The Arabidopsis homolog of Scc4/MAU2 is essential for embryogenesis. *Journal of Cell Science, 130*, 1051–1063.

Mirouze, M., Lieberman-Lazarovich, M., Aversano, R., Bucher, E., Nicolet, J., Reinders, J., & Paszkowski, J. (2012). Loss of DNA methylation affects the recombination landscape in Arabidopsis. *Proceedings of the National Academy of Sciences of the United States of America, 109*, 5880–5885.

Mohrmann, L., & Verrijzer, C. P. (2005). Composition and functional specificity of SWI2/SNF2 class chromatin remodeling complexes. *Biochimica et Biophysica Acta, 1681*, 59–73.

Musacchio, A., & Salmon, E. D. (2007). The spindle-assembly checkpoint in space and time. *Nature Structural & Molecular Biology, 8*, 379–393.

Nelms, B., & Walbot, V. (2019). Defining the developmental program leading to meiosis in maize. *Science, 364*, 52–56.

Nikalayevich, E., El Jailani, S., Dupre, A., Cladiere, D., Gryaznova, Y., Fosse, C., Buffin, E., Touati, S. A., & Wassmann, K. (2022). Aurora B/C-dependent phosphorylation promotes Rec8 cleavage in mammalian oocytes. *Current Biology, 32*, 2281–2290. e4.

Niu, B., Wang, L., Zhang, L., Ren, D., Ren, R., Copenhaver, G. P., Ma, H., & Wang, Y. (2015). Arabidopsis *cell division cycle 20.1* is required for normal meiotic spindle assembly and chromosome segregation. *The Plant Cell, 27*, 3367–3382.

Nonomura, K. I., Nakano, M., Murata, K., Miyoshi, K., Eiguchi, M., Miyao, A., Hirochika, H., & Kurata, N. (2004). An insertional mutation in the rice *PAIR2* gene, the ortholog of Arabidopsis *ASY1*, results in a defect in homologous chromosome pairing during meiosis. *Molecular Genetics and Genomics, 271*, 121–129.

Nonomura, K. I., Nakano, M., Eiguchi, M., Suzuki, T., & Kurata, N. (2006). PAIR2 is essential for homologous chromosome synapsis in rice meiosis I. *Journal of Cell Science, 119*, 217–225.

Nonomura, K., Morohoshi, A., Nakano, M., Eiguchi, M., Miyao, A., Hirochika, H., & Kurata, N. (2007). A germ cell specific gene of the *ARGONAUTE* family is essential for the progression of premeiotic mitosis and meiosis during sporogenesis in rice. *The Plant Cell, 19*, 2583–2594.

Olivier, M., Da Ines, O., Amiard, S., Serra, H., Goubely, C., White, C. I., & Gallego, M. E. (2016). The structure-specific endonucleases MUS81 and SEND1 are essential for telomere stability in Arabidopsis. *The Plant Cell, 28*, 74–86.

Orr, J. N., Waugh, R., & Colas, I. (2021). Ubiquitination in plant meiosis: Recent advances and high throughput methods. *Frontiers in Plant Science, 12*, 667314.

Page, S. L., & Hawley, R. S. (2004). The genetics and molecular biology of the synaptonemal complex. *Annual Review of Cell and Developmental Biology, 20*, 525–558.

Paques, F., & Haber, J. E. (1999). Multiple pathways of recombination induced by double-strand breaks in *Saccharomyces cerevisiae*. *Microbiology and Molecular Biology Reviews, 63*, 349–404.

Parish, R. W., & Li, S. F. (2010). Death of a tapetum: A programme of developmental altruism. *Plant Science, 178*, 73–89.

Parvanov, E. D., Petkov, P. M., & Paigen, K. (2010). Prdm9 controls activation of mammalian recombination hotspots. *Science, 327*, 835.

Penkner, A., Tang, L., Novatchkova, M., Ladurner, M., Fridkin, A., Gruenbaum, Y., Schweizer, D., Loidl, J., & Jantsch, V. (2007). The nuclear envelope protein Matefin/SUN-1 is required for homologous pairing in *C. elegans* meiosis. *Developmental Cell, 12*, 873–885.

Perrella, G., Consiglio, M. F., Aiese-Cigliano, R., Cremona, G., Sanchez-Moran, E., Barra, L., Errico, A., Bressan, R. A., Franklin, F. C., & Conicella, C. (2010). Histone hyperacetylation affects meiotic recombination and chromosome segregation in Arabidopsis. *The Plant Journal, 62*, 796–806.

Petrovska, B., Cenklova, V., Pochylova, Z., Kourova, H., Doskocilova, A., Plihal, O., Binarova, L., & Binarova, P. (2012). Plant Aurora kinases play a role in maintenance of primary meristems and control of endoreduplication. *The New Phytologist, 193*, 590–604.

Phadnis, N., Hyppa, R. W., & Smith, G. R. (2011). New and old ways to control meiotic recombination. *Trends in Genetics, 27*, 411–421.

Powers, N. R., Parvanov, E. D., Baker, C. L., Walker, M., Petkov, P. M., & Paigen, K. (2016). The meiotic recombination activator PRDM9 trimethylates both H3K36 and H3K4 at recombination hotspots *in vivo*. *PLoS Genetics, 12*, e1006146.

Pradillo, M., Lopez, E., Linacero, R., Romero, C., Cunado, N., Sanchez-Moran, E., & Santos, J. L. (2012). Together yes, but not coupled: New insights into the roles of RAD51 and DMC1 in plant meiotic recombination. *The Plant Journal, 69*, 921–933.

Quan, L., Xiao, R., Li, W., Oh, S. A., Kong, H., Ambrose, J. C., Malcos, J. L., Cyr, R., Twell, D., & Ma, H. (2008). Functional divergence of the duplicated *AtKIN14a* and *AtKIN14b* genes: Critical roles in Arabidopsis meiosis and gametophyte development. *The Plant Journal, 53*, 1013–1026.

Rao, H. B., Qiao, H., Bhatt, S. K., Bailey, L. R., Tran, H. D., Bourne, S. L., Qiu, W., Deshpande, A., Sharma, A. N., Beebout, C. J., Pezza, R. J., & Hunter, N. (2017). A SUMO-ubiquitin relay recruits proteasomes to chromosome axes to regulate meiotic recombination. *Science, 355*, 403–407.

Reddy, T. V., Kaur, J., Agashe, B., Sundaresan, V., & Siddiqi, I. (2003). The *DUET* gene is necessary for chromosome organization and progression during male meiosis in Arabidopsis and encodes a PHD finger protein. *Development, 130*, 5975–5987.

Ren, L., Zhao, T., Zhao, Y., Du, G., Yang, S., Mu, N., Tang, D., Shen, Y., Li, Y., & Cheng, Z. (2021). The E3 ubiquitin ligase DESYNAPSIS1 regulates synapsis and recombination in rice meiosis. *Cell Reports, 37*, 109941.

Richardson, D. N., Simmons, M. P., & Reddy, A. S. (2006). Comprehensive comparative analysis of kinesins in photosynthetic eukaryotes. *BMC Genomics, 7*, 18.

Riehs, N., Akimcheva, S., Puizina, J., Bulankova, P., Idol, R. A., Siroky, J., Schleiffer, A., Schweizer, D., Shippen, D. E., & Riha, K. (2008). Arabidopsis SMG7 protein is required for exit from meiosis. *Journal of Cell Science, 121*, 2208–2216.

Ronceret, A., Doutriaux, M. P., Golubovskaya, I. N., & Pawlowski, W. P. (2009). PHS1 regulates meiotic recombination and homologous chromosome pairing by controlling the transport of RAD50 to the nucleus. *Proceedings of the National Academy of Sciences of the United States of America, 106*, 20121–20126.

Sakuno, T., Tanaka, K., Hauf, S., & Watanabe, Y. (2011). Repositioning of aurora B promoted by chiasmata ensures sister chromatid mono-orientation in meiosis I. *Developmental Cell, 21*, 534–545.

Samanic, I., Simunic, J., Riha, K., & Puizina, J. (2013). Evidence for distinct functions of MRE11 in Arabidopsis meiosis. *PLoS One, 8*, e78760.

Sanders, P. M., Bui, A. Q., Weterings, K., McIntire, K. N., Hsu, Y. C., Lee, P. Y., Truong, M. T., Beals, T. P., & Goldberg, R. B. (1999). Anther developmental defects in *Arabidopsis thaliana* male-sterile mutants. *Sexual Plant Reproduction, 11*, 297–322.

San-Segundo, P. A., & Roeder, G. S. (1999). Pch2 links chromatin silencing to meiotic checkpoint control. *Cell, 97*, 313–324.

Scherthan, H. (2001). A bouquet makes ends meet. *Nature Reviews. Molecular Cell Biology, 2*, 621–627.

Schild-Prufert, K., Saito, T. T., Smolikov, S., Gu, Y., Hincapie, M., Hill, D. E., Vidal, M., McDonald, K., & Colaiacovo, M. P. (2011). Organization of the synaptonemal complex during meiosis in *Caenorhabditis elegans*. *Genetics, 189*, 411–421.

Schubert, V., Weissleder, A., Ali, H., Fuchs, J., Lermontova, I., Meister, A., & Schubert, I. (2009). Cohesin gene defects may impair sister chromatid alignment and genome stability in *Arabidopsis thaliana*. *Chromosoma, 118*, 591–605.

Schwartz, E. K., & Heyer, W. D. (2011). Processing of joint molecule intermediates by structure-selective endonucleases during homologous recombination in eukaryotes. *Chromosoma, 120*, 109–127.

Sebastian, J., Ravi, M., Andreuzza, S., Panoli, A. P., Marimuthu, M. P., & Siddiqi, I. (2009). The plant adherin AtSCC2 is required for embryogenesis and sister-chromatid cohesion during meiosis in Arabidopsis. *The Plant Journal, 59*, 1–13.

Seguela-Arnaud, M., Crismani, W., Larcheveque, C., Mazel, J., Froger, N., Choinard, S., Lemhemdi, A., Macaisne, N., Van Leene, J., Gevaert, K., De Jaeger, G., Chelysheva, L., & Mercier, R. (2015). Multiple mechanisms limit meiotic crossovers: TOP3α and two BLM homologs antagonize crossovers in parallel to FANCM. *Proceedings of the National Academy of Sciences of the United States of America, 112*, 4713–4718.

Serra, H., Lambing, C., Griffin, C. H., Topp, S. D., Nageswaran, D. C., Underwood, C. J., Ziolkowski, P. A., Seguela-Arnaud, M., Fernandes, J. B., Mercier, R., & Henderson, I. R. (2018). Massive crossover elevation via combination of *HEI10* and *recq4a recq4b* during Arabidopsis meiosis. *Proceedings of the National Academy of Sciences of the United States of America, 115*, 2437–2442.

Shah, S. S., Hartono, S., Piazza, A., Som, V., Wright, W., Chédin, F., & Heyer, W.-D. (2020). Rdh54/Tid1 inhibits Rad51-Rad54-mediated D-loop formation and limits D-loop length. *eLife, 9*, e59112.

Shao, T., Tang, D., Wang, K., Wang, M., Che, L., Qin, B., Yu, H., Li, M., Gu, M., & Cheng, Z. (2011). OsREC8 is essential for chromatid cohesion and metaphase I monopolar orientation in rice meiosis. *Plant Physiology, 156*, 1386–1396.

Sheehan, M. J., & Pawlowski, W. P. (2009). Live imaging of rapid chromosome movements in meiotic prophase I in maize. *Proceedings of the National Academy of Sciences of the United States of America, 106*, 20989–20994.

Shen, Y., Tang, D., Wang, K., Wang, M., Huang, J., Luo, W., Luo, Q., Hong, L., Li, M., & Cheng, Z. (2012). ZIP4 in homologous chromosome synapsis and crossover formation in rice meiosis. *Journal of Cell Science, 125*, 2581–2591.

Shi, W., Ji, J., Xue, Z., Zhang, F., Miao, Y., Yang, H., Tang, D., Du, G., Li, Y., Shen, Y., & Cheng, Z. (2021). PRD1, a homologous recombination initiation factor, is involved in spindle assembly in rice meiosis. *The New Phytologist, 230*, 585–600.

Shilo, S., Melamed-Bessudo, C., Dorone, Y., Barkai, N., & Levy, A. A. (2015). DNA crossover motifs associated with epigenetic modifications delineate open chromatin regions in Arabidopsis. *The Plant Cell, 27*, 2427–2436.

Shintomi, K., & Hirano, T. (2009). Releasing cohesin from chromosome arms in early mitosis: Opposing actions of Wapl-Pds5 and Sgo1. *Genes & Development, 23*, 2224–2236.

Shrestha, R. L., Conti, D., Tamura, N., Braun, D., Ramalingam, R. A., Cieslinski, K., Ries, J., & Draviam, V. M. (2017). Aurora-B kinase pathway controls the lateral to end-on conversion of kinetochore-microtubule attachments in human cells. *Nature Communications, 8*, 150.

Siddiqui, N. U., Stronghill, P. E., Dengler, R. E., Hasenkampf, C. A., & Riggs, C. D. (2003). Mutations in Arabidopsis condensin genes disrupt embryogenesis, meristem organization and segregation of homologous chromosomes during meiosis. *Development, 130*, 3283–3295.

Sidhu, G. K., Fang, C., Olson, M. A., Falque, M., Martin, O. C., & Pawlowski, W. P. (2015). Recombination patterns in maize reveal limits to crossover homeostasis. *Proceedings of the National Academy of Sciences of the United States of America, 112*, 15982–15987.

Smagulova, F., Gregoretti, I. V., Brick, K., Khil, P., Camerini-Otero, R. D., & Petukhova, G. V. (2011). Genome-wide analysis reveals novel molecular features of mouse recombination hotspots. *Nature, 472*, 375–378.

Smith, S. J., Osman, K., & Franklin, F. C. (2014). The condensin complexes play distinct roles to ensure normal chromosome morphogenesis during meiotic division in Arabidopsis. *The Plant Journal, 80*, 255–268.

Stacey, N. J., Kuromori, T., Azumi, Y., Roberts, G., Breuer, C., Wada, T., Maxwell, A., Roberts, K., & Sugimoto-Shirasu, K. (2006). Arabidopsis SPO11-2 functions with SPO11-1 in meiotic recombination. *The Plant Journal, 48*, 206–216.

Starr, D. A., & Fridolfsson, H. N. (2010). Interactions between nuclei and the cytoskeleton are mediated by SUN-KASH nuclear-envelope bridges. *Annual Review of Cell and Developmental Biology, 26*, 421–444.

Stroud, H., Do, T., Du, J., Zhong, X., Feng, S., Johnson, L., Patel, D. J., & Jacobsen, S. E. (2014). Non-CG methylation patterns shape the epigenetic landscape in Arabidopsis. *Nature Structural & Molecular Biology, 21*, 64–72.

Su, H., Cheng, Z., Huang, J., Lin, J., Copenhaver, G. P., Ma, H., & Wang, Y. (2017). Arabidopsis RAD51, RAD51C and XRCC3 proteins form a complex and facilitate RAD51 localization on chromosomes for meiotic recombination. *PLoS Genetics, 13*, e1006827.

Surcel, A., Zhou, X. F., Quan, L., & Ma, H. (2008). Long-term maintenance of stable copy number in the eukaryotic family: Origin of a vertebrate meiotic and fate of recent segmental duplicates. *Journal of Systematics and Evolution, 46*, 405–423.

Tang, Y., Yin, Z., Zeng, Y., Zhang, Q., Chen, L., He, Y., Lu, P., Ye, & Zhang, X. (2017). MTOPVIB interacts with AtPRD1 and plays important roles in formation of meiotic DNA double-strand breaks in Arabidopsis. *Scientific Reports, 7*, 10007.

Turinetto, V., & Giachino, C. (2015). Multiple facets of histone variant H2AX: A DNA double-strand-break marker with several biological functions. *Nucleic Acids Research, 43*, 2489–2498.

Uanschou, C., Siwiec, T., Pedrosa-Harand, A., Kerzendorfer, C., Sanchez-Moran, E., Novatchkova, M., Akimcheva, S., Woglar, A., Klein, F., & Schlögelhofer, P. (2007). A novel plant gene essential for meiosis is related to the human *CtIP* and the yeast *COM1/SAE2* gene. *The EMBO Journal, 26*, 5061–5070.

Uanschou, C., Ronceret, A., Von Harder, M., De Muyt, A., Vezon, D., Pereira, L., Chelysheva, L., Kobayashi, W., Kurumizaka, H., Schlogelhofer, P., & Grelon, M. (2013). Sufficient amounts of functional HOP2/MND1 complex promote interhomolog DNA repair but are dispensable for intersister DNA repair during meiosis in Arabidopsis. *The Plant Cell, 25*, 4924–4940.

Underwood, C. J., Choi, K., Lambing, C., Zhao, X., Serra, H., Borges, F., Simorowski, J., Ernst, E., Jacob, Y., Henderson, I. R., & Martienssen, R. A. (2018). Epigenetic activation of meiotic recombination near *Arabidopsis thaliana* centromeres via loss of H3K9me2 and non-CG DNA methylation. *Genome Research, 28*, 519–531.

Varas, J., Graumann, K., Osman, K., Pradillo, M., Evans, D. E., Santos, J. L., & Armstrong, S. J. (2015). Absence of SUN1 and SUN2 proteins in *Arabidopsis thaliana* leads to a delay in meiotic progression and defects in synapsis and recombination. *The Plant Journal, 81*, 329–346.

Vrielynck, N., Chambon, A., Vezon, D., Pereira, L., Chelysheva, L., De Muyt, A., Mezard, C., Mayer, C., & Grelon, M. (2016). A DNA topoisomerase VI-like complex initiates meiotic recombination. *Science, 351*, 939–943.

Vrielynck, N., Peuch, M., Durand, S., Lian, Q., Chambon, A., Hurel, A., Guerin, J., Guerois, R., Mercier, R., Grelon, M., & Mezard, C. (2023). SCEP1 and SCEP2 are two new components of the synaptonemal complex central element. *Nature Plants, 9*, 2016–2030.

Wang, Y., & Copenhaver, G. P. (2018). Meiotic recombination: Mixing it up in plants. *Annual Review of Plant Biology, 69*, 577–609.

Wang, Y., Wu, H., Liang, G., & Yang, M. (2004). Defects in nucleolar migration and synapsis in male prophaseI in the *ask1-1* mutant of Arabidopsis. *Sexual Plant Reproduction, 16*, 273–282.

Wang, M., Wang, K., Tang, D., Wei, C., Li, M., Shen, Y., Chi, Z., Gu, M., & Cheng, Z. (2010). The central element protein ZEP1 of the synaptonemal complex regulates the number of crossovers during meiosis in rice. *The Plant Cell, 22*, 417–430.

Wang, K., Wang, M., Tang, D., Shen, Y., Qin, B., Li, M., & Cheng, Z. (2011a). PAIR3, an axis-associated protein, is essential for the recruitment of recombination elements onto meiotic chromosomes in rice. *Molecular Biology of the Cell, 22*, 12–19.

Wang, M., Tang, D., Wang, K., Shen, Y., Qin, B., Miao, C., Li, M., & Cheng, Z. (2011b). OsSGO1 maintains synaptonemal complex stabilization in addition to protecting centromeric cohesion during rice meiosis. *The Plant Journal, 67*, 583–594.

Wang, K., Wang, M., Tang, D., Shen, Y., Miao, C., Hu, Q., Lu, T., & Cheng, Z. (2012a). The role of rice HEI10 in the formation of meiotic crossovers. *PLoS Genetics, 8*, e1002809.

Wang, M., Tang, D., Luo, Q., Jin, Y., Shen, Y., Wang, K., & Cheng, Z. (2012b). BRK1, a Bub1-related kinase, is essential for generating proper tension between homologous kinetochores at metaphase I of rice meiosis. *The Plant Cell, 24*, 4961–4973.

Wang, Y., Cheng, Z., & Ma, H. (2014). Meiosis: Interactions between homologous chromosomes. In *Cell biology* (pp. 1–34). Springer.

Wang, K., Wang, C., Liu, Q., Liu, W., & Fu, Y. (2015a). Increasing the genetic recombination frequency by partial loss of function of the synaptonemal complex in rice. *Molecular Plant, 8*, 1295–1298.

Wang, S., Zickler, D., Kleckner, N., & Zhang, L. (2015b). Meiotic crossover patterns: Obligatory crossover, interference and homeostasis in a single process. *Cell Cycle, 14*, 305–314.

Wang, C., Wang, Y., Cheng, Z., Zhao, Z., Chen, J., Sheng, P., Yu, Y., Ma, W., Duan, E., Wu, F., Liu, L., Qin, R., Zhang, X., Guo, X., Wang, J., Jiang, L., & Wan, J. (2016a). The role of OsMSH4 in male and female gamete development in rice meiosis. *Journal of Experimental Botany, 67*, 1447–1459.

Wang, H., Hu, Q., Tang, D., Liu, X., Du, G., Shen, Y., Li, Y., & Cheng, Z. (2016b). OsDMC1 is not required for homologous pairing in rice meiosis. *Plant Physiology, 171*, 230–241.

Wang, J., Niu, B., Huang, J., Wang, H., Yang, X., Dong, A., Makaroff, C., Ma, H., & Wang, Y. (2016c). The PHD finger protein MMD1/DUET ensures the progression of male meiotic chromosome condensation and directly regulates the expression of the condensin gene *CAP-D3*. *The Plant Cell, 28*, 1894–1909.

Wang, C., Higgins, J. D., He, Y., Lu, P., Zhang, D., & Liang, W. (2017a). Resolvase OsGEN1 mediates DNA repair by homologous recombination. *Plant Physiology, 173*, 1316–1329.

Wang, J., Blevins, T., Podicheti, R., Haag, J. R., Tan, E. H., Wang, F., & Pikaard, C. S. (2017b). Mutation of Arabidopsis *SMC4* identifies condensin as a corepressor of pericentromeric transposons and conditionally expressed genes. *Genes & Development, 31*, 1601–1614.

Wang, Y., Jiang, L., Zhang, T., Jing, J., & He, Y. (2018). *ZmCom1* is required for both mitotic and meiotic recombination in maize. *Frontiers in Plant Science, 9*, 1005.

Wang, C., Huang, J., Zhang, J., Wang, H., Han, Y., Copenhaver, G. P., Ma, H., & Wang, Y. (2019). The largest subunit of DNA polymerase delta is required for normal formation of meiotic type I crossovers. *Plant Physiology, 179*, 446–459.

Wang, H., Xu, W., Sun, Y., Lian, Q., Wang, C., Yu, C., He, C., Wang, J., Ma, H., Copenhaver, G. P., & Wang, Y. (2020a). The cohesin loader SCC2 contains a PHD finger that is required for meiosis in land plants. *PLoS Genetics, 16*, e1008849.

Wang, J., Yu, C., Zhang, S., Ye, J., Dai, H., Wang, H., Huang, J., Cao, X., Ma, J., Ma, H., & Wang, Y. (2020b). Cell-type-dependent histone demethylase specificity promotes meiotic chromosome condensation in Arabidopsis. *Nature Plants, 6*, 823–837.

Wang, C., Huang, J., Zhang, J., Yu, Y., Copenhaver, G. P., You, C., & Wang, Y. (2022a). DNA polymerase epsilon interacts with SUVH2/9 to repress the expression of genes associated with meiotic DSB hotspot in Arabidopsis. *Proceedings of the National Academy of Sciences of the United States of America, 119*, e2208441119.

Wang, Y., Wang, Y., Zang, J., Chen, H., & He, Y. (2022b). *ZmPRD1* is essential for double-strand break formation, but is not required for bipolar spindle assembly during maize meiosis. *Journal of Experimental Botany, 73*, 3386–3400.

Waterworth, W. M., Altun, C., Armstrong, S. J., Roberts, N., Dean, P. J., Young, K., Weil, C. F., Bray, C. M., & West, C. E. (2007). NBS1 is involved in DNA repair and plays a synergistic role with ATM in mediating meiotic homologous recombination in plants. *The Plant Journal, 52*, 41–52.

West, P. T., Li, Q., Ji, L., Eichten, S. R., Song, J., Vaughn, M. W., Schmitz, R. J., & Springer, N. M. (2014). Genomic distribution of H3K9me2 and DNA methylation in a maize genome. *PLoS One, 9*, e105267.

White, C. I., Wang, Y., Cheng, Z., Huang, J., Shi, Q., Hong, Y., Copenhaver, G. P., Gong, Z., & Ma, H. (2012). The DNA replication factor *RFC1* is required for interference-sensitive meiotic crossovers in *Arabidopsis thaliana*. *PLoS Genetics, 8*, e1003039.

Wijeratne, A. J., Chen, C., Zhang, W., Timofejeva, L., & Ma, H. (2006). The *Arabidopsis thaliana PARTING DANCERS* gene encoding a novel protein is required for normal meiotic homologous recombination. *Molecular Biology of the Cell, 17*, 1331–1343.

Wijnker, E., & Schnittger, A. (2013). Control of the meiotic cell division program in plants. *Plant Reproduction, 26*, 143–158.

Wijnker, E., Velikkakam James, G., Ding, J., Becker, F., Klasen, J. R., Rawat, V., Rowan, B. A., de Jong, D. F., de Snoo, C. B., Zapata, L., Huettel, B., de Jong, H., Ossowski, S., Weigel, D., Koornneef, M., Keurentjes, J. J., & Schneeberger, K. (2013). The genomic landscape of meiotic crossovers and gene conversions in *Arabidopsis thaliana*. *eLife, 2*, e01426.

Woo, H. R., Pontes, O., Pikaard, C. S., & Richards, E. J. (2007). VIM1, a methylcytosine-binding protein required for centromeric heterochromatinization. *Genes & Development, 21*, 267–277.

Wu, Z., Ji, J., Tang, D., Wang, H., Shen, Y., Shi, W., Li, Y., Tan, X., Cheng, Z., & Luo, Q. (2015). *OsSDS* is essential for DSB formation in rice meiosis. *Frontiers in Plant Science, 6*, 21.

Xu, J., Zhao, L., Xu, Y., Zhao, W., Sung, P., & Wang, H. W. (2017). Cryo-EM structures of human RAD51 recombinase filaments during catalysis of DNA-strand exchange. *Nature Structural & Molecular Biology, 24*, 40–46.

Xu, R. Y., Xu, J., Wang, L., Niu, B., Copenhaver, G. P., Ma, H., Zheng, B., & Wang, Y. (2019). The Arabidopsis anaphase-promoting complex/cyclosome subunit 8 is required for male meiosis. *The New Phytologist, 224*, 229–241.

Xu, W., Yu, Y., Jing, J., Wu, Z., Zhang, X., You, C., Ma, H., Copenhaver, G. P., He, Y., & Wang, Y. (2023). SCF^{RMF} mediates degradation of the meiosis-specific recombinase DMC1. *Nature Communications, 14*, 5044.

Xue, Z., Li, Y., Zhang, L., Shi, W., Zhang, C., Feng, M., Zhang, F., Tang, D., Yu, H., Gu, M., & Cheng, Z. (2016). OsMTOPVIB promotes meiotic DNA double-strand break formation in rice. *Molecular Plant, 9*, 1535–1538.

Xue, Z., Liu, C., Shi, W., Miao, Y., Shen, Y., Tang, D., Li, Y., You, A., Xu, Y., Chong, K., & Cheng, Z. (2019). OsMTOPVIB is required for meiotic bipolar spindle assembly. *Proceedings of the National Academy of Sciences of the United States of America, 116*, 15967–15972.

Yang, M., Hu, Y., Lodhi, M., McCombie, W. R., & Ma, H. (1999). The Arabidopsis *SKP1-LIKE1* gene is essential for male meiosis and may control homologue separation. *Proceedings of the National Academy of Sciences of the United States of America, 96*, 11416–11421.

Yang, X., Makaroff, C. A., & Ma, H. (2003a). The Arabidopsis *MALE MEIOCYTE DEATH1* gene encodes a PHD-finger protein that is required for male meiosis. *The Plant Cell, 15*, 1281–1295.

Yang, X. H., Makaroff, C. A., & Ma, H. (2003b). The Arabidopsis gene encodes a PHD-finger protein that is required for male meiosis. *The Plant Cell, 14*, 1281–1295.

Yang, X., Timofejeva, L., Ma, H., & Makaroff, C. A. (2006). The Arabidopsis SKP1 homolog ASK1 controls meiotic chromosome remodeling and release of chromatin from the nuclear membrane and nucleolus. *Journal of Cell Science, 119*, 3754–3763.

Yang, C., Hamamura, Y., Sofroni, K., Bower, F., Stolze, S. C., Nakagami, H., & Schnittger, A. (2019). SWITCH 1/DYAD is a WINGS APART-LIKE antagonist that maintains sister chromatid cohesion in meiosis. *Nature Communications, 10*, 1755.

Yang, C., Sofroni, K., Wijnker, E., Hamamura, Y., Carstens, L., Harashima, H., Stolze, S. C., Vezon, D., Chelysheva, L., Orban-Nemeth, Z., Pochon, G., Nakagami, H., Schlogelhofer, P., Grelon, M., & Schnittger, A. (2020). The Arabidopsis Cdk1/Cdk2 homolog CDKA;1 controls chromosome axis assembly during plant meiosis. *The EMBO Journal, 39*, e101625.

Yang, X., Song, M., Wang, Y., Tan, T., Tian, Z., Zhai, B., Yang, X., Tan, Y., Cao, Y., Dai, S., Wang, S., & Zhang, L. (2022). The ubiquitin-proteasome system regulates meiotic chromosome organization. *Proceedings of the National Academy of Sciences of the United States of America, 119*, e2106902119.

Yelina, N. E., Choi, K., Chelysheva, L., Macaulay, M., de Snoo, B., Wijnker, E., Miller, N., Drouaud, J., Grelon, M., Copenhaver, G. P., Mezard, C., Kelly, K. A., & Henderson, I. R. (2012). Epigenetic remodeling of meiotic crossover frequency in *Arabidopsis thaliana* DNA methyltransferase mutants. *PLoS Genetics, 8*, e1002844.

Yelina, N. E., Lambing, C., Hardcastle, T. J., Zhao, X., Santos, B., & Henderson, I. R. (2015). DNA methylation epigenetically silences crossover hot spots and controls chromosomal domains of meiotic recombination in Arabidopsis. *Genes & Development, 29*, 2183–2202.

Yu, H., Wang, M., Tang, D., Wang, K., Chen, F., Gong, Z., Gu, M., & Cheng, Z. (2010). OsSPO11-1 is essential for both homologous chromosome pairing and crossover formation in rice. *Chromosoma, 119*, 625–636.

Yuan, W., Li, X., Chang, Y., Wen, R., Chen, G., Zhang, Q., & Wu, C. (2009). Mutation of the rice gene *PAIR3* results in lack of bivalent formation in meiosis. *The Plant Journal, 59*, 303–315.

Yuan, L., Yang, X., Ellis, J. L., Fisher, N. M., & Makaroff, C. A. (2012). The Arabidopsis SYN3 cohesin protein is important for early meiotic events. *The Plant Journal, 71*, 147–160.

Zamariola, L., De Storme, N., Tiang, C. L., Armstrong, S. J., Franklin, F. C., & Geelen, D. (2013). SGO1 but not SGO2 is required for maintenance of centromere cohesion in *Arabidopsis thaliana* meiosis. *Plant Reproduction, 26*, 197–208.

Zamariola, L., De Storme, N., Vannerum, K., Vandepoele, K., Armstrong, S. J., Franklin, F. C., & Geelen, D. (2014). SHUGOSHINs and PATRONUS protect meiotic centromere cohesion in *Arabidopsis thaliana*. *The Plant Journal, 77*, 782–794.

Zeng, X., Yan, X., Yuan, R., Li, K., Wu, Y., Liu, F., Luo, J., Li, J., & Wu, G. (2016). Identification and analysis of *MS5*^{d}: A gene that affects Double-Strand Break (DSB) repair during meiosis I in *Brassica napus* microsporocytes. *Frontiers in Plant Science, 7*, 1966.

Zhang, C., Song, Y., Cheng, Z. H., Wang, Y. X., Zhu, J., Ma, H., Xu, L., & Yang, Z. N. (2012). The *Arabidopsis thaliana* DSB formation (*AtDFO*) gene is required for meiotic double-strand break formation. *The Plant Journal, 72*, 271–281.

Zhang, J., Pawlowski, W. P., & Han, F. (2013). Centromere pairing in early meiotic prophase requires active centromeres and precedes installation of the synaptonemal complex in maize. *The Plant Cell, 25*, 3900–3909.

Zhang, J., Zhang, B., Su, H., Birchler, J. A., & Han, F. (2014a). Molecular mechanisms of homologous chromosome pairing and segregation in plants. *Journal of Genetics and Genomics, 41*, 117–123.

Zhang, L., Tang, D., Luo, Q., Chen, X., Wang, H., Li, Y., & Cheng, Z. (2014b). Crossover formation during rice meiosis relies on interaction of OsMSH4 and OsMSH5. *Genetics, 198*, 1447–1456.

Zhang, F., Tang, D., Shen, Y., Xue, Z., Shi, W., Ren, L., Du, G., Li, Y., & Cheng, Z. (2017). The F-box protein ZYGO1 mediates bouquet formation to promote homologous pairing, synapsis, and recombination in rice meiosis. *The Plant Cell, 29*, 2597–2609.

Zhao, D., Yang, X., Quan, L., Timofejeva, L., Rigel, N. W., Ma, H., & Makaroff, C. A. (2006). *ASK1*, a *SKP1* homolog, is required for nuclear reorganization, presynaptic homolog juxtaposition and the proper distribution of cohesin during meiosis in Arabidopsis. *Plant Molecular Biology, 62*, 99–110.

Zheng, G., Kanchwala, M., Xing, C., & Yu, H. (2018). MCM2-7-dependent cohesin loading during S phase promotes sister-chromatid cohesion. *eLife, 7*, 33920.

Zhou, L., Han, J., Chen, Y., Wang, Y., & Liu, Y. G. (2017). *Bivalent formation 1*, a plant-conserved gene, encodes an OmpH/coiled-coil motif-containing protein required for meiotic recombination in rice. *Journal of Experimental Botany, 68*, 2163–2174.

Zickler, D., & Kleckner, N. (2015). Recombination, pairing, and synapsis of homologs during meiosis. *Cold Spring Harbor Perspectives in Biology, 7*, a016626.

Zickler, D., & Kleckner, N. (2016). A few of our favorite things: Pairing, the bouquet, crossover interference and evolution of meiosis. *Seminars in Cell & Developmental Biology, 54*, 135–148.

Zickler, D., Kleckner, N., Campbell, A., Anderson, W. W., & Jones, E. W. (1999). Meiotic chromosomes: Integrating structure and function. *Annual Review of Genetics, 33*, 603–754.

Ziolkowski, P. A., Underwood, C. J., Lambing, C., Martinez-Garcia, M., Lawrence, E. J., Ziolkowska, L., Griffin, C., Choi, K., Franklin, F. C. H., Martienssen, R. A., & Henderson, I. R. (2017). Natural variation and dosage of the HEI10 meiotic E3 ligase control Arabidopsis crossover recombination. *Genes & Development, 31*, 306–317.

Chapter 12
Pollen Development

De Ye

Abstract Pollen grain, also referred to as a male gametophyte, is a male reproductive structure that produces male gametes (also called sperm cells) and plays important roles in the delivery of the male gametes into female gametophytes during pollination and fertilization. Pollen development is initiated from the microspore generated by meiosis at the middle stages of anther development. The microspore undergoes an asymmetric pollen mitosis I (PMI) to form a bicellular pollen grain that consists of a vegetative cell and a generative cell. The generative cell further has a symmetric pollen mitosis II (PMII) to generate two sperm cells and form a tricellular pollen grain. As a result, the pollen grain contains two smaller sperm cells that are enclosed in a larger vegetative cell. The tricellular pollen grain is further dehydrated to develop into a mature pollen grain ready for pollination and fertilization. In this chapter, the molecular genetics of pollen development will be discussed, including the genes and their molecular mechanisms that are involved in pollen mitosis, cell specification, and pollen wall formation.

Keywords Fertilization · Gametogenesis · Sporogenesis · Pollen · Pollination

12.1 Introduction

Pollen is a common name of the male gametophytes in flowering plants. The individual male gametophyte also is referred to as a pollen grain. A mature pollen grain consists of two cells (a vegetative cell and a generative cell) or three cells (a vegetative cell and two sperm cells) in different species. Pollen development is to produce male gametes (sperm cells) and the structure for protection and transport of the sperm cells into the female gametophyte (embryo sac) for double fertilization. The pollen formation in angiosperms takes place within the stamens. The process

D. Ye (✉)
China Agricultural University, Beijing, China

Hainan University, Haikou, Hainan, China
e-mail: yede@cau.edu.cn

F. Chang et al. (eds.), *Regulation of Plant Development*,
https://doi.org/10.1007/978-981-95-7033-1_12

comprises five phases: (i) differentiation of the male sporogenous cells; (ii) meiosis; (iii) post-meiotic development of microspores; (iv) microspore mitosis; and (v) maturation of pollen grains. The first two are collectively called microsporogenesis, while the last three phases are together called microgametogenesis (Fig. 12.1). Therefore, pollen formation can be defined as two sequential stages:

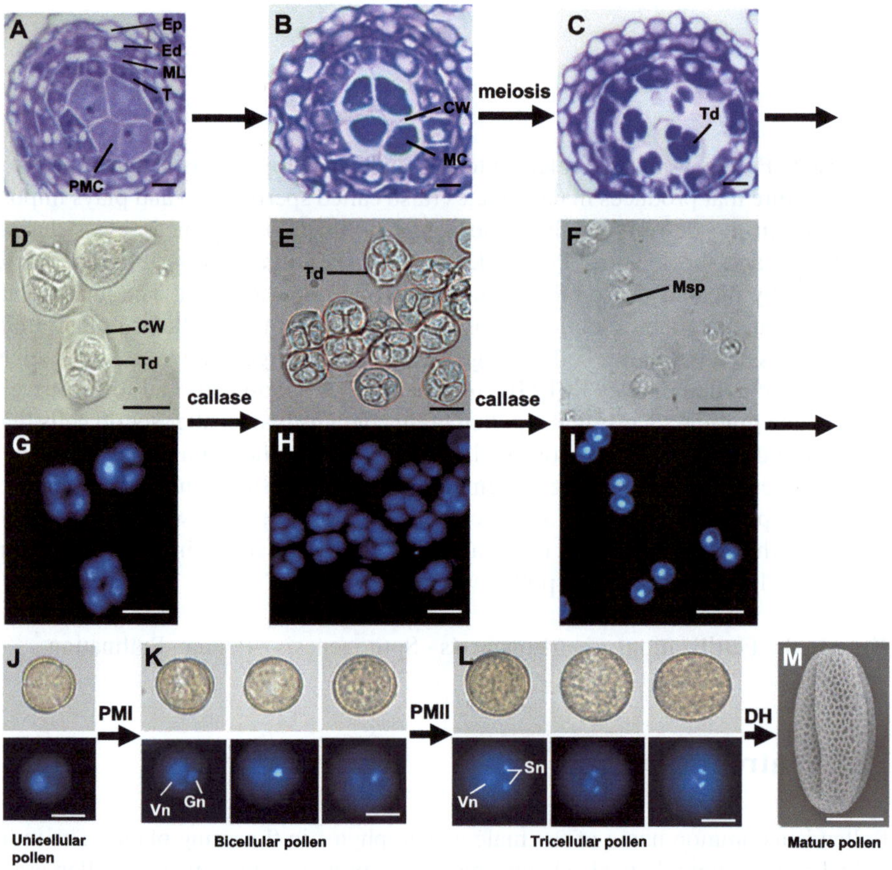

Fig. 12.1 Pollen development in Arabidopsis (Yang et al., 2003; Dou et al., 2011; Wang et al., 2011). (**a**) Pollen sac at the anther developmental stage 5, showing the PMCs. (**b**) Pollen sac at the anther developmental stage 6, in which the meiocytes are surrounded by the callose walls. (**c**) Pollen sac at the anther developmental stage 7, in which the tetrads of microspores are formed by meiosis. (**d**) Tetrads at the early tetrad stage. (**e**) Tetrads at the late tetrad stage. (**f**) Individual microspores released from the tetrads by callase. (**g–i**) Images of DAPI staining of (**d–f**), respectively. The bright spots show the nuclei stained by DAPI. (**j**) Unicellular pollen. (**k**) Bicellular pollen. (**l**) Tricellular pollen. The down panels in (**j–l**) are the images of the DAPI-stained pollen grains, showing the nuclei revealed by DAPI staining. (**m**) Mature pollen. CW callose walls, DH dehydration, Ed endothecium, Ep epidermis, Gn generative nucleate, Mc meiocytes, ML middle layer, PMCs pollen mother cells, Msp microspore, PMI pollen mitosis I, PMII pollen mitosis II, Sn sperm cell nucleates, T tapetum, Td tetrads, Vn vegetative nucleate. Bars = 10 μm in (**a–c**), 25 μm in (**d–i**), and 10 μm in (**j–m**)

microsporogenesis and microgametogenesis. Microsporogenesis has been described and discussed in the previous chapters for the anther development and meiosis. This chapter will focus on microgametogenesis or pollen development that produces male gametes for pollination and fertilization.

12.2 Microsporogenesis and Post-meiotic Development of the Microspores

Unlike animals in which the reproductive cells are determined early in embryogenesis, flowering plants do not have reproductive cells until they switch to the reproductive growth when they produce reproductive organs from a population of the undifferentiated stem cells in the floral meristem (Twell, 2011). The male sporogenous cells differentiate in the anther of stamen in flowers. At the early stages of the anther development, the diploid male sporogenous cell develops into pollen mother cell (PMC) (also referred to as microsporocyte) that is encased in a tapetum-derived callose wall and undergoes meiotic division, giving rise to a tetrad of haploid microspores surrounded by the callose walls (Fig. 12.1). The callose is then broken down, and the individual microspores are released into the anther locule. The free microspores serve as gametophyte initials and undergo post-meiotic development, including nuclear migration, cytoplasmic differentiation, and change of the vacuoles, leading to polarization of the microspores. The process is accompanied by the cell cycle control. The cell cycle control genes and related regulatory genes are expressed prolifically in the microspores at this stage. Loss of these genes, for example, the two RING finger E3 ligases, RHF1a and RHF2a, causes interruption of the pollen development (Liu et al., 2008).

12.3 Pollen Mitosis I and Specification of Germline and Vegetative Cells

Following the post-meiotic development, the polarized microspores undergo the first mitosis, also referred to as pollen mitosis I (PMI) (Fig. 12.1). The PMI is an asymmetric division that produces the larger vegetative cell and the smaller generative cell, forming a specified structure of "the smaller generative cell within the larger vegetative cell," also known as a bicellular pollen grain (Fig. 12.1). The vegetative cell then exits the cell cycle in the G1 phase and differentiates as a "supportive cell" that plays basic roles in the survival, protection, and transport of the male gametes (sperm cells). The generative cell serves as a germline cell that is functionally separated from the vegetative cell and functions as an initial of male gametes (Twell, 2011).

The asymmetric division in PMI immediately determines the differences between the vegetative cell and the generative cell, relying on the intrinsic factors as demonstrated by several lines of physiological and genetic evidence. Disturbing cell polarity using centrifugation or microtubule inhibitors can induce symmetric division of microspore, resulting in formation of two vegetative cells and failure in determination of germline cell fate. Mutations of the genes involved in gametophytic microtubule arrays, cytokinesis, and nuclear migration, for example, *GEMINI POLLEN 1* (*GEM1*) (Park et al., 1998), *GEM2* (Park et al., 2004), and *TWO-IN-ONE* (*TIO*) (Oh et al., 2005), cause failure of germ cell fate determination. These studies suggest that differentiation of generative cell relies on the intrinsic factors confined by the complete asymmetric division in PMI (Twell, 2011).

Genetic studies have uncovered two parallel pathways that regulate (1) maintenance of generative cell proliferation and (2) mitotic arrest of the vegetative cell (Twell, 2011). The A-type cyclin-dependent kinase (CDKA;1) plays important roles in progression of the S phase and entry to mitosis in a generative cell. The activity of CDKA;1 is regulated by the F-box-like 17 (FBL17) protein. In the generative cell, the FBL17 is transiently expressed and targets the CDK inhibitors KRP6 and KRP7 for proteasome-dependent degradation, enabling the generative cell to progress through the S phase (Kim et al., 2008) and entry of the generative cell to further mitosis. Furthermore, the KRP6 also is regulated by two RING finger E3 ligases, RHF1a and RHF2a, to promote progression through both microspore and the subsequent germ cell mitotic cycle. Therefore, RHF1a and RHF2a prevent inhibition of CDK activity in microspores and germ cells by reducing the level of KRP6 accumulated during meiosis (Liu et al., 2008).

Conversely, due to the absence of FBL17 in the vegetative cell, the persistence of KRP6 and KRP7 maintains inhibition of CDKA activity and vegetative cell cycle progression, thus enabling functional differentiation of the vegetative cell. Genetic studies identified the retinoblastoma-related (RBR) protein that negatively regulates proliferation of the vegetative cell dependent on CDKA;1 activity. Loss of RBR protein function results in an extra mitosis and generates two vegetative cells, leading to pollen with four sperm cells (Chen et al., 2009; Twell, 2011). Furthermore, introduction of a *cdka;1* mutant allele prevents cell proliferation in hypomorphic *rbr* pollen. Therefore, cooperation of RBR repression of the E2F-DP pathway and KRP-dependent CDKA;1 inhibition enforces cell cycle exit that enables determination of the vegetative cell fate.

12.4 Pollen Mitosis II and Specification of Sperm Cells

The further development of the generative cell involves tight control of cell cycle progression and integration of cell cycle control and sperm cell specification. In some species, for example, Arabidopsis (*Arabidopsis thaliana*), the generative cell undergoes an equal division, also referred to as the pollen mitosis II (PMII), to generate two sperm cells, forming a tricellular pollen grain (Fig. 12.1). In some species, for example, tobacco, the generative cell maintains not to divide until germination of the pollen grain. The sperm cells are produced by the equal division of the generative cell in the pollen tube.

A number of mutants have been described in Arabidopsis that produce bicellular pollen harboring a single germ cell due to failure of germ cell division (Twell, 2011). These mutants exhibit two types of characteristics: (1) the mutants that affect cell cycle progression of the generative cell, leading to differentiation of the generative cell to a diploid sperm-like germline cell that can fertilize egg cell or central cell; and (2) the mutants that affect both cell cycle progression and specification of the germline cell. The mutation in the chromatin remodeling factor CHROMATIN ASSEMBLY FACTOR 1 (CAF1) disrupts the PMII and results in formation of a proportion of pollen grains with single germ cell able to fertilize either the egg or the central cell. This indicates that CAF1 is important for the PMII, but not for sperm cell fate determination (Chen et al., 2008), suggesting that the specification of sperm cell differentiation is not dependent on normal cell cycle regulation (Chen et al., 2008; Ma & Sundaresan, 2010).

A group of pollen-expressed genes that regulate both cell cycle progression and sperm cell specification have been described, for example, the *DUO POLLEN 1* (*DUO1*) (Rotman et al., 2005). *DUO1* encodes a conserved R2R3 MYB protein that is specifically expressed in the male germline (Rotman et al., 2005). The *duo1* mutant is defective in the PMII, resulting in formation of pollen with a vegetative cell and a diploid sperm-like cell unable to fertilize either egg cell or central cell, unlike the *cdka;1* mutant sperm-like cell which can fertilize the egg cell or the central cell (Aw et al., 2010). This suggests that the *DUO1* is involved in the two parallel pathways that regulate the cell cycle progression of a germline cell or specification of a sperm cell. Thus, the *DUO1* is likely to participate in coordination of cell cycle progression control in germline cell and the sperm cell identity.

Loss of DUO1 suppresses germline expression of the G2/M regulator CYCLIN B1;1, indicating that failure of mitotic entry in *duo1* germ cells is related to germline expression of the CYCLIN B1;1 (Brownfield et al., 2009a). The mutations in the DUO1 also suppress the expression of several male germline-specific genes in the *duo1* germ cells, including that of AtGCS1 (HAP2), which encodes an essential sperm cell surface protein required for efficient pollen tube guidance and gamete fusion (Mori et al., 2006; von Besser et al., 2006). The ectopic expression of DUO1 can activate expression of those male germline-specific genes in vegetative tissue. However, the ectopic expression of DUO1 in vegetative tissues is not enough to induce expression of CYCLIN B1;1, indicating that a more complicate mechanism is involved in regulation of CYCLIN B1;1. This also suggests the difference of expression regulation for germline cell cycle entry from that for the other male germline-specific genes (Twell, 2011).

The nuclear protein DUO3 is identified also as being involved in male germ cell division and gamete specification in Arabidopsis (Brownfield et al., 2009b). DUO3 contains the domains related to Gonadless-4 (GON-4), a cell lineage regulator of gonadogenesis in *Caenorhabditis elegans* (Friedman et al., 2000). The DUO3 proteins exist widely in different plants and likely have conserved functions as demonstrated by the fact that the DUO3 from moss can restore germ cell division and specification in the Arabidopsis *duo3* mutants. DUO3 is expressed widely in different tissues, in part overlapping with the expression of DUO1 in germline cells. DUO3 is involved in both G2/M transition and the expression of a subset of DUO1-targeting genes. However, unlike DUO1, its role in the germline during G2/M transition acts independently of CYCB1;1 (Brownfield et al., 2009b). This suggests that DUO1 and DUO3 cooperate to integrate cell cycle progression and cell differentiation during male germline development.

12.5 Pollen Wall Formation

Pollen wall is the specialized extracellular cell wall of the vegetative cell, which has a highly organized multiple-layer structure surrounding outskirt of the pollen grain. Studies have revealed that pollen wall is required for protection of the male germline and the recognition of male with female. Defects in the pollen wall could result in male sterility (Shi et al. 2015a, b; Qiao et al., 2023; Wei & Ma, 2023).

12.5.1 Structure and Formation of Pollen Walls

The structures of pollen walls in different species vary largely, but they have a conserved fundamental structure that is commonly comprised of an outer layer (the exine) and an inner layer (the intine) (Heslop-Harrison, 1968; Blackmore et al., 2007; Qiao et al., 2023) (Fig. 12.2). The two layers have distinct components. The exine layer consists of a robust material called sporopollenin derived from somatic tapetal cells of anther. The intine layer consists of pectic polymers, celluloses, and hemicelluloses, which are synthesized by the microspores themselves (Hasegawa et al., 2000). In Arabidopsis, the exine consists of the outer sculpted layer, the sexine, and the inner simple layer, the nexine. The sexine has two specified structures: the tectum at the outermost edge and the bacula made up of radially directed rods (Paxson-Sowders et al., 1996; Ariizumi et al., 2003). The nexine also has two layers, the foot layer (nexine I) and the endexine (nexine II). The surface and cavities of the exine sculpture are covered with pollen coat materials. The intine also can be divided into two layers, exintine (facing the exterior) and the endintine (facing the interior) (Heslop-Harrison, 1968; Scott et al., 2004). Besides the two fundamental layers, a tryphine layer is also found in the Arabidopsis pollen wall. In addition, several precursor walls, such as PMC primary wall, callose wall, and primexine, which are successively synthesized and degraded, are also involved in pollen wall formation.

At the tetrad stage, the Arabidopsis microspore has the callose wall directly contacting to the plasma membrane (PM). The callose wall surrounding the microspore is proposed to act as the mold for primexine formation. The primexine is formed by

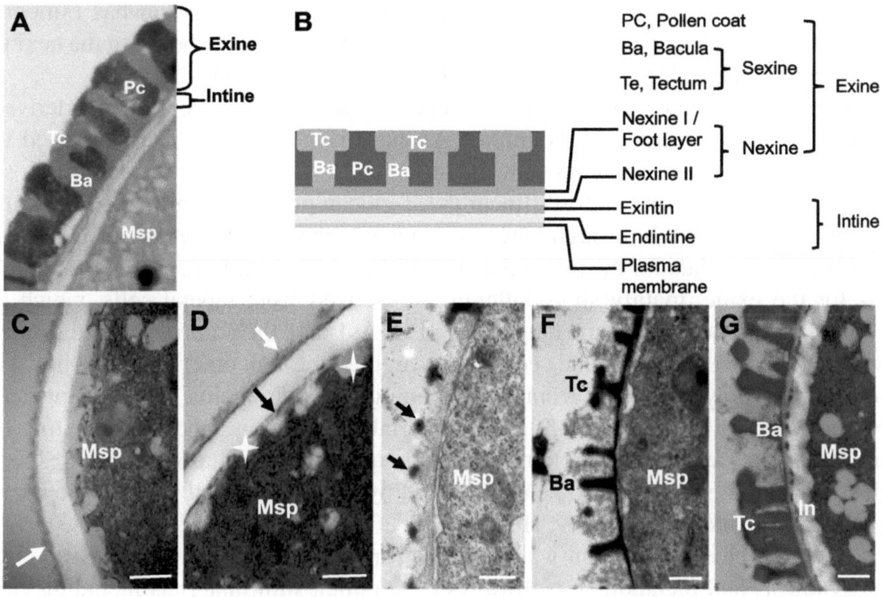

Fig. 12.2 Structures and formation process of the pollen walls in Arabidopsis (Dou et al., 2011; Shi et al. 2015a, b). (**a**) TEM image of mature pollen walls in Arabidopsis. (**b**) Schematic of the pollen wall structure in Arabidopsis. (**c**) TEM image of the Arabidopsis microspore at the early tetrad stage. The white arrow indicates the sporopollenin on the callose wall. (**d**) TEM image of the Arabidopsis microspore at the late tetrad stage, which shows the undulant plasma membrane surface and deposition of sporopollenin on the plasma membrane (black arrow) and the surface of the callose wall (white arrow). The white stars indicate the undulant structure of the plasma membrane. (**e**) TEM image of Arabidopsis pollen at the early unicellular microspore stage. The black arrows indicate the emerging probacula. (**f**) TEM image of the Arabidopsis pollen at the later unicellular microspore stage, which shows formation of the bacula and tectum. (**g**) TEM image of Arabidopsis pollen at the bicellular pollen stage, which shows nearly full formation of the sculptured exine. Ba bacula, In intine, Msp microspore, Ne nexine, TC tectum. Bars = 2 μm in (**c**) and (**d**), 1 μm in (**e**) and (**f**), and 0.5 μm in (**g**)

deposition of the glycocalyx-like fibrillar polysaccharide materials. Meanwhile, the PM changes to a wavy form (undulation) (Heslop-Harrison, 1968), which acts as the sporopollenin receptor. The successive deposition of the sporopollenin components in the primexine gives rise to formation of proexine with the probaculum and protectum. The probaculum grows and elongates after the primary and callose walls of the PMC are degraded by the callases secreted from the tapetum. While the microspore is released from the tetrad, the undulation structures on the PM disappear and the primexine is embedded in between the probacula. The mature exine structure is completed by the further polymerization of sporopollenin components. The spaces between the tectum and foot layers are filled with the materials from the programmed cell death (PCD)-degenerated tapetal cells to form the pollen coat. After the PMII, the integrated inner wall is formed as the intine (Owen & Markaroff, 1995; Paxson-Sowders et al., 1996; Ariizumi & Toriyama, 2011). In the monocot rice (*Oryza sativa*), compared to the Arabidopsis pollen, the rice pollen has a smooth and particulate exine patterning that has more inter-layer spaces between the nexine (foot layer) and sexine (Li & Zhang, 2010; Zhang et al., 2011).

The materials that constitute the different structures in the pollen wall are derived from different cell types of the anther. At the early stages, the materials used to construct the primary and callose walls of PMCs, mainly the pectins and cellulose in the primary wall and callose (linear β-1,3-glucan polymers) in the callose wall, are produced by PMCs through polysaccharide metabolism, while the fibrillar polysaccharide materials for the primexine are produced by the microspore. The materials for the exine, mainly sporopollenin, are derived from tapetal cells, which is composed of the long-chain fatty acids (LCFAs) and their derivatives, the fatty acids, and phenolic compounds covalently coupled through ether and ester linkage or aliphatic polymers with aromatic or conjugated side chains (Paxson-Sowders et al., 1996; Ahlers et al., 1999). The materials for exintine and the endintine in intine are the pectins and protein inclusions, and microfibrillar cellulose, respectively, which are proposed to be synthesized by the pollen itself (Paxson-Sowders et al., 1996; Ahlers et al., 1999; Huang et al., 2009).

Taken together, the pollen wall formation involves four main mechanisms: (1) biosynthesis and metabolism pathways of the pollen wall materials including callose, cellulose, pectins, sporopollenin, proteins, and other compounds; (2) translocation of the pollen wall materials from tapetum to microspores; (3) involvement of three cell types: PMC, tapetal cells, and gametophytic cell; and (4) expression regulation of the genes involved in pollen wall formation. The defects in the mechanisms could drastically affect pollen wall structures, resulting in pollen abortion or loss of male gametophytic functions, and leading to male sterility.

12.5.2 Genetics of Pollen Wall Formation

The pollen wall formation is controlled by a complex genetic network. Many genes have been identified as being involved in pollen wall formation (Jiang et al., 2013; Shi et al. 2015a, b).

12.5.2.1 Formation and Degradation of the Callose Walls

The callose wall is made of callose. Callose is a type of β-1,3-glucan. Callose is synthesized by callose synthase (CalS) and degraded by β-1,3-glucanases. Loss of the anther-expressed Arabidopsis CalS5s (Dong et al., 2005), GLUCAN SYNTHASE-LIKE 1(GSL1) and 12 (GSL12) (Enns et al., 2005), and the rice CalS5, GLUCAN SYNTHASE-LIKE 5 (GSL5) (Shi et al. 2015a, b) leads to defect in callose biosynthesis and exine patterning. Mutations in the a,β-1,3-glucanase-encoding genes *A6* in Arabidopsis, Brassicas, and rice cause defects in callose degradation during pollen development, indicating that the A6 gene is involved in callose degradation (Hird et al., 1993; Wan et al., 2011).

12.5.2.2 Primexine Formation and Microspore PM Undulations

Primexine formation is important for determination of pollen wall patterning through guides of PM undulation. Several membrane-related protein genes have been proposed as being involved in primexine formation and PM undulation. The mutation in the Arabidopsis *DEFECTIVE IN EXINE FORMATION 1* (*DEX1*), which encodes a membrane calcium-binding protein, delays primexine formation and significantly reduces levels of primexine and PM undulations, resulting in abnormal exine (Paxson-Sowders et al., 2001). The mutations in the Arabidopsis plastidic integral membrane protein gene, *NO EXINE FORMATION 1* (*NEF1*) (Ariizumi et al., 2004), and sugar transporter gene, *RUPTURED POLLEN GRAIN 1* (*RPG1*) (Sun et al., 2013), also exhibit the exine-defective phenotypes similar to those of *dex1*. Moreover, loss of the functional PM protein, NO PRIMEXINE AND PLASMA MEMBRANE UNDULATION (NPU), results in complete lack of primexine formation and PM undulations in the *npu* mutant pollen (Chang et al., 2012). These data suggest that the membrane-associated action is required for primexine formation. In addition, the de novo DNA methyltransferase gene *Exine Formation Defect* (*EFD*) influences the primexine patterning and is essential for exine formation, but does not influence the microspore PM undulation, indicating that the DNA modification is involved in primexine formation (Hu et al., 2012).

12.5.2.3 Biosynthesis of Sporopollenin

Sporopollenin is synthesized from the complex precursors related to many metabolism pathways of lipid, phenylpropanoid, polysaccharide, etc. Many genes and enzymes involved in the biosynthesis of sporopollenin and their precursors have been identified (Shi et al. 2015a, b). Loss or downregulation of the functions of the genes involved in the pathways often drastically affects the exine patterning. For example, loss of the ATP diphosphohydrolases (Apyrases) in the *apy6 apy7* mutant results in thinner exine with few bacula or tectum structures and random deposition of tryphine, leading to sticky pollen (Yang et al., 2013).

12.5.2.4 Tryphine Formation

The tryphine filled in the cavities of the pollen exine has complex components, including lipids, wax esters, flavonoids, hydroxycinnamoyl spermidine metabolites, and proteins (Quilichini et al., 2012a, b; Jessen et al., 2011; Shi et al. 2015a, b). The genes involved in the biosynthesis and transport of the components are required for tryphine formation. For example, the genes involved in biosynthesis and transport of the LCFAs have been proposed as being involved in tryphine formation. Mutations in *ECERIFERUM* (*CER*) genes, *CER1* (Aarts et al., 1995), *CER3*/*FLP1* (Ariizumi et al., 2003), and the three CER2-LIKE genes (Haslam et al., 2015) affect tryphine formation and result in abnormal exine patterning. The long-chain acyl-CoA synthetases

(LACS) 1 and 4 (Jessen et al., 2011) also have been proposed as having similar roles in tryphine formation and exine patterning. Furthermore, the phosphoserine phosphatase, which catalyzes the last step of the phosphorylated pathway of serine biosynthesis (PPSB), also is required for tryphine formation (Flores-Tornero et al., 2015).

12.5.2.5 Intine Development

The intine is made up of the pectins, protein inclusions, and microfibrillar cellulose, which are proposed as being produced mainly by the microspores. The Golgi-localized glycosyltransferase, GLYCOSYLTRANSFERASE 1 (OsGT1) (Moon et al., 2013), and the ARABINOKINASE-like protein, COLLAPSED ABNORMAL POLLEN 1 (CAP1), are proposed to be required for intine development in rice. The mutations in the OsGT1 or CAP1 disrupt intine formation (Ueda et al., 2013). The overexpression of a rhamnogalacturonan lyase or an endo-α-1,5-L-arabinanase, which act in removal of the side chains of pectic arabinan, results in failure of intine formation in the transgenic potato plants (Cankar et al., 2012). Furthermore, in *Brassica campestris*, silencing of the *PECTATE LYASE-LIKE* (*PLL*) genes, *BcPLL9* or *BcPLL10*, or putative arabinogalactan protein-encoding gene *BcMF8* drastically affects intine patterning with abnormal exintine, endintine, and tryphine (Jiang et al., 2012a, b; Lin et al., 2012). These data indicate that the pectins are important for intine formation.

12.6 Roles of the Tapetum in Pollen Development

The tapetum is a cell layer that surrounds the most-inner sporogenous cells and their descendant cells, namely, microspores or then pollen grains in the anther. Many genetic studies have proposed that the tapetum is required for pollen development. The genetic data are frequently reviewed for the latest understanding of anther development mechanisms (Ma & Sundaresan, 2010; Gómez et al., 2015; Qiao et al., 2023; Wei & Ma, 2023).

12.6.1 Determination of Tapetal Cell and Sporogenous Cell Fates

Characterization of the male-sterile mutants has proposed that interaction between the sporogenous cells and tapetal precursor cells influences the cell developmental fates of both tapetal precursor cells and sporogenous cells at the early anther developmental stages, which is regulated by a signaling pathway between the sporogenous cells and tapetal precursor cells. Several components have been found as being involved in the process in Arabidopsis, including the receptor-like protein kinases, EXCESS MALE SPOROCYTE1/EXCESS SPOROGENOUS CELLS (EMS1/EXS) and SERK1/ SERK2 (SOMATIC EMBRYO RECEPTOR KINASE1/2), which are expressed in tapetal precursor cells, and the secreted protein TAPETUM DETERMINANT1 (TPD1) expressed in sporogenous cells (Albrecht et al., 2005; Canales et al., 2002; Yang et al., 2003, 2005; Zhao et al., 2002). Loss of these proteins results in that the tapetal precursor cells develop into microsporocyte-like cells instead of the normal tapetal cells, while the male meiocytes failed to complete meiotic cytokinesis, leading to lack of tapetum and pollen in the mutant anthers (Albrecht et al., 2005; Canales et al., 2002; Yang et al., 2003, 2005; Zhao et al., 2002; also see the previous chapters for anther development and meiosis). However, it remains unclear whether the meiotic defect results from lack of tapetum.

12.6.2 Life and Death of Tapetum for Pollen Development

The tapetum functions and tapetal PCD are required for pollen development. A complex network has been proposed for regulation of tapetum function and tapetal PCD. The *DYT1* encodes a basic helix-loop-helix (bHLH) transcription factor. A mutation in *DYT1* does not affect the specification of tapetal cells but results in abnormal tapetum with enlarged vacuole during the anther meiotic stage, frequently leading to degeneration of microspores (Zhang et al., 2006). The rice ortholog, UNDEVELOPED TAPETUM1 (UDT1), also acts in a similar way to AtDYT1, indicating that the tapetum has the conserved roles in pollen development (Jung et al., 2005). The regulator gene *TAPETAL DEVELOPMENT and FUNCTION* (*TDF1/MYB35*) acts downstream of *DYT1* and upstream of *ABORTED MICROSPORES* (*AMS*) (Zhu et al., 2008, 2011). *AMS* encodes a putative bHLH-type transcription factor and is expressed at a lower level during the premeiotic stages and then at increased levels during the post-meiotic stages. The *AMS* regulates the expression of several tapetum genes that are associated with tapetum function and biosynthesis, including synthesis of lipidic and phenolic components essential for pollen wall patterning and flavonoids (Xu et al., 2010, 2012). The *ams* mutant has reduced microspore callose walls and has lack of sporopollenin secretion, leading to premature microspore degeneration (Xu et al., 2012). The rice *AMS* ortholog *TAPETUM DEGENERATION RETARDATION* (*TDR*) is involved in the tapetum function and PCD (Xu et al., 2010; Zhang et al., 2011). It regulates tapetal cell death via interaction with the bHLH transcription factor ETERNAL TAPETUM1 (EAT1), a positive regulator of the two aspartyl proteases OsAP25 and OsAP37, which direct tapetum PCD (Niu et al., 2013). These data indicated that both deformation and altered PCD of the tapetal cells affect pollen development, frequently leading to male sterility.

Many genetic data showed that defects in the tapetum development at the different stages all can affect pollen development, indicating that the roles of tapetum in pollen development cover all the developmental stages from tetrad to mature

(Gómez et al., 2015). At the tetrad stage, the a,β-1,3-glucanases for degradation of microspore callose wall are secreted from tapetal cells. Mutations in the genes encoding 1,3-glucanases interrupt the release of microspores from the tetrad. At the post-meiotic stages, tapetum participates in initiation of exine formation. Loss of the tapetum functions results in lack of pollen wall exine at the stage, leading to infertility of the pollen. At the further developmental stages, proper development of tapetum is required for the typical patterning of pollen exines and protection of pollen grains. Loss of the tapetum functions at the stages altered the patterning of pollen wall and results in immature degeneration or infertility of the pollen. Many genes have been identified as being expressed transiently in the tapetum at the different stages and involved in pollen wall formation. For example, *MALE STERILITY 1* (*MS1*), which encodes a plant homeodomain (PHD)-finger transcription factor, is expressed transiently in the tapetum from the callose breakdown to the free microspore stage. The *ms1* tapetal cells exhibit abnormal vacuolation with altered degeneration, resulting in unusual pollen wall composition with sticky minimal exine and leading to degeneration of immature pollen occurring soon after microspore release (Vizcay-Barrena & Wilson, 2006; Ito et al., 2007; Ariizumi & Toriyama, 2011). However, very little information has been known about the roles of tapetum in the pollen mitosis. Whether tapetum plays critical roles in pollen mitosis remains unclear.

12.6.3 Roles of Tapetum in Pollen Wall Formation

As mentioned above, the pollen exine is made from the materials secreted by tapetal cells. Therefore, the tapetal cells must equip three types of mechanisms for pollen wall formation: (1) biosynthesis of the exine wall materials, which is associated with many synthesis and metabolism pathways, especially the metabolism pathways for the sugar, lipid and flavonoid; (2) secretion and transport of the materials from tapetum to microspores; and (3) expression and functional regulation of the genes and proteins being involved in the biosynthesis and translocation of the pollen wall materials. Defects in any of the mechanisms may alter the structures and functions of the pollen walls, resulting in deformation and lack of the functions of pollen grains, and, as a result, leading to degeneration or infertility of pollen (Shi et al. 2015a, b).

The genetics of biosynthesis of the exine materials has been discussed above. For the translocation of pollen exine materials, several types of transporter and secretion-related genes have been identified, including ATP-binding cassette (ABC), lipid transfer protein (LTP), and multidrug and toxic efflux (MATE) proteins that are proposed responsible for the transport of sporopollenin precursors. The rice OsABCG15/PDA1 transports the lipidic precursors (Hu et al., 2010; Zhu et al., 2013), and its Arabidopsis ortholog AtABCG26 delivers both lipid precursors and polyketides for exine formation (Choi et al., 2011; Quilichini et al. 2012a, b). The

AtABCG9 and AtABCG31 are involved in the transport of steryl glycosides for tryphine deposition (Choi et al., 2012). The AtABCG1 and AtABCG16 transfer the lipidic precursors for nexine formation (Yadav et al., 2012). The Arabidopsis type III LTPs are involved in distribution of exine precursors. They also function as the exine components (Huang et al., 2013). The glycosylphosphatidylinositol (GPI)-anchored nonspecific LTPs (LTPGs) serve as the transporters of exine precursors (Edstam & Edqvist, 2012). Furthermore, the sugar transporter RPG1 for primexine deposition (Sun et al., 2013), the magnesium transporters (AtMGT4, AtMGT5, and AtMGT9) for pollen development (Li et al., 2015), and the UDP-GlcNAc and UDP-GalNAc transporter REPRESSOR OF CYTOKININ DEFICIENCY 1 (ROCK1) for sporopollenin and primexine formation (Niemann et al., 2015) have been identified as being required for pollen wall formation. In addition, different lepidic organelles also have been proposed as being involved in the translocation of the lipid molecules to microspores at the late pollen developmental stages in Arabidopsis and rice that have the secretory type of tapetum. For example, the orbicules (Ubisch bodies) located on the inner side of the tapetum facing the developing microspores transport the sporopollenin precursors in rice and other Poaceae plants (Zhu et al., 2013; Zhang et al., 2007). In the Arabidopsis and other Brassicaceae plants, the tapetal elaioplasts and tapetosomes service for transport of the tryphine components (Quilichini et al., 2012a, b; Xu et al., 2012; Tsou et al., 2015). All the data show that translocation of the pollen wall materials is associated with a batch of complex transportation tools.

Genetic data also suggest that the tapetal cells have a complex gene regulation network regulating the expression of the genes involved in biosynthesis and transportation of the pollen wall materials. The *AMS* and its rice ortholog *TDR* have been proposed as playing central roles in pollen wall formation. In Arabidopsis, the *AMS* regulates the expression of more than 20 downstream genes, including transcription factor gene *AtMYB103/MYB80*, *TRANSPOSABLE ELEMENT SILENCING* VIA *AT-HOOK* (*TEK*), and the genes involved in callose dissociation (*A6*), fatty acid hydroxylation (*CYP703A2*, *CYP704B1*, and *CYP86C3*), phenolic synthesis (*PKSB* and *TKPR1*), tryphine formation (extracellular lipase genes, *EXL*s, and glycine-rich protein genes, *GRP*s), and lipid transport (*ABCG26*) (Xu et al., 2010, 2012). However, in the rice, the *TDR* modulates the expression of the downstream genes, *OsC6* (Li et al., 2006; Zhang et al., 2010) and *CYP703A3* (Yang et al., 2012). The *AtMYB103/MYB80*, by acting downstream of *AMS*, regulates the expression of *MS1*, *A6*, *MALE-STERILE 2* (*MS2*), *Glyoxal oxidase 1* (*GLOX1*), and the pectin methylesterase-encoding gene *VANGUARD 1* (*VGD1*) (Jiang et al., 2005; Zhang et al., 2007; Phan et al., 2011). Furthermore, the GAMYB related to gibberellin (GA) signaling regulates pollen wall development by activating the expression of *CYP703A3* (Aya et al., 2009). Moreover, the *DYT1* also influences pollen wall formation by acting upstream of *AMS* and directly regulates the expression of *TDF1* (Gu et al., 2012). Taken together, the Arabidopsis *DYT1*, *TDF1*, *AMS*, *AtMYB103/MYB80/MS188*, and *MS1* form a gene expression regulation network regulating pollen wall formation (Zhu et al., 2011) (Fig. 12.3).

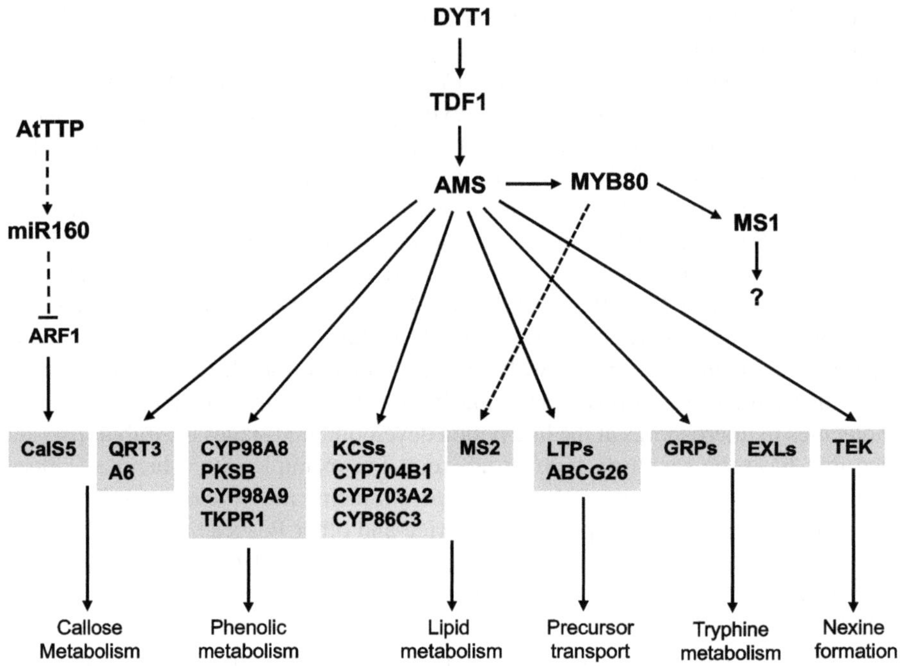

Fig. 12.3 Pollen wall formation is associated with multiple metabolism pathways in Arabidopsis (Shi et al. 2015a, b). The direct positive regulation is indicated by the solid and dotted arrows and supported by the promoter binding (solid arrows) and transcriptomic data (dotted arrows). The positive and negative actions are indicated by arrows and lines with bars, respectively

12.7 Epigenetics of Pollen Development

The difference between the vegetative cell and germ cells is reflected not only in their distinctive patterns of cell cycle control and cellular differentiation, but also in their nuclear chromatin "states" (Twell, 2011). Several lines of evidence suggest that during pollen development, the epigenetic regulation and small RNA pathways may play important roles in the male gametophytic cell fate determination and pollen wall formation (Shi et al. 2015a, b; Twell, 2011).

12.8 Perspectives

The regulation mechanisms of pollen development are proposed mainly based on the data from genetic dissection of the pollen development process. Although several genetic pathways that regulate pollen development have been proposed, the cell biological and cell physiological mechanisms of pollen development remain poorly understood. First, the microspores after meiosis differentiate cellular polarity to prepare for the asymmetric PMI. There must be a complex cellular mechanism that regulates the asymmetric distribution and organization of cellular components during the post-meiotic development of the microspores. The mechanisms that regulate cell cycle control and checkpoint of cytokinesis also are important for the asymmetric division, especially how the cellular polarity of microspores influences the checkpoint of cytokinesis to ensure the asymmetric division. Second, the mechanisms that determinate specification and fates of the vegetative cell and generative cell remain poorly understood. Third, although several transcription factors and cell cycle control proteins have been identified, more data are required to illustrate the mechanisms that regulate cell cycle control and determinate specification of male germline. Fourth, the pollen walls are synthesized by two types of pathways, namely, microspore pathways which are involved in intine formation, and tapetum pathways which are responsible for exintine formation. Whether there exists a mechanism that coordinates the microspore development and tapetum development to synergize the pollen wall formation remains being illustrated.

References

Aarts, M. G., Keijzer, C. J., Stiekema, W. J., & Pereira, A. (1995). Molecular characterization of the CER1 gene of Arabidopsis involved in epicuticular wax biosynthesis and pollen fertility. *Plant Cell, 7*, 2115–2127.

Ahlers, F., Thom, I., Lambert, J., Kuckuk, R., & Wiermann, R. (1999). 1H NMR analysis of sporopollenin from *Typha angustifolia*. *Phytochemistry, 50*, 1095–1098.

Albrecht, C., Russinova, E., Hecht, V., Baaijens, E., & de Vries, S. (2005). The *Arabidopsis thaliana* SOMATIC EMBRYOGENESIS RECEPTOR-LIKE KINASES1 and 2 control male sporogenesis. *Plant Cell, 17*, 3337–3349.

Ariizumi, T., & Toriyama, K. (2011). Genetic regulation of sporopollenin synthesis and pollen exine development. *Annual Review of Plant Biology, 62*, 437–460.

Ariizumi, T., Hatakeyama, K., Hinata, K., Sato, S., Kato, T., Tabata, S., & Toriyama, K. (2003). A novel male-sterile mutant of *Arabidopsis thaliana*, *faceless pollen-1*, produces pollen with a smooth surface and an acetolysis-sensitive exine. *Plant Molecular Biology, 53*, 107–116.

Ariizumi, T., Hatakeyama, K., Hinata, K., Inatsugi, R., Nishida, I., Sato, S., Kato, T., Tabata, S., & Toriyama, K. (2004). Disruption of the novel plant protein NEF1 affects lipid accumulation in the plastids of the tapetum and exine formation of pollen, resulting in male sterility in *Arabidopsis thaliana*. *Plant Journal, 39*, 170–181.

Aw, S. J., Hamamura, Y., Chen, Z., Schnittger, A., & Berger, F. (2010). Sperm entry is sufficient to trigger division of the central cell but the paternal genome is required for endosperm development in Arabidopsis. *Development, 137*, 2683–2690.

Aya, K., Ueguchi-Tanaka, M., Kondo, M., Hamada, K., Yano, K., Nishimura, M., & Matsuoka, M. (2009). Gibberellin modulates anther development in rice *via* the transcriptional regulation of *GAMYB*. *Plant Cell, 21*, 1253–1272.

Blackmore, S., Wortley, A. H., Skvarla, J. J., & Rowley, J. R. (2007). Pollen wall development in flowering plants. *New Phytologist, 174*, 483–498.

Brownfield, L., Hafidh, S., Borg, M., Sidorova, A., Mori, T., & Twell, D. (2009a). A plant germline-specific integrator of sperm specification and cell cycle progression. *PLoS Genetics, 5*, e1000430.

Brownfield, L., Hafidh, S., Durbarry, A., Khatab, H., Sidorova, A., Doerner, P., & Twell, D. (2009b). Arabidopsis DUO POLLEN3 is a key regulator of male germline development and embryogenesis. *Plant Cell, 21*, 1940–1956.

Canales, C., Bhatt, A. M., Scott, R., & Dickinson, H. (2002). EXS, a putative LRR receptor kinase, regulates male germline cell number and tapetal identity and promotes seed development in Arabidopsis. *Current Biology, 12*, 1718–1727.

Cankar, K., Kortstee, A., Toonen, M. A., Wolters-Arts, M., Houbein, R., Mariani, C., Ulvskov, P., Jorgensen, B., Schols, H. A., Visser, R. G., & Trindade, L. M. (2012). Pectic arabinan side chains are essential for pollen cell wall integrity during pollen development. *Plant Biotechnology Journal, 12*, 492–502.

Chang, H. S., Zhang, C., Chang, Y. H., Zhu, J., Xu, X. F., Shi, Z. H., Zhang, X. L., Xu, L., Huang, H., Zhang, S., & Yang, Z. N. (2012). No primexine and plasma membrane undulation is essential for primexine deposition and plasma membrane undulation during microsporogenesis in Arabidopsis. *Plant Physiology, 158*, 264–272.

Chen, Z., Tan, J. L. H., Ingouff, M., Sundaresan, V., & Berger, F. (2008). Chromatin assembly factor 1 regulates the cell cycle but not cell fate during male gametogenesis in Arabidopsis thaliana. *Development, 135*, 65–73.

Chen, Z., Hafidh, S., Poh, S. H., Twell, D., & Berger, F. (2009). Proliferation and cell fate establishment during Arabidopsis male gametogenesis depends on the retinoblastoma protein. *Proceedings of the National Academy of Sciences of the United States of America, 106*, 7257–7262.

Choi, H., Jin, J. Y., Choi, S., Hwang, J. U., Kim, Y. Y., Suh, M. C., & Lee, Y. (2011). An ABCG/WBC-type ABC transporter is essential for transport of sporopollenin precursors for exine formation in developing pollen. *Plant Journal, 65*, 181–193.

Choi, H., Ohyama, K., Kim, Y. Y., Jin, J. Y., Lee, S. B., Yamaoka, Y., Muranaka, T., Suh, M. C., Fujioka, S., & Lee, Y. (2012). The role of Arabidopsis ABCG9 and ABCG31 ATP binding cassette transporters in pollen fitness and the deposition of steryl glycosides on the pollen coat. *Plant Cell, 26*, 310–324.

Dong, X., Hong, Z., Sivaramakrishnan, M., Mahfouz, M., & Verma, D. P. (2005). Callose synthase (CalS5) is required for exine formation during microgametogenesis and for pollen viability in Arabidopsis. *Plant Journal, 42*, 315–328.

Dou, X. Y., Yang, K. Z., Zhang, Y., Wang, W., Liu, X. L., Chen, L. Q., Zhang, X. Q., & Ye, D. (2011). WBC27, an adenosine tri-phosphate-binding cassette protein, controls pollen wall formation and patterning in Arabidopsis. *Journal of Integrative Plant Biology, 53*, 74–88. https://doi.org/10.1111/j.1744-7909.2010.01010.x

Edstam, M. M., & Edqvist, J. (2012). Involvement of GPI-anchored lipid transfer proteins in the development of seed coats and pollen in *Arabidopsis thaliana*. *Physiologia Plantarum, 152*, 32–42.

Enns, L. C., Kanaoka, M. M., Torii, K. U., Comai, L., Okada, K., & Cleland, R. E. (2005). Two callose synthases, GSL1 and GSL5, play an essential and redundant role in plant and pollen development and in fertility. *Plant Molecular Biology, 58*, 333–349.

Flores-Tornero, M., Anoman, A. D., Rosa-Téllez, S., & Ros, R. (2015). Lack of phosphoserine phosphatase activity alters pollen and tapetum development in *Arabidopsis thaliana*. *Plant Science, 235*, 81–88.

Friedman, L., Santa Anna-Arriola, S., Hodgkin, J., & Kimble, J. (2000). Gon-4, a cell lineage regulator required for gonadogenesis in *Caenorhabditis elegans*. *Developmental Biology, 228*, 350–362.

Gómez, J. F., Talle, B., & Wilson, Z. A. (2015). Anther and pollen development: A conserved developmental pathway. *Journal of Integrative Plant Biology, 57*, 876–891.

Gu, J. N., Zhu, J., Yu, Y., Teng, X. D., Lou, Y., Xu, X. F., Liu, J. L., & Yang, Z. N. (2012). DYT1 directly regulates the expression of TDF1 for tapetum development and pollen wall formation in Arabidopsis. *Plant Journal, 80*, 1005–1013.

Hasegawa, Y., Nakamura, S., Uheda, E., & Nakamura, N. (2000). Immunolocalization and possible roles of pectins during pollen growth and callose plug formation in angiosperms. *Grana, 39*, 46–55.

Haslam, T. M., Haslam, R., Thoraval, D., Pascal, S., Delude, C., Domergue, F., Fernández, A. M., Beaudoin, F., Napier, J. A., Kunst, L., & Joubès, J. (2015). ECERIFERUM2-LIKE proteins have unique biochemical and physiological functions in very-long-chain fatty acid elongation. *Plant Physiology, 167*, 682–692.

Heslop-Harrison, J. (1968). Pollen wall development. *Science, 161*, 230–237.

Hird, D. L., Worrall, D., Hodge, R., Smartt, S., Paul, W., & Scott, R. (1993). The anther-specific protein encoded by the *Brassica napus* and *Arabidopsis thaliana* A6 gene displays similarity to b-1,3-glucanases. *Plant Journal, 4*, 1023–1033.

Hu, L., Tan, H., Liang, W., & Zhang, D. (2010). The *post-meiotic deficient Anther1* (*PDA1*) gene is required for post-meiotic anther development in rice. *Journal of Genetics and Genomics, 37*, 37–46.

Hu, J., Wang, Z., Zhang, L., & Sun, M. X. (2012). The Arabidopsis Exine Formation Defect (EFD) gene is required for primexine patterning and is critical for pollen fertility. *New Phytologist, 203*, 120–154.

Huang, L., Cao, J., Zhang, A., Ye, Y., Zhang, Y., & Liu, T. (2009). The polygalacturonase gene BcMF2 from *Brassica campestris* is associated with intine development. *Journal of Experimental Botany, 60*, 301–313.

Huang, M.-D., Chen, T. L., & Huang, A. H. (2013). Abundant type III lipid transfer proteins in Arabidopsis tapetum are secreted to the locule and become a constituent of the pollen exine. *Plant Physiology, 163*, 1218–1229.

Ito, T., Nagata, N., Yoshiba, Y., Ohme-Takagi, M., Ma, H., & Shinozaki, K. (2007). Arabidopsis *MALE STERILITY1* encodes a PHD-type transcription factor and regulates pollen and tapetum development. *Plant Cell, 19*, 3549–3562.

Jessen, D., et al. (2011). Combined activity of LACS1 and LACS4 is required for proper pollen coat formation in Arabidopsis. *Plant Journal, 68*, 715–726.

Jiang, L., Yang, S. L., Xie, L. F., Puah, C. S., Zhang, X. Q., Yang, W. C., Sundaresan, V., & Ye, D. (2005). VANGUARD1 encodes a pectin methylesterase that enhances pollen tube growth in the Arabidopsis style and transmitting tract. *Plant Cell, 17*, 584–596. https://doi.org/10.1105/tpc.104.027631

Jiang, J., Yao, L., Yu, Y., Liang, Y., Jiang, J., Ye, N., Miao, Y., & Cao, J. (2012a). PECTATE LYASE-LIKE 9 from *Brassica campestris* is associated with intine formation. *Plant Science, 229*, 66–75.

Jiang, J., Yao, L., Yu, Y., Lv, M., Miao, Y., & Cao, J. (2012b). PECTATE LYASE-LIKE 10 is associated with pollen wall development in *Brassica campestris*. *Journal of Integrative Plant Biology, 56*, 1095–1105.

Jiang, J., Zhang, Z., & Cao, J. (2013). Pollen wall development: The associated enzymes and metabolic pathways. *Plant Biology, 15*, 249–263. https://doi.org/10.1111/j.1238-8677.2012.00706.x

Jung, K. H., Han, M. J., Lee, Y. S., Kim, Y. W., Hwang, I., Kim, M. J., Kim, Y. K., Nahm, B. H., & An, G. (2005). Rice undeveloped tapetum1 is a major regulator of early tapetum development. *Plant Cell, 17*, 2705–2722.

Kim, H. J., Oh, S. A., Brownfield, L., Hong, S. H., Ryu, H., Hwang, I., Twell, D., & Nam, H. G. (2008). Control of plant germline proliferation by SCF^{FBL17} degradation of cell cycle inhibitors. *Nature, 455*, 1134–1137.

Li, H., & Zhang, D. (2010). Biosynthesis of anther cuticle and pollen exine in rice. *Plant Signaling and Behavior, 5*, 1121–1123.

Li, N., Zhang, D. S., Liu, H. S., Yin, C. S., Li, X. X., Liang, W. Q., Yuan, Z., Xu, B., Chu, H. W., Wang, J., Wen, T. Q., Huang, H., Luo, D., Ma, H., & Zhang, D. B. (2006). The rice tapetum degeneration retardation gene is required for tapetum degradation and anther development. *Plant Cell, 18*, 2999–3012.

Li, J., Huang, Y., Tan, H., Yang, X., Tian, L., Luan, S., Chen, L., & Li, D. (2015). An endoplasmic reticulum magnesium transporter is essential for pollen development in Arabidopsis. *Plant Science, 231*, 212–220.

Lin, S., Dong, H., Zhang, F., Qiu, L., Wang, F., Cao, J., & Huang, L. (2012). *BcMF8*, a putative arabinogalactan protein encoding gene, contributes to pollen wall development, aperture formation and pollen tube growth in *Brassica campestris*. *Annals of Botany, 113*, 777–788.

Liu, J., Zhang, Y., Qin, G., Tsuge, T., Sakaguchi, N., Luo, G., Sun, K., Shi, D., Aki, S., Zheng, N., Aoyama, T., Oka, A., Yang, W., Umeda, M., Xie, Q., Gu, H., & Qu, L. J. (2008). Targeted degradation of the cyclin dependent kinase inhibitor ICK4/KRP6 by RING-type E3 ligases is essential for mitotic cell cycle progression during Arabidopsis gametogenesis. *Plant Cell, 20*, 1538–1554.

Ma, H., & Sundaresan, V. (2010). Development of flowering plant gametophytes. *Current Topics in Developmental Biology, 91*, 379–412.

Moon, S., Kim, S. R., Zhao, G., Yi, J., Yoo, Y., Jin, P., Lee, S. W., Jung, K. H., Zhang, D., & An, G. (2013). Rice glycosyltransferase 1 encodes a glycosyltransferase essential for pollen wall formation. *Plant Physiology, 161*, 663–675.

Mori, T., Kuroiwa, H., Higashiyama, T., & Kuroiwa, T. (2006). GENERATIVE CELL SPECIFIC 1 is essential for angiosperm fertilization. *Nature Cell Biology, 8*, 64–71.

Niemann, M. C., Bartrina, I., Ashikov, A., Weber, H., Novák, O., Spíchal, L., Strnad, M., Strasser, R., Bakker, H., Schmülling, T., & Werner, T. (2015). Arabidopsis ROCK1 transports UDP-GlcNAc/UDP-GalNAc and regulates ER protein quality control and cytokinin activity. *Proceedings of the National Academy of Sciences of the United States of America, 112*, 291–296.

Niu, N., Liang, W., Yang, X., Jin, W., Wilson, Z. A., Hu, J., & Zhang, D. (2013). EAT1 promotes tapetal cell death by regulating aspartic proteases during male reproductive development in rice. *Nature Communications, 4*, 1245.

Oh, S. A., Johnson, A., Smertenko, A., Rahman, D., Park, S. K., Hussey, P. J., & Twell, D. (2005). A divergent cellular role for the FUSED kinase family in the plant-specific cytokinetic phragmoplast. *Current Biology, 15*, 2107–2111.

Owen, H. A., & Markaroff, C. A. (1995). Ultrastructure of microsporogenesis and microgametogenesis in *Arabidopsis thaliana* (L.) Heynh. ecotype Wassilewskija (*Brassicaceae*). *Protoplasma, 185*, 7–21.

Park, S. K., Howden, R., & Twell, D. (1998). The *Arabidopsis thaliana* gametophytic mutation *gemini pollen 1* disrupts microspore polarity, division asymmetry and pollen cell fate. *Development, 125*, 3789–3799.

Park, S. K., Rahman, D., Oh, S. A., & Twell, D. (2004). *Gemini pollen 2*, a male and female gametophytic cytokinesis defective mutation. *Sexual Plant Reproduction, 17*, 63–70.

Paxson-Sowders, D., Owen, H., & Makaroff, C. (1996). A comparative ultrastructural analysis of exine pattern development in wild-type Arabidopsis and a mutant defective in pattern formation. *Protoplasma, 198*, 53–65.

Paxson-Sowders, D. M., Dodrill, C. H., Owen, H. A., & Makaroff, C. A. (2001). DEX1, a novel plant protein, is required for exine pattern formation during pollen development in Arabidopsis. *Plant Physiology, 127*, 1739–1749.

Phan, H. A., Iacuone, S., Li, S. F., & Parish, R. W. (2011). The MYB80 transcription factor is required for pollen development and the regulation of tapetal programmed cell death in *Arabidopsis thaliana*. *Plant Cell, 23*, 2209–2224.

Qiao, Y., Hou, B., & Qi, X. (2023). Biosynthesis and transport of pollen coat precursors in angiosperms. *Nature Plants, 9*, 9864–876.

Quilichini, T. D., Douglas, C. J., & Samuels, A. L. (2012a). New views of tapetum ultrastructure and pollen exine development in *Arabidopsis thaliana*. *Annals of Botany, 112*, 1189–1201.

Quilichini, T. D., Samuels, A. L., & Douglas, C. J. (2012b). ABCG26-mediated polyketide trafficking and hydroxycinnamoyl spermidines contribute to pollen wall exine formation in Arabidopsis. *Plant Cell, 26*, 4483–4498.

Rotman, N., Durbarry, A., Wardle, A., Yang, W. C., Chaboud, A., Faure, J. E., Berger, F., & Twell, D. (2005). A novel class of MYB factors controls sperm-cell formation in plants. *Current Biology, 15*, 244–248.

Scott, R. J., Spielman, M., & Dickinson, H. (2004). Stamen structure and function. *The Plant Cell, 16*(Suppl. 1), S46–S60.

Shi, J., Cui, M., Yang, L., Kim, Y. J., & Zhang, D. (2015a). Genetic and biochemical mechanisms of pollen wall development. *Trends in Plant Science, 20*, 741–753.

Shi, X., Sun, X., Zhang, Z., Feng, D., Zhang, Q., Han, L., Wu, J., & Lu, T. (2015b). GLUCAN SYNTHASE-LIKE 5 (GSL5) plays an essential role in male fertility by regulating callose metabolism during microsporogenesis in rice. *Plant Cell Physiology, 56*, 497–509.

Sun, M. X., Huang, X. Y., Yang, J., Guan, Y. F., & Yang, Z. N. (2013). Arabidopsis RPG1 is important for primexine deposition and functions redundantly with RPG2 for plant fertility at the late reproductive stage. *Plant Reproduction, 26*, 83–91.

Tsou, C. H., Cheng, P. C., Tseng, C. M., Yen, H. J., Fu, Y. L., You, T. R., & Walden, D. B. (2015). Anther development of maize (Zea mays) and longstamen rice (Oryza longistaminata) revealed by cryo-SEM, with foci on locular dehydration and pollen arrangement. *Plant Reproduction, 28*, 47–60.

Twell, D. (2011). Male gametogenesis and germline specification in flowering plants sex plant. *Reproduction, 24*, 129–160.

Ueda, K., Yoshimura, F., Miyao, A., Hirochika, H., Nonomura, K., & Wabiko, H. (2013). *COLLAPSED ABNORMAL POLLEN 1* gene encoding the arabinokinase-like protein is involved in pollen development in rice. *Plant Physiology, 162*, 858–871.

Vizcay-Barrena, G., & Wilson, Z. A. (2006). Altered tapetal PCD and pollen wall development in the Arabidopsis ms1 mutant. *Journal of Experimental Botany, 57*, 2709–2717.

von Besser, K., Frank, A. C., Johnson, M. A., & Preuss, D. (2006). Arabidopsis HAP2 (GCS1) is a sperm-specific gene required for pollen tube guidance and fertilization. *Development, 133*, 4761–4769.

Wan, L., Zha, W., Cheng, X., Liu, C., Lv, L., Liu, C., Wang, Z., Du, B., Chen, R., Zhu, L., & He, G. (2011). A rice b-1,3-glucanase gene Osg1 is required for callose degradation in pollen development. *Planta, 233*, 309–323.

Wang, W., Wang, L., Chen, C., Xiong, G., Tan, X. Y., Yang, K. Z., Wang, Z. C., Zhou, Y., Ye, D., & Chen, L. Q. (2011). Arabidopsis CSLD1 and CSLD4 are required for cellulose deposition and normal growth of pollen tubes. *Journal of Experimental Botany, 62*, 5161–5177. https://doi.org/10.1093/jxb/err221

Wei, S., & Ma, L. (2023). Comprehensive insight into tapetum-mediated pollen development in *Arabidopsis thaliana*. *Cells, 12*, 247. https://doi.org/10.3390/cells12020247

Xu, J., Yang, C., Yuan, Z., Zhang, D., Gondwe, M. Y., Ding, Z., Liang, W., Zhang, D., & Wilson, Z. A. (2010). The ABORTED MICROSPORES regulatory network is required for postmeiotic male reproductive development in *Arabidopsis thaliana*. *Plant Cell, 22*, 91–107.

Xu, J., Ding, Z., Vizcay-Barrena, G., Shi, J., Liang, W., Yuan, Z., Werck-Reichhart, D., Schreiber, L., Wilson, Z. A., & Zhang, D. (2012). ABORTED MICROSPORES acts as a master regulator of pollen wall formation in Arabidopsis. *Plant Cell, 26*, 1544–1556.

Yadav, V., Molina, I., Ranathunge, K., Castillo, I. Q., Rothstein, S. J., & Reed, J. W. (2012). ABCG transporters are required for suberin and pollen wall extracellular barriers in Arabidopsis. *Plant Cell, 26*, 3569–3588.

Yang, S. L., Xie, L. F., Mao, H. Z., Puah, C. S., Yang, W. C., Jiang, L. X., Sundaresan, V., & Ye, D. (2003). Tapetum determinant1 is required for cell specialization in the Arabidopsis anther. *Plant Cell, 15*, 2792–2804.

Yang, S. L., Jiang, L. X., Puah, C. S., Xie, L. F., Zhang, X. Q., Chen, L. Q., Yang, W. C., & Ye, D. (2005). Overexpression of *TAPETUM DETERMINANT1* alters the cell fates in the Arabidopsis carpel and tapetum via genetic interaction with excess microsporocytes1/extra sporogenous cells. *Plant Physiology, 139*, 186–191.

Yang, X., Wu, D., Shi, J., He, Y., Pinot, F., Grausem, B., Yin, C., Zhu, L., Chen, M., Luo, Z., Liang, W., & Zhang, D. (2012). Rice CYP703A3, a cytochrome P450 hydroxylase, is essential for development of anther cuticle and pollen exine. *Journal of Integrative Plant Biology, 56*, 979–994.

Yang, J., Wu, J., Romanovicz, D., Clark, G., & Roux, S. J. (2013). Co-regulation of exine wall patterning, pollen fertility and anther dehiscence by Arabidopsis Apyrases 6 and 7. *Plant Physiology and Biochemistry, 69*, 62–73.

Zhang, W., Sun, Y., Timofejeva, L., Chen, C., Grossniklaus, U., & Ma, H. (2006). Regulation of Arabidopsis tapetum development and function by *DYSFUNCTIONAL TAPETUM1* (*DYT1*) encoding a putative bHLH transcription factor. *Development, 133*, 3085–3095.

Zhang, Z. B., Zhu, J., Gao, J. F., Wang, C., Li, H., Li, H., Zhang, H. Q., Zhang, S., Wang, D. M., Wang, Q. X., Huang, H., Xia, H. J., & Yang, Z. N. (2007). Transcription factor AtMYB103 is required for anther development by regulating tapetum development, callose dissolution and exine formation in Arabidopsis. *Plant Journal, 52*, 528–538.

Zhang, D., Liang, W., Yin, C., Zong, J., Gu, F., & Zhang, D. (2010). OsC6, encoding a lipid transfer protein, is required for postmeiotic anther development in rice. *Plant Physiology, 154*, 129–162.

Zhang, D., Luo, X., & Zhu, L. (2011). Cytological analysis and genetic control of rice anther development. *Journal of Genetics and Genomics, 38*, 379–390.

Zhao, D. Z., Wang, G. F., Speal, B., & Ma, H. (2002). The *EXCESS MICROSPOROCYTES1* gene encodes a putative leucine-rich repeat receptor protein kinase that controls somatic and reproductive cell fates in the Arabidopsis anther. *Genes and Development, 16*, 2021–2031.

Zhu, J., Chen, H., Li, H., Gao, J. F., Jiang, H., Wang, C., Guan, Y. F., & Yang, Z. N. (2008). Defective in Tapetal development and function 1 is essential for anther development and tapetal function for microspore maturation in Arabidopsis. *Plant Journal, 55*, 266–277.

Zhu, J., Lou, Y., Xu, X. F., & Yang, Z. N. (2011). A genetic pathway for tapetum development and function in Arabidopsis. *Journal of Integrative Plant Biology, 53*, 892–900.

Zhu, L., Shi, J., Zhao, G., Zhang, & Liang, W. (2013). Post-meiotic deficient anther1 (PDA1) encodes an ABC transporter required for the development of anther cuticle and pollen exine in rice. *Journal of Plant Biology, 56*, 59–68.

Chapter 13
Female Gametophyte Development

Dong-Qiao Shi, Yu Luo, and Wei-Cai Yang

Abstract The life cycle of plants alternates between a diploid sporophytic and a haploid gametophytic generation. In contrast to human and animal which set aside a specialized cell lineage (germline) for reproduction in early embryogenesis, the germline of plant is formed de novo. Female gametophyte, or embryo sac, is a multicellular structure hosted in the ovule. *Polygonum*-type embryo sac is the most prevalent female gametophyte that occurs in about 70% of angiosperms. Plant sexual reproduction initiates from differentiation of somatic cell to archesporal cell. In majority of angiosperms, only a single archesporial cell is produced from hypodermal cell of the ovule primordium to a megaspore mother cell which produces four megaspores after meiosis. One megaspore becomes functional megaspore that forms a large eight-nucleate coenocytic cell after three rounds of nuclear mitosis. Subsequently, nuclear migration and fusion, and cellularization occur, leading to the formation of a mature female gametophyte with three antipodals, one central cell, one egg cell, and two synergids. Female gametophyte development undergoes a complicated and elaborately tuned process including cell fate specification, cell division, and differentiation. There are great interests to investigate ovule and female gametophyte development not only for crop improvement, but also for answers to fundamental biological questions. Breakthroughs have been made in understanding molecular mechanisms underlying germline specification, functional megaspore selection, nuclear division, migration and fusion, and cell fate determination of the gametophyte.

Plant reproductive development is a concerted work of complex regulatory mechanisms. This chapter provides a detailed overview of advances in genetic regulation and molecular pathways leading to female gametophyte development. The origin and evolution of the *Polygonum*-type female gametogenesis are discussed briefly.

Keywords Female gametophyte · Megaspore · Meiosis · Egg apparatus

D.-Q. Shi · Y. Luo · W.-C. Yang (✉)
Institute of Genetics and Developmental Biology, The University of Chinese Academy of Sciences, Chinese Academy of Sciences, Beijing, China
e-mail: wcyang@genetics.ac.cn

F. Chang et al. (eds.), *Regulation of Plant Development*,
https://doi.org/10.1007/978-981-95-7033-1_13

13.1 Introduction

The angiosperm female gametophyte (FG), which is also named as embryo sac (ES), or megagametophyte, develops within the ovule in pistil (Fig. 13.1a, b). There are mainly three types of embryo sacs in respect to the variation of meiosis, selection of functional megaspore(s) [FM(s)], mitosis and cellularization, namely monosporic (*Polygonum*), bisporic (*Alisma*), and tetrasporic (*Drusa*) (Yadegari & Drews, 2004). The *Polygonum*-type ES and its development have been studied extensively since about 70% of angiosperms (including model plants *Arabidopsis*, maize, and rice) follow this type (Fig. 13.1c–h). The FG is a highly organized seven-celled structure composed of four cell types within the ovule. FG development consists of two sequential processes, namely megasporogenesis and megagametogenesis. The development of FG in *Arabidopsis* is of the *Polygonum*-type and serves as a model for genetic studies. Megasporogenesis begins from the differentiation of archesporial cell (AC) and ends with the formation of functional megaspore, while megagametogenesis covers the development from FM to a mature seven-celled FG. A hypodermal cell at the tip of ovule primordium differentiates to archesporial cell initially, which develops directly into megaspore mother cell (MMC) (Fig. 13.1c). Subsequently, the diploid MMC undergoes a rapid growth phase and proceeds meiosis to produce four haploid megaspores. Only the chalazal-most megaspore

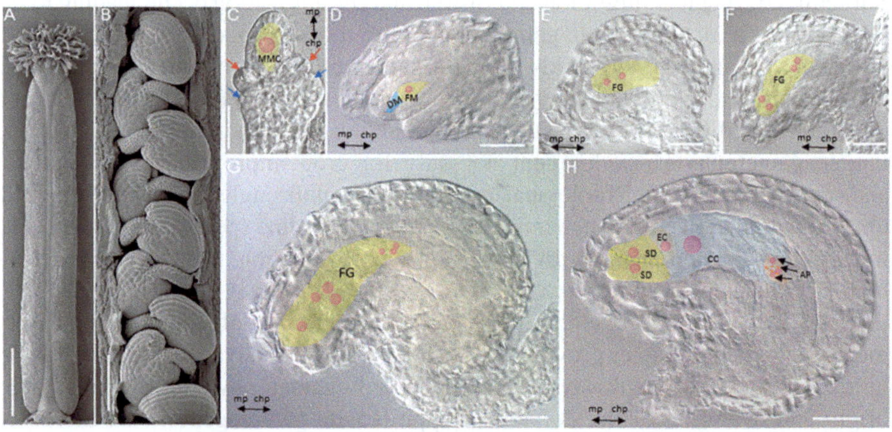

Fig. 13.1 Development of Female Gametophyte in *Arabidopsis thaliana*. (**a, b**) *Arabidopsis* pistil (**a**) and the inside ovules (**b**) under electron scanning microscope; (**c**) MMC and integument primordia. The inner and outer integument primordia are shown with red and yellow arrows respectively. Bar = 600 μm in (**a**) and 120 μm in (**b**). (**d**) FM (yellow) and degenerated megaspores (DM, blue) at FG1 stage. (**e**) A two-nucleate FG at FG3 stage. (**f**) A FG4-staged FG with four nuclei. (**g**) A FG of late FG5 stage with eight nuclei (one nucleus at mycropylar end is out of focus). (**h**) A mature FG with three antipodals (AP), one central cell (CC), one egg cell (EG), and two synergids (SD). The cytoplasm and nucleus are false-colored yellow and red in (**c**)–(**g**), respectively. The micropylar pole (mp) is upward and chalazal pole (chp) is downward in (**c**). The mp is leftward and chp is rightward in (**d**)–(**h**). Bars = 20 μm in (**c**)–(**h**)

becomes the FM; the other three megaspores undergo programmed cell death (PCD) and degeneration (Fig. 13.1d). During megagametogenesis, the FM grows rapidly and undergoes three rounds of nuclear division to form a syncytium of eight nuclei, with four nuclei at the chalazal end and the other four at the micropylar end (Fig. 13.1d–g). Then, nuclear migration, cellularization, and nuclear fusion take place successively. As a sequel of above events, three small cells at the chalazal end differentiate into antipodal, three cells close to the micropyle differentiate into two synergids and one egg, the middle large cell with a big vacuole and two nuclei originated from each pole forms the central cell. The two polar nuclei migrate close to each other and fuse to give rise to a diploid nucleus in the central cell (Fig. 13.1h). The three antipodals often degenerate during embryo sac maturation which results in the formation of a four-celled embryo sac (also called a female germ unit, FGU) at fertilization in some species.

13.2 Megasporogenesis and Megagametogenesis

13.2.1 Megaspore Mother Cell

13.2.1.1 Cytological and Molecular Characters of Megaspore Mother Cell

Archesporial cell is the initial cell that adopts germline cell fate transiting from somatic to reproductive lineage. Archesporial cell originates from a subepidermal cell at the distal end of an ovule primordium; however, it gradually becomes distinctive from other cells by its big size, condensed cytoplasm, and conspicuous nucleus and vacuole (Schneitz et al., 1995; Yang et al., 1999). This cell expands extensively in volume to become MMC, a unique cell characterized as extremely large size, condensed cytoplasm, and very big nucleus (Webb & Gunning, 1990; Ouedraogo et al., 2023) (Fig. 13.1c). The MMC is committed to proceed meiosis, instead of mitosis, to produce haploid megaspores; therefore, MMC is also called female meiocyte or megasporocyte, in which mitosis is transited to meiosis.

In *Arabidopsis*, MMC is a polarized cell with unequally distributed organelles. Mitochondria and plastids are rich in the chalazal region, while endoplasmic reticulum, small vacuoles, and dictyosomes are prominent in the micropylar region (Bajon et al., 1999). The cytoplasmic polarity in MMC is thought to contribute to the selection of FM from the tetrad after meiosis. Plasmodesmata are abundant in the chalazal cell wall, indicating active material exchange between the MMC and neighboring cells (Bajon et al., 1999). However, in some species, such as *Sedum hispanicum* and *Eleocharis sellowiana*, no obvious asymmetric distribution, or opposite polarity of above organelles has been found in the MMC, although the development of the FG is *Polygonum*-type (Brzezicka & Kozieradzka-Kiszkurno, 2018; Rocha et al., 2014). Compared to cytological observation from diverse species, transcriptome profiling and functional study on genes are more informative to understand the mechanism underlying cell fate transition and FM determination. Zhao et al. identified 862 genes expressed in the MMC by comparing the expression profile of the wild-type and mutant lacking MMC (Zhao et al., 2014). Gene ontology analysis indicates that these genes are involved in biogenesis, metabolism, structure, and transporter activities. Further analysis suggests that they encode mitochondrial carriers, cosyl transferases, DNA-binding proteins (WRC and HMG-box), and proteins involved in calmodulin-binding and protein-protein interaction (QLQ and WD40) (Zhao et al., 2014). Cell-specific transcriptome analysis also suggests that as far as the expression profile, MMC is closer to the sporophytic nucellar cells than to the haploid gametophytic cells. 796 genes are expressed predominantly in MMC and are categorized as translational regulation, ribosome biogenesis and structure, metabolic functions, transport processes, structural constituent of chromatin, and ATP-dependent helicase activity (Schmidt et al., 2011). All these suggest that female germline specification is likely an orchestration of multiple regulatory networks. By study of the transcriptome dynamics of young rice ovules at three stages (AC, MMC, and FM), Zhao et al. (2020) compared gene expression profiling and found that 5274 genes are preferentially expressed in ovules during megasporogenesis comparatively to mature female gametophyte. Benefited from high-throughput single-cell sequencing, 16,872 single-cell transcriptomics of *Arabidopsis* MMC and its neighboring cells have been acquired and analyzed (Hou et al., 2021). At three developmental stages during MMC specification, expression files of the germline-associated cell clusters, the germline subclusters, and the adjacent somatic cell subclusters show heterogeneity among the cell types in ovule primordia. The germline subcluster is enriched with "gene expression", "macromolecule biosynthetic and metabolic processes" from AC to MMC2 stage, and "chromosome organization" and "microtubule-based process" at MMC2 in accordance to the material and morphological preparation during germline specification. The reconstruction of pseudotime trajectory of female germline-associated cell clusters is no doubt a valuable reference for more insights of female germline specification (Hou et al., 2021).

The three-dimensional (3D) ovule primordium images of *Arabidopsis* (Hernandez-Lagana et al., 2016) and *Zea mays* (Ouedraogo et al., 2023) indicate that spatially confined signals are pivotal for MMC specification. During ovule primordium growth, the layer 2 (L2) cells, or the hypodermal cells of the enlarged niche, offer a pool of MMC candidates. The central-most cell of this layer turns to a nondividing and expanding state, acquiring an anisotropic, trapezoidal shape subsequently (Hernandez-Lagana et al., 2016). It is suggested that there is morphogenetic feedback between the singular MMC and surrounding somatic cells. In *katanin* mutant that lacks the microtubule-severing protein KATANIN, geometry of ovule primordium is altered, resulting in the singleness of spore mother cell (SMC) being affected and ectopic SMC candidates appearing (Hernandez-Lagana et al., 2016). This is consistent with the point that geometry and spatial signals contribute to MMC determination (Hernandez-Lagana et al., 2016). How the biochemical and mechanical signals coordinate during the cell identity specification is still an open question to be answered (Varapparambath et al., 2022).

Genes and pathways have been identified and reported to be involved in the acquisition and restriction of reproductive fate. Existing data indicate that the somatic-to-reproductive transition (SRT) is an integrative result of transcriptional, translational, and epigenetic regulation. She and coworkers investigated nuclear organization, presence of histone variants, and histone modification of MMC and FM. Their results showed that drastic reprogramming of epigenetic landscape takes place in developing MMC (She et al., 2013). MMC specification is accompanied by the establishment of a highly distinct chromatin configuration greatly different from that of the nucellar cells. In differentiating MMC, heterochromatin content and number of chromo centers are reduced, and histone variants are dynamically exchanged (She et al., 2013). Histone marks that characterize the permissive state of chromatin are increased, while those associated with repressive chromatin function are diminished. However, marks such as H3K16ac and RNA polymerase II, indicative of active transcription, are also reduced (She et al., 2013). These data imply that MMC has an open, permissive chromatin state with low transcriptional activity (She et al., 2013). Furthermore, the selected FM shares similar epigenetic resetting with the MMC, such as chromatin decondensation, reduced heterochromatin content, and low transcriptional activity (She et al., 2013). It seems that, just as in germline of animals, the SRT transition in plants induces obvious epigenetic reprogramming. Undoubtedly, chromatin remodeling and chromosome dynamics are vital for MMC specification. Presumably, the meiosis or SRT will fail if epigenetic reprogramming is disrupted in the MMC.

13.2.1.2 Mechanisms Underlying Specification of Megaspore Mother Cell

MMC formation is under transcriptional, post-translational, and epigenetic controls at diverse levels. At transcription level, several key players were identified. *SPOROCYTELESS* [*SPL*, also known as *NOZZLE* (*NZZ*)] is a key transcription regulator that controls both male and female germline formation (Yang et al., 1999; Schiefthaler et al., 1999). *SPL* is specifically expressed in archesporial cell, young MMC, and distal nucellar cells during MMC development (Schiefthaler et al., 1999; Yang et al., 1999). SPL is a central player connecting with complex networks. It was shown that SPL acts as a transcription repressor by interacting with TCP/TPL proteins that might recruit components involved in DNA methylation and histone modification (Chen et al., 2014; Wei et al., 2015). Study on DOMAINS REARRANGED METHYLASEs (DRM) methyltransferase double mutant *drm1drm2* demonstrates that RNA-dependent DNA methylation (RdDM) pathway is required to confine *SPL* expression and limit to a single MMC (Mendes et al., 2020). Furthermore, *SPL* may regulate auxin homeostasis by repressing the transcription of *YUC2* and *YUC6* and participate in lateral organ morphogenesis (Li et al., 2008), and is also critical for cytokinin-induced *PIN1* expression in ovules (Bencivenga et al., 2012). Recent study shows that *SPL* is genetically coupled with miR160-associating auxin signaling for acquisition of MMC fate (Huang et al., 2022). Fine-tuning of spatial expression of *AUXIN RESPONSE FACTOR17* (*ARF17*) by miR60 is critical for MMC specification. *ARF17* genetically interacts with *SPL*, and the miR160-*ARF17* axis controls establishment of auxin maximum via defining the expression domain of PIN1 (Huang et al., 2022). It is demonstrated that *SPL* cooperates with auxin distribution during MMC specification (Huang et al., 2022). The transcription of *SPL* is directly activated by the floral organ identity regulator *AGAMOUS* (*AG*) (Ito et al., 2004), suggesting that *SPL* is a key player in the transition from floral organ identity gene to germline development. *SPL* is also epigenetically regulated by *SET DOMAIN GROUP2* (*SDG2*), an important player primarily responsible for H3K4me3 deposition (Berr et al., 2010). ChIP analysis indicates that H3K4me3 level of *SPL* is dramatically reduced in *sdg2* mutant, that is consistent with the downregulation of *SPL* expression. On protein level, SPL is quite unstable and its stability is enhanced via phosphorylation by protein kinases MPK3 and MPK6 (Zhao et al., 2017a).

WUSCHEL (*WUS*), a key transcription regulator critical for maintaining stem cell and shoot apical meristem, also functions in germline formation and ovule development (Groβ-Hardt et al., 2002; Groβ-Hardt & Laux, 2003; Lieber et al., 2011). During ovule development, *WUS* is expressed exclusively in distal domain of the nucellus including archesporial cell and MMC (Groβ-Hardt et al., 2002), that overlaps with *SPL* expression domain (Yang et al., 1999). Using genetic and conditional rescue approach, Lieber et al. created ovules lacking WUS activity (Lieber et al., 2011). Their data showed that about 9%–12% *wus-1* ovules do not form MMC. Subepidermal cells at the distal nucellus become smaller than a typical MMC; however, they resemble the parenchyma-like neighboring somatic cells instead. While in the remaining ovules, MMC is formed but stopped after meiosis, to produce arrested and aborted ES (Groβ-Hardt et al., 2002; Lieber et al., 2011). This suggests that *WUS* is required for germ cell formation although the underlying mechanism is not known. Using a *35S:WUS-GR* inducible system and subtractive screen, Lieber et al. identified a downstream gene *WINDHOSE1* (*WIH1*). WIH1 and its homolog WIH2 act redundantly to control MMC formation since neither *wih1* nor *wih2* is sterile (Lieber et al., 2011). In *wih1wih2* double mutant, about 34% developing ovules contain no MMC but parenchyma-like cells instead. Interestingly, a significant portion of the remaining ovules contains a tracheid-like structure in place of the ES. Further analysis found that *WIH1* acts genetically in the same pathway with *TORNADO2* (*TRN2*) during ovule development (Lieber et al., 2011). Genetic analysis also suggests that *WIH* and *TRN* interact not at transcription level but more likely through protein-protein interaction. As *WUS* has been previously shown to act downstream of *SPL* during ovule development (Skinner et al., 2004), these data demonstrate that the *SPL-WUS-WIH-TRN* pathway is critical for female germline formation in *Arabidopsis* (Wang & Ma, 2011).

WIH1 encodes a peptide of 82 amino acid residues and has two homologs in *Arabidopsis* (Lieber et al., 2011). WIH1 protein contains a large N-terminal domain with six GYPP motifs and a conserved C-terminal WIH domain. WIH2 and WIH3 are longer than WIH1 and contain two additional GYPP-repeats (Lieber et al., 2011). GYPP repeats have been shown to bind membrane phospholipid that likely helps to recruit proteins to the membrane via the GYPP-containing domain (Lieber et al., 2011). Interestingly, TRN2 encodes a tetraspanin-type integral membrane protein. Since tetraspanins can interact with each other and cluster with other proteins to form tetraspanin-enriched microdomain in animal cells, it is possible that WIH and TRN together create a membrane microdomain for cell-cell signaling during germline formation in plants (Lieber et al., 2011).

On the other hand, accumulation of WUS in nucellus, but not in MMC, is directly under transcriptional control of *RBR1* (*RETINOBLASTOMA-RELATED 1*) (Zhao et al., 2017b; Jiang & Zheng, 2022). *RBR1* is the orthologue of human tumor suppressor *Retinoblastoma* (*Rb*) in *Arabidopsis*. *RBR1* is a cell cycle repressor that plays key roles in many aspects of plant growth and development (Ebel et al., 2004; Harashima & Sugimoto, 2016). As a transcriptional repressor, RBR1 directly binds to *WUS* promoter (Zhao et al., 2017b). RBR1 itself is inactivated by CYCLIN-DEPENDENT KINASE A-1 (CDKA-1), which is in turn inhibited by KIP-RELATED PROTEIN (KRPs), as a result, repression on *WUS* is removed in *rbr1* and *krp* triple mutants, leading to ectopic expression of *WUS* and failure of transition from mitosis to meiosis in the MMC, and supernumerary MMCs in *rbr1* homozygote or *krp* triple mutants (Zhao et al., 2017b). These indicate that *WUS* repression is necessary for MMC formation, and downregulation of *WUS* is critical for entry into meiosis in *Arabidopsis*. The excessive MMCs in *rbr1* or *krp* triple mutants are likely originated from one or more mitotic divisions instead of meiosis of an already specified MMC (Zhao et al., 2017b; Jiang & Zheng, 2022). These suggest that, similarly to that in animals, entry into meiosis is somehow associated with the loss of stem cell fate in plants.

MMC formation is regulated epigenetically at multiple layers (Jiang & Zheng, 2022). The small RNA (sRNA) pathways play key roles in MMC specification through repression of somatic cells to acquire germline cell fate (Aslam et al., 2022). In plants and animals, sRNA is involved in posttranscriptional regulation, chromatin organization, and remodeling of epigenetic landscape. The spread of silencing effect mediated by sRNA can be operated between cells or in a nucleotide sequence-specific manner at long distance. Various proteins associated with silencing effects on DNA and chromatin through sRNA, such as AGONAUTEs (AGOs), RNA-DEPENDENT RNA POLYMERASEs (RDRs), TEX1/THO3 components, have been reported to be involved in MMC specification, meiosis, and FM determination. AGO proteins can bind microRNA (miRNA) or short interfering RNA (siRNA) to cleave endogenous mRNA. In addition, AGO members can also mediate target silencing through a broad spectrum of actions such as transcript cleavage, translational repression, methylation, and subsequent chromatin modifications (Ghildiyal & Zamore, 2009). The ten AGO proteins of *Arabidopsis* are grouped into three distinct clades: AGO1/AGO5/AGO10, AGO2/AGO3/AGO7 (Vaucheret, 2008). AGO4/6/8/9 and AGO1/10 are more conserved than members of other clades (Nonomura et al., 2007; Rodriguez-Leal et al., 2016). Multiple AGO proteins have been shown to play roles in plant germ cell development on the RNA-mediated silencing system (Nonomura et al., 2007). AGO9, involved in RdDM, preferentially binds 24-nt sRNAs derived from transposable elements (TEs) (Olmedo-Monfil et al., 2010). AGO9 is only present in companion somatic cells including epidermal layer or integuments, but not found in MMC or FG; therefore, AGO9 controls MMC commitment through a non-cell autonomous sRNA-dependent pathway (Olmedo-Monfil et al., 2010). Compared to a single MMC per ovule primordium in the wild-type, more than one (up to six cells with conspicuous nucleus) MMC-like cells are found at variable frequencies in *ago9* mutants in a dosage-dependent manner (Olmedo-Monfil et al., 2010). However, among the enlarged cells, only one cell proceeds meiosis and produces haploid megaspores, the other MMC-like cells are able to differentiate into gametocytes and initiate gametogenesis directly, rather than undergo meiosis (Olmedo-Monfil et al., 2010). It is reported that AGO10 spatially sequesters the *miR165/6* to promote its degradation by SMALL RNA DEGRADING NUCLEASE1 (SDN1) and SDN2 for stem cell maintenance in plants (Yu et al., 2017). In ovule, *AGO10* is activated by Swi2/Snf2-Related1 (SWR1) and ERECTA at the chalaza (Cai et al., 2021). AGO10, coupling with SWR1 and the ER-MPK-signaling pathway, sequesters *miR398* to ensure the *miR398* target genes *AGL51/52/78* regulate the communication between the FG and surrounding integument tissues (Cai et al., 2021). Accordingly, the spatiotemporal control of *miR398* biogenesis is essential for proper ovule morphogenesis (Cai et al., 2021). Mutations affecting other AGO members, AGO4, AGO6, or AGO8, lead to abnormal premeiotic ovules with excessive MMCs too (Hernandez-Lagana et al., 2016). It indicates that all members of the AGO4 clade play an important role during the SRT process (Hernandez-Lagana et al., 2016). Consistently, mutations in *RDR6* and *SUPPRESSOR OF GENE SILENCING 3* (*SGS3*), two genes involved in sRNA function, also cause similar phenotype as *ago9* (Olmedo-Monfil et al., 2010). According to data from natural variations of *Arabidopsis*, the number of female gamete precursors is influenced by the activity of AGO9 and RDR6 in different ecotypes (Rodriguez-Leal et al., 2015). Maize mutant *Dominant nonreduction 4* (*Dnr4*) shows defects in chromatin condensation and spindle formation during diakinesis, metaphase I and II, results in abnormal tetrads that include triads, dyads, and microspores with multiple nuclei subsequently, which can lead to unreduced gametophytes (Singh et al., 2011). Interestingly, the *Mu* insertion in *Dnr4* locates in locus *AGO104*, a homolog of *AGO9* controlling the specification of MMC in *Arabidopsis* (Olmedo-Monfil et al., 2010; Singh et al., 2011). It indicates that maize AGO104 shares functional conservation with *Arabidopsis* AGO9 in preventing the adjacent cells of the MMC from adopting germ cell fate (Singh et al., 2011). Very similar to AGO9 in *Arabidopsis*, AGO104 is also expressed in the nucellar cells surrounding the female meiocyte (Singh et al., 2011). In *ago104* mutant, transcriptome changes significantly compared to the wild-type, suggesting that many genes are regulated through the small RNA pathway during MMC specification in maize (Singh et al., 2011). MEL1 (MEIOSIS ARRESTED AT LEPTOTENE 1) is a distinct AGO protein that shows high identity with rice and *Arabidopsis* AGO1 in their PAZ and PIWI domains (Nonomura et al., 2007). MEL1 is localized in cytoplasm of MMC. The *mel1* mutation causes failure of MMC formation and ES development in some ovules (Nonomura et al., 2007). All these data indicate that function of AGO-mediated sRNA pathways is crucial for the specification of the female germ cell.

Except the AGO pathway, the RDR pathway is also important for germ cell development. In *j66*, an enhancer mutation of *rdr6*, a large portion of developing ovules contains two MMCs (44.5% in *j66* vs. 15.8% in *rdr6*) (Su et al., 2017). Similar to other mutants with multiple MMC-like cells, the supernumerary MMC phenotype in *j66* did not affect ovule fertility (Su et al., 2017). Mutation in *j66* was mapped to *TEX1* gene that expressed in the epidermal layer of developing ovules and later concentrated in the FM, central cell, and antipodal, until diminishing along the micropylar-chalazal axis (Su et al., 2017). TEX1 is a component of the THO/TREX complex that is conserved in human, *Drosophila*, yeast, and plant (Masuda et al., 2005; Strässer et al., 2002; Yelina et al., 2010). THO complex functions in transcription elongation, while TREX complex has a conserved role in coupling transcription to mRNA export (Strässer et al., 2002). TREX complex is composed of mRNA export factors SUB2 and YRA1 as well as the THO complex in yeast (Chavez et al., 2000; Strässer et al., 2002). In plants, AtTHO1 (AtHPR1), AtTHO2, AtTHO3 (AtTEX1), AtTHO5, AtTHO6, and AtTHO7 are identified from the THO complex (Yelina et al., 2010). The plant THO/TREX mRNA export complex is reported to be involved in transgene and endogenous siRNA biosynthesis (Yelina et al., 2010). The THO/TREX complexes are responsible to transport *trans*-acting siRNA (tasiRNA) precursors transcribed from *TAS* genes from the nucleus to the subcellular location to process into siRNA (Su et al., 2017). In plants, tasiRNA is a class of endogenous secondary siRNAs produced through the action of RDR6 upon microRNA-mediated cleavage of non-coding *TAS* transcripts (Kumakura et al., 2006). There are four *TAS* family genes (TAS1–4) in *Arabidopsis* (Su et al., 2017). Mutation of *tex1* downregulates level of *TAS3*-derived tasiRNA, tasiR-ARF. The tasiR-ARF targets the ARF family genes *ARF2*, *ARF3*, and *ARF4*. Decrease of tasiR-ARF results in derepression of the target *ARF3* (Su et al., 2017). Suppression of *ARF3* in lateral epidermal cells is important for preventing excessive MMC formation (Su et al., 2017). In line with the phenotype of *tex1*, mutations in components of *THO*, *TAS3* members, and *ARF3* lead to the formation of multiple MMCs as well (Su et al., 2017). Later in Su et al. (2020) study, it is confirmed that loss-of-function *MIR390* or *AGO7* leads to supernumerary MMC-like cells. It is plausible that the THO complex, TAS3 and ARF3 pathway, and sRNA mobility of tasiR-ARF along the mediolateral axis contribute to MMC specification by patterning auxin distribution in the ovule primordium (Su et al., 2017, 2020). In addition, RNA helicases, MNEME (MEM) which is significantly enriched in the MMC and UAP56 (U2AF65-associated protein), also take part in restricting germline fate (Schmidt et al., 2011). The ATP-dependent RNA helicase UAP56 binds both RNA and dsDNA, and interacts directly with mRNA export factors MOSI and ALY2 in *Arabidopsis* (Kammel et al., 2013), suggesting that it might act via the THO/TREX complex in siRNA biogenesis. It is not known whether MEM has similar activity as UAP56. It will be of interest to explore how MEM regulates MMC specification.

Third, DNA methylation pathway is also critical for specification and meiosis of MMC. DMT102 and DMT103 are two maize DNA methyltransferases that are homologous to *Arabidopsis* CHROMOMETHYLASEs (CMTs) and DOMAINS REARRANGED METHYLTRANSFERASE (DRMs) proteins (Garcia-Aguilar et al., 2010). DMT102 is required for cytosine methylation at CNG sites and it is likely involved in a maintenance function, while DMT103 is a close homolog of DRM1 and/or 2 (Papa et al., 2001; Makarevitch et al., 2007; Garcia-Aguilar et al., 2010). Loss of function or knock-down expression alleles produce unreduced male gametophytes, multiple egg cells or embryo sacs with supernumerary gametes, and abnormal antipodals (Garcia-Aguilar et al., 2010). Both genes are active during early reproductive stages, and their expression is confined in a small population of the nucellus during megasporogenesis: *DMT102* is expressed in both MMC and the neighboring nucellar cells, while *DMT103* is only detected in the epidermal layer and integuments of the developing ovule (Garcia-Aguilar et al., 2010). Furthermore, the repressive mark H3K9me2 and permissive mark H3K9ac in *dmt102*, sexual B73 plants, and apomictic 38C plants are analyzed and compared (Garcia-Aguilar et al., 2010). The results showed that, reversely to the enhanced H3K9me2 signal in the egg cell of the wild-type B73, H3K9me2 is obviously reduced in the *dmt102::Mu* egg cell; however, a hyperacetylated state of H3K9 in archesporial domain of *dmt102::Mu* ovule mimics that in 38C (Garcia-Aguilar et al., 2010). This implies that DNA methylation mechanism in germ cell and surrounding sporophytic cells are essential for meiosis of MMC, differentiation of apomixis, and sexual pathway.

In addition to above mechanisms underlying MMC determination, pathways by receptor kinases are also reported to control the limitation of MMC in rice ovule. *MSP1* (*MULTIPLE SPOROCYTE 1*) encodes a putative leucine-rich repeat (LRR) receptor-like protein kinase in rice (Nonomura et al., 2003). The *msp1* mutation results in excessive male and female sporocytes, suggesting that *MSP1* plays important roles in restricting the number of cells to adopt germ cell fate and initiate meiosis (Nonomura et al., 2003). Similarly, *MSP1* is also expressed in the surrounding nucellar cells, but not in the sporocyte at all (Nonomura et al., 2003). However, the extra MMCs in *msp1* can enter meiosis and mitosis subsequently; curiously, a small number of ovules can give rise to twin embryos, and even twin shoots later (Nonomura et al., 2003). OsTDL1A is a close homolog of *Arabidopsis* TPD1 (TAPETUM DETERMINANT 1) that acts as a small ligand of LRR receptor kinase EXS/EMS1 to repress germ cell fate in tapetum in anthers (Jia et al., 2008; Ma, 2005; Zhang & Yang, 2014). OsTDL1A is detected to physically interact with LRR domain of MSP1, and its expression overlaps with that of *MSP1* in both ovule and anther (Zhao et al., 2008; Nonomura et al., 2003). Multiple MMCs are produced in *OsTDL1A-RNAi* plants, which means that OsTDL1A-MSP1 signaling pathway is critical for MMC specification in rice as that in *Arabidopsis* anther (Zhao et al., 2008). Similarly, loss of *MULTIPLE ARCHESPORIAL CELLS 1* (*MAC1*), an orthologue of *OsTDL1A* also results in extra archesporial cells in both ovule and anther (Sheridan et al., 1996, 1999). In *mac1* ovules, several hypodermal cells develop into extra archesporial cells which can proceed with meiosis and produce several surviving FMs (Sheridan et al., 1996). Consequently, more than one embryo sac is formed in each *mac1* ovule (Sheridan et al., 1996, 1999; Wang et al., 2012). Interestingly, unlike in rice and maize, loss of either *TPD1* or *EXS/EMS1* leads to switch of tapetal fate to male germ cell fate but has no effect on female germline development in *Arabidopsis* (Canales et al., 2002; Yang et al., 2003; Zhao et al., 2002). Cai et al.

(2022, 2023) dissect the role of brassinosteroid (BR) signaling in MMC specification in *Arabidopsis*. Genes of BR biogenesis are active only in sporophytic cells of ovule, and consistently, the BR signaling components are distributed in those somatic cells too (Cai et al., 2022). Excessive MMC-like cells appear both in the BR-deficient mutants and in the BR-insensitive ones. BRASSINAZOL-RESISTANT 1 (BZR1) transcription factor, the key player of BR signaling, directly targets *WRKY23* which is exclusively expressed in the sub-epidermal somatic cells surrounding MMC during megasporogenesis (Cai et al., 2023). It means that BR signaling prohibits the L2 cells flanking MMC from acquiring germline fate (Cai et al., 2022). In accordance with BR gradient in the distal nucellus and the maximum BR in L1 cells of ovule primordia (Cai et al., 2022), the associate EPIDERMAL PATTERNING FACTOR-like (EPFL) peptide ligands are expressed in L1 and control specification of MMC with their ERECTA family (ERf) receptor kinases in *Arabidopsis* (Cai et al., 2023). The joining of EPFL-ERf signaling to BR network in MMC specification is highlighted based on this report (Franco, 2023). Under the BR signaling hierarchy cooperated by redundant ligand-receptor complex, the specification of the unique MMC is ensured. It is indicated that BZR1 family members play downstream of EPFL-ER ligand–receptor pairs (Cai et al., 2023). Depending on EPFL-ERf signaling, BZR1 directly targets to *NSN1*, which encodes a nucleolar GTP-binding protein that genetically interacts with WRKY23 to prevent generation of multiple MMC-like cells (Cai et al., 2023).

It is interesting that quite a few mutants with multiple MMCs in the developing ovule have no defect in fertility because they can encompass or overcome the abnormal development at the MMC stage, to reach a successful female gametogenesis. Even for *rbr1* or triple *krp* mutants (Zhao et al., 2017b), the supernumerary MMCs caused by over mitosis are able to enter meiosis and mitosis sequentially during gametogenesis, until the formation of functional ES to guide pollen tube. Does this mean that MMC specification defects are likely endurable for subsequent steps, or there are some backup pathways to remedy the abnormalities?

13.2.1.3 Meiosis Initiation of Megaspore Mother Cell

Meiosis is a specialized cellular division required for sexual reproduction among most eukaryotes. Germ cell specification is indispensable to meiosis, so meiosis initiation and determination of MMC are two correlated events; it is difficult to isolate one from the other, although they are presented as two sections in this chapter. Chapter for meiosis in this book is recommended for more detailed concepts and advances of meiosis. As meiosis is a special cell division taken by germ cells, it is inevitable that loss of function of genes involved in this process will cause meiosis failure in both sexes. Comparison of transcriptomes from female and male meiocytes reveals that 96.9% MMC genes are also expressed in male meiocytes (Schmidt et al., 2011; Yang et al., 2010; Zhao et al., 2014). To date, most mutations showing direct effects on meiosis are for both male and female, so that PMCs are often selected as materials for observation; the knowledge about clues enabling MMC to cease mitotic cell cycle and transit to meiosis is still very limited. Despite mutations for meiosis in both sexes, such as *DMC1* (*DISRUPTED MEIOTIC cDNA1*) (Couteau et al., 1999), *RAD51* (Franklin et al., 1999), *DETERMINATE, INFERTILE1* (*DIF1*) (Bhatt et al., 1999), *SYN3* (Yuan et al., 2012), and other genes in *Arabidopsis*, rice, and maize, few mutations have been identified to date that specifically affect megasporogenesis, or show different phenotypes for male and female meiosis. However, the switch of somatic mitosis to sexual meiosis is undoubtedly coordinated with development of reproductive organs. It is suggestive that the mechanisms controlling initiation of meiosis are sexually diverse, so mutants specifically defective in meiosis of MMC are particularly valuable for dissection of female regulation on meiosis.

Arabidopsis allele *swi1.1* is a female-specific mutation defective in meiotic switch and performs mitotic-like division in MMC (Motamayor et al., 2000). There are 59% of developing ovules containing two megasporocyte-like cells due to a single equational mitotic cell division of MMC in the *swi1.1* primordium (Motamayor et al., 2000). Whereas allele *swi1.2* is abnormal in sister chromatid cohesion and synapsis during male meiosis besides the female defect (Mercier et al., 2001). Almost at the same time, a female-specific meiotic mutant *dyad* was identified (Siddiqi et al., 2000). In majority of *dyad* ovules, only one equational division is proceeded in morphologically normal MMC and results in diploid cells of dyad, rather than haploid tetrad from a typical meiosis (Siddiqi et al., 2000; Agashe et al., 2002). These two cells persist in *dyad* embryo sac until later stage of ovule development, but only the chalazal one expresses FM-specific markers (Motamayor et al., 2000; Ravi et al., 2008). Statistical data show that MMCs of about 4% *dyad* ovules cannot enter meiosis (Siddiqi et al., 2000); meanwhile, about 0.24% products from equational division of MMCs are functional gametes, thus these fertile, apomeiotic, and diploid female gametes can be fertilized by haploid sperm to produce triploid embryos and plants (Ravi et al., 2008). Interestingly, *SWI1* (*SWITCH1*) gene is identical to *DYAD*, a regulator of meiotic chromosome organization that is expressed in female and male meiocytes (Agashe et al., 2002; Mercier et al., 2001; Ravi et al., 2008). *Arabidopsis SWI1/DYAD*, maize *AM1*, and rice *OsAM1* are three homologs required for meiosis initiation (Mercier et al., 2001; Motamayor et al., 2000). Maize gene *AMEIOTIC1* (*AM1*) directs the transition from mitosis to meiosis and plays essential roles through meiotic prophase in maize (Pawlowski et al., 2009). Phenotype of maize *am1* alleles was reported as early as in 1970s (Palmer, 1971). Meiosis in *am1* homozygotes never occurs; instead, it is substituted by one or several rounds of synchronized mitotic cell division(s) (Golubovskaya et al., 1993; Pawlowski et al., 2009). Chromosome behavior in *am1* meiocytes is similar to that of mitosis, as it leads to degeneration of the germ cell (Golubovskaya et al., 1993; Palmer, 1971; Pawlowski et al., 2009). The *AM1* gene regulates the transition through a novel leptotene–zygotene checkpoint. It means that AM1 exists as foci in early prophase I meiocytes (Pawlowski et al., 2009). Analysis of transcriptomes indicates that *AM1* is a predominant regulator required for all meiotic processes in a manner epistatic over other genes for the initiation of meiosis in maize (Golubovskaya et al., 1993; Nan et al., 2011; Pawlowski et al.,

2009). Proteins encoded by *SWI1/DYAD*, *AM1*, and *OsAM1* are plant-specific coil-coil proteins, which suggests that meiosis in plant germ cell is initiated by a novel plant-specific mechanism (Che et al., 2011). SWI1/DYAD, OsAM1, and AM1 are all nuclear proteins that bind to chromatin in early meiosis prophase I in meiocyte (Che et al., 2011; Mercier et al., 2001, 2003; Pawlowski et al., 2009). It is suggested that meiosis initiation is a conserved process regulated by conserved factors in plants, or even in other eukaryotic organisms. However, it is still not clear why some mutations of genes for meiosis affect megasporogenesis specifically.

Intriguingly *ACTIN-RELATED PROTEIN6* (*ARP6*) regulates expression of many meiosis-related genes such as *DMC1*, *ASY1* (*ASYNAPTIC1*), *SWI1/DYAD*, and *MND1* (*MEIOTIC NUCLEAR DIVISION PROTEIN1*) during megasporogenesis (Qin et al., 2014). ARP6 is a subunit of the SWR1 (SICK WITH RAT8) ATP-dependent chromatin remodeling complex which is unique in substituting H2A.Z-H2B dimers for H2A-H2B (Mizuguchi et al., 2004). In *arp6-1* and *arp6-2*, ovule primordium growth and MMC differentiation are not interrupted, except that prophase I of meiosis is defective (Qin et al., 2014). Meiotic recombination-associated genes are upregulated in *arp6* developing ovules, indicating an upstream and repressive effect of *ARP6* on those genes (Qin et al., 2014). It is shown that ARP6 mediates the deposition of H2A.Z at locus of *DMC1*, so ARP6 acts as a dual regulator of *DMC1* by repressing *DMC1* in nonsporogenous cells, and activating *DMC1* expression specifically in megasporocytes after initiation of meiosis (Qin et al., 2014). It is possible that SWR1 complex functions as a developmental sensor of crosstalk between the MMC and surrounding somatic cells to ensure that initiation of meiosis happens only in germ cell (Qin et al., 2014). Mutation in gene encoding cytochrome P450 KLUH (KLU) results in a very similar phenotype to *arp6* mutant: chromosome pairing and organization during meiosis are perturbed in *klu* female germ cells (Zhao et al., 2014). Expression of *DMC1* in MMC is dependent on *KLU* in a non-cell-autonomous manner. *KLU* is preferentially expressed in the inner integument; therefore, it may produce a mobile signal other than phytohormone to regulate MMC meiosis (Adamski et al., 2009; Anastasiou et al., 2007; Eriksson et al., 2010; Zhao et al., 2014). In *arp6 klu* double mutant, more than one MMC is observed per developing ovule, meaning that *KLU* genetically interacts with *ARP6* to control MMC specification (Zhao et al., 2018). In *arp6 klu* ovule primordia, *WRKY28* that preferentially expressed in the hypodermal somatic cell surrounding the MMC is downregulated (Zhao et al., 2018). Loss of *WRKY28* phenocopies *arp6 klu* double mutant phenotype in ovule primordia and produces MMC-like cells as well (Zhao et al., 2018). ChIP result obtained by using H2A.Z antibody indicates that ARP6 can deposit H2A.Z onto *WRKY28* in a KLU-dependent manner (Zhao et al., 2018). In total, ARP6 and KLU control *WRKY28* expression in hypodermal cells neighboring MMC by depositing H2A.Z to *WRKY28*, to prevent somatic hypodermal cells from acquiring MMC fate (Zhao et al., 2018).

13.2.2 Female Gametogenesis

13.2.2.1 Selection of Functional Megaspore

As the meiotic products of MMC, four megaspores are arranged as a "T" shape, or linearly. Subsequently, only the proximal megaspore becomes the functional one to proceed the following mitosis during embryo sac development. The other three megaspores at distal end undergo PCD and degeneration, in comparison with the surviving chalazal-most megaspore which is enlarged and obviously distinguished from nucellar cells. However, why do different megaspores have different fate? What genes are involved in the selection of FM, or death of non-functional megaspores? These questions have not been answered until now.

The formation of the FM is crucial for the establishment of the gametophytic generation, it marks the initiation of megagametogenesis. Determination and selection of FM are under complicated and programmed regulation. Cytological studies suggest that the developmental cues mediating the selection of the FM may be present in MMC before meiosis. More mitochondria and plastids are located in the chalazal region of the MMC, whereas rough ER and vacuoles are more abundant at the micropylar portion (Bajon et al., 1999). It is not known whether the asymmetric distribution of cellular organelles in MMC and asymmetric division play a role. The megaspores at different position can receive different amount and/or types of cellular contents and organelles after meiosis (Medina et al., 1981; Papini et al., 2011; Bajon et al., 1999). Alternatively, position information and intercellular communications via, for example, plasmodesmata with neighboring cells might play important roles.

Cell wall composition varies considerably between cell types during early ovule development. Rapid wall modifications and changes of cell wall thickness take place during the specification and development of MMC and FM (Bajon et al., 1999; Coimbra et al., 2007; Rafinska & Bednarska, 2011; Rodkiewicz, 1970). The modifications known to date include callose deposition, plasmodesmata construction, and accumulation of some special proteins. Transient incorporation of callose (β-1,3-glucan polymer) is involved in pathogen response, cell plate stability, and pollen wall formation (Pirselova & Matusikova, 2013; Thiele et al., 2009; Underwood, 2012). Dynamic callose deposition in cell walls of MMC proceeding meiosis (Fig. 13.2a) is a common feature of angiosperms with the monosporic and bisporic type of FG development (Bajon et al., 1999). In MMC which undergoes apomixis rather than meiosis, callose is reduced or completely absent. Large amount of callose accumulates at the early stage of meiotic prophase I (Fig. 13.2b) and disappears initially from the FM (Fig. 13.2c) and subsequently from non-functional megaspores upon degeneration (Bajon et al., 1999). When stained with aniline blue, very distinctive fluorescent signal of callose is observed in cell walls, especially at the distal pole of the MMC and cell plates between megaspores (Bajon et al., 1999). The permeability of the cell wall containing callose is likely weaker than that of cellulose cell walls, so that callose is regarded as an envelope or a molecular filter

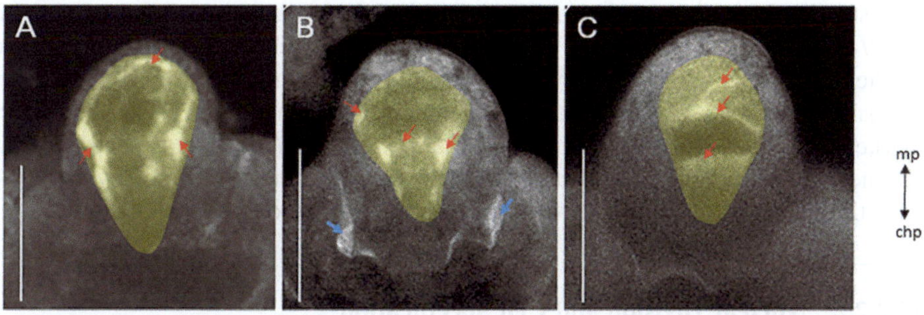

Fig. 13.2 Callose accumulation in the walls of MMC and megaspores. (**a**) Callose is detected by aniline blue staining in MMC at meiosis initiation. (**b**) Callose is visible in MMC at meiosis I (red arrows) and the cells at inner integument cells (blue arrows) surrounding nucellus. (**c**) Callose accumulation in cell walls (red arrows) in a T-shaped tetrad. The micropylar pole (mp) is upward and chalazal pole (chp) is downward in (**a**)–(**c**). Bars = 20 μm

which isolates the germ cell from surrounding cells, to ensure that meiosis is proceeded in this special reproductive cell, rather than in other somatic cells (Bajon et al., 1999). After meiosis, callose locates at the side wall of non-functional megaspores, as well as the transversal walls between the tetrad, but not in the cell walls between FM and its adjacent nucellar cells (Bajon et al., 1999). The FM is thus separated from other three non-functional megaspores by callose (Bajon et al., 1999). Numerous plasmodesmata are observed at the lateral and chalazal end, but not at the micropylar end of the surviving tear-drop-shaped FM (Bajon et al., 1999). This dynamic pattern of callose deposition and plasmadesmata is thought to allow FM development and make sure that the degeneration only occurs in non-functional megaspores.

Deposition of arabinogalactan-proteins (AGPs) is also characterized in reproductive cells. AGPs are glycosylated hydroxyproline-rich glycoproteins that are ubiquitous in plants (Showalter, 2001; Leszczuk et al., 2019). Large amount of AGPs exist in cell walls, plasma membranes, and extracellular secretions (Showalter, 2001). However, the type and depositing pattern of specific AGP sugar epitopes are regarded as molecular markers in *Arabidopsis* sexual reproduction (Coimbra et al., 2007; Leszczuk et al., 2019). AGPs are specifically accumulated in or around reproductive cells, and this group of proteins plays an important role during the alternation between sporophytic and gametophytic transitions (Coimbra et al., 2007; Leszczuk et al., 2019). Immunostaining with the monoclonal antibodies to AGPs shows that they are present on the surface of MMC, megaspores, ES wall, synergids, and the filiform apparatus at the micropylar end (Coimbra et al., 2007). It was reported that the classical AGP18 exerts an active regulation over the selection and survival of megaspores in *Arabidopsis* (Acosta-Garcia & Vielle-Calzada, 2004). AGP18 is localized in the nucellar cells covering MMC at the distal end, but not in MMC at all. Later after meiosis, AGP18 is detected in FM and abundant in the cellular edges adjacent to the FM (Acosta-Garcia & Vielle-Calzada, 2004). In mature ES, AGP18 is expressed specifically in the central cell, egg, and synergid

(Acosta-Garcia & Vielle-Calzada, 2004). The FM does not initiate differentiation in *AGP18-RNAi* ovules but persists in the nucellus of defective ovules (Acosta-Garcia & Vielle-Calzada, 2004). However, two or more surviving megaspores from one meiosis are visible, and these additional megaspores can acquire the FM-specific markers but cannot induce FM differentiation in *AGP18*-overexpressing ovules (Demesa-Arevalo & Vielle-Calzada, 2013). Taken together, *AGP18* positively regulates the selection and viability of FM during early gametogenesis.

13.2.2.2 Nuclear Division and Cell Specification During Megagametogenesis

The enlarged and differentiated FM initiates megagametogenesis and undergoes three rounds of free-nuclear mitosis, eventually gives rise to a large syncytium with a big central vacuole and eight nuclei. It is visualized that nuclear divisions occur along the micropylar-chalazal (M-C) axis by living dynamics of FG development (Susaki et al., 2021). The eight-nucleate coenocytic stage is a short stage preceding nuclear migration and cytokinesis. Initially the eight nuclei locate at two poles, with four at chalazal end and the other four at micropylar end. The two polar nuclei from two poles of the syncytium respectively move toward each other along cytoplasm bridge for fusion. While the polar nuclei migrate and fuse linearly, the edge of the cell plate forms (Susaki et al., 2021). The synergid-specific marker, MYB98pro::GFP-MYB98, is initiated just after cellularization in the synergid, egg, and central cells and eventually confined to the synergid cells (Susaki et al., 2021). It means that cell fates are determined immediately after cellularization, and the ES cellularizes simultaneously to produce seven cells of four cell types, namely, three antipodals, one central cell, one egg, and two synergids patterned in order along the micropylar-chalazal axis (Susaki et al., 2021; Christensen et al., 1997; Murgia et al., 1993; Song et al., 2014). Megagametogenesis in *Arabidopsis* is arbitrarily divided into seven continuous developmental stages from FG1 to FG8 that are widely accepted for FG study (Christensen et al., 1997).

Benefited by abundant mutants defective in FG development, studies of functional genetics in this field have been booming in recent 25 years. A large number of genes involved in nuclear mitosis, cell fate determination, and/or female guidance for pollen tube have been identified. However, it should be noted that many genes are multifunctional, so that pleiotropic defects are caused even by single mutations of these players.

Mitotic Progression

Megagametogenesis initiates from mitotic division of FM. The mitosis of FM is limited to nuclear division, so that the mitotic progression of megaspore results in a syncytium containing eight nuclei from three rounds of mitosis. This progression is under the control of pathways specific to germ line or sexual reproduction, as well as common factors involved in cell cycle in sporophytic tissues. The haploid nature of the FM suggests that genes essential for cellular functions would also be required for embryo sac development. Paralleled with sRNA regulatory action during MMC specification, somatic sRNA pathways are also required to promote the initiation of megaspore mitosis. *AGO5* gene that encodes a putative effector involved in miRNA-directed target cleavage in sRNA silencing pathways, functions as an essential player at this transition (Oliver et al., 2014; Tucker et al., 2012). Mutant *ago5-4* produced a truncated AGO5 protein with PAZ domain but lacking the MID-domain and catalytic PIWI-domain never initiates mitosis of the functional megaspore and becomes arrested at FG1 stage (Tucker et al., 2012). Interestingly, *AGO5* is not expressed in the megaspore, but predominantly active in surrounding somatic cells of ovule (Tucker et al., 2012). Similar phenotype was also observed in *pWUS:AGO5-4* transgenic ovules (Tucker et al., 2012). This suggests clearly that AGO5 expressed in neighboring somatic cells has an impact on FM development but the underlying mechanism is not clear.

There are several other mutants reported to be defective in mitosis initiation of FM. In *agl23-1* mutant, the FM is formed in a small portion of the mutant ovules manifested by the expression of *pFM1:GUS* marker but arrested at FM stage (Colombo et al., 2008). This suggests that *AGL23* plays a role in initiating FM development. Compared to restricted *AGO5* expression in surrounding somatic cells, *AGL23* is expressed in FM and throughout FG development, but not in the archesporial stage (Colombo et al., 2008). However, the FM expression of *AGL23* was not observed using genomic translational GFP fusion (Bemer et al., 2010). It is not known whether this discrepancy is caused by the promoter or genomic region used by the two approaches; therefore, its role in FM needs more investigation. Isoforms of secretion-associated Ras-related GTPase 1 (Sar1) for the assembly and disassembly of the coat protein complex II (COPII) coat mediate mitosis of FM redundantly in *Arabidopsis* (Liang et al., 2023). Another mutation that affects FM development is *cyp85a1* (Pérez-España et al., 2011). CYP85A1 and CYP85A2 are members of CYP85A family which catalyze oxidative reactions in the biosynthesis of active forms of BRs (Pérez-España et al., 2011). *cyp85a1* plants are not distinctive from wild-type plants in vegetative growth, but ovule development is arrested at FM stage (Pérez-España et al., 2011). *pCYP85A1::GUS* transformants showed that it is expressed from the one-nucleate to two-nucleate stage of female gametogenesis, and prevalent throughout female gametophyte development (Pérez-España et al., 2011). This suggests that BR biosynthesis plays a role in early embryo sac development, although the underlying mechanisms are far from elucidation.

Interestingly, RNA processing and ribosome biogenesis seem fundamental for controlling the mitotic progression during megagametogenesis as mutations in genes involved in the processes often arrest embryo sacs at two-, four-, or eight-nucleate stages (Shi & Yang, 2011). These include *slow walker* (*swa*) mutants, *gfa1/clo/vaj*, *maa3*, and *clps3*. *SWA1* is the first gene identified as a player essential for 18S pre-RNA splicing and 18S rRNA biogenesis during female gametogenesis, whereas *SWA2* is involved in pre-ribosome biogenesis and nucleolar export (Li et al., 2009; Liu et al., 2010; Shi et al., 2005). *GAMETOPHYTIC*

FACTOR1/CLOTHO/VAJRA(*GFA1/CLO/VAJ*), homologous to the yeast U5 snRNP spliceosomal component Snu114p for spliceosome dynamics, and MAGATAMA3 (MAA3), a plant homolog of the yeast Sen1 that is necessary to RNA metabolism in many aspects, cause delay or arrest of mitotic progression (Coury et al., 2007; Liu et al., 2009; Shimizu et al., 2008). Mutants of five putative protein co-factors for ribosome biogenesis in *Arabidopsis*, namely, RRP5, PWP2, NOB1, ENP1, and NOC4, show strong transmission distortion through female gametophyte (Missbach et al., 2013). It will be interesting to reveal why the RNA biogenesis and ribosomal function that are general to cell activity show sex priority on gametophyte development. It is plausible to speculate that the rapid expansion of embryo sac, the increased number of mitotic nuclear divisions, and/or cellularization processes that are completed in very short time, may require extremely active processing of rRNA, tRNA, and mRNA, and highly efficient ribosomal metabolism simultaneously (Missbach et al., 2013; Shi & Yang, 2011).

Antipodals

The three small antipodals locate at the pocket of chalazal end in the ES. Knowledge about their function has been largely limited. Antipodals are regarded as accessory cells that are dispensable for fertilization (Yang et al., 2010). It is hypothesized that antipodal cells are backup of germ cells in case of gametic failure (Groβ-Hardt et al., 2007). It is evidenced that antipodals can adopt a central cell fate in mutants for splicing factors (Groβ-Hardt et al., 2007; Moll et al., 2008), or express egg-specific markers when auxin synthesis gene is overexpressed in the ES of *Arabidopsis* (Pagnussat et al., 2009). It was previously reported that *Arabidopsis* antipodals undergo degeneration shortly before fertilization (Christensen et al., 1997; Murgia et al., 1993). The lifespan of antipodals has been linked to mitochondrial integrity in the central cell (Kägi et al., 2010). However, the observation of Song et al. (2014) using fluorescent markers showed that antipodals persist in mature embryo sac and even at early stage of young seed, so it is concluded that *Arabidopsis* mature ES for fertilization is a seven-celled structure at FG7 stage, but not a four-celled one (Christensen et al., 1997; Murgia et al., 1993). The life span of antipodals varies in angiosperms. In *Tofieldia glutinosa*, the uninucleate antipodals proceed free-nuclear divisions after the initial cellularization to generate binucleate antipodal cells (Holloway & Friedman, 2008). In cereals such as rice and maize, antipodals are multi- or uni-nucleate, they continue to proliferate to a cluster as many as several hundred cells (Diboll & Larson, 1966; Nonomura et al., 2003). The proliferative antipodals are thought to function as nutrient transfer cells from the maternal sporophyte to the young seed (Evans & Grossniklaus, 2009). In addition, finger-like cell wall projections from antipodals are present in some species (Drews & Koltunow, 2011; Huang & Russell, 1992). These observations suggest that the antipodal cells indeed have function to be discovered. Antipodal proliferation correlates with auxin biogenesis in the maize ES (Chettoor & Evans, 2015). The cessation of antipodal division is coupled with loss of expression of *DR5* and *PIN1* in the antipodal in maize leaf polarity mutant *Laxmidrib1* (Chettoor & Evans, 2015).

Central Cell

As one of the female gametes, the central cell is fertilized by a sperm to initiate endosperm development and nourish the developing embryo. Morphologically, the central cell and egg apparatus (synergids and egg cell) are highly polarized cells. There is a large central vacuole in central cell, which pushes central nucleus to the micropylar end of the cell. In some species, *Arabidopsis*, for example, the two haploid polar nuclei from opposite poles fuse to form a big homodiploid (2n) secondary nucleus with a dense nucleolus in central cell during maturation of the ES, however, in other species, such as rice or wheat, there are two separate polar nuclei, instead of a fused central nucleus, in the central cell before fertilization, a sperm fuses with the polar nucleus proximal to egg and subsequently fuse with the other polar nucleus, so the fusion of polar nuclei will not occur before fertilization. The cell walls vary between the central cell and adjacent cells, for example, the wall between central cell and the egg is not continuous in some places, in addition, there are also some unknown materials accumulated discontinuously in cell wall between the two gametes (Fig. 13.3a). The cell wall is thin, or entirely absent between the central cell and the synergid, compared to a little thicker wall and abundant plasmodesmata between the central cell and the antipodal (Mansfield et al., 1991) (Fig. 13.3b) The complex cytoplasm is full of well-differentiated organelles such as rough ER,

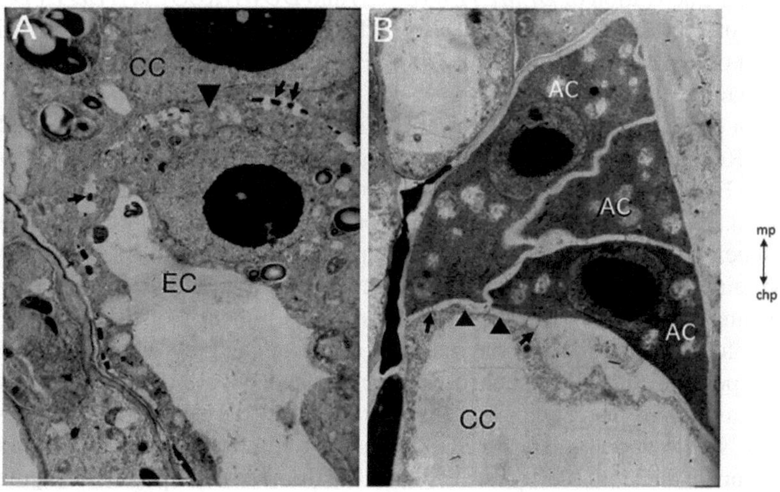

Fig. 13.3 Cell walls between female gametophyte cells. (**a**) Cell wall between central cell (CC) and egg cell (EC) is not continuous (arrow head). There are some unknown materials (arrows) accumulated discontinuously in cell wall. (**b**) Cell wall between central cell (CC) and antipodal cells (AC) is thick (arrow head). There are plasmodesmata (arrows) in cell wall. Bars = 20 μm

ribosomes, mitochondria, as well as plastids containing large starch granules in *Arabidopsis* (Mansfield et al., 1991). This implies that the central cell has high metabolic activity.

There is intercellular communications between central cell and other FG cells, the antipodals, the egg cell, and the synergids (Li & Yang, 2020). As one of the female gametes for double fertilization, central cell shows potential contribution for pollen tube guidance and polyspermy block (Chen et al., 2007; Yang et al., 2010; Zhang et al., 2020; Nagahara et al., 2021). The molecular mechanism underlying specification of the central cell is poorly understood. Genetic studies revealed that type I MADS proteins AGAMOUS-LIKE 80 (AGL80) and AGL61/DIANA are associated with central cell specification (Bemer et al., 2008; Portereiko et al., 2006a; Steffen et al., 2008; Zhang et al., 2020). Loss of function of *AGL80* and *AGL61* results in misspecification of the central cell (Bemer et al., 2008; Steffen et al., 2008; Zhang et al., 2020), reduction in size of vacuole and nucleolus, unfused polar nuclei, and absence of endosperm after fertilization (Bemer et al., 2008; Steffen et al., 2008). It seems that AGL80 plays more effect on central cell differentiation after polar nuclei fusion (Zhang et al., 2020). Further molecular analysis indicates that genes expressed in the antipodal and synergid are ectopically activated in the dysfunctional central cell of *agl61* or *agl80* ovules (Bemer et al., 2008; Steffen et al., 2008; Zhang et al., 2020). This suggests that AGL61 and AGL80 inhibit central cell from expressing genes active in the antipodal or synergid (Bemer et al., 2008; Steffen et al., 2008; Zhang et al., 2020). Both *AGL80* and *AGL61* are transcription factors expressed exclusively in the central cell and endosperm. It is suggested that AGL80 mediates regulation of downstream genes in manner of AGL80-AGL61 heterodimer or AGL80-AGL80 homodimer in the central cell, for example, the activation of *DEMETER* (*DME*) and *DOWNREGULATED IN DIF1 46* (*DD46*), or repression of *DD1*, *DD3* or *MYB98* that specifically expressed in the antipodal or synergids (Bemer et al., 2008; Portereiko et al., 2006a, b; Steffen et al., 2008; Zhang et al., 2020).

Fusion of polar nuclei in the central cell is similar to yeast nuclear fusion during mating (Rose, 1996). The yeast Ig binding protein (BiP), a molecular chaperone Hsp70 in ER, is essential to the nuclear membrane fusion (Normington et al., 1989; Rose et al., 1989). There are three *BiP* genes in *Arabidopsis*. BiP-deficient mutants show specific defect in polar nuclear fusion in the embryo sac. It is suggested that BiP proteins have conserved function as a nuclear fusion protein in plant and yeast (Maruyama et al., 2010). It is surprising that the dysfunction of *BiP* does not interfere double fertilization, since the fusion of sperm and egg nuclei, and the fusion of another sperm nucleus with an unfused polar nucleus, occur normally, to produce an embryo, and endosperm composed of diploid nuclei divided by nucleus from the fusion of sperm and a polar nucleus, and haploid nuclei from the unfertilized and unfused polar nucleus (Maruyama et al., 2010). It means that pollen tube guidance, double fertilization, and first several rounds of nuclear division in endosperm are independent of polar nuclear fusion, although the seed is aborted possibly because of abnormal development of embryo and endosperm (Maruyama et al., 2010). *Arabidopsis* gamete expressed 1 (GEX1) is a membrane protein that is conserved among plant species as a putative karyogamy factor. Mutant *gex1* shows unfused polar nuclei in mature FG, and it means that GEX1 is necessary for the nuclear fusion in *Arabidopsis* reproduction (Nishikawa et al., 2020). Armadillo BTB *Arabidopsis* protein 1 (ABAP1), the one reported to control proliferative cell divisions during leaf growth in *Arabidopsis*, can interact with different transcription factors in the course of plant reproduction (Cabral et al., 2021). It is suggested that the fusion of polar nuclei in FG is regulated by interaction of ABAP1 and ARIA-interacting Double AP2-domain (ADAP) (Cabral et al., 2021). There are more mutants showing unfused polar nuclei, such as *lachesis* (*lis*) (Groβ-Hardt et al., 2007) and *clotho/gametophytic factor 1* (*clo*) (Moll et al., 2008; Völz et al., 2012), *nuclear fusion defective 1* (*nfd1*) (Portereiko et al., 2006a, b), *gametophytic factor 2* (*gfa2*) (Christensen et al., 2002), *rbr* (Johnston et al., 2010), *maa3* (Shimizu et al., 2008; Shimizu & Okada, 2000), *Athemn* (Pratibha et al., 2017), and *glauce* (*glc*) (Leshem et al., 2012). These mutants exhibit diverse extent of defects in polar nuclear fusion and central cell function. Molecular studies on these target genes demonstrate that central cell development, specification, and polar nuclear fusion are controlled by multiple genes that are involved in mitochondrial function, pre-mRNA splicing, and specification of other gametophytic cells.

Besides functioning as one of female gametes to fertilize with sperm and as precursor of endosperm, the central cell is also a vital cell for pollen tube guidance. Mutant *maa3* that shows unfused polar nuclei or delay of central cell maturation, is incapable of pollen tube guidance or inhibition of polytubey (Shimizu & Okada, 2000; Shimizu et al., 2008). *CENTRAL CELL GUIDANCE* (*CCG*) is specifically expressed in the central cell. Mutant *ccg* has morphologically normal embryo sac, but fails in micropylar pollen tube guidance. However, the tube attraction defect of *ccg* mutant can be rescued by the central cell-specific expression of *CCG*, indicating the indispensable role of the central cell in pollen tube guidance (Chen et al., 2007). CCG is a nuclear protein with a conserved N-terminal zinc ribbon domain and interacts with another central cell-specific protein CBP1 (CCG-BINDING PROTEIN 1) to form a transcription regulatory complex, CCG-CBP1 complex (Li et al., 2015). It is assumed that the expression of genes involved in pollen tube guidance within the central cell is modulated by CCG-CBP1 complex through recruiting the Mediator and RNA polymerase II machinery to AGAMOUS-like (AGL) transcription factors (Li et al., 2015).

There is a group of FIS-like genes identified to prevent onset of central cell nuclear proliferation or autonomous initiation of endosperm proliferation independent of fertilization. The FERTILIZATION-INDEPENDENT SEED (FIS) Polycomb Repressive Complex 2 (FIS-PRC2) represses the activation of target genes by establishing histone modification (such as H3K27me3), chromatin remodeling, or DNA methylation for epigenetic regulation of central cell/endosperm proliferation (Baroux et al., 2006; Gehring et al., 2006; Jullien et al., 2006; Köhler et al., 2003a, b; Makarevich et al., 2006, 2008; Schmidt et al., 2013; Villar et al., 2009). The core components of FIS-RPC2 are MEDEA (MEA), FIS2, FERTILIZATION INDEPENDENT ENDOSPERM (FIE) and MULTICOPY SUPPRESSOR OF IRA1 (MSI1), mutations in any of the *FIS* class genes show maternal-effect seed abortion (Grossniklaus et al., 2001) or fertilization-independent endosperm proliferation (Chaudhury et al., 1997; Grossniklaus et al., 1998; Köhler et al., 2003a; b; Ohad et al., 1996; Ohad et al., 1999). Interestingly, the cell cycle repressor *RBR* also takes part in this regulatory pathway of central cell proliferation (Johnston et al., 2008). However, it seems that *RBR* plays roles in cell differentiation independently of its function in cell cycle. As in mutant *rbr* which shows unrestricted nuclear proliferation in egg apparatus, central cell region, or other cells in the ES, the phenotype of autonomous central cell proliferation is similar to that caused by mutation of above *PRC2* genes. MSI1-RBR binds *MET1* promoter and controls heterochromatin maintenance (Ach et al., 1997; Johnston et al., 2008; Köhler et al., 2003a, b). MET1 is a regulator of *FIS2* gene, while expression of *FIS2* is lost in *rbr* embryo sacs. It is revealed that RBR regulates expression of *PRC2* genes through MET1-dependent DNA methylation (Johnston et al., 2008). Together, RBR and PRC2 form a dynamic and reciprocally repressive loop to mediate cell proliferation and specification during embryo sac development. In view of interaction with PcG complexes, plant RBR is functionally conserved with its animal orthologue pRb which associates with varied chromatin-modifying factors (Flowers et al., 2010; Kotake et al., 2007; Kuwabara & Gruissem, 2014; Talluri & Dick, 2012; Zhang et al., 2000).

Egg

The egg cell is broad at chalazal end and tapered toward the micropylar end (Mansfield et al., 1991). It shows an opposite polarity to the neighboring central cell and the synergid, since the conspicuous vacuole of egg cell occupies the micropylar and central part of the cell, the cytoplasm and nucleus lie chalazally and proximal to central nucleus. The close proximity of the female gamete nuclei (the central cell and egg nuclei) and the discontinuous cell wall bordering central cell and egg are favorable for reception of two sperms during double fertilization. The egg cell and central cell are female gametes to accept sperm cells; therefore, the fertilized egg and central cell are precursors of embryo and endosperm respectively. However, only DNA from egg cell can be inherited by next generation, thus egg cell is regarded as the "true" female gamete that contributes genetically to the offspring (Sprunck & Gross-Hardt, 2011). Fertilization of egg cell means the transition from gametophyte generation to sporophyte generation. It is inevitable that cell differentiation is accompanied by cellular reprogramming at the physiological, cytological, transcriptomic, and the chromatin level. Differentiation of female gametophytic cells can be recognized as the reprogramming and establishing of cell-specific transcriptome and epigenetic state. The two female gametes show global dimorphism in chromatin and transcriptional states consequently (Baroux & Autran, 2015; Pillot et al., 2010). Totally, data from epigenetic markers and transcriptomes indicate that the egg cell is featured with repressive chromatin and a relative quiescent state of transcription globally, whereas the central cell shows depletion of epigenetic markers in chromatin and active transcription status (Baroux & Autran, 2015). Expression of *CHROMOMETHYLASE 3* (*CMT3*) induces transposon silencing and silencing of euchromatic domains in egg cell. It suggests that the siRNAs possibly act in the egg cell-specific silencing, since *CMT3* acts downstream of siRNAs in somatic cells in the RdDM pathway (Cao et al., 2013; Pillot et al., 2010). Aided by laser-assisted microdissection, populations of different female gametophytic cell types are acquired. Transcriptome analysis shows that predominant expression of genes encoding the double-stranded RNA-binding factors, factors necessary for RdDM, and groups of RNA metabolism, protein degradation, and cell cycle control in the egg, suggesting again that active sRNA pathways are functioning in the female gamete (Wuest et al., 2010). The RWP-RK domain-containing (RKD) transcription factors have been focused for their function in germ specification of land plants (Boavida & Becker, 2016; Koi et al., 2016; Koszegi et al., 2011; Rovekamp et al., 2016; Wuest et al., 2010). RWP-RK family can be divided into two subfamilies, the NIN-like and the RKD families (Koszegi et al., 2011). RKD homologs are prevalent and conserved among land plants, but they are not found in animals. In wheat, TaRKD1, 2 are enriched in cDNA library from egg cells (Koszegi et al., 2011). The *MpRKD* gene of *Marchantia polymorpha* is specifically active in developing egg and sperm precursors, consistently, egg cell cannot be formed in loss-of-function mutant *mprkd* (Koi et al., 2016). Importantly, the *MpRKD* gene functions during the transition from gametophytic generation to sporophytic generation; it can maintain the transcriptional quiescence and prevent proliferation of the egg cell before fertilization (Rovekamp et al., 2016). It is suggested that RKD proteins have fundamental association with differentiation and maintenance of the egg cell in non-flowering plants (Rovekamp et al., 2016). In *Arabidopsis*, *AtRKD1* to *AtRKD4* are mainly expressed in reproductive organs, *AtRKD1* and *AtRKD2* are expressed in egg apparatus and the egg cell, respectively (Koszegi et al., 2011). In addition, ectopic expression of *AtRKD1, 2* can induce expression of egg-specific markers and more importantly, overexpression of *AtRKD2* confers an egg-like transcriptome to the seedling, although there is no detectable phenotype for loss of function of these two genes (Koszegi et al., 2011). The high similarity of expression patterns in egg cell from a broad spectrum of angiosperm species implies the conservation of roles that RKDs take in egg differentiation or function (Koszegi et al., 2011).

Synergids

The egg apparatus consists of two identical synergids and an egg cell. The polarity of synergids is opposite to the egg. Although synergid cells do not contribute directly to the seeds, they are required for pollen tube attraction and discharge of pollen tube content (Escobar-Restrepo et al., 2007; Higashiyama et al., 2001; Kasahara et al., 2005; Márton et al., 2005; Okuda et al., 2009; Rotman et al., 2003). The cell wall boundary at the chalazal end of the egg and synergids has disappeared, the absence or discontinuous cell wall facilitates the migration of sperm from the burst tube tip to the female gametes. Another specialization associated with synergid function is the filiform apparatus (FA), a highly specialized, proliferated, and thickened cell wall at the micropylar pole of the two synergids. The FA shows a unique structure of finger-like wall projections extending deep into the synergid cytoplasm. The invasion of FA enlarges the surface area of the plasma membrane in this region, and it is proposed that synergids are specialized secretory cells (Dresselhaus et al., 2016; Higashiyama, 2002; Higashiyama & Yang, 2017). The FA facilitates secretion of small peptides fundamental for pollen tube attraction and reception. Investigations indicate that the secretory organelles of synergids are abundantly accumulated nearby the FA, furthermore, the pollen tube attractants, such as LUREs and other small Cysteine-rich Proteins (CRPs) are localized in the FA (Amien et al., 2010; Cordts et al., 2001; Higashiyama et al., 2006; Huck et al., 2003; Márton et al., 2005; Takeuchi & Higashiyama, 2012; Okuda et al., 2009; Russell, 1992). It has long been assumed that the FA is the site through which the pollen tube enters the synergid (Drews & Koltunow, 2011; Kessler & Grossniklaus, 2011). However, based on observation of pollen tube behavior, Leshem et al. hypothesized that the arriving pollen tube grows beyond the FA to a certain zone of the receptive synergid, rather than growing through the FA to enter the synergid and release sperms (Leshem et al., 2013). The direct proof that the main function of the synergid to attract pollen tubes is from the investigation of laser ablation of individual gametophytic cells. *Torenia* has naked and protruded embryo sac out of ovule. Higashiyama et al. ablated female gametophytic cells and checked the pollen tube guidance with a semi in vitro guidance assay in *Torenia*. Their data showed that only the synergid cell is necessary and sufficient for attraction of pollen tubes in *Torenia* (Higashiyama et al., 2001). Synergids are widely accepted as one of the most important female gametophytic cells for pollen tube guidance, although the central cell (Chen et al., 2007) and the egg cell (Márton et al., 2012) are reported crucial to pollen tube attractant in *Arabidopsis* as well. EGG APPARATUS1 (EA1), a small diffusible peptide that expressed in egg apparatus of maize (*Zea mays*), functions as the attractant for pollen tube (Márton et al., 2005, 2012). Loss-of-function mutant *myb98* exhibits defects in pollen tube guidance (Kasahara et al., 2005). In *myb98* ovules, the FA lacks extensive invaginations, but just shows simple thickening of cell wall at micropylar end of the synergid. *MYB98*, encoding a transcription factor, is exclusively expressed in the synergid, it controls differentiation of the synergid by positively regulating a battery of synergid-expressed genes encoding filiform apparatus-localized proteins (Punwani et al., 2007, 2008). Notably, in the 83 downstream genes of *MYB98*, there are 48 genes encoding CRPs (Punwani et al., 2008). The pollen tube attractants TfLUREs and AtLUREs are also CRPs of the defensin-like subfamily that are abundantly and predominantly expressed in the synergid cell of *T. fournieri* and *Arabidopsis*, respectively (Okuda et al., 2009; Takeuchi & Higashiyama, 2012). Consistently, the AtLURE1 to AtLURE5 are members of the CRP810 subclade which are regulated by *MYB98* (Jones-Rhoades et al., 2007; Punwani et al., 2008). Several other genes that involved in synergid function are identified in addition to *MYB98* and *LUREs*, including *FERONIA/SIRÈNE* encoding a plasma membrane receptor-like kinase (Escobar-Restrepo et al., 2007; Huck et al., 2003; Rotman et al., 2003), *LORELEI* coding a small plant-specific putative glucosylphosphatidylinositol-anchored protein (GAP) (Capron et al., 2008) and

GFA2 encoding a J domain–containing protein required for mitochondrial function (Christensen et al., 2002). These genes are expressed in synergids predominantly or exclusively. Mutation of above players results in dysfunction of the synergid manifested by loss of pollen tube guidance or overgrowth in the synergid, failure of tube rupture, polytubey, or persistence of synergid. These indicate there is a complicated networks controlling synergid differentiation and functioning.

Cell Fate Specification in Embryo Sac

It is still puzzling how the eight nuclei from a single MMC acquire different cell fates during female gametogenesis. The female gametophyte is few-celled but featured with oriented mitosis and different cell types, to date, it has become an ideal model system for the studies of cell specification, cell-cell communication, and PCD (Tekleyohans et al., 2017), although clues are very limited.

After three rounds of nuclear division, an eight-nucleate coenocyte is formed. Subsequently, cell walls are produced and cells undergo differentiation or specification. It was proposed that there should be a transition after stage FG5 (eight-nucleate), at this transition, the nuclear division is ceased, cell wall formation is initiated, and cell expansion and differentiation are activated; however, the molecular pathways regulating the transition are largely unknown. In double mutant *myb64 myb119*, the embryo sac fails to exit coenocytic development (FG5 stage), or initiate FG5 transition, resulting in absence or ectopic expression of cell-specific markers, defective cell differentiation, and gametophytic polarity. It has not been elucidated how these two nuclear proteins are involved in FG5 transition, except that *MYB119* is transcriptionally regulated by the histidine kinase CKI1 (CYTOKININ INDEPENDENT 1), an activator of the cytokinin signaling pathway required for cell specification in FG (Rabiger & Drews, 2013; Wang et al., 2010; Yuan et al., 2016).

It is proposed that all cells in the ES have potential to differentiate to each other, the accessory cells have the tendency to specify to gametic cells (the egg and central cell), given that the lateral non-cell-autonomous inhibition mechanism is eliminated. This speculation is corroborated by studies on *Arabidopsis* mutants *lis* (synergids and central cell adopt egg cell fate) (Groβ-Hardt et al., 2007), *cki* (excessive egg cells arise on expense of antipodals and central cell) (Yuan et al., 2016), *eostre* and *amp1* (synergids are misspecified to egg cell) (Kong et al., 2015; Pagnussat et al., 2007), and maize *ZmEAL1-RNAi* plant (antipodals adopt a central cell fate) (Krohn et al., 2012). The additional gametic cells of above mutants can express cell-specific markers; furthermore, some are even functional for fertilization. *CKI1* encodes an *Arabidopsis* histidine kinase (AHK) (Deng et al., 2010; Yuan et al., 2016). CKI1 can activate the cytokinin two-component signaling (TCS) pathway, although it cannot bind cytokinin. *CKI1* is specifically expressed in the four chalazal nuclei at FG5, later in antipodals and the central cell. It is hypothesized that CKI1 may act in cooperation with an "X factor" at the micropylar pole, with auxin and its downstream targets acting as candidates for "X factor" (Pagnussat et al.,

2009), or CKI1 alone determines the central cell fate (Liu et al., 2017; Liu et al., 2018). *CKI1* promotes central cell specification and restricts egg cell fate. Loss of CKI1 function results in ectopic expression of egg cell markers in the central cell and antipodal, or in up to six nuclei in the embryo sac. Dual endosperms are formed when additional central cell-like cells resulted from ectopic *CKI* expression are fertilized. Interestingly, ectopic *CKI1* expression in root cells can even activate central cell-specific genes and downstream *AHPs* (*Arabidopsis His Phosphotransfer proteins*) for cell specification in the FG (Deng et al., 2010; Yuan et al., 2016). AHPs are responsible to transfer phosphorylation signals from *Arabidopsis* histidine kinases (AHKs) to *Arabidopsis* response regulators (ARRs) in TCS (Liu et al., 2017). *AHP1*, *AHP2*, *AHP3*, and *AHP5* are expressed in central cell. Mutant with triple mutations in *AHP2*, *AHP3*, and *AHP5* phenocopies the *cki1* mutant that shows loss of central cell and antipodal cell fates and gain of egg cell or synergid cell attributes (Liu et al., 2017). However, analysis of *ahk2 ahk3 ahk4* triple mutants indicates that *AHK2-AHK4* are not necessary for ES development (Riefler et al., 2006). It is indicated that CKI1 takes roles through downstream factors AHP2, AHP3, and AHP5 (Liu et al., 2017). A recent publication revealed that the type-B ARR10, 12, and 18 (ARR10/12/18) are localized at the chalazal pole, and the central cell identity is also lost in *arr10 12 18* plants (Zhu et al., 2022). From the genetic analysis, it is concluded that ARR10/12/18 function downstream of CKI1–AHP2/3/5 as core factors to determine cell fate of the FG (Zhu et al., 2022) and the CKI1-ARRs specify the central cell in *Arabidopsis* (Osnato, 2022).

It is speculated that once the cell fate is determined, the egg cell acts as a main organizer of cell specification in the ES. It sounds reasonable from investigation of *ZmEAL1* (Krohn et al., 2012) and *LIS* (Völz et al., 2012). *ZmEAL1* encodes an EA1-like protein (Márton et al., 2005; Uebler et al., 2015) that is exclusively expressed in the egg. The ZmEAL1-containing vesicles are secreted in egg and directly targeted to the chalazal pole of the egg. However, the deficiency of ZmEAL1 from *RNAi* plants results in upto three extra secondary central cells derived from the antipodal cell cluster. Whereas *LIS* codes for the *Arabidopsis* homolog of the pre-mRNA splicing factor PRP4 and is expressed dynamically in the gametophytic cells: it is expressed at all stages of female gametophyte development; however, shortly after cellularization, its expression is hardly detectable in accessory cells, in comparison to the upregulation in gametic cells (Groβ-Hardt et al., 2007). Expression of the egg cell marker expands to the synergid and the central cell in *lis* mutant (Groβ-Hardt et al., 2007). Synergids, central cell, and antipodals are affected when LIS is specifically reduced in the egg cell (Völz et al., 2012). It is likely that the egg is a spatial headquarter harboring cell fate factors which are able to function in long distance in the embryo sac.

Data from genetic studies suggest that there is an association of nuclear position and cell specification in the ES. It is believed that nucleus will adopt cell fate according to its position in the embryo sac (Sun et al., 2021). For example, two cells show marker of egg identity in vitro or in vivo when nuclear positions are altered artificially by chemical to inhibit actin polymerization or in dominant negative mutant *DN-ACT8*, conclusively, cell fate of egg apparatus is likely determined by final position of nucleus (Sun et al., 2021). Mutation in WYRD, a putative plant orthologue of the inner centromere protein (INCENP) which is implicated in the control of chromosome segregation and cytokinesis in yeast and animals, causes faulty position of a nucleus at the chalazal end of the synergid cell, but not at the micropylar end as in the wild-type (Kirioukhova et al., 2011). Interestingly, the cell with aberrant nuclear position acquires an egg cell fate in *wyrd* (Kirioukhova et al., 2011). Mutations affecting *RBR* result in additional nuclear divisions during female gametogenesis and supernumerary nuclei in the embryo sac. Some nuclei differentiate to extra egg or synergid nucleus in accordance to their position in the embryo sac, moreover, twin embryos are produced after fertilization (Ebel et al., 2004; Ingouff et al., 2009). Maize *indeterminate gametophyte 1* (*ig1*) mutant has multiple synergids, eggs, and central cells because of additional mitosis either in the FG before cellularization or in individual micropylar cells after cellularization (Evans, 2007; Guo et al., 2004; Huang & Sheridan, 1996). The cell fate and function of the egg and central cell during fertilization are corresponding to their cellular position in *ig1* embryo sacs (Evans, 2007; Guo et al., 2004; Huang & Sheridan, 1996). For example, the extra central cells close to micropylar end or near the synergid can be fertilized, while those locating at the chalazal end cannot. It is suggested that the fate of the central cell is closely correlated with its location in the embryo sac (Evans, 2007; Guo et al., 2004; Huang & Sheridan, 1996). Consistently, in *eostre* embryo sacs, only the synergid nucleus locating at the chalazal end of the cell adopts egg cell fate, while the micropylar nucleus does not (Pagnussat et al., 2007).

According to the above data, it is reasonable to suppose that there may be a positional mechanism and some cell fate determinants that control gametophytic cell fate during megagametophyte development (Pagnussat et al., 2007; Sun et al., 2021). The fate of cells at different position or compartment is determined in accordance to the concentration or gradient of these clues. It is well accepted that auxin gradients and cytokinin status established in the sporophytic nucellar tissue are critical for patterning of ovules (Bencivenga et al., 2012; Kamiuchi et al., 2014; Marsch-Martinez & de Folter, 2016; Skinner & Sundaresan, 2018; Zúñiga-Mayo et al., 2023; Fu et al., 2023). Many studies reveal that positional information generated from context-dependent auxin homeostasis implicates cell specification during embryogenesis, organogenesis, vascular differentiation, and phyllotaxis patterning (Benkova et al., 2003; Bhatia et al., 2016; Friml et al., 2003; Mironova et al., 2017; Weijers et al., 2006; Sun et al., 2021). For the establishment of the embryo sac polarity, auxin is proposed as a potential candidate controlling specification of gametophytic cells as well (Bencivenga et al., 2011; Pagnussat et al., 2009; Panoli et al., 2015; Sundaresan & Alandete-Saez, 2010; Sun et al., 2021). Auxin distribution was investigated with *DR5* promoter and reporter markers during ovule and megagametogenesis. It was shown that the DR5::GFP signal is strong in nucellus at FG1 stage, subsequently the signal can be detected inside the embryo sac from stage FG3, later the distribution of auxin is polarized to the micropylar pole at FG4, and the maximum signal is present at micropylar end in the eight-nucleate embryo sac at FG5, until less polarized at FG6 (Pagnussat et al., 2009; Sundaresan & Alandete-Saez, 2010). Downregulation of *ARF* genes results in the loss of synergid cell identity and acquirement of egg cell fate in the three micropylar cells (Pagnussat et al., 2009; Sundaresan & Alandete-Saez, 2010). Whereas overexpression of *YUCCA* (*YUC*) genes encoding key enzymes in auxin biosynthesis leads to the expression of synergid-specific marker in multiple cells, including those that should be specified to egg cell, central cell, and even antipodals at the chalazal end of embryo sac (Pagnussat et al., 2009). Consistently, auxin signaling mutants have an additional egg-like cell from one of two synergids (Pagnussat et al., 2009). It means that cell fate is changed consistently with disruption of auxin gradient in the embryo sac. It was proposed that a gradient of auxin in the embryo sac directs the fate of each gametophytic cell in respect to its position (Pagnussat et al., 2009; Sundaresan & Alandete-Saez, 2010). Auxin is generated from sporophytic nucellus at the early stage of megagametogenesis, subsequently auxin source is produced within the gametophyte at the micropylar pole during embryo sac development, then auxin gradient is established as the maximum accumulation at micropylar end to promote synergid fate (Pagnussat et al., 2009), probably also as the cue of egg determination (Sun et al., 2021). This discrepancy is most likely caused by the cell-specific markers used in the two studies. Consistently to auxin action for FG development and cellularization, genes for auxin synthesis (*YUC* and *TAA*/*TAR*) and auxin carriers (*AUX1*, *LAX1*, *PIN1*) are expressed in ovule and the developing ES (Panoli et al., 2015). In *yuc1* and *yuc2* mutants, there are more than one cell expressing the egg cell marker in FG, indicating the misspecification of micropylar cells resulted from loss of *YUC1* and *YUC2* functions (Panoli et al., 2015). Mitotic divisions, cell expansion, and nuclear migration during ES development are affected when YUC8 is deficient (Panoli et al., 2015). Whereas overexpression of *YUC1* for high levels of auxin inside the ES activates the expression of micropylar cell type markers in chalazal cells (Panoli et al., 2015). On the other hand, auxin dynamics is largely dependent on the activity of auxin carrier proteins PINs. The expression of *PIN1* in the sporophytic tissue of the ovule proves critical to megagametogenesis. Downregulation of *PIN1* leads to megagametophyte arrest at FG1 stage (Ceccato et al., 2013). In the study of maize ES, it is reported that the cluster of antipodals is a site of auxin accumulation (Chettoor & Evans, 2015). Reduced or absent antipodal proliferation in *Laxmidrib1* mutants is associated with loss of auxin signaling in the antipodal cluster (Chettoor & Evans, 2015). PIN1a localization suggests that auxin transport is conducted directionally from the central cell toward the chalazal tip cluster, and continued from antipodal cells to the surrounding sporophytic tissue (Chettoor & Evans, 2015). Given the strong activity of PIN1a in the antipodal of maize (Chettoor & Evans, 2015), it seems that auxin accumulation in maize ES is opposite to the situation of *Arabidopsis*, the auxin signaling is active and highly concentrated in maize antipodals (Chettoor & Evans, 2015). Furthermore, in Lituiev et al. report, neither the expression of auxin reporters was found inside the ES, nor robust auxin gradient could be deduced from their mathematical models (Lituiev et al., 2013). Does this mean diverse mechanisms underlying FG patterning? Or, "a clear candidate for a cell-fate determinant" is waiting for discovering (Schmidt et al., 2015)?

13.3 Female Gametogenesis in Ancient Lineages of Angiosperms

Recent consensus of the earliest-diverging lineages of angiosperms has raised reconsideration of the origin and early developmental evolution of the angiosperm seven-celled/eight-nucleate (*Polygonum*-type) female gametophyte (Figs. 13.1h and 13.4a). The above megasporogenesis and megagametogenesis are the typical developmental process of the *Polygonum*-type which is taken by about 70% of angiosperms. The evolutionary origin of *Polygonum*-type is still mysterious until now. It had been thought that the *Polygonum*-type embryo sac was also general for the basal-most clades of flowering plants, the "ANITA" grade angiosperms which comprise eight families: *Amborellaceae*, *Nymphaeaceae*, *Hydatellaceae*, *Cabombaceae*, *Austrobaileyaceae*, *Trimeniaceae*, *Schisandraceae*, and *Illiciaceae* (Tobe et al., 2000). However, reports show that three types, *Amborella* type, *Nuphar/Schisandra*-type (Fig. 13.4b), and *Polygonum*-type female gametophytes are exhibited in the most ancient extant lineages of flowering plants (Bachelier & Friedman, 2011; Friedman, 2006; Friedman & Ryerson, 2009; Friedman & Williams, 2004; Friedman et al., 2003; Williams & Friedman, 2002, 2004; Rudall et al., 2008). Based on the observation, *Amborella* type and

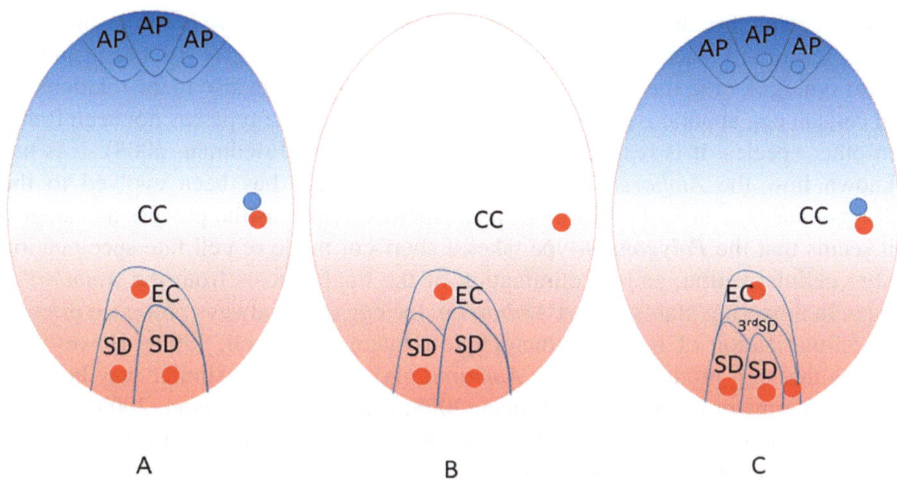

Fig. 13.4 FG pattern of *Polygonum*-type, Nuphar/Schisandra-type, and *Amborella*-type. According to model of module and monosporic FG patterning of *Polygonum*-type, Nuphar/Schisandra-type, and *Amborella*-type (Friedman & Williams, 2003; Friedman, 2008; Friedman & Ryerson, 2009), the *Polygonum*-type FG (**a**) and *Amborella*-type FG (**c**) are composed of two modules from micropylar end (red) and chalazal end (blue); however, the four-nucleate Nuphar/Schisandra-type FG develops from one module at the micropylar end (blue). The four-celled egg apparatus in *Amborella*-type FG (**c**) is made of two synergids (SD), and the additional division-produced egg (EC) and the third synergids (3rd SD). The nucleus is shown with red or blue circles. The chalazal end is upward and micropylar end is downward in (**a**)–(**c**). CC, central cell; AP, antipodal

Nuphar/Schisandra-type of mature embryo sac structure and developmental pattern are characterized in the extant ancient lineages of angiosperms (Friedman & Ryerson, 2009). It indicates that the mature female gametophyte of *Amborella* contains eight cells and nine nuclei (Friedman, 2006) (Fig. 13.4c), whereas members of the *Nymphaeales* and *Austrobaileyales* are identified as four-celled/four-nucleate embryo sac (Friedman & Williams, 2004; Williams & Friedman, 2002).

The development of female gametophyte in *Amborella trichopoda* parallels that of *Polygonum*-type from MMC to cellularization after eight-nucleate syncytium. Thus it was demonstrated that *Amborella* followed the typical *Polygonum*-type previously (Tobe et al., 2000). From Friedman's observation (2008), the seven-celled, eight-nucleate embryo sac is not the mature form of *Amborella*, but the penultimate stage of the ontogeny. A subsequent mitosis occurs and one cell at micropylar domain divides into two cells, one is the third synergid, and the other one is egg cell (Fig. 13.4c). Therefore, the mature *Amborella* female gametophyte is composed of eight cells with nine nuclei, making a pattern as: three antipodals at the chalazal end, a central cell with two polar nuclei in the middle, three synergids and one egg at the micropylar end (Friedman, 2008). So the endosperm from fusion of haploid sperm and diploid central cell is triploid as well. However, the four-celled egg apparatus is a definitive feature of *Amborella*. Furthermore, the egg cell is lineal to the third synergid in the *Amborella*-type, whereas in the *Polygonum*-type embryo sac, the egg nucleus and the polar nucleus from micropylar domain are daughters of a same nucleus (Friedman, 2008). Megagamtogenesis of *Ambrella* is unique, not only in the terminal asynchronous division of a micropylar cell to give rise to egg and the third synergid, or in the patterning of gametophytic cells, but also in the relationship of cells in egg apparatus (Friedman, 2008). The *Amborella*-type has not been found in other species, it is restricted to *A. trichopoda* only (Friedman, 2008). It is not known how the *Amborella*-type of the basal lineage has been evolved to the *Polygonum*-type in vast majority of angiosperms. As far as the process and stages, it seems that the *Polygonum*-type takes a short-cut mode of cell fate specification after cellularization, and an elimination of the final mitosis from the *Amborella*-type. In flowering plants except *Amborella*, the egg cell is always produced directly at cellularization of female gametophyte (Huang & Russell, 1992). However, it seems that egg production in the *Amborella*-type shares the similar way how the gymnosperms produce egg (Friedman, 2008; Friedman & Ryerson, 2009). Unlike the angiosperms, gymnosperms never produce egg cell directly, but form a mother cell in the archegonium, subsequently the egg mother cell divides into two cells, the egg and ventral canal cell (Friedman & Ryerson, 2009). The cell in *Amborella* that undergoes the final division to produce an egg cell and an additional synergid is considered homologous to the egg mother cell in gymnosperm archegonia (Friedman, 2008). If so, does it mean that the *Amborella*-type female gametophyte is a transitional structure from gymnosperm to angiosperm megagametogenesis?

The *Nuphar/Schisandra*-type female gametogenesis is taken by members of the *Nymphaeales* and *Austrobaileyales*. There is always only one MMC that appears in every ovule, except several species that have multiple MMCs per ovule, such as *Trimenia moorei*, an extant member of the ancient angiosperm clade *Austrobaileyales* that has as many as twenty megasporocytes differentiated and later several female gametophytes present in each ovule (Bachelier & Friedman, 2011; Endress & Sampson, 1983; Williams & Friedman, 2004). Only a single megaspore (the chalazal most one) from MMC meiosis survives to proceed gametogenesis in various ancient angiosperms of *Nymphaeales* and *Austrobaileyales*. Megasporogenesis in the *Nuphar/Schisandra*-type is monosporic too (Friedman, 2008; Friedman & Williams, 2003), similar to that of the *Polygonum*-type. However, the two types differ from megagametogenesis. At one-nucleate stage of the *Nuphar/Schisandra*-type megagametogenesis, the nucleus is specially positioned at the micropylar end of the embryo sac, both nuclei from the first mitosis stay at the micropylar pole, rather than being separated to two poles (Friedman, 2008; Friedman & Williams, 2003). Another feature is that only two rounds of nuclear mitosis happen in the *Nuphar/Schisandra*-type, instead of three rounds of mitosis in the *Polygonum*-type, leading to four nuclei in the micropylar domain, then one nucleus migrates to the central zone of the embryo sac to become the central nucleus, while the other three nuclei at the micropylar end undergo cellularization and differentiation to form the egg apparatus with two synergids and one egg cell (Friedman, 2008; Friedman & Williams, 2003). Consequently, the mature embryo sac of the *Nuphar/Schisandra*-type has four cells/four nuclei (Williams & Friedman, 2002). It means that there is no antipodal in this type of embryo sac; meanwhile, each cell has one nucleus (Friedman, 2008; Friedman & Williams, 2003). This four-celled structure is different from the four-celled (after degeneration of the antipodal) embryo sac of the *Polygonum*-type, because in the *Nuphar/Schisandra*-type the central cell is a haploid, uninucleate gamete that contains only one, instead of two polar nuclei as in the *Polygonum*-type (Friedman, 2008; Friedman & Williams, 2003; Rudall et al., 2008). It means that endosperm of these ancestor lineages is diploid, but not triploid (Friedman, 2008; Friedman & Williams, 2003; Rudall et al., 2008). Embryo in *Nymphaeales* is minute, the endosperm development is minimal, and embryo-nourishing reserves are stored in a maternal tissue from nucellus, the perisperm (Losada et al., 2017; Povilus et al., 2015). The diploid biparental endosperm is prevalent in early angiosperms; it is assumed as a basal pattern from which the triploid endosperm is evolved (Williams & Friedman, 2002). Consistently, as the female gametophyte, the four-celled/four-nucleate *Nuphar/Schisandra*-type is likely the ancestral origin of seven-celled/eight-nucleate *Polygonum-type* (Williams & Friedman, 2002). This point is supported by the identification of four-celled/four-nucleate female gametophyte in *Illicium* and other *Austrobaileyales* (Endress & Sampson, 1983; Friedman et al., 2003; Williams & Friedman, 2004). The ontogeny of female gametophyte in *Austrobaileyales* is very similar to that in *Nymphaeales*. However, the diploid endosperm of *Nymphaeales* is entirely devoid or minute of reserves, the maternal nucellar-derived perisperm acts as embryo-nourishing tissue (Floyd & Friedman, 2001; Friedman & Bachelier, 2013). In the mature seeds of *Austrobaileyales* plant *T. moorei*, no perisperm is found, while the endosperm contains small starch grains, significant quantities of protein bodies, and lipid droplets, which are primary nutrient-storing and embryo-nourishing tissue in some species (Floyd & Friedman, 2001; Friedman & Bachelier, 2013). Molecular analyses demonstrate that *Austrobaileyales* is a clade sister to all angiosperms above the basal grade. Group *Austrobaileyales* includes monocots, eumagnoliids, and eudicots (Soltis et al., 2000; Zanis et al., 2002). A comparative analysis based on more than 250 embryological studies of the earliest diverging lineages within monocots and eudicots, the eumagnoliid clade, *Ceratophyllaceae* and *Chloranthaceae*, indicates that the common ancestor of angiosperms above the basal grade possessed a monosporic, seven-celled and eight-nucleate female gametophyte, while taxa with bisporic or tetrasporic development occur only in nested positions (Williams & Friedman, 2004). The existence of four-celled/four-nucleate female gametophyte and acquirement of endosperm-storing function in *Austrobaileyales* are suggestive of *Nuphar/Schisandra*-type as ancestral pattern (Williams & Friedman, 2004). It is not known how and when the female gametophyte transits from four-celled/four-nucleate type to seven-celled/eight-nucleate type, or the endosperm transits from diploid to triploid.

It seems that the first nuclear mitosis at one-nucleate stage in the developing embryo sac is the key step to generate fundamental difference between the four-celled/four-nucleate and the seven-celled/eight-nucleate female gametophytes. In condition of the *Nuphar/Schisandra*-type, both daughter nuclei from the first mitosis are located at the micropylar domain, they take the same pole in the embryo sac, whereas in the *Polygonum*-type, these twin nuclei migrate to the chalazal and the micropylar domain respectively, they occupy two poles in the embryo sac (Friedman, 2008; Williams & Friedman, 2002). It is supposed that the nucleus and its circumambient cytoplasm at each pole compose a module, four nuclei are produced after two mitoses in each module, then three uninucleate cells are adjacent to each other, and the fourth nucleus migrates to the central region (Friedman, 2008; Friedman & Williams, 2003; Friedman & Ryerson, 2009). Thus it can be assumed that only one module has been formed in the *Nuphar/Schisandra*-type embryo sac, since the two nuclei from the first nuclear mitosis are not separated to two poles but positioned at the same domain, the micropylar domain (Friedman, 2008). A single developmental module in embryo sac of *Nymphaeales* and *Austrobaileyales* produces a three-celled micropylar egg apparatus and a central cell with one polar nucleus (Fig. 13.4b). Whereas in the *Polygonum*-type condition, two nuclei from the first mitosis migrate to chalazal end and micropylar end, respectively, thus two developmental modules formed, and four nuclei are generated after two rounds of mitosis in each module (Fig. 13.4a). The micropylar module produces three nuclei for egg apparatus and a polar nucleus that moves to the central region, meanwhile, the chalazal module produces three antipodal nuclei and a polar nucleus that migrates to the central region too (Friedman, 2008; Friedman & Ryerson, 2009; Friedman & Williams, 2003). It means that two polar nuclei are born from chalazal module and micropylar module respectively. Similarly, the *Amborella*-type can be regarded as FG with two modules as well (Fig. 13.4c). It sounds reasonable to consume that the generation of one or two modules in the female gametophyte depends on the positioning of the first two free nuclei. It remains elusive what factors are involved in nuclear localization or migration at this stage yet. It is possible that some clues control the times of mitosis as two, or nucleus number as four in each module, so that the modular unit can reach a stable status with four nuclei before cellularization.

13.4 Open Questions to Explore

The FG, or ES, is a highly specified sexual reproductive structure. The archesporial cell is originated from somatic cell, but is unique to take a sexual way. It has been recognized that the developmental process of FG is nicely orchestrated with many regulatory networks, from the determination of MMC, the selection of FM and degeneration of non-functional megaspores, the nuclear division in the syncytium, the migration and fusion of polar nuclei, to the specification of each cell in the FG. Brilliant light has been shed on our understanding about mechanisms underlying MMC specification, meiosis initiation and the cell fate determination of female gametes, due to results from reverse and forward genetics. It is suggestive that MMC is specified under an integrative effect of small RNA, DNA methylation, transcription regulation, and protein modification. Cell-cell communication during MMC development indicates that development of the germ cell is not isolated from somatic cell, conversely, germ cell specification relies on molecular signals from surrounding nucellar cells. Successful meiosis of MMC does not only require genes common for both male and female, but also some unknown female-specific factors essential for megaspore production. It is likely that these sex-dependent meiotic genes function prior to the group of genes general to MMC and PMC. Almost no molecular clue has been acquired about the PCD of non-FMs until now. When is PCD initiated in non-FMs? How is FM selected? Whether any of the other three megaspores can adopt fate of FM when the chalazal most one is killed? However, no key for above questions is available presently. Furthermore, no mutant defective at this process has been reported formally to date.

Megagametogenesis initiates from free nuclear mitosis of the FM. It is interesting that only nuclear division, rather than complete cell division, occurs during FG mitosis. It is not clear how FG skips cytoplasm division and cell plate formation during this process. Unfortunately, mutant that undergoes cellularization earlier than eight-nucleate stage has not been discovered until now either. The nuclear division in FG is under control of genes for mitosis, as well as players that take roles in ribosome processing and RNA biogenesis. It means that as a giant cell with multiple nuclei, the developing FG may require large amount of proteins and abundant energy to continue its development and execute its function. As we know, FG, even as well as MMC in some species, is polarized. The polarity is fundamental to differentiate the nuclei and cells. Is the polarity of FG established de novo, or inherited from MMC? It is demonstrated that gametophytic cell fate is determined according to nuclear position in the FG. Whether the polarity is formed because of the distribution of auxin is still under debate. It seems that the polarity is maintained throughout the whole process of development. How is auxin gradient stabilized in such a dynamic cytosol environment? All these questions are to be clarified.

Acknowledgments This work was supported by National Natural Science Foundation of China Project Grant 32130032.

References

Ach, R. A., Taranto, P., & Gruissem, W. (1997). A conserved family of WD-40 proteins binds to the retinoblastoma protein in both plants and animals. *Plant Cell, 9*, 1595–1606.

Acosta-Garcia, G., & Vielle-Calzada, J. P. (2004). A classical arabinogalactan protein is essential for the initiation of female gametogenesis in *Arabidopsis*. *Plant Cell, 16*, 2614–2628.

Adamski, N. M., Anastasiou, E., Eriksson, S., O'Neill, C. M., & Lenhard, M. (2009). Local maternal control of seed size by KLUH/CYP78A5-dependent growth signaling. *Proceedings of the National Academy of Sciences USA, 106*, 20115–20120.

Agashe, B., Prasad, C. K., & Siddiqi, I. (2002). Identification and analysis of *DYAD*: A gene required for meiotic chromosome organisation and female meiotic progression in *Arabidopsis*. *Development, 129*, 3935–3943.

Amien, S., Kliwer, I., Márton, M. L., et al. (2010). Defensin-like ZmES4 mediates pollen tube burst in maize via opening of the potassium channel KZM1. *PLoS Biology, 8*, e1000388.

Anastasiou, E., Kenz, S., Gerstung, M., et al. (2007). Control of plant organ size by KLUH/CYP78A5-dependent intercellular signaling. *Developmental Cell, 13*, 843–856.

Aslam, M., Fakher, B., & Qin, Y. (2022). Big role of small RNAs in female gametophyte development. *International Journal of Molecular Sciences, 23*, 1979.

Bachelier, J. B., & Friedman, W. E. (2011). Female gamete competition in an ancient angiosperm lineage. *Proceedings of the National Academy of Sciences USA, 108*, 12360–12365.

Bajon, C., Horlow, C., Motamayor, J. C., Sauvanet, A., & Robert, D. (1999). Megasporogenesis in *Arabidopsis thaliana* L.: An ultrastructural study. *Sexual Plant Reproduction, 12*, 99–109.

Baroux, C., & Autran, D. (2015). Chromatin dynamics during cellular differentiation in the female reproductive lineage of flowering plants. *The Plant Journal, 83*, 160–176.

Baroux, U., Gagliardini, V., Page, D. R., & Grossniklaus, U. (2006). Dynamic regulatory interactions of Polycomb group genes: MEDEA autoregulation is required for imprinted gene expression in *Arabidopsis*. *Genes & Development, 20*, 1081–1086.

Bemer, M., Wolters-Arts, M., Grossniklaus, U., & Angenent, G. C. (2008). The MADS domain protein DIANA acts together with AGAMOUS-LIKE80 to specify the central cell in *Arabidopsis* ovules. *Plant Cell, 20*, 2088–2101.

Bemer, M., Heijmans, K., Airoldi, C., Davies, B., & Angenent, G. C. (2010). An atlas of type I MADS box gene expression during female gametophyte and seed development in *Arabidopsis*. *Plant Physiology, 154*, 287–300.

Bencivenga, S., Colombo, L., & Masiero, S. (2011). Cross talk between the sporophyte and the megagametophyte during ovule development. *Sexual Plant Reproduction, 24*, 113–121.

Bencivenga, S., Simonini, S., Benkova, E., & Colombo, L. (2012). The transcription factors BEL1 and SPL are required for cytokinin and auxin signaling during ovule development in *Arabidopsis*. *Plant Cell, 24*, 2886–2897.

Benkova, E., Michniewicz, M., Sauer, M., et al. (2003). Local, efflux-dependent auxin gradients as a common module for plant organ formation. *Cell, 115*, 591–602.

Berr, A., McCallum, E. J., Menard, R., et al. (2010). *Arabidopsis SET DOMAIN GROUP2* is required for H3K4 trimethylation and is crucial for both sporophyte and gametophyte development. *Plant Cell, 22*, 3232–3248.

Bhatia, N., Bozorg, B., Larsson, A., Ohno, C., Jonsson, H., & Heisler, M. G. (2016). Auxin acts through MONOPTEROS to regulate plant cell polarity and pattern phyllotaxis. *Current Biology, 26*, 3202–3208.

Bhatt, A. M., Lister, C., Page, T., et al. (1999). The *DIF1* gene of *Arabidopsis* is required for meiotic chromosome segregation and belongs to the REC8/RAD21 cohesin gene family. *The Plant Journal, 19*, 463–472.

Boavida, L. C., & Becker, J. D. (2016). Plant evolution: What does it take to be an egg? *Current Biology, 26*, R527–R530.

Brzezicka, E., & Kozieradzka-Kiszkurno, M. (2018). Ultrastructural and cytochemical aspects of female gametophyte development in *Sedum hispanicum* L. (*Crassulaceae*). *Protoplasma, 255*, 247–261.

Cabral, L. M., Masuda, H. P., Ballesteros, H. F., et al. (2021). ABAP1 plays a role in the differentiation of male and female gametes in *Arabidopsis thaliana. Frontiers in Plant Science, 12*, 642758.

Cai, H., Liu, L., Zhang, M., et al. (2021). Spatiotemporal control of miR398 biogenesis, via chromatin remodeling and kinase signaling, ensures proper ovule development. *Plant Cell, 33*, 1530–1553.

Cai, H., Liu, L., Huang, Y., et al. (2022). Brassinosteroid signaling regulates female germline specification in *Arabidopsis. Current Biology, 32*, 1102–1114.

Cai, H., Huang, Y., Liu, L., et al. (2023). Signaling by the EPFL-ERECTA family coordinates female germline specification through the BZR1 family in *Arabidopsis. Plant Cell, 35*, 1455–1473.

Canales, C., Bhatt, A. M., Scott, R., & Dickinson, H. (2002). EXS, a putative LRR receptor kinase, regulates male germline cell number and tapetal identity and promotes seed development in *Arabidopsis. Current Biology, 12*, 1718–1727.

Cao, X. F., Aufsatz, W., Zilberman, D., et al. (2013). Role of the DRM and CMT3 methyltransferases in RNA-directed DNA methylation. *Current Biology, 13*, 2212–2217.

Capron, A., Gourgues, M., Neiva, L. S., et al. (2008). Maternal control of male-gamete delivery in *Arabidopsis* involves a putative GPI-anchored protein encoded by the *LORELEI* gene. *Plant Cell, 20*, 3038–3049.

Ceccato, L., Masiero, S., Roy, D. S., et al. (2013). Maternal control of PIN1 is required for female gametophyte development in Arabidopsis. *PLoS ONE, 8*, e66148.

Chaudhury, A. M., Ming, L., Miller, C., Craig, S., Dennis, E. S., & Peacock, W. J. (1997). Fertilization-independent seed development in Arabidopsis thaliana. *Proceedings of the National Academy of Sciences USA, 94*, 4223–4228.

Chavez, S., Beilharz, T., Rondon, A. G., et al. (2000). A protein complex containing Tho2, Hprl, Mftl and a novel protein, Thp2, connects transcription elongation with mitotic recombination in *Saccharomyces cerevisiae. The EMBO Journal, 19*, 5824–5834.

Che, L., Tang, D., Wang, K., et al. (2011). *OsAM1* is required for leptotene-zygotene transition in rice. *Cell Research, 21*, 654–665.

Chen, Y. H., Li, H. J., Shi, D. Q., et al. (2007). The central cell plays a critical role in pollen tube guidance in *Arabidopsis. Plant Cell, 19*, 3563–3577.

Chen, G. H., Sun, J. Y., Liu, M., et al. (2014). SPOROCYTELESS is a novel embryophyte-specific transcription repressor that interacts with TPL and TCP proteins in *Arabidopsis. Journal of Genetics and Genomics, 41*, 617–625.

Chettoor, A. M., & Evans, M. M. (2015). Correlation between a loss of auxin signaling and a loss of proliferation in maize antipodal cells. *Frontiers in Plant Science, 6*, 187.

Christensen, C. A., King, E. J., Jordan, J. R., & Drews, G. N. (1997). Megagametogenesis in *Arabidopsis* wild type and the *Gf* mutant. *Sexual Plant Reproduction, 10*, 49–64.

Christensen, C. A., Gorsich, S. W., Brown, R. H., et al. (2002). Mitochondrial GFA2 is required for synergid cell death in *Arabidopsis. Plant Cell, 14*, 2215–2232.

Coimbra, S., Almeida, J., Junqueira, V., Costa, M. L., & Pereira, L. G. (2007). Arabinogalactan proteins as molecular markers in *Arabidopsis thaliana* sexual reproduction. *Journal of Experimental Botany, 58*, 4027–4035.

Colombo, M., Masiero, S., Vanzulli, S., Lardelli, P., Kater, M. M., & Colombo, L. (2008). AGL23, a type I MADS-box gene that controls female gametophyte and embryo development in *Arabidopsis*. *The Plant Journal, 54*, 1037–1048.

Cordts, S., Bantin, J., Wittich, P. E., Kranz, E., Lorz, H., & Dresselhaus, T. (2001). *ZmES* genes encode peptides with structural homology to defensins and are specifically expressed in the female gametophyte of maize. *The Plant Journal, 25*, 103–114.

Coury, D. A., Zhang, C. Q., Ko, A., et al. (2007). Segregation distortion in *Arabidopsis gametophytic factor 1* (*gfa1*) mutants is caused by a deficiency of an essential RNA splicing factor. *Sexual Plant Reproduction, 20*, 87–97.

Couteau, F., Belzile, F., Horlow, C., Grandjean, O., Vezon, D., & Doutriaux, M. P. (1999). Random chromosome segregation without meiotic arrest in both male and female meiocytes of a dmc1 mutant of *Arabidopsis*. *Plant Cell, 11*, 1623–1634.

Demesa-Arevalo, E., & Vielle-Calzada, J. P. (2013). The classical arabinogalactan protein AGP18 mediates megaspore selection in *Arabidopsis*. *Plant Cell, 25*, 1274–1287.

Deng, Y., Dong, H. L., Mu, J. Y., et al. (2010). *Arabidopsis* histidine kinase CKI1 acts upstream of HISTIDINE PHOSPHOTRANSFER PROTEINS to regulate female gametophyte development and vegetative growth. *Plant Cell, 22*, 1232–1248.

Diboll, A. G., & Larson, D. A. (1966). An electron microscopic study of the mature megagametophyte in *Zea mays*. *American Journal of Botany, 53*, 391–402.

Dresselhaus, T., Sprunck, S., & Wessel, G. M. (2016). Fertilization mechanisms in flowering plants. *Current Biology, 26*, R125–R139.

Drews, G. N., & Koltunow, A. M. (2011). The female gametophyte. *The Arabidopsis Book, 9*, e0155.

Ebel, C., Mariconti, L., & Gruissem, W. (2004). Plant retinoblastoma homologues control nuclear proliferation in the female gametophyte. *Nature, 429*, 776–780.

Endress, P. K., & Sampson, F. B. (1983). Floral structure and relationships of the *Trimeniaceae* (*Laurales*). *J Arnold Arboretum, 64*, 447–473.

Eriksson, S., Stransfeld, L., Adamski, N. M., Breuninger, H., & Lenhard, M. (2010). KLUH/CYP78A5-dependent growth signaling coordinates floral organ growth in *Arabidopsis*. *Current Biology, 20*, 527–532.

Escobar-Restrepo, J. M., Huck, N., Kessler, S., et al. (2007). The FERONIA receptor-like kinase mediates male-female interactions during pollen tube reception. *Science, 317*, 656–660.

Evans, M. M. (2007). The indeterminate gametophyte1 gene of maize encodes a LOB domain protein required for embryo sac and leaf development. *Plant Cell, 19*, 46–62.

Evans, M. M. S., & Grossniklaus, U. (2009). The maize megagametophyte. In *Handbook of maize: Its biology* (pp. 79–104). Springer.

Flowers, S., Beck, G. R., & Moran, E. (2010). Transcriptional activation by pRB and its coordination with SWI/SNF recruitment. *Cancer Research, 70*, 8282–8287.

Floyd, S. K., & Friedman, W. E. (2001). Developmental evolution of endosperm in basal angiosperms: Evidence from Amborella (*Amborellaceae*), Nuphar (*Nymphaeaceae*), and Illicium (*Illiciaceae*). *Plant Systematics and Evolution, 228*, 153–169.

Franco, J. (2023). BZR1, you have an invite: EPFL-ERECTA wants to join your female germline specification network. *Plant Cell, 35*, 1298–1299.

Franklin, A. E., McElver, J., Sunjevaric, I., Rothstein, R., Bowen, B., & Cande, W. Z. (1999). Three-dimensional microscopy of the Rad51 recombination protein during meiotic prophase. *Plant Cell, 11*, 809–824.

Friedman, W. E. (2006). Embryological evidence for developmental lability during early angiosperm evolution. *Nature, 441*, 337–340.

Friedman, W. E. (2008). *Hydatellaceae* are water lilies with gymnospermous tendencies. *Nature, 453*, 94–97.

Friedman, W. E., & Bachelier, J. B. (2013). Seed development in *Trimenia* (*Trimeniaceae*) and its bearing on the evolution of embryo-nourishing strategies in early flowering plant lineages. *American Journal of Botany, 100*, 906–915.

Friedman, W. E., & Ryerson, K. C. (2009). Reconstructing the ancestral female gametophyte of angiosperms: Insights from Amborella and other ancient lineages of flowering plants. *American Journal of Botany, 96*, 129–143.
Friedman, W. E., & Williams, J. H. (2003). Modularity of the angiosperm female gametophyte and its bearing on the early evolution of endosperm in flowering plants. *Evolution, 57*, 216–230.
Friedman, W. E., & Williams, J. H. (2004). Developmental evolution of the sexual process in ancient flowering plant lineages. *Plant Cell, 16*, S119–S132.
Friedman, W. E., Gallup, W. N., & Williams, J. H. (2003). Female gametophyte development in Kadsura: Implications for *Schisandraceae*, *Austrobaileyales*, and the early evolution of flowering plants. *International Journal of Plant Sciences, 164*, S293–S305.
Friml, J., Vieten, A., Sauer, M., et al. (2003). Efflux-dependent auxin gradients establish the apical-basal axis of *Arabidopsis*. *Nature, 426*, 147–153.
Fu, Y., Zhang, H., Ma, Y., Li, C., Zhang, K., & Liu, X. (2023). A model worker: Multifaceted modulation of AUXIN RESPONSE FACTOR3 orchestrates plant reproductive phases. *Frontiers in Plant Science, 14*, 1123059.
Garcia-Aguilar, M., Michaud, C., Leblanc, O., & Grimanelli, D. (2010). Inactivation of a DNA methylation pathway in maize reproductive organs results in apomixis-like phenotypes. *Plant Cell, 22*, 3249–3267.
Gehring, M., Huh, J. H., Hsieh, T. F., et al. (2006). DEMETER DNA glycosylase establishes MEDEA polycomb gene self-imprinting by allele-specific demethylation. *Cell, 124*, 495–506.
Ghildiyal, M., & Zamore, P. D. (2009). Small silencing RNAs: An expanding universe. *Nature Reviews. Genetics, 10*, 94–108.
Golubovskaya, I., Grebennikova, Z. K., Avalkina, N. A., & Sheridan, W. F. (1993). The role of the ameiotic1 gene in the initiation of meiosis and in subsequent meiotic events in maize. *Genetics, 135*, 1151–1166.
Grossniklaus, U., Vielle-Calzada, J. P., Hoeppner, M. A., & Gagliano, W. B. (1998). Maternal control of embryogenesis by medea, a Polycomb group gene in *Arabidopsis*. *Science, 280*, 446–450.
Grossniklaus, U., Spillane, C., Page, D. R., & Kohler, C. (2001). Genomic imprinting and seed development: Endosperm formation with and without sex. *Current Opinion in Plant Biology, 4*, 21–27.
Groβ-Hardt, R., & Laux, T. (2003). Stem cell regulation in the shoot meristem. *Journal of Cell Science, 116*, 1659–1666.
Groβ-Hardt, R., Lenhard, M., & Laux, T. (2002). *WUSCHEL* signaling functions in interregional communication during *Arabidopsis* ovule development. *Genes & Development, 16*, 1129–1138.
Groβ-Hardt, R., Kagi, C., Baumann, N., et al. (2007). *LACHESIS* restricts gametic cell fate in the female gametophyte of *Arabidopsis*. *PLoS Biology, 5*, e47.
Guo, F., Huang, B. Q., Han, Y., & Zee, S. Y. (2004). Fertilization in maize *indeterminate gametophyte1* mutant. *Protoplasma, 223*, 111–120.
Harashima, H., & Sugimoto, K. (2016). Integration of developmental and environmental signals into cell proliferation and differentiation through RETINOBLASTOMA-RELATED 1. *Current Opinion in Plant Biology, 29*, 95–103.
Hernandez-Lagana, E., Rodriguez-Leal, D., Lua, J., & Vielle-Calzada, J. P. (2016). A multigenic network of ARGONAUTE4 clade members controls early megaspore formation in *Arabidopsis*. *Genetics, 204*, 1045–1056.
Higashiyama, T. (2002). The synergid cell: Attractor and acceptor of the pollen tube for double fertilization. *Journal of Plant Research, 115*, 149–160.
Higashiyama, T., & Yang, W. C. (2017). Gametophytic pollen tube guidance: Attractant peptides, gametic controls, and receptors. *Plant Physiology, 173*, 112–121.
Higashiyama, T., Yabe, S., Sasaki, N., et al. (2001). Pollen tube attraction by the synergid cell. *Science, 293*, 1480–1483.
Higashiyama, T., Inatsugi, R., Sakamoto, S., et al. (2006). Species preferentiality of the pollen tube attractant derived from the synergid cell of *Torenia fournieri*. *Plant Physiology, 142*, 481–491.

Holloway, S. J., & Friedman, W. E. (2008). Embryological features of *Tofieldia glutinosa* and their bearing on the early diversification of monocotyledonous plants. *Annals of Botany, 102*, 167–182.

Hou, Z., Liu, Y., & Zhang, M. (2021). High-throughput single-cell transcriptomics reveals the female germline differentiation trajectory in *Arabidopsis thaliana*. *Communications Biology, 4*, 1149.

Huang, B. Q., & Russell, S. D. (1992). Female germ unit—Organization, isolation, and function. *International Review of Cytology, 140*, 233–293.

Huang, B. Q., & Sheridan, W. F. (1996). Embryo sac development in the maize indeterminate gametophyte1 mutant: Abnormal nuclear behavior and defective microtubule organization. *Plant Cell, 8*, 1391–1407.

Huang, J., Zhao, L., Malik, S., et al. (2022). Specification of female germline by microRNA orchestrated auxin signaling in *Arabidopsis*. *Nature Communications, 13*, 6960.

Huck, N., Moore, J. M., Federer, M., & Grossniklaus, U. (2003). The *Arabidopsis* mutant *feronia* disrupts the female gametophytic control of pollen tube reception. *Development, 130*, 2149–2159.

Ingouff, M., Sakata, T., Li, J., Sprunck, S., Dresselhaus, T., & Berger, F. (2009). The two male gametes share equal ability to fertilize the egg cell in *Arabidopsis thaliana*. *Current Biology, 19*, R19–R20.

Ito, T., Wellmer, F., Yu, H., et al. (2004). The homeotic protein AGAMOUS controls microsporogenesis by regulation of SPOROCYTELESS. *Nature, 430*, 356–360.

Jia, G., Liu, X., Owen, H. A., et al. (2008). Signaling of cell fate determination by the TPD1 small protein and EMS1 receptor kinase. *Proceedings of the National Academy of Sciences, 105*, 2220–2225.

Jiang, T., & Zheng, B. (2022). Epigenetic regulation of megaspore mother cell formation. *Frontiers in Plant Science, 12*, 826871.

Johnston, A. J., Matveeva, E., Kirioukhova, O., Grossniklaus, U., & Gruissem, W. (2008). A dynamic reciprocal RBR-PRC2 regulatory circuit controls *Arabidopsis* gametophyte development. *Current Biology, 18*, 1680–1686.

Johnston, A. J., Kirioukhova, O., Barrell, P. J., et al. (2010). Dosage-sensitive function of retinoblastoma related and convergent epigenetic control are required during the *Arabidopsis* life cycle. *PLoS Genetics, 6*, e1000988.

Jones-Rhoades, M. W., Borevitz, J. O., & Preuss, D. (2007). Genome-wide expression profiling of the *Arabidopsis* female gametophyte identifies families of small, secreted proteins. *PLoS Genetics, 3*, 1848–1861.

Jullien, P. E., Katz, A., Oliva, M., Ohad, N., & Berger, F. (2006). Polycomb group complexes self-regulate imprinting of the polycomb group gene *MEDEA* in *Arabidopsis*. *Current Biology, 16*, 486–492.

Kägi, C., Stierhof, Y. D., & Groß-Hardt, R. (2010). The gametic central cell of *Arabidopsis* determines the lifespan of adjacent accessory cells. *Proceedings of the National Academy of Sciences, 107*, 22350–22355.

Kamiuchi, Y., Yamamoto, K., Furutani, M., Tasaka, M., & Aida, M. (2014). The *CUC1* and *CUC2* genes promote carpel margin meristem formation during *Arabidopsis* gynoecium development. *Frontiers in Plant Science, 5*, 165.

Kammel, C., Thomaier, M., Sorensen, B. B., et al. (2013). *Arabidopsis* DEAD-Box RNA helicase UAP56 interacts with both RNA and DNA as well as with mRNA export factors. *PLoS ONE, 8*, e60644.

Kasahara, R. D., Portereiko, M. F., Sandaklie-Nikolova, L., Rabiger, D. S., & Drews, G. N. (2005). MYB98 is required for pollen tube guidance and synergid cell differentiation in *Arabidopsis*. *Plant Cell, 17*, 2981–2992.

Kessler, S. A., & Grossniklaus, U. (2011). She's the boss: Signaling in pollen tube reception. *Current Opinion in Plant Biology, 14*, 622–627.

Kirioukhova, O., Johnston, A. J., Kleen, D., et al. (2011). Female gametophytic cell specification and seed development require the function of the putative *Arabidopsis* INCENP ortholog WYRD. *Development, 138*, 3409–3420.

Köhler, C., Hennig, L., Bouveret, R., Gheyselinck, J., Grossniklaus, U., & Gruissem, W. (2003a). *Arabidopsis* MSI1 is a component of the MEA/FIE Polycomb group complex and required for seed development. *The EMBO Journal, 22*, 4804–4814.

Köhler, C., Hennig, L., Spillane, C., Pien, S., Gruissem, W., & Grossniklaus, U. (2003b). The Polycomb-group protein MEDEA regulates seed development by controlling expression of the MADS-box gene PHERES1. *Genes & Development, 17*, 1540–1553.

Koi, S., Hisanaga, T., Sato, K., et al. (2016). An evolutionarily conserved plant RKD factor controls germ cell differentiation. *Current Biology, 26*, 1775–1781.

Kong, J., Lau, S., & Jürgens, G. (2015). Twin plants from supernumerary egg cells in *Arabidopsis*. *Current Biology, 25*, 225–230.

Koszegi, D., Johnston, A. J., Rutten, T., et al. (2011). Members of the RKD transcription factor family induce an egg cell-like gene expression program. *The Plant Journal, 67*, 280–291.

Kotake, Y., Cao, R., Viatour, P., Sage, J., Zhang, Y., & Xiong, Y. (2007). pRB family proteins are required for H3K27 trimethylation and Polycomb repression complexes binding to and silencing p16INK4alpha tumor suppressor gene. *Genes & Development, 21*, 49–54.

Krohn, N. G., Lausser, A., Juranic, M., & Dresselhaus, T. (2012). Egg cell signaling by the secreted peptide ZmEAL1 controls antipodal cell fate. *Developmental Cell, 23*, 219–225.

Kumakura, N., Takeda, A., Fujioka, Y., Motose, H., Takano, R., & Watanabe, Y. (2006). SGS3 and RDR6 interact and colocalize in cytoplasmic SGS3/RDR6-bodies. *FEBS Letters, 583*, 1261–1266.

Kuwabara, A., & Gruissem, W. (2014). *Arabidopsis* Retinoblastoma-related and Polycomb group proteins: Cooperation during plant cell differentiation and development. *Journal of Experimental Botany, 65*, 2667–2676.

Leshem, Y., Johnson, C., Wuest, S. E., et al. (2012). Molecular characterization of the *glauce* mutant: A central cell-specific function is required for double fertilization in *Arabidopsis*. *Plant Cell, 24*, 3264–3277.

Leshem, Y., Johnson, C., & Sundaresan, V. (2013). Pollen tube entry into the synergid cell of *Arabidopsis* is observed at a site distinct from the filiform apparatus. *Plant Reproduction, 26*, 93–99.

Leszczuk, A., Szczuka, E., & Zdunek, A. (2019). Arabinogalactan proteins: Distribution during the development of male and female gametophytes. *Plant Physiology and Biochemistry, 135*, 9–18.

Li, H. J., & Yang, W. C. (2020). Central cell in flowering plants: Specification, signaling, and evolution. *Frontiers in Plant Science, 11*, 590307.

Li, L. C., Qin, G. J., Tsuge, T., et al. (2008). *SPOROCYTELESS* modulates *YUCCA* expression to regulate the development of lateral organs in *Arabidopsis*. *The New Phytologist, 179*, 751–764.

Li, N., Yuan, L., Liu, N., et al. (2009). SLOW WALKER2, a NOC1/MAK21 homologue, is essential for coordinated cell cycle progression during female gametophyte development in *Arabidopsis*. *Plant Physiology, 151*, 1486–1497.

Li, H. J., Zhu, S. S., Zhang, M. X., et al. (2015). *Arabidopsis* CBP1 is a novel regulator of transcription initiation in central cell-mediated pollen tube guidance. *Plant Cell, 27*, 2880–2893.

Liang, X., Li, S. W., Wang, J. L., Zhao, H. M., Li, S., & Zhang, Y. (2023). *Arabidopsis* Sar1 isoforms play redundant roles in female gametophytic development. *Plant Reproduction, 36*, 349–354.

Lieber, D., Lora, J., Schrempp, S., Lenhard, M., & Laux, T. (2011). *Arabidopsis WIH1* and *WIH2* genes act in the transition from somatic to reproductive cell fate. *Current Biology, 21*, 1009–1017.

Lituiev, D. S., Krohn, N. G., Muller, B., et al. (2013). Theoretical and experimental evidence indicates that there is no detectable auxin gradient in the angiosperm female gametophyte. *Development, 140*, 4544–4553.

Liu, M., Yuan, L., Liu, N. Y., Shi, D. Q., Liu, J., & Yang, W. C. (2009). *GAMETOPHYTIC FACTOR 1*, involved in pre-mRNA splicing, is essential for megagametogenesis and embryogenesis in *Arabidopsis*. *Journal of Integrative Plant Biology, 51*, 261–2711.

Liu, M., Shi, D. Q., Yuan, L., Liu, J., & Yang, W. C. (2010). *SLOW WALKER3*, encoding a putative DEAD-box RNA helicase, is essential for female gametogenesis in *Arabidopsis*. *Journal of Integrative Plant Biology, 52*, 817–828.

Liu, Z., Yuan, L., Song, X., Yu, X., & Sundaresan, V. (2017). AHP2, AHP3, and AHP5 act downstream of CKI1 in *Arabidopsis* female gametophyte development. *Journal of Experimental Botany, 68*, 3365–3373.

Liu, Z., Yuan, L., Sundaresan, V., & Yu, X. (2018). *Arabidopsis* CKI1 mediated two-component signaling in the specification of female gametophyte. *Plant Signaling & Behavior, 13*, e1469360.

Losada, J. M., Bachelier, J. B., & Friedman, W. E. (2017). Prolonged embryogenesis in *Austrobaileya scandens* (Austrobaileyaceae): Its ecological and evolutionary significance. *The New Phytologist, 215*, 851–864.

Ma, H. (2005). Molecular genetic analyses of microsporogenesis and microgametogenesis in flowering plants. *Annual Review of Plant Biology, 56*, 393–434.

Makarevich, G., Leroy, O., Akinci, U., et al. (2006). Different Polycomb group complexes regulate common target genes in *Arabidopsis*. *EMBO Reports, 7*, 947–952.

Makarevich, G., Villar, C. B. R., Erilova, A., & Kohler, C. (2008). Mechanism of PHERES1 imprinting in *Arabidopsis*. *Journal of Cell Science, 121*, 906–912.

Makarevitch, I., Stupar, R. M., Iniguez, A. L., et al. (2007). Natural variation for alleles under epigenetic control by the maize chromomethylase zmet2. *Genetics, 177*, 749–760.

Mansfield, S. G., Briarty, L. G., & Erni, S. (1991). Early embryogenesis in *Arabidopsis Thaliana*. 1. The mature embryo sac. *Canadian Journal of Botany, 69*, 447–460.

Marsch-Martinez, N., & de Folter, S. (2016). Hormonal control of the development of the gynoecium. *Current Opinion in Plant Biology, 29*, 104–114.

Márton, M. L., Cordts, S., Broadhvest, J., & Dresselhaus, T. (2005). Micropylar pollen tube guidance by egg apparatus 1 of maize. *Science, 307*, 573–576.

Márton, M. L., Fastner, A., Uebler, S., & Dresselhaus, T. (2012). Overcoming hybridization barriers by the secretion of the maize pollen tube attractant ZmEA1 from *Arabidopsis* ovules. *Current Biology, 22*, 1194–1198.

Maruyama, D., Endo, T., & Nishikawa, S. (2010). BiP-mediated polar nuclei fusion is essential for the regulation of endosperm nuclei proliferation in *Arabidopsis thaliana*. *Proceedings of the National Academy of Sciences USA, 107*, 1684–1689.

Masuda, S., Das, R., Cheng, H., Hurt, E., Dorman, N., & Reed, R. (2005). Recruitment of the human TREX complex to mRNA during splicing. *Genes & Development, 19*, 1512–1517.

Medina, F. J., Risueno, M. C., & Rodriguez-Garsia, M. I. (1981). Evolution of the cytoplasmic organelles during female meiosis in *Pisum sativum* L. *Planta, 151*, 215–225.

Mendes, M. A., Petrella, R., Cucinotta, M., et al. (2020). The RNA-dependent DNA methylation pathway is required to restrict *SPOROCYTELESS/NOZZLE* expression to specify a single female germ cell precursor in *Arabidopsis*. *Development, 147*, dev194274.

Mercier, R., Vezon, D., Bullier, E., Motamayor, J. C., Sellier, A., Lefèvre, F., Pelletier, G., & Horlow, C. (2001). SWITCH1 (SWI1): A novel protein required for the establishment of sister chromatid cohesion and for bivalent formation at meiosis. *Genes & Development, 15*, 1859–1871.

Mercier, R., Armstrong, S. J., Horlow, C., Jackson, N., Makaroff, C. A., Vezon, D., Pelletier, G., Jones, G. H., & Franklin, F. C. H. (2003). The meiotic protein SWI1 is required for axial element formation and recombination initiation in *Arabidopsis*. *Development, 130*, 3297–3308.

Mironova, V., Teale, W., Shahriari, M., Dawson, J., & Palme, K. (2017). The systems biology of auxin in developing embryos. *Trends in Plant Science, 22*, 225–235.

Missbach, S., Weis, B. L., Martin, R., Simm, S., Bohnsack, M. T., & Schleiff, E. (2013). 40S ribosome biogenesis co-factors are essential for gametophyte and embryo development. *PLoS ONE, 8*, e54084.

Mizuguchi, G., Shen, X., Landry, J., Wu, W. H., Sen, S., & Wu, C. (2004). ATP-driven exchange of histone H2AZ variant catalyzed by SWR1 chromatin remodeling complex. *Science, 303*, 343–348.

Moll, C., von Lyncker, L., Zimmermann, S., et al. (2008). CLO/GFA1 and ATO are novel regulators of gametic cell fate in plants. *The Plant Journal, 56*, 913–921.

Motamayor, J. C., Vezon, D., Bajon, C., et al. (2000). Switch (swi1), an *Arabidopsis thaliana* mutant affected in the female meiotic switch. *Sexual Plant Reproduction, 12*, 209–218.

Murgia, M., Huang, B. Q., Tucker, S. C., & Musgrave, M. E. (1993). Embryo sac lacking antipodal cells in *Arabidopsis thaliana* (Brassicaceae). *American Journal of Botany, 80*, 824–838.

Nagahara, S., Takeuchi, H., & Higashiyama, T. (2021). Polyspermy block in the central cell during double fertilization of *Arabidopsis thaliana. Frontiers in Plant Science, 11*, 588700.

Nan, G. L., Ronceret, A., Wang, R. C., Fernandes, J. F., Cande, W. Z., & Walbot, V. (2011). Global transcriptome analysis of two *ameiotic1* alleles in maize anthers: Defining steps in meiotic entry and progression through prophase I. *BMC Plant Biology, 11*, 120.

Nishikawa, S. I., Yamaguchi, Y., Suzuki, C., et al. (2020). *Arabidopsis* GEX1 is a nuclear membrane protein of gametes required for nuclear fusion during reproduction. *Frontiers in Plant Science, 11*, 548032.

Nonomura, K. I., Miyoshi, K., Eiguchi, M., et al. (2003). The *MSP1* gene is necessary to restrict the number of cells entering into male and female sporogenesis and to initiate anther wall formation in rice. *Plant Cell, 15*, 1728–1739.

Nonomura, K. I., Morohoshi, A., Nakano, M., et al. (2007). A germ cell-specific gene of the *ARGONAUTE* family is essential for the progression of premeiotic mitosis and meiosis during sporogenesis in rice. *Plant Cell, 19*, 2583–2594.

Normington, K., Kohno, K., Kozutsumi, Y., Gething, M. J., & Sambrook, J. S. (1989). *S. cerevisiae* encodes an essential protein homologous in sequence and function to mammalian BiP. *Cell, 57*, 1223–1236.

Ohad, N., Margossian, L., Hsu, Y. C., Williams, C., Repetti, P., & Fischer, R. L. (1996). A mutation that allows endosperm development without fertilization. *Proceedings of the National Academy of Sciences USA, 93*, 5319–5324.

Ohad, N., Yadegari, R., Margossian, L., et al. (1999). Mutations in FIE, a WD polycomb group gene, allow endosperm development without fertilization. *Plant Cell, 11*, 407–415.

Okuda, S., Tsutsui, H., Shiina, K., et al. (2009). Defensin-like polypeptide LUREs are pollen tube attractants secreted from synergid cells. *Nature, 458*, 357–361.

Oliver, C., Santos, J. L., & Pradillo, M. (2014). On the role of some ARGONAUTE proteins in meiosis and DNA repair in *Arabidopsis thaliana. Frontiers in Plant Science, 5*, 177.

Olmedo-Monfil, V., Duran-Figueroa, N., Arteaga-Vazquez, M., et al. (2010). Control of female gamete formation by a small RNA pathway in *Arabidopsis. Nature, 464*, 628–632.

Osnato, M. (2022). CKI1-ARRs specify the central cell in *Arabidopsis. Plant Cell, 34*, 4665–4666.

Ouedraogo, I., Lartaud, M., Baroux, C., et al. (2023). 3D cellular morphometrics of ovule primordium development in *Zea mays* reveal differential division and growth dynamics specifying megaspore mother cell singleness. *Frontiers in Plant Science, 14*, 1174171.

Pagnussat, G. C., Yu, H. J., & Sundaresan, V. (2007). Cell-fate switch of synergid to egg cell in *Arabidopsis eostre* mutant embryo sacs arises from misexpression of the BEL1-like homeodomain gene BLH1. *Plant Cell, 19*, 3578–3592.

Pagnussat, G. C., Alandete-Saez, M., Bowman, J. L., & Sundaresan, V. (2009). Auxin-dependent patterning and gamete specification in the *Arabidopsis* female gametophyte. *Science, 324*, 1684–1689.

Palmer, R. G. (1971). Cytological studies of ameiotic and normal maize with reference to premeiotic pairing. *Chromosoma, 35*, 233–246.

Panoli, A., Martin, M. V., Alandete-Saez, M., et al. (2015). Auxin import and local auxin biosynthesis are required for mitotic divisions, cell expansion and cell specification during female gametophyte development in *Arabidopsis thaliana. PLoS ONE, 10*, e0126164.

Papa, C. M., Springer, N. M., Muszynski, M. G., Meeley, R., & Kaeppler, S. M. (2001). Maize chromomethylase Zea methyltransferase2 is required for CpNpG methylation. *Plant Cell, 13*, 1919–1928.

Papini, A., Mosti, S., Milocani, E., Tani, G., Di Falco, P., & Brighigna, L. (2011). Megasporogenesis and programmed cell death in *Tillandsia* (Bromeliaceae). *Protoplasma, 248*, 651–662.

Pawlowski, W. P., Wang, C. J., Golubovskaya, I. N., et al. (2009). Maize AMEIOTIC1 is essential for multiple early meiotic processes and likely required for the initiation of meiosis. *Proceedings of the National Academy of Sciences USA, 106*, 3603–3608.

Pérez-España, V. H., Sánchez-León, N., & Vielle-Calzada, J.-P. (2011). CYP85A1 is required for the initiation of female gametogenesis in *Arabidopsis thaliana*. *Plant Signaling & Behavior, 6*, 321–326.

Pillot, M., Baroux, C., Vazquez, M. A., et al. (2010). Embryo and endosperm inherit distinct chromatin and transcriptional states from the female gametes in *Arabidopsis*. *Plant Cell, 22*, 307–320.

Pirselova, B., & Matusikova, I. (2013). Callose: The plant cell wall polysaccharide with multiple biological functions. *Acta Physiologiae Plantarum, 35*, 635–644.

Portereiko, M. F., Lloyd, A., Steffen, J. G., Punwani, J. A., Otsuga, D., & Drews, G. N. (2006a). *AGL80* is required for central cell and endosperm development in *Arabidopsis*. *Plant Cell, 18*, 1862–1872.

Portereiko, M. F., Sandaklie-Nikolova, L., Lloyd, A., Dever, C. A., Otsuga, D., & Drews, G. N. (2006b). Nuclear fusion defective1 encodes the *Arabidopsis* RPL21M protein and is required for karyogamy during female gametophyte development and fertilization. *Plant Physiology, 141*, 957–965.

Povilus, R. A., Losada, J. M., & Friedman, W. E. (2015). Floral biology and ovule and seed ontogeny of *Nymphaea thermarum*, a water lily at the brink of extinction with potential as a model system for basal angiosperms. *Annals of Botany, 115*, 211–226.

Pratibha, P., Singh, S. K., Srinivasan, R., Bhat, S. R., & Sreenivasulu, Y. (2017). Gametophyte development needs mitochondrial coproporphyrinogen III oxidase function. *Plant Physiology, 174*, 258–275.

Punwani, J. A., Rabiger, D. S., & Drews, G. N. (2007). MYB98 positively regulates a battery of synergid-expressed genes encoding filiform apparatus-localized proteins. *Plant Cell, 19*, 2557–2568.

Punwani, J. A., Rabiger, D. S., Lloyd, A., & Drews, G. N. (2008). The *MYB98* subcircuit of the synergid gene regulatory network includes genes directly and indirectly regulated by MYB98. *The Plant Journal, 55*, 406–414.

Qin, Y., Zhao, L., Skaggs, M. I., et al. (2014). ACTIN-RELATED PROTEIN6 regulates female meiosis by modulating meiotic gene expression in *Arabidopsis*. *Plant Cell, 26*, 1612–1628.

Rabiger, D. S., & Drews, G. N. (2013). *MYB64* and *MYB119* are required for cellularization and differentiation during female gametogenesis in *Arabidopsis thaliana*. *PLoS Genetics, 9*, e1003783.

Rafinska, K., & Bednarska, E. (2011). Localisation pattern of homogalacturonan and arabinogalactan proteins in developing ovules of the gymnosperm plant *Larix decidua* Mill. *Sexual Plant Reproduction, 24*, 75–87.

Ravi, M., Marimuthu, M. P., & Siddiqi, I. (2008). Gamete formation without meiosis in *Arabidopsis*. *Nature, 451*, 1121–1124.

Riefler, M., Novak, O., Strnad, M., & Schmulling, T. (2006). *Arabidopsis* cytokinin receptor mutants reveal functions in shoot growth, leaf senescence, seed size, germination, root development, and cytokinin metabolism. *Plant Cell, 18*, 40–54.

Rocha, D. M., Andrade, C. G. T. J., & Vanzela, A. L. L. (2014). Female gametophyte development in Eleocharis sellowiana (CYPERACEAE): A species with atypical nuclear staining and polarity of the megaspore mother cell. *Plant Systematics and Evolution, 301*, 1287–1292.

Rodkiewicz, B. (1970). Callose in cell walls during megasporogenesis in angiosperms. *Planta, 93*, 39–47.

Rodriguez-Leal, D., Leon-Martinez, G., Abad-Vivero, U., & Vielle-Calzada, J. P. (2015). Natural variation in epigenetic pathways affects the specification of female gamete precursors in *Arabidopsis*. *Plant Cell, 27*, 1034–1045.

Rodriguez-Leal, D., Castillo-Cobian, A., Rodriguez-Arevalo, I., & Vielle-Calzada, J. P. (2016). A primary sequence analysis of the ARGONAUTE protein family in plants. *Frontiers in Plant Science, 7*, 1347.

Rose, M. D. (1996). Nuclear fusion in the yeast *Saccharomyces cerevisiae*. *Annual Review of Cell and Developmental Biology, 12*, 663–695.

Rose, M. D., Misra, L. M., & Vogel, J. P. (1989). Kar2, a karyogamy gene, is the yeast homolog of the mammalian Bip/Grp78 gene. *Cell, 57*, 1211–1221.

Rotman, N., Rozier, F., Boavida, L., Dumas, C., Berger, F., & Faure, J. E. (2003). Female control of male gamete delivery during fertilization in *Arabidopsis thaliana*. *Current Biology, 13*, 432–436.

Rovekamp, M., Bowman, J. L., & Grossniklaus, U. (2016). Marchantia MpRKD regulates the gametophyte-sporophyte transition by keeping egg cells quiescent in the absence of fertilization. *Current Biology, 26*, 1782–1789.

Rudall, P. J., Remizowa, M. V., Beer, A. S., et al. (2008). Comparative ovule and megagametophyte development in hydatellaceae and water lilies reveal a mosaic of features among the earliest angiosperms. *Annals of Botany, 101*, 941–956.

Russell, S. D. (1992). Double fertilization. *International Review of Cytology, 140*, 357–387.

Schiefthaler, U., Balasubramanian, S., Sieber, P., Chevalier, D., Wisman, E., & Schneitz, K. (1999). Molecular analysis of NOZZLE, a gene involved in pattern formation and early sporogenesis during sex organ development in *Arabidopsis thaliana*. *Proceedings of the National Academy of Sciences USA, 96*, 11664–11669.

Schmidt, A., Wuest, S. E., Vijverberg, K., Baroux, C., Kleen, D., & Grossniklaus, U. (2011). Transcriptome analysis of the *Arabidopsis* megaspore mother cell uncovers the importance of RNA helicases for plant germline development. *PLoS Biology, 9*, e1001155.

Schmidt, A., Wohrmann, H. J. P., Raissig, M. T., et al. (2013). The Polycomb group protein MEDEA and the DNA methyltransferase MET1 interact to repress autonomous endosperm development in *Arabidopsis*. *The Plant Journal, 73*, 776–787.

Schmidt, A., Schmid, M. W., & Grossniklaus, U. (2015). Plant germline formation: Common concepts and developmental flexibility in sexual and asexual reproduction. *Development, 142*, 229–241.

Schneitz, K., Hulskamp, M., & Pruitt, R. E. (1995). Wild-type ovule development in *Arabidopsis thaliana*—A light-microscope study of cleared whole-mount tissue. *The Plant Journal, 7*, 731–749.

She, W., Grimanelli, D., Rutowicz, K., et al. (2013). Chromatin reprogramming during the somatic-to-reproductive cell fate transition in plants. *Development, 140*, 4008–4019.

Sheridan, W. F., Avalkina, N. A., Shamrov, I. I., Batygina, T. B., & Golubovskaya, I. N. (1996). The *mac1* gene: Controlling the commitment to the meiotic pathway in maize. *Genetics, 142*, 1009–1020.

Sheridan, W. F., Golubeva, E. A., Abrhamova, L. I., & Golubovskaya, I. N. (1999). The *mac1* mutation alters the developmental fate of the hypodermal cells and their cellular progeny in the maize anther. *Genetics, 153*, 933–941.

Shi, D. Q., & Yang, W. C. (2011). Ovule development in *Arabidopsis*: Progress and challenge. *Current Opinion in Plant Biology, 14*, 74–80.

Shi, D. Q., Liu, J., Xiang, Y. H., Ye, D., Sundaresan, V., & Yang, W. C. (2005). *SLOW WALKER1*, essential for gametogenesis in *Arabidopsis*, encodes a WD40 protein involved in 18S ribosomal RNA biogenesis. *Plant Cell, 17*, 2340–2354.

Shimizu, K. K., & Okada, K. (2000). Attractive and repulsive interactions between female and male gametophytes in *Arabidopsis* pollen tube guidance. *Development, 127*, 4511–4518.

Shimizu, K. K., Ito, T., Ishiguro, S., & Okada, K. (2008). *MAA3* (*MAGATAMA3*) helicase gene is required for female gametophyte development and pollen tube guidance in *Arabidopsis thaliana*. *Plant and Cell Physiology, 49*, 1478–1483.
Showalter, A. M. (2001). Arabinogalactan-proteins: Structure, expression and function. *Cellular and Molecular Life Sciences, 58*, 1399–1417.
Siddiqi, I., Ganesh, G., Grossniklaus, U., & Subbiah, V. (2000). The *dyad* gene is required for progression through female meiosis in *Arabidopsis*. *Development, 127*, 197–207.
Singh, M., Goel, S., Meeley, R. B., et al. (2011). Production of viable gametes without meiosis in maize deficient for an ARGONAUTE protein. *Plant Cell, 23*, 443–458.
Skinner, D. J., & Sundaresan, V. (2018). Recent advances in understanding female gametophyte development. *F1000Research, 7*, F1000 Faculty Rev-804.
Skinner, D. J., Hill, T. A., & Gasser, C. S. (2004). Regulation of ovule development. *Plant Cell, 16*, S32–S45.
Soltis, D. E., Soltis, P. S., Chase, M. W., et al. (2000). Angiosperm phylogeny inferred from 18S rDNA, *rbcL*, and *atpB* sequences. *Botanical Journal of the Linnean Society, 133*, 381–461.
Song, X., Yuan, L., & Sundaresan, V. (2014). Antipodal cells persist through fertilization in the female gametophyte of *Arabidopsis*. *Plant Reproduction, 27*, 197–203.
Sprunck, S., & Gross-Hardt, R. (2011). Nuclear behavior, cell polarity, and cell specification in the female gametophyte. *Sexual Plant Reproduction, 24*, 123–136.
Steffen, J. G., Kang, I. H., Portereiko, M. F., Lloyd, A., & Drews, G. N. (2008). AGL61 interacts with AGL80 and is required for central cell development in Arabidopsis. *Plant Physiology, 148*, 259–268.
Strässer, K., Masuda, S., Mason, P., et al. (2002). TREX is a conserved complex coupling transcription with messenger RNA export. *Nature, 417*, 304–308.
Su, Z. X., Zhao, L. H., Zhao, Y. Y., et al. (2017). The THO complex non-cell-autonomously represses female germline specification through the TAS3-ARF3 module. *Current Biology, 27*, 1597–1609.
Su, Z., Wang, N., & Hou, Z. (2020). Regulation of female germline specification via small RNA mobility in *Arabidopsis*. *Plant Cell, 32*, 2842–2854.
Sun, Y., Wang, X., Pan, L., et al. (2021). Plant egg cell fate determination depends on its exact position in female gametophyte. *Proceedings of the National Academy of Sciences USA, 118*, e2017488118.
Sundaresan, V., & Alandete-Saez, M. (2010). Pattern formation in miniature: The female gametophyte of flowering plants. *Development, 137*, 179–189.
Susaki, D., Suzuki, T., Maruyama, D., Ueda, M., Higashiyama, T., & Kurihara, D. (2021). Dynamics of the cell fate specifications during female gametophyte development in *Arabidopsis*. *PLoS Biology, 19*, e3001123.
Takeuchi, H., & Higashiyama, T. (2012). A species-specific cluster of defensin-like genes encodes diffusible pollen tube attractants in *Arabidopsis*. *PLoS Biology, 10*, e1001449.
Talluri, S., & Dick, F. A. (2012). Regulation of transcription and chromatin structure by pRB: Here, there and everywhere. *Cell Cycle, 11*, 3189–3198.
Tekleyohans, D. G., Nakel, T., & Gross-Hardt, R. (2017). Patterning the female gametophyte of flowering plants. *Plant Physiology, 173*, 122–129.
Thiele, K., Wanner, G., Kindzierski, V., et al. (2009). The timely deposition of callose is essential for cytokinesis in *Arabidopsis*. *The Plant Journal, 58*, 13–26.
Tobe, H., Jaffre, T., & Raven, P. H. (2000). Embryology of Amborella (*Amborellaceae*): Descriptions and polarity of character states. *Journal of Plant Research, 113*, 271–280.
Tucker, M. R., Okada, T., Hu, Y., Scholefield, A., Taylor, J. M., & Koltunow, A. M. (2012). Somatic small RNA pathways promote the mitotic events of megagametogenesis during female reproductive development in *Arabidopsis*. *Development, 139*, 1399–1404.
Uebler, S., Márton, M. L., & Dresselhaus, T. (2015). Classification of EA1-box proteins and new insights into their role during reproduction in grasses. *Plant Reproduction, 28*, 183–197.

Underwood, W. (2012). The plant cell wall: A dynamic barrier against pathogen invasion. *Frontiers in Plant Science, 3*, 85.

Varapparambath, V., Mathew, M. M., Shanmukhan, A. P., et al. (2022). Mechanical conflict caused by a cell-wall-loosening enzyme activates de novo shoot regeneration. *Developmental Cell, 57*, 2063–2080.

Vaucheret, H. (2008). Plant ARGONAUTES. *Trends in Plant Science, 13*, 350–358.

Villar, C. B. R., Erilova, A., Makarevich, G., Trosch, R., & Kohler, C. (2009). Control of *PHERES1* imprinting in *Arabidopsis* by direct tandem repeats. *Molecular Plant, 2*, 654–660.

Völz, R., von Lyncker, L., Baumann, N., Dresselhaus, T., Sprunck, S., & Gross-Hardt, R. (2012). *LACHESIS*-dependent egg-cell signaling regulates the development of female gametophytic cells. *Development, 139*, 498–502.

Wang, Y., & Ma, H. (2011). Development: A pathway to plant female germ cells. *Current Biology, 21*, R477.

Wang, D., Zhang, C. Q., Hearn, D. J., et al. (2010). Identification of transcription-factor genes expressed in the *Arabidopsis* female gametophyte. *BMC Plant Biology, 10*, 110.

Wang, C. J. R., Nan, G. L., Kelliher, T., et al. (2012). Maize *multiple archesporial cells 1* (*mac1*), an ortholog of rice *TDL1A*, modulates cell proliferation and identity in early anther development. *Development, 139*, 2594–2603.

Webb, M. C., & Gunning, B. E. S. (1990). Embryo sac development in *Arabidopsis thaliana* megasporogenesis, including the microtubular cytoskeleton. *Sexual Plant Reproduction, 3*, 244–256.

Wei, B., Zhang, J., Pang, C., et al. (2015). The molecular mechanism of SPOROCYTELESS/NOZZLE in controlling *Arabidopsis* ovule development. *Cell Research, 25*, 121–134.

Weijers, D., Schlereth, A., Ehrismann, J. S., Schwank, G., Kientz, M., & Jurgens, G. (2006). Auxin triggers transient local signaling for cell specification in *Arabidopsis* embryogenesis. *Developmental Cell, 10*, 265–270.

Williams, J. H., & Friedman, W. E. (2002). Identification of diploid endosperm in an early angiosperm lineage. *Nature, 415*, 522–526.

Williams, J. H., & Friedman, W. E. (2004). The four-celled female gametophyte of Illicium (*Illiciaceae*; *Austrobaileyales*): Implications for understanding the origin and early evolution of monocots, eumagnoliids, and eudicots. *American Journal of Botany, 91*, 332–351.

Wuest, S. E., Vijverberg, K., Schmidt, A., et al. (2010). *Arabidopsis* female gametophyte gene expression map reveals similarities between plant and animal gametes. *Current Biology, 20*, 506–512.

Yadegari, R., & Drews, G. N. (2004). Female gametophyte development. *Plant Cell, 16*, S133–S141.

Yang, W. C., Ye, D., Xu, J., & Sundaresan, V. (1999). The *SPOROCYTELESS* gene of *Arabidopsis* is required for initiation of sporogenesis and encodes a novel nuclear protein. *Genes & Development, 13*, 2108–2117.

Yang, S. L., Xie, L. F., Mao, H. Z., et al. (2003). Tapetum determinant1 is required for cell specialization in the *Arabidopsis* anther. *Plant Cell, 15*, 2792–2804.

Yang, W. C., Shi, D. Q., & Chen, Y. H. (2010). Female gametophyte development in flowering plants. *Annual Review of Plant Biology, 61*, 89–108.

Yelina, N. E., Smith, L. M., Jones, A. M. E., Patel, K., Kelly, K. A., & Baulcombe, D. C. (2010). Putative *Arabidopsis* THO/TREX mRNA export complex is involved in transgene and endogenous siRNA biosynthesis. *Proceedings of the National Academy of Sciences USA, 107*(13948–13), 953.

Yu, Y., Ji, L., Le, B. H., et al. (2017). ARGONAUTE10 promotes the degradation of miR165/6 through the SDN1 and SDN2 exonucleases in *Arabidopsis*. *PLoS Biology, 15*, e2001272.

Yuan, L., Yang, X. H., Ellis, J. L., Fisher, N. M., & Makaroff, C. A. (2012). The *Arabidopsis* SYN3 cohesin protein is important for early meiotic events. *The Plant Journal, 71*, 147–160.

Yuan, L., Liu, Z. N., Song, X. Y., Johnson, C., Yu, X. L., & Sundaresan, V. (2016). The CKI1 histidine kinase specifies the female gametic precursor of the endosperm. *Developmental Cell, 37*, 34–46.

Zanis, M. J., Soltis, D. E., Soltis, P. S., Mathews, S., & Donoghue, M. J. (2002). The root of the angiosperms revisited. *Proceedings of the National Academy of Sciences USA, 99*, 6848–6853.

Zhang, D., & Yang, L. (2014). Specification of tapetum and microsporocyte cells within the anther. *Current Opinion in Plant Biology, 17*, 49–55.

Zhang, H. S., Gavin, M., Dahiya, A., et al. (2000). Exit from G1 and S phase of the cell cycle is regulated by repressor complexes containing HDAC-Rb-hSWI/SNF and Rb-hSWI/SNF. *Cell, 101*, 79–89.

Zhang, M. X., Zhu, S. S., Xu, Y. C., Guo, Y. L., Yang, W. C., & Li, H. J. (2020). Transcriptional repression specifies the central cell for double fertilization. *Proceedings of the National Academy of Sciences USA, 117*, 6231–6236.

Zhao, D. Z., Wang, G. F., Speal, B., & Ma, H. (2002). The *EXCESS MICROSPOROCYTES1* gene encodes a putative leucine-rich repeat receptor protein kinase that controls somatic and reproductive cell fates in the *Arabidopsis* anther. *Genes & Development, 16*, 2021–2031.

Zhao, X., de Palma, J., Oane, R., et al. (2008). OsTDL1A binds to the LRR domain of rice receptor kinase MSP1, and is required to limit sporocyte numbers. *The Plant Journal, 54*, 375–387.

Zhao, L., He, J., Cai, H., et al. (2014). Comparative expression profiling reveals gene functions in female meiosis and gametophyte development in *Arabidopsis*. *The Plant Journal, 80*, 615–628.

Zhao, F., Zheng, Y. F., Zeng, T., et al. (2017a). Phosphorylation of SPOROCYTELESS/NOZZLE by the MPK3/6 kinase is required for anther development. *Plant Physiology, 173*, 2265–2277.

Zhao, X., Bramsiepe, J., Van Durme, M., et al. (2017b). RETINOBLASTOMA RELATED1 mediates germline entry in *Arabidopsis*. *Science, 356*, 378–385.

Zhao, L., Cai, H., Su, Z., et al. (2018). *KLU* suppresses megasporocyte cell fate through SWR1-mediated activation of *WRKY28* expression in *Arabidopsis*. *Proceedings of the National Academy of Sciences USA, 115*, E526–E535.

Zhao, H., Guo, M., Yan, M., Cheng, H., et al. (2020). Comparative expression profiling reveals genes involved in megasporogenesis. *Plant Physiology, 182*, 2006–2024.

Zhu, M., Tao, L., Zhang, J., et al. (2022). The type-B response regulators ARR10, ARR12, and ARR18 specify the central cell in *Arabidopsis*. *Plant Cell, 34*, 4714–4737.

Zúñiga-Mayo, V. M., Durán-Medina, Y., Marsch-Martínez, N., & de Folter, S. (2023). Hormones and flower development in *Arabidopsis*. *Methods in Molecular Biology, 2686*, 111–127.

Chapter 14
Pollination and Fertilization

Sheng Zhong, Zijun Lan, Zengxiang Ge, and Li-Jia Qu

Abstract In order to cope with arid terrestrial environments, angiosperms have evolved a unique fertilization way—siphonogamy. The success of siphonogamy requires several prerequisites, including the normal development of the female gametophyte and male gametophyte (pollen). Appropriate pollination methods ensure the successful encounter between pollen and the stigma. After pollen lands on the stigma, pollen/pollen tube interacts with different tissues and cells of the pistil, completing a series of male-female communications. The smooth progress of these interactions ensures that the pollen tube can enter the female gametophyte, burst, and release sperm cells to complete the double fertilization. In this chapter, we provide an overview of pollination and the interactions between male and female, comprehensively summarizing the factors involved in these processes.

Keywords Sexual reproduction · Siphonogamy · Pollination · Male-female interaction · Double fertilization

S. Zhong · Z. Lan
State Key Laboratory for Protein and Plant Gene Research, Peking-Tsinghua Center for Life Sciences, New Cornerstone Science Laboratory, Beijing, China

Z. Ge
State Key Laboratory for Protein and Plant Gene Research, Peking-Tsinghua Center for Life Sciences, New Cornerstone Science Laboratory, Beijing, China

Institute of Science and Technology Austria (ISTA) I IST, Klosterneuburg, Austria

L.-J. Qu (✉)
State Key Laboratory for Protein and Plant Gene Research, Peking-Tsinghua Center for Life Sciences, New Cornerstone Science Laboratory, Beijing, China

College of Life Sciences, Peking University, Beijing, People's Republic of China
e-mail: qulj@pku.edu.cn

F. Chang et al. (eds.), *Regulation of Plant Development*,
https://doi.org/10.1007/978-981-95-7033-1_14

14.1 Pollination

Pollination in plants refers to the transfer of pollen grains, often by an insect or the wind, from the male parts of a flower to the stigma of the plant gynoecium. Plant pollination is the crucial step in the sexual reproduction of flowering plants since all plants must undergo pollination before the seed or fruit set. If pollen is transferred from the anther to the stigma of the same flower or between different flowers on the same plant, the process is called self-pollination, whereas cross-pollination occurs when the pollen grains pollinate ovule gametes with discrepant genetic material. Various intermediary agents, such as animals, insects, wind, or water, can serve as pollinators of seed plants. For this reason, plant pollination can be divided into two groups: biotic pollination, the transfer of which is achieved by insects (entomophily), birds (ornithophily), or bats (chiropterophily), and abiotic pollination, including wind (anemophily) and water (hydrophily) agents (Fig. 14.1).

14.1.1 How Pollination Is Studied

Pollination is one of the most important links between plants and animals in ecosystems. Studies of pollination can help us understand not only the process of sexual reproduction of plants but also the application of pollination in biodiversity conservation, agriculture, and forestry production practices. Studies of the interactions between plants and pollinators have been conducted at the species or community level.

14.1.1.1 Pollination at the Species Level

Examinations of pollination at the species level mainly focus on the attraction and reciprocity relationship between flowers and pollinators. Darwin first studied pollination at the species level. After finishing "The Origin of Species," he began to

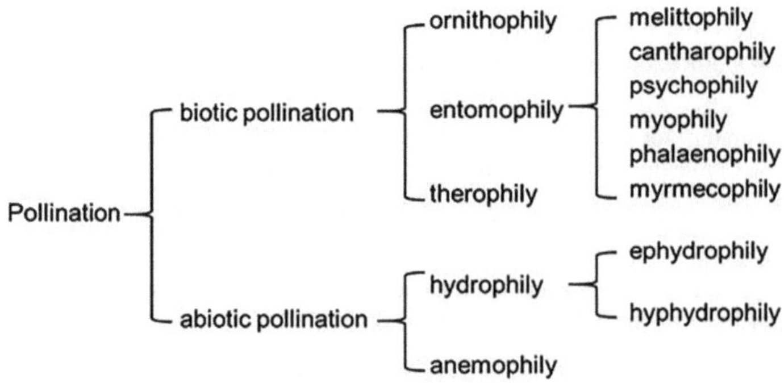

Fig. 14.1 Types of pollination

propose concrete studies to clarify his theory of evolution by describing, for the first time, the adaptation and differentiation among species based on the pollination and fertilization mechanism of *Orchidaceae* (Darwin, 1862). The native white orchid *Angraecum sesquipedale* in Madagascar, which has a spur length of 10–11 inches, was a classic example described in Darwin's work. Darwin found nectar at the end of the spur and therefore predicted that, in Madagascar, a moth must exist with a particularly long tongue to collect that nectar and pollinate the orchid. This prediction was found to be true 41 years later when a moth was discovered with an 11-inch tongue. The moth was named *Xanthopan morganii praedicta* (Rothschild & Jordan, 1903) in memory of Darwin's prediction and was finally confirmed to be the pollinator of the white orchids more than 90 years later (Wasserthal, 1997). To verify that the evolution of these flowers could be due to selection by these pollinators, Nilsson reduced the tube length of the spur and found that it reduced the removal of pollen grains and acceptance by the stigmas, reducing pollen transmissibility and plant fecundity (Nilsson, 1988); thus, this result supported Darwin's conjecture.

Fossil records indicate that angiosperm and insect species underwent rapid differentiation nearly simultaneously after the Late Cretaceous (Herrera & Pellmyr, 2002). Pollinator isolation is thought to be one of the major drivers for the differentiation of angiosperms, and flower features are thought to be shaped by the most frequent and effective pollinators in a region (Stebbins, 1970). The attraction of different pollinator groups by plants with different combinations of floral features is called pollination syndrome. For this reason, the length of the nectar spur in different species of *Aquilegia* is obviously differentiated via pollination by bees, hummingbirds, or moths (Huang, 2014). Phylogenetic analysis indicates that the differentiation of pollinators occurs significantly earlier than that of plants, thus suggesting that the evolution of the spur length in *Aquilegia* is an adaptation to the differentiation of pollinators (Whittall & Hodges, 2007). In other cases, flowers of different plant species pollinated by birds are generally red or orange, with anthers and stigmas that are prominent, odorless, and exude large amounts of thick, concealed nectar. Similarly, flowers pollinated by bats are usually dull white or green, open in the dusk or at night, produce exposed nectar, and have an obvious smell (Table 14.1).

14.1.1.2 Pollination at the Community Level

Pollination surveys at the species level are mostly focused on a single or few species as a research unit and ignore reticulation among species. However, plants are often not independent, and various links exist between different visitors and/or with other flowering plants in the community. For example, multiple visitors perform pollination for the same or similar types of floral plants, while the same pollinators usually feed on a variety of plants. Therefore, studying the interactions between plants and pollinators, pollinators and pollinators, and plants and plants is helpful to elucidate the evolution of the interaction between pollinators and plants.

Table 14.1 Classical pollination syndromes

Syndrome	Melittophily	Cantharophily	Psychophily	Myophily	Phalaenophily	Ornithophily	Chiropterophily
Pollinator	Bees	Beetles	Butterflies	Flies	Most moths	Birds	Bats
Timing of anthesis	Dawn, day	Day or night	Day	Usually day	Dusk, night	Day	Dusk, night, nigth only
Main colors	Pink/purple/blue, white, yellow	Cream, green, usually dull	Red, orange, yellow, mauve	White, yellow, greenish	Cream, yellow, greenish	Red, orange	Dull white, dull beige/green
Nectar guides	Yes	No	Maybe	No	No	No	No
Scent	Moderate, usually sweet	Strong, fruity or fermenting	Slight to moderate, sweet	Usually mild, not sweet	Fairly strong, sweet	Usually none	Strong, fruity, or fermenting
Shape	Bilateral or radial, exposed tube	Radial, flat, or bowl-shaped	Small, long tube, often en mass	Radial, flat; or flat inflorescence	Usually radial, moderate tube	Bilateral or radial, medium tube	Bilateral or radial, bowl or brush
Nectar shape	Exposed or concealed	Exposed	Concealed	Usually exposed	Concealed	Concealed	Usually exposed

The tendency of some plants, such as *Ficus*, *Orchidaceae*, and *Euphorbiaceae*, to select for one or a few particular pollinator/pollinators and the tendency of the pollinator and plant to maintain a "one-to-one" relationship is called ecological specialization. *Ficus* is one of the largest genera of angiosperms and is especially dominant in the tropical rainforest. The interaction between *Ficus* and its unique pollinator fig wasp (*Agaonidae-Chalcidoidea*) is one of the most specific symbiotic systems known to date and has been considered the ideal material for the study of symbiotic evolution, cospeciation, and other symbioses. The most typical feature of *Ficus* is the hypanthium inflorescence, which allows only the wasp, with its small body size, to enter for pollination. These wasps lay their eggs only in the female flowers of *Ficus*. The eggs then develop into an insect gall, and the young live in the gall and feed on the flower (Herre et al., 2008).

In contrast with ecological specialization, however, many plants are pollinated by various types of visitors or a pollinator that is able to visit many plants; this is called ecological generalization. Pollination generalization usually includes mutual complementation of pollinators for the plant, which ensures fertilization success. However, pollination also frequently causes competition for the pollen type, quantity, and quality among the various flowers, species, or visitors. For example, pollen grains from the same species transported by different pollinators will promote pollination efficiency, whereas if too many grains from different species are deposited on the same stigma by pollinators, heterogeneous pollen interference will occur, which can also have negative impacts on the pollen exporters. By contrast, some plants, especially when they share the same flowering period (for both pollen importers and exporters), will have adaptations to attract the pollinators, while pollinators would simultaneously also be competing for the same flower or plant. Quantitative analysis of the visiting frequency, pollen removal and acceptance, and residence time on a single flower of three common bees on *Hedysarum laeve* suggests that *Anthophora albifronella* is the most effective pollinator for the plant (Meng et al., 2007). *Besleria triflora* pollinated by hummingbirds has both bisexual flowers and staminate flowers with the pistil aborted. The staminate flowers could not be pollen exporters, but these flowers are more conspicuous, which attracts visitors, thus also raising the pollination efficiency of the bisexual flowers (Podolsky, 1992).

In addition, the relationship between plants and pollinators does not remain unchanged over time. Some plants or pollinators may develop a new mutualism by adapting to temporal and spatial variability. A classic case is that some plants lose their original pollinators after invading a new habitat and quickly establish a new pollination relationship with local visitors. For example, *Impatiens glandulifera*, originally distributed in the Himalayas in Asia, is pollinated by native bumblebees after its introduction to Europe and North America as ornamental flowers. *Impatiens glandulifera* was found to attract local bumblebees (mainly *Bombus pascuorum*) as pollinators due to the higher concentration of sugars secreted in its nectar than in that of the native plants of Wurzburg, Germany, resulting in a significant decrease in the fertilization of the native plant *Stachys palustris*, which shares the same pollinator (Chittka & Schürkens, 2001). In contrast with *Impatiens glandulifera* above, some invasive species could also benefit native species through pollination. For example, the exotic plant *Lythraceae* helps the indigenous species to attract more pollinators to their flowers and thus increases their seed production (Silva et al., 2013). Plants that bloom at the same time are assumed to attract more pollinators when their population density is low or if their flowering time is short, which can promote the likelihood of their survival in turn.

14.1.2 Self-Pollination and Cross-Pollination

Reproductive systems in flowering plants vary greatly among many species, but two processes are essential for all angiosperms: pollination (the transfer of pollen grains to the stigmas) and fertilization (cell fusion between sperm and egg cells or central cells). Pollination is divided into various types in the literature. If a single flower is viewed as the research unit, pollination can be classified by whether it occurs within a single flower or between different flowers. If the research unit is focused on the individual plant, pollination can be classified by whether it occurs within the same plant or between different plants. In many cases, plants may have multiple flowers; thus, pollination falls into three groups: (1) pollination within a single flower itself; (2) pollination between different flowers of the same plant; and (3) pollination between different plants. However, pollination is more frequently divided into self-pollination and cross-pollination.

Self-pollination is one of the two major types of pollination, in which grains are transferred from the anthers to the adjacent stigmas of a single flower (term 1 above) or from one flower to another on the same plant (term 2 above). Self-pollination can be further grouped into two categories: cleistogamy (also called "closed marriage"), which means that the mature pollen grains directly germinate to produce the pollen tubes in the pollen sac of non-opening flowers and then penetrate into the pistil for fertilization, and chasmogamy, or "open marriage," which is the opposite of cleistogamy as it occurs in opening flowers. Cleistogamy is the most typical self-pollination and is usually found in barley, peas, beans, and peanuts. Since the development of petals, nectar, and large amounts of pollen is not necessary, cleistogamy is more efficient in producing seeds than chasmogamy, which causes cleistogamy to be particularly useful for seed production under, for example, unfavorable or adverse conditions (Clavijo & Jimenez, 1993). In contrast to cleistogamy, chasmogamous flowers, which are open for insect pollination, are generally able to outcross in nature.

Self-pollination is traditionally considered an ancestral form of pollination, and two hypotheses are widely accepted to explain the evolution of self-pollination. The first is reproduction assurance, in which self-pollination is used to ensure seed production and the propagation of offspring when plants lack the possibility of cross-pollination (see below) due to the absence of pollinators, such as insects, wind, or water. For plants to survive successfully in the new habitat or under unfavorable conditions, the first and most important task is to complete pollination and fertilization. Therefore, self-pollination, especially cleistogamy, is the most effective way by reducing the dependence on any biotic or abiotic agents, and self-pollination is generally considered an adaptive mechanism of pioneer plants to ensure fast dispersal and robustness by sacrificing genetic diversity and reducing energy consumption (Baker, 1955). If self-pollinating plants could also offer pollen to other individuals but at a reduced fertilization efficiency, self-pollination would increase male fitness, providing the plants with more advantages in reproduction, that is, an automatic selection advantage (Holsinger, 2000). In this hypothesis, self-pollination is undoubtedly superior to cross-pollination. Self-pollination secures high reproduction rates in the early stages of invasion into a new habitat, but the genotypes suitable for survival will further be rapidly replicated by self-pollination, thus allowing the plant to adapt (Willmer et al., 2000).

Some disadvantages of self-pollination exist; for instance, it restricts gene flow within plant populations and reduces the rate of genetic recombination and genetic heterogeneity. Self-pollination also induces mutation accumulation and renders the plant less flexible in adapting to a changing environment. To compensate for these disadvantages, some plants evolve cross-pollination, which occurs in the majority of flowering plants, in which pollen is transferred to the stigmas of another individual plant, usually by animals, insects, or wind. Plants with cross-pollination are usually divided into three types based on their floral features: (1) Dioecious plants, in which the male and female flowers are on different plants, such as spinach, ginkgo, and papaya. The outcrossing rate of these plants is almost 100%. (2) Monoecious plants, with male and female flowers on the same plant. Male flowers of maize are on the top of the plants, while the female are on the central leaf axils; (3) Monoclinous flowers, with the male and female parts on the same flower; generally, they are self-incompatible or have asynchronous spatiotemporal development of the stamens and pistils on the same flower. Self-incompatibility (SI) refers to the failure of the processes of pollen germination, pollen tube growth, ovule fertilization, or embryo development when pollen lands on the stigma of the same plant or another plant with a similar genetic background; consequently, no seeds are produced. Self-incompatibility is the most important and efficient means to prevent self-pollination, promote the generation of new genotypes, and ensure genetic diversity, which are all conducive to evolution and adaptation to the environment.

The individual genotype of the offspring of cross-pollinated plants is usually heterozygous, with no completely similar individuals among the population. If cross-pollination occurs between different inbred lines, the offspring will present improved or increased function of some biological qualities compared to the parents; this process is called heterosis, also known as hybrid vigor or outbreeding enhancement. For example, cross-pollination in normally self-pollinated crops also generates higher yields, better fruit quality, and reduced response to stress. Heterosis has been widely used in crops and has significantly contributed to solving the world's food problem since it was reported for the first time in rice in 1926 (Jones, 1926). Although some scientists have proposed hypotheses, such as the genetic dominance hypothesis (Bruce, 1910) and the genetic overdominance hypothesis (Shull, 1908), the genetic mechanism of heterosis still lacks a uniform and clear explanation.

14.1.3 Entomophilous Pollination

There are approximately 1.4 million species described, with insects and angiosperms accounting for approximately 1/2 and 1/6, respectively. The high diversity of these two taxa is thought to be related to their complex symbiotic relationship (Wilson, 1992). In general, some insects feed on various parts of plants, such as the leaves, seeds, and pollen grains, and can also reside in some parts of plants. At the same time, the majority of flowering plants partially or strictly rely on insects for pollination, seed dispersal, and protection from other plant-based natural enemies, and some even prey upon insects. Among the diverse interspecific interactions between plants and insects, the mutualism between plants and pollinators has been used as a model system to study species adaptation and diversification since Charles Darwin. By visiting flowers, insects incidentally transfer pollen grains from one flower to another of the same or another plant, performing entomophilous pollination, the most common and typical biotic pollination, including almost 80% of angiosperms. Fossil studies reveal that entomophily was critical for some plants during the Early and Middle Cretaceous (Crane et al., 1995).

Various types of insects, such as bees (melittophily), butterflies (psychophily), moths (phalaenophily), flies (myophily and sapromyophily), and beetles (cantharophily), serve as the pollinators for plants. Melittophily, including approximately 20,000 species of bees, is the largest and most important group in entomophily. Some plants rely strictly on bees as pollinators because the anthers release pollen grains internally, where they can only be released by buzz pollination, also known as "sonication." Visits to the flowers by diverse types of bees, such as honeybees, bumblebees, and orchid bees, result in pollen grains being attached to the mouthparts, body hair, back, and feet of bees, distributing them for pollination. To attract the visitors, melittophilous flowers are frequently variable in size, shape, and color, frequently with bright petals, usually in blue or yellow, together with obvious nectar and a strong scent. As rewards, melittophilous flowers usually offer amounts of nectar and/or pollen to the visitors. Psychophily and phalaenophily are also very common in nature and are similar to melittophily since both of them rely on visual and/or olfactory signals to sense flowers. Flowers pollinated by moths or butterflies are usually scented and tend to be large and showy, with red, white, yellow, or orange colors. For instance, the typical phalaenophily is white and releases a strong, sweet scent to attract moths to visit in the evening, at night, or in the early morning. Because pollen grains cannot be digested by these insects (with some exceptions), more nectar is offered than pollen, and the nectaries are usually hidden in narrow tubes or spurs, which are easy for the visitors to reach using their long tongues. Compared to bees, butterflies, and moths, flies tend to be more important pollinators in high-altitude and high-latitude systems since there are fewer insect species. Myophilous flowers are typically purple, violet, blue, or white, with no obvious scent, while sapromyophilous flowers, which are less common than myophilous flowers, often mimic the odor of dead animals. However, cantharophilous flowers are generally white or dark in color, often with a fruity scent, because beetles have a good sense of smell. Beetles often eat pollen and bite petals, leaves, and other parts of the flowers.

Once a pollination relationship between plants and insects is established, the most basic study of entomophilous pollination will focus on two aspects: how floral plants attract visitors and how pollinators detect signals in a mutualism. Plants have evolved a variety of features to attract insects to visit (Table 14.1), and pollinators visiting flowers also have a variety of adaptations; some insects and plants have even developed an "obligate mutualism," in which they depend on one another.

To attract insects, entomophilous flowers have evolved various striking colors and/or color patterns, which are common for many angiosperms, and these visual features cause the flowers to be more conspicuous and attractive than other tissues or plants. When insects approach a plant, they would first sense the color, then land on the flowers and encounter the anthers and pollen grains. Different types of insects will be attracted by different colors or color patterns caused by the accumulation of different anthocyanins. For example, myophilous insects like white, yellow, and greenish flowers, whereas psychophilous flowers are often pink/purple/blue, white, or yellow to attract butterflies. Different color patterns, such as the red and black lines on the petals of *Canna indica*, are also typical visual signals to catch the attention of pollinators and guide them to the stamens and pistils.

Compared to color signals, olfactory signals of flowers have also long been among the most effective and common ways to attract insects (Andersson & Dobson, 2003) because the visitors to flowers, especially insects, often have specialized abilities in detecting and responding to floral odors. The main components of flower olfactory signals usually include monoterpenes, sesquiterpenes, phenols, simple alcohols, and ketones. The floral odor of a given species generally varies in chemical composition, but with only one or a few components accounting for a large proportion of the odor. Pollinators can accurately identify the specific scent of a plant based on its composition. Therefore, specific visitors can recognize specific floral scents: for instance, the odor of cantharophilous flowers is strong, fruity, or fermenting, while that of myophilous flowers is usually mild and not too sweet.

As rewards for pollination, some secondary products, such as nectar, pollen grains, oils, gums, and resin, are often offered to pollinators by plants. Among these products, nectar and pollen grains are the most common foods for insects. Nectar, mostly secreted by a special nectary that is usually located in the base of the pistil, is easily produced by plants and handled expediently by insects as well. The classic example of feeding on nectar is the honeybee, which is also famous for honey production. Honeybees visit a variety of flower types, both for sipping nectar and for collecting and distributing pollen grains. Nectar mainly contains three types of sugar: sucrose, glucose, and fructose, but the nectar composition varies among plant species, which affects which pollinators are attracted. Insects with long tongues frequently prefer medium concentrations of nectar, while short-tongued insects prefer more concentrated nectar (Richards, 1978). In addition, plants produce much more pollen, which is highly rich in starches, sugars, minerals, and free amino acids, than is necessary for fertilization. For example, each *Tilfa* flower contains 43,000 pollen grains, and those of *Typha* contain 170 million grains. Therefore, these abundant and nutritious grains are welcoming to many insects as well as other animals. Moreover, the morphological features, such as the color, shape, and size, and the chemical components of the grains, as well as the nectar, are also key factors in attracting specialized pollinators.

To attract insects, some plants even use deception strategies, and pollination cheating by visitors has also been widely studied. A plant can not only mimic an interspecific or intersexual flower by developing a pseudoflower, pseudonectar, pseudoanther, or pseudopollen but also mimic objects other than flowers, such as other animals, using aggressive mimicry or pseudocopulation mimicry, which would attract insects attempting to attack or mate to transport pollen grains. Some non-rewarding plants, which constitute a large proportion of angiosperms, especially some orchids, attempt to benefit from pollination without paying the normal cost of pollen grains or nectar. These orchids are always nectarless and usually deceive pollinators by resembling the flowers of other rewarding plants. For example, *Eulophia zeyheriana* develops flowers similar to those of *Wahlenbergia cuspidata* in spectrum and morphology, thus attracting the visitors of *Wahlenbergia cuspidata* for pollination (Peter & Johnson, 2008). Moreover, the mirror orchid *Ophrys speculum* is pollinated by bees of the genus *Andrena* in the Mediterranean. The special color and shape of the flower lip of this plant resemble those of female bees and also release a pheromone-like compound that is usually secreted by female bees. This mimicry attracts the male bees for pollen grain dispersal without providing pollen or nectar.

The pollination relationship between plants and insects is a mutualistic interaction commonly used to study species adaptation and diversification resulting from long-term co-evolution. The purposes of insect visits are highly variable: some insects feed on nectar and pollen, such as beetles, butterfly moths, and flies, while some collect them as food for their offspring; for example, social bees and bumblebees collect nectar mainly for feeding offspring as well as for their own consumption. In this case, pollinators have evolved a variety of strategies, such as specialized body structures to sense signals to collect, distribute, and deposit pollen grains. On the one hand, they have specialized external structures suitable for collecting grains. To reach nectaries in different positions in flowers easily, different pollinators have evolved different types of mouthparts, with specialized lengths and/or widths. For instance, the large body size and numerous body hairs of some bumblebees and all bees are adapted to carry pollen grains. The nectaries of some flowers are usually hidden in narrow spurs, where they are easily reached by moths and butterflies using their long tongues. On the other hand, some insects have specialized sense organs, such as olfactory and spectrum organs, to sense their preferred flowers. The most classic example is olfactory sensitivity, which is widely distributed in various types of biotic pollination. The olfactory organs of insects cause them to be highly sensitive to floral odors, secreted hormones, and other volatiles and guide them to specific flowers.

14.1.4 Anemophilous Pollination

Anemophilous pollination is when pollen is distributed to female tissues randomly by wind. Anemophilous pollination is the dominant type of abiotic pollination, found in the majority of gymnosperms and 10% of angiosperms. For example, several species, including the grasses (*Poaceae*) and sedges (*Cyperaceae*), strictly rely on wind for pollination. Compared with biotic pollination, anemophily is usually considered a wasteful process for most plants. To this end, species using anemophilous pollination have gradually gained, through long-term evolutionary processes, many morphological adaptations to increase pollen dispersal and capture. Some of these flower features are as follows:

1. Anemophilous flowers are commonly unisexual, whether in monoecious or dioecious species, and are often temporally or spatially separated from the flowers of the opposite sex to avoid a stigma becoming entirely covered with its own pollen, especially given the high output of pollen required by anemophily; that is, this adaptation not only prevents self-fertilization but also increases outcrossing. Flowers of anemophilous species are generally grouped as inflorescences, such as catkins, spikes, and umbels, which tend to be more condensed than flowers in zoophilous plants. The flower is also often found above or outside the leaves, possibly providing easier access to wind currents. The filaments are usually very long, also allowing the anthers to extend or be exposed outside the canopy, enabling easy release of pollen grains into the air. The perianth of anemophilous plants is always reduced or absent; if present, it is, in most cases, green to dark brown to reddish in color, with no protective or attractive functions because they are unnecessary.
2. To ensure pollination, anemophilous plants tend to produce and release as many pollen grains as possible since the efficiency of anemophily is very low for most plants. For example, a corn plant produces approximately five million pollen grains, with each inflorescence producing approximately 160,000 grains. The pollen is fairly small, usually 20–40 μm on average, and low in water content rather than sticky due to the thin or entirely absent pollenkitt. This pollen grain structure will help them to avoid clumping when they are being distributed in the wind. To reduce air drag and increase buoyancy, the surfaces of the pollen grains are relatively unsculptured and often contain some air-filled bladders, especially in conifers, which provide a larger surface area without increasing weight appreciably. Grains also have a reduction in the number and/or size of apertures, which probably helps to reduce water loss during the long journey in the wind.
3. Pistils of anemophilous flowers often evolve various special features that differ from most zoophilous plants to increase the capture efficiency of the pollen grains from the air. On the one hand, the number of ovules in the pistil is often much lower than that of biotically pollinated plants, with often just one ovule in many species, meaning only a single seed will set. Hence, they usually have a very high pollen/ovule (P/O) ratio. The P/O ratio of the *Pinaceae* is one million, while that of *European hazelnuts* reaches 2.5 million. On the other hand, the

stigmas are distinctly larger, with finely divided or expanded forms, both of which increase the ability of the plant to "capture" pollen. For example, stigmas of grasses are often plumose or featherlike, and cattails are often found to have brush-like stigmas.

4. Anemophilous plants generally have a distinct flowering season, or the anthers do not open unless the weather is favorable, that is, warm and dry. Commonly, the flowering onset is statistically earlier than that of biotically pollinated plants, both in terms of the seasonal timing and of the daily timing of the release of grains from the anthers. For example, most angiosperm trees bloom in early spring when the foliage is still very sparse and thus does not disturb mass air flow and pollen movement, rather than in late summer when the leaf cover is decreasing.

In general, anemophily is considered a less efficient method of pollination, but some studies show that it is not as ineffective as imagined and that its efficiency is even higher than that of entomophilous pollination under some conditions or in certain plant groups (Friedman & Barrett, 2009). Some adaptations contribute to this improved efficiency: the floral and accessory structures not only act to filter pollen grains in the air, removing unsuitable pollen and dust, but also could facilitate the flow of the pollen of the same species to the stigma. In addition, many anemophilous plants often grow in patches or clusters, which reduces the distance the pollen grains must travel (Niklas, 1985).

Anemophily and entomophily are not strictly separate, and some species even share the two pollination systems at the same time. For example, several species of *Plantago*, usually considered pollinated by wind under most circumstances, are also frequently pollinated by honeybees and flies. Some entomophilous plants, such as *Gnetum* and *Aconitum*, with attractive features, including colors and smell, have been observed to attract insects and were previously considered entomophilous, but recent studies have shown them to be anemophilous (Gong et al., 2016; Wang et al., 2016a). Anemophilous pollination is widely considered to be derived from entomophilous pollination. Anemophily is considered to have evolved independently in many groups of angiosperms, and the evolution of anemophilous pollination from entomophilous pollination has occurred at least 65 times (Friedman, 2011; Friedman & Barrett, 2009). Although the mechanism is still unknown, plants could easily adapt to the wind vector through the evolution of some relatively simple flower features. However, the shift from anemophily to entomophily is rarely found in nature because entomophilous plants often require more complex floral features associated with attraction, adaptation, and compensation to the visiting insect (Huang et al., 2013).

14.1.5 Hydrophilous Pollination

Abiotic pollination is traditionally divided into two categories: anemophilous pollination (wind-mediated, see Sect. 14.1.4) and hydrophilous pollination (water-mediated, in which pollen is distributed by the flow of water, particularly rivers and streams). Compared to anemophily, hydrophily is much rarer, constituting less than 2% of all abiotic pollinations, and often occurs in aquatic species. Hydrophilous plants often share some crucial features: the flowers tend to produce many pollen grains, but the P/O ratio is quite low; the sizes of the pollen grains are highly variable, and the shapes are generally elongate; compared with terrestrial pollen grains, hydrophilous grains are much more rigid, often with a reduced exine and increased apertures, and more clumped together; finally, the hydrophilous plants often tend to have large, erect, and rigid stigmas, thus increasing the possibility of contact with pollen grains.

Hydrophilous species usually fall into two groups: epihydrophily and hypohydrophily. Ephydrophily refers to pollination occurring on the surface of water. Pollen grains on the water's surface are either dry or wet. If dry, the pollen itself or the pollen along with the flower or the anther is spread across the surface of the water before touching the stigma, whereas wet pollen falls from the flower, directly floats across the water's surface, and moves across the surface film before pollination. For example, the male flowers of the dioecious plant, *Vallisneria spiralis*, grow on slender stems, while the female flowers are at the base of the plant. The male flowers rise to the water's surface when they bloom and are released and drift around, driven by wind and water flow, with the pollen itself not touching the water at all. Simultaneously, the female flowers rapidly expand, emerge out of the water, and let the stigma touch the male flowers. Then, the female flowers withdraw back under the water to finish pollination and fertilization. Subsequently, a seed will develop, germinate after it becomes mature, and develop into a new plant underwater.

Hypohydrophily is a type of pollination that occurs underwater and often involves the maturation, release, movement, and capture of pollen grains, anthers, or whole flowers under or through water. Therefore, successful pollination is related to the structure of the male flowers, such as the pollen grains, as well as the female flowers. On the one hand, the grains frequently have unique structures, such as an extremely thin exine or a lipid coating on the outside, to ensure that the pollen or flowers are not destroyed underwater. For example, the pollen of *Callitriche autumnalis* does not have an outer wall but is covered with lipids, which insulate the pollen from the water. On the other hand, flowers of some aquatic species have reduced perianths and exposed anthers, with large stigmas and few ovules—in many cases, only one for many species. The simplest case of hypohydrophily is the *Ceratophyllum*, whose male and female flowers in the leaf axil are very tiny. The female flower only contains a single ovary and has an oblique and bent style, with the tip pointing upward. The pollination of *Ceratophyllum* occurs in standing water when the pollen sinks after the stamens dehisce.

14.2 Fertilization

14.2.1 Female Gametogenesis

The female gametophyte is a multicellular structure that consists of four different types of cells, playing a crucial role in various stages of the reproduction processes, including pollen tube guidance and reception, and finally develops into the

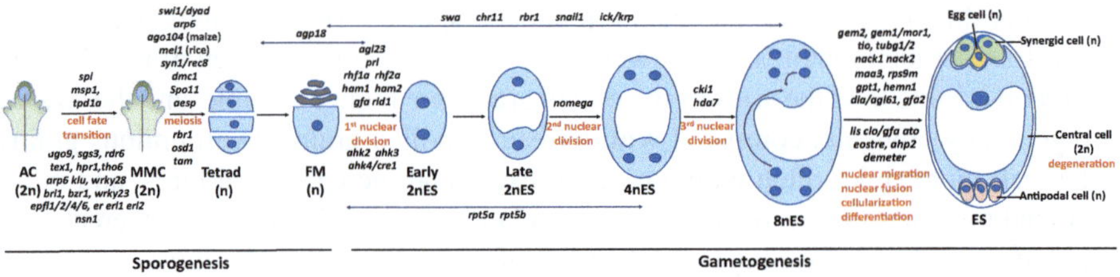

Fig. 14.2 Schematic drawing depicting the sequential development of female gametophyte in *Arabidopsis*. The mutants and genes involved in the meiotic and mitotic cell-cycle regulation are indicated. AC archesporial cell, MMC megaspore mother cell, FM functional megaspore, ES embryo sac

embryo and endosperm after double fertilization. Female gametophyte development can be divided into two phases: megasporogenesis and megagametogenesis (Yadegari & Drews, 2004; Yang et al., 2010). Megasporogenesis is the stage preceding female gametophyte development, where a specialized archesporial cell differentiates into a mature megaspore. During megagametogenesis, the functional megaspore undergoes three rounds of karyokinesis, and, after cellularization, differentiates to form an embryo sac containing seven cells with eight nuclei (Fig. 14.2). In recent decades, numerous genes related to the regulatory networks in female gametophyte development have been identified.

14.2.1.1 Megasporogenesis—Cell Fate Transition

During megasporogenesis, the archesporial cell, originating from the nucellus tip located in the hypodermal cell, initially differentiates into a megaspore mother cell (MMC). The diploid MMC undergoes meiosis to produce four haploid megaspores. Finally, the three megaspores in the micropylar region undergo programmed cell death, while the megaspore located in the chalaza becomes the mature functional megaspore. Cell fate transition always plays a pivotal role in development. During megasporogenesis, the genes associated with the transition from somatic cell to germline are limited. SPOROCYTELESS/NOZZLE (SPL/NZZ) was identified as relevant to sporogenesis in *Arabidopsis thaliana* years ago. In the spl mutant, the archesporial cell fails to differentiate into the megasporocyte and microsporocyte in both the female and male organs (Schiefthaler et al., 1999; Yang et al., 1999b). A recent study indicates that the homolog of SPL in rice, OsSPL, also participates in meiotic fate acquisition in sexual reproduction, suggesting the conservation of the SPL-mediated signaling pathway in angiosperms (Ren et al., 2018). SPL functions as an adaptor-like transcriptional repressor and interacts with the C-terminal binding protein TOPLESS/TOPLESS-RELATED (TPL/TPR) to inhibit its N-terminal EAR motif recruited transcription factor TEOSINTE BRANCHED1/CYCLOIDEA/PCF (TCP) family members (Chen et al., 2014; Wei et al., 2015).

Another significant scientific query regarding megasporogenesis involves the control of somatic cells entering the germline. Several studies indicate that genes expressed in the region surrounding the germline cells play an important role in this process. In maize, the loss of MULTIPLE ARCHESPORIAL CELLS 1 (MAC1) results in an excessive number of archesporial cells (Wang et al., 2012). In rice, mutants of MULTIPLE SPOROCYTE 1 (MSP1) possess more than one sporocyte, likely due to the overproduction of archesporial cells. MSP1 is a leucine-rich repeat receptor-like kinase (LRR-RLK) expressed in nucellar cells surrounding the germline cells (Nonomura et al., 2003). MSP1 interacts with the ligand OsTDL1A, having a similar expression pattern to that of MSP1, and regulates the female germline fate (Zhao et al., 2008). The small RNA pathway has also been shown to participate in regulating female germline specification. ARGONAUTE 9 (AGO9) cleaves endogenous mRNA using its processed miRNA or siRNA and is expressed in the somatic companion cells of the germline cell. Loss of AGO9 leads to the formation of multiple MMCs (Olmedo-Monfil et al., 2010). Both SUPPRESSOR OF GENE SILENCING 3 (SGS3) and RNA-DEPENDENT RNA POLYMERASE 6 (RDR6) function in siRNA production. Mutants of SGS3 and RDR6 show similar phenotypes to those of *ago9* (Olmedo-Monfil et al., 2010). These results suggest that the movement of sRNAs from somatic companion cells plays an essential role in controlling the specification of the megasporocyte. The TEX1, HPR1, and THO6-formed THO/TREX complex participates in ta-siRNA biogenesis. Knocking out any of these three genes results in multiple MMCs (Su et al., 2017b). The protein expression pattern shows that TEX1 is specifically expressed in the epidermal cell layer of ovules in the megasporogenesis stage. The THO/TREX complex could repress ARF3 expression in the epidermal cell layer through one of the complexes transporting products of TAS3-derived ta-siRNA. ARF is thought to repress an unknown factor transported to the adjacent cells of MMC to restrict their somatic fate. Thus, in *tex1*, *hpr1*, and *tho6* mutants, ARF could not be repressed since no ta-siRNA is produced. The unknown factor repressed by ARF could no longer be expressed to repress the adjacent cells. Meanwhile, the double mutant of *tex1 rdr6–11* shows a much more severely affected phenotype than either of the single mutants, indicating that TEX1 and RDR6 have functional redundancy (Su et al., 2017b). Additionally, chromatin remodeling has been identified as controlling germline specification. ARP6 is a subunit component of the chromatin-remodeling SWR1 complex (Qin et al., 2014b). Recent work shows that ARP6 is genetically interacting with Cytochrome P450 KLUH/CYP78A5 (KLU), which is expressed in the inner integument primordia. The double mutant *arp6 klu* has multiple MMC-like cells. RNA-seq analysis identifies that the transcription factor WRKY28 works downstream of the SWR1 complex and KLU. Interestingly, WRKY28 is specifically expressed in a few somatic cells surrounding the MMC. Since the *wrky28* mutant exhibits a similar phenotype to that of *arp6 klu*, KLU-SWR1-WRKY28 is thought to mediate a pathway controlling the germline cell fate specification (Zhao et al., 2018). However, how the inner integument located in KLU regulates the MMC surrounding the cell expressing SWR1 remains to be identified in the future.

Plant hormone brassinosteroid (BR) also plays an important role in the specialization process of MMC. A BR gradient is established in the distal nucellus of ovule primordia, with its maximum concentration found in the epidermal L1 layer cells. This BR-induced microenvironment is crucial for restricting the fate of female germline cells to a single L2 layer nucellus cell, achieved through the activation of the BRI1–BZR1–WRKY23 pathway (Cai et al., 2022). Additionally, EPFL-ERF ligand–receptor pairs operate upstream of the BZR1 family, orchestrating the regulation of female germline specification by activating the nucleolar GTP-binding protein, NUCLEOSTEMIN-LIKE 1 (NSN1) (Cai et al., 2023). It is now crucial to elucidate how NSN1 coordinates with WRKY23 to suppress the formation of MMC in ovule primordia.

14.2.1.2 Megasporogenesis—Functional Megaspore Formation

Once the MMC is formed, it undergoes two rounds of meiosis to generate a functional megaspore, and numerous genes related to meiosis are involved in this phase. Firstly, genes regulating chromosome structure, such as ARP6, play a role in MMC meiosis. In *Arabidopsis*, the *arp6* mutant exhibits a defect in the first round of meiosis, suggesting its role in histone methylation and acetylation, potentially controlling the temporal expression of meiosis-related genes. DMC1, a recombinase in meiosis, is repressed by ARP6 in both MMC and its surrounding cells. However, during MMC meiosis, ARP6 selectively activates DMC1 expression only within the MMC itself (Qin et al., 2014b). SWITCH1(SWI)/DYAD involved in chromatid cohesion and chromosome structure in *Arabidopsis*, is expressed in both the megasporocyte and microsporocyte. Mutants in *AtSWI* or its homolog in maize, *ZmAM1*, undergo mitosis instead of meiosis (Mercier et al., 2001; Pawlowski et al., 2009). The AGO family members MEL1 and AGO104 in maize also contribute to chromatin structure, and mutations in these genes lead to developmental arrest due to the failure of MMC meiosis (Nonomura et al., 2007; Singh et al., 2011).

Secondly, factors regulating chromosome behavior, including synapsis, crossing-over, and sister chromatid cohesion, are crucial for MMC meiosis. In *Arabidopsis*, the SPO11 family members, SPO11-1 and SPO11-2, initiate meiotic recombination and are necessary for synapsis, with mutants exhibiting female gametophyte developmental defects (Grelon et al., 2001; Stacey et al., 2006). The SYN protein family, homologous to Rec8p/Rad21p, is essential for sister chromatid cohesion, and the *syn1/rec8* knockout mutant shows defects in chromosome segregation, leading to aberrant MMC meiosis. AESP, encoding a separase in *Arabidopsis*, contributes to sister chromatid segregation, and its mutation results in aberrant meiosis in MMC (Liu & Makaroff, 2006). Additionally, the cell-cycle regulator RBR1, expressed in MMC, inhibits E2F and controls the cell cycle in the G1 stage. The *rbr1* mutant undergoes mitosis instead of meiosis, leading to multiple MMC, and RBR1 inhibits WUS to initiate MMC meiosis (Ebel et al., 2004; Johnston et al., 2008; Zhao et al., 2017).

In angiosperms, only the chalaza, among the four haploid megaspores, survives to become a functional megaspore, with the other three degenerating rapidly. The selection of the functional megaspore is position-dependent, as observed in mutants like *swi/dyad*, where two unreduced diploid megaspores are formed, but only the chalazal megaspore expresses the megaspore marker, indicating position-dependent selection of the functional megaspore.

14.2.1.3 Megagametogenesis—The Initiation of Megaspores

The megagametogenesis phase initiates from the functional megaspore. Once the mature megaspore is established, it undergoes rapid enlargement, filling the space vacated by the three degenerated megaspores. The single nucleus within the megaspore undergoes the first round of karyokinesis. Subsequently, the central vacuole develops, separating the two daughter nuclei, which move to the two poles. These nuclei at each pole undergo a second round of karyokinesis, resulting in a 4n + 4n structure. One of the four nuclei from each pole migrates to the central position, fusing to create a diploid central nucleus. The central vacuole expands, accentuating the distinction between the poles. As the polar nuclei migrate, cellularization is established, forming a seven-celled structure with eight nuclei (Yadegari & Drews, 2004; Yang et al., 2010). The seven cells include two synergid cells, one egg cell, one large central cell, and three antipodal cells.

Factors involved in the initiation of megaspores are limited. AtAGP18 (Arabinogalactan protein 18) encodes a classic arabinogalactan protein in *Arabidopsis* expressed in the female germline, especially in the functional megaspore. The *agp18* mutant exhibits a failure in the enlargement and division of the functional megaspore, suggesting a crucial role for proteoglycan in functional megaspore development (Acosta-Garcia & Vielle-Calzada, 2004; Demesa-Arevalo & Vielle-Calzada, 2013). AGL23, a type I MAD-box protein, is implicated in the first division of the megaspore nucleus, with the *agl23* mutant showing a blockage in this division. This suggests that AGL23-mediated transcriptional regulation is essential for the first division of the functional megaspore (Colombo et al., 2008).

14.2.1.4 Megagametogenesis—Three Rounds of Karyokinesis of Megaspore

During the successive mitotic processes that follow, genes associated with the cell cycle, as well as those involved in chromosome modification and structure, regulate the megaspore mitosis. PROLIFERA (PRL), encoding a homolog of yeast and mammalian MCM proteins, is crucial for initiating DNA replication, and its mutation arrests the embryo sac at the one-nucleate stage (Springer et al., 2000). INTERACTOR/INHIBITOR OF CYCLIN-DEPENDENT KINASE (ICK)/KIP-RELATED PROTEINS (KRP), inhibitors of CDK, are expressed in the megaspore and embryo sac, and their septuple mutant produces multiple embryo sacs in one ovule. RING-H2 GROUP F1A (RHF1A) and RHF2A, functionally redundant RING-finger proteins, regulate mitotic cell-cycle progression through the degradation of their substrate ICK/KRP6, leading to mitotic arrest at interphase in the FG1 stage in *rhf1a rhf2a* mutants (Liu et al., 2008). ANAPHASE-PROMOTING COMPLEX/CYCLOSOME (APC/C), an E3 ubiquitin ligase, triggers cyclin degradation during both mitosis and meiosis, and the knockout of NOMEGA, encoding an APC/C subunit, halts embryo sac growth at the two-nucleate stage (Kwee & Sundaresan, 2003).

In addition to the cell cycle, genes involved in chromosome modification and structure, such as HAM1/HAM2 (histone acetylases), HDA7 (a member of the REDUCED POTASSIUM DEPENDENCY3 subfamily of HISTONE DEACETYLASE), and CHR11 (an ATP-dependent chromatin-remodeling factor), participate in megaspore mitosis (Cigliano et al., 2013; Huanca-Mamani et al., 2005; Latrasse et al., 2008). Mutations in these genes result in mitotic defects during female gametophyte development. Some ribosome-related factors also function in mitosis, with SLOW WALKER (SWA), involved in 18 s-RNA processing, displaying a slower cycle of female gametophyte development and ultimately producing an aborted female gametophyte (Liu et al., 2010). Loss of *SNAIL1*, which encodes a large-subunit homolog of the ribosome, leads to delayed gametophyte development, although it forms a mature embryo sac with pollen tube attraction activity, suggesting unaffected synergid cell fate (Hao et al., 2017). Additionally, other signaling pathways, such as AHK2/3/4 (cytokinin receptors), CKI1 (downstream of the cytokinin pathway), NUWA (an imprinted gene regulating mitochondria function), and ROOT INITIATION DEFECTIVE 1 (RID) (an RNA helicase regulating female gametophyte development), all play roles in mitosis (He et al., 2017; Liu et al., 2017; Yuan et al., 2016; Zhu et al., 2016).

Upon the completion of the final mitosis, one nucleus from each pole migrates and fuses to form the diploid secondary nucleus. Research on nuclei migration is very limited. Only a few mutants show some nuclei mispositioning phenotypes. TUBGs and NACK1/2 (see below) are factors essential for cellularization. Their mutants show nuclei mispositioning, suggesting that the two genes might also be involved in nucleus migration. Nucleus fusion studies are more extensive, revealing that MAGATAMA 3 (MAA3), GLUCOSE 6-PHOSPHATE/PHOSPHATE TRANSLOCATOR 1 (GPT1), and DIANA (DIA)/AGL61 mutants exhibit migrated but unfused nuclei, resulting in a fertilization defect of the central cell (Niewiadomski et al., 2005; Shimizu et al., 2008; Steffen et al., 2008). Mitochondria-related genes, including RPS9M (a mitochondrial ribosome protein), HEMN1 (encoding a coproporphyrinogen III oxidase in mitochondria), and GFA2 (encoding a chaperone in mitochondria), are important for nuclear fusion, with mutations in these genes causing defects in this process (Lu et al., 2017; Pratibha et al., 2017; Christensen et al., 2002).

14.2.1.5 Megagametogenesis—Cellularization of Gametophytic Cells

When nuclei undergo migration and fusion, cellularization within the syncytial embryo sac becomes an important process for gametophytic cell fate specification. The temporal control of cellularization in female gametophytes is evident in *hadad* (*hdd*) ovules, where cellularization occurs prematurely after the first or second gametophytic division, resulting in arrested embryo sacs and eventual degeneration (Moore et al., 1997). Spatial control of cellularization is also observed, particularly in the case of GEMINI POLLEN 2 (GEM2), which functions in both male and female gametophyte development. Female gametophytes in *gem2* mutants fail to complete cellularization, leading to abnormal nuclei distribution in the embryo sac. Simultaneously, defective boundaries between nuclei at the micropylar end cause their fusion into one or two large nuclei (Park et al., 2004).

The involvement of both microtubules and phragmoplast is essential for cellularization during female gametophyte development. GEM1/MOR1, a microtubule-associated protein, and TIO, a phragmoplast-associated protein localized to the midline of the nascent phragmoplast, play essential roles. Mutants of GEM1/MOR1 or TIO exhibit similar phenotypes to *gem2* mutants. TIO is specifically required for cell plate expansion during cellularization, emphasizing its importance in this process (Oh et al., 2005).

The positioning of the cell plate and phragmoplast also contributes to cellularization. Loss of the two γ-tubulin genes, TUBG1 and TUBG2, in *Arabidopsis* results in aberrant spindle and phragmoplast structures. The *tubg1 tubg2* mutant displays an incorrect number and aberrant positioning of nuclei in the gametophytes (Pastuglia et al., 2006). In addition, NPK1-ACTIVATING KINESIN 1 (NACK1) and NACK2, which activate the MAPKK kinase NPK1, are implicated in cellularization. Mutants of these genes exhibit mispositioning of nuclei, leading to the formation of nonfunctional gametophytes (Jin et al., 2002; Krysan et al., 2002; Nishihama et al., 2001; Tanaka et al., 2004).

14.2.1.6 Megagametogenesis—The Polarity of Embryo Sac

The mature embryo sac is a polarized structure, with egg cells and synergid cells located at the micropylar end and antipodal cells positioned at the chalazal end. This polarity arises from the intricate processes of nuclear division, migration, vacuole expansion, and cellularization. Notably, the post-cellularization destiny of nuclei at FG5 marks their final positions, emphasizing the critical role of nuclear positioning rather than lineages in egg cell specification (Christensen et al., 1997; Susaki et al., 2021). Disrupting the actin cytoskeleton before cellularization and impairing nuclear migration lead to the formation of two functional egg cells, reinforcing the significance of nuclear positions in cell fate determination (Sun et al., 2021).

Distinct patterns of gene expression during this process contribute to positional cues in gametic cell specification. For example, DEMETER, a key regulator of gene imprinting in endosperm development, exhibits specific expression in the micropylar region. It is initially expressed in the polar nuclei and synergid nuclei before cellularization and later becomes restricted to the central cells. In contrast, FIS2, a downstream gene of DME, is exclusively expressed in polar nuclei (Choi et al., 2002). These polar expressions suggest the presence of positional cues in the specification of gametic cells.

Several signal pathways also play roles in cell fate determination. LACHESIS (LIS), CLOTHO (CLO)/GFA1, and ATROPOS (ATO), components of the RNA splicing machinery, are primarily expressed in egg cells and central cells. Mutations in these genes result in synergid cells or central cells adopting an egg-like structure, generating multiple egg cells. Thus, their expression may induce lateral inhibition from the egg cell (Gross-Hardt et al., 2007; Moll et al., 2008). Additionally, auxin and cytokinin pathways are crucial for cell specification, as disrupting normal auxin synthesis or distribution patterns impairs egg cell specification (Pagnussat et al., 2009; Yuan et al., 2016; Sun et al., 2021).

In the two-component cytokinin signaling pathway, five *Arabidopsis* Histidine Phosphotransfer protein genes (AHPs) (AHP1–AHP5) exist. Recent studies suggest that AHP2, AHP3, and AHP5 act as major downstream factors of CKI1 and are expressed in the central cell. Furthermore, ARR10/12/18 function downstream of CKI1–AHP2/3/5. In the *ahp2 ahp3 ahp5/+* and *arr10 arr12 arr18/+* triple mutant ovules, central and antipodal cells lose their fates and acquire attributes of egg cells or synergid cells (Cheng & Kieber, 2013; Deng et al., 2010a; Zhu et al., 2022).

14.2.2 Male Gametogenesis

In angiosperms, male gametophyte development takes place in stamens and occurs synchronously with female gametophyte development. The process involves two sequential phases: microsporogenesis and microgametogenesis (Borg et al., 2009). During microsporogenesis, the archesporial cell differentiates into a diploid POLLEN MOTHER CELL (PMC), which then undergoes meiosis to produce tetrads of haploid microspores. The release of distinct unicellular microspores from the tetrads marks the completion of this phase. Subsequently, during microgametogenesis, the microspore undergoes two rounds of mitosis to form the male germ unit (MGU), establishing a physical association between the sperm cells and the vegetative nucleus (Borg et al., 2009) (Fig. 14.3). Numerous genes have been identified in recent years to play important roles in male gametogenesis, with some genes functioning in both female and male gametogenesis due to shared processes in development.

14.2.2.1 Microsporogenesis—Cell Fate Transition

A cell fate transition from the archesporial cell to the PMC mirrors processes in female gametophyte development, with several factors playing key roles in both. For instance, SPL, expressed in anthers, is essential for male germline development,

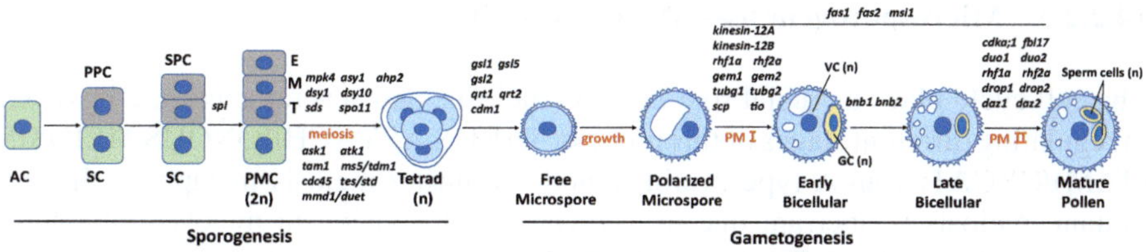

Fig. 14.3 Schematic drawing depicting the sequential development of trinucleate pollen. The mutants and genes involved in the meiotic and mitotic cell-cycle regulation are indicated. AC archesporial cell, SC sporogenous cell, PPC primary parietal cell, SPC secondary parietal cell, PMC pollen mother cell, E endothecium, M middle layer, T tapetum, VC vegetative cell, GC generative cell, PM I pollen mitosis I, PM II pollen mitosis II

as its loss results in a complete absence of male germline (Schiefthaler et al., 1999; Yang et al., 1999b). In rice, the TPD1-EMS1/EXS1 mediated signal pathway shares similarities with the mechanism of TDLA1-MSP1, which controls male sporocyte fate (Huang et al., 2016).

14.2.2.2 Microsporogenesis—Pollen Mother Cell (PMC) Formation

The PMC undergoes two rounds of meiosis to produce a tetrad, and gene functions controlling chromosome behavior and the cell cycle contribute to this process. SYN1/DIF1, SPO11, and DMC1, known for their roles in female gametogenesis meiosis, also participate in PMC meiosis, and mutants of these genes exhibit defects in PMC meiosis (Bai et al., 1999; Bhatt et al., 1999). Additionally, RAD50 and MRE11, involved in SPO11-mediated DNA double-strand breaks, are essential for meiosis in male gametogenesis. RAD51 functions in chromosome pairing, synapsis, and chromosome recombination, with mutants such as *asy1*, *ahp2*, and *sds* showing defects in synapsis and reduced fertility (Armstrong et al., 2002; Azumi et al., 2002; Hollingsworth et al., 1990; Peirson et al., 1996; Ross et al., 1997). In *Arabidopsis*, ASYNAPTIC1 (ASY1) is a homolog of HOP1 in yeast, which is localized to the axis-related chromatin (Armstrong et al., 2002; Hollingsworth et al., 1990). In *asy1*, homologous chromosomes cannot undergo synapsis at meiosis prophase I. AHP2, encoding a protein similar to yeast MEU13P and HOP2, is involved in autosyndetic pairing, and its loss results in defects in pairing and bivalent formation during meiosis. SOLO DANCERS (SDS), which encodes a protein with a cyclin-like domain, affects association, recombination, and bivalent formation in MMC (Azumi et al., 2002). In *Arabidopsis thaliana*, ARABIDOPSIS SKP1-LIKE 1 (ASK1), a homolog of human and yeast S-PHASE KINASE ASSOCIATED PROTEIN 1 (SKP1), participates in ubiquitin-mediated protein degradation, with the *ask1* mutant showing abnormal chromosome segregation during meiosis anaphase I and producing nuclear-defective microspores (Yang et al., 1999a).

14.2.2.3 Microsporogenesis—Meiosis of PMC

The cell cycle plays a crucial role in PMC meiosis, and several genes are involved in regulating different stages of this process. TARDY ASYNCHRONOUS MEIOSIS (TAM)/CYCA1;2, an A-type cyclin, functions during meiosis metaphase. In *tam* mutant, meiosis is aberrant due to asynchronized division in meiosis metaphase within the PMC. MS5/TDM1/POLLENLESS3 encodes a protein similar to the SYNAPTONEMAL COMPLEX (SCPs) and cell-cycle regulatory proteins (Chaudhury et al., 1994; Glover et al., 1998; Ross et al., 1997; Sanders et al., 1999). After meiosis II, the MMC in *ms5* undergoes an additional division, leading to abnormal pollen development (Glover et al., 1998). MMD1/DUET may control the cell cycle of meiosis, as observed in the *mmd1/duet* mutant, where early chromosomes exhibit normal behavior, but from diakinesis onward, cytosol constriction occurs, chromosomes break, and cells die before the completion of the second cytokinesis (Reddy et al., 2003; Yang et al., 2003b). AtCDC45, a homolog of the yeast cell-cycle regulatory protein in *Arabidopsis*, is essential for PMC meiosis. Knocking out CDC45 prevents the PMC from producing tetrads (Stevens et al., 2004).

Furthermore, microtubules and their associated proteins are essential for proper meiosis. KATA/ATK1 encodes a kinesin-like protein involved in spindle formation during meiosis. The *kata/atk1* mutant, identified through transposon tagging, exhibits reduced fertility (Chen et al., 2002a). During metaphase I of the PMC in the *kata/atk1* mutant, chromosomes fail to align properly on the equatorial plate, leading to defective chromosome segregation in anaphase I. Abnormal spindle formation is implicated as the cause of the meiosis defect in the PMCs. TES (TETRASPORE)/STD (STUD) encodes a kinesin protein, and in the *tes/std* mutant, karyokinesis is normal, but cytokinesis does not proceed normally during meiosis, resulting in the production of huge spores with four nuclei. These oversized spores can continue development, undergo two rounds of mitosis, and eventually produce large pollen grains with up to eight sperm cells each (Hulskamp et al., 1997; Spielman et al., 1997; Yang et al., 2003a).

14.2.2.4 Microsporogenesis—Synthesis and Degeneration of Callose

Plasmodesmata facilitate communication between each PMC, potentially transferring signaling molecules that synchronize the initiation of meiosis in PMCs within the same anther sac. Concurrently, as a PMC enters meiosis, gradual callose deposition occurs on its cell wall, coinciding with the interruption of plasmodesmata connections. The callose, comprised of β-1,3-glucan, is considered essential for tetrad formation. Researchers think that this callose wall, easily synthesized and degraded, is a prerequisite for meiosis, providing sugar for the pollen outer wall and preventing cell adhesion during microspore separation. Mutants with defective callose synthesis, particularly in the GLUCAN SYNTHASE-LIKE (GSLs) family responsible for callose synthesis, exhibit pollen abortion. For instance, *gsl1gsl5* mutants show abortive pollen due to callose synthesis defects, while *gsl2* mutants display improper exine sculpting, leading to pollen abortion (Dong et al., 2005; Enns et al., 2005; Shi et al., 2015, 2016). Callose synthesis may impact microspore cytokinesis by influencing cell plate formation, with MPK4, a kinase in the MPK family involved in the MAPKK signaling pathway, playing an important role. In the *mpk4* mutant, MMCs develop normally but fail to form intersporal callose walls post-meiosis, preventing completion of meiotic cytokinesis (Zeng et al., 2011).

Meanwhile, callose degradation is temporally regulated, and premature degradation influences pollen activity. Callase, the enzyme responsible for callose degradation, is produced by the tapetum after tetrad formation. Tapetum integrity is crucial for callose secretion, influencing pollen development. CALLOSE DEFECTIVE MICROSPORE1 (CDM1), encoding a tandem CCCH-type zinc finger protein highly expressed in tapetum at stages 4–7, is essential for microspore development. In *cdm1*, abnormal callose formation leads to rapid microspore degeneration. CDM1 potentially regulates callose metabolism through genes involved in callose synthesis (e.g., CALLOSE SYNTHASE 5 and 12) and callose-related genes (e.g., A6 and MYB80) (Lu et al., 2014).

14.2.2.5 Microgametogenesis—Mitosis of Microspores

During microgametogenesis, free microspores undergo rapid enlargement, accompanied by the formation of a single large vacuole (Borg et al., 2009; Owen & Makaroff, 1995; Yamamoto et al., 2003). The formation of the large vacuole eventually causes the microspore nucleus to migrate to a peripheral position against the cell wall. Following this, polarized microspores undergo two rounds of mitosis to mature into pollen. The first mitosis, known as pollen mitosis I (PMI), is an asymmetric division resulting in two distinct daughter cells: a large vegetative cell and a small generative cell, separated by the microspore cell wall. The generative cell migrates into the vegetative cell, forming a distinctive "cell-within-a-cell" structure (Hafidh et al., 2016; Russell, 1996; Russell & Jones, 2015). The engulfing process concludes with the degradation of the hemispherical callose wall that initially separated the vegetative and germ cells. The importance of the asymmetric division at PMI lies in correct cellular patterning in male gametophyte development due to the unique gene expression profiles of the daughter cells (Borg et al., 2009).

In *Arabidopsis*, the *sidecar* pollen (*scp*) mutant is the first reported male gametophytic mutant, displaying abnormal microspore division and cellular patterns. In *scp*, microspores undergo symmetrical division instead of the typical asymmetric division, resulting in both daughter cells acquiring the vegetative cell fate (Chen & McCormick, 1996). Nuclear division and cytokinesis, specifically the establishment of a cell plate and/or cell wall, are crucial events during PMI (De Storme & Geelen, 2013). GEM2, known for its role in cellularization of the female gametophyte, also functions in the first asymmetric division of the microspore. In *gem2*, mis-positioned cell plates and abolished cytokinesis lead to both daughter cells possessing the vegetative cell fate (Park et al., 2004). Microtubules play a vital role in the first asymmetric division, with mutants of γ-tubulin genes TUBG1 and TUBG2 exhibiting abnormal polarity post-PMI (Pastuglia et al., 2006). GEM1/MOR1, a microtubule-associated protein, is essential for establishing microspore polarity by stimulating spindle growth and myoblast microtubulins in interphase (Park et al., 1998; Twell et al., 2002; Whittington et al., 2001). In *gem1* mutants, some microspores exhibit symmetry. In tobacco, the mutant of TMBP2000, which is the homolog of GEM1/MOR1, shows a similar phenotype. The phragmoplast midline localized TIO is crucial for the first asymmetric division (Oh et al., 2005, 2014). The *tio* mutant displays cytokinesis defects, resulting in microspore arrest in the bi-nucleate form. Two functionally redundant microtubule motor kinesins, PAKRP1/Kinesin-12A and PAKRP1L/Kinesin-12B, coordinate with TIO to regulate mitosis (Lee et al., 2007). In *kinesin-12a/kinesin-12b* mutants, dividing microspores exhibit microtubule organization defects and form an antiparallel microtubule array between the nuclei. Importantly, in *gem1*, *tio*, and *kinesin-12a/kinesin-12b* mutants, karyokinesis is normal, suggesting that microtubulins might not function in nucleus division. At the early bicellular stage, the germ cell migrates into the vegetative nuclei to form the "cell-within-a-cell" structure. Recent studies emphasize the significance of germ cell migration into the vegetative nuclei for male gametophyte development. In *Arabidopsis thaliana*, BNB1/2, encoding the VIIIa subfamily of bHLH transcription factors, is involved. In *bnb1/bnb2* mutants, the microspore divides to form a two-nucleate pollen, but the germ cell remains within the pollen wall boundaries instead of migrating into the vegetative nucleus. Finally, the mature pollen contains only one or no nucleus, highlighting the importance of migration (Yamaoka et al., 2018).

At the late bicellular stage, the large vegetative cell remains in the G1 stage, accumulating proteins essential for rapid pollen tube growth (Borg et al., 2009; Hafidh et al., 2016). Simultaneously, the germ cell undergoes chromatin condensation and initiates the second mitosis (PMII) to generate twin sperm cells. PMII can occur in either the pollen grain or the pollen tube, resulting in two types of mature pollen: bi-nucleate pollen and trinucleate pollen (Brewbaker, 1967).

Numerous genes play important roles in the PMII stage, including components of the canonical cell-cycle machinery. In *Arabidopsis*, T-DNA insertion in CDKA;1 leads to mitosis failure in PMII and delayed DNA synthesis in the (S) phase of the cell cycle, resulting in the presence of only one sperm-like cell in *cdka;1* mutant (Iwakawa et al., 2006; Nowack et al., 2005). F-box-like 17 (FBL17), an F-box protein, interacts with SKP1 and CUL1 to form the SKP1-CUL1-F-box protein (SCF) E3 ubiquitin ligase complex (Petroski & Deshaies, 2005; Smalle & Vierstra, 2004). FBL17, transiently expressed in the later stage of PMI, degrades KRP6/7, an inhibitor of CDK. The *fbl17* mutant exhibits a phenotype similar to that of *cdka;1* (Gusti et al., 2009; Iwakawa et al., 2006; Kim et al., 2008; Nowack et al., 2005). RHF1A and RHF2A, mentioned earlier, participate in male gametophyte mitosis through the same mechanism as in the female gametophyte. The reassembly of nucleosomes and chromatin is another critical event in PMII. Members of the CHROMATIN ASSEMBLY FACTOR-1 (CAF-1) signal pathway, FAS1, FAS2, and MSI1, are essential for both PMI and PMII. Mutants of these genes display defects in PMI or PMII, emphasizing the significance of the CAF-1 signal pathway (Chen et al.,

2008). Additionally, some transcription factors and their target genes play important roles in PMII. DUO POLLEN 1 (DUO1), encoding an R2R3-MYB transcription factor specifically expressed in germ and sperm cells, is essential for germ cell entry into PMII. In the *duo1* mutant, mitosis is arrested at the G2-M phase, resulting in the formation of only one sperm-like cell in the pollen (Durbarry et al., 2005; Rotman et al., 2005). DAZ1 and DAZ2 are target genes of DUO3, encoding the C2H2-type zinc finger protein. The daz1 daz2 mutant exhibits a phenotype similar to that of duo1, indicating their essential role as target genes of DUO1 for mitosis (Borg et al., 2014). Throughout the developmental stages, nuclear stability is very important. bHLH transcription factors DEFECTIVE REGION OF POLLEN 1 (DROP1) and DROP2 may function in this process. While PMI can occur in most *drop1 drop2* mutants, the germline cell or both the vegetative and germline cells gradually degenerate. In *drop1 drop2* mutants, mature pollen possesses only one vegetative nucleus or no nucleus, underscoring the importance of these two transcription factors in maintaining nuclear stability during microgametogenesis (Zhang et al., 2017a).

14.2.3 Different Strategies for Sperm to Meet Female Gamete in Plants

Evolutionarily, the living environments of plants shifted from water to terrestrial habitats. This substantial change necessitated the development of new structures, including vasculature, roots, and guard cells. In response to these environmental shifts and to ensure successful fertilization, plants also adapted their reproductive system structures, employing different strategies to facilitate the meeting of sperm cells with female gamete.

Aquatic plants and basal land plants inhabit moist environments conducive to the mobility of their sperm cells. Comparable to animals, seedless plants produce motile sperm cells that exhibit chemotaxis, allowing them to perceive female cues and navigate toward the caulonema. Recent research in the basal land plant *Physcomitrella patens* highlights sperm chemotaxis, revealing the GLUTAMATE RECEPTOR-LIKE (GLR) family's role as non-selective Ca^{2+} permeable channels regulating cytoplasmic Ca^{2+}. Mutants of the GLR genes, *glr1/2*, in *Physcomitrella patens* show defects in sperm cell targeting of female reproductive organs (Ortiz-Ramirez et al., 2017). Thus, GLR channels play an essential role in sperm chemotaxis in this basal land plant. MADS-box transcription factors, known for their involvement in various developmental processes in higher plants, also participate in motile flagella formation of moss *Physcomitrella patens* sperm cells (Koshimizu et al., 2018). However, this biological function is lost with the loss of flagella in higher plant sperm cells.

Unlike seedless plants, the sperm cells of most seed plants are non-motile, with the exception of gymnosperms, Cycadales, and Ginkgoales. Seed plants have evolved a new fertilization pathway called siphonogamy. This process relies on a novel structure, the pollen tube, which transports immotile sperm cells to the female gametophyte for fertilization. The mutant of two bHLH transcription factors, *drop1 and drop2*, results in a pollen tube without sperm cells but with normal pollen tube development. In siphonogamy, sperm cells become passive cargo within the pollen tube (Zhang et al., 2017a).

14.2.4 Pollen Adheres to the Stigma and Hydrates to Germinate

After landing on the stigma, pollen grains undergo a series of processes: pollen capture, adhesion, hydration, and germination (Fig. 14.4a). The interaction between the male gametophyte (pollen tube) and the female gametophyte (pistil) initiates at the stigma, making the stigma a pivotal site for the "first" male-female interaction. In angiosperms, stigmas are categorized into two types: "dry" and "wet" (Heslopharrison, 1982). Extensively studied plants with "wet" stigmas include *Petunia*, *Nicotiana*, and *Lilium*, while well-studied "dry" stigma plants include *Brassica*, Raphanus, *Arabidopsis*, and *Secale*. Each step of the pollen-stigma interaction differs distinctly in "dry" and "wet" stigma plants.

14.2.4.1 Pollen Capture and Adhesion

The pollen-stigma interaction commences with rapid pollen capture, immediately followed by physical adhesion between the pollen and papilla cells of the stigma. In plants with "wet" stigmas, pollen capture and adhesion are considered passive and

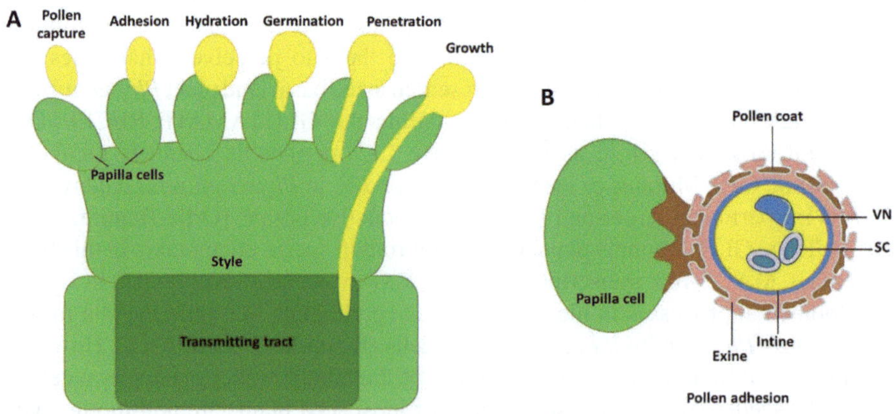

Fig. 14.4 Schematic diagram of the interaction between pollen/pollen tube and the stigma. (**a**) Schematic diagram of the successive interactions between pollen/pollen tube and the stigma at various steps. (**b**) Schematic diagram of the pollen structure and status during the adhesion process. VN vegetative nucleus, SC sperm cell

nonspecific processes (Dumas et al., 1984; Sanchez et al., 2004), relying on the surface tension of the exudate on the stigma. These exudates are identified as lipids and carbohydrate-rich substances (Dumas et al., 1984; Sanchez et al., 2004; Wolters-Arts et al., 1998). In contrast to "wet" stigma plants, pollen capture and adhesion are intricate in "dry" stigma plants. These two processes have been studied separately. For pollen capture, exine is necessary, while a pollen coat is not. The eceriferum mutant *cer6-2*, unable to form a pollen coat, shows normal pollen capture between pollen and stigma (Fiebig et al., 2000). Mutants such as less adherent pollen 1 (*lap1*), *male sterility 2* (*ms2*), and *faceless pollen-1* (*flp1*), exhibiting abnormal exine development, demonstrate significantly reduced pollen capture forces between pollen and stigma (Zinkl & Preuss, 2000).

After capture by the stigma, pollen adheres to the papilla cells. During this time, the pollen coat mobilizes, flowing into the contact point with the stigma. Subsequently, the pollen coat protrudes from the bacula on the exine, binding tightly with the papilla cell (Elleman & Dickinson, 1986) (Fig. 14.4b). Evidence also suggests that adhesion in "dry" stigma plants is species-specific (Zinkl et al., 1999). When attaching to the papilla cell, lipids, carbohydrates, and proteins from both pollen and stigma mix. The stigma cell wall contains two pairs of S (self-incompatibility) protein family members, and their binding proteins, S-Locus-Related-1 proteins (SLR1)—SLR-BP1 and S-Locus Glycoprotein (SLG)—PCP-A1, participate in this process (Doughty et al., 1998; Kandasamy et al., 1989; Lalonde et al., 1989; Luu et al., 1997, 1999; Takayama et al., 2000; Umbach et al., 1990). When SLR1 expression is reduced by transgenic suppression or stigmas are treated with SLG monoclonal antibodies, pollen-stigma adhesion is significantly reduced (Luu et al., 1997, 1999).

14.2.4.2 Pollen Hydration

After adhering to the stigma, the pollen undergoes hydration, and this process differs in "wet" and "dry" stigma plants. In "wet" stigma plants, hydration is a passive process, akin to pollen adhesion. Exudates from "wet" stigma plants supply sufficient water for pollen hydration (Swanson et al., 2004), and lipids in the stigma exudates are crucial for this process. The ectopic expression of the cytotoxic gene STIG10-barnease in *Nicotiana tabacum*, resulting in the transgenic plants' inability to produce triacylglyceride-rich lipids, led to impaired pollen hydration (Goldman et al., 1994).

In contrast, pollen hydration in "dry" stigma plants is a highly regulated process. Lipids and proteins in the pollen coat, along with proteins in the stigma, collectively contribute to pollen hydration. Mutants in the CER gene family, unable to form a pollen coat due to a defect in long-chain lipid synthesis, exhibit in vivo pollen hydration defects, leading to male sterility. Interestingly, these mutants can hydrate either in vitro or under high-humidity conditions (Hulskamp et al., 1995). Additionally, when the cer6 mutant is pollinated with wild-type pollen, the hydration defect in *cer6* can be rescued—a phenomenon termed the pollen mentor effect

(Hulskamp et al., 1995). Proteins in the pollen coat, such as GLYCINE-RICH OLEOSIN PROTEINS (GRPs), specifically GRP17, which are abundant in the pollen coat, play a role in pollen hydration. Knocking out *GRP17* results in significantly reduced pollen hydration, indicating its involvement in the process. EXL4, a lyolipase expressed in the pollen coat, collaborates with GRP17 to regulate pollen hydration by affecting water transfer from the stigma to the pollen (Updegraff et al., 2009). Proteins located in the stigma also play an important role in pollen hydration in "dry" stigma plants. The endocytosis complex subunits SEC3, SEC5, SEC6, SEC8, SEC10, SEC15, and EXO84, as well as PIP4, are reported to participate in pollen hydration (Safavian et al., 2015; Chapman & Goring, 2011). Ca^{2+} is essential for pollen hydration, with dynamic distribution observed during pollen hydration (Iwano et al., 2004). The pollen-specific gene MSL8, a mechanosensitive channel, is required for pollen to survive the hypoosmotic shock of rehydration, indicating the involvement of osmotic forces in pollen hydration (Hamilton et al., 2015). Stigma reactive oxygen species (ROS) levels also play an essential regulatory role in pollen hydration, with the modulation orchestrated by a redox-controlled gate mediated by peptides and receptors in *Arabidopsis* (Liu et al., 2021). The deposition of wild-type pollen suppresses stigmatic ROS, easing the stigmatic gate and facilitating hydration. In unpollinated stigmas, high ROS levels result from the binding of peptides RALF23/RALF33 to receptors from the Catharanthus roseus RLK1-LIKE (CrRLK1L) family, including FERONIA, ANJEA, and the co-receptor LLG1. This interaction activates RBOHD, leading to ROS accumulation in the stigma, inhibiting pollen hydration. PCP-B class pollen coat proteins (PCPs) from the pollen compete with RALF23/33 for receptor binding, reducing pistil ROS levels and promoting pollen hydration (Liu et al., 2021).

14.2.4.3 Pollen Germination

After pollen completes hydration, it undergoes germination to form a tube and then invades the style. In *Arabidopsis*, pollen germination is a rapid process in vivo, whereas it is slower on a medium in vitro. This observation suggests that some factors of the stigma promote pollen germination in vivo. Small molecule sulfinylated azadecalins have been detected in *Arabidopsis* pistils through mass spectrometry analysis and can stimulate *Arabidopsis* pollen germination in vitro (Qin & Yang, 2011). The pollen wall is considered essential for pollen germination. SEC31B, a component of the secretory COPII complex functioning in ER-Golgi vesicle trafficking, is crucial for normal primexine development and exhibits a significant reduction in pollen germination in the *sec31b* mutant. Pollen-expressed LEUCINE-RICH REPEAT EXTENSINS (LRXs) are important for maintaining pollen wall integrity and also contribute to pollen germination in *Arabidopsis* (Wang et al., 2018). The dynamics of the actin cytoskeleton play an important role in pollen germination. The participation of the subclass III ACTIN-DEPOLYMERIZING FACTORs (ADFs) member ADF5 in bundling actin filaments is critical for pollen germination (Zhu et al., 2017).

14.2.4.4 Pollen Tube Penetration into the Stigmatic Papilla

Upon pollen germination, the pollen tube penetrates into the cell wall of the papilla cell, navigating through the loosened cell wall (Kandasamy et al., 1994; Riglet et al., 2020; Lan et al., 2023a). In *Arabidopsis*, a "lock-and-key model" mediated by RAPID ALKANIZATION FACTOR (RALF) peptides and CrRLK1L receptors has been reported to govern the pollen tube penetration process (Lan et al., 2023a). The stigmatic gate is effectively locked by members of the CrRLK1L family, including FER-ANJ/HERCULES1/CURVY1, four self-secreted small peptides sRALF (RALF1/22/23/33), and cell wall proteins LRX3/4/5. Pollen grains carrying the "keys" pRALFs (RALF10/11/12/13/25/26/30) are allowed to penetrate the papilla cell, as pRALFs can competitively bind to the receptors, displacing sRALFs, and thereby unlocking the stigmatic gate. In contrast, the pollen tubes of the *ralf sept* (a mutant with simultaneous mutations in all seven pRALFs) are unable to penetrate the stigmas. However, mutant stigmas lacking the "lock" components, such as the stigmas of *fer*, *cvy1 anj herk1*, *ralf quad* (a mutant with simultaneous mutations in all four sRALFs), and *lrx345*, facilitate the penetration of *ralf sept* pollen tubes that lack the corresponding "keys." Although some components are shared between the stigmatic gate during pollen tube penetration and pollen hydration, such as the autocrine peptides RALF23/33 and the receptors FER-ANJ, it appears that the stigmatic ROS levels are not involved in regulating the process of pollen tube penetration, which is evident as the reduction of stigmatic ROS levels does not facilitate the penetration of *ralf sept* pollen tubes through the papilla cells (Lan et al., 2023a). The synthetic pRALF peptides can functionally substitute for the mentor pollen observed in the mentor effect. Treatment of *Arabidopsis* stigma with synthesized pRALF peptides not only promotes the penetration of *ralf sept* pollen through the stigma but, more importantly, also facilitates the penetration of interspecies pollen tubes through the *Arabidopsis* stigma, which is analogous to the facilitation of interspecies pollen tube penetration in the presence of *Arabidopsis* pollen during mixed-pollination events (Lan et al., 2023a).

14.2.4.5 Pollen Tube Elongation in Style

Once pollen germinates, the pollen tube first needs to invade the stigma and then grow into the style. This process is also quite different in "wet" and "dry" stigma plants. In "wet" stigma plants, the stigma lacks a cuticle. The cuticle is a defense against various stresses, such as pathogen attack and environmental changes. Therefore, without a cuticle, the pollen tube can easily invade the stigma after it germinates. In "wet" stigma plants with a close-type style, pollen tube growth depends on many enzymes from both the pollen and stigma. Pectinesterases (Jiang et al., 2005; Mu et al., 1994; Tian et al., 2006), polygalacturonases (Kim et al., 2006), and glucanases (Kotake et al., 2000) can loosen the extracellular matrix (ECM) between the papilla cells or modify the cell wall of the pollen tube. Unlike the "wet" stigma plants, the pollen tube must invade the stigma through the cuticle layer on the papilla cell in the "dry" stigma plants, *Arabidopsis* and *Brassica*. Intine-located cutinase is responsible for degenerating the cuticle on the papilla cells. Since cutinase is one type of serine esterase, the pollen tube of *Brassica napus* L. could not invade the papilla cell when the stigma was pretreated with serine esterase inhibitors (Hiscock et al., 1994). In *Brassica*, the pollen tube will grow to the bottom of the papilla cell between the cell wall and cell membrane when it breaks through the cuticle (Elleman et al., 1992). However, the pollen tube of *Arabidopsis* invades the cuticle, cell wall, and plasma membrane of the papilla cell, then grows to the bottom of the papilla cell (Elleman et al., 1992). Various enzymes, such as pectinesterases, pectin lyases, polygalacturonases, and xylanases, play an important role during this process (Suen et al., 2003; Suen & Huang, 2007). In *Arabidopsis*, a few proteins have been found to participate in pollen tube growth in the styles. Hydroxyproline O-arabinosyltransferases (HPATs) are involved in pollen tube growth in both the papilla cell and the style in *Arabidopsis*, and three HPAT genes exist. The pollen tube growth of *hpat1/hpat3* double mutant is arrested either on the papilla cell or in the style. HPAT proteins may regulate the synthesis of the pollen tube cell wall by affecting the stability of its extensin (Ogawa-Ohnishi et al., 2013). AtTOD1 codes a Golgi-localized alkaline ceramidase in *Arabidopsis*. The *tod1* pollen tube shows a growth defect because of the higher turgor in the pollen tube (Chen et al., 2015). The AtVPS41-mediated late stage of the endocytic pathway is very essential for pollen tube growth in the style. The growth of *vps41* pollen tube was seriously blocked in the pistil. Thus, the male transmission efficiency was reduced to nearly zero (Hao et al., 2016). O-FUCOSYLTRANSFERASE1 (AtOFT1) also participates in the penetration of the pollen tube through the stigma-style interface, suggesting that protein O-glycosylation events play an important role in pollen-pistil interaction (Smith et al., 2018).

Additionally, some biological processes and signaling pathways play an essential role in multiple steps of the pollen-pistil interaction. EXO70A1, which is a component of the endocytosis complex, participates in pollen adhesion, hydration, and pollen tube growth on the papilla cell. The *exo70a1* exhibits a significantly reduced adhesion force between pollen and stigma. When the *exo70a1* stigma was pollinated with wild-type pollen, only partial pollen hydration occurred. Pollen tube growth on the papilla cell also showed a defect in *exo70a1* (Samuel et al., 2009). Thus, EXO70A1-mediated endocytosis is important during pollen-stigma interaction. RLK and its ligand-mediated signaling pathway also participate in pollen-stigma interaction in tomato (Kaothien et al., 2005; Muschietti et al., 1998). The pollen tube-specific expressed peptide LATE ANTHER TOMATO 52 (LAT52) contributes to pollen hydration, germination, and in vitro pollen tube growth and specifically interacts with the pollen tube-located RLK LePRK2 (Twell et al., 1989). LAT52-LePRK2 may form a complex to regulate pollen tube growth (Tang et al., 2002). Recently, four secreted cysteine-rich peptides PCP-Bα/β/γ/σ have been identified in the pollen coat in Cruciferae. The *pcp-bα/β/γ* mutant pollen exhibits defects in pollen adhesion, pollen hydration, and pollen tube growth in vivo (Wang et al., 2017); see Table 14.2 and Fig. 14.4.

Table 14.2 Factors involved in the regulation of pollen-stigma recognition and pollen tube growth

Protein	Description	Species	Function	Reference
LAP1	/	*A. thaliana*	Pollen capture	Zinkl & Preuss (2000)
MS2	Fatty acyl carrier protein reductase	*A. thaliana*	Pollen capture	Aarts et al. (1997)
FLP1	Synthesis of the components of tryphine, sporopollenin of exine	*A. thaliana*	Pollen capture	Ariizumi et al. (2003)
SLR1 & SLR-BP1	SLR1: S-locus related 1 (SLR1) protein SLR-BP1: a defensin-like class A pollen coat protein	*B. campestris*	Pollen stigma adhesion	Luu et al. (1999) and Takayama et al. (2000)
SLG & PCP-A1	SLG: S-Locus Glycoprotein PCP A1: a defensin like class A pollen coat protein	*B. oleracea*	Pollen stigma adhesion	Luu et al. (1997) and Doughty et al. (1998)
CER1/3/6	Putative fatty acid reductase responsible for aldehyde formation	*A. thaliana*	Pollen hydration	Fiebig et al. (2000)
GRP17	Pollen coat oleosin	*A. thaliana*	Pollen hydration	Mayfield et al. (2001)
EXL4	Lyolipase	*A. thaliana*	Pollen hydration	Updegraff et al. (2009)
EXO84	Endocytosis complex subunit	*A. thaliana*	Pollen hydration	Safavian et al. (2015)
MSL8	Mechnosensitive channel	*A. thaliana*	Pollen hydration	Halmilton et al. (2015)
ADF5	Subclass III ADFs	*A. thaliana*	Pollen germination	Zhu et al. (2017)
SEC31B	COPII protein SECRETORY31B	*A. thaliana*	Pollen germination	Zhao et al. (2016)
HPATs	Hydroxyproline O-arabinosyltransferases	*A. thaliana*	PT penetration	Ogawa-Ohnishi et al. (2013)
AtTOD1	Alkaline ceramidase	*A. thaliana*	PT penetration	Chen et al. (2015)
AtVPS41	Vacuolar protein sorting	*A. thaliana*	PT penetration	Hao et al. (2016)
AtOFT1	O-lucosyltransferase1	*A. thaliana*	PT penetration	Smith et al. (2018)
EXO70A1	Component of endocytosis complex	*A. thaliana*	Pollen adhesion, hydration and PT growth	Samuel et al. (2009)

(continued)

Table 14.2 (continued)

Protein	Description	Species	Function	Reference
LRX3/4/5	Leucine-rich repeat extensins	*A. thaliana*	PT penetration	Lan et al. (2023a)
LRX8/9/10/11		*A. thaliana*	PT growth	Wang et al. (2018)
FER	*Catharanthus roseus* RLK1-LIKE family	*A. thaliana*	Pollen hydration and PT penetration	Liu et al. (2021) and Lan et al. (2023a)
ANJ		*A. thaliana*	Pollen hydration and PT penetration	Liu et al. (2021) and Lan et al. (2023a)
HERK1		*A. thaliana*	PT penetration	Lan et al. (2023a)
CVY1		*A. thaliana*	PT penetration	Lan et al. (2023a)
LLG1	LRE-like GPI-anchored protein	*A. thaliana*		Liu et al. (2021)
RALF1/22/23/33	Rapid alkanization factor family	*A. thaliana*	Pollen hydration and PT penetration	Liu et al. (2021) and Lan et al. (2023a)
RALF1/22		*A. thaliana*	PT penetration	Lan et al. (2023a)
RALF10/11/12/13/25/26/30		*A. thaliana*	PT penetration	Lan et al. (2023a)
RBOHD	NADPH/respiratory burst oxidase protein	*A. thaliana*	Pollen hydration	Liu et al. (2021)
PCP-Bα/β/γ/σ	Cysteine-rich peptides	*A. thaliana*	Pollen adhesion, pollen hydration, and PT growth	Wang et al. (2017) and Liu et al. (2021)
LAT52	A cysteine-rich extracellular protein	*S. lycopersicum*	Pollen adhesion, pollen germination and PT growth	Twell et al. (1989)
LePRK2	LRR RLK	*S. lycopersicum*	Pollen adhesion, pollen germination and PT growth	Twell et al. (1989)

(continued)

Table 14.2 (continued)

Protein	Description		Species	Function	Reference
CSLD1/4	Cellulose synthase-like D		*A. thaliana*	PT growth	Bernal et al. (2008) and Wang et al. (2011)
CSLA7	Cellulose synthase-like putative glycosyltransferase		*A. thaliana*	PT growth	Goubet et al. (2003)
VGD1	Pectin methylesterase		*A. thaliana*	PT growth	Jiang et al. (2005)
PME48			*A. thaliana*	Pollen germination, PT growth	Leroux et al. (2015)
AtPPME1	Pollen-specific pectin methylesterase		*A. thaliana*	PT growth	Tian et al. (2006)
NtPPME1			*N. tabacum*	PT growth	Bosch et al. (2005)
AtPMEI2	Pectin methylesterase inhibitor		*A. thaliana*	PT growth	Rockel et al. (2008)
AtPMEI1			*A. thaliana*	PT growth	Rockel et al. (2008)
ZmPMEI1			*Z. mays*	PT growth	Woriedh et al. (2013)
BoPMEI1			*B. oleracea*	PT growth	Zhang et al. (2010)
FH3	Actin-binding protein	Nucleating protein	*A. thaliana*	PT growth	Ye et al. (2009)
FH5			*N. tabacum*	PT growth	Cheung et al. (2010)
LIFH1			*L. brownii*	PT growth	Li et al. (2017)
ADF5/7/9		Lateral binding protein	*A. thaliana*	Pollen germination and PT growth	Zhu et al. (2013)
FIM4/5		Cross-linking protein	*A. thaliana*	Pollen germination and PT growth	Su et al. (2017a)
VLN5			*A. thaliana*	PT growth	Zheng et al. (2013) and Zhu et al. (2017)
MAP18	Microtubulin binding protein		*A. thaliana*	PT growth	Zhu et al. (2013) and Kato et al. (2010)
MDP25			*A. thaliana*	PT growth	Qin et al. (2014a)

(continued)

Table 14.2 (continued)

Protein	Description	Species	Function	Reference
SEC3/5/6/8	Exocyst complex	*A. thaliana*	PT growth	Cole et al. (2005), Hala et al. (2008), and Kulich et al. (2010)
EXO70		*A. thaliana*	PT growth	Bloch et al. (2016)
AtVps11		*A. thaliana*	PT growth	Tan et al. (2017)
POK/AtVPS52		*A. thaliana*	PT growth	Lobstein et al. (2004) and Guermonprez et al. (2008)
COG3/8	Oligomeric Golgi (COG) complex	*A. thaliana*	PT growth	Tan et al. (2016)
ALA3	Aminophospholipid ATPase	*A. thaliana*	PT growth	Zhou et al. (2020)
ROP1/RIC3/RIC4	Small GTPases	*A. thaliana*	Pollen germination and PT growth	Lin et al. (1996), Li et al. (1998), and Zheng & Yang (2000b)
ROP3/ROP5		*A. thaliana*	Functionally redundant to ROP1	Li et al. (1998)
ROP9/Rac7		*A. thaliana*	PT growth	Cheung et al. (2003)
Rac11/2		*A. thaliana*	PT growth	Fu et al. (2001) and Kost et al. (1999)
REN1		*A. thaliana*	Inhibition of ROP1	Hwang et al. (2008)
GDI1/GDI2a/GDI2b		*A. thaliana*	PT growth	Feng et al. (2016)
RabA4d		*A. thaliana*	PT growth	Szumlanski and Nielsen (2009)
NtRab2/11b		*N. tabacum*	PT growth	Cheung et al. (2002), de Graaf et al. (2005), Haizel et al. (1995), and Hepler et al. (2001)

(continued)

Table 14.2 (continued)

Protein	Description		Species	Function	Reference
CNGC18	Ca^{2+} channel	Calcium-permeable cyclic nucleotide-gated channel	*A. thaliana*	PT growth guidance	Frietsch et al. (2007)
OsCNGC13			*O. sativa*	PT growth	Chaiwongsar et al. (2009)
CNGC2			*A. thaliana*	PT growth	Xu et al. (2017)
ACA9		Ca^{2+}-ATPase	*A. thaliana*	PT growth and rupture	Schiott et al. (2004)
CBL1/9		Calcium sensor	*A. thaliana*	Pollen germination and PT growth	Mähs et al. (2013)
CBL2/3			*A. thaliana*	Pollen germination and PT growth	Steinhorst et al. (2015)
CML24/25		Calmodulin-like protein	*A. thaliana*	Pollen germination and PT growth	Yang et al. (2014) and Wang et al. (2015)
CIPK12/19		Calcineurin B-like protein-interacting protein kinase	*A. thaliana*	PT growth and polarity	Steinhorst et al. (2015) and Zhou et al. (2015)
CPK17/34/2/20/24/11		Calcineurin-dependent protein kinases	*A. thaliana*	PT growth	Myers et al. (2009), Gutermuth et al. (2013), and Zhao et al. (2013)
SIPK	K^+ channel	Salicylic acid-induced protein kinase	*A. thaliana*	Pollen germination and PT growth	Mouline et al. (2002)
SLAC1	S-type anion channel		*A. thaliana*	PT growth	Vahisalu et al. (2008)
SLAH3	S-type anion channel		*A. thaliana*	PT growth	Gutermuth et al. (2013, 2018)
ALMT12/13/14	R-type anion channel Al-activated malate transporter		*A. thaliana*	PT growth	
MGP2	Sialyltransferase-like protein		*A. thaliana*	Pollen germination and PT growth	Deng et al. (2010b)
MGP4			*A. thaliana*	PT and root growth	Liu et al. (2011)

(continued)

Table 14.2 (continued)

Protein	Description		Species	Function	Reference
SETH1/2	GPI anchor biosynthetic pathway		*A. thaliana*	Pollen germination and PT growth	Lalanne et al. (2004)
GAUT13/14	Galacturonosyltransferase		*A. thaliana*	PT growth	Wang et al. (2013)
HPAT1/2/3	Hyp O-arabinosyltransferase		*A. thaliana*	PT growth	Ogawa-Ohnishi et al. (2013)
HMGB15	ARID-HMG DNA-binding protein		*A. thaliana*	PT growth	Xia et al. (2014)
PIP5K4/5/6	Phosphoinositides		*A. thaliana*	PT growth and polarity	Ischebeck et al. (2008, 2010) and Sousa et al. (2008)
PIP5K10/11			*N. tabacum*	PT growth	Ischebeck et al. (2011)
PIKβ1/PIKβ2			*A. thaliana*	PT growth	Szumlanski and Nielsen (2009)
MPK6/PtdIns(4,5)P2			*A. thaliana*	PT growth	Hempel et al. (2017)
Pet PLC1	Secondary messengers	Phospholipase C	*P. infanta*	PT growth	Dowd et al. (2006)
PLC2			*A. thaliana*	PT growth	Dowd et al. (2006)
NtPLC3			*N. tabacum*	PT growth	Helling et al. (2006)
Phospholipase D		Phospholipase D	*N. tabacum*	PT growth	Potocký et al. (2003)
PA		Phosphatidic acid	*N. tabacum*	PT growth	Potocký et al. (2003)
PSiP		cAMP production	*Z. mays*	PT morphology growth	Moutinho et al. (2001)
CLE45	Small peptide		*A. thaliana*	PT growth	Endo et al. (2013)
LeSTIG1			*S. lycopersicum*	PT growth	Tang et al. (2004)
LePRK	LRR-RLK		*S. lycopersicum*	PT growth	Tang et al. (2002, 2004)
RPK1/3/6			*A. thaliana*	PT growth	Takeuchi and Higashiyama (2016)

14.2.5 Pollen Tube Growth Toward the Embryo Sac

The pollen tube, a novel structure formed through siphonogamy, embarks on a lengthy journey from the papilla cell to the embryo sac. The pollen tube exhibits polar growth, which is also found in root hair, neurons, and fungal hyphae. The pollen tube is divided into four zones: apical dome, apical flank, subapical, and shank (Cheung & Wu, 2008). The apical region and flank contain numerous vesicles and a high calcium concentration but lack organelles. In contrast, the subapical region features an active Golgi apparatus, abundant endoplasmic reticulum (ER), ribosomes, and mitochondria. Vesicles from the Golgi apparatus are transported to the apical region by actin-polymerized microfilaments, subsequently fusing with the plasma membrane at the tip. This fusion releases contents to synthesize the new cell wall, ultimately promoting pollen tube elongation.

Numerous genes and signaling pathways, particularly those related to cell wall components, cytoskeleton, endomembrane trafficking system, and ion concentration balance, have been identified in the polar growth of pollen tubes. Additionally, factors in papilla cells, style, and the transmitting tract contribute to the directional growth of the pollen tube, referred to as preovular pollen tube guidance, which will be discussed in the subsequent section.

14.2.5.1 Cell Wall Components

Maintaining pollen tube integrity is essential due to the high inner turgor pressure during growth, where a robust cell wall plays a vital role. The composition of the cell wall varies in each sub-region of the pollen tube. The tip of the pollen tube features a single-layered cell wall composed of cellulose and esterified pectin, while the shank region contains two layers: an outer coating mainly composed of hemicellulose and de-esterified pectins, and an inner sheath composed of cellulose and callose. These distinct compositions are essential for pollen tube growth. Although cellulose makes up a small proportion of the cell wall, enzymes related to cellulose synthesis, such as CELLULOSE SYNTHASE-LIKED 1 (CSLD1) and CSLD4, are important for pollen tube growth. Pollen tubes of *csld1* and *csld4* exhibit an enlarged tip and are arrested in tip growth (Wang et al., 2011). CELLULOSE SYNTHASE-LIKE PUTATIVE GLYCOSYLTRANSFERASE 7 (CSLA7) in *Arabidopsis*, considered a polysaccharide β-glycosyltransferase, also plays a significant role in pollen tube growth. In *Arabidopsis*, *csla7* exhibits slower pollen tube growth, indicating the essential role of CSLA7-mediated polysaccharide synthesis in pollen tube wall integrity (Goubet et al., 2003). The rapidly growing pollen tube tip, characterized by a thin cell wall, continuously secretes pectin to sustain growth. Pectin secretion is de-esterified by PECTIN METHYLENE ESTERASE (PME), with the de-esterified pectin remaining in the subapical and shank regions. PME and its inhibitor (PMEI) collaboratively regulate pectin distribution, and mutants of PME and PMEI exhibit precocious bursting of the pollen tube (Jiang et al., 2005; Woriedh et al., 2013). VANGUARD1 (VGD1), the first reported pectin methylesterase with a biological function in the pollen tube, is found in both the plasma membrane and cell wall. In *vgd1* mutants, pollen tube growth is significantly reduced in the style and transmitting tract, accompanied by decreased pectin methylesterase activity in the pollen tube (Jiang et al., 2005). Another pectin methylesterase, PPME1, specifically expressed in pollen, influences pollen tube growth and morphology (Tian et al., 2006). In *Arabidopsis*, PMEI2 is the inhibitor of PPME1. Interestingly, compared with the whole tube wall distribution of PPME1, PMEI2 is specifically located at the pollen tube tip. Therefore, the secreted esterified pectin could only be de-esterified in the subapical region and shank, achieving fine regulation of pollen tube growth (Rockel et al., 2008). In addition, AtPME48 and ZmPMEI1 also participate in pollen tube growth in *Arabidopsis* and maize (Leroux et al., 2015; Woriedh et al., 2013).

14.2.5.2 Cytoskeleton

The kinetics analysis reveals that pollen tube growth is oscillatory. Consequently, feedback mechanisms must be in place to achieve precise spatiotemporal regulation of pollen tube growth. The dynamic changes in microfilament/F-actin play an important role in this fine regulation. The assembly and biological function of microfilaments depend on ACTIN-BINDING PROTEINS (ABPs), which encompass nucleating protein, monomer actin-binding protein, capping protein, severing protein, lateral binding protein, and cross-linking protein. Most ABP mutants exhibit abnormal pollen tube growth, underscoring the importance of these genes. Studies have highlighted the significance of nucleating protein, lateral binding protein, and cross-linking protein in pollen tube growth.

Nucleating proteins, such as formins, directly bind to the elongating end of F-actin, enhancing filament elongation by protecting the growing end from capping proteins (Kovar & Pollard, 2004; Romero et al., 2004). In *Arabidopsis*, ARABIDOPSIS THALIANA FORMIN 3 (AFH3) is responsible for actin cable formation, influencing pollen tube morphology upon overexpression or knockdown (Ye et al., 2009). FORMIN HOMOLOGY-5 (FH5) in tobacco facilitates subapical actin structure assembly and provides actin filaments for vesicular trafficking to the apical dome, with FH5 knockdown resulting in a curved pollen tube (Cheung et al., 2010). In Lily, the class I formin member LIFH1 is localized in both apical vesicles and plasma membrane. The apical vesicles located in LIFH1 have been shown to initiate actin polymerization and coordinate the interaction between the actin fringe and exocytic vesicle trafficking to regulate pollen tube tip growth (Li et al., 2017).

ACTIN-DEPOLYMERIZING FACTOR (ADF) is a lateral binding protein that binds with F-actin, leading to depolymerization and impeding F-actin reassembly. In *Arabidopsis*, ADF7 co-localizes with F-actin in the shank, and in *adf7*, the speed of depolymerization decreases, resulting in slower pollen tube growth (Zheng et al., 2013).

Fimbrins and villins serve as cross-linking proteins with two F-actin-binding sites, allowing for F-actin cross-linking. In *Arabidopsis*, the loss of FIM4 and FIM5 affects pollen germination and tube growth, generating distinct actin structures in vitro (Su et al., 2017a). VILLINS (VLNs) are another cross-linking factor; in *vln2/vln5*, F-actin exhibits irregular distribution and less bundling in the pollen tube, affecting the actin fringe/collar formation (Qu et al., 2013). A newly discovered cross-linking protein, CROLIN1, stabilizes F-actin structure, contributing to pollen tube growth (Jia et al., 2013).

In addition to ABPs, several microtubulin binding proteins have been identified with the ability to bind F-actin, participating in pollen tube growth. In *Arabidopsis*, MICROTUBULE-ASSOCIATED PROTEIN18 (MAP18) has MF-severing activity, dependent on Ca^{2+}, impacting pollen tube morphology (Kato et al., 2010; Zhu et al., 2013). MICROTUBULE-DESTABILIZING PROTEIN25 (MDP25) in the subapical region, regulated by Ca^{2+}, negatively regulates MF, with *mdp25* showing faster pollen tube growth (Qin et al., 2014a).

14.2.5.3 Endomembrane Trafficking

The rapid and continuous growth of the pollen tube tip is dependent on polarized endocytosis and exocytosis. Golgi-derived secretory vesicles are transported to the tip, where they fuse with the plasma membrane, supplying cell wall constituents and regulators (such as ion transporters) for tip growth. Exocytosis, in many cell types, is primarily mediated by a conserved tethering complex known as the exocyst, comprising eight subunits: SEC3, SEC5, SEC6, SEC8, SEC10, SEC15, EXO70, and EXO84 (Rybak et al., 2014; Schekman & Novick, 2004). Among these, SEC3, SEC5, SEC6, SEC8, and EXO70 are implicated in pollen tube growth (Bloch et al., 2016; Cole et al., 2005; Hala et al., 2008; Kulich et al., 2010). The exo70c2 mutant exhibits an enhanced growth rate and decreased thickness of the tip cell wall, leading to tip bursts (Synek et al., 2017). As exocytosis cargos exceed the requirements for pollen tube growth, endocytosis is necessary for their reinternalization in the tube tip (Cheung & Wu, 2008). In yeast, Vps52p/SAC2, a subunit of the GARP/VFT complex, mediates vesicle transport from pre-vacuolar compartments (PVC) to the Golgi apparatus. The *Arabidopsis* homolog POKY POLLEN TUBE (POK) mutation results in shortened pollen tubes (Lobstein et al., 2004). Additionally, Golgi-localized proteins, COG3 and COG8, forming a COG complex to modulate vesicular trafficking, are identified as crucial factors for pollen tube growth (Tan et al., 2016).

Proteins from the small GTPase superfamily play important roles in vesicle transport within the cell, with subfamilies including Ras, Rho, Rab, ARF, and Ran (Kahn et al., 2006). Plants feature a Rho GTPase subfamily, RAC/ROP GTPase, whose members regulate polarized pollen tube growth through well-organized conversion between active and inactive forms and restricted distribution (Bloch et al., 2005; Kost, 2008; Nibau et al., 2006). AtROP1, specifically expressed in pollen tubes, functions redundantly with AtROP3 and AtROP5 (Li et al., 1998; Lin et al.,

1996; Zheng & Yang, 2000a). Constant expression of AtROP1 causes depolarization of pollen tubes, resembling the phenotype of ren1 knockout mutants. REN1 inactivates Rho by promoting GTP hydrolysis, while its dominant-negative mutants exhibit aberrant pollen tube growth (Hwang et al., 2008). In *Arabidopsis*, AGC1.5 subfamily of cytoplasmic kinases interacts with RopGEFs via their catalytic PRONE domain and phosphorylates RopGEFs, restricting the localization of ROP-GTP during pollen tube growth. Loss of AGC1.5 and AGC1.7 functions resulted in the mistargeting of active ROPs and defective events downstream of ROP signaling in pollen tubes (Li et al., 2018). In tobacco, pollen tube tip growth is regulated by the ROP GTPase-RAC5 and its effector RISAP, with RISAP overexpression interfering with apical membrane traffic and blocking tip growth (Stephan et al., 2014). GDI2, the GDP dissociation inhibitor 2, negatively regulates the dissociation of ROP from the membrane in the ROP signaling pathway (Hwang et al., 2010; Klahre et al., 2006). In *Arabidopsis*, GDI1, GDI2a, and GDI2b maintain the same conserved functions as GDI2 (Feng et al., 2016). Downstream of the ROP1 GTPase, the RIC4 and RIC3 pathways promote filamentous actin (F-actin) assembly and disassembly, respectively (Chen et al., 2002b; Gu et al., 2005; Lee et al., 2008; Wu et al., 2001). Another family of RAS-related small GTPases, Rabs, also regulates vesicular trafficking at the pollen tube tip. Tobacco NtRab2, predominantly expressed in pollen, functions in the secretory pathway between the endoplasmic reticulum and the Golgi in elongating pollen tubes. Dominant-negative mutations in NtRab2 proteins block the delivery of proteins to normal locations and inhibit pollen tube growth (Cheung et al., 2002). In *Arabidopsis*, RABA4D is expressed in a pollen-specific manner, and the disruption of RABA4D results in bulged pollen tubes with a reduced rate of growth in vitro, possibly due to altered deposition of some cell wall components (Szumlanski & Nielsen, 2009). Tip-localized PS established by AMINOPHOSPHOLIPID ATPASE 3 (ALA3) is crucial for Rab GTPase-mediated vesicle targeting/trafficking and polar growth of pollen tubes. Loss of ALA3 function resulted in loss of polar localization of apical PS, leading to a marked decrease in YFP-RabA4b/RFP-RabA4d- and FM4–64-labeled vesicles from the inverted-cone zone of the pollen tube tip and an obvious increase in pollen tube width (Zhou et al., 2020).

14.2.5.4 Ion Concentration Balance

A polarized ion gradient and ion flow exist in the pollen tube tip, and they are essential for establishing the growth of the polarized pollen tube (Hepler et al., 2001; Taylor & Hepler, 1997). Four major ions, Ca^{2+}, H^+, K^+, and Cl^-, play important roles in pollen tube elongation. Ca^{2+} is considered a key regulator of pollen tube germination and growth and is a necessary component for in vitro pollen tube germination in the medium (Boavida & McCormick, 2007; Steinhorst & Kudla, 2013). Disruption of the Ca^{2+} gradient results in the obvious inhibition of pollen tube growth (Obermeyer & Weisenseel, 1991; Picton & Steer, 1983; Pierson et al., 1994). Calcium channels, pumps, and exchange proteins must cooperate to ensure Ca^{2+}

flow. CNGC, a cyclic nucleotide-gated channel located in the plasma membrane of pollen tubes, facilitates Ca^{2+} influx when activated. In rice and *Arabidopsis*, OsCNGC13 and AtCNGC2 are crucial for style and pistil development, respectively, and they function in pollen tube growth by affecting ECM secretion (Chaiwongsar et al., 2009; Xu et al., 2017). ACA9 (Auto-inhibited Ca^{2+} ATPases) encodes a plasma membrane (PM) Ca^{2+}-ATPase necessary for pollen tube growth and the rupture of the pollen tube in the synergid cell (Schiott et al., 2004). Furthermore, many downstream factors in the calcium signaling pathway are essential for pollen tube growth by receiving and transferring Ca^{2+} signals. Various types of proteins can interact with intracellular calcium to transfer signals, such as Ca^{2+} sensors, calmodulin (CAM), calmodulin-like protein (CML), Ca^{2+}-dependent protein kinases (CPKs), calcineurin B-like proteins (CBLs), and calcineurin B-like interacting protein kinases (CIPKs). Two closely related tonoplast-localized Ca^{2+} sensor proteins, CBL2 and CBL3, act as vacuole-located Ca^{2+} sensors, forming a complex with CBL-Interacting Protein Kinase 12 (CIPK12) to regulate pollen tube polar growth by recruiting cytosol to the vacuole membrane (Steinhorst et al., 2015). Additionally, CML24, CML25, CPK24, CPK11, CPK17, CPK34, and CIPK19 have been reported to enable the rapid growth of pollen tubes (Myers et al., 2009; Steinhorst et al., 2015; Wang et al., 2015; Yang et al., 2014; Zhao et al., 2013; Zhou et al., 2015).

Distinct polarized potassium ion (K^+), hydrogen ion (H^+), and anion fluxes are also essential for pollen tube polar growth. Since K^+ signaling pathways are crucial for pollen tube rupture, they will be introduced in detail in the chapter on pollen tube reception. Meanwhile, auto-inhibited plasma membrane proton (H^+) ATPase (AHA) plays an important role in pollen tube growth; *aha6/8/9* show reduced cytosolic pH and delayed pollen tube growth (Hoffmann et al., 2020). Anion fluxes present a peculiar case. On the one hand, the uptake of anions into the PTs aids rapid osmosis; on the other hand, tip efflux of anions strongly correlates with PT growth rate (Gutermuth et al., 2013, 2018; Zonia et al., 2001). SLAH3 and SLAC1 are S-type (slow deactivating) anion channels, and ALMT12/13/14 are R-type (rapid deactivating) anion channels. Pollen tube growth in *slac1-3*, *slah3-1*, *almt12-1*, and *almt13-1* was impaired, similar to *cpk2-1*, *cpk20-1*, and *cpk6-1*. Both S- and R-type anion channels regulate PT growth depending on CPK2, CPK20, and CPK6 (Gutermuth et al., 2018).

14.2.5.5 Other Factors

Meanwhile, various enzymes have been identified as being involved in pollen tube growth. MALE GAMETOPHYTE DEFECTIVE 2 (MGP2) is a Golgi apparatus-located sialyltransferase-like protein. Both pollen germination and pollen tube growth are impaired in the *mgp2* mutant (Deng et al., 2010b). MGP4 is a rhamnogalacturonan II xylosyltransferase, and the loss of MGP4 results in defective pollen tube growth (Liu et al., 2011). The Golgi apparatus-located proteins GAUT13 and GAUT14 are two functionally redundant galacturonyl transferases that may participate in pectin biosynthesis in the pollen tube cell wall (Wang et al., 2013). MAT3 encodes methionine adenosyltransferase. Knocking out MAT3 leads to slower pollen tube growth (Chen et al., 2016). HPAT is a Hyp O-arabinosyltransferase. Pollen tube growth is arrested in the stigma or style in the *hapt1*, *hapt2*, and *hapt3* triple mutant, suggesting that these enzymes are essential for the reproduction process (MacAlister et al., 2016). AtHMGB15 belongs to the plant-specific ARID-HMGDNA binding protein family, which has conserved ARID and HMG-box domains. In *athmgb15*, fertility and male transmission efficiency are abnormal since pollen tube growth has been impaired (Xia et al., 2014). SETH1 and SETH2 participate in the first step of GPI synthesis. Callose is aberrant in *seth1/seth2*, resulting in abnormal pollen germination and pollen tube growth (Lalanne et al., 2004).

In addition, some receptors and ligands are involved in pollen tube growth. LeSTIG1 and LePRK1/2 are reported to promote pollen tube growth in the stigma in tomato (Kaothien et al., 2005; Tang et al., 2004). In *Arabidopsis*, PRK1/3/6 not only participate in pollen tube micropylar guidance but are also essential for pollen tube growth by interacting with ROPGEFs to activate the downstream ROP1 signaling pathway (Takeuchi & Higashiyama, 2016).

14.2.6 Pollen Tube Guidance Involves Communication with Female Tissues

In flowering plants, the female gametophyte is spatially distant from the transmitting tract. To transport the two immotile sperm cells and ensure they meet with the female gametes, the continuously growing pollen tube must be precisely guided to enter the female gametophyte—a process known as pollen tube guidance.

14.2.6.1 Preovular Pollen Tube Guidance

Pollen tube guidance encompasses preovular and ovular guidance (Higashiyama & Takeuchi, 2015) (Fig. 14.5). Preovular guidance directs compatible pollen tube growth toward the ovary and initiates when the pollen tube grows into the stigma. Various factors contributing to preovular guidance have been identified in many species. For instance, chemocyanin, a small cell wall peptide expressed in the Lily stigma, induces pollen tube chemotropism (Kim et al., 2003). The lipid transfer protein subfamily member STIGMA/STYLAR CYSTEINE-ADHESION (SCA) enhances the chemotropic activity (Mollet et al., 2000; Park et al., 2000), potentially facilitating chemocyanin penetration into the pollen tube. In tomato, STIGMA-SPECIFIC PROTEIN 1 (STIG1) interacts with the pollen tube-located receptor kinase LePRK2, promoting pollen tube growth in vivo and in vitro (Huang et al., 2014; Tang et al., 2004). In *Brassica*, compatibility factors like GLYOXALASE 1 (GLO1) promote pollen tube growth (Sankaranarayanan et al., 2015). Specialized cells in the transmitting tract, such as the arabinogalactan protein TRANSMITTING TRACT TISSUE-SPECIFIC (TTS) in tobacco, stimulate pollen tube growth and attract the pollen tube in vitro (Cheung et al., 1995; Wu et al., 2000). The C2H2/C2HC zinc finger transcription factor NTT (NO TRANSMITTING TRACT) regulates the extracellular matrix composition. In the *ntt* mutant, the pollen tube fails to reach the end of the pistil, resulting in seed abortion in the lower half of the ovary (Crawford et al., 2007). Additionally, the amino acid D-ser activates GLUTAMATE RECEPTORS (GLRs), participating in pollen tube directional growth by increasing the intercellular Ca^{2+} concentration (Michard et al., 2011).

14.2.6.2 Ovular Guidance in Dicots—Funicular Guidance

Ovular guidance garners much more attention than preovular guidance and differs between dicots and monocots due to their distinct anatomies. In dicots like *A. thaliana*, ovular guidance involves two steps: funicular guidance, attracting pollen tubes in the transmitting tract to penetrate the septum and climb onto the funiculus, and micropylar guidance, where pollen tubes respond to attractants produced by the FG, targeting and entering the micropyle (Lausser & Dresselhaus, 2010) (Fig. 14.5). Several molecules have been identified to function in funicular guidance. CATION/PROTON EXCHANGERS 21/23 (CHX21/23), located in the endoplasmic reticulum (ER), serve as potassium transporters. The *chx21/23* mutant pollen tube exhibits normal growth in the transmitting tract but fails to penetrate the septum (Lu et al., 2011). However, the precise mechanism through which potassium transporters regulate funicular guidance remains unknown. Two mitogen-activated protein kinases, MPK3 and MPK6, also contribute to funicular guidance (Guan et al., 2014). The *mpk3 mpk6* mutant pollen tube displays a defect in funicular guidance.

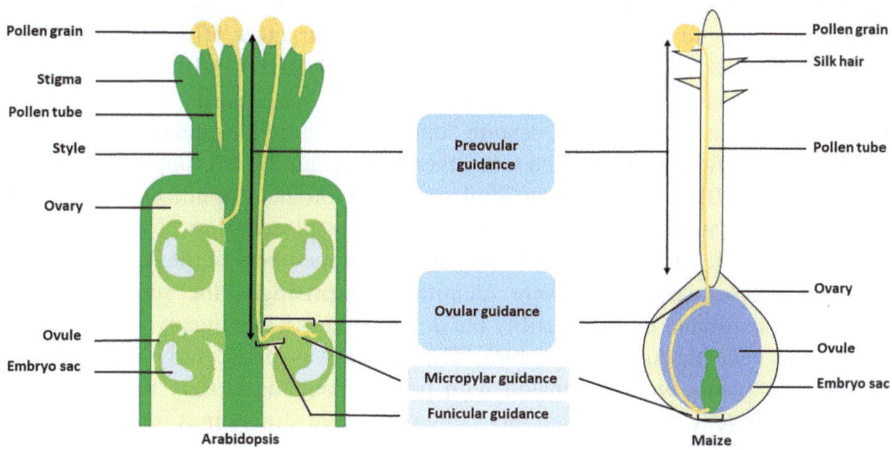

Fig. 14.5 Schematic diagram of pollen tube guidance process in the dicot *Arabidopsis* and monocot maize

Interestingly, a semi-in vivo ovule targeting assay reveals that micropylar guidance in *mpk3 mpk6* appears normal, suggesting that there are distinct molecular mechanisms for funicular guidance and micropylar guidance. GPI-anchored proteins, such as COBL10, are also involved in funicular guidance. COBL10, present in the pollen tube tip and punctate vesicles, influences cell wall organization, affecting pollen tube growth. The tip localization of COBL10 proves critical for cell wall constituent deposition, impacting both funicular and micropylar guidance (Li et al., 2013).

14.2.6.3 Ovular Guidance in Dicots—Micropylar Guidance

After the pollen tube climbs onto the funiculus, it continues growing and is attracted to enter the micropyle. In *T. fournieri*, synergid cells play a pivotal role in micropylar guidance. Laser ablation experiments, facilitated by the naked embryo sac, reveal that ovules lose pollen tube attraction only when both synergid cells are eliminated, indicating synergid cells likely produce micropylar guidance attractants (Higashiyama et al., 2001). In *Arabidopsis*, synergid cells expressed R2R3-MYB transcription factor MYB98, which is crucial for micropylar guidance. *myb98* ovules exhibit abnormal filiform apparatus (FA) structure and a defect in micropylar guidance. While multiple pollen tubes may climb onto the funiculus, most fail to penetrate the micropyle, suggesting the downstream genes of MYB98 may act as micropylar guidance attractants (Kasahara et al., 2005).

Cysteine-rich peptides (CRPs) are identified as participants in numerous signaling pathways (Ge et al., 2017; Sugano et al., 2010). In *T. fournieri*, CRPs are abundantly expressed in isolated synergid cells. The CRP TfLUREs are predominantly expressed at the micropylar surface of the FA. More importantly, TfLURE1/2/3 have been verified to possess pollen tube attraction activity directly using newly established attraction method (Okuda et al., 2009). Additionally, the ovular factor METHYL-GLUCURONOSYL ARABINOGALACTAN (AMOR) induces pollen tube competency to respond to the ovular guidance attractant TfLUREs (Mizukami et al., 2016). In *A. thaliana*, five CRPs known as AtLURE1.1-AtLURE1.5, possessing pollen tube attraction activity, are secreted from synergid cells and diffuse to the funicular surface. Interestingly, despite abnormal micropylar guidance behavior in AtLURE1s RNAi mutants, pollen tubes eventually enter the micropyle (Takeuchi & Higashiyama, 2012). These results imply that the biological significance of the pollen tube attractants AtLURE1s might extend beyond ensuring plant fertility. In 2019, AtLURE1.7 and AtLURE1.8, two additional members of the AtLURE1 family with species specificity, were identified. Surprisingly, the septuple mutant *atlure1null* exhibited normal fertility but demonstrated a significant delay in pollen tube emergence through the septum. This delay was particularly evident when *A. thaliana* pistils were pollinated with pollen from the closely related species *A. lyrata*. While *A. lyrata* pollen tubes emerged more slowly in *A. thaliana* pistils, the absence of AtLURE1s did not further impair the emergence of *A. lyrata* pollen tubes in *atlure1null* pistils. This suggests that species-specific AtLURE1s function as conspecific pollen precedence factors, accelerating conspecific pollen tube penetration through the septum for reproductive isolation (Zhong et al., 2019).

Receptor-like kinases (RLKs) have been identified as receptors for many CRPs in various signaling pathways. The receptor-like cytoplasmic kinases (RLCKs) LOST IN POLLEN TUBE GUIDANCE 1 and 2 (LIP1/2) without extracellular domains are the first receptor complex components responding to AtLURE1s. Both LIP1 and LIP2, located in the plasma membrane and cytoplasm of the pollen tube, contribute to micropylar pollen tube guidance, as evidenced by reduced male transmission efficiency and reduced attraction ratio toward the AtLUREs in the *lip1 lip2* mutant (Liu et al., 2013). The pollen-specific receptor-like kinase PRK6 interacts with LIP1 and LIP2 and is identified as the receptor for AtLURE1s, as evidenced by the complete loss of response to AtLURE1s in the *prk6* pollen tube. Structural evidence indicates AtLURE1s bind directly with the juxtamembrane domain of PRK6 (Zhang et al., 2017b). Interestingly, PRK6, along with its family members PRK1 and PRK3, also contributes to pollen tube growth. This suggests that these RLKs may play a role in coordinating the processes of both pollen tube growth and guidance. Recent studies reveal the involvement of the ROP1 Rho GTPase in an exocytosis-coordinated mechanism linking tip growth and pollen tube guidance (Luo et al., 2017). Moreover, LRR-RLKs MDIS1-INTERACTING RECEPTOR-LIKE KINASE 1/2 (MIK1/2) and MALE DISCOVERER 1/2 (MDIS1/2) form a heteromer serving as another receptor for AtLURE1s (Wang et al., 2016b). Mutants of *mik1/2* and *mdis1/2* show reduced efficiency in pollen tube attraction toward AtLURE1.

In addition to AtLURE1s, nonspecies-specific attractants called XIUQIUs have been identified, attracting pollen tubes from closely related *Arabidopsis* species. These substances may enhance genomic preservation in species facing extreme environmental conditions. XIUQIUs, along with AtLURE1s, collectively regulate plant fertility, offering insights into the working model of pollen tube attractants in fertility regulation. Furthermore, the discovery of other attractants like TICKETs implies that the molecular regulatory mechanisms governing pollen tube guidance are more complex than initially imagined (Meng et al., 2019).

Although the ablation of the central cell did not result in a micropylar guidance defect in *T. fournieri* (Higashiyama et al., 2001), the central cell is essential for micropylar guidance in *Arabidopsis*. CENTRAL CELL GUIDANCE (CCG) is a nuclear protein that might act as a transcription factor and is specifically expressed in the central cell. The ovule of the *ccg* mutant undergoes normal embryo sac development but loses micropyle attraction activity, leading to severe sterility (Chen et al., 2007). Another central cell-located protein, CCG BINDING PROTEIN 1 (CBP1), interacts with CCG to recruit the Mediator complex and RNA Pol II. Transcriptome analysis reveals downregulation of several genes, including MYB98, in *ccg* ovules, suggesting that the CCG-CBP1 complex may function in micropylar guidance through the MYB98 signaling pathway (Li et al., 2015b).

In addition, several other factors, including enzymes, Ca^{2+} channels, and GPI-anchor-related proteins, have been identified to participate in micropylar guidance. SUMO E3 LIGASE1 (SIZ1), expressed in the female gametophyte, regulates both funicular and micropylar guidance (Ling et al., 2012). N-ACETYLGLUTAMATE KINASE (NAGK), catalyzing the second step of arginine biosynthesis, is expressed in both the female and male gametophytes (Huang et al., 2017). The nagk mutant exhibits normal female gametophyte morphology but is defective in micropylar guidance due to dysfunctional synergid cells. ABNORMAL POLLEN TUBE GUIDANCE 1 (APTG1), a putative mannosyltransferase, regulates GPI anchor biogenesis, and its mutant pollen tube shows a defect in micropylar guidance (Dai et al., 2014). CYCLIC NUCLEOTIDE-GATED CHANNEL 18 (CNGC18), a Ca^{2+} channel on the pollen tube, exhibits defective pollen tube guidance due to abnormal Ca^{2+} oscillation in the pollen tube tip. COBL10-related proteins also play a role in micropylar guidance (Chang et al., 2007). MLO5/9/15 are essential for the plasma membrane localization of CNGC18. In the *mlo5 mlo9* mutant, CNGC18 fails to be recruited to the cell membrane, resulting in the failure to decode the ovular attraction signal (Meng et al., 2020). Since GPI-anchored COBL10 is a substrate of APTG1, the subcellular localization of COBL10 is affected in aptg1. ASPARTIC PROTEASE A36 and A39 co-localize with COBL10 on the ER (Gao et al., 2017), but the relationship between the two proteins remains unclear. The a36 a39 pollen tube has a compromised micropylar guidance. The mutant pollen tube of the ER-located protein POLLEN DEFECTIVE IN GUIDANCE 1 (POD1) exhibits a micropylar guidance defect. POD1 interacts with the ER chaperone CALRETICULIN 3 (CRT3), involved in the folding of membrane receptors, suggesting that POD1 may participate in folding the receptors for micropylar guidance (Li et al., 2011).

14.2.6.4 Ovular Guidance in Monocots

Ovular guidance in monocots differs from that in dicots due to distinct ovular anatomy. Only micropylar guidance occurs in monocots (Lausser & Dresselhaus, 2010) (Fig. 14.5). For example, in maize, the transmitting tract ends close to the upper ovary. Thus, the pollen tube enters the ovary cavity by breaking through the inner epidermis of the ovary wall. However, no funiculus structure exists in maize. The pollen tube grows toward the micropylar region because the special anatomical structure of the ovary provides a growth direction clue. The 94-amino acid peptide, Zea mays EGG APPARATUS 1 (ZmEA1), exclusively expressed in the maize egg apparatus, has been identified to regulate micropylar guidance. The ZmEA1 RNAi mutant exhibits a severe micropylar pollen tube guidance defect and significantly reduced seed set (Marton et al., 2005).

To date, many factors have been identified to function in pollen tube guidance (Fig. 14.6). However, some questions remain. For example, how do the two sets of receptors of AtLURE1s cooperatively regulate AtLURE1-mediated pollen tube guidance? What is the molecular mechanism by which members of the PRK family simultaneously regulate pollen tube growth and guidance? What are the other pollen tube attractants? How do different pollen tube attractants collectively control pollen tube guidance? These interesting scientific questions are worthy of future study.

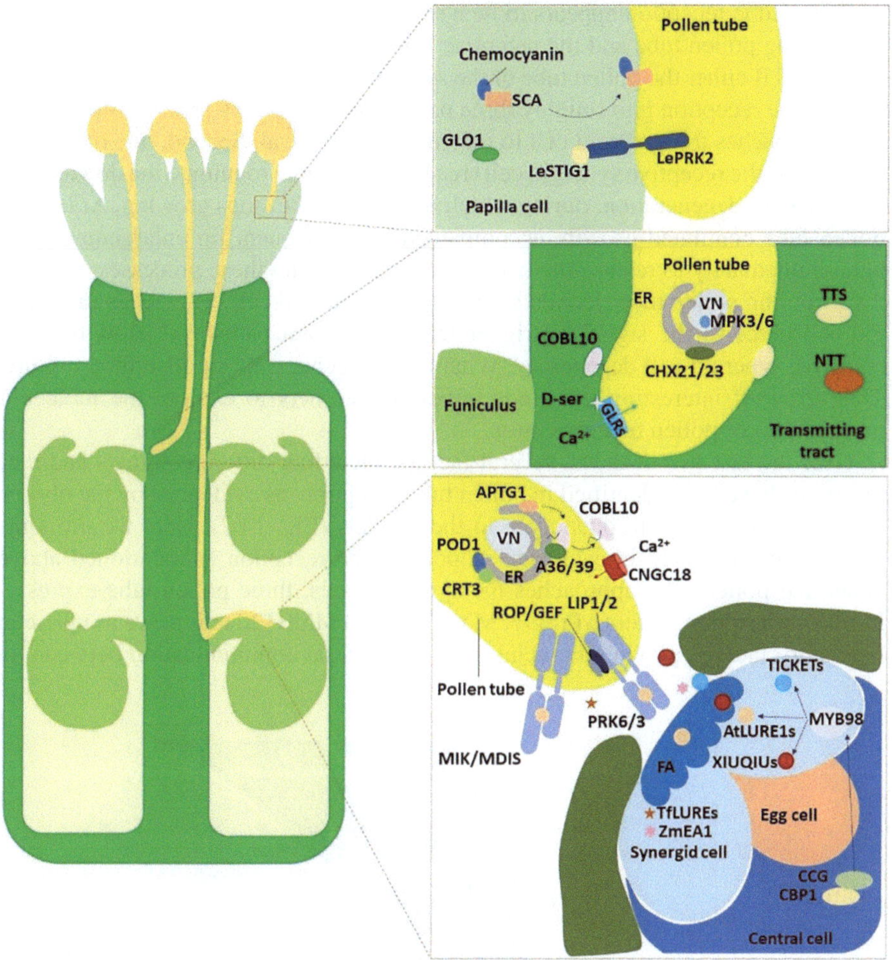

Fig. 14.6 Genes involved in different processes of pollen tube guidance

14.2.7 Male-Female Communications for Pollen Tube Reception

After entering the micropyle, the pollen tube continues growing and approaches the synergid cell. This process lasts approximately 30–60 minutes (Iwano et al., 2012). Notably, recent evidence shows that the pollen tube does not enter the synergid cell through FA as previously predicted but enters the embryo sac at a more distant site by growing around or along the synergid cell (Leshem et al., 2013). After approaching the synergid cells, many male-female communications occur between the pollen tube and the receptive synergid cell. These complicated communications between the two cells are responsible for triggering the pollen tube rupture and the degeneration of the receptive synergid cell. This process is called pollen tube reception.

Pollen tube reception appears to be a series of greetings and responses between the arriving pollen tube and the receptive synergid cell. The reception could not be completed if either the pollen tube or the synergid cell were defective. We divide the pollen tube reception into roughly three parts. First, the pollen tube secretes signals as it approaches the synergid cell to inform it of its arrival. Second, one of the synergid cells (the receptive synergid cell) receives the signal to initiate the downstream pathways for degeneration, during which the pollen tube stops growing. At last, the pollen tube continued growth, accompanied by subsequent tip enlargement, then burst immediately to release the two sperm cells. During these processes, what are the incoming pollen tube-derived signals to the synergid cell? How does the synergid cell perceive the signal to trigger the downstream pathways? How does the receptive synergid cell degenerate? What is the mechanism of pollen tube rupture? This series of interesting questions impels researchers to explore the molecular mechanisms of pollen tube reception.

Over the last two decades, several factors, including different signals and their receptors, have been identified in pollen tube reception, as evidenced by the observable pollen tube growth phenotype in the mutants (Fig. 14.7). Here we introduce these factors involved in each step of pollen tube reception as mentioned above. When the pollen tube approaches the synergid cells, three pollen tube-expressed MYB transcription factors MYB97, MYB101, and MYB120 may serve as upstream regulatory factors, playing a crucial role in the initial communication between the

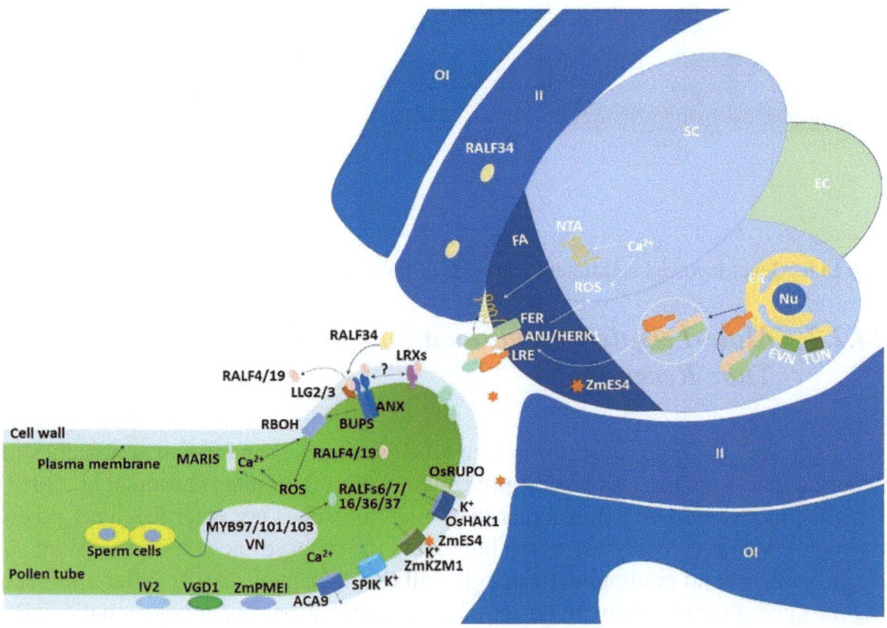

Fig. 14.7 Genes involved in different processes of pollen tube reception. VN vegetative nucleus, OI outer integument, II inner integument, FA filiform apparatus, SC synergid cell, EC egg cell

pollen tube and the receptive synergid cell. The triple mutant *myb97*, *myb101*, and *myb120* exhibit an extraordinarily high percentage of pollen tube overgrowth, resulting in severe abortion (Leydon et al., 2013; Liang et al., 2013). Transcriptome analysis shows that several CRP genes are regulated by these three transcription factors (Leydon et al., 2017). Among them, five RALF peptides RALF6/7/16/36/37 were identified as male signals regulating the pollen tube reception, as evidenced by the inability of pollen tubes from the *ralf* quintuple mutant to burst in wild-type (Zhong et al., 2022).

The CrRLK1L family member FERONIA/ SIRÈNE, which is expressed and localized in the synergid cell, is the first factor identified to be involved in pollen tube reception. When the pollen tube reaches the synergid cell, the RALF peptides interact with FER and two other CrRLK family members, ANJ and HERK1, to activate the downstream signaling pathways. In *fer* and *anj herk1* ovules, a high proportion of pollen tubes exhibit an overgrowth phenotype, indicating the significance of this peptide/receptor-mediated signaling pathway in pollen tube reception (Escobar-Restrepo et al., 2007; Huck et al., 2003; Rotman et al., 2003; Galindo-Trigo et al., 2020). Until now, several factors have been found to be involved in the FER-mediated pollen tube reception. First, post-translational modification is important for the biological function of FER. The extracellular domain of FER is glycosylated. The uridine diphosphate (UDP)-glycosyltransferase superfamily protein TURAN (TUN) and dolichol kinase EVAN (EVN) are required for N-glycosylation of FER in the ER. The *tun* and *evn* mutants display the same pollen tube overgrowth phenotype as *fer* (Lindner et al., 2015). Both LORELEI (LRE) and EARLY NODULIN-LIKE PROTEINS (ENODLs/ENs) are GPI-anchored proteins participating in pollen tube reception (Liu et al., 2016b; Tsukamoto et al., 2010; Hou et al., 2016). LRE and EN14/15 are asymmetrically distributed at the synergid cells and accumulate at the FA when the pollen tube arrives. Although LRE lacks the intracellular domain, it is considered a co-receptor in the FER signaling pathway. The proportion of pollen tube overgrowth observed in *lre* mutants, comparable to that in fer mutants, also underscores its importance in this signaling pathway. EN14 could interact with the extracellular domain of FER. The EN quintuple mutant exhibits pollen tube overgrowth, suggesting that ENs may participate in the FER signaling pathway to regulate pollen tube reception. LLG1 is the closely related LRE-like-GPI-anchored protein 1. LLG1 and LRE interact with FER and seem to participate in facilitating the localization of FER to the cell membrane and stabilizing the FER-receptor complex at the cell surface (Li et al., 2015a). Plant-specific MILDEW-RESISTANCE LOCUS O (MLO) family member AtMLO7/NORTIA (NTA) is the first identified downstream protein of FER. The pollen tube in the *nta* ovule also exhibited pollen tube overgrowth. Live imaging of pollen tube reception has revealed that the polar accumulation of NTA at the filiform apparatus in response to signals from an arriving pollen tube depends on the FER receptor. Artificial delivery of NTA to the filiform apparatus is sufficient to bypass the FER signaling pathway and to promote interspecific pollen tube reception (Ju et al., 2021). The arrival of the pollen tube is reported to induce the formation of a trimeric complex of FER-LRE-NTA, facilitating the relocation of NTA from the cytosol to the cell membrane. This, in turn, results in alterations to the dynamics of synergid cytoplasmic Ca^{2+} (Iwano et al., 2012; Denninger et al., 2014; Ngo et al., 2014; Gao et al., 2022). Meanwhile, NTA also has a calmodulin (CaM)-binding domain. The previously mentioned Ca^{2+} oscillations in the synergid cell may activate CaM, which can interact with the CaM-binding domain of NTA and lead to the polar localization of NTA (Kessler et al., 2010). In addition, the peroxisomes ABSTINENCE BY MUTUAL CONSENT (AMC) is expressed in both the pollen tube and synergid cells. The *amc* mutant shows pollen tube overgrowth. The seed set rate of self-pollinated *fer+/– amc+/–* is lower than that of the *fer+/–* or *amc+/–* single mutant, suggesting that the FER and AMC pathways are at least partially independent (Boisson-Dernier et al., 2008).

The downstream events of the FER-mediated signal pathway seem to trigger two important events: the degeneration of the receptive synergid cell and the initiation of pollen tube burst. The exact sequential order of the two key events in pollen tube reception remains unclear. Image and movie evidence shows that the pollen tube bursts when it occupies the space of the degenerated synergid cell, implying a location effect for the pollen tube burst. The best explanation for this result is that the two sperm cells require the pollen tube to burst at a certain location for the gametes successfully meet. Little is known regarding the molecular mechanism of the degeneration of the receptive synergid cell, but Ca^{2+} may play an important role in it. When the pollen tube arrives, pollen tube apex-induced Ca^{2+} oscillation has been detected in the synergid cells, which seems to be a degeneration signal.

All the events described above lead to the eventual pollen tube burst. A number of studies have demonstrated that both the "female" produced factors and "male" signal pathways contribute to pollen tube burst by affecting the integrity of the pollen tube. On the one hand, the female factor FER-LRE signaling complex has been identified to control the production of reactive oxygen species (ROS), especially hydroxyl free radicals, when the pollen tube arrives (Duan et al., 2014; Li et al., 2015a). In vitro assays show that the pollen tube bursts by an ROS process. When the "female" produced ROS were inhibited, the pollen tube could be guided into the ovule without rupture. Thus, the "female" produced ROS are required for pollen tube burst. Notably, ROS-triggered pollen tube burst is dependent on Ca^{2+}. The ratio of the pollen tube burst is decreased when Ca^{2+} concentration is reduced (Duan et al., 2014).

Studies of the male factors determining pollen tube burst have progressed in recent years. ANX1/2 (ANXUR1/2) are CrRLK1L protein family members that are highly and specifically expressed in the pollen and pollen tube. *Anx1/anx2* pollen bursts very quickly in vivo and in vitro (Boisson-Dernier et al., 2009; Miyazaki et al., 2009). This precocious burst phenotype is attributed to the breaking of the cell wall integrity. However, whether this precocious pollen tube burst is comparable with the pollen tube burst in the embryo sac is not clear, nor is it clear why the pollen tube maintains its integrity during the whole pistil journey until it meets the synergid cell. Last year, a number of studies explored this interesting scientific question. Another pollen tube-expressed CrRLK1L family members Buhda's Paper Seal 1 (BUPS1) and 2, which interact with ANX1/2 to control pollen tube integrity and pollen tube burst by perceiving different RALF peptides at different times. RAPID

ALKANIZATION FACTOR (RALF) family members RALF4/19 and RALF34 are identified to be the mobile ligands of the BUPS/ANX complex with the opposite biological function. All *bups1/2*, *anx1/2*, and *ralf4/19* mutants show precocious pollen tube bursting. RALF4/19 can interact with the BUPS/ANX complex to trigger the downstream pathway to maintain the pollen tube integrity, which ensures the pollen tube reaches the ovule without premature rupture. RALF34 is another peptide expressed by the female gametophyte. In vitro, it can compete with RALF4/19 and trigger pollen tube burst. When the pollen tube arrives at the synergid cell, RALF34 expressed by the female gametophyte will compete with RALF4/19 to interact with BUPS/ANXs (Ge et al., 2017). This newly formed receptor-ligand interaction will suspend the BUPS/ANXs-RALF4/19 triggered pollen tube integrity, maintaining the pathway and rapidly trigger pollen tube burst. The BUPS/ANXs-mediated pollen tube burst pathway may be conserved or partially conserved in different species. In rice, the homolog of the BUPS protein RUPTURED POLLEN TUBE (RUPO) also participates in pollen tube burst (Liu et al., 2016a). In maize (Zea mays), ZmRALF2/3 also regulates cell wall integrity during pollen tube growth by interacting with the pollen-specific CrRLK1L receptor kinases Z. mays FERONIA-like (ZmFERL) 4/7/9, LORELEI-like glycosylphosphatidylinositol-anchor (LLG) proteins Z. mays LLG 1 and 2 (ZmLLG1/2), and Z. mays pollen extension-like (PEX) cell wall proteins ZmPEX2/4 (Zhou et al., 2023). The NADPH oxidases RBOHH and RBOHJ may act as the downstream of ANXs to regulate the integrity of the pollen tube (Boisson-Dernier et al., 2013; Lassig et al., 2014). In the wild-type, overexpressed ANXs could inhibit pollen germination, while in *rbohh rbohj*, they could not. RBOHH and RBOHJ may function in pollen tube rupture by regulating the ROS level. In the *rbohh rbohj* mutant, ROS are significantly reduced, suggesting that the pollen tube itself requires a certain amount of ROS to maintain its integrity, and the "female"-produced ROS and "male"-produced ROS do have opposite biological functions. Compared with the "female"-produced ROS, the "male"-produced ROS EW also correlated with Ca^{2+}. First, both RBOHH and RBOHJ oxidases contain two Ca^{2+} binding motifs, EF-hand motifs. The activity of these two oxidases is Ca^{2+} dependent. Therefore, RBOHH- and RBOHJ-dependent ROS are also related to Ca^{2+}. In *rbohh rbohj*, both Ca^{2+} balance and ROS level are affected (Boisson-Dernier et al., 2013; Kaya et al., 2015; Lassig et al., 2014). Second, ROS functions by modifying proteins and may possibly regulate Ca^{2+} concentration by affecting the Ca^{2+} channel. In addition, EMS mutagenesis with *anx1 anx2* revealed that the RLCK-VIII subfamily member MRI (MARIS) also participates in the regulation of pollen tube integrity. In MRI, the pollen tube exhibits a precocious burst (Boisson-Dernier et al., 2015; Liao et al., 2016). Genetic evidence shows that MRI is the downstream of ANXs. When overexpressing MRI^{R240C} (R240 is the activity site of MRI) in *anx1 anx2*, the ratio of the pollen tube burst will be partially impacted. In addition to the BUPS/ANXs-mediated signaling pathway, the LRX family member LRX8-11 plays a role in maintaining pollen tube integrity (Fabrice et al., 2018; Mecchia et al., 2017; Sede et al., 2018; Wang et al., 2018). LRX protein family members have an N-terminal leucine-rich repeat domain and a C-terminal extensin domain, and LRXs directly affect the synthesis of extensins in the pollen tube and function in regulating the content and distribution of polysaccharides on the cell wall. The loss-of-function mutants of these extensins and polysaccharides exhibit defects in pollen growth and pollen tube integrity. LRX8-11 are expressed in pollen and located on the pollen wall and also interact with RALF4. Triple mutants of different combinations of the LRX8-11 genes exhibit an abolished ability of exogenous RALF4 to repress pollen tube growth, indicating that LRX8-11 are in the RALF4 signaling pathway (Mecchia et al., 2017). According to the report, the LRX8-RALF4 complex specifically interacts with demethylesterified pectins in a charge-dependent manner through RALF4's polycationic surface. This interaction between LRX8-RALF4 and pectin exerts a condensing effect, organizing the cell wall's polymers into a reticulated network crucial for wall integrity and expansion (Moussu et al., 2023). In summary, RALF4 plays a dual role: on one hand, they interact with CrRLK1L receptors as signaling molecules, and on the other hand, it interacts with cell wall proteins and pectins as a structural factor. Further research is needed to understand how RALFs coordinate these two functions.

In addition, the cell wall constitution of pectin-related genes and ion-related pathways is important for pollen tube growth and integrity. In *Arabidopsis*, pectin methylesterase VGD1 is located in the cell wall and cytosol of the pollen tube. The *vgd1* mutant pollen tube exhibits precocious pollen tube burst, suggesting that VGD1 participates in maintaining pollen tube integrity (Jiang et al., 2005). In maize, pectin methylesterase inhibitor PMEI could induce pollen tube burst, indicating the importance of pectin (Woriedh et al., 2013). In addition to Ca^{2+}, K^+ is also important for pollen tube reception in different species. AtSPIK and ZmKZM1 encode the K^+ channel in *Arabidopsis* and maize, respectively. In *Arabidopsis*, SPIK could induce the influx of K^+ (Mouline et al., 2002). In the *spik* mutant, the influx of K^+ is reduced, exhibiting defects in pollen germination and pollen tube growth. In maize, ZmKZM1 cooperates with the defense-like genes EMBRYO SAC1-4 (ES1-4), inducing pollen tube burst. When pollen tubes are treated with ZmES4, ZmKZM1 will be activated to induce the influx of K^+. Consistently, the mutant of ES1-4 exhibits the pollen tube overgrowth phenotype (Amien et al., 2010). In rice, the K^+ transporter OsHAK1/19/20 directly interacts with the above-mentioned receptor RUPO, and *rupo* mutant pollen exhibits high accumulation of K^+. In rice, K^+ could also induce precocious pollen tube bursts in vitro (Liu et al., 2016a). The pollen tube-expressed putative methylesterase IV2 is also reported to participate in pollen tube reception by maintaining the pollen tube cell wall (Rutley & Twell, 2015).

Although numerous female and male factors have been found in pollen tube reception, some unresolved questions remain. For example, why must RALF4/19 bind with both the BUPS/ANX complex and LRX? Does the BUPS/ANX complex interact with LRX to regulate both pollen tube burst and pollen tube integrity together? What is the relationship between the exact sequential order of synergid cell degeneration and pollen tube burst? Why does the pollen tube not burst when it arrives at the micropyle since RALF34 is expressed from the inner integument? How does RALF34 compete with RALF4/19, depending on the concentration or on more stable binding amino acids? Structural analysis is likely required for the future study of these questions (Fig. 14.7).

14.2.8 Double Fertilization Requires Successful Male-Female Gamete Interaction

The processes described above are orchestrated to ensure that the male gametes can successfully meet with the female gametes. Upon the pollen tube rupture, two sperm cells are promptly delivered into the embryo sac, where each sperm cell fuses with either the egg cell or the central cell in a process known as double fertilization. Successful double fertilization hinges on the successful fusion of the male and female gametes. Despite the rapid occurrence of gamete fusion, it involves multiple steps: gamete attachment and gamete fusion (plasmogamy and karyogamy) (Dresselhaus et al., 2016).

In *Arabidopsis*, the two sperm cells within a pollen grain appear isomorphic, suggesting no discernible preference for fusion with either the egg cell or the central cell. In contrast, the two female gametes exhibit dimorphism, displaying distinct preferences during fertilization. Typically, the fusion between sperm and egg occurs first, followed by sperm-central cell fusion a few minutes later (Denninger et al., 2014; Hamamura et al., 2014). Using the cell-cycle defective mutant *cdka;1*, which produces a single sperm-like cell while retaining fertilization competence, revealed a pronounced preference for fertilization of the egg cell over the central cell (Li et al., 2022). Furthermore, two egg cell-specific aspartic endopeptidases, ECS1 and ECS2, have been identified to contribute to the preferential fertilization of the egg cell (Jiang et al., 2022).

During gamete recognition, the male gametes are essential for maintaining sperm cell integrity. In the *chx17/18/19* mutant, where three sperm cell endomembrane-associated cation/H^+ exchangers (CHXs) were mutated, the sperm cells disappear after release. The disrupted osmotic balance in *chx17/18/19* sperm cells leads to their rupture within the embryo sac, indicating a potential interplay between turgor pressure and the integrity of the sperm cell surface (Chen et al., 2022).

14.2.8.1 Gamete Attachment

In *Arabidopsis*, GAMETE EXPRESSED 2 (GEX2) plays a role in gamete attachment. GEX2 is a single transmembrane protein located on the sperm cell surface, which contains an extracellular domain with a filamin repeat domain similar to the immunoglobulin-like (Ig-like) domain involved in gamete interaction in algae and mammals (Inoue et al., 2005; Misamore et al., 2003; Mori et al., 2014). GEX2 participates in gamete attachment between sperm cells and the two female gametes. In the *gex2* mutant, sperm cells cannot attach to either the egg cell or the central cell, resulting in single fertilization and failed double fertilization. Unlike fusogen, the ratio of attached sperm cells to female gametes is significantly lower with polysaccharide-digesting enzyme treatment, suggesting that GEX2-mediated gamete attachment is important for fertilization (Mori et al., 2014).

A family of cysteine-rich peptides, EGG CELL1s (EC1s), has been reported to play a pivotal role in attachment. EC1s were identified to accumulate in the storage vesicles of the egg cell before fertilization. Upon delivery of sperm cells into the embryo sac, EC1-GFP fusion proteins are rapidly secreted out of the egg cell, indicating that exocytosis of EC1s is triggered by the incoming sperm cells. Ovules lacking EC1s exhibit a significant defect and delayed fusion with the female gametes, emphasizing the essential role of EC1s in facilitating the two fusion events (Sprunck et al., 2012). In *ec1-RNAi* ovules, a high percentage of paired sperm cells were neither attached to the egg cell nor the central cell after protoplasting, underscoring the crucial role of EC1s in the attachment process (Cyprys et al., 2019).

14.2.8.2 Gamete Fusion

The last step of double fertilization is gamete fusion, consisting of two processes: plasmogamy and karyogamy. HAPLESS2/GENERATIVE CELL SPECIFIC1 (HAP2/GCS1) is a conserved protein participating in plasmogamy across eukaryotes. In *Arabidopsis*, HAP2/GCS1 is specifically expressed in the endomembrane system and plasma membrane of sperm cells. It is a single transmembrane protein with a conserved HAP2/GCS1 domain and a histidine-rich carboxy-terminal. Sperm cells lacking HAP2/GCS1 can attach to the egg cell and the central cell but cannot fuse (Mori et al., 2006, 2014; von Besser et al., 2006). Monomeric HAP2/GCS1 trimerizes upon interaction with lipid bilayers and three β sheet-rich domains during fertilization, forming a "hairpin" -like structure similar to viral class II fusion proteins (Doms, 2017; Fédry et al., 2017; Feng et al., 2018; Gibbons et al., 2003).

The fine-tuning regulation, including the timing and localization of fusogens, is very important for fusion fate. AtHAP2/GCS1 relocates from the endomembrane compartment to the plasma membrane in response to the secretion of egg cell-expressed cysteine-rich EGG CELL 1s (EC1s) (Sprunck et al., 2012). EC1-induced translocation of membrane proteins in sperm cells seems to be universal among angiosperms. Other putative membrane fusion proteins, DUF679 membrane proteins 8/9 (DMP8/9), are also translocated to the surface of sperm cells when EC1s are applied exogenously to male gametes. The translocation of AtHAP2/GCS1 triggered by EC1s also relies on DMP8/9, whose functions are conserved across seed plants (Wang et al., 2022). Sperm cells from the *dmp8/dmp9* mutant exhibit a preference for fertilizing the central cell over the egg cell, indicating that DMP8/9 play a more substantial role in regulating sperm-egg fusion (Cyprys et al., 2019).

When plasmogamy between male and female gametes is complete, the gamete nuclei migrate to the fusion site for karyogamy. This process relies on the cell cytoskeleton (Dunkley et al., 2022; Fatema et al., 2019). F-actin has been shown to play an essential role in gamete nuclei migration for karyogamy, while microtubules are not necessary (Kawashima & Berger, 2015; Kawashima et al., 2014).

To date, the understanding of gamete fusion in plants is limited (Fig. 14.8). Numerous interesting scientific questions remain unclear at each step of gamete interaction. Known genes participating in gamete interaction shed light on us. For

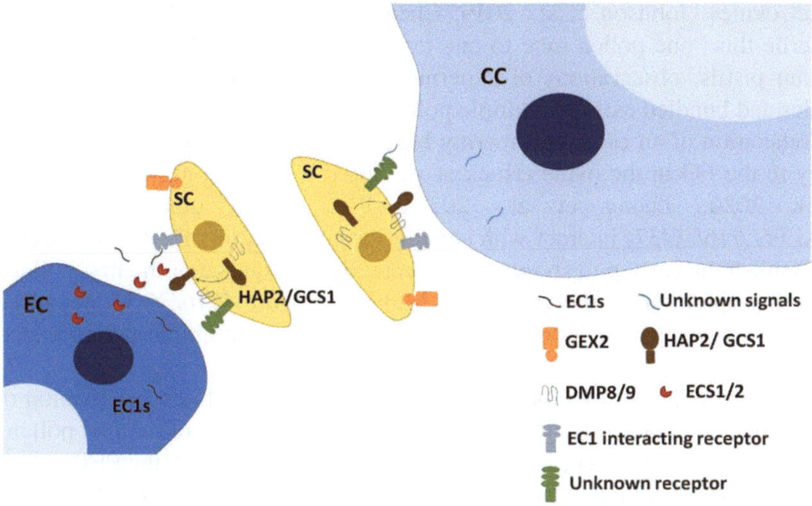

Fig. 14.8 Genes involved in different processes of gamete fusion. EC egg cell, CC central cell, SC sperm cell

example, since EC1s are essential for attachment and EC1 mediates the translocation of the fusogen HAP2/GCS1, the biological mechanism of EC1s warrants investigation. Additionally, considering that EC1s are peptides specifically expressed in the egg cell, does the central cell express EC1s-like peptides to participate in sperm activation? If so, what is the receptor on the sperm cell that interacts with these EC1s and EC1s-like peptides? Another crucial question is the identification of the partner protein of GEX2 on female gametes in *Arabidopsis*. Furthermore, what is the regulation mechanism of the relocation of HAP2/GCS1 to the plasma membrane of sperm cells, and how do DMP8/9 regulate the relocation of GCS1/HAP2? These intriguing questions are worthy of further exploration (Fig. 14.8).

14.2.9 Polytubey Block and Fertilization Recovery

Polyspermy, the entry of multiple sperm cells into a single egg, can lead to chromosomal anomalies in the zygote, resulting in severe consequences in animals (van der Ven et al., 1985). In plants, including angiosperms, polyspermy is generally considered disadvantageous (Spielman & Scott, 2008; Bianchi & Wright, 2016; Tekleyohans et al., 2017). While animals employ barriers within the female reproductive tract to prevent polyspermy (Bianchi & Wright, 2016), angiosperms have evolved multiple mechanisms to prevent the entry of multiple pollen tubes into the same ovule and thereby avert polyspermy.

During pollen tube growth in the pistil, a strict one-to-one relationship is established between pollen tubes and ovules as pollen tubes emerge from the septum to target ovules (Johnson et al., 2019; Cheung et al., 2022). Multiple mechanisms underlie this "one pollen tube to one ovule" phenomenon. In loss-of-function *fer* mutant pistils, observations of supernumerary pollen tube penetration of single ovules and bundled exits of multiple pollen tubes from the transmitting track reveal the relaxation of an early polyspermy block at the septum and a local block at the entry to the FG in the ovule (Huck et al., 2003; Rotman et al., 2003; Duan et al., 2014, 2020; Zhong et al., 2022). Five pollen tube-secreted RALFs (RALF6/7/16/36/37) interact with septum-located receptors FER/ANJEA/HERK1 to establish the early polyspermy block upon the emergence of the first pollen tube from the septum (Zhong et al., 2022). Multiplexed *ralf6/7/16/36/37* mutant pollen tubes fail to trigger the block, resulting in the entry of multiple pollen tubes into single ovules.

The local polyspermy block, primarily achieved through the inactivation of the pollen tube attractant AtLURE1s, is activated upon the entry of the first pollen tube into the ovule (Duan et al., 2020; Yu et al., 2021). On one hand, the FER-controlled block engages in signaling events with FER, pectin, and nitric oxide (NO) triggered by pollen tube arrival at the filiform apparatus. NO mediates a dual mechanism to disengage AtLURE1-mediated pollen tube attraction, inhibiting both the activity of the already-secreted attractant and preventing its further secretion from the synergids (Duan et al., 2020). In addition to the synergid cell, the egg cell also initiates the polyspermy block mechanism upon fusion with a sperm cell by secreting two aspartic proteases, ECS1 and ECS2, which cleave and deactivate any residual attractants (Yu et al., 2021).

If double fertilization by the first-emerged pollen tube fails, as observed in mutants such as *cdka;1*, *duo1*, *duo3*, and *hap2*, a phenotype resembling "polytubey" is observed. This phenomenon is termed "fertilization recovery" (Kasahara et al., 2012, 2013; Cheung et al., 2022). Fertilization recovery allows the plant to attract additional pollen tubes to a penetrated but not fertilized ovule, providing more opportunities to produce offspring.

What triggers fertilization recovery after fertilization failure? Firstly, the bursting of the first-arriving pollen tube in the FG is accompanied by the relaxation of the septum polyspermy block, as the secretion of the pollen RALF 6, 7, 16, 36, and 37 ceases. In the absence of pollen RALFs for maintenance, the septum block re-opens, allowing additional pollen tubes to exit and target already penetrated ovules (Zhong et al., 2022). Since there are two synergid cells in the embryo sac, after the initial fertilization attempt fails, the receptive synergid cell undergoes degradation (Maruyama & Higashiyama, 2016), and the embryo sac could continue to secrete attractants to attract additional pollen tubes, relying on the persistent synergid cell. In addition to synergid cell-mediated fertilization recovery, the central cell also plays a crucial role and directly participates in fertilization recovery. When double fertilization proves unsuccessful, and both synergids degenerate following the entry of the pollen tube, the central cell initiates an additional response by secreting pollen tube attractants, namely, SALVAGER1 and SALVAGER2 (Meng et al., 2023; Dresselhaus & van der Linde, 2023; Lan et al., 2023b). In *myb98* mutant ovules lacking functional synergids, an upregulation is observed in SALVAGERs and their upstream transcription factor CCG. The peptide SALVAGERs, initially retained within the central cell before fertilization, are selectively secreted toward the micropyle (Meng et al., 2023). This central cell-controlled fertilization recovery mechanism, in contrast to the transient impact of synergid-based fertilization recovery, assumes the role of the primary backup system for ensuring fertilization success (Dresselhaus & van der Linde, 2023).

It is worth noting that the preservation of the fertilization recovery mechanism in plants poses a risk to the one pollen tube to one ovule relationship between pollen tubes and ovules. Therefore, plants carefully regulate both the establishment of this one-to-one relationship and the decision to initiate fertilization recovery to ensure the highest efficiency in offspring production.

References

Aarts, M. G., Hodge, R., Kalantidis, K., Florack, D., Wilson, Z. A., Mulligan, B. J., Stiekema, W. J., Scott, R., & Pereira, A. (1997). The Arabidopsis MALE STERILITY 2 protein shares similarity with reductases in elongation/condensation complexes. *The Plant Journal, 12*, 615–623.

Acosta-Garcia, G., & Vielle-Calzada, J. P. (2004). A classical arabinogalactan protein is essential for the initiation of female gametogenesis in *Arabidopsis*. *Plant Cell, 16*, 2614–2628.

Amien, S., Kliwer, I., Marton, M. L., Debener, T., Geiger, D., Becker, D., & Dresselhaus, T. (2010). Defensin-like ZmES4 mediates pollen tube burst in maize via opening of the potassium channel KZM1. *PLoS Biology, 8*, e1000388.

Andersson, S., & Dobson, H. E. (2003). Behavioral foraging responses by the butterfly Heliconius melpomene to Lantana camara floral scent. *Journal of Chemical Ecology, 29*, 2303–2318.

Ariizumi, T., Hatakeyama, K., Hinata, K., Sato, S., Kato, T., Tabata, S., & Toriyama, K. (2003). A novel male-sterile mutant of Arabidopsis thaliana, faceless pollen-1, produces pollen with a smooth surface and an acetolysis-sensitive exine. *Plant Molecular Biology, 53*,107–116.

Armstrong, S. J., Caryl, A. P., Jones, G. H., & Franklin, F. C. H. (2002). Asy1, a protein required for melotic chromosome synapsis, localizes to axis-associated chromatin in *Arabidopsis* and *Brassica*. *Journal of Cell Science, 115*, 3645–3655.

Azumi, Y., Liu, D. H., Zhao, D. Z., Li, W. X., Wang, G. F., Hu, Y., & Ma, H. (2002). Homolog interaction during meiotic prophase I in *Arabidopsis* requires the SOLO DANCERS gene encoding a novel cyclin-like protein. *The EMBO Journal, 21*, 3081–3095.

Bai, X. F., Peirson, B. N., Dong, F. G., Xue, C., & Makaroff, C. A. (1999). Isolation and characterization of SYN1, a RAD21-like gene essential for meiosis in *Arabidopsis*. *Plant Cell, 11*, 417–430.

Baker, H. (1955). Self-compatibility and establishment after 'long-distance' dispersal. *Evolution, 9*, 347–349.

Bernal, A. J., Yoo, C. M., Mutwil, M., Jensen, J. K., Hou, G., Blaukopf, C., Sørensen, I., Blancaflor, E. B., Scheller, H. V., & Willats, W. G. (2008). Functional analysis of the cellulose synthase-like genes CSLD1, CSLD2, and CSLD4 in tip-growing Arabidopsis cells. *Plant Physiology, 148*, 1238–1253.

Bhatt, A. M., Lister, C., Page, T., Fransz, P., Findlay, K., Jones, G. H., Dickinson, H. G., & Dean, C. (1999). The DIF1 gene of Arabidopsis is required for meiotic chromosome segregation and belongs to the REC8/RAD21 cohesin gene family. *The Plant Journal, 19*, 463–472.

Bianchi, E., & Wright, G. J. (2016). Sperm meets egg: The genetics of mammalian fertilization. *Annual Review of Genetics, 50*, 93–111.

Bloch, D., Lavy, M., Efrat, Y., Efroni, I., Bracha-Drori, K., Abu-Abied, M., Sadot, E., & Yalovsky, S. (2005). Ectopic expression of an activated RAC in *Arabidopsis* disrupts membrane cycling. *Molecular Biology of the Cell, 16*, 1913–1927.

Bloch, D., Pleskot, R., Pejchar, P., Potocky, M., Trpkosova, P., Cwiklik, L., Vukasinovic, N., Sternberg, H., Yalovsky, S., & Zarsky, V. (2016). Exocyst SEC3 and phosphoinositides define sites of exocytosis in pollen tube initiation and growth. *Plant Physiology, 172*, 980–1002.

Boavida, L. C., & McCormick, S. (2007). Temperature as a determinant factor for increased and reproducible in vitro pollen germination in *Arabidopsis thaliana*. *The Plant Journal, 52*, 570–582.

Boisson-Dernier, A., Frietsch, S., Kim, T. H., Dizon, M. B., & Schroeder, J. I. (2008). The peroxin loss-of-function mutation *abstinence by mutual consent* disrupts male-female gametophyte recognition. *Current Biology, 18*, 63–68.

Boisson-Dernier, A., Roy, S., Kritsas, K., Grobei, M. A., Jaciubek, M., Schroeder, J. I., & Grossniklaus, U. (2009). Disruption of the pollen-expressed FERONIA homologs ANXUR1 and ANXUR2 triggers pollen tube discharge. *Development, 136*, 3279–3288.

Boisson-Dernier, A., Lituiev, D. S., Nestorova, A., Franck, C. M., Thirugnanarajah, S., & Grossniklaus, U. (2013). ANXUR receptor-like kinases coordinate cell wall integrity with growth at the pollen tube tip via NADPH oxidases. *PLoS Biology, 11*, e1001719.

Boisson-Dernier, A., Franck, C. M., Lituiev, D. S., & Grossniklaus, U. (2015). Receptor-like cytoplasmic kinase MARIS functions downstream of CrRLK1L-dependent signaling during tip growth. *Proceedings of the National Academy of Sciences of the United States of America, 112*, 12211–12216.

Borg, M., Brownfield, L., & Twell, D. (2009). Male gametophyte development: A molecular perspective. *Journal of Experimental Botany, 60*, 1465–1478.

Borg, M., Rutley, N., Kagale, S., Hamamura, Y., Gherghinoiu, M., Kumar, S., Sari, U., Esparza-Franco, M. A., Sakamoto, W., Rozwadowski, K., et al. (2014). An EAR-dependent regulatory module promotes male germ cell division and sperm fertility in *Arabidopsis*. *Plant Cell, 26*, 2098–2113.

Bosch, M., Cheung, A. Y., & Hepler, P. K. (2005). Pectin methylesterase, a regulator of pollen tube growth. *Plant Physiology, 138*, 1334–1346.

Brewbaker, J. L. (1967). Distribution and phylogenetic significance of binucleate and trinucleate pollen grains in angiosperms. *American Journal of Botany, 54*, 10691083.

Bruce, A. (1910). The mendelian theory of heredity and the augmentation of vigor. *Science, 32*, 627–628.

Cai, H., Liu, L., Huang, Y., Zhu, W., Qi, J., Xi, X., Aslam, M., Dresselhaus, T., & Qin, Y. (2022). Brassinosteroid signaling regulates female germline specification in *Arabidopsis*. *Current Biology, 32*, 1102–1114.e1105.

Cai, H., Huang, Y., Liu, L., Zhang, M., Chai, M., Xi, X., Aslam, M., Wang, L., Ma, S., Su, H., Liu, K., Tian, Y., Zhu, W., Qi, J., Dresselhaus, T., & Qin, Y. (2023). Signaling by the EPFL-ERECTA family coordinates female germline specification through the BZR1 family in *Arabidopsis*. *The Plant Cell, 35*, 1455–1473.

Chaiwongsar, S., Strohm, A. K., Roe, J. R., Godiwalla, R. Y., & Chan, C. W. M. (2009). A cyclic nucleotide-gated channel is necessary for optimum fertility in high-calcium environments. *The New Phytologist, 183*, 76–87.

Chang, F., Yan, A., Zhao, L. N., Wu, W. H., & Yang, Z. B. (2007). A putative calcium-permeable cyclic nucleotide-gated channel, CNGC18, regulates polarized pollen tube growth. *Journal of Integrative Plant Biology, 49*, 1261–1270.

Chapman, L. A., & Goring, D. R. (2011). Misregulation of phosphoinositides in *Arabidopsis thaliana* decreases pollen hydration and maternal fertility. *Sexual Plant Reproduction, 24*, 319–326.

Chaudhury, A. M., Lavithis, M., Taylor, P. E., Craig, S., Singh, M. B., Signer, E. R., Knox, R. B., & Dennis, E. S. (1994). Genetic control of male fertility in *Arabidopsis thaliana*: Structural-analysis of premeiotic developmental mutants. *Sexual Plant Reproduction, 7*, 17–28.

Chen, Y. C. S., & McCormick, S. (1996). Sidecar pollen, an *Arabidopsis thaliana* male gametophytic mutant with aberrant cell divisions during pollen development. *Development, 122*, 3243–3253.
Chen, C. B., Marcus, A., Li, W. X., Hu, Y., Calzada, J. P. V., Grossniklaus, U., Cyr, R. J., & Ma, H. (2002a). The *Arabidopsis* ATK1 gene is required for spindle morphogenesis in male meiosis. *Development, 129*, 2401–2409.
Chen, C. Y., Wong, E. I., Vidali, L., Estavillo, A., Hepler, P. K., Wu, H. M., & Cheung, A. Y. (2002b). The regulation of actin organization by actin-depolymerizing factor in elongating pollen tubes. *Plant Cell, 14*, 2175–2190.
Chen, Y. H., Li, H. J., Shi, D. Q., Yuan, L., Liu, J., Sreenivasan, R., Baskar, R., Grossniklaus, U., & Yang, W. C. (2007). The central cell plays a critical role in pollen tube guidance in *Arabidopsis*. *Plant Cell, 19*, 3563–3577.
Chen, Z., Hui, J. T. L., Ingouff, M., Sundaresan, V., & Berger, F. (2008). Chromatin assembly factor 1 regulates the cell cycle but not cell fate during male gametogenesis in *Arabidopsis thaliana*. *Development, 135*, 65–73.
Chen, G. H., Sun, J. Y., Liu, M., Liu, J., & Yang, W. C. (2014). SPOROCYTELESS is a novel embryophyte-specific transcription repressor that interacts with TPL and TCP proteins in *Arabidopsis*. *Journal of Genetics and Genomics, 41*, 617–625.
Chen, L. Y., Shi, D. Q., Zhang, W. J., Tang, Z. S., Liu, J., & Yang, W. C. (2015). The *Arabidopsis* alkaline ceramidase TOD1 is a key turgor pressure regulator in plant cells. *Nature Communications, 6*, 6030.
Chen, Y., Zou, T., & McCormick, S. (2016). S-adenosylmethionine synthetase 3 is important for pollen tube growth. *Plant Physiology, 172*, 244–253.
Chen, S.-Y., Wang, L., Jia, P.-F., Yang, W.-C., Sze, H., & Li, H.-J. (2022). Osmoregulation determines sperm cell geometry and integrity for double fertilization in flowering plants. *Molecular Plant, 15*, 1488–1496.
Cheung, A. Y., Chen, C. Y., Tao, L. Z., et al. (2003). Regulation of pollen tube growth by Rac-like GTPases. *Journal of Experimental Botany, 54*, 73–81.
Cheng, C. Y., & Kieber, J. J. (2013). The role of cytokinin in ovule development in *Arabidopsis*. *Plant Signaling & Behavior, 8*, e23393.
Cheung, A. Y., & Wu, H. M. (2008). Structural and signaling networks for the polar cell growth machinery in pollen tubes. *Annual Review of Plant Biology, 59*, 547–572.
Cheung, A. Y., Wang, H., & Wu, H. M. (1995). A floral transmitting tissue-specific glycoprotein attracts pollen tubes and stimulates their growth. *Cell, 82*, 383–393.
Cheung, A. Y., Chen, C. Y. H., Glaven, R. H., de Graaf, B. H. J., Vidali, L., Hepler, P. K., & Wu, H. M. (2002). Rab2 GTPase regulates vesicle trafficking between the endoplasmic reticulum and the Golgi bodies and is important to pollen tube growth. *Plant Cell, 14*, 945–962.
Cheung, A. Y., Niroomand, S., Zou, Y. J., & Wu, H. M. (2010). A transmembrane formin nucleates subapical actin assembly and controls tip-focused growth in pollen tubes. *Proceedings of the National Academy of Sciences of the United States of America, 107*, 16390–16395.
Cheung, A. Y., Duan, Q., Li, C., James Liu, M.-C., & Wu, H.-M. (2022). Pollen-pistil interactions: It takes two to tangle but a molecular cast of many to deliver. *Current Opinion in Plant Biology, 69*, 102279.
Chittka, L., & Schürkens, S. (2001). Successful invasion of a floral market. *Nature, 411*, 653.
Choi, Y. H., Gehring, M., Johnson, L., Hannon, M., Harada, J. J., Goldberg, R. B., Jacobsen, S. E., & Fischer, R. L. (2002). DEMETER, a DNA glycosylase domain protein, is required for endosperm gene imprinting and seed viability in *Arabidopsis*. *Cell, 110*, 33–42.
Christensen, C. A., King, E. J., Jordan, J. R., & Drews, G. N. (1997). Megagametogenesis in *Arabidopsis* wild type and the Gf mutant. *Sexual Plant Reproduction, 10*, 49–64.
Christensen, C. A., Gorsich, S. W., Brown, R. H., Jones, L. G., Brown, J., Shaw, J. M., & Drews, G. N. (2002). Mitochondrial GFA2 is required for synergid cell death in *Arabidopsis*. *Plant Cell, 14*, 2215–2232.

Cigliano, R. A., Cremona, G., Paparo, R., Termolino, P., Perrella, G., Gutzat, R., Consiglio, M. F., & Conicella, C. (2013). Histone deacetylase AtHDA7 is required for female gametophyte and embryo development in *Arabidopsis*. *Plant Physiology, 163*, 431–440.

Clavijo, E., & Jimenez, M. (1993). Cleistogamy and chasmogamy in Ceratocapnos heterocarpa (Fumariaceae). *International Journal of Plant Sciences, 154*, 325–333.

Cole, R. A., Synek, L., Zarsky, V., & Fowler, J. E. (2005). SEC8, a subunit of the putative Arabidopsis exocyst complex, facilitates pollen germination and competitive pollen tube growth. *Plant Physiology, 138*, 2005–2018.

Colombo, M., Masiero, S., Vanzulli, S., Lardelli, P., Kater, M. M., & Colombo, L. (2008). AGL23, a type I MADS-box gene that controls female gametophyte and embryo development in *Arabidopsis*. *The Plant Journal, 54*, 1037–1048.

Crane, P. R., Friis, E. M., & Pedersen, K. R. (1995). The origin and early diversification of angiosperms. *Nature, 374*, 27–33.

Crawford, B. C. W., Ditta, G., & Yanofsky, M. F. (2007). The NTT gene is required for transmitting-tract development in carpels of *Arabidopsis thaliana*. *Current Biology, 17*, 1101–1108.

Cyprys, P., Lindemeier, M., & Sprunck, S. (2019). Gamete fusion is facilitated by two sperm cell expressed DUF679 membrane proteins. *Nature Plants, 5*, 253–257.

Dai, X. R., Gao, X. Q., Chen, G. H., Tang, L. L., Wang, H., & Zhang, X. S. (2014). ABNORMAL POLLEN TUBE GUIDANCE1, an endoplasmic reticulum-localized mannosyltransferase homolog of GLYCOSYLPHOSPHATIDYLINOSITOL10 in yeast and PHOSPHATIDYLINOSITOL GLYCAN ANCHOR BIOSYNTHESIS B in human, is required for *Arabidopsis* pollen tube micropylar guidance and embryo development. *Plant Physiology, 165*, 1544–1556.

Darwin, C. (1862). *On the various contrivances by which british and foreign orchids are fertilised by insects*. John Murray.

de Graaf, B. H., Cheung, A. Y., Andreyeva, T., Levasseur, K., Kieliszewski, M., & Wu, H. M. (2005). Rab11 GTPase-regulated membrane trafficking is crucial for tip-focused pollen tube growth in tobacco. *The Plant Cell, 17*, 2564–2579.

De Storme, N., & Geelen, D. (2013). Cytokinesis in plant male meiosis. *Plant Signaling & Behavior, 8*, e23394.

Demesa-Arevalo, E., & Vielle-Calzada, J. P. (2013). The classical arabinogalactan protein AGP18 mediates megaspore selection in *Arabidopsis*. *Plant Cell, 25*, 1274–1287.

Deng, Y., Dong, H. L., Mu, J. Y., Ren, B., Zheng, B. L., Ji, Z. D., Yang, W. C., Liang, Y., & Zuo, J. R. (2010a). Arabidopsis histidine kinase CKI1 acts upstream of HISTIDINE PHOSPHOTRANSFER PROTEINS to regulate female gametophyte development and vegetative growth. *Plant Cell, 22*, 1232–1248.

Deng, Y., Wang, W., Li, W. Q., Xia, C. A., Liao, H. Z., Zhang, X. Q., & Ye, D. (2010b). MALE GAMETOPHYTE DEFECTIVE 2, encoding a sialyltransferase-like protein, is required for normal pollen germination and pollen tube growth in *Arabidopsis*. *Journal of Integrative Plant Biology, 52*, 829–843.

Denninger, P., Bleckmann, A., Lausser, A., Vogler, F., Ott, T., Ehrhardt, D. W., Frommer, W. B., Sprunck, S., Dresselhaus, T., & Grossmann, G. (2014). Male-female communication triggers calcium signatures during fertilization in *Arabidopsis*. *Nature Communications, 5*, 4645.

Doms, R. W. (2017). What came first—The virus or the egg? *Cell, 168*, 755–757.

Dong, X. Y., Hong, Z. L., Sivaramakrishnan, M., Mahfouz, M., & Verma, D. P. S. (2005). Callose synthase (CalS5) is required for exine formation during microgametogenesis and for pollen viability in *Arabidopsis*. *The Plant Journal, 42*, 315–328.

Doughty, J., Dixon, S., Hiscock, S. J., Willis, A. C., Parkin, I. A. P., & Dickinson, H. G. (1998). PCP-A1, a defensin-like *Brassica* pollen coat protein that binds the S locus glycoprotein, is the product of gametophytic gene expression. *Plant Cell, 10*, 1333–1347.

Dowd, P. E., Coursol, S., Skirpan, A. L., Kao, T. H., Gilroy, S. (2006). Petunia Phospholipase C1 Is Involved in Pollen Tube Growth. *The Plant Cell, 18*, 1438–1453.

Dresselhaus, T., & van der Linde, K. (2023). Plant reproduction: Fertilization SALvaged by the central cell. *Current Biology, 33*, R1013–R1015.

Dresselhaus, T., Sprunck, S., & Wessel, G. M. (2016). Fertilization mechanisms in flowering plants. *Current Biology, 26*, R125–R139.

Duan, Q. H., Kita, D., Johnson, E. A., Aggarwal, M., Gates, L., Wu, H. M., & Cheung, A. Y. (2014). Reactive oxygen species mediate pollen tube rupture to release sperm for fertilization in *Arabidopsis*. *Nature Communications, 5*, 3129.

Duan, Q., Liu, M.-C. J., Kita, D., Jordan, S. S., Yeh, F.-L. J., Yvon, R., Carpenter, H., Federico, A. N., Garcia-Valencia, L. E., Eyles, S. J., et al. (2020). FERONIA controls pectin- and nitric oxide-mediated male-female interaction. *Nature, 579*, 561–566.

Dumas, C., Knox, R. B., & Gaude, T. (1984). Pollen pistil recognition – New concepts from electron-microscopy and cyto-chemistry. *International Review of Cytology, 90*, 239–272.

Dunkley, S., Scheffler, K., & Mogessie, B. (2022). Cytoskeletal form and function in mammalian oocytes and zygotes. *Current Opinion in Cell Biology, 75*, 102073.

Durbarry, A., Vizir, I., & Twell, D. (2005). Male germ line development in *Arabidopsis*. duo pollen mutants reveal gametophytic regulators of generative cell cycle progression. *Plant Physiology, 137*, 297–307.

Ebel, C., Mariconti, L., & Gruissem, W. (2004). Plant retinoblastoma homologues control nuclear proliferation in the female gametophyte. *Nature, 429*, 776–780.

Elleman, C. J., & Dickinson, H. G. (1986). Pollen stigma interactions in Brassica.4. Structural reorganization in the pollen grains during hydration. *Journal of Cell Science, 80*, 141–157.

Elleman, C. J., Franklintong, V., & Dickinson, H. G. (1992). Pollination in species with dry stigmas – The nature of the early stigmatic response and the pathway taken by pollen tubes. *The New Phytologist, 121*, 413–424.

Endo, S., Shinohara, H., Matsubayashi, Y., Fukuda, H. (2013). A novel pollen-pistil interaction conferring high-temperature tolerance during reproduction via CLE45 signaling. *Current Biology, 23*, 1670–1676.

Enns, L. C., Kanaoka, M. M., Torii, K. U., Comai, L., Okada, K., & Cleland, R. E. (2005). Two callose synthases, GSL1 and GSL5, play an essential and redundant role in plant and pollen development and in fertility. *Plant Molecular Biology, 58*, 333–349.

Escobar-Restrepo, J. M., Huck, N., Kessler, S., Gagliardini, V., Gheyselinck, J., Yang, W. C., & Grossniklaus, U. (2007). The FERONIA receptor-like kinase mediates male-female interactions during pollen tube reception. *Science, 317*, 656–660.

Fabrice, T. N., Vogler, H., Draeger, C., Munglani, G., Gupta, S., Herger, A. G., Knox, P., Grossniklaus, U., & Ringli, C. (2018). LRX proteins play a crucial role in pollen grain and pollen tube cell wall development. *Plant Physiology, 176*, 1981–1992.

Fatema, U., Ali, M. F., Hu, Z., Clark, A. J., & Kawashima, T. (2019). Gamete nuclear migration in animals and plants. *Frontiers in Plant Science, 10*, 517.

Fédry, J., Liu, Y., Péhau-Arnaudet, G., Pei, J., Li, W., Tortorici, M. A., Traincard, F., Meola, A., Bricogne, G., Grishin, N. V., et al. (2017). The ancient gamete fusogen HAP2 is a eukaryotic class II fusion protein. *Cell, 168*, 904–915.e910.

Feng, Q. N., Kang, H., Song, S. J., Ge, F. R., Zhang, Y. L., Li, E., Li, S., & Zhang, Y. (2016). Arabidopsis RhoGDIs are critical for cellular homeostasis of pollen tubes. *Plant Physiology, 170*, 841–856.

Feng, J., Dong, X., Pinello, J., Zhang, J., Lu, C., Iacob, R. E., Engen, J. R., Snell, W. J., & Springer, T. A. (2018). Fusion surface structure, function, and dynamics of gamete fusogen HAP2. *Elife, 7*, e39772.

Fiebig, A., Mayfield, J. A., Miley, N. L., Chau, S., Fischer, R. L., & Preuss, D. (2000). Alterations in CER6, a gene identical to CUT1, differentially affect long-chain lipid content on the surface of pollen and stems. *Plant Cell, 12*, 2001–2008.

Frietsch, S., Wang, Y. F., Sladek, C., Poulsen, L. R., Romanowsky, S. M., Schroeder, J. I., & Harper, J. F. (2007). A cyclic nucleotide-gated channel is essential for polarized tip growth of pollen. *Proceedings of the National Academy of Sciences of the United States of America, 104*, 14531–14536.

Friedman, J. (2011). Gone with the wind: Understanding evolutionary transitions between wind and animal pollination in the angiosperms. *The New Phytologist, 191*, 911.

Friedman, J., & Barrett, S. (2009). Wind of change: New insights on the ecology and evolution of pollination and mating in wind-pollinated plants. *Annals of Botany, 103*, 1515–1527.

Fu, Y., Wu, G., & Yang, Z. (2001). Rop GTPase-dependent dynamics of tip-localized F-actin controls tip growth in pollen tubes. *The Journal of Cell Biology, 152*, 1019–1032.

Galindo-Trigo, S., Blanco-Touriñán, N., DeFalco, T. A., Wells, E. S., Gray, J. E., Zipfel, C., & Smith, L. M. (2020). 1393 Cr RLK1L receptor-like kinases HERK1 and ANJEA are female determinants of pollen 1394 tube reception. *EMBO Reports, 21*, e48466.

Gao, H., Zhang, Y. H., Wang, W. L., Zhao, K. K., Liu, C. M., Bai, L., Li, R., & Guo, Y. (2017). Two membrane-anchored aspartic proteases contribute to pollen and ovule development. *The New Phytologist, 173*, 219–239.

Gao, Q., Wang, C., Xi, Y., Shao, Q., Li, L., & Luan, S. (2022). A receptor-channel trio conducts Ca^{2+} signalling for pollen tube reception. *Nature, 607*, 534–539.

Ge, Z., Bergonci, T., Zhao, Y., Zou, Y., Du, S., Liu, M. C., Luo, X., Ruan, H., Garcia-Valencia, L. E., Zhong, S., et al. (2017). *Arabidopsis* pollen tube integrity and sperm release are regulated by RALF-mediated signaling. *Science, 358*, 1596–1600.

Gibbons, D. L., Erk, I., Reilly, B., Navaza, J., Kielian, M., Rey, F. A., & Lepault, J. (2003). Visualization of the target-membrane-inserted fusion protein of semliki forest virus by combined electron microscopy and crystallography. *Cell, 114*, 573–583.

Glover, J., Grelon, M., Craig, S., Chaudhury, A., & Dennis, E. (1998). Cloning and characterization of MS5 from Arabidopsis: A gene critical in male meiosis. *The Plant Journal, 15*, 345–356.

Goldman, M. H. S., Goldberg, R. B., & Mariani, C. (1994). Female sterile tobacco plants are produced by stigma-specific cell ablation. *The EMBO Journal, 13*, 2976–2984.

Gong, Y., Yang, M., Vamosi, J. C., Yang, H. M., Mu, W. X., Li, J. K., & Wan, T. (2016). Wind or insect pollination? Ambophily in a subtropical gymnosperm Gnetum parvifolium (Gnetales). *Plant Species Biology, 31*, 272–279.

Goubet, F., Misrahi, A., Park, S. K., Zhang, Z. N., Twell, D., & Dupree, P. (2003). AtCSLA7, a cellulose synthase-like putative glycosyltransferase, is important for pollen tube growth and embryogenesis in *Arabidopsis*. *The New Phytologist, 131*, 547–557.

Grelon, M., Vezon, D., Gendrot, G., & Pelletier, G. (2001). AtSPO11-1 is necessary for efficient meiotic recombination in plants. *The EMBO Journal, 20*, 589–600.

Gross-Hardt, R., Kagi, C., Baumann, N., Moore, J. M., Baskar, R., Gagliano, W. B., Jurgens, G., & Grossniklaus, U. (2007). LACHESIS restricts gametic cell fate in the female gametophyte of *Arabidopsis*. *PLoS Biology, 5*, 494–500.

Gu, Y., Fu, Y., Dowd, P., Li, S. D., Vernoud, V., Gilroy, S., & Yang, Z. B. (2005). A Rho family GTPase controls actin dynamics and tip growth via two counteracting downstream pathways in pollen tubes. *The Journal of Cell Biology, 169*, 127–138.

Guan, Y. F., Lu, J. P., Xu, J., McClure, B., & Zhang, S. Q. (2014). Two mitogen-activated protein kinases, MPK3 and MPK6, are required for funicular guidance of pollen tubes in *Arabidopsis*. *The New Phytologist, 165*, 528–533.

Guermonprez, H., Smertenko, A., Crosnier, M. T., Durandet, M., Vrielynck, N., Guerche, P., Hussey, P. J., Satiat-Jeunemaitre, B., & Bonhomme, S. (2008). The POK/AtVPS52 protein localizes to several distinct post-Golgi compartments in sporophytic and gametophytic cells. *Journal of Experimental Botany, 59*, 3087–3098.

Gusti, A., Baumberger, N., Nowack, M., Pusch, S., Eisler, H., Potuschak, T., De Veylder, L., Schnittger, A., & Genschik, P. (2009). The *Arabidopsis thaliana* F-Box protein FBL17 is essential for progression through the second mitosis during pollen development. *PLoS One, 4*, e4780.

Gutermuth, T., Lassig, R., Portes, M. T., Maierhofer, T., Romeis, T., Borst, J. W., Hedrich, R., Feijo, J. A., & Konrad, K. R. (2013). Pollen tube growth regulation by free anions depends

on the interaction between the anion channel SLAH3 and calcium-dependent protein kinases CPK2 and CPK20. *Plant Cell, 25*, 4525–4543.
Gutermuth, T., Herbell, S., Lassig, R., Brosche, M., Romeis, T., Feijo, J. A., Hedrich, R., & Konrad, K. R. (2018). Tip-localized Ca2+-permeable channels control pollen tube growth via kinase-dependent R- and S-type anion channel regulation. *The New Phytologist, 218*, 1089–1105.
Hafidh, S., Fila, J., & Honys, D. (2016). Male gametophyte development and function in angiosperms: A general concept. *Plant Reproduction, 29*, 31–51.
Hala, M., Cole, R., Synek, L., Drdova, E., Pecenkova, T., Nordheim, A., Lamkemeyer, T., Madlung, J., Hochholdinger, F., Fowler, J. E., et al. (2008). An exocyst complex functions in plant cell growth in *Arabidopsis* and tobacco. *Plant Cell, 20*, 1330–1345.
Hamamura, Y., Nishimaki, M., Takeuchi, H., Geitmann, A., Kurihara, D., & Higashiyama, T. (2014). Live imaging of calcium spikes during double fertilization in *Arabidopsis*. *Nature Communications, 5*, 4722.
Hamilton, E. S., Jensen, G. S., Maksaev, G., Katims, A., Sherp, A. M., & Haswell, E. S. (2015). Mechanosensitive channel MSL8 regulates osmotic forces during pollen hydration and germination. *Science, 350*, 438–441.
Hao, L. H., Liu, J. J., Zhong, S., Gu, H. Y., & Qu, L. J. (2016). AtVPS41-mediated endocytic pathway is essential for pollen tube-stigma interaction in *Arabidopsis*. *Proceedings of the National Academy of Sciences of the United States of America, 113*, 6307–6312.
Hao, L. H., Wei, X. L., Zhu, J. L., Shi, J., Liu, J. J., Gu, H. Y., Tsuge, T., & Qu, L. J. (2017). SNAIL1 is essential for female gametogenesis in *Arabidopsis thaliana*. *Journal of Integrative Plant Biology, 59*, 629–641.
Haizel, T., Merkle, T., Turck, F., & Nagy, F. (1995). Characterization of membrane-bound small GTP-binding proteins from Nicotiana tabacum. *Plant Physiology, 108*, 59–67.
He, S., Sun, Y., Yang, Q., Zhang, X. Y., Huang, Q. P., Zhao, P., Sun, M. X., Liu, J. J., Qian, W. Q., Qin, G. J., et al. (2017). A novel imprinted gene NUWA controls mitochondrial function in early seed development in *Arabidopsis*. *PLoS Genetics, 13*, e1006553.
Helling, D., Possart, A., Cottier, S., Klahre, U., & Kost, B. (2006). Pollen tube tip growth depends on plasma membrane polarization mediated by tobacco PLC3 activity and endocytic membrane recycling. *The Plant Cell, 18*, 3519–3534.
Hempel, F., Stenzel, I., Heilmann, M., Krishnamoorthy, P., Menzel, W., Golbik, R., Helm, S., Dobritzsch, D., Baginsky, S., Lee, J., Hoehenwarter, W., & Heilmann, I. (2017). MAPKs influence pollen tube growth by controlling the formation of phosphatidylinositol 4,5-bisphosphate in an apical plasma membrane domain. *The Plant Cell, 29*, 3030–3050.
Hepler, P. K., Vidali, L., & Cheung, A. Y. (2001). Polarized cell growth in higher plants. *Annual Review of Cell and Developmental Biology, an17*, 159–187.
Herre, E. A., Jandér, K. C., & Machado, C. A. (2008). Evolutionary ecology of figs and their associates: Recent progress and outstanding puzzles. *Annual Review of Ecology and Systematics, 39*, 439–458.
Herrera, C. M., & Pellmyr, O. (2002). *Plant–animal interactions: An evolutionary approach*. Blackwell Science.
Heslopharrison, J. (1982). Pollen-stigma interaction and cross-incompatibility in the grasses. *Science, 215*, 1358–1364.
Higashiyama, T., & Takeuchi, H. (2015). The mechanism and key molecules involved in pollen tube guidance. *Annual Review of Plant Biology, 66*, 393–413.
Higashiyama, T., Yabe, S., Sasaki, N., Nishimura, Y., Miyagishima, S., Kuroiwa, H., & Kuroiwa, T. (2001). Pollen tube attraction by the synergid cell. *Science, 293*, 1480–1483.
Hiscock, S. J., Dewey, F. M., Coleman, J. O. D., & Dickinson, H. G. (1994). Identification and localization of an active cutinase in the pollen of *Brassica napus L. Planta, 193*, 377–384.
Hoffmann, R. D., Portes, M. T., Olsen, L. I., Damineli, D. S. C., Hayashi, M., Nunes, C. O., Pedersen, J. T., Lima, P. T., Campos, C., Feijó, J. A., & Palmgren, M. (2020). Plasma membrane H^+-ATPases sustain pollen tube growth and fertilization. *Nature Communications, 11*, 2395.

Hollingsworth, N. M., Goetsch, L., & Byers, B. (1990). The Hop1 gene encodes a meiosis-specific component of yeast chromosomes. *Cell, 61*, 73–84.

Holsinger, K. (2000). Reproductive systems and evolution in vascular plants. *Proceedings of National Academy of Sciences of United States of America, 97*, 7037.

Hou, Y. N., Guo, X. Y., Cyprys, P., Zhang, Y., Bleckmann, A., Cai, L., Huang, Q. P., Luo, Y., Gu, H. Y., Dresselhaus, T., et al. (2016). Maternal ENODLs are required for pollen tube reception in *Arabidopsis*. *Current Biology, 26*, 2343–2350.

Huanca-Mamani, W., Garcia-Aguilar, M., Leon-Martinez, G., Grossniklaus, U., & Vielle-Calzada, J. P. (2005). CHR11, a chromatin-remodeling factor essential for nuclear proliferation during female gametogenesis in Arabidopsis thaliana. *Proceedings of the National Academy of Sciences of the United States of America, 102*, 17231–17236.

Huang, S. (2014). Most effective pollinator principle of floral evolution: Evidence and query. *Chinese Bulletin of Life Sciences, 26*, 118–124.

Huang, S., Xiong, Y., & Barrett, S. (2013). Experimental evidence of insect pollination in *Juncaceae*, a primarily wind-pollinated family. *International Journal of Plant Sciences, 174*, 1219–1228.

Huang, W. J., Liu, H. K., McCormick, S., & Tang, W. H. (2014). Tomato pistil factor STIG1 promotes in vivo pollen tube growth by binding to phosphatidylinositol 3-phosphate and the extracellular domain of the pollen receptor kinase LePRK2. *Plant Cell, 26*, 2505–2523.

Huang, J., Zhang, T. Y., Linstroth, L., Tillman, Z., Otegui, M. S., Owen, H. A., & Zhao, D. Z. (2016). Control of anther cell differentiation by the small protein ligand TPD1 and its receptor EMS1 in *Arabidopsis*. *PLoS Genetics, 12*, e1006147.

Huang, J., Chen, D., Yan, H. L., Xie, F., Yu, Y., Zhang, L. Y., Sun, M. X., & Peng, X. B. (2017). Acetylglutamate kinase is required for both gametophyte function and embryo development in *Arabidopsis thaliana*. *Journal of Integrative Plant Biology, 59*, 642–656.

Huck, N., Moore, J. M., Federer, M., & Grossniklaus, U. (2003). The A*rabidopsis* mutant feronia disrupts the female gametophytic control of pollen tube reception. *Development, 130*, 2149–2159.

Hulskamp, M., Kopczak, S. D., Horejsi, T. F., Kihl, B. K., & Pruitt, R. E. (1995). Identification of genes required for pollen-stigma recognition in *Arabidopsist thaliana*. *The Plant Journal, 8*, 703–714.

Hulskamp, M., Parekh, N. S., Grini, P., Schneitz, K., Zimmermann, I., Lolle, S. J., & Pruitt, R. E. (1997). The STUD gene is required for male-specific cytokinesis after telophase II of meiosis in *Arabidopsis thaliana*. *Developmental Biology, 187*, 114–124.

Hwang, J. U., Vernoud, V., Szumlanski, A., Nielsen, E., & Yang, Z. B. (2008). A tip-localized RhoGAP controls cell polarity by globally inhibiting Rho GTPase at the cell apex. *Current Biology, 18*, 1907–1916.

Hwang, J. U., Wu, G., Yan, A., Lee, Y. J., Grierson, C. S., & Yang, Z. B. (2010). Pollen-tube tip growth requires a balance of lateral propagation and global inhibition of Rho-family GTPase activity. *Journal of Cell Science, 123*, 340–350.

Inoue, N., Ikawa, M., Isotani, A., & Okabe, M. (2005). The immunoglobulin superfamily protein Izumo is required for sperm to fuse with eggs. *Nature, 434*, 234–238.

Ischebeck, T., Vu, L. H., Jin, X., Stenzel, I., Löfke, C., & Heilmann, I. (2010). Functional cooperativity of enzymes of phosphoinositide conversion according to synergistic effects on pectin secretion in tobacco pollen tubes. *Molecular Plant, 3*, 870–881.

Ischebeck, T., Stenzel, I., & Heilmann, I. (2008). Type B phosphatidylinositol-4-phosphate 5-kinases mediate Arabidopsis and Nicotiana tabacum pollen tube growth by regulating apical pectin secretion. *The Plant Cell, 20*, 3312–3330.

Ischebeck, T., Stenzel, I., Hempel, F., Jin, X., Mosblech, A., & Heilmann, I. (2011). Phosphatidylinositol-4,5-bisphosphate influences Nt-Rac5-mediated cell expansion in pollen tubes of Nicotiana tabacum. *The Plant Journal, 65*, 453–468.

Iwakawa, H., Shinmyo, A., & Sekine, M. (2006). *Arabidopsis* CDKA;1, a cdc2 homologue, controls proliferation of generative cells in male gametogenesis. *The Plant Journal, 45*, 819–831.

Iwano, M., Shiba, H., Miwa, T., Che, F. S., Takayama, S., Nagai, T., Miyawaki, A., & Isogai, A. (2004). Ca^{2+} dynamics in a pollen grain and papilla cell during pollination of *Arabidopsis*. *Plant Physiology, 136*, 3562–3571.

Iwano, M., Ngo, Q. A., Entani, T., Shiba, H., Nagai, T., Miyawaki, A., Isogai, A., Grossniklaus, U., & Takayama, S. (2012). Cytoplasmic Ca^{2+} changes dynamically during the interaction of the pollen tube with synergid cells. *Development, 139*, 4202–4209.

Jia, H. L., Li, J. S., Zhu, J. G., Fan, T. T., Qian, D., Zhou, Y. L., Wang, J. J., Ren, H. Y., Xiang, Y., & An, L. Z. (2013). *Arabidopsis* CROLIN1, a novel plant actin-binding protein, functions in cross-linking and stabilizing actin filaments. *The Journal of Biological Chemistry, 288*, 32277–32288.

Jiang, L. X., Yang, S. L., Xie, L. F., Puah, C. S., Zhang, X. Q., Yang, W. C., Sundaresan, V., & Ye, D. (2005). VANGUARD1 encodes a pectin methylesterase that enhances pollen tube growth in the *Arabidopsis* style and transmitting tract. *Plant Cell, 17*, 584–596.

Jiang, J., Stührwohldt, N., Liu, T., Huang, Q., Li, L., Zhang, L., Gu, H., Fan, L., Zhong, S., Schaller, A., et al. (2022). Egg cell-secreted aspartic proteases ECS1/2 promote gamete attachment to prioritize the fertilization of egg cells over central cells in *Arabidopsis*. *Journal of Integrative Plant Biology, 64*, 2047–2059.

Jin, H. L., Axtell, M. J., Dahlbeck, D., Ekwenna, O., Zhang, S. Q., Staskawicz, B., & Baker, B. (2002). NPK1, an MEKK1-like mitogen-activated protein kinase kinase kinase, regulates innate immunity and development in plants. *Developmental Cell, 3*, 291–297.

Johnson, M. A., Harper, J. F., & Palanivelu, R. (2019). A fruitful journey: Pollen tube navigation from germination to fertilization. *Annual Review of Plant Biology, 70*, 809–837.

Johnston, A. J., Matveeva, E., Kirioukhova, O., Grossniklaus, U., & Gruissem, W. (2008). A dynamic reciprocal RBR-PRC2 regulatory circuit controls *Arabidopsis* gametophyte development. *Current Biology, 18*, 1680–1686.

Jones, J. W. (1926). Hybrid vigor in rice. *Agronomy Journal, 18*, 423–428.

Ju, Y., Yuan, J., Jones, D. S., Zhang, W. W., Staiger, C. J., & Kessler, S. A. (2021). Polarized NORTIA accumulation in response to pollen tube arrival at synergids promotes fertilization. *Developmental Cell, 56*, 2938–2951.

Kahn, R. A., Cherfils, J., Elias, M., Lovering, R. C., Munro, S., & Schurmann, A. (2006). Nomenclature for the human Arf family of GTP-binding proteins: ARF, ARL, and SAR proteins. *The Journal of Cell Biology, 172*, 645–650.

Kandasamy, M. K., Paolillo, D. J., Faraday, C. D., Nasrallah, J. B., & Nasrallah, M. E. (1989). The S-locus specific glycoproteins of Brassica accumulate in the cell-wall of developing stigma Papillae. *Developmental Biology, 134*, 462–472.

Kandasamy, M. K., Nasrallah, J. B., & Nasrallah, M. E. (1994). Pollen-pistil interactions and developmental regulation of pollen-tube growth in Arabidopsis. *Development, 120*, 3405–3418.

Kaothien, P., Ok, S. H., Shuai, B., Wengier, D., Cotter, R., Kelley, D., Kiriakopolos, S., Muschietti, J., & McCormick, S. (2005). Kinase partner protein interacts with the LePRK1 and LePRK2 receptor kinases and plays a role in polarized pollen tube growth. *The Plant Journal, 42*, 492–503.

Kasahara, R. D., Portereiko, M. F., Sandaklie-Nikolova, L., Rabiger, D. S., & Drews, G. N. (2005). MYB98 is required for pollen tube guidance and synergid cell differentiation in *Arabidopsis*. *Plant Cell, 17*, 2981–2992.

Kasahara, R. D., Maruyama, D., Hamamura, Y., Sakakibara, T., Twell, D., & Higashiyama, T. (2012). Fertilization recovery after defective sperm cell release in *Arabidopsis*. *Current Biology, 22*, 1084–1089.

Kasahara, R. D., Maruyama, D., & Higashiyama, T. (2013). Fertilization recovery system is dependent on the number of pollen grains for efficient reproduction in plants. *Plant Signaling & Behavior, 8*, e23690.

Kato, M., Nagasaki-Takeuchi, N., Ide, Y., & Maeshima, M. (2010). An *Arabidopsis* hydrophilic Ca^{2+}-binding protein with a PEVK-rich domain, PCaP2, is associated with the plasma mem-

brane and interacts with calmodulin and phosphatidylinositol phosphates. *Plant & Cell Physiology, 51*, 366–379.
Kawashima, T., & Berger, F. (2015). The central cell nuclear position at the micropylar end is maintained by the balance of F-actin dynamics, but dispensable for karyogamy in *Arabidopsis*. *Plant Reproduction, 28*, 103–110.
Kawashima, T., Maruyama, D., Shagirov, M., Li, J., Hamamura, Y., Yelagandula, R., Toyama, Y., & Berger, F. (2014). Dynamic F-actin movement is essential for fertilization in Arabidopsis thaliana. *Elife, 3*, e04501.
Kaya, H., Iwano, M., Takeda, S., Kanaoka, M. M., Kimura, S., Abe, M., & Kuchitsu, K. (2015). Apoplastic ROS production upon pollination by RbohH and RbohJ in *Arabidopsis*. *Plant Signaling & Behavior, 10*, e989050.
Kessler, S. A., Shimosato-Asano, H., Keinath, N. F., Wuest, S. E., Ingram, G., Panstruga, R., & Grossniklaus, U. (2010). Conserved molecular components for pollen tube reception and fungal invasion. *Science, 330*, 968–971.
Kim, S., Mollet, J. C., Dong, J., Zhang, K. L., Park, S. Y., & Lord, E. M. (2003). Chemocyanin, a small basic protein from the lily stigma, induces pollen tube chemotropism. *Proceedings of the National Academy of Sciences of the United States of America, 100*, 16125–16130.
Kim, J., Shiu, S. H., Thoma, S., Li, W. H., & Patterson, S. E. (2006). Patterns of expansion and expression divergence in the plant polygalacturonase gene family. *Genome Biology, 7*, R87.
Kim, H. J., Oh, S. A., Brownfield, L., Hong, S. H., Ryu, H., Hwang, I., Twell, D., & Nam, H. G. (2008). Control of plant germline proliferation by SCF(FBL17) degradation of cell cycle inhibitors. *Nature, 455*, 1134–U1114.
Klahre, U., Becker, C., Schmitt, A. C., & Kost, B. (2006). Nt-RhoGDI2 regulates Rac/Rop signaling and polar cell growth in tobacco pollen tubes. *The Plant Journal, 46*, 1018–1031.
Koshimizu, S., Kofuji, R., Sasaki-Sekimoto, Y., Kikkawa, M., Shimojima, M., Ohta, H., Shigenobu, S., Kabeya, Y., Hiwatashi, Y., Tamada, Y., et al. (2018). Physcomitrella MADS-box genes regulate water supply and sperm movement for fertilization. *Nature Plants, 4*, 36–45.
Kost, B., Lemichez, E., Spielhofer, P., Hong, Y., Tolias, K., Carpenter, C., Chua, N. H. (1999). Rac homologues and compartmentalized phosphatidylinositol 4, 5-bisphosphate act in a common pathway to regulate polar pollen tube growth. *The Journal of Cell Biology, 145*, 317–330.
Kost, B. (2008). Spatial control of Rho (Rac-Rop) signaling in tip-growing plant cells. *Trends in Cell Biology, 18*, 119–127.
Kotake, T., Li, Y. Q., Takahashi, M., & Sakurai, N. (2000). Characterization and function of wall-bound exo-beta-glucanases of *Lilium longiflorum* pollen tubes. *Sexual Plant Reproduction, 13*, 1–9.
Kovar, D. R., & Pollard, T. D. (2004). Insertional assembly of actin filament barbed ends in association with formins produces piconewton forces. *Proceedings of the National Academy of Sciences of the United States of America, 101*, 14725–14730.
Krysan, P. J., Jester, P. J., Gottwald, J. R., & Sussman, M. R. (2002). An *Arabidopsis* mitogen-activated protein kinase kinase kinase gene family encodes essential positive regulators of cytokinesis. *Plant Cell, 14*, 1109–1120.
Kulich, I., Cole, R., Drdova, E., Cvrckova, F., Soukup, A., Fowler, J., & Zarsky, V. (2010). *Arabidopsis* exocyst subunits SEC8 and EXO70A1 and exocyst interactor ROH1 are involved in the localized deposition of seed coat pectin. *The New Phytologist, 188*, 615–625.
Kwee, H. S., & Sundaresan, V. (2003). The NOMEGA gene required for female gametophyte development encodes the putative APC6/CDC16 component of the anaphase promoting complex in *Arabidopsis*. *The Plant Journal, 36*, 853–866.
Lalanne, E., Honys, D., Johnson, A., Borner, G. H. H., Lilley, K. S., Dupree, P., Grossniklaus, U., & Twell, D. (2004). SETH1 and SETH2, two components of the glycosylphosphatidylinositol anchor biosynthetic pathway, are required for pollen germination and tube growth in *Arabidopsis*. *Plant Cell, 16*, 229–240.

Lalonde, B. A., Nasrallah, M. E., Dwyer, K. G., Chen, C. H., Barlow, B., & Nasrallah, J. B. (1989). A highly conserved Brassica gene with homology to the S-locus-specific glycoprotein structural gene. *Plant Cell, 1*, 249–258.

Lan, Z., Song, Z., Wang, Z., Li, L., Liu, Y., Zhi, S., Wang, R., Wang, J., Li, Q., Bleckmann, A., et al. (2023a). Antagonistic RALF peptides control an intergeneric hybridization barrier on *Brassicaceae* stigmas. *Cell, 186*, 4773–4787.e12.

Lan, Z., Song, Z., Zhong, S., & Qu, L.-J. (2023b). The central cell: Another opportunity for fertilization recovery in plants. *Seed Biology, 2*, 22.

Lassig, R., Gutermuth, T., Bey, T. D., Konrad, K. R., & Romeis, T. (2014). Pollen tube NAD(P)H oxidases act as a speed control to dampen growth rate oscillations during polarized cell growth. *The Plant Journal, 78*, 94–106.

Latrasse, D., Benhamed, M., Henry, Y., Domenichini, S., Kim, W., Zhou, D. X., & Delarue, M. (2008). The MYST histone acetyltransferases are essential for gametophyte development in *Arabidopsis*. *BMC Plant Biology, 8*, 121.

Lausser, A., & Dresselhaus, T. (2010). Sporophytic control of pollen tube growth and guidance in grasses. *Biochemical Society Transactions, 38*, 631–634.

Lee, Y. R. J., Li, Y., & Liu, B. (2007). Two *Arabidopsis* phragmoplast-associated kinesins play a critical role in cytokinesis during male gametogenesis. *Plant Cell, 19*, 2595–2605.

Lee, Y. J., Szumlanski, A., Nielsen, E., & Yang, Z. B. (2008). Rho-GTPase-dependent filamentous actin dynamics coordinate vesicle targeting and exocytosis during tip growth. *The Journal of Cell Biology, 181*, 1155–1168.

Leroux, C., Bouton, S., Kiefer-Meyer, M. C., Fabrice, T. N., Mareck, A., Guenin, S., Fournet, F., Ringli, C., Pelloux, J., Driouich, A., et al. (2015). PECTIN METHYLESTERASE48 is involved in Arabidopsis pollen grain germination. *Plant Physiology, 167*, 367–380.

Leshem, Y., Johnson, C., & Sundaresan, V. (2013). Pollen tube entry into the synergid cell of Arabidopsis is observed at a site distinct from the filiform apparatus. *Plant Reproduction, 26*, 93–99.

Leydon, A. R., Beale, K. M., Woroniecka, K., Castner, E., Chen, J., Horgan, C., Palanivelu, R., & Johnson, M. A. (2013). Three MYB transcription factors control pollen tube differentiation required for sperm release. *Current Biology, 23*, 1209–1214.

Leydon, A. R., Weinreb, C., Venable, E., Reinders, A., Ward, J. M., & Johnson, M. A. (2017). The molecular dialog between flowering plant reproductive partners defined by SNP-informed RNA-sequencing. *Plant Cell, 29*, 984–1006.

Li, H., Wu, G., Ware, D., Davis, K. R., & Yang, Z. B. (1998). *Arabidopsis* Rho-related GTPases: Differential gene expression in pollen and polar localization in fission yeast. *Plant Physiology, 118*, 407–417.

Li, H. J., Xue, Y., Jia, D. J., Wang, T., Shi, D. Q., Liu, J., Cui, F., Xie, Q., Ye, D., & Yang, W. C. (2011). POD1 regulates pollen tube guidance in response to micropylar female signaling and acts in early embryo patterning in *Arabidopsis*. *Plant Cell, 23*, 3288–3302.

Li, S., Ge, F. R., Xu, M., Zhao, X. Y., Huang, G. Q., Zhou, L. Z., Wang, J. G., Kombrink, A., McCormick, S., Zhang, X. S., et al. (2013). *Arabidopsis* COBRA-LIKE 10, a GPI-anchored protein, mediates directional growth of pollen tubes. *The Plant Journal, 74*, 486–497.

Li, C., Yeh, F. L., Cheung, A. Y., Duan, Q., Kita, D., Liu, M. C., Maman, J., Luu, E. J., Wu, B. W., Gates, L., et al. (2015a). Glycosylphosphatidylinositol-anchored proteins as chaperones and co-receptors for FERONIA receptor kinase signaling in *Arabidopsis*. *Elife, 4*, e06587.

Li, H. J., Zhu, S. S., Zhang, M. X., Wang, T., Liang, L., Xue, Y., Shi, D. Q., Liu, J., & Yang, W. C. (2015b). *Arabidopsis* CBP1 is a novel regulator of transcription initiation in central cell-mediated pollen tube guidance. *Plant Cell, 27*, 2880–2893.

Li, S. W., Dong, H. J., Pei, W. K., Liu, C. N., Zhang, S., Sun, T. T., Xue, X. H., & Ren, H. Y. (2017). LlFH1-mediated interaction between actin fringe and exocytic vesicles is involved in pollen tube tip growth. *The New Phytologist, 214*, 745–761.

Li, E., Cui, Y., Ge, F.-R., Chai, S., Zhang, W.-T., Feng, Q.-N., Jiang, L., Li, S., & Zhang, Y. (2018). AGC1.5 kinase phosphorylates RopGEFs to control pollen tube growth. *Molecular Plant, 11*, 1198–1209.

Li, L., Hou, S., Xiang, W., Song, Z., Wang, Y., Zhang, L., Li, J., Gu, H., Dong, J., Dresselhaus, T., et al. (2022). The egg cell is preferentially fertilized in *Arabidopsis* double fertilization. *Journal of Integrative Plant Biology, 64*, 2039–2046.

Liang, Y., Tan, Z. M., Zhu, L., Niu, Q. K., Zhou, J. J., Li, M., Chen, L. Q., Zhang, X. Q., & Ye, D. (2013). MYB97, MYB101 and MYB120 function as male factors that control pollen tube-synergid interaction in *Arabidopsis thaliana* fertilization. *PLoS Genetics, 9*, e1003933.

Liao, H. Z., Zhu, M. M., Cui, H. H., Du, X. Y., Tang, Y., Chen, L. Q., Ye, D., & Zhang, X. Q. (2016). MARIS plays important roles in Arabidopsis pollen tube and root hair growth. *Journal of Integrative Plant Biology, 58*, 927–940.

Lin, Y. K., Wang, Y. L., Zhu, J. K., & Yang, Z. B. (1996). Localization of a Rho GTPase implies a role in tip growth and movement of the generative cell in pollen tubes. *Plant Cell, 8*, 293–303.

Lindner, H., Kessler, S. A., Muller, L. M., Shimosato-Asano, H., Boisson-Dernier, A., & Grossniklaus, U. (2015). TURAN and EVAN mediate pollen tube reception in *Arabidopsis* synergids through protein glycosylation. *PLoS Biology, 13*, e1002139.

Ling, Y., Zhang, C., Chen, T., Hao, H., Liu, P., Bressan, R. A., Hasegawa, P. M., Jin, J. B., & Lin, J. (2012). Mutation in SUMO E3 ligase, SIZ1, disrupts the mature female gametophyte in *Arabidopsis*. *PLoS One, 7*, e29470.

Liu, Z., & Makaroff, C. A. (2006). *Arabidopsis* separase AESP is essential for embryo development and the release of cohesin during meiosis. *Plant Cell, 18*, 1213–1225.

Liu, J. J., Zhang, Y. Y., Qin, G. J., Tsuge, T., Sakaguchi, N., Luo, G., Sun, K. T., Shi, D. Q., Aki, S., Zheng, N. Y., et al. (2008). Targeted degradation of the cyclin-dependent kinase inhibitor ICK4/KRP6 by RING-type E3 ligases is essential for mitotic cell cycle progression during Arabidopsis gametogenesis. *Plant Cell, 20*, 1538–1554.

Liu, M., Shi, D. Q., Yuan, L., Liu, J., & Yang, W. C. (2010). SLOW WALKER3, encoding a putative DEAD-box RNA helicase, is essential for female gametogenesis in *Arabidopsis*. *Journal of Integrative Plant Biology, 52*, 817–828.

Liu, X. L., Liu, L. F., Niu, Q. K., Xia, C. A., Yang, K. Z., Li, R., Chen, L. Q., Zhang, X. Q., Zhou, Y. H., & Ye, D. (2011). MALE GAMETOPHYTE DEFECTIVE 4 encodes a rhamnogalacturonan II xylosyltransferase and is important for growth of pollen tubes and roots in *Arabidopsis*. *The Plant Journal, 65*, 647–660.

Liu, J. J., Zhong, S., Guo, X. Y., Hao, L. H., Wei, X. L., Huang, Q. P., Hou, Y. N., Shi, J., Wang, C. Y., Gu, H. Y., et al. (2013). Membrane-bound RLCKs LIP1 and LIP2 are essential male factors controlling male-female attraction in *Arabidopsis*. *Current Biology, 23*, 993–998.

Liu, L. T., Zheng, C. H., Kuang, B. J., Wei, L. Q., Yan, L. F., & Wang, T. (2016a). Receptor-like kinase RUPO interacts with potassium transporters to regulate pollen tube growth and integrity in rice. *PLoS Genetics, 12*, e1006085.

Liu, X. L., Castro, C., Wang, Y. B., Noble, J., Ponvert, N., Bundy, M., Hoel, C., Shpak, E., & Palanivelu, R. (2016b). The role of LORELEI in pollen tube reception at the Interface of the synergid cell and pollen tube requires the modified eight-cysteine motif and the receptor-like kinase FERONIA. *Plant Cell, 28*, 1035–1052.

Liu, Z., Yuan, L., Song, X., Yu, X., & Sundaresan, V. (2017). AHP2, AHP3, and AHP5 act downstream of CKI1 in *Arabidopsis* female gametophyte development. *Journal of Experimental Botany, 68*, 3365–3373.

Liu, C., Shen, L., Xiao, Y., Vyshedsky, D., Peng, C., Sun, X., Liu, Z., Cheng, L., Zhang, H., Han, Z., et al. (2021). Pollen PCP-B peptides unlock a stigma peptide-receptor kinase gating mechanism for pollination. *Science, 372*, 171–175.

Lobstein, E., Guyon, A., Ferault, M., Twell, D., Pelletier, G., & Bonhomme, S. (2004). The putative *Arabidopsis* homolog of yeast Vps52p is required for pollen tube elongation, localizes to Golgi, and might be involved in vesicle trafficking. *Plant Physiology, 135*, 1480–1490.

Lu, Y. X., Chanroj, S., Zulkifli, L., Johnson, M. A., Uozumi, N., Cheung, A., & Sze, H. (2011). Pollen tubes lacking a pair of K^+ transporters fail to target ovules in *Arabidopsis*. *Plant Cell, 23*, 81–93.

Lu, P. L., Chai, M. F., Yang, J. E., Ning, G., Wang, G. L., & Ma, H. (2014). The Arabidopsis CALLOSE DEFECTIVE MICROSPORE1 gene is required for male fertility through regulating callose metabolism during microsporogenesis. *Plant Physiology, 164*, 1893–1904.

Lu, C. Q., Yu, F., Tian, L. F., Huang, X. Y., Tan, H., Xie, Z. J., Hao, X. H., Li, D. P., Luan, S., & Chen, L. B. (2017). RPS9M, a mitochondrial ribosomal protein, is essential for central cell maturation and endosperm development in *Arabidopsis*. *Frontiers in Plant Science, 8*, 2171.

Luo, N., Yan, A., Liu, G., Guo, J. Z., Rong, D. Y., Kanaoka, M. M., Xiao, Z., Xu, G. S., Higashiyama, T., Cui, X. P., et al. (2017). Exocytosis-coordinated mechanisms for tip growth underlie pollen tube growth guidance. *Nature Communications, 8*, 1687.

Luu, D. T., Heizmann, P., Dumas, C., Trick, M., & Cappadocia, M. (1997). Involvement of SLR1 genes in pollen adhesion to the stigmatic surface in *Brassicaceae*. *Sexual Plant Reproduction, 10*, 227–235.

Luu, D. T., Marty-Mazars, D., Trick, M., Dumas, C., & Heizmann, P. (1999). Pollen-stigma adhesion in Brassica spp involves SLG and SLR1 glycoproteins. *Plant Cell, 11*, 251–262.

MacAlister, C. A., Ortiz-Ramirez, C., Becker, J. D., Feijo, J. A., & Lippman, Z. B. (2016). Hydroxyproline O-arabinosyltransferase mutants oppositely alter tip growth in *Arabidopsis thaliana* and Physcomitrella patens. *The Plant Journal, 85*, 193–208.

Mähs, A., Steinhorst, L., Han, J. P., Shen, L. K., Wang, Y., & Kudla, J. (2013). The calcineurin B-like Ca2+ sensors CBL1 and CBL9 function in pollen germination and pollen tube growth in Arabidopsis. *Molecular Plant, 6*, 1149–1162.

Marton, M. L., Cordts, S., Broadhvest, J., & Dresselhaus, T. (2005). Micropylar pollen tube guidance by egg apparatus 1 of maize. *Science, 307*, 573–576.

Maruyama, D., & Higashiyama, T. (2016). The end of temptation: The elimination of persistent synergid cell identity. *Current Opinion in Plant Biology, 34*, 122–126.

Mayfield, J. A., & Preuss, D. (2000). Rapid initiation of *Arabidopsis* pollination requires the oleosin-domain protein GRP17. *Nature Cell Biology, 2*, 128–130.

Mayfield, J. A., Fiebig, A., Johnstone, S. E., & Preuss, D. (2001). Gene families from the *Arabidopsis* thaliana pollen coat proteome. *Science, 292*, 2482–2485.

Mecchia, M. A., Santos-Fernandez, G., Duss, N. N., Somoza, S. C., Boisson-Dernier, A., Gagliardini, V., Martinez-Bernardini, A., Fabrice, T. N., Ringli, C., Muschietti, J. P., et al. (2017). RALF4/19 peptides interact with LRX proteins to control pollen tube growth in *Arabidopsis*. *Science, 358*, 1600–1603.

Meng, Y., Xu, H., Chen, X., & Cai, Q. (2007). Pollination efficiency of the main bee pollinators of Hedysarum laeve, a legume in Mu Us Sandland, Inner Mongolia. *Biodiversity Science, 15*, 633–638.

Meng, J. G., Zhang, M. X., Yang, W. C., & Li, H. J. (2019). TICKET attracts pollen tubes and mediates reproductive isolation between relative species in *Brassicaceae*. *Science China Life Sciences, 62*, 1413–1419.

Meng, J. G., Liang, L., Jia, P. F., Wang, Y. C., Li, H. J., & Yang, W. C. (2020). Integration of ovular signals and exocytosis of a Ca^{2+} channel by MLOs in pollen tube guidance. *Nature Plants, 6*, 143–153.

Meng, J. G., Xu, Y. J., Wang, W. Q., Yang, F., Chen, S. Y., Jia, P. F., Yang, W. C., & Li, H. J. (2023). Central-cell produced attractants control fertilization recovery. *Cell, 186*, 3593–3605.e12.

Mercier, R., Vezon, D., Bullier, E., Motamayor, J. C., Sellier, A., Lefevre, F., Pelletier, G., & Horlow, C. (2001). SWITCH1 (SWI1): A novel protein required for the establishment of sister chromatid cohesion and for bivalent formation at meiosis. *Genes & Development, 15*, 1859–1871.

Michard, E., Lima, P. T., Borges, F., Silva, A. C., Portes, M. T., Carvalho, J. E., Gilliham, M., Liu, L. H., Obermeyer, G., & Feijo, J. A. (2011). Glutamate receptor-like genes form Ca^{2+} channels in pollen tubes and are regulated by pistil D-serine. *Science, 332*, 434–437.

Misamore, M. J., Gupta, S., & Snell, W. J. (2003). The Chlamydomonas Fus1 protein is present on the mating type plus fusion organelle and required for a critical membrane adhesion event during fusion with minus gametes. *Molecular Biology of the Cell, 14*, 2530–2542.

Miyazaki, S., Murata, T., Sakurai-Ozato, N., Kubo, M., Demura, T., Fukuda, H., & Hasebe, M. (2009). ANXUR1 and 2, sister genes to FERONIA/SIRENE, are male factors for coordinated fertilization. *Current Biology, 19*, 1327–1331.

Mizukami, A. G., Inatsugi, R., Jiao, J., Kotake, T., Kuwata, K., Ootani, K., Okuda, S., Sankaranarayanan, S., Sato, Y., Maruyama, D., et al. (2016). The AMOR arabinogalactan sugar chain induces pollen-tube competency to respond to ovular guidance. *Current Biology, 26*, 1091–1097.

Moll, C., von Lyncker, L., Zimmermann, S., Kagi, C., Baumann, N., Twell, D., Grossniklaus, U., & Gross-Hardt, R. (2008). CLO/GFA1 and ATO are novel regulators of gametic cell fate in plants. *The Plant Journal, 56*, 913–921.

Mollet, J. C., Park, S. Y., Nothnagel, E. A., & Lord, E. M. (2000). A lily stylar pectin is necessary for pollen tube adhesion to an *in vitro* stylar matrix. *Plant Cell, 12*, 1737–1749.

Moore, J. M., Calzada, J. P., Gagliano, W., & Grossniklaus, U. (1997). Genetic characterization of hadad, a mutant disrupting female gametogenesis in *Arabidopsis thaliana*. *Cold Spring Harbor Symposia on Quantitative Biology, 62*, 35–47.

Mori, T., Kuroiwa, H., Higashiyama, T., & Kuroiwa, T. (2006). GENERATIVE CELL SPECIFIC 1 is essential for angiosperm fertilization. *Nature Cell Biology, 8*, 64–71.

Mori, T., Igawa, T., Tamiya, G., Miyagishima, S. Y., & Berger, F. (2014). Gamete attachment requires GEX2 for successful fertilization in *Arabidopsis*. *Current Biology, 24*, 170–175.

Mouline, K., Very, A. A., Gaymard, F., Boucherez, J., Pilot, G., Devic, M., Bouchez, D., Thibaud, J. B., & Sentenac, H. (2002). Pollen tube development and competitive ability are impaired by disruption of a Shaker K^+ channel in *Arabidopsis*. *Genes & Development, 16*, 339–350.

Moussu, S., Lee, H. K., Haas, K. T., Broyart, C., Rathgeb, U., De Bellis, D., Levasseur, T., Schoenaers, S., Fernandez, G. S., Grossniklaus, U., et al. (2023). Plant cell wall patterning and expansion mediated by protein-peptide-polysaccharide interaction. *Science, 382*, 719–725.

Moutinho, A., Hussey, P. J., Trewavas, A. J., Malhó, R. (2001). cAMP acts as a second messenger in pollen tube growth and reorientation. *Proceedings of the National Academy of Sciences of the United States of America, 98*, 10481–10486.

Mu, J. H., Stains, J. P., & Kao, T. H. (1994). Characterization of a pollen-expressed gene encoding a putative pectin esterase of Petunia inflata. *Plant Molecular Biology, 25*, 539–544.

Muschietti, J., Eyal, Y., & McCormick, S. (1998). Pollen tube localization implies a role in pollen-pistil interactions for the tomato receptor-like protein kinases LePRK1 and LePRK2. *Plant Cell, 10*, 319–330.

Myers, C., Romanowsky, S. M., Barron, Y. D., Garg, S., Azuse, C. L., Curran, A., Davis, R. M., Hatton, J., Harmon, A. C., & Harper, J. F. (2009). Calcium-dependent protein kinases regulate polarized tip growth in pollen tubes. *The Plant Journal, 59*, 528–539.

Ngo, Q. A., Vogler, H., Lituiev, D. S., Nestorova, A., & Grossniklaus, U. (2014). A calcium dialog mediated by the FERONIA signal transduction pathway controls plant sperm delivery. *Developmental Cell, 29*, 491–500.

Nibau, C., Wu, H. M., & Cheung, A. Y. (2006). RAC/ROP GTPases: 'Hubs' for signal integration and diversification in plants. *Trends in Plant Science, 11*, 309–315.

Niewiadomski, P., Knappe, S., Geimer, S., Fischer, K., Schulz, B., Unte, U. S., Rosso, M. G., Ache, P., Flugge, U. I., & Schneider, A. (2005). The *Arabidopsis* plastidic glucose 6-phosphate/phosphate translocator GPT1 is essential for pollen maturation and embryo sac development. *Plant Cell, 17*, 760–775.

Niklas, K. J. (1985). The aerodynamics of wind pollination. *Botanical Review, 51*, 328–386.

Nilsson, L. A. (1988). The evolution of flowers with deep corolla tubes. *Nature, 334*, 147–149.

Nishihama, R., Ishikawa, M., Araki, S., Soyano, T., Asada, T., & Machida, Y. (2001). The NPK1 mitogen-activated protein kinase kinase kinase is a regulator of cell-plate formation in plant cytokinesis. *Genes & Development, 15*, 352–363.

Nonomura, K., Miyoshi, K., Eiguchi, M., Suzuki, T., Miyao, A., Hirochika, H., & Kurata, N. (2003). The MSP1 gene is necessary to restrict the number of cells entering into male and female sporogenesis and to initiate anther wall formation in rice. *Plant Cell, 15*, 1728–1739.

Nonomura, K. I., Morohoshi, A., Nakano, M., Eiguchi, M., Miyao, A., Hirochika, H., & Kurata, N. (2007). A germ cell-specific gene of the ARGONAUTE family is essential for the progression of premeiotic mitosis and meiosis during sporogenesis in rice. *Plant Cell, 19*, 2583–2594.

Nowack, M. K., Grini, P. E., Jakoby, M. J., Lafos, M., Koncz, C., & Schnittger, A. (2005). Isolation and characterization of a mutant of the *Arabidopsis* cdc2 homolog CDKA;1. *Computational Biochemical Physiology A, 141*, S322–S323.

Obermeyer, G., & Weisenseel, M. H. (1991). Calcium-channel blocker and calmodulin antagonists affect the gradient of free calcium-ions in Lily pollen tubes. *European Journal Cell Biology, 56*, 319–327.

Ogawa-Ohnishi, M., Matsushita, W., & Matsubayashi, Y. (2013). Identification of three hydroxyproline O-arabinosyltransferases in *Arabidopsis thaliana. Nature Chemical Biology, 9*, 726–730.

Oh, S. A., Johnson, A., Smertenko, A., Rahman, D., Park, S. K., Hussey, P. J., & Twell, D. (2005). A divergent cellular role for the FUSED kinase family in the plant-specific cytokinetic phragmoplast. *Current Biology, 15*, 2107–2111.

Oh, S. A., Bourdon, V., Dickinson, H. G., Twell, D., & Park, S. K. (2014). *Arabidopsis* Fused kinase TWO-IN-ONE dominantly inhibits male meiotic cytokinesis. *Plant Reproduction, 27*, 7–17.

Okuda, S., Tsutsui, H., Shiina, K., Sprunck, S., Takeuchi, H., Yui, R., Kasahara, R. D., Hamamura, Y., Mizukami, A., Susaki, D., et al. (2009). Defensin-like polypeptide LUREs are pollen tube attractants secreted from synergid cells. *Nature, 458*, 357–361.

Olmedo-Monfil, V., Duran-Figueroa, N., Arteaga-Vazquez, M., Demesa-Arevalo, E., Autran, D., Grimanelli, D., Slotkin, R. K., Martienssen, R. A., & Vielle-Calzada, J. P. (2010). Control of female gamete formation by a small RNA pathway in *Arabidopsis. Nature, 464*, 628–U200.

Ortiz-Ramirez, C., Michard, E., Simon, A. A., Damineli, D. S. C., Hernandez-Coronado, M., Becker, J. D., & Feijo, J. A. (2017). GLUTAMATE RECEPTOR-LIKE channels are essential for chemotaxis and reproduction in mosses. *Nature, 549*, 91.

Owen, H. A., & Makaroff, C. A. (1995). Ultrastructure of microsporogenesis and microgametogenesis in *Arabidopsis thaliana* (L) Heynh Ecotype Wassilewskija (Brassicaceae). *Protoplasma, 185*, 7–21.

Pagnussat, G. C., Alandete-Saez, M., Bowman, J. L., & Sundaresan, V. (2009). Auxin-dependent patterning and gamete specification in the *Arabidopsis* female gametophyte. *Science, 324*, 1684–1689.

Park, S. K., Howden, R., & Twell, D. (1998). The Arabidopsis thaliana gametophytic mutation gemini pollen1 disrupts microspore polarity, division asymmetry and pollen cell fate. *Development, 125*, 3789–3799.

Park, S. Y., Jauh, G. Y., Mollet, J. C., Eckard, K. J., Nothnagel, E. A., Walling, L. L., & Lord, E. M. (2000). A lipid transfer-like protein is necessary for lily pollen tube adhesion to an in vitro stylar matrix. *Plant Cell, 12*, 151–163.

Park, S. K., Rahman, D., Oh, S. A., & Twell, D. (2004). Gemini pollen 2, a male and female gametophytic cytokinesis defective mutation. *Sexual Plant Reproduction, 17*, 63–70.

Pastuglia, M., Azimzadeh, J., Goussot, M., Camilleri, C., Belcram, K., Evrard, J. L., Schmit, A. C., Guerche, P., & Bouchez, D. (2006). Gamma-tubulin is essential for microtubule organization and development in Arabidopsis. *The Plant Cell, 18*, 1412–1425. (未在tubulin的工作中引用此文).

Pawlowski, W. P., Wang, C. J. R., Golubovskaya, I. N., Szymaniak, J. M., Shi, L., Hamant, O., Zhu, T., Harper, L., Sheridan, W. F., & Cande, W. Z. (2009). Maize AMEIOTIC1 is essential for multiple early meiotic processes and likely required for the initiation of meiosis. *Proceedings of the National Academy of Sciences of the United States of America, 106*, 3603–3608.

Peirson, B. N., Owen, H. A., Feldmann, K. A., & Makaroff, C. A. (1996). Characterization of three male-sterile mutants of *Arabidopsis thaliana* exhibiting alterations in meiosis. *Sexual Plant Reproduction, 9*, 1–16.

Peter, C. I., & Johnson, S. D. (2008). Mimics and magnets: The importance of color and ecological facilitation in floral deception. *Ecology, 89*, 1583–1595.

Petroski, M. D., & Deshaies, R. J. (2005). Mechanism of lysine 48-linked ubiquitin-chain synthesis by the cullin-RING ubiquitin-ligase complex SCF-Cdc34. *Cell, 123*, 1107–1120.

Picton, J. M., & Steer, M. W. (1983). Evidence for the role of Ca^{2+} ions in tip extension in pollen tubes. *Protoplasma, 115*, 11–17.

Pierson, E. S., Miller, D. D., Callaham, D. A., Shipley, A. M., Rivers, B. A., Cresti, M., & Hepler, P. K. (1994). Pollen-tube growth is coupled to the extracellular calcium-ion flux and the intracellular calcium gradient – Effect of bapta-type buffers and hypertonic media. *Plant Cell, 6*, 1815–1828.

Podolsky, R. (1992). Strange floral attractors: Pollinator attraction and the evolution of plant sexual systems. *Science, 258*, 791–793.

Potocký, M., Eliás, M., Profotová, B., Novotná, Z., Valentová, O., Zárský, V. (2003). Phosphatidic acid produced by phospholipase D is required for tobacco pollen tube growth. *Planta, 217*, 122–130.

Pratibha, P., Singh, S. K., Srinivasan, R., Bhat, S. R., & Sreenivasulu, Y. (2017). Gametophyte development needs mitochondrial coproporphyrinogen III oxidase function1. *Plant Physiology, 174*, 258–275.

Qin, Y., & Yang, Z. B. A. (2011). Rapid tip growth: Insights from pollen tubes. *Seminars in Cell & Developmental Biology, 22*, 816–824.

Qin, T., Liu, X. M., Li, J. J., Sun, J. B., Song, L. N., & Mao, T. L. (2014a). *Arabidopsis* microtubule-destabilizing protein 25 functions in pollen tube growth by severing actin filaments. *Plant Cell, 26*, 325–339.

Qin, Y., Zhao, L. H., Skaggs, M. I., Andreuzza, S., Tsukamoto, T., Panoli, A., Wallace, K. N., Smith, S., Siddiqi, I., Yang, Z. B., et al. (2014b). ACTIN-RELATED PROTEIN6 regulates female meiosis by modulating meiotic gene expression in *Arabidopsis*. *The Plant Cell, 26*, 1612–1628.

Qu, X. L., Zhang, H., Xie, Y. R., Wang, J., Chen, N. Z., & Huang, S. J. (2013). *Arabidopsis* Villins promote actin turnover at pollen tube tips and facilitate the construction of actin collars. *Plant Cell, 25*, 1803–1817.

Reddy, T. V., Kaur, J., Agashe, B., Sundaresan, V., & Siddiqi, I. (2003). The DUET gene is necessary for chromosome organization and progression during male meiosis in *Arabidopsis* and encodes a PHD finger protein. *Development, 130*, 5975–5987.

Ren, L. J., Tang, D., Zhao, T. T., Zhang, F. F., Liu, C. Z., Xue, Z. H., Shi, W. Q., Du, G. J., Shen, Y., Li, Y. F., et al. (2018). OsSPL regulates meiotic fate acquisition in rice. *The New Phytologist, 218*, 789–803.

Richards, A. J. (1978). *The pollination of flowers by insects*. Academic Press for the Linnean Society of London.

Riglet, L., Rozier, F., Kodera, C., Bovio, S., Sechet, J., Fobis-Loisy, I., & Gaude, T. (2020). KATANIN-dependent mechanical properties of the stigmatic cell wall mediate the pollen tube path in *Arabidopsis*. *Elife, 9*, e57282.

Rockel, N., Wolf, S., Kost, B., Rausch, T., & Greiner, S. (2008). Elaborate spatial patterning of cell-wall PME and PMEI at the pollen tube tip involves PMEI endocytosis, and reflects the distribution of esterified and de-esterified pectins. *The Plant Journal, 53*, 133–143.

Romero, S., Le Clainche, C., Didry, D., Egile, C., Pantaloni, D., & Carlier, M. T. (2004). Formin is a processive motor that requires profilin to accelerate actin assembly and associated ATP hydrolysis. *Molecular Biology of the Cell, 15*, 267a.

Ross, K. J., Fransz, P., Armstrong, S. J., Vizir, I., Mulligan, B., Franklin, F. C. H., & Jones, G. H. (1997). Cytological characterization of four meiotic mutants of Arabidopsis isolated from T-DNA-transformed lines. *Chromosome Research, 5*, 551–559.

Rothschild, W., & Jordan, K. (1903). A revision of the lepidopterous family Sphingidae. *Novitates Zoologicae, 9*, 972.

Rotman, N., Rozier, F., Boavida, L., Dumas, C., Berger, F., & Faure, J. E. (2003). Female control of male gamete delivery during fertilization in *Arabidopsis thaliana*. *Current Biology, 13*, 432–436.

Rotman, N., Durbarry, A., Wardle, A., Yang, W. C., Chaboud, A., Faure, J. E., Berger, F., & Twell, D. (2005). A novel class of MYB factors controls sperm-cell formation in plants. *Current Biology, 15*, 244–248.

Russell, S. D. (1996). Attraction and transport of male gametes for fertilization. *Sexual Plant Reproduction, 9*, 337–342.

Russell, S. D., & Jones, D. S. (2015). The male germline of angiosperms: Repertoire of an inconspicuous but important cell lineage. *Frontiers in Plant Science, 6*, 173.

Rutley, N., & Twell, D. (2015). A decade of pollen transcriptomics. *Plant Reproduction, 28*, 73–89.

Rybak, K., Steiner, A., Synek, L., Klaeger, S., Kulich, I., Facher, E., Wanner, G., Kuster, B., Zarsky, V., Persson, S., et al. (2014). Plant cytokinesis is orchestrated by the sequential action of the TRAPPII and exocyst tethering complexes. *Developmental Cell, 29*, 607–620.

Safavian, D., Zayed, Y., Indriolo, E., Chapman, L., Ahmed, A., & Goring, D. R. (2015). RNA silencing of exocyst genes in the stigma impairs the acceptance of compatible pollen in *Arabidopsis*. *Plant Physiology, 169*, 2526–2538.

Samuel, M. A., Chong, Y. T., Haasen, K. E., Aldea-Brydges, M. G., Stone, S. L., & Goring, D. R. (2009). Cellular pathways regulating responses to compatible and self-incompatible pollen in *Brassica* and *Arabidopsis* stigmas intersect at Exo70A1, a putative component of the exocyst complex. *Plant Cell, 21*, 2655–2671.

Sanchez, A. M., Bosch, M., Bots, M., Nieuwland, J., Feron, R., & Mariani, C. (2004). Pistil factors controlling pollination. *Plant Cell, 16*, S98–S106.

Sanders, P. M., Bui, A. Q., Weterings, K., McIntire, K. N., Hsu, Y. C., Lee, P. Y., Truong, M. T., Beals, T. P., & Goldberg, R. B. (1999). Anther developmental defects in *Arabidopsis thaliana* male-sterile mutants. *Sexual Plant Reproduction, 11*, 297–322.

Sankaranarayanan, S., Jamshed, M., & Samuel, M. A. (2015). Degradation of glyoxalase I in Brassica napus stigma leads to self-incompatibility response. *Nature Plants, 1*, 15185.

Schekman, R., & Novick, P. (2004). 23 genes, 23 years later. *Cell, 116*, S13–S15, 11 p following S19.

Schiefthaler, U., Balasubramanian, S., Sieber, P., Chevalier, D., Wisman, E., & Schneitz, K. (1999). Molecular analysis of NOZZLE, a gene involved in pattern formation and early sporogenesis during sex organ development in *Arabidopsis thaliana*. *Proceedings of the National Academy of Sciences of the United States of America, 96*, 11664–11669.

Schiott, M., Romanowsky, S. M., Baekgaard, L., Jakobsen, M. K., Palmgren, M. G., & Harper, J. F. (2004). A plant plasma membrane Ca^{2+} pump is required for normal pollen tube growth and fertilization. *Proceedings of the National Academy of Sciences of the United States of America, 101*, 9502–9507.

Sede, A. R., Borassi, C., Wengier, D. L., Mecchia, M. A., Estevez, J. M., & Muschietti, J. P. (2018). *Arabidopsis* pollen extensins LRX are required for cell wall integrity during pollen tube growth. *FEBS Letters, 592*, 233–243.

Shi, X., Sun, X. H., Zhang, Z. G., Feng, D., Zhang, Q., Han, L. D., Wu, J. X., & Lu, T. G. (2015). GLUCAN SYNTHASE-LIKE 5 (GSL5) plays an essential role in male fertility by regulating callose metabolism during microsporogenesis in rice. *Plant & Cell Physiology, 56*, 497–509.

Shi, X., Han, X., & Lu, T. G. (2016). Callose synthesis during reproductive development in monocotyledonous and dicotyledonous plants. *Plant Signaling & Behavior, 11*, e1062196.

Shimizu, K. K., Ito, T., Ishiguro, S., & Okada, K. (2008). MAA3 (MAGATAMA3) helicase gene is required for female gametophyte development and pollen tube guidance in *Arabidopsis thaliana*. *Plant & Cell Physiology, 49*, 1478–1483.

Shull, G. (1908). The composition of a field of maize. *Journal of Heredity, os-4*, 296.

Silva, E., King, V., Jake, L., & Sargent, R. (2013). Evidence for pollen limitation of a native plant in invaded communities. *Oecologia, 172*, 469–476.

Singh, M., Goel, S., Meeley, R. B., Dantec, C., Parrinello, H., Michaud, C., Leblanc, O., & Grimanelli, D. (2011). Production of viable gametes without meiosis in maize deficient for an ARGONAUTE protein. *Plant Cell, 23*, 443–458.

Smalle, J., & Vierstra, R. D. (2004). The ubiquitin 26S proteasome proteolytic pathway. *Annual Review of Plant Biology, 55*, 555–590.

Smith, D. K., Jones, D. M., Lau, J. B. R., Cruz, E. R., Brown, E., Harper, J. F., & Wallace, I. S. (2018). A putative protein O-fucosyltransferase facilitates pollen tube penetration through the stigma-style interface. *Plant Physiology, 176*, 2804–2818.

Sousa, E., Kost, B., & Malho, R. (2008). Arabidopsis phosphatidylinositol-4-monophosphate 5-kinase 4 regulates pollen tube growth and polarity by modulating membrane recycling. *The Plant Cell, 20*, 3050–3064.

Spielman, M., & Scott, R. (2008). Polyspermy barriers in plants: From preventing to promoting fertilization. *Sexual Plant Reproduction, 21*, 53–65.

Spielman, M., Preuss, D., Li, F. L., Browne, W. E., Scott, R. J., & Dickinson, H. G. (1997). TETRASPORE is required for male meiotic cytokinesis in *Arabidopsis thaliana*. *Development, 124*, 2645–2657.

Springer, P. S., Holding, D. R., Groover, A., Yordan, C., & Martienssen, R. A. (2000). The essential Mcm7 protein PROLIFERA is localized to the nucleus of dividing cells during the G(1) phase and is required maternally for early *Arabidopsis* development. *Development, 127*, 1815–1822.

Sprunck, S., Rademacher, S., Vogler, F., Gheyselinck, J., Grossniklaus, U., & Dresselhaus, T. (2012). Egg cell-secreted EC1 triggers sperm cell activation during double fertilization. *Science, 338*, 1093–1097.

Stacey, N. J., Kuromori, T., Azumi, Y., Roberts, G., Breuer, C., Wada, T., Maxwell, A., Roberts, K., & Sugimoto-Shirasu, K. (2006). *Arabidopsis* SPO11-2 functions with SPO11-1 in meiotic recombination. *The Plant Journal, 48*, 206–216.

Stebbins, G. L. (1970). Adaptive radiation of reproductive characteristics in angiosperms, I: Pollination mechanisms. *Annual Review of Ecology and Systematics, 1*, 307–326.

Steffen, J. G., Kang, I. H., Portereiko, M. F., Lloyd, A., & Drews, G. N. (2008). AGL61 interacts with AGL80 and is required for central cell development in *Arabidopsis*. *Plant Physiology, 148*, 259–268.

Steinhorst, L., & Kudla, J. (2013). Calcium and reactive oxygen species rule the waves of signaling. *Plant Physiology, 163*, 471–485.

Steinhorst, L., Mahs, A., Ischebeck, T., Zhang, C. X., Zhang, X. X., Arendt, S., Schultke, S., Heilmann, I., & Kudla, J. (2015). Vacuolar CBL-CIPK12 Ca^{2+}-sensor-kinase complexes are required for polarized pollen tube growth. *Current Biology, 25*, 1475–1482.

Stephan, O., Cottier, S., Fahlen, S., Montes-Rodriguez, A., Sun, J., Eklund, D. M., Klahre, U., & Kost, B. (2014). RISAP is a TGN-associated RAC5 effector regulating membrane traffic during polar cell growth in tobacco. *Plant Cell, 26*, 4426–4447.

Stevens, R., Grelon, M., Vezon, D., Oh, J. S., Meyer, P., Perennes, C., Domenichini, S., & Bergounioux, C. (2004). A CDC45 homolog in *Arabidopsis* is essential for meiosis, as shown by RNA interference-induced gene silencing. *Plant Cell, 16*, 99–113.

Su, H., Feng, H. L., Chao, X. T., Ding, X., Nan, Q., Wen, C. X., Liu, H. D., Xiang, Y., & Liu, W. (2017a). Fimbrins 4 and 5 act synergistically during polarized pollen tube growth to ensure fertility in *Arabidopsis*. *Plant and Cell Physiology, 58*, 2006–2016.

Su, Z. X., Zhao, L. H., Zhao, Y. Y., Li, S. F., Won, S., Cai, H. Y., Wang, L. L., Li, Z. F., Chen, P. J., Qin, Y., et al. (2017b). The THO complex non-cell-autonomously represses female germline specification through the TAS3-ARF3 module. *Current Biology, 27*, 1597–1609.

Suen, D. F., Wu, S. S. H., Chang, H. C., Dhugga, K. S., & Huang, A. H. C. (2003). Cell wall reactive proteins in the coat and wall of maize pollen: Potential role in pollen tube growth on the stigma and through the style. *Journal of Biological Chemistry, 278*, 43672–43681.

Suen, D. F., & Huang, A. H. C. (2007). Maize pollen coat xylanase facilitates pollen tube penetration into silk during sexual reproduction. *Journal of Biological Chemistry, 282,* 625–636.

Sugano, S. S., Shimada, T., Imai, Y., Okawa, K., Tamai, A., Mori, M., & Hara-Nishimura, I. (2010). Stomagen positively regulates stomatal density in Arabidopsis. *Nature, 463*, 241–U130.

Sun, Y., Wang, X., Pan, L., Xie, F., Dai, B., Sun, M., & Peng, X. (2021). Plant egg cell fate determinationdepends on its exact position in female gametophyte. *Proceedings of the National Academy of Sciences of the United States of America, 118*, e2017488118.

Susaki, D., Suzuki, T., Maruyama, D., Ueda, M., Higashiyama, T., & Kurihara, D. (2021). Dynamics of the cell fate specifications during female gametophyte development in *Arabidopsis. PLoS Biology, 19*, e3001123.

Swanson, R., Edlund, A. F., & Preuss, D. (2004). Species specificity in pollen-pistil interactions. *Annual Review of Genetics, 38*, 793–818.

Synek, L., Vukasinovic, N., Kulich, I., Hala, M., Aldorfova, K., Fendrych, M., & Zarsky, V. (2017). EXO70C2 is a key regulatory factor for optimal tip growth of pollen. *Plant Physiology, 174*, 223–240.

Szumlanski, A. L., & Nielsen, E. (2009). The Rab GTPase RabA4d regulates pollen tube tip growth in *Arabidopsis thaliana. Plant Cell, 21*, 526–544.

Takayama, S., Shiba, H., Iwano, M., Asano, K., Hara, M., Che, F. S., Watanabe, M., Hinata, K., & Isogai, A. (2000). Isolation and characterization of pollen coat proteins of *Brassica campestris* that interact with S locus-related glycoprotein 1 involved in pollen-stigma adhesion. *Proceedings of the National Academy of Sciences of the United States of America, 97*, 3765–3770.

Takeuchi, H., & Higashiyama, T. (2012). A species-specific cluster of defensin-like genes encodes diffusible pollen tube attractants in *Arabidopsis. PLoS Biology, 10*, e1001449.

Takeuchi, H., & Higashiyama, T. (2016). Tip-localized receptors control pollen tube growth and LURE sensing in *Arabidopsis. Nature, 531*, 245–248.

Tan, X. Y., Cao, K., Liu, F., Li, Y. X., Li, P. X., Gao, C. J., Ding, Y., Lan, Z. Y., Shi, Z. X., & Rui, Q. C., et al. (2016). *Arabidopsis* COG complex subunits COG3 and COG8 modulate Golgi morphology, vesicle trafficking homeostasis and are essential for pollen tube growth. *PLoS Genetics, 12*, e1006140.

Tan, X., Wei, J., Li, B., Wang, M., & Bao, Y. (2017). AtVps11 is essential for vacuole biogenesis in embryo and participates in pollen tube growth in Arabidopsis. *Biochemical and Biophysical Research Communications, 491*, 794–799.

Tanaka, H., Ishikawa, M., Kitamura, S., Takahashi, Y., Soyano, T., Machida, C., & Machida, Y. (2004). The AtNACK1/HINKEL and STUD/TETRASPORE/AtNACK2 genes, which encode functionally redundant kinesins, are essential for cytokinesis in *Arabidopsis. Genes to Cells, 9*, 1199–1211.

Tang, W. H., Ezcurra, I., Muschietti, J., & McCormick, S. (2002). A cysteine-rich extracellular protein, LAT52, interacts with the extracellular domain of the pollen receptor kinase LePRK2. *Plant Cell, 14*, 2277–2287.

Tang, W. H., Kelley, D., Ezcurra, I., Cotter, R., & McCormick, S. (2004). LeSTIG1, an extracellular binding partner for the pollen receptor kinases LePRK1 and LePRK2, promotes pollen tube growth in vitro. *The Plant Journal, 39*, 343–353.

Taylor, L. P., & Hepler, P. K. (1997). Pollen germination and tube growth. *Annual Review of Plant Physiology, 48*, 461–491.

Tekleyohans, D. G., Mao, Y., Kägi, C., Stierhof, Y.-D., & Groß-Hardt, R. (2017). Polyspermy barriers: A plant perspective. *Current Opinion in Plant Biology, 35*, 131–137.

Tian, G. W., Chen, M. H., Zaltsman, A., & Citovsky, V. (2006). Pollen-specific pectin methylesterase involved in pollen tube growth. *Developmental Biology, 294*, 83–91.

Tsukamoto, T., Qin, Y., Huang, Y., Dunatunga, D., & Palanivelu, R. (2010). A role for LORELEI, a putative glycosylphosphatidylinositol-anchored protein, in *Arabidopsis thaliana* double fertilization and early seed development. *The Plant Journal, 62*, 571–588.

Twell, D., Wing, R., Yamaguchi, J., & Mccormick, S. (1989). Isolation and expression of an anther-specific gene from tomato. *Molecular Genetics and Genomics, 217*, 240–245.

Twell, D., Park, S. K., Hawkins, T. J., Schubert, D., Schmidt, R., Smertenko, A., & Hussey, P. J. (2002). MOR1/GEM1 has an essential role in the plant-specific cytokinetic phragmoplast. *Nature Cell Biology, 4*, 711–714.

Umbach, A. L., Lalonde, B. A., Kandasamy, M. K., Nasrallah, J. B., & Nasrallah, M. E. (1990). Immunodetection of protein glycoforms encoded by 2 independent genes of the self-incompatibility multigene family of *Brassica*. *Plant Physiology, 93*, 739–747.

Updegraff, E. P., Zhao, F., & Preuss, D. (2009). The extracellular lipase EXL4 is required for efficient hydration of *Arabidopsis* pollen. *Sexual Plant Reproduction, 22*, 197–204.

Vahisalu, T., Kollist, H., Wang, Y. F., Nishimura, N., Chan, W. Y., Valerio, G., Lamminmäki, A., Brosché, M., Moldau, H., Desikan, R., Schroeder, J. I., & Kangasjärvi, J. (2008). SLAC1 is required for plant guard cell S-type anion channel function in stomatal signalling. *Nature, 452*, 487–491.

van der Ven, H. H., Al-Hasani, S., Diedrich, K., Hamerich, U., Lehmann, F., & Krebs, D. (1985). Polyspermy in *in vitro* fertilization of human oocytes: Frequency and possible causes. *Annals of the New York Academy of Sciences, 442*, 88–95.

von Besser, K., Frank, A. C., Johnson, M. A., & Preuss, D. (2006). *Arabidopsis* HAP2 (GCS1) is a sperm-specific gene required for pollen tube guidance and fertilization. *Development, 133*, 4761–4769.

Wang, W., Wang, L., Chen, C., Xiong, G. Y., Tan, X. Y., Yang, K. Z., Wang, Z. C., Zhou, Y. H., Ye, D., & Chen, L. Q. (2011). *Arabidopsis* CSLD1 and CSLD4 are required for cellulose deposition and normal growth of pollen tubes. *Journal of Experimental Botany, 62*, 5161–5177.

Wang, C. J. R., Nan, G. L., Kelliher, T., Timofejeva, L., Vernoud, V., Golubovskaya, I. N., Harper, L., Egger, R., Walbot, V., & Cande, W. Z. (2012). Maize multiple archesporial cells 1 (mac1), an ortholog of rice TDL1A, modulates cell proliferation and identity in early anther development. *Development, 139*, 2594–2603.

Wang, L., Wang, W., Wang, Y. Q., Liu, Y. Y., Wang, J. X., Zhang, X. Q., Ye, D., & Chen, L. Q. (2013). Arabidopsis galacturonosyltransferase (GAUT) 13 and GAUT14 have redundant functions in pollen tube growth. *Molecular Plant, 6*, 1131–1148.

Wang, S. S., Diao, W. Z., Yang, X., Qiao, Z., Wang, M., Acharya, B. R., & Zhang, W. (2015). *Arabidopsis thaliana* CML25 mediates the Ca^{2+} regulation of K^+ transmembrane trafficking during pollen germination and tube elongation. *Plant, Cell & Environment, 38*, 2372–2386.

Wang, L., Zhang, C., Yang, M., Zhang, G., Zhang, Z., Yang, Y., & Duan, Y. (2016a). Intensified wind pollination mediated by pollen dimorphism after range expansion in an ambophilous biennial Aconitum gymnandrum. *Ecology and Evolution, 7*, 541–549.

Wang, T., Liang, L., Xue, Y., Jia, P. F., Chen, W., Zhang, M. X., Wang, Y. C., Li, H. J., & Yang, W. C. (2016b). A receptor heteromer mediates the male perception of female attractants in plants. *Nature, 536*, 360–360.

Wang, L. D., Clarke, L. A., Eason, R. J., Parker, C. C., Qi, B. X., Scott, R. J., & Doughty, J. (2017). PCP-B class pollen coat proteins are key regulators of the hydration checkpoint in *Arabidopsis thaliana* pollen-stigma interactions. *The New Phytologist, 213*, 764–777.

Wang, X., Wang, K., Yin, G., Liu, X., Liu, M., Cao, N., Duan, Y., Gao, H., Wang, W., & Ge, W., et al. (2018). Pollen-expressed leucine-rich repeat extensins are essential for pollen germination and growth. *Plant Physiology, 176*, 1993–2006.

Wang, W., Xiong, H., Cyprys, P., Malka, R., Flores-Tornero, M., Zhao, P., Peng, X., Sprunck, S., & Sun, M. (2022). DMP8 and 9 regulate HAP2/GCS1 trafficking for the timely acquisition of sperm fusion competence. *Proceedings of the National Academy of Sciences of the United States of America, 119*, e2207608119.

Wasserthal, L. (1997). The pollinators of the Malagasy star orchids Angraecum sesquipedale, A. sororium and A. compactum and the evolution of extremely long spurs by pollinator shift. *Botanica Acta, 110*, 343–359.

Wei, B. Y., Zhang, J. Z., Pang, C. X., Yu, H., Guo, D. S., Jiang, H., Ding, M. X., Chen, Z. Y., Tao, Q., Gu, H. Y., et al. (2015). The molecular mechanism of SPOROCYTELESS/NOZZLE in controlling *Arabidopsis* ovule development. *Cell Research, 25*, 121–134.

Whittall, J. B., & Hodges, S. A. (2007). Pollinator shifts drive increasingly long nectar spurs in columbine flowers. *Nature, 447*, 706–709.

Whittington, A. T., Vugrek, O., Wei, K. J., Hasenbein, N. G., Sugimoto, K., Rashbrooke, M. C., & Wasteneys, G. O. (2001). MOR1 is essential for organizing cortical microtubules in plants. *Nature, 411*, 610–613.

Willmer, P., Stone, G., & Johnston, I. (2000). *Environmental physiology of animals*. Blackwell.

Wilson, E. O. (1992). *The diversity of life*. W. W. Norton & Co.

Wolters-Arts, M., Lush, W. M., & Mariani, C. (1998). Lipids are required for directional pollen-tube growth. *Nature, 392*, 818–821.

Woriedh, M., Wolf, S., Marton, M. L., Hinze, A., Gahrtz, M., Becker, D., & Dresselhaus, T. (2013). External application of gametophyte-specific ZmPMEI1 induces pollen tube burst in maize. *Plant Reproduction, 26*, 255–266.

Wu, H. M., Wong, E., Ogdahl, J., & Cheung, A. Y. (2000). A pollen tube growth-promoting arabinogalactan protein from nicotiana alata is similar to the tobacco TTS protein. *Plant Journal, 22*, 165–176.

Wu, G., Gu, Y., Li, S. D., & Yang, Z. B. (2001). A genome-wide analysis of Arabidopsis Rop-interactive CRIB motif-containing proteins that act as Rop GTPase targets. *Plant Cell, 13*, 2841–2856.

Xia, C., Wang, Y. J., Liang, Y., Niu, Q. K., Tan, X. Y., Chu, L. C., Chen, L. Q., Zhang, X. Q., & Ye, D. (2014). The ARID-HMG DNA-binding protein AtHMGB15 is required for pollen tube growth in *Arabidopsis thaliana*. *The Plant Journal, 79*, 741–756.

Xu, Y., Yang, J., Wang, Y. H., Wang, J. C., Yu, Y., Long, Y., Wang, Y. L., Zhang, H., Ren, Y. L., Chen, J., et al. (2017). OsCNGC13 promotes seed-setting rate by facilitating pollen tube growth in stylar tissues. *PLoS Genetics, 13*, e1006906.

Yadegari, R., & Drews, G. N. (2004). Female gametophyte development. *Plant Cell, 16*, S133–S141.

Yamamoto, Y., Nishimura, M., Hara-Nishimura, I., & Noguchi, T. (2003). Behavior of vacuoles during microspore and pollen development in *Arabidopsis thaliana*. *Plant and Cell Physiology, 44*, 1192–1201.

Yamaoka, S., Nishihama, R., Yoshitake, Y., Ishida, S., Inoue, K., Saito, M., Okahashi, K., Bao, H., Nishida, H., Yamaguchi, K., et al. (2018). Generative cell specification requires transcription factors evolutionarily conserved in land plants. *Current Biology, 28*, 479–486.

Yang, M., Hu, Y., Lodhi, M., McCombie, W. R., & Ma, H. (1999a). The *Arabidopsis* SKP1-LIKE1 gene is essential for male meiosis and may control homologue separation. *Proceedings of the National Academy of Sciences of the United States of America, 96*, 11416–11421.

Yang, W. C., Ye, D., Xu, J., & Sundaresan, V. (1999b). The SPOROCYTELESS gene of *Arabidopsis* is required for initiation of sporogenesis and encodes a novel nuclear protein. *Genes & Development, 13*, 2108–2117.

Yang, C. Y., Spielman, M., Coles, J. P., Li, Y., Ghelani, S., Bourdon, V., Brown, R. C., Lemmon, B. E., Scott, R. J., & Dickinson, H. G. (2003a). TETRASPORE encodes a kinesin required for male meiotic cytokinesis in *Arabidopsis*. *The Plant Journal, 34*, 229–240.

Yang, X. H., Makaroff, C. A., & Ma, H. (2003b). The *Arabidopsis* MALE MEIOCYTE DEATH1 gene encodes a PHD-finger protein that is required for male meiosis. *Plant Cell, 15*, 1281–1295.

Yang, W. C., Shi, D. Q., & Chen, Y. H. (2010). Female gametophyte development in flowering plants. *Annual Review of Plant Biology, 61*, 89–108.

Yang, X., Wang, S. S., Wang, M., Qiao, Z., Bao, C. C., & Zhang, W. (2014). *Arabidopsis thaliana* calmodulin-like protein CML24 regulates pollen tube growth by modulating the actin cytoskeleton and controlling the cytosolic Ca^{2+} concentration. *Plant Molecular Biology, 86*, 225–236.

Ye, J. R., Zheng, Y. Y., Yan, A., Chen, N. Z., Wang, Z. K., Huang, S. J., & Yang, Z. B. (2009). *Arabidopsis* formin3 directs the formation of actin cables and polarized growth in pollen tubes. *Plant Cell, 21*, 3868–3884.

Yu, X., Zhang, X., Zhao, P., Peng, X., Chen, H., Bleckmann, A., Bazhenova, A., Shi, C., Dresselhaus, T., & Sun, M. (2021). Fertilized egg cells secrete endopeptidases to avoid polytubey. *Nature, 592*, 433–437.

Yuan, L., Liu, Z., Song, X., Johnson, C., Yu, X., & Sundaresan, V. (2016). The CKI1 histidine kinase specifies the female gametic precursor of the endosperm. *Developmental Cell, 37*, 34–46.

Zeng, Q. N., Chen, J. G., & Ellis, B. E. (2011). AtMPK4 is required for male-specific meiotic cytokinesis in *Arabidopsis*. *The Plant Journal, 67*, 895–906.

Zhang, G. Y., Feng, J., & Wu, J., et al. (2010). BoPMEI1, a pollen-specific pectin methylesterase inhibitor, has an essential role in pollen tube growth. *Planta, 231*, 1323–1334.

Zhang, J., Huang, Q. P., Zhong, S., Bleckmann, A., Huang, J. Y., Guo, X. Y., Lin, Q., Gu, H. Y., Dong, J., Dresselhaus, T., et al. (2017a). Sperm cells are passive cargo of the pollen tube in plant fertilization. *Nature Plants, 3*, 17079.

Zhang, X., Liu, W., Nagae, T. T., Takeuchi, H., Zhang, H., Han, Z., Higashiyama, T., & Chai, J. (2017b). Structural basis for receptor recognition of pollen tube attraction peptides. *Nature Communications, 8*, 1331.

Zhao, X., de Palma, J., Oane, R., Gamuyao, R., Luo, M., Chaudhury, A., Herve, P., Xue, Q., & Bennett, J. (2008). OsTDL1A binds to the LRR domain of rice receptor kinase MSP1, and is required to limit sporocyte numbers. *The Plant Journal, 54*, 375–387.

Zhao, L. N., Shen, L. K., Zhang, W. Z., Zhang, W., Wang, Y., & Wu, W. H. (2013). Ca^{2+}-dependent protein kinase11 and 24 modulate the activity of the inward rectifying K^+ channels in *Arabidopsis* pollen tubes. *Plant Cell, 25*, 649–661.

Zhao, B. C., Shi, H. D., Wang, W. L., Liu, X. Y., Gao, H., Wang, X. X., Zhang, Y. H., Yang, M. D., Li, R., & Guo, Y. (2016). Secretory COPII Protein SEC31B Is Required for Pollen Wall Development. *Plant Physiology, 172*, 1625–1642.

Zhao, X. A., Bramsiepe, J., Van Durme, M., Komaki, S., Prusicki, M. A., Maruyama, D., Forner, J., Medzihradszky, A., Wijnker, E., Harashima, H., et al. (2017). RETINOBLASTOMA RELATED1 mediates germline entry in *Arabidopsis*. *Science, 356*, eaaf6532.

Zhao, L. H., Cai, H. Y., Su, Z. X., Wang, L. L., Huang, X. Y., Zhang, M., Chen, P. J., Dai, X. Z., Zhao, H. M., Palanivelu, R., et al. (2018). KLU suppresses megasporocyte cell fate through SWR1-mediated activation of WRKY28 expression in *Arabidopsis*. *Proceedings of the National Academy of Sciences of the United States of America, 115*, E526–E535.

Zheng, Z. L., & Yang, Z. B. (2000a). The Rop GTPase switch turns on polar growth in pollen. *Trends in Plant Science, 5*, 298–303.

Zheng, Z. L., & Yang, Z. (2000b). The Rop GTPase: an emerging signaling switch in plants. *Plant Molecular Biology, 44*, 1–9.

Zheng, Y. Y., Xie, Y. R., Jiang, Y. X., Qu, X. L., & Huang, S. J. (2013). Arabidopsis ACTIN-DEPOLYMERIZING FACTOR7 severs actin filaments and regulates actin cable turnover to pomote normal pollen tube growth. *Plant Cell, 25*, 3405–3423.

Zhong, S., Liu, M. L., Wang, Z. J., Huang, Q. P., Hou, S. Y., Xu, Y. C., Ge, Z. X., Song, Z. H., Huang, J. Y., Qiu, X. Y., et al. (2019). Cysteine-rich peptides promote interspecific genetic isolation in *Arabidopsis*. *Science, 364*, eaau9564.

Zhong, S., Li, L., Wang, Z., Ge, Z., Li, Q., Bleckmann, A., Wang, J., Song, Z., Shi, Y., Liu, T., et al. (2022). RALF peptide signaling controls the polytubey block in *Arabidopsis*. *Science, 375*, 290–296.

Zhou, L. M., Lan, W. Z., Chen, B. Q., Fang, W., & Luan, S. (2015). A calcium sensor-regulated protein kinase, CALCINEURIN B-LIKE PROTEIN-INTERACTING PROTEIN KINASE19, is required for pollen tube growth and polarity. *Plant Physiology, 167*, 1351–U1395.

Zhou, Y., Yang, Y., Niu, Y., Fan, T., Qian, D., Luo, C., Shi, Y., Li, S., An, L., & Xiang, Y. (2020). The tip-localized phosphatidylserine established by *Arabidopsis* ALA3 is crucial for Rab GTPase-mediated vesicle trafficking and pollen tube growth. *Plant Cell, 32*, 3170–3187.

Zhou, L.-Z., Wang, L., Chen, X., Ge, Z., Mergner, J., Li, X., Küster, B., Längst, G., Qu, L.-J., & Dresselhaus, T. (2023). The RALF signaling pathway regulates cell wall integrity during pollen tube growth in maize. *The Plant Cell, 36*, koad324.

Zhu, L., Zhang, Y., Kang, E. F., Xu, Q. Y., Wang, M. Y., Rui, Y., Liu, B. Q., Yuan, M., & Fu, Y. (2013). MAP18 regulates the direction of pollen tube growth in *Arabidopsis* by modulating F-actin organization. *Plant Cell, 25*, 851–867.

Zhu, D. Z., Zhao, X. F., Liu, C. Z., Ma, F. F., Wang, F., Gao, X. Q., & Zhang, X. S. (2016). Interaction between RNA helicase ROOT INITIATION DEFECTIVE 1 and GAMETOPHYTIC FACTOR 1 is involved in female gametophyte development in *Arabidopsis*. *Journal of Experimental Botany, 67*, 5757–5768.

Zhu, J. G., Nan, Q., Qin, T., Qian, D., Mao, T. L., Yuan, S. J., Wu, X. R., Niu, Y., Bai, Q. F., An, L. Z., et al. (2017). Higher-ordered actin structures remodeled by *Arabidopsis* ACTIN-DEPOLYMERIZING FACTOR5 are important for pollen germination and pollen tube growth. *Molecular Plant, 10*, 1065–1081.

Zhu, M., Tao, L., Zhang, J., Liu, R., Tian, H., Hu, C., Zhu, Y., Li, M., Wei, Z., Yi, J., et al. (2022). The type-B response regulators ARR10, ARR12, and ARR18 specify the central cell in *Arabidopsis*. *Plant Cell, 34*, 4714–4737.

Zinkl, G. M., & Preuss, D. (2000). Dissecting *Arabidopsis* pollen-stigma interactions reveals novel mechanisms that confer mating specificity. *Annals of Botany, 85*, 15–21.

Zinkl, G. M., Zwiebel, B. I., Grier, D. G., & Preuss, D. (1999). Pollen-stigma adhesion in *Arabidopsis*: A species-specific interaction mediated by lipophilic molecules in the pollen exine. *Development, 126*, 5431–5440.

Zonia, L., Cordeiro, S., & Feijo, J. A. (2001). Ion dynamics and hydrodynamics in the regulation of pollen tube growth. *Sexual Plant Reproduction, 14*, 111–116.

Chapter 15
Male Fertility and Hybrid Sterility in Plants

Yongyao Xie, Huiwu Tang, Zhe Zhao, Yao-Guang Liu, and Letian Chen

Abstract In higher plants, pollen carries the male gametes. Male sterility, in which a plant generates abnormal anthers or aborted (inviable) pollen but normal embryo sacs, enables out-crossing to produce hybrids for the utilization of heterosis in agriculture. However, hybrid sterility also hinders crop breeding; for example, interspecific and intersubspecific hybrids always show various forms of sterility. This reproductive isolation prevents the production of hybrids with increased grain yields in key crops such as rice (*Oryza sativa*). In this chapter, we summarize the current understanding of the genetic and molecular bases of cytoplasmic male sterility, fertility restoration, and environment-sensitive genic male sterility in rice model systems. We also discuss how studies of these systems have shed light on cytoplasmic–nuclear interactions and plant–environment interactions. In addition, we also present new advances in understanding the molecular mechanisms of major loci for hybrid sterility in rice and highlight new strategies for overcoming reproductive barriers to better utilize heterosis in interspecific and intersubspecific hybrid rice.

Keywords Genic male sterility · Environment-sensitive genic male sterility · Cytoplasmic male sterility · Hybrid sterility

15.1 Introduction

In plants, male fertility has important implications for agriculture, as the use of male-sterile lines has enabled the widespread production of high-yield hybrid varieties. Development of pollen in higher plants occurs in the anthers and involves the formation of stamen primordia, early anther development, meiosis, and post-meiotic

Y. Xie · H. Tang · Z. Zhao · Y.-G. Liu · L. Chen (✉)
State Key Laboratory for Conservation and Utilization of Subtropical Agro-Bioresources, College of Agriculture, South China Agricultural University, Guangzhou, China

College of Life Sciences, South China Agricultural University, Guangzhou, China
e-mail: yyxie@scau.edu.cn; ygliu@scau.edu.cn; lotichen@scau.edu.cn

F. Chang et al. (eds.), *Regulation of Plant Development*,
https://doi.org/10.1007/978-981-95-7033-1_15

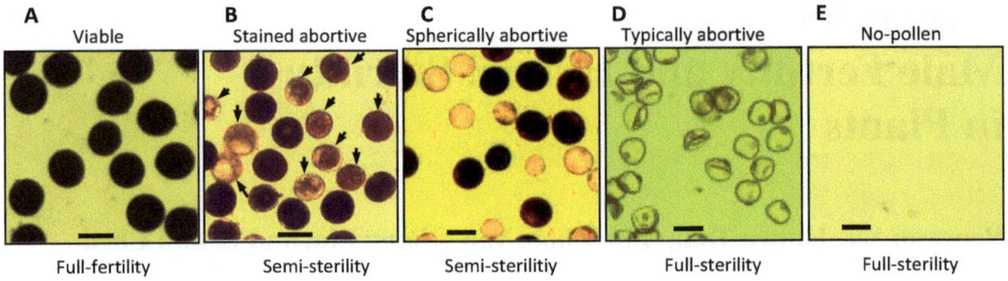

Fig. 15.1 Major cellular morphological phenotypes of pollen fertility (viability) for normal and sterile rice. The viable pollen grains in normal fertile lines are evenly stained by I_2-KI and are black and round (**a**). Stained abortive pollen grains are round, partially stained, and smaller than the fertile pollen (**b**). Spherically abortive pollen grains are small, round, empty, and unstainable, because they fail to accumulate starch (**c**). Typically, abortive pollen grains are empty, shrunken, and unstainable because of no starch (**d**). No-pollen male sterility is an extreme defect in pollen development, in which the anthers produce no pollen (**e**). Bars = 50 μm

microspore development. Defects in these developmental events caused by the dysfunction of nuclear genes may lead to genic male sterility (GMS), producing nonviable pollen of various types (Fig. 15.1). Most GMS mutants are difficult to use in hybrid seed production because they lack the maintainer lines required for propagation of the male-sterile lines. However, in environment-sensitive genic male sterility (EGMS) mutants, genes and environmental factors such as photoperiod and/or temperature control GMS synergistically. Under permissive conditions (such as in lower temperatures or short days), the EGMS lines are male fertile and can be self-pollinated, but under restrictive conditions (such as higher temperatures or long days), the EGMS lines are male sterile. This facilitates hybrid seed production in the two-line system, which involves the EGMS line as the female parent and another inbred line as the male parent. EGMS lines can be used as sterile lines or maintainer lines under different environmental conditions, thus simplifying seed production in two-line hybrid breeding.

In contrast to GMS, defects in mitochondria caused by incompatible interactions between specific cytoplasmic and nuclear genes can cause maternally inherited cytoplasmic male sterility (CMS), which occurs in many flowering plants. For hybrid breeding, CMS lines can be maintained by crossing to maintainer lines, and male fertility can be rescued by nuclear *restorer of fertility* (*Rf*) genes. CMS/*Rf* systems are vital germplasm for three-line hybrid breeding in crops; in this breeding scheme, the CMS line will be used as the female parent, and the maintainer and restorer lines are used as the male parents for maintaining the CMS line and producing hybrid seeds, respectively. In addition to their utility for hybrid seed production, CMS/*Rf* systems provide good molecular models for studying cytoplasmic–nuclear interactions in plants. Currently, work in rice has revealed the molecular bases of six major CMS systems and five EGMS systems.

CMS and EGMS lines are valuable genetic resources for the utilization of heterosis, but hybrid sterility (HS), in which the hybrids from interspecific and intersubspecific crosses show various forms of sterility, can hinder the production of hybrids with increased grain yields. HS is a major form of postzygotic reproductive isolation, which can help maintain species identity in plants. Genetic analyses indicated that two major models may control HS in rice: the one-locus sporogametophytic interaction model (one-locus model) and the duplicate gametophytic lethal model (two-locus model) (Oka, 1974). In the one-locus model, the interaction between functionally divergent alleles within the same locus can lead to a gain-of-function effect for allele-specific gamete abortion, resulting in distorted segregation of the alleles in the progeny. In the two-locus model, both parents are normal because each of them carries only one of two non-allelic loci affecting gamete development; however, when the two loci combine in a gamete, their interaction results in sterility in the hybrids.

This chapter focuses on the genetic and molecular characteristics of the isolated CMS/*Rf* and EGMS genes in rice. We also present new advances in understanding the molecular mechanisms of seven major HS loci in rice and highlight new strategies for overcoming reproductive barriers to utilize heterosis in interspecific and intersubspecific hybrid rice breeding programs.

15.2 Pollen Development and Male Sterility

Pollen is produced in anthers, and the development of mature anthers and viable pollen, which is mainly controlled by nuclear genes, involves three major steps: early anther development, meiosis, and post-meiotic pollen development. These steps involve interactions between the sporophytic tissues that produce the anther wall cells, including the tapetum, and the gametophytic cells that produce the pollen. The molecular mechanisms of pollen development in rice and *Arabidopsis* have been intensively studied, and the major processes are outlined below (for more detailed information, please refer to Chap. 9):

Meristem formation: The formation of the inflorescence meristem and floral meristem is a vital early process in reproductive development in flowering plants. Conserved pathways involving the *CLAVATA* and *WUSCHEL* genes strictly regulate the development of the inflorescence meristem and floral meristem (Guo & Liu, 2012; Muller et al., 2006; Schoof et al., 2000). Works in *Arabidopsis* have defined the ABCDE model for the determination of floral organs; in this model, the B-, C-, and E-class genes regulate stamen identity (Li et al., 2011a, b; Ma, 2005; Muller et al., 2006). Dysfunction in the genes that control the formation of stamen primordia generally results in abnormal reproductive organs (Li et al., 2011a, b; Ma, 2005; Muller et al., 2006).

Stamen cell differentiation: The stamen primordial cells differentiate into four concentric somatic (sporophytic) cell layers: the epidermis, endothecium, middle layer, and tapetum, which surround the pollen mother cells (PMCs), which undergo meiosis to generate haploid microspores and finally form the gametophytic cells, the pollen grains (Chang et al., 2011; Guo & Liu, 2012). These processes are mainly controlled by highly conserved genes involved in DNA replication, meiotic sister-chromatid cohesion, pairing, synapsis, meiotic recombination, spindle organization, and chromosome segregation; these genes include *EXCESS MICROSPOROCYTES1/EXTRA SPOROGENOUS CELLS* and *MULTIPLE SPOROCYTE1* in flowering plants (Canales et al., 2002; Chang et al., 2011; Guo & Liu, 2012; Jia et al., 2008; Nonomura et al., 2003; Yang et al., 2003, 2005; Zhao et al., 2002, 2008).

Pollen wall formation: The mature pollen wall consists of the outer exine, the inner intine, and the tryphine; proper wall formation is essential for the protection and release of the microspore. The biosynthesis of the exine (mainly sporopollenin) is regulated by ATP-diphosphohydrolase, cytosolic acyl-CoA-binding proteins, and UDP-arabinopyranose mutases (Ariizumi & Toriyama, 2011; Jiang et al., 2013); the formation of the intine and tryphine is associated with ECERIFERUM (CER), CER-LIKE proteins, long-chain acyl-CoA synthetases, Golgi-localized glycosyltransferase, and ARABINOKINASE-like proteins (Aarts et al., 1995; Ariizumi et al., 2003; Haslam et al., 2015; Jessen et al., 2011; Moon et al., 2013; Ueda et al., 2013).

A large number of male sterility (*ms*) mutants, which harbor defects in any of the above processes, have been identified in many monocots and dicots, especially in the model plants rice (*Oryza sativa* L.) and *Arabidopsis thaliana*. Most *ms* mutants are caused by a dysfunction of nuclear genes, and this GMS is further described in Chap. 9.

15.3 Cytoplasmic Male Sterility in Plants

CMS traits can be classified into sporophytic and gametophytic types, in which the sterility and restoration of male gametes depend on the actions of the CMS and restorer genes in sporophytic cells (anther, tapetum, etc.) or the gametophytic cells, respectively (Table 15.1). To date, several types of CMS systems have been identified in rice, including Boro II-type CMS (CMS-BT) derived from an *indica* rice cultivar Chinsurah Boro II, Hong-Lian-type CMS (CMS-HL) from red-awned wild rice, wild-abortive-type CMS (CMS-WA) from wild rice, CMS-LD from Lead rice, CMS-CW from Chinese wild rice, and CMS-FA from wild rice. Among these CMS systems, CMS-BT, CMS-HL, CMS-LD, and CMS-CW function in a gametophytic manner, and CMS-WA functions in a sporophytic manner (Table 15.1).

Origins and Characteristics of CMS Genes CMS is determined by mitochondrial genes and inherited maternally in higher plants (Chase, 2007). Due to their importance in agricultural crops and in basic research, plant CMS systems have been intensively studied. A number of mitochondrial genes determining CMS have been cloned and functionally studied in crops, including several in rice such as *orf79* (for CMS-BT), *orfH79* (*orf79* for CMS-HL), *orfL79* (*orf79* for CMS-LD), *WA352* (for CMS-WA), *FA182* (for CMS-FA) (Fig. 15.2) (Kazama et al., 2016; Luo

Table 15.1 Major CMS/*Rf* systems in rice

CMS type	CMS gene	Mode of CMS	*Rf* gene	Location	Mechanism of *Rf*	References
CMS-BT (G)	*orf79*	Cytotoxicity	*Rf1a* (PPR) *Rf1b* (PPR)	Chr.10 Chr.10	RNA level RNA level	Wang et al. (2006) and Komori et al. (2004)
CMS-HL (G)	*orfH79*	Energy deficiency	*Rf5* (PPR) *Rf6* (PPR)	Chr.10 Chr.8	RNA level RNA level	Huang et al. (2015), Wang et al. (2013), and Hu et al. (2012)
CMS-LD (G)	*orfL79*	UK	*Rf2*(GRP)	Chr.2	RNA level	Kazama et al. (2016) and Itabashi et al. (2011)
CMS-CW (G)	UK	UK	*Rf17*(ACPS)	Chr.4	UK	Fujii and Toriyama (2009)
CMS-WA (S)	*WA352*	Aberrant PCD	*Rf3* (UK) *Rf4* (PPR)	Chr.1 Chr.10	UK RNA level	Tang et al. (2017), Tang et al. (2014), Luo et al. (2013), and Zhang et al. (1997)
CMS-RT102 (S)	*orf352*	Aberrant PCD	*Rf3* (UK) *Rf4* (PPR)	Chr.1 Chr.10	UK RNA level	Okazaki et al. (2013)
CMS-FA (S)	*FA182*	Cytotoxicity	*Rf19* (PPR)	Chr.10	RNA level	Jiang et al. (2022)

S sporophytic, *G* gametophytic, *PPR* pentatricopeptide repeat, *GRP* glycine-rich protein, *ACPS* acyl-carrier protein synthase, *UK* gene/product or mechanism is unknown

et al., 2013; Jiang et al., 2022; Peng et al., 2010; Wang et al., 2006), *orf239* in common bean (*Phaseolus vulgaris*) (He et al., 1996), *orf456* in chili pepper (*Capsicum annuum*) (Kim et al., 2007), *orf129* in sugar beet (*Beta vulgaris*) (Yamamoto et al., 2008), *orf220* in mustard (*Brassica juncea*) (Yang et al., 2010), and *orf288* in oilseed rape (*Brassica napus*) (Jing et al., 2012).

Sequence analyses of cloned CMS genes have revealed that CMS genes generally consist of a chimeric open reading frame (ORF) derived from multiple origins. Most CMS genes contain fragments of essential mitochondrial genes for components of the electron transfer chain, such as *atp6*, *atp8*, and *cox1*. For example, in rice, the CMS gene *orf79* for CMS-BT and its variants *orfH79* (*orf79* of CMS-HL) and *orfL79* (*orf79* of CMS-LD) encode small chimeric peptides consisting of a partial sequence homologous to *cytochrome oxidase1* (*COX1*).

The CMS genes tend to be co-transcribed with essential mitochondrial genes and may cause cytotoxicity, energy deficiency, or aberrant programmed cell death, and may be involved in retrograde (mitochondria-to-nucleus) regulation (Chen & Liu, 2014). In CMS-BT lines, the mitochondrial genome contains two duplicated copies of the *atp6* gene, and an additional ORF, called *orf79*, is present 206 bp downstream of the CMS-BT-specific *atp6* (B-*atp6*) (Akagi et al., 1994). The *orf79* and B-*atp6* genes are co-transcribed, producing a bi-cistronic mRNA of about 2000

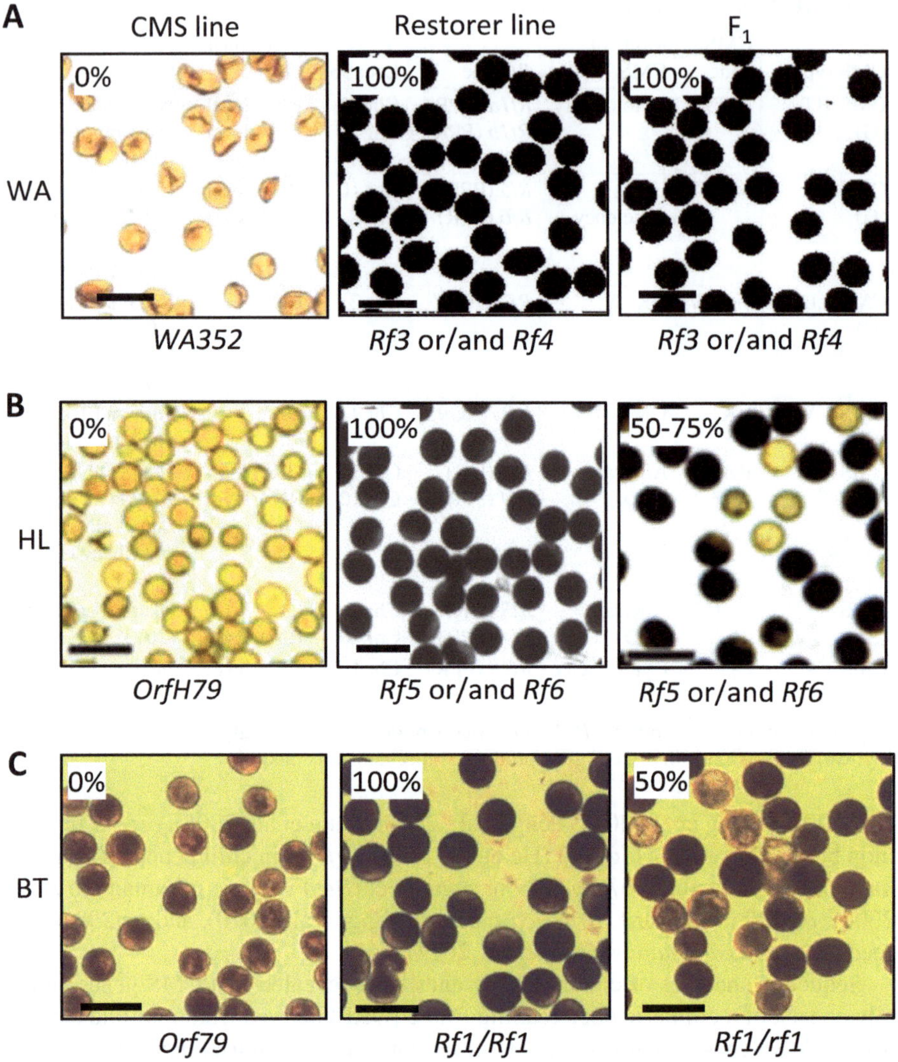

Fig. 15.2 Pollen fertility of major CMS/*Rf* systems. (**a**) The sterile pollen grains of the CMS-BT line are round and lightly-stained by I_2-KI. CMS-BT is a gametophytic-type CMS: the gametes carrying the *orfH79* CMS gene will abort. Therefore, the plants are semi-sterile in the F_1 generation, as half the grains are fertile and half are aborted. Bars = 100 μm. (**b**) The sterile pollen grains of the CMS-HL line are empty, round, and unstainable. CMS-HL is a gametophytic-type CMS and can be restored by two *Rf* genes. The pollen fertility varies from 50–75% in the F_1 generation. Bar = 100 μm. (**c**) The sterile pollen grains of the CMS-WA line are empty, shrunken, and unstainable. CMS-WA is a sporophytic-type CMS. Therefore, the pollen grains are fertile in the F_1 generation. Bars = 100 μm

nucleotides. The protein product of *orf79*, ORF79, specifically accumulates in the microspores and may be cytotoxic to the microspores depending on its C-terminus (Table 15.1) (Wang et al., 2006). In FA-CMS lines, a mitochondrial chimeric gene *FA182* confers male sterility due to that the protein product FA182 may also be cytotoxic to the microspores (Jiang et al., 2022). In CMS-HL lines, *orfH79* is also located downstream of the extra *atp6* gene and is co-transcribed with *atp6* (Table 15.1). The *atp6-orfH79* bi-cistronic transcript possesses differences in the intergenic sequence compared with B-*atp6-orf79*, but the ORFs of *orfH79* and *orf79* share high similarity (98% nucleotide sequence identity) (Peng et al., 2010). In CMS-HL lines, ORFH79 preferentially accumulates in pollen and suppresses the enzyme activity of the electron transport chain (ETC) complex III in the mitochondria via interaction with p61, a subunit of ETC complex III, resulting in increased production of reactive oxygen species (ROS) and reduced levels of ATP, ultimately causing pollen abortion (Peng et al., 2010; Wan et al., 2007; Wang et al., 2013).

CMS-LD lines contain a variant of the CMS-BT B-*atp6-orf79*, called L-*atp6-orf79*, with a 4-bp insertion in the 5′ untranslated region of *orf79* and a single-nucleotide polymorphism (SNP) in the coding region (Itabashi et al., 2009). However, because of the lower expression level of L-*atp6-orf79*, the pollen viability of CMS-LD lines is higher than that of CMS-BT lines (Kazama et al., 2016). This molecular evidence indicates that B-*atp6-orf79*, *atp6-orfH79*, and L-*atp6-orf79* might be derived from a common origin, and their intergenic sequences and the ORFs have diverged during evolution.

Moreover, the chimeric ORFs of CMS genes usually encode transmembrane proteins (Chen & Liu, 2014). For example, the CMS gene *WA352* for CMS-WA and its variant *orf352* for RT102-type CMS (CMS-RT102) comprise four mitochondrial genomic segments and encode a transmembrane protein with 352 aa (Luo et al., 2013; Okazaki et al., 2013). *WA352* is located downstream of the mitochondrial gene *rpl5*, and of the three *WA352*-containing transcripts, one is co-transcribed with *rpl5* and the other two are transcribed independently (Liu et al. 2007a, b; Luo et al., 2013). In CMS-WA lines, *WA352* is constitutively expressed in all tested tissues, but the WA352 protein accumulates only in the tapetum of anthers at the microspore mother cell (MMC) stage. WA352 interacts with the nuclear-encoded mitochondrial transmembrane protein COX11, a subunit of the mitochondrial complex IV, and impairs the function of COX11 in peroxide metabolism, thereby causing a burst of ROS and a release of cytochrome *c* from the mitochondria to the cytosol, which induces premature programmed cell death in the tapetum and consequent male sterility (Fig. 15.2; Zhang et al., 2025; Luo et al., 2013).

To understand the evolutionary origin of the chimeric ORF *WA352*, Tang et al. (2017) screened 808 accessions from 15 different *Oryza* species and discovered 10 new recombinant structures related to the *WA352* and *orf352*. These structures might have been generated by multiple rearrangements among conserved mitochondrial sequences, including a common sequence named cs1 that encodes the COX11-interaction domain, accompanied by substoichiometric shifting and sequence changes. Transformation assays using some of these recombinant structures showed that two structures (*WA352b* and *WA314*) can act as new, functional CMS genes, while two others (*orf367* and *orf356*) are nonfunctional CMS-like genes, called protogenes (Tang et al., 2017). Consistent with their biological role in CMS, all of these functional CMS proteins interact with COX11, but the proteins of the protogenes do not, due to the presence of special sequence domains that repress the interaction with COX11 (Tang et al., 2017). This evidence indicates that the functional CMS genes might have first occurred as nonfunctional CMS-like protogenes, and then they further evolved into CMS genes via a series of sequence recombinations and nucleotide variations to acquire their ability of the COX11 interaction.

Fertility Restoration of CMS The pollen fertility of CMS plants can be restored by nuclear-encoded *Restorer of fertility* (*Rf*) genes, which encode mitochondria-targeted products that inhibit the function of CMS genes. Depending on the different mechanisms of these *Rf* genes, the restoration of fertility in CMS plants may be achieved at the genomic DNA level, transcriptional level, or protein level (Chen & Liu, 2014).

To date, at least 12 *Rf* genes have been identified in eight plant species. Nine of these *Rf* genes encode pentatricopeptide repeat (PPR) proteins; these PPR-encoding genes are *Rf-PPR592* in petunia (*Petunia hybrida*) (Bentolila et al., 2002); *Rfk1* in radish (Koizuka et al., 2003); *Rf1a* (*Rf5*), *Rf1b*, *Rf4*, *Rf6*, and *Rf19* in rice (Hu et al., 2012; Huang et al., 2015; Jiang et al., 2022; Komori et al., 2004; Tang et al., 2014; Wang et al., 2006); *Rf1* in sorghum (*Sorghum bicolor*) (Klein et al., 2005); and *Rfo* and *Rfp* in oilseed rape (Liu et al., 2016; Uyttewaal et al., 2008). PPR proteins belong to a large family of organelle-targeting proteins that are involved in various post-transcriptional processes, including mRNA editing, splicing, cleavage, degradation, and translation (Barkan & Small, 2014). PPR proteins are classified into P and PLS types based on the structure of the PPR motif. The P-type PPR proteins only contain the 35-aa repeats, while the PLS-type PPR proteins contain the PPR repeats plus extra domain(s) at the C-terminal. Most PPR-type RF proteins process the mRNAs of CMS genes through post-transcriptional mechanisms, thereby suppressing CMS (Chen & Liu, 2014).

Current three-line hybrid rice breeding methods mainly depend on the CMS-BT, CMS-WA, CMS-HL, and CMS-FA systems and their corresponding *Rf* genes. Five *Rf* genes (*Rf1a*/*Rf5*, *Rf1b*, *Rf4*, *Rf6*, and *Rf19*) for these CMS systems are PPR-type *Rf* genes (Fujii & Toriyama, 2009; Hu et al., 2012; Huang et al., 2015; Itabashi et al., 2011; Jiang et al., 2022; Kazama et al., 2016; Komori et al., 2004; Luo et al., 2013; Okazaki et al., 2013; Peng et al., 2010; Tang et al., 2014, 2017; Wang et al., 2006). Studies of the underlying molecular mechanisms of these *Rf* genes have enhanced our understanding of the restoration of fertility in CMS plants as well as the molecular basis underlying cytoplasmic–nuclear interactions in plants.

Examination of *Rf* gene conservation and function has also enhanced our understanding of how CMS/Rf systems evolve. Interestingly, the CMS-related PPR genes tend to cluster in the chromosomal regions. For example, ten PPR genes are located on chromosome 10, of which three (*Rf1a*/*Rf5*, *Rf1b*, and *Rf4*) function as the *Rf* genes for CMS-BT, CMS-HL, or CMS-WA (Fig. 15.1). RF1A/RF5 and RF1B share 70% identity and restore fertility in CMS-BT systems via different post-transcriptional mechanisms (Table 15.1): RF1A cleaves the B-*atp6-orf79* co-transcript into two fragments (approximately 1500 and 450 nucleotides), which damages the ribosome-binding site at the 5'-UTR (untranslated region) of *orf79*. Therefore, the processed *orf79* transcript cannot be translated into the cytotoxic protein (Kazama et al., 2008; Wang et al., 2006). However, RF1B promotes complete degradation of the B-*atp6-orf79* mRNA. Moreover, when *Rf1a* and *Rf1b* coexist in a genome, RF1A is epistatic over RF1B in the processing of the *orf79* mRNA (Wang et al., 2006). Given that RF1A also promotes the C-to-U editing of the *atp6* mRNA, the role of *Rf1a* for restoration of fertility might be a new gain-of-function alteration of this gene (Wang et al., 2006). *Rf5* for CMS-HL is the same gene as *Rf1a* for CMS-BT, but *Rf5* uses a different mechanism for processing the *atp6-orfH79* mRNA (Fig. 15.2; Table 15.1; Hu et al., 2012). In the CMS-BT/*Rf* system, RF1A binds directly to the B-*atp6-orf79* mRNA for its cleavage (Kazama et al., 2008); however, RF5 interacts with the glycine-rich protein GRP162 to form a restoration of fertility complex, which functions in cleaving the *atp6-orfH79* transcript (Hu et al., 2012; Huang et al., 2015; Qin et al., 2016). These different modes of action between CMS-BT and CMS-HL may be due to the differences in the sequences within the intergenic regions of B-*atp6-orf79* and *atp6-orfH79*. The fertility in CMS-FA systems can be restored by only a single *Rf* gene, *Rf19*, located on 10. *Rf19* also encodes a mitochondria-targeted PPR protein, which mediates the cleavage of FA182 transcripts, thus restoring male fertility. Intriguingly, *Rf19* shares a common ancestor with *Rf1A*/*Rf5* (Jiang et al., 2022).

The restoration of fertility in the CMS-HL system is also controlled by another restorer gene, *Rf6*, on chromosome 8 (Table 15.1) (Huang et al., 2015). *Rf6* encodes a mitochondrial-localized PPR protein, which interacts with a hexokinase protein OsHXK6 to cleave the *atp6-orfH79* mRNA at nucleotide 1238 (Hu et al., 2012; Huang et al., 2015; Qin et al., 2016). Furthermore, duplication of the PPR motif 3 is crucial for the function of RF6 in the processing of mRNA (Huang et al., 2015).

Notably, the restorer gene *Rf2* for CMS-LD on chromosome 2 encodes a mitochondria-localized protein with a 152-aa glycine-rich domain (GRP), similar to GRP162 required for CMS-HL (Itabashi et al., 2011). *Rf2* is preferentially expressed in developing and mature anthers, peaking at the tricellular pollen stage (Itabashi et al., 2011). An SNP (T to C) in the recessive *rf2* allele in non-restorer lines causes an I78T amino acid variation. RF2 restores fertility in the CMS-LD system by reducing the abundance of the L-*atp6-orf79* transcript, suggesting that RF2 may use a different mechanism in post-transcriptional processing of the L-*atp6-orf79* transcript from that of the PPR RF1A for processing B-*atp6-orf79* in CMS-BT (Kazama et al., 2016). In the presence of *Rf2*, the amount of ORFL79 and ORF79 decreased to approximately zero in CMS-LD lines and 25% in CMS-BT lines, respectively. Therefore, *Rf2* fully restores the fertility of CMS-LD rice but only partially rescues the fertility of CMS-BT rice (Itabashi et al., 2009; Kazama et al., 2016). These results indicate that the levels of ORF79 and its variants may be regulated by overlapping restoration of fertility complexes with different processing efficiencies for the specific target mRNAs of CMS-BT, CMS-HL, and CMS-LD lines.

The fertility in CMS-WA systems can be rescued by two major *Rf* genes, *Rf3* and *Rf4*, located on chromosomes 1 and 10, respectively (Ahmadikhah & Karlov, 2006; Zhang et al., 1997; Zhang et al., 2002). *Rf4* is a member of the PPR cluster that also contains *Rf1a/Rf5* and *Rf1b* (Luo et al., 2013; Tang et al., 2014). *Rf4* encodes a mitochondria-targeted PPR protein with 782 amino acids and 18 PPR motifs, with a very similar structure to that of RF1A. RF4 restores fertility in CMS-WA systems by reducing the abundance of the *rpl5-WA352* and *WA352* transcripts, similar to the role of RF1B in the B-*atp6-orf79* mRNA in CMS-BT. Furthermore, the fertility restoration ability of *RF4* is dependent on the copy number variation of *RF4* (Zhao et al., 2023). It was found that varieties with a two-copy *RF4* haplotype harbor stronger fertility restoration ability. However, the *rpl5-WA352* and *WA352* transcripts are not affected by the other restorer gene *Rf3*, but WA352 is absent at the MMC stage of *Rf3*-restored plants, suggesting that *Rf3* probably functions to WA352 at the translational or post-translational level (Luo et al., 2013). Only three non-PPR-type *Rf* genes have been identified and characterized: *Rf2* for CMS-T in maize (*Zea mays*) was the first identified *Rf* gene and encodes a mitochondria-localized protein with aldehyde dehydrogenase activity (Cui et al., 1996). *Rf2* for rice CMS-LD encodes a mitochondria-localized protein containing a glycine-rich domain (GRP) (Itabashi et al., 2011), and *Rf17* for rice CMS-CW encodes a mitochondria-localized protein similar to acyl-carrier protein synthase (ACPS) (Fujii & Toriyama, 2005, 2009). CMS-CW is derived from recurrent backcrossing of a Chinese wild rice (*O. rufipogon*) with a *japonica* rice Fujisaka 5 (Fujii & Toriyama, 2005). Unlike other CMS in rice, the pollen grains of CMS-CW lines are morphologically normal, but they cannot germinate on the stigma. Although the causal gene for CMS-CW has not been identified, the restorer gene of CMS-CW, *Rf17*, has been isolated on chromosome 4. *Rf17* encodes a mitochondrial-localized acyl-carrier protein synthase (ACPS) of 178 aa, named *RETROGRADE-REGULATED MALE STERILITY* (*RMS*), which functions in a gametophytic manner via an unknown mechanism (Fujii & Toriyama, 2005, 2009). The transcript level of *RMS* in mature anthers of fertile rice is approximately 25% of that in the CMS-CW line, suggesting that the expression of *RMS* is differentially regulated by an unknown retrograde (mitochondria-to-nucleus) signal from the CMS and normal mitotypes. Knockdown of *RMS* in CMS-CW lines restores pollen fertility, whereas overexpression of *RMS* in fertile rice with normal cytoplasm induces pollen abortion. Sequence analysis found that *Rf17* contains an SNP at 2286-bp upstream of the promoter region compared with *rf17*, indicating that this SNP may be involved in suppression of *RMS* in response to retrograde signaling in the fertility-restored line.

15.4 Environment-Sensitive Genic Male Sterility in Plants

In EGMS, interactions between genetic loci and environmental factors such as temperature and/or photoperiod produce reversible male sterility (Chen & Liu, 2014). EGMS traits are classified into photoperiod-sensitive genic male sterility (PGMS),

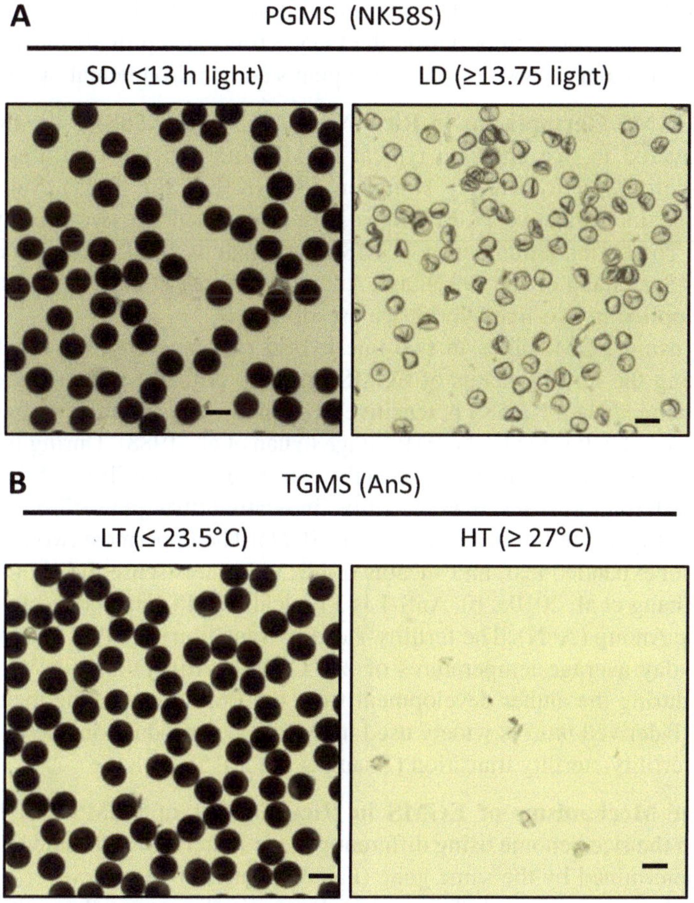

Fig. 15.3 Pollen fertility for major EGMS traits. (**a**) The pollen fertility of the PGMS line is normal under short-day conditions (≤13 h light), and the plants become male sterile under long-day conditions (≥13.75 light). Bars = 30 mm. (**b**) The pollen fertility of the TGMS line is normal under lower temperatures than 23.5 °C, but the plants become male sterile when the temperature is higher than 27 °C. Bars = 30 mm

temperature-sensitive genic male sterility (TGMS), or photoperiod- and temperature-sensitive genic male sterility (P/TGMS). PGMS and TGMS lines show normal male fertility under short-day (SD) conditions and low temperature, respectively, and become male sterile under long-day (LD) conditions and high temperatures (Fig. 15.3). Conversely, some EGMS germplasms are male fertile under LD or high-temperature conditions but male sterile under SD or low-temperature conditions and are thus called reverse PGMS (rPGMS) and reverse TGMS (rTGMS) lines, respectively. Either PGMS/TGMS or rPGMS/rTGMS lines can serve as the male-sterile lines for hybrid seed production under restrictive temperature and/or photoperiod conditions and can self-pollinate under permissive conditions. Therefore, the use of EGMS lines can simplify hybrid seed production by a two-line system. Besides their agronomic importance, the EGMS traits also provide ideal systems to study the molecular interactions between plants and environmental factors.

Typical EGMS Germplasms in Rice Nongken 58S (NK58S) is the first EGMS line discovered in rice and is a typical PGMS line derived from a spontaneous mutant of the *japonica* variety Nongken 58 (NK58) (Shi, 1985). Subsequently, NK58S and its derivative PGMS lines have been used for two-line hybrid rice breeding. Thereafter, a number of TGMS lines, such as 5460S (Yang et al., 1989), HengnongS-1 (Jiang et al., 1992), and AnongS-1 (Deng et al., 1999), were identified and applied in two-line hybrid rice breeding.

When using EGMS lines in two-line hybrid rice breeding, the conditions for transitioning the fertility status of EGMS lines are critical for hybrid seed quality (Fan & Zhang, 2018). NK58S is sensitive to environmental conditions from the stamen primordial cell stage to the PMC stage (Yuan et al., 1988). During these developmental stages, treatment with red light causes pollen sterility, while treatment with far-red light rescues male fertility, indicating that phytochromes may be involved in this PGMS trait (Chen & Xiao, 1991). The uppermost two leaves, especially the unexpanded leaf, and sheaths of NK58S plants sense the photoperiod for PGMS (Zhang et al. 2010a, b). AnS-1 is a typical TGMS line derived from normal *indica* rice Anong (AnN). The fertility-sterility transition of AnS-1 is controlled by threshold day-average temperatures of 24 °C (full fertility) and 26 °C (complete sterility) during the anther development from the PMC stage to the meiosis stage. The AnS-1-derived *tms5* is widely used in hybrid seed production due to the stability of its fertility-sterility transition (Zhang et al., 2015).

Molecular Mechanisms of EGMS in Rice Dozens of EGMS loci have been mapped in the rice genome using different genetic materials, and some EGMS traits may be determined by the same gene (Fan & Zhang, 2018). However, so far only five genes have been cloned (Table 15.2). Interestingly, among these characterized EGMS genes, two widely used loci (*Pms1* and *pms3* from NK58S) are involved in non-coding RNAs (ncRNAs).

The PGMS trait of NK58S is mainly controlled by the semi-dominant *Pms1* and recessive *pms3* loci, with the effects of the two loci depending on the genetic

Table 15.2 Major EGMS systems in rice

EGMS type	Locus	EGMS gene	Location	Function	References
PGMS	*pms1* *pms3* *p/ tms12-1*	*PMS1T* *LDMAR* *osa-smR5864m*	Chr. 7 Chr. 12	phasiRNA lnc RNA or small RNA	Fan et al. (2016), Ding et al. (2012a, b), and Zhou et al. (2012)
TGMS	*tms5* *ugp1*	*TMS5* UGP1	Chr. 2 Chr. 9	RNase Z UGPase	Zhou et al. (2014) and Chen et al. (2007)
rPGMS	*csa*	*CSA*	Chr. 1	MYB TF	Zhang et al. (2010a, b) and Zhang et al. (2013)

backgrounds of the parents in the crosses (Mei et al., 1999). The *pms3*-mediated PGMS trait in NK58S is caused by a single-base C-to-G nucleotide substitution in the locus on chromosome 12 (Ding, Lu, et al., 2012a). The *pms3* locus encodes a long ncRNA (lncRNA) referred to as *long-day-specific male-fertility-associated RNA* (*LDMAR*). *LDMAR* is highly expressed in young panicles, and the abundance is affected by LD conditions in NK58 but not in NK58S, suggesting that abundant levels of *LDMAR* are essential for male fertility (Ding et al., 2012a). The C-to-G mutation in *LDMAR* may change its secondary structure, leading to male sterility under LD conditions (Ding et al., 2012a).

Interestingly, when the *pms3* locus of the *japonica* NK58S line is introgressed into the genetic backgrounds of *indica* (*O. sativa* ssp. *indica*) varieties, such as Peiai 64 (PA), the resultant Peiai 64S (PA64S) plant becomes a P/TGMS line whose fertility transition is governed mainly by temperature but is enhanced by photoperiod (Zhou et al., 2012). Molecular cloning showed that *p/tms12-1*, the major locus conferring P/TGMS in PA64S, is the same gene as *pms3* (Zhou et al., 2012). The *p/tms12-1*(*pms3*) locus in PA64S also produces the *LDMAR* lncRNA, which is processed into a 21-nucleotide small RNA, *osa-smR5864m*. In PA64S, *osa-smR5864m* harbors the C-to-G SNP compared with the wild-type *osa-smR5864w*. Similar to the pattern of *LDMAR*, *osa-smR5864w* is preferentially expressed at higher levels, compared with *osa-smR5864m*, in young panicles under LD conditions or at high temperature. LncRNAs usually serve as precursors for miRNA- or small-interfering RNA-like small RNAs (siRNA) that can regulate gene expression (Taft et al., 2010). Therefore, *LDMAR* may act by generating the 21-nt *osa-smR5864m* small RNA and thus increasing the methylation of the *LDMAR* promoter, leading to suppression of *LDMAR* and causing premature tapetal PCD in anthers under LD conditions (Ding, Shen, et al., 2012b). The *pms3* (*p/tms12-1*)-generated small RNA may regulate certain uncharacterized target gene(s) to control T/PGMS. It would be interesting to understand whether *osa-smR5864* and Argonaute proteins contribute to changes in the levels of DNA methylation of these target gene(s).

In addition to the recessive *pms3* locus, the PGMS trait of NK58S is also controlled by the semi-dominant *Pms1* locus on chromosome 7 (Zhang et al., 1994). The transcript of *Pms1* (*PMS1T*) in NK58S has a dominant function in causing PGMS, while normal lines harbor a dysfunctional *pms1* (Fan et al., 2016). Compared with the transcript (1388 nt) of the wild-type lines, *PMS1T* (1453 nt) contains a SNP and a 65-nt insertion. *PMS1T* is expressed at higher levels under SD conditions than under LD conditions in the young panicles and anthers in NK58S. Furthermore, overexpression of the NK58S *PMS1T* in normal, fertile lines produced partial male sterility under LD conditions, while silencing of *PMS1T* in NK58S partially rescued male fertility under LD conditions (Fan et al., 2016).

In plants, microRNAs such as miRNA2118 or miR2275 trigger the cleavage of transcripts of precursor RNAs; then dicer-like protein (DCL) processes the transcript to produce 21-nt or 24-nt phased small-interfering RNAs (phasiRNAs) (Johnson et al., 2009). PhasiRNAs are associated with male sterility in rice and maize (Liu et al. 2007a, b; Zhai et al., 2015). Interestingly, in NK58S, *PMS1T* is cleaved by miRNA2118 to produce a number of 21-nt phasiRNAs in anthers under

LD conditions. Thus, the lncRNAs and their smRNA and phasiRNA products may serve as important regulators of gamete development and P/TGMS in rice and other plants (Guo & Liu, 2017). However, little is known about the targets of these small RNA molecules.

Distinct from the lncRNA-mediated PGMS caused by *Pms1* and *pms3* (*p/tms12-1*), those caused by *tms5*, *tms9*, *tms10,* and *Ugp1*, the rTGMS caused by *ago1d*, the rPGMS caused by *carbon-starved anther* (*csa*), and the PGMS caused by *csa2* are mediated by protein-coding genes (Chen et al., 2007; Shi et al., 2022; Yu et al., 2017; Wang et al., 2021; Wu et al., 2022; Zhang et al., 2013; Zhou et al., 2014). The *tms5* locus is located on the short arm of chromosome 2 (Deng et al., 1999; Zhou et al., 2014). Molecular cloning shows that the *TMS5* gene in wild-type rice encodes a short form of ribonuclease Z (RNase Z^S). AnN, a parental/sister line of AnS-1, contains a single-base mutation of *G*CG to *T*CG in a site of the coding region of *TMS5*. Within this same site, a second nucleotide mutation, T*C*G to T*A*G, occurred in AnS-1, producing a premature stop codon and causing the loss of function of the gene (Zhou et al., 2014). The wild-type RNase Z^S protein localizes in the cytoplasm and harbors endonuclease activity. Therefore, the wild-type RNase Z^S does not function in tRNA processing, in contrast to other nucleus-localized RNase Z^S proteins for tRNA processing; rather, this cytoplasmic RNase Z^S can cleave the mRNAs of $Ub_{L40}1$, $Ub_{L40}2$, and $Ub_{L40}4$, which encode the conserved ubiquitin-60S ribosomal L40 family protein Ub_{L40}. Transgenic overexpression of the Ub_{L40} genes in rice produced male sterility, suggesting that overaccumulation of Ub_{L40} is harmful for male development (Zhou et al., 2014).

Although the three *UbL40* genes are temperature-insensitive in rice lines carrying *TMS5*, the Ub_{L40} transcripts in the young panicles of AnS-1 carrying the non-functional *tms5* gene are more abundant at restrictive temperatures (high temperatures) than at permissive temperatures (low temperatures). Zhou et al. proposed a working model for the molecular mechanism of the RNase Z^S-mediated TGMS system. The relatively lower levels of Ub_{L40} expression are critical for normal male development. At low temperature, the expression of Ub_{L40} is maintained at a low level, and the *TMS5*-containing and wild-type rice lines are male fertile. At high temperatures (restrictive temperatures), Ub_{L40} genes are highly expressed in anthers, but in *TMS5*-containing wild-type plants, RNase Z^S processes the Ub_{L40} mRNAs and thus maintains their low levels. However, *TMS5*-containing plants accumulate excessive amounts of Ub_{L40} mRNAs and their protein products due to the dysfunction of RNase Z^S, resulting in abnormal anther development and male sterility (Zhou et al., 2014). Based on this molecular understanding of *tms5*-mediated TGMS, knockout of *TMS5* by CRISPR/Cas9 gene editing can generate artificial *tms5* alleles for TGMS. This technology can rapidly create new TGMS lines, such as converting the maintainer lines of the three-line hybrid rice system to TGMS lines for two-line hybrid rice breeding (Zhou et al., 2016).

In plants, callose is a major component of pollen cell walls and plays important roles in male microsporogenesis (Dong et al., 2005; Gibeaut, 2000; Tucker et al., 2001). Sugars, such as sucrose, are hydrolyzed by invertase into hexose monomers via the glycogenesis pathway, and pollen cells take up these hexoses through specific monosaccharide transporters (Truernit et al., 1999; Ylstra et al., 1998). In the pollen cells, hexoses are converted into glucose-1-phosphate, which is then converted into UDP-glucose by UDP-glucose pyrophosphorylase (UGPase). The resulting UDP-glucose is a precursor for the biosynthesis of callose for cell wall formation (Kleczkowski et al., 2004). Defects in essential genes for carbohydrate metabolism and transport usually affect pollen cell wall development, and in rice, their mutants exhibit male sterility. For example, *Ugp1* encodes a UDP-glucose pyrophosphorylase (UGPase) that converts glucose-1-phosphate and UTP to UDP-glucose, which is required for callose deposition and proper pollen development; loss of function of *Ugp1* causes pollen abortion. Surprisingly, a *Ugp1*-cosuppressing line derived from a transgenic line carrying a *Ubi* promoter-driven overexpression construct acquired a novel TGMS phenotype at high temperature but became male fertile when the temperature was lowered to around 21 °C. This occurred because the *Ugp1*-cosuppressing line had aberrant unspliced, intron-containing transcripts of the overexpressed transgene, while the endogenous *Ugp1* was completely silenced, resulting in male sterility at higher temperatures. However, at lower temperatures, the *Ugp1* transcripts were properly processed into mature RNAs for translation in the florets, resulting in a reversion from male sterility to fertility (Chen et al., 2007).

After that, several TGMS or rTGMS genes have been cloned. The *TMS9* gene encodes a PHD finger transcription factor. An SNP (T to C) in the coding sequence of *tms9* causes one amino substitution (Leu to Pro), which contributes to thermosensitive genic male sterility in the rice variety Hengnong S-1 (Wu et al., 2022). The abundance and transcriptional activity of TMS9 protein are regulated in a temperature-dependent manner. At high temperatures, the accumulation and activity of TMS9 protein are significantly lower than those of TMS9, thus leading to defective anther development. *TMS10* gene encodes a leucine-rich repeat receptor-like kinase (LRR-RLK) (Yu et al., 2017). *TMS10* gene is essential for both tapetal development and pollen wall formation. *TMS10* gene has a homologous gene, *TMS10-Like* (*TMS10L*). They play redundant roles for the maintenance of normal anther development only under low temperature. Therefore, dysfunction of *TMS10* gene causes male sterility only under high temperature. The first rTGMS gene *ARGONAUTE 1d* (*AGO1d*) unites with miR2118 and miR2275 to regulate pha-siRNA biogenesis in anther wall cells under low temperatures (Shi et al., 2022). Hence, disruption of *AGO1d* displays male sterility at lower temperatures, showing rTGMS phenotype.

The *CSA* gene encodes an R2R3 MYB transcription factor that regulates the expression of the monosaccharide transporter gene *OsMST8* to alter sugar partitioning from vegetative organs to reproductive organs (Zhang et al. 2010a, b). The expression of *OsMST8* is reduced in the *csa* mutant, which is caused by a single-base deletion and a G-to-A transition in the first exon of *CSA*; this mutation may abolish the regulation of the monosaccharide transporter responsible for the uptake of sugar monomers into floral organs. Thus, sugar accumulates in the leaves but is deficient in the floral organs, leading to abnormal pollen development and male sterility. However, the sterility of the *csa* mutant can be restored under LD conditions, probably due to functional complementation by two other day-length sensitive MYB homologs. The *csa* locus has been introgressed into several *japonica* and *indica* varieties to produce rPGMS lines for hybrid rice production (Zhang et al., 2013). Also, a recent study proposed that a brassinosteroid-signaling factor, rice BRASSINAZOLE-RESISTANT1 (OsBZR1), promotes *CSA* expression; suppression of *OsBZR1* expression causes partial male sterility independent of day length (Zhu et al., 2015). Nevertheless, the dysfunction of the homologous gene of the CSA gene, CSA2, displays PGMS, meaning that CSA and CSA2 may not redundantly function in anther development under different photoperiod conditions (Wang et al., 2021).

15.5 Reproductive Isolation and Hybrid Sterility in Plants

Reproductive isolation is a critical evolutionary mechanism for speciation (Baack et al., 2015; Bomblies, 2010; Rieseberg & Blackman, 2010; Schumer et al., 2015). Prezygotic and postzygotic reproductive isolation are two major reproductive barriers that prevent gene flow among different species/subspecies before or after fertilization and zygote formation, respectively (Ouyang & Zhang, 2013). In plants, different mating behaviors can also result in prezygotic reproductive isolation, which prevents the formation of hybrid zygotes (Bomblies, 2010). For instance, in *Antirrhinum*, distinct pollinators recognize various flower colors for reproduction. Variations in two *MYB* genes are involved in these flower color shifts, which affect pollinator visitation, thus inducing selective mating and reproductive isolation (Rausher, 2007). Postzygotic reproductive isolation occurs after mating, where the hybrids and their offspring exhibit hybrid incompatibility (Scopece et al., 2008). Postzygotic reproductive isolation occurs in three forms: hybrid sterility (HS), hybrid necrosis/weakness, and hybrid lethality in the F_1, F_2, or later generations (Ouyang & Zhang, 2013; Scopece et al., 2008). HS is the most common postzygotic reproductive isolation that occurs in plant hybrids, which grow normally at the vegetative stage but fail to produce viable pollen or embryo sacs at the reproductive stage, resulting in a low seed-setting rate (Lai et al., 2005; Ouyang et al., 2010; Skrede et al., 2008; Sweigart et al., 2006). Hybrid necrosis/weakness is mainly caused by immune response-related genes and has been characterized in tomato (*Solanum lycopersicum*), lettuce (*Lactuca sativa*), *Arabidopsis*, and rice (Alcazar et al., 2009; Bomblies, 2010; Chen et al., 2014; Jeuken et al., 2009; Kruger et al., 2002). Hybrid lethality leads to the death of the hybrid zygotes or embryos (Bikard et al., 2009). HS between species and subspecies (*indica* and *japonica*) in rice is the best-known postzygotic reproductive isolation in plants. Therefore, this chapter will focus on recent advances in rice HS to examine the molecular mechanisms of postzygotic reproductive isolation in plants.

Genetic Models for Rice Hybrid Sterility Cultivated rice includes two species, Asian rice (*Oryza sativa*) and African rice (*O. glaberrima*) (Oka, 1974). Asian rice further includes two major subspecies, *indica* and *japonica* (Liu et al., 1996). Although African rice is mostly grown only in Africa, it has some superior traits for tolerance to stresses such as heat, drought, disease, and aluminum toxicity (Brar & Khush, 1997; Chen & Liu, 2016). The inter-species hybrids of Asian and African rice or inter-subspecies hybrids of *indica* and *japonica* rice show strong hybrid vigor in biomass and stress tolerance, but they usually have severe HS, which hinders grain yield (Linares, 2002; Orr & Presgraves, 2000).

Because HS is a substantial barrier for the utilization of heterosis in hybrid rice, much work has focused on studying the molecular basis of HS in rice. To date, at least 50 HS loci affecting male and/or female fertility in rice inter-(sub)specific hybrids have been identified, and eight genes related to HS have been cloned (Table 15.3). HS mainly occurs in male gametes (hybrid male sterility, HMS), occasionally occurs in female gametes, and sometimes involves both the male and female gametes (Chen et al., 2008; Xie et al., 2017a; Yang et al., 2012). Herein, we summarize the underlying mechanisms of known rice HS genes.

The Dobzhansky-Muller model, based on genetic data from animal studies, supports the notion that postzygotic reproductive isolation is caused by a harmful interaction between functionally divergent genes or gene alleles in hybrids (Presgraves, 2010). In rice, two classical models have been proposed to explain HS. The first model, the duplicate gametic lethal model (also called the two-locus model), assumes that HS is controlled by two non-allelic loci essential for the development of male and/or female gametes. One parent carries the *A1/A1//a2/a2* genotype, and the other parent carries the *a1/a1//A2/A2* genotype. Both parents are normally fertile. However, the hybrid offspring with the *A1/a1//A2/a2* genotype derived from the cross of these two parents are partially sterile because some gametes harbor the *a1//a2* alleles together and lack *A1//A2* (Oka, 1957, 1974). Some loci, such as *S27/S28* and *DPL1/DPL2*, have been identified in rice and act consistently with the duplicate gametic lethal model (Mizuta et al., 2010; Yamagata et al., 2010).

The other model is the one-locus sporogametophytic interaction model (also called the one-locus model), in which gamete abortion is controlled by a single locus. The two parents are normally fertile when they carry the homozygous *AA* or *aa* genotype, but gametes carrying *a* from heterozygous *A/a* plants are abortive, leading to a distorted segregation ratio in the F_2 population (Kitamura, 1962). So far, the causal genes for ten rice loci (*Sa*, *S5*, *Sc*, *HSA1*, *S7*, *S1*, *qHMS7*, *ESA1*, *pf12/RHS12/Se,* and *qHMS1*) that show HS consistent with the one-locus model have been cloned (Fig. 15.4, Table 15.3).

Molecular Mechanisms of the One-Locus Model in Rice *Sa* confers HMS and is the first identified complex HS locus, consisting of two adjacent, functionally related genes (Long et al., 2008). The Taichung 65 (T65) variety carrying *Sa*$^-$/*Sa*$^-$ and the near-isogenic line (NIL) E4 carrying *Sa*$^+$/*Sa*$^+$ (here we define the *japonica* allele as "-" and the *indica* alleles as "+") show full male fertility, but the hybrid F_1 plants (carrying *Sa*$^+$/*Sa*$^-$) derived from the cross of T65 and E4 exhibit male semi-sterility, as male gametes carrying *Sa*$^-$ selectively abort (Long et al., 2008). The *Sa* locus was fine-mapped to a 10-kb region, which contains two genes named *SaF* and

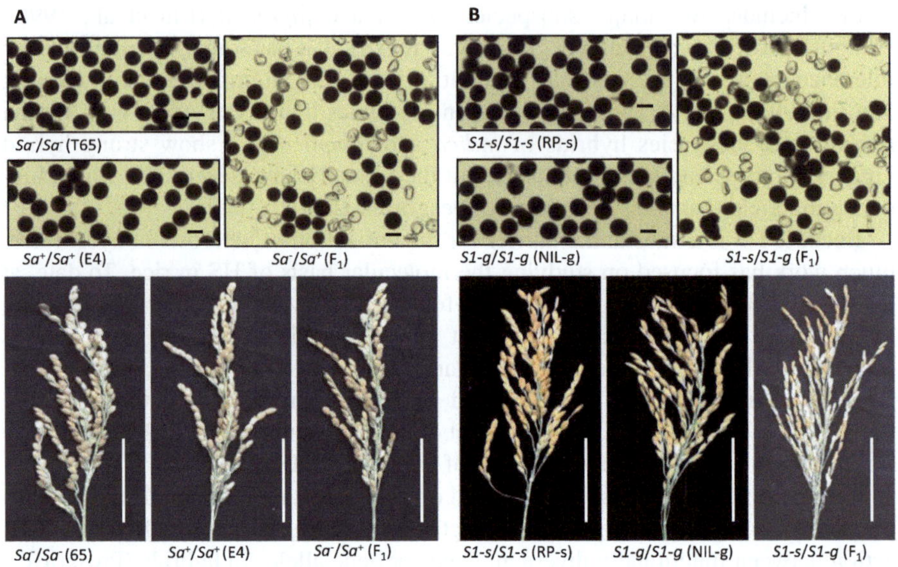

Fig. 15.4 Pollen and spikelet fertility for major hybrid sterility. (**a**) The pollen and spikelets of *indica* ($Sa^{+}Sa^{+}$) and *japonica* ($Sa^{-}Sa^{-}$) plants are fertile, but the pollen of F_1 ($Sa^{+}Sa^{-}$) is semi-sterile, as the pollen carrying the Sa^{+} allele is fertile, but the pollen grains carrying the Sa^{-} allele are aborted. However, the spikelet fertility is normal, as half the amount of viable pollen is sufficient for pollination in the spikelet. Bars = 30 mm (pollen); Bars = 5 cm (spikelet). (**b**) The pollen and spikelets of African rice (*S1-g S1-g*) and *japonica* rice (*S1-j S1-j*) are fertile, but the pollen of the F_1 (*S1-g S1-j*) is semi-sterile, as the pollen carrying *S1-g* is fertile but the pollen carrying *S1-j* is aborted. However, the spikelet fertility is low due to a poor seed-setting rate because *S1* causes both male and female sterility in the spikelet. Bars = 30 mm (pollen); Bars = 5 cm (spikelet)

SaM. The *japonica* allele has the SaF^{-} SaM^{-} haplotype, while the *indica* allele has the SaF^{+} SaM^{+} haplotype. SaM^{+} encodes a small E3 ligase-like protein, while SaM^{-} encodes a C-terminally truncated version of SaM^{+} due to a G-to-T mutation of SaM^{+} that alters the splicing in the fifth intron. SaF^{+} encodes a plant-specific F-box protein. SaF^{-} differs from SaF^{+} by an amino acid substitution (Phe-to-Ser). In Sa^{-} heterozygotes, SaF^{+} selectively associates with SaM^{-} and indirectly with SaM^{+}. This three-component complex results in selective abortion of Sa^{-}-carrying pollen, while the recombinants possessing SaM^{+}/SaM^{-} but lacking SaF^{+}, or harboring $SaF^{+}SaM^{-}$ without SaM^{+}, are fully fertile. Therefore, a two-gene/three-component interaction model has been proposed for *Sa*-mediated HMS. In this model, *Sa*-mediated HMS is determined by the SaF^{+}-SaM^{-}-SaM^{+} complex, resulting in the selective abortion of pollen carrying SaM^{-}. In fact, knocking down or knocking out SaF^{+} and SaM^{+} by RNA interference (RNAi) or CRISPR/Cas9 can rescue male fertility in the hybrids (Xie et al., 2017b). These studies revealed, for the first time, that this HS system, which acts genetically as a single locus, is in fact governed by closely linked genes. Furthermore, Li et al. (2025) revealed that the SaF^{+}-SaM^{+} complex acts as pollen killer, while another closely-linked gene *SaFL* functions as pollen protector and temperature switch of reproductive isolation.

Table 15.3 Major HS genes in rice

HS type	Locus	HS gene	Location	Protein(s)	References
Duplicate gametophytic lethal model	*S27/S28*	*mtRPL27*	Chr. 8 & Chr. 4	Mitochondrial ribosomal protein L27	Yamagata et al. (2010)
	DPL1/2	*DPL1*, *DPL2*	Chr. 1 & Chr. 11	Putative small protein	Mizuta et al. (2010)
One-locus sporogametophytic interaction model	*Sa*	*SaF*, *SaM*, *SaFL*	Chr.1	F-box protein SUMO E3 ligase, F-box like protein	Long et al. (2008), Li et al. (2025)
	S5	*ORF3*, *ORF4*, *ORF5*	Chr.6	Hsp 70 Transmembrane protein, Aspartic protease	Chen et al. (2008), and Yang et al. (2012)
	S7	*ORF3*^{ai}	Chr.7	TPR protein	Yu et al. (2016)
	S1	*OgTPR1* *SSP* *S1A4*	Chr.6	Trypsin-like peptidase Trypsin-like peptidase Unknown protein	Xie et al. (2017b), Koide et al. (2018), and Xie et al. (2019)
	Sc	*Sc-j*	Chr.3	DUF1618 protein	Shen et al. (2017)
	HSA1	*HSA1a* *HSA1b*	Chr.12	DUF1618 protein	Kubo et al. (2016)
	qHMS7	*ORF2* *ORF3*	Chr.7	Ribosome-inactivating protein Mitochondrial targeting protein	Yu et al. (2018)
	ESA1	*ESA1*	Chr.1	Armadillo repeat protein	Hou et al. (2019)
	Pf12/ RHS12/ Se	*iORF3* *iORF4*	Chr.12	RNase H domain protein Ank repeat protein	Zhou et al. (2023), Wang et al. (2023a, b), and Wang et al. (2023a, b)
	qHMS1	*HPT* *HPA*	Chr.1	DEAD-like helicase F-box protein	You et al. (2023)

Sequence analysis of the complex *Sa* locus in *Oryza* species found that the *SaF*$^{+}$*SaM*$^{+}$ haplotype is a primitive form present in ancient wild rice, while *SaF*$^{-}$*SaM*$^{-}$ is a diverged form derived from the first form by a nucleotide change in *SaF*$^{+}$, followed by a second mutation in *SaM*$^{+}$, suggesting a stepwise divergence of the complex locus (Long et al., 2008). The intermediate haplotype *SaF*$^{-}$*SaM*$^{+}$ found in some *O. rufipogon* and *O. sativa* populations is a neutral allele form, which can be used to overcome HMS in hybrid breeding (Xie et al. 2017a, b).

The *Sc* locus is another major locus on chromosome 3 for *indica–japonica* HMS with abortion of pollen grains containing the *japonica* allele *Sc-j* in F_1 plants (Yang et al., 2004). This locus was delimited to an 8.6-kb region in *japonica* rice; this region contains the *Sc-j* gene, which encodes a DUF1618 domain-containing protein, a member of a monocot-specific family (Wang et al., 2014). By contrast, in *indica* rice cultivars, the *Sc-i* allele variants have complex and variable genomic structures with two or three tandemly duplicated segments (28.2 kb or 28.4 kb in length), which contain *Sc-j*-homologous genes (*Sc-ia*, *Sc-ib1*, and *Sc-ib2*) that are controlled by a promoter distinct from that of *Sc-j* (Shen et al., 2017). Due to the structural variation in the promoter region and the copy number variation, the expression patterns of the *Sc-j* and *Sc-i* alleles are completely different in *japonica* and *indica* rice. *Sc-j* is specifically expressed at a low level in early-bicellular-stage pollen, while *Sc-i* is highly expressed in various sporophytic tissues and pollen during different stages. Moreover, knockdown of *Sc-j* by expression of an antisense transgene in *japonica* rice resulted in pollen sterility, indicating that *Sc-j* is required for male gametophyte development. In the *Sc-jSc-i* hybrids, the strong expression of *Sc-i* in sporophytic cells causes suppression of *Sc-j* expression in pollen, thus leading to selective abortion of *Sc-j*-pollen. However, reduction of the *Sc-i* dosage by knocking out one or two of the three *Sc-i* copies by CRISPR/Cas9 rescues *Sc-j* expression and restores male fertility in the hybrids. These results suggest that selective suppression of *Sc-j* by *Sc-i*, called "allelic suppression," may involve transcriptional-repressive epigenetic regulation of the *Sc-j* promoter, and this epiallele may be retained through meiosis in the pollen grains (Shen et al., 2017).

The structural changes in the promoter sequence in *Sc-i* not only alter its gene expression pattern and levels but might also allow the *Sc-i* paralogs to escape repression by their own products. Thus, the *Sc-i* allele variants have evolved into a selfish speciation gene in this HMS system. This study revealed that the new type of combinatorial genomic variation involving structural changes and copy number variation has an important genetic effect on hybrid incompatibility (Shen et al., 2017).

The *S5* locus is a major locus producing female semi-sterility in *indica–japonica* hybrids, and *S5* has three types of alleles: an *indica* allele (*S5-i*), a *japonica* allele (*S5-j*), and a neutral allele (*S5-n*) (Ikehashi & Araki, 1986). All homozygotes and the heterozygotes carrying *S5-n* (*S5-i/S5-n* or *S5-j/S5-n*) are fertile, but the heterozygotes carrying *S5-i/S5-j* show selective abortion of the embryo sacs carrying *S5-j* (Ikehashi & Araki, 1986). The *S5* locus was mapped on chromosome 6 within a 24.8-kb region containing three putative genes, *ORF3*, *ORF4*, and *ORF5*, which function as a trigenic reproductive isolation system (the wild-type form is defined as "+" and the mutant form is defined as "-") (Chen et al., 2008; Yang et al., 2012). Sequence analysis showed that the *indica S5* locus carries *ORF3+*, *ORF4-*, and *ORF5+*, and the *japonica S5* locus carries *ORF3-*, *ORF4+*, and *ORF5-*. Wide-compatible varieties, which show no HS with *japonica* or *indica*, carry neutral alleles, *S5-n*, with the combinations *ORF3+*, *ORF4+*, *ORF5n* or *ORF3-*, *ORF4-*, *ORF5n*.

In the *S5* locus, the *indica ORF3+* encodes an abundant heat shock protein, HSP70, and the *japonica ORF3-* encodes an HSP70 protein that is truncated at the

C-terminus due to a 13-bp deletion. Nevertheless, both ORF3+ and ORF3- localize in the endoplasmic reticulum (ER), which indicates that the truncation at the C-terminus of ORF3- does not change its subcellular localization but may alter its function. The *japonica ORF4+* encodes an unknown protein with a transmembrane domain at the C-terminus. The *indica ORF4-* harbors an 11-bp deletion in the coding sequence, resulting in loss of the transmembrane domain and thus alteration of the protein's localization. The *indica ORF5+* encodes an aspartic protease that is specifically expressed in ovule tissues. The *japonica ORF5-* allele has two SNPs compared with ORF5+, leading to a Phe-to-Leu amino acid substitution, which results in lower stability and enzymatic activity. However, neutral *ORF5n* has a large deletion in the coding region, which results in a loss of the signal peptide in the N-terminus. Therefore, *indica* ORF5+ and *japonica* ORF5- are localized in the cell wall, but ORF5n is localized in the cytoplasm.

Based on these genetic and molecular data, a killer–protector model was proposed to explain *S5*-mediated female sterility in rice. In this model, ORF5+ may be secreted in the extracellular domain and function as a gamete killer via its aspartic protease activity and the co-action of the helper protein ORF4+; the killer–helper complex triggers endoplasmic reticulum (ER) stress in the embryo sac, leading to abnormal PCD and female sterility. In *indica* rice, the female gametes containing *ORF3+* can eliminate the ER stress by the normal function of the protector ORF3 + .

Subsequent studies revealed that two sets of three tightly linked genes, termed *Ospara3-5* on chromosome 5 and *Os345-like* on chromosome 12, have similarity to the functional trigenic *ORFs3-5* (Ouyang et al., 2016). More interestingly, the intergenic regions of *Ospara3-5* and *ORFs3-5* exhibit high similarity and harbor multiple repetitive sequences of the RPO_OS *Helitron* transposon elements. *ORFs3-5* on chromosome 6 might have originated from a duplication and insertion of the ancestral *Ospara3-5* fragment on chromosome 5 via movement of the *Helitron* transposon. Moreover, most wild rice species harbor the sequences *ORF3 + ORF4 + ORF5+*, suggesting that this haplotype may be the ancestral proto-alleles, which could neutralize the killing and protecting processes. Since *ORF4+* and *ORF5+* function together as a killer complex, and *ORF3+* plays a protector role for the gametes, mutations of the protector *ORF3+* likely happened after the loss-of-function mutations in the killer genes because of natural survival selection on the gametes. Mutations in *ORF4+* or *ORF5+* resulted in the haplotype *ORF3 + ORF4-ORF5+* or *ORF3 + ORF4 + ORF5-*, and mutations in *ORF3+* occurred later, producing the haplotypes *ORF3-ORF4-ORF5+* or *ORF3-ORF4 + ORF5-*. Eventually, typical *indica* varieties evolved carrying the haplotype *ORF3 + ORF4-ORF5+* and typical *japonica* varieties evolved carrying the haplotype *ORF3-ORF4 + ORF5-* (Ouyang et al., 2016).

This killer–protector system has been also identified in other loci such as *S1*, *qHMS7*, *pf12/RHS12/Se*, and *qHMS1*. The *S1* locus is an important locus for interspecific HS in hybrids between Asian rice (*O. sativa*) and African rice (*O. glaberrima*) and affects pollen and embryo-sac fertility (Xie et al., 2017a). The homozygotes with the *O. glaberrima S1* locus allele (*S1-g*) or the *O. sativa* allele (*S1-s*) have normal fertility, but heterozygotes exhibit a reduction (about 50%) in fertility of both male and female gametes, resulting in a low transmission of the *S1-s* allele in the hybrid progenies. Koide et al. (2008) proposed that this *S1*-mediated HS acts as a sex-independent transmission ratio distortion (siTRD) system. With two different NILs carrying the *S1-g* allele in the *indica* and *japonica* backgrounds, a genetic study revealed that the *S1* locus is composed of two linked chromosomal sites that are responsible for the male and female sterilities. A modifier for enhancing the *S1*-mediated female sterility was identified and mapped to a 40-kb region (Koide et al., 2008). Garavito et al. (2010) mapped the *S1* locus in a 27.8-kb region that also contains a component for HMS defined by Koide et al. (2008). However, a study indicated that a mapped 55-kb region of *S1-g* derived from an African rice variety can control both male and female sterility (Xie et al., 2017a). Sequence and expression analysis showed that the 55-kb region of the *S1* locus has only two candidate genes, *ORF6* and *ORF7*, present in common in *O. sativa* and *O. glaberrima*. In *O. glaberrima*, *ORF6* encodes a protein (OgTPR1) containing two trypsin-like peptidase domains and a ribosome biogenesis regulatory domain. *ORF7* encodes a putative F-box protein; both *ORF6* and *ORF7* are expressed in anthers and panicles. Knockout of *OgORF6* (*OgTPR1*) and *OgORF7* in NIL (*S1-g*) by CRISPR/Cas9 did not affect male or female fertility. However, the hybrid plants from crossing the *ogtpr1* mutant with a *japonica* line showed normal male and female fertility, but the hybrids with the *ogorf7* mutant retained the HS phenotype. These results indicate that OgTPR1 is not essential for gametogenesis but is required for the *S1*-mediated pollen and embryo-sac HS (Xie et al., 2017a). After that, Koide and his colleagues found that an African rice lineage-specific gene *SSP* is also required for *S1*-mediated HS (Koide et al., 2018). We further found that another lineage-specific gene *S1A4* adjacent to *OgTPR1* (renamed as *S1TPR*) cooperates with *S1TPR* and *SSP* to form a killer complex for triggering a sterility signal in sporophytic meiotic cells. In addition, *S1TPR* also functions as a gametophytic protector against the sterility effect to rescue the gametes with the African rice allele (Xie et al., 2019). So far, many selfish genetic elements in African rice such as *S1* locus conferring HS have been identified. On the other hand, some selfish genetic elements in Asian rice such as *S58* locus have been found (Feng et al., 2023). Furthermore, cloning the corresponding genes for these loci will be greatly beneficial for overcoming interspecific HS in African-Asian hybrids.

The *qHMS7* and *qHMS1* loci are mapped on chromosomes 7 and 1, which cause pollen abortion between *O. sativa* and *O. meridionalis*, respectively. In the *qHMS7* locus, the pollens carrying the *O. meridionalis* allele are selectively aborted in a heterozygous status (Yu et al., 2018). The *O. sativa* allele contains two adjacent genes, *ORF2* and *ORF3*. *ORF2* encodes a ribosome-inactivating protein (RIP) domain-containing protein and functions as a toxic factor conferring pollen abortion. *ORF3* encodes a grass family-specific protein localized in mitochondria and acts as an antidote against the toxic factor in a gametophytic manner (Yu et al., 2018). In the *qHMS1* locus, the pollens carrying *O. sativa* allele are selectively killed in hybrids (You et al., 2023). Similar to the *qHMS7* locus, the *qHMS1* locus encodes a toxin-antidote system containing two genes *HPT* and *HPA* in the *O. meridionalis* allele. *HPT* encodes a DEAD-like helicase as a hybrid pollen toxin causing pollen abortion. *HPA* encodes an F-box protein as a hybrid pollen antidote rescuing the pollen in a gametophytic manner (You et al., 2023).

The *pf12/RHS12/Se* locus is also a major HMS locus on chromosome 12 between *indica* and *japonica* rice (Zhou et al., 2023; Wang et al., 2023a, b). In the *pf12/RHS12/Se* locus, the pollens with *japonica* allele are selectively aborted in *indica–japonica* hybrids (Zhou et al., 2023, Wang et al., 2023a, b). The *indica* allele harbors two tightly linked genes, *iORF3/DUYAO and iORF4/JIEYAO*, whereas the *japonica* allele lacks these two genes. *DUYAO* encodes a mitochondrion-targeted protein, which causes cell death by disrupting the function of OsCOX11. *JIEYAO* encodes a protein which degrades DUYAO through an autophagosome pathway (Wang et al., 2023a); thereby, the pollens with *JIEYAO* are rescued and preferentially transmitted in *indica–japonica* hybrids.

The *S7* locus, located on chromosome 7, controls hybrid female sterility of an Aus-type variety (one ecotype of *indica* rice) and *japonica/indica* rice (Ikehashi & Araki, 1987; Yanagihara et al., 1992). Female gametes with the *$S7^j$* allele from *japonica* or the *$S7^i$* allele from *indica* are aborted in the *Aus-japonica/indica* hybrids, whereas female gametes carrying the *$S7^{ai}$* allele from Aus survive. The *S7* locus was delimited to a 139-kb region in the centromere region of chromosome 7 containing 16 putative genes (Yu et al., 2016). Among them, the pistil-expressed *ORF3* is the candidate gene for the hybrid female sterility trait and encodes a tetratricopeptide repeat (TPR) protein. Knockdown of *ORF3* by RNAi rescues spikelet fertility and eliminates the distorted transmission of *$S7^{ai}$* in heterozygotes (Yu et al., 2016).

The *HSA1* locus, located on chromosome 12, causes F_1 and F_2 hybrid female sterility in *indica–japonica* hybrids (Kubo et al., 2016). The *HSA1* locus has two tightly linked genes, *HSA1a* and *HSA1b*. The *japonica* allele contains *$HSA1a^j$-$HSA1b^j$,* and the *indica* allele contains *$HSA1a^{is}$-$HSA1b^{is}$*. *$HSA1a^j$* encodes a plant-specific domain of unknown function protein (DUF1618), while *$HSA1a^{is}$* encodes a truncated protein due to two base-pair deletions in the coding region. *$HSA1b^{is}$* encodes an unknown protein with some similarity to a nucleotide-binding protein, while *$HSA1b^j$* encodes a truncated protein due to one base-pair insertion in the coding region. The indica homozygous alleles are introgressed into japonica, which causes female gamete abortion. Intriguingly, a recombinant haplotype *$HSA1a^j$-$HSA1b^{is}$* leads to semi-sterility in the heterozygotes with *$HSA1a^{is}$-$HSA1b^{is}$*, suggesting that *HSA1* may also cause sterility in F_1 hybrids (Kubo et al., 2016).

The *ESA1* locus has been mapped on chromosome 1, conferring embryo-sac semi-sterility between *O. rufipogon* and *O. sativa* (Hou et al., 2019). Unlike the other one-locus model-associated HS loci, the *ESA1* locus does not cause segregation distortion in F_2 population. The *O. rufipogon* allele introgression into the *O. sativa* results in embryo-sac semi-sterility. *ESA1* encodes an armadillo repeat domain protein targeting the nuclear membrane. An SNP (T to C) between *O. rufipogon* and *O. sativa* causes a premature stop codon in *O. rufipogon*-derived *ESA1*. The truncated protein product may associate with an unknown *O. sativa*-derived factor, causing harmful interaction for embryo-sac development (Hou et al., 2019).

These studies on the *Sa*, *Sc*, *S5*, *HSA1*, *S1*, *qHMS7*, *pf12/RHS12/Se,* and *qHMS1* loci revealed that one-locus model-associated HS generally involves a complex locus comprising multiple genes, and hybrid incompatibility involves diverse molecular genetic mechanisms.

Molecular Mechanisms of the Two-Locus Model in Rice Unlike the one-locus model-associated HS, rice *S27/S28*-, *DOPPELGANGER1* (*DPL1*)/*DPL2*-, and *DUPLICATED GAMETOPHYTIC STERILITY 1* (*DGS1*)/*DGS2*-mediated HS act consistently with the two-locus model (duplicate gametic lethal model).

In *S27/S28*-mediated HS, pollen sterility in the F_1 hybrids between T65 and a NIL carrying introgressed chromosomal segments from the wild rice (*O. glumaepatula*) is caused by an epistatic interaction between two loci, *S27* on chromosome 8 and *S28* on chromosome 4 (Sobrizal & Yoshimura, 2001, 2002). T65 carries *S27-T65*$^{+/+}$*//S28-T65*$^{S/S}$, and the NIL possesses *S27-glum*$^{S/S}$*//S28-glum*$^{+/+}$. In the F_1 hybrids, meiotic recombination of the chromosomes produces about 25% sterile pollen grains carrying *S27-glum*$^{S/S}$*//S28-T65*$^{S/S}$. In recombinants with heterozygous *S27* and homozygous *S28-T65*, pollen grains carrying the *S27-glum*^{S} allele are sterile, while in recombinants with heterozygous *S28* and homozygous *S27-glum*^{S}, pollen grains carrying the *S28-T65*^{S} allele are sterile; thus, both types of plants have pollen semi-sterility phenotypes.

S27 and *S28* were mapped to 134.9-kb and 1.68-Mb regions, respectively. The region of the *S28-glum*$^{+}$ allele harbors a UDP-glycosyltransferase gene (*UDPGT*) and a mitochondrial ribosomal protein L27 gene (*mtRPL27a*), whereas the *S28-T65*^{S} allele has no *UDPGT*, and the *mtPRL27* gene is silent due to loss of promoter activity. The region of *S27-T65*$^{+}$ allele harbors *UDPGT* and two tandem-duplicated versions of *mtRPL27*, *mtRPL27a* and *mtRPL27b*; these two genes are absent in *S27-glum*^{S}. Further genomic complementation assays confirmed that *mtRPL27*, but not *UDPGT*, rescues the fertility of pollen carrying *S27-glum*^{s}*/S28-T65*^{s}. Dysfunctional *mtRPL27* may cause a defect in protein synthesis in mitochondria, resulting in abnormal respiratory activity and sterile pollen (Yamagata et al., 2010). Comparison of genomic sequences in wild and cultivated rice suggested that the duplicated segments at *S27* on chromosome 8 were derived from *S28* on chromosome 4 (Presgraves, 2010). The segmental genomic duplication carrying *mtRPL27* at *S27* occurred before the divergence of the wild rice species with the AA genome, and the duplicated segments at *S27* spread in each species, except *O. glumaepatula*.

Similar to the *S27/S28* locus pair, the genetically interacting locus pair *DPL1/DPL2* and *DGS1/DGS2*, fitting the duplicate gametic lethal model, was also detected in *indica–japonica* and *O. nivara–O.sativa* F_1 hybrids, respectively (Mizuta et al., 2010; Nguyen et al., 2017). *DPL1* (located on chromosome 1) and *DPL2* (located on chromosome 6) are homologous genes derived from transposition/duplication. However, independent disruptions of *DPL1* and *DPL2* by sequence variations occurred in *indica* and *japonica* rice, respectively, during their evolution, resulting in only a single functional *DPL1* or *DPL2* gene located on the different chromosomes in each subspecies. *DPL1* and *DPL2* encode highly conserved, plant-specific small proteins essential for pollen germination. Therefore, in *indica–japonica* hybrids, the ~25% pollen grains that lack functional *DPL1* and *DPL2* alleles are sterile. *DGS1* (located on chromosome 4) and *DGS2* (located on chromosome 7) encode a putative DNA-dependent RNA polymerase (RNAP) III subunit C4 (RPC4) which is required for male gamete development. The *O. nivara* and *O.sativa* alleles only harbor a single functional *DGS1* or *DGS2* gene located on the different chromosomes, respectively, due to mutations that occurred during their evolution. The gametes without functional *DGS1* and *DGS2* are sterile in *O. nivara-O.sativa* hybrids.

Similar two-locus systems have also been reported in other organisms, such as the chromosome-transposed male fertility gene *JYAlpha* in *Drosophila melanogaster–D. simulans* hybrids (Masly et al., 2006). Therefore, the hybrid incompatibility caused by the gene transposition/duplication-reciprocal absent/loss-of-function mutation mechanism may be common for postzygotic reproductive isolation in different organisms.

Strategies for Overcoming HS in Hybrid Rice Breeding Programs The development of hybrid rice varieties has significantly increased grain yield since the 1970s. Yuan proposed three steps for hybrid rice development (Yuan, 1987). The first step is the utilization of intra-subspecific crosses (such as *indica–indica* or *japonica–japonica*). The second step is the utilization of intersubspecific crosses, such as *indica–japonica* crosses. The final step is the utilization of interspecific crosses, such as *sativa–glaberrima* crosses. Current hybrid rice is mainly derived from intra-subspecific crosses, especially *indica* lines. Several varieties from *indica–japonica* crosses, such as Yongyou 6 and Chunyou 6, have been cultivated by using sterile lines of *japonica* rice and restorer lines of *indica* carrying wide-compatible neutral alleles, which exhibit much stronger hybrid vigor (Wu et al., 2007; Xu et al., 2010). The ultimate goal for the investigation of HS is to break down HS in order to utilize the strong hybrid vigor in distant hybrids. However, there are still many challenges in controlling the fertility of distant hybrids.

Three strategies have been developed to overcome HS in rice hybrid breeding. Neutral alleles of certain HS loci were identified in some wide-compatibility rice lines (Kitamura, 1962; Ikehashi & Araki, 1986). These neutral alleles can be introgressed into the parents for intersubspecific hybrid breeding. However, very few identified neutral alleles are currently available, thus limiting their utility in breeding programs. A second option for this purpose is the creation of *indica*-compatible *japonica* lines by introgressing *indica* alleles of HS loci into the *japonica* genetic background. A recent study showed that this strategy works well for overcoming *indica–japonica* HS by building up such *indica*-compatible *japonica* lines with multiple *indica* HS alleles (Guo et al., 2016). The downside is that this approach is time-consuming and laborious due to the need for successive backcrosses. A third option is creating artificial hybrid-compatible lines by molecular approaches. With the increasing cloning of HS-related genes, understanding of their molecular mechanisms, and the development of CRISPR/Cas9-based genome editing technologies, our studies have demonstrated that artificial neutral lines for *Sa*, *Sc*, and *S1* loci can be generated by functional knockout or suppression of HS-related genes if such genes are not required for gamete development (Xie et al., 2017a, b), or the reduced gene dosage does not affect pollen fertility as shown in *Sc* (Shen et al., 2017).

Therefore, the molecular approach enables us to create artificial neutral alleles of multiple HS loci in elite rice varieties and rapid breeding of hybrid-compatible lines in the future.

15.6 Concluding Remarks

Major advances in our understanding of the molecular mechanisms of CMS and EGMS have been made using rice models. These traits are controlled by interactions between nuclear and cytoplasmic genes or genes and the environment. The CMS or EGMS traits in other plant species, especially in major crops such as wheat, maize, and soybean, need to be further studied. HS, caused by the interactions of multiple genes in one locus or between two non-allelic loci, is a major obstacle for the utilization of hybrid vigor in crops. The isolation of HS genes has enabled the generation of artificial hybrid-compatible lines using genome editing technologies. Therefore, advances in our knowledge on the molecular basis of CMS, EGMS, and HS will accelerate and broaden the utilization of distant hybrid vigor in crops.

Acknowledgments This work was supported by the grants from the National Natural Science Foundation of China (32030080, 31921004) and the Open Competition Program of the Top Ten Critical Priorities of Agricultural Science and Technology Innovation for the 14th Five-Year Plan of Guangdong Province (2022SDZG05) to LC, and the Guangdong Natural Science Funds for Distinguished Young Scholars (2021B1515020089) to YX. We are grateful to Dr. Jun Hu for the photos of CMS-HL.

References

Aarts, M. G., Keijzer, C. J., Stiekema, W. J., & Pereira, A. (1995). Molecular characterization of the *CER1* gene of arabidopsis involved in epicuticular wax biosynthesis and pollen fertility. *The Plant Cell, 7*, 2115–2127.

Ahmadikhah, A., & Karlov, G. I. (2006). Molecular mapping of the fertility-restoration gene *Rf4* for WA-cytoplasmic male sterility in rice. *Plant Breeding, 125*, 363–367.

Akagi, H., Sakamoto, M., Shinjyo, C., Shimada, H., & Fujimura, T. (1994). A unique sequence located downstream from the rice mitochondrial *atp6* may cause male sterility. *Current Genetics, 25*, 52–58.

Alcazar, R., Garcia, A. V., Parker, J. E., & Reymond, M. (2009). Incremental steps toward incompatibility revealed by Arabidopsis epistatic interactions modulating salicylic acid pathway activation. *Proceedings of the National Academy of Sciences of the United States of America, 106*, 334–339.

Ariizumi, T., & Toriyama, K. (2011). Genetic regulation of sporopollenin synthesis and pollen exine development. *Annual Review of Plant Biology, 62*, 437–460.

Ariizumi, T., Hatakeyama, K., Hinata, K., Sato, S., Kato, T., Tabata, S., & Toriyama, K. (2003). A novel male-sterile mutant of *Arabidopsis thaliana, faceless pollen-1*, produces pollen with a smooth surface and an acetolysis-sensitive exine. *Plant Molecular Biology, 53*, 107–116.

Baack, E., Melo, M. C., Rieseberg, L. H., & Ortiz-Barrientos, D. (2015). The origins of reproductive isolation in plants. *The New Phytologist, 207*, 968–984.

Barkan, A., & Small, I. (2014). Pentatricopeptide repeat proteins in plants. *Annual Review of Plant Biology, 65*, 415.

Bentolila, S., Alfonso, A. A., & Hanson, M. R. (2002). A pentatricopeptide repeat-containing gene restores fertility to cytoplasmic male-sterile plants. *Proceedings of the National Academy of Sciences of the United States of America, 99*, 10887–10892.

Bikard, D., Patel, D., Le Mette, C., Giorgi, V., Camilleri, C., Bennett, M. J., & Loudet, O. (2009). Divergent evolution of duplicate genes leads to genetic incompatibilities within *A. thaliana*. *Science, 323*, 623–626.

Bomblies, K. (2010). Doomed lovers: Mechanisms of isolation and incompatibility in plants. *Annual Review of Plant Biology, 61*, 109–124.

Brar, D. S., & Khush, G. S. (1997). Alien introgression in rice. *Plant Molecular Biology, 35*, 35–47.

Canales, C., Bhatt, A. M., Scott, R., & Dickinson, H. (2002). *EXS*, a putative LRR receptor kinase, regulates male germline cell number and tapetal identity and promotes seed development in *Arabidopsis*. *Current Biology, 12*, 1718–1727.

Chang, F., Wang, Y., Wang, S., & Ma, H. (2011). Molecular control of microsporogenesis in *Arabidopsis*. *Current Opinion in Plant Biology, 14*, 66–73.

Chase, C. D. (2007). Cytoplasmic male sterility: A window to the world of plant mitochondrial-nuclear interactions. *Trends in Genetics, 23*, 81–90.

Chen, L., & Liu, Y. G. (2014). Male sterility and fertility restoration in crops. *Annual Review of Plant Biology, 65*, 579–606.

Chen, L., & Liu, Y. G. (2016). Discovery, utilization and molecular mechanisms of CMS-WA in rice. *Chinese Science Bulletin, 35*, 3804–3812.

Chen, K., & Xiao, Y. (1991). Effect of night-break with different wavelengths light on fertility in HPGMR (in Chinese). *Journal of Wuhan University (Natural Science Edition)*, 127–128.

Chen, R., Zhao, X., Shao, Z., Wei, Z., Wang, Y., Zhu, L., Zhao, J., Sun, M., He, R., & He, G. (2007). Rice UDP-glucose pyrophosphorylase1 is essential for pollen callose deposition and its cosuppression results in a new type of thermosensitive genic male sterility. *The Plant Cell, 19*, 847–861.

Chen, J., Ding, J., Ouyang, Y., Du, H., Yang, J., Cheng, K., Zhao, J., Qiu, S., Zhang, X., & Yao, J. (2008). A triallelic system of *S5* is a major regulator of the reproductive barrier and compatibility of indica-japonica hybrids in rice. *Proceedings of the National Academy of Sciences of the United States of America, 105*, 11436–11441.

Chen, C., Chen, H., Lin, Y. S., Shen, J. B., Shan, J. X., Qi, P., Shi, M., Zhu, M. Z., Huang, X. H., Feng, Q., Han, B., Jiang, L., Gao, J. P., & Lin, H. X. (2014). A two-locus interaction causes interspecific hybrid weakness in rice. *Nature Communications, 5*, 3357.

Cui, X. Q., Wise, R. P., & Schnable, P. S. (1996). The *rf2* nuclear restorer gene of male-sterile T-cytoplasm maize. *Science, 272*, 1334.

Deng, H., Shu, F., & Yuan, D. (1999). An overview of research and utilization of Annong S1 (in Chinese). *Hybrid Rice, 14*, 1–3.

Ding, J., Lu, Q., Ouyang, Y., Mao, H., Zhang, P., Yao, J., Xu, C., Li, X., Xiao, J., & Zhang, Q. (2012a). A long noncoding RNA regulates photoperiod-sensitive male sterility, an essential component of hybrid rice. *Proceedings of the National Academy of Sciences of the United States of America, 109*, 2654–2659.

Ding, J., Shen, J., Mao, H., Xie, W., Li, X., & Zhang, Q. (2012b). RNA-directed DNA methylation is involved in regulating photoperiod-sensitive male sterility in rice. *Molecular Plant, 5*, 1210–1216.

Dong, X., Hong, Z., Sivaramakrishnan, M., Mahfouz, M., & Verma, D. P. (2005). Callose synthase (CalS5) is required for exine formation during microgametogenesis and for pollen viability in *Arabidopsis*. *The Plant Journal, 42*, 315–328.

Fan,Y., & Zhang, Q. (2018). Genetic and molecular characterization of photoperiod and thermo-sensitive male sterility in rice. *Plant Reproduction, 31*, 3–14.

Fan, Y., Yang, J., Mathioni, S. M., Yu, J., Shen, J., Yang, X., Wang, L., Zhang, Q., Cai, Z., Xu, C., Li, X., Xiao, J., Meyers, B. C., & Zhang, Q. (2016). *PMS1T*, producing phased small-interfering RNAs, regulates photoperiod-sensitive male sterility in rice. *Proceedings of the National Academy of Sciences of the United States of America, 113*, 15144–15149.

Feng, Y., Tang, J., Liu, R., Liu, Y.-G., Chen, L., & Xie, Y. (2023). Characterization and fine-mapping of a new Asian rice selfish genetic locus *S58* in Asian–African rice hybrids. *Theoretical and Applied Genetics, 136*, 87.

Fujii, S., & Toriyama, K. (2005). Molecular mapping of the fertility restorer gene for ms-CW-type cytoplasmic male sterility of rice. *Theoretical and Applied Genetics, 111*, 696–701.

Fujii, S., & Toriyama, K. (2009). Suppressed expression of *RETROGRADE-REGULATED MALE STERILITY* restores pollen fertility in cytoplasmic male sterile rice plants. *Proceedings of the National Academy of Sciences of the United States of America, 106*, 9513–9518.

Garavito, A., Guyot, R., Lozano, J., Gavory, F., Samain, S., Panaud, O., Tohme, J., Ghesquière, A., & Lorieux, M. (2010). A genetic model for the female sterility barrier between Asian and African cultivated rice species. *Genetics, 185*, 1425–1440.

Gibeaut, D. (2000). Nucleotide sugars and glycosyltransferases for synthesis of cell wall matrix polysaccharides. *Plant Physiology and Biochemistry, 38*, 69–80.

Guo, J. X., & Liu, Y. G. (2012). Molecular control of male reproductive development and pollen fertility in rice. *Journal of Integrative Plant Biology, 54*, 967–978.

Guo, J., & Liu, Y. G. (2017). Long non-coding RNAs play an important role in regulating photoperiod- and temperature-sensitive male sterility in rice. *Science China Life Sciences, 60*, 443–444.

Guo, J., Xu, X., Li, W., Zhu, W., Zhu, H., Liu, Z., Luan, X., Dai, Z., Liu, G., Zhang, Z., Zeng, R., Tang, G., Fu, X., Wang, S., & Zhang, G. (2016). Overcoming inter-subspecific hybrid sterility in rice by developing *indica*-compatible *japonica* lines. *Scientific Reports, 6*, 26878.

Haslam, T. M., Haslam, R., Thoraval, D., Pascal, S., Delude, C., Domergue, F., Fernandez, A. M., Beaudoin, F., Napier, J. A., Kunst, L., & Joubes, J. (2015). ECERIFERUM2-LIKE proteins have unique biochemical and physiological functions in very-long-chain fatty acid elongation. *Plant Physiology, 167*, 682–692.

He, S., Abad, A. R., Gelvin, S. B., & Mackenzie, S. A. (1996). A cytoplasmic male sterility-associated mitochondrial protein causes pollen disruption in transgenic tobacco. *Proceedings of the National Academy of Sciences of the United States of America, 93*, 11763–11768.

Hou, J., Cao, C., Ruan, Y., Deng, Y., Liu, Y., Zhang, K., Tan, L., Zhu, Z., Cai, H., Liu, F., Sun, H., Gu, P., Sun, C., & Fu, Y. (2019). *ESA1* is involved in embryo sac abortion in interspecific hybrid progeny of rice. *Plant Physiology, 180*, 356–366.

Hu, J., Wang, K., Huang, W., Liu, G., Gao, Y., Wang, J., Huang, Q., Ji, Y., Qin, X., Wan, L., Zhu, R., Li, S., Yang, D., & Zhu, Y. (2012). The rice pentatricopeptide repeat protein RF5 restores fertility in Hong-Lian cytoplasmic male-sterile lines via a complex with the glycine-rich protein GRP162. *Plant Cell, 24*, 109–122.

Huang, W., Yu, C., Hu, J., Wang, L., Dan, Z., Zhou, W., He, C., Zeng, Y., Yao, G., Qi, J., Zhang, Z., Zhu, R., Chen, X., & Zhu, Y. (2015). Pentatricopeptide-repeat family protein RF6 functions with hexokinase 6 to rescue rice cytoplasmic male sterility. *Proceedings of the National Academy of Sciences of the United States of America, 112*, 14984–14989.

Ikehashi, H., & Araki, H. (1986). Genetics of F_1 sterility in remote crosses in rice. In IRRI (Ed.), Rice genetics proceedings of the first rice genetics symposium *Rice genetics* (pp. 119–130). IRRI.

Ikehashi, H., & Araki, H. (1987). Screening and genetic analysis of wide-compatibility in F_1 hybrids of distant crosses in rice, *Oryza sativa* L. *Technical Bulletin of the Tropical Agriculture Research Center, 12*, 231–241.

Itabashi, E., Kazama, T., & Toriyama, K. (2009). Characterization of cytoplasmic male sterility of rice with Lead Rice cytoplasm in comparison with that with Chinsurah Boro II cytoplasm. *Plant Cell Reports, 28*, 233–239.

Itabashi, E., Iwata, N., Sota, F., Kazama, T., & Toriyama, K. (2011). The fertility restorer gene, *Rf2*, for Lead Rice-type cytoplasmic male sterility of rice encodes a mitochondrial glycine-rich protein. *The Plant Journal, 65*, 359–367.

Jessen, D., Olbrich, A., Knufer, J., Kruger, A., Hoppert, M., Polle, A., & Fulda, M. (2011). Combined activity of LACS1 and LACS4 is required for proper pollen coat formation in *Arabidopsis*. *The Plant Journal, 68*, 715–726.

Jeuken, M. J., Zhang, N. W., McHale, L. K., Pelgrom, K., den Boer, E., Lindhout, P., Michelmore, R. W., Visser, R. G., & Niks, R. E. (2009). Rin4 causes hybrid necrosis and race-specific resistance in an interspecific lettuce hybrid. *The Plant Cell, 21*, 3368–3378.

Jia, G., Liu, X., Owen, H. A., & Zhao, D. (2008). Signaling of cell fate determination by the TPD1 small protein and EMS1 receptor kinase. *Proceedings of the National Academy of Sciences of the United States of America, 105*, 2220–2225.

Jiang, Z., Xu, Q., & Dong, Y. (1992). Studies on the fertility changes and heredity in Hengnong S-1 (*indica* type) male-sterile rice (in Chinese). *Crop Research, 6*, 12–14.

Jiang, J., Zhang, Z., & Cao, J. (2013). Pollen wall development: The associated enzymes and metabolic pathways. *Plant Biology, 15*, 249–263.

Jiang, H., Lu, Q., Qiu, S., Yu, H., Wang, Z., Yu, Z., Lu, Y., Wang, L., Xia, F., Wu, Y., Li, F., Zhang, Q., Liu, F., Song, D., Ma, C., Ding, Q., Zhang, X., Zhang, L., Zhang, X., Li, X., Zhang, J., Xiao, J., Li, X., Wang, N., Ouyang, Y., Zhou, F., & Zhang, Q. (2022). Fujian cytoplasmic male sterility and the fertility restorer gene *OsRf19* provide a promising breeding system for hybrid rice. *Proceedings of the National Academy of Sciences of the United States of America, 119*, e2208759119.

Jing, B., Heng, S., Tong, D., Wan, Z., Fu, T., Tu, J., Ma, C., Yi, B., Wen, J., & Shen, J. (2012). A male sterility-associated cytotoxic protein ORF288 in *Brassica juncea* causes aborted pollen development. *Journal of Experimental Botany, 63*, 1285–1295.

Johnson, C., Kasprzewska, A., Tennessen, K., Fernandes, J., Nan, G. L., Walbot, V., Sundaresan, V., Vance, V., & Bowman, L. H. (2009). Clusters and superclusters of phased small RNAs in the developing inflorescence of rice. *Genome Research, 19*, 1429–1440.

Kazama, T., Nakamura, T., Watanabe, M., Sugita, M., & Toriyama, K. (2008). Suppression mechanism of mitochondrial ORF79 accumulation by Rf1 protein in BT-type cytoplasmic male sterile rice. *The Plant Journal, 55*, 619–628.

Kazama, T., Itabashi, E., Fujii, S., Nakamura, T., & Toriyama, K. (2016). Mitochondrial *ORF79* levels determine pollen abortion in cytoplasmic male sterile rice. *The Plant Journal, 85*, 707–716.

Kim, D. H., Kang, J. G., & Kim, B. (2007). Isolation and characterization of the cytoplasmic male sterility-associated *orf456* gene of chili pepper (*Capsicum annuum* L.). *Plant Molecular Biology, 63*, 519–532.

Kitamura, E. (1962). Genetics studies on sterility observed in hybrids between distantly related varieties of rice (*Oryza sativa* L.). *Bulletin of the Chugoku Agricultural Experiment Station Series A, 8*, 141–205.

Kleczkowski, L. A., Geisler, M., Ciereszko, I., & Johansson, H. (2004). UDP-glucose pyrophosphorylase. An old protein with new tricks. *Plant Physiology, 134*, 912–918.

Klein, R. R., Klein, P. E., Mullet, J. E., Minx, P., Rooney, W. L., & Schertz, K. F. (2005). Fertility restorer locus *Rf1* of *sorghum* (*Sorghum bicolor* L.) encodes a pentatricopeptide repeat protein not present in the colinear region of rice chromosome 12. *Theoretical and Applied Genetics, 111*, 994–1012.

Koide, Y., Onishi, K., Nishimoto, D., Baruah, A. R., Kanazawa, A., & Sano, Y. (2008). Sex-independent transmission ratio distortion system responsible for reproductive barriers between Asian and African rice species. *The New Phytologist, 179*, 888–900.

Koide, Y., Ogino, A., Yoshikawa, T., Kitashima, Y., Saito, N., Kanaoka, Y., Onishi, K., Yoshitake, Y., Tsukiyama, T., Saito, H., Teraishi, M., Yamagata, Y., Uemura, A., Takagi, H., Hayashi, Y., Abe, T., Fukuta, Y., Okumoto, Y., & Kanazawa, A. (2018). Lineage-specific gene acquisition or loss is involved in interspecific hybrid sterility in rice. *Proceedings of the National Academy of Sciences of the United States of America, 115*, E1955–E1962.

Koizuka, N., Imai, R., Fujimoto, H., Hayakawa, T., Kimura, Y., Kohno-Murase, J., Sakai, T., Kawasaki, S., & Imamura, J. (2003). Genetic characterization of a pentatricopeptide repeat protein gene, *orf687*, that restores fertility in the cytoplasmic male-sterile *Kosena radish*. *The Plant Journal, 34*, 407–415.

Komori, T., Ohta, S., Murai, N., Takakura, Y., Kuraya, Y., Suzuki, S., Hiei, Y., Imaseki, H., & Nitta, N. (2004). Map-based cloning of a fertility restorer gene, *Rf-1*, in rice (Oryza sativa L.). *The Plant Journal, 37*, 315–325.

Kruger, J., Thomas, C. M., Golstein, C., Dixon, M. S., Smoker, M., Tang, S., Mulder, L., & Jones, J. D. (2002). A tomato cysteine protease required for Cf-2-dependent disease resistance and suppression of autonecrosis. *Science, 296*, 744–747.

Kubo, T., Takashi, T., Ashikari, M., Yoshimura, A., & Kurata, N. (2016). Two tightly linked genes at the *hsa1* locus cause both F_1 and F_2 hybrid sterility in rice. *Molecular Plant, 9*, 221–232.

Lai, Z., Nakazato, T., Salmaso, M., Burke, J. M., Tang, S., Knapp, S. J., & Rieseberg, L. H. (2005). Extensive chromosomal repatterning and the evolution of sterility barriers in hybrid sunflower species. *Genetics, 171*, 291–303.

Li, H., Liang, W., Hu, Y., Zhu, L., Yin, C., Xu, J., Dreni, L., Kater, M. M., & Zhang, D. (2011a). Rice *MADS6* interacts with the floral homeotic genes *SUPERWOMAN1*, *MADS3*, *MADS58*, *MADS13*, and *DROOPING LEAF* in specifying floral organ identities and meristem fate. *The Plant Cell, 23*, 2536–2552.

Li, H., Liang, W., Yin, C., Zhu, L., & Zhang, D. (2011b). Genetic interaction of OsMADS3, DROOPING LEAF, and OsMADS13 in specifying rice floral organ identities and meristem determinacy. *Plant Physiology, 156*, 263–274.

Li, G., Zhang, Y., Yu, H., Xie, Y., Luo, H., Wang Y., Tang, J., Zhang, J., Xie, X., Zong, W., Liu, K., Wang, X., Long, Y., Song, Q., Wu, Z., Liu, Y-G., & Chen, L. (2025). A tripartite pollen killer–protector system confers temperature-sensitive inter-subspecific reproductive isolation in rice. *Molecular Plant*. https://doi.org/10.1016/j.molp.2025.12.008

Linares, O. F. (2002). African rice (*Oryza glaberrima*): History and future potential. *Proceedings of the National Academy of Sciences of the United States of America, 99*, 16360–16365.

Liu, K. D., Zhou, Z. Q., Xu, C. G., Zhang, Q., & Maroof, M. (1996). An analysis of hybrid sterility in rice using a diallel cross of 21 parents involving *indica*, *japonica* and wide compatibility varieties. *Euphytica, 90*, 275–280.

Liu, B., Chen, Z., Song, X., Liu, C., Cui, X., Zhao, X., Fang, J., Xu, W., Zhang, H., Wang, X., Chu, C., Deng, X., Xue, Y., & Cao, X. (2007a). *Oryza sativa* dicer-like4 reveals a key role for small interfering RNA silencing in plant development. *The Plant Cell, 19*, 2705–2718.

Liu, Z. L., Hong, X., Guo, J. X., & Liu, Y. G. (2007b). Structural and expressional variations of the mitochondrial genome conferring the wild abortive type of cytoplasmic male sterility in rice. *Journal of Integrative Plant Biology, 49*, 908–914.

Liu, Z., Yang, Z., Wang, X., Li, K., An, H., Liu, J., Yang, G., Fu, T., Yi, B., & Hong, D. (2016). A mitochondria-targeted PPR protein restores pol cytoplasmic male sterility by reducing orf224 transcript levels in *Oilseed Rape*. *Molecular Plant, 9*, 1082–1084.

Long, Y., Zhao, L., Niu, B., Su, J., Wu, H., Chen, Y., Zhang, Q., Guo, J., Zhuang, C., & Mei, M. (2008). Hybrid male sterility in rice controlled by interaction between divergent alleles of two adjacent genes. *Proceedings of the National Academy of Sciences of the United States of America, 105*, 18871–18876.

Luo, D., Xu, H., Liu, Z., Guo, J., Li, H., Chen, L., Fang, C., Zhang, Q., Bai, M., Yao, N., Wu, H., Wu, H., Ji, C., Zheng, H., Chen, Y., Ye, S., Li, X., Zhao, X., Li, R., & Liu, Y. (2013). A detrimental mitochondrial-nuclear interaction causes cytoplasmic male sterility in rice. *Nature Genetics, 45*, 573–577.

Ma, H. (2005). Molecular genetic analyses of microsporogenesis and microgametogenesis in flowering plants. *Annual Review of Plant Biology, 56*, 393–434.

Masly, J. P., Jones, C. D., Noor, M. A., Locke, J., & Orr, H. A. (2006). Gene transposition as a cause of hybrid sterility in *Drosophila*. *Science, 313*, 1448–1450.

Mei, M., Dai, X., Xu, C., & Zhang, Q. (1999). Mapping and genetic analysis of the genes for photoperiod-sensitive genic male sterility in rice using the original mutant Nongken 58S. *Crop Science, 39*, 1711–1715.

Mizuta, Y., Harushima, Y., & Kurata, N. (2010). Rice pollen hybrid incompatibility caused by reciprocal gene loss of duplicated genes. *Proceedings of the National Academy of Sciences of the United States of America, 107*, 20417–20422.

Moon, S., Kim, S. R., Zhao, G., Yi, J., Yoo, Y., Jin, P., Lee, S. W., Jung, K. H., Zhang, D., & An, G. (2013). Rice *glycosyltransferase1* encodes a glycosyltransferase essential for pollen wall formation. *Plant Physiology, 161*, 663–675.

Muller, R., Borghi, L., Kwiatkowska, D., Laufs, P., & Simon, R. (2006). Dynamic and compensatory responses of *Arabidopsis* shoot and floral meristems to CLV3 signaling. *The Plant Cell, 18*, 1188–1198.

Nguyen, G. N., Yamagata, Y., Shigematsu, Y., Watanabe, M., Miyazaki, Y., Doi, K., Tashiro, K., Kuhara, S., Kanamori, H., Wu, J., Matsumoto, T., Yasui, H., & Yoshimura, A. (2017). Duplication and loss of function of genes encoding RNA Polymerase III subunit C4 causes hybrid incompatibility in rice. *G3 (Bethesda, Md.), 7*, 2565–2575.

Nonomura, K., Miyoshi, K., Eiguchi, M., Suzuki, T., Miyao, A., Hirochika, H., & Kurata, N. (2003). The *MSP1* gene is necessary to restrict the number of cells entering into male and female sporogenesis and to initiate anther wall formation in rice. *The Plant Cell, 15*, 1728–1739.

Oka, H. I. (1957). Genetic analysis for the sterility of hybrids between distantly related varieties of cultivated rice. *Japanese Journal of Breeding. Journal of Genetics, 55*, 397–409.

Oka, H. (1974). Analysis of genes controlling f(1) sterility in rice by the use of isogenic lines. *Genetics, 77*, 521–534.

Okazaki, M., Kazama, T., Murata, H., Motomura, K., & Toriyama, K. (2013). Whole mitochondrial genome sequencing and transcriptional analysis to uncover an RT102-type cytoplasmic male sterility-associated candidate gene derived from *Oryza rufipogon*. *Plant & Cell Physiology, 54*, 1560–1568.

Orr, H. A., & Presgraves, D. C. (2000). Speciation by postzygotic isolation: Forces, genes and molecules. *BioEssays, 22*, 1085–1094.

Ouyang, Y., & Zhang, Q. (2013). Understanding reproductive isolation based on the rice model. *Annual Review of Plant Biology, 64*, 111–135.

Ouyang, Y., Liu, Y. G., & Zhang, Q. (2010). Hybrid sterility in plant: Stories from rice. *Current Opinion in Plant Biology, 13*, 186–192.

Ouyang, Y., Li, G., Mi, J., Xu, C., Du, H., Zhang, C., Xie, W., Li, X., Xiao, J., Song, H., & Zhang, Q. (2016). Origination and establishment of a trigenic reproductive isolation system in rice. *Molecular Plant, 9*, 1542–1545.

Peng, X., Wang, K., Hu, C., Zhu, Y., & Wang, T. (2010). The mitochondrial gene *orfH79* plays a critical role in impairing both male gametophyte development and root growth in CMS-Honglian rice. *BMC Plant Biology, 10*, 125.

Presgraves, D. C. (2010). Darwin and the origin of interspecific genetic incompatibilities. *The American Naturalist, 176*(Suppl 1), S45.

Qin, X., Huang, Q., Xiao, H., Zhang, Q., Ni, C., Xu, Y., Liu, G., Yang, D., Zhu, Y., & Hu, J. (2016). The rice DUF1620-containing and WD40-like repeat protein is required for the assembly of the restoration of fertility complex. *The New Phytologist, 210*, 934–945.

Rausher, M. D. (2007). Comment on "Evolutionary paths underlying flower color variation in Antirrhinum". *Science, 315*(461), 461.

Rieseberg, L. H., & Blackman, B. K. (2010). Speciation genes in plants. *Annals of Botany, 106*, 439–455.

Schoof, H., Lenhard, M., Haecker, A., Mayer, K. F., Jurgens, G., & Laux, T. (2000). The stem cell population of *Arabidopsis* shoot meristems in maintained by a regulatory loop between the *CLAVATA* and *WUSCHEL* genes. *Cell, 100*, 635–644.

Schumer, M., Cui, R., Rosenthal, G. G., & Andolfatto, P. (2015). Reproductive isolation of hybrid populations driven by genetic incompatibilities. *PLoS Genetics, 11*, e1005041.

Scopece, G., Widmer, A., & Cozzolino, S. (2008). Evolution of postzygotic reproductive isolation in a guild of deceptive orchids. *The American Naturalist, 171*, 315–326.

Shen, R., Wang, L., Liu, X., Wu, J., Jin, W., Zhao, X., Xie, X., Zhu, Q., Tang, H., Li, Q., Chen, L., & Liu, Y. G. (2017). Genomic structural variation-mediated allelic suppression causes hybrid male sterility in rice. *Nature Communications, 8*, 1310.

Shi, M. (1985). The discovery and preliminary studies of the photoperiod-sensitive recessive male sterile rice (*Oryza sativa* L. subsp. *japonica*) (in Chinese). *Scientia Agricultura Sinica, 18*, 44–48.

Shi, C., Zhang, J., Wu, B., Jouni, R., Yu, C., Meyers, B. C., Liang, W., & Fei, Q. (2022). Temperature-sensitive male sterility in rice determined by the roles of *AGO1d* in reproductive phasiRNA biogenesis and function. *The New Phytologist, 236*, 1529–1544.

Skrede, I., Brochmann, C., Borgen, L., & Rieseberg, L. H. (2008). Genetics of intrinsic postzygotic isolation in a circumpolar plant species, *Draba nivalis* (*Brassicaceae*). *Evolution, 62*, 1840–1851.

Sobrizal, Y. M., & Yoshimura, A. (2002). Mapping of pollen semi-sterility gene, S28 (t), on rice chromosome 4. *Rice Genetics Newsletter, 19*, 80–82.

Sobrizal, Y. M., & Yoshimura, A. (2001). Mapping of a gene for pollen semi-sterility on chromosome 8 of rice. *Rice Genetics Newsletter, 18*, 59–61.

Sweigart, A. L., Fishman, L., & Willis, J. H. (2006). A simple genetic incompatibility causes hybrid male sterility in *mimulus*. *Genetics, 172*, 2465–2479.

Taft, R. J., Pang, K. C., Mercer, T. R., Dinger, M., & Mattick, J. S. (2010). Non-coding RNAs: Regulators of disease. *The Journal of Pathology, 220*, 126–139.

Tang, H., Luo, D., Zhou, D., Zhang, Q., Tian, D., Zheng, X., Chen, L., & Liu, Y. G. (2014). The rice restorer *Rf4* for wild-abortive cytoplasmic male sterility encodes a mitochondrial-localized PPR protein that functions in reduction of WA352 transcripts. *Molecular Plant, 7*, 1497–1500.

Tang, H., Zheng, X., Li, C., Xie, X., Chen, Y., Chen, L., Zhao, X., Zheng, H., Zhou, J., Ye, S., Guo, J., & Liu, Y. G. (2017). Multi-step formation, evolution, and functionalization of new cytoplasmic male sterility genes in the plant mitochondrial genomes. *Cell Research, 27*, 130–146.

Truernit, E., Stadler, R., Baier, K., & Sauer, N. (1999). A male gametophyte-specific monosaccharide transporter in *Arabidopsis*. *The Plant Journal, 17*, 191–201.

Tucker, M. R., Paech, N. A., Willemse, M. T., & Koltunow, A. M. (2001). Dynamics of callose deposition and beta-1,3-glucanase expression during reproductive events in sexual and apomictic Hieracium. *Planta, 212*, 487–498.

Ueda, K., Yoshimura, F., Miyao, A., Hirochika, H., Nonomura, K., & Wabiko, H. (2013). *Collapsed abnormal pollen1* gene encoding the Arabinokinase-like protein is involved in pollen development in rice. *Plant Physiology, 162*, 858–871.

Uyttewaal, M., Arnal, N., Quadrado, M., Martin-Canadell, A., Vrielynck, N., Hiard, S., Gherbi, H., Bendahmane, A., Budar, F., & Mireau, H. (2008). Characterization of *Raphanus sativus* pentatricopeptide repeat proteins encoded by the fertility restorer locus for ogura cytoplasmic male sterility. *The Plant Cell, 20*, 3331–3345.

Wan, C., Li, S., Wen, L., Kong, J., Wang, K., & Zhu, Y. (2007). Damage of oxidative stress on mitochondria during microspores development in Honglian CMS line of rice. *Plant Cell Reports, 26*, 373.

Wang, Z., Zou, Y., Li, X., Zhang, Q., Chen, L., Wu, H., Su, D., Chen, Y., Guo, J., Luo, D., Long, Y., Zhong, Y., & Liu, Y. G. (2006). Cytoplasmic male sterility of rice with boro II cytoplasm is caused by a cytotoxic peptide and is restored by two related PPR motif genes *via* distinct modes of mRNA silencing. *The Plant Cell, 18*, 676–687.

Wang, K., Gao, F., Ji, Y., Liu, Y., Dan, Z., Yang, P., Zhu, Y., & Li, S. (2013). ORFH79 impairs mitochondrial function *via* interaction with a subunit of electron transport chain complex III in Honglian cytoplasmic male sterile rice. *The New Phytologist, 198*, 408–418.

Wang, L., Shen, R., Chen, L. T., & Liu, Y. G. (2014). Characterization of a novel *DUF1618* gene family in rice. *Journal of Integrative Plant Biology, 56*, 151–158.

Wang, D., Li, J., Sun, L., Hu, Y., Yu, J., Wang, C., Zhang, F., Hou, H., Liang, W., & Zhang, D. (2021). Two rice MYB transcription factors maintain male fertility in response to photoperiod by modulating sugar partitioning. *The New Phytologist, 231*, 1612–1629.

Wang, C., Wang, J., Lu, J., Xiong, Y., Zhao, Z., Yu, X., Zheng, X., Li, J., Lin, Q., Ren, Y., Hu, Y., He, X., Li, C., Zeng, Y., Miao, R., Guo, M., Zhang, B., Zhu, Y., Zhang, Y., Tang, W., Wang, Y., Hao, B., Wang, Q., Cheng, S., He, X., Yao, B., Gao, J., Zhu, X., Yu, H., Wang, Y., Sun, Y., Zhou, C., Dong, H., Ma, X., Guo, X., Liu, X., Tian, Y., Liu, S., Wang, C., Cheng, Z., Jiang, L., Zhou, J., Guo, H., Jiang, L., Tao, D., Chai, J., Zhang, W., Wang, H., Wu, C., & Wan, J. (2023a). A natural gene drive system confers reproductive isolation in rice. *Cell, 186*, 3577–3592.

Wang, D., Wang, H., Xu, X., Wang, M., Wang, Y., Chen, H., Ping, F., Zhong, H., Mu, Z., Xie, W., Li, X., Feng, J., Zhang, M., Fan, Z., Yang, T., Zhao, J., Liu, B., Ruan, Y., Zhang, G., Liu, C., & Liu, Z. (2023b). Two complementary genes in a presence-absence variation contribute to *indica-japonica* reproductive isolation in rice. *Nature Communications, 14*, 4531.

Wu, M., Lin, J., Song, X., Ruan, G., & Li, C. (2007). Breeding of *indica-japonica* hybrid rice combination Chunyou 58 with super high yield (in Chinese). *Hybrid Rice, 22*, 17–19.

Wu, L., Jing, X., Zhang, B., Chen, S., Xu, R., Duan, P., Zou, Huang, S., Zhou, T., An, C., Luo, Y., & Li, Y. (2022). A natural allele of *OsMS1* responds to temperature changes and confers thermosensitive genic male sterility. *Nature Communications, 13*, 1–15.

Xie, Y., Xu, P., Huang, J., Ma, S., Xie, X., Tao, D., Chen, L., & Liu, Y. G. (2017a). Interspecific hybrid sterility in rice is mediated by *OgTPR1* at the *S1* locus encoding a peptidase-like protein. *Molecular Plant, 10*, 1137–1140.

Xie, Y., Niu, B., Long, Y., Li, G., Tang, J., Zhang, Y., Ren, D., Liu, Y. G., & Chen, L. (2017b). Suppression or knockout of *SaF/SaM* overcomes the *Sa*-mediated hybrid male sterility in rice. *Journal of Integrative Plant Biology, 59*, 669–679.

Xie, Y., Tang, J., Xie, X., Li, X., Huang, J., Fei, Y., Han, J., Chen, S., Tang, H., Zhao, X., Tao, D., Xu, P., Liu, Y. G., & Chen, L. (2019). An asymmetric allelic interaction drives allele transmission bias in interspecific rice hybrids. *Nature Communications, 10*, 2501.

Xu, D., Wang, X., Ma, R., Yu, S., Zhu, L., Ouyang, Y., & Jin, Q. (2010). Analysis on physiological properties of the heavy panicletype of *indica-japonica* inter-subspecific hybrid rice Yongyou 6 (in Chinese). *Scientia Agricultura Sinica, 43*, 4796–4804.

Yamagata, Y., Yamamoto, E., Aya, K., Win, K. T., Doi, K., Ito, T., Kanamori, H., Wu, J. Z., Matsumoto, T., & Matsuoka, M. (2010). Mitochondrial gene in the nuclear genome induces reproductive barrier in rice. *Proceedings of the National Academy of Sciences of the United States of America, 107*, 1494–1499.

Yamamoto, M. P., Shinada, H., Onodera, Y., Komaki, C., & Mikami, T. (2008). A male sterility-associated mitochondrial protein in wild beets causes pollen disruption in transgenic plants. *The Plant Journal, 54*, 1027–1036.

Yanagihara, S., Kato, H., & Ikehashi, H. (1992). A new locus for multiple alleles causing hybrid sterility between an *aus* variety and *javanica* varieties in rice (*Oryza sativa* L.). *Japanese Journal of Breeding, 42*, 793–801.

Yang, R., Li, W., Wang, N., Liang, K., & Chen, Q. (1989). Discovery and preliminary study on *indica* photosensitive genic male-sterile germplasm 5460ps (in Chinese). *Chinese Journal of Rice Science, 3*, 47–48.

Yang, S. L., Xie, L. F., Mao, H. Z., Puah, C. S., Yang, W. C., Jiang, L., Sundaresan, V., & Ye, D. (2003). Tapetum determinant1 is required for cell specialization in the *Arabidopsis* anther. *The Plant Cell, 15*, 2792–2804.

Yang, C., Chen, Z., Zhuang, C., Mei, M., & Liu, Y. (2004). Genetic and physical fine-mapping of the *Sc* locus conferring *indica-japonica* hybrid sterility in rice (*Oryza sativa* L.). *Chinese Science Bulletin, 49*, 1718–1721.

Yang, S. L., Jiang, L., Puah, C. S., Xie, L. F., Zhang, X. Q., Chen, L. Q., Yang, W. C., & Ye, D. (2005). Overexpression of *TAPETUM DETERMINANT1* alters the cell fates in the *Arabidopsis* carpel and tapetum *via* genetic interaction with *excess microsporocytes1/extra sporogenous cells. Plant Physiology, 139*, 186–191.

Yang, J. H., Liu, X. Y., Yang, X. D., & Zhang, M. F. (2010). Mitochondrially-targeted expression of a cytoplasmic male sterility-associated *orf220* gene causes male sterility in *Brassica juncea. BMC Plant Biology, 10*, 231.

Yang, J., Zhao, X., Cheng, K., Du, H., Ouyang, Y., Chen, J., Qiu, S., Huang, J., Jiang, Y., Jiang, L., Ding, J., Wang, J., Xu, C., Li, X., & Zhang, Q. (2012). A killer-protector system regulates both hybrid sterility and segregation distortion in rice. *Science, 337*, 1336–1340.

Ylstra, B., Garrido, D., Busscher, J., & van Tunen, A. J. (1998). Hexose transport in growing petunia pollen tubes and characterization of a pollen-specific, putative monosaccharide transporter. *Plant Physiology, 118*, 297–304.

You, S., Zhao, Z., Yu, X., Zhu, S., Wang, J., Lei, D., Zhou, J., Li, J., Chen, H., Xiao, Y., Chen, W., Wang, Q., Lu, J., Chen, K., Zhou, C., Zhang, X., Cheng, Z., Guo, X., Ren, Y., Zheng, X., Liu, S., Liu, X., Tian, Y., Jiang, L., Tao, D., Wu, C., & Wan, J. (2023). A toxin-antidote system contributes to interspecific reproductive isolation in rice. *Nature Communications, 14*, 7528.

Yu, Y., Zhao, Z., Shi, Y., Tian, H., Liu, L., Bian, X., Xu, Y., Zheng, X., Gan, L., Shen, Y., Wang, C., Yu, X., Wang, C., Zhang, X., Guo, X., Wang, J., Ikehashi, H., Jiang, L., & Wan, J. (2016). Hybrid sterility in rice (*Oryza Sativa* L.) involves the tetratricopeptide repeat domain containing protein. *Genetics, 203*, 1439–1451.

Yu, J., Han, J., Kim, Y. J., Song, M., Yang, Z., He, Y., Fu, R., Luo, Z., Hu, J., Liang, W., & Zhang, D. (2017). Two rice receptorlike kinases maintain male fertility under changing temperatures. *Proceedings of the National Academy of Sciences of the United States of America, 114*, 12327–12332.

Yu, X., Zhao, Z., Zheng, X., Zhou, J., Kong, W., Wang, P., Bai, W., Zheng, H., Zhang, H., Li, J., Liu, J., Wang, Q., Zhang, L., Liu, K., Yu, Y., Guo, X., Wang, J., Lin, Q., Wu, F., Ren, Y., Zhu, S., Zhang, X., Cheng, Z., Lei, C., Liu, S., Liu, X., Tian, Y., Jiang, L., Ge, S., Wu, C., Tao, D., Wang, H., & Wan, J. (2018). A selfish genetic element confers non-Mendelian inheritance in rice. *Science, 360*, 1130–1132.

Yuan, L. (1987). A tentative plan for the breeding of hybrid rice (in Chinese). *Hybrid Rice, 1*, 1–3.

Yuan, S., Zhang, Z., & Xu, C. (1988). Studies on the critical stage of fertility change induced by light and its phase development in HPGMR (in Chinese). *Acta Agronomica Sinica, 14*, 7–13.

Zhai, J., Zhang, H., Arikit, S., Huang, K., Nan, G. L., Walbot, V., & Meyers, B. C. (2015). Spatiotemporally dynamic, cell-type-dependent premeiotic and meiotic phasiRNAs in maize anthers. *Proceedings of the National Academy of Sciences of the United States of America, 112*, 3146–3151.

Zhang, Q., Shen, B. Z., Dai, X. K., Mei, M. H., Saghai, M. M., & Li, Z. B. (1994). Using bulked extremes and recessive class to map genes for photoperiod-sensitive genic male sterility in rice. *Proceedings of the National Academy of Sciences of the United States of America, 91*, 8675–8679.

Zhang, G., Lu, Y., Bharaj, T. S., Virmani, S. S., & Huang, N. (1997). Mapping of the *Rf-3* nuclear fertility-restoring gene for WA cytoplasmic male sterility in rice using RAPD and RFLP markers. *Theoretical and Applied Genetics, 94*, 27–33.

Zhang, Q. Y., Liu, Y. G., Zhang, G. Q., & Mei, M. T. (2002). Molecular mapping of the fertility restorer gene *Rf-4* for WA cytoplasmic male sterility in rice. *Acta Genetica Sinica, 29*, 1001–1004.

Zhang, P., Ding, J., & Zhang, Q. (2010a). The leaves and sites for inducing fertility change by photoperiod in photoperiod-sensitive genic male sterile rice (in Chinese). *Molecular Plant Breeding, 8*, 641–646.

Zhang, H., Liang, W., Yang, X., Luo, X., Jiang, N., Ma, H., & Zhang, D. (2010b). *Carbon starved anther* encodes a MYB domain protein that regulates sugar partitioning required for rice pollen development. *The Plant Cell, 22*, 672–689.

Zhang, H., Xu, C., He, Y., Zong, J., Yang, X., Si, H., Sun, Z., Hu, J., Liang, W., & Zhang, D. (2013). Mutation in *CSA* creates a new photoperiod-sensitive genic male sterile line applicable for hybrid rice seed production. *Proceedings of the National Academy of Sciences of the United States of America, 110*, 76–81.

Zhang, H., Chen, X., Huang, J., E, Z., Gong, J., & Shu, Q. (2015). Identification and transition analysis of Photo-/Thermo-sensitive genic male sterile genes in two-line hybrid rice in China. *Scientia Agricultura Sinica, 48*, 1–9.

Zhang, Z., Ding, Z., Feng, X., Huang, J., Peng, X., Xiao, Y., Zong, W., Zhao, Z., Liu, Y-G., Xie, Y., & Chen, L. (2025). Ubiquitin-mediated degradation restricts spatiotemporal accumulation of the cytoplasmic male sterility protein WA352 to anthers in rice. *Proceedings of the National Academy of Sciences, 122*, e2504381122.

Zhao, D. Z., Wang, G. F., Speal, B., & Ma, H. (2002). The *excess microsporocytes1* gene encodes a putative leucine-rich repeat receptor protein kinase that controls somatic and reproductive cell fates in the *Arabidopsis* anther. *Genes & Development, 16*, 2021–2031.

Zhao, X., de Palma, J., Oane, R., Gamuyao, R., Luo, M., Chaudhury, A., Herve, P., Xue, Q., & Bennett, J. (2008). OsTDL1A binds to the LRR domain of rice receptor kinase MSP1, and is required to limit sporocyte numbers. *The Plant Journal, 54*, 375–387.

Zhao, Z., Ding, Z., Huang, J., Meng, H., Zhang, Z., Gou, X., Tang, H., Xie, X., Ping, J., Xiao, F., Liu, Y. G., Xie, Y., & Chen, L. (2023). Copy number variation of the restorer *Rf4* underlies human selection of three-line hybrid rice breeding. *Nature Communications, 14*, 7333.

Zhou, H., Liu, Q., Li, J., Jiang, D., Zhou, L., Wu, P., Lu, S., Li, F., Zhu, L., Liu, Z., Chen, L., Liu, Y. G., & Zhuang, C. (2012). Photoperiod- and thermo-sensitive genic male sterility in rice are caused by a point mutation in a novel noncoding RNA that produces a small RNA. *Cell Research, 22*, 649–660.

Zhou, H., Zhou, M., Yang, Y., Li, J., Zhu, L., Jiang, D., Dong, J., Liu, Q., Gu, L., & Zhou, L. (2014). RNase Z(S1) processes UbL40 mRNAs and controls thermosensitive genic male sterility in rice. *Nature Communications, 5*, 4884.

Zhou, H., He, M., Li, J., Chen, L., Huang, Z., Zheng, S., Zhu, L., Ni, E., Jiang, D., Zhao, B., & Zhuang, C. (2016). Development of commercial thermo-sensitive genic male sterile rice accelerates hybrid rice breeding using the CRISPR/Cas9-mediated *TMS5* editing system. *Scientific Reports, 6*, 37395.

Zhou, P., Wang, Z., Zhu, X., Tang, Y., Ye, L., Yu, H., Li, Y., Zhang, N., Liu, T., Wang, T., Wu, Y., Cao, D., Chen, Y., Li, X., Zhang, Q., Xiao, J., Yu, S., Zhang, Q., Mi, J., & Ouyang, Y. (2023). A minimal genome design to maximally guarantee fertile inter-subspecific hybrid rice. *Molecular Plant, 16*, 726–738.

Zhu, X., Liang, W., Cui, X., Chen, M., Yin, C., Luo, Z., Zhu, J., Lucas, W. J., Wang, Z., & Zhang, D. (2015). Brassinosteroids promote development of rice pollen grains and seeds by triggering expression of Carbon Starved Anther, a MYB domain protein. *The Plant Journal, 82*, 570–581.

Chapter 16
Genetic and Molecular Regulation of Self-Incompatibility in Flowering Plants

Hong Zhao, Yanzhai Song, Yue Zhang, and Yongbiao Xue

Abstract Self-incompatibility (SI) is an intraspecific reproductive barrier widely found in flowering plants to prevent self-fertilization but to promote outcrossing. SI is usually controlled by a single polymorphic *S*-locus. Since the *S*-loci from several families encode different molecular components regulating SI, they are typically classified into five types (type-1 ~ −5) with type-1 originated from the common ancestor of eudicots and type-2 ~ −5 evolved following the loss of the type-1 *S*. Both self- and non-self-recognition systems are associated with SI. In self-recognition systems, pollen protein Prp S (*Papver rhoeas* pollen S) interacts with its cognate stigma Prs S (*Papver rhoeas* stigmatic S) to initiate a signaling pathway leading to programmed self-pollen death (Papaveraceae, type-3), or pollen protein SCR/SP11 (*S*-locus cystein-rich/S Pollen 11) interacts with its cognate stigma SRK (*S*-locus Receptor-like Kinase) leading to destruction of proposed pollen compatibility factors resulting in self-pollen rejection (Brassicaceae, type-2). In non-self-recognition system found in the Solanaceae, Plantaginaceae, Rosaceae, and Rutaceae (type-1), pollen-expressed multiple SLFs (*S*-locus F-box) form functional SCF complexes to degrade non-self S-RNases allowing cross-pollen growth. Recently, in dimorphic *Primula* (type-4) and *Turnera* (type-5), the *S*-locus encodes several hemizygous genes to control SI, but little is known about their roles in self-pollen rejection. Together, these findings indicate that flowering plants have adopted multiple mechanisms to inhibit self-pollen growth but allow cross-pollen fertilization. In this chapter, we highlight our knowledge about the genetic and molecular regulation of SI responses in flowering plants and several future challenges in addressing their detailed molecular mechanisms and evolution.

Keywords SI · Heterostyly · Papaveraceae · Brassicaceae · Solanaceae · Rosaceae · Plantaginaceae · Rutaceae · *Primula* · S-RNase · SLF · SCR/SP11 · SRK · PrpS · PrsS

H. Zhao · Y. Song · Y. Zhang · Y. Xue (✉)
Institute of Genetics and Developmental Biology and Beijing Institute of Genomics, Chinese Academy of Sciences, Beijing, China
e-mail: ybxue@genetics.ac.cn

F. Chang et al. (eds.), *Regulation of Plant Development*,
https://doi.org/10.1007/978-981-95-7033-1_16

16.1 Introduction

Angiosperms, also known as the flowering plants, are the most abundant and diverse group in the plant kingdom, and most of them have a bisexual flower. Hermaphrodite flowers have the strong tendency to self-fertilize, but this would lead to inbreeding depression. In many species, self-incompatibility (SI) is adopted to prevent self-fertilization and maintain genetic diversity (de Nettancourt, 2001; Franklin-Tong, 2008). As an intraspecific reproductive barrier, the SI system is controlled by rejection of self-pollen and acceptance of pollen from the plants unrelated genetically. In most cases, the recognition between pollen and pistil is determined by a single multi-allelic *S*-locus encoding the pollen *S* and the pistil *S* (Takayama & Isogai, 2005). The SI systems are generally classified into gametophytic SI (GSI) or sporophytic SI (SSI). This is based on whether the pollen behavior in SI response is determined by its own pollen haploid genotype or the diploid genotype of mother plant. According to the molecular differences, five types of SI have been defined recently, including type-1 controlled by the pistil *S S-RNase* and pollen *S SLF*s commonly spread in Plantaginaceae, Solanaceae, Rosaceae, and Rutaceae, type-2 by *SRK* and *SCR* in Brassicaceae, type-3 by *PrsS* and *PrpS* in Papaveraceae, type-4 by *CYP-GLO2-KFB-CCM-PUM* in Primulaceae, type-5 by *TsSPH1-TsYUC6-TsBAHD* in Turneraceae, with type-1 characterized as a non-self-recognition system but types 2–5 self-ones (Zhao et al., 2022). Furthermore, remarkable progresses have been made in their origin and evolutionary mechanisms recently. Among them, type-1 SI possessed a single origin in the most recent common ancestor (MRCA) of eudicots and types 2–5 dynamically evolved following its losses (Igić & Kohn, 2001; Liang et al., 2020; Ramanauskas & Igić, 2017; Vieira et al., 2008; Xue et al., 1996; Zhao et al., 2022; Zhu et al., 2023). In this chapter, we highlight recent major progress in the genetic and molecular control of several SI species in angiosperms (Fig. 16.1).

16.2 *Papaver* SI

SI in *Papaver rhoeas* has the features of GSI, which occurs when the pistil of a plant is not able to receive pollen of the same plant related genetically but that of any other plant of the same species unrelated genetically. In this species, SI is controlled by a single polymorphic *S*-locus containing one pistil *S* and one pollen *S*. The self-recognition system, when the pollen S protein encoded by an *S*-haplotype interacts with the pistil S protein encoded by the same *S*-haplotype, triggers self-incompatible response resulting in inhibition of self-pollen (Fig. 16.2). Recent studies of SI in *Papaver* have focused mainly on the signal and mechanism related to inhibition of incompatible pollen tubes, including a calcium signal pathway, phosphorylation of a soluble inorganic pyrophosphatase, depolymerization of actin cytoskeleton, and programmed cell death (PCD).

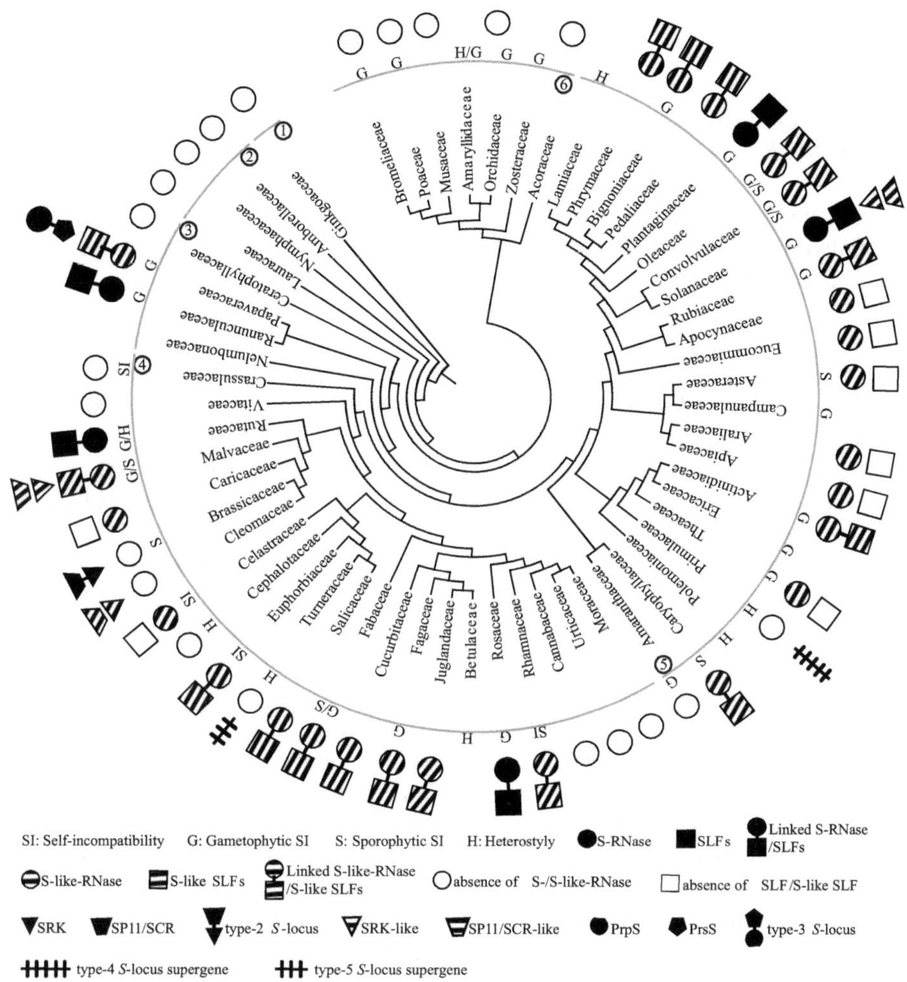

Fig. 16.1 Distribution of SI systems and their associated molecules in angiosperms. (Modified from Zhao et al., 2022). The phylogenetic family tree is derived from TimeTree website and APG (Angiosperm Phylogeny Group classification) IV. Numbers 1–6 marked with black circles indicate the gymnosperms (Ginkgoaceae), the broad families of the angiosperms, and the Lauraceae of Magnolia, monocots, basal dicots, superrosids, and superasterids, respectively. G, S, and H are gametophytic, sporophytic, and heterostyly SI. The symbols depicting the *S* genes and *S*-like genes are shown on the bottom

16.2.1 Genetics of SI in Papaver

P. rhoeas (the field poppy) is a species possessing SI, first reported by Hoffman in 1878 (East, 1940). The pistil of *P. rhoeas* has no style and is composed of a cap-shaped stigma containing 10 to 14 rays of papillae. The pollen released from the anther lands on the dry stigma surface and forms a visible "foot." In the "foot," the

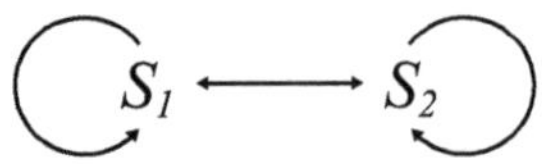

Fig. 16.2 Self- and non-self-recognition SI systems. Arrowed circles and a line with two arrows indicate self (S_1 or S_2 haplotype) and non-self (S_1 and S_2 haplotypes) recognition, respectively

secretion of the pollen coat stimulates the cuticle separated from papilla cell wall. After hydration and germination of the pollen, the pollen tubes penetrate the cuticle beneath the "foot" and grow into large subcuticular space below the cuticle along the direction of the stigmatic ray (Elleman et al., 1992). In this process, the incompatible pollen is inhibited after germination and during the growth of pollen tubes, whereas the compatible pollen normally germinates and grows. Because of no style in *Papaver*, SI response and incompatible pollen inhibition occur on the stigmatic surface very rapidly.

Genetic studies on *Papaver* SI began by M. Lawrence in the 1970s. Early studies showed that most plants possessing GSI have binuclear pollen and that reciprocal pollinations between members of families of these plants are successful (Brewbaker, 1957). Moreover, other studies suggested that inhibition of incompatible pollen occurs on stigma for most plants that have SSI. It is conflicting that plants in *Papaver* have both gametophytic and sporophytic features. In order to solve this problem, Lawrence used aniline blue staining to investigate progenies of reciprocal pollinations between individuals in families of *P. rhoeas* and found that SI in *Papaver* is controlled by a single gene (Lawrence, 1975). In addition, half-compatible pollinations containing compatible and incompatible pollen tubes were identified in reciprocal pollinations, which suggest that the SI system of *Papaver* is of gametophytic instead of sporophytic nature (Lawrence, 1975). In the study of population genetics in *Papaver*, he investigated the cross-pollination phenotypes between full-sib families made from several natural populations and classified these pollinations into incompatible, half-compatible, and full-compatible (Lawrence & O'Donnell, 1981; Lawrence et al., 1993). These results indicated that a large number of *S*-haplotypes appear in the artificial families and that the frequencies of *S*-alleles are significantly unequal.

16.2.2 The Pistil S *Determinant in* Papaver

The pistil S protein candidate of *P. rhoeas* was identified by Frankin-Tong's group in 1994. Based on in vitro bioassays, crude stigma extracts of *P. rhoeas* carrying an *S*-haplotype can inhibit growth of pollen tubes containing the same *S*-haplotype and were separated into two components using purification of ConA, including ConA-bound and ConA-unbound fractions. Only ConA-bound fraction has inhibition activity for incompatible pollen (Franklin-Tong et al., 1988, 1989). This suggests that the pistil S protein is a glycoprotein and expressed in stigma. Through isoelectric focusing gel (IEF), a less abundant glycoprotein isolated from ConA-bound fraction was a candidate for the pistil S protein because of stigma-specificity and inhibitory activity. Subsequently, Foote et al. (1994) isolated the stigmatic proteins from over 260 plants of *P. rhoeas* and reported that they co-segregate with the S_1 allele (Foote et al., 1994). The IEF results revealed that the stigmatic proteins have two isoforms with pIs of 7.55 and 6.9, which were designated S_{1a} and S_{1b}, respectively. After SDS-PAGE for the two isoforms, it was clearly visible that S_{1a} protein was separated into two bands with masses of 16.7 and 14.7 kDa, while masses of the two bands of S_{1b} protein were 16.8 and 14.8 kDa. N-terminal 12–17 amino acids of the four proteins were analyzed by N-terminal protein sequencing and found to be identical. Using an oligonucleotide probe based on the N-terminal sequence, the pistil S_1 was cloned by screening cDNA library and had a 417 bp open reading frame encoding a 14.5 kDa protein of 120 amino acids (Foote et al., 1994). The sequence analysis revealed that the stigmatic S_1 protein has several features, including putative signal peptide, potential N-glycosylation site, glycine at the cleavage site, and a central hydrophobic region. These results demonstrate that it is a secreted and hydrophobic glycoprotein. Subsequently, several pistil *S* genes from different *S*-haplotypes of *Papaver* were identified, which showed a high degree of amino acid sequence polymorphism (51.3%–63.7%) among proteins encoded by these genes sharing similar secondary structures (Kurup et al., 1998; Walker et al., 1996).

The pistil *S* gene in *Papaver* is of stigma-specific expression and a single copy, while it has no highly homologous gene in other species. Using *Arabidopsis* genomic sequence, however, it was found that the *Papaver* pistil *S* gene has low homology to a large gene family of *Arabidopsis* named *SPH* (*S-protein homolog*) (Ride et al., 1999). Additionally, the *Papaver SI-like* (*PSIL*) genes in *Arabidopsis* have homology to the *Papaver* pistil S_1 gene (Jones-Rhoades et al., 2007). To investigate the function of the pistil *S* genes, because of the difficulty of genetic transformation in *Papaver*, they were expressed in *Escherichia coli,* producing the recombinant protein. Through in vitro bioassay, the growing pollen tubes were treated with stigmatic extracts or recombinant stigmatic S protein. It is certain that the recombinant proteins specifically inhibit the growth of incompatible pollen tubes of the same *S*-haplotype, whereas the heated recombinant protein has no inhibitory activity (Foote et al., 1994). This suggests that these genes play a role of the pistil *S* determinant in Papaver and that the inhibitory activity is determined by the stigmatic S protein rather than glycosylation.

The secondary structure of the stigmatic S protein was predicted to be composed of six β strands with two α helices and seven hydrophilic loops located at the surface. Based on the predictions, two studies analyzed the amino acid residues of the stigmatic S proteins that are crucial for the *S*-haplotype-specific recognition and inhibition of incompatible pollen (Jordan et al., 1999; Kakeda et al., 1998). Using site-directed mutagenesis, mutant derivatives of the stigmatic S protein were expressed in *E. coli* and the inhibitory activity for incompatible pollen was investigated (Kakeda et al., 1998). The experiments revealed that variable and conserved amino acid residues of stigmatic S protein in loop 6 predicted at hydrophilic surface have a critical function for the recognition and inhibition of incompatible pollen. The pistil *S* gene of *P. rhoeas* was named as *PrsS* (*Papaver rhoeas* stigmatic *S* determinant).

16.2.3 The Pollen S *Determinant in* Papaver

After the identification of the pistil *S* determinant, it has taken a long time to isolate the pollen *S* determinant in *Papaver*. A 70–120 kDa glycoprotein expressed specifically in mature pollen of *P. rhoeas*, designed SBP (S protein-binding protein), was initially identified through western ligand blotting and located on the pollen plasma membrane (Hearn et al., 1996). Nevertheless, an *S*-allele specificity, which is the feature of the pollen S protein, does not appear in SBP, whereas SBP can bind with the stigmatic S proteins of several S-haplotypes. Thus, it was suggested that SBP could not be the authentic pollen S receptor but an accessory receptor (Hearn et al., 1996). Subsequently, in order to test the function of the interaction between SBP and Prs S, several mutant derivatives of Prs S_1 containing the mutations of amino acid residue in loops 2 and 6 weakened the interaction between SBP and Prs S and reduced the inhibition of incompatible pollen, which indicates that the role of SBP binding with Prs S is very essential for the rejection of self-pollen (Jordan et al., 1999). Based on these studies, it was suggested that SBP is the key pollen *S* component rather than the pollen *S* determinant.

Based on the linkage of the pollen *S* and pistil *S*, through the screening of a cosmid genomic DNA library of *P. rhoeas*, an open reading frame that is 475 bp from the pistil S_1 gene was found by sequencing of a 42 kb clone containing the S_1-locus (Wheeler et al., 2009). This gene, named *Prp* S_1 (*Papaver rhoeas* pollen S_1), has a CDS region of 579 bp encoding a 20.5 kDa protein and displays the characteristic of pollen-specific expression, while *Prp* S_3 and *Prp* S_8 were cloned from RNA of S_3S_8 *P. rhoeas*. The analysis of *Prp S* using Southern blotting indicated that this gene is single copy in genome, and that *Prp* S_3 and *Prp* S_8 are allelic to *Prp* S_1. Based on amino acid sequence analyses of the Prp S proteins between different *S*-haplotypes, it has been shown that these proteins have a high level of polymorphism, which is similar to Prs S. Additionally, the prediction of 3–5 transmembrane helices implied that the pollen S protein could be a transmembrane protein, which was determined by immunolocalization. Through the investigation of two full-sib families carrying the different *S*-haplotypes, both *Prs S* and *Prp S* containing the same *S*-allele are closely linked (Wheeler et al., 2009).

In order to test the biological activity of *Prp S*, the silencing expression of *Prp S* via the antisense oligonucleotide approach results in a strong reduction of the inhibition of incompatible pollen tubes through in vitro bioassay (Wheeler et al., 2009). Thus, it was suggested that *Prp S* is the pollen *S* determinant of *Papaver* and is involved in the *S*-allelic-specific recognition with Prs S and activation of the SI response. Moreover, when *Prp S* of *P. rhoeas* was transformed to a self-compatible plant of *Arabidopsis thaliana* that lost SI system, the transgenic *A. thaliana* pollen expressing Prp S protein was inhibited by the *Papaver* recombinant Prs S of the same *S*-haplotype using in vitro bioassay, which indicates that the pollen-specific expression of *Prp S* in *A. thaliana* induces the response and downstream signal of SI in other species (de Graaf et al., 2012). Thus, the result implies a possibility that utility of *Prp S* and *Prs S* might reconstruct SI in species possessing no SI system. To examine this hypothesis, the *S* determinants (*Prp S* and *Prs S*) of *P. rhoeas* were co-expressed in a compatible *A. thaliana* plant. The transgenic plants containing an artificial *S*-locus had no seed after self-pollination and displayed strong SI (Lin et al., 2015). This suggests that the transgenic stigma expressing the Prs S of an *S*-haplotype specifically rejects the transgenic pollen carrying cognate Prp S, because of the recognition and interaction of cognate Prs S and Prp S in vivo. Meanwhile, it is found that the *Papaver S* determinants have a function of inducing SI in vivo and an ability of establishing SI system in compatible divergent species, which provides a method for cross-breeding.

16.2.4 Ca^{2+} Signaling in Papaver *SI*

As a second messenger, Ca^{2+}-mediated signal transduction pathway regulates various biological processes in plants, particularly in pollen germination and pollen tube growth. Moreover, because of the characteristics of the stigmatic S protein, a small secreted protein, it was speculated that the stigmatic S protein could be a ligand with regard to the signaling of pollen tubes under SI stimulation. Thus, there have been several studies involved in Ca^{2+} signaling induced by SI response in *Papaver*. To analyze $[Ca^{2+}]_i$ (cytosolic free calcium) change in pollen tubes during SI response using Ca^{2+}-sensitive dye, one study revealed that a transient rise of the $[Ca^{2+}]_i$ was detectable by treatment of incompatible stigmatic S glycoproteins in pollen tubes, whereas the level of $[Ca^{2+}]_i$ had no change in pollen tubes after addition of either compatible or heated incompatible stigmatic S glycoproteins (Franklin-Tong et al., 1993). This result suggested that the $[Ca^{2+}]_i$ rise is specifically accompanied by the incompatible reaction and involved in the inhibition of incompatible pollen tube growth. Furthermore, to test this proposal, two experiments showed that the recombinant stigmatic S protein induces the elevation of $[Ca^{2+}]_i$ in the incompatible pollen tubes, and that artificial increasing in the level of $[Ca^{2+}]_i$ led to the inhibition of incompatible pollen tubes growth, both of which indicate that calcium signaling mediates SI response in *Papaver* (Franklin-Tong et al., 1993, 1995). In addition, Ca^{2+} influx emerged by the increase of intracellular free Ca^{2+} appears in the shank of incompatible pollen tubes nearby the nuclear region, though it occurs in the tip of normally growing pollen tubes (Franklin-Tong et al., 1997, 2002). Therefore, it is obvious that Ca^{2+} influx stimulated by SI response comes from extracellular Ca^{2+}. Since Ca^{2+} influx could act as a downstream signal of SI response produced by a ligand-receptor interaction between the *S* determinants, it was speculated that Prp S as the pollen S protein plays a role of a channel in the shank of pollen tubes. Subsequently, a recent study, using whole cell patch clamping of pollen protoplasts, showed that Prs S triggers a conductance in incompatible pollen grain plasma membrane but not in compatible pollen grains, which suggests that SI reaction in *Papaver* induces a nonspecific cation conductance permeable to Ca^{2+} and K^+ (Wu et al., 2011). It is proposed that the conductance activated by SI may be produced by a nonspecific cation channel. In summary, it is clear that Ca^{2+}-dependent signal transduction pathway mediates SI in *Papaver*, which is an early event in SI response.

16.2.5 Protein Phosphorylation in Papaver *SI*

Generally, activation of Ca^{2+}-dependent signal transduction pathway may change protein kinase activity and subsequently produce the phosphorylation of its substrates. To test SI-induced protein phosphorylation in incompatible pollen tubes of *P. rhoeas*, a study revealed that a phosphoprotein with a molecular mass of 26 kDa was accumulated when pollen was treated by incompatible stigmatic S protein but not when pollen was treated by compatible or heated incompatible stigmatic S protein. The protein named Pr-p26.1 was located in soluble fractions through high-speed differential centrifugation and phosphorylated in an *S*-dependent manner (Rudd et al., 1996). The study suggested that the phosphorylation of Pr-p26.1 involves in SI response in *P. rhoeas*. A decade later, two genes, designed as *Pr-p26.1a* and *Pr-p26.1b*, were cloned through the complementary DNA encoding Pr-p26.1 (de Graaf et al., 2006). Both amino acid sequences of Pr-p26.1a and Pr-p26.1b have 78% identity and high homology to Family I sPPases (soluble inorganic pyrophosphatases) that possess a function of hydrolysis of inorganic pyrophosphate (PPi) to provide a driving force maintaining metabolism. Pr-p26.1a and Pr-p26.1b were specifically expressed in mature pollen using western blotting and detected in cytosolic through immunolocalization and green fluorescent protein (GFP) fusion protein. Moreover, sPPase activity of both recombinant Pr-p26.1a and Pr-p26.1b is Mg^{2+}-dependent and inhibited by Ca^{2+} and phosphorylation. Thus, it is proposed that sPPase activity of Pr-p26.1 is inhibited by Ca^{2+}-dependent phosphorylation under SI induction. To confirm the function of Pr-p26.1 in pollen tubes, using antisense oligonucleotide method to silence the expression of Pr-p26.1, the inhibition of pollen tube growth was observed when pollen was treated with antisense oligonucleotides related to *Pr-p26.1a* and *Pr-p26.1b*, but not when pollen was treated with the corresponding sense oligonucleotides. This showed that Pr-p26.1 plays a key role of pollen tube growth. Because of biochemical function of sPPase, to test whether sPPase activity of Pr-p26.1 is involved in SI stimulation, the level of PPi was detected after SI stimulation. As expected, the concentration of PPi in incompatible pollen tubes was significantly increased by SI stimulation, which is the metabolic association between sPPase activity and SI response. Pr-p26.1 as an early target of the SI-induced Ca^{2+} signal is critical for SI mechanism. Thus, the study on the phosphorylation of Pr-p26.1 contributes to the regulation mechanism of SI.

Recently, one study made great progress with regard to the phosphorylation of Pr-p26.1 (Eaves et al., 2017). The phosphorylation sites of Pr-p26.1a/b were mapped by liquid chromatographic tandem mass spectrometry (LC-MS) when Pr-p26.1a/b was catalyzed by endogenous pollen kinase or recombinant Ca^{2+}-dependent protein kinases (CPKs). Meanwhile, the results showed that Pr-p26.1a/b were substrates of three *P. rhoeas* CPKs expressed in pollen, including PrCPK14, PrCPK6/26, and PrCPK17/34. To investigate whether the phosphorylation sites of Pr-p26.1a/b are involved in the biological event of SI, this study determined if phosphomimic substitutions of the phosphorylated amino acid alter the sPPase activities. The data revealed that these phosphomimic mutants could not alter the sPPase activities under standard conditions in vitro. Nevertheless, sPPase activities of the phosphomimic mutants were reduced significantly and differentially at low pH and in the presence of Ca^{2+} and H_2O_2. Several studies suggested that SI response induced Ca^{2+} and ROS signaling and eventually triggered programmed cell death (PCD) involved in dramatic acidification of incompatible pollen tube cytosol (Franklin-Tong et al., 1993; Wilkins et al., 2011, 2015). Therefore, it was indirectly demonstrated that the key phosphorylation sites of Pr-p26.1 are crucial for modulating its sPPase activity and related to the biological events of SI reaction.

Additionally, another phosphoprotein of *P. rhoeas* pollen was found at 240 and 400 s after SI stimulation, whose phosphorylation was increased in incompatible pollen tubes but not in compatible pollen tubes (Rudd et al., 1997). The phosphoprotein with a molecular mass of 68 kDa, designated as p68, was phosphorylated at the latter stage of SI response. In details, different from the phosphorylation of p26.1 enhancing within 90 s of SI induction and further increasing with 400 s, the phosphorylation of p68 was strikingly increased between 240 and 400 s of the addition of stigmatic S protein. Moreover, p68 was phosphorylated by Ca^{2+}-independent manner under SI response. There are two key suggestions for the SI-induced protein phosphorylation. On the one hand, the phosphorylation of p68 appears to occur downstream of p26.1 in SI-mediated signaling. On the other hand, two different signal transduction pathways, Ca^{2+}-dependent and Ca^{2+}-independent signaling, may mediate the protein phosphorylation in the initial and later stages of SI response, respectively.

16.2.6 Cytoskeleton Dynamics in Papaver *SI*

Cytoskeleton in eukaryotes commonly contains three types: microfilaments, microtubules, and intermediate filaments. They are the main target of signal transduction in plant cells and play an important role in pollen tube growth. In *P. rhoeas*, actin cytoskeleton after SI induction changes dramatically in incompatible pollen tubes but not in compatible pollen tubes, which was divided into several distinct phases (Geitmann et al., 2000). First, within 1 min of SI induction, the subapical actin mesh of pollen tube loses its original structure and is relocated in the apex. In this process, the actin bundles are rapidly disintegrated. Subsequently, the actin filaments in the shank of pollen tube become diffused and marginalized; meanwhile, small punctate foci composed of small actin structure begin to appear. Then, the longitudinal actin bundles fragment and are invisible with large punctate foci forming gradually. Finally, at approximately 3 h after SI induction, punctate foci of actin are distributed in the entire pollen tube. At this point, actin cytoskeleton of incompatible pollen tubes is rearranged, which suggests that actin cytoskeleton in incompatible pollen tubes occurs during SI response (Geitmann et al., 2000). In addition, an experiment, the inhibition of pollen tube growth in nonspecific manner, cannot give rise to the rearrangements of actin cytoskeleton in pollen tubes, which demonstrates that the rearrangements as a consequence of SI induction are not resulting from the inhibition of pollen tube growth (Geitmann et al., 2000).

To measure alterations of F-actin during the SI stimulation quantitatively, one study revealed that F-actin depolymerization was sustained for at least 1 h under the SI response (Snowman et al., 2002). Furthermore, artificial increasing of $[Ca^{2+}]_i$ results in the depolymerization of F-actin in pollen tubes, which indicates that actin cytoskeleton is a target of Ca^{2+} signal pathway triggered by SI response. In this study, profilin, an actin-binding protein (ABP), was analyzed to examine whether it involves in actin depolymerization in *P. rhoeas*. The results showed that the Ca^{2+}-dependent monomeric actin sequestering activity of profilin was not sufficient for SI-induced actin depolymerization, though it might contribute to the depolymerization (Snowman et al., 2002). Profilin is an abundant small protein that can form a complex with actin monomers and thought to mediate the organization of actin cytoskeleton (Sun et al., 1995). Moreover, profilin of *P. rhoeas* gives rise to the phosphorylation of pollen proteins in vitro, which suggests that it may change protein kinase or phosphatase activity and acts as a signal regulating pollen tube growth (Clarke et al., 1998). Collectively, it was proposed that profilin requires the cooperation of other ABPs to accomplish actin depolymerization in pollen. PrABP80 identified from *P. rhoeas* pollen, a Ca^{2+}-regulated 80 kDa ABP involved in gelsolin, has the ability to nucleate actin filaments and enhance profilin-mediated actin depolymerization through blocking onto actin filament ends (Huang et al., 2004).

Previous studies lack evidence showing what influences the formation of the punctate actin foci. To solve this question, Poulter et al. (2010) analyzed and quantified the formation of incompatible pollen tubes of *P. rhoeas* and found that the formation requires actin polymerization and is co-localized with two intracellular ABPs, which are actin-depolymerizing factor (ADF) and cyclase-associated protein (CAP). Nevertheless, profilin and fimbrin are not correlated with the actin foci (Poulter et al., 2010). ADF has the ability to bind to both G- and F-actin in order to depolymerize F-actin and is related to F-actin treadmilling (Andrianantoandro & Pollard, 2006; dos Remedios et al., 2003). CAP as an interactor of adenylyl cyclase can combine with ADP-actin monomers in yeast (Field et al., 1990; Mattila et al., 2004). These data suggest that ABPs play an important role of Ca^{2+}-mediated actin depolymerization under SI response.

After the presence of SI-induced alterations of actin cytoskeleton, a question to ask is what would happen to the incompatible pollen tube after actin depolymerization and whether the SI-stimulated actin changes are involved in the inhibition of incompatible pollen tube growth. Previous study reported that PCD is triggered by SI response in incompatible pollen tubes of *P. rhoeas* and also reported that this process is mediated by caspase-3-like activity (Thomas & Franklin-Tong, 2004). Thus, to investigate the association between actin depolymerization and PCD cascade induced by SI, using jasplakinolide (Jasp) and latrunculin B (Lat B) to stabilize and depolymerize the actin of *P. rhoeas* pollen tube, the results revealed that the addition of Jasp to stabilize actin filament in incompatible pollen alleviates SI-induced PCD effect, which suggests that SI-induced actin depolymerization is an upstream signal of the PCD activation. Lat B-induced actin depolymerization triggers DNA fragmentation that is a target of PCD cascade, which is mediated by caspase-3-like activity. These data demonstrated that the alterations of actin filament dynamics are crucial for initiating PCD in pollen of *P. rhoeas* (Thomas et al., 2006).

As the actin cytoskeleton was tested to possess the critical function in SI of *P. rhoeas*, the microtubule, another cytoskeleton component, was studied in SI-mediated signal transduction in *P. rhoeas*. The study indicated that the microtubule cytoskeleton undergoes rapid and obvious depolymerization in incompatible pollen tubes and is invisible within 1 min of SI induction, demonstrating that the pollen microtubule is also an early target of the SI signal, so as the actin (Poulter et al., 2008). However, different from the actin, the microtubule cytoskeleton has no reorganization in the latter of depolymerization under SI stimulation (Poulter et al., 2008). Significantly, artificial alterations of the actin cytoskeleton containing stabilization or depolymerization through the responding drugs trigger the same as the alterations in microtubule cytoskeleton. But the microtubule depolymerization is not required for actin changes. These data suggest that the microtubule depolymerization is a consequence of the SI-induced actin depolymerization and that actin is an upstream signal of microtubules. It is clear that the microtubule disruption is different from the actin depolymerization during SI response. Subsequently, the disruption of microtubule dynamics alone is not sufficient to trigger PCD in SI, whereas it is required for PCD process (Poulter et al., 2008).

16.2.7 PCD in Papaver *SI*

Strikingly, a key point of *Papaver* SI response is the eventual cell fate of pollen during SI induction, which is related to the inhibition or death of incompatible pollen tubes. In *P. rhoeas*, the SI response triggers PCD to inhibit incompatible pollen tubes (Thomas & Franklin-Tong, 2004). Generally, PCD is detected by several features, including DNA fragmentation, cell acidification, caspase activation, and leakage of cytochrome c from the mitochondria into the cytosol. The study on SI-induced PCD in *Papaver* was initiated by Jordan et al. (2000). Through FragEL assay, he found that nuclear DNA fragmentation was detected in incompatible pollen tubes but not in compatible pollen tubes. This result indicates that the SI response mediates the nuclear DNA fragmentation. Elevation of $[Ca^{2+}]_i$ through mastoparan also triggers nuclear DNA fragmentation in pollen tubes, which implicates that Ca^{2+} signaling is involved in this phenomenon. According to these data, it is supposed that PCD might be relative to SI response because of DNA fragmentation as a hallmark feature of PCD (Jordan et al., 2000). Subsequently, another study confirmed that SI triggers PCD in *Papaver* incompatible pollen tubes (Thomas & Franklin-Tong, 2004). Based on caspase-3 as a vital enzyme mediating PCD process, the nuclear DNA fragmentation was significantly inhibited by the pretreatment of the caspase-3 inhibitor (Ac-DEVD-CHO, DEVD) in incompatible pollen tubes before SI induction, but not in incompatible pollen tubes induced by SI alone. Meanwhile, the inhibition of incompatible pollen tubes under SI response was weakened by the pretreatment of DEVD. These data suggest that the *Papaver* caspase-3-like activity was activated by SI reaction and involved in the inhibition of incompatible pollen tubes. Moreover, another hallmark feature of PCD, cytochrome c leakage from the mitochondria into the cytosol, occurred in incompatible pollen tubes after SI induction. To investigate the involvement of Ca^{2+} signaling and cytochrome c leakage under SI-mediation, mastoparan stimulated the elevation of $[Ca^{2+}]_i$ in *Papaver* pollen tubes, resulting in cytochrome c leakage into the cytosol. The study also revealed that SI-induced pollen protein extract cleaved a PARP (poly ADP-ribose polymerase) which is a substrate of caspase-3 activity, and that the PARP-cleavage activity was alleviated by DEVD and triggered by increase of $[Ca^{2+}]_i$ in pollen. The above data in the study provide solid evidence that PCD is triggered by SI in *Papaver* (Thomas & Franklin-Tong, 2004). Additionally, it was proved that DEVDase (caspase-3-like) activity was triggered by SI using a fluorescent substrate of caspase-3 (AC-DEVD-AMC) and located in the cytosol, vegetative nucleus, and generative cell nucleus after SI stimulation (Bosch & Franklin-Tong, 2007).

Through AMC-based peptide caspase substrates, besides DEVDase, two caspase-like activities induced by SI were found in incompatible pollen tubes, including LEVDase (caspase-4) and VEIDase (caspase-6) (Bosch & Franklin-Tong, 2007). The temporal profile of VEIDase is similar to that of DEVDase, for which significant activity was detected 3 h after SI stimulation and peaked at 5 h. However, the LEVDase activity was continuously increased until 8 h after SI stimulation, although the early increase is slower than DEVDase and VEIDase. These results suggest that DEVDase and VEIDase activities were activated early in SI response, whereas the LEVDase activity was stimulated later. To test the association between SI-induced PCD and caspase-like containing LEVDase or VEIDase, using AC-LEVD-CHO or AC-VEID-CHO to inhibit LEVDase or VEIDase activity, respectively, SI-induced DNA fragmentation was alleviated by AC-VEID-CHO but not by AC-LEVD-CHO, which shows that VEIDase plays a key role in SI-induced PCD. Thus, there are two caspase-like activities involved in SI-mediated PCD, including DEVDase and VEIDase (Bosch & Franklin-Tong, 2007).

In animal cells, it is thought that cytosol acidification usually occurs in PCD process. Based on this phenomenon, to determine whether SI-induced PCD results in cytosol acidification in the incompatible pollen tubes of *P. rhoeas*, the intracellular cytosolic pH ($[pH]_i$) was measured in normally growing or incompatible pollen tubes. The result indicates that $[pH]_i$ fell from pH 6.9 to pH 5.5 at 1–4 h after SI induction, which is in accordance with the optimal pH of the DEVDase and VEIDase activities in vitro (Bosch & Franklin-Tong, 2007). Subsequently, further study on cytosol acidification in SI-stimulated PCD revealed that the cytosol acidification is a dramatic and rapid process during SI response (Wilkins et al., 2015). The cytosolic pH drops from 6.9 to 6.4 within 10 min of SI stimulation and reaches 5.5 at 60 min. To test the function of SI-induced cytosolic acidification, the study used propionic acid to change artificially [pH]i of the pollen tube. The data indicate that alteration of pollen tube [pH]i is relative to several process of PCD induced by SI response, including alteration of DEVDase activity, formation of F-actin foci, colocalization of two ABPs with F-actin foci, the sPPase activity, and vacuolar alteration (Wilkins et al., 2015).

In *P. rhoeas*, besides the above descriptions, there are several biological affairs with regard to SI-induced PCD process. It is known that MAPK (mitogen-activated protein kinase) of plants has an important ability to respond to extracellular signals and is involved in stress responses (Tena et al., 2001; Zhang & Klessig, 2001). An MBP (myelin basic protein) kinase, named p56 because of its 56 kDa molecular weight, was identified in *P. rhoeas* pollen growing in vitro and activated in incompatible pollen under SI stimulation. Additionally, p56 can cross-react with AtMPK3 antibody. These data indicate that p56 has MAPK activity and is an authentic MAPK (Li et al., 2007; Rudd et al., 2003). Then, a study investigated the association between the MAPK and SI-induced PCD process (Li et al., 2007). The MAPK inhibitor U0126, which inhibits p56 activity and pollen tube growth, rescues the viability of incompatible pollen tube lost by SI stimulation. These results demonstrate that p56 involves in the cell death of incompatible pollen tube. U0126 also can inhibit DEVDase activity and DNA fragmentation, which implicates the involvement of the MAPK in signaling during SI-induced PCD. Recently, *PrMPK9-1* expressed in pollen of *P. rhoeas* was identified and cloned that encodes p56-MAPK and belongs to threonine-aspartate-tyrosine (TDY) MAPK (Chai et al., 2017). Antisense oligonucleotides for PrMPK9-1 can alleviate the SI-specific inhibition of pollen tube and inhibit the formation of actin punctate foci during SI response, which implicates that PrMPK9-1 is crucial for SI response and signaling network of actin and caspase3-like activation during SI-induced PCD process. In addition, another study found that SI induces the rapid increases of ROS and NO which are downstream of the Ca^{2+} elevation. ROS and NO signal activates SI-induced DEVDase activity and stimulates the formation of actin punctate foci mediated by SI response, which indicates that both signals are upstream of PCD events during SI response (Wilkins et al., 2011).

16.2.8 A Model for **Papaver** *SI*

In an incompatible response, the interaction between the pollen and pistil *S* determinants triggers initially K^+ and Ca^{2+} influx in the cytosol of the pollen tube during the early SI response (Fig. 16.3a). Then, the rapid increase of $[Ca^{2+}]_i$ directly mediates several downstream signals, including inhibition of Prp26.1 sPPase activity and

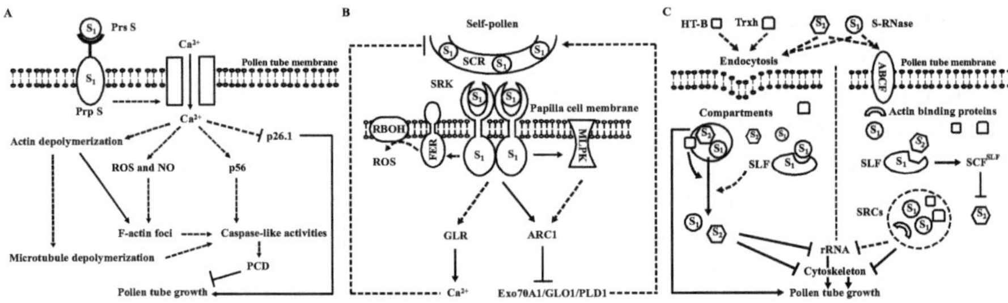

Fig. 16.3 Signaling pathways of three major SI systems in angiosperms. (**a**) SI of *Papaver rhoeas*. Prs S_1 reciprocally recognizes and interacts with Prp S_1, triggering a Ca^{2+} influx and further stimulating a series of downstream responses including protein phosphorylation, ROS accumulation, actin depolymerization, and PCD of self-pollen. (**b**) SI of Brassicaceae. The pollen coat-localized S_1-SCR penetrates the papilla cell wall and specifically binds to the extracellular domain of S_1-SRK, triggering several signaling cascades leading to self-pollen rejection. (**c**) S-RNase-based SI. S-RNases enter into pollen tubes without *S*-haplotype specificity. S_1-SLFs can recognize S_1- and S_2-RNase in the cytosol, forming SCF^{SLF} complexes with CUL1-P and SSK1, leading to the ubiquitination and degradation of non-self S_2-RNase (degradation model), and the survived S_1-RNase can form the S-RNase condensates (SRCs) resulting in self-pollen inhibition (condensation model), or promoting the destruction of compartments leading to S-RNase release to exert cytotoxicity (compartmentalization model)

actin depolymerization, resulting in the inhibition of incompatible pollen tube tip growth. Later in the SI reaction, the PCD process is initiated by several signaling events. A process, which is the depolymerization of F-actin forms punctate foci, is stimulated by ABPs and signals of NO and ROS. The depolymerization of F-actin also mediates microtubule depolymerization. Cytoskeleton depolymerization is involved in caspase-like activity to trigger PCD. Another event with regard to PCD, p56 is phosphorylated by the elevation of $[Ca^{2+}]_i$, which activates its MAPK activity to change the DEVDase activity. The DEVDase, caspase-3 in *P. rhoeas*, plays a crucial role in mediating the PCD reaction, eventually resulting in cell death of the incompatible pollen tube. In the PCD process, there are three main events in incompatible pollen tube, including DNA fragmentation, cytochrome c leakage from mitochondria to cytosol, and cytosolic acidification. The cytosolic acidification is not only a marker of PCD but also a mediator to activate caspase-3 activity. Thus, these signaling events constitute a signal network to induce PCD response and inhibition of incompatible pollen tubes, which leads to unsuccessful self-fertilization.

16.3 Brassicaceae SI

16.3.1 Genetics of Brassicaceae SI

The Brassicaceae family is characterized by the SSI self-recognition system (Fig. 16.2). The earliest report of Brassicaceae SI appeared in 1920 by Stout, who observed it in radish (Stout, 1920). During the study of the SI trait of *Brassica oleracea*, Kakizaki (1930) proposed a genetic model indicating that the *T*-locus functioned as alleles with the *S*-locus, and the *S*-locus was epistatic over the *T*-locus. However, the *T*-locus was controversial. In the 1950s, Bateman (1955) proposed the sporophytic model through working on *Iberis amara* and interpreting previous results from *Raphanus*. He concluded that the SI phenotype of Brassicaceae was controlled by a single *S* (Sterility) locus with multiple alleles, each of which represented a different mating specificity, and the pollen SI phenotype was determined by the diploid genotype of the pollen-producing parent. This mode of genetic control had been later confirmed in several other taxa, including *Cardemine*, *Capsella*, *Raphanus raphanistrum*, *Brassica oleracea* (Bateman, 1955; Sampson, 1957, 1964; Thompson & Howard, 1959), and more recently in *Arabidopsis lyrata* (Bechsgaard et al., 2004; Charlesworth et al., 2003; Kusaba et al., 2001; Schierup et al., 2001). As new alleles increase chances of compatibility, *S* alleles increase frequently in order to reach equilibrium. As a result, the number of the *S*-locus variants in a particular species can be quite large, for example, 22 *S* alleles in *Iberis* (Bateman, 1955), 34 in *Raphanus* (Sampson, 1957), ~100 in *Brassica rapa* (Nou et al., 1993), and more than 50 in *Brassica oleracea* (Ockendon, 1982). In addition, the *S*-locus alleles were found to exhibit complex genetic interactions in all species examined in Brassicaceae, not only in stigma but also in pollen, including dominance/recessive interactions or co-dominance relationships (Sampson, 1964; Thompson & Taylor, 1966).

16.3.2 The Pistil S *Determinant in Brassicaceae*

The *S*-locus-specific glycoproteins (SLSG), or the S-locus glycoproteins (SLG), were the first *S*-locus-specific proteins discovered in the plant kingdom. *SLG* was found in most *S*-haplotypes of *Brassica* and *Raphanus*, except for *Arabidopsis* (Kusaba et al., 2001) and *Capsella* (Nasrallah et al., 2007). Multiple strategies had been carried out in order to obtain the *S*-specific products, initially beginning with the detection of the *S*-allele-specific antigen through immunological technique in cabbage, *B. oleracea* (Wallace, 1967). Furthermore, isoelectric focusing (IEF), a technique used for separating different molecules by differences in their isoelectric point (pI), had made great contribution afterwards (Muntoni et al., 1970; Nishio et al., 1977). Through IEF analysis, SLG exhibited polymorphic molecular weight and charge in different *S*-haplotypes, and the polymorphism of SLG was found co-segregated with *S*-alleles with no exception not only in parents but also in genetic crosses F_1 and F_2 populations (Nasrallah et al., 1972; Nishio & Hinata, 1978). Simultaneously, *SLG* was identified through differential screening of a stigma cDNA library, and the corresponding genomic DNA sequences were determined (Lalonde et al., 1989; Nasrallah et al., 1987, 1988). In addition to *SLG*, some SLG-like proteins expressed by *S*-locus-related (*SLR*) genes in the stigma of *Brassica* had also been discovered (Boyes et al., 1991; Cock et al., 1995; Isogai et al., 1988). SLG is primarily synthesized in papilla cells on the stigma surface, then secreted into papillar cell walls and accumulates there. The expression level of SLG increases along with pistil maturing prior to florescence and peaks after the acquisition of the pollen-stigma incompatible response (Nasrallah et al., 1985). Except for the highly polymorphic regions and the conserved protein regions, several conserved carbohydrate chains are present in different SLGs (Takayama et al., 1986a, b).

The second *S*-locus-linked gene discovered afterwards from stigma is the *S* receptor kinase (*SRK*) gene, which is expressed specifically in papilla cells like *SLG* (Watanabe et al., 1994), but at a level nearly 200-fold lower than SLG (Stein et al., 1991). Through analysis of the DNA flanking sequence of *SLG* homologous genomic clones, researchers had found sequences of high similarity with the known protein kinase gene, *ZmPK1*, which was the first receptor-like protein kinase gene identified in plants. Restriction fragment length polymorphism (RFLP) analysis of F_2 population revealed that SRK was genetically linked to the *S*-locus. Consistent with their *S*-haplotypes, *SRK* genes are highly polymorphic between haplotypes and co-segregated with *S*-alleles. The corresponding genomic DNA and cDNA sequence of SRK predicted that it was a transmembrane receptor kinase containing three domains, including the extracellular S (for *S*-locus) domain, a single-pass transmembrane domain, and a serine/threonine-type protein kinase catalytic domain (Stein et al., 1991). The S domain of SRK is about 90% homologous to SLG within a haplotype (Stein et al., 1991). The receptor-like structures of SRK and SLG suggest that they function as the receptor of the pollen ligand, thus potentially distinguishing self-pollen and non-self-pollen and stimulating the downstream response for SI.

In order to confirm the roles of these *S* locus stigma-specific proteins in SI recognition reaction, several early studies had focused on screening self-compatible *Brassica* strains. Some self-compatible phenotypes resulted from downregulated RNA levels of SLG or nonexpressed transcripts of SRK were reported successively (Nasrallah et al., 1992, 1994), which had implied the essential function of SLG and SRK in SI indirectly. But these findings are all circumstantial evidence, the most direct ways to investigate the roles of these *S*-locus-linked genes in SI response are transgenic approaches including gain-of-function and loss-of-function assays. In a loss-of-function assay, both the transcription level and translation level of *SLG* and *SRK* greatly reduced by introducing antisense-*SLG* into self-incompatible *B. rapa* and rendered the transgenic plants become self-compatible. The breakdown of SI was only due to the alteration of stigma rather than the pollen, which served as direct evidence that SLG and SRK play important roles during stigma-pollen recognition as stigma factors (Shiba et al., 1995, 2000; Takasaki et al., 2001). In a gain-of-function experiment, *SLG* or *SRK* of a different *S* haplotype was transformed into *Brassica* plants and further determined whether the transgenic plants acquired *S* specificity of the transgene. However, the transformed *SLG* or *SRK* caused a remarkable reduction in the expression level of the endogenous *S*-locus and related genes and rendered the transgenic plants turn from SI to SC (Conner et al., 1997; Stahl et al., 1998). This phenomenon was described as homology-dependent gene silencing or co-suppression (Conner et al., 1997). In order to avoid the negative effect of co-suppression, *SRK* and *SLG* (like SRK^{28} and SLG^{28}) were separately transformed into self-incompatible plants of *B. rapa* which contain highly different *SLG* and *SRK* genes (like SRK^{60} and SLG^{60}) (Takasaki et al., 2000). As a result, it was the expression of *SRK* alone, rather than *SLG*, conferred transgenic plants the ability to reject self-pollen that belongs to the same haplotype with the transgene. Simultaneously, other researchers had also achieved the similar conclusion through gain-of-function approaches by transforming the self-compatible *Brassica napus cv.* Westar (Cui et al., 2000; Shiba et al., 2001). Although *SLG* exhibits some characteristics such as high-level stigma-specific expression, polymorphism, and tight linkage with the *S*-locus, it does not seem to be involved in the *S* haplotype-specific recognition. Nevertheless, some evidence showed that SLG can facilitate the recognition process of SI possibly through forming a complex with the S domain of its cognate SRK to enhance the combination of SRK and a pollen ligand. This role of SLG is dependent on the amino-acid identity of SLG and the S domain of SRK, the higher identity, the stronger ability to form the complex, thus resulting in the stronger rejection (Takasaki et al., 2000). However, the enhancing effect of SLG in SI discovered by Takasaki et al. (2000) was not further verified by other studies (Shiba et al., 2001). Thus, several gain-of-function experiments have demonstrated that SRK is the sole female determinant of *S* haplotype specificity in *Brassica*. However, the precise function of SLG is still unclear.

16.3.3 The Pollen S Determinant in Brassicaceae

Identification of the pollen determinant of *S*-haplotype specificity is undoubtedly a difficult task. First, the establishment of in vitro pollination bioassay system provided a strong hint to this exploration. In the assay, the coating-removed pollen grains were supplemented with pollen coat proteins (PCPs) isolated from self-pollen or cross-pollen, then using them to pretreat stigma by micromanipulation (Doughty et al., 1993). This system revealed that PCPs could induce an *S*-haplotype-specific SI response in stigma. Furthermore, through analysis of PCPs fractions, a ≤10 kDa molecule fraction enriched with basic cysteine-rich PCP-A-like (PCP-class A, one of which was known to bind stigma-specific components of the *S* locus) proteins was proved to contain male *S* determinant (Stephenson et al., 1997). Several studies revealed that there are a number of genes existed around SLG and SRK (Boyes & Nasrallah, 1995; Boyes et al., 1997; Yu et al., 1996). With the hypothesis that SI of *Brassica* was activated through a receptor-ligand recognition, the *S* locus should contain one or more genes involved in the SI reaction such as a gene encoding pollen ligand for SRK. As there are too many repeat sequences of *Brassica* genome, it was difficult to conduct chromosome walking (Suzuki et al., 1997). In order to study the genomic structure of the *S* locus of *Brassica* in detail, the primary task was to clone a large fragment containing the intact *S* locus. Although up to 1 Mb DNA could be cloned by YAC vector, it was difficult to purify the inserted DNA from yeast genome (Suzuki et al., 1997). With the advantages of P1-derived artificial chromosomes (PACs) cloning vector (Ioannou et al., 1994), a 74-kb genomic region containing both *SLG* and *SRK* in the S_9 haplotype of *B. rapa* was successfully cloned through direct cloning of the completely digested DNA fragment into PAC vector (Suzuki et al., 1997). Using this cloned *S*-locus, a potential candidate cysteine-rich PCP-A-like protein gene, designated as *SP11* (*S* locus protein 11), was identified by DNA sequence analysis and immature anther cDNA library screening. *SP11*, as the proposed pollen ligand gene, is located in an A + T-rich region just downstream of SRK_9 and also shows anther-specific expression and a tight linkage to the *S*-locus (Suzuki et al., 1999). Subsequently, Takayama et al. (2000) identified *SP11* genes in additional *S* haplotypes, including S_8, S_{12}, and S_{52} of *B. campestris* and further characterized the genes. They showed that *SP11* could strictly fulfill several criteria of the pollen *S* determinant including *S*-haplotype-associated polymorphism, sporophytic control of SI, and inducing *S*-haplotype-specific SI response of the stigma (Takayama et al., 2000). At the same time, through sequence analysis of the region between *SLG* and *SRK* in S_8 haplotype of *B. rapa* and screening of S_8 anther cDNA library, Schopfer et al. (1999) also discovered an anther-specific, cysteine-rich, and single-copy protein gene named the *S*-locus cysteine-rich (*SCR*) protein gene. The DNA sequence of *SCR* is consisted of two exons and one intron as large as 4.1 kb (Schopfer et al., 1999). The first exon encodes a signal peptide, and the second encodes a small cysteine-rich protein. Analysis of the F_2 population and DNA gel blot proved that *SCR* was co-segregated with its cognate *SRK* and the *S* haplotype. Gain-of-function and loss-of-function experiments had further ascertained the function of *SCR* in *Brassica* SI. Besides, the sequence of SCR_8 is completely identical to $SP11_8$, indicating that *SCR* is the same as *SP11* (Takayama et al., 2000). In *Brassica*, *SP11/SCR* alleles are highly polymorphic, which exhibit more than 60% amino-acid sequence divergence (Kusaba et al., 2001; Schopfer et al., 1999; Watanabe et al., 2000). Furthermore, the expression pattern varies in different *SP11/SCR* alleles. As expected for sporophytic control of SI, most *SP11/SCR* alleles are expressed in anther tapetum, then their products are secreted and incorporated into the outer pollen coat at the late developmental stages, while some alleles showed gametophytic expression in microspores (Kachroo et al., 2001; Schopfer et al., 1999; Shiba et al., 2001; Takayama et al., 2001).

16.3.4 The S-Haplotype-Specific Interaction of SRK and SCR

As the sole female determinant of *S* haplotype specificity, SRK confers stigma the capacity of discriminating between self-pollen and non-self-pollen, which are determined by the small basic cysteine-rich pollen coat protein, *SP11/SCR*. In order to maintain their specific interaction, *SRK* and *SCR* genes must co-evolve, consistent with their high polymorphisms. The interaction of SRK and SCR in a haplotype-dependent manner is a milestone for understanding the molecular mechanism of Brassicaceae SI (Kemp & Doughty, 2007).

First, several studies had proved the physical interaction between SP11/SCR and SRK. Using recombinant proteins purified from *Nicotiana* leaves and bacteria,

Kachroo et al. (2001) demonstrated that SCR could directly interact with the ectodomain of SRK (eSRK) in an *S*-haplotype-specific manner without the participation of any other components of stigma and pollen. Cross-linking experiments using ^{25}I-S_8-SP11 proved that S_8-SP11 (stabilized by four intramolecular disulfide bonds) could bind its cognate SRK_8 with high affinity ($K_d = 0.7$ nM) on the stigma membrane and further induce autophosphorylation of SRK_8, which had also been discovered after self-pollination in vivo (Cabrillac et al., 2001; Takayama et al., 2001). Except for SRK, cross-linking experiments also revealed another 65 kDa protein that strongly binds to ^{25}I-S_8-SP11. Subsequent researches proved that this 65 kDa protein is a truncated SRK (tSRK) containing ectodomain, transmembrane domain, and part of juxtamembrane domain of SRK_8 (Shimosato et al., 2007). Besides, although both the intact SRK and artificial membrane-anchored SRK can bind SP11 with high affinity, the soluble eSRK as well as SLG exhibits no high affinity. When eSRK was artificially dimerized, its high affinity with SP11 appeared. These results indicated that membrane-anchored SRK dimerization could provide a primed condition for rapid recruitment and activation of the receptor on ligand binding. Actually, Giranton et al. (2000) had already found the existence of constitutive homooligomers of stigma membrane-anchored SRK in the absence of ligand, while eSRK and SLG were mostly monomeric.

The modular structure of eSRK shows that it contains three kinds of domain, locating in the N-terminus are two lectin-like domains, then followed by an epidermal growth factor (EGF)-like domain and a C-terminal PAN-APPLE domain (Shiu & Bleecker, 2001). Furthermore, yeast-two-hybrid studies demonstrated that SRK dimerization, a crucial prerequisite for high-affinity interaction between SRK and SCR, is mediated by two EFG-like domains and PAN-APPLE domain in eSRK (Naithani et al., 2007). Through sequence comparison of eSRK variants, four hypervariable (hv) regions, designated as hvI (within the second lectin-like domain), hvII, hvIII (within the EGF-like domain), and C-terminal variable region (CVR) (within the PAN-APPLE domain) were discovered, which were predicted as potential determinants of SI specificity (Kusaba et al., 1997; Sato et al., 2002). Based on studies of a few variants, residues of hvI and hvII rather than hvIII were proposed to be more likely required for self-SCR recognition (Kemp & Doughty, 2007), but it was still unclear whether these regions are generally required for all the SRK variants to recognize their cognate SCR. Kemp and Doughty (2007) suggested that SCR bound to SRK through a two-step process in which a haplotype-specific activation of SRK was mediated by a generic low-affinity binding of SCR to the hv region of eSRK or the recruitment of other accessory molecules in order to acquire a tighter interaction between SCR and membrane-anchored SRK.

SP11/SCR variants share a higher polymorphism than SRKs. For most SCR variants, including SCRs from *Brassica* class-I S-haplotypes and *A. lyrata*, only eight cysteine residues (C1 through C8) are typically conserved. However, NMR spectroscopy studies for SCR_8 of *B. rapa* and the subsequent modeling for SCRs from some other variants suggested that these small proteins of different alleles shared a similar 3D structure like defensins except for loop 1. Loop 1 is an hv region among allelic SP11s, residues of which are likely to be important for allele-specific interactions with SRKs (Mishima et al., 2003). Consistently, studies of chimeric SP11 genes demonstrated that C2-C3 and C5-C6 regions are more likely to be responsible for recognition specificity (Chookajorn et al., 2004; Sato et al., 2003). Furthermore, it is worth noting that docking of ligand lacks correlation with a stable binding actually resulting in receptor activation, and they were suggested to be distinct events (Chookajorn et al., 2004).

Recently, the crystal structure of eSRK9-SCR9 complex from *B. rapa* has been solved (Ma et al., 2016). It suggested that SCR binding could induce homodimerization of eSRK, forming a 2:2 eSRK:SCR heterotetramer through mediating the direct interaction of two SRK and two independent SCR molecules. Nearly all three hv regions of eSRK contribute to the recognition event. For SCR, the eSRK-interacting residues scatter along all the length, but most of them are concentrated in C5-C6 region that interacts with hvI and hvII of eSRK. Through modeling analysis of other eSRK-SCR pairs from *A. lyata*, a stable 2:2 tetrameric complexes with a similar recognition mechanism like eSRK-SCR from *B. rapa* generally existed. Besides, further structure analysis indicated that compared to self SCR, there were different interfaces between SRK and non-self SCR, which provided the direct molecular evidence for discrimination of self and non-self SCR by eSRK. Moreover, Murase et al. (2020) showed that the binding free energies are most stable for cognate eSRK-SP11 combinations, and the interaction modes of eSRK and SP11 are different between intra- and inter-subgroup.

16.3.5 Early SI Responses in Brassicaceae

The initiation of pollen-stigma recognition is completed by pollen capture and adhesion, which is mediated by the surface properties of both stigma and pollen grain (Zinkl et al., 1999). For the stigma of Brassicaceae, the proteinaceous pellicle coating the stigmatic papillae plays a crucial role in pollen adhesion and pollen-stigma recognition (Mattsson et al., 1974; Stead et al., 1980). Stigma-specific proteins like SLG and S-locus related-1 (SLR) are suggested to positively mediate this event through binding to PCPs. While in the case of pollen grain, its exine mediates initial binding, and pollen coat composed of lipids and proteins is responsible for pollen adhesion (Elleman et al., 1992; Zinkl et al., 1999). Pollen adhesion is a highly selective process. Once a compatible response initiated after pollen capture, the pollen coat moves toward the pollen-stigma interface, forming a meniscus-shaped "foot" and a mixture of lipids and proteins from both interfaces of pollen and stigma (Elleman & Dickinson, 1990; Preuss et al., 1993). Although after self-pollination, pollen capture occurs rapidly and pollen adhesion is also initiated, the "foot" disappears and poor adhesion can be found. Furthermore, as Brassicaceae possesses dry-type stigma, pollen hydration is dependent on water transferred from stigma, which is controlled by the unique passage formed by a suitable lipid-interface occurring during pollen adhesion (Dickinson, 1995; Elleman et al., 1992). But in SI, the lipid-interface may not form and as a result the water passage is damaged, thus resulting in the inhibition of pollen hydration. Although sometimes self-pollen can hydrate due to the humid environment, it cannot further germinate a pollen tube, or the pollen tube cannot penetrate into the papilla surface. This can be attributed to the destruction of enzymes secreted by pollen or stigma leading to the breakdown of the papilla surface and the signals responsible for pollen tube growth, such as GABA that is identified in *Arabidopsis* (Hiscock et al., 2002; Palanivelu et al., 2003).

16.3.6 Downstream SI Signaling Events in Brassicaceae

After self-pollination, the pollen ligand SP11 interacts specifically with the extracellular domain of its cognate SRK and activates its intracellular kinase domain (Cabrillac et al., 2001; Takayama et al., 2001). Like other receptor-like kinases (RLKs), activation of SRK will stimulate a series of phosphorylation cascades downstream of SRK to respond to the extracellular signals like self-pollination. However, our understanding of the downstream signaling pathway of SRK is relatively limited, and so far, several factors have been found to be involved in the signal transduction of SRK, including the *M* locus protein kinase (MLPK), ARM-repeat-containing1 (ARC1), Exo70A1, glyoxalase (GLO1), and phospholipase alpha 1 (PLD1).

The first key molecule that regulates the *Brassica* SI is the *M* locus protein kinase (MLPK), which was discovered through positional cloning of a modifier (*m*) mutation of SI in the *B. rapa* yellow sarson variety (Murase et al., 2004). Here, the *M* locus is independent of and epistatic to the *S* locus, and the *mm* recessive mutants exhibit a complete absence of SI in the stigma rather than in the pollen. *MLPK*, composed of seven introns and eight exons, encodes a 404-amino acid protein kinase that belongs to the receptor-like cytoplasmic kinase VII (RLCK VII) subfamily. Like SRK, MLPK is also a membrane-anchored serine/threonine protein kinase that positively regulates SI signaling. Transient expression of MLPK in the papillae of *B. rapa* var. yellow sarson could completely restore SI phenotype (Murase et al., 2004). Besides, two isoforms that only differ in N terminus had been identified in *Brassica* MLPK and its *Arabidopsis* ortholog, which are produced by alternative transcription initiation sites (Kakita et al., 2007b). These two isoforms are both targeted to the plasma membrane, but through different membrane targeting motifs, either by an N-terminal hydrophobic region of one isoform or by myristoylation motif in the other isoform (Kakita et al., 2007b). Membrane localization was indispensable for both isoforms to independently complement the homozygote mutation (Kakita et al., 2007b), which was speculated to be associated with direct interaction of SRK and MLPK in the plasma membrane, as demonstrated through a bimolecular fluorescence complementation assay in tobacco protoplasts, to transduce SI signaling (Kakita et al., 2007b). Furthermore, in vitro phosphorylation assay showed that MLPK could be efficiently phosphorylated by SRK, altogether suggested that following the recognition between SRK and SP11/SCR and SRK activation, MLPK is activated by SRK and functions in a complex with SRK to induce downstream signaling events (Kakita et al., 2007a).

One receptor kinase can always interact with a number of substrates, leading to the activation of multiple intracellular signaling pathways (Cohen et al., 1995), and SRK is no exception. Several interacting partners for SRK kinase domain have been identified, including ARC1, two thioredoxin *h* proteins (THL1 and THL2) (Bower et al., 1996), the kinase-associated protein phosphatase (KAPP), a sorting nexin, and calmodulin (Vanoosthuyse et al., 2003). Among them, only KAPP and ARC1 show phosphorylation-dependent interaction with SRK, while the others can interact with both the inactive and active kinase domain of SRK (Bower et al., 1996; Gu et al., 1998; Vanoosthuyse et al., 2003). Besides, thioredoxin *h* proteins were demonstrated to be an inhibitor of SRK in the absence of SI to keep SRK inactive (Cabrillac et al., 2001), while KAPP plays a general negative regulation role for multiple activated receptor kinases (Braun et al., 1997). Additionally, the expression of ARC1 is restricted in the stigma papillae (Gu et al., 1998). All these information implicates that ARC1 is more likely to be a downstream effector following SRK activation in response to self-pollination.

ARC1 contains a novel N-terminal domain (UND) conserved in lots of U-box/ARM proteins, a U-box in the center, and an ARM repeat domain in the C-terminal half of the protein (Samuel et al., 2006). Through ARM repeat domain, ARC1 interacts with phosphorylated SRK kinase domain and in turn phosphorylated by SRK (Gu et al., 1998). MLPK is also capable of phosphorylating the ARM-repeat domain of ARC1 in vitro, and the phosphorylation efficiency is higher than that of SRK (Samuel et al., 2008). In order to explore the relationship between ARC1 and SI, Stone et al. (1999) proved a positive role of ARC1 in self-pollen rejection, since suppression of ARC1 mRNA level in self-incompatible *B. napus* W1 line led to incomplete breakdown of SI, but SP11/SCR and SRK were functional (Stone et al., 1999). Furthermore, in vitro studies showed that ARC1 possesses U-box-dependent E3 ligase activity, thus defined as an E3 ubiquitin ligase, which can tag a substrate protein with ubiquitin and lead to proteolysis by the 26S proteasome (Stone et al., 2003). Although ARC1 exhibited several subcellular localizations via different motifs when expressed in tobacco BY-2 suppression-cultured cells including the nucleus, cytosol, and proteasome/COP9 signalosome (CSN), it is localized to the ER-associated proteasome only in the presence of a constitutively active SRK as well as its U-box (Stone et al., 2003). Additionally, MLPK could also affect the re-localization of ARC1 to the proteasome, consistent with the previous hypothesis that MLPK functions in a complex with SRK to activate downstream events (Samuel et al., 2008). Given that the levels of ubiquitinated proteins increased after self-incompatible pollinations and decreased in the ARC1 antisense-suppressed pistils, ARC1 is likely to be responsible for ubiquitination of compatible regulators and their subsequent degradation during SI response (Stone et al., 2003). Pre-treatments of pistils with proteasome inhibitor MG132 also broke down the SI response, indicating that SI is dependent on proteasome degradation pathway (Stone et al., 2003). These results suggested that self-pollen rejection is likely to be a direct result from re-localization of ARC1 to proteasome through activated SRK and MLPK to target several downstream regulators for degradation. The identification of the ubiquitination pathway implicated in this process has absolutely provided an important insight into downstream events after initial pollen recognition, although the targets of ARC1 still remain largely unclear.

The putative targets of ARC1 were hypothesized to be factors promoting the compatible pollination and fertilization (Stone et al., 2003). In order to identify the substrate proteins of ARC1, a screen of *B. napus* pistil cDNA yeast two-hybrid library using the N-terminus of ARC1 was performed, and BnExo70A1 was isolated as an interactor of ARC1 (Samuel et al., 2009). In vitro assays further confirmed the interaction between Exo70A1 and ARC1 and the ubiquitination of Exo70A1 by ARC1. Amino acid sequence comparison showed that BnExo70A1 is similar to a conserved eukaryotic protein, Exo70, a subunit of exocyst complex that is well known in yeast and mammalian as a regulator of specialized secretory events, such as polarized exocytosis (Cole & Fowler, 2006). In plants, Exo70 is also belong to a large gene family, and Exo70 mutants exhibit various defects in cell morphogenesis, root hair elongation, pollen germination, and polar growth (Cole & Fowler, 2006).

Transgenic experiments directly confirmed that in both *Brassica* and *A. thaliana*, loss of Exo70A1 in stigmas led to rejection of compatible pollen via a similar mechanism as the incompatible pollen (Samuel et al., 2009). What is more, overexpression of Exo70A1 resulted in partial breakdown of SI (Samuel et al., 2009). These genetic results are consistent with the predictions that Exo70A1 plays a positive role in pollen adhesion, pollen hydration, and penetration in cross-pollination. Previous electron microscopic studies had identified distinct endomembrane system changes following compatible or self-incompatible pollination (Elleman & Dickinson, 1996). When stimulated by compatible pollen coat, the outer cell wall of the stigma papilla rapidly expands towards the pollen attachment site, with plasma membrane vesiculation as well as ER and Golgi structures assembling in the expanded region (Elleman & Dickinson, 1996). The vesicles in this region were proposed to be essential for pollen hydration and pollen tube penetration as they may carry hydration factors and cell wall loosening enzymes (Elleman & Dickinson, 1996). In addition to vesicles, the vacuolar network localized in the *Brassica* papilla is also involved in compatible pollination through actively transporting water, calcium, ion, and other factors essential for normal pollen germination into pollen grains, while the vacuolar network is disrupted in self-incompatible pollination, leading to pollen arrest (Iwano et al., 2007). Based on the conserved functions of Exo70A1 in regulating polarized secretion, Exo70A1 was predicted to promote water transfer for pollen hydration and enzyme secretion for pollen tube penetration in *Brassica* through marking pollen-stigma interface for polarized exocytosis during compatible pollination. But in self-incompatible pollination, ExoA1 is ubiquitinated and inactivated by ARC1, thus disrupting the polarized secretion as well as pollen hydration and pollen tube penetration thereafter. However, since it is unclear whether these subcellular events are the cause or the consequence of compatible or self-incompatible pollination, the exact role of Exo70A1 remains to be further confirmed (Zhang et al., 2009).

In addition to Exo70A1, GLO1 and PLD1 have been identified as two other targets of ARC1 through 2D-difference gel electrophoresis (DIGE) (Samuel et al.,

2011; Sankaranarayanan et al., 2015, 2017; Scandola & Samuel, 2019). GLO1 serves as the rate-limiting enzyme of the glyoxalase pathway, which is closely associated with stress tolerance, particularly in methylglyoxal (MG) detoxification. However, following self-incompatible pollination, GLO1 is modified by MG, resulting in more efficient ubiquitination of it by ARC1 as well as its proteasomal degradation. Consequently, the accumulated MG, a cytosolic by-product of glycolysis, can modify and disrupt other compatibility factors such as Exo70A1, ultimately leading to self-pollen rejection. To validate the genetic function of GLO1, both RNAi and overexpression assays have been conducted using self-incompatible *B. napus*, resulting in a reduction of cross-seed sets and a partial breakdown of SSI, respectively, thus supporting a positive role of GLO1 during compatible response. Similarly, as PLD1 is indispensable for phosphatidic acid (PA) production to facilitate endocytosis critical for pollen germination, its degradation mediated by ARC1 also leads to SI.

Actin dynamics associated with vacuolar network regulation are also involved in compatible and self-incompatible (Iwano et al., 2007). After cross-pollination, actin polymerization and actin bundles formation occur at the apical region of the papilla cells. They are indispensable for both pollen hydration and germination as they can indirectly control transports of water and ions necessary for pollen hydration and germination through regulating vacuolar network structure and targeting vacuoles to the pollen attachment site. In contrast, during SI response, actin reorganization, likely depolymerization, occurs, thus disrupting the dynamics of the vacuolar network and blocking the support of necessities for pollen hydration and germination. In a word, cross-pollination initiates actin cytoskeleton changes and is in turn initiated by actin polymerization. Actin dynamics are determined by the action of different actin binding proteins (ABPs). Cytosolic calcium concentration regulates the activity of some ABPs, and the increases of cytosolic calcium concentration are always linked to and precede the actin rearrangement. Additionally, recent studies showed that calcium ion influx mediated by glutamate receptor-like channel (GLR) is required for and sufficient to induce SI response leading to inhibition of pollen hydration in Brassicaceae (Iwano et al., 2015). Compared to deceased ROS levels following compatible pollination, self-recognition of cognate SRK and SCR can also trigger a rapid elevation of reactive oxygen species (ROS) to inhibit pollen germination and penetration in stigma papilla cells (Zhang et al., 2021). Although FERONIA (FER) receptor kinase homolog and Rac/Rop guanosine triphosphatase (GTPase) were proved to control this process, it is still unclear how the signaling module is specifically hyperactivated during SI as well as its relationship with the downstream response.

16.3.7 Physiological and Environmental Features of Brassicaceae SI

Several studies showed that SI in Brassicaceae can be overcome by a series of physiological and environmental conditions, including plant age (Horisaki & Niikura, 2008; Ockendon, 1978), chemical treatment such as ether, KOH, and NaCl

(Monteiro et al., 1988; Tatebe, 1964), and high temperature (Matsubara, 1980; Okazaki & Hinata, 1987). Among them, CO_2 gas (3–5%) treatment is the most effective way to overcome SI, and it has been used for decades to obtain a great number of inbred seeds of cruciferous vegetables (Lao et al., 2014). Lines of Brassicaceae always display different degrees of CO_2 sensitivity (Lao et al., 2014; Nakanishii et al., 1969). Various early studies had led to different conclusions on dominant or recessive genetic control of high CO_2 sensitivity based on different materials and different genetic experiment designs (Ji et al., 2007; Niikura & Matsuura, 2000). In some inbred lines of *B. rapa*, CO_2 sensitivity is a quantitative trait controlled by more than one gene and two major QTLs, BrSIO1 and BrSIO2, have been identified as controlling high CO_2 sensitivity (Lao et al., 2014). These genes are not linked with the *S*-locus and are different from QTLs associated with SI stability (Hatakeyama et al., 2010). For CO_2-sensitive *B. rapa* cv. Hiratsuka, a shunken and distorted papilla cell surface had been observed, which may contribute to overcome SI (Lee et al., 2001). CO_2 sensitivity is predicted to be controlled by genes expressed in pistil, but pretreating non-pollinated pistil with high CO_2 gas had no effects on SI breakdown, indicating that physiological process stimulated by pollination is essential for SI to overcome (Lao et al., 2014). Under high CO_2 conditions, substantial Ca^{2+} accumulation at the pollen-stigma interface was specifically observed in CO_2-sensitive lines. As a second messenger, calcium is predicted to be related to secretory events associated with papilla cell wall loosening as described above and high concentration of Ca^{2+} is also necessary for activating pectinase to loosen the papilla cell wall as well as for keeping the pollen tube rigid enough not to burst (Hepler & Winship, 2010). Cytosolic calcium increases have been observed in both *Arabidopsis* and compatible pollinations in *Brassica* (Dearnaley et al., 1997; Elleman & Dickinson, 1999; Iwano et al., 2004). All the information suggests that CO_2 treatment induces a compatible reaction and Ca^{2+} may be an essential factor involved in SI breakdown. Thus, the mechanism causing overcome of SI by CO_2 remains largely unclear, further studies mainly focused on exploring new effective methods or identifying major genetic locus controlling SI overcome will be helpful for both better understanding of SI signaling pathway and hybrid breeding.

16.3.8 Dominance Relationship in Brassicaceae SI

As Brassicaceae belong to SSI system, both the female *S* determinant *SRK* and the male *S* determinant *SCR/SP11* are sporophytically expressed, and the SI phenotypes of the stigma and the pollen in an *S*-heterozygote are determined by the two *S* haplotypes it carries. Dominance relationship is a key characteristic of SSI in *Brassica*. Both *SRK* and *SCR/SP11* independently exhibit a dominance hierarchy that determines which of the two *S* haplotypes of an *S*-heterozygote is expressed at the SI phenotypic level (Durand et al., 2014). There are two classes of *S*-alleles in the Brassicaceae, including Class I alleles displaying co-dominance in expression and Class II alleles exhibiting as recessive when associated with Class I *S*-alleles

(Billiard et al., 2006; Hatakeyama et al., 1998). When individuals contain both classes of alleles, self-fertilization occurs with a high chance. In addition to dominance and recessiveness, different haplotypes can also exhibit co-dominance, linear dominance, or non-linear dominance relationship (Bateman, 1955; Hatakeyama et al., 1998; Thompson & Taylor, 1966). In the stigma, the dominance relationships are determined by *SRK* itself rather than its relative transcript level (Hatakeyama et al., 2001), while in the pollen, dominance relationships are regulated at the mRNA level of SCR/SP11 (Shiba et al., 2002).

Dominance or recessiveness occurs more frequently in the pollen than in the stigma. Phylogenetic analysis uncovered that the class II *SP11s* form a distinct group separated from class I *SP11s*. The transcripts of *SP11s* from pollen-recessive *S*-haplotypes have never been identified through homology-based cloning, as they are greatly diminished in the anther tapetum of the *S* heterozygotes compared with pollen-dominant *S*-haplotypes (Shiba et al., 2002). The promoter sequence and expression patterns of Class II SP11s are also different from the Class I. Shiba et al., 2006 revealed that 5′ promoter sequences of recessive SP11 alleles are specifically methylated in tapetum before the initiation of *SP11* transcription, leading to tissue-specific monoallelic de novo DNA methylation involved in gene silencing of recessive *SP11s* and SI phenotype determination. Tarutani et al. (2010) showed that this allele-specific methylation is controlled by *trans*-acting small non-coding RNA (sRNA) located in the flanking region of dominant *SP11* alleles. These sRNAs are specifically expressed in the anther tapetum, similar to the promoter sequences of the recessive *SP11s* alleles and are processed into 24-nucleotide sRNA, designated as *SP11* methylation inducer (*Smi*) (Tarutani et al., 2010). *Smi* can repress transcription of the recessive *SP11s* through triggering the methylation of their promoter. Based on the discovery of *Smi*, Durand et al. (2014) further studied the sRNAs and their targets, revealing the evolution of how at least 17 sRNAs-producing loci and their multiple target sequences collectively control the dominance hierarchy in an SI *Arabidopsis* species. Furthermore, Kakizaki et al. (2006) discovered that the intergenic regions of recessive *SP11* genes contain several short repetitive sequences highly similar to the target methylation region of the other recessive *SP11* alleles, which may also be involved in regulation of dominance relationships.

16.3.9 Evolutionary Switches from SI to Self-Fertility in Brassicaceae

The switch from outbreeding to inbreeding has occurred frequently in flowering plants. Since self-fertility is associated with inbreeding depression and represents dead ends of plant lineages evolved from outbreeding ancestors (Barrett, 2002, 2003; Jain, 1976; Stebbins, 1957), this switch is proposed to be mainly caused by pollinator scarcity or population bottlenecks (Jarne & Charlesworth, 1993).

Nevertheless, inbreeding provides several benefits, including reproductive assurance and increased potential for colonization. In some cases, only when the benefits of reproductive assurance outweigh inbreeding depression, inbreeding depression can be overcome and self-fertile taxa can be established (Porcher & Lande, 2005).

Similar to other flowering plants, switches to inbreeding have also occurred frequently in Brassicaceae (Boggs et al., 2009). *A. thaliana* is a predominantly self-compatible species derived from a self-incompatible ancestor, while its closely related congeners possess obligate SI (Abbott & Gomes, 1989; Bechsgaard et al., 2006). Despite many studies on the genetic basis of loss of SI in *A. thaliana*, the mechanism switches from outbreeding to self-fertility remain poorly understood in a long time. All the mutations that lead to breakdown of SI can be classified into two groups. One type includes mutations disrupting the genetic linkage of matched SRK and SCR alleles or disrupting the integrity of SRK and SCR, causing the stigma-specific and pollen-specific loss of SI, respectively (Goring et al., 1993; Naithani et al., 2007; Nasrallah et al., 1992; Okamoto et al., 2007). However, the recombination between SRK and SCR is rare and there is a strong selection for maintaining their tight linkage (Boggs et al., 2009). But on the other hand, multiple independent mutations of SRK and SCR in *A. thaliana* have been discovered through analysis of the *S*-locus genes in self-compatible species or strains (Kusaba et al., 2001; Nasrallah et al., 2007). All *A. thaliana* accessions analyzed so far possess a non-functional *S* locus, called pseudo-*S* (*ΨS*) carrying inactivating mutations in *SRK* or *SCR* rather than recombination between *SRK* and *SCR* (Tang et al., 2007). Tsuchimatsu et al. (2010) revealed a disruptive 213 bp inversion in the *SCR* gene in 95% of European *A. thaliana* accessions while *SRK* is still functional. The prevalence of this mutation within SCR that leads to loss of SI also further confirmed the predictions that disruption of the male recognition component is more advantageous than the female component under conditions of selfing (Busch & Schoen, 2008; Uyenoyama et al., 2001).

The other kind, modifier-loci mutations, contains inactivation of the indispensable SI signaling molecules or mutations reducing the expression of the *S*-locus recognition genes, such as *PUB8*, a female modifier gene adjacent to the *S*-locus, which causes reduced *SRK* transcript levels at later developmental stages of stigma (Liu et al., 2007a). The polymorphism of the promoter region of *PUB8* also contributes to the variation in the strength of loss of SI. For example, when pairs of *SRK* and *SCR* from self-incompatible *A. lyrata* heterologously transformed into *A. thaliana*, some accessions showed developmentally stable SI while the others exhibit SI only in young flowers (Boggs et al., 2009; Liu et al., 2007a; Nasrallah & Nasrallah, 2002). Thus, it is still controversial about which mutation(s) predominantly contributed to the loss of SI and how *A. thaliana* become self-fertile. The transition to self-fertility in *A. thaliana* is induced by at least two independent mutations of the *S*-locus, and an unstable allele at a locus unlinked with the *S* locus is also associated with self-fertility (Boggs et al., 2009).

16.3.10 A Model for Brassicaceae SI

The SI response in Brassicaceae is based on a phosphorylation and ubiquitin-mediated degradation pathway, which inhibits the secretion of factors required by compatible pollination (Fig. 16.3b). Once self-pollinated, SI reaction occurs rapidly on the pollen-stigma interface. The discrimination between self-pollen and non-self pollen is based on the activity of two highly polymorphic proteins encoded by the tightly linked *S*-locus genes: *SRK* and *SCR*. When stigmas are pollinated with pollen derived from the same plant or from plants expressing the same *S* haplotype, pollen coat-localized SCR penetrates papilla cell wall and specifically binds to the extracellular domain of SRK, which induces homodimerization and autophosphorylation of SRK, thus triggering several signaling cascades leading to self-pollen rejection. SRK can phosphorylate and activate MLPK, an essential positive signaling intermediate localizing to papilla cell membrane. MLPK could interact directly with SRK and may function in a complex with SRK to further promote the phosphorylation-mediated recruitment of ARC1 E3 ligase. Both SRK and MLPK can affect the re-localization of ARC1 to proteasome, leading to the ubiquitination and degradation of specific proteins that function either as inhibitors of SI or as factors required for successful pollen tube growth. Typically, three substrates, including EXO70A1, GLO1, and PLD1, have been proved to be ubiquitinated by ARC1 during incompatible pollination. Exo70A1, a putative component of the exocyst complex, is predicted to promote water transfer for pollen hydration and enzyme secretion for pollen tube penetration. PLD1, a key factor contributing to PA production, is also hypothesized to promote exocytosis necessary for pollen germination. GLO1, an essential enzyme in the glyoxalase pathway, plays a role methylglyoxal (MG) detoxification. These substrates are vital for compatible pollen growth but are inactivated by the ubiquitin-proteasome pathway following incompatible pollination. Furthermore, actin reorganization, likely depolymerization, is also stimulated by self-pollen on the pollen-stigma interface, disrupting the dynamics of the vacuolar network and blocking the support of water, calcium, ion, and other factors essential for pollen hydration and germination, rendering the pollination incompatible. Additionally, the initial self-recognition of SRK and SCR can trigger a Ca^{2+} influx mediated by glutamate receptor-like channel (GLR) and a ROS burst controlled by the FERONIA-Rac/Rop-Rboh module, thus leading to self-pollen rejection by the inhibiting pollen hydration.

16.4 Solanaceae-Type SI

16.4.1 Genetics of Solanaceae-Type SI

SI studies in Solanaceae began more than a century ago. East and Mangelsdorf (1925) found that SI is controlled by a poly-allelic *S*-locus, and totally 25 multiple alleles were identified in *Nicotiana* groups in this study. Besides, they found that in the pollination of heterozygous the *S*-allele of compatibility pollen should be different with either *S*-alleles of the style (East & Mangelsdorf, 1925). Further studies showed that the Plantaginaceae, Rosaceae, and Rutaceae share the similar type of SI to Solanaceae. SI of these species is controlled by the paternal gametophyte known as GSI (Gruber, 1930; Lewis, 1942) and adopts a non-self recognition system (Fig. 16.2).

16.4.2 The Pistil S *Determinant*

A set of approaches were adopted to identify this factor in the 1980s. Bredemeijer and Blaas (1981) first isolated the pistil *S*-specific protein in *N. alata* by isoelectric focusing experiments. They found a dense broad band (S_2-band) that was presented in the pistil protein of both S_2S_2 and S_2S_3 genotypes, but this band is absent in pollen protein of same genotypes and pistil protein of S_3S_3 genotype. Further analysis of F_1 progenies showed that S_2-band is always present in mature pistils that contain S_2 genotype. And the S_3- and S_6-bands have the same performance thus S_2-, S_3-, and S_6-bands are identified to represent different alleles of the pistil *S*-specific determinant gene. The cDNA of S_2-protein was then cloned and sequenced, and this glycoprotein is confirmed as the *S*-specific determinant of pistil of *Nicotiana* (Bredemeijer & Blaas, 1981; McClure et al., 1989). By analyzing the protein sequences of the *S*-proteins, McClure et al. (1989) found that part of the amino acid sequence of the protein is similar to the conserved region of ribonuclease T2 of *Aspergillus oryzae,* and *S*-proteins do have ribonuclease activity. S-RNases are then successively isolated in *Petunia*, *Pyrus*, *Malus*, *Prunus,* and *Antirrhinum* (Bošković & Tobutt, 1996; Broothaerts et al., 1990, 1991, 1995; Sassa et al., 1997; Xue et al., 1996). Phylogenetic analysis showed that *S-RNase* genes belonging to Class III T2 *RNases* have a single origin from the most recent common ancestor (MRCA) of eudicots (Xue et al., 1996; Igić & Kohn, 2001; Ramanauskas & Igić, 2017; Zhao et al., 2022). The protein sequences of S-RNase show high polymorphisms between different *S*-haplotypes. The sequence similarity of the S-RNase in Solanaceae is about 38% ~ 98% (McCubbin & Kao, 2000). The expression of S-RNase occurs in stigma and throughout the secretory tissue and is high in mature styles but weak in immature ones, showing a developmentally regulated expression profile and a correlation with self-incompatibility (Cornish et al., 1987).

All the characteristics of *S-RNase* suggest that this gene is the candidate of the pistil *S* determinant, and the transformation experiments of gain-of-function and loss-of-function provided direct evidence. A 3.6 kb DNA fragment containing *PiS*$_3$-*RNase* and its own promoter was introduced into S_1S_2 in *Petunia inflata*. Pollination analysis showed that four of eighty transgenic plants can completely reject the pollen of S_3S_3 and get no seed set, while other transgenic plants got fewer seeds than fully compatible pollinations. Fast protein liquid chromatography analysis showed that the protein expression level of S_3-RNase is higher in completely incompatible plants. By contrast, when the antisense sequence of *PiS*$_3$-*RNase* was introduced into

S_2S_3 to inhibit the expression level of S_3-RNase, five of the thirty transgenic plants lost the ability to reject S_3 pollen (Lee et al., 1994). The genetic function of *NaS*$_{A2}$-*RNase* of *N. alata* was verified by the similar transformation experiments. A construct containing the *NaS*$_{A2}$-*RNase* gene driven by the *Chip* promoter was introduced into $S_{C10}S_{C10}$, and transgenic plants produced the ability to completely or partially reject S_{A2} pollen (Murfett et al., 1994).

In addition, style part mutants of Rosaceae were studied in the 1990s, which verified the genetic function of S-RNase. S_4-*RNase* of a spontaneous compatible mutant from incompatible *P. serotine* with the genotype of S_2S_4 was unexpressed in the style, which caused the style losing the ability to reject S_4 pollen, and this mutant has no effect on pollen function (Sassa et al., 1997). The expression of S-RNase in 13 different cultivars of *P. dulcis* with known *S*-genotypes was analyzed by two-dimensional electrophoresis PAGE (2D-PAGE) and western blot. These experiments showed that the S_c-RNase point of "Jeffries" ($S_c^mS_d$) which is an SC mutant of "Nonpareil" (S_cS_d) is absent in the 2D-PAGE gel of style proteins (Tao et al., 1997). These studies confirmed *S-RNase* as the pistil *S*-specific determinant in Rosaceae.

To examine the pistil *S*-specific determinant, researchers attempted to identify recognition domains of S-RNase. Amino acid sequence alignment showed that S-RNases of Solanaceae and Plantaginaceae species contain five conserved domains C1-C5 and two hypervariable domains Hva and Hvb while the C4 domain of S-RNases of Plantaginaceae species has a poor identity with that of Solanaceae. The S-RNases found in Rosaceae species contain four conserved domains (C1, C2, C3, C5) that are similar to those of Solanaceae, as well as a distinct RC4 domain specific to Rosaceae. Besides, this kind of S-RNase only contains one hypervariable region called the RHV, which corresponds to the Hva domain of other species (Ioerger et al., 1991; Ishimizu et al., 1998; Xue et al., 1996). All of the *S-RNase* genes have an intron located inside the Hva domain of the gene or the RHv domain of Rosaceae species; however, there is still a special case that the *S-RNase* genes of *Prunoideae* have an extra intron located between the signal peptide and the C1 domain.

Hva and Hvb domains of S-RNases were inferred to participate in the specific recognition. There are four different amino acid residues between the Hva domains of *S. chacoense* S_{11}-RNase and S_{13}-RNase. When the different amino acids of S_{11}-RNase were exchanged for those of S_{13}-RNase, the mutant S_{11}-RNase lost the ability to reject S_{11} pollen and possessed the ability to reject S_{13} pollen (Matton et al., 1997, 1999). However, the domain swapping experiments in *Nicotiana* showed that all mutant *S-RNases* lost the ability to reject self-pollen (Zurek et al., 1997). In summary, hypervariable regions play an important role in the discrimination of self- and non-self-pollen but other parts of S-RNase sequence are also necessary.

S-RNase is a glycol protein. To study the function of carbohydrate chain of S-RNase, Karunanandaa et al. (1994) mutated the glycosylation site of *PiS*$_3$-*RNase* and introduced it into S_1S_2. The transgenic plants still obtain the ability to reject S_3 pollen, which demonstrates that the glycosylation site does not take part in the pollen recognition and rejection function of *PiS*$_3$-*RNase* (Karunanandaa et al., 1994). Furthermore, a recent study in *S. chacoense* showed that the number of carbohydrate chains of S-RNases may influence pollen rejection thresholds of S-RNase (Liu et al., 2008).

16.4.3 The Pollen S *Determinant*

Several early studies have shown a genetic feature of the pollen *S* known as competitive interaction. Lewis (1942) found that allotetraploids of *Pyrus*, *Petunia,* and *Solanum* are compatible, and only heterozygous diploid pollen grains are compatible due to so-called competitive interaction (Crane & Lewis, 1942; Lewis, 1942). Eventually, Brewbaker and Natarajan (1960) found that an extra centric fragment makes the SC pollen possess another *S*-allele in the haplotype pollen (Brewbaker & Natarajan, 1960). Nevertheless, it has taken a long time to identify the pollen *S* gene in this SI system. In *P. inflata*, using RNA differential display experiments, 10 cDNA markers were identified that are tightly linked to *S-RNase* genes and showed restriction fragment length polymorphism (RFLP) among different *S* genotypes (McCubbin et al., 2000). Unfortunately, none of these genes were identified as the pollen *S* gene.

Lai et al. (2002) cloned a gene that is tightly linked to the *S-RNase* gene in a bacterial artificial chromosome (BAC) library of a self-incompatible *Antirrhinum* line with S_2S_4 genotypes. Through sequencing the S_2BAC containing *AhS*$_2$*-RNase*, 11 other genes were identified, and among them *gene11* located 9 kb away from the S_2*-RNase* gene has an uninterrupted ORF of 376 amino acids and was expressed in both immature and mature anthers. Analysis of the sequence showed that *gene11* encodes a conserved F-box domain at the N-terminal and three conserved domains C1-C3 located outside the F-box. But the regions between the conserved domains exhibit divergence. According to these characteristics, it was named as the *A. hispanicum S locus F-box-*S_2 (Ah*SLF-*S_2). The amino acid sequence of *AhSLF-*S_2 has 97.5% identity with its allele *AhSLF-*S_2 and *AhSLF-*S_2 can only be specifically detected in S_2-containing lines, which indicated that *AhSLF-*S_2 is polymorphic and allele-specific. *AhSLF-*S_2 is shown to be specifically expressed in pollen via RT-PCR and RNA in situ hybridization analysis. Although without direct evidence, the characteristics of *AhSLF-*S_2 totally agree with those of the pollen *S* gene so that this gene was determined to be the candidate for the pollen *S*-specific determinant gene (Lai et al., 2002).

To demonstrate its genetic function, Qiao et al. (2004a) introduced a transformation-competent artificial chromosome (TAC) that contains *AhSLF-*S_2 and *AhS*$_2$*-RNase* into SI *P. hybrida* S_3S_3 to examine the molecular genetic function of *AhSLF-*S_2. *AhSLF-*S_2 has a specific expression in pollen of *P. hybrida* S_3S_3. After self-pollination, all four of the transgenic plants got capsules with seeds, indicating the breakdown of SI. Wild-type *P. hybrida* S_3S_3 plants pollinated with transgenic pollen also set seeds. As a control, researchers also pollinated S_3S_3 and transgenic plants with wild-type S_3S_3 pollen, and both got no capsule. These pollination experiments ensured that it is the transgene that causes the SC phenotype of pollen (Qiao et al., 2004a). Similarly, Sijacic et al. (2004) verified the function of *PiSLF2* by introducing it into self-incompatible *P. inflata* S_1S_2, S_1S_3, and S_1S_1. These experiments indicated that *SLF* is the pollen *S*-specific determinant gene of Plantaginaceae and Solanaceae species (Qiao et al., 2004a; Sijacic et al., 2004).

By sequencing and segregation analyses of S_2TAC, S_5TACa, S_4TAC, and S_1TAC vectors of *Antirrhinum*, Zhou et al. (2003) found different numbers of other F-box genes linked with different haplotypes of *S-RNase* besides *AhSLF,* and these genes are all pollen specifically expressed with the characteristics of pollen *S*-specific determinant (Zhou et al., 2003). Six *PiSLF*-like (*PiSLFL*) genes were also identified at the *S*-locus of *P. inflata* (Hua et al., 2007).

There is only one *S-RNase* gene at the *S*-locus of Solanaceae, Plantaginaceae, Rosaceae, and Rutaceae species, while the number of *SLF* genes is more than one and variable in different species. Besides, the allelic diversity of *SLF* genes is much lower than that of S-RNase genes. Kubo et al. (2010) cloned 40 *SLF*s in different haplotypes of *P. inflata* using 5′- and 3′-RACE, and these *SLF*s were separated into six different types by phylogenetic analysis. In order to verify the in vivo function of them, they carried out a series of transgenic experiments. When S_7-*SLF1* gene was separately introduced into S_5S_7, S_7S_9, S_7S_{11}, and S_5S_{19}, only the SI of S_7S_9 was disrupted, though the expression of S_7-*SLF1* is at a similar level in different transgenic plants. This result suggested that S_7-*SLF1* could not disrupt the SI of S_5-, S_7-, S_{11}-, and S_{19}-pollen. When S_7-*SLF2* was introduced into S_5S_{11}, S_5S_9, S_5S_{19}, S_7S_9, and S_7S_{11}, the SI of all the transgenic plants was disrupted. Genotype analysis of progenies showed that there are no S_5S_5 and S_7S_7 progenies, which suggested that only the SI of S_9-, S_{11}-, and S_{19}-pollen was disrupted by S_7-*SLF2*. In conclusion, S_7-*SLF1* can cause competitive interaction in S_9-pollen but cannot in S_5-, S_7-, S_{11}-, and S_{19}-pollen while S_7-*SLF2* can cause competitive interaction in S_9-, S_{11}-, and S_{19}-pollen but cannot in S_7- and S_5-pollen. So both *SLF*s can only recognize and detoxify part haplotypes of S-RNases. They proposed a collaborative recognition model in which the *S*-locus of one haplotype comprises multiple types of *SLF* genes. Each type of SLFs can specifically recognize a portion of S-RNases, while these SLFs altogether collaboratively recognize S-RNases from all haplotypes except the self (Kubo et al., 2010).

16.4.4 The Pollen S *Determinant in Rosaceae*

By sequencing a 70 kb region contains *S-RNase* of *Prunus dulcis*, two homologs of *AhSLF-S₂* were identified, which are the *SFB* (*S haplotype-specific F-box*) and *SLF*. The similarity between different alleles of *SFB* genes was 68.4 to 76.4% while that of *SLF* genes was 95.7%. Considering the low sequence diversity of *SLF* genes, they were not regarded as the pollen *S*-specific determinant genes of *P. dulcis*, but the high polymorphism among *SFB* alleles was comparable to that of *S-RNase* alleles. Besides, RT-PCR revealed that *SFB*s are specifically expressed in pollen, and some *SLF*s are expressed both in pollen and style. These features of *SFB* make it the candidate for the *S*-determinant of *P. dulcis* (Ushijima et al., 2003). The *S*-loci of S_1- and S_7-haplotypes of *P. mume* were also sequenced, in which six *F-box* genes were identified. Similarly, based on the expression and sequence analyses, *SLFs*/*SFBs* were suggested to be the pollen *S*-specific determinant genes (Entani et al., 2003).

Ushijima et al. (2004) showed that in *PPM*s of *P. avium*, $SFB^{4'}$ and $SFB^{f'}$ were mutated. A 4 bp deletion of SFB^4 causes a frame-shift, which could eventually lead to a lack of two hypervariable domains of products, and a 6.8 kb insertion in the middle of the SFB^f coding region causes loss of the C terminal of the transcripts (Ushijima et al., 2004). *SFB* mutations of S_3- and S_5-haplotypes were then discovered in other *P. avium* SC mutants (Marchese et al., 2007; Sonneveld et al., 2005). Though *SFB*s rather than *SLFL*s were suggested as candidates of the pollen *S*-specific determinant of *Prunus* species, however, there is still different evidence suggesting otherwise. A 358 bp insertion of SFB_C was discovered in an SC *PPM* of *P. mume* S_2S_C. Pollination analysis showed that the self-pollinated progenies of S_2S_C contain S_2S_2, S_2S_C, and S_CS_C individuals, which suggested that both S_2- and S_c-pollen grains are self-compatible; thus, there might be a mutation site unlinked to *S-RNase* causing the SC of pollen (Vilanova et al., 2006). A mutation located at chromosome 3 of *P. avium* causes disruption of the SI of pollen (Zuriaga et al., 2013). Recent studies in *P. avium* and *P. persica* proposed that SLFLs may play as general inhibitors of all S-RNase (Matsumoto & Tao, 2016; Chen et al., 2018a). Thus, the reason leading to the SC of *Prunus* pollen is still unclear, and whether the *S*-specific determinants of *Prunus* are *SFB*s or *SFB*s and *SLFL*s remains to be further examined.

Cheng et al. cloned *SLF1*, *SLF2,* and six *SLF-like* genes linked with *S-RNase* in *Malus* × *domestica* (Cheng et al., 2006). By analyzing the *S*-locus of *Malus* × *domestica* and *P. pyrifolia*, two *F-box* genes at a 317 kb region of the apple S_9-haplotype and six pollen specifically expressed *F-box* genes in *P. pyrifolia* S_4S_5 were identified. These genes were named as the *S* locus *F-box* brothers (*SFBB*) (Sassa et al., 2007). All of these *SFBB* genes are polymorphic and tightly linked to *S-RNase*s and are specifically expressed in pollen, and their features meet those of the pollen *S*-specific determinant genes. Besides, no *PPM* caused by the mutation of *SFBB*s has been discovered in *Maloideae* species up to now.

16.4.5 S-Haplotype-Independent S-RNase Uptake into Pollen Tubes

An immunocytochemical labeling experiment of pollen tubes growing in styles showed that S_{11}-RNase of *S. chacoense* can be observed in both incompatible S_{11} pollen tubes and compatible S_{12} pollen tubes (Luu et al., 2000). It is clear that S-RNases are absorbed by pollen tubes without discrimination of *S*-haplotype, while it is unclear whether S-RNase uptake is mediated by membrane transporters or endocytosis. Goldraij et al. (2006) found that S-RNase is located in ribbon-like vacuoles of pollen tubes, which suggested S-RNase may uptake by endocytosis but no direct evidence and details have been identified. On the other hand, researchers have identified several factors that may participate in the transportation of S-RNase.

The ATP-binding cassette family is a large integral membrane protein family of both eukaryotes and prokaryotes that participates in the transportation of various substances, from ions to macromolecules across membranes using hydrolysis of ATP to power the process. A recent study in *Malus domestica* found a protein belonging to group F of the ATP-binding cassette family (MdABCF) can interact with S-RNase and assist in the transportation of S-RNase non-selectively by coordinating with the cytoskeleton in pollen tubes. Suppression of *PiABCF* in *Petunia inflata* causes breakdown of SI. According to this study, ABCF plays an important role in transportation of S-RNase both in the Rosaceae and Solanaceae species (Meng et al., 2014a). Another study in apple found that F-actin and Golgi vesicles participate in the transportation of S-RNase. Golgi apparatus mainly functions in the secretory, the vacuolar trafficking, and the endocytic pathway, which suggested that S-RNase may be transported by endocytosis (Meng et al., 2014b). Thus, ABCF may coordinate with the cytoskeleton to regulate the endocytosis of S-RNase into pollen tubes.

More recently, MdHT1 (hexose transporter 1) is a pollen-expressed hexose transporter of apple. The expression level of *MdHT1* is lower in an SC apple cultivar *CAU-1* (S_1S_9) than that in an SI cultivar *"Fuji"* (S_1S_9), and the *S*-specific determinant genes are all normal in *CAU-1*. Reciprocal cross between *CAU-1* and *"Fuji"* showed that progenies can be obtained only when CAU-1 serves as a pollen donor, which suggested that it is the pollen of CAU-1 that lost the SI function. Biochemical analysis showed that MdHT1 can interact with S-RNase. Less S-RNase was absorbed by pollen tubes of *CAU-1* than *"Fuji"* in vitro when they were treated with S_1- and S_9-RNase, which may attenuate their inhibition to CAU-1 pollen tubes. Inhibition of *MdHT1* in *"Fuji"* pollen tube arrests the inhibition of pollen tube growth by self S-RNase. Hence, MdHT1 is a pollen factor of apple that may participate in the absorption of S-RNase, but the exact function of this factor is unknown (Li et al., 2016c).

16.4.6 Cytotoxicity Mechanism of S-RNase

It is considered that the ribonuclease activity of S-RNase may play a toxicity role in inhibiting pollen tube growth, and rRNAs in incompatible pollen tubes are degraded (McClure et al., 1990). Alignment between S-RNase and T2 RNase identified the conserved domains that are associated with the ribonuclease activity located in the C2 and C3 domains. Further analysis found that two conserved histidine residues, both in S-RNase and fungal ribonucleases T2, are essential for the ribonuclease activity (Huang et al., 1994). A point mutation in C2 domain making the conserved His-33 residue change into an Asn (asparaginate) causes the loss of ribonuclease activity of Sc-RNase and eventually leads to SC of *Sc* allele in *L. peruvianum* (Royo et al., 1994). This study first proposed the probable relationship between ribonuclease activity and the pollen rejection function of S-RNase. Huang et al. (1994) created a mutant *Pi*S_3-RNase by changing the His-93 residue, which is also essential for ribonuclease activity, to an Asp and introduced the mutant *S_3-RNase* into S_1S_2. The transformed plant failed to reject S_3 pollen (Huang et al., 1994). Recently, Torres-Rodríguez et al. (2020) found that the ribonuclease activity of S_{C10}-RNase could be significantly enhanced when its conserved Cys155-Cys185 disulfide bond was reduced by *Nicotiana alata* thioredoxin type h (NaTrxh). Mutation of NaTrxh can suppress the *S*-specific pollen rejection, which may be due to the decreased ribonuclease activity of S-RNase. These studies showed that the ribonuclease activity is necessary for the cytotoxicity of S-RNase while the study of Gray et al. (1991) found a contrary result. The ribonuclease activity of S-RNase was lost by heat treatment, but the ability to inhibit the elongation of self-pollen tubes by the treated S-RNase was enhanced (Gray et al., 1991). A style grafting experiment showed that incompatible reaction of pollen tubes was fully reversed by grafting the incompatible styles onto compatible ones, which could not happen if RNAs of pollen tubes were degraded (Lush & Clarke, 1997).

As we know, S-RNase is T2 ribonuclease; it is reported that human RNASET2 and *Aspergillus* ACTIBIND can interact with actin (Leeuw et al., 2007). An Y2H assay showed that the *P. avium* S-RNase can interact with actin, and the binding of S-RNase to actin may disrupt the actin dynamics then arrest the growth of pollen tubes (Matsumoto & Tao, 2012). The eukaryotic elongation factor 1 (eEF1) was first identified from pollen proteins of *S. chacoense*. Pull-down experiments showed that this protein can interact with S-RNase in vitro. Besides, S-RNase can enhance the interaction between eEF1A and actin which causes actin-forming bundles. An immune-gold standard test showed the colocalization of S_{11}-RNase and eEF1A in vivo. The study implicated that eEF1A appears to help S-RNase arrest the elongation of pollen tubes through interacting with actin. These results showed that S-RNase can potentially disrupt actin dynamics or inhibit the depolymerization of actin via eEF1A to arrest the growth of pollen tubes (Soulard et al., 2014). Chen et al. (2018b) showed that *Pyrus bretschneideri* PbrS-RNase can directly interact with PbrActin1, leading to actin depolymerization and PCD in self-pollen tubes. But phospholipase D (PbrPLD1)-derived PA can protect actin from depolymerization in early SI response until PbrS-RNase accumulates to a threshold, causing cessation of pollen tube growth.

Ca^{2+} is an important factor to regulate the growth of pollen tubes and plays an essential role in *Papaver* SI. S-RNase also arrests pollen tube growth by inhibiting the concentration of Ca^{2+} in the cytosol of the pollen tube. Qu et al. (2017) found that self S-RNase can inhibit the activity of Ca^{2+}-permeable channel by decreasing the activity of phospholipase C (PLC) in the membrane of pollen tubes in *Pyrus*. PLC regulates the hydrolysis of PIP2 into inositol 1, 4, 5-trisphosphate (IP3) and diacylglycerol (DG). IP3 can interact with self S-RNase rather than non-self S-RNases and as a secondary messenger, IP3 regulates the concentration of Ca^{2+} which then influences the growth of pollen tubes. Adding either self S-RNase or inhibitors of PLC into culture medium can decrease the concentration of Ca^{2+} and then inhibit the growth of pollen tubes of *Pyrus*. Consistent with the polymorphism of *S-RNase* genes, *PLC* also showed polymorphism in *Pyrus* of different *S*-genotypes. These discoveries suggested that the activity of PLC is inhibited by self S-RNase, which ultimately inhibits self-pollen tube growth (Qu et al., 2017). Calcineurin B subunit protein (CBL) is a calcium sensor protein family which can interact with CBL-interacting protein kinases (CIPKs) and form a CBL-CIPK complex to regulate calcium signaling. *MdCBL5* is highly expressed in pollen tubes of apples and participates in the regulation of pollen tube growth. When pollen tubes were treated with antisense oligonucleotide of *MdCBL5* in vitro, the growth of pollen tubes was inhibited and the concentration of Ca^{2+} was decreased compared with the control. qRT-PCR showed that the expression of *MdCBL5* can be inhibited by S-RNase treatment, so S-RNase controls the expression of *MdCBL5* to regulate the growth of pollen tubes by altering the concentration of Ca^{2+} in the tip of pollen tubes (Gu et al., 2015).

A study in *P. pyrifolia* showed that the S-RNase induced alterations of mitochondria in SI pollen tubes, which may ultimately lead to programmed cell death (PCD) (Liu et al., 2007b). In addition, a study in *Nicotiana* identified a factor NaSIPP (*N. alata* SI Pollen Protein) that may participate in SI by interacting with NaStEP and is specifically expressed in mature pollen. NaSIPP belongs to the MPC (Mitochondrial Phosphate Carrier) subfamily, which can collapse mitochondrial membrane potential by increasing the mitochondrial permeability then the ATP synthesis stops and leads to a lack of energy. NaStEP and NaSIPP interact in vivo and together locate in mitochondria and cause the opening of NaSIPP, resulting in an arrest of the pollen tube growth (Jiménez-Durán et al., 2013). Recently, Tang et al. (2023) revealed that S-RNase signaling can induce acetylation of *P. bretschneideri* inorganic pyrophosphatase 5 (PbrPPA5), further repressing the expression of the pectin methylesterase (PME) gene *PbrPME44* and resulting in swelling at self pollen tube tips.

16.4.7 Detoxification of S-RNase by the Ubiquitin Proteasome Pathway

It appears that S-RNase acts as a cytotoxin to suppress the elongation of self-pollen tube. Qiao et al. (2004b) found that the C-terminal (AhSLF2-C) rather than the N-terminal (AhSLF2-N) of AhSLF2 interacts with S-RNase without discrimination of *S*-haplotypes in pull-down and yeast two-hybrid assays. Co-immunoprecipitation experiments showed that AhSLF2 interacts with S-RNase in vivo (Qiao et al., 2004b). Hua et al. (2007) compared sequences between PiSLF alleles and identified two variable regions, Va and Vb. Based on the sequence alignment, SLF was separated into three functional domains (FD1, FD2, and FD3). A domain swapping method was adopted to detect interactions between S-RNase and chimeric SLFs, which showed that FD2 plays a major role in interacting with non-self S-RNase, and FD1 and FD3 are important for the recognition of non-self S-RNases. As Va and Vb were separately included in FD1 and FD3, variable regions of SLFs may play a vital role in recognition (Hua et al., 2007).

An Y2H analysis showed that the C-terminal region of PhS_3-SLF1 and PhS_{3L}-SLF1 and the HVa and HVb regions of S-RNase play dominant roles in recognition, consistent with the former studies. Analysis of the structure, surface hydrophobicity, and electrostatic potentials of these SLFs and S-RNases showed that the recognition relationships of SLF and S-RNase are consistent with the electronic characters of the C-terminal domain of SLF and the HVa and HVb regions of S-RNase. To identify the in vivo function of different domains of SLF, six domain-swapping chimeric *SLF* genes (L-3-3, 3-L-L, 3-L-3, L-3-L, 3–3-L, and L-L-3) were introduced into PhS_3S_{3L}. Chimeric genes of 3-L-L, 3-L-3, L-3-L, or 3–3-L disrupted the SI of S_3 pollen from PhS_3S_{3L}, which suggested that the second and third domains of S_{3L}-*SLF1* play a dominant role in S_3-RNase recognition. Through analyzing the selective pressure of each amino acid site of type 1 SLFs, the 293 site was identified as a positive selection site at which S_3-SLF1 has a histidine acid (H) and S_{3L}-SLF1 a glutamic acid (E). Exchange of the amino acid residue at the 293 site causes S_3-*SLF1* (H293E) to be able to disrupt the SI of S_3 pollen in the transgenic plant PhS_3S_{3L}. Considering the electronic characters of H and E are opposite under neutral conditions, the point mutation changed the surface electronic characters of S_3-SLF1 that ultimately makes it possess the ability to recognize S_3-RNase. In conclusion, surface electronic characters may play a key role in the recognition process between SLFs and non-self S-RNases, and according to this discovery, a "like charges repel and unlike charges attract" model was proposed (Li et al., 2017).

AhSLF can interact with Cullin1-like protein in vivo, which suggests AhSLF function as a component of $SCF^{AhSLF\text{-}S2}$ complex. AhSSK1 (*Antirrhinum hispanicum* SKP1-like1) was first isolated by a yeast two-hybrid screening against the cDNA library of pollen in which AhSLF serves as the bait. *AhSSK1* is specifically expressed in pollen without allelic specificity and unlinked to the *S*-locus (Huang et al., 2006). As the genetic function of *AhSSK1* cannot be verified by transgene, a homolog of *AhSSK1* was cloned in *P. hybrid* by homology-based cloning and was named as *PhSSK1* (*P. hybrid SLF-interacting Skp1-like1*). *PhSSK1* is also specifically expressed in pollen and is not an *S*-haplotype-specific gene either. An RNAi construct was introduced into SI *P. hybrida* $S_{3L}S_{3L}$ to test the in vivo function of *PhSSK1*. Compared with wild-type $S_{3L}S_{3L}$, transgenic lines got fewer seeds when pollens were used to pollinate S_VS_V and S_IS_I, and such phenotypes were inherited to T_1 plants. These experiments indicated that suppression of *PhSSK1* weakens the cross-compatibility of pollen. Yeast two hybrid and pull down assays showed that SSK1 can interact with SLF in vitro (Zhao et al., 2010). Homologs of *SSK1* have been identified from *P. inflata* and Rosaceae species, and the interaction between SSK1 and corresponding SLFs is also verified in these species. It is noteworthy that SSK1 of *P. avium* can interact with both SFBs and SLFLs (Matsumoto et al., 2012; Meng et al., 2011; Xu et al., 2013). IP-MS analysis showed that PiSSK1 can interact with 17 PiSLFs in vivo. In addition, CUL1 and Rbx1 were also identified in the products of Co-IP (Li et al., 2016b). Based on these studies, *SSK1* is an essential factor in CPC response by forming an SCF^{SLF} complex with SLF and Cullin1.

Eight homologous genes of *CUL* were identified from *P. hybrida,* and phylogenetic analysis with *AhCUL1* genes showed that there are five *CUL1* genes. Among them only *CUL1-P* specifically expresses in pollens and anthers. IP-MS and pull-down experiments indicated that among five CUL1s only CUL1-P interacts with SSK1. Artificial microRNAs (*amiRNA*) were introduced into *P. hybrida* to separately suppress the expressions of *CUL1-P* and *CUL1-G*. Only the suppression of CUL1-P causes reduction of seed sets during compatible pollination, which directly confirmed the genetic function of CUL1-P. So CUL1-P is the specific CUL1 for SCF^{SLF} complex (Kubo et al., 2016).

As is mentioned above, SLF can directly bind to SSK1 and interact with CUL1-P via SSK1. Entani et al. (2014) used FLAG-affinity beads to immunoprecipitate the protein extracts of *S_7-SLF2-FLAG*-transformed pollen of *P. inflata* and S_7-SLF2-FLAG, CUL1, PhSSK1, and PhRbx1 are all detectable in the products. The Co-IP products were used as E3 to ubiquitinate self- and non-self S-RNase in vitro. The results showed that the S_9-RNase and S_{11}-RNase are ubiquitinated, but S_7-RNase is not, which verified the function of SCF^{SLF} as an E3 ligase to ubiquitinate non-self S-RNase. The reduction of the level of ubiquitinated S_9-RNase is detectable in pollen extracts and such reduction can be inhibited by the addition of 100 uM MG132 into the extracts, which indicates that the ubiquitinated S-RNases are degraded by 26S proteasome (Entani et al., 2014). Treating compatible pollination of *Antirrhinum* with MG132 caused blocking of compatibility, indicating that non-self S-RNase is also degraded by the 26S proteasome pathway in *Antirrhinum* (Qiao et al., 2004b). Immunogold labeling and subcellular fractionation assays revealed that PhSLF and PhSSK1 are co-localized in the cytosol of pollen tubes. And S-RNase signals are also detectable in the matrix of the cytoplasm of pollen tubes no matter in compatible or incompatible pollination, and S-RNases are degraded in compatible pollen tubes (Liu et al., 2014). Taken together, SLF functions as a component of SCF^{SLF} complex to recognize self and non-self S-RNase in the cytoplasm of the pollen tube, and only non-self S-RNases are ubiquitinated for 26S proteasome-mediated degradation.

Genetic studies have identified *SFB* as the pollen-specific determinant of *Prunoideae,* while biochemical studies found that both SFB and SLFL can interact with SSK1. Recent studies in *P. avium* and *P. persica* proposed that SLFLs may play as general inhibitors, as they can interact with SSK1, CUL1 and all haplotypes of S-RNase including self in vitro, while SFB prevents self S-RNase from being degraded (Matsumoto & Tao, 2016). Nevertheless, Zhao et al. (2022) found that SFB and SLFL transgene can also disrupt pollen SI on the basis of competitive interaction in *P. hybrida,* suggesting that SFB and SLFL have a similar function like SLFs, consistent with their single origin in the MRCA of eudicots.

Though it seems clear that S-RNase is ubiquitinated by SCF^{SLF} complex, SBP1 (S-RNase-binding protein1) was identified by a yeast two-hybrid method screening against cDNA library of *P. hybrida* pollen with the Hv region of S-RNase as the bait. Analyses of the protein sequence and the expression pattern of SBP1 showed that this protein is a Ring-finger and is generally expressed in all organs. Later studies found that SBP1 can interact with PiSLF, S-RNase, PiCUL1, and E2 in *P.*

hybrida. And NaSBP1 was found interacting with Na120 kDa in *N. alata*. Though SBP1 can interact with different factors of SI response, the function of this protein is still unclear. As SBP1 can interact with S-RNase and E2, it may play as an E3, but no direct evidence has been found so far (Lee et al., 2008; Sims & Ordanic, 2001).

16.4.8 Detoxification of S-RNase by Compartmentalization

Goldraij et al. (2006) reported that S-RNase is compartmentalized in the SC pollen tube of *Nicotiana,* and the cytotoxicity of S-RNases could not be exerted until they were released from compartments (Goldraij et al., 2006). A small protein named HT identified, initially by differential hybridization experiments, is highly expressed in *N. alata* styles and stigmas and absent in *N. plumbaginifolia.* Sequence analysis showed that this 101-residue protein contains a secretion domain and an asparagine-rich domain (McClure et al., 1999). *HT* genes were also isolated from tomato and were divided into two groups, *HT-A* and *HT-B,* by a phylogenetic analysis. To test the function of different *HT* genes, an antisense RNA construct of *HT-A* and an RNAi construct of *HT-B* were separately transformed into an SI species, *S. chacoense.* Pollination and expression analyses showed that only suppression of *HT-B* led to failure of self-pollen rejection of pistil, which indicated that only HT-B participates in SI (Kondo et al., 2002). *HT-A* and *HT-B* genes were then cloned from tomato species (*L. esculentum* and *L. peruvianum*) and *P. inflata*. HT-B can be absorbed by both self- and non-self-pollen tubes and located on the membrane of vacuoles. Analysis of style extracts showed that the protein levels of HT-B drop to 3–25% of unpollinated controls after compatible pollination but only drop 50% in incompatible pollination. In *P. inflata*, only a $<$ 1000-fold reduction causes partial breakdown of SI (Raul et al., 2009). It is now clear that HT-B is an essential factor of SI and a small amount of HT-B is sufficient. Based on these studies, researchers supposed that HT-B was an important factor maintaining the stability of vacuoles that contain S-RNase. *N. alata* Stigma-Expressed Protein NaStEP is a Kunitz-type proteinase inhibitor that can stabilize HT-B in incompatible pollen tubes. Studies showed that NaStEP also can be absorbed by pollen tubes regardless of the *S*-haplotypes of the pollen. Expression detection of *NaStEP* in SC and SI *Nicotiana* showed that this gene is unexpressed in SC species. Suppressing the expression of *NaStEP* in pistil of SI species caused partially SC. HT-B was almost undetectable in *NaStEP*-suppressed $S_{105}S_0$ hybrid pollinated with S_{105} pollen 36 h post-pollination, but when SI *N. alata* $S_{105}S_{105}$ was challenged with S_{105} pollen, HT-B was stabilized (Jiménez-Durán et al., 2013).

Thus, S-RNase appears to be compartmentalized in vacuoles in pollen tubes, and these vacuoles can be maintained in compatible pollen tubes with S-RNase that cannot be released as HT-B is degraded by unknown mechanisms. By contrast, HT-B is stabilized by NaStEP in incompatibility and causes the disruption of vacuoles and release of S-RNase. Though HT-B is an important factor causing SI for Solanaceae, especially *Nicotiana* and *Solanum* species, its detailed functions are still unclear, and no *HT-B* gene has been identified in the Plantaginaceae, Rosaceae, and Rutaceae.

16.4.9 *Models for Solanaceae-Type SI*

16.4.9.1 S-RNase Degradation Model

The degradation model has been proposed since the first SLF was identified as the pollen *S*-specific determinant, considering it possesses an F-box domain (Lai et al., 2002). With the successive discovery of other components, the degradation model was further established. Recently, SSK1 and CUL1-P were identified as the specific factors of SCF^{SLF} complex (Huang et al., 2006; Kubo et al., 2016; Zhao et al., 2010), and SCF^{SLF} complex has been proved capable of ubiquitinating S-RNase in vitro (Entani et al., 2014). Besides, researchers have identified that multiple SLFs encoded by an *S* locus can collaboratively recognize all the non-self S-RNases except for the self-ones (Kubo et al., 2010). Furthermore, Zhao et al. (2021) revealed that non-self S-RNase is mainly polyubiquitinated through K48 linkages by SCF^{SLF} at three regions comprising six ubiquitination sites. These studies further support the S-RNase degradation model (Fig. 16.3c). In this model, when non-self S-RNases enter into pollen tubes, SLFs form SCF^{SLF} complexes with CUL1-P, SSK1, and Rbx1 to recognize and ubiquitinate S-RNases, leading to their degradation by 26S proteasomes as well as pollen tube growth. But when self S-RNases are absorbed into pollen tubes, no SCF^{SLF} can interact with them so that they exert the cytotoxicity to stop the growth of pollen tubes.

16.4.9.2 S-RNase Compartmentalization Model

HT-B was identified as an essential factor of SI in *Nicotiana* and supposed to break the stability of vacuoles containing S-RNases (McClure et al., 1999). 120 Kda, NaStEP, and NaSIPP were also proved to be involved in SI response of Solanaceae species (Cruz-Zamora et al., 2020; García-Valencia et al., 2017; Goldraij et al., 2006; Hancock et al., 2005; Jiménez-Durán et al., 2013). Based on these studies, the S-RNase compartmentalization model was proposed (Fig. 16.3c). This model proposed that S-RNases are absorbed into pollen tubes without discrimination of *S*-haplotypes. HT-Bs are absorbed together with S-RNases and locate on the surface of ribbon-like compartments. After endocytosis, a few of S-RNases are released from the compartments and recognized with SLFs to determine their fate. The interaction between non-self S-RNases and SLFs promotes a decrease of HT-B so that compartments maintain stability and most S-RNases cannot be released. Interaction between self S-RNases and SLFs has no effect on HT-B, and HT-B stabilizing by NaStEP during this process. Consequently, the compartments are disrupted by HT-B and S-RNases are released to exert their cytotoxicity to inhibit the growth of pollen tubes. On the other hand, S-RNases may activate the interaction between NaStEP and NaSIPP, thereby contributing to the arrest of pollen tube growth. Nevertheless, the function and degradation mechanisms of HT-B remain to be elucidated.

16.4.9.3 S-RNase Condensation Model

Recently, Tian et al. (2023) found that S-RNases can form S-RNase condensates (SRCs) in the self-pollen tube cytoplasm through its intrinsically disordered region (IDR)-mediated phase separation, and their disruption breaks SI in self-incompatible *P. hybrida*, thus proposing a condensation model (Fig. 16.3c) showing that SRCs are required for self-pollen rejection. In this model, pollen compatibility factors SLF and SSK1 restrict, whereas the pistil SI factors HT-B and Trxh promote the formation of SRCs. A reduced state promotes S-RNase condensation ensuing self-pollen inhibition, but an oxidative state can reduce SRC formation resulting in SC. Furthermore, the SRCs formed in self-pollen tubes mainly assemble and enrich ABPs, such as PhABRACL, a promoter of actin polymerization. Therefore, actin filament organizations are disrupted to form actin foci in the self but not the non-self-pollen tubes, resulting in the self-pollen rejection and cross-pollen acceptance. This finding revealed a new mode of S-RNase action in angiosperms.

16.5 Dimorphic SI

In hermaphrodite flower plants, there are several species possessing heterostyly, which is found in at least 28 angiosperm families. Heterostyly means that hermaphrodite individuals have different morphological types or "morph" of flowers in the same species. The differences focus mainly on style, length, and anther height. Generally, heterostyly has distyly and tristyly, containing two (short and long morphs) and three (short, mid, and long morphs) morphs, respectively. In distylous plants, the short morph has a short style and high anthers. By contrast, the long morph has a long style and low anthers. It is interesting that pollinations between anthers and stigmas possessing the same height can produce progeny. Thus, the floral morphs are linked to heteromorphic SI.

Darwin (1877) first reported the phenomena of heterostyly and heteromorphic SI in his book "The different forms of flowers on plants of the same species," noting the role of heterostyly in facilitating cross-pollinations through animal pollinators (Darwin, 1877). Since then, being plants of dimorphic flowers, *Primula* gradually became a model system for studying heterostyly. In *Primula*, heterostyly is controlled by the *S*-locus having two alleles, termed as dominant *S* (short style) and recessive *s* (long style) (Shang et al., 2025). Short and long morphs are described as thrum and pin by Darwin and determined by *Ss* and *ss* genotypes, respectively. An early genetic model for heterostyly in *Primula*, defined as supergene model, is established by classic genetic experiments. The *S*-locus is composed of three tightly linked genes (supergene), including *G*, *P*, and *A* (Bateson & Gregory, 1905; Bridges, 1914; Darlington, 1931; De Winton & Haldane, 1935; Ernst, 1936). *G* determines style length and its incompatibility, *P* pollen size and its incompatibility, and *A* anther height.

A recent study has revealed that the *S* haplotype of the short morph encompasses a genomic sequence of 455 kb. However, the central region of this sequence, spanning 278 kb, is absent in the *s* haplotype of the long morph. Thus, it is suggested that the short morph phenotype is determined by the 278 kb fragment comprising five tightly linked genes, including CCM^T, GLO^T, CYP^T, PUM^T, and KFB^T (Huu et al., 2016, 2020; Li et al., 2016a). In addition, short morph (*S/s*) has hemizygosity of the *S* locus, resulting in the absence of recombination of the *S* locus because of lacking homologous sequences between *S* and *s* haplotypes. Through analyses of expression and function of the *S* locus gene, it is found that null mutation of GLO^T in the short homostyle is linked to loss of anther height, whereas style length and pollen size are changed. Analyses also found that nonsense mutation of CYP^T in the long homostyle is associated with loss of short style, whereas anther elevation and pollen size are changed. These results suggest that GLO^T and CYP^T could control anther elevation and style length, respectively. The mechanism underlying style length of heterostyly in *Primula* is explained by another study (Nguyen et al., 2016). A cytochrome P450 family gene, defined as *CYP734A50* (CYP^T), was identified by RNA-seq-based differential expression analysis and encodes an enzyme that degrades brassinosteroid (BR) promoting plant cell elongation. To test the function of *CYP734A50*, expression of this gene in short morph was reduced by virus-induced gene silencing (VIGS) and resulted in the formation of a longer style. Consistently, BR has a high level in long morph but is undetectable in short morph (Huu et al., 2022). When short morph was treated by brassinolide (BL) in vitro, longer style occurred. These data demonstrate that the style length phenotype of distyly in *Primula* is controlled by *CYP734A50*, and the short style is resulted from BR degradation by CYP734A50. In Turneraceae, the *S* locus possesses three tightly linked genes, including *Turnera subulata TsBAHD*, specifically expressed in pistils; *TsYUC6* in anthers and *TsSPH1* in both filaments and anthers (Matzke et al., 2020; Shore et al., 2019). Although the pistil gene *TsBAHD* encodes a protein containing the conserved domain found in BAHD acyltransferases that differ from that of *Primula*, it can also inactivate BR leading to short styles (Matzke et al., 2021). Through genomic analyses of *Linum*, a 260-kb region unique to thrum individuals were uncovered, which also contains a pistil-expressed candidate gene controlling style length (Gutiérrez-Valencia et al., 2022).

16.6 Perspectives

Although much progress has been made in understanding the genetic and molecular regulation of SI systems in angiosperms, there are still far more questions than answers. In self-recognition Brassicaceae and Papaveraceae SI systems, the specific interaction between the female and male *S* determinants from the same *S* haplotype is largely based on the ligand-receptor recognition modes. Solanaceae, Plantaginaceae, Rosaceae, and Rutaceae employ non-self-recognition SI systems based on cytosolic protein-protein interactions. Extensive studies have shown the specific interaction regions mediating recognition between S-RNases and SLFs or SFBs. Physical forces based on electrostatic potentials of SLF appear to contribute to the pollen *S* specificity through a mechanism of "like charges repel and unlike charges attract" between SLFs and S-RNases. Nevertheless, many questions remain to be answered, including whether a threshold value of electrostatic potentials determines discrimination of self- and non-self-pollen, are there any other physical forces involved in specific interaction between SLFs and S-RNase. To further examine the non-self-recognition mechanism, a crystal structure of SLF bound to S-RNase would be essential.

Despite a long history of studying SI systems in angiosperms, we are still some distance from understanding how different SI systems arose and evolved. During angiosperm evolution, loss and regain of SI have frequently occurred under the fluctuating pressure of inbreeding and outcrossing. Although self-fertility is associated with inbreeding depression, it provides several benefits for species survivals. Loss of SI seems to be an optimal result from weighing pros and cons. Recent studies have revealed that duplication of type-1 *S*-locus, mutation, and inactivation of SI component and deletion of the entire *S*-locus directly lead to loss of SI, further dissecting the mechanisms of transition from SI to self-fertility would also help us discover new factors involved in SI and speciation. Notably, although type-1 SI present in Plantaginaceae, Solanaceae, Rosaceae, and Rutaceae, type-2 in Brassicaceae, type-3 in Papaveraceae, type-4 in Primulaceae, and type-5 in Turneraceae have been elucidated, the mechanisms underlying SI of other families in angiosperms particularly the top four known ones as Asteraceae, Orchidaceae, Fabaceae, and Poaceae are still mysterious. Further studies of other SI systems are expected to shed light on the origin and evolution of angiosperm SI as well as its relationship with the rapid expansions of angiosperms.

Acknowledgments This work was supported by grants from Ministry of Science and Technology (2013CB945102) and National Natural Science Foundation of China (31401045).

References

Abbott, R. J., & Gomes, M. F. (1989). Population genetic structure and outcrossing rate of Arabidopsis thaliana (L.) Heynh. *Heredity, 62*, 411–418.

Andrianantoandro, E., & Pollard, T. D. (2006). Mechanism of actin filament turnover by severing and nucleation at different concentrations of ADF/cofilin. *Molecular Cell, 24*, 13–23.

Barrett, S. C. H. (2002). The evolution of plant sexual diversity. *Nature Reviews. Genetics, 3*, 274–284.

Barrett, S. C. H. (2003). Mating strategies in flowering plants: The outcrossing-selfing paradigm and beyond. *Philosophical Transactions of the Royal Society of London. Series B, Biological Sciences, 358*, 991–1004.

Bateman, A. J. (1955). SI systems in angiosperms. (III). Cruciferae. *Heredity, 9*, 52–68.

Bateson, W., & Gregory, R. P. (1905). On the inheritance of heterostylism in *Primula*. *Proceedings of the Royal Society of London B, 76*, 581–586.

Bechsgaard, J., Bataillon, T., & Schierup, M. H. (2004). Uneven segregation of sporophytic SI alleles in *Arabidopsis lyrata. Journal of Evolutionary Biology, 17*, 554–561.
Bechsgaard, J., Castric, V., Charlesworth, D., Vekemans, X., & Schierup, M. H. (2006). The transition to self-compatibility in *Arabidopsis thaliana* and evolution within *S*-haplotypes over 10 Myr. *Molecular Biology and Evolution, 23*, 1741–1750.
Billiard, S., Castric, V., & Vekemans, X. (2006). A general model to explore complex dominance patterns in plant sporophytic SI systems. *Genetics, 175*, 1351–1369.
Boggs, N. A., Nasrallah, J. B., & Nasrallah, M. E. (2009). Independent *S*-locus mutations caused self-fertility in *Arabidopsis thaliana. PLoS Genetics, 5*, e1000426.
Bosch, M., & Franklin-Tong, V. E. (2007). Temporal and spatial activation of caspase-like enzymes induced by SI in *Papaver* pollen. *Proceedings. National Academy of Sciences. United States of America, 104*, 18327–18332.
Bošković, R., & Tobutt, K. R. (1996). Correlation of stylar ribonuclease zymograms with incompatibility alleles in sweet cherry. *Euphytica, 90*, 245–250.
Bower, M. S., Matias, D. D., Fernandescarvalho, E., Mazzurco, M., Gu, T., Rothstein, S. J., & Goring, D. R. (1996). Two members of the thioredoxin-h family interact with the kinase domain of a *Brassica S* locus receptor kinase. *Plant Cell, 8*, 1641–1650.
Boyes, D. C., & Nasrallah, J. B. (1995). An anther-specific gene encoded by an *S* locus haplotype of *Brassica* produces complementary and differentially regulated transcripts. *The Plant Cell, 7*, 1283–1294.
Boyes, D. C., Chen, C., Tantikanjana, T., Esch, J. J., & Nasrallah, J. B. (1991). Isolation of a second s-locus-related cDNA from *Brassica oleracea*: Genetic relationships between the *S* locus and two related loci. *Genetics, 127*, 221–228.
Boyes, D. C., Nasrallah, M. E., Vrebalov, J., & Nasrallah, J. B. (1997). The SI (*S*) haplotypes of *Brassica* contain highly divergent and rearranged sequences of ancient origin. *The Plant Cell, 9*, 237–247.
Braun, D. M., Stone, J. M., & Walker, J. C. (1997). Interaction of the maize and *Arabidopsis* kinase interaction domains with a subset of receptor-like protein kinases: Implications for transmembrane signaling in plants. *The Plant Journal, 12*, 83–95.
Bredemeijer, G. M. M., & Blaas, J. (1981). *S*-specific proteins in styles of self-incompatible *Nicotiana alata. Theoretical and Applied Genetics, 59*, 185–190.
Brewbaker, J. L. (1957). Pollen cytology and SI in plants. *The Journal of Heredity, 48*, 271–277.
Brewbaker, J. L., & Natarajan, A. T. (1960). Centric fragments and pollen-part mutation of incompatibility alleles in *Petunia. Genetics, 45*, 699–704.
Bridges, C. B. (1914). The chromosome hypothesis of linkage applied to cases in sweetpeas and *Primula. The American Naturalist, 48*, 524–534.
Broothaerts, W. J., van Laere, A., Witters, R., Preaux, G., Decock, B., van Damme, J., & Vendrig, J. C. (1990). Purification and N-terminal sequencing of style glycoproteins associated with SI in *Petunia hybrida. Plant Molecular Biology, 14*, 93–102.
Broothaerts, W., Vanvinckenroye, P., Decock, B., Damme, J. V., & Vendrig, J. C. (1991). *Petunia hybrida* S-proteins: Ribonuclease activity and the role of their glycan side chains in self-incompatibility. *Sexual Plant Reproduction, 4*, 258–266.
Broothaerts, W., Janssens, G. A., Proost, P., & Broekaert, W. F. (1995). cDNA cloning and molecular analysis of two SI alleles from apple. *Plant Molecular Biology, 27*, 499–511.
Busch, J. W., & Schoen, D. J. (2008). The evolution of SI when mates are limiting. *Trends in Plant Science, 13*, 128–136.
Cabrillac, D., Cock, J. M., Dumas, C., & Gaude, T. (2001). The *S*-locus receptor kinase is inhibited by thioredoxins and activated by pollen coat proteins. *Nature, 410*, 220–223.
Chai, L., Tudor, R. L., & Poulter, N. S. (2017). MAP kinase PrMPK9-1 contributes to the SI response. *Plant Physiology, 174*, 1226–1237.
Charlesworth, D., Mable, B. K., Schierup, M. H., Bartolomé, C., & Awadalla, P. (2003). Diversity and linkage of genes in the SI gene family in *Arabidopsis lyrata. Genetics, 164*, 1519–1535.

Chen, Q., Meng, D., Li, W., Gu, Z., Yuan, H., Duan, X., Yang, Q., Li, Y., & Li, T. (2018a). SLFL genes participate in the ubiquitination and degradation of S-RNase in self-compatible chinese peach. *Frontiers in Plant Science, 9*, 227.

Chen, J., Wang, P., de Graaf, B. H. J., Zhang, H., Jiao, H., Tang, C., Zhang, S., & Wu, J. (2018b). Phosphatidic acid counteracts S-RNase signaling in pollen by stabilizing the actin cytoskeleton. *The Plant Cell, 30*, 1023–1039.

Cheng, J., Han, Z., Xu, X., & Li, T. (2006). Isolation and identification of the pollen-expressed polymorphic F-box genes linked to the *S*-locus in apple (*Malus* x *domestica*). *Sexual Plant Reproduction, 19*, 175–183.

Chookajorn, T., Kachroo, A., Ripoll, D. R., Clark, A. G., & Nasrallah, J. B. (2004). Specificity determinants and diversification of the Brassica SI pollen ligand. *Proceedings. National Academy of Sciences. United States of America, 101*, 911–917.

Clarke, S. R., Staiger, C. J., Gibbon, B. C., & Franklin-Tong, V. E. (1998). A potential signaling role for profilin in pollen of *Papaver rhoeas*. *The Plant Cell, 10*, 967–979.

Cock, J. M., Stanchev, B., Delorme, V., Croy, R. R., & Dumas, C. (1995). SLR3: A modified receptor kinase gene that has been adapted to encode a putative secreted glycoprotein similar to the S locus glycoprotein. *Molecular & General Genetics, 248*, 151–161.

Cohen, G. B., Ren, R., & Baltimore, D. (1995). Modular binding domains in signal transduction. *Cell, 80*, 237–248.

Cole, R. A., & Fowler, J. E. (2006). Polarized growth: Maintaining focus on the tip. *Current Opinion in Plant Biology, 9*, 579–588.

Conner, J. A., Tantikanjana, T., Stein, J. C., Kandasamy, M. K., Nasrallah, J. B., & Nasrallah, M. E. (1997). Transgene-induced silencing of *S*-locus genes and related genes in *Brassica*. *The Plant Journal, 11*, 809–823.

Cornish, E. C., Pettitt, J. M., Bonig, I., & Clarke, A. E. (1987). Developmentally controlled expression of a gene associated with self-incompatibility in *Nicotiana alata*. *Nature, 326*, 99–102.

Crane, M. B., & Lewis, D. (1942). Genetical studies in pears. III Incompatibility and sterility. *Journal of Genetics, 43*, 31–49.

Cruz-Zamora, Y., Nájera-Torres, E., Noriega-Navarro, R., Torres-Rodríguez, M.D., Bernal-Gracida, L.A., García-Valdés, J., Juárez-Díaz, J.A., & Cruz-García, F. (2020). NaStEP an essential protein for self-incompatibility in *Nicotiana* performs a dual activity as a proteinase inhibitor and as a voltage-dependent channel blocker. *Plant Physiology and Biochemistry, 151*, 352–361.

Cui, Y., Bi, Y., Brugiere, N., Arnoldo, M., & Rothstein, S. J. (2000). The *S* locus glycoprotein and the *S* receptor kinase are sufficient for self-pollen rejection in *Brassica*. *Proceedings. National Academy of Sciences. United States of America, 97*, 3713–3717.

Darlington, C. D. (1931). Meiosis in diploid and tetraploid *Primula sinensis*. *Journal of Genetics, 24*, 65–95.

Darwin, C. (1877). *The different forms of flowers on plants of the same species*. Murray.

De Graaf, B. H., Rudd, J. J., Wheeler, M. J., Perry, R. M., Bell, E. M., Osman, K., Franklin, F. C., & Franklin-Tong, V. E. (2006). SI in *Papaver* targets soluble inorganic pyrophosphatases in pollen. *Nature, 444*, 490–493.

De Graaf, B. H., Vatovec, S., Juarez-Diaz, J. A., Chai, L., Kooblall, K., Wilkins, K. A., Zou, H., Forbes, T., Franklin, F. C., & Franklin-Tong, V. E. (2012). The *Papaver* SI pollen S-determinant, PrpS, functions in *Arabidopsis thaliana*. *Current Biology, 22*, 154–159.

de Nettancourt, D. (2001). *Incompatibility and incongruity in wild and cultivated plants*. Springer.

De Winton, D., & Haldane, J. B. S. (1935). The genetics of *Primula sinensis*. III. Linkage in the diploid. *Journal of Genetics, 31*, 67–100.

Dearnaley, J. D., Levina, N. N., Lew, R. R., Heath, I. B., & Goring, D. R. (1997). Interrelationships between cytoplasmic Ca^{2+} peaks, pollen hydration and plasma membrane conductances during compatible and incompatible pollinations of *Brassica napus* papillae. *Plant & Cell Physiology, 38*, 985–999.

Dickinson, H. G. (1995). Dry stigmas, water and SI in *Brassica*. *Sexual Plant Reproduction, 8*, 1–10.

dos Remedios, C. G., Chhabra, D., Kekic, M., Dedova, I. V., Tsubakihara, M., Berry, D. A., & Nosworthy, N. J. (2003). Actin binding proteins: Regulation of cytoskeletal microfilaments. *Physiological Reviews, 83*, 433–473.

Doughty, J., Hedderson, F., McCubbin, A., & Dickinson, H. (1993). Interaction between a coating-borne peptide of the Brassica pollen grain and stigmatic S (self-incompatibility)-locus-specific glycoproteins. *Proceedings. National Academy of Sciences. United States of America, 90*, 467.

Durand, E., Meheust, R., Soucaze, M., Goubet, P. M., Gallina, S., Poux, C., Fobisloisy, I., Guillon, E., Gaude, T., & Sarazin, A. (2014). Dominance hierarchy arising from the evolution of a complex small RNA regulatory network. *Science, 346*, 1200–1205.

East, E. M. (1940). The distribution of self-sterility in the flowering plants. *Proceedings. American Philosophical Society, 82*, 448–518.

East, E. M., & Mangelsdorf, A. J. (1925). A new interpretation of the hereditary behavior of self-sterile plants. *Proceedings. National Academy of Sciences. United States of America, 11*, 166–171.

Eaves, D. J., Haque, T., Tudor, R. L., Barron, Y., Zampronio, C. G., Cotton, N. P. J., de Graaf, B. H., White, S. A., Cooper, H. J., Franklin, F. C., Harper, J. F., & Franklin-Tong, V. E. (2017). Identification of phosphorylation sites altering pollen soluble inorganic pyrophosphatase activity. *Plant Physiology, 173*, 1606–1616.

Elleman, C. J., & Dickinson, H. G. (1990). The role of the exine coating in pollen-stigma interactions in *Brassica oleracea* L. *The New Phytologist, 114*, 511–518.

Elleman, C. J., & Dickinson, H. G. (1996). Identification of pollen components regulating pollination-specific responses in the stigmatic papillae of *Brassica oleracea*. *The New Phytologist, 133*, 197–205.

Elleman, C. J., & Dickinson, H. G. (1999). Commonalities between pollen/stigma and host/pathogen interactions: Calcium accumulation during stigmatic penetration by *Brassica oleracea* pollen tubes. *Sexual Plant Reproduction, 12*, 194–202.

Elleman, C. J., Franklintong, V. E., & Dickinson, H. G. (1992). Pollination in species with dry stigmas: The nature of the early stigmatic response and the pathway taken by pollen tubes. *The New Phytologist, 121*, 413–424.

Entani, T., Iwano, M., Shiba, H., Che, F. S., Isogai, A., & Takayama, S. (2003). Comparative analysis of the SI (*S*-) locus region of *Prunus mume*: Identification of a pollen-expressed F-box gene with allelic diversity. *Genes to Cells, 8*, 203–213.

Entani, T., Kubo, K., Isogai, S., Fukao, Y., Shirakawa, M., Isogai, A., & Takayama, S. (2014). Ubiquitin-proteasome-mediated degradation of S-RNase in a solanaceous cross-compatibility reaction. *The Plant Journal, 78*, 1014–1021.

Ernst, A. (1936). Weitere untersuchungen zur phänanalyse zum fertilitätsproblem und zur genetik heterostyler primeln. II. *Primula hortensis*. *Archiv der Julius Klaus-Stiftung für Vererbungsforschung, Sozialanthropologie und Rassenhygiene, 11*, 1–280.

Field, J., Vojtek, A., Ballester, R., Bolger, G., Colicelli, J., Ferguson, K., Gerst, J., Kataoka, T., Michaeli, T., Powers, S., et al. (1990). Cloning and characterization of CAP, the *S. Cerevisiae* gene encoding the 70 kd adenylyl cyclase-associated protein. *Cell, 61*, 319–327.

Foote, H. C., Ride, J. P., Franklin-Tong, V. E., Walker, E. A., Lawrence, M. J., & Franklin, F. C. (1994). Cloning and expression of a distinctive class of SI (*S*) gene from Papaver rhoeas L. *Proceedings. National Academy of Sciences. United States of America, 91*, 2265–2269.

Franklin-Tong, V. E. (2008). *Self-incompatibility in flowering plants*. Springer.

Franklin-Tong, V. E., Lawrence, M. J., & Franklin, F. C. H. (1988). An *in vitro* bioassay for the stigmatic product of the SI gene in *Papaver rhoeas* L. *The New Phytologist, 110*, 109–118.

Franklin-Tong, V. E., Ruuth, E., Marmey, P., Lawrence, M. J., & Franklin, F. C. H. (1989). Characterization of a stigmatic component from *Papaver rhoeas* L. which exhibits the specific activity of a SI (*S*-) gene product. *The New Phytologist, 112*, 307–315.

Franklin-Tong, V. E., Ride, J. P., Read, N. D., Trewavas, A. J., & Franklin, F. C. H. (1993). The SI response in *Papaver rhoeas* is mediated by cytosolic free calcium. *The Plant Journal, 4*, 163–177.

Franklin-Tong, V. E., Ride, J. P., & Franklin, F. C. H. (1995). Recombinant stigmatic SI (S-) protein elicits a Ca^{2+} transient in pollen of *Papaver rhoeas*. *The Plant Journal, 8*, 299–307.

Franklin-Tong, V. E., Hackett, G., & Hepler, P. K. (1997). Ratio-imaging of Ca^{2+} in the SI response in pollen tubes of *Papaver rhoeas*. *The Plant Journal, 12*, 1375–1386.

Franklin-Tong, V. E., Holdaway-Clarke, T. L., Straatman, K. R., Kunkel, J. G., & Hepler, P. K. (2002). Involvement of extracellular calcium influx in the SI response of *Papaver rhoeas*. *The Plant Journal, 29*, 333–345.

García-Valencia, L. E., Bravo-Alberto, C. E., Wu, H. M., Rodríguez-Sotres, R., Cheung, A. Y., & Cruz-García, F. (2017). SIPP, a novel mitochondrial phosphate carrier, mediates in self-incompatibility. *Plant Physiology, 175*, 1105–1120.

Geitmann, A., Snowman, B. N., Emons, A. M., & Franklin-Tong, V. E. (2000). Alterations in the actin cytoskeleton of pollen tubes are induced by the SI reaction in *Papaver rhoeas*. *The Plant Cell, 12*, 1239–1251.

Giranton, J. L., Dumas, C., Cock, J. M., & Gaude, T. (2000). The integral membrane *S*-locus receptor kinase of *Brassica* has serine/threonine kinase activity in a membranous environment and spontaneously forms oligomers in planta. *Proceedings. National Academy of Sciences. United States of America, 97*, 3759–3764.

Goldraij, A., Kondo, K., Lee, C. B., Hancock, C. N., Sivaguru, M., Vazquez-Santana, S., Kim, S., Phillips, T. E., Cruz-Garcia, F., & McClure, B. (2006). Compartmentalization of S-RNase and HT-B degradation in self-incompatible *Nicotiana*. *Nature, 439*, 805–810.

Goring, D. R., Glavin, T. L., Schafer, U., & Rothstein, S. J. (1993). An S receptor kinase gene in self-compatible *Brassica napus* has a 1-bp deletion. *The Plant Cell, 5*, 531.

Gray, J. E., McClure, B. A., Bönig, I., Anderson, M. A., & Clarke, A. E. (1991). Action of the style product of the SI gene of *Nicotiana alata* (S-RNase) on *in vitro*-grown pollen tubes. *The Plant Cell, 3*, 271.

Gruber, F. (1930). *Über Selbstssterilitat und Selbstfertilitatbe Antirrhinum.*

Gu, T., Mazzurco, M., Sulaman, W., Matias, D. D., & Goring, D. R. (1998). Binding of an arm repeat protein to the kinase domain of the *S*-locus receptor kinase. *Proceedings of the National Academy of Sciences, 95*, 382–387.

Gu, Z., Meng, D., Yang, Q., Yuan, H., Wang, A., Li, W., Chen, Q., Zhang, Y., Wang, D., & Li, T. (2015). A CBL gene, MdCBL5, controls the calcium signal and influences pollen tube growth in apple. *Tree Genetics & Genomes, 11*, 1–11.

Gutiérrez-Valencia, J., Fracassetti, M., Berdan, E. L., Bunikis, I., Soler, L., Dainat, J., Kutschera, V. E., Losvik, A., Désamoré, A., Hughes, P. W., Foroozani, A., Laenen, B., Pesquet, E., Abdelaziz, M., Pettersson, O. V., Nystedt, B., Brennan, A. C., Arroyo, J., & Slotte, T. (2022). Genomic analyses of the *Linum* distyly supergene reveal convergent evolution at the molecular level. *Current Biology, 32*, 4360–4371.

Hancock, C., Kent, L., & McClure, B. A. (2005). The stylar 120 kDa glycoprotein is required for S-specific pollen rejection in *Nicotiana*. *The Plant Journal, 43*, 716.

Hatakeyama, K., Watanabe, M., Takasaki, T., Ojima, K., & Hinata, K. (1998). Dominance relationships between S-alleles in self-incompatible *Brassica campestris* L. *Heredity, 80*, 241–247.

Hatakeyama, K., Takasaki, T., Suzuki, G., Nishio, T., Watanabe, M., Isogai, A., & Hinata, K. (2001). The S receptor kinase gene determines dominance relationships in stigma expression of SI in Brassica. *The Plant Journal, 26*, 69.

Hatakeyama, K., Horisaki, A., Niikura, S., Narusaka, Y., Abe, H., Yoshiaki, H., Ishida, M., Fukuoka, H., & Matsumoto, S. (2010). Mapping of quantitative trait loci for high level of SI in *Brassica rapa* L. *Genome, 53*, 257–265.

Hearn, M. J., Franklin, F. C. H., & Ride, J. P. (1996). Identification of a membrane glycoprotein in pollen of *Papaver rhoeas* which binds stigmatic SI (S-) proteins. *The Plant Journal, 9*, 467–475.

Hepler, P. K., & Winship, L. J. (2010). Calcium at the cell wall-cytoplast interface. *Journal of Integrative Plant Biology, 52*, 147–160.
Hiscock, S. J., Bown, D., Gurr, S. J., & Dickinson, H. G. (2002). Serine esterases are required for pollen tube penetration of the stigma in *Brassica. Sexual Plant Reproduction, 15*, 65–74.
Horisaki, A., & Niikura, S. (2008). Developmental and environmental factors affecting level of SI response in *Brassica rapa* L. *Sexual Plant Reproduction, 21*, 123–132.
Hua, Z., Meng, X., & Kao, T. H. (2007). Comparison of *Petunia inflata S*-locus F-Box protein (*Pi SLF*) with Pi SLF like proteins reveals its unique function in S-RNase based SI. *The Plant Cell, 19*, 3593–3609.
Huang, S., Lee, H. S., Karunanandaa, B., & Kao, T. H. (1994). Ribonuclease activity of *Petunia inflata* S proteins is essential for rejection of self-pollen. *The Plant Cell, 6*, 1021–1028.
Huang, S., Blanchoin, L., Chaudhry, F., Franklin-Tong, V. E., & Staiger, C. J. (2004). A gelsolin-like protein from *Papaver rhoeas* pollen (PrABP80) stimulates calcium-regulated severing and depolymerization of actin filaments. *The Journal of Biological Chemistry, 279*, 23364–23375.
Huang, J., Zhao, L., Yang, Q., & Xue, Y. (2006). AhSSK1, a novel SKP1-like protein that interacts with the S-locus F-box protein SLF. *The Plant Journal, 46*, 780–793.
Huu, C. N., Kappel, C., Keller, B., Sicard, A., Takebayashi, Y., Breuninger, H., Nowak, M. D., Bäurle, I., Himmelbach, A., Burkart, M., Ebbing-Lohaus, T., Sakakibara, H., Altschmied, L., Conti, E., & Lenhard, M. (2016). Presence versus absence of CYP734A50 underlies the style-length dimorphism in primroses. *Elife, 5*, e17956.
Huu, C. N., Keller, B., Conti, E., Kappel, C., & Lenhard, M. (2020). Supergene evolution via step-wise duplications and neofunctionalization of a floral-organ identity gene. *Proceedings of the National Academy of Sciences of the United States of America, 117*, 23148–23157.
Huu, C. N., Plaschil, S., Himmelbach, A., Kappel, C., & Lenhard, M. (2022). Female self-incompatibility type in heterostylous *Primula* is determined by the brassinosteroid-inactivating cytochrome P450 CYP734A50. *Current Biology, 32*, 671–676.
Igić, B., & Kohn, J. R. (2001). Evolutionary relationships among SI RNases. *Proceedings. National Academy of Sciences. United States of America, 98*, 13167–13171.
Ioannou, P. A., Amemiya, C. T., Garnes, J., Kroisel, P. M., Shizuya, H., Chen, C., Batzer, M. A., & De Jong, P. J. (1994). A new bacteriophage P1-derived vector for the propagation of large human DNA fragments. *Nature Genetics, 6*, 84–89.
Ioerger, T. R., Gohlke, J. R., Xu, B., & Kao, T. H. (1991). Primary structural features of the self-incompatibility protein in Solanaceae. *Sexual Plant Reproduction, 4*, 81–87.
Ishimizu, T., Shinkawa, T., Sakiyama, F., & Norioka, S. (1998). Primary structural features of rosaceous S-RNases associated with gametophytic SI. *Plant Molecular Biology, 37*, 931–941.
Isogai, A., Takayama, S., Shiozawa, H., Tsukamoto, C., Kanbara, T., Hinata, K., Okazaki, K., & Suzuki, A. (1988). Existence of a common glycoprotein homologous to S-glycoproteins in two self-incompatible homozygotes of *Brassica campestris. Plant & Cell Physiology, 29*, 1331–1336.
Iwano, M., Shiba, H., Miwa, T., Che, F., Takayama, S., Nagai, T., Miyawaki, A., & Isogai, A. (2004). Ca^{2+} dynamics in a pollen grain and papilla cell during pollination of *Arabidopsis. Plant Physiology, 136*, 3562–3571.
Iwano, M., Shiba, H., Matoba, K., Miwa, T., Funato, M., Entani, T., Nakayama, P., Shimosato, H., Takaoka, A., & Isogai, A. (2007). Actin dynamics in papilla cells of *Brassica rapa* during self- and cross-pollination. *Plant Physiology, 144*, 72.
Iwano, M., Ito, K., Fujii, S., Kakita, M., Asanoshimosato, H., Igarashi, M., Kaothiennakayama, P., Entani, T., Kanatani, A., & Takehisa, M. (2015). Erratum: Calcium signalling mediates SI response in the Brassicaceae, *1*, 15128.
Jain, S. K. (1976). The evolution of inbreeding in plants. *Annual Review of Ecology, Evolution, and Systematics, 7*, 469–495.
Jarne, P., & Charlesworth, D. (1993). The evolution of the selfing rate in functionally hermaphrodite plants and animals. *Annual Review of Ecology, Evolution, and Systematics, 24*, 441–466.

Ji, Y. H., Gothandam, K. M., Baek, N. K., Wang, G., & Chung, Y. Y. (2007). Dominance relationship between two self-incompatible *Brassica campestris* haplotypes in response to CO_2 gas. *Journal Of Plant Biology, 50*, 161–166.

Jiménez-Durán, K., McClure, B., Garcíacampusano, F., Rodríguezsotres, R., Cisneros, J., Busot, G., & Cruz-García, F. (2013). NaStEP: A proteinase inhibitor essential to SI and a positive regulator of HT-B stability in *Nicotiana alata* pollen tubes. *Plant Physiology, 161*, 97.

Jones-Rhoades, M. W., Borevitz, J. O., & Preuss, D. (2007). Genome-wide expression profiling of the Arabidopsis female gametophyte identifies families of small, secreted proteins. *PLoS Genetics, 3*, 1848–1861.

Jordan, N. D., Kakeda, K., Conner, A., Ride, J. P., Franklin-Tong, V. E., & Franklin, F. C. (1999). S-protein mutants indicate a functional role for SBP in the SI reaction of *Papaver rhoeas*. *The Plant Journal, 20*, 119–125.

Jordan, N. D., Franklin, F. C., & Franklin-Tong, V. E. (2000). Evidence for DNA fragmentation triggered in the SI response in pollen of *Papaver rhoeas*. *The Plant Journal, 23*, 471–479.

Kachroo, A., Schopfer, C. R., Nasrallah, M. E., & Nasrallah, J. B. (2001). Allele-specific receptor-ligand interactions in *brassica* SI. *Science, 293*, 1824–1826.

Kakeda, K., Jordan, N. D., Conner, A., Ride, J. P., Franklin-Tong, V. E., & Franklin, F. C. (1998). Identification of residues in a hydrophilic loop of the *Papaver rhoeas* S protein that play a crucial role in recognition of incompatible pollen. *The Plant Cell, 10*, 1723–1732.

Kakita, M., Shimosato, H., Murase, K., Isogai, A., & Takayama, S. (2007a). Direct interaction between S-locus receptor kinase and M-locus protein kinase involved in *Brassica* SI signaling. *Plant Biotechnology, 24*, 185–190.

Kakita, M., Murase, K., Iwano, M., Matsumoto, T., Watanabe, M., Shiba, H., Isogai, A., & Takayama, S. (2007b). Two distinct forms of M-locus protein kinase localize to the plasma membrane and interact directly with S-locus receptor kinase to transduce SI signaling in *Brassica rapa*. *The Plant Cell, 19*, 3961–3973.

Kakizaki, Y. (1930). Studies on the genetics and physiology of self and cross incompatibility in the common cabbage (*Brassica oleracea* L. var. capitata L). *Journal of Botany, 5*, 133–208.

Kakizaki, T., Takada, Y., Fujioka, T., Suzuki, G., Satta, Y., Shiba, H., Isogai, A., Takayama, S., & Watanabe, M. (2006). Comparative analysis of the S-intergenic region in class-II *S* haplotypes of self-incompatible *Brassica rapa* (syn. campestris). *Genes & Genetic Systems, 81*, 63–67.

Karunanandaa, B., Huang, S., & Kao, T. (1994). Carbohydrate moiety of the *Petunia inflata* S_3 protein is not required for SI interactions between pollen and pistil. *The Plant Cell, 6*, 1933–1940.

Kemp, B. P., & Doughty, J. (2007). S cysteine-rich (SCR) binding domain analysis of the *Brassica* self-incompatibility S-locus receptor kinase. *The New Phytologist, 175*, 619–629.

Kondo, K., Yamamoto, M., Matton, D. P., Sato, T., Hirai, M., Norioka, S., Hattori, T., & Kowyama, Y. (2002). Cultivated tomato has defects in both S-RNase and HT genes required for stylar function of SI. *The Plant Journal, 29*, 627.

Kubo, K., Entani, T., Takara, A., Wang, N., Fields, A. M., Hua, Z., Toyoda, M., Kawashima, S., Ando, T., Isogai, A., et al. (2010). Collaborative non-self recognition system in S-RNase-based SI. *Science, 330*, 796–799.

Kubo, K. I., Tsukahara, M., Fujii, S., Murase, K., Wada, Y., Entani, T., Iwano, M., & Takayama, S. (2016). Cullin1-P is an essential component of non-self recognition system in SI in *Petunia*. *Plant and Cell Physiology, 57*, pcw152.

Kurup, S., Ride, J. P., Jordan, N., Fletcher, G., Franklin-Tong, V. E., & Franklin, F. C. H. (1998). Identification and cloning of related SI *S*-genes in *Papaver rhoeas* and *Papaver nudicaule*. *Sexual Plant Reproduction, 11*, 192–198.

Kusaba, M., Nishio, T., Satta, Y., Hinata, K., & Ockendon, D. J. (1997). Striking sequence similarity in inter- and intra-specific comparisons of class I SLG alleles from *Brassica oleracea* and *Brassica campestris*: Implications for the evolution and recognition mechanism. *Proceedings of the National Academy of Sciences of the United States of America, 94*, 7673–7678.

Kusaba, M., Dwyer, K., Hendershot, J., Vrebalov, J., Nasrallah, J. B., & Nasrallah, M. E. (2001). SI in the genus *Arabidopsis*: Characterization of the *S* locus in the outcrossing *A. lyrata* and its autogamous relative *A. thaliana*. *The Plant Cell, 13*, 627–643.

Lai, Z., Ma, W., Han, B., Liang, L., Zhang, Y., Hong, G., & Xue, Y. (2002). An F-box gene linked to the SI (*S*) locus of *Antirrhinum* is expressed specifically in pollen and tapetum. *Plant Molecular Biology, 50*, 29–42.

Lalonde, B. A., Nasrallah, M. E., Dwyer, K. G., Chen, C. H., Barlow, B., & Nasrallah, J. B. (1989). A highly conserved *Brassica* gene with homology to the *S*-locus-specific glycoprotein structural gene. *The Plant Cell, 1*, 249.

Lao, X., Suwabe, K., Niikura, S., Kakita, M., Iwano, M., & Takayama, S. (2014). Physiological and genetic analysis of CO_2-induced breakdown of SI in *Brassica rapa*. *Journal of Experimental Botany, 65*, 939.

Lawrence, M. J. (1975). The genetics of SI in *Papaver rhoeas*. *Proceedings of the Royal Society of London. Series B, 188*, 275–285.

Lawrence, M. J., & O'Donnell, S. (1981). The population genetics of the SI polymorphism in *Papaver Rhoeas* III. The number and frequency of *S* alleles in two further natural populations R-102 and R-104. *Heredity, 47*, 53–62.

Lawrence, M. J., Lane, M. D., O'Donnell, S., & Franklin-Tong, V. E. (1993). The population genetics of the SI polymorphism in *Papaver Rhoeas* V. Cross-classification of the *S*-alleles of samples from 3 natural-populations. *Heredity, 71*, 581–590.

Lee, H. S., Huang, S., & Kao, T. (1994). S proteins control rejection of incompatible pollen in *Petunia inflata*. *Nature, 367*, 560–563.

Lee, S. H., Hong, M. Y., Kim, S., Lee, J. S., Kim, B. D., Min, B. H., Baek, N. K., & Chung, Y. Y. (2001). Controlling SI by CO_2 gas treatment in *Brassica campestris*: Structural alteration of papillae cell and differential gene expression by increased CO_2 gas. *Molecules and Cells, 11*, 186–191.

Lee, C. B., Swatek, K. N., & McClure, B. (2008). Pollen proteins bind to the C-terminal domain of *Nicotiana alata* pistil arabinogalactan proteins. *The Journal of Biological Chemistry, 283*, 26965–26973.

Leeuw, M. D., Roiz, L., Smirnoff, P., Schwartz, B., Shoseyov, O., & Almog, O. (2007). Binding assay and preliminary X-ray crystallographic analysis of ACTIBIND, a protein with anticarcinogenic and antiangiogenic activities. *Acta Crystallographica. Section F, Structural Biology and Crystallization Communications, 63*, 716–719.

Lewis, D. (1942). *Competiotion and dominance of incompatibility alleles in diploid pollen*.

Li, S., Samaj, J., & Franklin-Tong, V. E. (2007). A mitogen-activated protein kinase signals to programmed cell death induced by SI in *Papaver* pollen. *Plant Physiology, 145*, 236–245.

Li, J., Cocker, J. M., Wright, J., Webster, M. A., & McMullan, M. (2016a). Genetic architecture and evolution of the *S* locus supergene in *Primula vulgaris*. *Nature Plants, 2*, 16188.

Li, S., Williams, J. S., Sun, P., & Kao, T. H. (2016b). All 17 types of S-locus F-box proteins of S_2 - and S_3 -haplotypes of *Petunia inflata* are assembled into similar SCF complexes with specific function in SI. *The Plant Journal, 87*, 606–616.

Li, W., Yang, Q., Gu, Z., Wu, C., Meng, D., Yu, J., Chen, Q., Li, Y., Yuan, H., Wang, D., & Li, T. (2016c). Molecular and genetic characterization of a self-compatible apple cultivar, 'CAU-1'. *Plant Science, 252*, 162–175.

Li, J., Zhang, Y., Song, Y., Zhang, H., Fan, J., Li, Q., Zhang, D., & Xue, Y. (2017). Electrostatic potentials of the S-locus F-box proteins contribute to the pollen *S* specificity in SI in *Petunia hybrida*. *The Plant Journal, 89*, 45–57.

Liang, M., Cao, Z., Zhu, A., Liu, Y., Tao, M., Yang, H., Xu, Q., Jr., Wang, S., Liu, J., Li, Y., Chen, C., Xie, Z., Deng, C., Ye, J., Guo, W., Xu, Q., Xia, R., Larkin, R. M., Deng, X., Bosch, M., Franklin-Tong, V. E., & Chai, L. (2020). Evolution of self-compatibility by a mutant S_m*-RNase* in *Citrus*. *Nature Plants, 6*, 131–142.

Lin, Z., Eaves, D. J., Sanchez-Moran, E., Franklin, F. C., & Franklin-Tong, V. E. (2015). The *Papaver rhoeas* S determinants confer SI to *Arabidopsis thaliana* in planta. *Science (New York, N.Y.), 350*, 684–687.

Liu, P., Shermanbroyles, S., Nasrallah, M. E., & Nasrallah, J. B. (2007a). A cryptic modifier causing transient SI in *Arabidopsis thaliana*. *Current Biology, 17*, 734–740.

Liu, Z. Q., Xu, G. H., & Zhang, S. L. (2007b). *Pyrus pyrifolia* stylar S-RNase induces alterations in the actin cytoskeleton in self-pollen and tubes *in vitro*. *Protoplasma, 232*, 61.
Liu, B., Morse, D., & Cappadocia, M. (2008). Glycosylation of S-RNases may influence pollen rejection thresholds in *Solanum chacoense*. *Journal of Experimental Botany, 59*, 545–552.
Liu, W., Fan, J., Li, J., Song, Y., Li, Q., Zhang, Y. E., & Xue, Y. (2014). SCF^{SLF}-mediated cytosolic degradation of S-RNase is required for cross-pollen compatibility in S-RNase-based SI in *Petunia hybrida*. *Frontiers in Genetics, 5*, 228.
Lush, W. M., & Clarke, A. E. (1997). Observations of pollen tube growth in *Nicotiana alata* and their implications for the mechanism of SI. *Sexual Plant Reproduction, 10*, 27–35.
Luu, D.-T., Qin, X., Morse, D., & Mario. (2000). S-RNase uptake by compatible pollen tubes in gametophytic SI. *Nature, 407*, 649–651.
Ma, R., Han, Z., Hu, Z., Lin, G., Gong, X., Zhang, H., Nasrallah, J. B., & Chai, J. (2016). Structural basis for specific self-incompatibility response in *Brassica*. *Cell Research, 26*, 1320–1329.
Marchese, A., Boskovic, R. I., Caruso, T., Raimondo, A., Cutuli, M., & Tobutt, K. R. (2007). A new self-compatibility haplotype in the sweet cherry 'Kronio', S5', attributable to a pollen-part mutation in the SFB gene. *Journal of Experimental Botany, 58*, 4347–4356.
Matsubara, S. (1980). Overcoming SI in *Raphanus sativus* L. with high temperature. *Journal of the American Society for Horticultural, 105*, 842–846.
Matsumoto, D., & Tao, R. (2012). Isolation of pollen-expressed actin as a candidate protein interacting with S-RNase in *Prunus avium* L. *Journal of the Japanese Society for Horticultural Science, 81*, 41–47.
Matsumoto, D., & Tao, R. (2016). Distinct self-recognition in the *Prunus* S-RNase-based gametophytic SI system. *The Horticulture Journal, 85*, 289–305.
Matsumoto, D., Yamane, H., Abe, K., & Tao, R. (2012). Identification of a Skp1-like protein interacting with SFB, the pollen *S* determinant of the gametophytic SI in *Prunus*. *Plant Physiology, 159*, 1252.
Mattila, P., Quintero-Monzon, O., Kugler, J., Moseley, J., Almo, S., Lappalainen, P., & Goode, B. (2004). A high-affinity interaction with ADP-actin monomers underlies the mechanism and *in vivo* function of Srv2/cyclase-associated protein. *Molecular Biology of the Cell, 15*, 5158–5171.
Matton, D. P., Maes, O., Laublin, G., Xike, Q., Bertrand, C., Morse, D., & Cappadocia, M. (1997). Hypervariable domains of SI RNases mediate allele-specific pollen recognition. *The Plant Cell, 9*, 1757–1766.
Matton, D. P., Luu, D. T., Xike, Q., Laublin, G., O'Brien, M., Maes, O., Morse, D., & Cappadocia, M. (1999). Production of an S-RNase with dual specificity suggests a novel hypothesis for the generation of new *S* alleles. *The Plant Cell, 11*, 2087–2097.
Mattsson, O., Knox, R. B., Heslopharrison, J. S., & Heslopharrison, Y. (1974). Protein pellicle of stigmatic papillae as a probable recognition site in incompatibility reactions. *Nature, 247*, 298–300.
Matzke, C. M., Shore, J. S., Neff, M. M., & McCubbin, A. G. (2020). The *Turnera* style *S*-Locus gene *TsBAHD* possesses brassinosteroid-inactivating activity when expressed in *Arabidopsis thaliana*. *Plants, 9*, 1566.
Matzke, C. M., Hamam, H. J., Henning, P. M., Dougherty, K., Shore, J. S., Neff, M. M., & McCubbin, A. G. (2021). Pistil mating type and morphology are mediated by the brassinosteroid inactivating activity of the *S*-Locus gene *BAHD* in heterostylous *Turnera* species. *International Journal of Molecular Sciences, 22*, 10603.
McClure, B. A., Haring, V., Ebert, P. R., Anderson, M. A., Simpson, R. J., Sakiyama, F., & Clarke, A. E. (1989). Style SI gene products of *Nicotiana alata* are ribonucleases. *Nature, 342*, 955–957.
McClure, B. A., Gray, J. E., Anderson, M. A., & Clarke, A. E. (1990). Self-incompatibility in *Nicotiana alata* involves degradation of pollen rRNA. *Nature, 347*, 757–760.
McClure, B., Mou, B., Canevascini, S., & Bernatzky, R. (1999). A small asparagine-rich protein required for S-allele-specific pollen rejection in *Nicotiana*. *Proceedings of the National Academy of Sciences of the United States of America, 96*, 13548–13553.

McCubbin, A. G., & Kao, T. (2000). Molecular recognition and response in pollen and pistil interactions. *Annual Review of Cell and Developmental Biology, 16*, 333–364.
McCubbin, A. G., Wang, X., & Kao, T. H. (2000). Identification of SI (*S*-) locus linked pollen cDNA markers in *Petunia inflata*. *Genome, 43*, 619–627.
Meng, X., Sun, P., & Kao, T. H. (2011). S-RNase-based SI in *Petunia inflata*. *Annals of Botany, 108*, 637–646.
Meng, D., Gu, Z., Li, W., Wang, A., Yuan, H., Yang, Q., & Li, T. (2014a). Apple MdABCF assists in the transportation of S-RNase into pollen tubes. *The Plant Journal, 78*, 990–1002.
Meng, D., Gu, Z., Yuan, H., Wang, A., Li, W., Yang, Q., Zhu, Y., & Li, T. (2014b). The microtubule cytoskeleton and pollen tube golgi vesicle system are required for *in vitro* S-RNase internalization and gametic SI in apple. *Plant & Cell Physiology, 55*, 977–989.
Mishima, M., Takayama, S., Sasaki, K., Jee, J., Kojima, C., Isogai, A., & Shirakawa, M. (2003). Structure of the male determinant factor for *Brassica* SI. *The Journal of Biological Chemistry, 278*, 36389–36395.
Monteiro, A. A., Gabelman, W. H., & Williams, P. H. (1988). Use of sodium chloride solution to overcome SI in *Brassica campestris*. *HortScience, 23*, 876–877.
Muntoni, E., Canaparo, R., Della, P. C., Serpe, L., Casale, F., Barbera, S., Romano, P., Zara, G. P., & Eandi, M. (1970). SI proteins in plants: Detection, genetics, and possible mode of action. *Heredity, 25*, 23–27.
Murase, K., Shiba, H., Iwano, M., Che, F., Watanabe, M., Isogai, A., & Takayama, S. (2004). A membrane-anchored protein kinase involved in *Brassic*a SI signaling. *Science, 303*, 1516–1519.
Murase, K., Moriwaki, Y., Mori, T., Liu, X., Masaka, C., Takada, Y., Maesaki, R., Mishima, M., Fujii, S., Hirano, Y., Kawabe, Z., Nagata, K., Terada, T., Suzuki, G., Watanabe, M., Shimizu, K., Hakoshima, T., & Takayama, S. (2020). Mechanism of self/nonself-discrimination in *Brassica* self-incompatibility. *Nature Communications, 11*, 4916.
Murfett, J., Atherton, T. L., Mou, B., Gasser, C. S., & McClure, B. A. (1994). S-RNase expressed in transgenic *Nicotiana* causes *S*-allele-specific pollen rejection. *Nature, 367*, 563–566.
Naithani, S., Chookajorn, T., Ripoll, D. R., & Nasrallah, J. B. (2007). Structural modules for receptor dimerization in the S-locus receptor kinase extracellular domain. *Proceedings. National Academy of Sciences. United States of America, 104*, 12211–12216.
Nakanishii, T., Esashi, Y., & Hinata, K. (1969). Control of SI by CO_2 gas in *Brassica*. *Plant & Cell Physiology, 10*, 351–359.
Nasrallah, M. E., & Nasrallah, J. B. (2002). Generation of self-incompatible *Arabidopsis thaliana* by transfer of two *S* locus genes from *A. lyrata*. *Science, 297*, 247–249.
Nasrallah, M. E., Wallace, D. H., & Savo, R. M. (1972). Genotype, protein, phenotype relationships in SI of *Brassica*. *Genetical Research, 20*, 151–160.
Nasrallah, J. B., Kao, T. H., Goldberg, M. L., & Nasrallah, M. E. (1985). A cDNA clone encoding an S-locus-specific glycoprotein from *Brassica oleracea*. *Nature, 318*, 263–267.
Nasrallah, J. B., Kao, T. H., Chen, C. H., Goldberg, M. L., & Nasrallah, M. E. (1987). Amino-acid sequence of glycoproteins encoded by three alleles of the *S* locus of *Brassica oleracea*. *Nature, 326*, 617–619.
Nasrallah, J. B., Yu, S., & Nasrallah, M. E. (1988). SI genes of *Brassica oleracea*: Expression, isolation, and structure. *Proceedings. National Academy of Sciences. United States of America, 85*, 5551–5555.
Nasrallah, M. E., Kandasamy, M. K., & Nasrallah, J. B. (1992). A genetically defined trans-acting locus regulates *S*-locus function in *Brassica*. *The Plant Journal, 2*, 497–506.
Nasrallah, J. B., Rundle, S. J., & Nasrallah, M. E. (1994). Genetic evidence for the requirement of the *Brassica* S-locus receptor kinase gene in the self-incompatibility response. *The Plant Journal, 5*, 373–384.
Nasrallah, J. B., Liu, P., Sherman-Broyles, S., Schmidt, R., & Nasrallah, M. E. (2007). Epigenetic mechanisms for breakdown of SI in interspecific hybrids. *Genetics, 175*, 1965–1973.
Nguyen, H. C., Christian, K., Barbara, K., Adrien, S., Yumiko, T., Holger, B., Nowak, M. D., Isabel, B., Axel, H., & Michael, B. (2016). Presence versus absence of CYP734A50 underlies the style-length dimorphism in primroses. *Elife, 5*.

Niikura, S., & Matsuura, S. (2000). Genetic analysis of the reaction level of SI to a 4% CO_2 gas treatment in the radish (*Raphanus sativus* L.). *Theoretical and Applied Genetics, 101*, 1189–1193.

Nishio, T., & Hinata, K. (1978). Stigma proteins in self-incompatible brassica campestris l. and self-compatible relatives, with special reference to s-allele specificity. *Japanese Journal of Genetics, 53*, 27–33.

Nishio, T., Hinata, K., Nishio, T., & Hinata, K. (1977). Analysis of S-specific proteins in stigma of *Brassica oleracea* L. by isoelectric focusing. *Heredity, 38*, 391–396.

Nou, I. S., Watanabe, M., Isogai, A., & Hinata, K. (1993). Comparison of *S*-alleles and S-glycoproteins between two wild populations of *Brassica campestris* in Turkey and Japan. *Sexual Plant Reproduction, 6*, 79–86.

Ockendon, D. J. (1978). Effect of hexane and humidity on SI in *Brassica oleracea. Theoretical and Applied Genetics, 52*, 113–117.

Ockendon, D. J. (1982). An *S*-allele survey of cabbage (*Brassica oleracea* var. Capitata). *Euphytica, 31*, 325–331.

Okamoto, S., Odashima, M., Fujimoto, R., Sato, Y., Kitashiba, H., & Nishio, T. (2007). Self-compatibility in *Brassica napus* is caused by independent mutations in *S*-locus genes. *The Plant Journal, 50*, 391–400.

Okazaki, K., & Hinata, K. (1987). Repressing the expression of SI in crucifers by short-term high temperature treatment. *Theoretical and Applied Genetics, 73*, 496.

Palanivelu, R., Brass, L., Edlund, A. F., & Preuss, D. (2003). Pollen tube growth and guidance is regulated by POP2, an *Arabidopsis* gene that controls GABA levels. *Cell, 114*, 47–59.

Porcher, E., & Lande, R. (2005). Loss of gametophytic si with evolution of inbreeding depression. *Evolution, 59*, 46–60.

Poulter, N. S., Vatovec, S., & Franklin-Tong, V. E. (2008). Microtubules are a target for SI signaling in *Papaver* pollen. *Plant Physiology, 146*, 1358–1367.

Poulter, N. S., Staiger, C. J., Rappoport, J. Z., & Franklin-Tong, V. E. (2010). Actin-binding proteins implicated in the formation of the punctate actin foci stimulated by the SI response in *Papaver. Plant Physiology, 152*, 1274–1283.

Preuss, D., Lemieux, B., Yen, G., & Davis, R. W. (1993). A conditional sterile mutation eliminates surface components from *Arabidopsis* pollen and disrupts cell signaling during fertilization. *Genes & Development, 7*, 974–985.

Qiao, H., Wang, F., Zhao, L., Zhou, J., Lai, Z., Zhang, Y., Robbins, T. P., & Xue, Y. (2004a). The F-box protein AhSLF-S_2 controls the pollen function of S-RNase-based self-incompatibility. *The Plant Cell, 16*, 2307–2322.

Qiao, H., Wang, H., Zhao, L., Zhou, J., Huang, J., Zhang, Y., & Xue, Y. (2004b). The F-box protein AhSLF-S_2 physically interacts with S-RNases that may be inhibited by the ubiquitin/26S proteasome pathway of protein degradation during compatible pollination in *Antirrhinum. The Plant Cell, 16*, 582–595.

Qu, H., Guan, Y., Wang, Y., & Zhang, S. (2017). PLC-mediated signaling pathway in pollen tubes regulates the gametophytic SI of *Pyrus* species. *Frontiers in Plant Science, 8*, 1164.

Ramanauskas, K., & Igić, B. (2017). The evolutionary history of plant T2/S-type ribonucleases. *PeerJ, 5*, e3790.

Raul, P. A., Koichiro, U., Takato, K., & Hidenori, S. (2009). Identification and functional analysis of pistil SI factor HT-B of *petunia. Journal of Experimental Botany, 60*, 1309–1318.

Ride, J. P., Davies, E. M., Franklin, F. C., & Marshall, D. F. (1999). Analysis of *Arabidopsis* genome sequence reveals a large new gene family in plants. *Plant Molecular Biology, 39*, 927–932.

Royo, J., Kunz, C., Kowyama, Y., Anderson, M., Clarke, A. E., & Newbigin, E. (1994). Loss of a histidine residue at the active site of S-locus ribonuclease is associated with self-compatibility in *Lycopersicon peruvianum. Proceedings of the National Academy of Sciences of the United States of America, 91*, 6511–6514.

Rudd, J. J., Franklin, F., Lord, J. M., & Franklin-Tong, V. E. (1996). Increased phosphorylation of a 26-kD pollen protein is induced by the SI response in *Papaver rhoeas*. *The Plant Cell, 8*, 713–724.
Rudd, J. J., Franklin, F. C. H., & Franklin-Tong, V. E. (1997). Ca^{2+}-independent phosphorylation of a 68 kDa pollen protein is stimulated by the SI response in *Papaver rhoeas*. *The Plant Journal, 12*, 507–514.
Rudd, J. J., Osman, K., Franklin, F. C., & Franklin-Tong, V. E. (2003). Activation of a putative MAP kinase in pollen is stimulated by the SI (SI) response. *FEBS Letters, 547*, 223–227.
Sampson, D. R. (1957). The genetics of SI in the radish. *Canadian Journal of Botany, 36*, 39–56.
Sampson, D. R. (1964). A one-locus SI system in *Raphanus raphanistrum*. *Genome, 6*, 435–445.
Samuel, M. A., Salt, J. N., Shiu, S., & Goring, D. R. (2006). Multifunctional arm repeat domains in plants. *International Review of Cytology, 253*, 1–26.
Samuel, M. A., Mudgil, Y., Salt, J. N., Delmas, F., Ramachandran, S., Chilelli, A., & Goring, D. R. (2008). Interactions between the S-domain receptor kinases and AtPUB-ARM E3 ubiquitin ligases suggest a conserved signaling pathway in *Arabidopsis*. *Plant Physiology, 147*, 2084.
Samuel, M. A., Chong, Y. T., Haasen, K. E., Aldeabrydges, M. G., Stone, S. L., & Goring, D. R. (2009). Cellular pathways regulating responses to compatible and self-incompatible pollen in *Brassica* and *Arabidopsis* stigmas intersect at Exo70A1, a putative component of the exocyst complex. *The Plant Cell, 21*, 2655.
Samuel, M. A., Tang, W., Jamshed, M., Northey, J., Patel, D., Smith, D., Siu, K. W., Muench, D. G., Wang, Z. Y., & Goring, D. R. (2011). Proteomic analysis of *Brassica* stigmatic proteins following the self-incompatibility reaction reveals a role for microtubule dynamics during pollen responses. *Molecular and Cellular Proteomics, 10*, M111.011338.
Sankaranarayanan, S., Jamshed, M., & Samuel, M. A. (2015). Degradation of glyoxalase I in *Brassica napus* stigma leads to self-incompatibility response. *Nature Plants, 1*, 15185.
Sankaranarayanan, S., Jamshed, M., Kumar, A., Skori, L., Scandola, S., Wang, T., Spiegel, D., & Samuel, M. A. (2017). Glyoxalase goes green: The expanding roles of glyoxalase in plants. *International Journal of Molecular Sciences, 18*, 898.
Sassa, H., Hirano, H., Nishio, T., & Koba, T. (1997). Style-specific self-compatible mutation caused by deletion of the S-RNase gene in Japanese pear (*Pyrus serotina*). *The Plant Journal, 12*, 223–227.
Sassa, H., Kakui, H., Miyamoto, M., Suzuki, Y., Hanada, T., Ushijima, K., Kusaba, M., Hirano, H., & Koba, T. (2007). S locus F-box brothers: Multiple and pollen-specific F-box genes with *S* haplotype-specific polymorphisms in apple and Japanese pear. *Genetics, 175*, 1869.
Sato, K., Nishio, T., Kimura, R., Kusaba, M., Suzuki, T., Hatakeyama, K., Ockendon, D. J., & Satta, Y. (2002). Coevolution of the *S*-locus genes *SRK*, *SLG* and *SP11/SCR* in *Brassica oleracea* and *B. rapa*. *Genetics, 162*, 931–940.
Sato, Y., Fujimoto, R., Toriyama, K., & Nishio, T. (2003). Commonality of self-recognition specificity of *S* haplotypes between *Brassica oleracea* and *Brassica rapa*. *Plant Molecular Biology, 52*, 617–626.
Scandola, S., & Samuel, M. A. (2019). A flower-specific phospholipase D is a stigmaticcompatibility factor targeted by the self-incompatibility response in *Brassica napus*. *Current Biology, 29*, 506–512.
Schierup, M. H., Mable, B. K., Awadalla, P., & Charlesworth, D. (2001). Identification and characterization of a polymorphic receptor kinase gene linked to the SI locus of *Arabidopsis lyrata*. *Genetics, 158*, 387–399.
Schopfer, C. R., Nasrallah, M. E., & Nasrallah, J. B. (1999). The male determinant of SI in *Brassica*. *Science, 286*, 1697–1700.
Shang, L., Gad, K., & Lenhard, M. (2025). Converging on long and short: the genetics, molecular biology and evolution of heterostyly. *Current Opinion in Plant Biology, 85*, 102731.
Shiba, H., Hinata, K., Suzuki, A., & Isogai, A. (1995). Breakdown of SI in *Brassica* by the antisense RNA of the *SLG* gene. *Proceedings of the Japan Academy Series B, 71*, 81–83.

Shiba, H., Kimura, N., Takayama, S., Hinata, K., Suzuki, A., & Isogai, A. (2000). Alteration of the SI phenotype in *Brassica* by transformation of the antisense *SLG* gene. *Bioscience Biotechnology and Biochemistry, 64*, 1016–1024.

Shiba, H., Takayama, S., Iwano, M., Shimosato, H., Funato, M., Nakagawa, T., Che, F., Suzuki, G., Watanabe, M., & Hinata, K. (2001). A pollen coat protein, SP11/SCR, determines the pollen *S*-specificity in the SI of *Brassica* species. *Plant Physiology, 125*, 2095–2103.

Shiba, H., Iwano, M., Entani, T., Ishimoto, K., Shimosato, H., Che, F., Satta, Y., Ito, A., Takada, Y., & Watanabe, M. (2002). The dominance of alleles controlling SI in *Brassica* pollen is regulated at the RNA level. *The Plant Cell, 14*, 491–504.

Shiba, H., Kakizaki, T., Iwano, M., Tarutani, Y., Watanabe, M., Isogai, A., & Takayama, S. (2006). Dominance relationships between SI alleles controlled by DNA methylation. *Nature Genetics, 38*, 297–299.

Shimosato, H., Yokota, N., Shiba, H., Iwano, M., Entani, T., Che, F., Watanabe, M., Isogai, A., & Takayama, S. (2007). Characterization of the SP11/SCR high-affinity binding site involved in self/nonself recognition in *Brassica* SI. *The Plant Cell, 19*, 107–117.

Shiu, S. H., & Bleecker, A. B. (2001). Receptor-like kinases from *Arabidopsis* form a monophyletic gene family related to animal receptor kinases. *Proceedings of the National Academy of Sciences of the United States of America, 98*, 10763–10768.

Shore, J. S., Hamam, H. J., Chafe, P. D. J., Labonne, J. D. J., Henning, P. M., & McCubbin, A. G. (2019). The long and short of the *S*-locus in *Turnera* (Passifloraceae). *The New Phytologist, 224*, 1316–1329.

Sijacic, P., Wang, X., Skirpan, A. L., Wang, Y., Dowd, P. E., McCubbin, A. G., Huang, S., & Kao, T. H. (2004). Identification of the pollen determinant of S-RNase-mediated SI. *Nature, 429*, 302–305.

Sims, T. L., & Ordanic, M. (2001). Identification of a S-ribonuclease-binding protein in *Petunia hybrida*. *Plant Molecular Biology, 47*, 771.

Snowman, B. N., Kovar, D. R., Shevchenko, G., Franklin-Tong, V. E., & Staiger, C. J. (2002). Signal-mediated depolymerization of actin in pollen during the SI response. *The Plant Cell, 14*, 2613–2626.

Sonneveld, T., Tobutt, K. R., Vaughan, S. P., & Robbins, T. P. (2005). Loss of pollen-*S* function in two self-compatible selections of *Prunus avium* is associated with deletion/mutation of an *S* haplotype-specific F-box gene. *The Plant Cell, 17*, 37–51.

Soulard, J., Boivin, N., Morse, D., & Cappadocia, M. (2014). eEF1A is an S-RNase binding factor in self-incompatible *Solanum chacoense*. *PLoS One, 9*, e90206.

Stahl, R., Arnoldo, M. A., Glavin, T. L., Goring, D. R., & Rothstein, S. J. (1998). The SI phenotype in *Brassica* is altered by the transformation of a mutant *S* locus receptor kinase. *The Plant Cell, 10*, 209–218.

Stead, A. D., Roberts, I. N., & Dickinson, H. G. (1980). Pollen-stigma interaction in *Brassica oleracea*: The role of stigmatic proteins in pollen grain adhesion. *Journal of Cell Science, 42*, 417–423.

Stebbins, G. L. (1957). Self fertilization and population variability in the higher plants. *The American Naturalist, 91*, 337–354.

Stein, J. C., Howlett, B., Boyes, D. C., Nasrallah, M. E., & Nasrallah, J. B. (1991). Molecular cloning of a putative receptor protein kinase gene encoded at the SI locus of *Brassica oleracea*. *Proceedings of the National Academy of Sciences of the United States of America, 88*, 8816–8820.

Stephenson, A. G., Doughty, J., Dixon, S., Elleman, C., Hiscock, S. J., & Dickinson, H. G. (1997). The male determinant of SI in *Brassica oleracea* is located in the pollen coating. *The Plant Journal, 12*, 1351–1359.

Stone, S. L., Arnoldo, M., & Goring, D. R. (1999). A breakdown of *Brassica* SI in ARC1 antisense transgenic plants. *Science, 286*, 1729–1731.

Stone, S. L., Anderson, E. M., Mullen, R. T., & Goring, D. R. (2003). ARC1 is an E3 ubiquitin ligase and promotes the ubiquitination of proteins during the rejection of self-incompatible *Brassica* pollen. *The Plant Cell, 15*, 885–898.

Stout, A. B. (1920). Further experimental studies on SI in hermaphrodite plants. *Journal of Genetics, 9*, 85–129.

Sun, H. Q., Kwiatkowska, K., & Yin, H. L. (1995). Actin monomer binding proteins. *Current Opinion in Cell Biology, 7*, 102–110.

Suzuki, G., Watanabe, M., Toriyama, K., Isogai, A., & Hinata, K. (1997). Direct cloning of the *brassica S* locus by using a P1-derived artificial chromosome (PAC) vector. *Gene, 199*, 133–137.

Suzuki, G., Kai, N., Hirose, T., Fukui, K., Nishio, T., Takayama, S., Isogai, A., Watanabe, M., & Hinata, K. (1999). Genomic Organization of the *S* locus: Identification and characterization of genes in *SLG* / *SRK* region of S_9 haplotype of *Brassica campestris* (syn. rapa). *Genetics, 153*, 391–400.

Takasaki, T., Hatakeyama, K., Suzuki, G., Watanabe, M., Isogai, A., & Hinata, K. (2000). The S receptor kinase determines SI in *Brassica* stigma. *Nature, 403*, 913–916.

Takasaki, T., Hatakeyama, K., Watanabe, M., Toriyama, K., & Hinata, K. (2001). Homology-dependent suppression of stigma phenotype by an antisense *S*-locus glycoprotein (SLG) gene in *Brassica rapa* L. *Breeding Science, 51*, 89–94.

Takayama, S., Isogai, A., Tsukamoto, C., Ueda, Y., Hinata, K., Okazaki, K., & Suzuki, A. (1986a). Isolation and some characterization of S-locus-specific glycoproteins associated with SI in *Brassica campestris*. *Journal of the Agricultural Chemical Society of Japan, 50*, 1365–1367.

Takayama, S., Isogai, A., Tsukamoto, C., Ueda, Y., Hinata, K., Okazaki, K., Koseki, K., & Suzuki, A. (1986b). Structure of carbohydrate chains of S-glycoproteins in *Brassica campestris* associated with SI. *Agricultural and Biological Chemistry, 50*, 1673–1676.

Takayama, S., Shiba, H., Iwano, M., Shimosato, H., Che, F., Kai, N., Watanabe, M., Suzuki, G., Hinata, K., & Isogai, A. (2000). The pollen determinant of SI in *Brassica campestris*. *Proceedings. National Academy of Sciences. United States of America, 97*, 1920–1925.

Takayama, S., Shimosato, H., Shiba, H., Funato, M., Che, F., Watanabe, M., Iwano, M., & Isogai, A. (2001). Direct ligand-receptor complex interaction controls *Brassica* SI. *Nature, 413*, 534–538.

Takayama, S., & Isogai, A. (2005) Self-incompatibility in plants. *Annual Review of Plant Biology, 56*, 467–489.

Tang, C., Toomajian, C., Shermanbroyles, S., Plagnol, V., Guo, Y. L., Hu, T. T., Clark, R. M., Nasrallah, J. B., Weigel, D., & Nordborg, M. (2007). The evolution of selfing in *Arabidopsis thaliana*. *Science, 317*, 1070–1072.

Tang, C., Wang, P., Zhu, X., Qi, K., Xie, Z., Zhang, H., Li, X., Gao, H., Gu, T., Gu, C., Li, S., de Graaf, B. H. J., Zhang, S., & Wu, J. (2023). Acetylation of inorganic pyrophosphatase by S-RNase signaling induces pollen tube tip swelling by repressing pectin methylesterase. *The Plant Cell, 35*, 3544–3565.

Tao, R., Yamane, H., Sassa, H., Mori, H., Gradziel, T. M., Dandekar, A. M., & Sugiura, A. (1997). Identification of stylar RNases associated with gametophytic SI in almond (*Prunus dulcis*). *Plant & Cell Physiology, 38*, 304–311.

Tarutani, Y., Shiba, H., Iwano, M., Kakizaki, T., Suzuki, G., Watanabe, M., Isogai, A., & Takayama, S. (2010). Trans-acting small RNA determines dominance relationships in *Brassica* SI. *Nature, 466*, 983–986.

Tatebe, T. (1964). Studies on the physiological mechanism of SI in Japanese radish. I. *Journal of the Japanese Society for Horticultural Science, 33*, 62–66.

Tena, G., Asai, T., Chiu, W. L., & Sheen, J. (2001). Plant mitogen-activated protein kinase signaling cascades. *Current Opinion in Plant Biology, 4*, 392–400.

Thomas, S. G., & Franklin-Tong, V. E. (2004). SI triggers programmed cell death in *Papaver* pollen. *Nature, 429*, 305–309.

Thomas, S. G., Huang, S., Li, S., Staiger, C. J., & Franklin-Tong, V. E. (2006). Actin depolymerization is sufficient to induce programmed cell death in self-incompatible pollen. *The Journal of Cell Biology, 174*, 221–229.

Thompson, K. F., & Howard, H. W. (1959). SI in marrow-stem kale, *Brassica oleracea* var. Acephala. *Journal Of Genetics, 56*, 325–340.

Thompson, K. F., & Taylor, J. P. (1966). Non-linear dominance relationships between *S* alleles. *Heredity, 21*, 345–362.
Tian, H., Zhang, H., Huang, H., Zhang, Y., & Xue, Y. (2023). Phase separation of S-RNase promotes self-incompatibility in *Petunia hybrida. Journal of Integrative Plant Biology, 66*, 986–1006.
Torres-Rodríguez, M. D., Cruz-Zamora, Y., Juárez-Díaz, J. A., Mooney, B., McClure, B. A., & Cruz-García, F. (2020). NaTrxh is an essential protein for pollen rejection in *Nicotiana* by increasing S-RNase activity. *The Plant Journal, 103*, 1304–1317.
Tsuchimatsu, T., Suwabe, K., Shimizuinatsugi, R., Isokawa, S., Pavlidis, P., Städler, T., Suzuki, G., Takayama, S., Watanabe, M., & Shimizu, K. K. (2010). Evolution of self-compatibility in *Arabidopsis* by a mutation in the male specificity gene. *Nature, 464*, 1342.
Ushijima, K., Sassa, H., Dandekar, A. M., Gradziel, T. M., Tao, R., & Hirano, H. (2003). Structural and transcriptional analysis of the self-incompatibility locus of almond: Identification of a pollen-expressed F-box gene with haplotype-specific polymorphism. *The Plant Cell, 15*, 771–781.
Ushijima, K., Yamane, H., Watari, A., Kakehi, E., Ikeda, K., Hauck, N. R., Iezzoni, A. F., & Tao, R. (2004). The *S* haplotype-specific F-box protein gene,*SFB*, is defective in self-compatible haplotypes of *Prunus avium* and *P. mume. The Plant Journal, 39*, 573–586.
Uyenoyama, M. K., Zhang, Y., & Newbigin, E. (2001). On the origin of SI haplotypes: Transition through self-compatible intermediates. *Genetics, 157*, 1805–1817.
Vanoosthuyse, V., Tichtinsky, G., Dumas, C., Gaude, T., & Cock, J. M. (2003). Interaction of calmodulin, a sorting nexin and kinase-associated protein phosphatase with the *Brassica oleracea S* locus receptor kinase. *Plant Physiology, 133*, 919–929.
Vieira, J., Fonseca, N. A., & Vieira, C. P. (2008). An S-RNase-based gametophytic SI system evolved only once in eudicots. *Journal of Molecular Evolution, 67*, 179.
Vilanova, S., Badenes, M. L., Burgos, L., Martinez-Calvo, J., Llacer, G., & Romero, C. (2006). Self-compatibility of two apricot selections is associated with two pollen-part mutations of different nature. *Plant Physiology, 142*, 629–641.
Walker, E. A., Ride, J. P., Kurup, S., Franklin-Tong, V. E., Lawrence, M. J., & Franklin, F. C. H. (1996). Molecular analysis of two functional homologues of the $S_{(3)}$ allele of the *Papaver rhoeas* self-incompatibilty gene isolated from different populations. *Plant Molecular Biology, 30*, 983–994.
Wallace, D. H. (1967). Immunogenetics of SI in *Brassica oleracea* L.1. *Heredity, 22*, 519–527.
Watanabe, M., Takasaki, T., Toriyama, K., Yamakawa, S., Isogai, A., Suzuki, A., & Hinata, K. (1994). A high degree of homology exists between the protein encoded by *SLG* and the S receptor domain encoded by *SRK* in self-incompatible *Brassica campestris* L. *Plant & Cell Physiology, 35*, 1221–1229.
Watanabe, M., Ito, A., Takada, Y., Ninomiya, C., Kakizaki, T., Takahata, Y., Hatakeyama, K., Hinata, K., Suzuki, G., & Takasaki, T. (2000). Highly divergent sequences of the pollen SI (*S*) gene in class-I *S* haplotypes of *Brassica campestris* (syn. rapa) L. *FEBS Letters, 473*, 139–144.
Wheeler, M. J., de Graaf, B. H. J., Hadjiosif, N., Perry, R. M., Poulter, N. S., Osman, K., Vatovec, S., Harper, A., Franklin, F. C. H., & Franklin-Tong, V. E. (2009). Identification of the pollen SI determinant in *Papaver rhoeas. Nature, 459*, 992–995.
Wilkins, K. A., Bancroft, J., Bosch, M., Ings, J., Smirnoff, N., & Franklin-Tong, V. E. (2011). ROS and NO mediate actin reorganization and programmed cell death in the SI response of *Papaver. Plant Physiology, 156*, 404–416.
Wilkins, K. A., Bosch, M., Haque, T., Teng, N., Poulter, N. S., & Franklin-Tong, V. E. (2015). SI-induced programmed cell death in field poppy pollen involves dramatic acidification of the incompatible pollen tube cytosol. *Plant Physiology, 167*, 766–779.
Wu, J., Wang, S., Gu, Y., Zhang, S., Publicover, S. J., & Franklin-Tong, V. E. (2011). SI in *Papaver rhoeas* activates nonspecific cation conductance permeable to Ca^{2+} and K^{+}. *Plant Physiology, 155*, 963–973.
Xu, C., Li, M., Wu, J., Guo, H., Li, Q., Zhang, Y., Chai, J., Li, T., & Xue, Y. (2013). Identification of a canonical SCF^{SLF} complex involved in S-RNase-based SI of *Pyrus* (Rosaceae). *Plant Molecular Biology, 81*, 245.

Xue, Y., Carpenter, R., Dickinson, H. G., & Coen, E. S. (1996). Origin of allelic diversity in *Antirrhinum S* locus RNases. *Plant Cell, 8*, 805–814.

Yu, K., Schafer, U., Glavin, T. L., Goring, D. R., & Rothstein, S. J. (1996). Molecular characterization of the *S* locus in two self-incompatible *Brassica napus* lines. *The Plant Cell, 8*, 2369–2380.

Zhang, S., & Klessig, D. F. (2001). MAPK cascades in plant defense signaling. *Trends in Plant Science, 6*, 520–527.

Zhang, Y., Zhao, Z., & Xue, Y. (2009). Roles of proteolysis in plant SI. *Annual Review of Plant Biology, 60*, 21.

Zhang, L., Huang, J., Su, S., Wei, X., Yang, L., Zhao, H., Yu, J., Wang, J., Hui, J., Hao, S., Song, S., Cao, Y., Wang, M., Zhang, X., Zhao, Y., Wang, Z., Zeng, W., Wu, H. M., Yuan, Y., Zhang, X., Cheung, A. Y., & Duan, Q. (2021). FERONIA receptor kinase-regulated reactive oxygen species mediate self-incompatibility in *Brassica rapa*. *Current Biology, 31*, 3004–3016.

Zhao, L., Huang, J., Zhao, Z., Li, Q., Sims, T. L., & Xue, Y. (2010). The Skp1-like protein SSK1 is required for cross-pollen compatibility in S-RNase-based SI. *The Plant Journal, 62*, 52–63.

Zhao, H., Song, Y., Li, J., Zhang, Y., Huang, H., Li, Q., Zhang, Y., & Xue, Y. (2021). Primary restriction of S-RNase cytotoxicity by a stepwise ubiquitination and degradation pathway in *Petunia hybrida*. *The New Phytologist, 231*, 1249–1264.

Zhao, H., Zhang, Y., Zhang, H., Song, Y., Zhao, F., Zhang, Y. E., Zhu, S., Zhang, H., Zhou, Z., Guo, H., Li, M., Li, J., Gao, Q., Han, Q., Huang, H., Copsey, L., Li, Q., Chen, H., Coen, E., Zhang, Y. J., & Xue, Y. (2022). Origin, loss, and regain of self-incompatibility in angiosperms. *Plant Cell, 34*, 579–596.

Zhou, J., Wang, F., Ma, W., Zhang, Y., Han, B., & Xue, Y. (2003). Structural and transcriptional analysis of S-locus F-box genes in *Antirrhinum*. *Sexual Plant Reproduction, 16*, 165–177.

Zhu, S., Zhang, Y., Copsy, L., Han, Q., Zheng, D., Coen, E., & Xue, Y. (2023). The snapdragon genomes reveal the evolutionary dynamics of the *S*-locus supergene. *Molecular Biology and Evolution, 40*, msad080.

Zinkl, G. M., Zwiebel, B. I., Grier, D. G., & Preuss, D. (1999). Pollen-stigma adhesion in Arabidopsis: A species-specific interaction mediated by lipophilic molecules in the pollen exine. *Development, 126*, 5431–5440.

Zurek, D. M., Mou, B., Beecher, B., & McClure, B. (1997). Exchanging sequence domains between S-RNases from *Nicotiana alata* disrupts pollen recognition. *The Plant Journal, 11*, 797–808.

Zuriaga, E., Munoz-Sanz, J. V., Molina, L., Gisbert, A. D., Badenes, M. L., & Romero, C. (2013). An *S*-locus independent pollen factor confers self-compatibility in 'Katy' apricot. *PLoS One, 8*, e53947.

Chapter 17
A Reconsideration of Sex: From the Perspective of the Sexual Reproduction Cycle

Shu-Nong Bai

Abstract Sexual reproduction is indispensable for most eukaryotes. However, the definition of "sex" has remained controversial for centuries. Since the early 1930s, unisexual flowers have been used to investigate sex differentiation or determination in plants, but all unisexual flowers investigated to date result from prevention of sex cell differentiation in either the stamen or carpel, not link to the sex cells as originally proposed by Robbins and Pearson in 1933. To solve the problem, the concept of the "sexual reproduction cycle" (SRC) has been developed. The SRC represents a modified cell cycle that integrates changes of surrounding components, "sex" refers to heterogametes in the SRC (Fig. 17.1). Multicellular organisms are derived via the interpolation of multicellular structures (MCSs) into the two (diploid and haploid) intervals of SRC (Fig. 17.2). Therefore, sex differentiation occurs only in multicellular organisms and ensures heterogametogenesis. In plants, only differentiation of archegonia and antheridia fits the definition and could be considered as true sex differentiation, while differentiation of heterosporangia (e.g., stamens and ovules in angiosperms), is a form of pseudo sex differentiation. Finally, a third sex-related phenomenon, sexual behavior, has evolved in multicellular organisms to facilitate SRC completion. In plants, sexual behaviors include pollen tube growth, the attraction of pollinators to ensure that gametes meet, and the promotion of outcrossing. Unisexual flower development is one of the mechanisms of outcross promotion. The SRC-based comparison of plants with other eukaryotes provides an alternative reference framework for understanding the general principles of sex differentiation in all eukaryotes.

Keywords Eukaryotes · Sexual reproductive cycle · Sex · Sex differentiation · Sex behavior

---

S.-N. Bai (✉)
College of Life Sciences, Peking University, Beijing, China
e-mail: shunongb@pku.edu.cn

F. Chang et al. (eds.), *Regulation of Plant Development*,
https://doi.org/10.1007/978-981-95-7033-1_17

17.1 Overview of the Current Understanding of Sex Differentiation in Plants

Sex has always been a hot topic in human society, and it is much more compelling to the general public than to academic researchers. However, many notions about sex are vague and controversial, not only to the general public but also in academic circles. Three essential questions remain elusive: what is sex, what is sex differentiation, and what is the ultimate function of sex. These three questions are so fundamental that without clear answers, there would be no solid basis for solving other controversies.

Given the importance of these three questions, I will first summarize the current mainstream understanding of issues regarding sex and analyze the problems inherent in the current explanations. Then I will describe the concept of the sexual reproduction cycle (SRC) and an SRC-based framework for understanding issues related to sex. Finally, I attempt to provide answers to the three questions posed above, using this new framework.

An important aspect of the view described in this chapter is that it puts plants and animals in the same position when comparing sexual issues. Current notions about sex are mainly derived from observations and investigations on animals, and the concepts and logic used to explain sex-related processes in plants were generally borrowed from animal studies. I hope that by providing a more equal comparison between animals and plants, this chapter improves our understanding of sex-related phenomena as a whole.

17.1.1 Mainstream Concepts About Plant Sex and Sex Determination

Information about mainstream concepts of sex-related phenomena in plants can be obtained through three resources: representative reviews, classic books, and websites. The most influential reviews include those from Dellaporta and Calderon-Urrea (1993), Charlesworth (2002), Tanurdzic and Banks (2004), Ming et al. (2011), Diggle et al. (2011), and Bachtrog et al. (2014). Fewer books focus on the topic of sex-related phenomena in plants. While the earliest book is Carl Linnaeus (Carl von Linne)'s *A Dissertation on the Sexes of Plants* (1786), Yampolsky's *Distribution of Sex Forms in the Phanerogamic Flora* (1922) is a classic, information-rich book that is widely cited. To the best of my knowledge, the current understanding of sex-related phenomena in plants was shaped by Robbins and Pearson's book *Sex in the Plant World* (1933). In recent years, mainstream concepts about plant sex have been summarized in two books, *Sex Determination in Plants*, edited by Ainsworth (1999), and *The Evolution of Sex Determination*, by Beukboom and Perrin (2014). Websites are also powerful tools for obtaining information about any topic of interest. It is convenient to access these websites to compare very divergent concepts.

Based on the currently accessible information, the mainstream concepts on plant sex can be summarized as follows. First, studies on plant sex have mainly focused on flowering plants, although a few studies have focused on sexual phenomena in non-flowering plants (Atallah & Banks, 2015; Pannell, 2017). Second, the definition of plant sex was mainly derived from Robbins and Pearson (1933). In their book, the authors reasoned that

> An organism, or an organ, or even a trait, is one sex or the other simply because it is associated with the production of the sex cells of that sex. We human beings are male because we produce sperms, or female because we produce eggs! Beards and beardlessness are human male and female characters because of their association with the production of sperms and eggs, respectively. So, in plants, a stamen is a male organ because within it are produced the pollen grains, which in turn are the sperm-producing plants. And a flower or a plant is male if it bears only stamens. Likewise, a pistil is a female organ, and a flower or a plant is female if it bears only pistils.

Third, it is widely recognized that the majority (approximately 90%) of flowering plants are "hermaphrodites." Sex differentiation and determination have mainly been studied using unisexual flowers. How can the knowledge about the minority of species, i.e., those with unisexual flowers, help us to understand the process of sex differentiation if most plants are hermaphrodites? According to Ainsworth (1999), "The determination of sex in plants is thus a key developmental decision which leads to the suppression of the program of development of either the male or female organs. This intriguing group of plants has gone about unisexuality in many different ways with the developmental blocks occurring across the spectrum of organ initiation and development. For this reason, unisexual plants offer us the opportunity of investigating plant development in a way which is complementary to the study of flower development in hermaphrodite species." It is interesting that he did not mention sex at all in regard to the plants with unisexual flowers.

Under such a conceptual framework, studies on plant sex have mainly focused on how unisexual flowers develop. Before the 1950s, these studies mainly involved descriptive analyses of the developmental processes of unisexual flowers, such as cucumber (Heimlich, 1927), searching for genetic control mechanism determining the transition from hermaphrodites to unisexual (reviewed by Harkes and Leebens-Mack, 2017), and investigations of the distribution of plant species that produce unisexual flowers (Yampolsky, 1922). Since the late 1950s, two different lines of investigation have emerged: physiological analysis of the effects of hormones, such as ethylene, and the environment, such as temperature, on unisexual flower development (Heslop-Harrison, 1956, 1957; Witter & Bokovac, 1958; McMurray & Miller, 1968; Iwahori et al., 1969; Shannon & Guardia, 1969; Rudich et al., 1969); and genetic analysis of genes involved in the determination of unisexual flowers, starting from studies of cucumber (Galun, 1961). While the roles of floral organ identity genes in the determination of unisexual flower development were examined during the mid-1990s (Hardenack et al., 1994; Ainsworth et al., 1995; Sheppard et al., 2000; Kater et al., 2001), hormonal or so-called sex genes and other regulatory components were demonstrated to play crucial roles (see the next section).

The concept of sex chromosomes has also been applied to investigations of plant sex. Since sex chromosomes were first discovered in animals in the late nineteenth century and in moss in the early twentieth century (review see Bowman, 2016), a correlation between unisexual flowers and heterochromosomes has been reported in plants (Westergaard, 1958). Although such a correlation is not a universal character of plants, even among dioecious species, this type of research has a high profile among studies of sex differentiation in plants (Charlesworth, 2002, 2015, 2016; Ming et al., 2011).

17.1.2 Current Progress on the Molecular Mechanisms of Unisexual Flower Development

Sex Determination in Plants, edited by Ainsworth (1999), offers a summary of progress in our understanding of unisexual flower development in eight plant species, including maize (*Zea mays*), cucumber (*Cucumis sativus*), asparagus (*Asparagus officinalis*), kiwifruit (*Actinidia deliciosa*), white campion (*Silene latifolia*), sorrel (*Rumex acetosa*), common hop (*Humulus lupulus*), and wild yam (*Dioscorea tokoro*). Since that time, some significant progress has been made in this field, which I briefly summarize below.

17.1.2.1 Maize

Maize has unisexual flowers, and the regulatory mechanisms underlying floral morphogenesis have been well studied at the molecular level. According to Robbins and Pearson (1933), the unisexual flowers of maize were first recognized by Logan in the mid-eighteenth century. Although the underlying genetic and physiological factors were analyzed during the early twentieth century (Heslop-Harrison, 1960; reviewed by Coe et al., 1989), a systematic observation of flower morphogenesis was first reported by Cheng et al. (1983). Since 1993, several genes related to unisexual flower development have been cloned in maize, including *TASSELSEED2* (*TS2*, DeLong et al., 1993), *ANTHER EAR1* (*AN1*, Bensen et al., 1995), *TS4* (Chuck et al., 2007), *TS1* (Acosta et al., 2009), *SILKLESS1* (*SK1*, Hayward et al., 2016), and *TS5* (Wang et al., 2020).

Current understanding of the regulation of unisexual flower development in maize mainly focuses on how male flowers are formed, i.e., the arrest of pistil development during early floral development. Programmed cell death is involved in this process (Calderon-Urrea & Dellaporta, 1999), and jasmonic acid signaling (*TS1*, Acosta et al., 2009; *TS5*, Wang et al., 2022) and brassinosteroids (*SK1*, Hartwig et al., 2011) are required for its regulation. In contrast to the regulation of pistil differentiation in tassels, i.e., male flowers, little is known about how female flower development is regulated. Also, little is known about how the female inflorescence, the corn, is positioned in the axils of lower leaves, whereas the male inflorescence, the tassel, is terminally positioned.

17.1.2.2 Cucumber and Melon

Cucumber and other cucurbit plants represent another group of well-studied plants with unisexual flowers. Perl-Treves (1999; Perl-Treves & Rajagopalan, 2006) systematically summarized the history of unisexual flower research in cucumber. Bai and Xu (2013) reviewed recent progress based on the work carried out in their own laboratory. However, more information has since become available, not only on the regulatory mechanisms of unisexual flower development in cucumber but also on this process in other cucurbit plants.

The first report on morphological observation of unisexual cucumber flowers in an English-language journal was from Heimlich (1927). Since the late 1950s, three approaches have been taken to mechanistically investigate the formation of these flowers, including morphological, physiological, and genetic studies. While Atsmon and Galun (1960) provided the first morphological description of unisexual flower development in cucumber, Goffinet (1990) and Malepszy and Niemirowicz-Szczytt (1991) refined this description, and Bai et al. (2004) proposed a standard for describing developmental processes in unisexual cucumber flowers. This standard has been widely used in subsequent investigations (http://apps.webofknowledge.com/Search.do?product=WOS&SID=2FzJawmDfVkwiRhIG38&search_mode=CitedRefIndex&prID=d02003aa-77c2-4803-9142-86cf147ea4cb).

Physiological investigations of unisexual flower development in cucumber began in the late 1950s (Heslop-Harrison, 1957; Witter & Bokovac, 1958). Such investigations included analyses of effects of various internal and external factors on the ratio of male to female flowers, e.g., daylength and temperature; specifically, ethylene was shown to play a key role in female flower development (reviews see Perl-Treves, 1999; Perl-Treves & Rajagopalan, 2006; Bai & Xu, 2013).

Genetic analysis of flower development in cucumber began in the 1960s. Galun (1961) and Kubicki (1969a, b, c) proposed a criterion to distinguish the genetic effects on unisexual flower development, i.e., the position of the first female flower that is initiated. Using this criterion to investigate segregation, they proposed that among the genes found to affect this criterion, three major genes are *F*, *M*, and *A*, which are required for female flower development (summarized by Perl-Treves, 1999). Following the demonstration of the cosegregation of the *F* locus with a known ethylene biosynthesis gene, *CsACS1-G* (*Cucumis sativus 1-aminocycloprop ane-1-carboxylic acid synthase1-G*, Trebitsh et al., 1997; Mibus & Tatlioglu, 2004; Zhang et al., 2020a, b), the *M* gene was found to encode CsACS2 (Boualem et al., 2009; Li et al., 2009), and the *A* gene was found to encode CsACS11 (Boualem et al., 2015). In addition to the enzymes encoded by *F*, *M*, and *A*, *CsACO2* (*Cucumis sativus ACC oxidase 2*) was recently identified as a new player in the regulation of unisexual flower differentiation (Chen et al., 2016).

Two genes controlling female flower development in melon (*Cucumis melo*), *A* and *G*, have been identified as *CmACS7* (a homolog of *CsACS2*) and *CmWIP* (encoding a transcriptional regulator), respectively (Boualem et al., 2008; Martin et al., 2009). Recently, Boualem et al. (2015) determined that *CmACS11* is a homolog of the cucumber *A* gene *CsACS11*, which encodes an enzyme involved in regulating unisexual flower development together with CmACS7 and CmWIP. Further investigation revealed that *CRABS CLAW* (*CRC*) is a target of CmWIP together with TOPLESS for suppression of carpel development during male flower development (Zhang et al., 2022).

It is clear that physiological and genetic approaches to investigate the regulation of unisexual flower development in cucurbits have been plausibly merged: the increase in ethylene production leading to female flower development is satisfactorily explained by the genetic model of interactions between genes encoding ethylene synthesis enzymes (CsACS2, CsACS11, CsACO2, and the protein encoded in the *CsACS1-G* locus in cucumber; CmACS7 and CmACS11 in melon). Some new findings provided detailed regulatory information, such as a protein (CmWIP), together with other genes required for organ development. However, only by these findings, it is not yet clear how these ethylene production components function in an organ-specific manner and why high ethylene levels preferentially lead to female flower development.

As summarized in Bai and Xu (2013), the regulatory mechanism of unisexual flower development was explored using a third approach: developmental analysis. It has been demonstrated repeatedly with various tools (from analysis under a dissecting microscope to SEM) that the initiation of both stamen and carpel primordia occurs during early flower development in cucumber (Heimlich, 1927; Atsmon & Galun, 1960; Goffinet, 1990; Malepszy & Niemirowicz-Szczytt, 1991; Bai et al., 2004). Based on this finding, it was reasoned that the formation of unisexual flowers results from the inhibition of stamen or carpel organ formation during floral bud differentiation. Thus, an important question to ask about the regulation of unisexual flower development is what and how changes occur that lead to the inappropriate stamen or carpel organ formation rather than how an entire floral bud becomes male or female.

Following previous findings of a correlation between primordial anther-specific DNA damage and the inhibition of stamen development in female flowers (Hao et al., 2003; Gu et al., 2011) and of stamen-specific downregulation of *CsETR1* expression (Wang et al., 2010), it was demonstrated that the B-class MADS box protein CsAP3 binds to and activates the *CsETR1* promoter and may be responsible for its stamen-specific downregulation during female flower development (Sun et al., 2016). Interestingly, this work uncovered in the C-terminal region of CsAP3 a GV repeat (found in a wide range of species with a so far unknown function) that is required for the transcriptional activation of the *CsETR1* promoter. Moreover, this GV repeat is unique to the AP3 homologs of cucurbits and is not found in other known AP3 homologs (Sun et al., 2016). This finding implies that the stamen-specific regulation of ethylene perception in cucumber may be a conserved mechanism during unisexual flower development in cucurbits.

Another interesting finding is that some pre-miRNAs, such as 396a, 319a, and 171b, are expressed at low levels in male flowers but at high levels in female and perfect flowers, suggesting that microRNA might be involved in the organ-preferential regulation of the ethylene gene expression (Sun et al., 2010). Based on such an observation, it was proposed that the organ-preferential regulation of ethylene gene expression might result initially from particular stress responses for ensuring carpel development, and the ethylene-induced inhibition of stamen development is a by-product, that selected by the advantage of promotion of outcross. Such a hypothesis was supported by the fact that all the ethylene produced genes are preferentially during early carpel primordial development (Zhang et al., 2020a, b)

17.1.2.3 Papaya

Papaya is an interesting plant in terms of unisexual flower development. Papaya varieties can be dioecious or trioecious, with three types of flowers produced: male, female, and perfect flowers (Ming et al., 2007). As summarized in Bai and Xu (2013), systematic morphological and molecular analyses have been carried out in papaya to reveal the regulatory mechanism of unisexual flower formation, or as those in the field term it, sex determination. Since then, some progress has been made in analyzing genomic variation and in sequencing the loci responsible for unisexual flower development in papaya (Gschwend et al., 2012; Wang et al., 2012; Na et al., 2014; VanBuren et al., 2015; VanBuren et al., 2016). It should be quite interesting to learn the functions of the genes encoded by these loci.

17.1.2.4 Other Plants

In recent years, in addition to maize and cucumber, significant progress has been made in understanding the mechanisms underlying unisexual flower development in asparagus and kiwifruit, among the eight plant species described in the book edited by Ainsworth (1999) (please refer to the first paragraph of Sect. 17.2.1). While some progress has been made in the remaining four species (Leite Montalvao et al., 2021), some findings have been well recognized regarding unisexual flower development in some species not in the book, such as melon, persimmon (*Diospyros lotus*), banyan tree (*Ficus microcarpa* L. and *F. hispida*), and grapevine (*Vitis vinifera* L.).

In 2017, Harkess et al. reported that the unisexual flower of asparagus is caused by particular miRNAs that are differentially localized in a heterochromosome the called young Y-chromosome, adding another example of the role of miRNA in the regulation of unisexual flower development. Then, they reported that hemizygosity underlies the loss of recombination between the genes that, respectively, suppresses female organogenesis (*SOFF/DUF247*) and promotes male function (*aspTDF1*) (Harkess et al., 2020). Murase et al. (2017) found that a Myb transcription factor is also involved in the unisexual flower development.

The regulatory mechanism of unisexual flower development in kiwifruit was uncovered recently as well. Akagi et al. (2018, 2019) reported that a type-C cytokinin response regulator and a fasciclin-like gene play key roles.

The progress of the regulatory mechanism of unisexual flower development in melon has been mentioned above. In contrast to its previously less known situation, through a joint effort between researchers at UC-Davis and Kyoto University, Akagi et al. (2014) identified the *OGI* (Japanese for "male tree") gene from male persimmon trees and demonstrated that both *OGI* and *MeGI* (Japanese for "female tree") are homologs of the barley *VRS1* gene, which encodes a homeodomain-leucine zipper I-class protein. Unlike *MeGI*, the *OGI* mRNA includes a unique 21 bp small RNA that interferes with the expression of *MeGI*. On the other hand, overexpressing *MeGI* in transgenic *Arabidopsis* and tobacco plants does not affect carpel development but causes a male-sterile phenotype. These results suggest that the small RNA encoded by *OGI* interferes with *MeGI* expression in plants bearing the "Y-chromosome" and inhibits carpel development, leading to the production of male flowers, whereas the expression of *MeGI* in female plants inhibits male fertility. This work (for the first time) revealed a mechanism by which small RNAs play a role in regulating unisexual flower development.

It is interesting that comparative analysis of two *Ficus* genomes has identified an *AGAMOUS* paralog, *FhAG2*, specific for male flowers, as a candidate gene for unisexual flower development (Zhang et al., 2020a, b). The regulatory mechanism underlying the unisexual flower development of grapes still remains elusive. While domesticated grape varieties are hermaphroditic and derived from the dioecious wild species, and there is a particular genome region determining the unisexual flower development, different reports proposed different mechanisms. Massoneet et al. (2020) suggested the *VviINP1* gene as a candidate gene defective in a male-sterile mutant and a potential female-fertility function associated with the transcription factor *VviYABBY3*; on the other hand, Iocco-Corena et al. (2021) suggested that the VviPLATZ1 transcription factor is a key candidate for female flower development. Furthermore, Zou et al. (2021) suggested that multiple independent recombination events led to hermaphroditism in grapes.

17.1.3 Controversial Issues in the Study of Plant Sex Differentiation Using Unisexual Flowers

While the dominant concept about plant sex differentiation came from Robbins and Pearson (1933), Juraze and Banks (1998) had proposed a different definition of this. According to Robbins and Pearson, "a flower or a plant is male if it bears only stamens. Likewise, a pistil is a female organ, and a flower or a plant is female if it bears only pistils." The logic behind this statement is that stamens and pistils bear different sex cells. Juraze and Banks stated that, "sex determination is the developmental decision that occurs during the plant life cycle that leads to the differentiation of the two organs or cells that produce the two gametes." At first glance, these two statements appear to be quite similar. However, if one reads carefully, there are some significant differences between them: first, Robbins and Pearson's statement does not clearly indicate how to define "bears," primordia or functional organs, whereas Juraze and Banks's statement emphasizes the "decision" that leads to the differentiation to gamete production. Second, Robbins and Pearson's statement does not include nonflowering plants, but Juraze and Banks's statement does. Third, Robbins and Pearson's statement implies that only unisexual flowers exhibit sex differentiation, whereas Juraze and Banks's statement is more inclusive. It is understandable that during their time, Robbins and Pearson had little morphological and developmental information available to provide more precise criteria for defining sex. However, it is obviously inappropriate to exclude nonflowering plants and plants with perfect flowers from the scope of sex differentiation.

It is intriguing why unisexual flowers are still used as a criterion to define sex in the investigation of the regulatory mechanism of sex differentiation, even though a wealth of data indicates that in many plants bearing unisexual flowers, both male and female organ primordia are initiated in floral buds (Dellaporta and Calderon-Urrea 1993; Ainsworth 1999). To solve the problem, it is worthwhile to analyze why the concept proposed by Robbins and Pearson more than 90 years ago still dominates the study of plant sex differentiation even though so many observations are not in line with it and are therefore intentionally neglected.

17.1.3.1 Comparisons with Animals

In the definition of plant sex proposed by Robbins and Pearson, one can find that they used human sexuality as a reference. Historically, the world we described has been a human-centered one. Along with the accumulating information acquired from the exploration of unknown nature over history, we have gradually realized that humans are not the center of the world, and the characteristics of humans are no longer used as standards to describe and understand the characteristics of other organisms. From this perspective, whether it is proper to use human sexuality as a reference to describe or define plant sex seems need to be reexamined.

First, animals (including humans) are mainly unitary and dioecious organisms. In general, although with some exceptions, each individual bears only one type of gonad, and the germline localized in the gonad produces one type of gamete. By contrast, plants are mainly clonal organisms. A plant is similar to a coral and is not comparable to an individual unitary organism (details discussed in Sect. 17.2; Bai, 1999, 2015b, 2016, 2017a, 2019). Similar to those of animals, a plant life cycle begins with a zygote and ends with two different types of gametes (after the fusion of the two gametes, the new zygote starts a new life cycle). The difference between animals and plants is that by forming branches, MCSs (multicellular structures) in plants can multiply into numerous developmental units, which are observed as multiple flowers. This is why C. H. Waddington (1966), who coined the term "epigenetics," pointed out that "each branch meristem gives rise to a whole new cycle of growth and development." From this perspective, it is not proper to compare a plant colony to an animal individual in terms of sex differentiation.

Second, any multicellular species has to complete its life cycle in order to pass its genetic information to the next generation for sustainable survival. The core of a life cycle is the "sexual reproduction cycle (SRC)," which emerged from unicellular eukaryotes (details discussed in Sect. 17.2.1; Bai 2015a). Heterogametogenesis is an indispensable event in the SRC, which is why heterogametes are required to complete the life cycles of all eukaryotes, regardless of whether they are unicellular or multicellular. In animals, each dioecious individual is essentially a carrier of one particular type of gamete. In plants, including bryophytes, pteridophytes, and spermatophytes, regardless of the existence of flowers and multiple branches and the methods used for gamete production, every population must bear both male and female gametes. It is difficult to imagine that only plants that bear unisexual flowers, which account for ~10% of angiosperms, undergo sex differentiation while the other 90% of angiosperms and other non-flowering plants do not. If this inference stands, there would be no reason to consider the mechanism of unisexual flower development to be the mechanism of sex differentiation in plants.

Third, human sexuality is primarily determined by sex chromosomes. However, sex chromosomes are not a universal mechanism of sex differentiation in the animal kingdom, not to mention plant species (Bai & Xu, 2013). Detailed investigation revealed that the genes linked with sex chromosomes mainly affect gonad differentiation (Gilbert, 2010; Beukeboom & Perrin, 2014). The differentiation of the germline into particular gametes is not directly determined by genes linked with sex chromosomes but is indirectly affected by differentiated gonads (Casper & Van Doren, 2006; Swain, 2006; Jemc, 2011). Plants utilize a completely different morphogenetic strategy from animals, i.e., no differentiation of gonads and germlines. Referring to heterochromosomes that are correlated with unisexual flowers as "sex chromosomes" will easily lead to the notion that they play a similar role to animal sex chromosomes in sex determination or differentiation, but this is not true.

17.1.3.2 To Lead to or Prevent Sex Cell Formation

Although Robbins and Pearson's definition dominated the field for more than 90 years, John Coulter (1914) proposed a different theory about plant sex, which was not mutually exclusive from that of Robbins and Pearson. At the conclusion of his book, *Evolution of Sex in Plants*, he stated that "the impression one obtains of sexuality as a method of reproduction is that it represents protoplasts engaged in reproduction under peculiar difficulties that do not obtain in reproduction by spores or by vegetative multiplication, and that its significance lies in the fact that it makes organic evolution more rapid and far more varied." The experimental system Coulter used to investigate sex differentiation was a green alga. The use of this experimental system allowed him to pay attention to the conditions under which vegetative cells are converted into reproductive cells. Coulter's view on plant sex is consistent with the proposal made by August Weismann (1889) that the purpose of sex is to generate genetic variation (Winther, 2001), especially under stressful environments.

In Coulter's book, he clearly indicated that in his view, the origin of the gamete is the same as the origin of sex, and sex differentiation is the differentiation of gametes that can be recognized as male and female. This could be interpreted as meaning that sex differentiation, regardless of the level of complexity (organ or individual), should lead to gamete formation. From this perspective, Robbins and Pearson's correlation of sex cells and organs is somewhat consistent with Coulter's view of sex. However, the manner in which they refer to sex, i.e., that a flower or a plant bears only one sex organ, blurs the key issue of whether unisexual flowers or plants result from the inhibition of the formation of inappropriate sex organs.

If unisexual flowers result from the inhibition of one type of sex organ, this should not be considered to represent a sex differentiation mechanism, as inhibition of one type of "sex organ" does not lead to the corresponding gametogenesis; instead, it prevents (a subset of) the process. Unfortunately, not enough attention has been paid to this subtle difference.

17.1.3.3 A Bird-Nest Puzzle

Many unisexual flowers investigated to date, including flowers from maize, cucurbits, white campion, asparagus, wild yam, kiwi fruit, *Fragaria*, persimmon, grape, and banyan (Dellaporta & Calderon-Urrea, 1993; Ainsworth, 1999; Bai & Xu, 2013; Akagi et al., 2014, 2018; Harkess et al., 2017; Massonnet et al., 2020; Zhang et al., 2020a, b), have both stamen and carpel primordia during the early stage of floral bud formation. While female papaya flowers have no trace of stamen primordia, Ronse De Craene et al. (2011) reported feminization in this species ranging from the fusion of stamen tissue to the gynoecium or carpeloidy of antepetalous stamens, pointing to the existence of stamen primordia during early floral bud development. Among these plants, all known mechanisms for unisexual flower development are centered on the inhibition of inappropriate organ development. For example, male flowers in maize result from the inhibition of carpel differentiation, whereas female flowers in cucumber and persimmon result from the inhibition of stamen development via completely different molecular mechanisms.

Based on the previous findings of my colleagues and I, we questioned the dominant concept of using unisexual flowers to investigate regulatory mechanisms of sex differentiation by addressing the "Bird-nest puzzle" (Bai & Xu, 2012). We reasoned that, as plants are multicellular organisms, plant sex cells are produced in particular organs, i.e., the haploid antheridium/archegonium or equivalent structures embraced in the diploid stamens/ovules. Therefore, these organs can be seen as analogous to the nest, with the sex cells as eggs. While a bird needs a nest for laying and hatching eggs, the bird—not the nest—lays the eggs. Although birds build their own nests in most cases, can we try to understand how birds lay and hatch eggs by investigating how the nest is built or destroyed? Of the mechanisms of unisexual flower formation elucidated to date, all describe the inhibition of inappropriate organ development, which can be viewed as analogous to the destruction of the nest, and they have nothing to do with the development of appropriate organs that leads to sex cell differentiation. For example, little is known about how proper carpel development is regulated in female maize flowers and how proper stamen development is regulated in male cucumber flowers. As mentioned in Sect. 17.1.1, Ainsworth was indeed aware of such a situation. Unfortunately, this awareness did not receive the attention it deserved.

Taken together, perhaps the major reason for the dominance of using unisexual flowers to define sex differentiation is the neglect of the logical conflicts in the conceptual framework while focusing on experimental details.

17.2 Sexual Reproduction Cycle (SRC) and Plant Morphogenesis 123 (PM123)

If one is convinced that the unisexual flower is not an issue of sex differentiation, it would be natural to ask the following questions: (1) Does sex occur in plants? and (2) What is sex differentiation in plants?

According to Robbins and Pearson (1933), Rudolph Jakob Gamerarius, in his letter written in 1694, provided the first experimental evidence that seed production requires anthers. In 1736, Linnaeus introduced his "sexual system" for classifying plants. The first question is no longer an issue, as the occurrence of sex in plants is evidenced by the existence of two types of gametes in all major types of organisms categorized as plants, from bryophytes to pteridophytes to spermatophytes. The question then could be revised as this: what (precisely) does "sex" in plants refer to?

It is not easy to answer the revised question. Since sex in plants has traditionally been viewed in a conceptual framework built on observations and investigations in animals, to properly answer the revised question, one must first know how sex is defined in animals. One must then find a way to properly compare the process of the completion of the life cycle in animals vs. plants.

Table 17.1 lists definitions and descriptions of sex found in various authoritative resources. It is bewildering that even in animals, the meanings of "sex" vary significantly and are not biologically equivalent. These meanings can include denoting a general phenomenon, i.e., members of species, being divided into two groups; an event, i.e., combining genes from two different individuals or the occurrence of meiosis; or a composite process that includes multiple events. Given that the meaning of "sex" is ambiguous even in animals, one wonders how sex can be clearly defined in plants.

In terms of comparing plants and animals, I have already pointed out that a plant is a colony of multiple developmental units (Bai, 1999, 2015b, 2016, 2017a, 2019; Waddington, 1966) and as such is not comparable to an individual of a unitary animal organism. Is there then a proper conceptual framework for comparing animals and plants in terms of clarifying the sex issue?

To clarify what precisely "sex" denotes and to properly compare sex differentiation in animals and plants, new concepts and reference frameworks are needed.

17.2.1 The SRC and What a "Generation" Really Means

While the meaning of sex is ambiguous, most researchers can agree on one thing: sexual reproduction has been occurring since the advent of unicellular eukaryotes (refer to the resources listed in Table 17.1). However, the classic references regarding sex place the emphasis on two different aspects of the function of sex: reproduction and genetic variation. Regarding the first, Berrill wrote, "Because the life span of all individual forms of life, from microbes to men, is limited, the first concern of any particular population is to produce successors (https://www.britannica.com/science/sex#toc29374)." Therefore, sex or sexual reproduction is viewed as simply one way to produce successors. By contrast, August Weismann (1889) pointed out that the purpose of sex is to generate genetic variation. Such a viewpoint was echoed by Gilbert (Table 17.1). What does "reproduction" mean in this context, and how is it associated with genetic variation?

We recently proposed the concept of the SRC (Bai & Xu, 2013; Bai, 2015a). According to this concept, the SRC is a modified cell cycle in which three events, meiosis, heterogametogenesis, and fertilization, are integrated. This concept clarifies the relationship between "reproduction" and increased genetic variation.

First, regarding the change in cell number, the SRC is essentially a cell cycle, meaning that through the SRC, one cell becomes two. It is correct to say that "any particular population must produce successors" to maintain its survival. However, according to D'Arcy Thompson (1917), cell division is merely a physical process equivalent to a soap bubble: along with the increase in contents inside of a cell, which is driven by the positive feedback characteristics of a living system, the cell may divide to release the increasing surface tension and to maintain a proper surface/volume ratio. Similarly, upon an increase in inner pressure, a soap bubble may divide into two. However, unlike a soap bubble, in which the number of resulting bubbles that form is uncertain, cell division usually results in one cell becoming two. While "proliferation" was traditionally used for describing such a process, the increase of cell number is just a passive outcome, not an intentional one. There is no doubt that along with evolution, the regulation of this process has gradually become more complicated to ensure robustness and has ended up as the cell cycle we know of today.

It should be clarified that the term "cell cycle" should be understood here only as describing the resumption of the internal status of cellular structures and functions. The continuous cell division process in which one cell becomes two is simply a linear and unidirectional process resulting in an increase of cell number, occurring in both prokaryotes and eukaryotes (shown in yellow-lined round-cornered rectangles in Fig. 17.1).

Table 17.1 Definition or description of sex in authorized resources

Types	Years	Definition or description	Authors	Resource
Encyclopedia	2017	Sex, the sum of features by which members of species can be divided into two groups—male and female—that complement each other reproductively	N.J. Berrill	https://www.britannica.com/science/sex#toc29374
	2017	Organisms of many species are specialized into male and female varieties, each known as a sex		http://en.wikipedia.org/wiki/Sex
Textbooks	2005	Sexual reproduction is the creation of offspring by the fusion of haploid gametes to form a zygote, which is diploid	Campbell and Reece	*Biology* 7th ed.
	2000	It should be note that sex and reproduction are two distinct and separable processes. Reproduction involves the creation of new individuals; sex involves the combining of genes from two different individual into new arrangements	S. F. Gilbert	*Developmental Biology* 6th ed.
Monograph	1982	Sex is a composite process in the course of which genomes are diversified by a type of nuclear division called meiosis, and by type of nuclear fusion called syngamy, or fertilization Sex and reproduction are quite distinct processes: sex is a change in the state of cells or individuals, whilst reproduction is a change in their number	G. Bell	*The Masterpiece of Nature:* *The evolution and genetics of sexuality*
	1983	*Fisher, 1930: No practical biologist interested in sexual reproduction would be led to work out the detailed consequences experience by organism having three or more sexes, yet what else should he do if he wishes to understand why the sexes are, in fact, always two?* Sex is defined as gender, male or female Sex development refers collectively to the various molecular, genetic, and physiological processes that produce a male or a female from a zygote of a given genotype and parents in a given environment	J. Bull	*Evolution of Sex Determining Mechanisms*
	2014	Sex is defined by the occurrence of meiosis	Beukeboom and Perrin	*The Evolution of Sex Determination*

Review articles	2002	True sex – syngamy, nuclear fusion and meiosis – is found only in eukaryotes	Cavalier-Smith	Origins of the machinery of recombination and sex
	2013	The core features of sexual reproduction involve: (i) ploidy changes from diploid to haploid to diploid states, (ii) the production of haploid mating partners or gametes from the diploid state via meiosis which recombines the two parental genomes to produce novel genotypes and halves the ploidy and (iii) cell-cell recognition between the mating partners or gametes followed by cell-cell fusion to generate the diploid zygote and complete the cycle	Heitman et al	

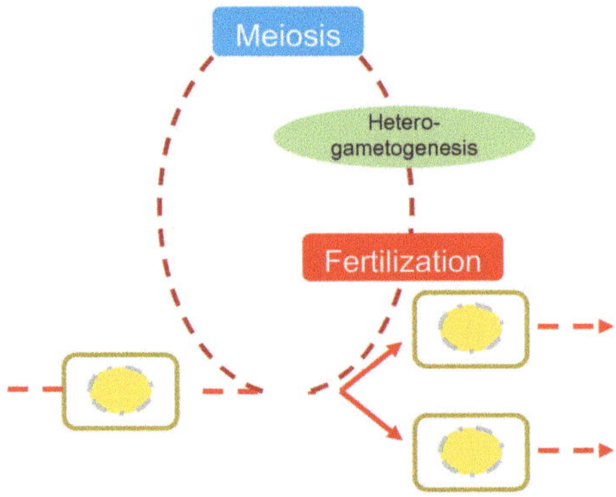

Fig. 17.1 A diagram of the modified cell cycle called the "sexual reproduction cycle," SRC. (**a**) Ordinary cell division, i.e., one cell becomes two. (**b**) SRC. The 3 rounded rectangles containing yellow ovals represent diploid cells. The red dashed line and arrows represent one diploid cell becoming 2 (a cell cycle). Dark red dashed curve represents a process, in which 3 biologic events, i.e., meiosis, fertilization, and heterogametogenesis, are integrated, and inserted into the cell cycle represented by two rounded rectangles (blue and red, respectively) and an oval (light green). Through the SRC, a diploid eukaryote can autonomously generate genetic variations and increase fitness to the unpredictably changed environment. (Reprinted from Bai 2016, by permission of Science Press)

Second, as described in Bai (2015a), after eukaryotes emerged, meiosis, heterogametogenesis, and fertilization independently arose during the linear and unidirectional cell division and were by chance integrated and selected as a programed process termed the SRC (between the one to two cells, there are three events inserted shown in Fig. 17.1). As pointed out by many authors, the number of genetic variations can autonomously increase during meiosis (DNA sequence itself can also change without meiosis in various ways, often at much lower frequencies; Long et al. 2013). From this perspective, the number of variations increases not via the entire SRC process but via meiosis, one of the three key events constituting this process.

While heterogametogenesis functions in harnessing variation and enhancing heterogeneity, fertilization functions in restoring diploidy and integrating adaptive variations into the next generation. The latter function of fertilization is easily understood because only the haploid, meiotically produced cells, which contain new variations, surviving to the completion of gametogenesis are able to meet and fertilize.

Third, although the SRC is essentially a process by which one cell becomes two, it differs from linear cell proliferation, as the two resulting cells produced by the SRC are no longer the same as the initial one due to genetic variation. That is why I proposed that only through the SRC can the concept of "generation" be rationally defined (Bai, 2015a). This adjusts the previous description of linear cell division into a cyclical process, meaning that through one SRC, along with each diploid cell becoming two to achieve cell proliferation, there are repeated cell divisions and fusions and an alternation between diploidy and haploidy, through which the genetic variations are increased and selected.

Using the concept of the SRC, it is possible to distinguish the innovative autonomous mechanism to the unpredictable changes of surrounding components from the ancestral process of cell division. While the final results of both processes are one cell becoming two, the former (SRC) is cyclical and results in different generations, and the latter is linear and results in the proliferation of cells in the same generation (in terms of genetic setting). If one considers the essential function or evolutionary advantage of sex to be the increase in genetic variation rather than a method for reproduction, one can focus on the three events, i.e., meiosis, heterogametogenesis, and fertilization, and no longer be concerned with cell proliferation.

17.2.2 Interpolation of MCSs

The above discussion is based on analysis of the life cycles of unicellular eukaryotes. Is this concept applicable to multicellular eukaryotes? Bai (2015a) proposed that the SRC provides a framework or backbone onto which MCS is interpolated (Fig. 17.2). The idea of inserting the MCS into the sexual reproductive process was originally proposed in the early twentieth century by the British botanist F.O. Bower (1935), although during that time, there was no concept of the SRC. If two facts are taken as the foundation, i.e., that a cell has to divide while it remains alive, and that each of the

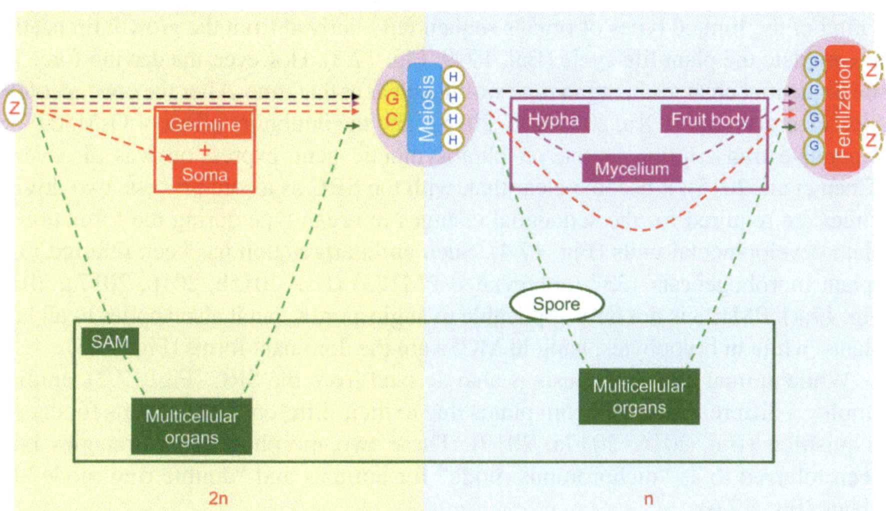

Fig. 17.2 Comparison of morphogenetic strategies of animals, fungi, and plants in the framework of SRC. A yellowish background indicates the diploid phase and bluish haploid phase. In the intervals between zygote and germ cells, multicellular structures of animals (red) and plants (green) are interpolated, while there are none in fungi (pink). In the intervals between meiotically produced cells and gametogenic cells, multicellular structures are interpolated for fungi and plants but not animals. (Reprinted from Bai 2015, by permission of Frontiers in Plant Sciences)

three events in the SRC is induced by stress (Bai, 2015a), it is easy to imagine cell division would occur during the intervals of the SRC between the formation of the zygote and meiotic cells and between the formation of meiotically produced cells and gametogenic cells. It is easy to imagine that these cells can be organized into an MCS via various mechanisms. Therefore, multicellular organisms emerged along with the SRC process. As shown in Fig. 17.2, it has been proposed that due to differences in their modes of energy acquisition, e.g., photoautotrophy in plants and heterotrophy in animals and fungi, MCSs can be interpolated into these intervals once in heterotrophic organisms (in diploid intervals in animals and haploid intervals in most fungi) and twice in photoautotrophic plants (in both intervals). From this perspective, there is a clear division of labor in which the SRC, a process that occurs at the unicellular level, is responsible for increasing genetic variation to help organisms to integrate unpredictable changes of surrounding components (stresses), whereas MCSs are responsible for increasing the capacity for integration of surrounding components, which are conventionally expressed as energy acquisition and internal/external environmental responses. Different organisms can have different ways to integrate the SRC and MCSs.

17.2.3 Double Ring Mode and PM123

Based on my investigation of the *Arabidopsis* mutant "*embryonic flower* (*emf*)," I proposed the concept of the "plant developmental unit (PDU)" to describe the combination of the limited types of organs sequentially derived from the growth tip needed to complete the plant life cycle (Bai, 1999; Fig. 17.3). However, the driving force for the sequential changes in organ type was not clear at that time. After the concept of the SRC emerged (Bai & Xu, 2013; Bai, 2015a) and the inhibitory effect of OsMADS58, a putative transcription factor, on photosynthetic gene expression was discovered (Chen et al., 2015), it became clear that, with the SRC as a core process, two driving forces are required for the sequential changes in organ type during the formation of plant developmental units (Fig. 17.4). Such an interpretation has been referred to as "plant morphogenesis 123" (abbreviated PM123) (Bai, 2015b, 2016, 2017a, 2019; Fig. 17.4). PM123 is not only applicable to angiosperms, but it also applies to all land plants, while in bryophytes, haploid MCSs are the dominant forms (Fig. 17.5).

While animal morphogenesis is also derived from the SRC (Fig. 17.2), animals employ a different strategy from plants due to their different mechanisms for energy acquisition (Bai, 2016, 2017a, 2019). These two morphogenetic strategies have been referred to as "dichotomous mode" for animals and "double ring mode" for plants (Fig. 17.6).

Thus, concepts such as SRC and PDU and conceptual frameworks such as PM123 have been introduced to dispel the ambiguity in the existing definitions of sex and the problematic comparison of an animal individual to a plant. Using these concepts as a reference point, we are now ready to move forward to address our three questions on sex, sex differentiation, and the ultimate function of sex not only in plants but also in animals.

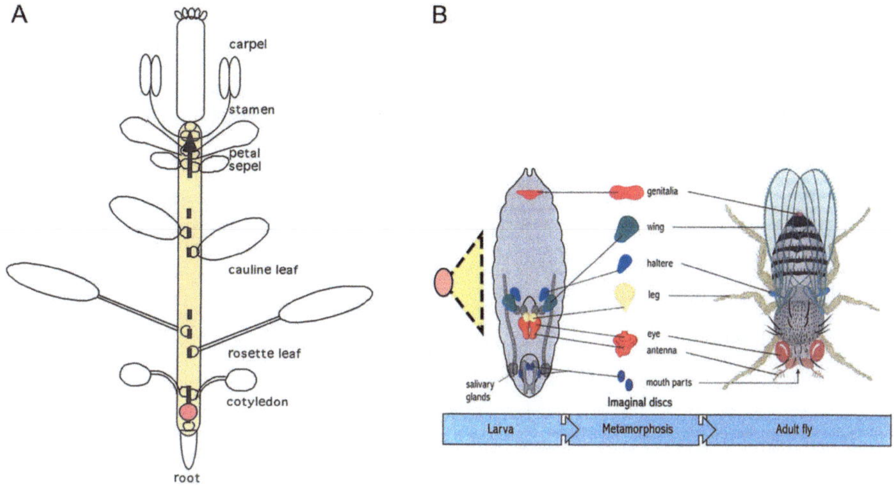

Fig. 17.3 Comparison of developmental units in plants and animals required for life-cycle completion. While a plant growth tip generated from a zygote (pink circle) can produce numerous lateral organs and branches, only seven organ types in *Arabidopsis* are present in a complete life cycle (**A**). The half circles along the dashed orange arrow represent organ primordia. By comparison, the basic structure required for Drosophila to complete its life cycle is the embryo, elaborated from a zygote to larva. Embryogenesis is represented by an orange-lined yellow triangle (**B**). Unlike animal individuals, which contain limited types and numbers of organs in a determined pattern, the functionally equivalent structure in plants is the imaginal unit shown in (**A**), referred to as a "developmental unit," rather than the whole plant. According to this perspective, the structural equivalent of an animal embryo would be the process represented as the yellow area in (**A**). Orange dashed lines in the yellow region indicate that the process is relatively open but ultimately limited. (**B**) Was modified from Figures 2–6 in Wolpert et al.'s (2007) edited Principles of Development. (This figure was modified from Figure 6 in Bai 1999)

17.3 Sex: Division, Heterogametes, and Heterogametogenesis

At the beginning of this chapter, I stated that while sex is a hot topic, many notions about sex are vague and controversial, and answers to the three essential questions about sex, sex differentiation, and the ultimate function of sex remain elusive. The above sections described the controversies regarding these three questions demonstrated that the development of unisexual flowers in angiosperms is not an issue of sex differentiation but is instead utilized as a mechanism for promoting outcrossing; furthermore, they highlighted that "sex" is a natural phenomenon that is not related to reproduction but is instead a mechanism for increasing genetic variation. In addition, I introduced the concept of the developmental unit to properly compare the life cycles of animals and plants regardless of their different morphogenetic strategies.

Traditionally, studies investigating natural phenomena begin with an inductive approach in which related samples from the environment are collected and their similarity analyzed to find the common principles behind these samples. This method has been quite successful for investigating many important concepts since

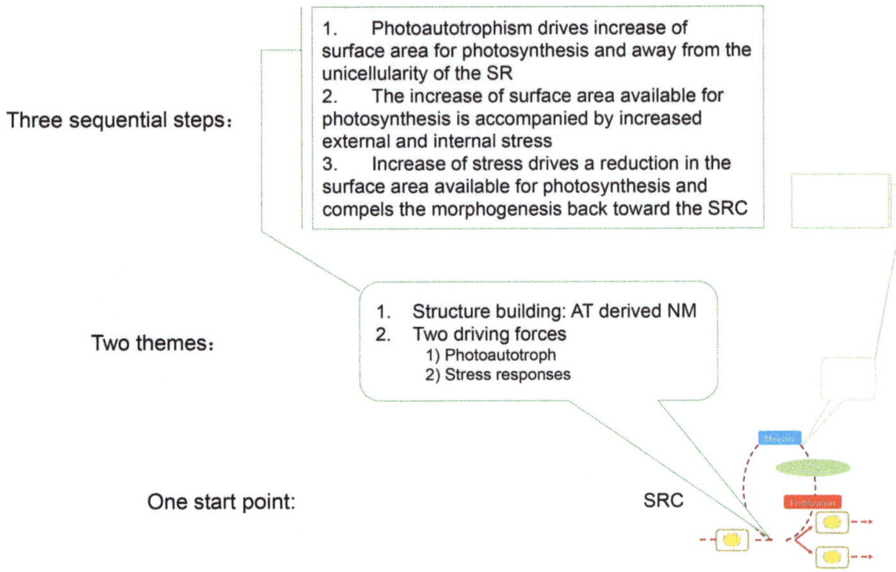

Fig. 17.4 A diagram of the PM123 theory. Two multicellular structures are interpolated into the two intervals (green framed for the diploid and light green framed for the haploid) during the SRC (the one of the PM123). Represented in the diploid phase, two themes (the two of the PM123) underlie the morphogenesis of the multicellular structures: structure building, the "axial tree" (AT) derived neo-modularization (NM), and two driving forces for sequential change of organ types, including photoautotroph and stress response. Three sequential steps (the Three of the PM123) are elaborated in the upper frame

the scientific revolution, such as "force," "energy," and "molecule." However, to investigate highly complex biological phenomena, such as sex, this method will inevitably face difficulties: surrounded by such a huge diversity after billions of years of evolution, from which perspective should one find essential things in common? To avoid such complications, here, I took an alternate approach to analyzing sex-related phenomena. Based on the LUCA (last universal cell ancestor; Darwin, 1859; Theobald, 2010) hypothesis, which fits well into the conceptual framework of evolutionary theory and the concept of the SRC, I discuss sex-related phenomena in the simplest system, unicellular eukaryotes, from a function-based perspective.

First, let us examine how the word "sex" originated. According to the open access online English dictionary www.dictionary.com, we found the following information: "*Origin*: 1350–1400; ME < L *sexus*, perh. akin to *secāre* to divide (see 'section'), meaning that the word sex originated from the Latin word *sexus*, which is akin to *secare*, meaning "divide." While such a definition perfectly matches the observation that humans and many other animals are dimorphic, which can be divided into two groups, it is not clear at what level such a division originated, i.e., the core question of what sex is still remains.

Clearly, when answering this question, the SRC as a whole should be ruled out; although through this process, one cell divides into two, this occurs during other cell

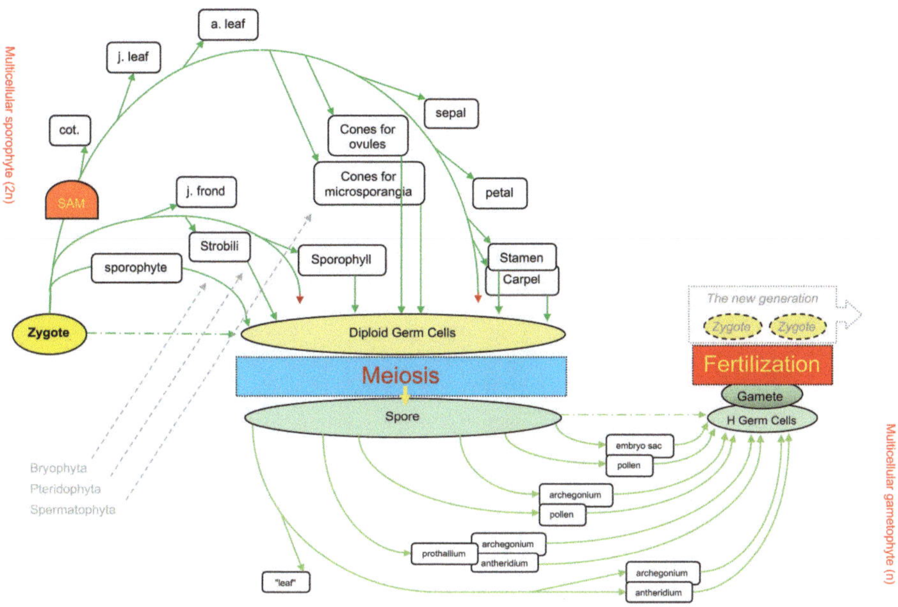

Fig. 17.5 Different levels of elaboration around the core processes in the life cycles of the three plant phyla. The sexual reproduction cycle (Bai, 2015a) from one zygote to the next generation's zygotes through meiosis and fertilization is the backbone of the life cycle for all three land plant groups: Bryophyta, Pteridophyta, and Spermatophyta. Green arrows show the differentiation of various organ types in the diploid phase, and light green for organs in the haploid phase. Dark red arrowheads indicate unlimited tip growth activity. cot., cotyledons; j. leaf, juvenile leaf (e.g., rosette leaves in *Arabidopsis*); a. leaf, adult leaf (e.g., cauline leaves in *Arabidopsis*). (This figure was revised from Fig. 1.9 in Bai and Xu 2013)

division processes as well. If the division of one cell into two were considered to be "sex," one would not be able to distinguish the SRC from other cell division processes.

Fertilization should not be referred to as "sex" either, as this event involves the fusing of two cells into one, which is functionally an opposite event from "dividing."

Meiosis has been considered to represent the core process of sex (Table 17.1 and the references therein). However, while meiosis is indeed almost always followed by sex differentiation in animals, such a correlation does not always occur in plants and unicellular eukaryotes. As summarized in Fig. 1.6 in Bai and Xu (2013), meiosis produces homospores in bryophytes and some ferns, and heterogametogenesis occurs in antheridia and archegonia of gametophytes derived from homospores in these plants. The relationship between meiosis and heterogametogenesis is more complicated in unicellular eukaryotes, or protists (detailed information refers to Shen, 1999). From this perspective, it is obviously not proper to define sex as meiosis simply because such a relationship cannot be universally applied to all eukaryotes.

The final event in the conceptual framework of the SRC is heterogametogenesis. To date, according to the available references, heterogametogenesis, i.e., the

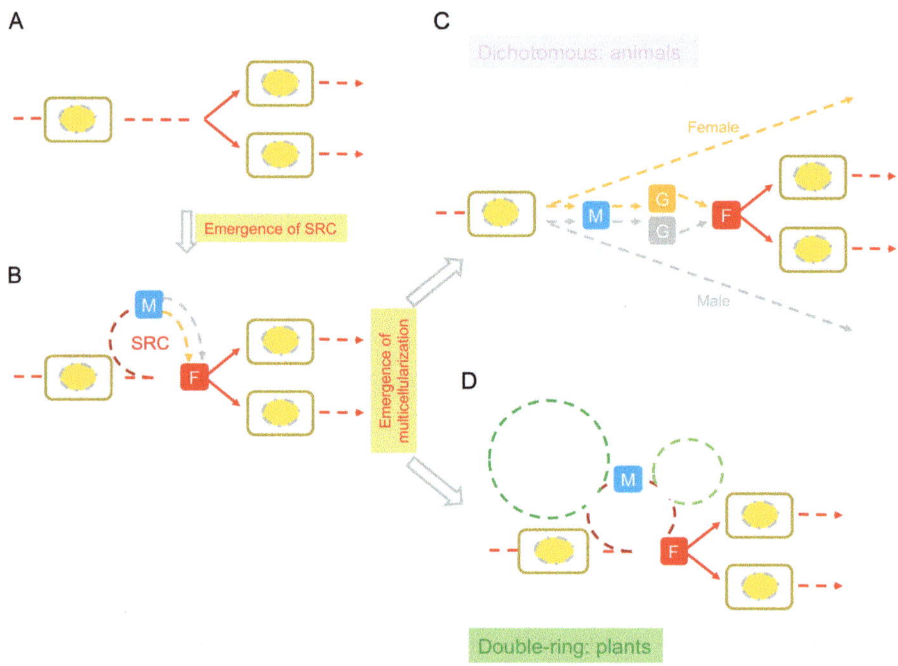

Fig. 17.6 Comparison of SRC-derived developmental modes in animals and plants. (**A**) Represents a cell cycle, i.e., one diploid cell becomes two cells through mitosis. (**B**) Represents the SRC (sexual reproduction cycle, Bai, 2015a). Two arrows (orange and gray) between M (meiosis) and F (fertilization) represent heterogametogenesis. (**C**) Represents the "dichotomous mode" for animal development. Orange and gray lines represent female and male soma and germlines, respectively, and orange and gray Gs represent female and male gametogenesis, respectively. (**D**) Represents the "double-ring mode" of plant development. On the SRC backbone, the green dashed ring on the left represents diploid multicellular structures composed of various types of organs; the light green ring on the right represents haploid multicellular structures composed of various types of organs. (This figure was modified from Fig. 16 in Bai 2016)

generation of different types of gametes, i.e., heterogametes, appears to be a universal phenomenon in eukaryotes, regardless of the huge diversity in gamete size, shape, number of mating types, and molecular mechanisms (Bai & Xu, 2013; Bai, 2015a). While genetic variations can be randomly generated through recombination in meiosis, all meiotically produced cells can ultimately lead to two or more types after heterogametogenesis. Since it is widely accepted that the ultimate difference between the "male" and "female" can be traced back to so-called sex cells, i.e., gametes, it can unambiguously be stated that the division of "sex cells" originates from heterogametogenesis and ends up at heterogametes. In other words, what the word "sex" truly means from a biological perspective is heterogametogenesis in terms of process, or heterogametes in terms of ultimate results. Therefore, the answer to the question "what is sex" is that sex refers to heterogametes, or heterogametogenesis.

The concept of the SRC clarifies the relationship of cell proliferation and the increase in genetic variation, and it solves the controversial issue of mixing the two events when defining sex. If the entire SRC is taken to represent a modified cell cycle, the problem of the "two-fold cost of sex (Williams, 1975; Maynard Smith, 1978)" no longer exists as well. Using heterogametes, or heterogametogenesis to define sex can solve another controversial issue of whether the genetic variation in prokaryotes should be considered to represent sex. If meiosis is used as the definition of sex, increase of genetic variation would represent the core feature of sex. This would imply that prokaryotes have sex. However, no particular differentiation process that is functionally equivalent to gamete production, the cyclical process required for cell division and fusion, or the alternation of diploid and haploid status have been reported in prokaryotes. It is clear that increasing genetic variation is universal in all cellularized living systems including both eukaryotes and prokaryotes. This is the process that helps the organisms to integrate unpredictable changes of surrounding components. By contrast, sex, i.e., heterogametes or heterogametogenesis, is unique to eukaryotes, resulting in the labeling of meiotically produced cells to harness variations while enhancing heterogeneity (Bai, 2015a). In other words, heterogametogenesis is an internal control system that balances randomly generated genetic variations.

17.4 Sex Differentiation: Mechanisms that Occur in Soma to Ensure Heterogametogenesis

The above section answered the question "what is sex." Now we move to the second question: What is "sex differentiation"?

Before elaborating on the answer presented in the title of this section, it is important to clarify an adjustment to the original description of the SRC (Bai, 2015a), which was described as the "integration of meiosis, sex differentiation, and fertilization." Based on the above discussion, the definition of the SRC should be updated as a modified cell cycle in which meiosis, heterogametogenesis, and fertilization have been inserted and integrated. When the concept of the SRC is applied to unicellular eukaryotes, the terms "sex differentiation" and "heterogametogenesis" are functionally equivalent and interchangeable. However, when the SRC concept is applied to multicellular eukaryotes, the term "sex differentiation" is no longer interchangeable with heterogametogenesis but an added novel process in which somatic cells are differentiated, accompanying the germ cell differentiation that ends up as heterogametogenesis. Thus, I propose the use of "sex differentiation" to refer to the mechanisms that occur in the soma to ensure heterogametogenesis in multicellular eukaryotes, while "sex" per se, based on its original meaning "to divide," refers to heterogametogenesis during the SRC at the unicellular level in both unicellular and multicellular eukaryotes.

The need to distinguish between these terms comes from the idea that multicellular organisms originated from the interpolation of MCSs into the intervals of the SRC (Figs. 17.2, 17.3, and 17.6). As discussed above, the advantages of having an MCS are enhancements of the integration of surrounding components. Under various selective pressures, numerous differentiation processes have occurred in MCSs, e.g., the differentiation of various types of tissues, organs, and systems. As mentioned above, the three events involved in the completion of the SRC, i.e., meiosis, heterogametogenesis, and fertilization, are induced by stress in unicellular eukaryotes. Therefore, in multicellular eukaryotes, some somatic cells differentiated to protect the cells that function as carriers of the SRC, and some somatic cells must be differentiated in sensing and responding to the changes of surrounding components to induce these SRC events. Thus, the somatic cell differentiations with the particular function of ensuring heterogametogenesis need to be specified as "sex differentiation."

However, in light of the overwhelming diversity in all multicellular eukaryotes from animals to plants, what types of differentiation in the soma can be properly referred to as "sex differentiation"? If the SRC is thought of as an ancestral and universal core process in all multicellular eukaryotes, from a function-based perspective, we can see that to ensure heterogametogenesis, two functions are essential: (1) to ensure the setting of a divergence point between the two sexes and (2) to ensure that heterogametogenesis can be fulfilled when these cells are surrounded by somatic cells.

Since animals and plants possess different morphogenetic strategies, to emphasize their functional similarity and strategic diversity, we will discuss these two processes in animals and plants separately.

17.4.1 Setting the Divergence Point

17.4.1.1 Setting the Divergence Point in Animals

From the perspective of SRC-derived morphogenesis, I proposed that animals possess a "dichotomous mode," whereas plants possess a "double ring mode" (Figs. 17.2, 17.6, and references therein). Fan and Bai (2002) reasoned that as animals acquire energy by fetching food, they have to stably differentiate MCSs suited for different functions, e.g., mobility, digestion, and sensation of surrounding components. To ensure the efficiency of the labor division of MCSs, cellular differentiation must be a stable or irreversible process. In this case, if all cells derived from the division of the zygote are consumed for integration of surrounding components, where do the cells that carry out the indispensable SRC come from? Based on current observations of animal morphogenetic strategies, it appears that the solution to this problem is to reserve a group of cells as germline cells in various ways during early embryogenesis and to use these cells as carriers to complete the SRC. This would explain the origin of the "dichotomous mode" in animal morphogenesis.

Where are these germline cells located in the animal's body? According to our current knowledge, except in some basal animals such as hydras (Loomis, 1954; Bosch & David, 1986; Littlefield, 1986; Nishimiya-Fujisawa & Sugiyama, 1995; Alexandrova et al., 2005), a majority of animals have evolved gonads into which germ cells migrate (Gilbert, 2010; Sommer & Bumbarger, 2012; Swain, 2006; Jemc, 2011; Wilhelm et al., 2013; Eggers et al., 2014). Under such a morphogenetic strategy, there would be at least two scenarios for setting the divergence point. One scenario is that the gonads remain neutral and allow the germ cells to differentiate. Later, when the individual matures, the germ cells would be induced into hetero-gametogenesis. Alternatively, the gonads could differentiate into one of two types during early embryogenesis and thereby force the germ cells within them to differentiate into only one type of gamete. While numerous scenarios might have arisen during evolution, the second scenario has survived in most animals.

In addition to the knowledge obtained from investigating the developmental biology of gonad initiation, the regulatory mechanisms at the molecular level revealed that the genes responsible for sex determination in vertebrates essentially determine whether the initiated gonad differentiates into an ovary or testis (Gilbert, 2010).

Taken together, an interesting scenario emerges: after zygotic division, the resulting cells are assigned into two different functional paths: some cells are destined to become germ cells, which carry out the SRC, while the remaining cells become somatic cells. Among somatic cells, some are committed to differentiate as gonads to host germ cells. During gonad differentiation, one decision has to be made in the majority of animals: to become an ovary or a testis. Such a decision determines not only the final shape, structure, and function of the gonad but also whether the migrated germ cells differentiate into eggs or sperm, as well as affecting secondary sex characteristics in mammals. Under such a scenario, it is obvious that the divergence point that leads to heterogametogenesis in animals is set during a certain stage of gonad development.

While the regulatory network for gonad differentiation appears to be conserved in various species ranging from reptiles to mammals (Janzen & Phillips, 2006), the mechanisms triggering this network vary. In mammals, for example, genes such as *SRY* are localized on the Y chromosome, and the setting of the divergence point for gonad differentiation is triggered by the distribution of the Y chromosome in the zygote (Wilhelm et al., 2013; Eggers et al., 2014). In the reptile red-eared turtle, an aromatase, rather than SRY, plays a key role in setting this divergence point in response to temperature, which triggers gonad differentiation into either an ovary or a testis (Ramsey & Crews, 2009). Comparing the mechanisms utilized in mammals and red-eared turtles, it is obvious that the former is more autonomous and robust against changes of surrounding components, while the latter also has some advantages, such as generating a proper sex ratio to adjust to variable temperature situations. From this perspective, one may imagine that so-called sex chromosomes were selected based on their advantage of ensuring an equal distribution of two types of individuals carrying different gametes during early embryogenesis to ultimately achieve a stable sex ratio in the offspring population. This might explain why sex chromosomes are present in some animal species but not in others. The systematic studies on the evolution of vertebrate sex chromosomes carried out by Graves (2016) provide favorable support to the speculation.

In summary, the first function of sex differentiation (i.e., divergence point setting) in animals is gonad-centered differentiation. Regardless of the detailed mechanisms, some somatic cells derived from division of the zygote differentiate into gonads to host germ cells and canalize the germ cells to differentiate into egg or sperm cells. During gonad differentiation, a decision has to be made to further differentiate into an ovary or a testis. During evolution, various mechanisms have emerged to stabilize the divergence point setting process to obtain a proper sex ratio in the offspring population. The emergence of sex chromosomes is one of the mechanisms that function in this process. However, sex chromosomes may affect differentiation of other somatic cells in addition to the gonad (Zhao et al., 2010). On the other hand, germ cells with particular sex chromosomes may autonomously regulate their own differentiation in addition to canalizing the gonads (Hempel et al., 2008; Murray et al., 2010; Rios-Rojas et al., 2015). Furthermore, hormones derived from gonads and/or determined by sex chromosomes collectively determine dioecy during individual morphogenesis in animals. All of the factors involved in gonad differentiation are interwoven, resulting in great diversity in the mechanisms regulating this process, which has long made the analysis quite confusing.

17.4.1.2 Setting the Divergence Point in Plants

In contrast to the dichotomous mode of animal morphogenesis, plant morphogenesis occurs in double-ring mode (Fig. 17.6). Since plants are photoautotrophic organisms, the most efficient pattern of morphogenesis of their MCS in terms of energy acquisition is to build up the maximum photosynthetic surface using the minimum amount of living materials. However, due to surface tension, the surface area of the MCS should be kept to a minimum. To achieve a maximum photosynthetic surface area, the rigid cellulosic cell wall has evolved to provide plant cells with mechanical support against surface tension. However, the presence of cell walls results in the immobility of the MCS in plants. To easily recover from attacks from herbivorous animals and to better utilize photosynthetic resources, plants have evolved the property of maintaining meristematic cells or easily reversing differentiated cells to meristematic cells, which can differentiate in various directions to help the plant survive in the changed surrounding components. Such a property has enabled the unique integration between the SRC and the MCS during plant morphogenesis: while all cells derived from division of the zygote and/or spore are committed to photoautotrophy in various ways, some somatic cells can be induced to differentiate into germ cells during later developmental stages under particular stress conditions (Figs. 17.2, 17.5; Bai, 2016, 2017a), a process known as the double-ring mode.

Based on the above analysis, it is clear that, unlike in animals, there is no cell lineage equivalent to animal germline cells in plants. The SRC is completed via the induction of germ cells from the two interpolated MCSs. Two issues should be clarified: (1) how are germ cells initiated from the somatic cells of the MCS during the two intervals of the SRC and (2) in which MCS should divergence point setting for heterogametogenesis be considered sex differentiation.

To clarify the first issue, during plant morphogenesis, the meiotic cells (diploid germ cells, Bai, 2017b) are initiated in the primordial sporangia in all land plants, and the gametogenic cells (haploid germ cells, Bai, 2017b) are initiated separately in primordial antheridia and archegonia. The traditional view is that reproductive organs (e.g., sporangia and antheridia/archegonia) differentiate first, followed by the initiation of germ cells. However, logically, it is reasonable to hypothesize that germ cells initiate first, and the surrounding cells are then recruited by the germ cells to ensure their differentiation (detailed elaboration see Bai, 2024). These two scenarios would lead to different theories on plant sex differentiation. However, little experimental data are available on gametophyte development in bryophytes and pteridophytes. More detailed investigations are needed to determine which of these scenarios is true.

To clarify the second issue, there is little doubt that heterogametogenesis occurs only after the interpolation of the MCS, i.e., gametophyte, into the second interval of the SRC. In bryophytes and some pteridophytes, homospores are generated; the divergence point leading to the differentiation of antheridia and archegonia (in which heterogametogenesis occurs) should be considered sex differentiation in plants. However, since spermatophytes produce heterospores in sporophytes, i.e., the MCSs interpolated into the first interval of the SRC, this phenomenon has complicated the interpretation of sex differentiation in plants in the past decades. This issue will be further discussed in Sect. 17.4.3.

17.4.2 Niche Establishment to Ensure Heterogametogenesis

One of the advantages of multicellular organisms is that their MCSs can provide protection and nutrition to germ cells regardless of whether the germline diverged from the soma during early embryogenesis in a continuous cell lineage (as in animals) or via separate initiations of two types, diploid and haploid, of germ cells from the two interpolated MCSs (as in plants). An obvious difference between animal gonad differentiation and plant antheridia/archegonia differentiation is that in animals, gonad initiation is spatiotemporally separated from germline initiation. During gonad differentiation, divergence occurs, causing the gonad to differentiate into an ovary or testis. By contrast, in plants, haploid germ cell and antheridia/archegonia initiation are spatiotemporally concurrent, so it is not yet clear where and how the divergence point is set. Given the diversity of mechanisms, three major processes in animals are involved in sex differentiation: differentiation of gonads (including setting of the divergence point), differentiation of the germline, and the interaction between the gonads and germline. By contrast, little is known about the antheridia/archegonia initiation and the haploid germ cell initiation and differentiation in plants, and whether and when the divergence point leading to heterogametogenesis is set in the haploid germ cells or in the surrounding cells remains unclear.

17.4.3 Pseudo Versus Real Sex Differentiation in Plants

One may question why the differentiation of antheridia/archegonia is referred to as sex differentiation in plants, whereas the divergence of stamens and ovules in angiosperms or micro- and mega-sporangia in some lycophytes and related structures in gymnosperms is not. The simple answer is that, according to the double ring mode of plant morphogenesis, stamens and ovules are essentially sporangia in which the diploid germ cells are committed to undergo meiosis, NOT GAMETOGENESIS. After meiosis, there is another round of MCS interpolation in which the antheridia and archegonia formation is initiated, together with the formation of haploid germ cells, starting heterogametogenesis.

However, in plants with heterospores, the divergence points are actually set in sporophytes. Regardless of the complexity of gametophytes, heterogametogenesis is canalized by the heterospores; for example, haploid germ cells generated from gametophytes derived from microspores differentiate into sperm cells, and those from megaspores differentiate into egg cells. How should such a situation be explained?

Dr. Xin Wang and I proposed a hypothesis on the origin and significance of heterospory in plant evolution (Wang & Bai, 2019). The main points are: (1) heterospory is derived from homospory (Bateman & Dimichele, 1994; Qiu et al., 2012). Homosporous plants such as bryophytes are gametophyte dominant, and they are monosporangiate, i.e., sporophytes possessing single capsules are a typical feature of these plants. In contrast, plants of heterospory are generally sporophyte dominant. (2) The larger size of sporophyte requires stronger physical support to against gravity. Gifford and Foster (1989) had suggested that the formation of the polymer lignin, part of secondary cell wall (SCW) thickening, is of paramount importance for vascular plants to support their large photosynthetic surface. Why did the SCW thickening mainly occur in the diploid sporophyte rather than the haploid gametophyte? Chen et al. (2012) found that the SCW thickening is a process that requires a dramatic transient increase in protein synthesis. Since diploidy may better meet the need for a transient increase in RNA for this increased protein synthesis, the finding of Chen et al. (2012) provided a piece of empirical evidence supporting the hypothesis proposed by Gifford and Foster on the transition from a gametophyte-dominant to a sporophyte-dominant life cycle. (3) In addition to the importance of diploidy in the emergence of SCW thickening and the sporophyte-dominant morphogenetic strategy, elaborate sporophytes provide more opportunities for sporangium/diploid germ cells to initiate than simpler structures do. Based on the view that a plant is a colony, each growth tip may terminate as a sporangium. Among the multiple growth tips and sporangia, there must be heterogeneity in nutritional allocation and stress reception. French (1972) observed the early termination of the increase in sporogenous cell number in megasporangia. Brooks (1973) reported that the application of Ethephon can affect the ratio of mega- vs. microsporangia. While more empirical data are needed to elucidate the mechanism of the origin of heterospory, heterogeneity of multiple sporangia in elaborate sporophytes might have evolved prior to (and may be a precondition for) heterospory.

If the above reasoning is valid, it will be clear that heterospory is just one of the results of the inevitable heterogeneity of elaborate sporophytes. How is heterogametogenesis in spermatophytes canalized into heterospory? Based on the SRC, the two interpolations of the MCSs, and the sessile properties of plants, our answer to this question is as follows.

Since gametes are produced in spatiotemporally separate locations, how the two gametes meet becomes a critical issue for the completion of the SRC. In bryophytes and homosporous pteridophytes, plants release their sperm into water, and the sperm swim to the archegonia for fertilization. Some heterosporous pteridophytes release both mega- and microspores, and the sperm swim to the female gametophytes for fertilization. However, when spermatophytes emerged, megagametophytes were retained in the aerial terminals of the sporophytes in ovules (Bateman & Dimichele, 1994; Herr, 1995; Bai et al., 2022). While this innovation affords adaptive advantages in terms of nutrient supply to and protection of the female gametophytes, as well as the female gametes (eggs), it generated a novel challenge, namely, how do the male gametes reach the aerially localized female parts and accomplish fertilization? It appears that the solution to this problem was to reduce the male gametophytes to make them mobile. Since the reduction of the male gametophytes left little room for antheridium differentiation, but a divergence point has been set in sporophytes (for whatever reason), the best option for spermatophytes to complete their SRC appears to be borrowing heterospory as a divergence point and to canalize heterogametogenesis into successive differentiation of heterospores. This borrowing has enabled spermatophytes to retain the advantage of receiving nutrition and protection from the sporophytes while ensuring that the gametes meet.

According to the above analysis, although heterospory originated from the heterogeneity of the elaborate sporophytes and has no direct causal relationship with heterogametogenesis, under the selective pressure of completing the SRC, heterogametogenesis in spermatophytes appears to be canalized with heterospory. In the other words, the divergence of mega- and microsporangia in spermatophytes has been borrowed to fulfill the role as the divergence point originally set in the gametophytes of bryophytes and some pteridophytes that leads to heterogametogenesis. To distinguish the two events, I refer to the divergence of mega- and microsporangia as "pseudo sex differentiation" and that leading to antheridia/archegonia differentiation as "real sex differentiation" in plants.

In light of the two sites for interpolation of MCSs in the SRC, more complicated situations in plant sex differentiation can occur. Figure 17.7 summarizes the meaning of sex and sex differentiation in unicellular eukaryotes and multicellular eukaryotes (animals and plants) from the perspective of the SRC.

17.5 Sexual Behavior: Ensuring Gamete Meeting and Enhancing Adaptive Trait Selection

So far, I have argued that "sex" should refer to heterogametes and heterogametogenesis from the perspective of the SRC. This definition is not only applicable to unicellular eukaryotes but also applies to multicellular eukaryotes. Also, I argued that "sex differentiation" should refer to mechanisms that occur in the soma to ensure heterogametogenesis, including divergence point setting and niche establishment. While the function of sex differentiation is the same in animals and plants, the detailed approaches for achieving this function vary. One issue remains elusive: Are some particular mechanisms required to ensure that gametes meet?

As mentioned above, in bryophytes and pteridophytes, sperm cells are released into water and swim to the archegonia for fertilization. In this case, there are no specific mechanisms to ensure that the gametes meet. In some animals, such as fishes, both males and females release their gametes into the water as well. However, for a significant number of animal and plant species, mechanisms have evolved to facilitate or ensure that the gametes meet. Regardless of how complicated these mechanisms are, the ultimate function is quite simple: to ensure that the gametes meet! From this perspective, I collectively refer to these mechanisms as "sexual behaviors." It is understandable that due to their different strategies for energy acquisition, the mobile or sessile properties of animals or plants lead to completely different sexual behaviors.

17.5.1 *Sexual Behaviors in Animals*

While sexual behaviors in animals are quite colorful and showy, they are actually quite simple in terms of functions: courtship (which occurs intersexually) and mating competition (which occurs intrasexually) (Fig. 17.7). The first function is mainly to ensure that gametes meet. The second function has evolved to enhance the selection for adaptive traits and originated as the second function of fertilization.

A notion is very popular that courtship was selected because parents want their genes to be passed onto their offspring. However, from the perspective of the SRC, the essential function of sexual reproduction (as discussed above) is merely to increase the rate of genetic variations and select the survival ones that have integrated unpredictable changes of surrounding components. The three key events comprising the SRC originated independently, driven by changes of surrounding components, and selected by its probability of survival capacity (Bai, 2015b, 2016). Accordingly, before the last step of the SRC (fertilization), courtship, is a mechanism used to ensure that gametes meet, and it is therefore driven by internal and external stresses. Courtship does not occur because parents want to pass their genomes to their offspring, but instead, without the completion of the SRC, the species may not be able to survive (and be observed by human beings) during the long path of evolution. A boy approaches a girl at age 15 simply because of hormonal drive. He does not do so at age 5, nor at age 10, simply because his hormones have not yet reached the appropriate levels. From this perspective, both sex and sexual behavior are simply biological phenomena, as simple as breathing air and fetching food. Unfortunately, these processes have been extremely and unnecessarily overemphasized and overexposed in human society.

The mating competition evolved from courtship. In unicellular eukaryotes, the SRC is driven by changes of surrounding components. Adaptive genetic variations generated during meiosis can be selected through fertilization. By contrast, the relationship between changes of surrounding components and SRC completion is much more complicated for multicellular organisms. In animals, the germ cells are well protected by the MCSs, and the effects of changes of surrounding components upon the SRC are realized through the structures and behaviors of individuals. To fulfill the SRC, the subjects are the two individuals carrying opposite sex cells. Therefore, it is the individuals, not the gametes, that sense and respond to the changes of surrounding components to complete the SRC. Both males and females play roles in courtship. As the diversified DNA pool shared in the population is the core for fitness, the more chances for the individuals carrying adaptive genes to get mated, the better the entire population integrates the unpredictable changes of internal and external components. Therefore, it is not because the male or female wants to pass their genes on, but only by the mating competition, the adaptive ones can be selected for making their contribution to the population. In this way, the genes of individuals strong enough to win the competition are selected during evolution. A paradoxical situation is that since the individuals take autonomy in completion of SRC, while the ultimate function of sexual behavior is to ensure the gamete meeting and enhancing the selection of adaptive traits, the pressure to win the mating competition may lead to the selection of some traits not adaptive to the survival of the winners in a complicated environment. Such a phenomenon, e.g., the ornate plumage of birds or the antlers of deer, was termed Fisherian runaway (https://en.wikipedia.org/wiki/Fisherian_runaway). Regardless of the maladaptive effects in Fisherian runaway, mating competitions introduce an effective mechanism to accelerate the accumulation of variations from the winners in a population and have become a highly effective evolutionary driving force (Manyuan Long, personal communication) in the animal kingdom. However, this is no longer a significant mechanism in human evolution, as humans have diverged along a novel evolutionary path driven by the capacity for cognition (Bai, 2021). From this perspective, the significance of sex and sexual behavior in human society and for human evolution has been greatly overestimated.

17.5.2 Sexual Behaviors in Plants

Considering their sessile properties, it might appear foolish to discuss sexual behavior in plants. However, if the functions of sexual behavior in ensuring gamete meeting and enhancing selection for adaptive traits are considered, there are, of course, many mechanisms that can or should be referred to as sexual behavior in plants.

The obvious mechanisms in plants used to ensure or facilitate the gametes meeting are pollen tube growth in spermatophytes and pollinator attraction in angiosperms. Unlike bryophytes and pteridophytes, which release sperm cells into water and let them swim to meet egg cells, in spermatophytes (including gymnosperms and angiosperms), the pollen tube functions as a carrier to deliver the male gametes to eggs. While the pollen tube itself may be derived from ancient filament structures

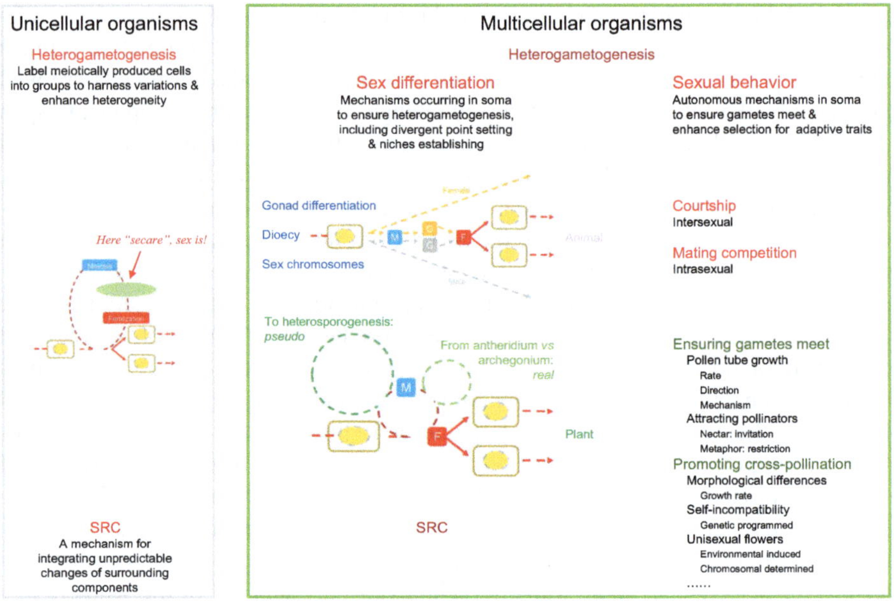

Fig. 17.7 A diagram of three key concepts of sex, sex differentiation, and sexual behavior from the perspective of SRC

such as protonema in mosses, this novel function significantly increases the chances that gametes meet.

Due to their sessile nature, pollination in plants is dependent on surrounding components. Nevertheless, diverse mechanisms have evolved to promote cross-pollination to maintain or enhance heterogeneity, which is indispensable resource for the adaptive trait selection, in the last step of the SRC. These mechanisms include the following: morphological differences, e.g., spatiotemporal separation of stamen and ovule differentiation (such as distyly and dichogamy); self-incompatibility; and the formation of unisexual flowers (Fig. 17.7).

It is worth noting that both diploid sporophytes and haploid gametophytes are involved in facilitating the gametes meeting and outcross promotion. Although there are no nerve-centered real behaviors, plants evolved as complicated systems to fulfill the function of sexual behaviors.

17.5.3 Unisexual Flowers in Basal Angiosperms

In the first section of this chapter, I argued that unisexual flower development is not an issue of sex differentiation but is instead one mechanism that promotes cross-pollination. This conclusion appears to be applicable only to advanced angiosperms, including magnoliids, monocots, and eudicots (Angiosperm Phylogeny Group,

APG, http://www.mobot.org/MOBOT/Research/APweb/welcome.html). The presence of unisexual flowers in basal angiosperms implies a different story. The rationale behind this notion is quite simple: while we propose that unisexual flowers evolved as a way to promote outcrossing, this process is thought to have occurred after perfect flowers emerged. However, from where and how did flowers evolve?

It is widely accepted that a perfect flower is a condensed shoot bearing both elaborated micro- and megasporangia, known as stamens and ovules, respectively (Gifford et al., 1989). However, in most heterosporous plants, such as some pteridophytes and gymnosperms, heterosporangia are spatiotemporally located on separate branches. How were the separately located heterosporangia integrated into a condensed shoot that functions as a unit we call the flower? If we hypothesize that this integration occurred in various ways (details see Bai, 2025), one can imagine that there might be some intermediate states that occurred during the integration of heterosporangia. If this is the case, there could be some types of dioecious and monoecious plants in which there no male or female organ primordium is initiated during any stage of floral bud formation, which would remain in today's basal angiosperms. From this perspective, while heterospory should not be referred to as real sex differentiation, as I have argued above, investigating unisexual flowers in basal angiosperms might play an indispensable role in solving a long-standing question put forth by Darwin: how did flowers evolve?

17.6 A New Frontier for Discovering Fundamental Biological Principles

Sex is a biological phenomenon that has existed long before human beings arose. Sex also accompanied human evolution and was recognized by our ancestors, since it obviously appeared at the very beginning of human development both ontogenetically and phylogenetically. However, as shown in Table 17.1, there is clear ambiguity about the meaning of sex. Why are we still unable to reach a rational consensus about this topic after so many years? Here, I introduce a new conceptual framework, derived from the SRC, to interpret the phenomenon of sex. Under this conceptual framework, the phenomenon of sex was dissected into three functionally different events or processes. The first is "sex", referring to heterogamete and heterogametogenesis. This event or process, as one of the three events of SRC, is originally emerged in unicellular eukaryotes and preserved in all multicellular eukaryotes. The second is "sex differentiation," referring to somatic cell differentiation to ensure heterogametogenesis. This event or process originated in multicellular eukaryotes and diversified in various forms into different kingdoms and species. The third is "sexual behavior," referring to somatic cell differentiations and behaviors to ensure gametes meeting and enhance autonomous selection of the adaptive variations. This event or process is originated in multicellular eukaryotes as well, and diversified even more than "sex differentiation" across the kingdoms and species. This new conceptual framework can logically solve long-lasting controversies regarding sex-related issues, such as reproduction vs. genetic variation, the twofold cost, and the relationship between meiosis and sex. I am addressing this topic not because of new experimental findings regarding some components that interact with each other at the molecular level, but because of the conceptual puzzling of a long-lasting concept that defies common sense, namely, the use of unisexual flower development to define sex differentiation in plants. Although the SRC-centered conceptual framework for sex-related issues and the morphogenetic strategies described here await more discussion and experimental evidence, some points underlying the unanticipated emergence of such a perspective are worth discussing.

Prior to the formulation of the concept of the SRC, one puzzling issue was whether there is an end to a plant's life cycle, and if there is, when or where? In the mainstream conceptual framework, plants have indeterminate developmental programs (Goldberg, 1988). The word "indeterminate" means "no end." If there is no end, when does the next generation begin? It is obvious that plant life cycles do have ends as well as beginnings, i.e., they start from zygotes and end at the gametes. When the two gametes meet and fuse into a new zygote, a new generation or life cycle (or for unicelluar eukaryotes, a new SRC) starts. It is also clear that each flowering plant produces numerous flowers and can continuously produce them for years. If the production of gametes that fuse for zygotes of new generation is used to define the end of the life cycle, should the branch that produces a flower be treated as a unit that completes the life cycle that is functionally equivalent to a pair of animal individuals (male and female)? The view that buds or shoots should be treated as developmental units and that a plant should be treated as a colony was proposed in the 1670s by founding fathers of modern botany such as Grew and Malpighi and was carried on by W. Goethe in the eighteenth century, Erasmus Darwin in the nineteenth century, and C.H. Waddington in the twentieth century (Arber, 1950; White, 1979; Bai, 2013, 2016, 2017a, 2019). One can find understandable reasons for the viewpoint in the eighteenth to nineteenth centuries to have diverged from that of the founding fathers (Bai, 2016). However, unless there is a paradigm shift, the current mainstream conceptual framework becomes solidified in the younger generations, and few spend time questioning its logic. Convenience outweighs the questioning of the logical inconsistency in the mainstream conceptual framework.

When I studied an *embryonic flower* mutant in Dr. Renee Sung's laboratory at UC-Berkeley in the early 1990s, I was puzzled by the observation that ALL plants described in the literature ultimately enter reproduction. If this is the case, what are the most highly conserved events that occur in all plant species regardless of their diversity in form? By comparing the life cycles of representative plants, I realized that only three events are essential in all plant species: fertilization, meiosis, and heterogametogenesis. Because of the three events, there are three core cell types in the life cycles of all plants: zygotes, meiotic cells and the resulting spores, and gametes (Fig. 17.5). Such a comparison established a new reference framework in which to view the plant life cycle and morphogenesis, leading to my formulation of concepts such as the PDU, the SRC, the integration of MCS and the SRC, and even new interpretations of why there are only two mating types in multicellular eukaryotes while there are often multiple gamete types in unicellular eukaryotes. All data required for such a comparison are available in the literature. The difference is simply a shift in perspective. Traditionally, researchers observe plants based on their appearances and investigate the concrete structures underlying their appearance, followed by observations at the molecular level, which I refer to as a "structure-based perspective." By contrast, the question of why all plants ultimately enter reproduction led me to focus on the abstract functions that enabled the emergence of a "function-based evolutionary perspective" (Bai, 2017a). It is easy to discuss previous paradigm shifts but difficult to judge ongoing paradigm shifts, perhaps because no one needs to take responsibility for a previous paradigm shift, as it already occurred, and anybody can discuss it. However, for on-going paradigms shifts, there is no guarantee of their success. Compared to staying in the mainstream, challenging traditional concepts is not only inconvenient, but it is also risky.

In the literature about sex in the animal kingdom, it appears that sex differentiation leading to heterogametogenesis always occurs prior to meiosis. Such an observation led Cavalier-Smith (2002) to strongly argue that sex is syngamy and meiosis. However, if we look at the plant kingdom, it is obvious that in all bryophytes and many pteridophytes, meiosis occurs prior to heterogametogenesis (Campbell & Reece, 2005; Raven et al., 2013; Bai & Xu, 2013). Such observations point to an inevitable dilemma: either the link of sex to meiosis is wrong, or one has to reject the common origin of meiosis in animals and plants, which appears to be undeniable (Bai, 2015a and reference therein). The failure to acknowledge the phenomenon that in many plants, meiosis occurs prior to heterogametogenesis appears not due to a lack of knowledge, but neglecting the existence of plants or not placing plants at the equal position with animals in terms of investigation of sex-related phenomena.

There is a saying that many biologists may have heard: "Ideas are cheap." When I first heard this saying, my immediate response was to ask the question, "what is expensive?" After about three decades, it appears to me that, in the current scientific society, an idea is cheap if it is empirically testable and leads to anticipated discoveries or is not empirically testable at all. New discoveries are always highly appreciated, but the conceptual framework and the underlying questions that led to the discovery received little attention. However, an idea is "expensive" if it sounds correct and is empirically testable but does not lead to any anticipated discoveries that get investment paid off under certain circumstances.

How does one determine if an idea is right or wrong? It depends on what type of questions an idea refers to. If questions such as how one protein interacts with another are asked, the answers to such questions could be tested only by experimental analysis. However, if questions such as what sex refers to or whether a plant developmental program is indeterminate are asked, the answers to such questions could be tested not only by experimental analysis, but (more importantly) by conceptual analysis of how and on what basis the "conventional concepts" were originally proposed and whether the paradigm should be shifted.

If the above analyses on plants, plant sex differentiation, and sex in general hold true, an attractive scenario would emerge. In the new scenario, research on plants will no longer be aimed solely at meeting human needs for food, energy, and the environment, but instead be expanded to focus on plants as a system to explore some fundamental biological principles shared with animals. It is widely accepted that all current organisms on Earth are derived from the LUCA, and there is likely a common ancestor for all eukaryotes. While experimental data suggest that the process of meiosis shares a common evolutionary origin and the cyclical life cycle is a common property of eukaryotes, how the two are integrated and whether and how multicellular eukaryotes were derived from unicellular eukaryotes that still retain this cyclical property remain unknown. The concept of the SRC provides a new frame of reference to think about these fundamental issues. In this new framework, I have proposed new definitions of sex, sex differentiation, and sexual behavior. I anticipate that based on the SRC, more controversial issues could be solved and crucial new questions could be addressed and investigated, opening up a new frontier for exploring nature. In short, it is time to give plants the dignity they deserve!

17.7 Conclusions

In this chapter, I clarified the current concept of plant sex differentiation, briefly introduced resources addressing this issue for those who are interested, and summarized the current progress in this field. After pointing out the controversial issues in the study of plant sex differentiation, I introduced a new conceptual framework for viewing plant morphogenesis and life cycle completion, which was derived from the recently proposed concept of the SRC. In this new conceptual framework, we refer to heterogametes and heterogametogenesis in the SRC as "sex," the mechanisms that occur in the soma to ensure heterogametogenesis as "sex differentiation," that to ensure gamete meeting and enhance selection of adaptive variations as "sexual behavior."

By examining long-lasting controversial issues in plant sex differentiation as well as sex-related issues in general, I analyzed the possible problems associated with the current thinking that may be responsible for these controversies. I propose that, to understand fundamental biological principles, a proper balance is needed between empirical investigation and conceptual reasoning.

When preparing a lecture for undergraduate students at an agricultural university, I realized that, throughout history, plants have been viewed from two different perspectives: humans are animals and plants are their food, and (from the perspective of an eighteenth century naturalist) plants are something unique that were created to satisfy our curiosity. However, there should be a third perspective: plants are indispensable systems for understanding fundamental biological principles that generally apply to multicellular eukaryotes including ourselves. From this perspective, plants and animals are all derived from the SRC through the interpolation of the MCSs into its intervals in different manners, resulting in different strategies for morphogenesis, i.e., the double-ring mode and the dichotomous mode. If this is the case, compelling new questions remain to be investigated, especially for plant biologists, that are far more challenging to human intelligence than simply improving crop production to meet human needs.

Acknowledgments I would cordially thank Prof. Hong Ma of Pennsylvania State University and Prof. Zhi-Hong Xu of Peking University for their invitation of contributing the chapter, thank Prof. Xin Wang of Nanjing Institute of Geology and Paleontology, Chinese Academy of Sciences for his inspiring insight on the origin of heterospory, and help in proof reading of the manuscript; Prof. Qi-Chang Fan of Peking University for sharing his deep thinking on sex in animals with me more than 20 years ago and his critical comments on the manuscript; Prof. Manyuan Long of University of Chicago for spending time with me to discuss sex-related issues during past 10 years, sharing his systematic thinking and new ideas on the origin of genes and sexual selection, and critical reading, very constructive comments and suggestions of the manuscript; Prof. Ray Ming of University of Illinois at Urbana-Champaign for his critical reading and kind comments and suggestions on the manuscript. I would also like to thank Prof. Guang-Yuan Rao for his kind comments and encouragement of the manuscript; Prof. Garland Allen for his comments on the manuscript and kind suggestions on revision; Prof. Hong Ma for his careful proof reading and helpful suggestions on the manuscript. I also want to thank the students of my graduate class "Plant Developmental Biology" of year 2017 to help me to read the manuscript and send me their feedback. Those who read the manuscript and discuss with me may or may not agree all my opinions. I take full responsibility of all mistakes in the chapter. Finally, I have to thank all the students and postdocs worked with me during their graduate studies in my lab. Without their hard work on cucumber unisexual flower development, I would never get in touch with the sticky issue and have something to say in this chapter.

References

Acosta, I. F., Laparra, H., Romero, S. P., Schmelz, E., Hamberg, M., Mottinger, J. P., Moreno, M. A., & Dellaporta, S. L. (2009). Tasselseed1 is a lipoxygenase affecting jasmonic acid signaling in sex determination of maize. *Science, 323*(5911), 262–265. https://doi.org/10.1126/science.1164645

Ainsworth, C. C. (1999). *Sex determination in plants*. BIOS Scientific Publishers.

Ainsworth, C., Crossley, S., Buchanan-Wollaston, V., Thangavelu, M., & Parker, J. (1995). Male and female flowers of the dioecious plant sorrel show different patterns of MADS box gene expression. *Plant Cell, 7*(10), 1583–1598. https://doi.org/10.1105/tpc.7.10.1583

Akagi, T., Henry, I. M., Tao, R., & Comai, L. (2014). A Y-chromosome-encoded small RNA acts as a sex determinant in persimmons. *Science, 346*(6209), 646–650. https://doi.org/10.1126/science.1257225

Akagi, T., Henry, I. M., Ohtani, H., Morimoto, T., Beppu, K., Kataoka, I., & Tao, R. (2018). A Y-encoded suppressor of feminization arose via lineage-specific duplication of a cytokinin response regulator in Kiwifruit. *Plant Cell, 30*(4), 780–795. https://doi.org/10.1105/tpc.17.00787

Akagi, T., Pilkington, S. M., Varkonyi-Gasic, E., Henry, I. M., Sugano, S. S., Sonoda, M., Firl, A., McNeilage, M. A., Douglas, M. J., Wang, T. C., Rebstock, R., Voogd, C., Datson, P., Allan, A. C., Beppu, K., Kataoka, I., & Tao, R. (2019). Two Y-chromosome-encoded genes determine sex in kiwifruit. *Nature Plants, 5*(8), 801–809.

Alexandrova, O., Schade, M., Bottger, A., & David, C. N. (2005). Oogenesis in Hydra: Nurse cells transfer cytoplasm directly to the growing oocyte. *Developmental Biology, 281*(1), 91–101. https://doi.org/10.1016/j.ydbio.2005.02.015

Arber, A. R. (1950). *The natural philosophy of plant form*. University Press.
Atallah, N. M., & Banks, J. A. (2015). Reproduction and the pheromonal regulation of sex type in fern gametophytes. *Frontiers in Plant Science, 6*, 100. https://doi.org/10.3389/fpls.2015.00100
Atsmon, D., & Galun, E. (1960). A morphogenetic study of staminate, pistillate and hermaphrodite flowers in Cucumis sativus (L.). *Phytomorphology, 10*, 110–115.
Bachtrog, D., Mank, J. E., Peichel, C. L., Kirkpatrick, M., Otto, S. P., Ashman, T. L., Hahn, M. W., Kitano, J., Mayrose, I., Ming, R., Perrin, N., Ross, L., Valenzuela, N., Vamosi, J. C., & Tree of Sex C. (2014). Sex determination: Why so many ways of doing it? *PLoS Biology, 12*(7), e1001899. https://doi.org/10.1371/journal.pbio.1001899
Bai, S. N. (1999). Phenomena, interpretation of the phenomena, and the developmental unit in plants. In C. S. Li (Ed.), *Advances in plant sciences (in Chinese)* (pp. 52–69). Higher Education Press.
Bai, S. N. (2013). Is a flower an organ? *Biology Teaching in University (Electronic Edition), 3*, 52–56.
Bai, S. N. (2015a). The concept of the sexual reproduction cycle and its evolutionary significance. *Frontiers in Plant Science, 6*, 11. https://doi.org/10.3389/fpls.2015.00011
Bai, S. N. (2015b). Plant developmental program: Sexual reproduction cycle derived "double ring". *Scientia Sinica Vitae, 45*, 811–819.
Bai, S. N. (2016). Make a new cloth for a grown body: From plant developmental unit to plant developmental program. *Annual Review of New Biology, 2015*, 73–116.
Bai, S. N. (2017a). Reconsideration of plant morphological traits: From a structure-based perspective to a function-based evolutionary perspective. *Frontiers in Plant Science, 8*, 345. https://doi.org/10.3389/fpls.2017.00345
Bai, S. N. (2017b). Two types of germ cells, the sexual reproduction cycle, and the double-ring mode of plant developmental program. *Plant Signaling & Behavior, 12*(7), e1320632. https://doi.org/10.1080/15592324.2017.1320632
Bai, S. N. (2019). Plant morphogenesis 123: A renaissance in modern botany? *Science China. Life Sciences, 62*(4), 453–466.
Bai, S. N. (2021). Where will human society go after the COVID-19 pandemic? – Thinking from the perspective of Bio-orientated concept framework. In B. Song (Ed.), *Out of the Anthropocene*. CITIC Press Group.
Bai, S. N. (2024). Two-way induction: How germ cells are differentiated in plants. *Discover Plants, 1*, 60. https://doi.org/10.1007/s44372-024-00066-3
Bai, S. N. (2025). On the plant developmental unit: from virtual concept to visual plantlet. *Plants* (Basel), PMID 39942958.
Bai, S. N., & Xu, Z. H. (2012). Bird-nest puzzle: Can the study of unisexual flowers such as cucumber solve the problem of plant sex determination? *Protoplasma, 249*(Suppl 2), S119–S123. https://doi.org/10.1007/s00709-012-0396-4
Bai, S. N., & Xu, Z. H. (2013). Unisexual cucumber flowers, sex and sex differentiation. *International Review of Cell and Molecular Biology, 304*, 1–55. https://doi.org/10.1016/B978-0-12-407696-9.00001-4
Bai, S. L., Peng, Y. B., Cui, J. X., Gu, H. T., Xu, L. Y., Li, Y. Q., Xu, Z. H., & Bai, S. N. (2004). Developmental analyses reveal early arrests of the spore-bearing parts of reproductive organs in unisexual flowers of cucumber (Cucumis sativus L.). *Planta, 220*(2), 230–240. https://doi.org/10.1007/s00425-004-1342-2
Bai, S. N., Rao, G. Y., & Yang, J. (2022). Origins of the seed: The "golden-trio hypothesis". *Frontiers in Plant Science, 13*, 965000. https://doi.org/10.3389/fpls.2022.96500
Bateman, R. M., & Dimichele, W. A. (1994). Heterospory – The most iterative key innovation in the evolutionary history of the plant kingdom. *Biological Reviews, 69*(3), 345–417. https://doi.org/10.1111/j.1469-185X.1994.tb01276.x
Bell, G. (1982). *The masterpiece of nature: The evolution and genetics of sexuality*. University of California Press.

Bensen, R. J., Johal, G. S., Crane, V. C., Tossberg, J. T., Schnable, P. S., Meeley, R. B., & Briggs, S. P. (1995). Cloning and characterization of the maize An1 gene. *The Plant Cell, 7*(1), 75–84. https://doi.org/10.1105/tpc.7.1.75

Beukeboom, L. W., & Perrin, N. (2014). *The evolution of sex determination* (1st ed.). Oxford University Press.

Bosch, T. C., & David, C. N. (1986). Male and female stem cells and sex reversal in Hydra polyps. *Proceedings of the National Academy of Sciences of the United States of America, 83*(24), 9478–9482.

Boualem, A., Fergany, M., Fernandez, R., Troadec, C., Martin, A., Morin, H., Sari, M. A., Collin, F., Flowers, J. M., Pitrat, M., Purugganan, M. D., Dogimont, C., & Bendahmane, A. (2008). A conserved mutation in an ethylene biosynthesis enzyme leads to andromonoecy in melons. *Science, 321*(5890), 836–838. https://doi.org/10.1126/science.1159023

Boualem, A., Troadec, C., Kovalski, I., Sari, M. A., Perl-Treves, R., & Bendahmane, A. (2009). A conserved ethylene biosynthesis enzyme leads to andromonoecy in two cucumis species. *PLoS One, 4*(7), e6144. https://doi.org/10.1371/journal.pone.0006144

Boualem, A., Troadec, C., Camps, C., Lemhemdi, A., Morin, H., Sari, M. A., Fraenkel-Zagouri, R., Kovalski, I., Dogimont, C., Perl-Treves, R., & Bendahmane, A. (2015). A cucurbit androecy gene reveals how unisexual flowers develop and dioecy emerges. *Science, 350*(6261), 688–691. https://doi.org/10.1126/science.aac8370

Bower, F. O. (1935). *Primitive land plants, also known as the Archegoniatae*. Macmillan.

Bowman, J. L. (2016). A brief history of marchantia from greece to genomics. *Plant & Cell Physiology, 57*(2), 210–229. https://doi.org/10.1093/pcp/pcv044

Brooks, K. E. (1973). Reproductive biology of Selaginella. 1. Determination of megasporangia by 2-Chloroethylphosphonic acid, an ethylene-releasing compound. *Plant Physiology, 51*(4), 718–722. https://doi.org/10.1104/Pp.51.4.718

Bull, J. J. (1983). Evolution of sex determining mechanisms. In *Evolution* (Vol. 1). Benjamin/Cummings Pub. Co., Advanced Book Program.

Calderon-Urrea, A., & Dellaporta, S. L. (1999). Cell death and cell protection genes determine the fate of pistils in maize. *Development, 126*(3), 435–441.

Campbell, N. A., & Reece, J. B. (2005). *Biology* (7th ed.). Pearson Benjamin Cummings.

Casper, A., & Van Doren, M. (2006). The control of sexual identity in the Drosophila germline. *Development, 133*(15), 2783–2791. https://doi.org/10.1242/dev.02415

Cavalier-Smith, T. (2002). Origins of the machinery of recombination and sex. *Heredity, 88*, 125–141. https://doi.org/10.1038/sj/hdy/6800034

Charlesworth, D. (2002). Plant sex determination and sex chromosomes. *Heredity, 88*(2), 94–101. https://doi.org/10.1038/sj.hdy.6800016

Charlesworth, D. (2015). Plant contributions to our understanding of sex chromosome evolution. *The New Phytologist, 208*(1), 52–65. https://doi.org/10.1111/nph.13497

Charlesworth, D. (2016). Plant sex chromosomes. *Annual Review of Plant Biology, 67*, 397–420. https://doi.org/10.1146/annurev-arplant-043015-111911

Chen, M. Q., Zhang, A. H., Zhang, Q., Zhang, B. C., Nan, J., Li, X., Liu, N., Qu, H., Lu, C. M., Sudmorgen, Z. Y. H., Xu, Z. H., & Bai, S. N. (2012). Arabidopsis NMD3 is required for nuclear export of 60S ribosomal subunits and affects secondary cell wall thickening. *PLoS One, 7*(4), e35904. https://doi.org/10.1371/journal.pone.0035904

Chen, R., Shen, L. P., Wang, D. H., Wang, F. G., Zeng, H. Y., Chen, Z. S., Peng, Y. B., Lin, Y. N., Tang, X., Deng, M. H., Yao, N., Luo, J. C., Xu, Z. H., & Bai, S. N. (2015). A gene expression profiling of early rice stamen development that reveals inhibition of photosynthetic genes by OsMADS58. *Molecular Plant, 8*(7), 1069–1089. https://doi.org/10.1016/j.molp.2015.02.004

Chen, H., Sun, J., Li, S., Cui, Q., Zhang, H., Xin, F., Wang, H., Lin, T., Gao, D., Wang, S., Li, X., Wang, D., Zhang, Z., Xu, Z., & Huang, S. (2016). An ACC oxidase gene essential for cucumber carpel development. *Molecular Plant, 9*(9), 1315–1327. https://doi.org/10.1016/j.molp.2016.06.018

Cheng, P. C., Greyson, R. L., & Walden, D. B. (1983). Organ initiation and the development of unisexual flowers in the tassel and ear of Zea mays. *American Journal of Botany, 70*, 450–462.
Chuck, G., Meeley, R., Irish, E., Sakai, H., & Hake, S. (2007). The maize tasselseed4 microRNA controls sex determination and meristem cell fate by targeting Tasselseed6/indeterminate spikelet1. *Nature Genetics, 39*(12), 1517–1521. https://doi.org/10.1038/ng.2007.20
Coe, E. H., Neuffer, M. G., & Hoisington, D. A. (1989). The genetics of corn. In G. F. Sprangue & J. Dudley (Eds.), *Corn and corn improvement* (3rd ed.). American Society of Agronomy.
Coulter, J. M. (1914). *The evolution of sex in plants* (The University of Chicago Science Series). The University of Chicago Press.
Darwin, C. (1859). *On the origin of species by means of natural selection*. J. Murray.
Dellaporta, S. L., & Calderon-Urrea, A. (1993). Sex determination in flowering plants. *The Plant Cell, 5*(10), 1241–1251. https://doi.org/10.1105/tpc.5.10.1241
DeLong, A., Calderon-Urrea, A., & Dellaporta, S. L. (1993). Sex determination gene TASSELSEED2 of maize encodes a short-chain alcohol dehydrogenase required for stage-specific floral organ abortion. *Cell, 74*(4), 757–768.
Diggle, P. K., Di Stilio, V. S., Gschwend, A. R., Golenberg, E. M., Moore, R. C., Russell, J. R., & Sinclair, J. P. (2011). Multiple developmental processes underlie sex differentiation in angiosperms. *Trends in Genetics: TIG, 27*(9), 368–376. https://doi.org/10.1016/j.tig.2011.05.003
Eggers, S., Ohnesorg, T., & Sinclair, A. (2014). Genetic regulation of mammalian gonad development. *Nature Reviews. Endocrinology, 10*(11), 673–683. https://doi.org/10.1038/nrendo.2014.163
Fan, Q. C., & Bai, S. N. (2002). *Principles of developmental biology*. Higher Education Press.
French, J. C. (1972). Dimensional correlations in developing Selaginella-sporangia. *American Journal of Botany, 59*(3), 224. https://doi.org/10.2307/2441421
Galun, E. (1961). Study of the inheritance of sex expression in the cucumber. *The Interaction of Major Genes with Modifying Genetic and Non-genetic Factors Genetica, 32*, 134–163.
Gifford, E. M., Foster, A. S., & Foster, A. S. (1989). *Morphology and evolution of vascular plants* (A Series of Books in Biology) (3rd ed.). W.H. Freeman and Co.
Gilbert, S. F. (2010). *Developmental biology* (9th ed.). Sinauer Associates.
Goffinet, M. C. (1990). Comparative ontogeny of male and female flowers of Cucumis sativus. In D. M. Bates, R. W. Bobinson, & C. Jeffrey (Eds.), *Biology and utilization of the cucurbitaceae* (pp. 288–304). Cornell University Press.
Goldberg, R. B. (1988). Plants: Novel developmental processes. *Science, 240*(4858), 1460–1467.
Graves, J. A. M. (2016). Evolution of vertebrate sex chromoomes and dosage compensation. *Nature Reviews Genetics, 17*(1), 33–46. https://doi.org/10.1038/nrg
Gschwend, A. R., Yu, Q., Tong, E. J., Zeng, F., Han, J., VanBuren, R., Aryal, R., Charlesworth, D., Moore, P. H., Paterson, A. H., & Ming, R. (2012). Rapid divergence and expansion of the X chromosome in papaya. *Proceedings of the National Academy of Sciences of the United States of America, 109*(34), 13716–13721. https://doi.org/10.1073/pnas.1121096109
Gu, H. T., Wang, D. H., Li, X., He, C. X., Xu, Z. H., & Bai, S. N. (2011). Characterization of an ethylene-inducible, calcium-dependent nuclease that is differentially expressed in cucumber flower development. *The New Phytologist, 192*(3), 590–600. https://doi.org/10.1111/j.1469-8137.2011.03825.x
Hao, Y. J., Wang, D. H., Peng, Y. B., Bai, S. L., Xu, L. Y., Li, Y. Q., Xu, Z. H., & Bai, S. N. (2003). DNA damage in the early primordial anther is closely correlated with stamen arrest in the female flower of cucumber (Cucumis sativus L.). *Planta, 217*(6), 888–895. https://doi.org/10.1007/s00425-003-1064-x
Hardenack, S., Ye, D., Saedler, H., & Grant, S. (1994). Comparison of MADS box gene expression in developing male and female flowers of the dioecious plant white campion. *The Plant Cell, 6*(12), 1775–1787. https://doi.org/10.1105/tpc.6.12.1775
Harkess, A., & Leebens-Mack, J. (2017). A century of sex determination in flowering plants. *The Journal of Heredity, 108*(1), 69–77.

Harkess, A., Zhou, J., Xu, C., Bowers, J. E., Van der Hulst, R., Ayyampalayam, S., Mercati, F., Riccardi, P., McKain, M. R., Kakrana, A., Tang, H., Ray, J., Groenendijk, J., Arikit, S., Mathioni, S. M., Nakano, M., Shan, H., Telgmann-Rauber, A., Kanno, A., Yue, Z., Chen, H., Li, W., Chen, Y., Xu, X., Zhang, Y., Luo, S., Chen, H., Gao, J., Mao, Z., Pires, J. C., Luo, M., Kudrna, D., Wing, R. A., Meyers, B. C., Yi, K., Kong, H., Lavrijsen, P., Sunseri, F., Falavigna, A., Ye, Y., Leebens-Mack, J. H., & Chen, G. (2017). The asparagus genome sheds light on the origin and evolution of a young Y chromosome. *Nature Communications, 8*(1), 1279. https://doi.org/10.1038/s41467-017-01064-8

Harkess, A., Huang, K., Van Der Hulst, R., Tissen, B., Caplan, J. L., Koppula, A., Batish, M., Meyers, B. C., & Leebens-Mack, J. (2020). Sex determination by two Y-linked genes in garden asparagus. *Plant Cell, 32*, 1790–1796. https://doi.org/10.1105/tpc.19.00859

Hartwig, T., Chuck, G. S., Fujioka, S., Klempien, A., Weizbauer, R., Potluri, D. P., Choe, S., Johal, G. S., & Schulz, B. (2011). Brassinosteroid control of sex determination in maize. *Proceedings of the National Academy of Sciences of the United States of America, 108*(49), 19814–19819. https://doi.org/10.1073/pnas.1108359108

Hayward, A. P., Moreno, M. A., Howard, T. P., 3rd, Hague, J., Nelson, K., Heffelfinger, C., Romero, S., Kausch, A. P., Glauser, G., Acosta, I. F., Mottinger, J. P., & Dellaporta, S. L. (2016). Control of sexuality by the sk1-encoded UDP-glycosyltransferase of maize. *Science Advances, 2*(10), e1600991. https://doi.org/10.1126/sciadv.1600991

Heimlich, L. F. (1927). The development and anatomy of the staminate flower of the cucumber. *American Journal of Botany, 14*, 227–236.

Heitman, J., Sun, S., & James, T. Y. (2013). Evolution of fungal sexual reproduction. *Mycologia, 105*(1), 1–27. https://doi.org/10.3852/12-253

Hempel, L. U., Kalamegham, R., Smith, J. E., 3rd, & Oliver, B. (2008). Drosophila germline sex determination: Integration of germline autonomous cues and somatic signals. *Current Topics in Developmental Biology, 83*, 109–150. https://doi.org/10.1016/S0070-2153(08)00404-3

Herr, J. M. (1995). The origin of the ovule. *American Journal of Botany, 82*(4), 547–564. https://doi.org/10.2307/2445703

Heslop-Harrison, J. (1956). Auxin and sexuality in Cannabis sativa. *Physiologia Plantarum, 9*, 588–597.

Heslop-Harrison, J. (1957). The experimental modification of sex expression in flowering plants. *Biological Reviews, 32*, 38–89.

Heslop-Harrison, J. (1960). The experimental control of sexuality and inflorescence structure in Zea mays L. *Proceedings of the Linnean Society of London, 172*, 108–124.

Iocco-Corena, P., Chaïb, J., Torregrosa, L., Mackenzie, D., Thomas, M. R., & Smith, H. M. (2021). VviPLATZ1 is a major factor that controls female flower morphology determination in grapevine. *Nature Communications, 12*(1), 6995.

Iwahori, S., Lyons, J. M., & Sims, W. L. (1969). Induced femaleness in cucumber by 2-chloroethanephosphonic acid. *Nature, 222*(5190), 271–272.

Janzen, F. J., & Phillips, P. C. (2006). Exploring the evolution of environmental sex determination, especially in reptiles. *Journal of Evolutionary Biology, 19*(6), 1775–1784. https://doi.org/10.1111/j.1420-9101.2006.01138.x

Jemc, J. C. (2011). Somatic gonadal cells: The supporting cast for the germline. *Genesis, 49*(10), 753–775. https://doi.org/10.1002/dvg.20784

Juarez, C., & Banks, J. A. (1998). Sex determination in plants. *Current Opinion in Plant Biology, 1*(1), 68–72.

Kater, M. M., Franken, J., Carney, K. J., Colombo, L., & Angenent, G. C. (2001). Sex determination in the monoecious species cucumber is confined to specific floral whorls. *The Plant Cell, 13*(3), 481–493.

Kubicki, B. (1969a). Comparative studies on sex determination in cucumber (Cucumis sativus L.) and muskmelon (Cucumis melo L.). *Genet Pol, 10*, 167–183.

Kubicki, B. (1969b). Investigations on sex determination in cucumber (Cucumis sativus L.). VII. Andromonoecious and hermaphroditism. *Genet Pol, 10*, 101–120.

Kubicki, B. (1969c). Investigations on sex determination n cucumber (Cucumis sativus L.). V. Genes controlling intensity of femaleness. *Genet Pol, 10*, 69–85.

Leite Montalvão, A. P., Kersten, B., Fladung, M., & Müller, N. A. (2021). The diversity and dynamics of sex determination in dioecious plants. *Frontiers in Plant Science, 11*, 580488. https://doi.org/10.3389/fpls.2020.580488

Li, Z., Huang, S., Liu, S., Pan, J., Zhang, Z., Tao, Q., Shi, Q., Jia, Z., Zhang, W., Chen, H., Si, L., Zhu, L., & Cai, R. (2009). Molecular isolation of the M gene suggests that a conserved-residue conversion induces the formation of bisexual flowers in cucumber plants. *Genetics, 182*(4), 1381–1385. https://doi.org/10.1534/genetics.109.104737

Linné, C. V., Smith, J. E., & Thomas Jefferson Library Collection Supplement (Library of Congress). (1786). *A dissertation on the sexes of plants*. Printed for the Author, and Sold by George Nicol ….

Littlefield, C. L. (1986). Sex determination in hydra: Control by a subpopulation of interstitial cells in Hydra oligactis males. *Developmental Biology, 117*(2), 428–434.

Long, M., VanKuren, N. W., Chen, S., & Vibranovski, M. D. (2013). New gene evolution: Little did we know. *Annual Review of Genetics, 47*, 307–333. https://doi.org/10.1146/annurev-genet-111212-133301

Loomis, W. F. (1954). Reversible induction of sexual differentiation in Hydra. *Science, 120*(3108), 145–146.

Malepszy, S., & Niemirowicz-Szczytt, K. (1991). Sex determination in cucumber (Cucumis sativus) as a model system for molecular biology. *Plant Science, 80*, 39–47.

Martin, A., Troadec, C., Boualem, A., Rajab, M., Fernandez, R., Morin, H., Pitrat, M., Dogimont, C., & Bendahmane, A. (2009). A transposon-induced epigenetic change leads to sex determination in melon. *Nature, 461*(7267), 1135–1138. https://doi.org/10.1038/nature08498

Massonnet, M., Cochetel, N., Minio, A., Vondras, A. M., Lin, J., Muyle, A., Garcia, J. F., Zhou, Y. F., Delledonne, M., Riaz, S., Figueroa-Balderas, R., Gaut, B. S., & Cantu, D. (2020). The genetic basis of sex determination in grapes. *Nature Communications, 11*, 2902. https://doi.org/10.1038/s41467-020-16700-z

Maynard Smith, J. (1978). *The evolution of sex*. Cambridge University Press.

McMurray, A. L., & Miller, C. H. (1968). Cucumber sex expression modified by 2-chloroethanephosphonic acid. *Science, 162*(3860), 1397–1398.

Mibus, H., & Tatlioglu, T. (2004). Molecular characterization and isolation of the F/f gene for femaleness in cucumber (Cucumis sativus L.). *TAG Theoretical and Applied Genetics Theoretische und Angewandte Genetik, 109*(8), 1669–1676. https://doi.org/10.1007/s00122-004-1793-7

Ming, R., Yu, Q., & Moore, P. H. (2007). Sex determination in papaya. *Seminars in Cell & Developmental Biology, 18*(3), 401–408. https://doi.org/10.1016/j.semcdb.2006.11.013

Ming, R., Bendahmane, A., & Renner, S. S. (2011). Sex chromosomes in land plants. *Annual Review of Plant Biology, 62*, 485–514. https://doi.org/10.1146/annurev-arplant-042110-103914

Murase, K., Shigenobu, S., Fujii, S., Ueda, K., Murata, T., Sakamoto, A., Wada, Y., Yamaguchi, K., Osakabe, Y., Osakabe, K., Kanno, A., Ozaki, Y., & Takayama, S. (2017). MYB transcription factor gene involved in sex determination in Asparagus officinalis. *Genes to Cells, 22*, 115–123. https://doi.org/10.1111/gtc.12453

Murray, S. M., Yang, S. Y., & Van Doren, M. (2010). Germ cell sex determination: A collaboration between soma and germline. *Current Opinion in Cell Biology, 22*(6), 722–729. https://doi.org/10.1016/j.ceb.2010.09.006

Na, J. K., Wang, J., & Ming, R. (2014). Accumulation of interspersed and sex-specific repeats in the non-recombining region of papaya sex chromosomes. *BMC Genomics, 15*, 335. https://doi.org/10.1186/1471-2164-15-335

Nishimiya-Fujisawa, C., & Sugiyama, T. (1995). Genetic analysis of developmental mechanisms in Hydra. XXII. Two types of female germ stem cells are present in a male strain of Hydra magnipapillata. *Developmental Biology, 172*(1), 324–336. https://doi.org/10.1006/dbio.1995.0026

Pannell, J. R. (2017). Plant sex determination. *Current Biology: CB, 27*(5), R191–R197. https://doi.org/10.1016/j.cub.2017.01.052

Perl-Treves, R. (1999). Male to female conversion along the cucumber shoot: Approaches to studying sex genes and flora development in Cucumis sativus. In C. Ainsworth (Ed.), *Sex determination in plants* (pp. 189–216). BIOS Scientific Publishers.

Perl-Treves, R., & Rajagopalan, P. A. (2006). Close, yet separate: Patterns of male and female floral development in monecious species. In C. Ainsworth (Ed.), *Flower development and manipulation* (pp. 117–146). Blackwell.

Qiu, Y. L., Taylor, A. B., & McManus, H. A. (2012). Evolution of the life cycle in land plants. *Journal of Systematics and Evolution, 50*(3), 171–194. https://doi.org/10.1111/j.1759-6831.2012.00188.x

Ramsey, M., & Crews, D. (2009). Steroid signaling and temperature-dependent sex determination-Reviewing the evidence for early action of estrogen during ovarian determination in turtles. *Seminars in Cell & Developmental Biology, 20*(3), 283–292. https://doi.org/10.1016/j.semcdb.2008.10.004

Raven, P. H., Evert, R. F., & Eichhorn, S. E. (2013). *Biology of plants* (8th ed.). W.H. Freeman and Company Publishers.

Rios-Rojas, C., Bowles, J., & Koopman, P. (2015). On the role of germ cells in mammalian gonad development: Quiet passengers or back-seat drivers? *Reproduction, 149*(4), R181–R191. https://doi.org/10.1530/Rep-14-0663

Robbins, W. W., & Pearson, H. M. (1933). *Sex in the plant world* (Appleton New World of Science Series). D. Appleton-Century Company, Incorporated.

Ronse De Craene, L., Trehin, C., Morel, P., & Negrutiu, I. (2011). Carpeloidy in flower evolution and diversification: A comparative study in Carica papaya and Arabidopsis thaliana. *Annals of Botany, 107*(9), 1453–1463. https://doi.org/10.1093/aob/mcr087

Rudich, J., Halevy, A. H., & Kedar, N. (1969). Increase in femaleness of three cucurbits by treatment with Ethrel, an ethylene releasing compound. *Planta, 86*(1), 69–76. https://doi.org/10.1007/BF00385305

Shannon, S., & Guardia, D. L. (1969). Sex expression and the production of ethylene induced by auxin in the cucumber (Cucumis sativum L.). *Nature, 223*, 186–186.

Shen, Y. F. (1999). *Protozoology*. Science Press.

Sheppard, L. A., Brunner, A. M., Krutovskii, K. V., Rottmann, W. H., Skinner, J. S., Vollmer, S. S., & Strauss, S. H. (2000). A DEFICIENS homolog from the dioecious tree black cottonwood is expressed in female and male floral meristems of the two-whorled, unisexual flowers. *Plant Physiology, 124*(2), 627–640.

Sommer, R. J., & Bumbarger, D. J. (2012). Nematode model systems in evolution and development. *WIREs Developmental Biology, 1*(3), 389–400. https://doi.org/10.1002/wdev.33

Sun, J. J., Li, F., Li, X., Liu, X. C., Rao, G. Y., Luo, J. C., Wang, D. H., Xu, Z. H., & Bai, S. N. (2010). Why is ethylene involved in selective promotion of female flower development in cucumber? *Plant Signaling & Behavior, 5*(8), 1052–1056.

Sun, J. J., Li, F., Wang, D. H., Liu, X. F., Li, X., Liu, N., Gu, H. T., Zou, C., Luo, J. C., He, C. X., Huang, S. W., Zhang, X. L., Xu, Z. H., & Bai, S. N. (2016). CsAP3: A cucumber homolog to Arabidopsis APETALA3 with novel characteristics. *Frontiers in Plant Science, 7*, 1181. https://doi.org/10.3389/fpls.2016.01181

Swain, A. (2006). Sex determination: Time for meiosis? The gonad decides. *Current Biology: CB, 16*(13), R507–R509. https://doi.org/10.1016/j.cub.2006.06.009

Tanurdzic, M., & Banks, J. A. (2004). Sex-determining mechanisms in land plants. *The Plant Cell, 16*(Suppl), S61–S71. https://doi.org/10.1105/tpc.016667

Theobald, D. L. (2010). A formal test of the theory of universal common ancestry. *Nature, 465*(7295), 219–222. https://doi.org/10.1038/nature09014

Thompson, D. A. W. (1917). *On growth and form*. University Press.

Trebitsh, T., Staub, J. E., & O'Neill, S. D. (1997). Identification of a 1-aminocyclopropane-1-carboxylic acid synthase gene linked to the female (F) locus that enhances female sex expression in cucumber. *Plant Physiology, 113*(3), 987–995.

VanBuren, R., Zeng, F., Chen, C., Zhang, J., Wai, C. M., Han, J., Aryal, R., Gschwend, A. R., Wang, J., Na, J. K., Huang, L., Zhang, L., Miao, W., Gou, J., Arro, J., Guyot, R., Moore, R. C.,

Wang, M. L., Zee, F., Charlesworth, D., Moore, P. H., Yu, Q., & Ming, R. (2015). Origin and domestication of papaya Yh chromosome. *Genome Research, 25*(4), 524–533. https://doi.org/10.1101/gr.183905.114

VanBuren, R., Wai, C. M., Zhang, J., Han, J., Arro, J., Lin, Z., Liao, Z., Yu, Q., Wang, M. L., Zee, F., Moore, R. C., Charlesworth, D., & Ming, R. (2016). Extremely low nucleotide diversity in the X-linked region of papaya caused by a strong selective sweep. *Genome Biology, 17*(1), 230. https://doi.org/10.1186/s13059-016-1095-9

Waddington, C. H. (1966). *Principles of development and differentiation* (Current Concepts in Biology Series). Macmillan.

Wang, X., & Bai, S. N. (2019). Key innovations in transition from homospory to heterospory. *Plant Signaling & Behavior, 14*(6), 1596010.

Wang, D. H., Li, F., Duan, Q. H., Han, T., Xu, Z. H., & Bai, S. N. (2010). Ethylene perception is involved in female cucumber flower development. *The Plant Journal, 61*(5), 862–872. https://doi.org/10.1111/j.1365-313X.2009.04114.x

Wang, J., Na, J. K., Yu, Q., Gschwend, A. R., Han, J., Zeng, F., Aryal, R., VanBuren, R., Murray, J. E., Zhang, W., Navajas-Perez, R., Feltus, F. A., Lemke, C., Tong, E. J., Chen, C., Wai, C. M., Singh, R., Wang, M. L., Min, X. J., Alam, M., Charlesworth, D., Moore, P. H., Jiang, J., Paterson, A. H., & Ming, R. (2012). Sequencing papaya X and Yh chromosomes reveals molecular basis of incipient sex chromosome evolution. *Proceedings of the National Academy of Sciences of the United States of America, 109*(34), 13710–13715. https://doi.org/10.1073/pnas.1207833109

Wang, F., Yuan, Z., Zhao, Z., Li, C., Zhang, X., Liang, H., Liu, Y., Xu, Q., & Liu, H. J. (2020). Tasselseed5 encodes a cytochrome C oxidase that functions in sex determination by affecting jasmonate catabolism in maize. *Journal of Integrative Plant Biology, 62*(2), 247–255.

Weismann, A. (1889). *Essays upon heredity*. Oxford Clarendon Press.

Westergaard, M. (1958). The mechanism of sex determination in dioecious flowering plants. *Advances in Genetics, 9*, 217–281.

White, J. (1979). Plant as a meta-population. *Annual Review of Ecology and Systematics, 10*, 109–145. https://doi.org/10.1146/annurev.es.10.110179.000545

Wilhelm, D., Yang, J. X., & Thomas, P. (2013). Mammalian sex determination and gonad development. *Current Topics in Developmental Biology, 106*, 89–121. https://doi.org/10.1016/B978-0-12-416021-7.00003-1

Williams, G. C. (1975). *Sex and evolution. Monographs in population biology* (Vol. 8). Princeton University Press.

Winther, R. G. (2001). August Weismann on germ-plasm variation. *Journal of the History of Biology, 34*(3), 517–555.

Witter, S. H., & Bokovac, M. J. (1958). The effect of gibberellins on economic crops. *Economic Botany, 12*, 213–255.

Wolpert, L., Jessell, T., Lawrence, P., Meyerowitz, E., Robertson, E., & Smith, J. (2007). *Principles of development* (3rd ed.). Oxford: Oxford University.

Yampolsky, C., & Yampolsky, H. (1922). *Distribution of sex forms in the phanerogamic flora*. Gebrüder Borntraeger.

Zhang, H. M., Li, S., Yang, L., Cai, G. H., Chen, H. M., Gao, D. L., Lin, T., Cui, Q. Z., Wang, D. H., Li, Z., Cai, R., Bai, S. N., Lucas, W., Huang, S. W., Zhang, Z. H., & Sun, J. J. (2020a). Gain-of-function ACS1G dominates in controlling female flower development in cucumber. *Plant Cell, 33*(2), 306–321.

Zhang, X. T., Wang, G., Zhang, S. C., Chen, S., Wang, Y. B., Wen, P., Ma, X. K., Shi, Y., Qi, R., Yang, Y., Liao, Z. Y., Lin, J., Lin, J. S., Xu, X. M., Chen, X. Q., Xu, X. D., Deng, F., Zhao, L. H., Lee, Y. L., Wang, R., Chen, X. Y., Lin, Y. R., Zhang, J. S., Tang, H. B., Chen, J., & Ming, R. (2020b). Genomes of the banyan tree and pollinator wasp provide insights into fig-wasp coevolution. *Cell, 183*, 875–889.

Zhang, S. Q., Tan, F. Q., Chung, C. H., Slavkovic, F., Devani, R. S., Troadec, C., Marcel, F., Morin, H., Camps, C., Roldan, M. V. G., Benhamed, M., Dogimont, C., Boualem, A., & Bendahmane,

A. (2022). The control of carpel determinacy pathway leads to sex determination in cucurbits. *Science, 378*(6619). https://doi.org/10.1126/science.add4250

Zhao, D., McBride, D., Nandi, S., McQueen, H. A., McGrew, M. J., Hocking, P. M., Lewis, P. D., Sang, H. M., & Clinton, M. (2010). Somatic sex identity is cell autonomous in the chicken. *Nature, 464*(7286), 237–U115. https://doi.org/10.1038/nature08852

Zou, C., Massonnet, M., Minio, A., Patel, S., Llaca, V., Karn, A., Gouker, F., Cadle-Davidson, L., Reisch, B., Fennell, A., Cantu, D., Sun, Q., & Londo, J. P. (2021). Multiple independent recombinations led to hermaphroditism in grapevine. *Proceedings of the National Academy of Sciences of the United States of America, 118*(15), e2023548118.

Chapter 18
Embryogenesis: The Establishment of a New Life in Plants

Lei Guo and Chun-Ming Liu

Abstract In both plants and animals, embryogenesis is the beginning of a new life, during which the basic body pattern is established. In flowering plants, embryogenesis starts with a single-celled diploid zygote, formed by the fusion of an egg and a sperm cell within the embryo sac during double fertilization. Another sperm cell carried along by the pollen tube fuses with the diploid central cell in the embryo sac to produce a triploid endosperm that nourishes the embryo during embryogenesis and, sometimes, seed germination as well. In embryo development, cell divisions and cell fate specifications occur orderly to produce elaborated architectural patterns of tissues and organs for a mature embryo. In this chapter, we give a general overview of key developmental events in plant embryogenesis, including polarity setup, zygote activation, tissue pattern formation, and organ primordial establishment, with attention to molecular mechanisms underlying these processes known to date.

Keywords Embryogenesis · Zygote activation · Tissue specification · Organ establishment · Meristem formation · Postembryonic development · *Arabidopsis*

18.1 Introduction

Embryogenesis, which starts after the fusion of an egg and a sperm cell, initiates a new life cycle in plants through successive cell divisions and differentiations. As the embryo develops, the basic body plan with one or two cotyledons is formed for monocotyledonous and dicotyledonous plants, respectively (Fig. 18.1). Mature

---

L. Guo
College of Horticultural Science and Engineering, Shandong Agricultural University, Tai'an, Shandong, China

C.-M. Liu (✉)
Institute of Botany, Chinese Academy of Sciences, Fragrant Hill, Beijing, China

School of Advanced Agricultural Sciences, Peking University, Beijing, China
e-mail: cmliu@ibcas.ac.cn

F. Chang et al. (eds.), *Regulation of Plant Development*,
https://doi.org/10.1007/978-981-95-7033-1_18

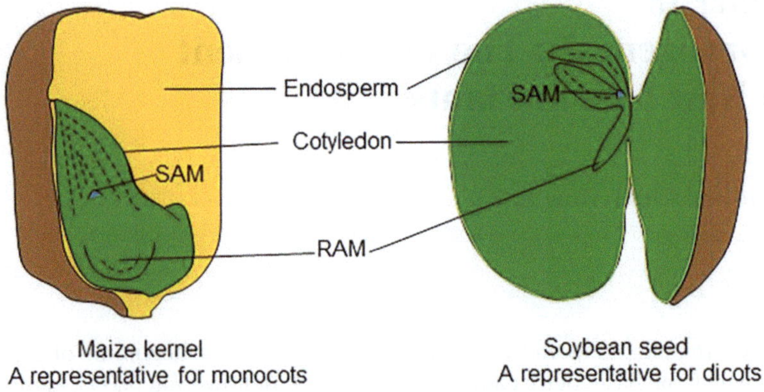

Fig. 18.1 Schematic illustration of a mature kernel of maize and a mature seed of soybean, representatives for monocotyledonous and dicotyledonous plants, respectively. Corresponding organs and tissues, including endosperm, cotyledon(s), SAM, and RAM, are color-coded and illustrated in the maize kernel and soybean seed. Note that the endosperm in mature soybean seed is hardly visible

seeds for most monocotyledonous plants, such as maize, have a large endosperm and a relatively small embryo (Fig. 18.1, left), whereas mature seeds for most dicotyledonous plants, such as soybeans, have a larger embryo with almost no endosperm (Fig. 18.1, right). During embryogenesis, the shoot apical meristem (SAM) and root apical meristem (RAM) are established. The former builds aboveground leaves, stems, and floral organs, while the latter forms the underground root systems after seed germination. In parallel, the basic tissue pattern—including tissues of protoderm/epidermis, cortex, endodermis, and pro-vasculature—is formed (Mansfield & Briarty, 1991). These organ primordia and tissue initials are the groundwork for postembryonic development (Johri, 1984; Maheshwari & Johri, 1950; Johri et al., 1992). Stem cells in the SAM and RAM continuously produce cells for new tissue specification and organ formation (van den Berg et al., 1997; Weigel & Jürgens, 2002).

This chapter focuses on summarizing recent progress achieved through molecular genetic studies of developmental events within plant embryogenesis, including zygote divisions, the establishment of tissue patterns, the formation of organ primordia, and the coordination of embryonic and postembryonic development. Most of these progresses stem from investigations in the model plant species *Arabidopsis*, unless otherwise specified. We provide a general overview of recent developments in this area. Readers are guided to read several review articles published in this field, each offering unique insights (Willemsen & Scheres, 2004; Lau et al., 2012; De Vries & Weijers, 2017; Dresselhaus & Jürgens, 2021; Zhao et al., 2022) and avoid redundancy by refraining from reiterating points that have been extensively covered in these articles. The chapter concludes with a discussion of prospects in plant embryogenesis research, highlighting key questions that remain to be addressed.

18.2 Divisions and Polarity Setup of the Zygote

In higher plants, double fertilization triggers and initiates both embryogenesis and endosperm development. Activation of the zygote occurring upon the fusion of an egg cell and a sperm cell is marked by successive events, including cell wall thickening, zygote elongation, and zygote divisions (Fig. 18.2). In *Arabidopsis*, after a pollen grain (the male gametophyte) lands on the stigma, it takes approximately 5 h to deliver two sperm cells in the pollen tube to the embryo sac (the female gametophyte), while the formation of a zygote takes another hour (Bleckmann et al., 2014; Dumas & Rogowsky, 2008; Guo et al., 2016).

As in unfertilized egg cells, the distribution of cellular components in zygotes remains asymmetrical in most flowering plants, with the nucleus located in the apical end and a large vacuole positioned in the basal end (Dumas & Rogowsky, 2008; Mansfield & Briarty, 1991; Zhao et al., 2011; Fig. 18.2). Detailed observation of cytoskeletal movement and vacuolar distribution during *Arabidopsis* zygote polarization demonstrated that dynamic rearrangement of cytoskeletal components and asymmetric vacuolar distribution is essential for nuclear migration and apical-basal

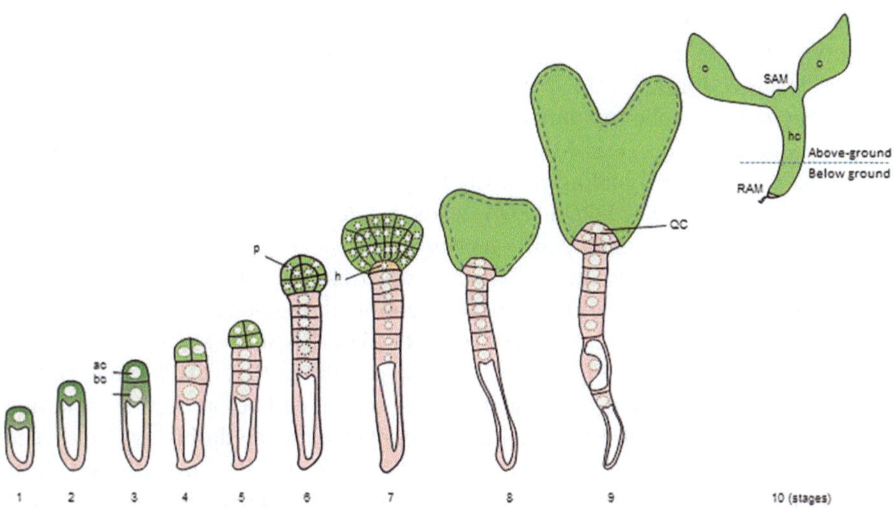

Fig. 18.2 The cell division patterns in embryogenesis of *Arabidopsis thaliana*. In *Arabidopsis*, cell divisions in embryogenesis follow a highly ordered pattern. The zygote at stage 1 elongates (stage 2) and undergoes an asymmetric cell division (stage 3) to produce a smaller apical cell (ac, green) and a larger basal cell (bc, pink). The ac undergoes a series of sequential divisions to produce 2-cell (stage 4), octant (stage 5), dermatogen (stage 6), heart-shaped (stages 7–8), and torpedo-shaped embryos (stage 9) and forms most parts of the seedling, including 2 cotyledons (c), a hypocotyl (hc), and shoot apical meristem (SAM, stage 10). By contrast, the bc undergoes several rounds of transversal divisions to produce a suspensor and a hypophysis cell (h). The h cell forms the quiescent center (QC) and columella cells of the root cap in the root apical meristem (RAM). Periclinal divisions occurred at the transition from the 16- to 32-cell stage, leading to the formation of the protoderm (p), which forms the epidermis in the plant body

axis establishment in zygotes (Kimata et al., 2016, 2019). Notably, in *Arabidopsis*, the zygote experiences a transient nonpolar period following fertilization, characterized by a centrally positioned nucleus and small vacuoles distributed throughout the cytoplasm (Drews & Koltunow, 2011; Mansfield et al., 1991). Prior to the division of the zygote, a shift in zygotic symmetry occurs, with the nucleus relocated to the apical region, and the zygote elongates along the apical-basal axis to reestablish a polarized pattern (Mansfield & Briarty, 1991; Faure et al., 2002; Fan et al., 2008; Ueda et al., 2011).

In the majority of flowering plant species, the initial division of the zygote is transversal, producing a smaller apical cell and a larger basal cell (Fig. 18.2). It is important to note that, despite this asymmetric division observed in most species, there are significant variabilities in the size ratio between the apical and basal cells (Sivaramakrishna, 2011; Dresselhaus & Jürgens, 2021). This asymmetrical division generally establishes apical-basal polarity, with the apical cell predominantly contributing to most of the embryo, while the basal cell primarily gives rise to the suspensor, a rudimentary organ responsible for positioning the embryo at the center of the embryo sac (Fig. 18.2; Johri, 1984). However, variations have been observed across different species, in which the basal cell may contribute partially to the formation of the embryo, and the apical cell also contributes partially to the formation of the suspensor (Johri, 1984, 1992; Maheshwari & Johri, 1950; Mansfield & Briarty, 1991). In some rare cases, the first division of the zygote is longitudinal (Maheshwari & Johri, 1950), showing the complexity of plant embryogenesis.

During the organelle repolarization, the zygote elongates about 3-fold along the apical-basal axis (Fig. 18.3; Mansfield & Briarty, 1991). A mitogen-activated protein kinase (MAPK) signaling pathway plays an important role in this process. A mutation in the MAPKK *YODA* (*YDA*) leads to defective zygote elongation, producing two equal-sized daughter cells—a phenotype also observed in the double mutant of two downstream MAP kinases, *mpk3 mpk6* (Lukowitz et al., 2004; Wang et al., 2007). The activation of *YDA* after fertilization is regulated by the membrane-associated receptor-like pseudokinase SHORT SUSPENSOR (SSP), with a paternal effect (Bayer et al., 2009) (Fig. 18.3). The small peptide EMBRYO SURROUNDING FACTOR 1 (ESF1), which accumulates in the central cell before fertilization, functions synergistically with SSP to regulate the YDA pathway during zygote division (Costa et al., 2014). Additionally, the RWP-RK family transcription factor GROUNDED (GRD) is required for the establishment of the early embryo polarity. Although GRD is not regulated by the YDA MAPK cascade, GRD signaling is YDA-dependent (Jeong et al., 2011). Mutations in a putative exo-polygalacturonase NIMNA (Babu et al., 2013) also affect the zygote elongation and subsequent zygote division. Like SSP, reciprocal crosses showed that NIMNA also acts on early embryo patterning with a mild paternal effect (Babu et al., 2013).

Another evidence of paternal contribution to zygote division and embryogenesis initiation is the paternally expressed *BABY BOOM1* (*BBM1*), an AP2/PLETHORA (PLT) family transcription factor (Fig. 18.3). The expression of *BBM1* occurs in sperm cells but is absent in unfertilized egg cells in both rice and *Arabidopsis* (Anderson et al., 2017; Khanday et al., 2019; Borg et al., 2020). When rice BBM1

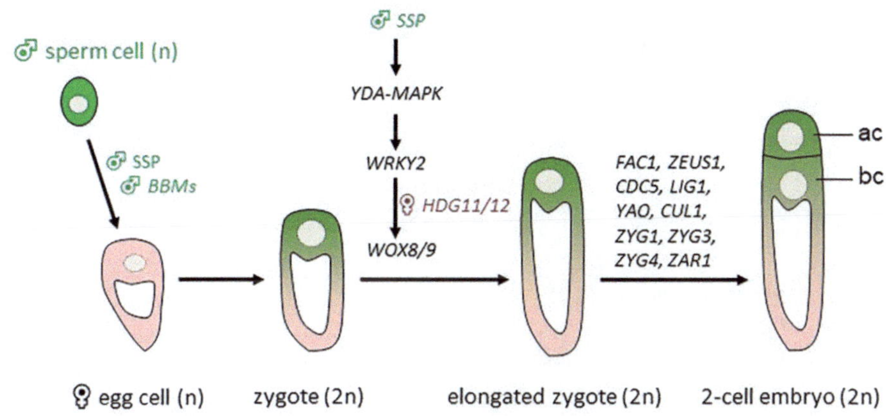

Fig. 18.3 Zygote elongation, repolarization, and the activation of the asymmetric cell division after the double fertilization. Upon the fusion of a sperm and an egg cell, *SSP*, as the transiently and paternally expressed gene, activates the mitogen-activated protein kinase YDA in the zygote. In addition, sperm-expressed *BBMs* activate the zygote to start embryogenesis. During zygote repolarization, the YDA signaling pathway activates the WRKY2 transcription factor through phosphorylation. Once activated, *WRKY2* induces expressions of *WOX8* and *WOX9*, along with maternally expressed *HDG11/12* transcriptional regulators, which collectively contribute to elongation and asymmetric division of the zygote. During asymmetric zygotic division, 10 genes—including *FAC1, ZEUS1, CDC1, LIG1, YAO, CUL1, ZYG1, ZYG3, ZYG4*, and *ZAR1*—have been showed to beessential for the activation of embryogenesis (ac apical cell, bc basal cell)

transcription was ectopically induced in an egg cell, embryogenesis occurred in the egg cell without fertilization, which further led to haploid progeny (Khanday et al., 2019).

During the course of zygote repolarization, a putative transcription factor, WRKY2, the phosphorylation of which was directly regulated by the MAPK signaling pathway, is also involved (Ueda et al., 2017). In *wrky2* mutants, zygotes elongate normally, but polarized organelle positioning fails, leading to defective zygote division and subsequently aberrant embryo development (Ueda et al., 2011). Because the defect could be rescued by overexpression of a *WUSCHEL RELATED HOMEOBOX 8* (*WOX8*) gene, suggesting a WRKY2-mediated de novo *WOX8* transcription in the zygote, which is essential for the repolarization of organelles and the formation of a normal embryo (Fig. 18.4). The WRKY2-mediated activation in the zygote was also applicable to the transcription of *WOX9*, the redundant gene of *WOX8*. Additionally, female-derived class IV HOMEODOMAIN GLABROUS (HDG) transcription factors *HDG11/12* are also required for the activation of *WOX8* transcription (Ueda et al., 2011, 2017). Considering the normal polarity setup in the *wrky2* egg cell, it seems that the regulation of zygote polarity is different from the polarity establishment in the egg cell (Ueda et al., 2011), although whether the WRKY2-WOX8 module regulates the repolarization of organelles in the zygote through the cytoskeleton or other cellular machinery is still not known.

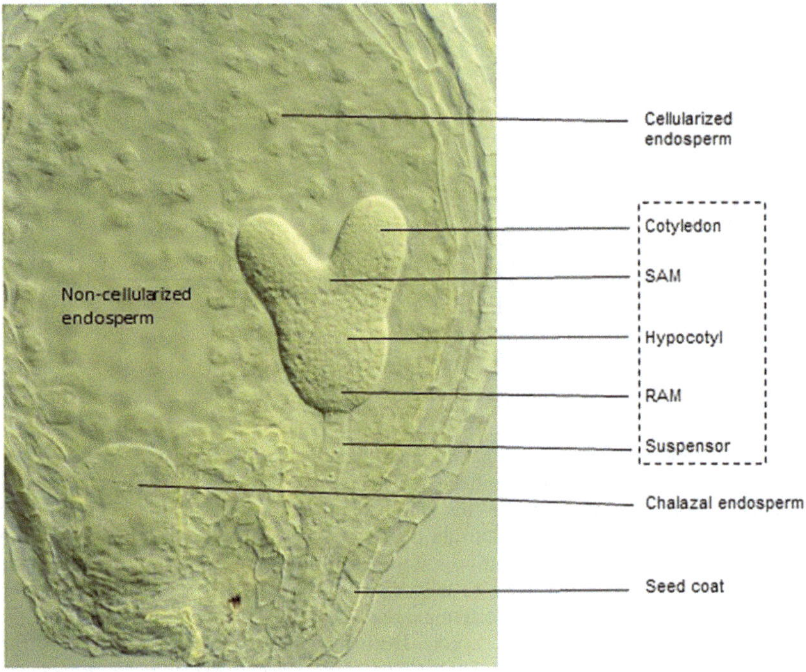

Fig. 18.4 The structure of an *Arabidopsis* seed with torpedo-shaped embryo. The torpedo-shaped embryo forms at the micropylar end of the embryo sac, with the suspensor connected to maternal tissues of the seed coat. The cellularization of the endosperm starts at the micropylar end and goes towards the chalazal end. The chalazal endosperm, located adjacent to the vascular bundle, facilitates the loading of sugars to the endosperm and embryo. The basic tissue structure of the embryo, including cotyledons, hypocotyl, hypophysis, suspensor, shoot apical meristem (SAM), and root apical meristem (RAM), is established at this stage

Although division of the zygote is a milestone event in early embryogenesis, the molecular mechanism underlying this process is still largely unknown. Due to the accumulation of maternal proteins and RNAs in the egg cell, zygote-lethal mutants are seldom found in animals. Most such embryo-lethal mutants in animals are able to carry out several rounds of mitotic divisions. In plants, however, several zygote-lethal mutants have been reported (Xu et al., 2005; Meinke et al., 2008; Yu et al., 2016; Guo et al., 2016; Yang et al., 2017). The *embryonic factor 1* (*fac1*) is the first zygote-lethal mutant identified in plants. Genetic analyses showed that the *FAC1* inheritance occurs in a typical Mendelian manner, suggesting that the presence of either maternal or paternal *FAC1* allele is sufficient for the initiation of zygotic embryogenesis. *FAC1* encodes an AMP deaminase and is expressed in both zygote and central cell right after fertilization. A *GUS* reporter construct carrying *FAC1* regulatory elements marks the cell lineage of somatic embryos, which can thus be used to follow somatic embryo development during in vitro embryogenesis (Xu et al., 2005) (Fig. 18.3).

Mutations of genes underlying cell cycle regulation often cause zygote-lethal phenotypes in plants. *ZEUS1* encodes a thymidylate kinase (AtTMPK) expressed during the mitotic G1/S transition stage. For example, the *zeus* mutant, with a T-DNA inserted in the AtTMPK, exhibited a zygote-lethal phenotype, suggesting the importance of DNA metabolism in the initiation of zygote division (Ronceret et al., 2008). Other cell cycle regulators, such as Myb-related CDC5 and DNA LIGASE I (AtLIG1), also play critical roles in zygote division, as mutations in either of them result in a zygote-lethal phenotype. AtLIG1 acts downstream of the DNA demethylase DEMETER (DME) to repair single-strand DNA breaks caused by DME-mediated demethylation (Andreuzza et al., 2010; Lin et al., 2007). Two genes that encode RNA processing proteins, WD-repeat containing protein Yao and tRNA ligase ZYGOTE-LETHAL 3 (ZYG3), are essential for the initiation of the zygote division in *Arabidopsis* (Li et al., 2010; Yang et al., 2017). Mutation of the *DEAD-BOX RNA HELICASE 27* that regulates microRNA biogenesis also led to zygote lethality and defective stem cell homeostasis in roots (Hou et al., 2021), suggesting a link between zygote activation and stem cell homeostasis in plants. The zygote-lethal phenotype of these mutants suggests that the mRNA and proteins of these genes involved are either not accumulated or not sufficiently accumulated in the zygote (Fig. 18.3).

Protein degradation is also important for the first division of the zygote. Mutations in either Cullin 1 or APC11, the subunits of the SCF (for SKP1, Cullin/CDC53, F-box protein) or Anaphase Promoting Complex/Cyclosome (APC/C) ubiquitin ligase complex, respectively, showed a zygote-lethal phenotype (Guo et al., 2016; Shen et al., 2002). As the smallest subunit of the APC/C, APC11/ZYG1 has ubiquitin ligase activity (Fig. 18.3; Guo et al., 2016), allowing for highly regulated protein degradation. Biochemical analyses showed that APC/C acts directly on cyclin B1;1 to promote the cyclin B1;1 degradation, promoting irreversible progression of the cycle. Remarkably, in addition to the zygote-lethal phenotype, the mutation in ZYG1/APC11 leads to asynchronized mitotic divisions in the coenocytic endosperm of *Arabidopsis*. Such a phenotype was also observed in multinucleate somatic cells formed when *APC11* expression is suppressed by microRNA, suggesting that APC/C may serve as an intracellular diffusible signal to coordinate cell cycle synchronizations in multinucleate cells (Guo et al., 2018).

ZYGOTIC ARREST 1 (ZAR1) is a receptor kinase known for regulating zygote divisions. Although the knockout mutant of *zar1* showed a very subtle phenotype, the mutant carrying a truncated ZAR1 protein exhibited a defective asymmetrical division in the zygote (Fig. 18.3). It is plausible that ZAR1 interacts with an unknown ligand to set up the division plane of the zygote (Yu et al., 2016).

The setup of apical-basal polarity in early plant embryos relies on auxin biosynthesis in the suspensor and polarized auxin transport from the suspensor to the embryo, guided primarily by PIN7. The local auxin production at the base of young embryos and the accompanying PIN7-mediated auxin flow toward the proembryo are required for the apical auxin maximum and the specification of the apical structures of the embryo. At the later stage of embryogenesis, the onset of localized apical auxin biosynthesis mediates PIN1 polarization, basal auxin maximum, and specification of the root pole (Robert et al., 2013). The role of the suspensor in the apical-basal polarity setup was supported by results from an in vitro microspore embryogenesis that mimics zygotic embryogenesis, which showed that the suspensor-like filament specifies the embryonic identity of the distal cell of the filament and later directs the apical-basal polarity establishment in the embryo. However, in vitro treatment of proembryos with or without suspensor does not change the polarity of the embryo nor the formation of SAM and RAM (Liu et al., 1993b), suggesting that the process of polarity setup is more complex than expected.

18.3 Establishment of Tissue Pattern

The zygote in *A. thaliana* undergoes an asymmetrical division to produce a smaller apical and a larger basal cell. The former contains dense cytoplasm, whereas the latter is elongated and highly vacuolated. These two cells differ not only in their cellular compositions but also in subsequent division patterns. The basal cell undergoes several rounds of transversal divisions to generate the extra-embryonic single-cell file suspensor, with seven to nine cells (Fig. 18.2). The roles of the filamentous structure are to anchor the embryo to maternal tissues, to push the developing embryo to the center of the embryo sac (Babu et al., 2013; Kawashima & Goldberg, 2010; Mansfield & Briarty, 1991), and to help the apical cell to set up the embryo polarity (Friml et al., 2003; Supena et al., 2008; Robert et al., 2013).

In contrast to the relatively simple division pattern in the suspensor, the apical and descendant cells undergo highly coordinated cell division patterns to form the embryo by changing division planes and by on-and-off regulations of cell cycles in different cells (Fig. 18.2). The most intriguing fact of the early plant embryogenesis is the capability of individual cells to reorient their division planes, allowing new cell layers to be formed and enabling tissue patterns to be established. For example, in *Arabidopsis*, the apical cell divides twice vertically, and then each of these progeny cells divides transversely to produce an 8-cell octant-staged embryo. Further divisions of these 8 cells are periclinal, producing a 16-cell embryo, with 8 external cells as the first protodermal tissue and 8 internal cells that produce the ground tissue and vascular bundles in the future. The divisions of these protodermal cells afterward are, without exception, anticlinal, to produce the epidermal tissue for the plant body. In contrast, divisions of the ground tissue cells are more complex, leading to the formation of organ primordia, including cotyledons, radicle, and meristematic tissues, and tissue initial cells with different cell fates (Fig. 18.4; Johri, 1984; Johri et al., 1992; Maheshwari & Johri, 1950; Mansfield & Briarty, 1991). It is anticipated that intercellular communications in plants occur effectively among cells in different cell lineages as regenerations and reorganizations occur quickly after ablations of individual cells (van den Berg et al., 1997).

In *Arabidopsis*, the basal cell also contributes partially to the formation of the embryo. The hypophysis, the uppermost suspensor cell, and its decedent cells form the quiescent center (QC) and columella cells of the root cap in the primary root

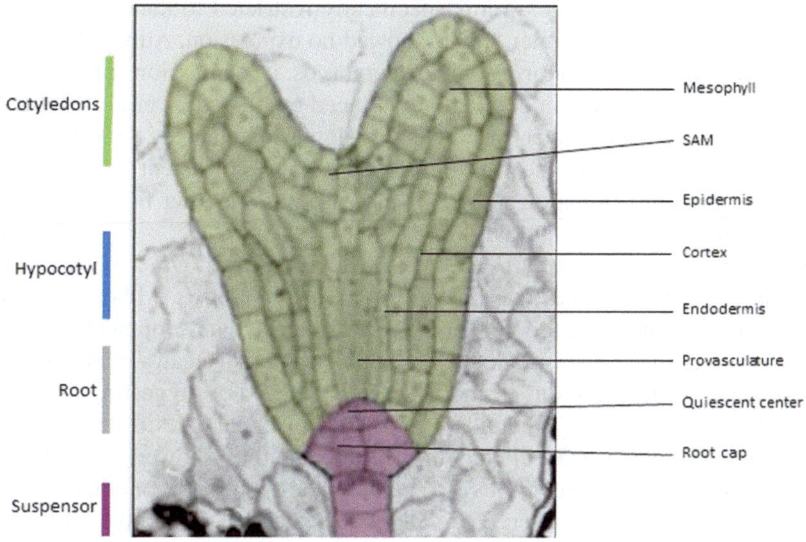

Fig. 18.5 Major tissues and organs are formed during plant embryogenesis. Transversal section of a torpedo-staged embryo of *Arabidopsis* to show different tissues formed in embryogenesis. Cells highlighted in green are derived from the apical cell, while cells highlighted in pinkish are from the basal cell after the first division of the zygote. Major tissues are labeled at the right, and major organs are labeled at the left. The cortex and endodermis together are called ground tissue

(Fig. 18.2). At the late-globular stage, the hypophysis divides transversally and asymmetrically to form an upper lens-shaped cell that forms the QC in the future and a lower basal cell that forms the columella cells of the root cap (Fig. 18.2). The QC, consisting mostly of four cells, is the stem cell organizing center of the root. Cells located adjacent to the QC are stem cells for different cell lineages and various tissues afterwards (Fig. 18.5; Scheres et al., 1994). The lower basal cell and its decedent cells located next to the QC are stem cells for the central root cap (Hamann et al., 1999; Scheres et al., 1994).

18.3.1 The Formation of Protoderm

After the periclinal divisions in the octant-staged embryo, eight outer cells form protoderm precursor cells that will differentiate into the epidermis and further maintain the integrity of the outer layer (called L1) of the plant body (Fig. 18.2; Scheres et al., 1994; Yoshida et al., 2014). The homeobox transcription factor of *A. thaliana* MERISTEM LAYER 1 (AtML1) is considered to be a master regulator in protoderm specification. *AtML1* starts to express in the apical cell after the first division of the zygote and continues to express in all decedent cells in proembryos (Lu et al., 1996; Sessions et al., 1999). Differential expression of the *AtML1* occurs at the

16-cell embryo, as the expression is specifically restricted to the outer protoderm layer thereafter, while the inner 8 cells showed no expression. Although no *AtML1* expression was observed in torpedo-staged embryos, its expression appears again in the epidermis of the SAM in cotyledonary-staged embryos (Lu et al., 1996). After seed germination, the L1-layer-specific expression is maintained in SAM and newly formed organs, but not in the well-developed organs, consistent with the role of AtML1 in epidermis initiation. *AtML1* is also used as a marker for the developing epidermis of the above-ground tissues.

As the close homolog of *AtML1*, *ROTODERMAL FACTOR 2* (*PDF2*) is also expressed exclusively in the outer layer of SAM in *Arabidopsis*. Although the single mutants of *atml1* and *pdf2* did not show visible defects in development, *atml1/pdf2* double mutants showed severe defects in embryogenesis (Ogawa et al., 2015) and epidermal cell differentiations in shoots (Abe et al., 2003), suggesting that AtML1 and PDF2 are functionally redundant and play a critical role in developing and maintaining the epidermal cell identity. Further studies showed that PDF2 and AtML1 may execute their functions by regulating the expressions of other epidermis-specific genes through a conserved cis-regulatory element named the L1 box (Abe et al., 2001, 2003; Ogawa et al., 2015; San-Bento et al., 2014). As expected, *AtML1* overexpression triggers the expression of these epidermis-specific genes, inducing the formation of stomatal cells and trichome-like cells in non-epidermal tissues (Peterson et al., 2013; Takada et al., 2013). *AtML1* orthologs in other species are also preferably expressed in protodermal cells (Ingram et al., 1999, 2000; Ito et al., 2002), indicating an evolutionarily conserved function in protoderm formation.

Other protoderm/epidermis-expressed genes reported so far include receptor-like kinases of *ARABIDOPSIS CRINKLY4* (*ACR4*) and *ABNORMAL LEAF SHAPE 2* (*ALE2*), a subtilase of *ABNORMAL LEAF SHAPE 1* (*ALE1*), and a small peptide-coding gene *CLE19* (Tanaka et al., 2001, 2002, 2007; Xu et al., 2015). Mutations in *ACR4* and *ALE2* disrupted expressions of both *PDF2* and *AtML1,* leading to defective protoderm differentiation (Tanaka et al., 2002, 2007). Genetic analyses suggest that *ACR4* and *ALE2* may also function redundantly in the same pathway as *AtML1* and *PDF2* to regulate protoderm specification (Tanaka et al., 2001, 2002, 2007). *CLE19* is expressed in protodermal cells in the upper portion of the embryo in *Arabidopsis* (Xu et al., 2015). In situ disruption of the CLE19 function by expression of the antagonistic $CLE19_{G6T}$ peptide led to defective cotyledon formation and delayed cellularization of the endosperm, suggesting that CLE19 may function in cotyledon formation and in signal transduction between embryo and endosperm (Xu et al., 2015). LRR family receptor kinases of *RPK2* and *TOADSTOOL 2* (*TOAD2*) have also been shown to be critical for epidermis formation, as the *rpk1/toad2* double mutant exhibited severe defects in cell organization and expressions of protoderm-specific markers in early embryos (Johnson et al., 2005; Nodine et al., 2007). It is known that the signal of small peptides is perceived by LRR family receptor kinases; whether CLE19 functions through any of these receptor kinases remains to be examined.

18.3.2 The Formation of Ground Tissue

Ground tissue, consisting of cortex and endodermis, is formed as the inner tissues of the embryo (Fig. 18.5). Roots in different plant species differ in diameter, which is correlated mainly with the number of cortex cell layers. For instance, the cortex in *Arabidopsis* has only one cell layer, while it may have more than 12 cell layers in rice roots. Variations in the number of cortex cell layers are also observed in different types of rice roots (Pauluzzi et al., 2012).

At the early globular stage, ground tissue precursor cells are formed from the longitudinal divisions of lower-tier inner cells. Subsequently, these ground tissue precursor cells and their daughter cells divide periclinally to form cortex and endodermis (Fig. 18.2; Petricka et al., 2009). Genetic studies have identified several factors involved in ground tissue specification and maintenance in *Arabidopsis* embryogenesis. The GRAS family transcription factors SHORT ROOT (SHR) and SCARECROW (SCR) are key regulators in ground tissue specification in embryonic roots (Di Laurenzio et al., 1996; Helariutta et al., 2000; Levesque et al., 2006; Nakajima et al., 2001). Although both *SHR* and *SCR* are expressed in early globular-stage embryos, defects in *shr* or *scr* mutants are only observed in early heart-shaped-stage embryos, with one single ground tissue layer formed in these mutants (Helariutta et al., 2000; Scheres et al., 1995; Wysocka-Diller et al., 2000). The expression of these two TFs is highly specific, with *SHR* in immature vasculature (Helariutta et al., 2000) and *SCR* in the endodermis (Di Laurenzio et al., 1996). The SHR protein can move through plasmodesmata from the vasculature into the cortex-endodermal initial cells and their daughter cells, where it interacts with SCR to promote asymmetric cell divisions and endodermal specification (Heidstra et al., 2004; Helariutta et al., 2000; Nakajima et al., 2001; Vaten et al., 2011).

A heat shock transcription factor, SCHIZORIZA (SCZ), is also found to specify the ground tissue identity. The *scz* mutant showed defects in radial patterning and the formation of ectopic root hairs from the cortex layer, suggesting that SCZ may participate in the SCR and SHR regulatory network (Mylona et al., 2002; Pernas et al., 2010; ten Hove et al., 2010).

18.3.3 The Formation of the Vasculature

In *Arabidopsis*, at the transition from the octant to the globular stage, formative divisions of the lower-tier inner cells generate not only four outer ground tissue precursor cells but also four inner vascular precursor cells. These vascular initial cells undergo several rounds of divisions to generate the pro-vasculature (Fig. 18.6), which consists of a central xylem axis and two phloem poles at each side of the xylem axis (De Rybel et al., 2016), although the formation of vascular bundles is only achieved after seed germination. Therefore, genes regulating phloem establishment (Ren et al., 2019) and xylem formation (Li et al., 2013) usually showed no defects in embryogenesis, even though they are expressed in very early-stage embryos (Ren et al., 2019). In recent years, fundamental progress has been made in understanding the initiation, specification, and formation of the vascular tissue.

The patterning of the vascular tissue is dependent on the auxin signaling pathway (Fig. 18.6). As the first identified transcription factor in the process, auxin-dependent transcription factor MONOPTEROS (MP) is essential for the formation of the vascular tissue in *Arabidopsis* (Hardtke & Berleth, 1998; Schlereth et al., 2010). A mutation in *MP* leads to defective formative divisions of vascular precursor cells and abnormal formation of an embryonic root (De Rybel et al., 2013; Hardtke & Berleth, 1998). Among those direct downstream targets of MP, the basic helix-loop-helix (bHLH) transcription factor TARGET OF MONOPTEROS 5 (TMO5) specifies the vascular precursor cells at the early globular stage (Schlereth et al., 2010). Biochemical analyses showed that TMO5 interacted with another bHLH transcription factor, LONESOME HIGHWAY (LHW), to form a heterodimeric complex (De Rybel et al., 2013; Ohashi-Ito et al., 2013). Mutations in *TMO5* and its homologs led to a reduction in vascular size due to defective periclinal divisions of vascular precursor cells (De Rybel et al., 2013; Schlereth et al., 2010). Similarly, *lhw* mutants also displayed severe defects in vascular tissue formation (Ohashi-Ito et al., 2013). Combined with the observations that ectopic co-expression of *TMO5* and *LHW* was able to induce periclinal divisions in other cell types, it has been proposed that TMO5 and LHW are the central regulators for promoting periclinal divisions in the formation of vascular tissue (Fig. 18.6).

Moreover, expressions of *TMO5* and *LHW* are regulated in a negative-feedback loop. Thermospermine synthase ACAULIS5 (ACL5) was able to antagonize TMO5-LHW activities through promoting the accumulation of two bHLH transcription factors of SAC51 and SACL3 as the *SUPPRESSOR OF ACAULIS5-LIKE* (*SACL*)

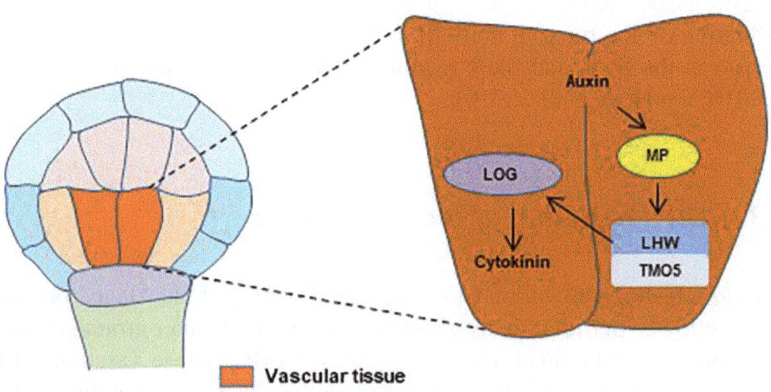

Fig. 18.6 The interplay between auxin and cytokinin during the initiation of the vascular tissue in the globular-stage embryo. In vascular precursor cells, auxin activates the expression of *MP*, which then initiates the expression of *TMO5*. TMO5 dimerizes with LHW to promote periclinal divisions by activating the expression of *LOG*, which encodes an enzyme in the final step of cytokinin biosynthesis

(Hanzawa et al., 2000; Vera-Sirera et al., 2015). The *acl5* mutant showed a dramatic increase in vascular bundle size (Hanzawa et al., 2010), and SAC51 and SACL3 are able to suppress the *acl5* phenotype in the formation of vascular bundles (Imai et al., 2006). SAC51 and SACL3 antagonize the TMO5-LHW dimer by interacting with LHW to suppress the over-proliferation of vascular cells caused by excess TMO5 and LHW activities (Vera-Sirera et al., 2015). In addition, SAC51 and SACL3 are potential targets of TMO5, and the TMO5-LHW dimer up-regulates the expressions of *ACL5*, *SAC51*, and *SACL3* in roots (Katayama et al., 2015; Vera-Sirera et al., 2015). At the globular stage, *TMO5* and *LHW* expressions overlap in four vascular initial cells and activate expressions of *LONELY GUY 3* and *4* (*LOG3/4*), enzymes involved in the final step of cytokinin biosynthesis (De Rybel et al., 2014; Kuroha et al., 2009; Ohashi-Ito et al., 2014). The TMO5/LHW complex triggers cytokinin biosynthesis by directly activating *LOG4* expression. Genetic analyses showed that the *log4* single mutant did not display any phenotype, but the *log1/2/3/4/5/7/8* septuple mutant exhibited severe defects in embryonic pro-vascular tissue and loss of the expression of a cytokinin response factor. Further, the defect in periclinal divisions in *lhw* and *tmo5/tmo5-like1* mutants can be rescued by the treatment with cytokinin (De Rybel et al., 2014). These findings demonstrate that the LOG-dependent cytokinin biosynthesis regulated by the TMO5-LHW dimer is essential for the formation of the vascular tissue (Fig. 18.6). Thus, the dynamic crosstalk between auxin and cytokinin is vital for the formation and development of embryonic vascular tissue in *Arabidopsis* (De Rybel et al., 2014; el-Showk et al., 2015; Muraro et al., 2014).

The *NO VEIN* (*NOV*) gene in *Arabidopsis*, encoding a novel plant-specific nuclear factor, is required for leaf vascular development, cellular patterning, and stem cell maintenance in the root meristem, and for cotyledon outgrowth and separation as well. Mutations in *nov* affected many aspects of auxin-dependent development without directly affecting auxin perception. NOV is required for provascular *PIN1* expression and region-specific expression of *PIN7* in leaf primordia, cell type-specific expressions of *PIN3*, *PIN4*, and *PIN7* in roots, and PIN2 polarity in the root cortex. *NOV* is specifically expressed in developing embryos, leaf primordia, and shoot and root apical meristems (Tsugeki et al., 2009). It seems that NOV regulates competence acquisition in embryos, allowing the establishment of an auxin-dependent cell specification and patterning.

18.4 The Establishment of Organ Primordia

Plant organogenesis is a highly regulated process, with continuous formations of new organs throughout their life cycles from SAM, RAM, and procambium in post-embryonic growth and development. The stem cell organizing center (SCOC) in the SAM and RAM is responsible for the maintenance of neighboring stem cells, which continuously provide initial cells for generations of new organs and tissues within these new organs. Without doubt, the establishment of SAM and RAM during embryogenesis is vital for postembryonic development (van den Berg et al., 1997; Weigel & Jürgens, 2002). During the transition from the octant stage to the dermatogen stage, the proembryo is separated into the upper- and lower-tier cells (Fig. 18.2, stages 5 and 6), generating aerial tissues from the upper tier and hypocotyl and root tissues from the lower tier during postembryonic development. The determination step of the primary SAM and RAM formations occurs at the 64-cell stage (Fig. 18.2, stage 7), when the inner upper-tier cells form the SAM initial, and the lower-tier cells and the uppermost cell in the suspensor (also called hypophysis) form the RAM initial (De Smet et al., 2010; Palovaara et al., 2016; ten Hove et al., 2015).

Auxin biosynthesis is critical for the establishment of the embryo pattern. IAA, the predominant form of auxin in plants, is synthesized in several pathways. The most important pathway in *Arabidopsis* is the tryptophan-dependent one involving two steps: first, the conversion from tryptophan to indole-3-pyruvate by Tryptophan Aminotransferases of *Arabidopsis* (TAAs), followed by oxidative decarboxylation by YUCCAs to produce IAA (Zhao et al., 2014). In the *Arabidopsis* genome, 5 *TAA* and 11 *YUCCA* genes have been identified. Mutations in *YUCCA1/4/10/11*—expressed in the upper regions of globular-stage embryos—led to the formation of embryos with mostly one cotyledon (Cheng et al., 2007), as observed in the *pin1* mutant (Liu et al., 1993a), suggesting a critical role of auxin in embryo patterning. It is also interesting to note that *YUCCA3*, *YUCCA4*, and *YUCCA9* are expressed in the suspensor of globular-stage embryos, and mutants of these genes showed defects in the upper regions of the embryo. Together with polar localization of PIN7 in suspensor cells (Friml et al., 2003; Robert et al., 2013), it is very likely that auxin biosynthesis and auxin polar transport together regulate the establishment of the apical-basal axis of the embryos.

18.4.1 Hypophysis Specification and RAM Formation

The RAM comprises the quiescent center (QC), specifically the SCOC, surrounded by stem cells and differentiating cells arranged in distinct cell files. The SCOC is responsible for the establishment and maintenance of the surrounding stem cell pool. As the precursor of the QC, the specification of lens-shaped hypophysis is the initiating step of RAM establishment in *Arabidopsis*. The expression of *WOX5*, a WUS family transcription factor, is expressed specifically in the hypophysis and QC and is required for the maintenance of the QC identity (Haecker et al., 2004; Pi et al., 2015; Sarkar et al., 2007; Scheres et al., 1994; Zhang et al., 2015). The specification of the hypophysis is regulated primarily by auxin signaling involving auxin biosynthesis, polar transport, perception, and response (De Smet et al., 2010). Although in many seed plants the presence of well-defined hypophysis and QC are not observed, a similar signaling machinery is evident (Hedman et al., 2013; Kamiya et al., 2003; Lim et al., 2005), indicating that the establishment of QC is due to its specific position within the root apical meristem rather than predetermined cell lineage (Palovaara et al., 2016).

Hypophysis presented a high level of auxin activity, indicated by the synthetic auxin response reporter constructs of *DR5::GUS* or *DR5::GFP*, due to auxin polar transport through polarized PIN1 protein localizations on cytoplasmic membranes (Friml et al., 2003). The localization of PIN1 protein at the basal membranes of provascular cells may have contributed to hypophysis specification by promoting the import of auxin into hypophysis (Lokerse & Weijers, 2009).

Furthermore, the transport of auxin into the hypophysis also depends on MP (Weijers et al., 2006), as embryos of *mp* mutants showed aberrant hypophysis specification (Hardtke & Berleth, 1998). Within vascular precursor cells, TMO7, one of the direct targets of MP, relocates to the hypophysis and regulates its division pattern. Down-regulation of *TMO7* showed aberrant hypophysis division, producing rootless seedlings (Schlereth et al., 2010). Mutations in another set of MP targets, the zinc finger family transcription factor NO TRANSMITTING TRACT (NTT) and its two paralogs WIP DOMAIN PROTEIN 4 (WIP4) and WIP5, also led to abnormal hypophysis divisions and rootless phenotypes (Crawford et al., 2015; Schlereth et al., 2010; Weijers et al., 2006), suggesting a critical role of NTT, WIP4, and WIP5 in hypophysis specification and division. In addition, other auxin response factors such as IAA10 and ARF9 are also key components in hypophysis specification (Rademacher et al., 2011, 2012).

Another set of genes underlying RAM establishment is the *PLTs*. *PLT1* and *PLT2* are expressed in the provascular and hypophysis cells. Expressions of these genes are responsive to auxin and dependent on MP activity but not directly regulated by MP (Aida et al., 2004; Galinha et al., 2007; Mahonen et al., 2014; Schlereth et al., 2010). *PLT3* and *BABY BOOM* (*BBM,* another paralog of PLT) are also expressed primarily in provascular and hypophysis cells from the heart-shaped stage onwards (Aida et al., 2004; Galinha et al., 2007). The *plt1/plt2/plt3/bbm* quadruple mutant showed aberrant RAM formation and rootless-seedling phenotypes (Galinha et al., 2007). These results together suggest that *PLT* genes regulate hypophysis specification and RAM establishment under the control of the positional signals generated by auxin accumulation. Their involvement adds further complexity to the molecular mechanisms orchestrating the intricate process of hypophysis specification within RAM.

Suspensor, which has been proposed to channel the maternal-to-embryo communication, is a terminal tissue that starts to enter programmed cell death (PCD) in late globular-stage embryos (Fig. 18.2). A recent study showed that a DELLA protein, NtCYS regulative factor 1 (NtCRF1), regulates the PCD of the suspensor. NtCRF1 binds to the promoter of NtCYS and regulates the suspensor PCD-switch module NtCYS-NtCP14 in response to gibberellin (GA). GA4 as a primary signal triggering suspensor PCD is generated in the micropylar endothelium by the transient activation of *NtGA3oxs* in the maternal tissue (Shi et al., 2019).

18.4.2 The Formation of SAM and Cotyledons

The formation of two cotyledons in a globular-stage embryo is regulated by auxin polar transport. Treatments of early globular-stage embryos of *Brassica juncea* (Indian mustard) in vitro with auxin polar transport inhibitors led to the formation of trumpet-shaped and radial symmetrical embryos with a fused cotyledon, instead of bilateral symmetrical embryos with two cotyledons. A similar phenotype was also observed when the *pin1* gene is mutated in *Arabidopsis* (Liu et al., 1993a), suggesting the critical importance of auxin polar transport in the formation of two cotyledons. In the globular-stage embryos, and in SAM as well, highly dynamic localizations of auxin efflux protein PIN1 direct the flow of auxin, allowing auxin to be transported and accumulated in cells from which new lateral organ primordia, such as cotyledons and leaves, are to be formed (Ni et al., 2001; Friml et al., 2003).

SAM, as a group of less vacuolated cells between two cotyledon primordia, is initiated at the late globular stage and formed at the torpedo-shaped stage (Barton & Poethig, 1993; Laux et al., 1996). Similar to *WOX5* in RAM, the funding member of the WOX family transcription factor, *WUSCHEL* (*WUS*), is recognized as a marker gene for SCOC in SAM and expressed first in four upper inner cells of dermatogen-staged embryos (Mayer et al., 1998). The *wus* mutant showed a reduced number of SAM stem cells in embryogenesis and further absence of SAM in seedlings, demonstrating the critical function of WUS in the maintenance of stem cells in SAM (Laux et al., 1996; Mayer et al., 1998). Mutations in *CLV3*, which encodes a small protein functioning as a peptide hormone, led to the formation of an enlarged SAM with an increased number of stem cells (Fletcher et al., 1999; Fiers et al., 2005). The CLV3 peptide hormone, produced in stem cells and perceived by the LRR family receptor kinase CLV1 and a receptor-like protein (without a kinase domain) CLV2, establishes a feedback regulatory loop with the WUS transcription factor in SCOC (Brand et al., 2000; for review, see Liu & Hu, 2010). Although mutants of either *WUS*, *CLV1*, *CLV2*, or *CLV3* caused severe defects in postembryonic development in SAM and/or inflorescence meristems, their phenotypes in embryogenesis were rather subtle, suggesting a major role of these genes in cell fate determination for postembryonic development instead of embryo patterning itself. The *KNOTTED1-like* homeobox (*KNOX*) gene *SHOOT MERISTEMLESS* (*STM*) was identified as another key regulator in SAM establishment during embryogenesis (Long et al., 1996). From the globular stage onwards, *STM* is expressed in cells destined for the formation of primary SAM and persists in all SAMs during postembryonic development (Long et al., 1996).

Two NAC family transcription factors, *CUP-SHAPED COTYLEDON 1* (*CUC1*) and *CUC2*, are expressed redundantly at the boundary of SAM and cotyledons and promote the expression of *STM* in the central domain. Conversely, STM downregulates expressions of *CUC1* and *CUC2* in a negative feedback regulatory loop and restricts expressions of *CUCs* in a ring of cells located at the upper tier of the embryo (Aida et al., 1997, 1999; Hibara et al., 2006; Tanaka et al., 2001; Vroemen et al., 2003). Single *cuc* mutants did not show a visible phenotype, whereas *cuc1/*

cuc2 double mutants displayed fused and cup-shaped cotyledons and a partial loss of SAM. *STM* expression was not detected in the *cuc1/cuc2* double mutant but was enhanced when *CUC1* was overexpressed (Aida et al., 1997, 1999; Tanaka et al., 2001). The *stm* mutant lacks a functional SAM due to ectopic expression of the MYB domain gene *ASYMMETRIC LEAVES1* (*AS1*), which is normally confined to organ primordia (Byrne et al., 2000). A mutation in *STM* also led to disturbance of *CUC2* expression during late embryogenesis (Aida et al., 1999) and resulted in a SAM phenotype similar to the *cuc1/cuc2* double mutant (Barton & Poethig, 1993). Thus, the coordination between *CUC* and *STM* defines the boundary between the SAM and the cotyledons.

In addition, an MYB family transcription factor, AS1, and a LOB family transcription factor, AS2, are another pair of factors involved in SAM establishment. AS1 and AS2 negatively regulate the function of STM in SAM. *AS1* and *AS2* are expressed asymmetrically in primordia of cotyledons and lateral organs. Mutations in either *AS1* or *AS2* led to a similar phenotype, and both could restore the SAM phenotype of the *stm* mutant (Byrne et al., 2000, 2002; Iwakawa et al., 2002; Ori et al., 2000). As such, STM appears to act antagonistically with AS1 and AS2 to regulate the SAM formation. JAGGED LATERAL ORGANS (JLO), a member of the LATERAL ORGAN BOUNDARY DOMAIN (LBD) family, is required for coordinated organ development in SAM and floral meristems. Loss of *JLO* function resulted in ectopic expression of the *STM* and another KNOX family gene, *BREVIPEDICELLUS* (*BP*), indicating that JLO restricts expression of these *KNOX* genes. JLO acts in a trimeric protein complex with AS2 and AS1 to suppress *BP* expression in lateral organs (Rast & Simon, 2012).

Several class III homeodomain-leucine zipper (HD-ZIP III) family transcription factors, REVOLUTA, PHABULOSA (PHB), PHAVOLUTA (PHV), CORONA (CNA), and ARABIDOPSIS THALIANA HOMEOBOX 8 (ATHB8), are also indispensable for the establishment of the SAM (Prigge et al., 2005). Although single mutants of these genes did not show any visible phenotype in SAM, different mutant combinations of these genes displayed varying defects in SAM establishment and/or cotyledon formation, indicating overlapping functions of these genes (Emery et al., 2003; Prigge et al., 2005). Further, a member of the ARGONAUTE (AGO) family protein, ZWILLE (ZLL), is required to positively regulate these *HD-ZIPIII* genes in SAM by restricting the accumulation of miR165/166 that are negative regulators of these *HD-ZIPIII* genes (Liu et al., 2009; Mallory & Vaucheret, 2006). Since expression of *ZLL* is weak in SAM and strong in adjacent vascular precursor cells, *ZLL* may function in a non-cell-autonomous manner to control the SAM establishment in early embryogenesis (Lynn et al., 1999; Moussian et al., 1998; Tucker et al., 2008).

Additionally, HD-ZIP III genes also exert their roles on SAM establishment by regulating *HD-ZIP II* genes such as *ATHB2*, *ATHB4*, *HOMEOBOX ARABIDOPSIS THALIANA 1* (*HAT1*), *HAT2*, and *HAT3*. Expressions of these *HD-ZIP II* genes overlap with those of *HD-ZIP III* genes during early embryogenesis. A *hat3/athb2/athb4* triple mutant showed defects in SAM, which are further enhanced by introduced *HD-ZIP III* mutants of either *phv*, *phb,* or *rev* (Turchi et al., 2013, 2015).

These *HD-ZIP II* genes are also targeted directly by REV (Brandt et al., 2012; Reinhart et al., 2013), suggesting that these *HD-ZIP II* and *III* family genes act in harmony in coordinating the refinement of SAM.

18.5 Future Prospective

Compared with mammals, mature embryos in plants are substantially less complex, as many tissues and organs in adult plants are not formed in embryos. However, embryogenesis in plants establishes the basic body plan, with the formation of an apical-basal organ pattern consisting of cotyledons, epicotyl/hypocotyl, SAM and RAM, and a radial tissue pattern consisting of epidermis, ground tissues, and provasculature. Such an embryo provides a foundation for postembryonic growth and development, allowing for lifelong organ and tissue formations over hundreds and thousands of years. In this chapter, we have provided a general overview of cell division, cell fate specification, and the formation of tissue and organ patterns during plant embryogenesis, mostly through molecular genetic studies in *Arabidopsis*. We also updated our most recent understanding of the genetic network underlying the initiation and establishment of tissue and organ patterns in *Arabidopsis* embryos. Having acquired this knowledge, we are still very far from a complete understanding of plant embryogenesis. Many key questions in the field remain to be addressed in the future.

1. Early embryogenesis in animals relies largely on maternal deposits, including proteins and transcripts. After fertilization, maternal factors are progressively cleared, while de novo expressions of the zygote genome are gradually activated. This process is defined as the maternal-to-zygotic transition (MZT). In plants, two models have been proposed regarding the extent to which maternal and paternal transcripts contribute to early embryogenesis. One suggests that maternal and paternal alleles contribute equally to the zygotic embryonic transcriptome (Ingouff et al., 2010; Meyer & Scholten, 2007; Nodine & Bartel, 2012; Weijers et al., 2001). Data from genetic analyses showed the presence of a relatively large number of zygote-lethal mutants (Liu & Meinke, 1998; Fan et al., 2008), and most embryo-lethal mutants are inherited in a Mendelian manner (Meinke et al., 2008), which appears to support this model. Another model proposes that, similar to the findings in animals, most transcripts in the zygote are derived from maternal alleles (Autran et al., 2011; Pillot et al., 2010). However, RNA sequencing performed in proembryos has revealed significant variations in maternal and paternal transcripts in hybrids made between different *Arabidopsis* ecotypes, indicating a conflict between equal parental contributions and primary maternal contributions (Del Toro-De Leon et al., 2014). Single-cell transcriptomic analyses in dissected gametes and proembryos may help to decipher these discrepancies (Del Toro-De Leon et al., 2016).

2. Although the highly ordered cell division and tissue patterning in *Arabidopsis* embryogenesis have been beneficial for image analyses (Gooh et al., 2015) and computational modeling, leading to the identification of numerous key factors involved in cell specification and tissue patterning by genetic analyses. It is as yet unclear how the division pattern of individual cells is determined. For instance, are cell plate orientation, directional cell elongations, and cell fate specification regulated by cell lineage, positional information, or both? Although data obtained to date support the suggestion that tissue and organ establishment during embryogenesis is dependent on auxin, cytokinin, and peptide hormones (Friml et al., 2003; Ohashi-Ito et al., 2014; Xu et al., 2015), functional elucidations on how these signal molecules act in harmony to execute cellular actions may help to answer these questions.
3. In many non-cruciferous plants, such as cotton and pea (Liu et al., 1995, 1996), a highly ordered cell division pattern is not present, which does not affect the formation of an elegant embryo. Thus, ordered cell division is not essential for the formation of plant embryos. Although studies of embryogenesis in these species are more challenging, they may provide better and more important answers for the machinery underlying plant embryogenesis.
4. In all angiosperm species, two types of embryos are formed, monocotyledonous and dicotyledonous embryos, with a SAM and a RAM. The morphologies of their SAM are very similar, while their postembryonic morphologies in leaves, stems, flowers, and fruits were very different. How these differences were determined and defined remains to be answered.

References

Abe, M., Takahashi, T., & Komeda, Y. (2001). Identification of a cis-regulatory element for L1 layer-specific gene expression, which is targeted by an L1-specific homeodomain protein. *The Plant Journal, 26*, 487–494.

Abe, M., Katsumata, H., Komeda, Y., & Takahashi, T. (2003). Regulation of shoot epidermal cell differentiation by a pair of homeodomain proteins in Arabidopsis. *Development, 130*, 635–643.

Aida, M., Ishida, T., Fukaki, H., Fujisawa, H., & Tasaka, M. (1997). Genes involved in organ separation in Arabidopsis: An analysis of the cup-shaped cotyledon mutant. *The Plant Cell, 9*, 841–857.

Aida, M., Ishida, T., & Tasaka, M. (1999). Shoot apical meristem and cotyledon formation during Arabidopsis embryogenesis: Interaction among the *CUP-SHAPED COTYLEDON* and *SHOOT MERISTEMLESS* genes. *Development, 126*, 1563–1570.

Aida, M., Beis, D., Heidstra, R., Willemsen, V., Blilou, I., Galinha, C., Nussaume, L., Noh, Y. S., Amasino, R., & Scheres, B. (2004). The *PLETHORA* genes mediate patterning of the *Arabidopsis* root stem cell niche. *Cell, 119*, 109–120.

Anderson, S. N., Johnson, C. S., Chesnut, J., Jones, D. S., Khanday, I., Woodhouse, M., Li, C., Conrad, L. J., Russell, S. D., & Sundaresan, V. (2017). The zygotic transition is initiated in unicellular plant zygotes with asymmetric activation of parental genomes. *Developmental Cell, 43*, 349–358.

Andreuzza, S., Li, J., Guitton, A. E., Faure, J. E., Casanova, S., Park, J. S., Choi, Y., Chen, Z., & Berger, F. (2010). DNA LIGASE I exerts a maternal effect on seed development in *Arabidopsis thaliana*. *Development, 137*, 73–81.

Autran, D., Baroux, C., Raissig, M. T., Lenormand, T., Wittig, M., Grob, S., Steimer, A., Barann, M., Klostermeier, U. C., Leblanc, O., et al. (2011). Maternal epigenetic pathways control parental contributions to Arabidopsis early embryogenesis. *Cell, 145*, 707–719.

Babu, Y., Musielak, T., Henschen, A., & Bayer, M. (2013). Suspensor length determines developmental progression of the embryo in Arabidopsis. *Plant Physiology, 162*, 1448–1458.

Barton, M. K., & Poethig, R. S. (1993). Formation of the shoot apical meristem in Arabidopsis thaliana: An analysis of development in the wild type and in the *shoot meristemless* mutant. *Development, 119*, 823–831.

Bayer, M., Nawy, T., Giglione, C., Galli, M., Meinnel, T., & Lukowitz, W. (2009). Paternal control of embryonic patterning in *Arabidopsis thaliana*. *Science, 323*, 1485–1488.

Bleckmann, A., Alter, S., & Dresselhaus, T. (2014). The beginning of a seed: Regulatory mechanisms of double fertilization. *Frontiers in Plant Science, 5*, 452.

Borg, M., Jacob, Y., Susaki, D., LeBlanc, C., Buendia, D., Axelsson, E., Kawashima, T., Voigt, P., Boavida, L., Becker, J., et al. (2020). Targeted reprogramming of H3K27me3 resets epigenetic memory in plant paternal chromatin. *Nature Cell Biology, 22*, 621–629.

Brand, U., Fletcher, J. C., Hobe, M., Meyerowitz, E. M., & Simon, R. (2000). Dependence of stem cell fate in Arabidopsis on a feedback loop regulated by CLV3 activity. *Science, 289*, 617–619.

Brandt, R., Salla-Martret, M., Bou-Torrent, J., Musielak, T., Stahl, M., Lanz, C., Ott, F., Schmid, M., Greb, T., Schwarz, M., et al. (2012). Genome-wide binding-site analysis of REVOLUTA reveals a link between leaf patterning and light-mediated growth responses. *The Plant Journal, 72*, 31–42.

Byrne, M. E., Barley, R., Curtis, M., Arroyo, J. M., Dunham, M., Hudson, A., & Martienssen, R. A. (2000). Asymmetric leaves1 mediates leaf patterning and stem cell function in *Arabidopsis*. *Nature, 408*, 967–971.

Byrne, M. E., Simorowski, J., & Martienssen, R. A. (2002). *ASYMMETRIC LEAVES1* reveals *knox* gene redundancy in *Arabidopsis*. *Development, 129*, 1957–1965.

Cheng, Y., Dai, X., & Zhao, Y. (2007). Auxin synthesized by the YUCCA Flavin monooxygenases is essential for embryogenesis and leaf formation in Arabidopsis. *Plant Cell, 19*, 2430–2439.

Costa, L. M., Marshall, E., Tesfaye, M., Silverstein, K. A., Mori, M., Umetsu, Y., Otterbach, S. L., Papareddy, R., Dickinson, H. G., Boutiller, K., et al. (2014). Central cell-derived peptides regulate early embryo patterning in flowering plants. *Science, 344*, 168–172.

Crawford, B. C., Sewell, J., Golembeski, G., Roshan, C., Long, J. A., & Yanofsky, M. F. (2015). Plant development. Genetic control of distal stem cell fate within root and embryonic meristems. *Science, 347*, 655–659.

De Rybel, B., Moller, B., Yoshida, S., Grabowicz, I., Barbier de Reuille, P., Boeren, S., Smith, R. S., Borst, J. W., & Weijers, D. (2013). A bHLH complex controls embryonic vascular tissue establishment and indeterminate growth in *Arabidopsis*. *Developmental Cell, 24*, 426–437.

De Rybel, B., Adibi, M., Breda, A. S., Wendrich, J. R., Smit, M. E., Novak, O., Yamaguchi, N., Yoshida, S., Van Isterdael, G., Palovaara, J., et al. (2014). Plant development. Integration of growth and patterning during vascular tissue formation in *Arabidopsis*. *Science, 345*, 1255215.

De Rybel, B., Mahonen, A. P., Helariutta, Y., & Weijers, D. (2016). Plant vascular development: From early specification to differentiation. *Nature Reviews Molecular Cell Biology, 17*, 30–40.

De Smet, I., Lau, S., Mayer, U., & Jurgens, G. (2010). Embryogenesis–The humble beginnings of plant life. *The Plant Journal, 61*, 959–970.

De Vries, S. C., & Weijers, D. (2017). Plant embryogenesis. *Current Biology, 27*, R870–R873.

Del Toro-De Leon, G., Garcia-Aguilar, M., & Gillmor, C. S. (2014). Non-equivalent contributions of maternal and paternal genomes to early plant embryogenesis. *Nature, 514*, 624–627.

Del Toro-De Leon, G., Lepe-Soltero, D., & Gillmor, C. S. (2016). Zygotic genome activation in isogenic and hybrid plant embryos. *Current Opinion in Plant Biology, 29*, 148–153.

Di Laurenzio, L., Wysocka-Diller, J., Malamy, J. E., Pysh, L., Helariutta, Y., Freshour, G., Hahn, M. G., Feldmann, K. A., & Benfey, P. N. (1996). The *SCARECROW* gene regulates an asymmetric cell division that is essential for generating the radial organization of the *Arabidopsis* root. *Cell, 86*, 423–433.

Dresselhaus, T., & Jürgens, G. (2021). Comparative embryogenesis in angiosperms: Activation and patterning of embryonic cell lineages. *Annual Review of Plant Biology, 72*, 641–676.

Drews, G. N., & Koltunow, A. M. (2011). The female gametophyte. *The Arabidopsis Book, 9*, e0155.

Dumas, C., & Rogowsky, P. (2008). Fertilization and early seed formation. *Comptes Rendus Biologies, 331*, 715–725.

el-Showk, S., Help-Rinta-Rahko, H., Blomster, T., Siligato, R., Maree, A. F., Mahonen, A. P., & Grieneisen, V. A. (2015). Parsimonious model of vascular patterning links transverse hormone fluxes to lateral root initiation: Auxin leads the way, while cytokinin levels out. *PLoS Computational Biology, 11*, e1004450.

Emery, J. F., Floyd, S. K., Alvarez, J., Eshed, Y., Hawker, N. P., Izhaki, A., Baum, S. F., & Bowman, J. L. (2003). Radial patterning of Arabidopsis shoots by class III *HD-ZIP* and *KANADI* genes. *Current Biology, 13*, 1768–1774.

Fan, Y. F., Jiang, L., Gong, H. Q., & Liu, C. M. (2008). Sexual reproduction in higher plants I: Fertilization and the initiation of zygotic program. *Journal of Integrative Plant Biology, 50*, 860–867.

Faure, J. E., Rotman, N., Fortune, P., & Dumas, C. (2002). Fertilization in *Arabidopsis thaliana* wild type: Developmental stages and time course. *The Plant Journal, 30*, 481–488.

Fiers, M., Golemiec, E., Xu, J., van der Geest, L., Heidstra, R., Stiekema, W., & Liu, C. M. (2005). The 14-amino acid CLV3, CLE19, and CLE40 peptides trigger consumption of the root meristem in Arabidopsis through a CLAVATA2-dependent pathway. *The Plant Cell, 9*, 2542–2553.

Fletcher, J. C., Brand, U., Running, M. P., Simon, R., & Meyerowitz, E. M. (1999). Signaling of cell fate decisions by CLAVATA3 in Arabidopsis shoot meristems. *Science, 283*, 1911–1914.

Friml, J., Vieten, A., Sauer, M., Weijers, D., Schwarz, H., Hamann, T., Offringa, R., & Jurgens, G. (2003). Efflux-dependent auxin gradients establish the apical-basal axis of Arabidopsis. *Nature, 426*, 147–153.

Galinha, C., Hofhuis, H., Luijten, M., Willemsen, V., Blilou, I., Heidstra, R., & Scheres, B. (2007). PLETHORA proteins as dose-dependent master regulators of Arabidopsis root development. *Nature, 449*, 1053–1057.

Gooh, K., Ueda, M., Aruga, K., Park, J., Arata, H., Higashiyama, T., & Kurihara, D. (2015). Live-cell imaging and optical manipulation of Arabidopsis early embryogenesis. *Developmental Cell, 34*, 242–251.

Guo, L., Jiang, L., Zhang, Y., Lu, X. L., Xie, Q., Weijers, D., & Liu, C. M. (2016). The anaphase-promoting complex initiates zygote division in *Arabidopsis* through degradation of cyclin B1. *The Plant Journal, 86*, 161–174.

Guo, L., Jiang, L., Lu, X. L., & Liu, C. M. (2018). ANAPHASE PROMOTING COMPLEX/CYCLOSOME-mediated cyclin B1 degradation is critical for cell cycle synchronization in syncytial endosperms. *Journal of Integrative Plant Biology, 60*, 448–454.

Haecker, A., Gross-Hardt, R., Geiges, B., Sarkar, A., Breuninger, H., Herrmann, M., & Laux, T. (2004). Expression dynamics of *WOX* genes mark cell fate decisions during early embryonic patterning in *Arabidopsis thaliana*. *Development, 131*, 657–668.

Hamann, T., Mayer, U., & Jurgens, G. (1999). The auxin-insensitive bodenlos mutation affects primary root formation and apical-basal patterning in the *Arabidopsis* embryo. *Development, 126*, 1387–1395.

Hanzawa, Y., Takahashi, T., Michael, A. J., Burtin, D., Long, D., Pineiro, M., Coupland, G., & Komeda, Y. (2000). ACAULIS5, an Arabidopsis gene required for stem elongation, encodes a spermine synthase. *The EMBO Journal, 19*, 4248–4256.

Hanzawa, Y., Takahashi, T. F., Komeda, Y., & Komeda, Y. (2010). ACL5: An Arabidopsis gene required for internodal elongation after flowering. *The Plant Journal, 12*, 863–874.

Hardtke, C. S., & Berleth, T. (1998). The Arabidopsis gene MONOPTEROS encodes a transcription factor mediating embryo axis formation and vascular development. *The EMBO Journal, 17*, 1405–1411.

Hedman, H., Zhu, T., von Arnold, S., & Sohlberg, J. J. (2013). Analysis of the *WUSCHEL-RELATED HOMEOBOX* gene family in the conifer picea abies reveals extensive conservation as well as dynamic patterns. *BMC Plant Biology, 13*, 89.

Heidstra, R., Welch, D., & Scheres, B. (2004). Mosaic analyses using marked activation and deletion clones dissect Arabidopsis SCARECROW action in asymmetric cell division. *Genes & Development, 18*, 1964–1969.

Helariutta, Y., Fukaki, H., Wysocka-Diller, J., Nakajima, K., Jung, J., Sena, G., Hauser, M. T., & Benfey, P. N. (2000). The *SHORT-ROOT* gene controls radial patterning of the Arabidopsis root through radial signaling. *Cell, 101*, 555–567.

Hibara, K., Karim, M. R., Takada, S., Taoka, K., Furutani, M., Aida, M., & Tasaka, M. (2006). Arabidopsis CUP-SHAPED COTYLEDON3 regulates postembryonic shoot meristem and organ boundary formation. *The Plant Cell, 18*, 2946–2957.

Hou, X. L., Chen, W. Q., Hou, Y. F., Gong, H. Q., Sun, J., Wang, Z., Zhao, H., Cao, X. F., Song, X. F., & Liu, C. M. (2021). DEAD-BOX RNA HELICASE 27 regulates microRNA biogenesis, zygote division and stem cell homeostasis. *The Plant Cell, 33*, 66–84.

Imai, A., Hanzawa, Y., Komura, M., Yamamoto, K. T., Komeda, Y., & Takahashi, T. (2006). The dwarf phenotype of the Arabidopsis acl5 mutant is suppressed by a mutation in an upstream ORF of a bHLH gene. *Development, 133*, 3575–3585.

Ingouff, M., Rademacher, S., Holec, S., Soljic, L., Xin, N., Readshaw, A., Foo, S. H., Lahouze, B., Sprunck, S., & Berger, F. (2010). Zygotic resetting of the HISTONE 3 variant repertoire participates in epigenetic reprogramming in Arabidopsis. *Current Biology, 20*, 2137–2143.

Ingram, G. C., Magnard, J. L., Vergne, P., Dumas, C., & Rogowsky, P. M. (1999). *ZmOCL1*, an *HDGL2* family homeobox gene, is expressed in the outer cell layer throughout maize development. *Plant Molecular Biology, 40*, 343–354.

Ingram, G. C., Boisnard-Lorig, C., Dumas, C., & Rogowsky, P. M. (2000). Expression patterns of genes encoding HD-ZipIV homeo domain proteins define specific domains in maize embryos and meristems. *The Plant Journal, 22*, 401–414.

Ito, M., Sentoku, N., Nishimura, A., Hong, S. K., Sato, Y., & Matsuoka, M. (2002). Position dependent expression of GL2-type homeobox gene, Roc1: Significance for protoderm differentiation and radial pattern formation in early rice embryogenesis. *The Plant Journal, 29*, 497–507.

Iwakawa, H., Ueno, Y., Semiarti, E., Onouchi, H., Kojima, S., Tsukaya, H., Hasebe, M., Soma, T., Ikezaki, M., Machida, C., et al. (2002). The *ASYMMETRIC LEAVES2* gene of Arabidopsis thaliana, required for formation of a symmetric flat leaf lamina, encodes a member of a novel family of proteins characterized by cysteine repeats and a leucine zipper. *Plant & Cell Physiology, 43*, 467–478.

Jeong, S., Palmer, T. M., & Lukowitz, W. (2011). The RWP-RK factor GROUNDED promotes embryonic polarity by facilitating YODA MAP kinase signaling. *Current Biology, 21*, 1268–1276.

Johnson, K. L., Degnan, K. A., Ross Walker, J., & Ingram, G. C. (2005). AtDEK1 is essential for specification of embryonic epidermal cell fate. *The Plant Journal, 44*, 114–127.

Johri, B. M. (1984). *Embryology of angiosperms*. Springer-Verlag.

Johri, B. M., Ambegaokar, K. B., & Srivastava, P. S. (1992). *Comparative embryology of angiosperms*. Springer Berlin Heidelberg.

Kamiya, N., Nagasaki, H., Morikami, A., Sato, Y., & Matsuoka, M. (2003). Isolation and characterization of a rice *WUSCHEL*-type homeobox gene that is specifically expressed in the central cells of a quiescent center in the root apical meristem. *The Plant Journal, 35*, 429–441.

Katayama, H., Iwamoto, K., Kariya, Y., Asakawa, T., Kan, T., Fukuda, H., & Ohashi-Ito, K. (2015). A negative feedback loop controlling bHLH complexes is involved in vascular cell division and differentiation in the root apical meristem. *Current Biology, 25*, 3144–3150.

Kawashima, T., & Goldberg, R. B. (2010). The suspensor: Not just suspending the embryo. *Trends in Plant Science, 15*, 23–30.

Khanday, I., Skinner, D., Yang, B., Mercier, R., & Sundaresan, V. (2019). A male-expressed rice embryogenic trigger redirected for asexual propagation through seeds. *Nature, 565*, 91–95.

Kimata, Y., Higaki, T., Kawashima, T., Kurihara, D., Sato, Y., Yamada, T., Hasezawa, S., Berger, F., Higashiyama, T., & Ueda, M. (2016). Cytoskeleton dynamics control the first asymmetric cell division in Arabidopsis zygote. *Proceedings of the National Academy of Sciences of the United States of America, 113*, 14157–14162.

Kimata, Y., Kato, T., Higaki, T., Kurihara, D., Yamada, T., Segami, S., Morita, M. T., Maeshima, M., Hasezawa, S., Higashiyama, T., et al. (2019). Polar vacuolar distribution is essential for accurate asymmetric division of Arabidopsis zygotes. *Proceedings of the National Academy of Sciences of the United States of America, 116*, 2338–2343.

Kuroha, T., Tokunaga, H., Kojima, M., Ueda, N., Ishida, T., Nagawa, S., Fukuda, H., Sugimoto, K., & Sakakibara, H. (2009). Functional analyses of LONELY GUY cytokinin-activating enzymes reveal the importance of the direct activation pathway in Arabidopsis. *The Plant Cell, 21*, 3152–3169.

Lau, S., Slane, D., Herud, O., Kong, J., & Jürgens, G. (2012). Early embryogenesis in flowering plants: Setting up the basic body pattern. *Annual Review of Plant Biology, 63*, 483–506.

Laux, T., Mayer, K. F., Berger, J., & Jurgens, G. (1996). The *WUSCHEL* gene is required for shoot and floral meristem integrity in Arabidopsis. *Development, 122*, 87–96.

Levesque, M. P., Vernoux, T., Busch, W., Cui, H., Wang, J. Y., Blilou, I., Hassan, H., Nakajima, K., Matsumoto, N., Lohmann, J. U., et al. (2006). Whole-genome analysis of the SHORT-ROOT developmental pathway in Arabidopsis. *PLoS Biology, 4*, e143.

Li, H. J., Liu, N. Y., Shi, D. Q., Liu, J., & Yang, W. C. (2010). YAO is a nucleolar WD40-repeat protein critical for embryogenesis and gametogenesis in Arabidopsis. *BMC Plant Biology, 10*, 169.

Li, S. P., Chen, M., Yu, D., Ren, S. C., Sun, S., Liu, L., Ketelaar, T., Emons, A. M., & Liu, C. M. (2013). EXO70A1-mediated vesicle trafficking is critical for tracheary element development in Arabidopsis. *The Plant Cell, 25*, 1774–1786.

Lim, J., Jung, J. W., Lim, C. E., Lee, M. H., Kim, B. J., Kim, M., Bruce, W. B., & Benfey, P. N. (2005). Conservation and diversification of SCARECROW in maize. *Plant Molecular Biology, 59*, 619–630.

Lin, Z., Yin, K., Zhu, D., Chen, Z., Gu, H., & Qu, L. J. (2007). AtCDC5 regulates the G2 to M transition of the cell cycle and is critical for the function of Arabidopsis shoot apical meristem. *Cell Research, 17*, 815–828.

Liu, C. M., & Hu, Y. (2010). Plant stem cells and their regulations in shoot apical meristem. *Frontiers of Biology, 5*, 417–423.

Liu, C. M., & Meinke, D. W. (1998). The *titan* mutants of Arabidopsis are disrupted in mitosis and cell cycle control during seed development. *The Plant Journal, 16*, 21–31.

Liu, C. M., Xu, Z. H., & Chua, N. H. (1993a). Auxin polar transport is essential for the establishment of bilateral symmetry during early plant embryogenesis. *The Plant Cell, 5*, 621–630.

Liu, C. M., Xu, Z. H., & Chua, N. H. (1993b). Proembryo culture: In vitro development of early globular-staged zygotic embryos of *Brassica juncea*. *The Plant Journal, 3*, 291–300.

Liu, C. M., Johnson, S., & Wang, T. L. (1995). Cyd, a mutant of pea that alters embryo morphology is defective in cytokinesis. *Developmental Genetics, 16*, 321–331.

Liu, C. M., Johnson, S., Hedley, C. L., & Wang, T. L. (1996). The generation of a legume embryo: Morphological and cellular defective in pea mutants. In T. L. Wang (Ed.), *Embryogenesis: The generation of a plant* (pp. 191–213). Taylor & Francis.

Liu, Q., Yao, X., Pi, L., Wang, H., Cui, X., & Huang, H. (2009). The *ARGONAUTE10* gene modulates shoot apical meristem maintenance and establishment of leaf polarity by repressing miR165/166 in Arabidopsis. *The Plant Journal, 58*, 27–40.

Lokerse, A. S., & Weijers, D. (2009). Auxin enters the matrix--assembly of response machineries for specific outputs. *Current Opinion in Plant Biology, 12*, 520–526.

Long, J. A., Moan, E. I., Medford, J. I., & Barton, M. K. (1996). A member of the KNOTTED class of homeodomain proteins encoded by the *STM* gene of *Arabidopsis*. *Nature, 379*, 66–69.
Lu, P., Porat, R., Nadeau, J. A., & O'Neill, S. D. (1996). Identification of a meristem L1 layer-specific gene in Arabidopsis that is expressed during embryonic pattern formation and defines a new class of homeobox genes. *The Plant Cell, 8*, 2155–2168.
Lukowitz, W., Roeder, A., Parmenter, D., & Somerville, C. (2004). A *MAPKK* kinase gene regulates extra-embryonic cell fate in *Arabidopsis*. *Cell, 116*, 109–119.
Lynn, K., Fernandez, A., Aida, M., Sedbrook, J., Tasaka, M., Masson, P., & Barton, M. K. (1999). The PINHEAD/ZWILLE gene acts pleiotropically in Arabidopsis development and has overlapping functions with the *ARGONAUTE1* gene. *Development, 126*, 469–481.
Maheshwari, P., & Johri, B. M. (1950). Development of the embryo sac, embryo and endosperm in *Helixanthera ligustrina* (wall.) dans. *Nature, 165*, 978–979.
Mahonen, A. P., Ten Tusscher, K., Siligato, R., Smetana, O., Diaz-Trivino, S., Salojarvi, J., Wachsman, G., Prasad, K., Heidstra, R., & Scheres, B. (2014). PLETHORA gradient formation mechanism separates auxin responses. *Nature, 515*, 125–129.
Mallory, A. C., & Vaucheret, H. (2006). Functions of microRNAs and related small RNAs in plants. *Nature Genetics, 38*(Suppl), S31–S36.
Mansfield, S. G., & Briarty, L. G. (1991). Early embryogenesis in *Arabidopsis thaliana*. II. The developing embryo. *Canadian Journal of Botany, 69*, 461–476.
Mansfield, S. G., Briarty, L. G., & Erni, S. (1991). Early embryogenesis in *Arabidopsis thaliana*. I. The mature embryo sac. *Canadian Journal of Botany, 69*, 447–460.
Mayer, K. F., Schoof, H., Haecker, A., Lenhard, M., Jurgens, G., & Laux, T. (1998). Role of WUSCHEL in regulating stem cell fate in the *Arabidopsis* shoot meristem. *Cell, 95*, 805–815.
Meinke, D., Muralla, R., Sweeney, C., & Dickerman, A. (2008). Identifying essential genes in Arabidopsis thaliana. *Trends in Plant Science, 13*, 483–491.
Meyer, S., & Scholten, S. (2007). Equivalent parental contribution to early plant zygotic development. *Current Biology, 17*, 1686–1691.
Moussian, B., Schoof, H., Haecker, A., Jurgens, G., & Laux, T. (1998). Role of the *ZWILLE* gene in the regulation of central shoot meristem cell fate during *Arabidopsis* embryogenesis. *The EMBO Journal, 17*, 1799–1809.
Muraro, D., Mellor, N., Pound, M. P., Help, H., Lucas, M., Chopard, J., Byrne, H. M., Godin, C., Hodgman, T. C., King, J. R., et al. (2014). Integration of hormonal signaling networks and mobile microRNAs is required for vascular patterning in Arabidopsis roots. *Proceedings of the National Academy of Sciences of the United States of America, 111*, 857–862.
Mylona, P., Linstead, P., Martienssen, R., & Dolan, L. (2002). SCHIZORIZA controls an asymmetric cell division and restricts epidermal identity in the Arabidopsis root. *Development, 129*, 4327–4334.
Nakajima, K., Sena, G., Nawy, T., & Benfey, P. N. (2001). Intercellular movement of the putative transcription factor SHR in root patterning. *Nature, 413*, 307–311.
Ni, D. A., Wang, L. J., Ding, C. H., & Xu, Z. H. (2001). Auxin distribution and transport during embryogenesis and seed germination of *Arabidopsis*. *Cell Research, 11*, 273–278.
Nodine, M. D., & Bartel, D. P. (2012). Maternal and paternal genomes contribute equally to the transcriptome of early plant embryos. *Nature, 482*, 94–97.
Nodine, M. D., Yadegari, R., & Tax, F. E. (2007). RPK1 and TOAD2 are two receptor-like kinases redundantly required for arabidopsis embryonic pattern formation. *Developmental Cell, 12*, 943–956.
Ogawa, E., Yamada, Y., Sezaki, N., Kosaka, S., Kondo, H., Kamata, N., Abe, M., Komeda, Y., & Takahashi, T. (2015). ATML1 and PDF2 play a redundant and essential role in Arabidopsis embryo development. *Plant & Cell Physiology, 56*, 1183–1192.
Ohashi-Ito, K., Matsukawa, M., & Fukuda, H. (2013). An atypical bHLH transcription factor regulates early xylem development downstream of auxin. *Plant & Cell Physiology, 54*, 398–405.
Ohashi-Ito, K., Saegusa, M., Iwamoto, K., Oda, Y., Katayama, H., Kojima, M., Sakakibara, H., & Fukuda, H. (2014). A bHLH complex activates vascular cell division via cytokinin action in root apical meristem. *Current Biology, 24*, 2053–2058.

Ori, N., Eshed, Y., Chuck, G., Bowman, J. L., & Hake, S. (2000). Mechanisms that control knox gene expression in the Arabidopsis shoot. *Development, 127*, 5523–5532.

Palovaara, J., de Zeeuw, T., & Weijers, D. (2016). Tissue and organ initiation in the plant embryo: A first time for everything. *Annual Review of Cell and Developmental Biology, 32*, 47–75.

Pauluzzi, G., Divol, F., Puig, J., Guiderdoni, E., Dievart, A., & Perin, C. (2012). Surfing along the root ground tissue gene network. *Developmental Biology, 365*, 14–22.

Pernas, M., Ryan, E., & Dolan, L. (2010). SCHIZORIZA controls tissue system complexity in plants. *Current Biology, 20*, 818–823.

Peterson, K. M., Shyu, C., Burr, C. A., Horst, R. J., Kanaoka, M. M., Omae, M., Sato, Y., & Torii, K. U. (2013). Arabidopsis homeodomain-leucine zipper IV proteins promote stomatal development and ectopically induce stomata beyond the epidermis. *Development, 140*, 1924–1935.

Petricka, J. J., Van Norman, J. M., & Benfey, P. N. (2009). Symmetry breaking in plants: Molecular mechanisms regulating asymmetric cell divisions in Arabidopsis. *Cold Spring Harbor Perspectives in Biology, 1*, a000497.

Pi, L., Aichinger, E., van der Graaff, E., Llavata-Peris, C. I., Weijers, D., Hennig, L., Groot, E., & Laux, T. (2015). Organizer-derived WOX5 signal maintains root columella stem cells through chromatin-mediated repression of *CDF4* expression. *Developmental Cell, 33*, 576–588.

Pillot, M., Baroux, C., Vazquez, M. A., Autran, D., Leblanc, O., Vielle-Calzada, J. P., Grossniklaus, U., & Grimanelli, D. (2010). Embryo and endosperm inherit distinct chromatin and transcriptional states from the female gametes in Arabidopsis. *The Plant Cell, 22*, 307–320.

Prigge, M. J., Otsuga, D., Alonso, J. M., Ecker, J. R., Drews, G. N., & Clark, S. E. (2005). Class III homeodomain-leucine zipper gene family members have overlapping, antagonistic, and distinct roles in Arabidopsis development. *The Plant Cell, 17*, 61–76.

Rademacher, E. H., Moller, B., Lokerse, A. S., Llavata-Peris, C. I., van den Berg, W., & Weijers, D. (2011). A cellular expression map of the Arabidopsis *AUXIN RESPONSE FACTOR* gene family. *The Plant Journal, 68*, 597–606.

Rademacher, E. H., Lokerse, A. S., Schlereth, A., Llavata-Peris, C. I., Bayer, M., Kientz, M., Freire Rios, A., Borst, J. W., Lukowitz, W., Jurgens, G., et al. (2012). Different auxin response machineries control distinct cell fates in the early plant embryo. *Developmental Cell, 22*, 211–222.

Rast, M., & Simon, R. (2012). Arabidopsis JAGGED LATERAL ORGANS acts with ASYMMETRIC LEAVES2 to coordinate KNOX and PIN expression in shoot and root meristems. *The Plant Cell, 24*, 2917–2933.

Reinhart, B. J., Liu, T., Newell, N. R., Magnani, E., Huang, T., Kerstetter, R., Michaels, S., & Barton, M. K. (2013). Establishing a framework for the Ad/abaxial regulatory network of Arabidopsis: Ascertaining targets of class III homeodomain leucine zipper and KANADI regulation. *The Plant Cell, 25*, 3228–3249.

Ren, S. C., Song, X. F., Chen, W. Q., Lu, R., Lucas, W. J., & Liu, C. M. (2019). CLE25 peptide regulates phloem initiation in Arabidopsis through a CLERK-CLV2 receptor complex. *Journal of Integrative Plant Biology, 61*, 1043–1061.

Robert, H. S., Grones, P., Stepanova, A. N., Robles, L. M., Lokerse, A. S., Alonso, J. M., Weijers, D., & Friml, J. (2013). Local auxin sources orient the apical-basal axis in *Arabidopsis* embryos. *Current Biology, 23*, 2506–2512.

Ronceret, A., Gadea-Vacas, J., Guilleminot, J., Lincker, F., Delorme, V., Lahmy, S., Pelletier, G., Chaboute, M. E., & Devic, M. (2008). The first zygotic division in *Arabidopsis* requires de novo transcription of thymidylate kinase. *The Plant Journal, 53*, 776–789.

San-Bento, R., Farcot, E., Galletti, R., Creff, A., & Ingram, G. (2014). Epidermal identity is maintained by cell-cell communication via a universally active feedback loop in *Arabidopsis thaliana*. *The Plant Journal, 77*, 46–58.

Sarkar, A. K., Luijten, M., Miyashima, S., Lenhard, M., Hashimoto, T., Nakajima, K., Scheres, B., Heidstra, R., & Laux, T. (2007). Conserved factors regulate signalling in Arabidopsis thaliana shoot and root stem cell organizers. *Nature, 446*, 811–814.

Scheres, B., Wolkenfelt, H., Willemsen, V., Terlouw, M., Lawson, E., Dean, C., & Weisbeek, P. (1994). Embryonic origin of the Arabidopsis primary root and root meristem initials. *Development, 120*, 2475–2487.

Scheres, B., Di Laurenzio, L., Willemsen, V., Hauser, M. T., Janmaat, K., Weisbeek, P., & Benfey, P. N. (1995). Mutations affecting the radial organisation of the Arabidopsis root display specific defects throughout the embryonic axis. *Development, 121*, 53–62.

Schlereth, A., Moller, B., Liu, W., Kientz, M., Flipse, J., Rademacher, E. H., Schmid, M., Jurgens, G., & Weijers, D. (2010). MONOPTEROS controls embryonic root initiation by regulating a mobile transcription factor. *Nature, 464*, 913–916.

Sessions, A., Weigel, D., & Yanofsky, M. F. (1999). The *Arabidopsis thaliana* MERISTEM LAYER 1 promoter specifies epidermal expression in meristems and young primordia. *The Plant Journal, 20*, 259–263.

Shen, W. H., Parmentier, Y., Hellmann, H., Lechner, E., Dong, A., Masson, J., Granier, F., Lepiniec, L., Estelle, M., & Genschik, P. (2002). Null mutation of AtCUL1 causes arrest in early embryogenesis in Arabidopsis. *Molecular Biology of the Cell, 13*, 1916–1928.

Shi, C., Luo, P., Du, Y. T., Chen, H., Huang, X., Cheng, T. H., Luo, A., Li, H. J., Yang, W. C., Zhao, P., & Sun, M. X. (2019). Maternal control of suspensor programmed cell death via gibberellin signaling. *Nature Communications, 10*, 3484.

Sivaramakrishna, D. (2011). Size relationships of apical cell and basal cell in two-celled embryos in angiosperms. *Canadian Journal of Botany, 56*, 1434–1438.

Supena, E. D., Winarto, B., Riksen, T., Dubas, E., van Lammeren, A., Offringa, R., Boutilier, K., & Custers, J. (2008). Regeneration of zygotic-like microspore-derived embryos suggests an important role for the suspensor in early embryo patterning. *Journal of Experimental Botany, 59*, 803–814.

Takada, S., Takada, N., & Yoshida, A. (2013). ATML1 promotes epidermal cell differentiation in *Arabidopsis* shoots. *Development, 140*, 1919–1923.

Tanaka, H., Onouchi, H., Kondo, M., Hara-Nishimura, I., Nishimura, M., Machida, C., & Machida, Y. (2001). A subtilisin-like serine protease is required for epidermal surface formation in *Arabidopsis* embryos and juvenile plants. *Development, 128*, 4681–4689.

Tanaka, H., Watanabe, M., Watanabe, D., Tanaka, T., Machida, C., & Machida, Y. (2002). ACR4, a putative receptor kinase gene of Arabidopsis thaliana, that is expressed in the outer cell layers of embryos and plants, is involved in proper embryogenesis. *Plant & Cell Physiology, 43*, 419–428.

Tanaka, H., Watanabe, M., Sasabe, M., Hiroe, T., Tanaka, T., Tsukaya, H., Ikezaki, M., Machida, C., & Machida, Y. (2007). Novel receptor-like kinase ALE2 controls shoot development by specifying epidermis in *Arabidopsis*. *Development, 134*, 1643–1652.

ten Hove, C. A., Willemsen, V., de Vries, W. J., van Dijken, A., Scheres, B., & Heidstra, R. (2010). SCHIZORIZA encodes a nuclear factor regulating asymmetry of stem cell divisions in the *Arabidopsis* root. *Current Biology, 20*, 452–457.

ten Hove, C. A., Lu, K. J., & Weijers, D. (2015). Building a plant: Cell fate specification in the early *Arabidopsis* embryo. *Development, 142*, 420–430.

Tsugeki, R., Ditengou, F. A., Sumi, Y., Teale, W., Palme, K., & Okada, K. (2009). NO VEIN mediates auxin-dependent specification and patterning in the *Arabidopsis* embryo, shoot, and root. *The Plant Cell, 21*, 3133–3151.

Tucker, M. R., Hinze, A., Tucker, E. J., Takada, S., Jurgens, G., & Laux, T. (2008). Vascular signalling mediated by ZWILLE potentiates WUSCHEL function during shoot meristem stem cell development in the *Arabidopsis* embryo. *Development, 135*, 2839–2843.

Turchi, L., Carabelli, M., Ruzza, V., Possenti, M., Sassi, M., Penalosa, A., Sessa, G., Salvi, S., Forte, V., Morelli, G., et al. (2013). Arabidopsis HD-zip II transcription factors control apical embryo development and meristem function. *Development, 140*, 2118–2129.

Turchi, L., Baima, S., Morelli, G., & Ruberti, I. (2015). Interplay of HD-zip II and III transcription factors in auxin-regulated plant development. *Journal of Experimental Botany, 66*, 5043–5053.

Ueda, M., Zhang, Z., & Laux, T. (2011). Transcriptional activation of Arabidopsis axis patterning genes WOX8/9 links zygote polarity to embryo development. *Developmental Cell, 20*, 264–270.

Ueda, M., Aichinger, E., Gong, W., Groot, E., Verstraeten, I., Vu, L. D., De Smet, I., Higashiyama, T., Umeda, M., & Laux, T. (2017). Transcriptional integration of paternal and maternal factors in the Arabidopsis zygote. *Genes and Development, 31*, 617–627.

van den Berg, C., Willemsen, V., Hendriks, G., Weisbeek, P., & Scheres, B. (1997). Short-range control of cell differentiation in the Arabidopsis root meristem. *Nature, 390*, 287–289.

Vaten, A., Dettmer, J., Wu, S., Stierhof, Y. D., Miyashima, S., Yadav, S. R., Roberts, C. J., Campilho, A., Bulone, V., Lichtenberger, R., et al. (2011). Callose biosynthesis regulates symplastic trafficking during root development. *Developmental Cell, 21*, 1144–1155.

Vera-Sirera, F., De Rybel, B., Urbez, C., Kouklas, E., Pesquera, M., Alvarez-Mahecha, J. C., Minguet, E. G., Tuominen, H., Carbonell, J., Borst, J. W., et al. (2015). A bHLH-based feedback loop restricts vascular cell proliferation in plants. *Developmental Cell, 35*, 432–443.

Vroemen, C. W., Mordhorst, A. P., Albrecht, C., Kwaaitaal, M. A., & de Vries, S. C. (2003). The CUP-SHAPED COTYLEDON3 gene is required for boundary and shoot meristem formation in Arabidopsis. *The Plant Cell, 15*, 1563–1577.

Wang, H., Ngwenyama, N., Liu, Y., Walker, J. C., & Zhang, S. (2007). Stomatal development and patterning are regulated by environmentally responsive mitogen-activated protein kinases in Arabidopsis. *The Plant Cell, 19*, 63–73.

Weigel, D., & Jürgens, G. (2002). Stem cells that make stems. *Nature, 415*, 751.

Weijers, D., Geldner, N., Offringa, R., & Jurgens, G. (2001). Seed development: Early paternal gene activity in *Arabidopsis*. *Nature, 414*, 709–710.

Weijers, D., Schlereth, A., Ehrismann, J. S., Schwank, G., Kientz, M., & Jurgens, G. (2006). Auxin triggers transient local signaling for cell specification in Arabidopsis embryogenesis. *Developmental Cell, 10*, 265–270.

Willemsen, V., & Scheres, B. (2004). Mechanisms of pattern formation in plant embryogenesis. *Annual Review of Genetics, 38*, 587–614.

Wysocka-Diller, J. W., Helariutta, Y., Fukaki, H., Malamy, J. E., & Benfey, P. N. (2000). Molecular analysis of SCARECROW function reveals a radial patterning mechanism common to root and shoot. *Development, 127*, 595–603.

Xu, J., Zhang, H. Y., Xie, C. H., Xue, H. W., Dijkhuis, P., & Liu, C. M. (2005). EMBRYONIC FACTOR 1 encodes an AMP deaminase and is essential for the zygote to embryo transition in Arabidopsis. *The Plant Journal, 42*, 743–756.

Xu, T. T., Ren, S. C., Song, X. F., & Liu, C. M. (2015). CLE19 expressed in the embryo regulates both cotyledon establishment and endosperm development in *Arabidopsis*. *Journal of Experimental Botany, 66*, 5217–5227.

Yang, K. J., Guo, L., Hou, X. L., Gong, H. Q., & Liu, C. M. (2017). *ZYGOTE-ARREST 3* that encodes the tRNA ligase is essential for zygote division in *Arabidopsis*. *Journal of Integrative Plant Biology, 59*, 680–692.

Yoshida, S., Barbier de Reuille, P., Lane, B., Bassel, G. W., Prusinkiewicz, P., Smith, R. S., & Weijers, D. (2014). Genetic control of plant development by overriding a geometric division rule. *Developmental Cell, 29*, 75–87.

Yu, T. Y., Shi, D. Q., Jia, P. F., Tang, J., Li, H. J., Liu, J., & Yang, W. C. (2016). The Arabidopsis receptor kinase ZAR1 is required for zygote asymmetric division and its daughter cell fate. *PLoS Genetics, 12*, e1005933.

Zhang, Y., Jiao, Y., Liu, Z., & Zhu, Y. X. (2015). ROW1 maintains quiescent centre identity by confining *WOX5* expression to specific cells. *Nature Communications, 6*, 6003.

Zhao, J., Xin, H., Qu, L., Ning, J., Peng, X., Yan, T., Ma, L., Li, S., & Sun, M. X. (2011). Dynamic changes of transcript profiles after fertilization are associated with de novo transcription and maternal elimination in tobacco zygote, and mark the onset of the maternal-to-zygotic transition. *The Plant Journal, 65*, 131–145.

Zhao, P., Shi, C., Wang, L., & Sun, M. X. (2022). The parental contributions to early plant embryogenesis and the concept of maternal-to-zygotic transition in plants. *Current Opinion in Plant Biology, 65*, 102144.

Zhao, Y. (2014). Auxin biosynthesis. *The Arabidopsis Book* 12e0173. https://doi.org/10.1199/tab.0173

Chapter 19
Endosperm Development

Wen-Hui Lin and Hong-Wei Xue

Abstract Seed development involves the formation of the embryo, endosperm, and seed coat. Endosperm development is not only crucial for embryo development and seed maturation (i.e., plant reproductive growth) but also affects grain yield and crop quality, especially in cereal crops such as rice, wheat, and maize. Elucidating the relevant regulatory mechanisms and genetic networks is important from a basic research perspective and may have implications for agricultural improvements. Studies have revealed the tightly regulated cytological development of the endosperm as well as the associated genetic and epigenetic controls. This chapter summarizes the cytological basis and genetic networks of endosperm initiation, cellularization, aleurone layer specification, accumulation of storage materials, and interactions between the embryo and endosperm.

Keywords Endosperm · Development · Aleurone layer · Imprinting · Starch

Because seeds represent the reproductive organs of higher plants, seed development is an important stage of the plant life cycle. The seeds provide nutrition during germination, and additionally, seed dormancy under stress conditions enables plants to adapt to the environment. Moreover, genetic recombination and variation during fertilization provide opportunities for natural selection. There are thousands of seed types in nature, with seeds produced by diverse plant species exhibiting considerable differences regarding size, color, internal structure, and chemical composition.

W.-H. Lin
School of Life Sciences and Biotechnology, Shanghai Jiao Tong University, Shanghai, China
e-mail: whlin@sjtu.edu.cn

H.-W. Xue (✉)
Guangdong Laboratory for Lingnan Modern Agriculture, Guangdong Basic Research Center of Excellence for Precise Breeding of Future Crops, College of Agriculture, South China Agricultural University, Guangzhou, China

Shanghai Collaborative Innovation Center of Agri-Seeds, Joint Center for Single Cell Biology, School of Agriculture and Biology, Shanghai Jiao Tong University, Shanghai, China
e-mail: xhw@scau.edu.cn

F. Chang et al. (eds.), *Regulation of Plant Development*,
https://doi.org/10.1007/978-981-95-7033-1_19

Anatomical analyses have revealed that most seeds have a common structure consisting of an embryo, endosperm, and seed coat, which develop from the zygote (fertilized egg), endosperm (fertilized polar nucleus), and integument, respectively.

19.1 Development and Cytological Basis of the Endosperm

The processes mediating endosperm development are similar in higher plants. Specifically, there is typically an initial mitotic stage and a subsequent cellularization process, after which nutrients accumulate. The later stages of endosperm development vary in different plant species. Some plants (e.g., dicotyledonous *Arabidopsis thaliana*) have a small endosperm in mature seeds, while other plants (e.g., most *Gramineae* species) have a large endosperm filled with nutrients (Fig. 19.1). These different strategies of nutrient accumulation are determined by different cell fates.

Mature seeds can be divided into two groups based on their endosperm characteristics: endosperm seeds and non-endosperm seeds. The endosperm of angiosperms forms as a result of nuclear division and cellularization. Nuclear division refers to a process in which the polar nucleus (primary endosperm nucleus, which is produced by the fusion of two polar nuclei with one sperm nucleus during fertilization) undergoes several rounds of mitosis without cytokinesis. The primary

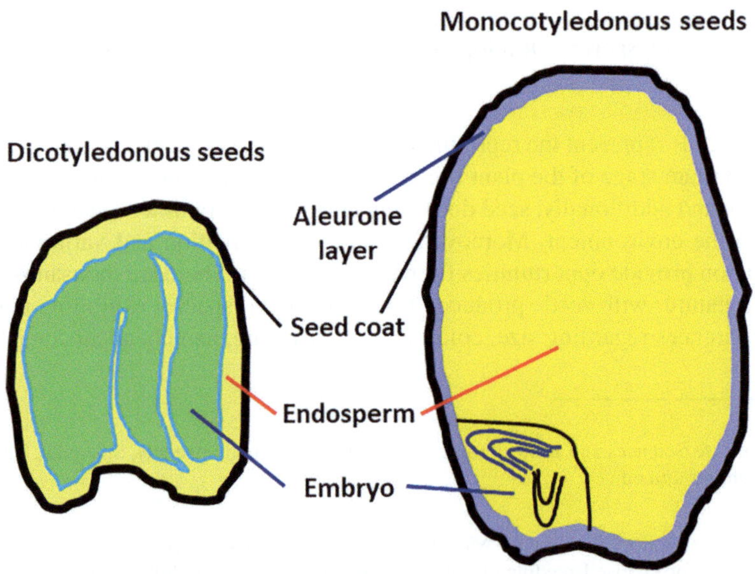

Fig. 19.1 Endosperms of most dicots and monocots. Along with seed development, endosperm of most dicotyledonous seeds (such as *Arabidopsis thaliana*) is consumed by embryo, which occupies most part of the mature seed; whereas endosperm of monocotyledonous seeds (such as rice) develops and keeps as a storage tissue to provide nutrients during seed germination and young seedling growth

endosperm nucleus is generally triploid. According to the cell division process (e.g., whether a cell wall is formed) during the early division of endosperm nuclei, the endosperm development of angiosperms can be divided into three types: nuclear, cellular, and helobial.

19.1.1 Nuclear Endosperm Development

Nuclear endosperm development is common among angiosperms, including *A. thaliana*, which is a model plant species, as well as crops such as corn, wheat, and rice (another model plant species). Nuclear endosperm development can be divided into two stages, namely the free nucleus stage and the cellularization stage. During the free nucleus stage, the endosperm nucleus divides, but there is no cell wall division. Many of the free nuclei disperse in the cytoplasm near the periphery of the embryo sac and form the coenocyte or syncytium. Finally, free nuclei and cytoplasm are pushed to the periphery of the embryo sac, and the endosperm sac is formed after the central vacuoles in the embryo sac expand.

After a rapid nuclear division, cell wall is generated around the free nucleus. Newly formed cells (nucleus with cell wall) gradually appear toward the central vacuole from the periphery, after which endosperm development enters the cellularization stage (Olsen, 2004). Before the cell wall is generated, the endoplasmic reticulum and Golgi vesicles among the free nuclei assume a plate-like arrangement and form the phragmoplast. The membranes of the endoplasmic reticulum and vesicles fuse to form the plasma membrane, after which the cell wall is formed. After cellularization, cell division occurs from the center to the periphery, and the proliferation of endosperm tissue depends on the continuous division and growth of endosperm outer cells.

The early stage of nuclear endosperm development is similar among different species, while the later stage varies considerably. In most dicotyledonous species (e.g., *A. thaliana*), the endosperm is absorbed by the embryo after cellularization. The internal space of the whole mature seed is basically occupied by the expanded embryo (primary nutrient storage organ), and the endosperm is very inconspicuous or completely degraded (Olsen, 2004). In contrast, in many monocotyledonous plant species, such as rice, the endosperm continues to grow. The centrally located cells expand to form the starchy endosperm, which accumulates starch and a small amount of proteins, which is followed by endoreduplication and an increase in genomic DNA content. The starchy endosperm occupies most of the space in mature rice seeds. During the later stages of endosperm development, programmed cell death is initiated, and all starchy endosperm cells die.

Aleurone cells are located at the periphery and mainly accumulate nutrients such as proteins, fatty acids, and vitamins. These cells are alive in mature seeds (Sabelli et al., 2009).

19.1.2 Cellular Endosperm Development

Cellular endosperm development is characterized by the formation of cell walls from the first division of the central nucleus, thus without a free nucleus stage. This type of endosperm development has been observed in *Solanum* species (*Solanaceae*) as well as in some *Crassulaceae*, *Labiatae*, and *Lagerstroemia* species.

19.1.3 Helobial Endosperm Development

Helobial endosperm development is an intermediate type of development between the nuclear and cellular types and mainly occurs in monocotyledonous plants. The primary endosperm nucleus moves to the chalazal end of the embryo sac. The first division of the chalazal end separates the embryo sac into two chambers, with the chamber close to the micropyle wider than that close to the chalazal end. The chamber close to the micropyle undergoes nuclear division followed by cellularization at a later stage. In the chamber close to the chalazal end, nuclei do not divide or divide only a few times and usually appear as free nuclei. Helobial endosperm development has been observed in plant species, including Alisma (*Rhizoma alismatis*), arrowhead (*Sagittaria sagittifolia* L.), and Grimmia (*Hosta ventricosa* (Salisb.) Stearn).

The endosperm is arranged along the anterior/posterior axis as well as according to radial symmetry. Moreover, the endosperm can be separated into three different regions along the axes. For example, in *A. thaliana*, part of the endosperm is close to the micropyle, part of it forms the peripheral endosperm (main part), and part of it is close to the chalazal end. Meanwhile, in maize, part of the endosperm is adjacent to the embryo, part of it forms the starchy endosperm (main part), and part of it is close to the basal endosperm transfer layer (BETL).

19.2 Genetic Basis of Endosperm Development

Analyses of various mutant resources, including populations resulting from insertional mutagenesis (Tos17, T-DNA, Ds, and dSpm) as well as populations mutagenized by chemical (ethyl methanesulfonate, methyl nitrosourea, diepoxybutane, and sodium azide) and physical (fast-neutron, γ-rays, and ion beam irradiation) means, have enabled the identification of mutants exhibiting altered endosperm development. The corresponding genes/regulators related to various processes have been functionally characterized (Fig. 19.2). Moreover, omics-based investigations have provided useful information for elucidating the endosperm development regulatory network.

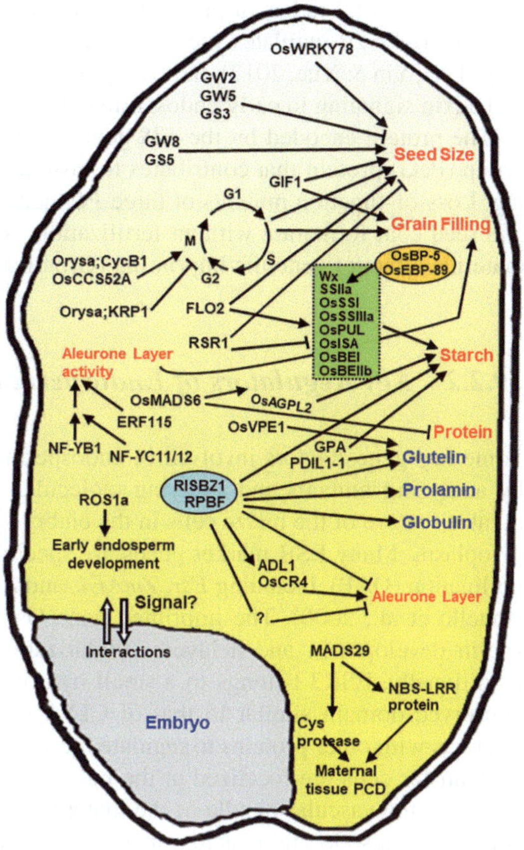

Fig. 19.2 Regulatory networks of rice endosperm development. Currently identified genes and related functions in endosperm development are summarized. (Modified from Zhou et al., 2013)

19.2.1 Transcriptional Regulation

Many genes are specifically expressed during endosperm development, and transcription factors are critical for regulating the expression patterns of these genes (summarized by Zhao et al., 2013).

For example, MADS-box transcription factors are involved in *A. thaliana* seed and endosperm development, while FEM111/AGAMOUS (AGL80) regulates the development of central cells and the subsequent endosperm cellularization. The central cells of loss-of-function mutants have small nuclei and vacuoles and do not develop any further. Additionally, *AGL61* is expressed in the central cells and endosperm, and the phenotype of the loss-of-function mutant is similar to that of the *agl80* mutant. Yeast two-hybrid analyses confirmed that AGL80 interacts with AGL61 to form heterodimers, which synergistically regulate the expression of downstream target genes and regulate the central cell development (Steffen et al., 2008). Expression of the rice MADS-box transcription factor gene, *OsMADS29*, which is specifically expressed in the nucellus, is induced after pollination or by an auxin treatment. Studies of plants with suppressed *OsMADS29* expression showed that OsMADS29 regulates the degradation of nuclei and the nucellar projection (Fig. 19.3, Yin & Xue, 2012), indicating the importance of transcriptional regulation and auxin signaling in early endosperm development.

The protein encoded by the *FIS* gene is highly conserved with the Polycomb-group (PcG) protein that contributes to histone methylation-mediated gene silencing. Loss-of-function mutants of three *FIS* genes undergo endosperm development and seed coat formation without fertilization. Additionally, *fis* mutants exhibit the maternal effect of genomic imprinting (Grossniklaus et al., 2001).

19.2.2 Key Regulators of Endosperm Development

Some key genes/factors involved in endosperm development have been identified by analyzing mutants and applying molecular genetics techniques (Fig. 19.2). A notable feature of the maize cells in the embryo surrounding region (ESR) is dense cytoplasm. Many ESR marker genes are specifically expressed at 5–20 days after pollination (DAP), including *Esr*, *ZmAE1*, and *ZmAE3* (Opsahl Ferstad et al., 1997; Bonello et al., 2000). The imprinted gene OsEMF2a induces autonomous endosperm development and delayed cellularization in rice (Tonosaki et al., 2021). Additionally, ESR3 belongs to a small hydrophilic protein family and contains a conserved domain similar to that of CLV3 (CLAVATA3). In *A. thaliana*, CLV3 interacts with other proteins to regulate meristem size.

Transfer cells are localized at the base of the endosperm, where they are connected to the vascular bundle of the maternal tissue. The main function of transfer cells is to transport amino acids, sucrose, and monosaccharides into the endosperm from the parental tissue. Characterizing the genes expressed at the early stage of endosperm transfer cells may be important for elucidating the differentiation of these cells. Barley *Endosperm1* (*End1*) is expressed in the polykaryon stage endosperm as well as in the differentiated transfer cells (Doan et al., 1996). The expression pattern of the *Zea mays* gene encoding MYB-related protein-1 (*ZmMRP-1*) is very similar to that of *End1*. ZmMRP-1 is considered to initiate the transient expression of some tissue-specific genes in endosperm transfer cells (Gómez et al., 2002). Maize mutant *defective kernel1* (*dek1*) lacks aleurone cells but contains normal endosperm transfer cells, indicating these cells are differentially regulated (Lid et al., 2002). Rice *GIF1*, which encodes an invertase that hydrolyzes sucrose into glucose and fructose, is specifically expressed in the dorsal vascular bundle and is important for grain filling (Wang et al., 2008).

A transcriptome study of rice seeds at four developmental stages revealed that among 2231 analyzed transcription factors, 1131 are transcribed in endosperm (74 are specifically transcribed in endosperm) and are involved in the regulation of nutrient storage and abscisic acid (ABA) response (Xue et al., 2012). Laser capture technology has been applied to analyze the transcriptome of the endosperm at 4 DAP, and results indicated that the expressed genes can be clustered into the

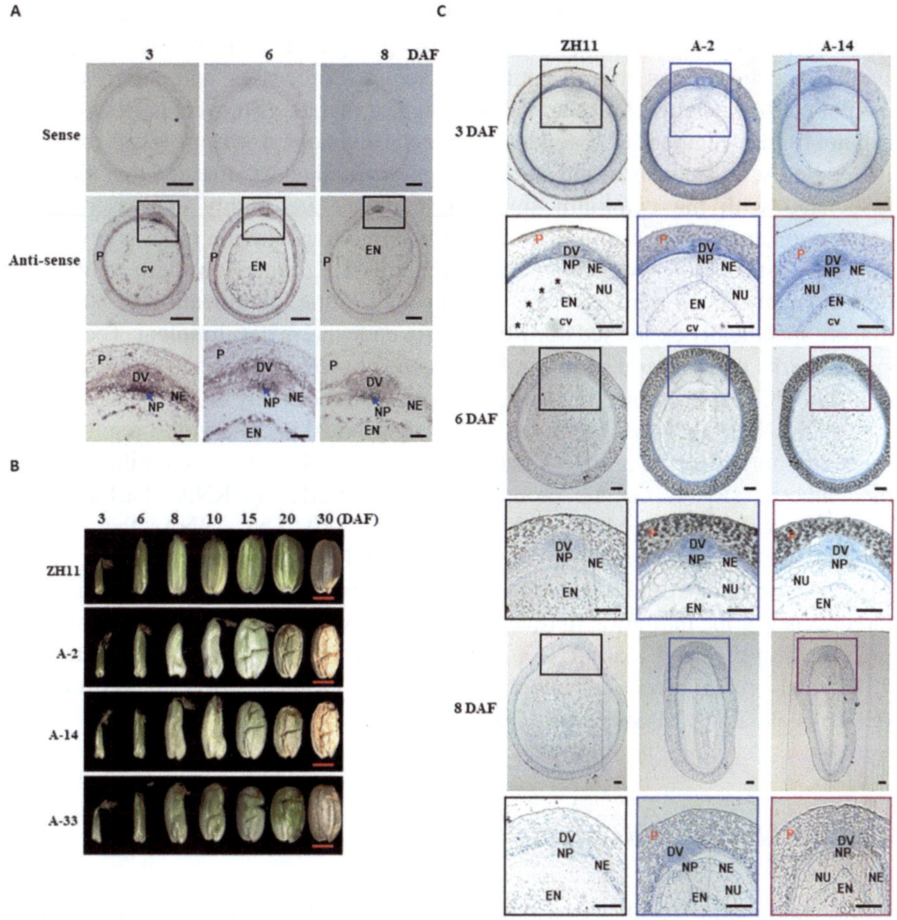

Fig. 19.3 MADS29 regulates the degradation of the nucellus and the nucellar projection during rice seed development. (Reprinted with permission from Yin and Xue et al., 2012). (**a**) MADS29 is preferentially expressed in reproductive tissues, especially in the nucellus and nucellar projection. The transverse sections of rice seeds at 3, 6, and 8 DAF were used for the RNA in situ hybridization analysis, revealing the strong expression of MADS29 in the nucellus and nucellar projection. The high expression level of MADS29 in the nucellar projection (squared regions) is enlarged (bottom panel). Bars = 200 mm (top and middle panels, 3, 6, 8 DAF), 50 mm (bottom panel). DV dorsal vascular bundle, EN endosperm, NE nucellar epidermis, NP nucellar projection, P pericarp, V vascular bundle. (**b**) Rice plants with suppressed MADS29 expression show abnormal seed development, and phenotypic observation revealed the shrunken seeds of A-MADS29 lines. Seeds at 3, 6, 8, 10, 15, 20, and 30 DAF were observed. Bar = 2 mm. (**c**) Transverse section analysis of the A-MADS29 and ZH11 seeds showed the non-degraded nucellus under suppressed MADS29 expression (bar = 100 mm). The squared regions were enlarged (bar = 50 mm), and stars highlighted the collapsed nucellus. NU nucellus, EN endosperm, NE nucellar epidermis, NP nucellar projection, DV dorsal vascular bundle, P pericarp, CV central vacuole. Seeds at 3, 6, and 8 DAF were observed

following groups based on the processes they affect: cell cycle regulation, DNA replication and modification, chromosome aggregation and assembling, protein synthesis, cytoskeleton and microtubule-associated processes, and energy metabolism of cells/organelles. Further analyses suggest that the cell cycle genes may contribute to the regulation of endosperm cell division at the polykaryon stage. Interestingly, an analysis of the plant hormone-related genes revealed that the genes associated with the cytokinin signaling pathway are highly expressed at the polykaryon stage, suggesting that cytokinin signal transduction plays a key role in the early stages of endosperm development. A comparison with other seed development-related transcriptome data resulted in the identification of 800 genes that are specifically expressed during the early endosperm development. The expression patterns of some transcription factor genes have been verified by quantitative real-time PCR or promoter-reporter gene fusion studies (Day et al., 2008).

Small RNAs (sRNAs), which are widely distributed in eukaryotes, regulate gene transcription and degradation (Valencia-Sanchez et al., 2006). According to their biosynthesis, sRNAs can be grouped as short interfering RNAs (siRNAs) and microRNAs (miRNAs; Mallory & Vaucheret, 2006). The siRNAs facilitate the splicing of target genes or help to modify DNA and histones to mediate gene silencing (Zilberman et al., 2003), while miRNAs mainly contribute to the splicing of target genes or the inhibition of translation (Mallory & Vaucheret, 2006). A previous investigation revealed that sRNAs are crucial for seed development (Chen et al., 2017) and are involved in genomic imprinting during endosperm development (Hsieh et al., 2009; Mosher et al., 2009).

19.2.3 Hormone Regulation

Hormone signaling regulates various aspects of plant development and plays important roles during seed development. Endosperm development involves cell differentiation and division, which are mainly regulated by auxin (Batista et al., 2019), cytokinin (Yang et al., 2002a, b, c), and brassinosteroids (BRs). When the developing seed starts to accumulate storage material, the hormone contents are also changed. For example, the ABA response and signaling are highly active in the endosperm at later developing stages (Xue et al., 2012).

OsYUC11-mediated auxin biosynthesis is essential for endosperm development in rice (Xu et al., 2021). However, increased auxin prevents endosperm cellularization in Arabidopsis (Batista et al., 2019). The application of kinetin, indole-3-acetic acid, or ABA to spikelets results in increased cell division and grain filling in the inferior spikelet of hybrid rice (Zhang et al., 2009). Leaf-derived ABA controls rice seed development in a temperature-dependent manner and is regulated by defective grain-filling 1 (DG1). The important transcription factors of starch synthesis involved in regulating storage material accumulation are inhibited in dg1 (Qin et al., 2021). Rice mutants that are deficient in BR biosynthesis or signaling (e.g., *d11*, *d2*, *brd1*, and *d61*) produce a relatively small embryo and endosperm (Yamamuro et al.,

2000; Yang et al., 2002a, b, c; Zhu et al., 2003a, b; Tanabe et al., 2005). Brassinosteroids influence the SHB1-MINI3-IKU2 pathway to regulate endosperm development and subsequently affect seed size via the transcription factor BZR1 (Jiang et al., 2013). In pea seeds, a lack of gibberellin (GA) stops seed growth before maturation. An analysis of the rice seed transcriptome at different developmental stages confirmed that 28 genes related to the ABA response and signaling are highly expressed, suggesting ABA has a key function during endosperm development (Xue et al., 2012).

Overexpression of a GA oxidase gene driven by an endosperm-specific promoter in the *A. thaliana* endosperm terminates seed development, suggesting that GA activity in the endosperm is essential for normal seed growth. There is complex crosstalk among various hormones, especially between ABA and GA, during seed and endosperm development. Although many hormones, such as auxin, ABA, ethylene, cytokinin, and BRs, are reportedly involved in the regulation of seed development, their effects on the regulation of endosperm development remain largely uncharacterized.

19.2.4 Omics-Based Analysis

Studies of transcription profiles may help to illustrate the transcriptional network associated with the regulation of endosperm development as well as the functions of transcription factors. A transcriptome investigation of the rice grain-filling stage by cDNA chips (Zhu et al., 2003a, b) unveiled the expression patterns of 21,000 genes. Of these genes, 269 exhibited upregulated expression during the grain-filling stage and consisted of promoters with the AACA element (i.e., *cis*-acting element for endosperm-specific expression). Thus, these genes may encode key factors influencing rice endosperm development. Transcription factors that potentially target the AACA element were predicted based on the results of a gene co-expression analysis. Additionally, the spatio-temporal expression patterns of genes involved in starch biosynthesis during the grain-filling stage were studied. Through a high-throughput microarray technique, protein–protein interactions in developing rice seeds were examined by yeast two-hybrid assays, and results indicated that many MADS family members interact with each other or with other proteins (e.g., DoF transcription factor) (Cooper et al., 2003). These findings help to explain the related crosstalk and regulatory network.

A transcriptome study of rice seeds at four developmental stages revealed that 474 genes differentially expressed in the endosperm are involved in various metabolic processes, including starch metabolism and fatty acid synthesis, as well as cellular organization (Xue et al., 2012). Additionally, experiments using cDNA chips to examine gene expression patterns during various seed developmental stages (from fertilization to maturation) indicated that homeobox genes are actively expressed, while an analysis of the *cis*-acting elements during seed/endosperm development suggested that DoF transcription factors may play an important role (Duan et al., 2005).

19.3 Aleurone Layer Cell Specification

Most monocotyledonous plants, such as cereals, contain two endosperm parts, namely, the central starchy endosperm and the peripheral aleurone layer. There are about 80,000–90,000 cells in the starchy endosperm of mature barley seeds (Cochrane & Duffus, 1983; Kvaale & Olsen, 1986) and about 60,000 cells in the starchy endosperm of wheat seeds (Chojecki et al., 1986). The programmed cell death of starchy endosperm during a late developmental stage (16 DAP in wheat seeds) was first observed in maize and wheat (Gibson et al., 1999). With the exception of barley (with three aleurone layers), all cereals have one aleurone layer in the endosperm. However, the aleurone in rice, maize, barley, and wheat has multiple cell layers near the main vascular bundle, which differentiate to form the transfer cells with trapped cell walls (except for rice). These transfer cells are believed to affect grain filling by transporting nutrients.

Investigations on the differentiation of the aleurone layer based on molecular genetics resulted in the identification of genes contributing to the differentiation of maize aleurone cells. The *CRINKLY4* (*CR4*, Becraft et al., 1996), *DEFECTIVE KERNEL 1* (*DEK1*, Lid et al., 2002), and *SAL1* (Shen et al., 2003) genes are involved in the regulation of cell fate determination and differentiation of aleurone cells during endosperm development. The *CR4* gene encodes a receptor-like kinase that is homologous to an animal tumor necrosis factor kinase. A loss-of-function mutation to *CR4* results in starchy endosperm cells replacing some of the aleurone cells. Meanwhile, *DEK1* encodes a protein that is similar to an animal calcium-dependent protein. The C-terminal domain of DEK1 has a calcium-activated protease domain that is active in vitro. Additionally, DEK1 contains a membrane-localization signal, 21 transmembrane motifs, and an extracellular ring structure that is considered to interact with extracellular molecules such as receptor proteins (Lid et al., 2002). The *dek1* mutant lacks aleurone cells. SAL1 is homologous to the human chromatin-modulating protein 1 and influences intracellular vesicle transport. Moreover, SAL1 is a conserved member of the vacuolar protein sorting gene E class. The maize *sal1* mutant exhibits increased amounts of aleurone cells (up to seven layers; Shen et al., 2003), implying that SAL1 is a negative regulator of aleurone layer determination. Similarly, the maize *thk1* mutant also produces multiple aleurone cell layers. The *THK1* gene may be a downstream target of DEK1 signaling involved in the regulation of aleurone cell differentiation (Yi et al., 2011). OsROS1-mediated DNA demethylation restricts the number of aleurone cell layers in rice and provides a way to improve the nutrition of rice (Liu et al., 2018).

Two transcription factors, RISBZ1 and RPBF, which control the storage proteins in rice seeds, regulate the differentiation of rice aleurone cells. Mutations to both of these transcription factors result in aleurone cell abnormalities, including the production of multiple cell layers and larger cells as well as disordered arrangements. Interestingly, the expression levels of *OsCR4*, *OsDEK1*, and *OsSAL1* decreased significantly in the double mutant, indicating the regulatory mechanism associated with aleurone cells is complex (Kawakatsu et al., 2009; Becraft & Yi, 2010).

Furthermore, a knockout of the aleurone-specific transcription factor gene *NF-YB1* by the CRISPR/Cas9 system or a knockdown by RNAi resulted in a dramatic decrease in the filling efficiency of rice endosperm. Furthermore, NF-YB1 controls grain filling by activating the expression of three genes encoding sucrose transporters (Bai et al., 2016) or through interacting with the ERF transcription factor to regulate genes whose promoter consists of the GCC-box and hence the multiple transport-related processes (Fig. 19.2, Xu et al., 2016).

19.4 Storage Products

Fresh weight and water content increase rapidly in developing seeds. Following the accumulation of storage materials, the seed dry weight increases. Most mature seeds contain at least two or three main storage materials, including carbohydrates (starch), triacylglycerols (fatty acids), and proteins, which are synthesized and stored in the dispersing organelles of the storage tissues during seed development.

During the milk stage and dough stage of the grain-filling period, nutrients transported from the dorsal vascular bundle are first offloaded to the nucellus and then diverted from nucellus cells. Some of the nutrients are transported to the endosperm directly through the back aleurone layers after going through the degraded nucellus layers. The remaining nutrients are transported to the periphery of the endosperm through the nucellus tissue and then transported to the central endosperm through the aleurone layers. Because the nutrients that fill the central part of the endosperm must be delivered by the dorsal vascular bundle through the nucellar epidermis and aleurone layers, it is difficult to efficiently fill the abdominal endosperm cells. Grains with this kind of endosperm have been named white belly rice. The central endosperm cells of the large grains of some rice cultivars are not filled well because they are far from the aleurone layer. These grains are called white core rice.

19.4.1 Synthesis and Storage of Fatty Acids

Triacylglycerols are the major constituent of the storage lipids in seeds. The main raw materials for the synthesis of lipids are phosphoglycerol and lipoyl coenzyme A. The fatty acids are synthesized in the seed endoplasmic reticulum, which expands partly because of the accumulation of lipids and proteins. Vesicles are then formed and detach from the endoplasmic reticulum. The vesicles fuse with each other and gradually expand to form an oil body containing the accumulated lipids.

The seed parts that accumulate lipids vary in different plant species. Lipids are accumulated in the cotyledons of soybean, peanut, and vegetables, while they are accumulated in the scutellum and aleurone layer in cereals. The starchy endosperm cells normally do not accumulate lipids. In oil plants, soluble sugars and starch are initially accumulated in the seeds, but their abundance rapidly decreases as the seeds develop. Meanwhile, seed weight and fat content increase. As seeds mature, the iodine value of fat increases while the acid value gradually decreases, indicating that saturated fatty acids are synthesized and then transformed into unsaturated fatty acids. The free fatty acids initially synthesized gradually form the complex oils during seed maturation.

19.4.2 Storage Proteins

Diverse types of proteins are synthesized at different seed development stages. Proteins related to embryo differentiation are synthesized during the early seed development, while storage-related proteins are synthesized during the mid-development stage. Synthesis of seed storage proteins is important for endosperm development because these proteins are the key components of the endosperm and are the main nitrogen source for developing seeds. The proteins synthesized during the late stages of seed development are mainly related to seed dormancy and anti-dehydration activities. Globulin is the primary type of storage protein in the endosperm.

Genes encoding storage proteins or their polypeptides are expressed under spatio-temporal regulation. The raw materials for storage protein synthesis, such as amino acids and amides, are derived from the vegetative organs. Amino acids are activated by ATP in the cytoplasm and then transferred to the granular endoplasmic reticulum by tRNAs. The amino acids are used to synthesize peptides in the ribosomes of the endoplasmic reticulum and finally as storage proteins after processing and modification, which accumulate in proteoplasts.

The seed protein contents are closely related to seed quality. Seeds with high protein contents are valuable as a source of nutrients. However, high protein contents in rice endosperm are negatively correlated with taste.

19.4.3 Starch and Development of Starch Granules

The sulcus (groove parts, the region close to where sucrose is transported) starts to accumulate starch after endosperm cells stop dividing. Starch granules are formed in the proplastids. The appearance of granules indicates the initiation of starch biosynthesis. The plastids containing starch granules split and proliferate through budding or by intermediate contraction and disconnection. As the plastids divide, starch granules and the attached enzymes that synthesize starch granules separate into daughter plastids. The starch granules in the daughter plastids increase in number and size, and the plastids become oval. Early stage plastids have obvious endometrial structures (thylakoids), which are distributed among the starch granules. The thylakoids are transformed to amyloid when the starch granules expand until the endometrial structures are full with starch granules.

Seed starch exists as amylose and amylopectin. The raw material for starch synthesis is ADP-glucose, which is derived from sucrose or hexose transported to the endosperm and cotyledon. The soluble sugar concentration gradually decreases, while the starch content continuously increases as seeds mature. The soluble sugar and starch contents of the caryopsis increase during the first few days after flowering. At approximately 10 days after flowering, the soluble sugar content starts to decrease, unlike the starch content, which continues to increase.

The starch in the amyloplast of the rice endosperm occurs as compounded starch, which develops from independent starch granules. An amyloplast transection full of starch granules exhibits a tortoise shell-like pattern. An analysis of the development of endosperm amyloplasts by electron microscopy suggests there are six developmental stages (i.e., plastid proliferation stage, plastid enlargement stage, starch accumulation stage, small amyloplast stage, amyloplast membrane expansion stage, and large amyloplast stage). When the endosperm cells are full of amyloid, the gap disappears, and the amyloplast and starch granules inside are squeezed to form an angular polyhedron. The starch granule structure comprises concentric rings or shells, with each ring representing the daily starch "growth." The biochemical basis of the "ring" formation involves the abrupt discontinuity of starch deposition density. However, the underlying mechanism remains unclear. Physiologically, these changes may be related to environmental conditions.

19.4.4 Regulation of Storage Proteins

Expression levels of genes encoding storage proteins exhibit distinct spatio-temporal patterns during endosperm development, which are mainly regulated at the transcript level (Qu & Takaiwa, 2004). There are some common regulatory sequences in the promoter region of genes specifically expressed in the endosperm, and analyses have identified some highly conserved *cis*-elements, such as P-box, GCN4, AACA, and ACGT motifs. Additionally, the transcription factors corresponding to these motifs have been identified. For example, the bZIP family transcription factor RISBZ1–5 specifically recognizes the GCN4 motif (TGAGTCA) and regulates the expression of genes related to the accumulation of glutelin, which is the main storage protein type in the endosperm. Interestingly, only RISBZ1 has been identified as a transcriptional activator, although the other members may help to regulate the glutelin accumulation by interacting with RISBZ1 to form a heterodimer (Onodera et al., 2001). Meanwhile, rice prolamin box-binding factor (RBPF), which is a Dof transcription activator, can bind to the P-box motif (TGTAAAG) in the promoter region of rice storage protein genes and activate the expression of prolamin- and glutelin-related genes. The interaction between RBPF and RISBZ1 has been confirmed using rice callus protoplasts, which revealed that their interaction enhances the expression of downstream genes but also induces the expression of downstream genes that are not activated by either RBPF or RISBZ1 alone (Yamamoto et al., 2006). In rice seeds, the downregulated expression of *RBPF* and *RISBZ1* leads to decreased expression levels of genes related to storage proteins. Additionally, the accumulation of starch and lipids decreases, and the aleurone develops with multiple cell layers, suggesting that the interaction between transcription factors is important for the development and storage of rice seeds.

The folding of endoplasmic reticulum storage proteins is mediated by BiP. *OsBiP1* is mainly expressed during rice grain maturation, and overexpression of *OsBiP1* in the endosperm decreases the accumulation of storage proteins and starch, resulting in crinkled grains and a floury endosperm (Yasuda et al., 2009). The small G protein OsRab5a is essential for the transport of glutelin precursors from Golgi bodies to the plant storage vacuole, and loss-of-function mutants present floury endosperms (Wang et al., 2010). Moreover, GAP3, which contains kelch repeats and forms a complex with OsRab5a, regulates the transport of glutelin precursors from the Golgi bodies to the plant storage vacuole. There is an increase in the abundance of glutelin precursors in the *gap3* mutant, resulting in the formation of floury endosperms (Ren et al., 2014). Furthermore, *Chalk5*, which is the major gene controlling chalkiness, encodes the proton-transferring pyrophosphatase located at the tonoplast membrane. Overexpression of *Chalk5* results in the increased endosperm chalkiness (Li et al., 2014).

The maize Dof transcription factor prolamin box-binding factor (PBF) interacts with the O2 protein (Vicente-Carbajosa et al., 1997). This transcription factor, along with its homologs BPBF and WPBP in barley and wheat, respectively, can bind to the P-box and regulate the expression of genes encoding seed storage proteins (Mena et al., 1998; Jakoby et al., 2002; Gualberti et al., 2002). The *A. thaliana* Dof transcription factor family also affects the seed germination by influencing gene expression responses to light and cold (Gualberti et al., 2002) or regulates the accumulation and metabolism of storage materials.

Abscisic acid promotes the accumulation of storage materials, especially proteins. The ABA level is highest during the most active stage in terms of seed expansion and the accumulation of storage materials during the normal seed development of wheat, soybean, and bean. The effects of exogenous ABA vary in different plant species and are not always positive. Abscisic acid does not affect the accumulation of globulin in the cultured pea embryo. However, application of exogenous ABA is slightly correlated with the accumulation of agglutinin in the barley embryo.

Abscisic acid also affects seed lipid metabolism. The biosyntheses of long-chain fatty acids in the *A. thaliana* seeds of ABA-related single and double mutants are similarly affected. The arachidic acid level in the seeds of the *abscisic acid insensitive3* (*abi3*) mutant was only a third of that of the wild-type control. The abundance of neutral lipids in the seeds of *aba tai3–1* double mutant is much less than that of wild-type or single mutant seeds, while the soluble carbohydrate and starch levels are higher. The effect of ABA on assimilate distribution may be due to the regulation of the lipid phthalidyl chain extension enzyme and lipase.

19.4.5 Regulation of Starch Biosynthesis

Starch is the main component of rice endosperm, and rice quality depends on the synthesis and accumulation of starch in seeds. Systemic genetic studies have resulted in the comprehensive characterization of the key enzymes participating in rice and corn endosperm starch biosynthesis, including sucrose synthase and invertase, ADP-glucose pyrophosphorylase, starch synthase, starch-branching enzyme, and starch-debranching enzyme. There are many isoforms of these enzymes with different activities or cellular localizations in different species, and most of these enzymes regulate starch synthesis via translational and post-translational modifications (Hannah & James, 2008). Additionally, starch synthesis is also affected at the transcriptional level by spatio-temporal, rhythmic, hormonal, and alternative splicing (AS) regulatory activities.

Starch properties are abnormal in transgenic lines with suppressed expression of the starch synthase genes *OsSSIIa* and *OsSSIIIa*, ultimately resulting in chalkiness (Zhang et al., 2011). A loss-of-function mutation of OsBEIIb (starch-branching enzyme) results in small, crinkled, and chalky rice grains (Tanaka et al., 2004). Seeds of double mutants with mutations to ISA1 (isoamylase) and PUL (starch-debranching enzyme) are also crinkled (Fujita et al., 2009). The plastid phosphorylase Pho1 is important for starch synthesis, and the *pho1* mutation results in altered starch synthesis and structure under low temperatures as well as the chalky grains (Satoh et al., 2008).

Expression of genes involved in starch synthesis is generally regulated at the transcriptional level. An investigation using *A. thaliana* revealed that starch synthesis is regulated by the co-expression of some starch synthesis-related genes in the same tissue. The expression levels of these genes are highly correlated in different tissues and cell types. Additionally, an analysis of their promoter regions revealed that the *cis*-element sequence [ATC][AC][CTG][ATC]AAAGN[AC][GCA][ATC] is present in 83.3% of these genes, indicating that the expression of these genes is regulated via common mechanisms (Tsai et al., 2009). Moreover, a similar mechanism may also regulate the expression of rice genes related to starch synthesis.

Studies on the transcriptional regulation of rice starch synthesis may provide important clues regarding the regulatory network associated with starch synthesis. However, there are still few reports of transcription factors regulating the genes related to starch synthesis. In rice, *GBSS* encodes a key enzyme involved in amylose synthesis, and its activity is controlled by the *Waxy* (*Wx*) locus, which also contributes to the synthesis of amylose in rice pollen and embryo sac. A bHLH transcription factor (OsBP-5) that binds to the *Wx* promoter was identified in a yeast one-hybrid assay. Although OsBP-5 cannot activate transcription, further analyses revealed that OsBP-5 interacts with the EREBP-type transcription factor OsEBP-89 to promote the expression of downstream genes. Downregulated *OsBP-5* expression results in decreased seed amylose contents, implying that the co-regulation of *Wx* expression by OsBP-5 and OsEBP-89 influences amylose synthesis (Zhu et al., 2003a, b). Additionally, the bZIP family REB protein binds to the promoter regions of *SBE1* and *Wx*. Because *SBE1* and *Wx* have very similar expression patterns at different stages of seed development, it is suggested that REB regulates their expression synergistically.

Fu and Xue (2010) identified factors regulating starch synthesis based on a gene co-expression analysis. They showed that genes related to starch synthesis can be clustered into the following two groups: class I genes are responsible for the starch synthesis in the endosperm, while class II genes are responsible for the temporary synthesis of starch in non-storage tissues. A subsequent analysis showed that 307 and 621 genes are co-expressed with the class I and II starch synthesis genes, respectively. One AP2/EREBP family transcription factor, rice starch regulator 1 (RSR1), is negatively correlated with the expression of the class I genes. Interestingly, the expression levels of all class I genes are upregulated in the loss-of-function mutant *rsr1*. The increased amylose content in *rsr1* seeds and the modifications of the amylopectin structure alter the morphology of starch granules and ultimately decrease the starch gelatinization temperature. Overexpression of *RSR1* enhances the starch gelatinization properties. Moreover, the *rsr1* mutant produces abnormally large seeds and increased yields.

Alternative splicing (AS) and single nucleotide polymorphisms (SNPs) affect transcript activities. The AS of *Wx* has been observed. There are two main *Wx* alleles, Wx^{a} and Wx^{b}, in rice *indica* and *japonica* varieties, respectively (Sano, 1984). The splicing efficiency of Wx^{b} is considerably lower than that of Wx^{a} because of one base mutation (T to G) in the splice site at the 5′ end of the first intron. This mutation results in the decreased accumulation of mature mRNA and GBSSI, which leads to a considerable decrease in the amylose content of *japonica* rice varieties (Wang et al., 1995; Hirano et al., 1998). The rice dull endosperm gene *DU1* encodes a Prp (pre-mRNA processing) family member, which specifically affects the splicing of Wx^{b}, thereby affecting starch biosynthesis (Zeng et al., 2007). Additionally, because of the presence of SNPs, there are at least four other *Wx* alleles (*wx*, Wx^{op}, Wx^{mq}, and Wx^{hp}). The additional 23-bp repeat in the second exon of *wx* leads to a premature termination of translation and the absence of GBSSI, resulting in an amylose deficiency in sticky rice (Wanchana et al., 2003).

Another gene related to starch synthesis, *AGPS2* (encoding the small subunit of AGPase), is also affected by AS. The precursor mRNA of *AGPS2* generates two mRNAs, *AGPS2a* and *AGPS2b*, after the AS of the first exon. AGPS2a contains a transit peptide associated with plastid localization and is mainly responsible for starch synthesis in leaves. In contrast, AGPS2b lacks a transit peptide and is located in the cytoplasm, where it is primarily responsible for starch synthesis in the endosperm (Ohdan et al., 2005).

In addition to the key enzymes for starch synthesis, the endosperm cells also regulate starch metabolism through a series of responsive reactions and specific gene expressions. Pyruvate phosphate double kinase (PPDKB) is an important regulator of carbon transportation in starch and fat biosynthesis. The loss-of-function mutant of PPDKB has white core floury endosperm (Kang et al., 2005). The *FLOURY ENDOSPERM2* (*FLO2*) gene encodes a protein containing a TPR domain, which regulates the expression of genes related to the synthesis of starch and storage proteins. Mutation of FLO2 results in a smaller grain and reduced starch quality (She et al., 2010). Transcriptional factor *MADS6* is highly expressed in rice flower and endosperm, and the starch content in the loss-of-function mutant *mads6* is reduced, as well as the seriously disturbed development of *mads6* (Zhang et al., 2015).

Plant hormones also affect starch synthesis. Abscisic acid and sucrose can synergistically induce the expression of *ADP-glucose pyrophosphorylase large subunit3* (Akihiro et al., 2005). Moreover, low GA concentrations promote starch accumulation in maize suspension cells, while high GA concentrations have the opposite effect. These observations indicate that GA helps to regulate starch synthesis, although the underlying mechanism remains unclear (Cao & Shannon, 1997). Overexpression of the ethylene receptor *ETR2* gene suppresses the expression of α-amylase gene *RAmy3D* and a monosaccharide transporter gene. These changes affect starch degradation and transport of sugar from the source to the sink (seed), leading to the accumulation of starch in rice internodes. The repressed expression of *ETR2* promotes the expression of *RAmy3D* and the monosaccharide transporter gene, thus enhancing starch accumulation in seeds (Wuriyanghan et al., 2009).

During seed development in pea (*Pisum* sp.), bean (*Phaseolus* sp.), maize, wheat, and rice, the stage with the highest cytokinin content is also the stage with the fastest embryo differentiation and endosperm development. High cytokinin levels of seed are consistent with the phenomenon of rapid cell division. Moreover, the zeatin and ribosylzeatin levels in the rice endosperm are closely correlated with the speed of cell division.

19.5 Interactions Between the Embryo and Endosperm

Interactions between the embryo and endosperm were initially observed a long time ago. Embryo and endosperm sizes vary in different plant species; however, the ratio of embryo size to endosperm size in a species is stable, indicating that the balance of embryo size to endosperm size is strictly regulated.

19.5.1 Regulatory Effect of the Endosperm on the Embryo

Current studies indicate that the endosperm has an important regulatory effect on embryo development. This effect is not limited to supplying nutrients to the embryo but is also closely related to embryo cell division and growth as well as epidermal formation. The endosperm tends to develop earlier (at the start of mitosis) than the embryo after a double fertilization in most angiosperms (Lersten & Carlson, 2004). In *A. thaliana*, a polar nucleus in a fertilized central cell divides into at least 16 free endosperm nuclei (Brown et al., 2003), while at least 4–8 free nuclei are formed in the maize endosperm (Randolph, 1936) before cellularization. These findings indicate that genes are more actively transcribed in the endosperm than in the embryo after fertilization (Grimanelli et al., 2005). Furthermore, early embryogenesis might depend on endosperm development. However, there is no conclusive evidence supporting these hypotheses.

Many genes affecting endosperm development and then embryo development have been identified through studies of mutants exhibiting deficient endosperm development. Large nuclei were observed in the embryo and endosperm of the *titan1* and *titan2* mutants because of the abnormal formation of microtubules and the microtubule-mediated continuous DNA duplication without cytokinesis (McElver et al., 2000; Tzafrir et al., 2002). Thus, the cell division in the endosperm and embryo may be regulated by a common mechanism (Kondou et al., 2008). In other mutants, such as *haiku1* (*iku1*) and *iku2*, endosperm cellularization occurs too early, and endosperm development is inhibited, which affects embryo proliferation and finally leads to the production of relatively small seeds (Garcia et al., 2003). The *A. thaliana* sucrose transporter gene *AtSUC5* is expressed specifically in the endosperm near the micropyle, thereby providing sucrose for early embryo development (Baud et al., 2005). Moreover, ZHOUPI (ZOU)/RGE1 is a transcription factor that is specifically expressed in endosperm, especially after the embryo reaches the heart stage. The endosperm of the loss-of-function mutant *zou* does not degrade, leading to a significantly smaller mature embryo and smaller/shriveled seeds compared with the wild-type embryo and seeds (Kondou et al., 2008; Yang et al., 2008), demonstrating that ZOU regulates embryogenesis by influencing endosperm development.

Rice *CycB1;1* is expressed in both endosperm and embryo. The mature seeds of transgenic plants with suppressed expression of *CycB1;1* by RNAi essentially lack an endosperm but contain an embryo with an enlarged volume. A previous study indicated that CycB1;1 influences embryo volume by regulating endosperm development (Guo et al., 2010). Meanwhile, *ENL1* encodes a member of the SNF2 helicase family (Hara et al., 2015). The mature seeds of rice mutant *enl1* contain a large embryo and no endosperm. Further analysis showed that endosperm development is abnormal when the embryo is at the syncytial stage.

The *GIANT EMBRYO* (*GE*) gene encodes the cytochrome P450 protein CYP78A13, which is expressed in the endosperm near the embryo. An enlarged embryo and small endosperm were observed in a GE loss-of-function mutant, while *GE* overexpression resulted in a relatively small embryo and enlarged endosperm (Nagasawa et al., 2013; Yang et al., 2013). These observations indicate that the expression patterns of *GE* gene in the embryo and endosperm are necessary and crucial for balancing embryo/endosperm sizes, and the embryo-endosperm interaction determines the balance of embryo/endosperm sizes (Nagasawa et al., 2013).

Endosperm development is believed to influence embryo development not just regarding nutrient supplies, but also in terms of embryo proliferation and growth as well as the formation of the epidermal layer. A seed contains three main parts (embryo, endosperm, and integument) as well as additional zygotic and maternal tissues. How these three main parts are coordinated and develop synergistically should be clarified in future studies.

19.5.2 Regulatory Effect of the Embryo on the Endosperm

How the embryo affects endosperm development has not been fully characterized. There is currently some evidence suggesting that the embryo positively regulates endosperm development. Some evidence has come from the studies on *fis* mutants. The endosperm of mutants of *FIS* family genes exhibits developmental defects (e.g., abnormal cellularization and an inability to mature). However, the endosperm can develop normally by using *cdka;1* pollen grains to pollinate *fis* mutants in the presence of a normal embryo, which implies that the embryo generates signals that promote the normal endosperm development (Aw et al., 2010). It remains unclear how the endosperm perceives signals from the embryo. The *CLE19* gene encodes a small extracellular peptide and is expressed exclusively in the embryo. Interestingly, the cellularization of the endosperm is severely delayed in response to suppressed *CLE19* expression, implying the embryo may secrete CLE19 to promote endosperm cellularization. Additionally, endosperm cellularization is unaffected if a dominant negative *CLE19* gene is expressed in the endosperm, suggesting CLE19 has a single effect (Xu et al., 2005).

In maize, *Esr* expression is undetectable in the endosperm if there is no embryo, indicating that the developing embryo provides signals that regulate *Esr* expression in endosperm. This represents strong evidence supporting the synergistic regulation of embryonic and endosperm development. Additionally, receptor-like kinases (RLKs) form one of the most important and largest receptor families in higher plants (Walker, 1994; Shiu & Bleecker, 2001). Endosperm development is regulated by signals from the embryo and integument. Receptors in the endosperm perceiving these signals are likely RLKs, although it is unclear which ones are responsible.

19.6 Parental Effects and Epigenetic Regulation During Early Endosperm Development

The parent-of-origin effect refers to the phenomenon in which the traits of the offspring are influenced by only one parent's genotype. In plants, the parent-of-origin effect is mainly observed during embryo and endosperm development, in which the effects of specific genes are influenced by only one parent. There are two types of parent-of-origin effects in plants (i.e., imprinting and non-imprinting effects) (Pagnussat et al., 2005).

The imprinting effect, which is also known as genomic imprinting, refers to the differential expression of a gene in different parents (Reik & Walter, 2001). The offspring specifically express an allele from one parent (Berger et al., 2006; Mosher & Melnyk, 2010). In mammals, genomic imprinting occurs mainly in the placenta, but in plants, it occurs primarily in the endosperm (Luo et al., 2011). Because the genes from the parents have the same DNA sequence, genomic imprinting is regarded as a kind of epigenetic regulation mediated by DNA and histone methylation (Gehring, 2013).

19.6.1 Parental Effects Due to Imprinting

Genomic imprinting can be divided into paternal and maternal imprinting. In angiosperms, imprinting is a genetic phenomenon specific to the endosperm, with important implications for the regulation of endosperm development and function. In plants, the epigenetic asymmetry of the parental chromosome is the basis of genetic imprinting, and this phenomenon only occurs in the endosperm. An earlier investigation revealed that the *A. thaliana* endosperm genome is actually demethylated, with highly methylated siRNAs at non-CG sites (Hsieh et al., 2009).

Imprinting by DNA methylation: DNA sugar kinase Demeter (DME) is a kind of DNA glycosyltransferase produced in the central cells. It specifically transfers 5-methylpyrimidine in DNA and is the main factor responsible for DNA demethylation in the central cells. Methylated imprinted genes in the central cells will be demethylated by DME. However, the paternal alleles in the endosperm, vegetative tissues, and pollen grains remain methylated, which determines the methylation difference between the parental alleles. In *A. thaliana*, in addition to the selective demethylation by DME, the methylation maintained by DNA methyltransferase MET1 in *Arabidopsis*, which is the main DNA methyltransferase and maintains the methylation of cytosine at CG sites, also contributes to the differential methylation of imprinted genes.

Imprinting by DNA methylation occurs mainly in *FIS*-class genes, including *FIE*/*FIS3*, *FIS2*, *MEA*/*FIS1*, *MSI1*/*IRA1*, and *SWN* (Grossniklaus et al., 1998; Guitton et al., 2004; Kiyosue et al., 1999; Köhler et al., 2003a; Luo et al., 1999; Ohad et al., 1999; Wang et al., 2006). Mutations to these genes result in the following two common phenotypes: (1) the unfertilized central cells can spontaneously initiate endosperm development without fertilization, and (2) the embryo and endosperm proliferate abnormally and are ultimately aborted after fertilization (embryo lethality by maternal imprinting). Most of these imprinted genes are regulated by MET1 and DME (Gehring et al., 2009; Hsieh et al., 2009; Kinoshita et al., 2004; Liu et al., 2010; Mosher & Melnyk, 2010).

The MET1 and DME enzymes are important for regulating the specific expression of imprinted genes in the endosperm. During the development of the female gametophyte, *MET1* expression is inhibited, and there is a decrease in the methylation of the maternal allele. Before fertilization, DME demethylates the maternal allele in the central cells, resulting in fewer methylated imprinted genes in the endosperm than in the embryo. Thus, the expression level of these genes is significantly higher in the endosperm.

Imprinting by histone methylation: Histone methylation also plays important roles in imprinting. Methyltransferase activity has been observed at the histone 3 lysine 27 (H3K27) site of the FIS-PcG complex because the MEA protein contains the SET domain (Grossniklaus et al., 1998; Kiyosue et al., 1999). The FIS-PcG complex continuously inhibits the silencing of the paternal *MEA* gene. A defective FIS-PcG complex will result in the silencing of the paternal *MEA* gene (i.e., non-imprinting of the paternal *MEA* gene) (Baroux et al., 2006; Gehring et al., 2006; Jullien et al., 2006). In the *A. thaliana* endosperm, *PHE1* is another imprinted gene. Unlike *MEA*, *FIS2*, *FWA*, and *MPC*, *PHE1* is a maternally imprinted gene, and the paternal allele is primarily expressed. PHE1 is associated with the histone methylation of the H3K27 site by the FIS-PcG complex (Köhler et al., 2003b).

Imprinting may also be due to siRNAs. The synthesis of siRNAs in *A. thaliana* mainly depends on RNA polymerase IV (homolog of RNA polymerase II). The accumulation of RNA polymerase IV-dependent siRNA mainly occurs during the development of the female gametophyte. The expression of these siRNAs in the developing endosperm is specific to the maternal tissue, demonstrating the connection/correlation between genetic imprinting and RNA silencing (Gehring et al., 2009; Hsieh et al., 2009; Mosher & Melnyk, 2010; Liu et al., 2010).

19.6.2 Identification of Imprinted Genes

Recent advances in genomics research have significantly accelerated the identification of new imprinted genes. Gehring et al. (2009) identified five novel imprinted genes in *A. thaliana* by comparing the endosperm and other tissues regarding DNA methylation. By analyzing the differences between the paternal and maternal transcripts in the transcriptomes of hybrid offspring, more than 200 candidate imprinted genes were identified in the *A. thaliana* endosperm (Gehring et al., 2011; Hsieh et al., 2011). Similar strategies were applied to detect 262 candidates in the rice endosperm and three candidates in the rice embryo. Transcriptional differences were verified for 56 endosperm imprinted genes and one embryo imprinted gene (Luo et al., 2011). However, the ratio of homologous imprinted genes in monocotyledonous and dicotyledonous plants is very low.

The PRC2 complex mediates the modification of the histone methylation at the H3K27 lysine residue (H3K27me3) in the endosperm and inhibits the expression of target genes. The *A. thaliana* subunits *PRC2*, *MEA*, *FIS2*, and fertilization independent endosperm (FIE) are the maternally imprinted genes in the endosperm. Mutations to these genes will lead to endosperm development without fertilization, serious abnormalities in cellularization, and the production of seeds unable to completely develop (Ohad et al., 1996; Figueiredo et al., 2015). The rice and maize homologs of *FIE*, *OsFIE1*, and *ZmFIE1*, respectively, are maternally imprinted genes. However, the phenotype of the corresponding mutants did not include the development of the endosperm without fertilization (Danilevskaya et al., 2003; Luo et al., 2009), implying there is a functional difference between monocotyledonous and dicotyledonous imprinted genes.

19.6.3 Epigenetically Regulated Imprinted Genes and Endosperm Development

Unlike the embryo, the endosperm is associated with genome-wide demethylation (mainly at CG sites) as well as the considerable methylation (some non-CG sites) of siRNA target sequences. The high degree of methylation in endosperm represents a highly specific RNAi-targeting process (Hsieh et al., 2009; Mosher & Melnyk, 2010). Methylation levels at CG sites are partially restored in *dme* mutants, indicating that the methylation at GC sites is specific in maternally imprinted genes. Endosperm demethylation accompanied by CHH transposon hypermethylation in the embryo indicates that endosperm demethylation induces some siRNA synthesis. The resulting siRNAs may be transferred to the embryo to increase transposon silencing (Hsieh et al., 2009). Gehring et al. discovered the demethylation of the large-scale repeat sequences, which results in the imprinting of genes, during *A. thaliana* seed development. On the basis of the methylation profiles of *A. thaliana* embryos and endosperm determined using the MeDIP-sequencing method, novel imprinted genes and endosperm development-regulated genes were identified following the comparison of the methylation of embryo and endosperm (Gehring et al., 2009).

Three *A. thaliana FIS* genes (*MEDEA/FIS1*, *FIS2*, and *FIS3/FIE*) have been identified. Of these genes, *MEA* and *FIS2* encode the main regulators of endosperm development and are imprinted in the endosperm. Specifically, *MEA* encodes a PcG protein with a SET domain, which enables the formation of FIS-PcG complexes consisting of the VEFS domain-containing protein FIS2, the WD40 domain-containing protein fertilization independent endosperm (FIE), and multicopy suppressor of IRA 1 (MSI1). The FIS complex is crucial for the development of female gametes in *A. thaliana*, and a mutation to this complex causes the spontaneous division of unfertilized central cells and abnormal development.

In the male gamete, *MEA* expression is silenced by the PcG complex, while *FIS2* expression is silenced by MET1. However, the relationship between the methylation process and MET1 function is unclear. During the late stages of female gamete development, DME demethylates H3K27 and activates the expression of *MEA*, which induces *FIS2* expression via the demethylation of *FIS2* at a 5′-upstream position. Additionally, FIS binds to the *PHE1* promoter; H3K27 methylation and *PHE1* demethylation ultimately result in the silencing of *PHE1* in the central cells of the female gamete. In the fertilized endosperm, the maternal *MEA* and *FIS2* alleles are expressed, which causes FIS to form a complex with FIE and MSI1 to maintain the silencing of the paternal *MEA* allele.

19.7 Prospects

The mechanism regulating endosperm development is not only a key part of plant reproductive development, but it is also closely related to crop yield and quality. Endosperm development is regulated by a complex network involving hormones, transcription factors, genetic and epigenetic regulation, and protein modifications, all of which can be affected by internal physiological signals and external factors.

Recently developed omics-based technologies have been very useful for studying the genetic and epigenetic regulation of endosperm development, which will provide important clues to enhance crop yield and quality. Additionally, research involving modeling/computational biology and systems biology will lead to the identification of additional regulators and reveal the regulatory network underlying endosperm development. Further analyses will likely characterize the cytological basis of early endosperm development, clarify how endosperm development is regulated through epigenetic control, uncover details regarding the genetic regulatory network associated with endosperm development, and elucidate the mechanism responsible for the differences in the endosperm fate in monocotyledonous and dicotyledonous plant species.

References

Akihiro, T., Mizuno, K., & Fujimura, T. (2005). Gene expression of ADP-glucose pyrophosphorylase and starch contents in rice cultured cells are cooperatively regulated by sucrose and ABA. *Plant and Cell Physiology, 46*(6), 937–946.

Aw, S. J., Hamamura, Y., Chen, Z., Schnittger, A., & Berger, F. (2010). Sperm entry is sufficient to trigger division of the central cell but the paternal genome is required for endosperm development in *Arabidopsis*. *Development, 137*(16), 2683–2690.

Bai, A. N., Lu, X. D., Li, D. Q., Liu, J. X., & Liu, C. M. (2016). NF-YB1-regulated expression of sucrose transporters in aleurone facilitates sugar loading to rice endosperm. *Cell Research, 26*(3), 384.

Baroux, C., Gagliardini, V., Page, D. R., & Grossniklaus, U. (2006). Dynamic regulatory interactions of Polycomb group genes: MEDEA autoregulation is required for imprinted gene expression in Arabidopsis. *Genes & Development, 20*(9), 1081–1086.

Batista, R. A., Figueiredo, D. D., Santos-González, J., & Köhler, C. (2019). Auxin regulates endosperm cellularization in Arabidopsis. *Genes & Development, 33*(7–8), 466–476.

Baud, S., Wuilleme, S., Lemoine, R., et al. (2005). The AtSUC5 sucrose transporter specifically expressed in the endosperm is involved in early seed development in *Arabidopsis*. *The Plant Journal, 43*(6), 824–836.

Becraft, P. W., & Yi, G. (2010). Regulation of aleurone development in cereal grains. *Journal of Experimental Botany, 62*(5), 1669–1675.

Becraft, P. W., Stinard, P. S., & McCarty, D. R. (1996). CRINKLY4: A TNFR-like receptor kinase involved in maize epidermal differentiation. *Science, 273*(2), 1406–1408.

Berger, F., Grini, P. E., & Schnittger, A. (2006). Endosperm: An integrator of seed growth and development. *Current Opinion in Plant Biology, 9*(6), 664–670.

Bonello, J. F., Opsahl-Ferstad, H. G., Perez, P., Dumas, C., & Rogowsky, P. M. (2000). Esr genes show different levels of expression in the same region of maize endosperm. *Gene, 246*(1), 219–227.

Brown, R. C., Lemmon, B. E., & Nguyen, H. (2003). Events during the first four rounds of mitosis establish three developmental domains in the syncytial endosperm of *Arabidopsis thaliana*. *Protoplasma, 222*(3–4), 167–174.

Cao, H., & Shannon, J. C. (1997). Effect of gibberellin on growth, protein secretion, and starch accumulation in maize endosperm suspension cells. *Journal of Plant Growth Regulation, 16*(3), 137–140.

Chen, J., Nolan, T. M., Ye, H., et al. (2017). *Arabidopsis* WRKY46, WRKY54, and WRKY70 transcription factors are involved in brassinosteroid-regulated plant growth and drought responses. *The Plant Cell, 29*(6), 1425–1439.

Chojecki, A. J. S., Gale, M. D., & Bayliss, M. W. (1986). The number and sizes of starch granules in the wheat endosperm, and their association with grain weight. *Annals of Botany, 58*(6), 819–831.

Cochrane, M. P., & Duffus, C. M. (1983). Endosperm cell number in cultivars of barley differing in grain weight. *Annals of Applied Biology, 102*(1), 177–181.

Cooper, B., Clarke, J. D., Budworth, P., et al. (2003). A network of rice genes associated with stress response and seed development. *Proceedings of the National Academy of Sciences of the United States of America, 100*, 4945–4950.

Danilevskaya, O. N., Hermon, P., Hantke, S., et al. (2003). Duplicated fie genes in maize expression pattern and imprinting suggest distinct functions. *The Plant Cell, 15*(2), 425–438.

Day, R. C., Herridge, R. P., Ambrose, B. A., & Macknight, R. C. (2008). Transcriptome analysis of proliferating *Arabidopsis* endosperm reveals biological implications for the control of syncytial division, cytokinin signaling, and gene expression regulation. *Plant Physiology, 148*(4), 1964–1984.

Doan, D. N. P., Linnestad, C., & Olsen, O. A. (1996). Isolation of molecular markers from the barley endosperm coenocyte and the surrounding nucellus cell layers. *Plant Molecular Biology, 31*(4), 877–886.

Duan, K., Luo, Y. H., Luo, D., Xu, Z. H., & Xue, H. W. (2005). New insights into the complex and coordinated transcriptional regulation networks underlying rice seed development through cDNA chip-based analysis. *Plant Molecular Biology, 57*, 785–804.

Figueiredo, D. D., Batista, R. A., Roszak, P. J., & Köhler, C. (2015). Auxin production couples endosperm development to fertilization. *Nature Plants, 1*, 15184.

Fu, F. F., & Xue, H. W. (2010). Coexpression analysis identifies Rice Starch Regulator1, a rice AP2/EREBP family transcription factor, as a novel rice starch biosynthesis regulator. *Plant Physiology, 154*(2), 927–938.

Fujita, N., Toyosawa, Y., Utsumi, Y., et al. (2009). Characterization of pullulanase (PUL)-deficient mutants of rice (*Oryza sativa* L.) and the function of PUL on starch biosynthesis in the developing rice endosperm. *Journal of Experimental Botany, 60*(3), 1009–1023.

Garcia, D., Saingery, V., Chambrier, P., et al. (2003). *Arabidopsis* haiku mutants reveal new controls of seed size by endosperm. *Plant Physiology, 131*(4), 1661–1670.

Gehring, M. (2013). Genomic imprinting: Insights from plants. *Annual Review of Genetics, 47*, 187–208.

Gehring, M., Huh, J. H., Hsieh, T. F., et al. (2006). DEMETER DNA glycosylase establishes MEDEA polycomb gene self-imprinting by allele-specific demethylation. *Cell, 124*(3), 495–506.

Gehring, M., Bubb, K. L., & Henikoff, S. (2009). Extensive demethylation of repetitive elements during seed development underlies gene imprinting. *Science, 324*(5933), 1447–1451.

Gehring, M., Missirian, V., & Henikoff, S. (2011). Genomic analysis of parent-of-origin allelic expression in Arabidopsis thaliana seeds. *PloS One, 6*(8), e23687.

Gibson, P. R., Nov, R., Fielding, M., et al. (1999). Relationship of hydrolase activities to epithelial cell turnover in distal colonic mucosa of normal rats. *Journal of Gastroenterology and Hepatology, 14*(9), 866–872.

G6mez, E., Royo, J., Guo, Y., Thompson, R., & Hueros, G. (2002). Establishment of cereal endosperm expression domains identification and properties of a maize transfer cell-specific transcription factor, ZmMRP-1. *The Plant Cell, 14*(3), 599–610.

Grimanelli, D., Perotti, E., Ramirez, J., & Leblanc, O. (2005). Timing of the maternal-to-zygotic transition during early seed development in maize. *The Plant Cell, 17*(4), 1061–1072.

Grossniklaus, U., Vielle-Calzada, J. P., Hoeppner, M. A., & Gagliano, W. B. (1998). Maternal control of embryogenesis by MEDEA, a polycomb group gene in *Arabidopsis*. *Science, 280*(5362), 446–450.

Grossniklaus, U., Nogler, G. A., & Van Dijk, P. J. (2001). How to avoid sex: The genetic control of gametophytic apomixis. *The Plant Cell, 13*, 1491–1498.

Gualberti, G., Papi, M., Bellucci, L., et al. (2002). Mutations in the Dof zinc finger genes DAG2 and DAG1 influence with opposite effects the germination of *Arabidopsis* seeds. *The Plant Cell, 14*, 1253–1263.

Guitton, A. E., Page, D. R., Chambrier, P., et al. (2004). Identification of new members of Fertilisation Independent Seed Polycomb Group pathway involved in the control of seed development in *Arabidopsis thaliana*. *Development, 131*(12), 2971–2981.

Guo, J., Wang, F., Song, J., Sun, W., & Zhang, X. S. (2010). The expression of Orysa; CycB1; 1 is essential for endosperm formation and causes embryo enlargement in rice. *Planta, 231*(2), 293.

Hara, T., Katoh, H., Ogawa, D., et al. (2015). Rice SNF2 family helicase ENL1 is essential for syncytial endosperm development. *The Plant Journal, 81*(1), 1–12.

Hannah, L. C., & James, M. (2008). The complexities of starch biosynthesis in cereal endosperms. *Current Opinion in Biotechnology, 19*(2), 160–165.

Hirano, H. Y., Eiguchi, M., & Sano, Y. (1998). A single base change altered the regulation of the Waxy gene at the posttranscriptional level during the domestication of rice. *Molecular Biology and Evolution, 15*, 978–987.

Hsieh, T. F., Ibarra, C. A., Silva, P., et al. (2009). Genome-wide demethylation of *Arabidopsis* endosperm. *Science, 324*, 1451–1454.

Hsieh, T. F., Shin, J., Uzawa, R., et al. (2011). Regulation of imprinted gene expression in *Arabidopsis* endosperm. *Proceedings of the National Academy of Sciences of the United States of America, 108*(5), 1755–1762.

Jakoby, M., Weisshaar, B., Dröge-Laser, W., et al. (2002). bZIP transcription factors in *Arabidopsis*. *Trends in Plant Science, 7*, 106–111.

Jiang, W. B., Huang, H. Y., Hu, Y. W., et al. (2013). Brassinosteroid regulates seed size and shape in *Arabidopsis*. *Plant Physiology, 162*, 1965–1977.

Jullien, P. E., Katz, A., Oliva, M., et al. (2006). Polycomb group complexes self-regulate imprinting of the Polycomb group gene MEDEA in Arabidopsis. *Current Biology, 16*(5), 486–492.

Kang, H. G., Park, S., Matsuoka, M., & An, G. (2005). White-core endosperm floury endosperm4 in rice is generated by knockout mutations in the C4-type pyruvate orthophosphate dikinase gene (OsPPDKB). *The Plant Journal, 42*(6), 901–911.

Kawakatsu, T., Yamamoto, M. P., Touno, S. M., Yasuda, H., & Takaiwa, F. (2009). Compensation and interaction between RISBZ1 and RPBF during grain filling in rice. *The Plant Journal, 59*(6), 908–920.

Kinoshita, T., Miura, A., Choi, Y., et al. (2004). One-way control of FWA imprinting in *Arabidopsis* endosperm by DNA methylation. *Science, 303*(5657), 521–523.

Kiyosue, T., Ohad, N., Yadegari, R., et al. (1999). Control of fertilization-independent endosperm development by the MEDEA polycomb gene in *Arabidopsis*. *Proceedings of the National Academy of Sciences of the United States of America, 96*(7), 4186–4191.

Köhler, C., Hennig, L., Bouveret, R., et al. (2003a). *Arabidopsis* MSI1 is a component of the MEA/FIE Polycomb group complex and required for seed development. *The EMBO Journal, 22*(18), 4804–4814.

Köhler, C., Hennig, L., Spillane, C., Pien, S., Gruissem, W., & Grossniklaus, U. (2003b). The Polycomb-group protein MEDEA regulates seed development by controlling expression of the MADS-box gene PHERES1. *Genes & Development, 17*(12), 1540–1553.

Kondou, Y., Nakazawa, M., Kawashima, M., et al. (2008). RETARDED GROWTH OF EMBRYO1, a new basic helix-loop-helix protein, expresses in endosperm to control EMBRYO growth. *Plant Physiology, 147*(4), 1924–1935.

Kvaale, A., & Olsen, A. (1986). Rates of cell division in developing barley endosperms. *Annals of Botany, 57*(6), 829–833.

Lersten, N. R., & Carlson, J. B. (2004). Reproductive morphology. In *Soybeans: Improvement, production, and uses* (pp. 15–57). American Society of Agronomy, Crop Science Society of America, Soil Science Society of America.

Li, Y., Fan, C., Xing, Y., et al. (2014). Chalk5 encodes a vacuolar H^+-translocating pyrophosphatase influencing grain chalkiness in rice. *Nature Genetics, 46*(4), 398–404.

Lid, S. E., Gruis, D., Jung, R., et al. (2002). The defective kernel 1 (dek1) gene required for aleurone cell development in the endosperm of maize grains encodes a membrane protein of the calpain gene superfamily. *Proceedings of the National Academy of Sciences of the United States of America, 99*(8), 5460–5465.

Liu, Y., Yan, Z., Chen, N., et al. (2010). Development and function of central cell in angiosperm female gametophyte. *Genesis, 48*(8), 466–478.
Liu, J., Wu, X., Yao, X., Yu, R., Larkin, P. J., & Liu, C. M. (2018). Mutations in the DNA demethylase OsROS1 result in a thickened aleurone and improved nutritional value in rice grains. *Proceedings of the National Academy of Sciences of the United States of America, 115*(44), 11327–11332.
Luo, M., Bilodeau, P., Koltunow, A., et al. (1999). Genes controlling fertilization-independent seed development in *Arabidopsis thaliana*. *Proceedings of the National Academy of Sciences of the United States of America, 96*(1), 296–301.
Luo, M., Platten, D., Chaudhury, A., Peacock, W. J., & Dennis, E. S. (2009). Expression, imprinting, and evolution of rice homologs of the polycomb group genes. *Molecular Plant, 2*(4), 711–723.
Luo, M., Taylor, J. M., Andrew, S., et al. (2011). A genome-wide survey of imprinted genes in rice seeds reveals imprinting primarily occurs in the endosperm. *PLoS Genetics, 7*(6), e1002125.
Mallory, A. C., & Vaucheret, H. (2006). Functions of microRNAs and related small RNAs in plants. *Nature Genetics, 38*, S31–S36.
McElver, J., Patton, D., Rumbaugh, M., et al. (2000). The TITAN5 gene of *Arabidopsis* encodes a protein related to the ADP ribosylation factor family of GTP binding proteins. *The Plant Cell, 12*(8), 1379–1392.
Mena, M., Vicente-Carbajosa, J., Schmidt, R. J., & Carbonero, P. (1998). An endosperm-specific DOF protein from barley, highly conserved in wheat, binds to and activates transcription from the prolamin-box of a native B-hordein promoter in barley endosperm. *The Plant Journal, 16*(1), 53–62.
Mosher, R. A., & Melnyk, C. W. (2010). siRNAs and DNA methylation: Seedy epigenetics. *Trends in Plant Science, 15*(4), 204–210.
Mosher, R. A., Melnyk, C. W., Kelly, K. A., et al. (2009). Uniparental expression of PolIV-dependent siRNAs in developing endosperm of *Arabidopsis*. *Nature, 460*(7252), 283–286.
Nagasawa, N., Hibara, K., Heppard, E. P., et al. (2013). GIANT EMBRYO encodes CYP78A13, required for proper size balance between embryo and endosperm in rice. *The Plant Journal, 75*(4), 592–605.
Ohad, N., Margossian, L., Hsu, Y. C., et al. (1996). A mutation that allows endosperm development without fertilization. *Proceedings of the National Academy of Sciences of the United States of America, 93*(11), 5319–5324.
Ohad, N., Yadegari, R., Margossian, L., et al. (1999). Mutations in FIE, a WD polycomb group gene, allow endosperm development without fertilization. *The Plant Cell, 11*(3), 407–416.
Ohdan, T., Francisco, P. B., Jr., Sawada, T., et al. (2005). Expression profiling of genes involved in starch synthesis in sink and source organs of rice. *Journal of Experimental Botany, 56*(422), 3229–3244.
Olsen, O. A. (2004). Nuclear endosperm development in cereals and *Arabidopsis thaliana*. *The Plant Cell, 16*(suppl 1), S214–S227.
Onodera, Y., Suzuki, A., Wu, C. Y., Washida, H., & Takaiwa, F. (2001). A rice functional transcriptional activator, RISBZ1, responsible for endosperm-specific expression of storage protein genes through GCN4 motif. *Journal of Biological Chemistry, 276*(17), 14139–14152.
Opsahl Ferstad, H. G., Deunff, E. L., Dumas, C., & Rogowsky, P. M. (1997). ZmEsr, a novel endosperm-specific gene expressed in a restricted region around the maize embryo. *The Plant Journal, 12*(1), 235–246.
Pagnussat, G. C., Yu, H. J., Ngo, Q. A., et al. (2005). Genetic and molecular identification of genes required for female gametophyte development and function in *Arabidopsis*. *Development, 132*(3), 603–614.
Qin, P., Zhang, G., Hu, B., Wu, J., Chen, W., Ren, Z., Liu, Y., Xie, J., Yuan, H., Tu, B., Ma, B., Wang, Y., Ye, L., Li, L., Xiang, C., & Li, S. (2021). Leaf-derived ABA regulates rice seed development via a transporter-mediated and temperature-sensitive mechanism. *Science Advances, 7*(3), eabc8873.

Qu, L. Q., & Takaiwa, F. (2004). Evaluation of tissue specificity and expression strength of rice seed component gene promoters in transgenic rice. *Plant Biotechnology Journal, 2*(2), 113–125.
Randolph, L. F. (1936). Developmental morphology of the caryopsis in maize. *Journal of Agricultural Research, 12*(53), 881–916.
Reik, W., & Walter, J. (2001). Genomic imprinting: Parental influence on the genome. *Nature Reviews Genetics, 2*(1), 21–32.
Ren, Y., Wang, Y., Liu, F., et al. (2014). GLUTELIN PRECURSOR ACCUMULATION3 encodes a regulator of post-Golgi vesicular traffic essential for vacuolar protein sorting in rice endosperm. *The Plant Cell, 26*(1), 410–425.
Sabelli, P. A., Hoerster, G., Lizarraga, L. E., et al. (2009). Positive regulation of minichromosome maintenance gene expression, DNA replication, and cell transformation by a plant retinoblastoma gene. *Proceedings of the National Academy of Sciences of the United States of America, 106*(10), 4042–4047.
Sano, Y. (1984). Differential regulation of waxy gene expression in rice endosperm. *Theoretical and Applied Genetics, 64*(5), 467–473.
Satoh, H., Shibahara, K., Tokunaga, T., et al. (2008). Mutation of the plastidial α-glucan phosphorylase gene in rice affects the synthesis and structure of starch in the endosperm. *The Plant Cell, 20*(7), 1833–1849.
She, K. C., Kusano, H., Koizumi, K., et al. (2010). A novel factor FLOURY ENDOSPERM2 is involved in regulation of rice grain size and starch quality. *The Plant Cell, 22*(10), 3280–3294.
Shen, B., Li, C., Min, Z., et al. (2003). Sal1 determines the number of aleurone cell layers in maize endosperm and encodes a class E vacuolar sorting protein. *Proceedings of the National Academy of Sciences of the United States of America, 100*(11), 6552–6557.
Shiu, S. H., & Bleecker, A. B. (2001). Receptor-like kinases from *Arabidopsis* form a monophyletic gene family related to animal receptor kinases. *Proceedings of the National Academy of Sciences of the United States of America, 98*(19), 10763–10768.
Steffen, J. G., Kang, I. H., Portereiko, M. F., Lloyd, A., & Drews, G. N. (2008). AGL61 interacts with AGL80 and is required for central cell development in *Arabidopsis*. *Plant Physiology, 148*, 259–268.
Tanabe, S., Ashikari, M., Fujioka, S., et al. (2005). A novel cytochrome P450 is implicated in brassinosteroid biosynthesis via the characterization of a rice dwarf mutant, dwarf11, with reduced seed length. *The Plant Cell, 17*(3), 776–790.
Tanaka, N., Fujita, N., Nishi, A., et al. (2004). The structure of starch can be manipulated by changing the expression levels of starch branching enzyme IIb in rice endosperm. *Plant Biotechnology Journal, 2*(6), 507–516.
Tonosaki, K., Ono, A., Kunisada, M., Nishino, M., Nagata, H., Sakamoto, S., Kijima, S. T., Furuumi, H., Nonomura, K. I., Sato, Y., Ohme-Takagi, M., Endo, M., Comai, L., Hatakeyama, K., Kawakatsu, T., & Kinoshita, T. (2021). Mutation of the imprinted gene *OsEMF2a* induces autonomous endosperm development and delayed cellularization in rice. *The Plant Cell, 33*(1), 85–103.
Tsai, H. L., Lue, W. L., Lu, K. J., et al. (2009). Starch synthesis in *Arabidopsis* is achieved by spatial cotranscription of core starch metabolism genes. *Plant Physiology, 151*(3), 1582–1595.
Tzafrir, I., McElver, J. A., Liu, C., et al. (2002). Diversity of TITAN functions in *Arabidopsis* seed development. *Plant Physiology, 128*(1), 38–51.
Valencia-Sanchez, M. A., Liu, J., Hannon, G. J., & Parker, R. (2006). Control of translation and mRNA degradation by miRNAs and siRNAs. *Genes & Development, 20*(5), 515–524.
Vicente-Carbajosa, J., Moose, S. P., Parsons, R. L., & Schmidt, R. J. (1997). A maize zinc-finger protein binds the prolamin box in zein gene promoters and interacts with the basic leucine zipper transcriptional activator Opaque2. *Proceedings of the National Academy of Sciences of the United States of America, 94*(14), 7685–7690.
Walker, J. C. (1994). Structure and function of the receptor-like protein kinases of higher plants. *Plant Molecular Biology, 26*(5), 1599–1609.

Wanchana, S., Toojinda, T., Tragoonrung, S., & Vanavichit, A. (2003). Duplicated coding sequence in the waxy allele of tropical glutinous rice (*Oryza sativa* L.). *Plant Science, 165*(6), 1193–1199.
Wang, Z. Y., Zheng, F. Q., Shen, G. Z., et al. (1995). The amylose content in rice endosperm is related to the post-transcriptional regulation of the Waxy gene. *The Plant Journal, 7*(4), 613–622.
Wang, D., Tyson, M. D., Jackson, S. S., & Yadegari, R. (2006). Partially redundant functions of two SET-domain polycomb-group proteins in controlling initiation of seed development in *Arabidopsis*. *Proceedings of the National Academy of Sciences of the United States of America, 103*(35), 13244–13249.
Wang, E., Wang, J., Zhu, X., et al. (2008). Control of rice grain-filling and yield by a gene with a potential signature of domestication. *Nature Genetics, 40*(11), 1370–1374.
Wang, Y., Ren, Y., Liu, X., et al. (2010). OsRab5a regulates endomembrane organization and storage protein trafficking in rice endosperm cells. *The Plant Journal, 64*(5), 812–824.
Wuriyanghan, H., Zhang, B., Cao, W. H., et al. (2009). The ethylene receptor ETR2 delays floral transition and affects starch accumulation in rice. *The Plant Cell, 21*(5), 1473–1494.
Xu, J., Zhang, H. Y., Xie, C. H., et al. (2005). EMBRYONIC FACTOR 1 encodes an AMP deaminase and is essential for the zygote to embryo transition in *Arabidopsis*. *The Plant Journal, 42*(5), 743–758.
Xu, J. J., Zhang, X. F., & Xue, H. W. (2016). Rice aleurone layer specific OsNF-YB1 regulates grain filling and endosperm development by interacting with an ERF transcription factor. *Journal of Experimental Botany, 67*(22), 6399–6411.
Xu, X., Zhang, D., Niu, B., Chen, C., Yun, Q., & Zhou, Y. (2021). OsYUC11-mediated auxin biosynthesis is essential for endosperm development of rice. *Plant Physiology, 185*(3), 934–950.
Xue, L. J., Zhang, J. J., & Xue, H. W. (2012). Genome-wide analysis of the complex transcriptional networks of rice developing seeds. *PLoS One, 7*, e31081.
Yamamoto, M. P., Onodera, Y., Touno, S. M., & Takaiwa, F. (2006). Synergism between RPBF Dof and RISBZ1 bZIP activators in the regulation of rice seed expression genes. *Plant Physiology, 141*(4), 1694–1707.
Yamamuro, C., Ihara, Y., Wu, X., et al. (2000). Loss of function of a rice brassinosteroid insensitive1 homolog prevents internode elongation and bending of the lamina joint. *The Plant Cell, 12*(9), 1591–1605.
Yang, H. J., Shen, H., Chen, L., et al. (2002a). The OsEBP-89 gene of rice encodes a putative EREBP transcription factor and is temporally expressed in developing endosperm and intercalary meristem. *Plant Molecular Biology, 50*(3), 379–391.
Yang, J. Y., Chung, M. C., Tu, C. Y., & Leu, W. M. (2002b). OSTF1: A HD-GL2 family homeobox gene is developmentally regulated during early embryogenesis in rice. *Plant Cell Physiology, 43*(6), 628–638.
Yang, J., Zhang, J., Huang, Z., et al. (2002c). Correlation of cytokinin levels in the endosperms and roots with cell number and cell division activity during endosperm development in rice. *Annals of Botany, 90*(3), 369–377.
Yang, S., Johnston, N., Talideh, E., et al. (2008). The endosperm-specific ZHOUPI gene of *Arabidopsis thaliana* regulates endosperm breakdown and embryonic epidermal development. *Development, 135*(21), 3501–3509.
Yang, W., Gao, M., Yin, X., et al. (2013). Control of rice embryo development, shoot apical meristem maintenance, and grain yield by a novel cytochrome p450. *Molecular Plant, 6*(6), 1945–1960.
Yasuda, H., Hirose, S., Kawakatsu, T., Wakasa, Y., & Takaiwa, F. (2009). Overexpression of BiP has inhibitory effects on the accumulation of seed storage proteins in endosperm cells of rice. *Plant and Cell Physiology, 50*(8), 1532–1543.
Yi, G., Lauter, A. M., Scott, M. P., & Becraft, P. W. (2011). The thick aleurone1 mutant defines a negative regulation of maize aleurone cell fate that functions downstream of defective kernel1. *Plant Physiology, 156*(4), 1826–1836.

Yin, L. L., & Xue, H. W. (2012). The MADS29 transcription factor regulates the degradation of the nucellus and the nucellar projection during rice seed development. *The Plant Cell, 24*(3), 1049–1065.
Zeng, D., Yan, M., Wang, Y., et al. (2007). Du1, encoding a novel Prp1 protein, regulates starch biosynthesis through affecting the splicing of Wxb pre-mRNAs in rice (*Oryza sativa* L.). *Plant Molecular Biology, 65*(4), 501–509.
Zhang, Y., Cao, G., Qu, L. J., & Gu, H. (2009). Involvement of an R2R3-MYB transcription factor gene AtMYB118 in embryogenesis in *Arabidopsis*. *Plant Cell Reports, 2*(3), 337–346.
Zhang, G., Cheng, Z., Zhang, X., et al. (2011). Double repression of soluble starch synthase genes SSIIa and SSIIIa in rice (*Oryza sativa* L.) uncovers interactive effects on the physicochemical properties of starch. *Genome, 54*(6), 448–459.
Zhang, Y., Du, L., Xu, R., et al. (2015). Transcription factors SOD7/NGAL2 and DPA4/NGAL3 act redundantly to regulate seed size by directly repressing KLU expression in *Arabidopsis thaliana*. *The Plant Cell, 27*(3), 620–632.
Zhao, P., Zhou, X. M., Zhang, L. Y., et al. (2013). A bipartite molecular module controls cell death activation in the basal cell lineage of plant embryos. *PLoS Biology, 11*, e1001655.
Zhou, S. R., Yin, L. L., & Xue, H. W. (2013). Functional genomics based understanding of rice endosperm development. *Current Opinion in Plant Biology, 16*, 236–246.
Zhu, T., Budworth, P., Chen, W. Q., et al. (2003a). Transcriptional control of nutrient partitioning during rice grain filling. *Plant Biotechnology Journal, 1*, 59–70.
Zhu, Y., Cai, X. L., Wang, Z. Y., & Hong, M. M. (2003b). An interaction between a MYC protein and an EREBP protein is involved in transcriptional regulation of the rice Wx gene. *Journal of Biological Chemistry, 278*(48), 47803–47811.
Zilberman, D., Cao, X., & Jacobsen, S. E. (2003). ARGONAUTE4 control of locus-specific siRNA accumulation and DNA and histone methylation. *Science, 299*(5607), 716–719.

Chapter 20
Genomic Imprinting in Plants

Binglian Zheng

Abstract Imprinting is a crucial phenomenon in the epigenetic regulation of flowering plants and mammals, where the preferential expression of alleles depends on their parental origin. Disruption of genomic imprinting usually causes abortion of early zygotic development in both plants and mammals. Genomic imprinting arises from extensive chromatin remodeling within the epigenome, typically involving clustered imprinting genes flanked by transposable elements. Consequently, the differential expression of alleles is closely linked to DNA methylation and histone modifications, mainly histone 3 lysine 27 tri-methylation (H3K27me3). In plants, imprinting predominantly occurs in the endosperm, a triploid placenta-like tissue that envelops and nurtures the embryo during seed development. Therefore, imprinted genes are involved in not only seed development but also the establishment of seed dormancy, an important adaptive trait for plant evolution. This chapter provides a comprehensive review of established knowledge and recent advancements in understanding the regulation of genomic imprinting, the key factors influencing these processes, and the biological functions associated with genomic imprinting.

Keywords Imprinting · DNA methylation · Histone methylation · Endosperm · Seed development

20.1 Introduction

Genomic imprinting is characterized by the preferential expression of alleles, determined by the parent of origin. Imprinted genes exhibit a predominant expression pattern from either the maternal or paternal allele, in stark contrast to non-imprinted genes, where both parental alleles are typically expressed. A gene primarily

B. Zheng (✉)
State Key Laboratory of Genetics and Development of Complex Phenotypes, Institute of Plant Biology, School of Life Sciences, Fudan University, Shanghai, China
e-mail: zhengbl@fudan.edu.cn

F. Chang et al. (eds.), *Regulation of Plant Development*,
https://doi.org/10.1007/978-981-95-7033-1_20

expressed from the maternally inherited allele is termed a maternally expressed imprinted gene (MEG), while a gene expressed from the paternally inherited allele is labeled as a paternally expressed imprinted gene (PEG). Genomic imprinting is recognized as a prototypical epigenetic phenomenon, as it involves the distinct expression of alleles with identical DNA sequences.

Breeders have noted the influence of parent-of-origin effects on plant and animal development for millennia. The outcome of crosses between donkeys and horses, resulting in a hinny when the male parent is a horse and a mule when the female parent is a horse, is a well-known example. In corn, specific alleles of the R and B genes yield fully colored kernels when inherited from one parent but not when inherited from the other parent. Arabidopsis seeds inheriting an excess of paternal or maternal genomes display reciprocal seed size phenotypes. A subset of parent-of-origin effects is attributed to imprinted gene expression. Imprinting can impact entire chromosomes, as demonstrated by paternal X-chromosome inactivation in most mammals, or more commonly, it may involve a single gene in plants.

While X-chromosome inactivation stands out as a familiar example of genomic imprinting in mammals, the initial demonstration of genomic imprinting was observed in the R gene in maize (Kermicle, 1970). It definitively clarified two key principles regarding plant genomic imprinting: (1) alleles are inherited in distinct forms from the male and female gametophytes; (2) this disparity impacts the endosperm but not the embryo. Subsequent studies have identified additional imprinted genes and elucidated the mechanisms of imprinting in plants (Huh et al., 2008; Ferguson-Smith, 2011). In contrast, the discovery of the first mammalian imprinted gene was reported much later (Bartolomei et al., 1991; Barlow et al., 1991; DeChiara et al., 1991). The identification of a differentially methylated imprinting control region at the H19-Igf2 locus supported the notion that differential DNA methylation played a crucial role in establishing or maintaining imprinted expression (Sasaki et al., 1992).

In contrast to the meiotic products, which are directly the sperm or the egg cell in animals, the haploid products of plant meiosis do not immediately differentiate into the gametes. Instead, they undergo multiple mitotic divisions to develop into multicellular gametophytes. The mature pollen, representing the male gametophyte, comprises two sperm cells and a vegetative nucleus. Upon germination, a pollen tube is formed, facilitating the delivery of twin sperm cells to the female gametophyte for double fertilization (Fig. 20.1; McCormick, 2004). In flowering plants, one sperm cell fuses with the egg to give rise to the embryo, while the other sperm cell fuses with the central cell to form the triploid endosperm (Fig. 20.1). The endosperm, akin to the mammalian placenta, serves as a nourishing tissue for the embryo. In flowering plants, imprinted gene expression is primarily restricted to the endosperm. Through genetic screening and genome-scale investigations, hundreds of imprinted genes have been identified in Arabidopsis, rice, and maize (Chaudhuri & Messing, 1994; Grossniklaus et al., 1998; Luo et al., 2000; Danilevskaya et al., 2003; Gutierrez-Marcos et al., 2004; Kohler et al., 2005; Kinoshita et al., 2007; Tiwari et al., 2008; Fitz Gerald et al., 2009; Gehring et al., 2011; Hsieh et al., 2011;

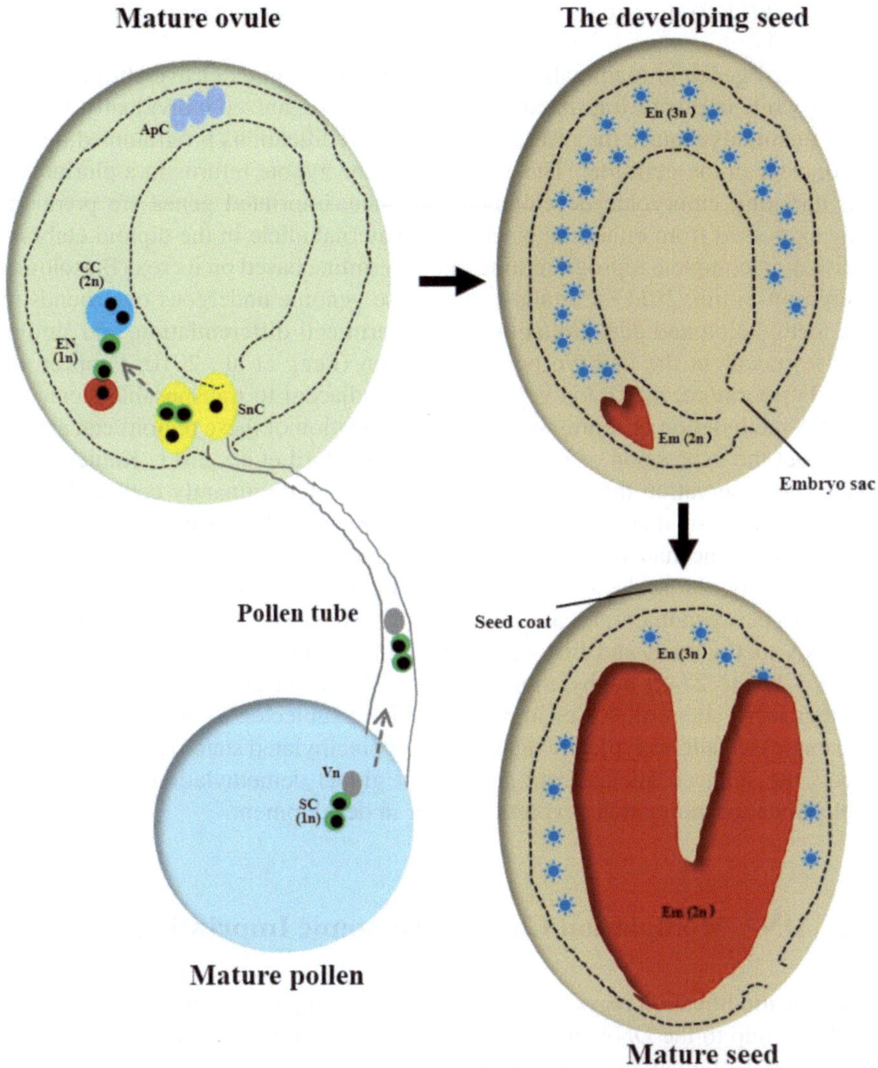

Fig. 20.1 Schematics depicting mature pollen, mature ovule, and seed development. Mature pollen consists of one vegetative nucleus (Vn) and twin sperm cells (SC). In contrast, a mature ovule contains the embryo sac, highlighted with a dashed line, which comprises four differentiated cell types: one egg cell (EC), one diploid (2n) central cell (CC), two synergid cells (SnC), and three antipodal cells (ApC). Upon fertilization, the egg cell is fertilized by one of the sperm delivered by the pollen tube, resulting in the production of a diploid embryo. Simultaneously, the fusion of the second sperm with the diploid central cell initiates triploid (3n) endosperm development. It is worth noting that Arabidopsis endosperm is transient and is consumed during seed development, leaving only one cell layer in the mature seed. The sporophytic tissue of the ovule develops into the seed coat after fertilization, resulting in a mature seed with three distinct parts and genetic constitutions: a diploid maternal seed coat, a diploid embryo, and an endosperm with two maternal and one paternal genome. This intricate process ensures the proper development and genetic makeup of seeds in Arabidopsis

Luo et al., 2011; Waters et al., 2011; Wolff et al., 2011; Zhang et al., 2011; Waters et al., 2013).

Imprinted genes in mammals are typically arranged in extensive chromosomal clusters, while in plants, imprinted genes seem to be organized individually. Unlike plants, mammals require the reset of epigenetic modifications programmed during development in the germline. This ensures that the zygote returns to a pluripotent state, initiating embryonic development. In mice, imprinted genes are predominantly expressed from either the paternal or maternal allele in the diploid embryo, necessitating complete reprogramming in the germline based on its sex (Bartolomei & Ferguson-Smith, 2011). Certainly, the mouse genome undergoes two rounds of DNA methylation and demethylation during germ cell differentiation, and similar processes occur in the embryo post-fertilization (Feng et al., 2010; Popp et al., 2010). Furthermore, imprinted genes are often adjacent to transposable elements, and DNA methylation or repressive histone methylation of these regions can extend and impact the expression of imprinted genes (Radford et al., 2011; Kohler et al., 2012). The observation that imprinting in plants is predominantly confined to the endosperm suggests that the initiation and preservation of imprinting undergo distinct regulatory mechanisms during the early stages of seed development.

Additionally, the molecular regulation of imprinted expression shows striking similarities between mammals and plants. Imprinted regions typically exhibit extensive DNA methylation and histone modifications, with a particular emphasis on histone 3 lysine 27 tri-methylation in plants (Kohler et al., 2012). During plant gametogenesis, silent alleles of imprinted loci are subjected to DNA methylation, while expressed alleles typically maintain a hypomethylated state. This methylation protects the silenced alleles from subsequent global demethylation, ensuring the establishment of monoallelic expression later in development.

20.2 DNA Methylation Controls Genomic Imprinting

Epigenetic modification plays a crucial role in regulating gene expression by adding a methyl group to the DNA molecule. In the context of genomic imprinting, DNA methylation is often associated with the silencing of one parental allele, leading to monoallelic expression of imprinted genes. This process occurs during gametogenesis, where specific regions of the genome are marked with methyl groups in a parent-of-origin-dependent manner. Understanding the dynamic nature of DNA methylation in the context of genomic imprinting provides valuable insights into the intricate mechanisms governing gene expression and development.

The regulation of imprinted gene expression in mammals is governed by crucial sequence elements known as imprinting control regions (ICRs). These regions extensively overlap with CpG islands and are subject to control through allele-specific DNA methylation. Allelic DNA methylation takes place during female gametogenesis. After fertilization, the DNA methylation pattern at imprinting control regions (ICRs) persists throughout development and is essential for the allelic expression of imprinted genes. In plants, cytosines in DNA can undergo methylation across all classes of sequence context, including CG, CHG, and CHH (where H = A, C, or T) (Law & Jacobsen, 2010). Arabidopsis maintains methylation states through three distinct pathways: CG methylation is maintained by MET1, a major DNA methyltransferase; CHG methylation is sustained by Chromomethylase 3 (CMT3), a plant-specific DNA methyltransferase; and CHH methylation is mainly controlled by two de novo DNA methyltransferases, Domains Rearranged Methyltransferase (DRM1 and DRM2) (Cao et al., 2003; Law & Jacobsen, 2010). Chromomethylase 2 (CMT2) is also responsible for the maintenance of a subset of CHH methylation independently of an RNA-directed DNA methylation pathway (Zemach et al., 2013). DNA methylation is typically linked to the silencing of genes and transposable elements (Jones, 2012). Consequently, the differential methylation of parental alleles can lead to imprinted expression, making it a recognized initial imprinting signal.

The link between DNA methylation and imprinting was first discovered in mammals (Bartolomei et al., 1993; Stoger et al., 1993; Ferguson-Smith et al., 1993) and in plants (Kinoshita et al., 2004). The regulatory importance of DNA methylation is substantiated by evidence showing that disruptions in methylation patterns at individual genes lead to the abolition of imprinting (Adams et al., 2000; Xiao et al., 2003; Kinoshita et al., 2004; Jullien et al., 2006; Makarevich et al., 2008; Hsieh et al., 2011; Vu et al., 2013). In contrast to that in the embryo, the global level of DNA methylation in the endosperm is much less, mainly due to low expression of DNA methyltransferases and high expression of DNA demethylases (Ibarra et al., 2012; Rodrigues et al., 2013; Waters et al., 2011; Zhang et al., 2014). In general, demethylation is the default status of the imprinted genes.

Genomic imprinting in plants involves the antagonistic activities of two DNA-modifying enzymes, MET1 and Demeter (DME), a DNA glycosylase that excises 5-methylcytosines from DNA (Gehring et al., 2006). MET1 had been identified as a key component of the imprinting machinery a long time ago in plants (Adams et al., 2000). DME was isolated as a mutant with a phenotype similar to that of MEA, a Polycomb Repressive Complex 2 (PRC2) component responsible for H3K27me3, consistent with its function as an activator of *MEA* (Choi et al., 2002; Guitton et al., 2004). Mutations in *DME* that disrupt the imprinting events resulted in abnormal endosperm development and seed abortion (Choi et al., 2002). DME is necessary to activate the maternal alleles of almost all imprinted genes by counteracting MET1-mediated DNA methylation in the central cell (Xiao et al., 2003; Kinoshita et al., 2004). The search for genetic suppressors of the *dme* mutant has identified several *met1* alleles, indicating that the antagonistic regulation between MET1 and DME is essential for the establishment of genomic imprinting in plants.

DME-mediated DNA demethylation has been consistently linked to imprinted gene expression in nearly all higher plants (Gehring et al., 2006; Hermon et al., 2007; Hsieh et al., 2011; Ibarra et al., 2012; Kinoshita et al., 2004; Rodrigues et al., 2013; Waters et al., 2011). The sites of demethylation that regulate imprinted genes are commonly transposable elements or other DNA repeats (Kinoshita et al., 2007; Gehring et al., 2009; Villar et al., 2009; Pignatta et al., 2014), indicating that imprinting formation by DNA demethylation evolved from a transposon-directed silencing mechanism (Kim and Zilberman 2014). Maternally and paternally expressed genes exhibit distinct patterns of DNA demethylation that are similar among Arabidopsis, maize, and rice (Ibarra et al., 2012; Rodrigues et al., 2013; Zhang et al., 2014), indicating that the regulatory mechanisms governing imprinted regions are highly conserved across 150 million years of evolution (Chaw et al., 2004).

The comprehensive maps of global DNA methylation during female gametophyte development (Ibarra et al., 2012) and male gametophyte development (Calarco et al., 2012) in Arabidopsis provide evidence that DNA hypomethylation is observed in both the central cell and the vegetative cell. This suggests that the companion cells undergo selective demethylation. In the vegetative nucleus, DNA methylation is specifically lost from the targets of DME, which encompass imprinted loci and recurrent epialleles. These loci accumulate corresponding small RNAs and undergo pre-methylation in the sperm cells (Calarco et al., 2012). Therefore, the reprogramming of DNA methylation during pollen development plays a crucial role in contributing to epigenetic inheritance, transposon silencing, and genomic imprinting. Similar to pollen, the central cell undergoes global DNA demethylation before fertilization (Ibarra et al., 2012). In Arabidopsis, rice, and maize, endosperm DNA demethylation is initiated in the central cell and requires DME (Park et al., 2016). DME is expressed in the central cell before fertilization (Gehring et al., 2006), implying that the demethylation patterns observed in the endosperm are inherited from the central cell. Additional evidence indicates that DNA demethylation in both the central cell and the endosperm is a passive process specifically mediated by DME (Ibarra et al., 2012). Therefore, these results conclusively demonstrate that demethylation in the companion cells reinforces the establishment of imprinting during plant reproductive development and likely contributes to the stable inheritance of imprinting across generations.

Given that DNA demethylation is differentially regulated between companion cells and germ cells, and considering the tight relationship between DME and the regulation of imprinted genes, a key question arises: How is the expression of DME regulated during plant reproductive development? Differential patterns of DNA methylation between the endosperm and the embryo result from the distinct expression of both DNA methyltransferases and demethylases in their respective cell types (Table 20.1). In pollen, MET1 is highly expressed in the sperm cells but not in the vegetative cell (Table 20.1; Borges et al., 2008; Li et al., 2017a). In contrast, DRM2 is specifically accumulated in the vegetative cell but not in the sperm cells (Table 20.1; Calarco et al., 2012). Unexpectedly, MET1 is very lowly expressed in both the central cell and the egg cell before fertilization (Jullien et al., 2008, 2012). After double fertilization, MET1 is primarily expressed in the embryo but absent in the endosperm (Jullien et al., 2012), suggesting that local hypomethylation in companion cells or tissues could be attributed, at least in part, to the absence of MET1 (Ibarra et al., 2012; Slotkin et al., 2009). Consistent with the higher level of CHG methylation in the egg cell, CMT3 levels are much higher in the egg cell than in the central cell (Jullien et al., 2012). DRM2 is predominantly expressed in the egg cell

Table 20.1 Expression of DNA methylation regulators in reproductive cells

Gene	Sperm cell	Vegetative cell	Central cell	Egg cell	Embryo	Endosperm
DME	down	up	high	N/A	N/A	high
MET1	up	absent	down	down	up	down
CMT3	N/A	N/A	N/A	down	up	down
DMR2	down	up	down	up	up	down
CMT2	N/A	N/A	N/A	down	up	down

rather than in the central cell (Jullien et al., 2012). DME is highly expressed in the companion cells, including the central cell and the vegetative cell, prior to fertilization (Fig. 20.2; Choi et al., 2002; Schoft et al., 2011). However, in rice endosperm, prevalent maternal-specific DNA hypomethylation occurs in the transcribed regions of active genes and is strongly correlated with paternally biased expression (Rodrigues et al., 2013). This suggests that demethylation, primarily influencing gene activity, may have evolved in at least some plant lineages.

20.3 H3K27me3 Controls Genomic Imprinting

In addition to DNA methylation's role in genomic imprinting, another well-known chromatin hallmark associated with imprinting in flowering plants is H3K27me3 (Kohler et al., 2005; Baroux et al., 2006; Gehring et al., 2006; Jullien et al., 2006; Makarevich et al., 2008; Raissig et al., 2013; Zhang et al., 2014). H3K27me3 marks are catalyzed by PRC2 in both plants and animals. In the Arabidopsis endosperm, the PRC2 complex is known as the FIS-PRC2 complex, consisting of four core subunits: MEA, FIS2, fertilization-independent endosperm (FIE), and MSI1 (Kohler et al., 2003). Removal of H3K27me3 by loss of function of the PcG proteins causes ectopic expression of imprinted genes (Ingouff et al., 2005). Remarkably, MEA, the first identified imprinted gene in Arabidopsis, itself encodes a PcG protein (Xiao et al., 2003). MEA self-regulates its own expression in the central cell and in the endosperm (Xiao et al., 2003). In the endosperm, the PRC2 complex involved in MEA imprinting also comprises the PcG proteins FIS2, FIE, and MSI1 (Kohler et al., 2003). A complex feedback loop, involving the maternal alleles of genes encoding the MEA–FIS2–FIE–MSI1 repressive Polycomb complex, maintains the silencing of the paternal allele of MEA. The presence of H3K27me3 at the promoter site of the paternally imprinted gene *PHE* relies on the activity of PcG complexes, which are active during both vegetative development and in the endosperm (Kohler et al., 2005).

Mutations in PRC2 genes disrupt the imprinting of many genes, with the majority of affected genes experiencing a disruption in their paternal expression (Hsieh et al., 2011; Wolff et al., 2011; Raissig et al., 2013). Characterizing the endosperm H3K27me3 profile and identifying specific target genes of the FIS-PRC2 complex, researchers demonstrated that the FIS-PRC2 complex has a subset of specific target

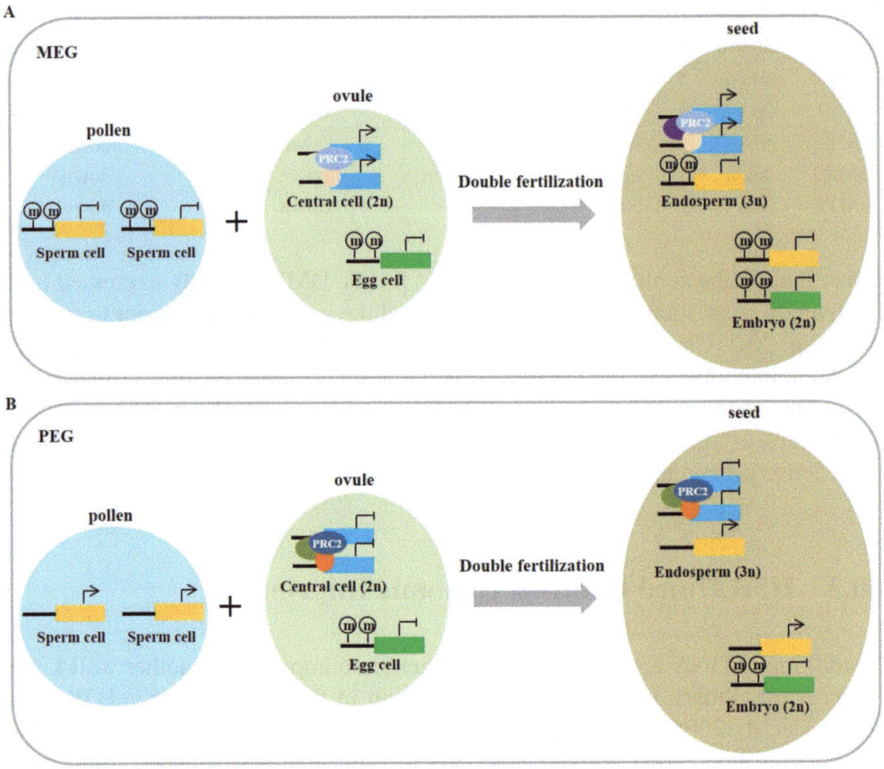

Fig. 20.2 The mechanisms governing imprinted gene expression in plants. (**a**) For maternally expressed genes (MEGs), DME demethylates the maternal allele in the central cell, resulting in specific expression. The PRC2 complex (indicated by light shading) monitors the expression level of the maternal allele without completely silencing it. (**b**) For paternally expressed genes (PEGs), the maternally inherited allele is extensively targeted by the PRC2 complex in the central cell, resulting in the repression of its expression. In contrast, the paternal allele of PEGs lacks DNA methylation in sperm and evades PRC2 targeting through an unknown mechanism in the endosperm after fertilization, thus maintaining active transcription

genes. This subset includes transposable elements and protein-coding genes that are depleted from DNA methylation and specifically targeted by the FIS-PRC2 complex in the endosperm (Weinhofer et al., 2010). It is well-known that PRC2-mediated silencing exhibits the requisite features, where PRC2-catalyzed H3K27me3 is inherited after DNA replication and subsequently recruits PRC2, establishing a self-reinforcing loop (Hansen et al., 2008; Margueron et al., 2009). Therefore, it is common for many imprinted genes to, in turn, promote DNA methylation or repressive histone modifications at components involved in these two processes (Gehring et al., 2011; Hsieh et al., 2011; Klosinska et al., 2016). In addition to H3K27me3, other histone modifications have also been found to be associated with the regulation of imprinting. At imprinting control regions (ICRs) in mice, for instance, H3K9me3 and H4K20me3 were detected in an allele-dependent manner (Delaval et al., 2007). In plants, H3K4me3 and H3K36me3 are associated with imprinting (Dong et al., 2017). Different imprinted loci in plants and mammals recruit specific epigenetic machinery to achieve allelic repression. This recruitment relies on the potential organization of repressive chromatin according to the nature of the DNA sequences in the vicinity of the imprinted genes.

This disruption is associated with H3K27me3 of maternally inherited alleles, which correlates with DNA hypomethylation (Fig. 20.2; Makarevich et al., 2008; Moreno-Romero et al., 2016). In the endosperm, PRC2 activity is generally anti-correlated with DNA methylation, particularly targeting transposons associated with DME-mediated demethylation (Weinhofer et al., 2010). However, despite many studies in both mammals and plants implicating DNA methylation as a significant hallmark of imprinting, it's worth noting that some imprinted genes regulated by PRC2 are not marked by DNA methylation (Hsieh et al., 2011; Wolff et al., 2011; Zhang et al., 2014). For example, the imprinting of an *MEA* transgene does not require DME or DNA methylation, as demonstrated by Wohrmann et al. in 2012. In this case, a 200-bp fragment named ICR is necessary and sufficient to mediate *MEA* activation and imprinted expression. Notably, imprinted *MEA* expression mediated by this ICR is independent of DME and MET1, consistent with the lack of any significant DNA methylation in this region (Wohrmann et al., 2012). This is the first example of an ICR without differential DNA methylation, suggesting that factors other than DME and MET1 are required for imprinting at the *MEA* locus. Of note, by characterizing the genome-wide map of allele-specific nucleosome occupancy, people have uncovered a mechanistic connection between chromatin organization and genomic imprinting (Dong et al., 2018).

20.4 Small RNA and Genomic Imprinting

Molecular mechanisms crucial for monoallelic expression include differential DNA methylation and differential targeting of PRC2 to maternal and paternal alleles. However, these processes do not account for all instances of imprinting, and it is likely that additional mechanisms remain to be discovered. Small interfering RNAs (siRNAs) regulate imprinted genes by their ability to initiate DNA methylation at transposable elements and DNA repeats through the RdDM pathway (Rodrigues & Zilberman, 2015). DNA methylation directed by small RNAs is required for imprinting at some loci that are demethylated in the central cell (Vu et al., 2013). Large populations of maternally biased 24-nt small RNAs were reported in the endosperm (Mosher et al., 2009), and some of these have been implicated in imprinted gene expression (Lu et al., 2012).

Moreover, small RNAs specifically inherited from the sperm cells were found to play an important role in the establishment of the triploid block, and the interruption of the delivery of paternal small RNA was able to rescue the triploid block (Martinez et al., 2018; Borges et al., 2018). Analysis of small RNA profiles indicated that endosperm 24-nt small RNAs exhibit reduced presence in transposable elements and are enriched in genes when compared with the embryo (Erdmann et al., 2017). Other investigations revealed an inverse relationship between the parent of origin of small RNAs and mRNAs, with genes more likely to be associated with maternally biased small RNAs than paternally biased ones (Erdmann et al., 2017). Disruption of the RdDM pathway causes a shift toward maternal allele mRNA expression for many genes (Erdmann et al., 2017). Furthermore, paternal inheritance of an RNA Pol IV mutation is sufficient to rescue seed abortion caused by excess paternal genome dosage (Erdmann et al., 2017; Borges et al., 2018; Martinez et al., 2018). Thus, paternally inherited small RNAs mediate the transcriptional balance between maternally and paternally inherited genomes in the endosperm and finally facilitate the establishment of imprinting during seed development. In the case of the triploid block, an increase in paternally inherited small RNAs disrupted the initiation and maintenance of imprinting, leading to aborted seeds. Despite intriguing correlations between small RNAs and imprinted genes, it remains unclear whether small RNAs associated with imprinted genes are a cause or a result of imprinting. Additionally, it's worth noting that most imprinted genes are not associated with imprinted small RNAs (Rodrigues et al., 2013; Pignatta et al., 2014; Xin et al., 2014). In conclusion, these epigenetic marks act as a heritable memory, ensuring that imprinted genes maintain their monoallelic expression patterns across generations. The interplay between DNA methylation, histone modifications, and small RNAs contributes to the precise regulation of imprinted genes during crucial developmental stages, such as embryogenesis and seed development in plants.

20.5 Known and Unknown Imprinted Genes in Plants

As mentioned above, the first identified genomically imprinted locus is the R1 gene in maize (Kermicle, 1970). Genome-wide RNA profiling led to the discovery of the second imprinted gene no-apical-meristem (NAM)-related protein 1 (*NRP1*) in maize (Table 20.2; Guo et al., 2003). *NPR1* encodes a transcription factor that was expressed only in the endosperm and regulated by gene-specific imprinting (Guo et al., 2003). Subsequently, the third imprinted gene *MEG1* was isolated in maize (Table 20.2), whose maternally inherited alleles are expressed specifically in the transfer layer of the endosperm (Gutierrez-Marcos et al., 2004). Notably, although the maternal allele of MEG1 is expressed at early stages of seed development, the paternal allele of MEG1 also becomes active at around 10 days after pollination (Gutierrez-Marcos et al., 2004), indicating that MEG1 represents a stage-specific imprinted gene. Other familiar imprinted genes in maize include ZmFIE1 and ZmFIE2 (Table 20.2; Danilevskaya et al., 2003), two homologs to the Arabidopsis imprinted gene fertilization-independent endosperm (FIE). The newly isolated maternally imprinted gene *FL3* is specifically expressed in starchy endosperm cells and regulated by genomic imprinting, whose function is involved in RNA polymerase III (Li et al., 2017b). Other imprinted genes in maize are not listed in detail.

Table 20.2 Familiar imprinted genes in arabidopsis and maize

Gene	Annotation	Parental	Species
MEA	PcG protein	maternal	Arabidopsis
FIS2	PcG protein	maternal	Arabidopsis
FIE	PcG protein	maternal	Arabidopsis
MSI	PcG protein	maternal	Arabidopsis
FWA	Transcription factor	maternal	Arabidopsis
PHE1	Transcription factor	paternal	Arabidopsis
FH5	Formin homolog	maternal	Arabidopsis
NUWA	A mitochondrial protein	maternal	Arabidopsis
NRP1	Transcription factor	maternal	Maize
NF-YC8	Transcription factor	paternal	Maize
MEG1	Cystein-rich protein	maternal	Maize
FIE1	PcG protein	maternal	Maize
FIE2	PcG protein	maternal	Maize
FL3	A novel protein	maternal	Maize

Overall, the function of most imprinted genes isolated in maize is currently unknown.

In contrast, the function of those imprinted loci in Arabidopsis, especially the fertilization-independent seed (*FIS*) genes, is significantly involved in seed development. The first identified imprinted gene, MEA (MEDEA), was isolated in a screen for gametophytic maternal effect mutations (Table 20.2; Grossniklaus et al., 1998). Fertilization-independent endosperm (FIE) and other *FIS* genes were identified in a screen for silique elongation in the absence of fertilization (Table 20.2; Ohad et al., 1999), and MSI1 was identified by looking for interacting proteins with FIE and MEA (Table 20.1; Kohler et al., 2003). Mechanistically, these FIS proteins form the Polycomb Repressive Complex (PRC2) to control the developmental timing of endosperm (Ingouff et al., 2005), in which they repress target genes that prevent endosperm proliferation before fertilization and also regulate target genes controlling cell proliferation after fertilization in the developing seed.

In Arabidopsis, there are several other functionally characterized imprinted genes, including *FWA*, *PHE (PHERES1)*, *FH5 (Formin Homolog 5)*, and *NUWA* (Table 20.2). FWA was initially isolated as a later flowering mutant (Koornneef et al., 1991), in which no nucleotide change but loss of DNA methylation in the promoter region was observed (Soppe et al., 2000). Subsequent studies showed that *FWA* is expressed in the central cell before fertilization and in the developing endosperm. Maternal expression is correlated with a loss of DNA methylation in the promoter region of the *FWA* gene (Table 20.2; Kinoshita et al., 2004). *PHERES1* encodes a transcription factor and is the first identified paternally imprinted gene in plants (Table 20.2; Kohler et al., 2005). *PHERES1* is expressed during early seed development, both in embryo and endosperm, and later its expression becomes restricted to the endosperm, where it is imprinted (Kohler et al., 2005). Loss of *NUWA* causes defective nuclear proliferation in early embryo and endosperm

(Table 20.2; He et al., 2017). Aiming to identify targets of the PcG proteins, the actin regulator formin Arabidopsis Formin Homologue 5 (AtFH5) was identified (Table 20.2). PcG activity silences the paternal allele of *AtFH5*, restricting its expression to the maternal allele (Fitz Gerald et al., 2009). *NUWA* is a mitochondria-localized maternally imprinted gene, and the de novo transcription of the maternal allele of *NUWA* starts from the zygote stage (He et al., 2017). Very few genes except the FIS proteins are commonly imprinted in both arabidopsis and maize, implying different control mechanisms underlying endosperm cellularization in monocots and dicots.

Indeed, various strategies have been employed to identify imprinted genes in plants, leveraging both experimental and computational approaches. Some key methodologies include: *(1) Expression pattern analysis*: identifying genes with endosperm-specific expression patterns similar to those of known imprinted genes. *(2) ChIP-seq and transcriptomic analysis*: utilizing Chromatin Immunoprecipitation sequencing (ChIP-seq) to identify targets of key regulators of imprinting and analyzing transcriptomic data; *(3) Differential DNA methylation analysis:* focusing on genes associated with differential DNA methylation between the embryo and endosperm, as DNA methylation is a key epigenetic mechanism in imprinting; *(4) Genome-wide polymorphism studies:* conducting genome-wide studies involving reciprocal crosses between different polymorphism ecotypes to discover genes with maternal and paternal origins of transcripts. These strategies collectively contribute to the comprehensive exploration of imprinted genes, enhancing our understanding of their roles in plant development and reproduction.

In the last method, two strains with known polymorphisms are crossed, and allele-specific gene expression is assayed by RNA sequencing in the resulting F1 progeny. It is critical to perform the cross in a reciprocal manner, meaning strain X female is crossed with strain Y male, and vice versa (strain Y female is crossed with strain X male). Imprinting can then be assessed for expressed genes that have at least one polymorphism in the transcribed sequence. These efforts yielded approximately hundreds of imprinted genes in Arabidopsis (Hsieh et al., 2011; Wolff et al., 2011), rice (Luo et al., 2011), maize (Waters et al., 2013), bean (Xu et al., 2014), and sorghum (Zhang et al., 2016) and endosperm, vastly expanding the imprinted gene universe. More importantly, this technology has enabled imprinted genes to be identified in those non-model species.

20.6 Functions of Genomic Imprinting in Plants

Although hundreds of imprinted genes have been identified in plants and much progress has been made in elucidating the epigenetic mechanisms of imprinting, the biological significance of genomic imprinting in plants remains mysterious. Why do diploids evolve a silencing system to give up their diploidy during reproduction? Current studies show that the biological functions of these imprinted genes are mostly embedded in two aspects: seed development and genome defense.

In mammals, imprinting has important roles in extra-embryonic development and function (Arima et al., 2006). Generally, imprinted genes are proposed to control nutrient transfer from the mother to the developing embryo (Constancia et al., 2004). Endosperm growth is enhanced by increased paternal dosage and negatively affected by increased maternal dosage. Indeed, when imprinting is disturbed by mutations in PcG genes, the duration of the syncytial endosperm developmental phase is prolonged, which enhances growth (Kiyosue et al., 1999; Ingouff et al., 2005). Genome-wide analysis of gene expression and DNA methylation in the rice endosperm shows that hundreds of imprinted genes were associated with grain yield quantitative trait loci. Most imprinted genes were expressed specifically in the endosperm. Disruption of their maternally imprinted expression of some genes increased the amount of small starch granules and reduced grain and embryo size, whereas mutation of their paternally imprinted expression of other loci reduced starch content and seed fertility. Therefore, both maternally and paternally imprinted genes in rice regulate nutrient metabolism and endosperm development, which optimize seed development and offspring fitness to facilitate parental-offspring co-adaptation. These imprinted genes could be used to improve the grain yield of rice and other cereal crops.

Polyploidization, a widespread phenomenon among plants, is considered a major speciation mechanism due to the reproductive barrier, also named the triploid block, caused by the malfunction of the endosperm. However, during the formation of polyploidy, there are ways to overcome the triploid block. Chromosome dosage has substantial effects on reproductive isolation and speciation in both plants and animals (Birchler & Veitia, 2012), and small RNAs in plants have been postulated to regulate dosage response via neighboring imprinted genes (Martienssen, 2010; Kohler et al., 2010). These findings show that miR845, a highly conserved microRNA in plants, targets the long terminal repeat (LTR) retrotransposons in pollen and triggers the accumulation of 21–22-nt small RNAs (epigenetically reactivated small RNAs, easiRNAs) in a dose-dependent fashion via RNA polymerase IV. Natural variation for miR845 may account for "endosperm balance" allowing the formation of triploid seeds (Borges et al., 2018). In another study, they further showed that depletion of paternal easiRNAs bypasses the triploid block in response to increased paternal ploidy, and paternal loss of RNA polymerase IV suppressed easiRNA biogenesis could rescue triploid seeds by restoring siRNA-directed DNA methylation at transposons, correlating with reduced expression of paternally expressed imprinted genes (Martinez et al., 2018). Thus paternally inherited siRNAs provide a unique means to monitor chromosome dosage and imprinting in the developing seed.

As mentioned above, DNA methylation is tightly related to imprinting, and especially CHH methylation is dynamically reprogrammed during gametogenesis and seed development. However, the role that CHH methylation plays in seed development remains elusive. By investigating Arabidopsis mutants defective in both RdDM and non-RdDM pathways, researchers found no detectable morphological defects during seed development, and the germination frequencies of these mutants were indistinguishable from wild-type (Lin et al., 2017). The simplest hypothesis for the role of dynamic CHH methylation in seed development may be a safeguarding mechanism to reinforce transposon silencing in seeds. The programmed increase of CHH methylation during seed development supports this hypothesis (Lin et al., 2017).

In addition, studies have identified imprinted genes associated with seed dormancy, and the parent-of-origin-specific expression of these genes can influence various aspects of seed biology, including germination timing and dormancy release. Many genes marked by a single H3K27me3 were found to be induced during seed germination (Sato et al., 2021). Mutants in the H3K27me3 demethylases REF6/ELF6 exhibit increased seed dormancy and decreased expression of those genes marked by a single H3K27me3 (Chen et al., 2020; Sato et al., 2021). Moreover, genes with single H3K27me3 show maternally biased expression patterns in dormant seeds rather than in non-dormant seeds (Piskurewicz et al., 2016; Sato et al., 2021). More interestingly, single H3K27me3 deposition is connected with the accumulation of CHH methylation on the paternal alleles, and CHH methylation can be induced by cold (Iwasaki et al., 2019). These data suggest that maternal-specific control of seed dormancy is mediated by both the removal of H3K27me3, causing activation of maternal alleles and the cold-induced silencing of the maternal alleles by establishment of CHH methylation. Therefore, understanding the molecular mechanisms underlying genomic imprinting can provide insights into the regulation of seed dormancy and contribute to the development of crops with improved germination characteristics.

20.7 Evolution of Genomic Imprinting in Plants

Although genomic imprinting is widely present from plants to mammals, why and how genomic imprinting has evolved still remains largely a mystery. The evolutionary forces that select for imprinted expression are still debated, but conflict between parental genomes in the developing offspring or dosage control may be important drivers of imprinting. In mammals, genomic imprinting is considered to have first appeared approximately 150 million years ago (Killian et al., 2001). In plants that arose about 140 million years ago, reproduction occurs by double fertilization, whereby one of the two sperm cells fertilizes the egg cell to produce the embryo, while the other sperm cell fertilizes the binucleate central cell to generate the triploid endosperm, which nourishes the plant embryo. Therefore, genomic imprinting has likely evolved independently in mammalian and plant lineages.

In both mammals and plants, the embryo remains connected to the mother inside the reproductive apparatus. Maternal support of embryogenesis is linked to differentiation of distinctive structures specialized for the transport of maternal nutrients to the embryo—the fetal placenta in mammals and the seed endosperm connected to the maternal seed coat in plants. Consequently, male and female parental investment in reproduction is extremely imbalanced. This imbalance has been considered as the parental conflict theory (Wilkins & Haig, 2003). The parental conflict hypothesis proposes that the interest of the father is to drive more maternal resources to his offspring, leading to greater fitness. By contrast, the maternal interest is to produce as many viable embryos as possible and, therefore, to share resources equally among embryos. Moreover, the mother has to balance resource allocation to the embryo and her own nutritional needs to ensure her survival. As a result, embryo growth would be enhanced by paternal regulators and inhibited by maternal regulators (Wilkins & Haig, 2003).

In plants, the parental conflict hypothesis was supported by observations of growth variations in seeds originating from crosses between parents of different ploidy (Scott et al., 1998). For example, increased dosage of the maternal genome causes a reduction of endosperm growth and of seed size. This effect could result from the increased dosage of active maternal alleles of imprinted growth inhibitors. A parallel phenomenon was noted in reciprocal crosses involving the wild-type ovules and pollen from the *met1* mutant plants (Adams et al., 2000). However, in most self-fertilizing plants, male and female gametogenesis takes place in the same flower. The prerequisite for the parental conflict theory is that plants must have separate sexes, and reproduction involves multiple "fathers." The existing imprinting status in self-fertilizing plants, such as Arabidopsis, suggests the conservation of an original mechanism acquired in ancestral flowering plants. Notably, gene imprinting is well conserved in mammals, yet its degree of conservation in plants is comparatively lower. Analysis of the genome-wide imprinting status of *Capsella rubella*, a species sharing a common recent ancestor with *Arabidopsis thaliana* approximately 10–14 million years ago, revealed that less than 30% of imprinted genes were conserved between the two species (Hatorangan et al., 2016). This suggests that genomic imprinting is a rapidly evolving phenomenon in plants.

Perspective

While imprinting is not universally present in all eukaryotes, the regulatory mechanism governing imprinting shows a high degree of conservation from plants to mammals. While maternal-specific DNA demethylation and PRC2-mediated histone methylation have been identified as general mechanisms controlling the distinct activation and silencing of maternal and paternal alleles for certain genes (Huh et al., 2008), the precise targeting specificity of DME and the mechanism by which the PRC2 complex is recruited to imprinted genes in plants remain unclear. Since then, hundreds of new imprinted genes have been discovered (Wolff et al., 2011), firmly establishing the generality of mechanisms governing imprinted gene regulation. Furthermore, why there is a need for preset DNA methylation followed by demethylation in plant imprinted genes remains a question. The regulation of some imprinted genes is not linked to DNA methylation, suggesting that the true trigger of imprinting might not be solely DNA methylation. Active crosstalk among different DNA methylations, histone modifications, and interactions between DNA and histone covalent modifications has been widely observed. How these parallel pathways coordinate in plant gametes to establish and/or maintain the inheritance of imprinting remains a key question. The plasticity of epigenetic marks during gametogenesis and early seed development, the biogenesis and functions of imprinting-related small RNAs, and the specificity of multiple DNA methyltransferases and histone modification machineries remain largely unclear. The role of DNA methylation and H3K27me3 events in seed biology across different plant species remains largely unexplored.

Furthermore, while the core mechanisms responsible for plant imprinting are fairly robust, our understanding of the evolution and biological significance of plant imprinting is still in its infancy. Considering the significance of imprinted loci in yield and seed development, it should be a priority to elucidate the biological functions of plant imprinted genes, despite very few of them currently having known functions.

References

Adams, S., Vinkenoog, R., Spielman, M., Dickinson, H. G., & Scott, R. J. (2000). Parent-of-origin effects on seed development in *Arabidopsis thaliana* require DNA methylation. *Development, 127*, 2493–2502.

Arima, T., Hata, K., Tanaka, S., Kusumi, M., Li, E., Kato, K., Shiota, K., Sasaki, H., & Wake, N. (2006). Loss of the maternal imprint in Dnmt3Lmat−/− mice leads to a differentiation defect in the extraembryonic tissue. *Developmental Biology, 297*, 361–373.

Barlow, D. P., Stoger, R., Herrmann, B. G., Saito, K., & Schweifer, N. (1991). The mouse insulin-like growth factor type-2 receptor is imprinted and closely linked to the Tme locus. *Nature, 349*, 84–87.

Baroux, C., Gagliardini, V., Page, D. R., & Grossniklaus, U. (2006). Dynamic regulatory interactions of Polycomb group genes: MEDEA autoregulation is required for imprinted gene expression in Arabidopsis. *Genes & Development, 20*, 1081–1086.

Bartolomei, M. S., & Ferguson-Smith, A. C. (2011). Mammalian genomic imprinting. *Cold Spring Harbor Perspectives in Biology, 3*, a002592.

Bartolomei, M. S., Zemel, S., & Tilghman, S. M. (1991). Parental imprinting of the mouse H19 gene. *Nature, 351*, 153–155.

Bartolomei, M. S., Webber, A. L., Brunkow, M. E., & Tilghman, S. M. (1993). Epigenetic mechanisms underlying the imprinting of the mouse H19 gene. *Genes & Development, 7*, 1663–1673.

Birchler, J. A., & Veitia, R. A. (2012). Gene balance hypothesis: Connecting issues of dosage sensitivity across biological disciplines. *Proceedings of the National Academy of Sciences of the United States of America, 109*, 14746–14753.

Borges, F., Gomes, G., Gardner, R., Moreno, N., McCormick, S., Feijo, J. A., & Becker, J. D. (2008). Comparative transcriptomics of Arabidopsis sperm cells. *Plant Physiology, 148*, 1168–1181.

Borges, F., Parent, J. S., van Ex, F., Wolff, P., Martinez, G., Kohler, C., & Martienssen, R. A. (2018). Transposon-derived small RNAs triggered by miR845 mediate genome dosage response in Arabidopsis. *Nature Genetics, 50*, 186–192.

Calarco, J. P., Borges, F., Donoghue, M. T., Van Ex, F., Jullien, P. E., Lopes, T., Gardner, R., Berger, F., Feijo, J. A., Becker, J. D., & Martienssen, R. A. (2012). Reprogramming of DNA methylation in pollen guides epigenetic inheritance via small RNA. *Cell, 151*, 194–205.

Cao, X., Aufsatz, W., Zilberman, D., Mette, M. F., Huang, M. S., Matzke, M., & Jacobsen, S. E. (2003). Role of the DRM and CMT3 methyltransferases in RNA-directed DNA methylation. *Current Biology, 13*, 2212–2217.

Chaudhuri, S., & Messing, J. (1994). Allele-specific parental imprinting of dzr1, a posttranscriptional regulator of zein accumulation. *Proceedings of the National Academy of Sciences of the United States of America, 91*, 4867–4871.

Chaw, S. M., Chang, C. C., Chen, H. L., & Li, W. H. (2004). Dating the monocot-dicot divergence and the origin of core eudicots using whole chloroplast genomes. *Journal of Molecular Evolution, 58*, 424–441.

Chen, H., Tong, J., Fu, W., Liang, Z., Ruan, J., Yu, Y., Song, X., Yuan, L., Xiao, L., Liu, J., et al. (2020). The H3K27me3 demethylase RELATIVE of early FLOWERING6 Suppresses seed dormancy by inducing abscisic acid catabolism. *Plant Physiology, 184*, 1969–1978.

Choi, Y., Gehring, M., Johnson, L., Hannon, M., Harada, J. J., Goldberg, R. B., Jacobsen, S. E., & Fischer, R. L. (2002). DEMETER, a DNA glycosylase domain protein, is required for endosperm gene imprinting and seed viability in Arabidopsis. *Cell, 110*, 33–42.

Constancia, M., Kelsey, G., & Reik, W. (2004). Resourceful imprinting. *Nature, 432*, 53–57.

Danilevskaya, O. N., Hermon, P., Hantke, S., Muszynski, M. G., Kollipara, K., & Ananiev, E. V. (2003). Duplicated fie genes in maize: Expression pattern and imprinting suggest distinct functions. *Plant Cell, 15*, 425–438.

DeChiara, T. M., Robertson, E. J., & Efstratiadis, A. (1991). Parental imprinting of the mouse insulin-like growth factor II gene. *Cell, 64*, 849–859.

Delaval, K., Govin, J., Cerqueira, F., Rousseaux, S., Khochbin, S., & Feil, R. (2007). Differential histone modifications mark mouse imprinting control regions during spermatogenesis. *The EMBO Journal, 26*, 720–729.

Dong, X., Zhang, M., Chen, J., Peng, L., Zhang, N., Wang, X., & Lai, J. (2017). Dynamic and antagonistic allele-specific epigenetic modifications controlling the expression of imprinted genes in maize endosperm. *Molecular Plant, 10*, 442–455.

Dong, X., Chen, J., Li, T., Li, E., Zhang, X., Zhang, M., Song, W., Zhao, H., & Lai, J. (2018). Parent-of-origin-dependent nucleosome organization correlates with genomic imprinting in maize. *Genome Research, 28*, 1020–1028.

Erdmann, R. M., Satyaki, P. R. V., Klosinska, M., & Gehring, M. (2017). A small RNA pathway mediates allelic dosage in endosperm. *Cell Reports, 21*, 3364–3372.

Feng, S., Jacobsen, S. E., & Reik, W. (2010). Epigenetic reprogramming in plant and animal development. *Science, 330*, 622–627.

Ferguson-Smith, A. C. (2011). Genomic imprinting: The emergence of an epigenetic paradigm. *Nature Reviews. Genetics, 12*, 565–575.

Ferguson-Smith, A. C., Sasaki, H., Cattanach, B. M., & Surani, M. A. (1993). Parental-origin-specific epigenetic modification of the mouse H19 gene. *Nature, 362*, 751–755.

Fitz Gerald, J. N., Hui, P. S., & Berger, F. (2009). Polycomb group-dependent imprinting of the actin regulator AtFH5 regulates morphogenesis in *Arabidopsis thaliana. Development, 136*, 3399–3404.

Gehring, M., Huh, J. H., Hsieh, T. F., Penterman, J., Choi, Y., Harada, J. J., Goldberg, R. B., & Fischer, R. L. (2006). DEMETER DNA glycosylase establishes MEDEA polycomb gene self-imprinting by allele-specific demethylation. *Cell, 124*, 495–506.

Gehring, M., Bubb, K. L., & Henikoff, S. (2009). Extensive demethylation of repetitive elements during seed development underlies gene imprinting. *Science, 324*, 1447–1451.

Gehring, M., Missirian, V., & Henikoff, S. (2011). Genomic analysis of parent-of-origin allelic expression in *Arabidopsis thaliana* seeds. *PLoS One, 6*, e23687.

Grossniklaus, U., Vielle-Calzada, J. P., Hoeppner, M. A., & Gagliano, W. B. (1998). Maternal control of embryogenesis by MEDEA, a polycomb group gene in Arabidopsis. *Science, 280*, 446–450.

Guitton, A. E., Page, D. R., Chambrier, P., Lionnet, C., Faure, J. E., Grossniklaus, U., & Berger, F. (2004). Identification of new members of Fertilisation Independent Seed Polycomb Group pathway involved in the control of seed development in *Arabidopsis thaliana. Development, 131*, 2971–2981.

Guo, M., Rupe, M. A., Danilevskaya, O. N., Yang, X., & Hu, Z. (2003). Genome-wide mRNA profiling reveals heterochronic allelic variation and a new imprinted gene in hybrid maize endosperm. *The Plant Journal, 36*, 30–44.

Gutierrez-Marcos, J. F., Costa, L. M., Biderre-Petit, C., Khbaya, B., O'Sullivan, D. M., Wormald, M., Perez, P., & Dickinson, H. G. (2004). maternally expressed gene1 is a novel maize endosperm transfer cell-specific gene with a maternal parent-of-origin pattern of expression. *Plant Cell, 16*, 1288–1301.

Hansen, K. H., Bracken, A. P., Pasini, D., Dietrich, N., Gehani, S. S., Monrad, A., Rappsilber, J., Lerdrup, M., & Helin, K. (2008). A model for transmission of the H3K27me3 epigenetic mark. *Nature Cell Biology, 10*, 1291–1300.

Hatorangan, M. R., Laenen, B., Steige, K. A., Slotte, T., & Kohler, C. (2016). Rapid evolution of genomic imprinting in two species of the Brassicaceae. *Plant Cell, 28*, 1815–1827.

He, S., Sun, Y., Yang, Q., Zhang, X., Huang, Q., Zhao, P., Sun, M., Liu, J., Qian, W., Qin, G., et al. (2017). A novel imprinted gene NUWA controls mitochondrial Function in early seed development in Arabidopsis. *PLoS Genetics, 13*, e1006553.

Hermon, P., Srilunchang, K. O., Zou, J., Dresselhaus, T., & Danilevskaya, O. N. (2007). Activation of the imprinted Polycomb Group Fie1 gene in maize endosperm requires demethylation of the maternal allele. *Plant Molecular Biology, 64*, 387–395.

Hsieh, T. F., Shin, J., Uzawa, R., Silva, P., Cohen, S., Bauer, M. J., Hashimoto, M., Kirkbride, R. C., Harada, J. J., Zilberman, D., & Fischer, R. L. (2011). Regulation of imprinted gene expression in Arabidopsis endosperm. *Proceedings of the National Academy of Sciences of the United States of America, 108*, 1755–1762.

Huh, J. H., Bauer, M. J., Hsieh, T. F., & Fischer, R. L. (2008). Cellular programming of plant gene imprinting. *Cell, 132*, 735–744.

Ibarra, C. A., Feng, X., Schoft, V. K., Hsieh, T. F., Uzawa, R., Rodrigues, J. A., Zemach, A., Chumak, N., Machlicova, A., Nishimura, T., et al. (2012). Active DNA demethylation in plant companion cells reinforces transposon methylation in gametes. *Science, 337*, 1360–1364.

Ingouff, M., Haseloff, J., & Berger, F. (2005). Polycomb group genes control developmental timing of endosperm. *The Plant Journal, 42*, 663–674.

Iwasaki, M., Hyvärinen, L., Piskurewicz, U., & Lopez-Molina, L. (2019). Non-canonical RNA-directed DNA methylation participates in maternal and environmental control of seed dormancy. *eLife, 8*, e37434.

Jones, P. A. (2012). Functions of DNA methylation: Islands, start sites, gene bodies and beyond. *Nature Reviews. Genetics, 13*, 484–492.

Jullien, P. E., Katz, A., Oliva, M., Ohad, N., & Berger, F. (2006). Polycomb group complexes self-regulate imprinting of the Polycomb group gene MEDEA in Arabidopsis. *Current Biology, 16*, 486–492.

Jullien, P. E., Mosquna, A., Ingouff, M., Sakata, T., Ohad, N., & Berger, F. (2008). Retinoblastoma and its binding partner MSI1 control imprinting in Arabidopsis. *PLoS Biology, 6*, e194.

Jullien, P. E., Susaki, D., Yelagandula, R., Higashiyama, T., & Berger, F. (2012). DNA methylation dynamics during sexual reproduction in *Arabidopsis thaliana*. *Current Biology, 22*, 1825–1830.

Kermicle, J. L. (1970). Dependence of the R-mottled aleurone phenotype in maize on mode of sexual transmission. *Genetics, 66*, 69–85.

Killian, J. K., Nolan, C. M., Wylie, A. A., Li, T., Vu, T. H., Hoffman, A. R., & Jirtle, R. L. (2001). Divergent evolution in M6P/IGF2R imprinting from the Jurassic to the Quaternary. *Human Molecular Genetics, 10*, 1721–1728.

Kim, M.Y., & Zilberman, D. (2014). DNA methylation as a system of plant genomic immunity. *Trends in Plant Science, 19*, 320–326.

Kinoshita, T., Miura, A., Choi, Y., Kinoshita, Y., Cao, X., Jacobsen, S. E., Fischer, R. L., & Kakutani, T. (2004). One-way control of FWA imprinting in Arabidopsis endosperm by DNA methylation. *Science, 303*, 521–523.

Kinoshita, Y., Saze, H., Kinoshita, T., Miura, A., Soppe, W. J., Koornneef, M., & Kakutani, T. (2007). Control of FWA gene silencing in *Arabidopsis thaliana* by SINE-related direct repeats. *The Plant Journal, 49*, 38–45.

Kiyosue, T., Ohad, N., Yadegari, R., Hannon, M., Dinneny, J., Wells, D., Katz, A., Margossian, L., Harada, J. J., Goldberg, R. B., & Fischer, R. L. (1999). Control of fertilization-independent endosperm development by the MEDEA polycomb gene in Arabidopsis. *Proceedings of the National Academy of Sciences of the United States of America, 96*, 4186–4191.

Klosinska, M., Picard, C. L., & Gehring, M. (2016). Conserved imprinting associated with unique epigenetic signatures in the Arabidopsis genus. *Nature Plants, 2*, 16145.

Kohler, C., Hennig, L., Bouveret, R., Gheyselinck, J., Grossniklaus, U., & Gruissem, W. (2003). Arabidopsis MSI1 is a component of the MEA/FIE Polycomb group complex and required for seed development. *The EMBO Journal, 22*, 4804–4814.

Kohler, C., Page, D. R., Gagliardini, V., & Grossniklaus, U. (2005). The Arabidopsis thaliana MEDEA Polycomb group protein controls expression of PHERES1 by parental imprinting. *Nature Genetics, 37*, 28–30.

Kohler, C., Mittelsten Scheid, O., & Erilova, A. (2010). The impact of the triploid block on the origin and evolution of polyploid plants. *Trends in Genetics, 26*, 142–148.

Kohler, C., Wolff, P., & Spillane, C. (2012). Epigenetic mechanisms underlying genomic imprinting in plants. *Annual Review of Plant Biology, 63*, 331–352.

Koornneef, M., Hanhart, C. J., & van der Veen, J. H. (1991). A genetic and physiological analysis of late flowering mutants in *Arabidopsis thaliana*. *Molecular & General Genetics, 229*, 57–66.

Law, J. A., & Jacobsen, S. E. (2010). Establishing, maintaining and modifying DNA methylation patterns in plants and animals. *Nature Reviews. Genetics, 11*, 204–220.

Li, L., Wu, W., Zhao, Y., & Zheng, B. (2017a). A reciprocal inhibition between ARID1 and MET1 in male and female gametes in Arabidopsis. *Journal of Integrative Plant Biology, 59*, 657–668.

Li, Q., Wang, J., Ye, J., Zheng, X., Xiang, X., Li, C., Fu, M., Wang, Q., Zhang, Z., & Wu, Y. (2017b). The maize imprinted gene Floury 3 encodes a PLATZ protein required for tRNA and 5S rRNA transcription through interaction with RNA polymerase III. *Plant Cell, 29*, 2661–2675.

Lin, J. Y., Le, B. H., Chen, M., Henry, K. F., Hur, J., Hsieh, T. F., Chen, P. Y., Pelletier, J. M., Pellegrini, M., Fischer, R. L., et al. (2017). Similarity between soybean and Arabidopsis seed methylomes and loss of non-CG methylation does not affect seed development. *Proceedings of the National Academy of Sciences of the United States of America, 114*, E9730–E9739.

Lu, J., Zhang, C., Baulcombe, D. C., & Chen, Z. J. (2012). Maternal siRNAs as regulators of parental genome imbalance and gene expression in endosperm of Arabidopsis seeds. *Proceedings of the National Academy of Sciences of the United States of America, 109*, 5529–5534.

Luo, M., Bilodeau, P., Dennis, E. S., Peacock, W. J., & Chaudhury, A. (2000). Expression and parent-of-origin effects for FIS2, MEA, and FIE in the endosperm and embryo of developing Arabidopsis seeds. *Proceedings of the National Academy of Sciences of the United States of America, 97*, 10637–10642.

Luo, M., Taylor, J. M., Spriggs, A., Zhang, H., Wu, X., Russell, S., Singh, M., & Koltunow, A. (2011). A genome-wide survey of imprinted genes in rice seeds reveals imprinting primarily occurs in the endosperm. *PLoS Genetics, 7*, e1002125.

Makarevich, G., Villar, C. B., Erilova, A., & Kohler, C. (2008). Mechanism of PHERES1 imprinting in Arabidopsis. *Journal of Cell Science, 121*, 906–912.

Margueron, R., Justin, N., Ohno, K., Sharpe, M. L., Son, J., Drury, W. J., 3rd, Voigt, P., Martin, S. R., Taylor, W. R., De Marco, V., et al. (2009). Role of the polycomb protein EED in the propagation of repressive histone marks. *Nature, 461*, 762–767.

Martienssen, R. A. (2010). Heterochromatin, small RNA and post-fertilization dysgenesis in allopolyploid and interploid hybrids of Arabidopsis. *The New Phytologist, 186*, 46–53.

Martinez, G., Wolff, P., Wang, Z., Moreno-Romero, J., Santos-Gonzalez, J., Conze, L. L., DeFraia, C., Slotkin, R. K., & Kohler, C. (2018). Paternal easiRNAs regulate parental genome dosage in Arabidopsis. *Nature Genetics, 50*, 193–198.

McCormick, S. (2004). Control of male gametophyte development. *Plant Cell, 16*(Suppl), S142–S153.

Moreno-Romero, J., Jiang, H., Santos-Gonzalez, J., & Kohler, C. (2016). Parental epigenetic asymmetry of PRC2-mediated histone modifications in the Arabidopsis endosperm. *The EMBO Journal, 35*, 1298–1311.

Mosher, R. A., Melnyk, C. W., Kelly, K. A., Dunn, R. M., Studholme, D. J., & Baulcombe, D. C. (2009). Uniparental expression of Pol IV-dependent siRNAs in developing endosperm of Arabidopsis. *Nature, 460*, 283–286.

Ohad, N., Yadegari, R., Margossian, L., Hannon, M., Michaeli, D., Harada, J. J., Goldberg, R. B., & Fischer, R. L. (1999). Mutations in FIE, a WD polycomb group gene, allow endosperm development without fertilization. *Plant Cell, 11*, 407–416.

Park, K., Kim, M. Y., Vickers, M., Park, J. S., Hyun, Y., Okamoto, T., Zilberman, D., Fischer, R. L., Feng, X., Choi, Y., & Scholten, S. (2016). DNA demethylation is initiated in the central cells of Arabidopsis and rice. *Proceedings of the National Academy of Sciences of the United States of America, 113*, 15138–15143.

Pignatta, D., Erdmann, R. M., Scheer, E., Picard, C. L., Bell, G. W., & Gehring, M. (2014). Natural epigenetic polymorphisms lead to intraspecific variation in Arabidopsis gene imprinting. *eLife, 3*, e03198.

Piskurewicz, U., Iwasaki, M., Susaki, D., Megies, C., Kinoshita, T., & Lopez-Molina, L. (2016). Dormancy-specific imprinting underlies maternal inheritance of seed dormancy in *Arabidopsis thaliana*. *eLife, 5*, e19573.

Popp, C., Dean, W., Feng, S., Cokus, S. J., Andrews, S., Pellegrini, M., Jacobsen, S. E., & Reik, W. (2010). Genome-wide erasure of DNA methylation in mouse primordial germ cells is affected by AID deficiency. *Nature, 463*, 1101–1105.

Radford, E. J., Ferron, S. R., & Ferguson-Smith, A. C. (2011). Genomic imprinting as an adaptative model of developmental plasticity. *FEBS Letters, 585*, 2059–2066.

Raissig, M. T., Bemer, M., Baroux, C., & Grossniklaus, U. (2013). Genomic imprinting in the Arabidopsis embryo is partly regulated by PRC2. *PLoS Genetics, 9*, e1003862.

Rodrigues, J. A., & Zilberman, D. (2015). Evolution and function of genomic imprinting in plants. *Genes & Development, 29*, 2517–2531.

Rodrigues, J. A., Ruan, R., Nishimura, T., Sharma, M. K., Sharma, R., Ronald, P. C., Fischer, R. L., & Zilberman, D. (2013). Imprinted expression of genes and small RNA is associated with localized hypomethylation of the maternal genome in rice endosperm. *Proceedings of the National Academy of Sciences of the United States of America, 110*, 7934–7939.

Sasaki, H., Jones, P. A., Chaillet, J. R., Ferguson-Smith, A. C., Barton, S. C., Reik, W., & Surani, M. A. (1992). Parental imprinting: Potentially active chromatin of the repressed maternal allele of the mouse insulin-like growth factor II (Igf2) gene. *Genes & Development, 6*, 1843–1856.

Sato, H., Santos-González, J., & Köhler, C. (2021). Combinations of maternal-specific repressive epigenetic marks in the endosperm control seed dormancy. *eLife, 10*, e64593.

Schoft, V. K., Chumak, N., Choi, Y., Hannon, M., Garcia-Aguilar, M., Machlicova, A., Slusarz, L., Mosiolek, M., Park, J. S., Park, G. T., et al. (2011). Function of the DEMETER DNA glycosylase in the *Arabidopsis thaliana* male gametophyte. *Proceedings of the National Academy of Sciences of the United States of America, 108*, 8042–8047.

Scott, R. J., Spielman, M., Bailey, J., & Dickinson, H. G. (1998). Parent-of-origin effects on seed development in *Arabidopsis thaliana. Development, 125*, 3329–3341.

Slotkin, R. K., Vaughn, M., Borges, F., Tanurdzic, M., Becker, J. D., Feijo, J. A., & Martienssen, R. A. (2009). Epigenetic reprogramming and small RNA silencing of transposable elements in pollen. *Cell, 136*, 461–472.

Soppe, W. J., Jacobsen, S. E., Alonso-Blanco, C., Jackson, J. P., Kakutani, T., Koornneef, M., & Peeters, A. J. (2000). The late flowering phenotype of *fwa* mutants is caused by gain-of-function epigenetic alleles of a homeodomain gene. *Molecular Cell, 6*, 791–802.

Stoger, R., Kubicka, P., Liu, C. G., Kafri, T., Razin, A., Cedar, H., & Barlow, D. P. (1993). Maternal-specific methylation of the imprinted mouse Igf2r locus identifies the expressed locus as carrying the imprinting signal. *Cell, 73*, 61–71.

Tiwari, S., Schulz, R., Ikeda, Y., Dytham, L., Bravo, J., Mathers, L., Spielman, M., Guzman, P., Oakey, R. J., Kinoshita, T., & Scott, R. J. (2008). MATERNALLY EXPRESSED PAB C-TERMINAL, a novel imprinted gene in Arabidopsis, encodes the conserved C-terminal domain of polyadenylate binding proteins. *Plant Cell, 20*, 2387–2398.

Villar, C. B., Erilova, A., Makarevich, G., Trosch, R., & Kohler, C. (2009). Control of PHERES1 imprinting in Arabidopsis by direct tandem repeats. *Molecular Plant, 2*, 654–660.

Vu, T. M., Nakamura, M., Calarco, J. P., Susaki, D., Lim, P. Q., Kinoshita, T., Higashiyama, T., Martienssen, R. A., & Berger, F. (2013). RNA-directed DNA methylation regulates parental genomic imprinting at several loci in Arabidopsis. *Development, 140*, 2953–2960.

Waters, A. J., Makarevitch, I., Eichten, S. R., Swanson-Wagner, R. A., Yeh, C. T., Xu, W., Schnable, P. S., Vaughn, M. W., Gehring, M., & Springer, N. M. (2011). Parent-of-origin effects on gene expression and DNA methylation in the maize endosperm. *Plant Cell, 23*, 4221–4233.

Waters, A. J., Bilinski, P., Eichten, S. R., Vaughn, M. W., Ross-Ibarra, J., Gehring, M., & Springer, N. M. (2013). Comprehensive analysis of imprinted genes in maize reveals allelic variation for imprinting and limited conservation with other species. *Proceedings of the National Academy of Sciences of the United States of America, 110*, 19639–19644.

Weinhofer, I., Hehenberger, E., Roszak, P., Hennig, L., & Kohler, C. (2010). H3K27me3 profiling of the endosperm implies exclusion of polycomb group protein targeting by DNA methylation. *PLoS Genetics, 6*, e1001152.

Wilkins, J. F., & Haig, D. (2003). What good is genomic imprinting: The function of parent-specific gene expression. *Nature Reviews. Genetics, 4*, 359–368.

Wohrmann, H. J., Gagliardini, V., Raissig, M. T., Wehrle, W., Arand, J., Schmidt, A., Tierling, S., Page, D. R., Schob, H., Walter, J., & Grossniklaus, U. (2012). Identification of a DNA methylation-independent imprinting control region at the Arabidopsis MEDEA locus. *Genes & Development, 26*, 1837–1850.

Wolff, P., Weinhofer, I., Seguin, J., Roszak, P., Beisel, C., Donoghue, M. T., Spillane, C., Nordborg, M., Rehmsmeier, M., & Kohler, C. (2011). High-resolution analysis of parent-of-origin allelic expression in the Arabidopsis endosperm. *PLoS Genetics, 7*, e1002126.

Xiao, W., Gehring, M., Choi, Y., Margossian, L., Pu, H., Harada, J. J., Goldberg, R. B., Pennell, R. I., & Fischer, R. L. (2003). Imprinting of the MEA Polycomb gene is controlled by antagonism between MET1 methyltransferase and DME glycosylase. *Developmental Cell, 5*, 891–901.

Xin, M., Yang, R., Yao, Y., Ma, C., Peng, H., Sun, Q., Wang, X., & Ni, Z. (2014). Dynamic parent-of-origin effects on small interfering RNA expression in the developing maize endosperm. *BMC Plant Biology, 14*, 192.

Xu, W., Dai, M., Li, F., & Liu, A. (2014). Genomic imprinting, methylation and parent-of-origin effects in reciprocal hybrid endosperm of castor bean. *Nucleic Acids Research, 42*, 6987–6998.

Zemach, A., Kim, M. Y., Hsieh, P. H., Coleman-Derr, D., Eshed-Williams, L., Thao, K., Harmer, S. L., & Zilberman, D. (2013). The Arabidopsis nucleosome remodeler DDM1 allows DNA methyltransferases to access H1-containing heterochromatin. *Cell, 153*, 193–205.

Zhang, M., Zhao, H., Xie, S., Chen, J., Xu, Y., Wang, K., Guan, H., Hu, X., Jiao, Y., Song, W., & Lai, J. (2011). Extensive, clustered parental imprinting of protein-coding and noncoding RNAs in developing maize endosperm. *Proceedings of the National Academy of Sciences of the United States of America, 108*, 20042–20047.

Zhang, M., Xie, S., Dong, X., Zhao, X., Zeng, B., Chen, J., Li, H., Yang, W., Zhao, H., Wang, G., et al. (2014). Genome-wide high resolution parental-specific DNA and histone methylation maps uncover patterns of imprinting regulation in maize. *Genome Research, 24*, 167–176.

Zhang, M., Li, N., He, W., Zhang, H., Yang, W., & Liu, B. (2016). Genome-wide screen of genes imprinted in Sorghum endosperm, and the roles of allelic differential cytosine methylation. *The Plant Journal, 85*, 424–436.

Chapter 21
Developmental Genetics and the Evolution of Endless Fruit Forms

Chunjing Song and Chaoying He

Abstract Fruit, as an evolutionary innovation, helps to protect and disperse seeds in angiosperms. To fulfill such functions, fruit forms have diversified to display numerous types and morphologies, including a diversity of shapes and sizes, and even novel forms. Developmental genetics has revealed how fruit organs initiate, how fruit forms develop, and how they have evolved. Here, we summarize the accumulated knowledge of the genetic basis underlying these developmental and evolutionary processes in model plants, such as *Arabidopsis*, rice, and tomato. The molecular basis underlying carpel organ identity and functionality seems to be evolutionarily conserved. In contrast, the developmental trajectories in post-initiation processes are too diverse to generalize a developmental genetic toolkit or propose a theoretical, unifying model for fruit formation that explains the transition of fruit types and morphogenesis. Developmental repatterning is considered to play a role in fruit diversification, which could be primarily accomplished through variation in gene expression over time, space, and/or dosage, thereby generating diverse morphologies. A comparison between closely related species to show developmental and genetic variation during fruit development is the key to addressing the mechanisms underlying developmental repatterning. The current knowledge on fruit patterning and repatterning from model plants is expansive. However, additional models are critically needed to better understand these general processes. Lastly, we argue that an evolutionary development biology (evo-devo) perspective could

C. Song
State Key Laboratory of Plant Diversity and Specialty Crops/State Key Laboratory of Systematic and Evolutionary Botany, Institute of Botany, Chinese Academy of Sciences, Beijing, China

College of Life Sciences, University of Chinese Academy of Sciences, Beijing, China

C. He (✉)
State Key Laboratory of Plant Diversity and Specialty Crops/State Key Laboratory of Systematic and Evolutionary Botany, Institute of Botany, Chinese Academy of Sciences, Beijing, China

College of Life Sciences, University of Chinese Academy of Sciences, Beijing, China

China National Botanical Garden, Beijing, China
e-mail: chaoying@ibcas.ac.cn

F. Chang et al. (eds.), *Regulation of Plant Development*,
https://doi.org/10.1007/978-981-95-7033-1_21

provide a multi-dimensional view to understand how fruits form and repattern, why fruits evolve, and further provide insights into crop design and breeding.

Keywords Carpel · Fruit · Developmental genetics · Evo-devo

21.1 Introduction

Fruits are the reproductive organs of angiosperms and thus contain mature seeds. There are two types of fruit according to the mechanism of formation. The narrow definition involves just the mature ovary, resulting in "true fruit."

Conversely, in the more general definition, fruit develops from the flower as a complex unity, and other floral organs that are exterior to the carpels may be involved in fruit formation, including the receptacle, calyx, corolla, and bracts, which form some of the fruit flesh (Roth, 1977). The latter fruit types are often called "false fruit." Fruit packages seeds and aids seed dispersal, thus facilitating the adaptation of angiosperms to environments. The diversification of fruit forms, as a result of plant adaptation to different environments, has caused angiosperms to diversify into the most prosperous phytogroup on the earth and has also allowed them to dominate most terrestrial ecosystems.

Regardless of the mechanism of fruit formation, the carpel-containing ovules develop into the primary and core components of fruit, while the final forms may be diverse (even including the origin of novel structures in some cases). Consequently, the organ identity and functionality of the carpel are instrumental toward producing a functional fruit. Upon pollination and reception of the fertilization signal, the ovules develop into mature seeds during fruit development, and this is the main determinant for carpel functionality. However, the diversity of fruit forms, which can be identified by types and morphologies, is required for seed dispersal. The molecular mechanisms underlying the determination of fruit types are poorly known. Fruit morphological features, including shape and size, are governed by different schemes of cell division and cell expansion, which involve gene products with various chemical properties and functions. Genetic modifications that occur in any of the involved genes likely alter the fruit developmental process, which is termed developmental repatterning, and result in the generation of different morphologies. A comparison between closely related species to show developmental and genetic variations during fruit development is an effective way to understand the mechanisms underlying developmental repatterning. Here, we have summarized the current body of knowledge on the genetics involved in these developmental and evolutionary processes of carpels and fruits. In particular, we focus on what is known of model plants, such as *Arabidopsis*, tomato, rice, and *Physalis*, thereby producing novel insights into crop design and breeding, which is particularly important since fruit is also a major source of human sustenance.

21.2 Regulation of Initiation of a Flower with a Fertile Carpel—The Fruit's Major Predecessor

LEAFY (*LFY*) is a plant-specific transcription factor gene that has a central role in blossoming and subsequent development (Schultz & Haughn, 1991; Siriwardana & Lamb, 2012) and is also hypothesized to play a crucial role in the flower initiation process. The loss of *NEEDLY (NLY)*, which is an ortholog of the *LFY* gene, causes an ectopic translocation of the ovule to the adaxial side of laminar microsporophylls. The consequent rolling of laminar microsporophyll to form ascidiate carpels thus plays a possible role in the origin of the flower and fruit in angiosperms (Frohlich & Parker, 2000; Specht & Bartlett, 2009; Moyroud et al., 2010; Silva et al., 2016). As summarized in Fig. 21.1a, FLOWERING LOCUS T (FT) and FLOWERING LOCUS D (FD) together activate the expression of *LFY* and *APETALA1* (*AP1*), which are required for floral meristem identity. They then act in a complex to activate the floral homeotic genes that control floral organ identity (Corbesier et al., 2007; Weigel et al., 1992). The above model has been well summarized as the floral organ identity ABC model, which was determined mainly via developmental genetics of the model plants *Arabidopsis thaliana* and *Antirrhinum majus* (Coen & Meyerowitz, 1991; Weigel & Meyerowitz, 1994). The floral model has been optimized to the widely accepted, current "ABCDE" model (Angenent & Colombo, 1996; Pelaz et al., 2000; Theissen, 2001), which posits that the development of a flower is generally controlled by at least five functional genes. The A and E function genes specify the identity of the sepals, while the A, B, and E function

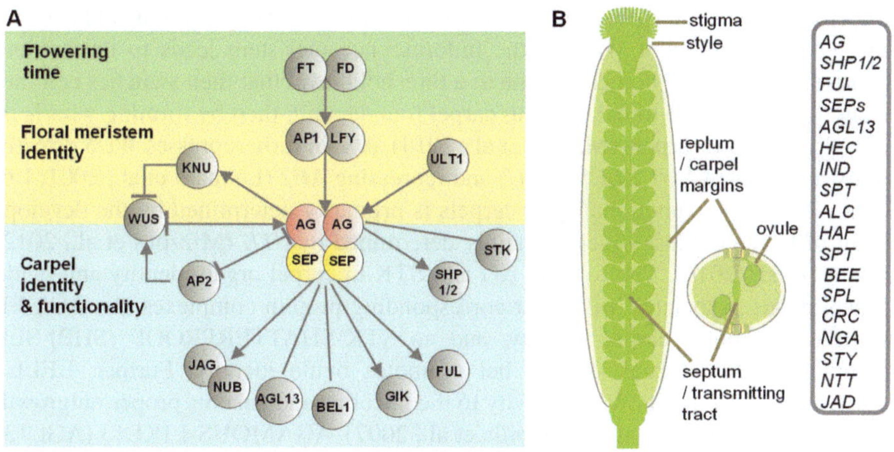

Fig. 21.1 Developmental regulation of *Arabidopsis* carpel. (**a**) Floral pattern and carpel organ identity regulatory network. The line with arrows indicates activation of gene expression, and the blocked line indicates gene repression. The adherence of two proteins, or lines, indicates protein-protein interactions. (**b**) The genetics underlying carpel formation. A schematic section of *Arabidopsis* carpel is shown, and the essential characterized genes are listed in the box to the side. (See text for additional details)

genes co-control the petal identity; the B, C, and E function genes determine stamen formation; the C and E function genes control carpel formation; and the D and E function genes are required for ovule formation. Consequently, the E function genes are indispensable for the development of all of the floral organs. In *Arabidopsis*, the A function genes include *AP1* and *APETALA2* (*AP2*), while the B function genes are *APETALA3* (*AP3*) and *PISTILLATA* (*PI*). The C, D, and E function genes are *AGAMOUS* (*AG*), *SEEDSTICK* (*STK*, formerly *AGAMOUS-LIKE11*), and *SEPALATA1/2/3/4* (*SEP1/2/3/4*, formerly known as *AGAMOUS-LIKE2/4/9/3*) respectively. These genes exclusively encode MADS-box regulatory proteins, with the exception of *AP2* (Bowman et al., 1991; Coen & Meyerowitz, 1991; Jofuku et al., 1994; Theissen, 2001).

The pistil is the female reproductive structure of a flower that is made up of a single carpel or a group of carpels fused together and is the central component of the flower. A carpel is usually differentiated into a stigma, style, and ovary, which enclose the ovules after fertilization and the development of seeds. The carpel is the main fruit predecessor, and its organogenesis thus lays the foundation for specific fruit forms. In *Arabidopsis*, *AG* is required for floral meristem termination (Krizek & Fletcher, 2005; Liu et al., 2014) and determines the organ identity of carpels (Bowman et al., 1989; Yanofsky et al., 1990). The above processes are noted by the observation that the mutation of *AG* leads to the replacement of the carpel by petals or a flower. The ULTRAPETALA 1 (ULT1) and LFY pathways act separately to regulate the identity and determinacy of the floral meristem. In addition, they independently induce *AG* expression in the center of the flower in order to terminate meristem activity (Engelhorn et al., 2014). WUSCHEL (WUS) is well known as a stem cell regulator, and it is also required for floral organ initiation because WUS directly activates *AG* by binding to the second intron of *AG* (Lohmann et al., 2001). Strong upregulation of *WUS* in the inflorescence meristem leads to the ectopic expression of the *AG* homeotic gene to a threshold level that then switches cell fate from floral meristem founder cell to carpel founder cell, thereby forming whorls of ectopically fused carpels (Mandel et al., 2014). AG, in turn, represses *WUS* directly or by activating *KNUCKLES* (*KNU*) and repressing *AP2* (Lenhard et al., 2001; Liu et al., 2014). The functionality of carpels is primarily determined by the development of fertile ovules, which is mainly determined by *STK* (Mizzotti et al., 2012; Ezquer et al., 2016). The roles of AG and STK in carpel organ identity and ovule functions are associated with their corresponding protein complexes. An AG-SEP complex promotes carpel identity, and an STK-SHATTERPROOF (SHP)-SEP complex inhibits carpel identity but promotes ovule identity. Further, BELL1 (BEL1)-AG-SEP limits WUS activity to the nucellus and ensures proper outgrowth of the ovule's integuments (Brambilla et al., 2007). AGAMOUS-LIKE13 (AGL13) is a putative E function protein that interacts with AG to form the AGL13-AG complex that regulates ovule development (Hsu et al., 2014). *JAGGED* (*JAG*) and *GIANT KILLER* (*GIK*) are also direct targets of AG. JAG, together with its paralog NUBBIN (NUB), controls stamen and carpel shape in *Arabidopsis* (Dinneny et al., 2004; Ohno et al., 2004; Gómez-Mena et al., 2005; Dinneny et al., 2006), while *GIK* regulates patterning and differentiation of reproductive organs (Ng et al., 2009).

The close homologs, AG, STK, and SEP, have been extensively characterized in various plants (Favaro et al., 2003; Acri-Nunes-Miranda & Mondragón-Palomino, 2014; Tani et al., 2009), and they are involved in regulating carpel induction, organ identity, functionality, and maturation of carpels and ovules. However, the specific molecular regulatory details require further investigation. Nevertheless, the above observations suggest that the developmental roles of these genes in carpel organ identity and functionality are basically conserved among angiosperms.

21.3 Developmental Genetics of Carpel in Producing Functional Fruits: The *Arabidopsis* Model

The carpel is a composite structure comprising ovary walls, placenta, and ovules, and its development has been thoroughly investigated in *Arabidopsis*. The *A. thaliana* gynoecium (the sum of all carpels) consists of two carpels that are fused along their entire margins, a style, a dry stigma, and the false septum, which is a specialized tissue derived from the carpel margins that harbors the transmitting tract (Reyes-Olalde et al., 2013; Fig. 21.1b). Genetic analysis has revealed some intermediate regulatory genes that control specific aspects of carpel development. Specifically, several transcription factor gene families, including the MADS-box, bHLH, YABBY, RAV, and zinc finger groups, were found to be associated with this developmental process and were further mapped on a schematic structure of *Arabidopsis* carpel (Fig. 21.1b).

The MADS-box gene *AG* not only regulates carpel organ identity but also controls earlier carpel development after organ initiation. Later in flower development when ovules are formed, *AG* is expressed in the funiculus and integuments, but the AG protein moves out of its expression domain and can be detected in the chalaza and nucellus (Urbanus et al., 2009). AG acts together with SEP3 and BELL1 (BEL1) to gradually restrict *WUS* expression to the nucellus, which then confines the proper site of integument development from the chalaza (Brambilla et al., 2007; Urbanus et al., 2009). AG activates *SPOROCYTELESS* (*SPL*), which controls sporogenesis in the carpel (Ito et al., 2004), and regulates the expression of *SUPERMAN* (*SUP*), which controls cell proliferation in carpel primordia (Sakai et al., 2000). The two MADS-box *SHATTERPROOF* genes (*SHP1* and *SHP2*) are required in the carpel margins for differentiation of the dehiscence zone, where the fruit later splits open to release the seeds (Liljegren et al., 2000). The heterodimer of MADS-domain protein AGL13-AG is involved in regulating ovule development (Hsu et al., 2014), and FUL-SEPs influences fruit development by mediating cell differentiation (Gu et al., 1998; de Folter et al., 2005).

The bHLH transcription factors, including *SPATULA* (*SPT*), *ALCATRAZ* (*ALC*), *HECATE1, 2, 3* (*HEC1*, *2*, and *3*, respectively), *INDEHISCENT* (*IND*), and *HALF FILLED* (*HAF*; also known as *CESTA* [*CES*]), are also involved in carpel development in *Arabidopsis*. *SPT* controls cell differentiation at the carpel margins and in the transmitting tract (the tissue that guides the growth of pollen tubes toward the ovules) (Alvarez & Smyth, 2002; Heisler et al., 2001). *SPT* is expressed in carpel margin tissues and in the pollen transmitting tract. The mutant *spt* lacks the transmitting tract, the gynoecium forms a medially flattened shape, and the carpel's stylar region is unfused (Heisler et al., 2001; Alvarez & Smyth, 2002). *ALC* is a close homolog of *SPT* that can fully restore an *alc* mutant (Groszmann et al., 2011), suggesting that *ALC* thus acts redundantly with *SPT* during early gynoecium development. Nonetheless, *ALC* has specialized essential roles in dehiscent zone development in the formation of the lignified separation layer that is required for the opening of fruit to release seeds (Rajani & Sundaresan, 2001; Groszmann et al., 2011). The *HEC1*, *2*, and *3* genes overlap in expression in the stigma, the transmitting tract, and the septum. Double mutants have defects in fertility and septum development, while the triple mutant is sterile and has no septum (Schuster et al., 2015). Of the three single *hec* mutants, only the *hec3* mutant results in obvious defects. The mutant does not have a seed abscission zone, and seeds therefore remain attached to the fruit, even after pod shattering, which severely affects seed dispersal. The specification of valve margin cells into separation and lignification layers is important for fruit dehiscence and seed dispersal, and *IND* is involved in this developmental process (Liljegren et al., 2004). Other regulatory modules that regulate the exact specification of the fruit valve, valve margin, and the replum include *FUL*, *REPLUMLESS* (*RPL*), *ALC*, *SHP1*, and *SHP2* (Liljegren et al., 2004). In addition, *IND* directly regulates *SPT* expression and cooperatively binds to target genes that are required for the regulation of auxin dynamics in carpel margin differentiation (Girin et al., 2011). *HAF*, also known as *CES*, acts as a positive regulator in brassinosteroid biosynthesis and is required for reproductive tract development (Poppenberger et al., 2011). The expression of *HAF* spans the reproductive tract in the zone of the fused septa and then expands to the center of the septum in all of the tissues that are related to the transmitting tract, including the stigma, style, ovary, and funiculi.

BRASSINOSTEROID ENHANCED EXPRESSION1 (*BEE1*) and *BEE3* are also bHLH members that share homology with *HAF* and are similarly expressed as *HAF* (Pires & Dolan, 2010). However, *BEE1* is also expressed in the upper part of the style and stigma, and *BEE3* is also expressed in the ovary transmitting tract (Crawford & Yanofsky, 2011). Many ovules of the *haf/bee1/bee3* triple mutant cannot be fertilized, especially in the lower parts of the ovary, due to defects in promoting pollen tube growth by reduction in the extracellular matrix and programmed cell death (Crawford & Yanofsky, 2011). The three genes are regulated by *HEC*, *STYLISH (STY)*, *NO TRANSMITTING TRACT* (*NTT*), and *AUXIN RESPONSE FACTOR6/8* (*ARF6/ARF8*) (Friedrichsen et al., 2002; Crawford et al., 2007; Crawford & Yanofsky, 2011).

The YABBY transcription factor gene *CRABS CLAW* (*CRC*) is also important for several aspects of carpel development. *Arabidopsis CRC* is expressed in the abaxial domain of the carpel and participates in directing the development of tissues derived from the abaxial side of the carpel primordium (Eshed et al., 1999). Compared with the wild-type, the *Arabidopsis CRC* mutant produces shorter, wider carpels that are apex fusion impaired and have no nectary glands, gynoecium, or fruits developed (Lee et al., 2005; Fourquin et al., 2005; Orashakova et al., 2009).

NGATHA (*NGA*) genes (*NGA1*, *2*, *3*, and *4*) encode transcription factors of the RAV family. *NGA2* and *3* are expressed in roots, shoot meristems, leaves, flowers, and seeds. In contrast, *NGA1* and *4* are expressed in the distal parts of leaves and floral organs after their identity determination processes have been completed (Swaminathan et al., 2008; Alvarez et al., 2009). The four *NGA* genes are redundantly expressed when the style and stigmas begin to differentiate in the apical region of the gynoecium (Alvarez et al., 2009; Trigueros et al., 2009). In the style, *NGA* and *STY* cooperatively activate the auxin biosynthesis genes *YUCCA2* and *YUCCA4* in order to regulate auxin gradients and the apical-basal patterning of the gynoecium, which may also incorporate *CRC* (Alvarez et al., 2009; Trigueros et al., 2009).

NTT encodes a transcription factor of the zinc finger family that is involved in the development of the transmitting tract and replum (Crawford et al., 2007). *NTT* is generally expressed in the septum, replum, funiculus, small domain in the ovule, the contact area between the two carpels where the septum will develop, and the replum of developing carpels and fruits (Crawford et al., 2007; Marsch-Martínez et al., 2014). The *ntt* mutant shows reduced fruit lengths and seed sets, especially in the ovules in the basal part of the ovary, wherein almost no pollen tube arrival occurs (Crawford et al., 2007). The role of NTT in replum development is mediated by binding to the promoter of *BREVIPEDICELLUS* (*BP*), which is an essential gene for replum development. The subsequent formation of a protein complex with SHOOT MERISTEMLESS (STM), REPLUMLESS (RPL), FUL, SHP1, SHP2, and other developmental regulators of the replum occurs (Marsch-Martínez et al., 2014). The *Arabidopsis* zinc finger proteins JAG and NUB control carpel shape (Dinneny et al., 2004; Ohno et al., 2004; Gómez-Mena et al., 2005; Dinneny et al., 2006), and thus likely also contribute to the determination of fruit shape.

21.4 Developmental Evolution of Fruit Endless Morphologies

The genetic regulators involved in carpel development appear to be conserved in seed plants (Pfannebecker et al., 2017). However, the diversity of flowers, especially the diversity of gynoecium, provides a foundation for the variety of fruit forms that exist. Fruit forms, broadly speaking, are associated with both fruit types and fruit morphologies. The developmental evolution of fruit types is poorly understood, while most research in this field has focused on fruit shapes and sizes (morphology). Several crop models have been established, such as in rice and tomato, and fruit size and shape variations generally result from repatterning of cell division and cell expansion in these model systems.

21.4.1 The Rice Model

The rice seed is a grain, which is a staple food that provides significant carbohydrate content for humans globally. The grain consists of the basal glume, the peripheral spikelet hull, the aleurone layer, the seed coat, and the endosperm with the embryo inside (Fig. 21.2a). The spikelet hull limits the cavity of the grain, and the endosperm fills most of the grain; these two components are crucial in determining grain size and shape in rice. Many genes can alter grain shape and growth (Li & Li, 2016). Here, we discuss only the major quantitative trait loci (QTLs) that have been identified for rice grain size (weight) and shape (width and length), their encoded proteins, genetic variation in regions, and functional modes (Fig. 21.2a; See Table 21.1 for additional details).

GRAIN SIZE 3 (*GS3*) and *DENSE AND ERECT PANICLE1* (*DEP1*) QTLs encode subunits of the G protein-coupled receptor, and negatively control grain size and panicle size. Further, the gain-of-function of *GS3* or *DEP1* in rice results in short grains (Fan et al., 2006; Mao et al., 2010; Huang et al., 2009). *GRAIN WIDTH AND WEIGHT 2* (*GW2*) encodes a primary QTL that dictates rice grain size and weight (Song et al., 2007). *GW2* is a close homolog of DA2, the E3 ubiquitin ligase, which is also involved in fruit and organ size regulation in *Arabidopsis* (Xia et al., 2013). *GW2* controls small grains by limiting the size of the spikelet hulls and negatively regulating cell division (Song et al., 2007). Moreover, *GW2* homologs in maize and wheat exhibit the conserved function of grain size control (Li et al., 2010; Su et al., 2011; Bednarek et al., 2012; Yang et al., 2012).

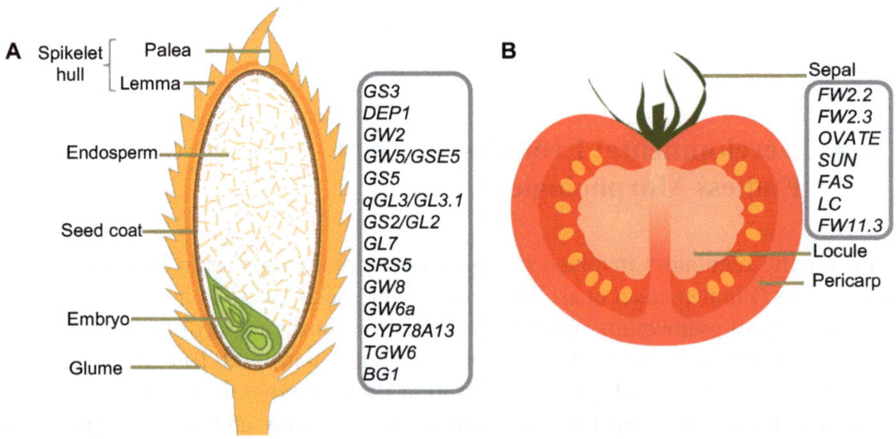

Fig. 21.2 Developmental genetics of fruit size and shape in rice and tomato. (**a**) Schematic section of a rice grain. Major QTLs that have been determined to control grain size and shape are indicated in the box to the side. (**b**) Schematic section of a tomato berry and the regulators that have been determined to control fruit size and shape. Further details of the listed QTLs and genes are summarized in Table 21.1. (See text and Table 21.1 for additional details)

Table 21.1 The main QTLs or genes controlling fruit morphology in rice, tomato, and *Physalis*

Species	QTL/gene	Protein	Allelic variation	Trait affected	Process affected	References
Rice	*GS3*	G protein	Coding region	Grain length/weight	Cell division	Mao et al. (2010)
	DEP1	G protein	Premature stop	Grain size/length	Cell division	Huang et al. (2009)
	GW2	E3 ubiquitin ligase	Premature stop	Grain width/weight	Cell division	Song et al. (2007)
	GSE5/GW5	IQD protein	Promoter	Grain width/weight	Cell division	Weng et al. (2008) and Duan et al. (2017)
	GS5	Serine carboxypeptidase	Promoter	Grain size	Cell division/cell expansion	Li et al. (2011) and Xu et al. (2015a)
	qGL3/GL3.1	Ser/Thr phosphatase	Coding region	Grain size/length	Cell division	Hu et al. (2012) and Qi et al. (2012)
	GS2/GL2	OsGRF4	Coding region	Grain size	Cell division/cell expansion	Hu et al. (2015), Duan et al. (2015) and Che et al. (2015)
	GLW7	OsSPL13	5′ UTR	Grain length	Cell expansion	Si et al. (2016)
	GW8	OsSPL16	Promoter	Grain size	Cell division/cell expansion	Wang et al. (2012a, b)
	GL7/GW7/SLG7	TON1 RECRUITING MOTIF protein	Regulatory region	Grain size	Cell division/cell expansion	Wang et al. (2015a, b, c) and Zhou et al. (2015)
	GW6a	Histone H4 acetyltransferase	Promoter	Grain size	Cell division	Song et al. (2015)
	GE	CYP78A13	Coding region	Grain size	Cell division/cell expansion	Nagasawa et al. (2013) and Xu et al. (2015b)
	TGW6	IAA-glucose hydrolase	Premature stop	Grain weight	Cell division	Ishimaru et al. (2013)

(continued)

Table 21.1 (continued)

Species	QTL/gene	Protein	Allelic variation	Trait affected	Process affected	References
Tomato	*FW2.2*	CNR protein	Promoter	Fruit size	Cell division	Frary et al. (2000) and Cong et al. (2002)
	FW3.2	Cytochrome P450	Promoter	Fruit size	Cell division	Chakrabarti et al. (2013)
	FW11.3	FAF-domain protein	Truncated protein	Fruit size	Cell expansion	Huang and van der Knaap (2011) and Mu et al. (2017)
	OVATE	Ovate family protein	Premature stop	Fruit shape	Cell division	Liu et al. (2002)
	SUN	IQD family of calmodulin-binding proteins	Transposon-mediated an increase in gene expression	Fruit shape	Cell division	Xiao et al. (2008) and Wu et al. (2011)
	FAS	YABBY-like protein	Intron	Locule number/ fruit shape/size	Cell division	Cong et al. (2008)
	LC	Putative WUSCHEL ortholog	Regulatory region	Locule number/ fruit shape/size	Cell division	Muños et al. (2011)
Physalis	*POS1*	CRF-AP2	Intron	Fruit size	Cell expansion	Wang et al. (2014)
	POS2	FW2.2 ortholog	Promoter	Fruit size	Cell division	Li and He (2015)
	MPF2	MADS-box protein	Promoter	Fruiting calyx size	Cell division/cell expansion	He and Saedler (2005)
	MPF3	MADS-box protein	Coding region	Calyx identity/ size	Cell division/cell expansion	Zhao et al. (2013)

SEED WIDTH ON CHROMOSOME 5 (*qSW5/GW5*) controls the grain size by limiting the spikelet hull size via repression of cell division (Shomura et al., 2008; Weng et al., 2008). *GSE5*, within the qSW5/GW5 locus, encodes a plasma membrane-associated protein that exhibits IQ domains. Loss of *GSE5* function results in wide and heavy grains, while the overexpression of *GSE5* results in narrow grains (Duan et al., 2017). *GRAIN SIZE 5* (*GS5*) encodes a putative serine carboxypeptidase that positively regulates cell proliferation and expansion in spikelet hulls. Elevation of *GS5* expression results in wide and heavy grains (Li et al., 2011; Xu et al., 2015a). *GRAIN LENGTH 3* (*qGL3*; also known as *GL3.1*) encodes a protein phosphatase that negatively controls cell division in spikelet hulls by dephosphorylating cyclin-T1;3 (CycT1;3) (Hu et al., 2012; Qi et al., 2012). *GRAIN SIZE* 2 (*GS2*; also known as *GL2*) encodes OsGRF4 (growth-regulating factor), which primarily functions in enlarging spikelet hulls by promoting cell expansion and slightly increasing cell proliferation (Hu et al., 2015; Duan et al., 2015; Che et al., 2015). The expression of *GS2/OsGRF4* is negatively regulated by OsmiR396 (Duan et al., 2015; Che et al., 2015). OsGSK2 (GSK3/SHAGGY-like kinase 2) is also a negative regulator of *GS2* and interacts with GS2 to reduce its transcriptional level (Che et al., 2015). OsGIF1/2/3 (GRF-interacting factors) are transcriptional coactivators of GS2, which physically interact with GS2, resulting in the overexpression of *OsGIF1*, thereby also producing large grain sizes (Duan et al., 2015).

GRAIN WEIGHT 6a (*GW6a*) is a grain weight QTL in rice and is a histone H4 acetyltransferase gene. The upregulation of *GW6a* results in increases in grain weight and yield by increasing cell numbers in spikelet hulls and promoting grain filling (Song et al., 2015). The rice *GIANT EMBRYO* (*GE*) gene encodes CYP78A13, which is required for proper size balancing between embryos and endosperms (Nagasawa et al., 2013). Variation in the coding region of *GE* influences grain size and yield in rice (Xu et al., 2015b). Meanwhile, *THOUSANDGRAIN WEIGHT* 6 (*TGW6*) encodes an IAA-glucosehydrolase, which negatively affects endosperm development and grain weight (Ishimaru et al., 2013). BIG GRAIN1 (BG1) regulates auxin transport, thereby promoting cell division and cell expansion in spikelet hulls and generating large grains (Liu et al., 2015).

The QTLs encoding SQUAMOSA PROMOTER BINDING PROTEIN-LIKE transcription factors (SPL) play important roles in determining grain shapes and sizes. Further, grain length and weight on chromosome 7 (*GLW7)* encodes OsSPL13, which positively controls grain length and weight by promoting cell elongation (Si et al., 2016). Further, *GLW7* promotes *SMALL AND ROUND SEED 5* (*SRS5*), whose loss-of-function mutant reduces grain length by regulating cell elongation of lemma cells (Si et al., 2016; Segami et al., 2012). In addition, *GRAIN WIDTH 8* (*GW8)* encodes another rice SPL transcription factor, OsSPL16, which positively affects cell division in the grain-width direction, thereby regulating grain size and weight (Wang et al., 2012a, b).

GRAIN LENGTH ON CHROMOSOME 7/GRAIN WIDTH 7/SLENDER GRAIN ON CHROMOSOME 7 (*GL7/GW7/SLG7*) encodes a TON1 RECRUITING MOTIF (TRM)-containing protein that is homologous to *Arabidopsis* LONGIFOLIA1/2. The main QTL for slender grains contributes to grain length by influencing spikelet hulls. *GL7/GW7/SLG7* longitudinally increases cell length and transversely decreases cell width (Wang et al., 2015a; Zhou et al., 2015). However, *GL7/GW7/SLG7* was shown to regulate cell proliferation in a positive manner in the longitudinal direction, while negatively affecting it in the transverse direction (Wang et al., 2015b). Both strategies could contribute to slender grains, despite the inconsistencies that require further investigation.

21.4.2 The Tomato Model

Fruit size and shape are pronounced in common solanaceous plants, such as tomatoes, eggplants, and peppers. The tomato model for berry development has been established owing to the considerable genetic resources and tools that have been developed for tomatoes. Unlike grains in rice, the tomato fruit shape is determined in the early periods of ovary development and is highly consistent with ovary shape (Houghtaling, 1935). It is also dependent on the pericarp differentiation of longitudinal growth and thickness growth. The same growth rate in various dimensions results in a round tomato, while different growth rates result in other shapes, classified as flat, ellipsoid, rectangular, heart, long, obovoid, and oxheart (Rodríguez et al., 2011). Tomato fruit size is also dependent on cell division within the ovary and is highly correlated with ovary size, in addition to cell expansion in the pericarp and locule number (Fig. 21.2b).

Many genes can participate in the regulation of these traits in tomatoes. For example, the pericarp cell number is linked to the expression of the MADS-box gene *TOMATO AGAMOUS* (*TAG*)-*LIKE1* (*TAGL1*), which is an ortholog of *Arabidopsis SHP* genes (Vrebalov et al., 2009). *TAGL1* is highly expressed in floral organs and in the tissues of young tomato fruits. Further, downregulating *TAGL1* leads to a reduction in pericarp thickness and an absence of the inner pericarp (Matas et al., 2011). However, these fruit size and shape traits are all quantitatively inherited and controlled by multiple genes. Thus far, seven main QTLs, including *OVATE*, *fasciated* (*FAS*), *SUN*, *locule number* (*LC*), *Fruit weight 2.2* (*FW2.2*), *Fruit weight 2.3* (*FW2.3*), and *Fruit weight 11.3* (*FW11.3*), have been cloned and determined to play an essential role in the domestication of tomatoes (Fig. 21.2b; See Table 21.1 for additional details). FW2.2, FW3.2, and FW11.3 mainly affect fruit size, while SUN and OVATE control long shapes, and FAS and LC control locule number (a major determinant of fruit size in tomato) in addition to flat shapes.

FW2.2 was the first cloned plant QTL (Frary et al., 2000) and encodes a cell number regulator (CNR) family protein (Guo & Simmons, 2011). This gene alone is responsible for approximately 30% of the variation in fruit weight (Frary et al., 2000). Modification of fruit size owing to *FW2.2* is caused by the alteration of the promoter region rather than protein sequence or structure alteration (Frary et al., 2000). *FW2.2* expression levels predominate in carpel and are negatively correlated with berry size. FW2.2 is a cell membrane-anchored protein that is plant-specific and fruit-specific and negatively regulates mitosis. Further, it is possibly involved in part of the cell cycle signal transduction pathway and is theorized to act with a subunit of casein kinase II (CKII) to affect the cell cycle (Cong et al., 2008). *FW3.2* encodes a KLUH orthologous protein that is a cytochrome P450 enzyme belonging to the CYP78A subfamily. *FW3.2* controls fruit weight and also has a minor role in regulating fruit shape (Chakrabarti et al., 2013). Therefore, it has also been referred to as *SlKLUH*. *FW3.2* regulates cell division by increasing cell number, resulting in the enlargement of the pericarp and septum tissue, and thereby increasing fruit weight. A single nucleotide polymorphism in the promoter of *SlKLUH* leads to increased gene expression and is highly associated with increased fruit mass (Chakrabarti et al., 2013). *FW11.3* is a locus that is close to *FAS* (Huang & van der Knaap, 2011). Very recently, this QTL, which was renamed as *Cell Size Regulator* (*CSR*), was found to encode a poorly characterized protein that controls tomato fruit weight via cell expansion of fruit pericarp. The deleted CSR allele, with a 194 amino-acid truncation, is a critical selected marker that was important in the transition from intermediate ancestors to the current domesticated tomato (Mu et al., 2017).

The *SUN* gene regulates elongation of fruit shape, and the encoded protein belongs to the IQD family (Xiao et al., 2008). Fruit shape is controlled by increasing cell division in the longitudinal direction and decreasing cell division in the transverse direction. The overexpression of *SUN* leads to shape changes in multiple organs, such as very elongated cotyledons and leaves, which then dramatically elongates the proximal ends of the ovaries and fruits (Xiao et al., 2008). Moreover, *SUN's* influence on fruit shape is independent of fruit mass and seed numbers (Wu et al., 2011). *OVATE* encodes a member of the ovate family proteins (OFPs) and controls the modification of round tomatoes to pear-shaped fruits. *OVATE* negatively controls fruit length, and a mutation that results in a premature stop codon in the second exon leads to pear-shaped tomatoes (Liu et al., 2002).

The number of tomato locules can vary greatly from two to ten, or even more, which affects the shape and size of the fruit. The trait is also controlled by multiple loci, while the primary phenotypic diversity can be explained by *FAS*, *LC*, and their interactions. *FAS* is considered to be the primary gene and exhibits the strongest effect of increasing locule numbers, and it can regulate the increase in locules from two to six and even higher (Muños et al., 2011). Consequently, *FAS* has played a key role in the domestication of modern tomatoes. *FAS* encodes a YABBY transcription factor that is expressed in the carpel and stamen early in development. The *FAS* gene coding region is not significantly different in different genetic lines (Cong et al., 2008), but gene expression can be altered in the high-locule-number lines via different mechanisms. These include the reduction of gene expression due to a 6-8 kb insertion mutation in the first intron of the *FAS* gene (Cong et al., 2008) or null expression due to the inversion of a 294-kb fragment at the first intron of the *FAS* gene (Huang & van der Knaap, 2011). *LC* increases the number of locules from two to four and thus has a weaker effect than *FAS* (Rodríguez et al., 2011). The QTL is located 1600 bp from the 3′ end of *WUS* in the noncoding region, and its biochemical nature requires further investigation. In addition, *Arabidopsis* WUS regulates stem cell numbers (Yadav et al., 2010), and the *WUS* homolog is therefore considered to be a candidate gene for *LC* (Muños et al., 2011). Molecular evolution analysis of *LC* alleles showed that the polymorphism of two nucleotides is accompanied by a decrease in the diversity of cultivars, thus indicating that SNPs are associated with increased locule numbers (Muños et al., 2011).

21.4.3 The New Model: Origin of Fruit Morphological Novelty and Repatterning

Fruit size substantially varies in *Physalis* (Wang et al., 2012a, b, 2014, 2015c), which is a genus that is closely related to *Solanum*. The two genera feature fleshy fruits as berries, and fruit size or yield is the prime breeding target in both species. *Physalis organ size 1* (*POS1*) and *POS2* are the two genes that control natural variation in berry size that have been characterized to date in *Physalis* (Table 21.1). Interestingly, both conservation and divergence in the genetic control of fruit size are indicated by comparisons of these two genes between *Solanum* and *Physalis*. *POS1* encodes a putative transcription factor with two cytokinin response factor (CRF)-AP2 (APETALA2) domains and positively controls cell expansion and thus fruit size in *Physalis philadelphica* (tomatillo). Copy variation of a 37-bp repeat in the first intron of *POS1* alleles is negatively correlated with the expression levels of this gene (Wang et al., 2014). However, the *POS1* ortholog in tomatoes seems to be associated with stress response (Shi et al., 2012; Wang et al., 2014), suggesting a divergence in genetic control over fruit size between *Solanum* and *Physalis*. *POS2*, which is also named *P. floridana cell number regulator 1* (*PfCNR1*), encodes a putative ortholog of *FW2.2*. The heterochronic expression levels of different *POS2* alleles in ovaries are negatively correlated with cell division activity and berry size variation among different *Physalis* species (Li & He, 2015). This resembles the observation for *FW2.2* in tomatoes (Frary et al., 2000) and indicates a conservation in gene functions in the fruit size regulatory networks between *Solanum* and *Physalis*.

Physalis has been investigated in greater detail due to its distinctive fruit morphology. Fruit from plants of the genus *Physalis* features a lantern-like accessory structure known as the Chinese lantern, or inflated calyx syndrome (ICS), that is not observed in *Solanum* (Fig. 21.3). Some understanding of the evolutionary development of Chinese lanterns has accumulated over the last decade (Fig. 21.3 and Table 21.1). The origin of the Chinese lantern is associated with the heterotopic expression of the *Physalis* MADS-box gene 2 (*MPF2*) in floral organs, whereas its ortholog in *Solanum tuberosum*, MADS-box gene 16 (*STMADS16*), is specifically expressed in vegetative tissues (He & Saedler, 2005). *MPF2* is apparently involved in the development of both the calyx and stamen (Fig. 21.3a). Heterotopic expression of *MPF2* in the floral organs of *Physalis* may result from variations in CArG-box regions in its promoter (He & Saedler, 2005; Zhao et al., 2013). Phenotypic variations in *MPF2* knockdowns further support the role of MPF2 in male fertility and fruiting calyx inflation (He & Saedler, 2005). Moreover, calyx identity is

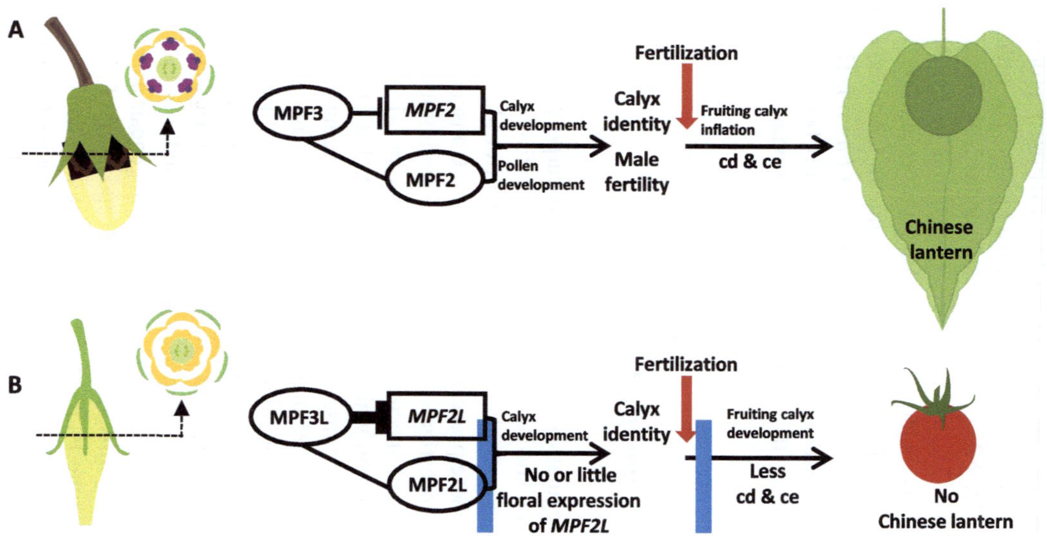

Fig. 21.3 The origin of the "Chinese lantern," a morphologically novel accessory to *Physalis* berry. (**a**) The MPF3-*MPF2*/MPF2 interactions involved in the development of the Chinese lantern in *Physalis*, where the fruiting calyx is inflated as the Chinese lantern. The line indicates the interaction of the two components: cd, cell division; ce, cell expansion. (**b**) Little to no activity of the MPF3-*MPF2*/MPF2-like interactions in the calyx of *Solanum*, where the fruiting calyx remains small in this genus. The *MPF2*-like (*MPF2L*) gene is strongly inhibited by the MPF3-like (MPF3L) regulator, and no or little expression of *MPF2L* is highlighted by the vertical blue line. In both **a** and **b**, the blocked line represents repression of gene expression. Arrows represent regulations. Proteins are presented in Roman numerals and ovals, while genes are identified in italics and rectangles

determined by the *Physalis* MADS-box gene 3 (*MPF3*), which is also involved in the control of Chinese lantern size. *MPF3* is specifically expressed in floral tissues and encodes a euAP1 MADS-domain protein that is primarily located in the nucleus. The protein interacts with MPF2 and some floral MADS-domain proteins in order to selectively bind variants of CArG-boxes in the *MPF2* promoter and repress *MPF2* expression in the calyx to low levels (Zhao et al., 2013), thereby generating Chinese lantern structures. In contrast, MPF3-like (MPF3L) proteins strongly restrict *MPF2*-like (*MPF2L*) expression in vegetative tissues. The little, or entirely absent, floral expression of *MPF2L* (i.e., calyx) results in the lack of Chinese lantern structures in *Solanum* (Fig. 21.3b). However, a mutation in the *MPF3L* gene *SlMADS-MC* leads to leaf-like enlarged sepals (Vrebalov et al., 2002). Therefore, the heterotopic expression of *MPF2* in floral organs due to CArG-box variation is the key to the origin of the Chinese lantern morphology in *Physalis* (He & Saedler, 2005; Zhao et al., 2013). Consequently, *Physalis* has currently become an additional model plant for understanding the developmental evolutionary genetics of fruit morphology.

21.4.4 Conservation Among Plant Models

While the developmental regulatory networks underlying carpel morphogenesis and development in the model species *A. thaliana* are relatively well understood (Ferrándiz et al., 1999; Reyes-Olalde et al., 2013; Robert et al., 2015), there has been less attention paid to these processes in other species. Consequently, scientists have been begun to further investigate the developmental evolution of fruit forms in some agronomically important species, such as tomatoes and rice. The molecular regulators of carpel development, which include *AG*, *SHP*, *ALC*, *CRC*, *HAF*, *HEC*, *IND*, *NGA*, and *SPT*, seem to be conserved among the most recent common ancestor of seed plants and *A. thaliana* (Pfannebecker et al., 2017). Indeed, some genetic regulators that have been functionally characterized are shared between plant species with large evolutionary distances between them, such as *Arabidopsis* and rice (Li & Li, 2016), or solanaceous species (Wang et al., 2015c). However, the degree of conservation lessens as additional models are investigated. For example, only a few conserved regulators of fruit development and morphogenesis are shared among *Arabidopsis*, rice, tomato, *Physalis*, or among a broad spectrum of plant species.

Members of the cytochrome P450 family are recruited to control organ size in various plant species. As mentioned above, *ORF6* in tomatoes (Chakrabarti et al., 2013) and *KLUH/CYP78A5* in *Arabidopsis* (Anastasiou et al., 2007; Adamski et al., 2009) control reproductive organ sizes. Other members of the cytochrome P450 subfamily also control organ and plant sizes, including *CYP78A9* and *CYP78A6* in *Arabidopsis*, and *CYP78A11* in rice (Fang et al., 2012; Ito & Meyerowitz, 2000; Miyoshi et al., 2004). *KLUH* (*CYP78A*) has also been found to regulate seed size in rice, wheat, and soybean (Ma et al., 2016; Xu et al., 2015b; Zhao et al., 2016), in addition to biomass and seed yield in maize (Sun et al., 2017). Moreover, *CaKLUH* is associated with fruit weights in peppers, suggesting that domestication-driven increases in fruit weight are, in part, derived from the KLUH signaling pathway (Chakrabarti et al., 2013). Two additional members of this subfamily, *CYP78A27* and *CYP78A28*, control colony size in moss (Katsumata et al., 2011). Thus, this gene family may play an ancient and conserved role in controlling organ size across plants.

Tomato *FW2.2* encodes a repressor of cell division, and this function is fulfilled by negatively regulating the expression of this gene, rather than through changes in protein structure (Frary et al., 2000; Cong et al., 2002). No information regarding this gene is available for *Arabidopsis*. However, its function seems to be conserved in many plant species that span dicots and monocots. In dicots, the *Pafw2.2-like* (*Persea Americana fruit weight 2.2-like*) gene in avocado has been proposed to play a conserved role as a negative regulator of fruit cell division (Dahan et al., 2010). *Prunus avium cell number regulator* (*PavCNR*) is a homolog of the tomato *FW2.2* and is a cell number regulator gene in *Prunus* species that is associated with fruit size in both sweet and sour cherries (De Franceschi et al., 2013). In *Physalis*, *POS2* (a *FW2.2*-like gene) regulates fruit size via the negative regulation of cell division, and it is associated with natural variation in berry sizes (Li & He, 2015). However, *GmFWL1* (*Glycine max fruit weight 2.2-like 1*) was found to be essential for soybean nodule organogenesis, resulting from effects on plant cell division (Libault et al., 2010). In monocots, the putative maize ortholog of tomato *FW2.2* is *cell number regulator1* (*CNR1*), which affects the size of entire plants and multi-organs by negatively regulating cell number (Guo et al., 2010). Further, *osfwl3* (*Oryza sativa fruit weight 2.2-like 3*) controls grain weights by negatively regulating cell division (Xu et al., 2013). Thus, *FW2.2*-like genes play a conserved role in cell division within different species. Furthermore, orthologs of FW2.2 in plants and the human oncogene *c-H-ras p21* have similar structures (Frary et al., 2000), which indicates an ancient origin of a FW2.2-associated cell cycle regulation pathway in plants and animals.

However, a large portion of the components in the fruit development regulatory networks among species, such as *Arabidopsis*, tomato, rice, and *Physalis*, do not overlap. This diversity in pathways might reflect the reason for the diverse and distinct fruit types and structures. Moreover, the number of unknown components involved in the development and evolution of fruit in each species is inestimable. This knowledge gap may suggest that large efforts should be invested in the field of fruit developmental genetics and that the conservation or divergence in genetic regulators that contribute to fruit evolution necessitates further investigation.

21.5 Future Research Highlights

21.5.1 Theoretical Models of Fruit Developmental Evolution

Fruits and their associated morphologies are complex and diverse. People are attracted to their amazing number of forms, which shows endless variation and demonstrates the power of evolution. However, this flower-derived organ is far from understood to the extent of its predecessor. There is no principle or model akin to the floral "ABC" model in fruit development. Roth (1977) attempted to describe and summarize fruit forms within all of the angiosperms by morphology alone, as there was no "general evolutionary system" for fruit development. Modern advancements in genetics and molecular biology have indicated that many genes control fruit morphology. Although the molecular genetics of fruit traits of model plants have been studied extensively, the complete network of fruit traits and the underlying molecular genetic basis are still very poorly understood. Moreover, as described above, conservation in genetic regulators of fruit development and morphology is poorly understood, and the interactions among the identified regulators are not fully understood as of yet. All of the above reasons significantly hinder the development of theoretical models for fruit developmental evolution.

21.5.2 The Developmental Evolution of Fruit Types

Fruits are divided into two main categories according to texture: dry and fleshy fruits. It is believed that the abundance of dispersal agents, in part, drives fruit evolution in particular habitats (Lorts et al., 2008). In arid and open habitats, such as in plains and prairies, the wind is continuous, and dry fruits prevail because they are easily dispersed by the wind. In places where animals that eat and disperse fruits are abundant, the majority of fruit genera produce fleshy fruits. However, the molecular mechanisms involved in the transition to different fruit types during evolution are not known, but despite this, the evolution of fruit types has already drawn considerable attention (Xiang et al., 2017). Unlike the conservation that has been demonstrated in determining carpel organ identity and development in angiosperms, little phylogenetic constraints have been observed for fruit type evolution among orders and families of angiosperm lineages (Lorts et al., 2008). Thus, the molecular switches for the transition of fruit types during evolution might be polyphyletic.

21.5.3 Exploiting New Systems for Fruit Evo-Devo Studies

Most current research is focused on *Arabidopsis*, rice, and tomato. In contrast to *Arabidopsis* silique, most crops produce grains and berries. Solanaceae includes some of the world's most economically and historically valuable crops that produce plentiful fleshy berries. Consequently, more solanaceous crops, including *Physalis* spp., eggplants, peppers, and wolfberry (*Lycium barbarum*), are invoked in fruit evo-devo studies (Wang et al., 2015a, b, c). Progress in understanding from these solanaceous crops will pave the way to understand the developmental genetics toolkit of solanaceous berries. Despite diversified fruit shape and size, the morphological novelty of fruit accessories is particularly fascinating (Wang et al., 2015a, b, c). The "Chinese lantern" is one accessory that has been elegantly investigated, but the origin of novelties in plants has been largely untouched. Moreover, the developmental regulation of the evolution of fruit color and functional secondary metabolite diversity also awaits investigation. Thus, understanding the nature of fruit form diversity requires multiple distinct systems, and particularly, genetically transformable systems, in order to understand fruit developmental evolution.

21.6 Concluding Remarks

Certain uncharacterized programs, also termed developmental trajectories, are presumed to determine fruit developmental patterns. Further, modifications that occur in developmental trajectories, which are processes of developmental repatterning, would result in additional fruit diversity. A large and accumulating body of evidence from studies of rice and solanaceous crops (Table 21.1) suggests that the evolution of fruit morphological variation and novelty can be attributed to genetic variation in the regulatory regions of preexisting genes. This observation further highlights the importance of gene expression variation in the evolution and domestication of certain species. Moreover, the comparison of closely related species or taxa suggests that the alteration of gene expression (i.e., heterotopy, heterochrony, or heterometry) may result from regulatory changes in either promoters or introns. Allelic variation in coding regions (heterotypy) also plays an evolutionary role in morphological variation. However, to identify simple rules of fruit developmental evolution and to explain the endless fruit forms are still vital and fascinating proposals. Evolutionary development biology (evo-devo) strategies that use a combination of multiple ideas and approaches from developmental biology, evolutionary biology, genetics, genomics, and bioinformatics are required in order to fully assess the nature of fruit morphological diversity. Moreover, we must exploit diverse plant systems to characterize fruit developmental repatterning events, dissect the developmental genetics underlying carpel and fruit development, and further reveal the molecular mechanisms underpinning the evolution of endless and beautiful fruit forms that encapsulate fruit types, fruit morphologies, and morphological novelties. A comparison between closely related species is an effective way to identify both developmental and genetic variations that occur in fruit development. The results and advancements from these investigations could then direct crop modification via breeding methodologies since some carpel developmental regulators have already been successfully used in the domestication or modification of some crops (Cong et al., 2008; Tiwari et al., 2011; Schiessl et al., 2015).

Acknowledgements This work was supported by a grant (31525003) from the National Natural Science Foundation of China to CYH. A large body of literature has emerged in this field, but we were not able to include all of it during the preparation of this review in 2017. We sincerely apologize to all of the colleagues whose work is not included and for our inability to update the literature after 2017.

References

Acri-Nunes-Miranda, R., & Mondragón-Palomino, M. (2014). Expression of paralogous *SEP-*, *FUL-*, *AG-* and *STK*-like MADS-box genes in wild-type and peloric *Phalaenopsis* flowers. *Front Plant Sci, 5*, 76.

Adamski, N. M., Anastasiou, E., Eriksson, S., O'Neill, C. M., & Lenhard, M. (2009). Local maternal control of seed size by KLUH/CYP78A5-dependent growth signaling. *Proc Natl Acad Sci USA, 106*, 20115–20120.

Alvarez, J., & Smyth, D. R. (2002). *CRABSCLAW* and *SPATULA* genes regulate growth and pattern formation during gynoecium development in *Arabidopsis thaliana*. *Int J Plant Sci, 163*, 17–41.

Alvarez, J. P., Goldshmidt, A., Efroni, I., Bowman, J. L., & Eshed, Y. (2009). The *NGATHA* distal organ development genes are essential for style specification in *Arabidopsis*. *Plant Cell, 21*, 1373–1393.

Anastasiou, E., Kenz, S., Gerstung, M., MacLean, D., Timmer, J., Fleck, C., & Lenhard, M. (2007). Control of plant organ size by KLUH/CYP78A5-dependent intercellular signaling. *Dev Cell, 13*, 843–856.

Angenent, G. C., & Colombo, L. (1996). Molecular control of ovule development. *Trends Plant Sci, 1*, 228–232.

Bednarek, J., Boulaflous, A., Girousse, C., Ravel, C., Tassy, C., Barret, P., Bouzidi, M. F., & Mouzeyar, S. (2012). Down-regulation of the *TaGW2* gene by RNA interference results in decreased grain size and weight in wheat. *J Exp Bot, 63*, 5945–5955.

Bowman, J. L., Smyth, D. R., & Meyerowitz, E. M. (1989). Genes directing flower development in *Arabidopsis*. *Plant Cell, 1*, 37–52.

Bowman, J. L., Smyth, D. R., & Meyerowitz, E. M. (1991). Genetic interactions among floral homeotic genes of *Arabidopsis*. *Development, 112*, 1–20.

Brambilla, V., Battaglia, R., Colombo, M., Masiero, S., Bencivenga, S., Kater, M. M., & Colombo, L. (2007). Genetic and molecular interactions between BELL1 and MADS box factors support ovule development in *Arabidopsis*. *Plant Cell, 19*, 2544–2556.

Chakrabarti, M., Zhang, N., Sauvage, C., Muños, S., Blanca, J., Cañizares, J., Diez, M. J., Schneider, R., Mazourek, M., McClead, J., Causse, M., & van der Knaap, E. (2013). A cytochrome P450 regulates a domestication trait in cultivated tomato. *Proc Natl Acad Sci USA, 110*, 17125–17130.

Che, R. H., Tong, H. N., Shi, B. H., Liu, Y. Q., Fang, S. R., Liu, D. P., Xiao, Y. H., Hu, B., Liu, L. C., Wang, H. R., Zhao, M. F., & Chu, C. C. (2015). Control of grain size and rice yield by *GL2*-mediated brassinosteroid responses. *Nat Plants, 2*, 15195.

Coen, E. S., & Meyerowitz, E. M. (1991). The war of the whorls: Genetic interactions controlling flower development. *Nature, 353*, 31–37.

Cong, B., Liu, J. P., & Tanksley, S. D. (2002). Natural alleles at a tomato fruit size quantitative trait locus differ by heterochronic regulatory mutations. *Proc Natl Acad Sci USA., 99*, 13606–13611.

Cong, B., Barrero, L. S., & Tanksley, S. D. (2008). Regulatory change in YABBY-like transcription factor led to evolution of extreme fruit size during tomato domestication. *Nat Genet, 40*, 800–804.

Corbesier, L., Vincent, C., Jang, S. H., Fornara, F., Fan, Q. Z., Searle, I., Giakountis, A., Farrona, S., Gissot, L., Turnbull, C., & Coupland, G. (2007). FT protein movement contributes to long-distance signaling in floral induction of *Arabidopsis*. *Science, 316*, 1030–1033.

Crawford, B. C. W., & Yanofsky, M. F. (2011). HALF FILLED promotes reproductive tract development and fertilization efficiency in *Arabidopsis thaliana*. *Development, 138*, 2999–3009.

Crawford, B. C. W., Ditta, G., & Yanofsky, M. F. (2007). The *NTT* gene is required for transmitting-tract development in carpels of *Arabidopsis thaliana*. *Curr Biol, 17*, 1101–1108.

Dahan, Y., Rosenfeld, R., Zadiranov, V., & Irihimovitch, V. (2010). A proposed conserved role for an avocado *FW2.2-like* gene as a negative regulator of fruit cell division. *Planta, 232*, 663–676.

de Folter, S., Immink, R. G., Kieffer, M., Parenicová, L., Henz, S. R., Weigel, D., Busscher, M., Kooiker, M., Colombo, L., Kater, M. M., Davies, B., & Angenent, G. C. (2005). Comprehensive interaction map of the *Arabidopsis* MADS box transcription factors. *Plant Cell, 17*, 1424–1433.

De Franceschi, P., Stegmeir, T., Cabrera, A., van der Knaap, E., Rosyara, U. R., Sebolt, A. M., Dondini, L., Dirlewanger, E., Quero-Garcia, J., Campoy, J. A., & Iezzoni, A. F. (2013). Cell number regulator genes in *Prunus* provide candidate genes for the control of fruit size in sweet and sour cherry. *Mol Breed, 32*, 311–326.

Dinneny, J. R., Yadegari, R., Fischer, R. L., Yanofsky, M. F., & Weigel, D. (2004). The role of *JAGGED* in shaping lateral organs. *Development, 131*, 1101–1110.

Dinneny, J. R., Weigel, D., & Yanofsky, M. F. (2006). *NUBBIN* and *JAGGED* define stamen and carpel shape in *Arabidopsis*. *Development, 133*, 1645–1655.

Duan, P. G., Ni, S., Wang, J. M., Zhang, B. L., Xu, R., Wang, Y. X., Chen, H. Q., Zhu, X. D., & Li, Y. H. (2015). Regulation of *OsGRF4* by *OsmiR396* controls gain size and yield in rice. *Nat Plants, 2*, 15203.

Duan, P. G., Xu, J. S., Zeng, D. L., Zhang, B. L., Geng, M. F., Zhang, G. Z., Huang, K., Huang, L. J., Xu, R., Ge, S., Qian, Q., & Li, Y. H. (2017). Natural variation in the promoter of *GSE5* contributes to grain size diversity in rice. *Mol Plant, 10*, 685–694.

Engelhorn, J., Moreau, F., Fletcher, J. C., & Carles, C. C. (2014). ULTRAPETALA1 and LEAFY pathways function independently in specifying identity and determinacy at the *Arabidopsis* floral meristem. *Ann Bot, 114*, 1497–1505.

Eshed, Y., Baum, S. F., & Bowman, J. L. (1999). Distinct mechanisms promote polarity establishment in carpels of *Arabidopsis*. *Cell, 99*, 199–209.

Ezquer, I., Mizzotti, C., Nguema-Ona, E., Gotté, M., Beauzamy, L., Viana, V. E., Dubrulle, N., Costa de Oliveira, A., Caporali, E., Koroney, A. S., Boudaoud, A., Driouich, A., & Colombo, L. (2016). The developmental regulator SEEDSTICK controls structural and mechanical properties of the *Arabidopsis* seed coat. *Plant Cell, 28*, 2478–2492.

Fan, C. H., Xing, Y. Z., Mao, H. L., Lu, T. T., Han, B., Xu, C. G., Li, X. H., & Zhang, Q. F. (2006). *GS3*, a major QTL for grain length and weight and minor QTL for grain width and thickness in rice, encodes a putative transmembrane protein. *Theor Appl Genet, 112*, 1164–1171.

Fang, W. J., Wang, Z. B., Cui, R. F., Li, J., & Li, Y. H. (2012). Maternal control of seed size by *EOD3/CYP78A6* in *Arabidopsis thaliana*. *Plant J, 70*, 929–939.

Favaro, R., Pinyopich, A., Battaglia, R., Kooiker, M., Borghi, L., Ditta, G., Yanofsky, M. F., Kater, M. M., & Colombo, L. (2003). MADS-box protein complexes control carpel and ovule development in *Arabidopsis*. *Plant Cell, 15*, 2603–2611.

Ferrándiz, C., Pelaz, S., & Yanofsky, M. F. (1999). Control of carpel and fruit development in *Arabidopsis*. *Annu Rev Biochem, 68*, 321–354.

Fourquin, C., Vinauger-Douard, M., Fogliani, B., Dumas, C., & Scutt, C. P. (2005). Evidence that *CRABS CLAW* and *TOUSLED* have conserved their roles in carpel development since the ancestor of the extant angiosperms. *Proc Natl Acad Sci USA, 102*, 4649–4654.

Frary, A., Nesbitt, T. C., Frary, A., Grandillo, S., Knaap, E., Cong, B., Liu, J. P., Meller, J., Elber, R., Alpert, K. B., & Tanksley, S. D. (2000). *fw2.2*: A quantitative trait locus key to the evolution of tomato fruit size. *Science, 289*, 85–88.

Friedrichsen, D. M., Nemhauser, J., Muramitsu, T., Maloof, J. N., Alonso, J., Ecker, J. R., Furuya, M., & Chory, J. (2002). Three redundant brassinosteroid early response genes encode putative bHLH transcription factors required for normal growth. *Genetics, 162*, 1445–1456.

Frohlich, M. W., & Parker, D. S. (2000). The mostly male theory of flower evolutionary origins: From genes to fossils. *Syst Bot, 25*, 155–170.

Girin, T., Paicu, T., Stephenson, P., Fuentes, S., Körner, E., O'Brien, M., Sorefan, K., Wood, T. A., Balanzá, V., Ferrándiz, C., Smyth, D. R., & Østergaard, L. (2011). INDEHISCENT and SPATULA interact to specify carpel and valve margin tissue and thus promote seed dispersal in *Arabidopsis*. *Plant Cell, 23*, 3641–3653.

Gómez-Mena, C., de Folter, S., Costa, M. M., Angenent, G. C., & Sablowski, R. (2005). Transcriptional program controlled by the floral homeotic gene *AGAMOUS* during early organogenesis. *Development, 132*, 429–438.

Groszmann, M., Paicu, T., Alvarez, J. P., Swain, S. M., & Smyth, D. R. (2011). *SPATULA* and *ALCATRAZ*, are partially redundant, functionally diverging bHLH genes required for *Arabidopsis* gynoecium and fruit development. *Plant J, 68*, 816–829.

Gu, Q., Ferrándiz, C., Yanofsky, M. F., & Martienssen, R. (1998). The *FRUITFULL* MADS-box gene mediates cell differentiation during *Arabidopsis* fruit development. *Development, 125*, 1509–1517.

Guo, M., Rupe, M. A., Dieter, J. A., Zou, J., Spielbauer, D., Duncan, K. E., Howard, R. J., Hou, Z., & Simmons, C. R. (2010). *Cell Number Regulator1* affects plant and organ size in maize: Implications for crop yield enhancement and heterosis. *Plant Cell, 22*, 1057–1073.

Guo, M., & Simmons, C. R. (2011). Cell number counts– the *fw2.2* and *CNR* genes and implications for controlling plant fruit and organ size. *Plant Sci, 181*, 1–7.

He, C. Y., & Saedler, H. (2005). Heterotopic expression of *MPF2* is the key to the evolution of the Chinese lantern of *Physalis*, a morphological novelty in Solanaceae. *Proc Natl Acad Sci USA, 102*, 5779–5784.

Heisler, M. G., Atkinson, A., Bylstra, Y. H., & Smyth, D. R. (2001). *SPATULA*, a gene that controls development of carpel margin tissues in *Arabidopsis*, encodes a bHLH protein. *Development, 128*, 1089–1098.

Houghtaling, H. B. (1935). A developmental analysis of size and shape in tomato fruits. *Bull Torrey Bot Club, 62*, 243–252.

Hsu, W. H., Yeh, T. J., Huang, K. Y., Li, J. Y., Chen, H. Y., & Yang, C. H. (2014). *AGAMOUS-LIKE13*, a putative ancestor for the E functional genes, specifies male and female gametophyte morphogenesis. *Plant J, 77*, 1–15.

Hu, Z. J., He, H. H., Zhang, S. Y., Sun, F., Xin, X. Y., Wang, W. X., Qian, X., Yang, J. S., & Luo, X. J. (2012). A kelch motif-containing serine/threonine protein phosphatase determines the large grain QTL trait in rice. *J Integr Plant Biol, 54*, 979–990.

Hu, J., Wang, Y. X., Fang, Y. X., Zeng, L. J., Xu, J., Yu, H. P., Shi, Z. Y., Pan, J. J., Zhang, D., Kang, S. J., Zhu, L., Dong, G. J., Guo, L. B., Zeng, D. L., Zhang, G. H., Xie, L. H., Xiong, G. S., Li, J. Y., & Qian, Q. (2015). A rare allele of *GS2* enhances grain size and grain yield in rice. *Mol Plant, 5*, 1455–1465.

Huang, Z. J., & van der Knaap, E. (2011). Tomato *fruit weight11.3* maps close to *fasciated* on the bottom of chromosome 11. *Theor Appl Genet, 123*, 465–474.

Huang, X. Z., Qian, Q., Liu, Z. B., Sun, H. Y., He, S. Y., Luo, D., Xia, G. M., Chu, C. C., Li, J. Y., & Fu, X. D. (2009). Natural variation at the *DEP1* locus enhances grain yield in rice. *Nat Genet, 41*, 494–497.

Ishimaru, K., Hirotsu, N., Madoka, Y., Murakami, N., Hara, N., Onodera, H., Kashiwagi, T., Ujiie, K., Shimizu, B., Onishi, A., Miyagawa, H., & Katoh, E. (2013). Loss of function of the IAA-glucose hydrolase gene *TGW6* enhances rice grain weight and increases yield. *Nat Genet, 45*, 707–711.

Ito, T., & Meyerowitz, E. M. (2000). Overexpression of a gene encoding a cytochrome P450, *CYP78A9*, induces large and seedless fruit in *Arabidopsis*. *Plant Cell, 12*, 1541–1550.

Ito, T., Wellmer, F., Yu, H., Das, P., Ito, N., Alves-Ferreira, M., Riechmann, J. L., & Meyerowitz, E. M. (2004). The homeotic protein AGAMOUS controls microsporogenesis by regulation of *SPOROCYTELESS*. *Nature, 430*, 356–360.

Jofuku, K. D., den Boer, B. G., Van Montagu, M., & Okamuro, J. K. (1994). Control of *Arabidopsis* flower and seed development by the homeotic gene *APETALA2*. *Plant Cell, 6*, 1211–1225.

Katsumata, T., Fukazawa, J., Magome, H., Jikumaru, Y., Kamiya, Y., Natsume, M., Kawaide, H., & Yamaguchi, S. (2011). Involvement of the CYP78A subfamily of cytochrome P450 monooxygenases in protonema growth and gametophore formation in the moss *Physcomitrella patens*. *Biosci Biotechnol Biochem, 75*, 331–336.

Krizek, B. A., & Fletcher, J. C. (2005). Molecular mechanisms of flower development: An armchair guide. *Nat Rev Genet, 6*, 688–698.

Lee, J. Y., Baum, S. F., Alvarez, J., Patel, A., Chitwood, D. H., & Bowman, J. L. (2005). Activation of *CRABS CLAW* in the nectaries and carpels of *Arabidopsis*. *Plant Cell, 17*, 25–36.

Lenhard, M., Bohnert, A., Jürgens, G., & Laux, T. (2001). Termination of stem cell maintenance in *Arabidopsis* floral meristems by interactions between WUSCHEL and AGAMOUS. *Cell, 105*, 805–814.

Li, Z. C., & He, C. Y. (2015). *Physalis floridana Cell Number Regulator1* encodes a cell membrane-anchored modulator of cell cycle and negatively controls fruit size. *J Exp Bot, 66*, 257–270.

Li, N., & Li, Y. H. (2016). Signaling pathways of seed size control in plants. *Curr Opin Plant Biol, 33*, 23–32.

Li, Q., Li, L., Yang, X. H., Warburton, M. L., Bai, G. H., Dai, J. R., Li, J. S., & Yan, J. B. (2010). Relationship, evolutionary fate and function of two maize co-orthologs of rice *GW2* associated with kernel size and weight. *BMC Plant Biol, 10*, 143.

Li, Y., Fan, C. C., Xing, Y. Z., Jiang, Y. H., Luo, L. J., Sun, L., Shao, D., Xu, C. J., Li, X. H., Xiao, J. H., He, Y. Q., & Zhang, Q. F. (2011). Natural variation in *GS5* plays an important role in regulating grain size and yield in rice. *Nat Genet, 43*, 1266–1269.

Libault, M., Zhang, X. C., Govindarajulu, M., Qiu, J., Ong, Y. T., Brechenmacher, L., Berg, R. H., Hurley-Sommer, A., Taylor, C. G., & Stacey, G. (2010). A member of the highly conserved *FWL* (tomato *FW2.2-like*) gene family is essential for soybean nodule organogenesis. *Plant J, 62*, 852–864.

Liljegren, S. J., Ditta, G. S., Eshed, Y., Savidge, B., Bowman, J. L., & Yanofsky, M. F. (2000). *SHATTERPROOF* MADS-box genes control seed dispersal in *Arabidopsis*. *Nature, 404*, 766–770.

Liljegren, S. J., Roeder, A. H., Kempin, S. A., Gremski, K., Ostergaard, L., Guimil, S., Reyes, D. K., & Yanofsky, M. F. (2004). Control of fruit patterning in *Arabidopsis* by *INDEHISCENT*. *Cell, 116*, 843–853.

Liu, J. P., Van Eck, J., Cong, B., & Tanksley, S. D. (2002). A new class of regulatory genes underlying the cause of pear-shaped tomato fruit. *Proc Natl Acad Sci USA, 99*, 13302–13306.

Liu, X. G., Dinh, T. T., Li, D. M., Shi, B. H., Li, Y. P., Cao, X. W., Guo, L., Pan, Y. Y., Jiao, Y. L., & Chen, X. M. (2014). *AUXIN RESPONSE FACTOR 3* integrates the functions of *AGAMOUS* and *APETALA2* in floral meristem determinacy. *Plant J, 80*, 629–641.

Liu, L. C., Tong, H. N., Xiao, Y. H., Che, R. H., Xu, F., Hu, B., Liang, C. Z., Chu, J. F., Li, J. Y., & Chu, C. C. (2015). Activation of *Big Grain1* significantly improves grain size by regulating auxin transport in rice. *Proc Natl Acad Sci USA, 112*, 11102–11107.

Lohmann, J. U., Hong, R. L., Hobe, M., Busch, M. A., Parcy, F., Simon, R., & Weigel, D. (2001). A molecular link between stem cell regulation and floral patterning in *Arabidopsis*. *Cell, 105*, 793–803.

Lorts, C. M., Briggeman, T., & Sang, T. (2008). Evolution of angiosperm fruit types and seed dispersal: A phylogenetic and ecological snapshot. *J Syst Evol, 46*, 396–404.

Ma, M., Zhao, H. X., Li, Z. J., Hu, S. W., Song, W. N., & Liu, X. L. (2016). *TaCYP78A5* regulates seed size in wheat (*Triticum aestivum*). *J Exp Bot, 67*, 1397–1410.

Mandel, T., Moreau, F., Kutsher, Y., Fletcher, J. C., Carles, C. C., & Eshed Williams, L. (2014). The ERECTA receptor kinase regulates *Arabidopsis* shoot apical meristem size, phyllotaxy and floral meristem identity. *Development, 141*, 830–841.

Mao, H. L., Sun, S. Y., Yao, J. L., Wang, C. R., Yu, S. B., Xu, C. G., Li, X. H., & Zhang, Q. F. (2010). Linking differential domain functions of the GS3 protein to natural variation of grain size in rice. *Proc Natl Acad Sci USA, 107*, 19579–19584.

Marsch-Martínez, N., Zuniga-Mayo, V. M., Herrera-Ubaldo, H., Ouwerkerk, P. B., Pablo-Villa, J., Lozano-Sotomayor, P., Greco, R., Ballester, P., Balanzá, V., Kuijt, S. J., Meijer, A. H., Pereira, A., Ferrándiz, C., & de Folter, S. (2014). The NTT transcription factor promotes replum development in *Arabidopsis* fruits. *Plant J, 80*, 69–81.

Matas, A. J., Yeats, T. H., Buda, G. J., Zheng, Y., Chatterjee, S., Tohge, T., Ponnala, L., Adato, A., Aharoni, A., Stark, R., Fernie, A. R., Fei, Z. J., Giovannoni, J. J., & Rose, J. K. (2011). Tissue- and cell-type specific transcriptome profiling of expanding tomato fruit provides insights into metabolic and regulatory specialization and cuticle formation. *Plant Cell, 23*, 3893–3910.

Miyoshi, K., Ahn, B. O., Kawakatsu, T., Ito, Y., Itoh, J., Nagato, Y., & Kurata, N. (2004). *PLASTOCHRON1*, a timekeeper of leaf initiation in rice, encodes cytochrome P450. *Proc Natl Acad Sci USA, 101*, 875–880.

Mizzotti, C., Mendes, M. A., Caporali, E., Schnittger, A., Kater, M. M., Battaglia, R., & Colombo, L. (2012). The MADS box genes *SEEDSTICK* and *ARABIDOPSIS* B_{sister} play a maternal role in fertilization and seed development. *Plant J, 70*, 409–420.

Moyroud, E., Elske, K., Monniaux, M., Koes, R., & Parcy, F. (2010). *LEAFY* blossoms. *Trends Plant Sci, 15*, 346–352.

Mu, Q., Huang, Z. J., Chakrabarti, M., Illa-Berenguer, E., Liu, X. X., Wang, Y. P., Ramos, A., & van der Knaap, E. (2017). Fruit weight is controlled by *Cell Size Regulator* encoding a novel protein that is expressed in maturing tomato fruits. *PLoS Genet, 13*, e1006930.

Muños, S., Ranc, N., Botton, E., Bérard, A., Rolland, S., Duffé, P., Carretero, Y., Le Paslier, M. C., Delalande, C., Bouzayen, M., Brunel, D., & Causse, M. (2011). Increase in tomato locule number is controlled by two single-nucleotide polymorphisms located near *WUSCHEL*. *Plant Physiol, 156*, 2244–2254.

Nagasawa, N., Hibara, K., Heppard, E. P., Vander Velden, K. A., Luck, S., Beatty, M., Nagato, Y., & Sakai, H. (2013). *GIANT EMBRYO* encodes CYP78A13, required for proper size balance between embryo and endosperm in rice. *Plant J, 75*, 592–605.

Ng, K. H., Yu, H., & Ito, T. (2009). AGAMOUS controls *GIANT KILLER*, a multifunctional chromatin modifier in reproductive organ patterning and differentiation. *PLoS Biol, 7*, e1000251.

Ohno, C. K., Reddy, G. V., Heisler, M. G., & Meyerowitz, E. M. (2004). The *Arabidopsis JAGGED* gene encodes a zinc finger protein that promotes leaf tissue development. *Development, 131*, 1111–1122.

Orashakova, S., Lange, M., Lange, S., Wege, S., & Becker, A. (2009). The *CRABS CLAW* ortholog from California poppy, (*Eschscholzia californica*, Papaveraceae), *EcCRC*, is involved in floral meristem termination, gynoecium differentiation, and ovule initiation. *Plant J, 58*, 682–693.

Pelaz, S., Ditta, G. S., Baumann, E., Wisman, E., & Yanofsky, M. F. (2000). B and C floral organ identity functions require *SEPALLATA* MADS-box genes. *Nature, 405*, 200–203.

Pfannebecker, K. C., Lange, M., Rupp, O., & Becker, A. (2017). Seed plant-specific gene lineages involved in carpel development. *Mol Biol Evol, 34*, 925–942.

Pires, N., & Dolan, L. (2010). Origin and diversification of basic-Helix-Loop-Helix proteins in plants. *Mol Bio Evol, 27*, 862–874.

Poppenberger, B., Rozhon, W., Khan, M., Husar, S., Adam, G., Luschnig, C., Fujioka, S., & Sieberer, T. (2011). CESTA, a positive regulator of brassinosteroid biosynthesis. *EMBO J, 30*, 1149–1161.

Qi, P., Lin, Y. S., Song, X. J., Shen, J. B., Huang, W., Shan, J. X., Zhu, M. Z., Jiang, L. W., Gao, J. P., & Lin, H. X. (2012). The novel quantitative trait locus *GL3.1* controls rice grain size and yield by regulating *Cyclin-T1;3*. *Cell Res, 22*, 1666–1680.

Rajani, S., & Sundaresan, V. (2001). The *Arabidopsis* myc/bHLH gene *ALCATRAZ* enables cell separation in fruit dehiscence. *Curr Biol, 11*, 1914–1922.

Reyes-Olalde, J. I., Zuñiga-Mayo, V. M., Chávez Montes, R. A., Marsch-Martínez, N., & de Folter, S. (2013). Inside the gynoecium: At the carpel margin. *Trends Plant Sci, 18*, 644–655.

Robert, H. S., CrhakKhaitova, L., Mroue, S., & Benková, E. (2015). The importance of localized auxin production for morphogenesis of reproductive organs and embryos in *Arabidopsis*. *J Exp Bot, 66*, 5029–5042.

Rodríguez, G. R., Muños, S., Anderson, C., Sim, S. C., Michel, A., Causse, M., Gardener, B. B., Francis, D., & van der Knaap, E. (2011). Distribution of *SUN*, *OVATE*, *LC*, and *FAS* in the tomato germplasm and the relationship to fruit shape diversity. *Plant Physiol, 156*, 275–285.

Roth, I. (1977). *Fruits of Angiosperms*. Gebrüder Borntraeger.

Sakai, H., Krizek, B. A., Jacobsen, S. E., & Meyerowitz, E. M. (2000). Regulation of *SUP* expression identifies multiple regulators involved in *Arabidopsis* floral meristem development. *Plant Cell, 12*, 1607–1618.

Schiessl, S., Iniguez-Luy, F., Qian, W., & Snowdon, R. J. (2015). Diverse regulatory factors associate with flowering time and yield responses in winter-type *Brassica napus*. *BMC Genomics, 16*, 737.

Schultz, E. A., & Haughn, G. W. (1991). *LEAFY*, a homeotic gene that regulates inflorescence development in *Arabidopsis*. *Plant Cell, 3*, 771–781.

Schuster, C., Gaillochet, C., & Lohmann, J. U. (2015). *Arabidopsis HECATE* genes function in phytohormone control during gynoecium development. *Development, 142*, 3340–3350.

Segami, S., Kono, I., Ando, T., Yano, M., Kitano, H., Miura, K., & Iwasaki, Y. (2012). *Small and round seed 5* gene encodes α-tubulin regulating seed cell elongation in rice. *Rice (NY), 5*, 4.

Shi, X. L., Gupta, S., & Rashotte, A. M. (2012). *Solanum lycopersicum cytokinin response factor* (*SlCRF*) genes: Characterization of CRF domain-containing *ERF* genes in tomato. *J Exp Bot, 63*, 973–982.

Shomura, A., Izawa, T., Ebana, K., Ebitani, T., Kanegae, H., Konishi, S., & Yano, M. (2008). Deletion in a gene associated with grain size increased yields during rice domestication. *Nat Genet, 40*, 1023–1028.

Si, L. Z., Chen, J. Y., Huang, X. H., Gong, H., Luo, J. H., Hou, Q. Q., Zhou, T. Y., Lu, T. T., Zhu, J. J., Shangguan, Y. Y., Chen, E. W., Gong, C. X., Zhao, Q., Jing, Y. F., Zhao, Y., Li, Y., Cui, L. L., Fan, D. L., Lu, Y. Q., Weng, Q. J., Wang, Y. C., Zhan, Q. L., Liu, K. Y., Wei, X. H., An, K., An, G., & Han, B. (2016). *OsSPL13* controls grain size in cultivated rice. *Nat Genet, 48*, 447–456.

Silva, C. S., Puranik, S., Round, A., Brennich, M., Jourdain, A., Parcy, F., Hugouvieux, V., & Zubieta, C. (2016). Evolution of the plant reproduction master regulators LFY and the MADS transcription factors: The role of protein structure in the evolutionary development of the flower. *Front Plant Sci, 6*, 1193.

Siriwardana, N. S., & Lamb, R. S. (2012). The poetry of reproduction: The role of *LEAFY* in *Arabidopsis thaliana* flower formation. *Int J Dev Biol, 56*, 207–221.

Song, X. J., Huang, W., Shi, M., Zhu, M. Z., & Lin, H. X. (2007). A QTL for rice grain width and weight encodes a previously unknown RING-type E3 ubiquitin ligase. *Nat Genet, 39*, 623–630.

Song, X. J., Kuroha, T., Ayano, M., Furuta, T., Nagai, K., Komeda, N., Segami, S., Miura, K., Ogawa, D., Kamura, T., Suzuki, T., Higashiyama, T., Yamasaki, M., Mori, H., Inukai, Y., Wu, J. Z., Kitano, H., Sakakibara, H., Jacobsen, S. E., & Ashikari, M. (2015). Rare allele of a previously unidentified histone H4 acetyltransferase enhances grain weight, yield, and plant biomass in rice. *Proc Natl Acad Sci USA, 112*, 76–81.

Specht, C. D., & Bartlett, M. E. (2009). Flower evolution: The origin and subsequent diversification of the angiosperm flower. *Annu Rev Ecol Evol Syst, 40*, 217–243.

Su, Z. Q., Hao, C. Y., Wang, L. F., Dong, Y. C., & Zhang, X. Y. (2011). Identification and development of a functional marker of *TaGW2* associated with grain weight in bread wheat (*Triticum aestivum* L.). *Theor Appl Genet, 122*, 211–223.

Sun, X. H., Cahill, J., Van Hautegem, T., Feys, K., Whipple, C., Novák, O., Delbare, S., Versteele, C., Demuynck, K., De Block, J., Storme, V., Claeys, H., Van Lijsebettens, M., Coussens, G., Ljung, K., De Vliegher, A., Muszynski, M., Inzé, D., & Nelissen, H. (2017). Altered expression of maize *PLASTOCHRON1* enhances biomass and seed yield by extending cell division duration. *Nat Commun, 8*, 14752.

Swaminathan, K., Peterson, K., & Jack, T. (2008). The plant B3 superfamily. *Trends Plant Sci, 13*, 647–655.
Tani, E., Polidoros, A. N., Flemetakis, E., Stedel, C., Kalloniati, C., Demetriou, K., Katinakis, P., & Tsaftaris, A. S. (2009). Characterization and expression analysis of *AGAMOUS*-like, *SEEDSTICK*-like, and *SEPALLATA*-like MADS-box genes in peach (*Prunus persica*) fruit. *Plant Physiol Biochem, 47*, 690–700.
Theissen, G. (2001). Development of floral organ identity: Stories from the MADS house. *Curr Opin Plant Biol, 4*, 75–85.
Tiwari, A., Vivian-Smith, A., Voorrips, R. E., Habets, M. E., Xue, L. B., Offringa, R., & Heuvelink, E. P. (2011). Parthenocarpic potential in *Capsicum annuum* L. is enhanced by carpelloid structures and controlled by a single recessive gene. *BMC Plant Biol, 11*, 143.
Trigueros, M., Navarrete-Gomez, M., Sato, S., Christensen, S. K., Pelaz, S., Weigel, D., Yanofsky, M. F., & Ferrandiz, C. (2009). The *NGATHA* genes direct style development in the *Arabidopsis* gynoecium. *Plant Cell, 21*, 1394–1409.
Urbanus, S. L., de Folter, S., Shchennikova, A. V., Kaufmann, K., Immink, R. G., & Angenent, G. C. (2009). *In planta* localisation patterns of MADS domain proteins during floral development in *Arabidopsis thaliana*. *BMC Plant Biol, 9*, 5.
Vrebalov, J., Ruezinsky, D., Padmanabhan, V., White, R., Medrano, D., Drake, R., Schuch, W., & Giovannoni, J. (2002). A MADS-box gene necessary for fruit ripening at the tomato ripening-inhibitor (*rin*) locus. *Science, 296*, 343–346.
Vrebalov, J., Pan, I. L., Arroyo, A. J., McQuinn, R., Chung, M., Poole, M., Rose, J., Seymour, G., Grandillo, S., Giovannoni, J., & Irish, V. F. (2009). Fleshy fruit expansion and ripening are regulated by the tomato *SHATTERPROOF* gene *TAGL1*. *Plant Cell, 21*, 3041–3062.
Wang, L., Li, Z. C., & He, C. Y. (2012a). Transcriptome-wide mining of the differentially expressed transcripts for natural variation of floral organ size in *Physalis philadelphica*. *J Exp Bot, 63*, 6457–6465.
Wang, S. K., Wu, K., Yuan, Q. B., Liu, X. Y., Liu, Z. B., Lin, X. Y., Zeng, R. Z., Zhu, H. T., Dong, G. J., Qian, Q., Zhang, G. Q., & Fu, X. D. (2012b). Control of grain size, shape and quality by *OsSPL16* in rice. *Nat Genet, 44*, 950–954.
Wang, L., He, L. L., Li, J., Zhao, J., Li, Z. C., & He, C. Y. (2014). Regulatory change at *Physalis organ size 1* correlates to natural variation in tomatillo reproductive organ size. *Nat Commun, 5*, 4271.
Wang, Y. X., Xiong, G. S., Hu, J., Jiang, L., Yu, H., Xu, J., Fang, Y. X., Zeng, L. J., Xu, E. B., Xu, J., Ye, W. J., Meng, X. B., Liu, R. F., Chen, H. Q., Jing, Y. H., Wang, Y. H., Zhu, X. D., Li, J. Y., & Qian, Q. (2015a). Copy number variation at the *GL7* locus contributes to grain size diversity in rice. *Nat Genet, 47*, 944–948.
Wang, S. K., Li, S., Liu, Q., Wu, K., Zhang, J. Q., Wang, S. S., Wang, Y., Chen, X. B., Zhang, Y., Gao, C. X., Wang, F., Huang, H. X., & Fu, X. D. (2015b). The OsSPL16-GW7 regulatory module determines grain shape and simultaneously improves rice yield and grain quality. *Nat Genet, 47*, 949–954.
Wang, L., Li, J., Zhao, J., & He, C. Y. (2015c). Evolutionary developmental genetics of fruit morphological variation within the Solanaceae. *Front Plant Sci, 6*, 248.
Weigel, D., & Meyerowitz, E. M. (1994). The ABCs of floral homeotic genes. *Cell, 78*, 203–209.
Weigel, D., Alvarez, J., Smyth, D. R., Yanofsky, M. F., & Meyerowitz, E. M. (1992). *LEAFY* controls floral meristem identity in *Arabidopsis*. *Cell, 69*, 843–859.
Weng, J. F., Gu, S. H., Wan, X. Y., Gao, H., Guo, T., Su, N., Lei, C. L., Zhang, X., Cheng, Z. J., Guo, X. P., Wang, J. L., Jiang, L., Zhai, H. Q., & Wan, J. M. (2008). Isolation and initial characterization of *GW5*, a major QTL associated with rice grain width and weight. *Cell Res, 18*, 1199–1209.
Wu, S., Xiao, H., Cabrera, A., Meulia, T., & van der Knaap, E. (2011). SUN regulates vegetative and reproductive organ shape by changing cell division patterns. *Plant Physiol, 157*, 1175–1186.

Xia, T., Li, N., Dumenil, J., Li, J., Kamenski, A., Bevan, M. W., Gao, F., & Li, Y. H. (2013). The ubiquitin receptor DA1 interacts with the E3 ubiquitin ligase DA2 to regulate seed and organ size in *Arabidopsis*. *Plant Cell, 25*, 3347–3359.
Xiang, Y. Z., Huang, C.-H., Hu, Y., Wen, J., Li, S. S., Yi, T. S., Chen, H. Y., Xiang, J., & Ma, H. (2017). Evolution of Rosaceae fruit types based on nuclear phylogeny in the context of geological times and genome duplication. *Mol Biol Evol, 34*, 262–281.
Xiao, H., Jiang, N., Schaffner, E., Stockinger, E. J., & van der Knaap, E. (2008). A retrotransposon-mediated gene duplication underlies morphological variation of tomato fruit. *Science, 319*, 1527–1530.
Xu, J., Xiong, W. T., Cao, B. B., Yan, T. Z., Luo, T., Fan, T. T., & Luo, M. Z. (2013). Molecular characterization and functional analysis of "*fruit-weight 2.2-like*" gene family in rice. *Planta, 238*, 643–655.
Xu, C. J., Liu, Y., Li, Y. B., Xu, X. D., Xu, C. G., Li, X. H., Xiao, J. H., & Zhang, Q. F. (2015a). Differential expression of *GS5* regulates grain size in rice. *J Exp Bot, 66*, 2611–2623.
Xu, F., Fang, J., Ou, S. J., Gao, S. P., Zhang, F. X., Du, L., Xiao, Y. H., Wang, H. R., Sun, X. H., Chu, J. F., Wang, G. D., & Chu, C. C. (2015b). Variations in *CYP78A13* coding region influence grain size and yield in rice. *Plant Cell Environ, 38*, 800–811.
Yadav, R. K., Tavakkoli, M., & Reddy, G. V. (2010). *WUSCHEL* mediates stem cell homeostasis by regulating stem cell number and patterns of cell division and differentiation of stem cell progenitors. *Development, 137*, 3581–3589.
Yang, Z. B., Bai, Z. Y., Li, X. L., Wang, P., Wu, Q. X., Yang, L., Li, L. Q., & Li, X. J. (2012). SNP identification and allelic-specific PCR markers development for *TaGW2*, a gene linked to wheat kernel weight. *Theor Appl Genet, 125*, 1057–1068.
Yanofsky, M. F., Ma, H., Bowman, J. L., Drews, G. N., Feldmann, K. A., & Meyerowitz, E. M. (1990). The protein encoded by the *Arabidopsis* homeotic gene *agamous* resembles transcription factors. *Nature, 346*, 35–39.
Zhao, J., Tian, Y., Zhang, J. S., Zhao, M., Gong, P. C., Riss, S., Saedler, R., & He, C. Y. (2013). The euAP1 protein MPF3 represses *MPF2* to specify floral calyx identity and displays crucial roles in 'Chinese lantern' development in *Physalis*. *Plant Cell, 25*, 2002–2021.
Zhao, B. T., Dai, A. H., Wei, H. C., Yang, S. X., Wang, B. S., Jiang, N., & Feng, X. Z. (2016). *Arabidopsis KLU* homologue *GmCYP78A72* regulates seed size in soybean. *Plant Mol Biol, 90*, 33–47.
Zhou, Y., Miao, J., Gu, H. Y., Peng, X. R., Leburu, M., Yuan, F. H., Gu, H. W., Gao, Y., Tao, Y. J., Zhu, J. Y., Gong, Z. Y., Yi, C. D., Gu, M. H., Yang, Z. F., & Liang, G. H. (2015). Natural variations in *SLG7* regulate grain shape in rice. *Genetics, 201*, 1591–1599.

MIX
Papier aus verantwortungsvollen Quellen
Paper from responsible sources
FSC® C105338

If you have any concerns about our products,
you can contact us on
ProductSafety@springernature.com

In case Publisher is established outside the EU,
the EU authorized representative is:
Springer Nature Customer Service Center GmbH
Europaplatz 3, 69115 Heidelberg, Germany

Printed by Libri Plureos GmbH
in Hamburg, Germany